Insights into Electrocardiograms with MCQs

Sitaram Mittal

Insights into Electrocardiograms with MCQs

Springer

Sitaram Mittal
Ex Senior professor and head of the department of cardiology
Jawaharlal Nehru Medical College
Ajmer, India

ISBN 978-981-99-0126-5 ISBN 978-981-99-0127-2 (eBook)
https://doi.org/10.1007/978-981-99-0127-2

This Springer imprint is published by the registered company Springer Nature Singapore Pte Ltd.
The registered company address is: 152 Beach Road, #21-01/04 Gateway East, Singapore 189721, Singapore

Dedicated

to

the memory of my parents

Shri Gouri Shankar ji Mittal

Smt. Ratan Devi Mittal

Foreword

While a medical student at Jawahar Lal Nehru Medical College in Ajmer (Rajasthan), I had heard of Dr. Sita Ram Mittal—his status as one of the earliest physicians to qualify as a cardiologist from Rajasthan and his dedication to medicine were both legendary. In your early years of training in medicine, finding a role model who you aspire to be like is both exciting and forever changes your life path. Dr Mittal was that person for me.

I have therefore had the pleasure of following Dr. Mittal's journey in medicine closely, from the days when he was training at Post Graduate Institute in Chandigarh to his retirement as Head of the Department of Cardiology and Principal of Jawahar Lal Nehru Medical College in Ajmer. It was always a privilege to see him knit together a patient's history, their findings on examination, and the interpretation of their electrocardiogram, always placing the electrocardiogram in the social and clinical context of the human being in front of him. He is a true master clinician with a wealth of experience in the advanced concepts of electrocardiography. He has inspired and trained multiple generations of cardiologists. There is no better person to compile a book on electrocardiography, enabling many more to benefit from a lifetime of clinical care and medical education.

Electrocardiography has evolved extensively since it was first described by Willem Einthoven, who used Lippman electrometers and then string galvanometers. Advances in recording technology with the use of filters and microchips, over the years, have improved the ease and quality of recordings and thus the value of this test in diagnosis of cardiac and even noncardiac conditions. Use of artificial intelligence will see more advancement in electrocardiography. Despite decades of advancement and development of imaging techniques including angiography, CT scans, MRI, and diagnostic electrophysiology, electrocardiography has continued to be of critical value in cardiac diagnosis.

The disciplines of interventional cardiology and invasive electrophysiology now complement the value of electrocardiography. For instance, we can now accurately locate the position of an atrioventricular accessory pathway or site of origin of premature beats using a simple electrocardiogram. This is the result of years of investigation correlating electrocardiographic findings with confirmed abnormalities. Clinical therapeutic decision making, to a large extent, is still based on electrocardiographic findings. The timely classification of acute coronary events, for example, and response to cardiac resynchronization therapy using QRS duration in heart failure patients are few obvious examples. We can diagnose various pacemaker malfunctions using electrocardiography, reminding us that the role of electrocardiography will continue to evolve.

A big advantage of electrocardiography is its low cost and availability at smaller medical offices, clinics, mobile units, and hospitals. The value of accurate knowledge of electrocardiography to a range of medical personnel including technicians, nurses, internists, and cardiologists is immense. It is no surprise that to be board certified in cardiology the American Board of Internal Medicine requires board-eligible general physicians to pass the electrocardiographic section separately.

Dr. Mittal has witnessed this evolving role of electrocardiography over the years and in this volume *Insights into Electrocardiograms with MCQs* has complied chapters exploring both traditional and contemporary concepts in electrocardiography for clinical diagnosis. The text is unique in both the simplicity of basic concepts presented and the broad range of topics covered. Electrocardiograms and graphic concepts are superbly illustrated and easy to follow. The inclusion of test questions is a practical addition and adds to the text's educational relevance. This book will be of immense value for a range of personnel including medical students, technologists, nurses, internists and general physicians, medical subspecialists, and cardiologists. Congratulations to Dr. Sita Ram Mittal on the compilation of this volume, and gratitude for his lifetime of service in this field.

Icahn School of Medicine at Mount Sinai Hospital
Al Sabah Cardiac Arrhythmia Institute,
Mount Sinai Morningside Hospital
New York, NY, USA

Davendra Mehta

Preface

While teaching undergraduates, postgraduate students in internal medicine, postgraduate students in pediatrics, and students of diploma course in cardiac instrument technology, I realized that the students felt that interpretation of electrocardiogram was difficult. Students also did not realize the clinical significance of various abnormalities. Students, therefore, did not feel interested in learning electrocardiography.

I felt that there was need of a book that could explain various electrocardiographic abnormalities with explanatory diagrams, representative electrocardiograms, and simultaneously explain the clinical relevance of various electrocardiographic abnormalities.

In this book, I have discussed abnormalities of various components of the electrocardiogram. Genesis of various abnormalities has been explained by simple and easy-to-understand diagrams. Sophisticated and controversial concepts have been avoided to keep the book interesting for clinicians. Concepts which help in correct understanding of various abnormalities in electrocardiograms are retained in spite of some disputes. Several electrocardiograms have been included to give a solid visual impression of various abnormalities. Differential diagnosis has been discussed with the help of simple diagrams and representative electrocardiograms. Attempt has been made to include the latest concepts in simple language and with easy-to-understand diagrams. Clinical relevance of various abnormalities is discussed to make the subject interesting.

Each chapter is followed by a summary for quick revision and several MCQs, which will help readers in self-evaluation. These MCQs will also help teachers and examiners in evaluating students.

Nowadays, the diagnosis of congenital heart disease is dependent on detailed echocardiographic evaluation. However, electrocardiograms are still an integral part of the diagnosis of congenital heart disease. Clinical examination combined with correct interpretation of the electrocardiogram can significantly narrow down the differential diagnosis. The last four chapters deal with clinic-electrocardiographic approach to the diagnosis of congenital heart diseases. In addition to summaries and MCQs, these chapters also have flowcharts for easy memorization of the approach. I hope these chapters will be useful for doctors planning to go for fellowship in pediatric cardiology.

This book will be of use for undergraduates, family physicians interested in interpreting electrocardiograms, postgraduate students in general medicine, students and teachers of diploma courses in electrocardiography and cardiac instrument technology, and staff working in emergency departments, intensive care units, and various cardiac laboratories.

I am grateful to CIMS Medica India Pvt Ltd for giving me "no objection certificate" to reproduce my articles published in the journal *Cardiology Today* in the form of a book with another publisher.

I am thankful to my younger sister Smt. Santosh Agrawal for constantly inspiring me to write this book. She keenly followed the progress and boosted my morale. She and her husband, Shri Suresh Chand ji Agrawal, have always prayed for my success and well-being.

I am also thankful to my children, their spouse, and my grandchildren for constantly looking after me and keeping the environment congenial for me to write the book.

I am thankful to the nursing staff who helped me in collection of these ECGs.

I am thankful to my computer operator Mr. Manoj Kumar Jain for typing, scanning, and editing the manuscripts to my satisfaction.

I am also thankful to the publishing team at Springer Nature for full cooperation.

There could be some errors or omissions. Some concepts could have changed during the period of writing of this book. I will be thankful for any suggestions for improvement and correction.

Ajmer, India Sitaram Mittal

Acknowledgments

Many thanks to my wife Smt. Shkuntla Mittal.

Disclaimer

The author has made all efforts to give clinically useful information. However, Medicine is a rapidly evolving branch, and it is humanly impossible to be perfect. An electrocardiogram is only one investigation. It should be interpreted in the context of clinical profile and other relevant investigations including electro-physiologic studies, serum enzymes, and echocardiography. The author would like to stress that electro-physiologic studies are more informative than trying to interpret a surface electrocardiogram. This book is targeted only to persons who are trying to learn the interpretation of electrocardiogram and does not provide any guidelines for the management of patients. Clinicians should depend on their knowledge, experience, and other sources of information including electro-physiologic studies and opinion of relevant experts. The author disclaims any responsibility for any error or omission in the contents of the book. The author also disclaims any responsibility for any harm caused by the use of information given in this book.

Contents

About the Author

Sitaram Mittal has retired from the post of Sr. Professor and Head of the Department of Cardiology and Principal of Jawahar Lal Nehru Medical college, Ajmer. He has published more than 470 journal articles, including over 250 in surface electrocardiography. He is the author of 37 book chapters and books such as *Echocardiographic Evaluation of Right Ventricular function* and *Echocardiographic Evaluation of Pericardial Diseases*. He received several honors from national and international societies, including the International Biographical Centre and the American Biographical Institute. He serves as the editorial board member of several cardiology journals. He is the section editor for "ECG section" of the journals *Cardiology Today*. He is a reviewer for several national and international journals in cardiology, including the most prestigious *Journal of Electrocardiology*. He was honored by the president of India for clinical research.

1 Technical Errors and Artefacts

Artefacts and technical errors in the recording of an electrocardiogram are not unusual. These can significantly hamper interpretation, especially if one is not conversant with these problems. Common patterns resulting from technical errors and artefacts are discussed.

1 Technical Errors

1.1 Wrong Placement of Leads

1. *Switching of limb leads.* It can produce error in interpretation of sequential ECGs [1]. Changes occurring in different limb leads due to wrong placement are shown in Fig. 1.1 Interchange of upper limb leads results in false impression of dextrocardia in limb leads (technical dextrocardia). Interchange of left upper limb and right lower limb electrodes produces zero potential in lead III. Interchange of right upper limb and right lower limb electrodes produces zero potential in lead II. Interchange between left upper limb and left lower limb electrodes produces an abnormal Q wave in lead III. Interchange of right upper limb and left lower limb electrodes produces abnormal Q waves in leads I and aVL; P wave inversion in leads II, III, and aVF; and positive P, QRS, and T in lead aVR.
2. *Switching of chest electrodes.* Total reversal of precordial electrodes can give wrong impression of dextrocardia in chest leads (Fig. 1.2a, a′). Interchange of right and left precordial electrodes can cause sudden disappearance of R resembling "reverse progression" (Fig. 1.2b, b′).
3. *Wrong placement of chest electrodes.* It is not infrequent. Higher placement of right precordial leads can result in appearance of "slow progression of r" and may be misinterpreted as old anterior infarction (Fig. 1.3a, a′). Sometimes this technical error can cause rsr' configuration with negative P wave in V_1 (Fig. 1.3b, b′). Wrong placement of posterior chest leads (V_7–V_9) can produce wrong impression of a posterior myocardial infarction (Fig. 1.3c, c′).
4. *Recording of chest leads without changing the machine switch to "V" lead.* This error occurs in single-channel recorders which are infrequently used nowadays. This error causes identical configuration of lead aVF and leads V_1–V_6 (Fig. 1.4a, b).
5. *Recording at 50 mm/s speed.* Recording at 50 mm/s speed causes broadening of all waves and intervals (Fig. 1.5a, b). It simulates bradycardia. However, in sinus bradycardia, the duration of waves and intervals remains normal.

S. Mittal, *Insights into Electrocardiograms with MCQs*,
https://doi.org/10.1007/978-981-99-0127-2_1

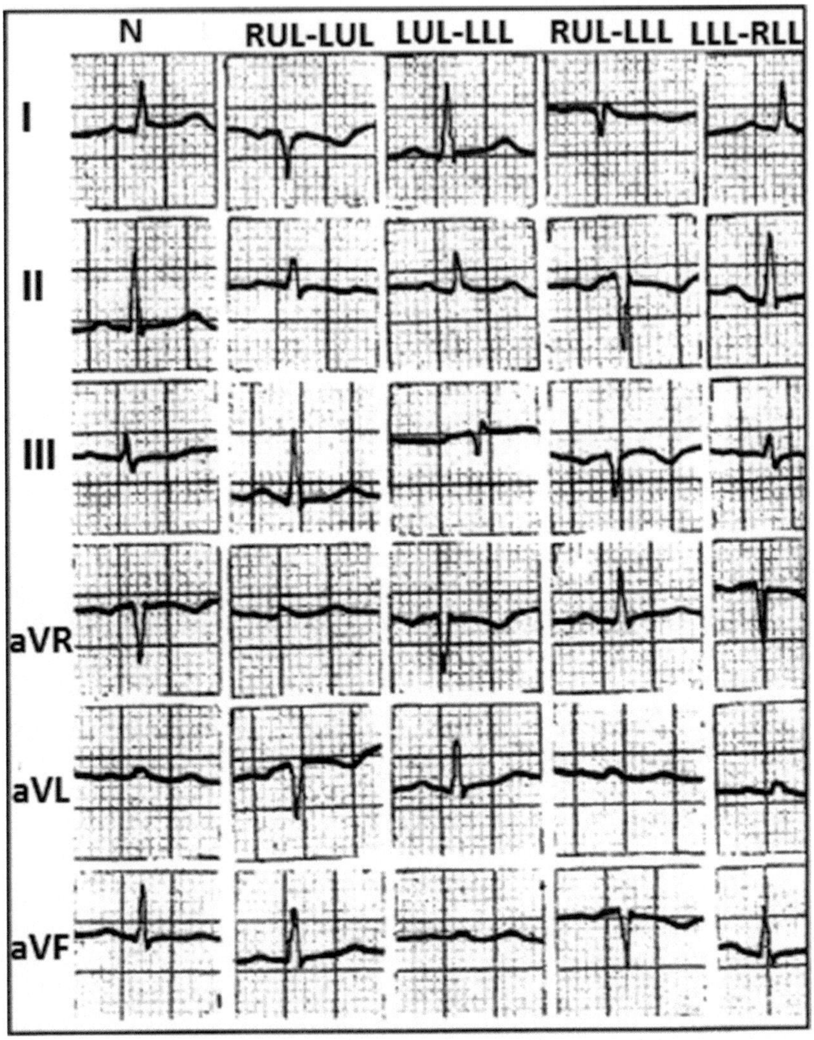

Fig. 1.1 Showing changes occurring in different limb leads due to wrong placement of limb electrodes. Abbreviation: *RUL* right upper limb, *LUL* left upper limb, *LLL* left lower limb, *RLL* right lower limb. (Reproduced with permission from: Mittal SR. Technical Errors and Artefacts. Cardiology Today 2011;15:81–85)

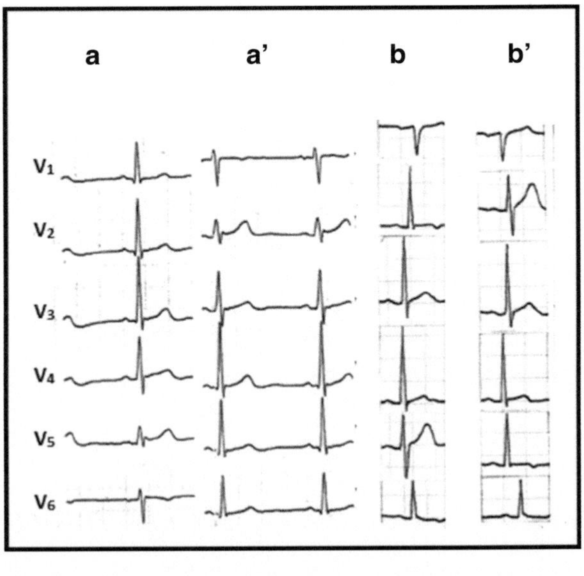

Fig. 1.2 (**a**) Electrocardiogram showing dextrocardia due to wrong placement of precordial leads. (**a′**) Electrocardiogram after correct placement of chest leads. (**b**) Electrocardiogram showing "reverse progression" due to wrong placement of chest leads. (**b′**) Electrocardiogram after correct placement of chest leads. (Reproduced with permission from: Mittal SR. Technical Errors and Artefacts. Cardiology Today 2011;15:81–85)

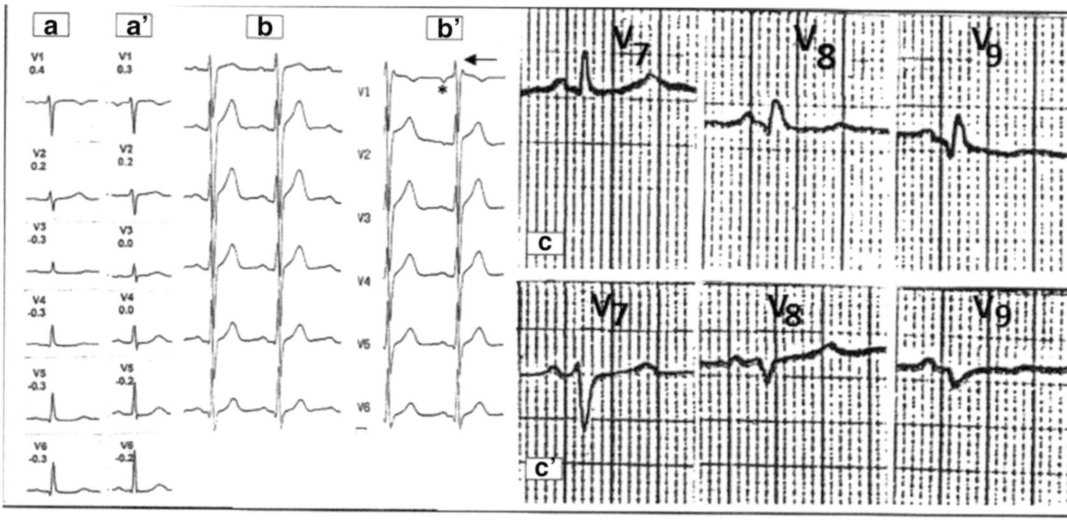

Fig. 1.3 (**a**) Showing normal precordial leads. (**a′**) Showing the impression of slow progression of r wave due to the high placement of chest leads. (**b**) Showing normal precordial pattern. (**b′**) Showing rsr' pattern (marked ←) with negative P wave in lead V_1 (marked *) due to high placement of chest leads. (**c**) Showing normal posterior leads. (**c′**) Showing false impression of posterior infarction due to high placement of electrodes. (Reproduced with permission from: Mittal SR. Technical Errors and Artefacts. Cardiology Today 2011;15:81–85)

Fig. 1.4 (**a**) Normal precordial leads. (**b**) Showing precordial leads resembling lead aVF due to failure of change of machine switch to "V" position. (Reproduced with permission from: Mittal SR. Technical Errors and Artefacts. Cardiology Today 2011;15:81–85)

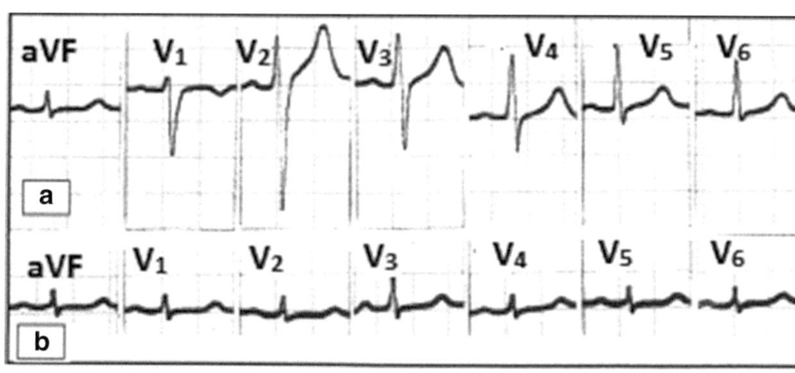

1.2 Poor Electrode Contact

This results in wandering baseline [2] (Fig. 1.6a), which can be confused with arrhythmia. However, careful observation and measurement show normal regular QRS spikes in the wandering baseline [3]. This can also result in an acute baseline shift (Fig. 1.6b, c) which creates problem in the interpretation of PR segment and ST segment.

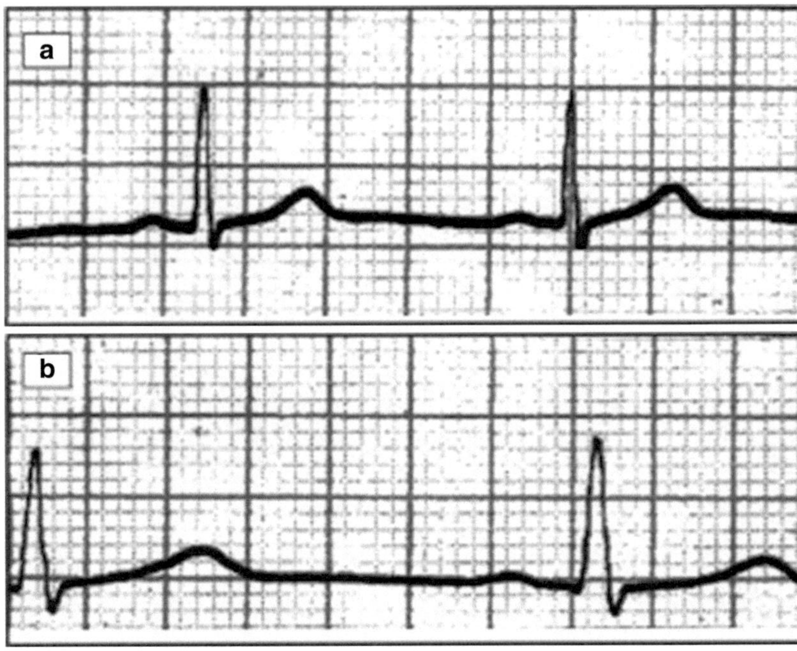

Fig. 1.5 (**a**) Showing ECG recorded at standard speed of 25 mm/s. (**b**) Showing ECG recorded at 50 mm/s speed. There is broadening of all waves and intervals. (Reproduced with permission from: Mittal SR. Technical Errors and Artefacts. Cardiology Today 2011;15:81–85)

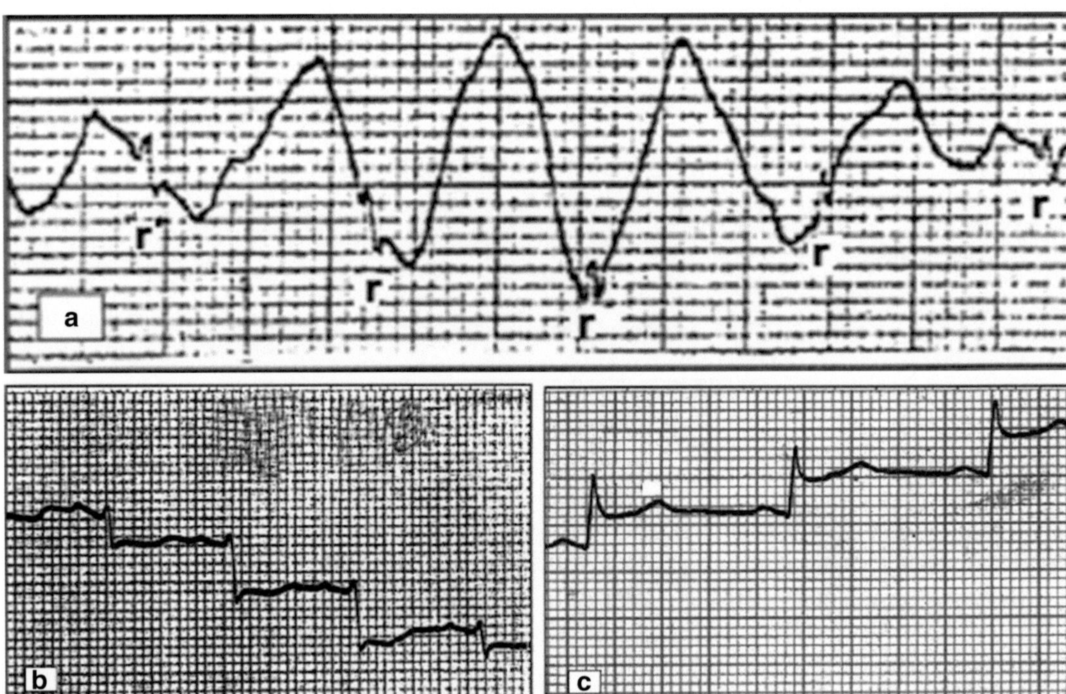

Fig. 1.6 (**a**) Showing wandering baseline due to poor electrode contact. Regular QRS (marked r) can be seen. (**b**, **c**) Showing acute baseline shift. (Reproduced with permission from: Mittal SR. Technical Errors and Artefacts. Cardiology Today 2011;15:81–85)

2 Artefacts

2.1 Related to Patient

Tremors, shivering, and hiccup can produce artefacts in the baseline that can be confused with atrial flutter (Fig. 1.7a–c). Such artefacts, however, are not present in all leads. Tremors produce artefacts in leads related to the limb having tremors. Hiccup produces disturbance in chest leads. Regular QRS spikes help in excluding a rhythm disorder. Limb tremors may disappear on command or during sleep [4] (Fig. 1.7a′).

2.2 Cable Malfunction

This error also occurs in single-channel recorders which are infrequently used nowadays. It can produce identical morphologies in left-sided chest leads (Fig. 1.8).

2.3 Electrical Noise

Alternating current interference can produce fine high-frequency vibrations in the baseline (Fig. 1.9). Such artefacts can also occur due to

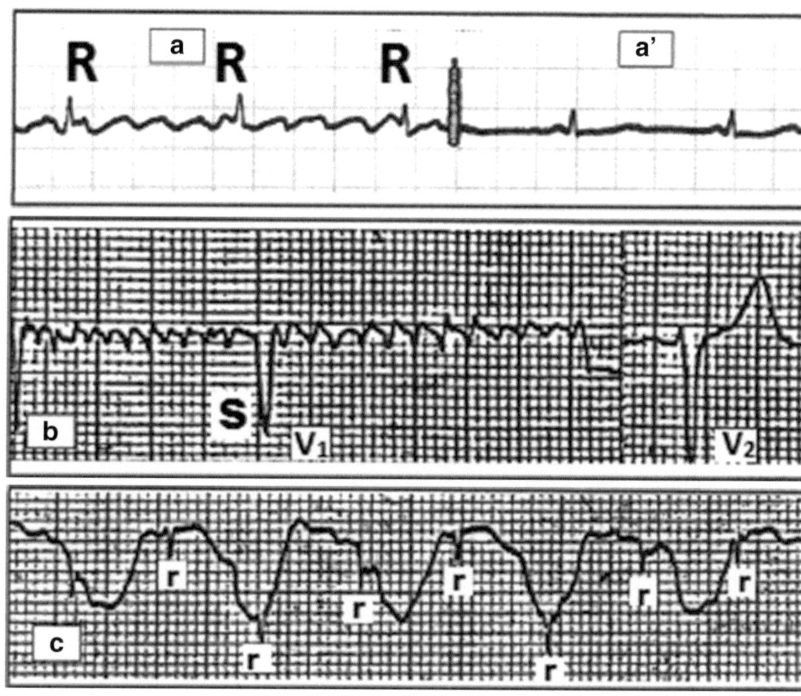

Fig. 1.7 (**a**) Showing effect of tremors on baseline. QRS (marked R) is regular. (**a′**) Showing disappearance of baseline vibration on stopping tremors. (**b**) Showing the effect of shivering on baseline. QRS (marked S) in regular. (**c**) Showing the effect of hiccup on baseline. Normal QRS (marked r) is occurring regularly. (Reproduced with permission from: Mittal SR. Technical Errors and Artefacts. Cardiology Today 2011;15:81–85)

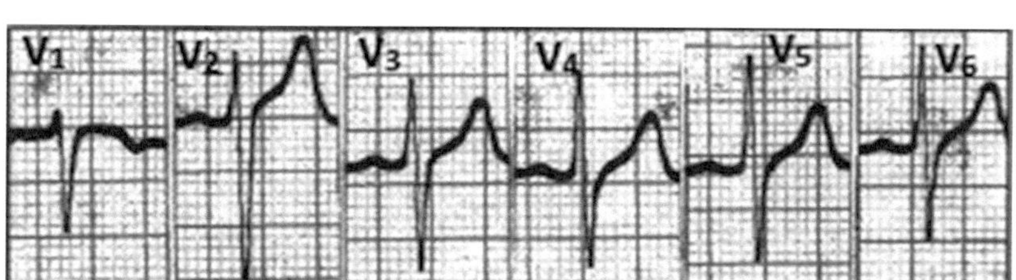

Fig. 1.8 Showing identical configuration in leads V_4–V_6. (Reproduced with permission from: Mittal SR. Technical Errors and Artefacts. Cardiology Today 2011;15:81–85)

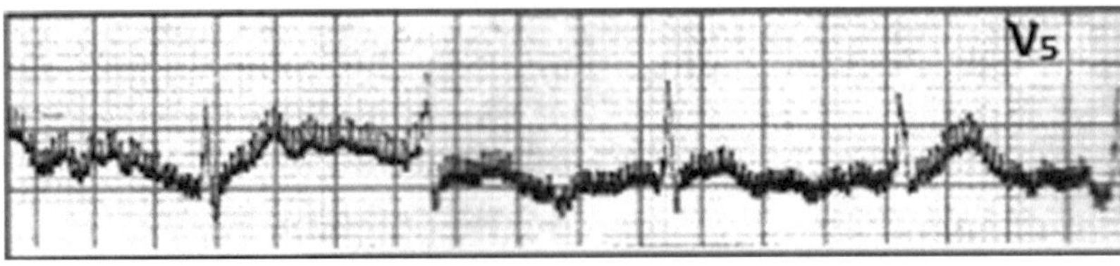

Fig. 1.9 Showing electrical noise. (Reproduced with permission from: Mittal SR. Technical Errors and Artefacts. Cardiology Today 2011;15:81–85)

improper earthing or improper contact of the neutral electrode (right lower limb electrode). Regular QRS helps in excluding an arrhythmia.

2.4 Artefacts Related to Motor Speed

Standard speed while recording an ECG is 25 mm/s. If the motor is defective, speed may change spontaneously. This may result in wrong impression of arrhythmia [3]. Slowing of the motor speed results in wrong impression of tachycardia (Fig. 1.10a, a′). It can be differentiated because tachycardia due to diminished motor speed squeezes all waves and intervals. This does not occur in tachyarrhythmias. Further, tachycardia produced by slow motor speed produces unphysiological RR interval (<140 ms). Increased speed results in false impression of bradycardia. However, unlike sinus bradycardia, all waves and intervals are widened (Fig. 1.10b). Changing speed can give wrong impression of atrial fibrillation (Fig. 1.10c). Motor problem produces changing width of waves and intervals as compared to fixed duration of QRS, T wave, and QT interval during atrial fibrillation. Transient retrograde motion of motor can produce curved QRS (Fig. 1.10d, d′) or splitting of the terminal portion of the QRS complex (Fig. 1.10e).

2.5 Artefacts Related to Stylus

1. *Artefacts related to stylus movement.*
   - *Standard.* Normal standard is of 10 mm/mv [4]. If the standard is more (>10 mm), QRS amplitude increases and can give a false impression of ventricular hypertrophy by voltage criteria. On the other hand, small standard (<10 mm) can give a false impression of low-voltage ECG (Fig. 1.11).
   - *High-frequency signals* [3] (Fig. 1.12). These can mimic pacing spikes and can result in wrong interpretation of ECG in patients with permanent pacemaker.
   - *Improper damping.* Normal standard is shown in Fig. 1.13a. It has sharp angles (marked → and ↑). Underdamping means when friction between the stylus and the ECG paper is less than normal. It is diagnosed when the standard shows spikes (Fig. 1.13b, marked ←). It can increase the height of the R wave and depth of the S wave. Overdamping means that the friction between the stylus and the ECG paper is more than normal. Overdamping results in loss of sharpness of the angles of the standard (Fig. 1.13c, marked →). It reduces sharpness of onset of the R wave and J point and produces false ST segment depression in leads with QRS having terminal S wave (Fig. 1.13d). It can also mask delta wave of preexcitation syndrome (Fig. 1.13e, f).
   - Restriction of movement of the stylus above or below a particular level results in cutting of R wave and/or S wave (Fig. 1.14a, b).
2. *Artefact related to transient loss of electrical current to the stylus.* This can result in a straight line with loss of all or some waves of ECG (Fig. 1.15a, b).
3. *Battery depletion.* It can change the size of various waves.

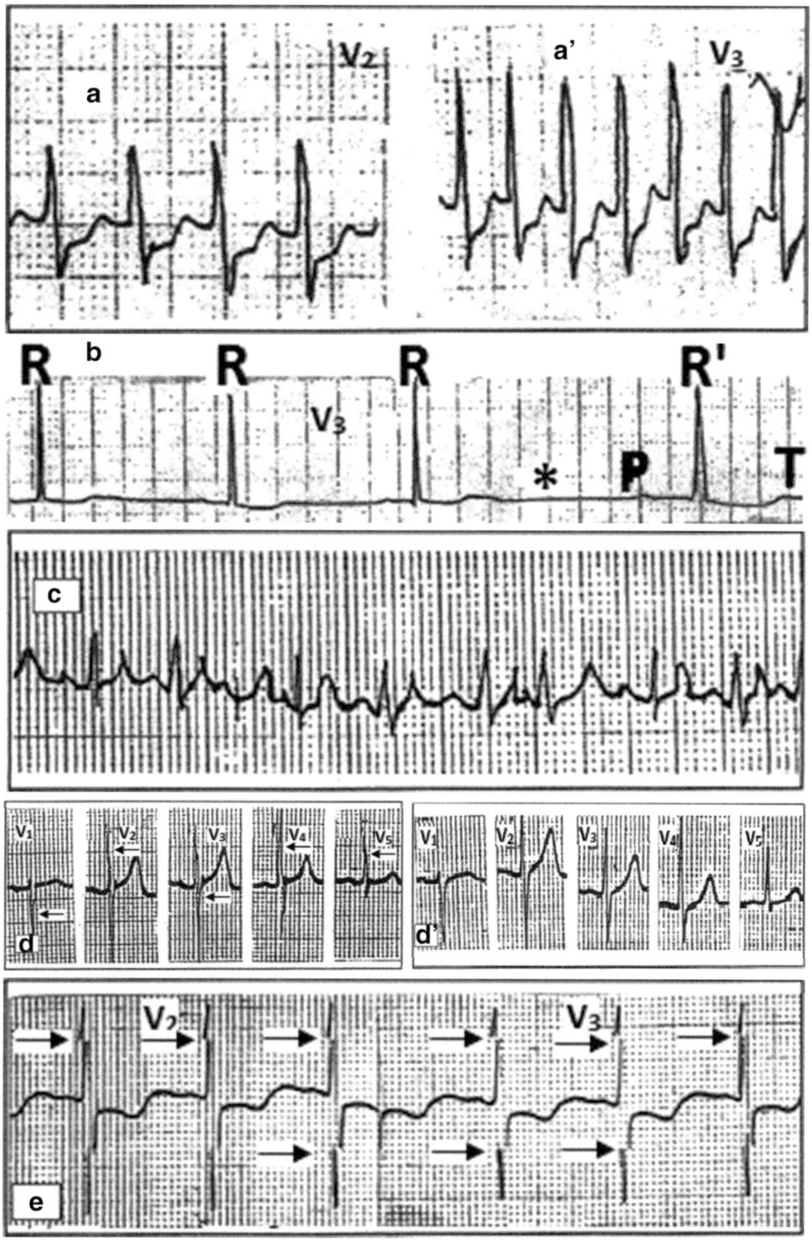

Fig. 1.10 (**a**) Showing ECG recorded at normal speed. (**a′**) Showing wrong impression of tachycardia due to slowing of the motor speed. (**b**) Showing sudden prolongation of RR interval (marked *), prolongation of PR interval, QRS duration and QT interval in fourth beat (marked R′) due to spontaneous increase in motor speed. (**c**) Showing effect of irregular motor speed. There is changing duration of P, QRS, and T waves with changing RR interval. (**d**) Showing curved QRS (marked ←). (**d′**) Showing normal QRS in ECG recorded from normal machine. (**e**) Showing splitting of QRS (marked →). (Reproduced with permission from: Mittal SR. Technical Errors and Artefacts. Cardiology Today 2011;15:81–85)

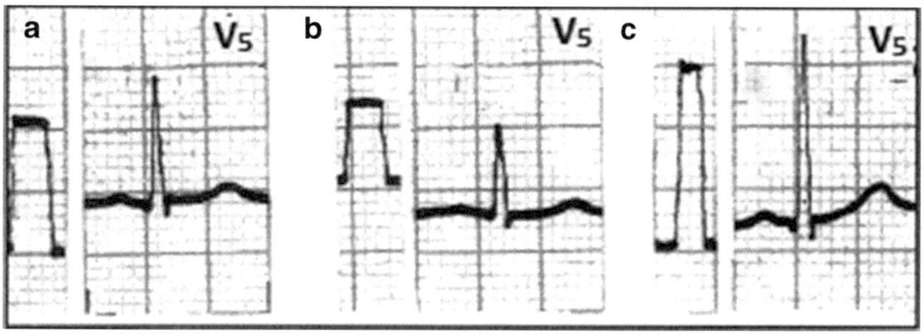

Fig. 1.11 Showing the effect of standards on amplitude of QRS. (**a**) Normal standard 10 mm/mv. (**b**) Small standard 6 mm/mv. (**c**) Large standard 15 mm/mv. (Reproduced with permission from: Mittal SR. Technical Errors and Artefacts. Cardiology Today 2011;15:81–85)

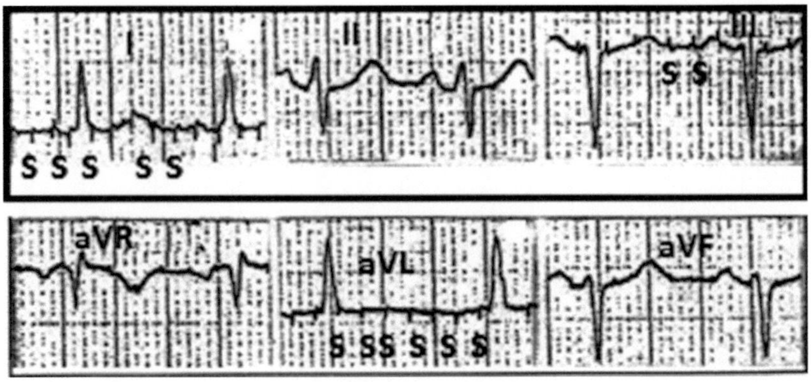

Fig. 1.12 Showing high-frequency signal artefacts (marked S) in leads I, III, and aVL. (Reproduced with permission from: Mittal SR. Technical Errors and Artefacts. Cardiology Today 2011;15:81–85)

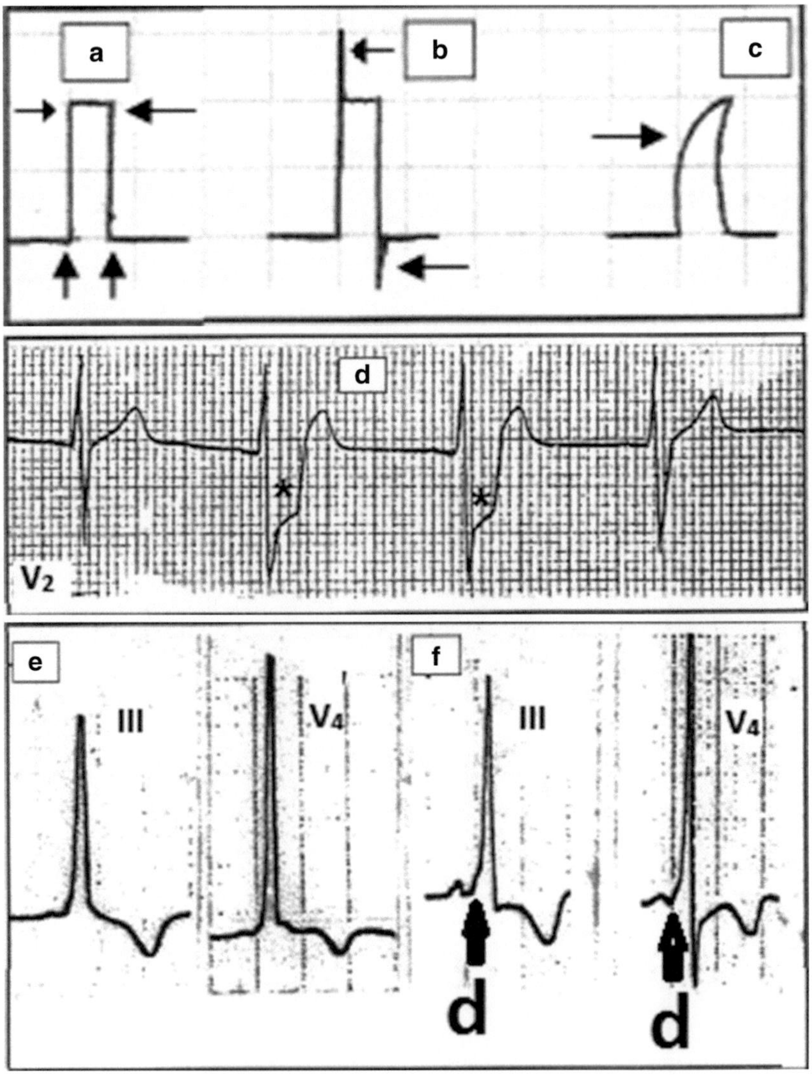

Fig. 1.13 (**a**) Normal standard. (**b**) Underdamping. (**c**) Overdamping. (**d**) False ST segment depression due to overdamping (marked *). (**e**) Apparently normal QRS due to overdamping. (**f**) Delta waves (marked as d) are clearly seen in ECG recorded from normal machine. (Reproduced with permission from: Mittal SR. Technical Errors and Artefacts. Cardiology Today 2011;15:81–85)

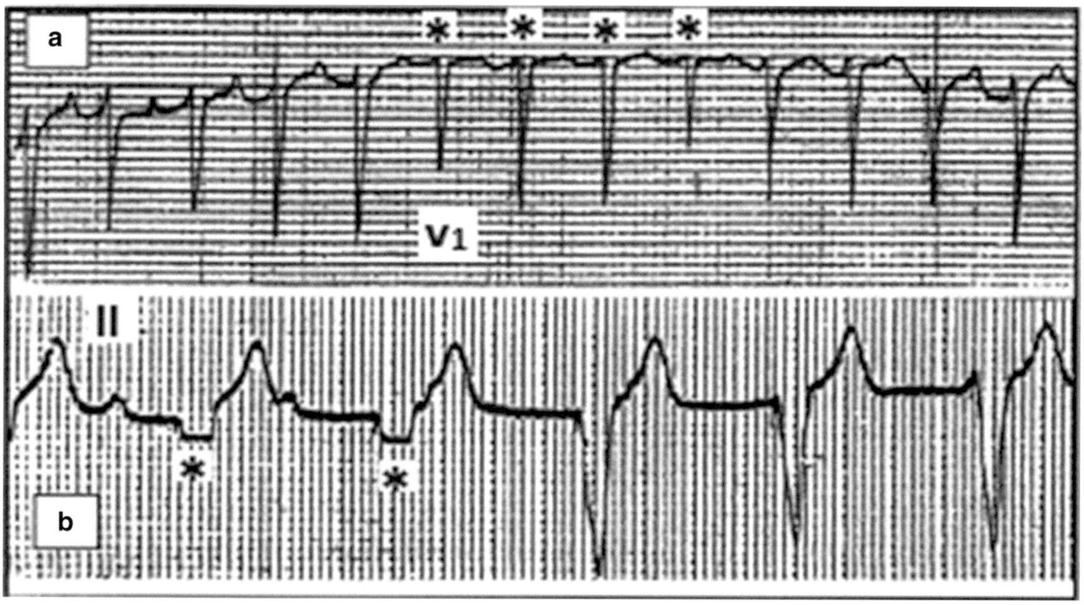

Fig. 1.14 (**a**) Restriction of movement above a particular level resulting in apparent loss of r wave in beats marked *. (**b**) Restriction of movement below a particular level resulting in cutting of S wave in leads marked *. (Reproduced with permission from: Mittal SR. Technical Errors and Artefacts. Cardiology Today 2011;15:81–85)

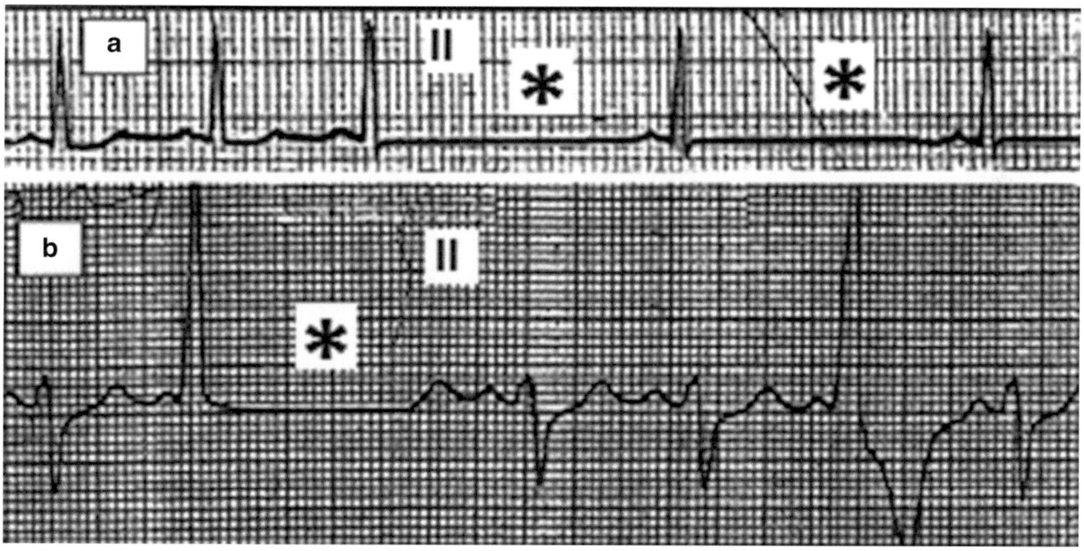

Fig. 1.15 (**a**) False impression of SA block due to transient loss of current to the stylus (marked *). Absence of T waves following QRS helps in correct diagnosis. (**b**) Loss of T wave and next QRS (marked *) following an ectopic due to transient loss of current to the stylus. (Reproduced with permission from: Mittal SR. Technical Errors and Artefacts. Cardiology Today 2011;15:81–85)

3 Computerized Interpretation

Currently available systems are not accurate enough to be relied on in the presence of complex abnormalities or in critical clinical environment [1]. No computer program can replace a skilled physician [4].

Placement of precordial leads is made with marked neglect of anatomical landmarks [5]. This is more common now than previously [4] and creates severe problems for computer interpretation of serial ECGs [6].

Summary

- Proximal placement of limb leads does not produce significant change in ECG.
- Interchange between right and left arm electrodes produces negative P wave and T wave in lead I and positive P wave and T wave in lead aVR (technical dextrocardia).
- Isoelectric line (zero potential) in lead II suggests wrong placement of right arm electrode on right leg.
- Isoelectric line in lead III suggests wrong placement of left arm electrode on right leg.
- Interchange between right and left leg electrodes does not produce significant change.
- Interchange between left upper limb and left lower limb electrodes produces abnormal Q wave in lead III.
- Interchange of right upper limb and left lower limb electrodes produces abnormal Q in leads I and aVL; P wave inversion in leads II, III, and aVF; and positive P, QRS, and T in lead aVR.
- Intercostal space below "sternal angle" is second intercostal space.
- High placement of lead V_1 can produce prominent negative component of the P wave and a secondary r'. This results in rsr' configuration in lead V_1.
- High placement of precordial leads may give false impression of poor progression of R wave.
- Interchange of leads V_1 and V_2 results in diminished amplitude of R wave and T wave inversion in lead V_2.
- Interchange of leads V_5 and V_6 results in increased voltage of R in V_6 than in V_5.
- Precordial lead placement over or below the breast in females does not produce significant change.
- Tremors can produce coarse undulations in baseline. These may resemble atrial flutter. However, these vibrations are confined to the lead attached to the involved limb and disappear during sleep.
- Coarse vibrations or noisy signals can simulate broad QRS tachycardia. Regular spikes of QRS help in differentiation.
- Electrical interference can produce high-frequency vibrations in the baseline. These may be confused with atrial fibrillation. QRS rhythm is, however, regular.
- Poor electrode contact and cable movement can produce a baseline shift.
- Erroneous recording at high speed or low speed can, respectively, simulate bradycardia or tachycardia. Such problems, however, affect duration of all components of the electrocardiogram, i.e., P wave, PR segment, QRS, and T wave.
- Underdamping results in increased amplitude of R wave or S wave.
- Overdamping results in loss of sharpness of onset of QRS and J point. It can produce false ST segment depression or elevation.

MCQs

Q1. Limb lead with green electrode is attached to:

(a) Right upper limb
(b) Left upper limb
(c) Left lower limb
(d) Right lower limb

Q2. Grounding electrode is attached to:

(a) Right upper limb
(b) Left upper limb
(c) Left lower limb
(d) Right lower limb

Q3. What is the effect on the electrocardiogram if the left lower limb electrode is placed on the left thigh?

(a) No change
(b) Right axis deviation
(c) T wave inversion in lead II
(d) Left axis deviation

Q4. Intercostal spaces are counted from:

(a) Suprasternal notch
(b) Sternal angle
(c) Clavicle
(d) Nipple

Q5. Intercostal space below the joint of manubrium sterni and body of the sternum is:

(a) First intercostal space
(b) Second intercostal space
(c) Third intercostal space
(d) Fourth intercostal space

Q6. Electrode of lead V_2 is placed over:

(a) Left half of the sternum in line of left 4th intercostal space
(b) Left parasternal region in left 4th intercostal space
(c) Left 4th intercostal space 2 inches lateral to left sternal border
(d) Left 4th intercostal space in mid-clavicular line

Q7. Electrode of lead V_3 is placed:

(a) Below lead V_2 electrode
(b) Lateral to lead V_2 electrode
(c) On the line joining lead V_2 and V_4 electrodes
(d) Medial to lead V_4 electrode

Q8. Electrode of lead V_6 is placed:

(a) In horizontal line from V_4 electrode
(b) In left 5th intercostal space
(c) In left 4th intercostal space
(d) In left 6th intercostal space

Q9. Electrode of lead V_5 is placed:

(a) In mid-clavicular line
(b) In anterior axillary line
(c) In mid-axillary line
(d) In posterior axillary line

Q10. Overdamping results in:

(a) Overshoot of the standardization signal
(b) Diminished amplitude of the standardization signal
(c) Increased amplitude of the standardization signal
(d) Rounding of the standardization signal

Q11. Gradual baseline shift can be due to:

(a) Patient movement
(b) Loss of electrode contact
(c) Muscular tremors
(d) Muscle tension

Q12. Vibrations in baseline can be due to:

(a) Unprepared skin
(b) Muscle tremor
(c) Selection of low-pass filter setting
(d) Cable movement

Q13. Respiration produces maximum change in:

(a) Lead I
(b) Lead III

(c) Lead aVR
(d) Lead aVL

Q14. Respiration can produce a beat to beat variation in:

(a) Heart rate
(b) Frontal plane QRS axis
(c) Frontal plane P wave axis
(d) Frontal plane T wave axis

Q15. Zero potential in lead III suggests:

(a) Interchange of left upper limb and right lower limb electrodes
(b) Interchange of right upper limb and right lower limb electrodes
(c) Interchange of right upper limb and left upper limb electrodes
(d) Interchange of right lower limb and left lower limb electrodes

Q16. Zero potential in lead II suggests:

(a) Interchange of right upper limb and right lower limb electrodes
(b) Interchange of left upper limb and left lower limb electrodes
(c) Interchange of right upper limb and left lower limb electrodes
(d) Interchange of left upper limb and right lower limb electrodes

Q17. Interchange of left upper limb and left lower limb electrodes produces:

(a) No change
(b) Abnormal Q wave in lead III
(c) Reversal of leads I and II
(d) Reversal of leads aVL and aVF

Q18. Interchange of right upper limb and right lower limb electrodes produces:

(a) Zero potential in lead II
(b) Abnormal Q wave in lead I
(c) Abnormal Q wave in lead aVL
(d) Positive R wave in lead aVR

Q19. Interchange of right upper limb and left lower limb electrode produces:

(a) Abnormal Q wave in leads I and II
(b) P wave inversion in leads II, III, and aVF
(c) Positive P, QRS, and T in lead aVR
(d) No change

Q20. Interchange of right upper limb and left upper limb electrodes produces:

(a) Negative P, QRS, and T in lead I
(b) Negative P, QRS, and T in lead aVL
(c) Positive P, QRS, and T in lead aVR
(d) No change in lead aVF

Q21. Interchanging of leads of right upper limb and left upper limb does not produce any change in lead:

(a) I
(b) aVR
(c) aVL
(d) aVF

Q22. Effects of interchanging of leads of right upper limb and left upper limb:

(a) P wave becomes positive in lead aVR
(b) T wave becomes inverted in lead aVL
(c) Change in horizontal plane QRS axis
(d) P wave in lead III becomes negative

Q23. Placing precordial electrodes one intercostal space higher can result in:

(a) Appearance of q wave in leads V_1 and V_2.
(b) Increase in the height of R wave in leads V_1 and V_2.
(c) Increase in the height of T wave in leads V_1 to V_4.
(d) Increase in amplitude of U wave in leads V_3, and V_4.

Q24. Placing chest electrodes, one intercostal space lower can result in:

(a) Decrease in the height of R waves in leads V_1 and V_2.
(b) Increase in the height of R waves in leads V_1 to V_3.
(c) Disappearance of q wave in leads V_1 and V_2.
(d) Decrease in amplitude of T wave in leads V_3 and V_4.

Q25. Placing chest electrode V_3 medial to the standard position can alter:

(a) R wave amplitude
(b) Frontal plane QRS axis
(c) Direction of depolarization
(d) U wave amplitude

Q26. Interchanging of leads of electrodes of lead V_1 and V_2:

(a) Decreases amplitude of QRS in lead V_2.
(b) Increases amplitude of T wave in lead V_2.
(c) Decreases amplitude of T wave in lead V_1.
(d) Decreases amplitude of QRS in lead V_1.

Answers

(1) c; (2) d; (3) a; (4) b; (5) b; (6) b; (7) c; (8) a; (9) b; (10) b, d; (11) a; (12) a, b; (13) b; (14) a, b; (15) a; (16) a; (17) b, c, d; (18) a, b, c, d; (19) a, b, c; (20) a, b, c, d; (21) d; (22) a, b; (23) a; (24) b, c; (25) a; (26) a

References

1. Mirvis DM, Goldberger AL. Electrocardiography. In: Libby P, Bonow RO, Mann DL, Zipes DP, editors. Braunwald's heart disease. Philadelphia: Saunders; 2008. p. 188.
2. Wagner GS, Lim TH. Recording the electrocardiogram. In: Wagner GS, editor. Marriot's practical electrocardiography. Philadelphia: Wolters Kulwer; 2008. p. 21–41.
3. Prystowsky EN, Padanilan BJ. Long term continuous electrocardiographic recording. In: Fuster V, O' Rourke RA, Walsh RA, Poole Wilson P, editors. Hurst's the heart. New York: McGraw Hill; 2008. p. 1058–61.
4. Castellanos A, Interian A Jr, Myerburg RJ. The resting electrocardiogram. In: Fuster V, O' Rourke RA, Walsh RA, Poole Wilson P, editors. Hurst's the heart. New York: McGraw Hill; 2008. p. 317.
5. Wegner W, Kligfreld P. Variability of precordial electrode placement during routine electrocardiography. J Electrocardiol. 1996;29:179–84.
6. Schijuenaars BJ, Kors JA, van Herpen G, et al. Effect of electrode positioning on ECG interpretation by a computer. J Electrocardiol. 1997;30:247–56.

2 Sinus Node Dysfunction

Normal cardiac depolarization starts from the sinoatrial node and traverses over intra-atrial conduction tissue to depolarize the atria. Depolarization of the sinoatrial node does not manifest on the surface electrocardiogram. Sinus node dysfunction is, therefore, evaluated indirectly by abnormalities in the rate and rhythm of the atrial depolarization. P waves of sinus origin are upright in lead II and slightly positive or biphasic in lead V_1.

1 Etiology

1. *Primary* sinoatrial node disease or intrinsic disease of the sinoatrial node:

   In elderly persons—ischemic heart disease and sclerodegeneration

   In younger persons—cardiomyopathy, myocarditis, amyloidosis, collagen disease, and surgery for congenital heart disease
2. Sinus node dysfunction *secondary* to causes other than structural disease of sinus node (extrinsic factors):

   Drugs—digitalis; beta-blockers; quinidine; procainamide; amiodarone; sotalol; diltiazem; verapamil; ivabradine; centrally acting sympatholytic antihypertensives, e.g., clonidine; and lithium

   Diseases—uremia, obstructive jaundice, myxedema, intracranial hypertension, hypokalemia, cervical or mediastinal tumors, gram-negative sepsis, and mental depression

   Vagotonic situations—trained athletes, hard manual workers, vomiting, vasovagal syncope, acute myocardial infarction, reperfusion due to thrombolytic agents, during sleep in elderly, carotid sinus hypersensitivity, and autonomic failure

   Sleep apnea.

   Some patients, especially elderly, may have a combination of extrinsic and intrinsic factors.

   In patients with primary sinoatrial node disease, the rest of the conduction system of the heart, e.g., intra-atrial conduction, the atrioventricular node, the His bundle, bundle branches, fascicles, and the Purkinje system, may be simultaneously involved (panconduction defect).

2 ECG Manifestations

2.1 Sinus Node Dysfunction

2.1.1 Failure of Impulse Generation [1]

1. *Sinus bradycardia*

   By definition, the P wave rate of less than 60/min is called sinus bradycardia (Fig. 2.1a–c). However rate as low as 50–60 beats per minute or even slower can be observed in

S. Mittal, *Insights into Electrocardiograms with MCQs*,
https://doi.org/10.1007/978-981-99-0127-2_2

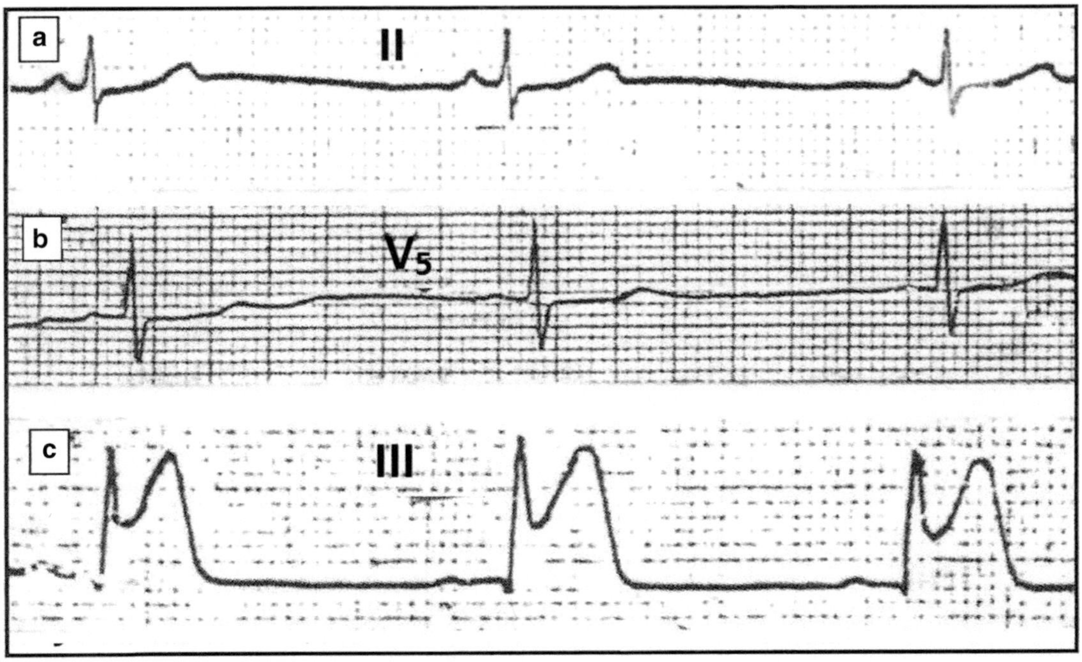

Fig. 2.1 (**a**) Lead II shows sinus bradycardia. (**b**) Lead V_5 showing sinus bradycardia with low voltage of QRS and T wave in a case of myxedema. (**c**) Lead III shows sinus bradycardia in a case of inferior myocardial infarction. (Reproduced with permission from: Mittal SR. Sinus Node Dysfunction. Cardiology Today 2011;15:221–224)

healthy persons during rest and sleep and does not necessarily indicate sinus node dysfunction. Rates as low as 40/min can be seen in training athletes during rest. Procedures like ocular pressure, carotid sinus massage, head-up tilt testing, and Valsalva maneuver affect sinus node rate through autonomic influence. Bradycardia during these procedures does not suggest intrinsic sinus node disease. A rate of less than 40/min without any apparent cause is certainly abnormal.

2. *Sinus arrhythmia*

   It is characterized by cyclic variation in sinus cycle length. Sinus arrhythmia without sinus bradycardia does not suggest sinus node dysfunction and needs further evaluation and correlation with symptoms. Respiratory sinus arrhythmia means an inspiratory increase in sinus rate and an expiratory decrease in sinus rate. It is frequent in children. Non-respiratory sinus arrhythmia is common in elderly but per se does not mean sinus node dysfunction.

3. *Wandering pacemaker*

   Because of the inability of the sinoatrial node to maintain an adequate rate, the pacemaker site gradually shifts from the sinoatrial node, across the atrium (abnormal P wave morphology), to the atrioventricular junction (P wave inverted in leads II, III, and aVF) and gradually back to the sinoatrial node. This phenomenon can occur in trained athletes (Fig. 2.2).

4. *Chronotropic incompetence*

   It means an inadequate increase in sinus rate during conditions like fever, intravenous injection of atropine, and adequate exercise. Inability of increase in sinus node activity to achieve at least 70% of age-predicted maximal heart rate during incremental dynamic exercise is considered abnormal and suggests sinus node dysfunction if there is no other reason for such a response (Fig. 2.3).

5. *Sinus pause*

   Irregular pauses of absence of P wave which are longer than the regular PP interval but are not exactly multiple of the regular

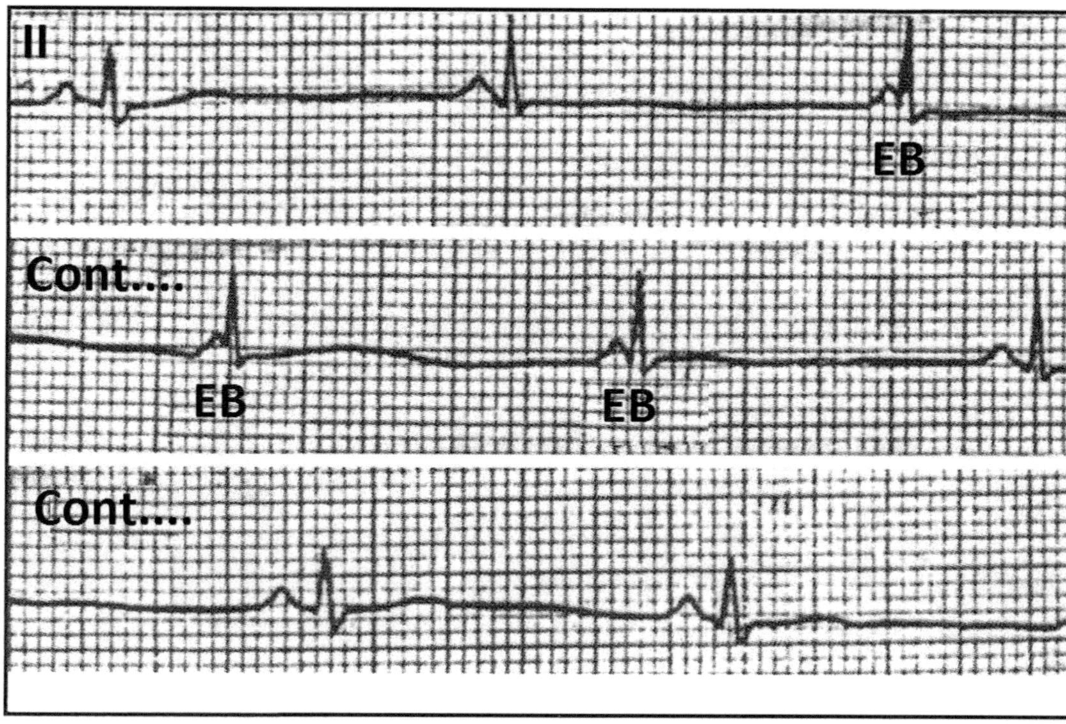

Fig. 2.2 Lead II showing sinus bradycardia with wandering pacemaker. First two beats show sinus rhythm. Third, fourth, and fifth beats are junctional escape beats (EB) with short PR interval because of delayed P wave. Sixth, seventh, and eighth beats are again sinus beats. (Reproduced with permission from: Mittal SR. Sinus Node Dysfunction. Cardiology Today 2011;15:221–224)

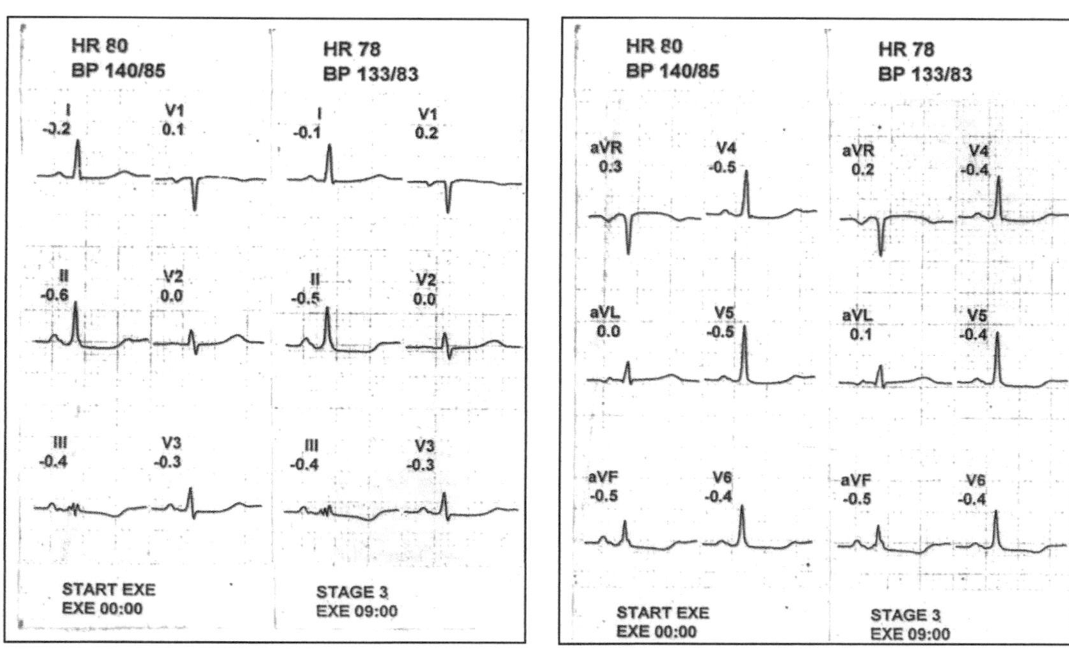

Fig. 2.3 Stress test recording showing failure of increase in heart rate from rest to stage 3. (Reproduced with permission from: Mittal SR. Sinus Node Dysfunction. Cardiology Today 2011;15:221–224)

sinus cycle length (PP interval) are called "sinus pause" [2] (Fig. 2.4a, b). Sinus pause of up to 2 s or more can be seen during sleep in adolescents and young adults. Pause for 2–3 s can occur during carotid sinus massage without meaning sinus node dysfunction. This results in sinus arrhythmia in the presence of sinus bradycardia. This type of sinus arrhythmia is not related to respiration (non-phasic sinus arrhythmia). Long sinus pauses can cause syncope. Sinus pause needs to be differentiated from pause produced by blocked atrial ectopic (Fig. 2.4c). Preceding T wave should be looked carefully for deformation by the blocked atrial ectopic.

A sinus pause of more than 3 s needs careful evaluation and correlation with symptoms.

6. *Increased sensitivity to suppression of impulse formation from drugs mentioned above (Fig. 2.5)*

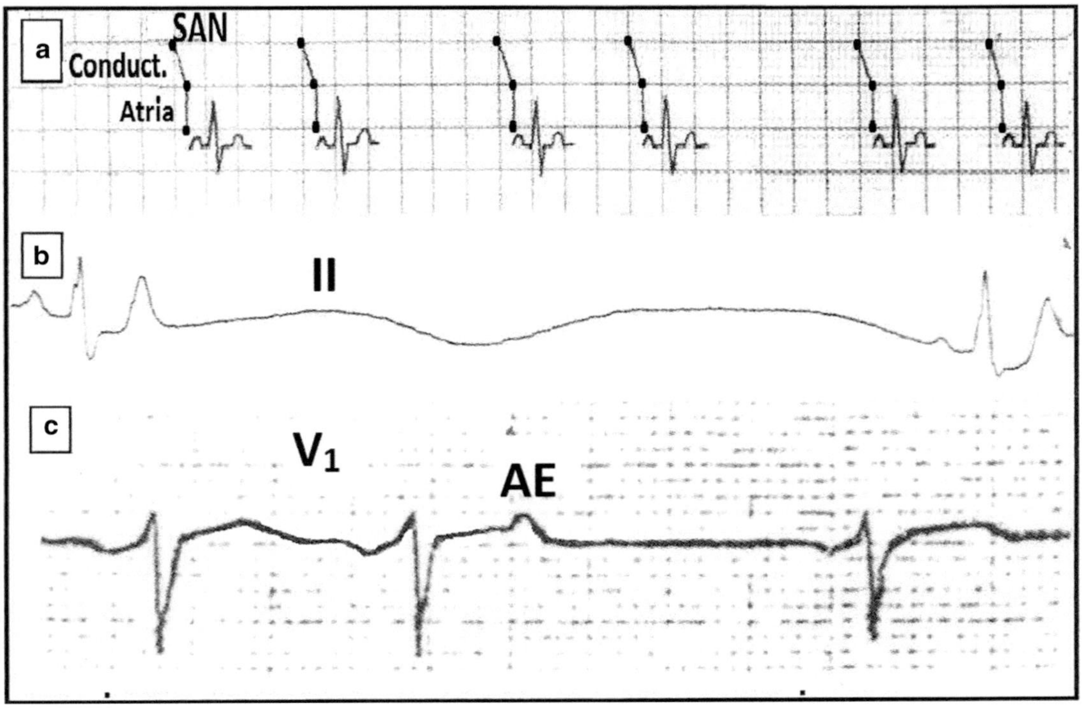

Fig. 2.4 (**a**) Ladder diagram showing the genesis of the ECG in sinus pause. (**b**) Lead II shows long sinus pause. It also shows prolonged PR interval due to concomitant atrioventricular nodal disease and broad QRS due to intraventricular conduction defect (pan-conduction defect). (**c**) Lead V_1 showing a sharp deflection in the T wave of second beat due to a blocked atrial ectopic (AE) followed by a pause. This pause should not be confused with sinus pause. (Reproduced with permission from: Mittal SR. Sinus Node Dysfunction. Cardiology Today 2011;15:221–224)

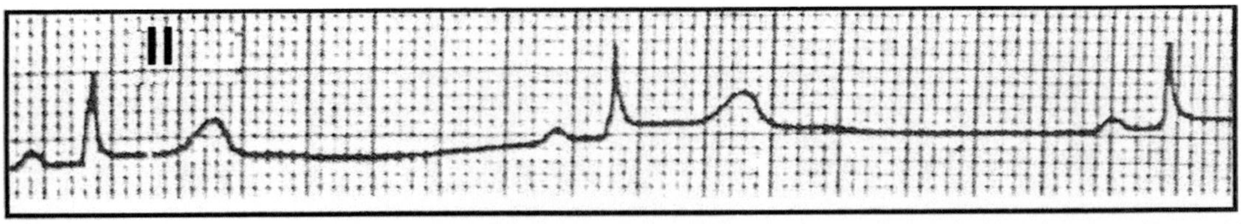

Fig. 2.5 Lead II shows marked sinus bradycardia following 10 mg of propranolol three times a day for control of anxiety. (Reproduced with permission from: Mittal SR. Sinus Node Dysfunction. Cardiology Today 2011;15:221–224)

7. *Long pause before appearance of sinus rhythm following:*
   - Supraventricular ectopic (Fig. 2.6a)
   - Atrioventricular nodal or ventricular ectopics with retrograde depolarization of the sinoatrial node (Fig. 2.6b)
   - Electric cardioversion for atrial flutter or fibrillation
   - Termination of atrioventricular nodal reentrant tachycardia by verapamil/diltiazem (Fig. 2.6c)

2.1.2 Failure of Impulse Transmission to Atria (Sinoatrial Exit Block)

The sinoatrial node consists of impulse generating cells in the center. Impulse generation in the sinoatrial node does not produce any manifestation on surface electrocardiogram. Impulse generating cells are surrounded by conducting cells, which conduct the impulse out to the atrial muscle and the internodal pathways. Involvement of these cells results in the sinoatrial exit block (Fig. 2.7). Like atrioventricular block, sinoatrial exit block can also be first-degree, second-degree, or complete sinoatrial exit block. First-degree and complete sinoatrial exit block cannot be diagnosed by surface ECG [3] (Fig. 2.7).

Second-degree sinoatrial exit block with Wenckebach phenomenon produces progressive shortening of the PP interval followed by the absence of one P wave. The pause is less than double of the previous PP cycle (Fig. 2.8a). Like atrioventricular block, the Wenckebach-type sinoatrial exit block can also produce bigeminy

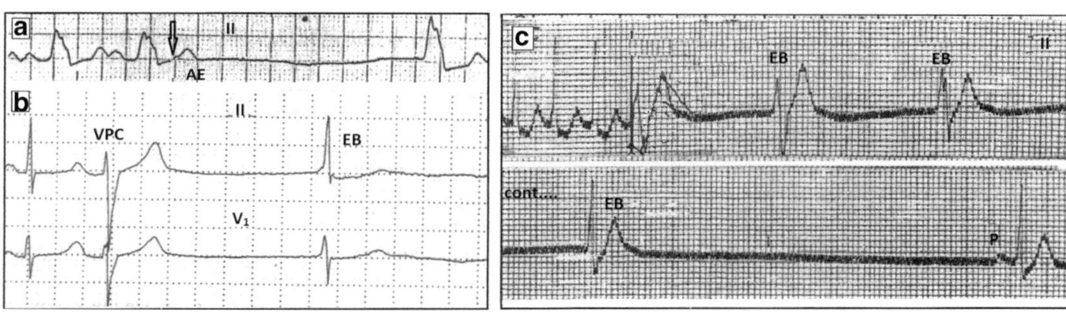

Fig. 2.6 (**a**) Lead II showing long sinus pause following an atrial ectopic (AE). Basal rhythm also shows prolonged PR interval and broad QRS. (**b**) Lead II and V_1 showing sinus pause and junctional escape rhythm (marked EB) after ventricular ectopic (VPC). (**c**) Lead II shows the termination of SVT with intravenous diltiazem. Termination of tachyarrhythmia is followed by three junctional escape beats (EB) and finally a sinus beat (P) after a long pause. (Reproduced with permission from: Mittal SR. Sinus Node Dysfunction Cardiology Today 2011;15:221–224)

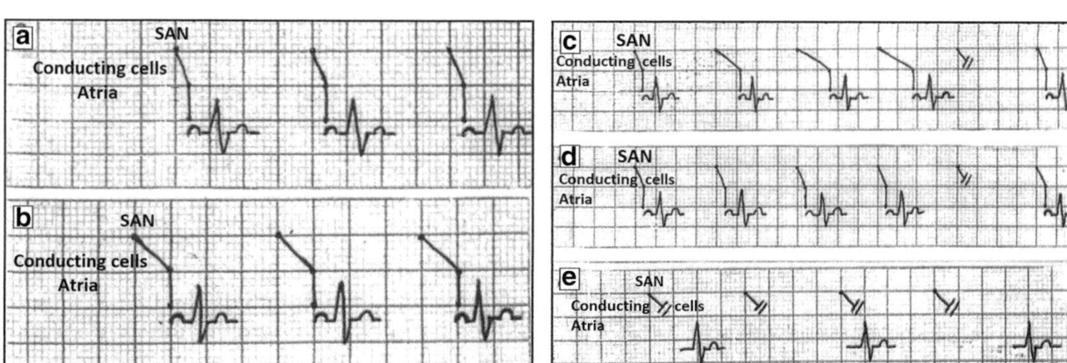

Fig. 2.7 Ladder diagram showing the genesis of ECG changes in the SA exit block. (**a**) Sinus rhythm. (**b**) 1° sinoatrial exit block. (**c**) Type I, second-degree sinoatrial exit block. (**d**) Type II, second-degree sinoatrial exit block. (**e**) Complete sinoatrial exit block. (Reproduced with permission from: Mittal SR. Sinus Node Dysfunction. Cardiology Today 2011;15:221–224)

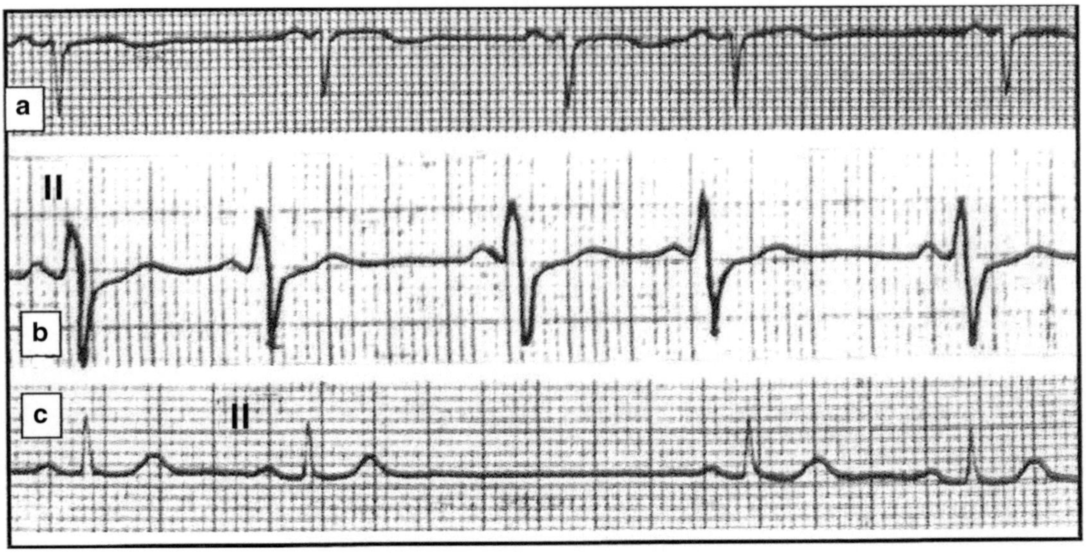

Fig. 2.8 (**a**) Lead II shows type I second-degree sinoatrial exit block. There is progressive shortening of PP interval followed by a pause which is less than twice of the preceding PP cycle. (**b**) Lead II shows bigeminy produced by type I second-degree sinoatrial exit block. It also shows left anterior hemiblock. (**c**) Lead II shows type II second-degree sinoatrial exit block. The pause produced by the dropped P wave is double of the preceding PP cycle. (Reproduced with permission from: Mittal SR. Sinus Node Dysfunction. Cardiology Today 2011;15:221–224)

(Fig. 2.8b). Second-degree sinoatrial exit block (Mobitz type II) results in regular PP interval with absence of one or more P waves. The pause so produced is exactly multiple of the regular PP cycle (Fig. 2.8c).

In patients with concomitant sinus arrhythmia, differentiation between sinus pause and sinoatrial exit block may be difficult.

2.2 Concomitant Intra-atrial Conduction Defect

Concomitant disease of the intra-atrial conduction tissue can be suspected by presence of any of the following findings:

(a) Abnormal P wave configuration—wide and notched or deformed P wave.

    Shape of P wave may change because of varying intra-atrial conduction (Fig. 2.9a).

(b) No or slow atrial escape rhythm during significant sinus bradycardia.

(c) Persistent atrial standstill—there is no spontaneous atrial activity. The ECG shows junctional bradycardia without any atrial activity (Fig. 2.9b). Such patients have a high risk of thromboembolism. It may be difficult to differentiate atrial standstill from fine atrial fibrillation.

(d) Hyperexcitability of atrial conduction tissue presenting as atrial fibrillation, atrial flutter, or atrial tachycardia. These rhythms may result in the following:

    - *Tachycardia-bradycardia syndrome.* It is an ill-sustained supraventricular tachyarrhythmia followed by long pause of asystole (Fig. 2.9c). This happens due to tachycardia-induced abnormal suppression of sinoatrial node and failure of lower pacemakers (e.g., atrioventricular node) to generate an escape rhythm. Long pause can produce syncope. Transient atrial fibrillation predisposes these persons to the risk of thromboembolism.
    - *Persistent atrial fibrillation, atrial flutter, or atrial tachycardia with slow ventricular rate.* It is due to concomitant atrioventricular nodal disease (Fig. 2.9d, e).

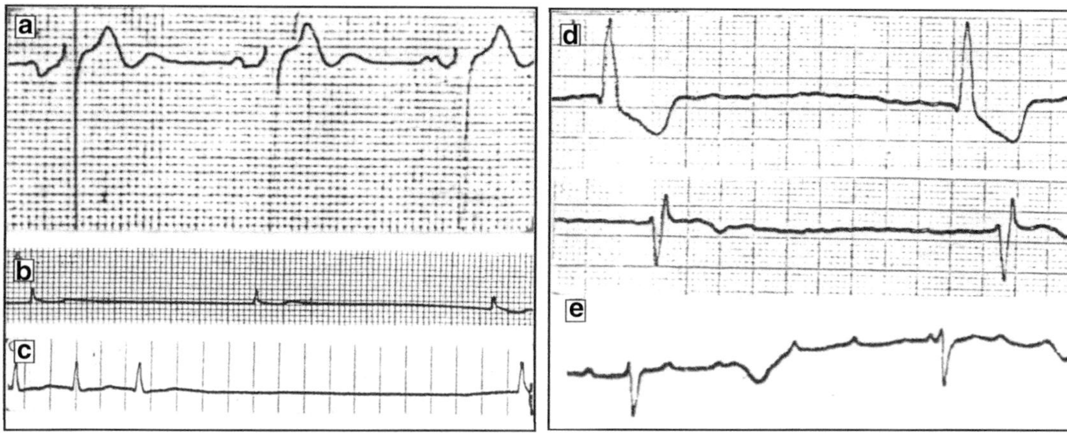

Fig. 2.9 (**a**) Lead V_2 showing changing configuration of P wave with bradycardia. (**b**) Lead II shows junctional bradycardia without any atrial activity (absence of any P wave). (**c**) Lead aVF showing three beats of atrial tachycardia followed by a long pause of asystole. (**d**) Leads II and V_1 showing atrial fibrillation with slow ventricular rate. (**e**) Lead V_1 showing atrial tachycardia with slow ventricular rate. QT interval is also prolonged. (Reproduced with permission from: Mittal SR. Sinus Node Dysfunction. Cardiology Today 2011;15:221–224)

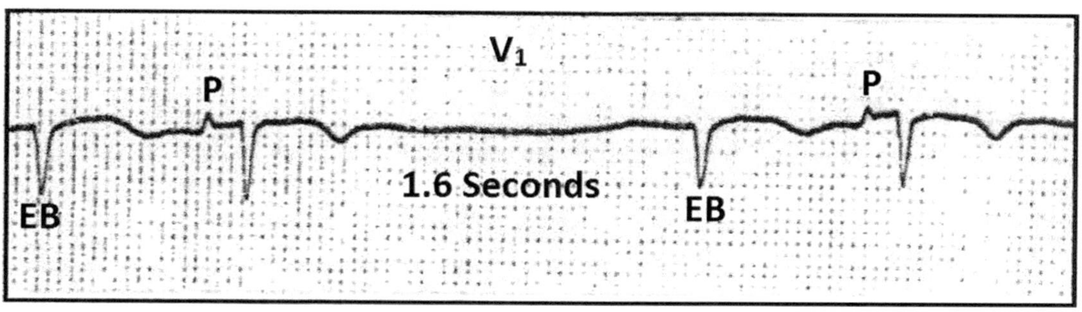

Fig. 2.10 Lead V_1 showing sinus bradycardia (P) with an alternate escape beat (EB). (Reproduced with permission from: Mittal SR. Sinus Node Dysfunction. Cardiology Today 2011;15:221–224)

2.3 Inadequate Subsidiary Pacemaker Activity

Normally a sinus pause of 1–2 s results in escape beat from a lower site capable of impulse formation. These escape beats can arise from the atria, the atrioventricular junction, or the ventricles. Failure of timely generation of escape beats from these subsidiary pacemaker sites occurs due to simultaneous involvement of the rest of the conduction system of the heart. Pause produced by failure of the lower pacemaker sites to produce escape beat timely can result in prolonged asystole with syncope.

2.4 Escape-Capture Bigeminy

Escape beat may be followed by a sinus beat if the next P wave falls after T wave of the escape beat. Depending on the sinus rate and rhythm, the sinus beat and escape beat may alternate producing escape-capture bigeminy (Fig. 2.10).

2.5 Concomitant Atrioventricular Block (Figs. 2.4b and 2.6a)

This may be first degree, second degree, or third degree. Atrioventricular block may be permanent or may manifest following a premature beat.

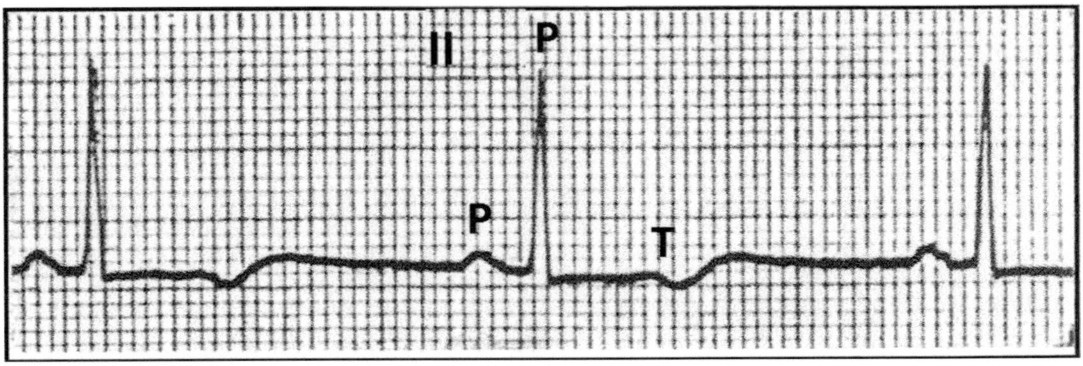

Fig. 2.11 Lead II shows sinus bradycardia with prolonged QT interval. (Reproduced with permission from: Mittal SR. Sinus Node Dysfunction. Cardiology Today 2011;15:221–224)

2.6 Concomitant Infra-His Conduction Defects (Figs. 2.4b, 2.6a, and 2.8b)

These can present as follows:

(a) Left anterior hemiblock
(b) Left posterior hemiblock
(c) Right bundle branch block
(d) Left bundle branch block

2.7 Repolarization Changes

Bradycardia may be accompanied by a prolonged QT interval (Fig. 2.11) and susceptibility to bradycardia-dependent ventricular tachyarrhythmia.

The term *sick sinus syndrome* should be used only when there are symptoms and signs that are definitely attributable to sinus node dysfunction. Prolonged electrocardiography recording and/or electrophysiologic study may be needed to confirm association of symptoms with electrocardiographic findings.

Summary

- Depolarization of the sinoatrial node does not manifest on surface electrocardiogram. The P wave is produced when the impulse comes out of the sinoatrial node and depolarizes the atria. Sinus node dysfunction on surface electrocardiogram is, therefore, evaluated indirectly by changes in rate and rhythm of the P waves of sinus origin.
- Sinus node dysfunction can be due to intrinsic disease of the sinoatrial node or may be secondary to extrinsic factors like drugs, systemic diseases, or vagotonia. Differentiation is important for management.
- Sinus rate between 45 and 60/min can be seen in healthy persons during sleep. Rates as low as 40/min can be present in trained athletes. These findings are due to increased vagal tone and do not suggest sinus node dysfunction. Procedures like ocular pressure, carotid sinus massage, head-up tilt testing, and Valsalva maneuver produce bradycardia due to autonomic influence and do not suggest intrinsic disease of the sinoatrial node. The unexplained sinus rate of 40/min or less needs further evaluation.
- Sinus arrhythmia without sinus bradycardia does not suggest sinus node dysfunction.
- Inability of the sinoatrial node to maintain an adequate rate can result in an intermittent shift of impulse generation to atria or atrioventricular junction (wandering pacemaker).
- Failure of adequate increase in sinus rate during dynamic exercise (chronotropic incompetence) suggests sinus node dysfunction, if there is no other reason to explain this finding.
- Absence of P wave for a duration longer than the regular PP interval and not having any mathematical relation with regular sinus cycle

length is called sinus pause. Sinus pause of up to 2 s or longer can occur in healthy persons during sleep. Sinus pause of more than 3 s needs further evaluation.

- First-degree sinoatrial exit block and complete sinoatrial exit block cannot be diagnosed by surface ECG.
- Progressive shortening of the PP interval followed by a pause which is less than double of the preceding PP cycle suggests Wenckebach-type (Mobitz type I) sinoatrial exit block.
- Absence of one or more P waves with a pause that is exactly multiple of the preceding PP interval suggests Mobitz type II sinoatrial exit block.
- Unduly long pause in the appearance of a sinus P wave after termination of supraventricular tachycardia, atrial flutter, or atrial fibrillation suggests sinus node dysfunction.
- Episodes of ill-sustained supraventricular tachycardia followed by a long pause (tachycardia-bradycardia syndrome) suggests sinus node dysfunction.
- Atrial flutter, atrial fibrillation, or atrial tachycardia with slow ventricular rate suggests concomitant disease of the atrioventricular node.
- The term "sick sinus syndrome" should be used only when there are symptoms and signs that are definitely attributable to sinus node dysfunction.

MCQs

Q1. What is the manifestation of activation of sinus node on the surface electrocardiogram?

(a) U wave
(b) U wave to P wave interval
(c) Initial part of P wave
(d) None

Q2. Which of the following does not produce sinus bradycardia?

(a) Salbutamol
(b) Beta-blockers
(c) Quinidine
(d) Digitalis

Q3. Which of the following is a common cause of sick sinus node in elderly?

(a) Coronary artery disease
(b) Amyloidosis
(c) Collagen diseases
(d) Sclerodegenerative process

Q4. Which of the following can produce sinus bradycardia even when the sinus node is normal?

(a) Orciprenaline
(b) Uremia
(c) Obstructive jaundice
(d) Increased parasympathetic drive

Q5. Ischemic involvement of the sinus node is less frequent in occlusion of:

(a) Left anterior descending coronary artery
(b) Proximal posterior descending coronary artery
(c) Proximal right coronary artery
(d) Proximal left circumflex coronary artery

Q6. Sinus bradycardia is diagnosed when the rate of sinus node is less than:

(a) 65/min
(b) 60/min
(c) 55/min
(d) 50/min

Q7. Sinus bradycardia is diagnosed when there are:

(a) P waves at a slow rate
(b) Narrow QRS at a slow rate without P wave
(c) Broad QRS at a slow rate without P wave
(d) Broad QRS at a slow rate with regular retrograde depolarization of atria

Q8. If there is no sinus beat, rhythm can arise from:

(a) Atria
(b) Atrioventricular node
(c) Ventricles
(d) Any of the above sites

Q9. When there is no sinus beat and impulse originates from a lower site, it is known as:

(a) Ectopic rhythm
(b) Premature beat
(c) Escape rhythm
(d) Capture beat

Q10. If there is a narrow QRS rhythm at a rate of 120/min and no P waves are visible, it suggests:

(a) Sinus node dysfunction
(b) Escape rhythm
(c) Accelerated rhythm
(d) None of above

Q11. Sinus bradycardia with low-voltage QRS and diffuse flattening of T waves suggests possibility of:

(a) Pericardial effusion
(b) Vagotonia
(c) Emphysema
(d) Myxedema

Q12. Sinus bradycardia with diffuse broad and inverted T waves and prolonged QT interval suggests possibility of:

(a) Increased intracranial pressure
(b) Well trained athlete
(c) Inferior myocardial infarction
(d) Neurocardiogenic syncope

Q13. Respiratory sinus arrhythmia:

(a) Means inspiratory increase in heart rate
(b) Means inspiratory decrease in heart rate
(c) Is seen in children
(d) Is common in elderly

Q14. When there is no P wave for a period which is multiple of preceding PP interval, it is known as:

(a) Sinus arrhythmia
(b) Sinoatrial exit block
(c) Sinus pause
(d) Sinus bradycardia

Q15. When there is no P wave for a period which is not multiple of preceding PP interval, it is called:

(a) Sinus arrhythmia
(b) Sinoatrial exit block
(c) Sinus pause
(d) Sinus bradycardia

Q16. Absence of P wave in the electrocardiogram can be due to:

(a) Sinus arrest
(b) Complete sinoatrial exit block
(c) Non-excitability of atrial muscle
(d) Block in atrioventricular node

Q17. Regular slow rate of P wave on electrocardiogram can be due to:

(a) Slow rate of sinus node
(b) Normal rate of sinus node with first-degree sinoatrial exit block
(c) Regular Mobitz type II sinoatrial exit block
(d) Sinus arrest

Q18. Progressive decrease in PP interval followed by a pause less than twice the preceding PP interval suggests:

(a) Sinus pause
(b) Mobitz type I, second-degree sinoatrial exit block
(c) Mobitz type II, second-degree sinoatrial exit block
(d) Sinus arrhythmia

Q19. Fixed PP interval followed by a pause which is twice the preceding PP interval suggests:

(a) Sinus pause
(b) Mobitz type I, second-degree sinoatrial block

(c) Mobitz type II, second-degree sinoatrial block
(d) Sinus arrhythmia

Q20. Complete absence of P wave for 2 s without any intervening QRS suggests:

(a) Sick sinus node
(b) Enhanced parasympathetic activity
(c) Atrioventricular block
(d) Diffuse disease of the cardiac conduction tissue

Q21. Lone atrial fibrillation with slow ventricular rate should suggest the possibility of:

(a) Sick sinus node
(b) Well-trained athlete
(c) Disease of atrial myocardium
(d) Diffuse disease of the cardiac conduction tissue

Q22. Which of the following suggests possibility of sick sinus?

(a) Resting heart rate of 56/min
(b) Resting heart rate of 50/min increasing to 70/min at the end of stage 4 of Bruce protocol
(c) Resting heart rate of 50/min increasing to 150/min at the end of the second stage of Bruce protocol
(d) Atrial fibrillation with a pause of 2 s

Q23. When the long pause between two P waves does not have a mathematical relation with preceding PP interval, it is called:

(a) First-degree sinoatrial exit block
(b) Complete sinoatrial exit block
(c) Second-degree sinoatrial exit block
(d) Sinus pause

Q24. Which of the following are not considered "sick sinus syndrome"?

(a) Sinus pause of 2 s followed by atrial fibrillation with fast ventricular rate
(b) Termination of paroxysm of atrioventricular nodal reentrant tachycardia followed by a sinus pause of 1 s
(c) Pause following a non-conducted atrial premature beat
(d) Pause following a ventricular premature beat

Q25. Which of the following suggests possibility of sinus node disease?

(a) Pulse rate dropping to 40/min from 70/min after 10 mg of propranolol
(b) Pulse rate dropping to 45/min during sleep
(c) Pulse rate of 45/min during episode of severe nausea
(d) Pulse rate of 45/min during carotid sinus massage

Q26. Which of the following does not suggest diagnosis of sinus node disease?

(a) Sinus bradycardia during release phase of Valsalva maneuver
(b) Sinus bradycardia during ocular pressure
(c) Sinus bradycardia in a case of increased intracranial pressure
(d) Sinus bradycardia during the table tilt test

Q27. Which of the following does not suggest sinus node disease?

(a) Failure of the sinus rate to increase after intravenous injection of atropine
(b) Atrioventricular junctional rhythm at a rate of 40/min
(c) Bradycardia during expiration
(d) Sinus bradycardia with atrioventricular block

Q28. Which of the following does not suggest sinus node disease?

(a) Atrial fibrillation with slow ventricular rate
(b) Atrial flutter with slow ventricular rate
(c) Atrial fibrillation with slow ventricular rate in a patient receiving Amiodarone
(d) Failure to resume sinus rhythm after electric cardioversion for atrial flutter

Q29. Which of the following does not suggest sinus node disease?

(a) Atrial rhythm at a rate of 50/min
(b) Junction rhythm at a rate of 40/min
(c) Pause of more than 2 s following an atrial premature beat
(d) Compensatory pause following a ventricular premature beat

Q30. In a case with sinus bradycardia at rest, if the rate of P wave suddenly doubles with effort or on injection of atropine, it suggests:

(a) First-degree sinoatrial exit block
(b) Second-degree sinoatrial exit block
(c) Complete sinoatrial exit block
(d) Atrioventricular block

Q31. If T wave preceding a pause in the PP cycle is deformed, it suggests:

(a) Sinoatrial exit block
(b) Sinus pause
(c) Blocked atrial ectopic
(d) Blocked ventricular ectopic

Q32. Which conditions cannot be diagnosed by surface ECG?

(a) First-degree sinoatrial exit block
(b) Complete sinoatrial exit block
(c) Sinus arrest
(d) Non-respiratory sinus arrhythmia

Answers

(1) d; (2) a; (3) d; (4) b, c, d; (5) a, b; (6) b; (7) a; (8) d; (9) c; (10) c; (11) d; (12) a; (13) a, c; (14) b; (15) c; (16) a, b, c; (17) a, c; (18) b; (19) c; (20) b, d; (21) d; (22) b, d; (23) d; (24) b, c, d; (25) a; (26) d; (27) c; (28) c; (29) d; (30) b; (31) c; (32) a, b, c

References

1. Vijayaraman P, Ellenbogen KA. Bradyarrhythmias and pacemakers. In: Fuster V, O' Rourke RA, Walsh RA, Poole Wilson P, editors. Hurst's the heart. New York: McGraw Hill; 2008. p. 1020–54.
2. Wagner GS, Wang TY. Decreased automaticity. In: Wagner GS, editor. Marriot's practical electrocardiography. Philadelphia: Wolters Kluwer; 2008. p. 391–9.
3. Schamroth L. Structural nodal disease: the so called sick sinus syndrome. In: Schamroth C, editor. An introduction to electrocardiography. Delhi: Oxford; 1990. p. 431–7.

3 The P Wave and the PR Segment

The P wave represents depolarization of the atria. Abnormalities in the P wave give information about the site of impulse generation and about the abnormalities of conduction within the atria. The PR segment extends from the end of the P wave to the beginning of the QRS complex. It represents conduction across the atrioventricular node and the His-Purkinje system.

1 Normal Sinus P Wave (Atrial Depolarization) [1]

The P wave is a composite deflection of the right and the left atrial depolarization. As sinoatrial node is situated in the right atrium, the right atrium depolarizes first, and the left atrium depolarizes later (Fig. 3.1a). However, in normal persons, the time difference in the depolarization of the two atria is not significant with the result that the P wave is smooth and rounded in lead II. Upstroke is formed by the right atrial depolarization, and downstroke is formed by the left atrial depolarization. In frontal plane the activation of the two atria is directed inferiorly and to the left toward the positive pole of lead II. Therefore the P wave is positive and most clear in lead II. In horizontal plane, the right atrial depolarization is anterior (toward lead V_1), and the left atrial depolarization moves posteriorly (away from lead V_1) (Fig. 3.1b). This results in a biphasic P wave in lead V_1. Initial deflection

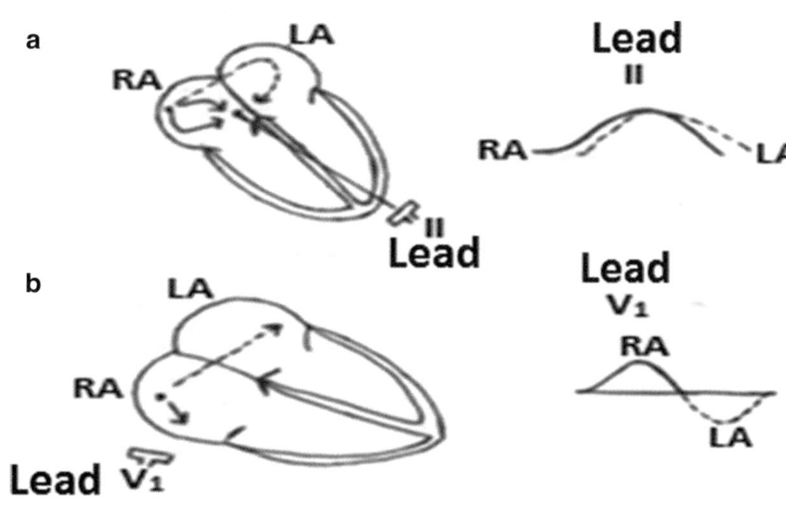

Fig. 3.1 (**a**) Atrial depolarization vector in the frontal plane. (**b**) Atrial depolarization vector in the horizontal plane. Abbreviations: *RA* right atrium, *LA* left atrium. (Reproduced with permission from: Mittal SR. P wave and The PR segment. Cardiology Today 2011;15:263–269)

S. Mittal, *Insights into Electrocardiograms with MCQs*,
https://doi.org/10.1007/978-981-99-0127-2_3

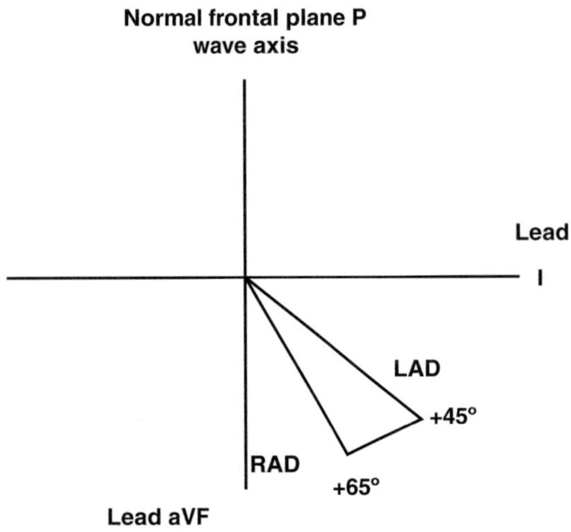

Fig. 3.2 Normal frontal plane P wave axis. Abbreviations: *LAD* left axis deviation, *RAD* right axis deviation. (Reproduced with permission from: Mittal SR. P wave and The PR segment. Cardiology Today 2011;15:263–269)

due to anteriorly directed right atrial depolarization is positive, and posteriorly directed left atrial depolarization produces terminal negative deflection.

Normal P wave parameters:

- Duration—less than 0.12 s, i.e., three small squares at the recording speed of 25 mm/s.
- Amplitude—less than 0.2 mV, i.e., two small squares at usual recording of 10 mm/mV.
- Frontal plane P wave axis—+45° to +65° (Fig. 3.2). Sinus tachycardia deviates frontal plane P wave axis to the right by about 10–15°.
- Inverted in lead aVR.
- Inverted or upright in lead aVL depending on the frontal plane P wave axis.
- In lead V_1 terminal negative deflection is less than 0.1 mV (one small square) in depth and less than 0.03 s (one small square) in width.

2 Atrial Repolarization (Ta Wave) [2]

It is a low-amplitude wave with a polarity against that of the P wave. It is usually not seen because of low amplitude and superimposition by the QRS complex. It may be visible during atrioventricular block (Fig. 3.3).

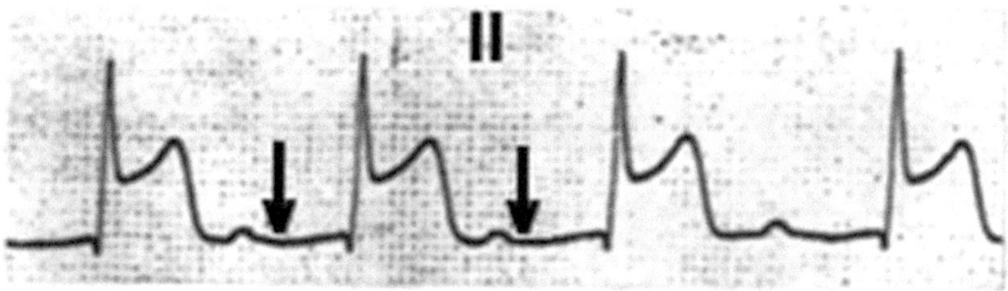

Fig. 3.3 Lead II in a case of inferior myocardial infarction and first-degree atrioventricular block showing the atrial repolarization wave (Ta wave) (arrow). (Reproduced with permission from: Mittal SR. P wave and The PR segment. Cardiology Today 2011;15:263–269)

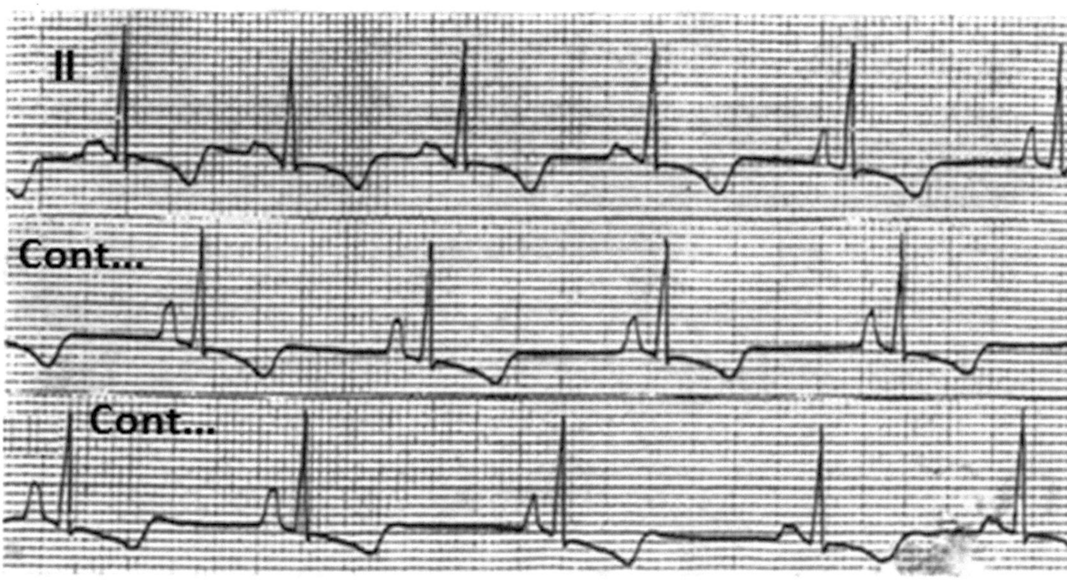

Fig. 3.4 Continuous strip of lead II shows M-shaped and tall peaked P waves in the same patient over a short period due to changing intra-atrial conduction defect. (Reproduced with permission from: Mittal SR. P wave and The PR segment. Cardiology Today 2011;15:263–269)

3 Atrial Depolarization Abnormalities [2, 3]

Depolarization abnormality is a better term than enlargement because such abnormalities can also occur in atrial overload and intra-atrial conduction defects (Fig. 3.4) without any echocardiographic evidence of atrial enlargement. Such abnormalities may also be transient during acute left ventricular failure or acute pulmonary embolism.

3.1 Left Atrial Depolarization Abnormality [2–4]

Left atrial depolarization abnormality manifests as follows:

- Frontal plane P wave axis of less than +45°.
- P wave duration of 0.12 s or more.
- Notched, bifid P wave (M-like appearance) with a distance of more than 0.04 s between two peaks (Fig. 3.5a) (sensitivity 15%, speci-

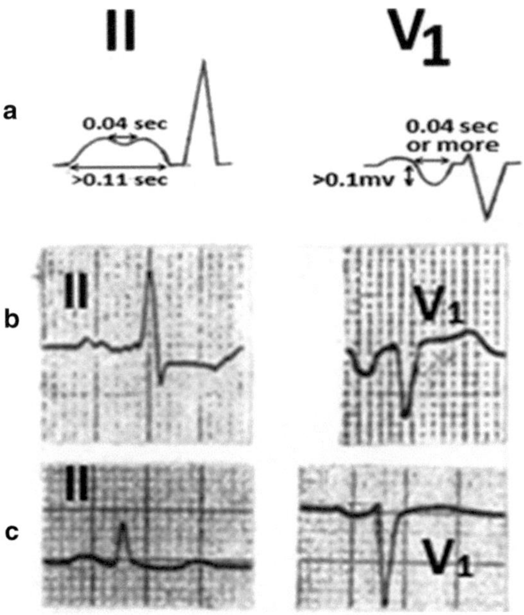

Fig. 3.5 (**a**) Diagram showing ECG features of the left atrial depolarization abnormalities. (**b**) Lead II shows M-shaped P wave. Lead V_1 shows deep and broad terminal negative deflection of the P wave. (**c**) Lead II shows only flattening of the top of the P wave, and lead V_1 shows increased duration of terminal negative deflection of the P wave. (Reproduced with permission from: Mittal SR. P wave and The PR segment. Cardiology Today 2011;15:263–269)

ficity 100%) in lead II, V_5, and V_6. As such P waves are frequently seen in mitral valve disease; they are also called "P mitrale."

- Negatively directed portion of P wave in lead V_1 is of 0.04 s or more in duration (Fig. 3.5b) (sensitivity 83%, specificity 80%) and 0.01 mV or more in amplitude (sensitivity 60%, specificity 93%). At times top P wave in lead II may not show a clear notch but only flattening (Fig. 3.5c). Similarly the negative terminal component in lead V_1 may not be deep but may be only broad (Fig. 3.5c).

3.2 Right Atrial Depolarization Abnormality [1, 3, 4]

It is characterized by the following:

- Rightward shift of the P wave axis to around +75°. Sinus tachycardia also tends to deviate the frontal plane P wave axis to right by about 10–15°.
- Peaked (pointed) P wave in leads II and aVF (Fig. 3.6a, b). There is no increase in duration of the P wave. Amplitude may not be more than 0.2 mV. Although amplitude criterion has high specificity, sensitivity is low. As such P waves are commonly seen in cases of cor pulmonale; they are also called "P pulmonale."
- In patients with gross right atrial enlargement, terminal negative deflection of P wave in lead V_1 may also be prominent, giving the wrong impression of concomitant left atrial enlargement. This is more likely to occur in patients with vertical heart as in emphysema. In vertical heart conventional lead V_1 remains higher in relation to the right atrium, with the result that the right atrial activation moves away from lead V_1 (Fig. 3.7a, b).
- However, unlike left atrial enlargement, the downstroke of P wave is rapid with an intrinsicoid deflection (time from beginning of the P wave to its peak) of less than 0.03 s. Also, there is no evidence of biatrial enlargement in lead II. Lead V_2 frequently shows a tall and pointed P wave (Fig. 3.8a, b).
- qR in lead V_1. In patients with gross right atrial enlargement with right ventricular displacement lead, lead V_1 may show a small q wave before a prominent R wave [4]

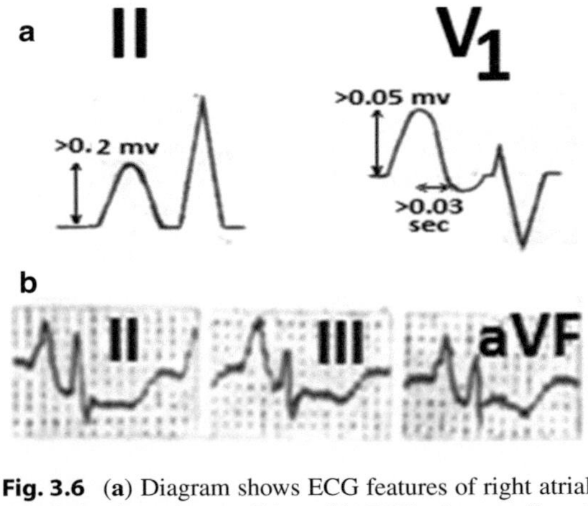

Fig. 3.6 (**a**) Diagram shows ECG features of right atrial depolarization abnormalities. (**b**) ECG shows tall and peaked P waves in leads II, III, and aVF. (Reproduced with permission from: Mittal SR. P wave and The PR segment. Cardiology Today 2011;15:263–269)

(Fig. 3.9a). This occurs because of the anatomical shift of RV by an enlarged and dilated RA with the result that lead V_1 records potential of the proximal basal region of RV (Fig. 3.9b).

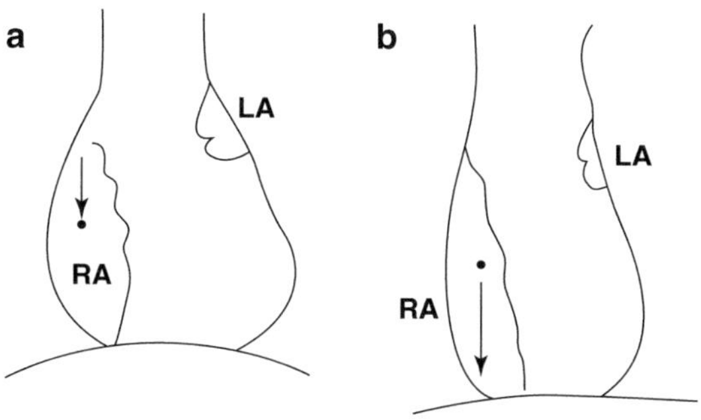

Fig. 3.7 Diagram showing relation of lead V_1 to right atrial depolarization in normal (**a**) and vertical heart (**b**). Abbreviations: *RA* right atrium, *LA* left atrium. (Reproduced with permission from: Mittal SR. P wave and The PR segment. Cardiology Today 2011;15:263–269)

3.3 Biatrial Depolarization Abnormality [4]

It manifests as follows:

– Lead II shows broad and bifid P wave with initial component taller than terminal component. Lead V_1 shows P wave with tall positive initial component followed by deep, wide, and delayed terminal component (Fig. 3.10).

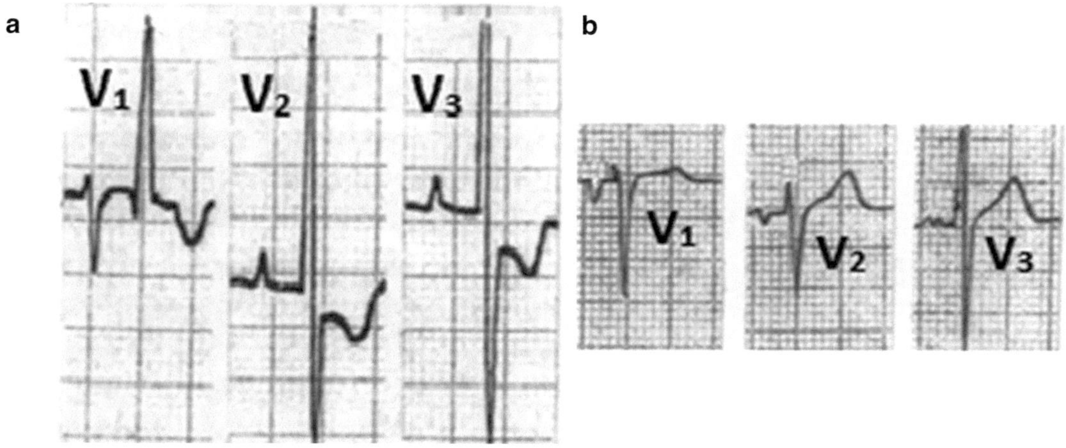

Fig. 3.8 (**a**) Leads V_1 to V_3 in right atrial depolarization abnormality. (**b**) Leads V_1 to V_3 in left atrial depolarization abnormality. (Reproduced with permission from: Mittal SR. P wave and The PR segment. Cardiology Today 2011;15:263–269)

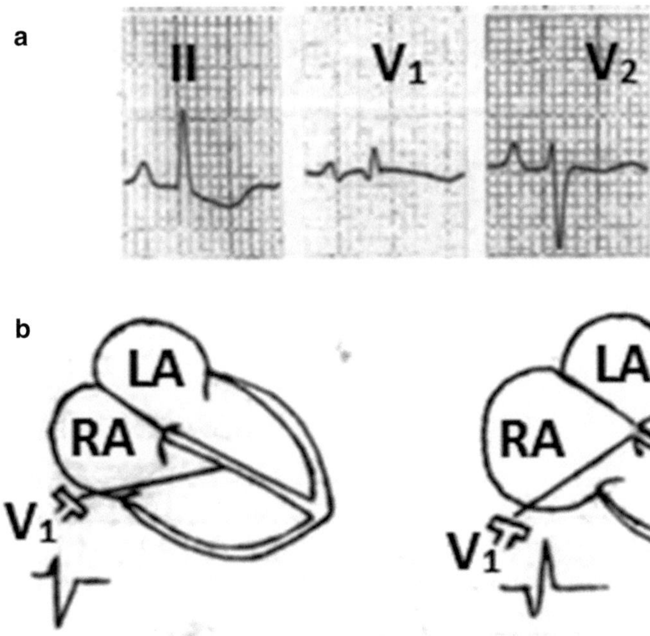

Fig. 3.9 (**a**) ECG shows qR in lead V_1. (**b**) Diagram shows effect of right atrial dilatation on lead V_1. Abbreviations: *RA* right atrium, *LA* left atrium. (Reproduced with permission from: Mittal SR. P wave and The PR segment. Cardiology Today 2011;15:263–269)

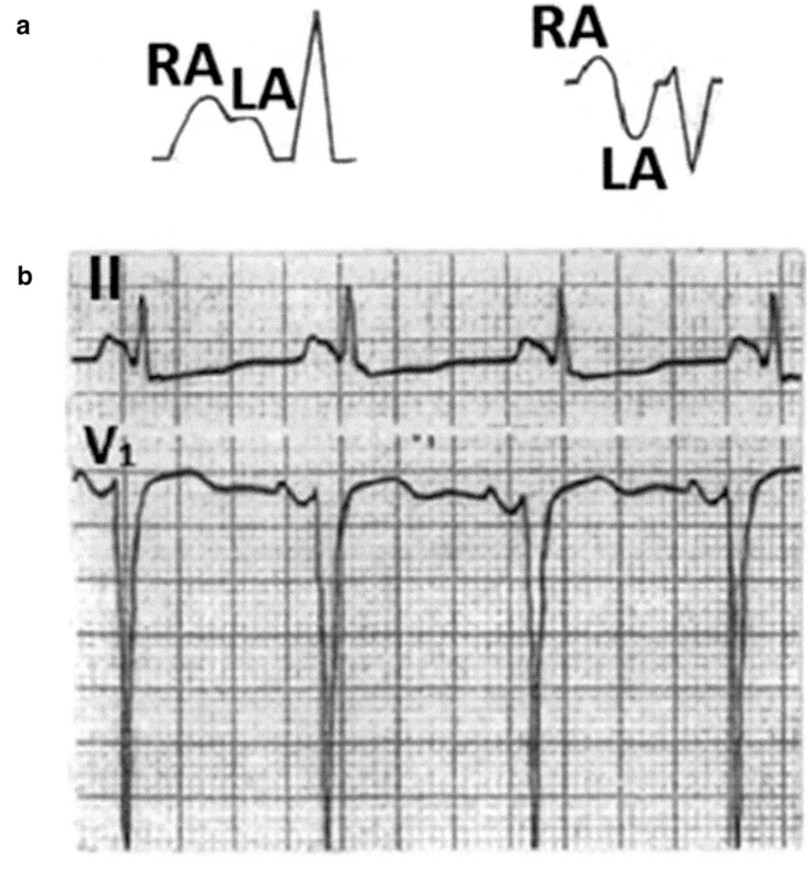

Fig. 3.10 (**a**) Diagram showing ECG features of biatrial enlargement. (**b**) ECG showing features of biatrial enlargement. Abbreviations: *RA* right atrium, *LA* left atrium. (Reproduced with permission from: Mittal SR. P wave and The PR segment. Cardiology Today 2011;15:263–269)

4 Interatrial Block

Partial interatrial conduction block results in delayed activation of the left atrium. This results in widening of P wave. P wave in lead II has two humps—first by depolarization of the right atrium and second by depolarization of the left atrium (Fig. 3.11a). This pattern resembles pattern produced by the left atrial enlargement. This pattern, therefore, suggests partial interatrial conduction block only if there is no other evidence of left atrial enlargement. P wave morphology in lead V_1 is, however, normal as opposed to prominent negative terminal deflection in cases with true left atrial enlargement.

In the presence of advanced interatrial block, the sinus impulse traverses through the right atrium up to the atrioventricular node. Subsequently the impulse traverses upward to depolarize the left atrium. This results in a biphasic P wave in lead II. Initial impulse traversing toward the atrioventricular node (toward positive pole of lead II) produces a positive deflection, whereas terminal retrograde conduction in the left atrium (away from positive pole of lead II) produces terminal negative deflection (Fig. 3.11b).

Interatrial block predisposes to atrial fibrillation, other supraventricular arrhythmias, left atrial thrombi, and systemic embolism.

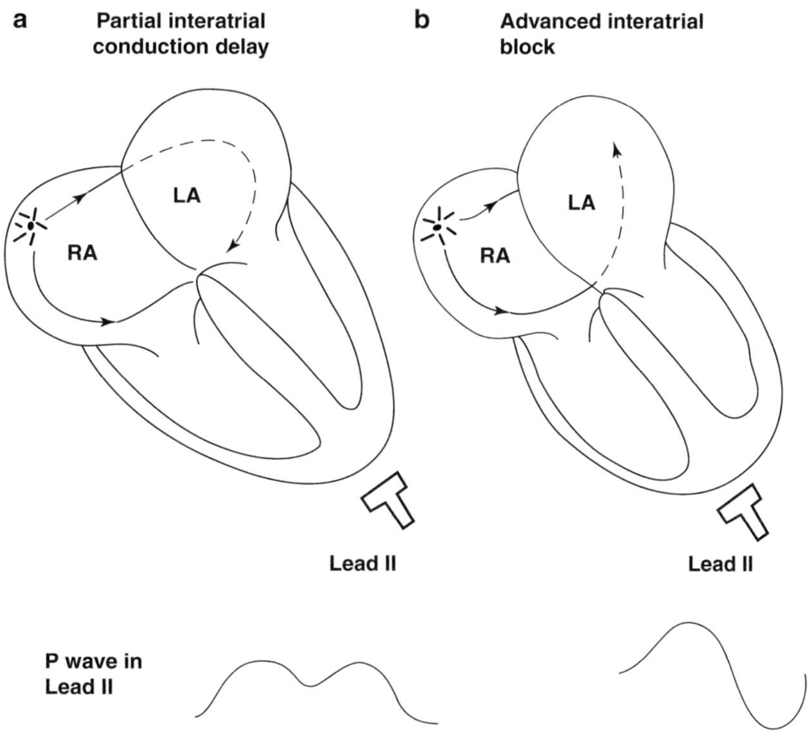

Fig. 3.11 (**a**) Showing effect of partial interatrial conduction delay on P wave of lead II. (**b**) Showing the effect of the advanced interatrial block on P wave of lead II. Abbreviations: *RA* right atrium, *LA* left atrium

5 Atrial Ectopics (Atrial Premature Beats, APB) [4, 5]

These are characterized by the following:

- Premature and abnormally appearing P wave (Fig. 3.12a, b). A very premature P wave may fall in the T wave of previous sinus beat and may be missed unless looked for very carefully. A little tall, more pointed, or slightly notched preceding T wave (Fig. 3.12c, d) as compared to T waves of other sinus beats should raise suspicion of the P wave of atrial ectopic merged in the T wave. Atrial ectopic arising in close proximity of sinoatrial node may produce a P wave similar to the sinus beats.
- Negative P wave in inferior leads II, III, and aVF suggests an origin in the lower atrium [2] (Fig. 3.13a). Left atrial origin of ectopic rhythm produces a P wave inversion in leads V_5 and V_6 in addition to inverted p wave in leads II, III, and aVF [4] (Fig. 3.13b). Uncommonly, left atrial rhythm may produce "dome and dart" appearance in lead V_1. There is initial slow upstroke (dome like) followed by a sharp upstroke (dart). Atrial ectopics could arise from more than two sites (multifocal) (Fig. 3.13c).
- PR interval of atrial ectopic beats is usually slightly longer than the PR interval of normal sinus beat due to relative refractoriness of the atrioventricular node (Fig. 3.14a). If atrial premature beat occurs so early that the atrioven-

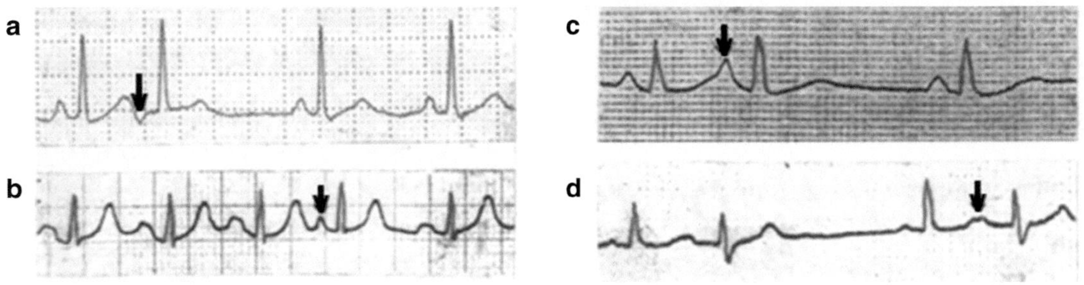

Fig. 3.12 (**a**, **b**) Atrial ectopic showing abnormal appearances of the P wave (arrow). (**c**) Atrial ectopic producing pointed T wave (arrow). (**d**) Atrial ectopic producing notched T wave (arrow). (Reproduced with permission from: Mittal SR. P wave and The PR segment. Cardiology Today 2011;15:263–269)

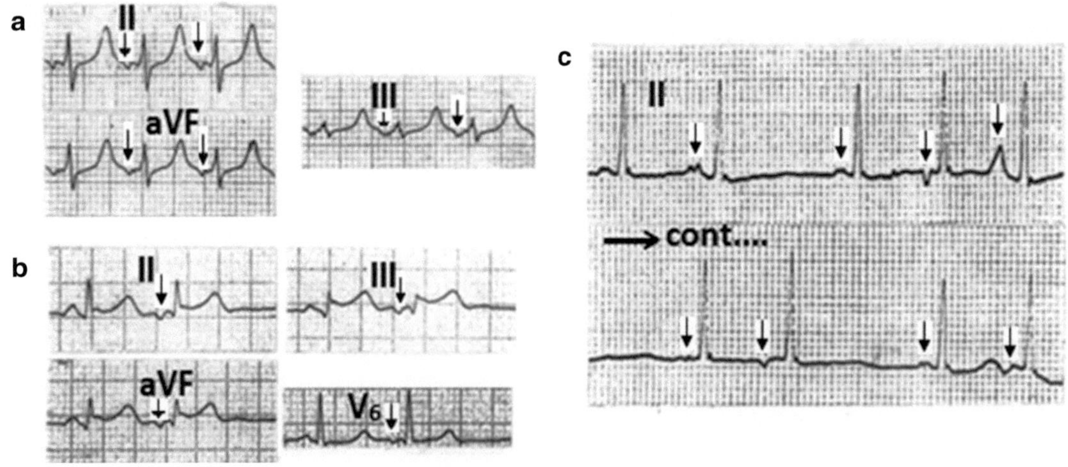

Fig. 3.13 (**a**) Low atrial rhythm (arrow). (**b**) Atrial ectopic arising from the left atrium (arrow). (**c**) Multifocal atrial ectopics (arrow). (Reproduced with permission from: Mittal SR. P wave and The PR segment. Cardiology Today 2011;15:263–269)

tricular node is in its absolute refractory period, atrial premature beat may not be conducted to the ventricles with the result that there is no QRS complex following the atrial premature beat (blocked or non-conducted atrial premature beat) (Fig. 3.14b). Non-conducted atrial premature beats may occur in a bigeminal pattern (Fig. 3.14c). ECG so produced may be confused with sinus bradycardia or sinoatrial exit block specially if the blocked atrial premature beat is merged in the preceding T wave and is not identified. If the atrial ectopic beat arises from low atrium, PR interval can be short.

- QRS complex following an atrial premature beat is usually similar to that of normal sinus beat because of normal conduction through ventricular conduction system (Fig. 3.12a, b). If the atrial premature beat is significantly premature or the refractory period of the bundle branches or fascicles is prolonged, atrial premature beat may find the ventricular conduction system in partial refractory state. This results in slow ventricular conduction through one or other bundle branch or fascicle resulting in abnormal QRS-aberrant conduction.
- Differentiation of atrial premature beat with aberrant ventricular conduction from premature ventricular beats [6].
  - Identification of a preceding P wave supports aberrant conduction. It may be difficult to identify if it falls in a preceding T wave.
  - Initial deflection of the QRS identical to sinus beat supports aberrancy (Fig. 3.12a, b).
  - If the broad QRS has right bundle branch block configuration, rsR' configuration in lead V_1 and qRS configuration in lead V_6 support aberrant conduction (Fig. 3.15a).

a

b

c

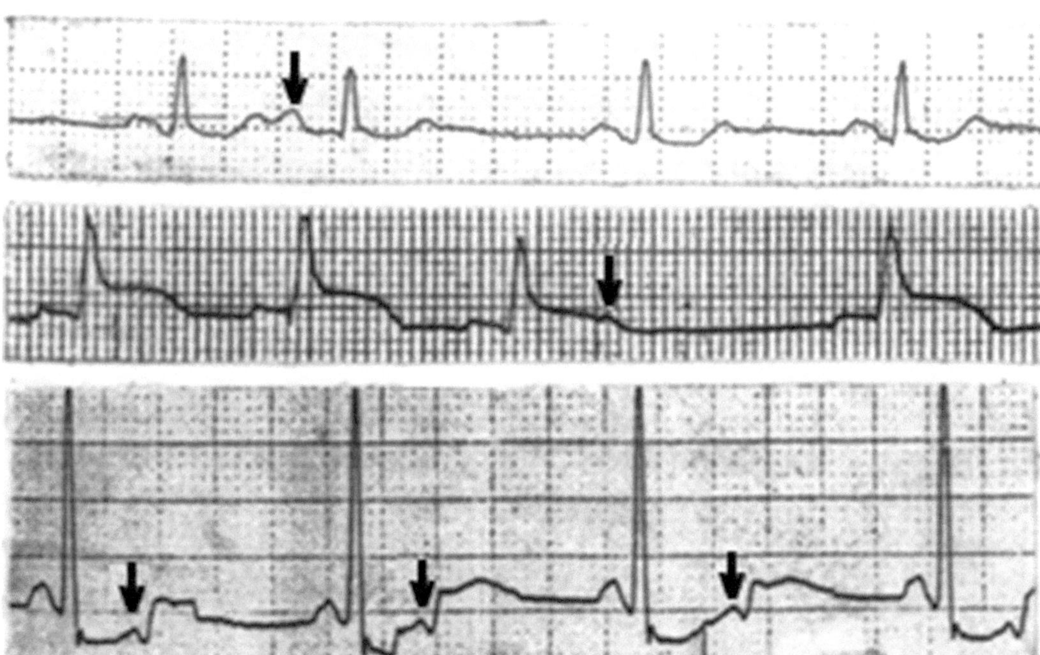

Fig. 3.14 (**a**) Atrial ectopic conducted with prolonged PR interval (arrow). (**b**) Non-conducted atrial premature beat (arrow). (**c**) Non-conducted atrial premature beats (arrows) occurring in bigeminal pattern. (Reproduced with permission from: Mittal SR. P wave and The PR segment. Cardiology Today 2011;15:263–269)

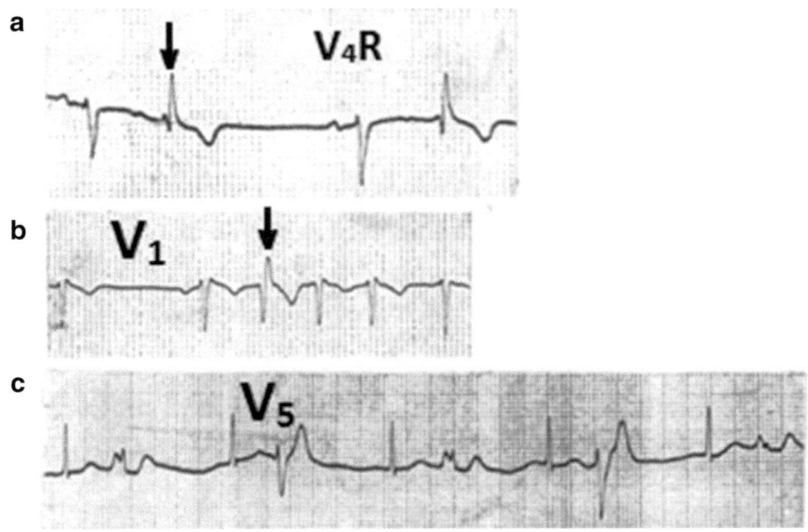

Fig. 3.15 (**a**) Lead V_4R showing atrial ectopic with rsR' pattern due to aberrant conduction (arrow). (**b**) Lead V_1 showing atrial ectopic with RBBB-type aberration in the second beat (arrow) in a cluster of four beats. (**c**) Lead V_5 showing junctional ectopics with alternate left and right bundle branch block pattern due to aberrancy. (Reproduced with permission from: Mittal SR. P wave and The PR segment. Cardiology Today 2011;15:263–269)

- Uncommonly aberrant conduction occurs over left bundle branch specially in patients with cardiac disease.
- If there is a cluster of ectopic beats, then broadening of QRS in the second beat of cluster supports aberrancy (Fig. 3.15b).
- Long QRS cycle followed by a short cardiac cycle with broad QRS (long-short sequence). This finding is not helpful as it can occur in aberrant conduction as well as in ventricular ectopic. However, absence of a long preceding cycle supports ventricular ectopic activity.
- In an irregular rhythm, if an even longer cycle, followed by an even shorter cycle is conducted with a normal QRS, broad QRS beats are suggestive of ventricular ectopics.
- Varying degrees of change in QRS depending on the length of the preceding RR cycle supports aberrancy.
- Alternating patterns of bundle branch block separated by single normal beat support possibility of bilateral aberrancy rather than ventricular ectopic arising alternately from the two the ventricles (Fig. 3.15c).

Aberrant conduction is rate dependent. Usually it is tachycardia dependent, i.e., it occurs with a short cycle length (Fig. 3.15a). Uncommonly, it can be bradycardia dependent.

6 Post-Ectopic Pause [4]

Pause following an atrial premature beat is longer than the normal PP cycle (Figs. 3.12a, b and 3.16a) because it includes the following:

- Time for the atrial impulse to enter the sinoatrial node and depolarize it
- Normal sinoatrial node cycle length
- Exit time for the impulse to come out of the sinoatrial node to the atrium

However, this pause is less than double of the normal PP cycle length (incomplete compensatory pause). This pause, however, may be variable due to several factors:

- Underlying sinoatrial node dysfunction. If the post-ectopic pause is longer than 25% of the

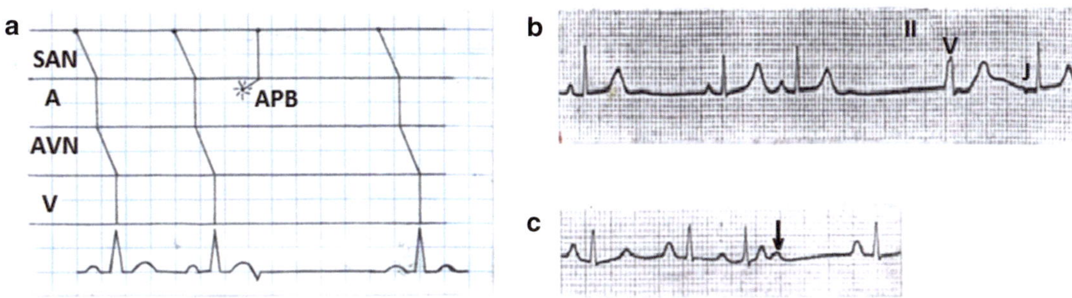

Fig. 3.16 (**a**) Ladder diagram showing the mechanism of incomplete compensatory pause in atrial ectopic. (**b**) Atrial ectopic beat (APB) followed by long sinus pause resulting in ventricular (V) and junctional (J) escape beats. (**c**) Atrial ectopic beat (APB) blocking the sinus P wave (arrow) resulting in a complete compensatory pause. (Reproduced with permission from: Mittal SR. P wave and The PR segment. Cardiology Today 2011;15:263–269)

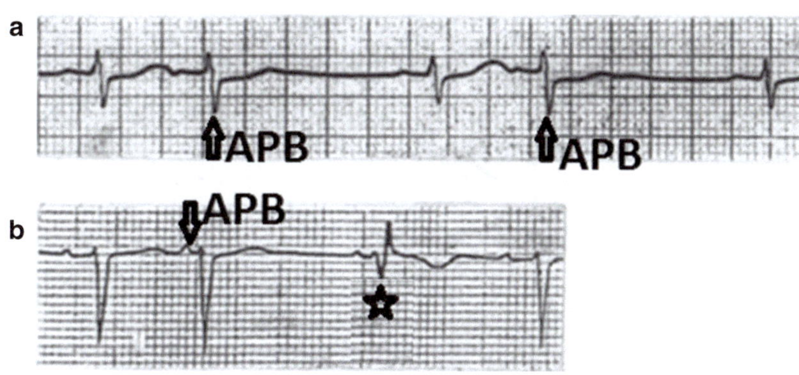

Fig. 3.17 (**a**) Atrial bigeminy (APB, atrial premature beats; marked as arrow). (**b**) Bradycardia-dependent aberrancy in post-ectopic beat (marked as *). (Reproduced with permission from: Mittal SR. P wave and The PR segment. Cardiology Today 2011;15:263–269)

original basal cycle, there is a probability of structural sinoatrial nodal disease (Fig. 3.16b).

- Temporary suppression of the sinoatrial node automaticity by atrial ectopic beat.
- Sinus arrhythmia.
- Exit block of next sinus P wave (Fig. 3.16c).
  - Atrial ectopics can occur after each sinus beat (bigeminy) (Fig. 3.17a) or after two sinuses beats (trigeminy). Frequent atrial ectopics can precede atrial tachyarrhythmias [5]. When the interval between sinus P wave and atrial ectopic beat is less than 50% of the preceding PP cycle, atrial premature beat is more likely to initiate a reentrant atrial tachyarrhythmia [5]. In such instance atrial premature beat forms the first beat of the atrial tachycardia/flutter or fibrillation. QRS following an atrial ectopic is mostly normal in configuration. Rarely the beat following compensatory pause may show broadening of QRS (bradycardia-dependent aberration) (Fig. 3.17b).
  - Atrial ectopics occur in conditions with atrial overload, e.g., mitral valve disease, tricuspid valve disease, and left and right ventricular dysfunction. They can also occur during infection, inflammation, thyrotoxicosis, tension states, or consumption of tobacco, alcohol, or caffeine. They can also occur without any apparent cause. Management is usually directed toward underlying cause. In the absence of underlying cardiac disease, beta-blockers and diltiazem may be helpful.

7 Junctional Premature Beats [5]

Premature beats arising from the atrioventricular junction also produce inverted P waves in leads II, III, and aVF. Relation of the P wave to the QRS depends on relative retrograde velocity of impulse to the atria and antegrade velocity to the ventricles. Faster retrograde velocity (common with premature beat arising in proximal part of the atrioventricular node) produces inverted P wave before the QRS (Fig. 3.18a). In premature beats with equal retrograde and antegrade velocity, the P wave is hidden in the QRS (Fig. 3.18b). If retrograde conduction time is longer than the antegrade conduction time to the ventricles, inverted P wave follows the QRS (Fig. 3.18c). This is common when ectopic focus is in the distal part of the atrioventricular node or in proximal part of the His bundle.

Junctional ectopic may or may not be able to penetrate the sinoatrial node. Sinus rhythm, therefore, may or may not be affected. Post-ectopic compensatory pause, therefore, may be incomplete or complete.

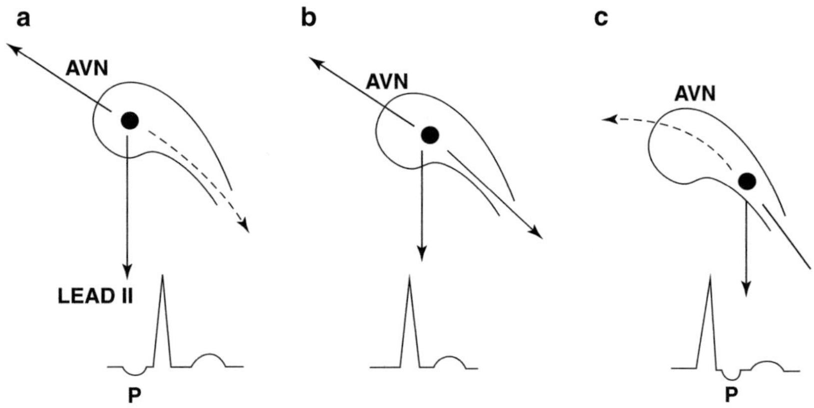

Fig. 3.18 (**a**) Junctional rhythm with inverted P wave preceding QRS. (**b**) Junctional rhythm with P wave hidden in QRS. (**c**) Junctional rhythm with inverted P wave following QRS. (Reproduced with permission from: Mittal SR. P wave and The PR segment. Cardiology Today 2011;15:263–269)

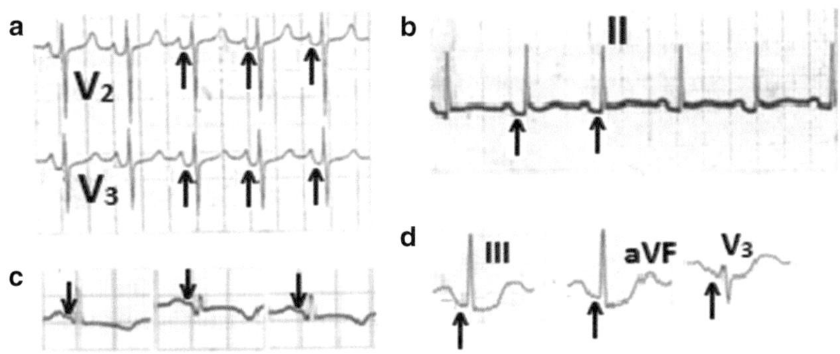

Fig. 3.19 (**a**) The PR segment depression in leads V_2 and V_3 (arrow). (**b**) The PR segment depression in leads II (arrow). (**c**) Showing elevation of the PR segment (arrow) in a case of inferior infarction. (**d**) Showing the effect of exercise on the PR segment (arrow). (Reproduced with permission from: Mittal SR. P wave and The PR segment. Cardiology Today 2011;15:263–269)

8 The PR Segment [2]

The segment between end of P wave and beginning of QRS is isoelectric. Acute pericarditis can be associated with depression of the PR segment in all leads (Fig. 3.19a, b) other than leads aVR and V_1. Absence of this finding, however, does not exclude the diagnosis. Rarely atrial infarction may be associated with elevation or depression of The PR segment (Fig. 3.19c) like the ST segment changes in ventricular infarction. Other findings like intra-atrial conduction defects or atrial tachyarrhythmia are more common in patients with atrial infarction. The PR segment deviation in a case of ventricular infarction is more likely to be due to concomitant pericarditis rather than atrial infarction. Sinus tachycardia during exercise testing is associated with down sloping PR segment which smoothly merges with rapid upsloping ST segment (Fig. 3.19d).

Summary

- P wave represents depolarization of the atria. Initial part is due to right atrial depolarization, and the terminal part is due to depolarization of the left atrium.
- Normal P wave has a duration of less than 0.12 s and amplitude of less than 0.2 mV. In lead V_1 terminal negative deflection is less than 0.01 mV in depth and less than 0.03 s in width.
- Atrial repolarization (Ta wave) is a low-amplitude wave with a polarity against the preceding P wave. Normally it is not seen as it falls in the QRS complex.
- Left atrial depolarization abnormality produces a notched, bifid P wave with a distance of more than 0.04 s between two peaks in lead II. Negative terminal deflection in lead V_1 is 0.04 s or more in duration and more than 0.01 mV in depth.
- Right atrial depolarization abnormality produces peaked P wave in lead II. Gross, right atrial enlargement can produce negative terminal deflection in lead V_1 without really meaning additional left atrial enlargement. Gross, right atrial enlargement can also produce qR configuration in lead V_1.
- Partial interatrial conduction block can also produce broad and bifid P wave in lead II. P wave in lead V_1 is, however, normal. Advanced interatrial block produces a broad and biphasic (+ve/−ve) P wave in lead II.
- Atrial ectopics produce P waves which are different in morphology as compared to the sinus P waves. PR interval is normal or mildly prolonged. Very premature atrial ectopics may fall in the preceding T wave producing mild deformation of the T wave. Negative P wave in leads II, III, and aVF with normal PR inter-

val suggest origin in low atrium. Left atrial rhythm can produce "dome and dart" appearance in lead V_1.

- In a premature beat with wide QRS, preceding P wave, and initial deflection of the QRS identical to sinus beat, rSR' configuration in lead V_1 and qRS pattern in lead V_6 support the possibility of supraventricular ectopic with aberrant conduction.
- Post-ectopic pause is not reliable in differentiating a supraventricular ectopic with aberration from a ventricular ectopic.
- Premature beats arising from atrioventricular junction produce inverted P waves in leads II, III, and aVF. The P wave may precede or follow the QRS or may fall within the QRS depending on the relative velocity of the antegrade and the retrograde conduction. The PR interval may be short.
- The PR segment represents conduction across the atrioventricular node and the His-Purkinje system. Normally it is isoelectric. It can be depressed in acute pericarditis except in leads aVR and V_1. Atrial infarction may be associated with elevation or depression of the PR segment. Sinus tachycardia during exercise testing is associated with down sloping the PR segment.

MCQs

Q1. Atrial extrasystoles with P wave inversion in lead I with normal PR interval suggest:

(a) Origin near sinus node
(b) Origin in the left atrium
(c) Origin in lower atrium
(d) Origin in high atrioventricular node

Q2. Atrial extrasystoles with P wave inversion in leads II, III, and aVF with normal PR interval suggests:

(a) Origin near sinus node
(b) Origin in the left atrium
(c) Origin in the posterior atrium
(d) Origin in the His bundle

Q3. Atrial extrasystoles with P wave inversion in leads II, III, and aVF with short PR interval suggests:

(a) Origin near sinus node
(b) Origin in the left atrium
(c) Origin in the lower atrium
(d) Origin in the His bundle

Q4. Ectopic beat arising within the sinus node:

(a) Occurs earlier than scheduled P wave
(b) Has configuration similar to other sinus P wave
(c) Has short PR interval
(d) Has normal QRS

Q5. A broad (duration of more than 0.12 s) and notched configuration of P wave in lead II with a normal configuration in lead V_1 suggests:

(a) Sinus rhythm with partial interatrial block
(b) Sinus rhythm with advanced interatrial block
(c) Ectopic atrial rhythm
(d) Ectopic junctional rhythm

Q6. P wave with duration of more than 0.12 s in lead II showing initial positive and terminal negative deflection suggests:

(a) Sinus rhythm with partial interatrial block
(b) Sinus rhythm with complete interatrial block
(c) Ectopic atrial rhythm
(d) Ectopic junctional rhythm

Q7. Atrial extrasystoles have:

(a) P wave configuration different from sinus P wave
(b) Normal PR interval
(c) Normal QRS
(d) Inverted T waves

Q8. Atrial extrasystole may not be followed by a QRS if:

(a) The atrioventricular node is fully repolarized

(b) The atrioventricular node is partially depolarized
(c) The atrioventricular node is fully depolarized
(d) The atrial extrasystole occurs on ascending limb of preceding T wave

Q9. An atrial extrasystole conducted with a prolonged PR interval and normal QRS suggests:

(a) Partial repolarization of the atrioventricular node
(b) Non-repolarization of the atrioventricular node
(c) Non-repolarization of the bundle branches
(d) Non-repolarization of the ventricular myocardium

Q10. An atrial extrasystole conducted with a normal PR interval and broad QRS suggests:

(a) Full repolarization of the atrioventricular node
(b) Partial repolarization of the atrioventricular node
(c) Non-repolarization of either bundle branch
(d) Full repolarization of bundle branches and Purkinje system

Q11. Compensatory pause following an atrial extrasystole depends on:

(a) Site of origin of atrial extrasystole
(b) Time taken for the impulse to enter the sinoatrial node
(c) Extent of suppression of the sinus node cycle
(d) Time taken by next sinus impulse to come out of sinus node

Q12. Compensatory pause following an atrial extrasystole suggests sinus node disease if compensatory pause is:

(a) Less than the preceding normal PP cycle length
(b) Equal to the preceding normal PP cycle length
(c) More than the preceding normal PP cycle length by more than 50%
(d) More than the preceding normal PP cycle length by more than 10%

Q13. A premature beat with inverted P wave in leads II, III, and aVF and short PR interval and normal QRS should suggest the possibility of:

(a) Ectopic focus in lower part of the atrioventricular node or His bundle
(b) Low atrial ectopic with preexcitation
(c) Left atrial ectopic with normal ventricular depolarization
(d) Ventricular ectopic with retrograde conduction

Q14. Deformation of the T wave, followed by a pause in the sinus rhythm suggests:

(a) Sinus pause
(b) Sinoatrial exit block
(c) Blocked atrial ectopic
(d) Blocked ventricular ectopic

Q15. Deformed T wave, followed by a broad QRS suggests:

(a) Blocked atrial ectopic
(b) Atrial ectopic with normal ventricular conduction
(c) Atrial ectopic with aberrant ventricular conduction
(d) Ventricular ectopic with retrograde conduction to atria

Q16. A broad QRS premature beat without any P wave and with a full compensatory pause suggests possibility of:

(a) Ventricular premature beat
(b) Ventricular premature beat with retrograde conduction
(c) Atrial premature beat with aberrant ventricular conduction
(d) Junctional premature beat with aberrant ventricular conduction

Q17. The interval from the sinus beat prior to a premature beat to the sinus beat following the premature beat is equal to the interval between two sinus beats. It suggests:

(a) Supraventricular premature beat not depolarizing sinus node
(b) Supraventricular premature beat depolarizing a normal sinus node
(c) Ventricular premature beat without retrograde conduction
(d) Ventricular premature beat with retrograde depolarization of the atria

Q18. An atrial premature beat:

(a) May not have a recognizable P wave
(b) May have a broad QRS not similar to that of the normal sinus beat
(c) May have a compensatory pause equal to two sinus cycles
(d) None of above

Q19. Atrial trigeminy means:

(a) One sinus beat followed by one atrial premature beat
(b) One sinus beat followed by two consecutive atrial premature beats
(c) One sinus beat followed by three consecutive atrial premature beats
(d) Two consecutive sinus beats followed by one atrial premature beat

Q20. Increased automaticity is the cause of atrial premature beat if:

(a) Interval between atrial premature beat and preceding sinus beat is constant
(b) Interval between atrial premature beat and preceding sinus beat is changing
(c) Interval between two atrial premature beats is constant or multiple of some interval
(d) Interval between two atrial premature beats is not constant nor multiple of some interval

Q21. In atrial bigeminy, if all atrial premature beats merge with the T wave and do not conduct to the ventricles, the rhythm is likely to be misdiagnosed as:

(a) Sinus bradycardia
(b) Sinus arrhythmia
(c) Atrioventricular block
(d) Junctional rhythm

Q22. An atrial premature beat occurring very early in cardiac cycle can:

(a) Prolong PR interval of the next sinus beat
(b) Produce aberrant intraventricular conduction of the next sinus beat
(c) Initiate a reentrant atrial tachyarrhythmia
(d) None of above

Q23. A premature beat arising in the atrioventricular node produces:

(a) Inverted P wave in leads II, III, and aVF
(b) Positive P wave in leads aVR and aVL
(c) Flat P wave in leads I and V_5
(d) Inverted P wave in lead V_1

Answers

(1) b; (2) c; (3) d; (4) a, b, d; (5) a; (6) b; (7) a, b, c; (8) c, d; (9) a; (10) a, c; (11) b, c, d; (12) c; (13) a; (14) c; (15) c; (16) a, d; (17) a, c; (18) a, b, c; (19) d; (20) b, c; (21) a; (22) a, b, c; (23) a, b, c

References

1. Wagner GS, Lim TH. Interpretation of normal electrocardiogram. In: Wagner GS, editor. Marriot's practical electrocardiography. Philadelphia: Wolters Kluwer; 2008. p. 43–70.
2. Mirvis DM, Goldberger AL. Electrocardiography. In: Libby P, Bonow RO, Mann DL, Zipes DR, editors. Braunwald's heart disease. Philadelphia: Saunders; 2008. p. 149–90.
3. Wagner GS, Lim TH. Chamber enlargement. In: Wagner GS, editor. Marriot's practical electrocardiography. Philadelphia: Wolters Kluwer; 2008. p. 71–96.

4. Schamroth L. The normal and abnormal P wave. In: Schamroth C, editor. An introduction to electrocardiography. France: Blackwell Science; 1982. p. 49–61.
5. Wagner GS, Wang TY. Premature beats. In: Wagner GS, editor. Marriot's practical electrocardiography. Philadelphia: Wolters Kluwer; 2008. p. 261–86.
6. Wagner GS, Wang TY. Ventricular versus supraventricular with aberrant conduction. In: Wagner GS, editor. Marriot's practical electrocardiography. Philadelphia: Wolters Kluwer; 2008. p. 371–89.

4 Narrow QRS Tachycardias

Tachycardia, in adults, is defined as a rate of 100 or more per minute [1]. Tachycardia can arise from any site in the heart. When it arises from a focus above the bundle branches and if the intraventricular conduction system is normal, QRS of the tachycardia is narrow (i.e., less than 0.10 s). All these tachycardias are grouped as "narrow QRS tachycardias." These include the following:

- Sinus tachycardia
- Inappropriate sinus tachycardia (IST)
- Sinoatrial node reentrant tachycardia (SNRT)
- Accelerated atrial rhythm
- Atrial tachycardia
- Multifocal atrial tachycardia (MAT) (chaotic atrial tachycardia)
- Accelerated atrioventricular nodal (junctional) rhythm
- Atrioventricular nodal reentrant tachycardia (AVNRT)
- Junctional ectopic tachycardia
- Orthodromic atrioventricular reciprocating (reentrant) tachycardia (AVRT) in patients with accessory pathways

Atrial flutter and atrial fibrillation also produce narrow QRS tachycardia. They are discussed separately.

The presence of persistent or rate-dependent intraventricular conduction delay can produce broad QRS in any of these tachycardias. Second beat at the onset of a narrow QRS tachycardia may be broad due to abnormal intraventricular conduction (aberrancy).

1 Sinus Tachycardia

Electrocardiographic features include the following:

- Rate rarely exceeds 160/min except in exercising adults. Age-predicted maximal heart rate in an individual during exercise is 220—age [2].
- The P wave is upright in lead II and inverted in lead aVR. P wave precedes each QRS. In marked tachycardia, P wave may fuse with the preceding T wave producing a notched, bifid, or peaked T wave (Fig. 4.1a). An abnormal configuration of P wave should suggest non-sinus origin of tachycardia.
- The PR interval is normal or may be slightly shorter due to accelerated atrioventricular nodal conduction secondary to sympathetic stimulation, which is usually the underlying cause (conditions like fever, anxiety, anger, exercise, thyrotoxicosis, anemia, and heart failure). Prolonged PR interval can be present if there is preexisting atrioventricular nodal disease or His bundle disease. Sinus tachycardia with prolongation of the PR interval is also an important feature of rheumatic carditis

S. Mittal, *Insights into Electrocardiograms with MCQs*,
https://doi.org/10.1007/978-981-99-0127-2_4

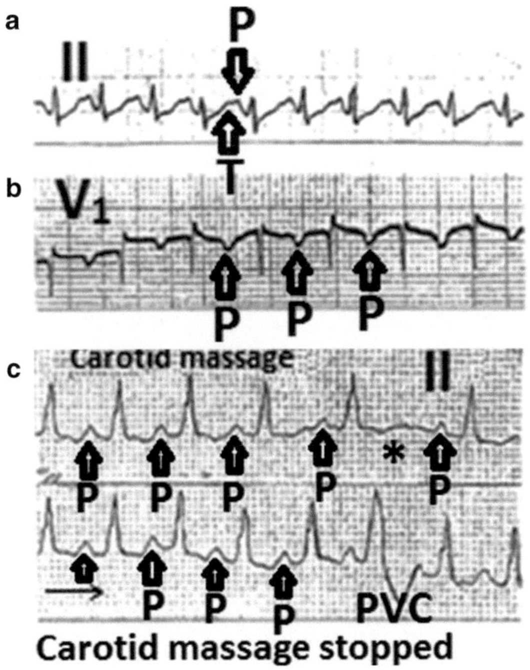

Fig. 4.1 (**a**) Lead II shows sinus tachycardia with fusion of T and P wave. (**b**) Lead V_1 showing sinus tachycardia, prominent negative P wave suggestive of left atrial depolarization abnormality and prolonged PR interval. (**c**) Lead II shows effect of carotid massage on sinus tachycardia. There is gradual slowing of tachycardia (marked *) with increase in heart rate on stopping carotid massage. *PVC* premature ventricular contraction. (Reproduced with permission from: Mittal SR. Narrow QRS Tachycardia. Cardiology Today 2012;16:82–89)

(Fig. 4.1b). Usually a prolonged PR interval should raise suspicion about another cause of tachycardia.

- Onset and termination are gradual.
- Carotid sinus massage produces a gradual slight slowing of tachycardia with gradual return to baseline on cessation of massage (Fig. 4.1c).

2 Inappropriate Sinus Tachycardia (IST) [3]

It is an incessant and symptomatic sinus tachycardia at rest or with minimal exertion. Heart rate may, however, markedly slow during sleep. It is usually seen in young females. Patients are disabled by persistent palpitation, dyspnea, chronic fatigue, and near syncope. Absence of causes of sinus tachycardia should raise suspicion of this entity. Patients require increased level of adrenergic activity for proper functioning and may not tolerate even minimal dose of beta-blockers. Clonidine or ivabradine may help some patients.

3 Sinus Node Reentrant Tachycardia (SNRT) [3]

A reentry circuit is established within the sinoatrial node (Fig. 4.2a). It stimulates the atria at a rapid rate. Electrocardiographically it resembles sinus tachycardia except that onset and termination may be sudden (Fig. 4.2b). Also, unlike sinus tachycardia, vagal maneuvers, adenosine, and verapamil may terminate this arrhythmia. Beta-blockers and calcium channel blockers may prevent recurrence.

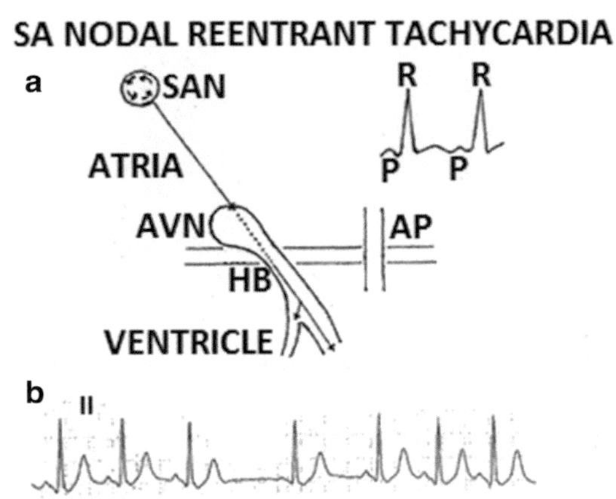

Fig. 4.2 (**a**) Diagram showing the genesis of SNRT. (**b**) Lead II shows sudden cessation and onset of SNRT. *SAN* sinoatrial node, *AVN* atrioventricular node, *HB* His bundle, *AP* accessory path. (Reproduced with permission from: Mittal SR. Narrow QRS Tachycardia. Cardiology Today 2012;16:82–89)

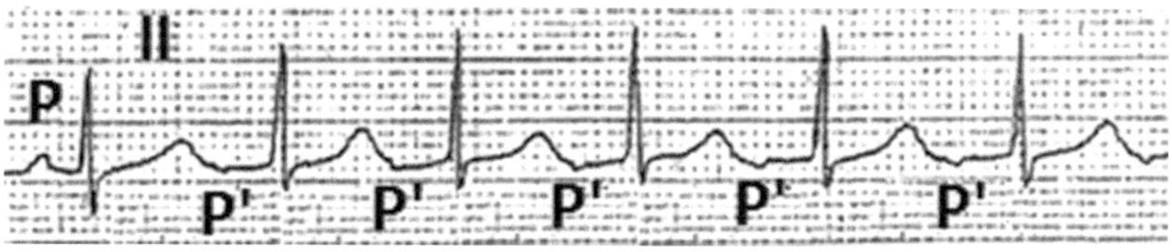

Fig. 4.3 Lead V_1 showing ECG features of accelerated atrial rhythm. (Reproduced with permission from: Mittal SR. Narrow QRS Tachycardia. Cardiology Today 2012;16:82–89)

4 Accelerated Atrial Rhythm [2]

Normally the rate of automaticity of atrial tissue is less than that of sinoatrial node. If the rate of automaticity of a focus in the atria increases above the rate of sinoatrial node, it takes over the control of the heart rate. It is characterized by the following:

- Usually rate is more than 60/min but less than 100/min. Uncommonly the rate may exceed 100/min and may qualify to be called a tachycardia. Even in such circumstances, rate does not exceed 130/min.
- Abnormal P wave morphology.
- Normal or prolonged PR interval (Fig. 4.3).

5 Atrial Tachycardia (Focal Atrial Tachycardia, Extra Systolic Atrial Tachycardia)

An abnormal focus of automaticity appears in either atrium (Fig. 4.4a). It discharges at a fast rate resulting in atrial tachycardia. It is characterized by the following:

- Sudden onset and cessation (Fig. 4.4b). Onset is with an atrial premature beat (Fig. 4.4b).
- Abnormal P wave morphology. However, if the focus is near sinoatrial node, P wave contour may be similar to the sinus P wave [1]. A positive or biphasic P wave in lead V_1 predicts a right atrial focus [1] (Fig. 4.4b), and negative P wave in lead V_1 predicts a left atrial focus [1] (Fig. 4.4c).
- Discrete P wave with a clear baseline rather than a sawtooth morphology is seen [2] in both the leads II and V_1 (Fig. 4.4b). Rate is around 200 ± 50/min.
- PR interval may be normal. If ectopic focus is very close to the atrioventricular node, PR interval may be short (Fig. 4.4b *first ectopic beat*). If tachycardia is rapid, the PR interval may be prolonged due to incomplete recovery of the atrioventricular node [4] (Fig. 4.4b). With prolongation of the PR interval, P wave may fall close to the previous QRS and may simulate atrioventricular reciprocating tachycardia (Fig. 4.4d, e).
- Atrial tachycardia may be associated with Wenckebach phenomena, i.e., progressive prolongation of the PR interval with failure of one P wave to conduct to the ventricles followed by normalization of the PR interval (Fig. 4.5a). Spontaneous long pause due to atrioventricular block should suggest the possibility of underlying atrioventricular nodal disease (Fig. 4.5b).
- QRS complexes are regular. However, at the onset, there may be warming up or gradual increase in rate [1].
- Decreasing atrioventricular nodal conduction (vagal maneuvers, diltiazem, verapamil, or adenosine) reduces ventricular rate but does not terminate atrial tachycardia [3] (Fig. 4.4d, e). Termination of a tachycardia by a P wave not conducted to the ventricles indicates atrioventricular node dependency of the tachycardia and excludes the possibility of an atrial tachycardia [3]. Atrial tachycardia is not affected by a premature ventricular beat

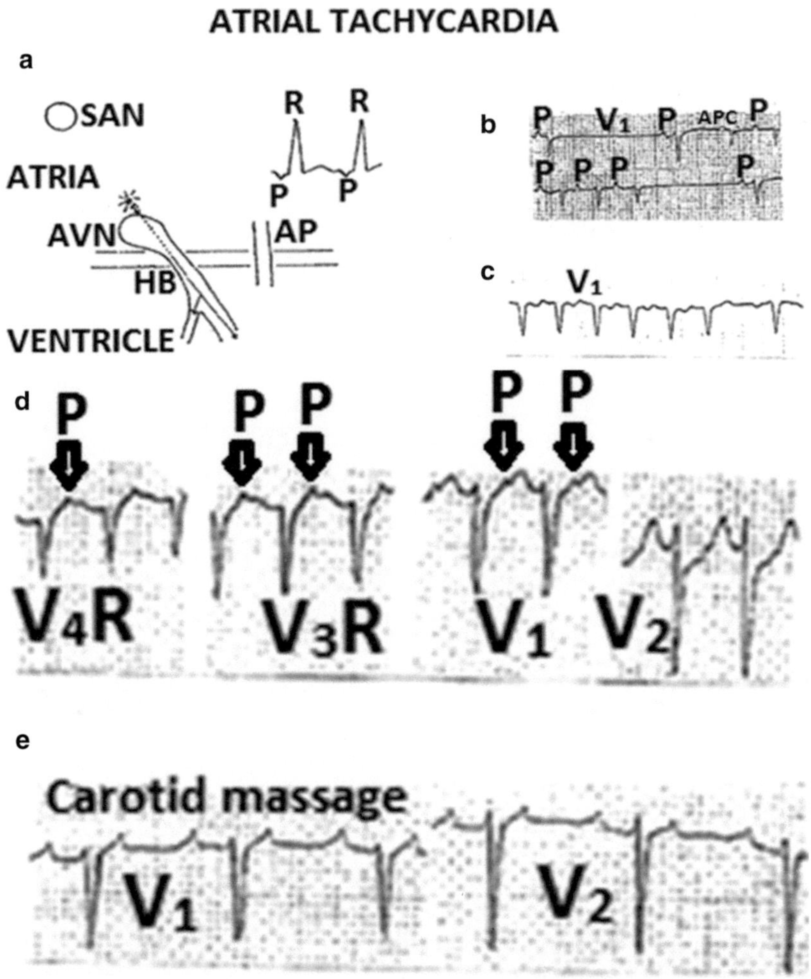

Fig. 4.4 (**a**) Diagram showing genesis of the atrial tachycardia. (**b**) Lead V_1 showing sudden onset and cessation of atrial tachycardia, onset with an atrial premature beat (APC), and positive P wave, suggesting right atrial focus, discrete P waves with clear baseline, and prolonged PR interval. (**c**) Lead V_1 showing atrial tachycardia with a negative P wave suggesting left atrial focus. *SAN* sinoatrial node, *AVN* atrioventricular node, *HB* His bundle, *AP* accessory path. (**d**) Leads V_4R to V_2 showing atrial tachycardia with the P wave close to the previous QRS (marked ↓). (**e**) Leads V_1 and V_2 showing atrial tachycardia with 2:1 atrioventricular block following carotid massage in the same patient as in (**d**). (Reproduced with permission from: Mittal SR. Narrow QRS Tachycardia. Cardiology Today 2012;16:82–89)

(Fig. 4.6a). Atrial tachycardia may be only short-lived and may stop spontaneously (Fig. 4.4b). Long pause after spontaneous reversal to sinus rhythm should raise suspicion of underlying sinus node disease (tachy-brady syndrome).

Atrial tachycardia may be due to an ectopic focus or a focal micro-reentry circuit. In ectopic atrial tachycardia initial and subsequent P waves are identical (Fig. 4.4b). If the initial P wave is different from the subsequent P waves, there is a possibility of micro-reentry following an atrial premature beat (Fig. 4.6b). Progressive reduction in rate of atrial tachycardia supports the possibility of reentry (Fig. 4.6c).

Atrial Tachycardia with Block If atrioventricular nodal conduction is slow due to preexisting atrioventricular nodal disease or drugs that delay

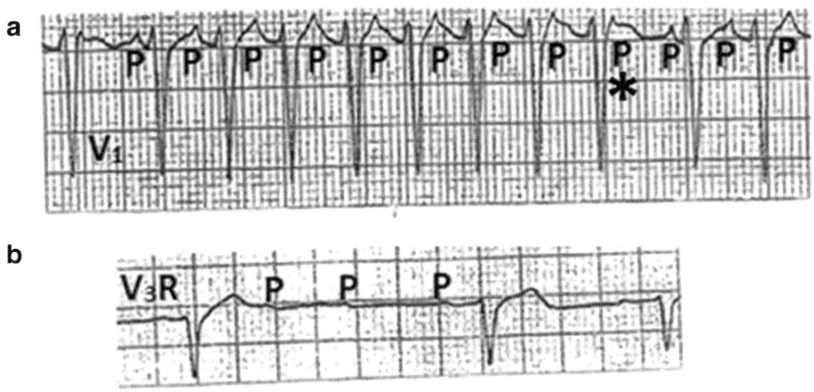

Fig. 4.5 (**a**) Lead V_1 showing atrial tachycardia with Wenckebach phenomena. There is progressive prolongation of the PR interval with failure of one P wave (marked star) to conduct to the ventricles. (There is no QRS after this P wave). (**b**) Lead V_3R showing atrial tachycardia with atrioventricular block. (Reproduced with permission from: Mittal SR. Narrow QRS Tachycardia. Cardiology Today 2012;16:82–89)

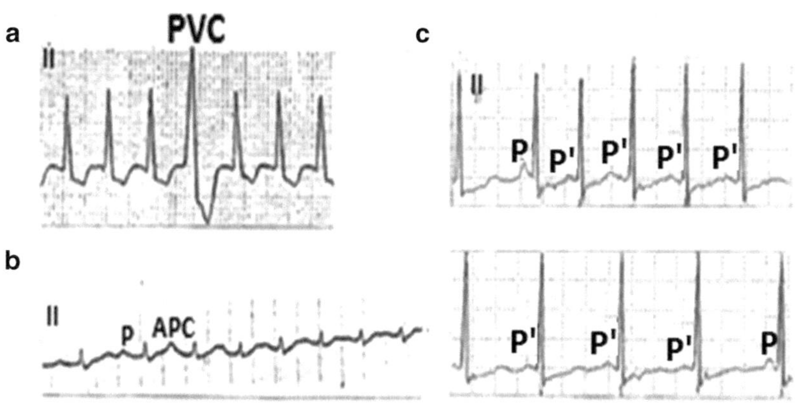

Fig. 4.6 (**a**) Lead II shows no effect of premature ventricular contraction on atrial tachycardia. (**b**) Lead II shows initiation of atrial tachycardia. The first P wave of the tachycardia (marked P) is different from subsequent P waves. (**c**) Lead II shows gradual slowing of atrial tachycardia (marked P′). (Reproduced with permission from: Mittal SR. Narrow QRS Tachycardia. Cardiology Today 2012;16:82–89)

atrioventricular conduction (e.g., beta-blockers, diltiazem, verapamil, or digoxin), there may be fixed (Fig. 4.7a, b), variable, or fluctuating atrioventricular block [5]. Irregular atrioventricular conduction may result in abnormal QRS due to abnormal conduction in the ventricles (aberrant intraventricular conduction) (Fig. 4.7c). Such a block does not terminate atrial tachycardia. Digoxin used to be a common culprit [2] specially in the setting of hypokalemia. Because of enhanced automaticity caused by digitalis, rate of ectopic atrial focus gradually increases. Concomitant parasympathetic effect of digitalis produces atrioventricular block [1]. The atrial rate progressively increases if digitalis toxicity is not suspected and digitalis is continued [2].

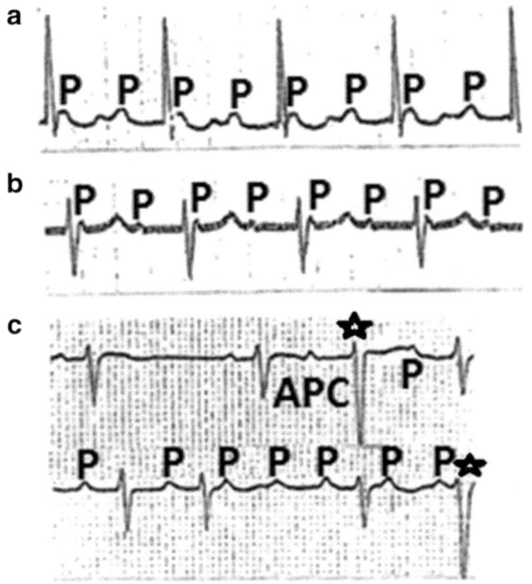

Fig. 4.7 (**a**) Lead II shows atrial tachycardia with 2:1 atrioventricular block. (**b**) Lead V_1 from the same patient as in (**a**). (**c**) Lead V_1 shows atrial tachycardia starting with an atrial premature beat (APC). There is irregular conduction to the ventricles. Third and eighth beats show aberrant intraventricular conduction (abnormal configuration of QRS marked as star) due to long-short cycle sequence. (Reproduced with permission from: Mittal SR. Narrow QRS Tachycardia. Cardiology Today 2012;16:82–89)

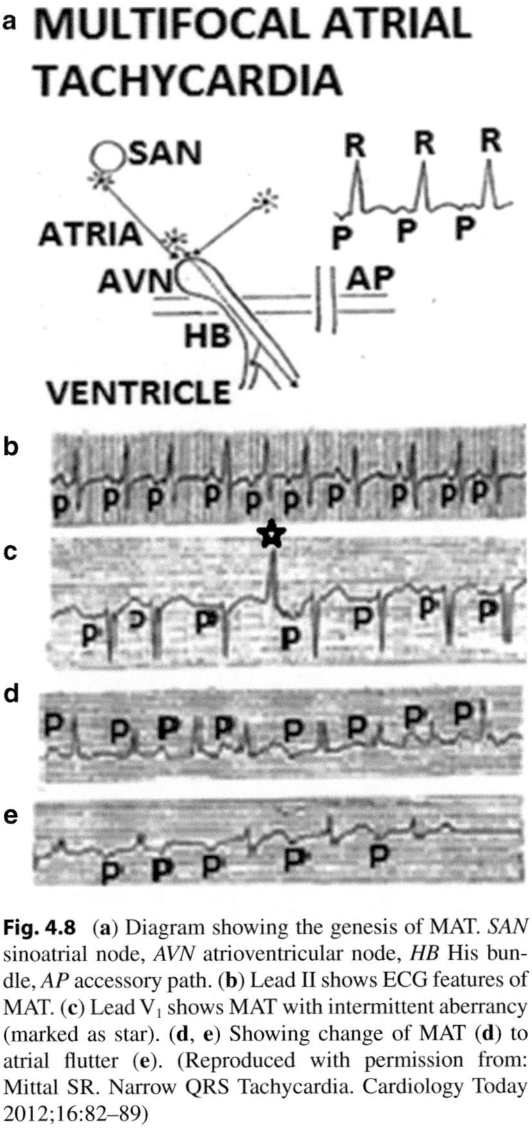

Fig. 4.8 (**a**) Diagram showing the genesis of MAT. *SAN* sinoatrial node, *AVN* atrioventricular node, *HB* His bundle, *AP* accessory path. (**b**) Lead II shows ECG features of MAT. (**c**) Lead V_1 shows MAT with intermittent aberrancy (marked as star). (**d**, **e**) Showing change of MAT (**d**) to atrial flutter (**e**). (Reproduced with permission from: Mittal SR. Narrow QRS Tachycardia. Cardiology Today 2012;16:82–89)

6 Multifocal Atrial Tachycardia (MAT, Chaotic Atrial Tachycardia)

Multiple foci of automaticity appear in different parts of the atria (Fig. 4.8a) and stimulate atria at a rapid rate. It produces multifocal atrial tachycardia.

It is characterized by the following.

Three or more different shapes of P wave due to changing atrial focus [3] (Fig. 4.8b). PP interval is totally irregular. The PR interval may vary due to the changing distance of the atrial ectopic focus from the atrioventricular node [2]. There is no atrioventricular block. QRS is normal in configuration but is irregularly irregular. Irregular rhythm can result in abnormal depolarization of the ventricles (aberrant conduction) producing abnormal QRS (Fig. 4.8c). It is almost always seen in the setting of decompensated cor pulmonale with hypoxia [2] or patients of diabetic ketoacidosis. It is usually preceded by atrial premature beats and may progress to atrial flutter (Fig. 4.8d, e) or fibrillation if not managed. Management of cor pulmonale usually results in gradual reversal to sinus rhythm. Antiarrhythmic agents are not needed. Theophyllines, beta-2 stimulants, and digitalis may aggravate this arrhythmia. Electric cardioversion is ineffective.

7 Accelerated Junctional Rhythm

Normally the rate of automaticity of atrioventricular junctional tissue is around 50/min. If the automaticity of the junctional tissue increases and exceeds that of sinoatrial node, it is called accelerated junctional rhythm.

Usually the rate is less than 100/min. Uncommonly the rate exceeds 100/min and may qualify to be a tachycardia (Fig. 4.9). Even in these situations, the rate usually does not exceed 130/min.

It is characterized by the following:

- Retrogradely conducted P waves (inverted in leads II, III, and aVF). Depending on relative velocity of retrograde conduction to the atria and antegrade conduction to the ventricles, P wave may merge with the QRS or may follow the QRS and fall on the ST segment.
- In some cases rate of sinoatrial node and accelerated junctional rhythm may be nearly similar. This may result in atrioventricular dissociation with normal P wave and QRS but changing P and QRS relation [2].
- PR interval may be short.
- Regular normal QRS at a rate of more than 60/min in the setting of atrial fibrillation is another manifestation of accelerated junctional rhythm. It is common with digitalis toxicity. Continuation of digitalis results in progressive acceleration of junctional rhythm [2].

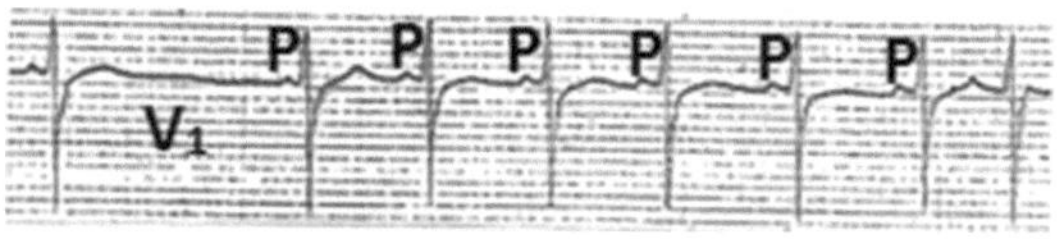

Fig. 4.9 Lead II showing accelerated junctional rhythm. P shows sinus P wave and P′ shows junctional P waves. (Reproduced with permission from: Mittal SR. Narrow QRS Tachycardia. Cardiology Today 2012;16:82–89)

8 Atrioventricular Nodal Reentrant Tachycardia (AVNRT)

A micro-reentrant circuit is established within the atrioventricular nodal tissue (Fig. 4.10a). It rapidly depolarizes the ventricles antegradely and atria retrogradely.

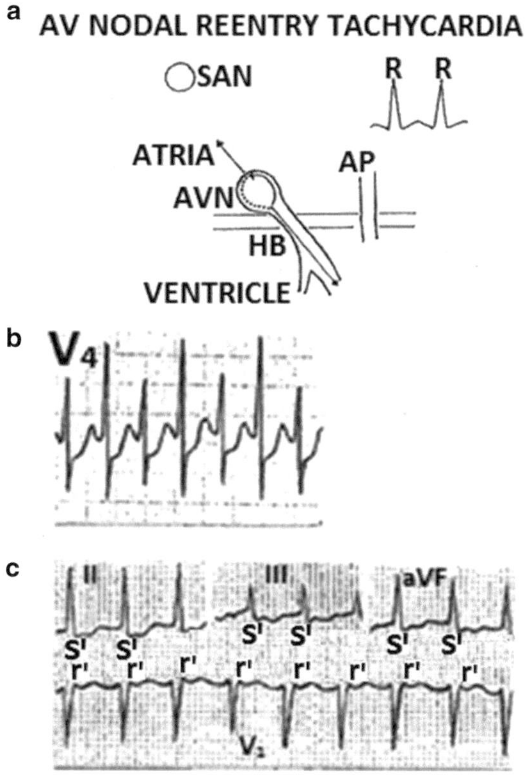

Fig. 4.10 (**a**) Diagram showing the mechanism of AVNRT. *SAN* sinoatrial node, *AVN* atrioventricular node, *HB* His bundle, *AP* accessory path. (**b**) Lead V_4 showing AVNRT without visible P wave, depression of the ST segment, and electrical alternans. (**c**) Leads II, III, and aVF showing "pseudo wave." Lead V_1 showing "pseudo wave." (Reproduced with permission from: Mittal SR. Narrow QRS Tachycardia. Cardiology Today 2012;16:82–89)

It is characterized by the following:

- Sudden onset and termination.
- Rate between 150 and 250/min [1].
- Negative P′ wave in lead II due to retrograde atrial activation. Position of the P′ wave in relation to the QRS depends on the relative velocity of conduction retrogradely to the atria and antegradely to the ventricles. The P wave is not visible in two thirds of cases as atrial activation occurs during ventricular activation [2] (Fig. 4.10b). When retrograde conduction to the atria is slower, inverted P′ wave may immediately follow the QRS (Fig. 4.10c). This produces a "pseudo s wave" in leads II, III, and aVF (Fig. 4.10c) or "pseudo r' wave" in lead V_1 (Fig. 4.10c).

It may be recognized only on comparison with the QRS complex in sinus rhythm. This occurs in nearly one third of the cases [2]. Only in very few cases the abnormal P′ wave may just precede the QRS [6]. Thus retrograde P wave in AVNRT typically deforms the initial or the terminal portion of the QRS. AVNRT is to be differentiated from the following:

- Sinus tachycardia—In sinus tachycardia, the P wave is usually visible separately or deforms the preceding T wave. Rate of sinus tachycardia is usually less than 150/min in a nonexercising adult. The rate of AVNRT is almost always more than 150/min [2]. Sinus tachycardia accelerates and decelerates gradually as opposed to sudden onset and termination of AVNRT.
- Atrial flutter with 2:1 atrioventricular conduction—Tachycardia rate, in an untreated patient of atrial flutter, is usually 150/min, and there is no isoelectric baseline. Carotid sinus compression (vagal maneuver) may help differentiation. In atrial tachycardia or atrial flutter, atrioventricular block is increased, resulting in slowing of the QRS and unmasking of the P′ wave of the atrial tachycardia or the flutter waves (Fig. 4.11a). Sinus tachycardia shows gradual slowing with separation of T and P waves. AVNRT is either not affected or suddenly terminates (Fig. 4.11b). Termination may be followed by junctional or ventricular escape beats before appearance of sinus rhythm (Fig. 4.11b). If carotid sinus compression does not yield diagnostic information, intravenous adenosine, beta-blocker, verapamil, or diltiazem may help.
- AVNRT can be of two types [6]:
  - Typical form (slow-fast)—Antegrade conduction is through the slow path, and retrograde conduction is via the fast pathway. It is initiated by an atrial premature beat which conducts with a prolonged PR interval. It is common. It may be congenital but usually manifests in the third or fourth decade. Diseases or drugs that impair atrioventricular nodal conduction help in initiation of this tachyarrhythmia. RR interval can shorten over the course of first few beats of AVNRT at the onset or lengthen during last few beats preceding termina-

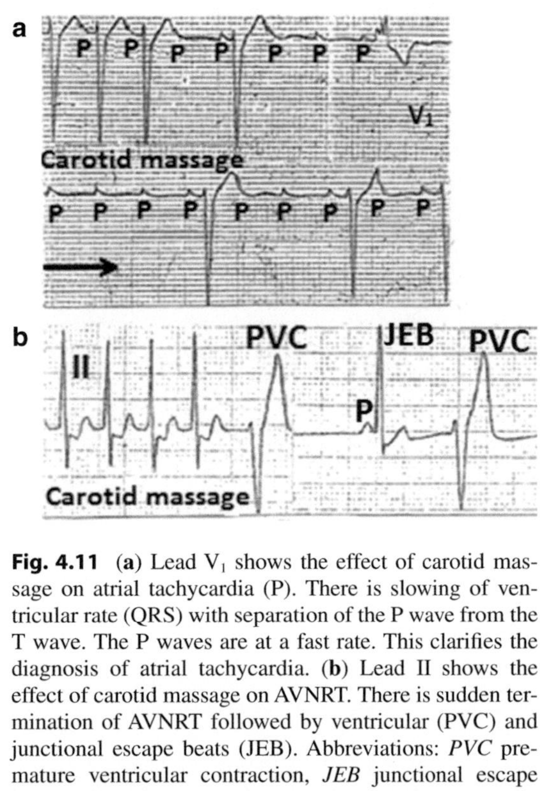

Fig. 4.11 (**a**) Lead V_1 shows the effect of carotid massage on atrial tachycardia (P). There is slowing of ventricular rate (QRS) with separation of the P wave from the T wave. The P waves are at a fast rate. This clarifies the diagnosis of atrial tachycardia. (**b**) Lead II shows the effect of carotid massage on AVNRT. There is sudden termination of AVNRT followed by ventricular (PVC) and junctional escape beats (JEB). Abbreviations: *PVC* premature ventricular contraction, *JEB* junctional escape beat. (Reproduced with permission from: Mittal SR. Narrow QRS Tachycardia. Cardiology Today 2012;16:82–89)

tion. It is usually caused by variation in the antegrade atrioventricular nodal conduction time.

  - Atypical form (fast-slow)—Antegrade conduction to the ventricles is via the fast pathway, and retrograde conduction to the atria is via the slow pathway. The P wave, therefore, falls clearly after the QRS [4]. RP′ interval (interval between the QRS complex and the P′ wave) is equal to or longer than the interval between the retrogradely conducted P′ wave and the next QRS (the P′-R interval). It is rare.
- Development of rate-dependent bundle branch block does not affect the rate of the tachycardia [4].
- There may be progressive prolongation of the P′-R interval, and failure of conduction of a P′ wave to the ventricles may result in termination of the tachycardia.

Other ECG findings that can be seen during or after AVNRT [4]:

- Significant depression of the ST segment (Fig. 4.10b). This finding itself does not indicate ischemia except in elderly with h/o angina during the tachyarrhythmia.
- Newly acquired T wave inversion immediately after termination of the tachycardia or within 6 h. These may persist for variable duration but do not suggest coronary artery disease.
- QRS alternans (Fig. 4.10b) or cycle length alternans can occur when tachycardia rate is very rapid [1].

9 Orthodromic Atrioventricular Reentrant Tachycardia Using Atrioventricular Bypass Tract

A macro-reentry circuit is established involving the atria-atrioventricular node-His bundle-bundle branch-Purkinje system and the accessory (extra) pathway connecting the atria and the ventricle (Fig. 4.12a). It is initiated by an atrial ectopic

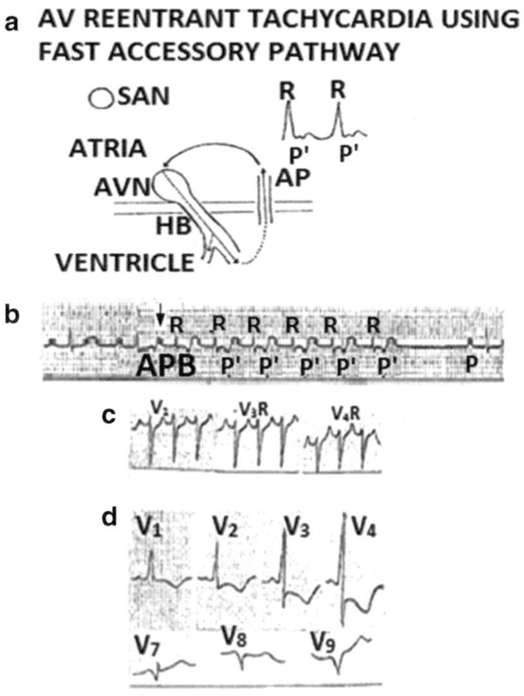

Fig. 4.12 (**a**) Diagram showing the mechanism of orthodromic AVRT. Abbreviation: *SAN* sinoatrial node, *AVN* atrioventricular node, *HB* His bundle, *AP* accessory path. (**b**) The ECG shows features of orthodromic AVRT. Atrial premature beat (APB is marked ↓) is followed by transient AVRT (marked P′). P′ wave is falling clearly after QRS (marked R). Tachycardia terminates due to failure of the fifth P′ wave to conduct to the ventricle (no QRS after the fifth P′ wave). (**c**) The ECG shows AVRT. (**d**) The ECG shows preexcitation during sinus rhythm in the same patient as in (**c**). (Reproduced with permission from: Mittal SR. Narrow QRS Tachycardia. Cardiology Today 2012;16:82–89)

(Fig. 4.12b). During tachycardia, there is antegrade conduction to the ventricles through the atrioventricular node and retrograde conduction to the atria via the bypass tract. Accessory pathway may (Fig. 4.12d) or may not be apparent (concealed) during the sinus rhythm. During tachycardia the impulse has to depolarize the whole ventricle before reaching the bypass tract which is situated along the atrioventricular groove. Then only, the impulse passes up along the bypass tract to depolarize the atria. Thus, ventricular depolarization is completed before atrial depolarization. Therefore, abnormal P wave always comes clearly after the QRS (Fig. 4.12b, c). Due to decreased QT interval during tachycar-

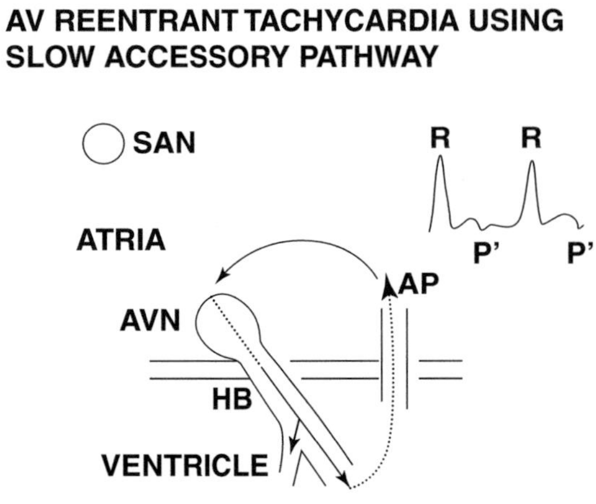

Fig. 4.13 Diagram showing the mechanism of orthodromic AVRT with slowly conducting accessory pathway. *SAN* sinoatrial node, *AVN* atrioventricular node, *HB* His bundle, *AP* accessory path. (Reproduced with permission from: Mittal SR. Narrow QRS Tachycardia. Cardiology Today 2012;16:82–89)

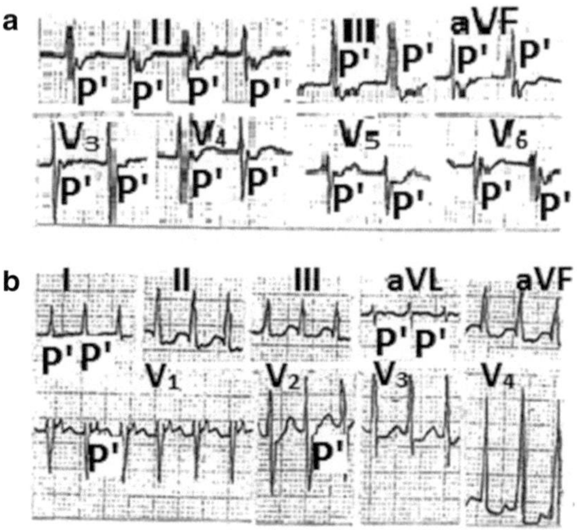

Fig. 4.14 (**a**) Showing AVRT with inverted P′ waves in leads II, III, and aVF and V_2 toV_6. (**b**) Showing AVRT with inverted P′ in the leads I and aVL. (Reproduced with permission from: Mittal SR. Narrow QRS Tachycardia. Cardiology Today 2012;16:82–89)

dia, the P′ wave may merge with the T wave and may be clear only in some leads.

Uncommonly, the retrograde conduction via the accessory path may be slow. In this situation the retrogradely conducted P′ wave occurs little farther from the preceding QRS (Fig. 4.13), as compared to the typical (common) orthodromic atrioventricular reentrant tachycardia [6].

In the majority of these cases, the accessory pathway is localized in the right posterior septum near the orifice of the coronary sinus with the result that right atrium is activated first [6]. This results in inverted P waves in leads II, III, and aVF as well as in leads V_2 to V_6 (Fig. 4.14a). Concealed accessory pathway, in most cases, is left sided (inserting in the left atrium) which makes the left atrium the first site of retrograde activation. P waves, during tachycardia, are therefore negative in lead I [1] (Fig. 4.14b).

Depression of the ST segment in leads V_3 to V_6 is common in left lateral accessory pathway [4]. Posteroseptal or posterior pathway can produce T wave inversion in the inferior leads [4]. Inverted or notched T wave in leads V_2 or V_3 with positive retrograde P′ wave in at least two inferior leads suggests possibility of anteroseptal pathway [4]. As tachycardia circuit involves one of the ventricles, appearance of rate-dependent bundle branch block in the ventricle with the accessory pathway results in slowing of the tachycardia rate [1]. Termination generally occurs in the antegrade direction, i.e., failure of the last retrograde P′ wave to conduct to the ventricle [1] (Fig. 4.12b).

10 Junctional Ectopic Tachycardia (His Bundle Ectopic Tachycardia)

An ectopic focus appears in the atrioventricular junction or the His bundle (Fig. 4.15). It depolarizes the ventricles at a fast rate. It is rapid, often incessant, and resistant to drugs. Onset is spontaneous without any atrial premature beat, and there may be warming up (gradual increase in rate at the beginning) and cooling down (gradual decrease in rate before termination). Tachycardia arising from any focus in the atrioventricular junction has narrow QRS because intraventricular conduction is normal.

Mostly the impulse fails to conduct retrogradely to the atria and there is atrioventricular dissociation (Fig. 4.16a). At times, there may be

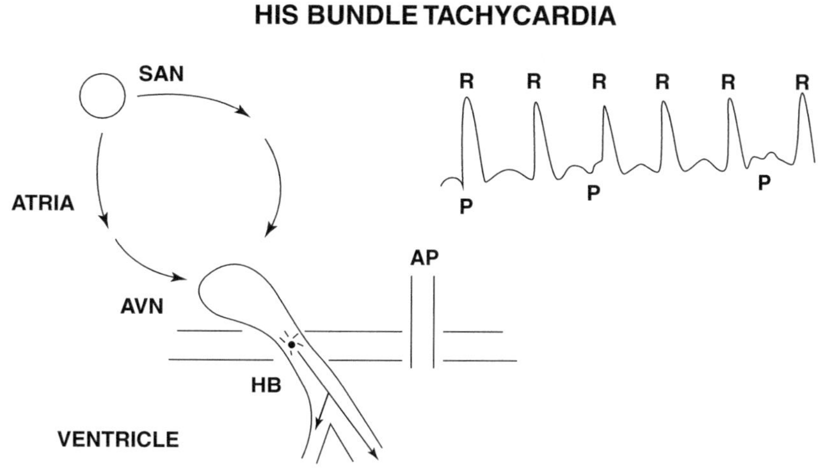

Fig. 4.15 Diagram showing genesis of the junctional ectopic tachycardia. *SAN* sinoatrial node, *AVN* atrioventricular node, *HB* His bundle, *AP* accessory path. (Reproduced with permission from: Mittal SR. Narrow QRS Tachycardia. Cardiology Today 2012;16:82–89)

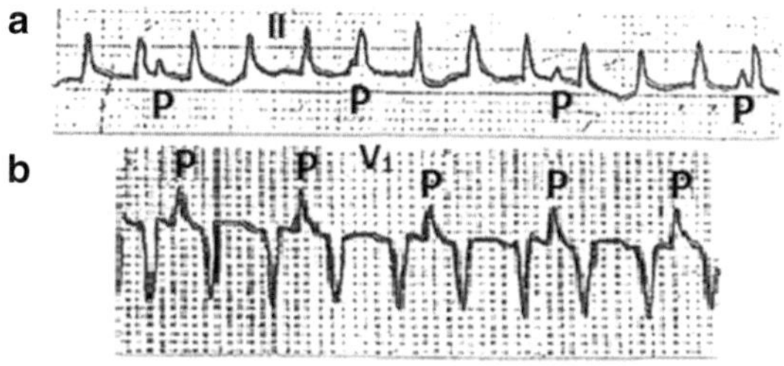

Fig. 4.16 (**a**) Lead II shows junctional ectopic tachycardia with atrioventricular dissociation. There is no association between the P waves and the QRS complexes. (**b**) Lead V_1 showing junctional ectopic tachycardia with 2:1 retrograde atrial conduction. Alternating QRS complex is followed by a P wave. (Reproduced with permission from: Mittal SR. Narrow QRS Tachycardia. Cardiology Today 2012;16:82–89)

irregular retrograde conduction to the atria (Fig. 4.16b). Ventricular premature beat may cause cessation of arrhythmia by depolarizing the abnormal focus.

Summary

- Sinus tachycardia has a gradual onset and termination. The rate varies from 100 to 180/min. Each QRS is preceded by a sinus P wave and a normal PR interval. Sinus arrhythmia supports the diagnosis of sinus tachycardia.
- Abnormal P wave preceding each QRS with a normal PR interval suggests the diagnosis of atrial tachycardia. A positive or biphasic P wave in lead V_1 suggests a left atrial focus. A negative P wave in lead V_1 suggests a right atrial focus. P wave nearly identical to the sinus P wave suggests a focus close to the sinoatrial node. Interval between P waves is isoelectric unlike atrial flutter waves. Ventricular rate is regular.
- If atrial rate is very fast or atrioventricular conduction is delayed, second-degree atrioventricular block can appear. This is called atrial tachycardia with block. It is usual in digitalis overdose.
- If there are three or more different contours of the P wave with changing PR interval, it is known as multifocal (chaotic) atrial tachycardia. It is common in decompensated cor pulmonale. Administration of theophylline or beta stimulants (e.g., salbutamol or terbutaline) could be contributory. Digitalis overdose is not a usual cause.
- Narrow QRS regular tachycardia with no apparent P waves or P waves deforming the initial or the terminal portion of the QRS complex producing a pseudo s wave in leads II, III, and aVF or pseudo r' wave in lead V_1 suggests typical (slow-fast) atrioventricular nodal reentrant tachycardia. Onset usually follows a premature atrial beat which conducts with a prolonged PR interval. The rate of the tachycardia may increase over the initial few beats or slow over a few beats before termination.

Termination may be followed by a pause because of tachycardia-dependent suppression of the sinoatrial node. Electrical alternans (change in the amplitude of alternate QRS) can occur if tachycardia is fast. Rate-dependent BBB developing during tachycardia does not change tachycardia cycle length.

- In atypical (fast-slow) atrioventricular nodal reentrant tachycardia, the P wave is inverted and falls clearly outside the QRS complex with the result that RP′ interval is equal to or longer than the interval between the P′ wave and next R wave (P′-R interval).
- Significant ST segment depression during AVNRT or T wave inversion appearing after tachycardia does not predict myocardial ischemia.
- During sinus rhythm, sudden prolongation of the PR interval, PR interval alternans, or two QRS complexes following a single P wave suggest dual atrioventricular nodal pathway.
- In a narrow QRS tachycardia, P wave occurring clearly after completion of QRS in the ST segment or early part of the T wave suggests atrioventricular reentrant tachycardia involving an accessory pathway. Negative P wave in lead I suggests left-sided accessory pathway. P wave inverted in leads II, III, and aVF suggests septal concealed accessory pathway.
- If during tachycardia rate-dependent bundle branch block develops in the ventricle in which the accessory pathway is located, the cycle length of the tachycardia lengthens.
- Atrioventricular dissociation or block during a narrow QRS tachycardia excludes possibility of AVNRT or involvement of accessory pathway.
- Electrocardiographic differential diagnosis of narrow QRS tachycardias can also be summarized as under.
  - Irregularly irregular tachycardia—atrial fibrillation, multifocal atrial tachycardia
  - Regularly irregular QRS (group beating)
  - Wenckebach periodicity or varying atrioventricular block, atrial rate 200 ± 50, isoelectric interval between p waves—atrial tachycardia
  - Sinoatrial sawtooth flutter waves without isoelectric interval—atrial flutter
  - P wave not visible—atrioventricular nodal reentry tachycardia
  - P wave deforming initial or terminal part of QRS—atrioventricular nodal reentry tachycardia
  - P wave clearly visible between two QRS

    Sinus configuration of P wave preceding each QRS—sinus tachycardia, inappropriate sinus tachycardia/sinoatrial nodal reentrant tachycardia

    Abnormal P wave—P wave is close to previous QRS, i.e., interval between QRS complex and P′ wave (RP′ interval is less than the distance between P′ wave and next QRS)—atrioventricular reciprocating tachycardia/atrial tachycardia with prolonged PR interval

    P wave close to the next QRS, i.e., distance between QRS and P′ wave is more than distance between P′ wave and next QRS—atypical atrioventricular nodal reentry tachycardia (fast-slow), atrial tachycardia/atrioventricular reciprocating tachycardia with slow-conducting accessory pathway

    Three or more different contours of the P wave with changing PR interval—chaotic atrial tachycardia

MCQs

Q1. The upper limit of normal automaticity of the Purkinje system is:

(a) 100 beats/min
(b) 60 beats/min
(c) 50 beats/min
(d) 40 beats/min

Q2. Tachyarrhythmias due to reentry mechanism show:

(a) Sudden onset with sudden termination
(b) Sudden onset with gradual termination

(c) Gradual onset with sudden termination
(d) Gradual onset with gradual termination

Q3. P wave rate of 150/min, P wave inverted in lead II, normal PR interval, and narrow QRS suggest:

(a) Sinus tachycardia
(b) Tachycardia arising from low atrium
(c) Junctional tachycardia
(d) Ventricular tachycardia

Q4. Tachycardia with normal P wave, normal PR interval, and a wide QRS suggests:

(a) Sinus tachycardia with preexcitation
(b) Sinus tachycardia with preexisting intraventricular conduction defect
(c) Sinus tachycardia with aberrant ventricular conduction
(d) Ventricular tachycardia

Q5. Tachycardia with discrete P waves with antegrade orientation, normal PR interval, and narrow QRS suggests:

(a) Sinus tachycardia
(b) Atrial tachycardia
(c) Typical AVNRT
(d) Atrioventricular reentry tachycardia

Q6. Which electrocardiographic findings suggest non-sinus origin of a tachycardia?

(a) Abnormal configuration of P waves
(b) Inverted P wave in lead II
(c) Inverted P wave in lead I
(d) Upright P wave in lead II

Q7. Abnormal P waves occurring regularly at a rate of 90/min with normal PR interval and normal QRS suggest:

(a) Accelerated atrial rhythm
(b) Atrial tachycardia
(c) AVNRT
(d) AVRT

Q8. Digitalis overdose with potassium depletion can cause:

(a) Sinus tachycardia
(b) Sinus node reentry tachycardia
(c) Atrial tachycardia
(d) Atrial tachycardia with block

Q9. Abnormal P waves of uniform morphology at a regular rate of 160/min with normal QRS at regular rate of 80/min suggest:

(a) Accelerated atrial rhythm
(b) Atrial tachycardia with block
(c) Accelerated junctional rhythm
(d) Accelerated ventricular rhythm

Q10. P waves at an irregular rate of 130/min, three or more different morphologies of the P waves, changing PR interval, and narrow QRS suggest:

(a) Accelerated junctional rhythm
(b) Atrial tachycardia with block
(c) Multifocal atrial tachycardia
(d) Atrioventricular reentry tachycardia

Q11. Chaotic atrial tachycardia:

(a) Occurs in digitalis overdose
(b) Occurs in chronic cor pulmonale
(c) Can progress to atrial fibrillation
(d) Can progress to ventricular tachycardia

Q12. Narrow QRS at a regular rate of 90/min without any accompanying P wave suggests:

(a) Accelerated atrial rhythm
(b) Accelerated junctional rhythm
(c) Atypical AVNRT
(d) Orthodromic AVRT

Q13. Narrow QRS at a regular rate 110/min with inverted P waves in the ST segment suggests:

(a) Sinus tachycardia

(b) Accelerated junctional rhythm
(c) Junctional tachycardia with retrograde conduction
(d) Typical AVNRT

Q14. Narrow QRS at a regular rate of 120/min with normal P waves at a rate of 80/min without any constant relation between P and QRS suggests:

(a) Sinus arrhythmia
(b) Junctional tachycardia
(c) Junctional tachycardia with atrioventricular dissociation
(d) Orthodromic AVRT

Q15. A narrow QRS tachycardia at a rate of 180/min without any P wave suggests:

(a) Typical AVNRT
(b) Atypical AVNRT
(c) Atrial tachycardia
(d) Sinus node reentry tachycardia

Q16. A narrow QRS tachycardia with r' in lead V_1 suggests:

(a) Typical AVNRT
(b) Atypical AVNRT
(c) Atrial tachycardia
(d) Sinus node reentry tachycardia

Q17. An absolutely regular narrow QRS tachycardia with normal P wave, normal PR interval, and normal QRS suggests:

(a) Sinus tachycardia
(b) Sinus node reentry tachycardia
(c) Typical atrioventricular nodal reentrant tachycardia
(d) Atrioventricular reentry tachycardia

Q18. A narrow QRS tachycardia with normal P waves, sinus arrhythmia, normal PR interval, and normal QRS suggests:

(a) Sinus tachycardia
(b) Sinus node reentry tachycardia
(c) Atypical AVNRT
(d) Atrioventricular reentry tachycardia

Q19. Typical AVNRT has:

(a) Pseudo r' in lead V_1
(b) Pseudo s in leads II, III, and aVF
(c) RP′ interval shorter than the P′R interval
(d) P′R interval shorter than RP′ interval

Q20. A narrow QRS tachycardia with clearly delineated inverted P waves in the leads II, III, and aVF and RP′ interval equal to or longer than the P′-R interval suggests:

(a) Atypical AVNRT
(b) Junctional tachycardia
(c) Atrial tachycardia
(d) Atrioventricular reentry tachycardia

Q21. A narrow QRS tachycardia changes into broad QRS tachycardia without change in tachycardia rate. It suggests:

(a) AVNRT with development of bundle branch block
(b) Change of supraventricular tachycardia into ventricular tachycardia
(c) Atrioventricular reentry tachycardia with development of bundle branch block on the side of the accessory pathway
(d) Atrioventricular reentry tachycardia with development of bundle branch block on the side opposite to the side of accessory pathway

Q22. Which of the following findings suggest a dual atrioventricular nodal pathway on ECG during sinus rhythm?

(a) Sudden prolongation of PR interval
(b) PR interval alternans

(c) Two QRS complexes in response to one P wave
(d) None of above

Q23. Which ECG changes during or after termination of AVNRT do not predict myocardial ischemia?

(a) Significant ST segment depression
(b) QRS alternans
(c) Newly acquired T wave inversions immediately after conversion of AVNRT to sinus rhythm
(d) All

Q24. Narrow QRS tachycardia with RP′ interval longer than the P′-R interval occurs in:

(a) Atypical AVNRT
(b) Atrial tachycardia
(c) Atrioventricular reentry tachycardia with a slow-conducting retrograde pathway
(d) Atrioventricular reentry tachycardia

Q25. Narrow QRS tachycardia with P′-R interval longer than RP′ interval is seen in:

(a) Sinus node reentry tachycardia
(b) Atrial tachycardia
(c) Typical AVNRT
(d) Atrioventricular reentry tachycardia

Q26. Narrow QRS tachycardia with inverted P waves in leads II, III, and aVF after the completion of the QRS complex in the ST segment or early in the T wave suggests:

(a) Orthodromic AVRT
(b) Antidromic AVRT
(c) Typical AVNRT
(d) Atrial tachycardia

Q27. Broad QRS tachycardia showing delta wave suggests:

(a) Orthodromic AVRT
(b) Antidromic AVRT
(c) Atypical AVNRT
(d) Sinus node reentry tachycardia

Answers

(1) d; (2) a; (3) b; (4) b, c; (5) a; (6) a, b, c; (7) a; (8) d; (9) b; (10) c; (11) b, c; (12) b; (13) c; (14) c; (15) a; (16) a; (17) b; (18) a; (19) a, b, c; (20) a, c; (21) a; (22) a, b, c; (23) a, b, c, d; (24) a, b, c; (25) c, d; (26) a; (27) a

References

1. Oligin JE, Zipes DP. Specific arrhythmias: diagnosis and treatment. In: Libby P, Bonow RO, Mann DL, Zipes DP, editors. Braunwald's heart disease. Philadelphia: Saunders; 2008. p. 863–922.
2. Wagner GS, Wang TY. Accelerated automaticity. In: Wagner GS, editor. Marriott's practical electrocardiography. Philadelphia: Wolters Kluwer; 2008. p. 288–300.
3. Prystowsky EN, Padanilam BJ, Waldo AL. Atrial fibrillation, atrial flutter and atrial tachycardia. In: Fuster V, Walsh RA, Harrington RA, editors. Hurst's the heart. New York: McGraw Hill; 2011. p. 963–81.
4. Calkins H. Supraventricular tachycardia: atrioventricular nodal reentry and Wolff-Parkinson White syndrome. In: Fuster V, Walsh RA, Harrington RA, editors. Hurst's the heart. New York: McGraw Hill; 2011. p. 987–1005.
5. Schamroth L. Ectopic atrial rhythm. In: Schamroth C, editor. An introduction to electrocardiography. France: Blackwell Science; 1982. p. 330–41.
6. Gilbert M, Wagner GS. Re-entrant junctional tachyarrhythmias. In: Wagner GS, editor. Marriott's practical electrocardiography. Philadelphia: Wolters Kluwer; 2008. p. 327–45.

5 Atrial Flutter

1 Mechanism

It is produced by a macro-reentry circuit (MR) usually involving the cavo-tricuspid isthmus in the right atrium [1] (Fig. 5.1).

It is initiated by a premature atrial contraction (PAC) (Fig. 5.2).

Uncommon varieties include left atrial flutter, double wave reentry flutter, and lower loop and upper loop reentry right atrial flutter [1]. Rarely rapid focal discharge may be responsible. It is uncommon in rheumatic heart disease. Hypertension, heart failure, ischemic heart disease (Fig. 5.3), cardiomyopathy, myocarditis, and thyrotoxicosis are common causes. Atrial flutter can develop late in patients after surgery for congenital heart disease or following ablation for atrial fibrillation. It may occur in elderly without apparent cause.

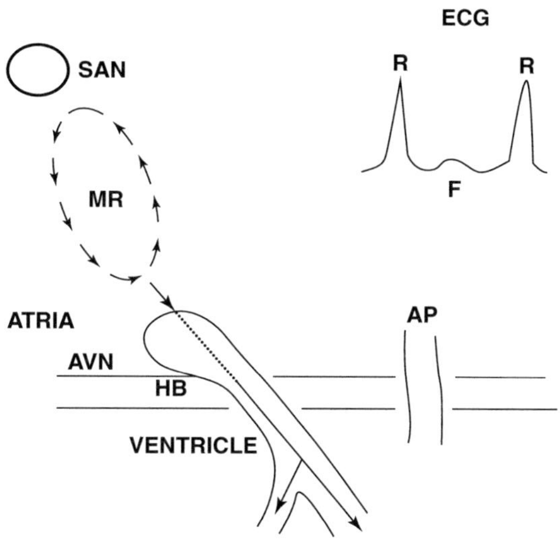

Fig. 5.1 Diagram showing mechanism of atrial flutter. *SAN* sinoatrial node, *AVN* atrioventricular node, *HB* His bundle, *AP* accessory pathway, *ECG* electrocardiogram, *F* flutter wave, *MR* macro-reentry. (Reproduced with permission from: Mittal SR. ECG of the month. Cardiology Today. 2012;16:134–137)

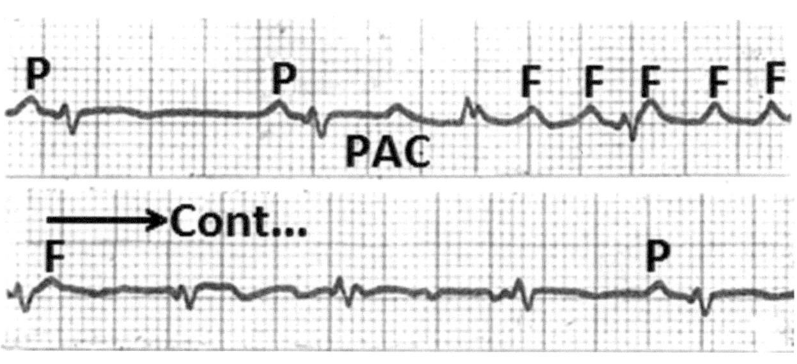

Fig. 5.2 Showing initiation of atrial flutter (F) by a premature atrial contraction (PAC) and spontaneous conversion to sinus rhythm. P—sinus P wave. (Reproduced with permission from: Mittal SR. ECG of the month. Cardiology Today. 2012;16:134–137)

S. Mittal, *Insights into Electrocardiograms with MCQs*,
https://doi.org/10.1007/978-981-99-0127-2_5

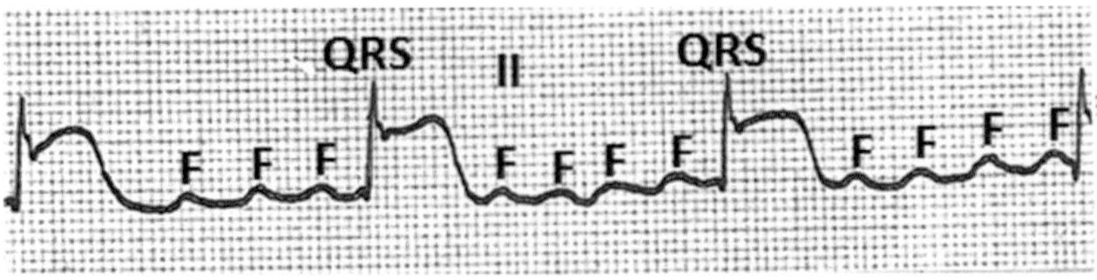

Fig. 5.3 Electrocardiogram shows atrial flutter in a case of inferior infarction. *F* atrial flutter wave. (Reproduced with permission from: Mittal SR. ECG of the month. Cardiology Today. 2012;16:134–137)

2 Atrial Waves

Flutter (F) waves:

(a) Rate is around 250–350/min [2]. It can be as slow as 200/min in patients treated with anti-arrhythmic drugs. Rarely, it can be as fast as 400/min (Fig. 5.4).

(b) Atrial flutter waves have regular sawtooth appearance without any isoelectric line (Fig. 5.5, lead II). These waves are best seen in leads II, III, aVF, and V_1 [2]. Rarely atrial flutter waves may be seen clearly only in the precordial leads V_1 to V_3 [3]. In precordial leads atrial flutter waves may mimic discrete P waves with isoelectric line [3] (Fig. 5.5). Rate and configuration in the inferior leads help differentiation from atrial tachycardia. Such discrete atrial flutter waves are seen especially in atria that have an extensive scar [1].

(c) Depending on the configuration of "F" wave in lead II, III, and aVF, atrial flutter can be as follows:
   - Type I (common, typical, counterclockwise)—Flutter waves are negative (inverted) in leads II, III, and aVF [2] (Fig. 5.6). The reentry wave travels down the right atrial free wall and upward along interatrial septum.
   - Type II (uncommon, reverse typical, clockwise)—Flutter waves are positive in leads II, III, and aVF [1] (Fig. 5.7). The reentry wave travels down the interatrial septum and upward along right atrial free wall.

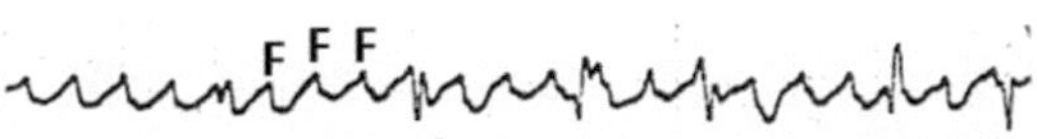

Fig. 5.4 Electrocardiogram shows fast atrial flutter (F). (Reproduced with permission from: Mittal SR. ECG of the month. Cardiology Today. 2012;16:134–137)

Atrial flutter not involving cavo-tricuspid isthmus may not have diagnostic contour, but rate between 250–350/min helps diagnosis. Positive flutter wave in lead I can suggest left atrial origin of atrial flutter. Morphology of the atrial flutter waves may change during the same episode of atrial flutter [2] (Fig. 5.8).

Atrial flutter may spontaneously change to atrial fibrillation [3]. Digitalis promotes conversion of atrial flutter to atrial fibrillation (Fig. 5.9a, b) by shortening the refractory period of the atrial myocardial cells [3]. As atrial flutter changes to atrial fibrillation, ventricular rate may change from rapid and regular to relatively slow and irregular [3]. Careful analysis of basal atrial rhythm helps in correct diagnosis.

Control of ventricular rate by drugs is easier in atrial fibrillation than in atrial flutter. Atrial flutter may spontaneously revert to sinus rhythm (Fig. 5.2). If sawtooth appearance shows some irregularity or distortion suggestive of atrial fibrillation, it is termed "atrial flutter-fibrillation" (Fig. 5.10). Most of these cases need to be managed as atrial fibrillation.

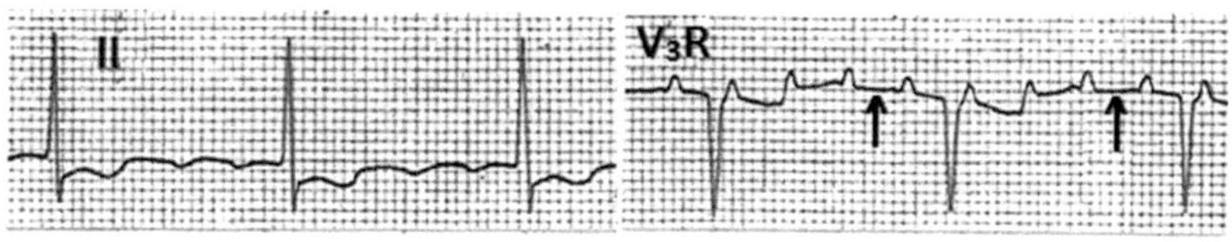

Fig. 5.5 Lead II shows atrial flutter. Lead V_1 from the same patient shows discrete P waves and isoelectric line (arrow). (Reproduced with permission from: Mittal SR. ECG of the month. Cardiology Today. 2012;16:134–137)

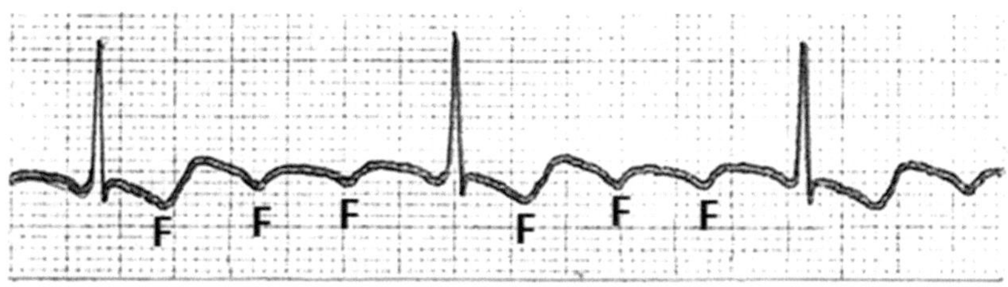

Fig. 5.6 Electrocardiogram is showing type I atrial flutter with negative "F" waves in lead II. (Reproduced with permission from: Mittal SR. ECG of the month. Cardiology Today. 2012;16:134–137)

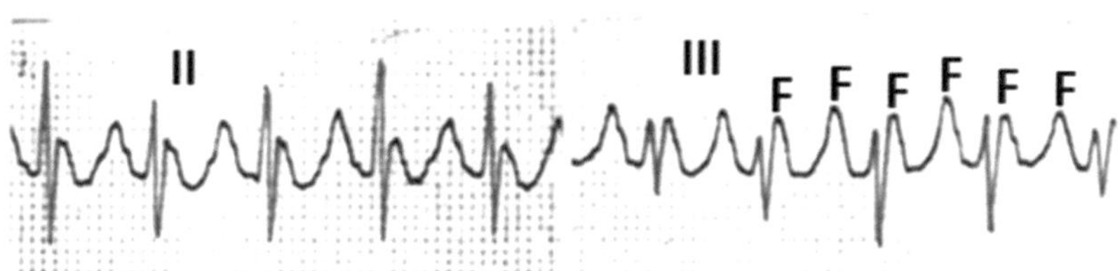

Fig. 5.7 Electrocardiogram is showing type II atrial flutter with positive "F" waves in leads II and III. (Reproduced with permission from: Mittal SR. ECG of the month. Cardiology Today. 2012;16:134–137)

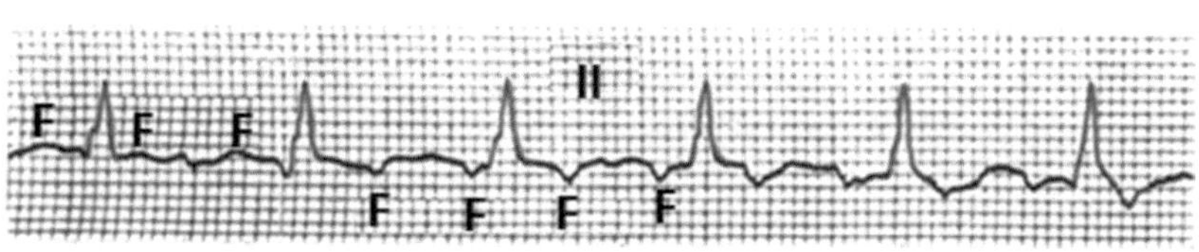

Fig. 5.8 Electrocardiogram shows the changing morphology of "F" waves. (Reproduced with permission from: Mittal SR. ECG of the month. Cardiology Today. 2012;16:134–137)

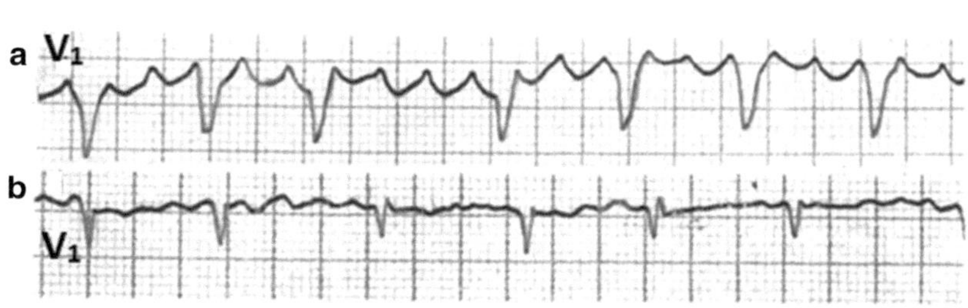

Fig. 5.9 Electrocardiogram is showing conversion of atrial flutter (**a**) to atrial fibrillation (**b**) with digoxin. (Reproduced with permission from: Mittal SR. ECG of the month. Cardiology Today. 2012;16:134–137)

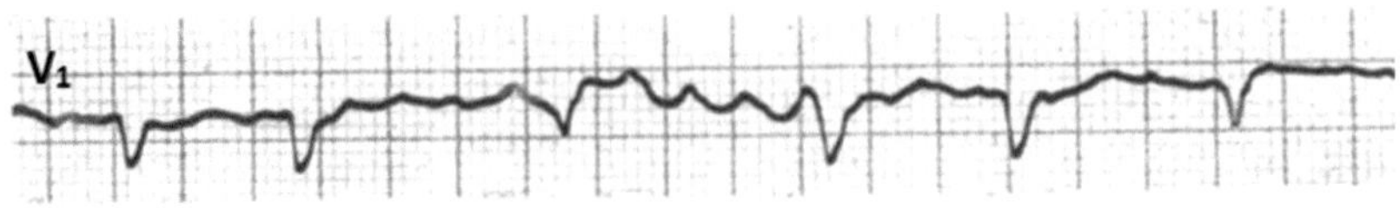

Fig. 5.10 Electrocardiogram is showing "atrial flutter-fibrillation." (Reproduced with permission from: Mittal SR. ECG of the month. Cardiology Today. 2012;16:134–137)

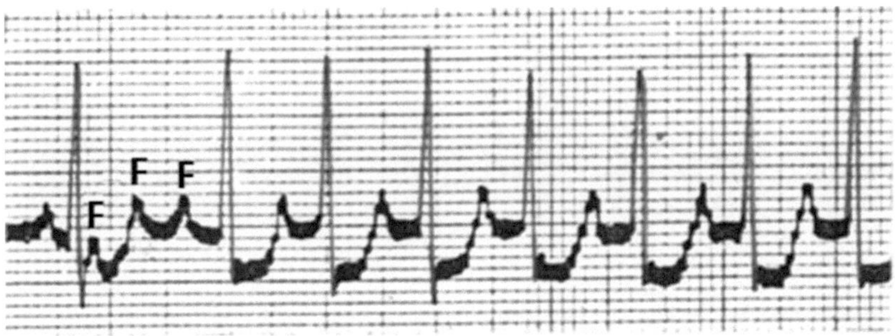

Fig. 5.11 Electrocardiogram is showing atrial flutter (marked F) (clear from initial beats) going into 2:1 conduction mimicking atrial or atrioventricular nodal reentry tachycardia. (Reproduced with permission from: Mittal SR. ECG of the month. Cardiology Today. 2012;16:134–137)

3 Atrioventricular Conduction and Ventricular Rate

(a) In atrial flutter, atrial rate is around 300/min. In normal persons, atrioventricular node can conduct up to 180 impulses per minute due to its long refractory period. Therefore, only alternate atrial flutter ("F") wave crosses the atrioventricular node to reach the ventricles (2:1 atrioventricular conduction). This results in classical ventricular rate of 150/min in an untreated patient [2]. In such an electrocardiogram, diagnosis of atrial flutter may be difficult because one of the "F" waves is partly or completely masked by the QRS and the other is masked by the T wave [3]. Such electrocardiogram is likely to be confused with atrial or atrioventricular nodal reentry tachycardia (Fig. 5.11).

Procedures or drugs that slow *atrioventricular* conduction can unmask the "F" wave. These include carotid compression (Fig. 5.12a, b) or intravenous diltiazem, adenosine, or beta-blockers.

1:1 atrioventricular conduction can occur in patients with an accessory pathway with short refractory period. This can cause regular wide QRS complexes without clear atrial activity and may lead to an erroneous diagnosis of ventricular tachycardia [3]. Blocking of the atrioventricular node does not help in diagnosis as antegrade conduction is over an accessory pathway. Electric cardioversion is effective in termination of such tachyarrhythmia. Presence of preexcitation on previous or subsequent ECG in sinus rhythm can help diagnosis. Hyperthyroidism and inherently fast atrioventricular nodal conduction may also produce 1:1 atrioventricular conduction (Fig. 5.13a). Carotid compression helps in unmasking atrial flutter in such cases (Fig. 5.13b). Concomitant occurrence of ventricular tachycardia with atrial flutter can also produce nearly similar ECG. Changing relation between "F" wave and QRS (atrioventricular dissociation) suggests diagnosis of atrial flutter with concomitant ventricular tachycardia.

(b) Atrial flutter rate is reduced by antiarrhythmic drugs that prolong the refractory period of the atrial muscle (quinidine, procainamide, flecainide) (Fig. 5.14). If the atrial

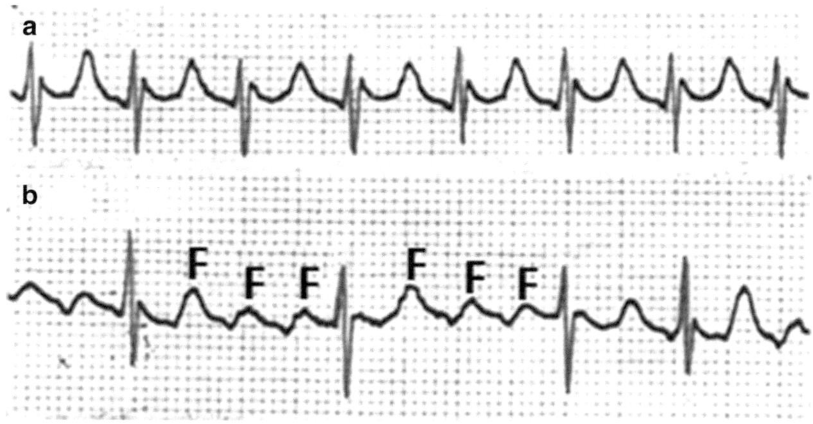

Fig. 5.12 Electrocardiogram is showing unmasking of atrial flutter with 2:1 atrioventricular conduction (**a**) by carotid massage (**b**). (Reproduced with permission from: Mittal SR. ECG of the month. Cardiology Today. 2012;16:134–137)

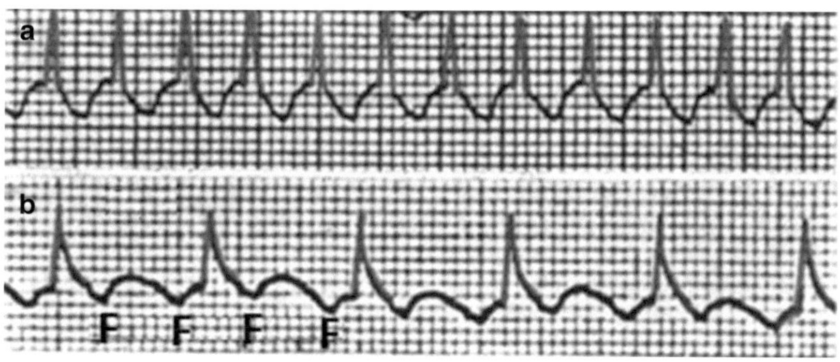

Fig. 5.13 Electrocardiogram is showing atrial flutter with 1:1 atrioventricular conduction (**a**) unmasked by decreasing atrioventricular conduction (**b**). *F* atrial flutter waves. (Reproduced with permission from: Mittal SR. ECG of the month. Cardiology Today. 2012;16:134–137)

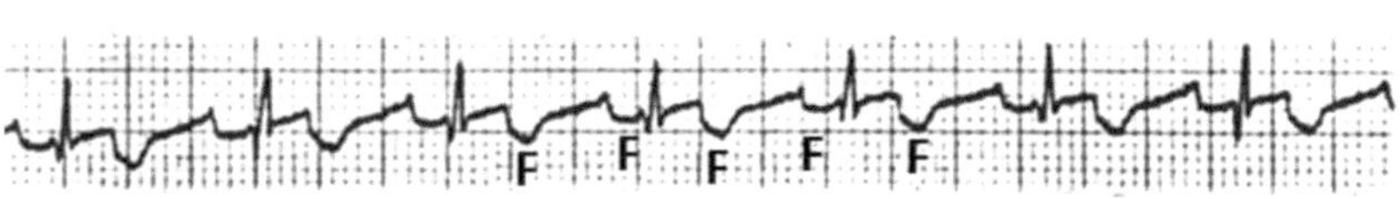

Fig. 5.14 Electrocardiogram is showing slow atrial flutter following treatment with class 1A antiarrhythmic. *F* atrial flutter waves. (Reproduced with permission from: Mittal SR. ECG of the month. Cardiology Today. 2012;16:134–137)

rate is reduced up to 200/min, each atrial impulse may reach the atrioventricular node after recovery of its refractory period and may cross the atrioventricular node (1:1 conduction). This results in an increase in the ventricular rate. Anticholinergic effect of these drugs also facilitates atrioventricular conduction. Therefore, in the management of atrial flutter, class 1A antiarrhythmic agents should be used in combination with drugs that block the atrioventricular conduction (digoxin, beta-blockers, verapamil) to avoid increase in atrioventricular conduction and ventricular rate.

(c) If there is slow atrioventricular conduction due to atrioventricular nodal disease or drugs (digoxin, verapamil, adenosine), atrioventricular ratio may increase from 2:1 to 3:1 or

4:1 with a corresponding ventricular rate of 100/min or 75/min. Atrioventricular conduction is most often even numbered, e.g., 2:1 or 4:1. Odd numbered atrioventricular ratios, e.g., 3:1 or 5:1, are uncommon (Fig. 5.15).

Carotid massage decreases the ventricular rate in stepwise fashion. Change in the atrioventricular conduction from 2:1 to 3:1 and 4:1 suddenly reduces ventricular rate from 150/min to 100/min and then to 75/min. On stopping the carotid massage, ventricular rate jumps back to 100/min and then to 150/min. There is no gradual change as seen in sinus tachycardia or atrial fibrillation.

(d) Atrioventricular block can produce the following:
   - Regularly irregular ventricular response [3]:
   - For example, alternating 2:1 and 4:1 conduction (Fig. 5.16). This occurs because of Wenckebach block and produces group beating of QRS [3] (Fig. 5.16).
   - Irregularly irregular ventricular response [3]:
   - For example, switching between different atrioventricular conduction ratios (Fig. 5.17).

Slow ventricular rate [3] with:

- A constant relation between atrial flutter waves and the QRS suggests second-degree atrioventricular block (Fig. 5.18).
- Changing relation between atrial flutter waves (F) and QRS complex suggests complete atrioventricular block (Fig. 5.19).

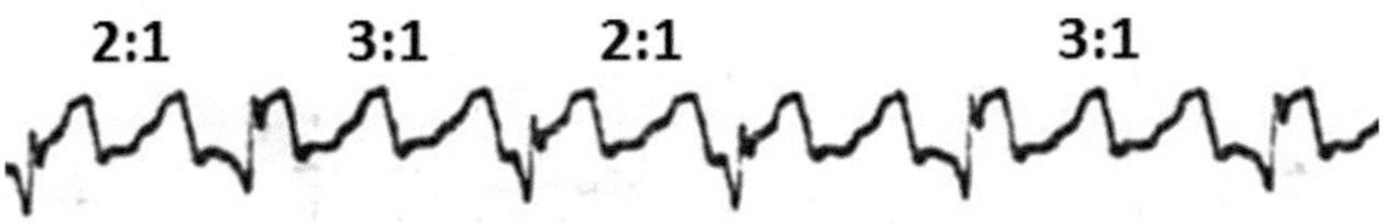

Fig. 5.15 Electrocardiogram is showing atrioventricular conduction changing between 2:1 and 3:1. (Reproduced with permission from: Mittal SR. ECG of the month. Cardiology Today. 2012;16:134–137)

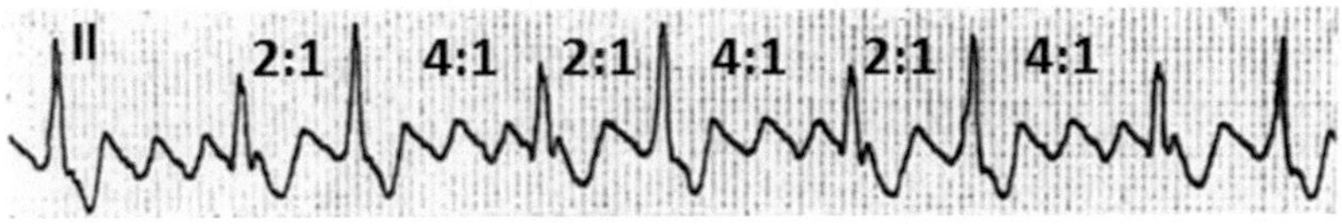

Fig. 5.16 Electrocardiogram showing group beating due to alternate 2:1 and 4:1 conduction. (Reproduced with permission from: Mittal SR. ECG of the month. Cardiology Today. 2012;16:134–137)

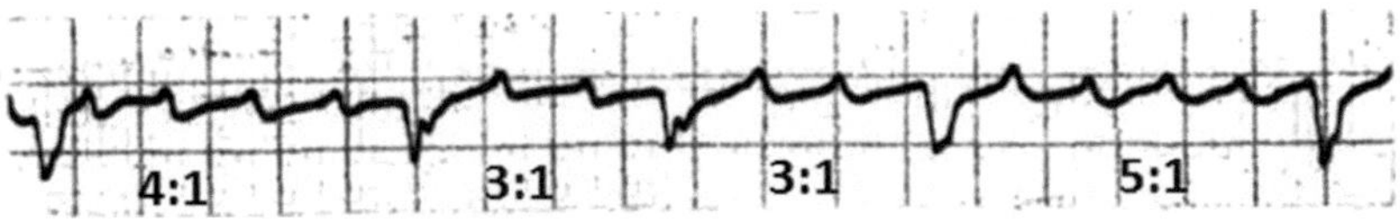

Fig. 5.17 Electrocardiogram is showing irregularly irregular ventricular rhythm due to changing atrioventricular conduction of the atrial flutter waves. (Reproduced with permission from: Mittal SR. ECG of the month. Cardiology Today. 2012;16:134–137)

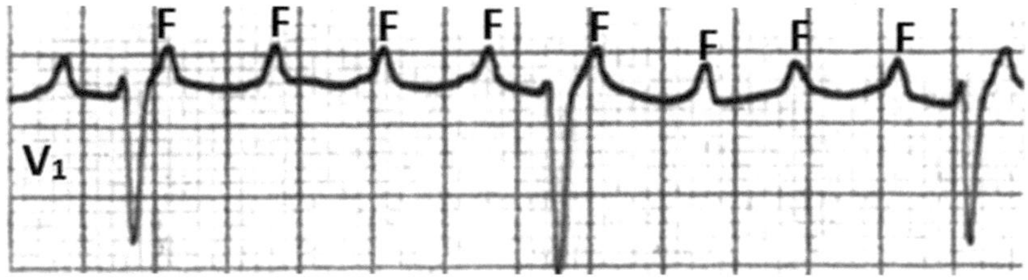

Fig. 5.18 Electrocardiogram is showing atrial flutter waves (F) with bradycardia and fixed interval between atrial flutter waves (F) and the QRS. (Reproduced with permission from: Mittal SR. ECG of the month. Cardiology Today. 2012;16:134–137)

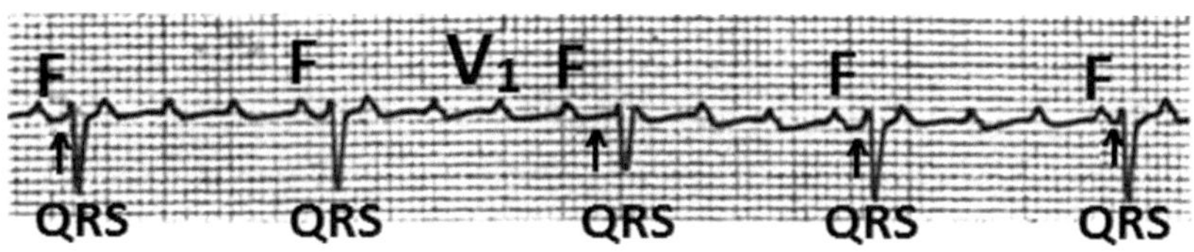

Fig. 5.19 Electrocardiogram is showing atrial flutter with bradycardia and changing interval (marked arrow) between the atrial flutter waves (marked F) and the QRS complex. (Reproduced with permission from: Mittal SR. ECG of the month. Cardiology Today. 2012;16:134–137)

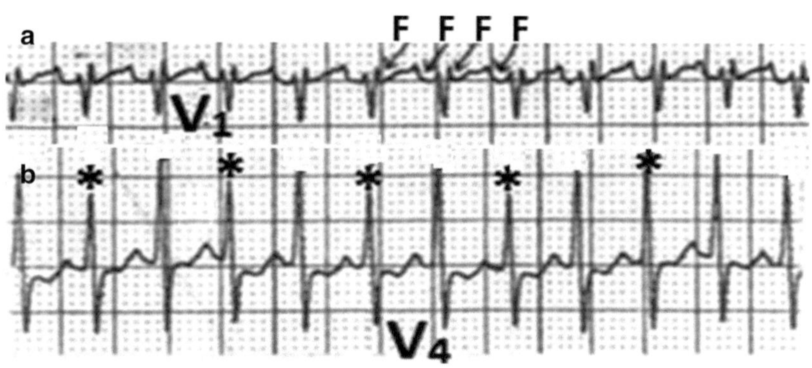

Fig. 5.20 (**a**) Lead V_1 is showing atrial flutter (F) (marked arrow) with 2:1 atrioventricular conduction. (**b**) Lead V_4 is showing changing amplitude of QRS in alternate beats (electrical alternans marked *). (Reproduced with permission from: Mittal SR. ECG of the month. Cardiology Today. 2012;16:134–137)

4 QRS Configuration

(a) Normally the QRS complex is narrow due to normal intraventricular conduction.
(b) Electrical alternans may be present at a fast rate (Fig. 5.20a, b).
(c) Atrial flutter with complete atrioventricular block can result in intermittent fusion of the atrial flutter waves ("F") and the QRS complex resulting in intermittent change in the configuration of QRX complex (Fig. 5.21).
(d) Regular broad QRS may be seen with:
   - Preexisting bundle branch block
   - 1:1 atrioventricular conduction with aberrant ventricular conduction
   - Antegrade conduction over an accessory pathway
   - Concomitant ventricular tachycardia

(e) Irregular broad QRS can occur with:
- Aberrant ventricular conduction—it is suggested by:
  - rSR' configuration in lead V_1 (Fig. 5.22a, b)
  - Broad QRS in short cycle following a long cycle
  - Initial QRS vector identical to narrow QRS
  - Change in the QRS configuration with change in the preceding cycle length
- Ventricular ectopics

They may have right bundle branch block configuration. If the QRS complex is upright in lead V_1, it is either uniphasic or biphasic. Initial deflection is not identical to that of narrow QRS. QRS configuration does not change with the length of the preceding cycle.

Fig. 5.21 Atrial flutter with complete atrioventricular block showing intermittent change in QRS configuration (R) due to fusion of the atrial flutter waves ("F") and the QRS complex (marked F+R). (Reproduced with permission from: Mittal SR. ECG of the month. Cardiology Today. 2012;16:134–137)

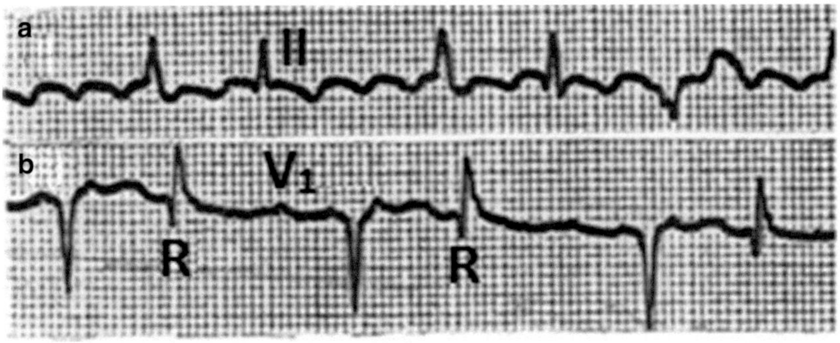

Fig. 5.22 (**a**) Lead II showing atrial flutter with alternating 2:1 and 4:1 conduction. (**b**) Lead V_1 showing aberrant conduction (marked R) in beats with short cycle length. (Reproduced with permission from: Mittal SR. ECG of the month. Cardiology Today. 2012;16:134–137)

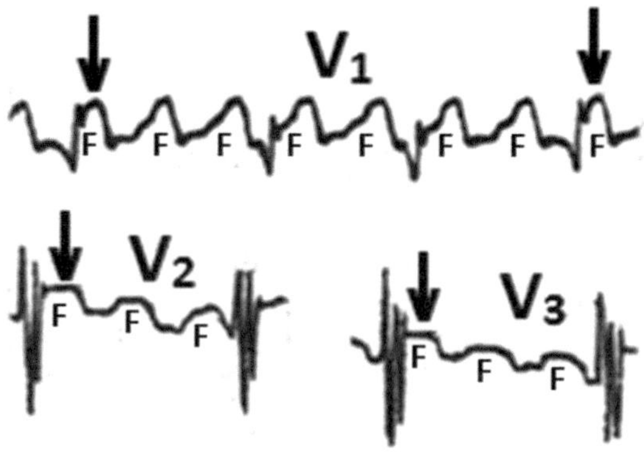

Fig. 5.23 Showing false impression of ST segment elevation due to superimposition of flutter waves (F) (arrow) in leads V_1, V_2, and V_3. (Reproduced with permission from: Mittal SR. ECG of the month. Cardiology Today. 2012;16:134–137)

5 ST Segment

Atrial flutter waves may give a false impression of ST segment deviation (Fig. 5.23).

Management Implications

Low-dose electric cardioversion is very effective in restoring sinus rhythm because it is easy to interrupt the macro-reentry circuit. For the same reason, catheter ablation is very effective. Adequate long-term control of ventricular rate with drugs is very difficult as compared to atrial fibrillation because the relatively powerful atrial flutter waves overcome drug-induced blockade of atrioventricular node.

Summary

- Atrial flutter is a macro-reentry tachycardia. Reentry circuit is mostly in the right atrium but rarely atrial flutter can arise in the left atrium.
- Atrial flutter waves appear as sawtooth waves best seen in lead II and/or lead V_1 without an isoelectric line. Usual rate of atrial flutter waves is between 250 and 350/min. Prominent flutter waves in lead I can suggest an origin in the left atrium. "Counterclockwise loop" of reentry produces negative flutter waves in lead II (typical atrial flutter). "Clockwise loop" (atypical atrial flutter) produces positive flutter waves in lead II.
- If the sawtooth appearance shows irregularity and/or rate is more than 400/min, it is labeled as atrial flutter-fibrillation and should be treated as atrial fibrillation.
- Usual atrioventricular conduction ratio is 2:1 with a ventricular rate around 150/min. At this ventricular rate, electrocardiographic diagnosis may be difficult because one flutter wave falls in the T wave and next flutter wave occurs with the QRS complex. Slowing atrioventricular conduction by carotid massage or drugs unmasks the atrial flutter waves.
- Carotid massage decreases the ventricular rate in stepwise fashion unlike the gradual reduction seen in sinus tachycardia or atrial fibrillation. Atrioventricular nodal reentrant tachycardia either remains unchanged or is terminated by carotid massage.
- The presence of accessory pathway with short refractory period can result in 1:1 atrioventricular conduction of the atrial flutter waves with precipitation of ventricular flutter/fibrillation.
- Irregular atrioventricular conduction of atrial flutter waves results in irregular ventricular rhythm. QRS with shorter RR interval may show aberrant ventricular conduction.
- Slow ventricular rate with changing relation between the flutter waves and the QRS complexes should suggest complete atrioventricular block.
- Electrical alternans (changing amplitude of QRS in alternating beats) can appear at fast ventricular rate.
- Prominent flutter waves can deform any part of the electrocardiogram.
- Adequate long-term control of ventricular rate with drugs is very difficult.

MCQs

Q1. Atrial flutter with 2:1 atrioventricular conduction can be differentiated from sinus tachycardia by:

(a) Sawtooth-like appearance of the atrial flutter waves
(b) Absence of any isoelectric interval
(c) Apparent prolongation of "PR" interval
(d) None of above

Q2. Regular tachycardia at a rate of 150/min without clear P waves could be:

(a) Atrial tachycardia with 2:1 atrioventricular conduction
(b) Atrial flutter with 2:1 atrioventricular conduction
(c) Sinus tachycardia
(d) Atrial fibrillation

Q3. Discrete P waves at a rate of 210/min and a ventricular rate of 70/min with isoelectric line between P waves suggest:

(a) Sinus tachycardia
(b) Atrial tachycardia with 3:1 atrioventricular conduction
(c) Atrial flutter with 3:1 atrioventricular conduction
(d) Atrial fibrillation

Q4. Sawtooth morphology of the "P" waves in lead II and V_1 suggests:

(a) Atrial tachycardia
(b) Atrial flutter
(c) Multifocal atrial tachycardia
(d) Left atrial rhythm

Q5. Which drugs can reduce the atrial rate in atrial flutter?

(a) Quinidine
(b) Procainamide
(c) Amiodarone
(d) Digitalis

Q6. Which drug decreases the refractory period of the atrial myocardial cell?

(a) Quinidine
(b) Procainamide
(c) Digitalis
(d) Amiodarone

Q7. In atrial flutter, ventricular rhythm can be:

(a) Regular
(b) Regularly irregular
(c) Irregularly irregular
(d) Any of above

Q8. In a case of atrial flutter with regular ventricular rhythm of 180/min, development of an irregular ventricular rhythm at a rate of around 120/min could be due to:

(a) Change of atrial flutter to atrial fibrillation
(b) Development of changing atrioventricular block
(c) New appearance of multiple ectopics
(d) Development of accelerated junctional rhythm

Q9. In a case of atrial flutter, ventricular rate can be:

(a) 250/min
(b) 200/min
(c) 150/min
(d) 100/min

Q10. In a case of tachycardia, carotid massage suddenly reduces ventricular rate from 180/min to 120/min. On stopping carotid massage, ventricular rate suddenly jumps back to 180/min. This response suggests:

(a) Suppression of the sinus node
(b) Termination of AVNRT to sinus rhythm
(c) Conversion of atrial flutter to atrial fibrillation
(d) Changing of atrioventricular block in atrial flutter

Q11. In a case of regular tachycardia, carotid massage produces gradual decrease in heart rate, and stopping carotid massage results in a gradual return to initial heart rate. This response suggests:

(a) Sinus tachycardia
(b) Antidromic AVRT
(c) Sinus node reentry tachycardia
(d) Changing atrioventricular block in atrial fibrillation

Q12. The upper limit of the number of atrial impulses that can be transmitted per minute through a normal atrioventricular node is:

(a) 300/min
(b) 250/min
(c) 200/min
(d) 150/min

Q13. What can increase the number of impulses that can be transmitted across a normal atrioventricular node?

(a) Exercise
(b) Sympathetic stimulation
(c) Parasympathetic stimulation
(d) None of above

Q14. Regularly irregular ventricular rhythm in a case of atrial flutter suggests:

(a) Changing rate of atrial flutter
(b) Changing atrioventricular conduction
(c) Two levels of atrioventricular block
(d) All

Q15. In atrial flutter, flutter waves are best seen in:

(a) Lead I
(b) Lead aVL
(c) Leads II, III, and aVF
(d) Leads V_1 and V_2

Q16. In atrial flutter, atrial rate is:

(a) 100–150/min
(b) 150–250/min
(c) 250–350/min
(d) 350–450/min

Q17. In atrial flutter, baseline:

(a) Is straight
(b) Shows regular coarse undulations
(c) Shows coarse irregular undulation
(d) Shows fine vibrations

Q18. Negative sawtooth flutter waves in leads II, III, and aVF suggest:

(a) Typical atrial flutter
(b) Reverse typical atrial flutter
(c) Left atrial flutter
(d) Double wave reentry atrial flutter

Q19. Atrial flutter is often initiated by:

(a) An atrial ectopic
(b) A junctional ectopic
(c) A ventricular ectopic
(d) Atrial fibrillation

Q20. Achieving adequate long-term control of ventricular rate in atrial flutter by drugs is:

(a) Very easy
(b) Easy
(c) Difficult
(d) Very difficult

Answers

(1) a, b, c; (2) b; (3) b; (4) b; (5) a, b, c; (6) c; (7) d; (8) a, b; (9) c, d; (10) d; (11) a; (12) d; (13) a, b; (14) b, c; (15) c, d; (16) c; (17) b; (18) a; (19) a, d; (20) d

References

1. Prystowsky EN, Padanilam BJ, Waldo AL. Atrial fibrillation, atrial flutter and atrial tachycardia. In: Fuster V, Walsh RA, Harrington RA, editors. Hurst's the heart. New York: McGraw Hill; 2011. p. 963–86.
2. Olgin JE, Zipes DP. Specific arrhythmias: diagnosis and treatment. In: Libby P, Bonow RO, Mann DL, Zipes DP, editors. Braunwald's heart disease. Philadelphia: Saunders; 2008. p. 863–922.
3. Wegner GS. Re-entry atrial tachyarrhythmias – the atrial flutter/fibrillation spectrum. In: Wagner GS, editor. Marriot's practical electrocardiography. Philadelphia: Wolters Kluwer; 2008. p. 301–26.

6 Atrial Fibrillation

1 Mechanism

It is caused by multiple micro-reentry circuits within the atria (Fig. 6.1) or a rapidly firing focus in the proximal portion of a pulmonary vein [1].

This arrhythmia can also be produced by a "focal" mechanism rather than reentry [2]. Atrial fibrillation promotes further atrial fibrillation due to electrical remodeling of the atria [1]. Atrial fibrillation in the absence of any known etiology is labeled "lone atrial fibrillation." Atrial fibrillation may be precipitated by an atrial premature beat (Fig. 6.2).

In other cases, atrial tachycardia, multifocal atrial tachycardia, or atrial flutter may degenerate into atrial fibrillation. It also forms a part of the spectrum of "sick sinus syndrome" (Fig. 6.3).

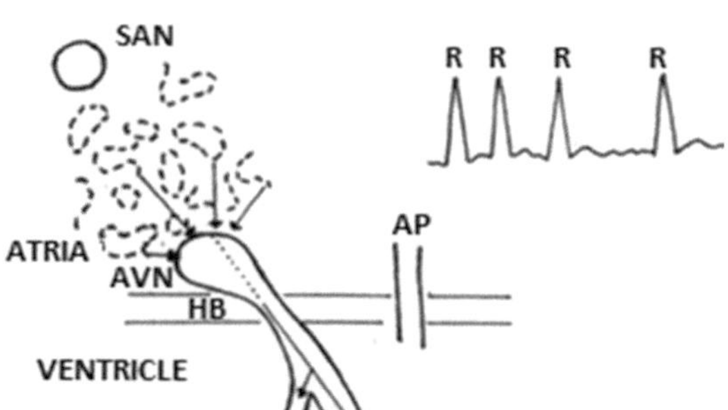

Fig. 6.1 Showing micro-reentry causing atrial fibrillation. *SAN* sinoatrial node, *AVN* atrioventricular node, *HB* His bundle, *AP* accessory pathway. (Reproduced with permission from: Mittal SR. Atrial fibrillation. Cardiology Today. 2012;14:174–177)

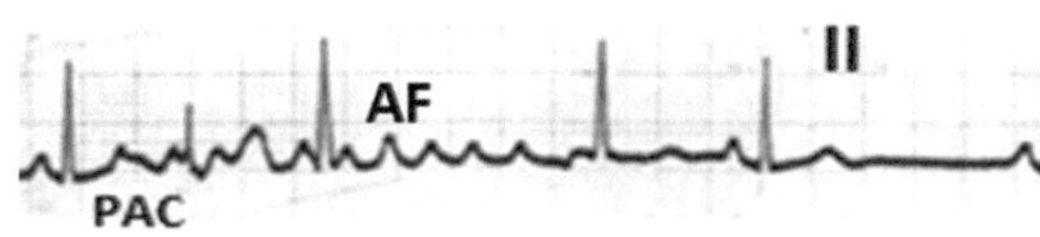

Fig. 6.2 Showing onset of transient atrial fibrillation (AF) with a premature atrial contraction (PAC). (Reproduced with permission from: Mittal SR. Atrial fibrillation. Cardiology Today. 2012;14:174–177)

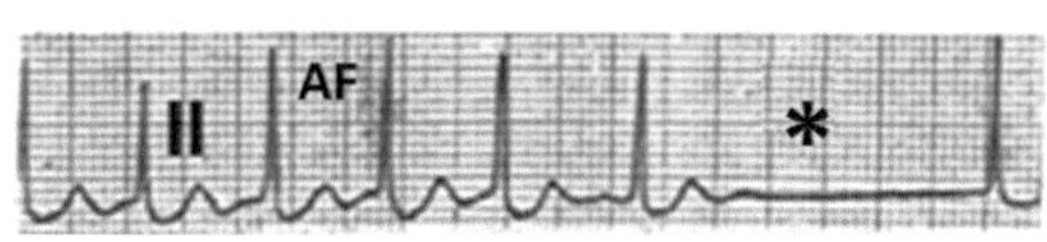

Fig. 6.3 Showing atrial fibrillation (AF) as a part of sick sinus syndrome. Fast ventricular rate is followed by a long pause (marked *). (Reproduced with permission from: Mittal SR. Atrial fibrillation. Cardiology Today. 2012;14:174–177)

In patients with sick sinus syndrome, management of atrial fibrillation with digoxin can produce longer pauses of bradycardia, and electric cardioversion can produce long period of asystole. Atrial fibrillation may be paroxysmal (transient) (Fig. 6.2) and may terminate spontaneously. Digoxin causes more frequent and longer paroxysms by decreasing the refractory period of the atrial musculature. Atrial fibrillation can also occur in the setting of acute myocardial infarction (Fig. 6.4). In the setting of acute myocardial infarction, it can be due to acute distension of the atria, concomitant atrial infarction, or pericarditis.

S. Mittal, *Insights into Electrocardiograms with MCQs*,
https://doi.org/10.1007/978-981-99-0127-2_6

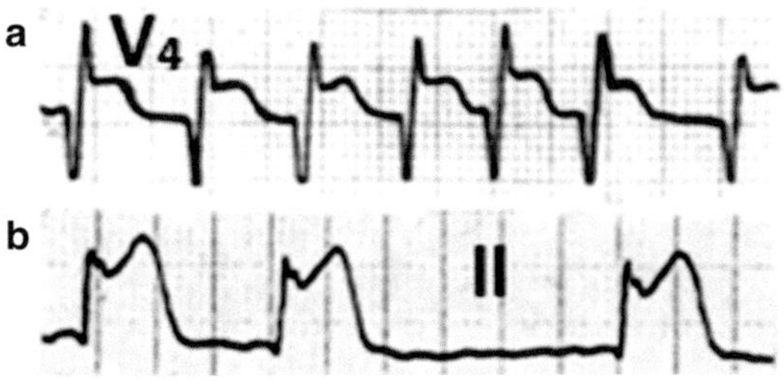

Fig. 6.4 (**a**) Atrial fibrillation in the setting of acute anterior myocardial infarction. (**b**) Atrial fibrillation in the setting of acute inferior myocardial infarction. (Reproduced with permission from: Mittal SR. Atrial fibrillation. Cardiology Today. 2012;14:174–177)

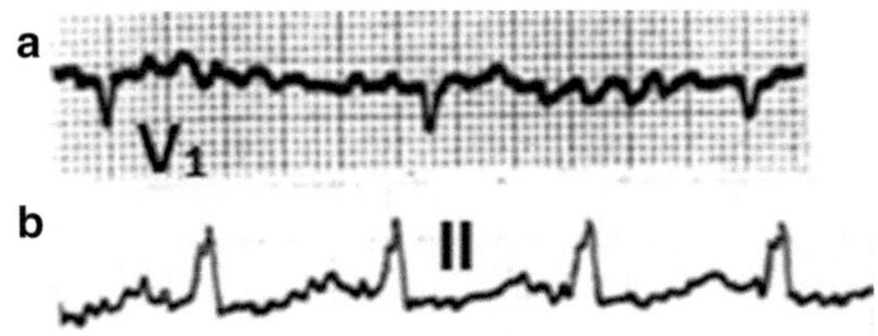

Fig. 6.5 (**a**) Showing undulating baseline in atrial fibrillation. (**b**) Showing vibrations due to artefacts. QRS complexes are regular. (Reproduced with permission from: Mittal SR. Atrial fibrillation. Cardiology Today. 2012;14:174–177)

2 Atrial Waves

Atrial activity appears as small baseline undulations of changing amplitude, morphology, and duration (Fig. 6.5a) at a rate varying between 350 and 600/min. These are called fibrillary waves ("f" waves). "f" waves should be differentiated from artefacts. Tremors produce vibrations localized to the leads which include the limb having tremors. Electrical disturbance and improper earthing produce very fast and sharp vibrations in the baseline. In both these conditions, the QRS complexes are regular (Fig. 6.5b) as opposed to irregularly irregular QRS complexes in atrial fibrillation.

When "f" waves are prominent, it is called coarse atrial fibrillation (Fig. 6.6).

When "f" waves are small or not visible, it is called fine atrial fibrillation (Fig. 6.6). This is common with long-standing atrial fibrillation. Usually "f" waves are prominent in leads II and V_1. Sometimes undulations are seen only in the right precordial leads (Fig. 6.7), and lead II may show a relatively straight line without any undulations.

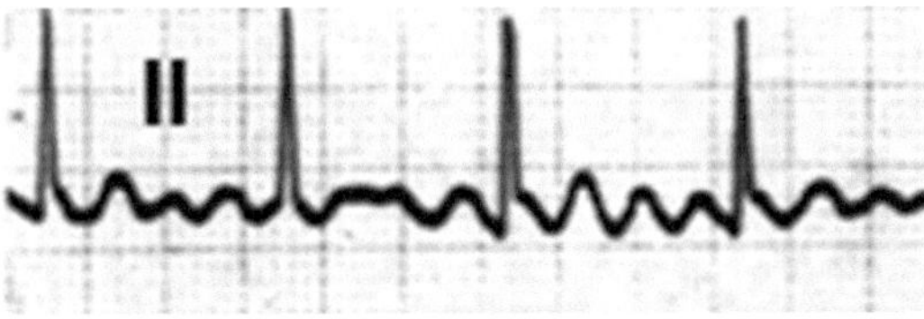

Fig. 6.6 Showing coarse atrial fibrillation. (Reproduced with permission from: Mittal SR. Atrial fibrillation. Cardiology Today. 2012;14:174–177)

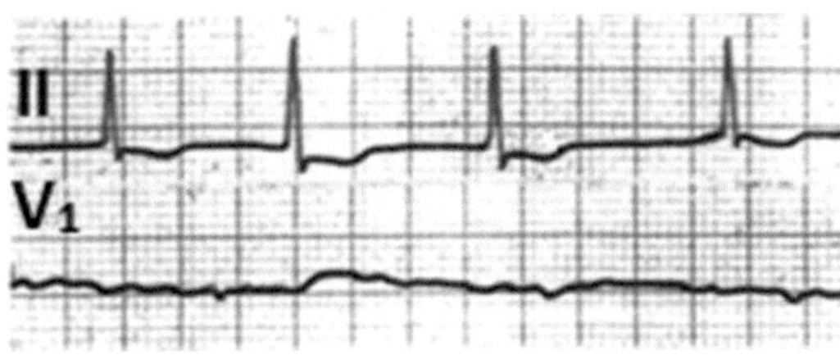

Fig. 6.7 Showing fine atrial fibrillation in lead II and undulations in lead V_1. (Reproduced with permission from: Mittal SR. Atrial fibrillation. Cardiology Today. 2012;14:174–177)

3 Atrioventricular Conduction and Ventricular Rate

(a) In atrial fibrillation, a large number of irregular atrial impulses (350–600/min) try to cross the atrioventricular node to reach the ventricles. Normal atrioventricular node can conduct only up to 180 impulses per minute due to its long refractory period. Therefore, only some impulses are able to cross the atrioventricular node. Other impulses penetrate the atrioventricular node for varying distance but are unable to cross it. These impulses, however, do depolarize the atrioventricular node for varying interval. This prevents penetration of subsequent impulses for variable interval resulting in irregularly irregular QRS. As atrial rate increases, the ventricular rate decreases because of increasing competition within the atrioventricular node [3]. At very fast ventricular rates, irregularity may be difficult to detect due to short RR interval. Carotid sinus massage can help in diagnosis by reducing the ventricular rate. Recording electrocardiogram at a higher speed may also help.

(b) If the atrioventricular node is normal, ventricular rate as high as 200/min may develop. Digoxin is helpful in controlling ventricular rate at rest by decreasing the atrioventricular conduction. It, however, poorly controls the ventricular rate during exercise and in other conditions of high sympathetic drive. Concomitant use of beta-blockers, verapamil, or diltiazem is needed for adequate control of ventricular rate in such situations.

(c) Ventricular rate of more than 250/min strongly suggests the possibility of preexcitation with antegrade conduction over an accessory pathway with short refractory period. It results in irregular broad QRS tachycardia (Fig. 6.8a, b).

It may be difficult to differentiate from ventricular tachycardia. Intermittent irregularity and slow QRS upstroke indicating delta wave may help in differentiation. Digitalis and verapamil are contraindicated in such situation as they increase conduction over the accessory pathway and can induce ventricular fibrillation [3]. Electric cardioversion is needed. Thyrotoxicosis, anxiety, sympathetic stimulation, bronchodilators, cor pulmonale (Fig. 6.9), hypoxia, and inherently fast atrioventricular conduction can also cause fast ventricular rate. However, in these situations, ventricular rate is usually less than 200/min, and there is no delta wave.

(d) Regular rhythm in the presence of atrial fibrillation:
  - Rate around 65–100/min suggests atrioventricular dissociation with accelerated junctional rhythm [3]. QRS is narrow if intraventricular conduction is normal. Patients with sick sinus syndrome may have slower junctional escape rhythm (Fig. 6.10).
  - Rate around 45/min with broad QRS suggests complete atrioventricular block with escape rhythm arising from either ventricle (Fig. 6.11).

Possibility of digitalis toxicity should be considered whenever there is regular rhythm with atrial fibrillation [3]. Continuation of digitalis results in acceleration of the junctional or ventricular escape rhythms (Fig. 6.12).

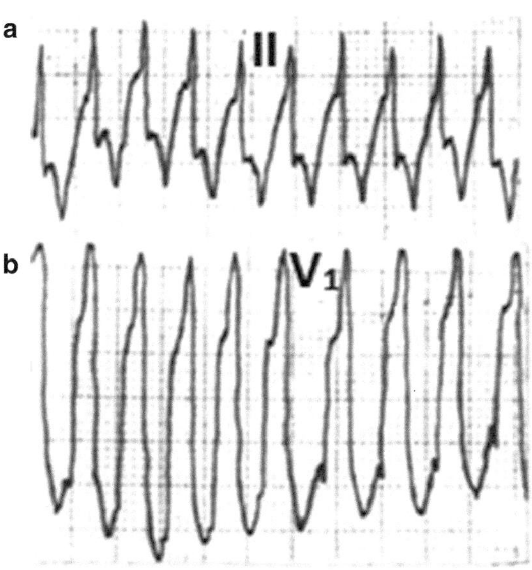

Fig. 6.8 (**a**, **b**) Lead II and V_1 showing broad QRS irregular tachycardia due to atrial fibrillation with antegrade conduction over an accessory pathway. (Reproduced with permission from: Mittal SR. Atrial fibrillation. Cardiology Today. 2012;14:174–177)

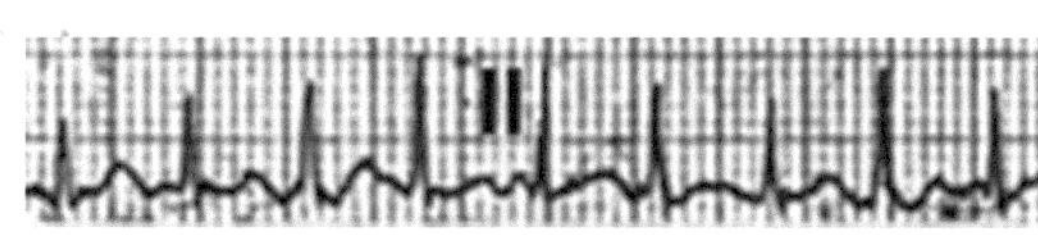

Fig. 6.9 Atrial fibrillation with fast ventricular rate in a case of cor pulmonale. (Reproduced with permission from: Mittal SR. Atrial fibrillation. Cardiology Today. 2012;14:174–177)

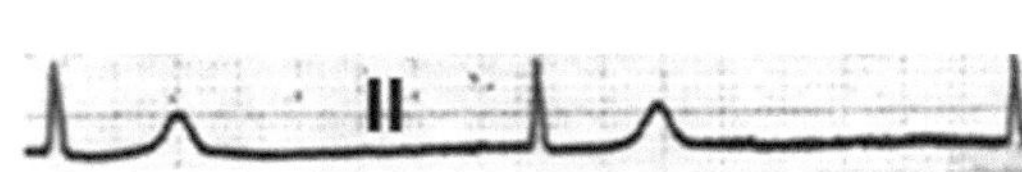

Fig. 6.10 Slow junctional escape rhythm in a case of atrial fibrillation due to sick sinus syndrome. There is no P wave and QRS is narrow. (Reproduced with permission from: Mittal SR. Atrial fibrillation. Cardiology Today. 2012;14:174–177)

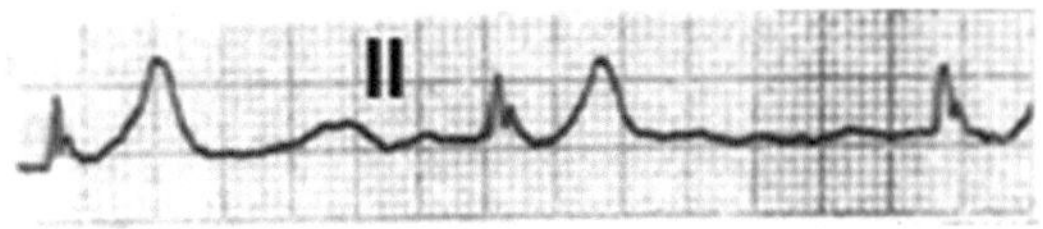

Fig. 6.11 Atrial fibrillation with complete atrioventricular block and ventricular escape rhythm. There are no P waves. QRS is broad regular and at a slow rate. (Reproduced with permission from: Mittal SR. Atrial fibrillation. Cardiology Today. 2012;14:174–177)

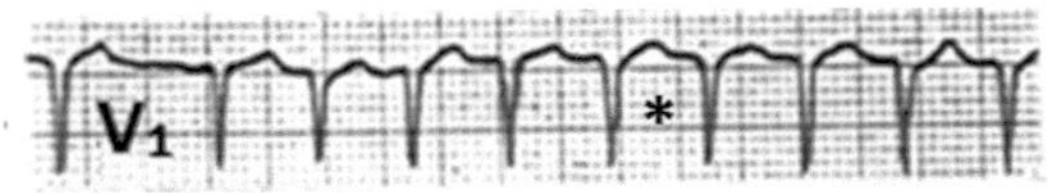

Fig. 6.12 Atrial fibrillation (absence of P wave) with episode of accelerated junctional rhythm (marked *) (regular narrow QRS at fast rate). (Reproduced with permission from: Mittal SR. Atrial fibrillation. Cardiology Today. 2012;14:174–177)

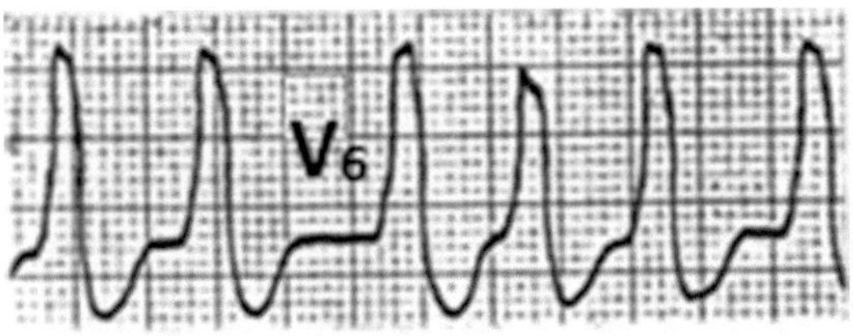

Fig. 6.13 Atrial fibrillation (absence of P wave) with pre-existing left bundle branch block (broad irregular QRS without any change in configuration irrespective of preceding cycle length). (Reproduced with permission from: Mittal SR. Atrial fibrillation. Cardiology Today. 2012;14:174–177)

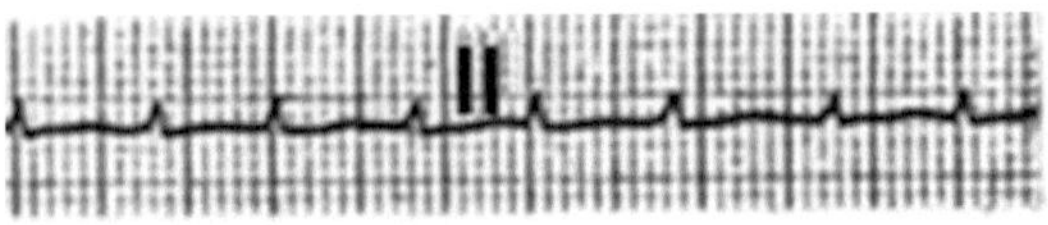

Fig. 6.14 Atrial fibrillation with low-voltage QRS due to pericardial constriction. (Reproduced with permission from: Mittal SR. Atrial fibrillation. Cardiology Today. 2012;14:174–177)

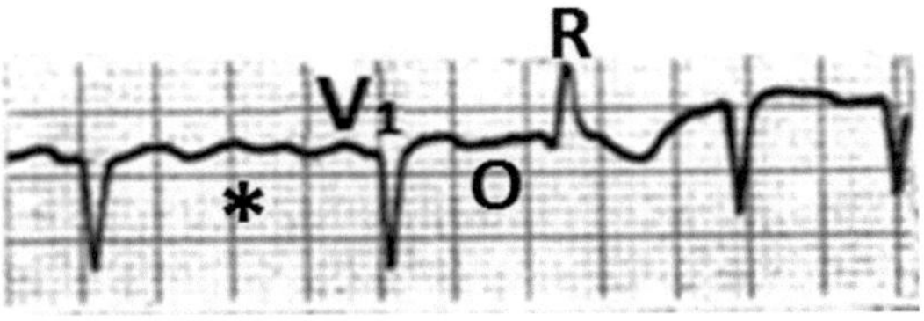

Fig. 6.15 Atrial fibrillation with aberrant conduction. Broad QRS (marked R) is seen in a short cycle (marked O) with a preceding long cycle (marked *). (Reproduced with permission from: Mittal SR. Atrial fibrillation. Cardiology Today. 2012;14:174–177)

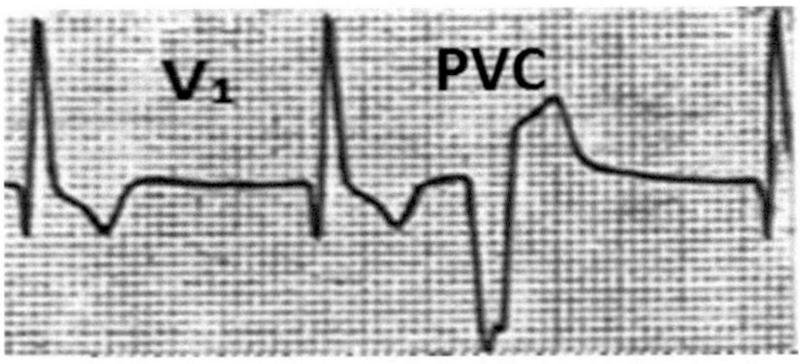

Fig. 6.16 Atrial fibrillation with ventricular premature beat (marked PVC). (Reproduced with permission from: Mittal SR. Atrial fibrillation. Cardiology Today. 2012;14:174–177)

4 QRS Configuration

(a) Usually QRS is narrow due to normal intraventricular conduction.

(b) It may be affected by underlying disease, e.g., rheumatic heart disease, hypertension, myocardial infarction, cardiomyopathy, electrolyte imbalance, bundle branch block (Fig. 6.13), or pericardial constriction (Fig. 6.14).

These conditions are also likely to affect the ST segment and the T wave:

(c) Irregular broad QRS (rate >250/min) with delta wave suggests antegrade conduction over an accessory pathway (Fig. 6.8a, b). Broad QRS in a short cycle following a long cycle suggests aberrant ventricular conduction (ventricular rate is usually <200/min).

(d) Regular broad QRS at slow rate is seen in complete atrioventricular block with escape ventricular rhythm.

(e) Intermittent broad QRS may be seen due to aberrant conduction in a short cycle following a long cycle (Fig. 6.15) or intermittent ventricular ectopics (Fig. 6.16).

Ventricular ectopics can also follow long-short cycle sequence, but as opposed to aberrant ventricular conduction, ventricular ectopics have the following differences.

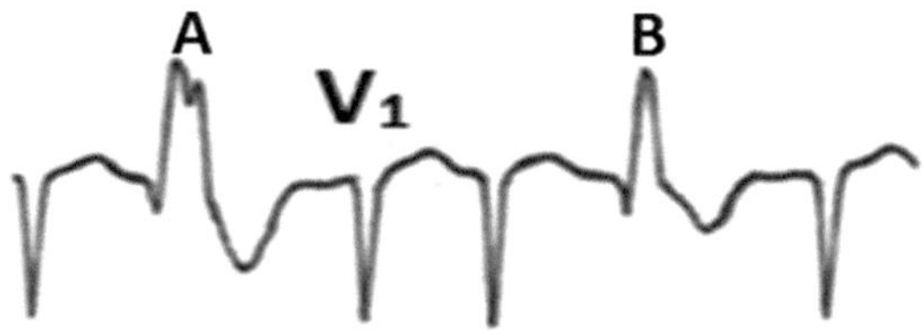

Fig. 6.17 Change in degree of aberrancy with change in preceding cycle length. Aberrantly conducted beat marked A is much broader than the beat marked B. (Reproduced with permission from: Mittal SR. Atrial fibrillation. Cardiology Today. 2012;14:174–177)

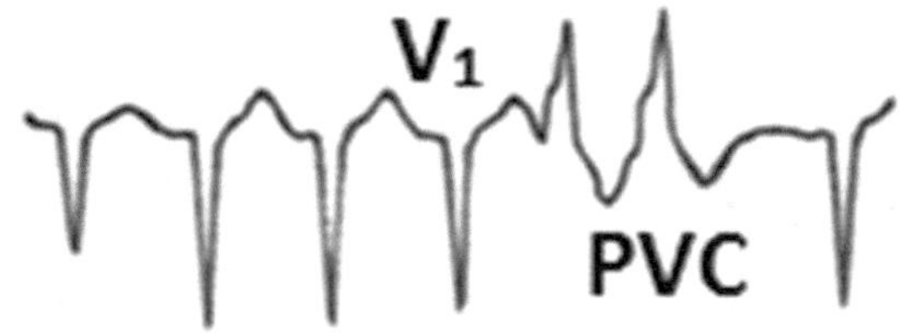

Fig. 6.19 Atrial fibrillation with ventricular premature beats showing biphasic QRS (rR') in lead V_1. (Reproduced with permission from: Mittal SR. Atrial fibrillation. Cardiology Today. 2012;14:174–177)

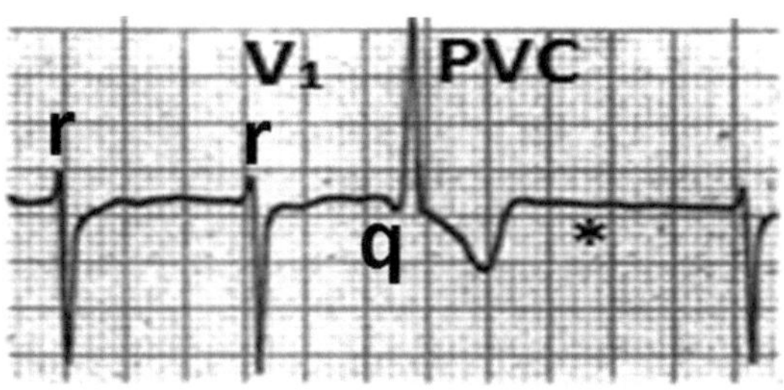

Fig. 6.18 Atrial fibrillation with ventricular ectopic showing difference in the initial QRS vector from that of narrow QRS complexes. In narrow QRS complexes, there is initial positive deflection (r) as compared to the initial negative deflection (q) in the premature ventricular contraction (PVC). Premature ventricular contraction is also followed by a compensatory pause (marked *). (Reproduced with permission from: Mittal SR. Atrial fibrillation. Cardiology Today. 2012;14:174–177)

- Ventricular ectopics have constant morphology which is not influenced by the cycle length. In aberrant ventricular conduction, QRS configuration changes with cycle length and heart rate (Fig. 6.17).
- In ventricular ectopic the initial QRS vector is opposite to that of the narrow QRS complexes (Fig. 6.18). In aberrant ventricular conduction, the initial vector of the broad QRS is similar to that of narrow QRS.
- Ventricular ectopics have a tendency for compensatory pause [4] (Fig. 6.18) due to concealed retrograde conduction of the ventricular extra systolic impulse into the atrioventricular node. This renders the atrioventricular node refractory to the ensuing impulses of atrial fibrillation. In aberrant ventricular conduction, there is no compensatory pause after the broad QRS.
- Ventricular ectopics have uniphasic or biphasic QRS complex (R or rR') in lead V_1 (Fig. 6.19) as opposed to triphasic (rsr') configuration in the aberrant ventricular conduction.

5 ST Segment and T Wave Changes

(a) These are influenced by the underlying disease or intraventricular conduction defect.

(b) Digitalis overdose causes ST segment depression with shortening of the QTc interval (Fig. 6.20).

Other ECG manifestations of digitalis overdose, which may be seen in a case of atrial fibrillation include accelerated junctional rhythm, atrioventricular block with bradycardia, and junctional or ventricular ectopics.

(c) Amiodarone can produce the following:
- Slow ventricular rate and prolongation of QT interval

(d) Hyperkalemia can cause tall tented T waves (Fig. 6.21). Hypokalemia can cause prolongation of QT interval.

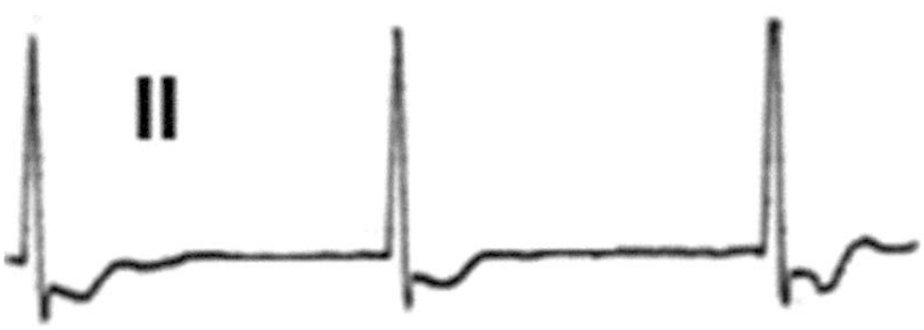

Fig. 6.20 Atrial fibrillation with ST segment depression and short QTc due to digitalis overdose. (Reproduced with permission from: Mittal SR. Atrial fibrillation. Cardiology Today. 2012;14:174–177)

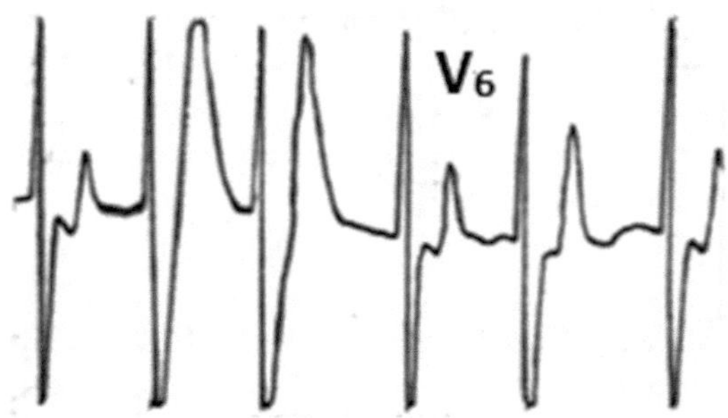

Fig. 6.21 Atrial fibrillation with tall tented T waves due to hyperkalemia. (Reproduced with permission from: Mittal SR. Atrial fibrillation. Cardiology Today. 2012;14:174–177)

Summary

- Atrial fibrillation is usually caused by multiple small micro-reentry circuits in the atria or a rapidly firing focus in the proximal portion of a pulmonary vein.
- Atrial dilatation and stretching of atrial muscles are the common culprits. Common causes are mitral valve disease, systemic hypertension, myocarditis, dilated cardiomyopathy, pulmonary thromboembolism, and thyrotoxicosis.
- Electrocardiogram is characterized by rapid (300–600/min) vibrations in the baseline that are of changing configuration. Fibrillary waves can be very small and baseline may appear straight.
- Conduction to the ventricles is irregularly irregular because of partial depolarization of the atrioventricular node by the rapidly firing atrial impulses. It may be difficult to appreciate the irregularity when QRS complexes occur at a fast rate.
- In untreated patients, ventricular rate varies from 100 to 160/min. Drugs blocking the atrioventricular node are effective in decreasing the atrioventricular conduction because atrial impulses are relatively weak as compared to those of atrial flutter.
- Fast regular rate should raise suspicion of accelerated junctional rhythm.
- Complete atrioventricular block can produce a slow and regular ventricular rate.
- Accessory pathway with short refractory period can conduct as many as 250–300 impulses per minute. This produces irregular broad QRS tachycardia. It may precipitate ventricular tachycardia or ventricular fibrillation.
- Persistent fast ventricular rate over a long period can result in tachycardia-induced cardiomyopathy.
- Long-short cycle sequence can produce intermittent broad QRS due to aberrant conduction in the ventricles.
- Slow ventricular rate in an untreated patient should raise suspicion of sick sinus syndrome.

MCQs

Q1. An electrocardiogram shows regular narrow QRS, at a rate of 75 beats/min with intervening straight baseline without any P wave. Possible diagnosis is:

(a) Junctional tachycardia with retrograde conduction
(b) Atrial fibrillation with controlled ventricular rate
(c) Accelerated junctional rhythm
(d) Atrial fibrillation with atrioventricular block and accelerated junctional rhythm

Q2. An electrocardiogram shows regular broad QRS at a rate of 50/min without any visible P wave. The possibilities are:

(a) Accelerated ventricular rhythm
(b) Accelerated ventricular rhythm with retrograde conduction
(c) Sinus bradycardia with accelerated ventricular rhythm
(d) Atrial fibrillation with atrioventricular block and accelerated ventricular rhythm

Q3. Atrial fibrillation with a regular ventricular rate of 90/min suggests:

(a) Digitalis underdose
(b) Digitalis overdose
(c) Overdose of beta blocker
(d) Amiodarone overdose

Q4. An electrocardiogram shows irregularly irregular ventricular rhythm without any atrial activity. What is the most likely diagnosis?

(a) Sinus arrhythmia
(b) Atrial tachycardia with changing atrioventricular conduction
(c) Atrial flutter with changing atrioventricular conduction
(d) Atrial fibrillation

Q5. An electrocardiogram shows irregularly irregular ventricular rhythm. Baseline shows vibrations varying from coarse to fine. What is the most likely diagnosis?

(a) Electrical vibrations
(b) Muscle tremors
(c) Lead movement
(d) Atrial fibrillation

Q6. Which drug decreases the refractory period of atrial myocardial cell?

(a) Quinidine
(b) Procainamide
(c) Flecainide
(d) Digitalis

Q7. In untreated atrial fibrillation, ventricular rhythm is:

(a) Regular and fast
(b) Regularly irregular
(c) Irregularly irregular
(d) Regular and slow

Q8. In a case of atrial fibrillation, development of a regular ventricular rhythm suggests:

(a) Conversion to atrial tachycardia
(b) Conversion to atrial flutter
(c) Increased atrioventricular conduction
(d) Accelerated junctional rhythm

Q9. Irregularly irregular ventricular rhythm could be due to:

(a) Atrial fibrillation
(b) Atrial tachycardia with changing atrioventricular conduction
(c) Atrial flutter with changing atrioventricular conduction
(d) All

Q10. The atrial rate in atrial fibrillation ranges between:

(a) 100–200/min
(b) 200–300/min
(c) 300–400/min
(d) 400–500/min

Q11. In a case of untreated atrial fibrillation, the number of atrial impulses that can be conducted to the ventricles is less than:

(a) 75/min
(b) 100/min
(c) 120/min
(d) 160/min

Q12. In atrial fibrillation, ventricular rate is irregular because of:

(a) Irregular rate of atria
(b) Atrioventricular block

(c) Atrioventricular dissociation
(d) Concealed conduction

Q13. Atrial fibrillation is caused by:

(a) Macro-reentry
(b) Multiple micro-reentry
(c) Increased automaticity
(d) All

Q14. Drugs are very effective in reducing ventricular rate in:

(a) Atrial flutter
(b) Atrial fibrillation
(c) Atrial tachycardia
(d) All

Q15. Irregular broad QRS tachycardia at a rate of 300 beats/min suggests:

(a) Atrial tachycardia with preexisting bundle branch block
(b) Atrial fibrillation with preexisting bundle branch block
(c) Ventricular tachycardia
(d) Atrial fibrillation with preexcitation

Q16. Digitalis can increase the ventricular rate in:

(a) Atrial flutter
(b) Atrial fibrillation
(c) Accelerated junctional rhythm
(d) Atrial fibrillation with preexcitation

Q17. Which drug decreases the refractory period of the accessory pathway?

(a) Digitalis
(b) Beta-blockers
(c) Verapamil
(d) Amiodarone

Q18. Atrial fibrillation with intermittent broad QRS suggests:

(a) Aberrant conduction
(b) Ventricular ectopic
(c) Preexcitation
(d) All

Q19. In atrial fibrillation, a wide QRS whose initial portion is identical to other narrow QRS complexes and is not followed by a pause suggests:

(a) Preexcitation
(b) Aberrant intraventricular conduction
(c) Ventricular ectopic
(d) All

Q20. In atrial fibrillation with occasional wide QRS complexes, increasing the dose of digitalis will reduce frequency of wide QRS in a case of:

(a) Aberrant intraventricular conduction
(b) Ventricular ectopic
(c) Preexcitation
(d) None

Answers

(1) d; (2) d; (3) b; (4) d; (5) d; (6) d; (7) c; (8) d; (9) d; (10) d; (11) d; (12) d; (13) b, c; (14) b; (15) d; (16) c, d; (17) a; (18) a, b; (19) b; (20) a

References

1. Prystowsky EN, Padanilam BJ, Waldo AL. Atrial fibrillation, atrial flutter and atrial tachycardia. In: Fuster V, Walsh RA, Harrington RA, editors. Hurst's the heart. New York: McGraw Hill; 2011. p. 963–86.
2. Olgin JE, Zipes DP. Specific arrhythmias: diagnosis and treatment. In: Libby P, Bonow RO, Mann DL, Zipes DP, editors. Braunwald's heart disease. Philadelphia: Saunders; 2008. p. 863–922.
3. Wegner GS. Re-entrant atrial tachyarrhythmias – the atrial flutter/fibrillation spectrum. In: Wagner GS, editor. Marriot's practical electrocardiography. Philadelphia: Wolters Kluwer; 2008. p. 301–26.
4. Schamroth L. Ectopic atrial rhythms. In: Schamroth C, editor. An introduction to electrocardiography. France: Blackwell Science; 1982. p. 330–41.

Short PR Interval, Ventricular Preexcitation

7

1 Introduction

The PR interval is the interval from the beginning of the P wave to the beginning of the QRS complex [1]. It includes the P wave and the PR segment [2]. The P wave is produced by atrial depolarization. Duration of the PR segment denotes the time taken for the electrical impulse to travel over the atrioventricular node, the His bundle, bundle branches, and the Purkinje system to reach the ventricular myocardium (Fig. 7.1). The atrioventricular node is usually the site for maximum delay in conduction. Normal range of the PR interval is from 0.12 to 0.20 s.

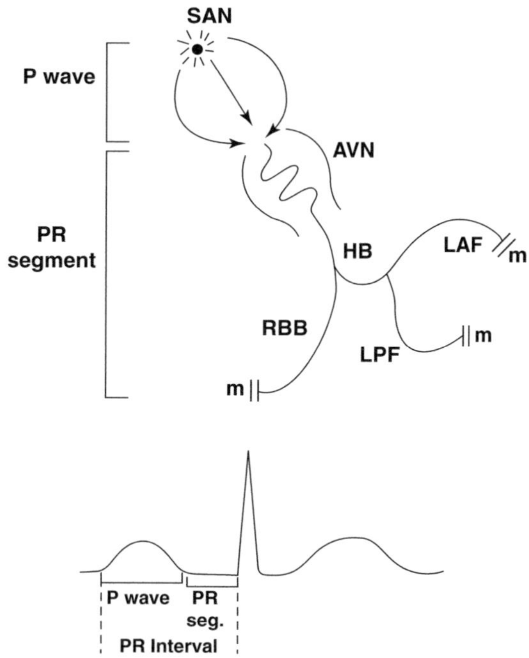

Fig. 7.1 Diagram showing components of the PR interval. *SAN* sinoatrial node, *AVN* atrioventricular node, *HB* His bundle, *LAF* left anterior fascicle, *LPF* left posterior fascicle, *RBB* right bundle branch, *m* myocardium. (Reproduced with permission from: Mittal SR. Short PR interval, ventricular preexcitation. Cardiology Today. 2013;17:43–48)

S. Mittal, *Insights into Electrocardiograms with MCQs*,
https://doi.org/10.1007/978-981-99-0127-2_7

2 Causes of the Short PR Interval

The PR interval can be shorter than normal in the following conditions:

1. *Impulse arising from the atrioventricular node (junctional rhythm) (Fig. 7.2a).* In this situation P wave is inverted in leads II, III, and aVF because of retrograde activation of the atria. The PR interval is short due to avoidance of time taken for the impulse to travel from the sinoatrial node to the atrioventricular node. Reduction in the atrioventricular nodal delay also contributes to the short PR interval. Representative ECG is shown in Fig. 7.2b.
2. *Rapid conduction over the atrioventricular node (hyperconductive atrioventricular node).* This may be congenital or acquired during increased sympathetic activity. P wave is normal, PR interval is short, and QRS is normal (Fig. 7.3a).
3. *Accessory (extra) pathway connecting lower atrium to the distal atrioventricular node (atrionodal pathway—James fibers) (Fig. 7.3b) or the His bundle (atrio-Hisian pathway—the Brechenmacher fibers) (Fig. 7.3c).* The P wave is normal. The PR interval is short as the impulse bypasses the atrioventricular node which is normally the site of maximum delay in cardiac conduction. QRS is normal because of normal depolarization of the ventricles.

   It is not possible to differentiate surface electrocardiograms of the hyperconductive atrioventricular node from those of atrionodal or atrio-Hisian bypass tracts. One example of short PR interval with normal QRS is shown in Fig. 7.3d.
4. *Ventricular preexcitation.* Sometimes there is an accessory pathway which connects the atria, the atrioventricular node, or the His bundle to the ventricles. Such pathways produce early excitation (preexcitation) of a part of the

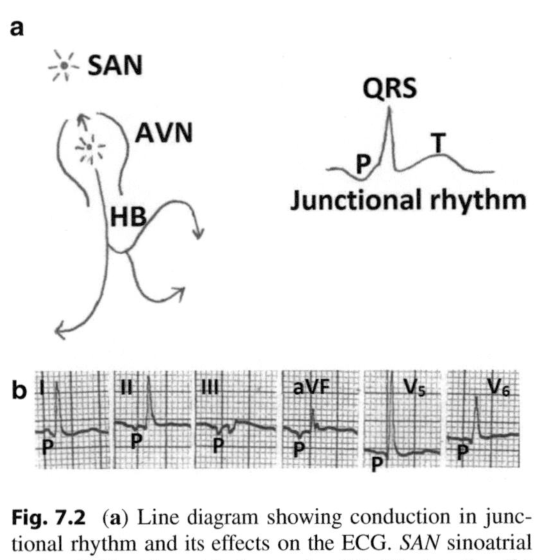

Fig. 7.2 (**a**) Line diagram showing conduction in junctional rhythm and its effects on the ECG. *SAN* sinoatrial node, *AVN* atrioventricular node, *HB* His bundle. (**b**) Electrocardiogram showing junctional rhythm with short PR interval. P wave is inverted in leads II, III, and aVF. (Reproduced with permission from: Mittal SR. Short PR interval, ventricular preexcitation. Cardiology Today. 2013;17:43–48)

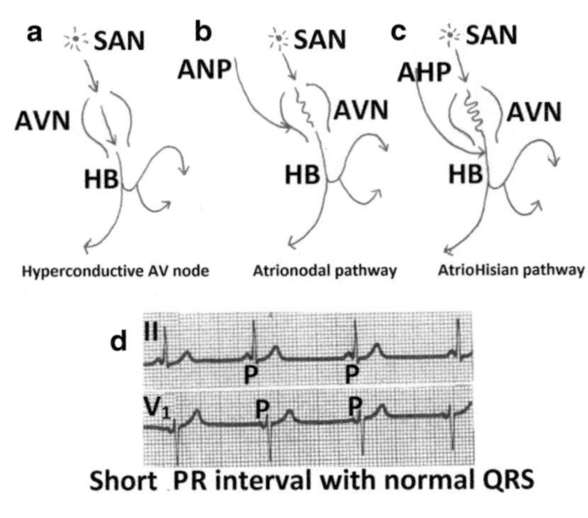

Fig. 7.3 Showing causes of short PR interval. (**a**) Diagram showing conduction and electrocardiogram in a hyperconductive atrioventricular node. (**b**) Diagram showing conduction and electrocardiogram in atrionodal pathway. (**c**) Diagram showing conduction and electrocardiogram in atrio-Hisian pathway. (**d**) ECG showing short PR interval with normal QRS. *SAN* sinoatrial node, *AVN* atrioventricular node, *HB* His bundle, *ANP* atrionodal pathway, *AHP* atrio-Hisian pathway. (Reproduced with permission from: Mittal SR. Short PR interval, ventricular preexcitation. Cardiology Today. 2013;17:43–48)

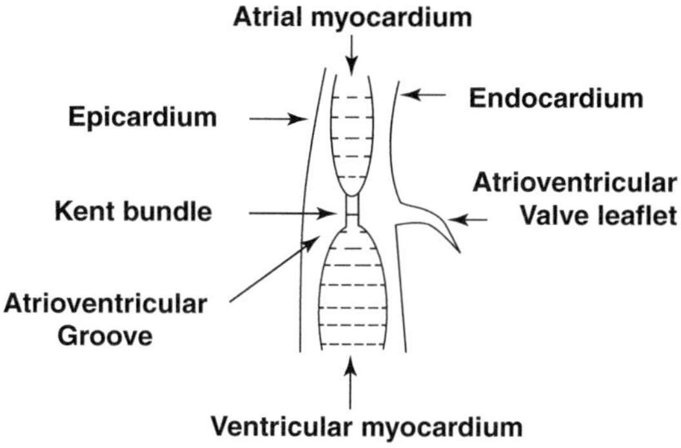

Fig. 7.4 Diagram showing the location of the Kent bundle

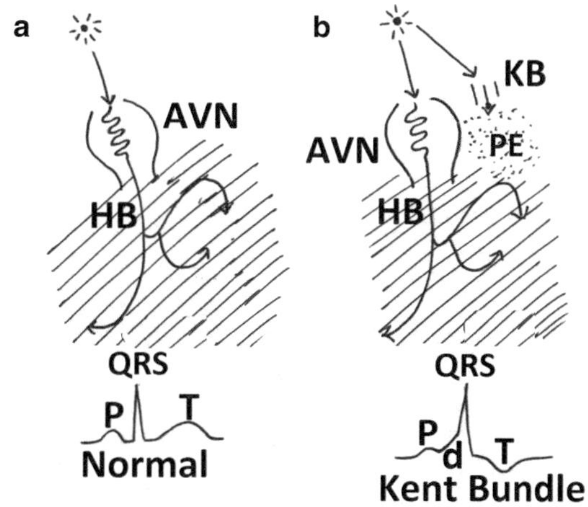

Fig. 7.5 Diagram showing normal conduction and conduction over the Kent bundle and its effects on the electrocardiogram. (**a**) Normal depolarization. (**b**) Preexcitation through the Kent bundle (KB). Abbreviations: *AVN* atrioventricular node, *HB* His bundle, *KB* Kent bundle, *d* delta wave, *PE* preexcited part of the ventricle. (Reproduced with permission from: Mittal SR. Short PR interval, ventricular preexcitation. Cardiology Today. 2013;17:43–48)

ventricle. Ventricular preexcitation can be due to the following:

(a) *Kent bundle.* It is an accessory muscular connection between the atria and the ventricles (Fig. 7.4).

Such an accessory pathway can conduct electrical impulse antegradely (from the atria to the ventricles), retrogradely (from the ventricle to the atria), or in both directions. When antegrade conduction over the accessory pathway is faster than that over the atrioventricular node—the His bundle pathway—it results in premature excitation (preexcitation) of the part of the ventricle that receives the accessory pathway. This results in early onset of QRS. QRS, therefore, becomes broad at the cost of shortening of the PR segment. PR interval is, therefore, short. Spread of impulse in the preexcited part of the ventricle is relatively slow due to conduction through the contractile myocardium which has minimal conducting fibers. This slow depolarization leads to slurring of the initial part of the QRS (the delta wave) (Fig. 7.5a, b).

(b) *Mahaim fiber (nodofascicular fibers).* This accessory pathway connects the distal part of the atrioventricular node or the His bundle to the ventricular myocardium. P wave is normal. PR interval is normal because of normal delay in the atrioventricular node conduction. There is a delta wave with broadening of the QRS complex due to preexcitation of a part of the ventricular myocardium (Fig. 7.6).

(c) *The atriofascicular fibers.* These fibers are long and connect the atria to the right bundle branch (Fig. 7.7). The P wave is normal. The PR interval may be short or may be normal because of delay in the long pathway. Preexcitation is present and generally results in left bundle branch block pattern.

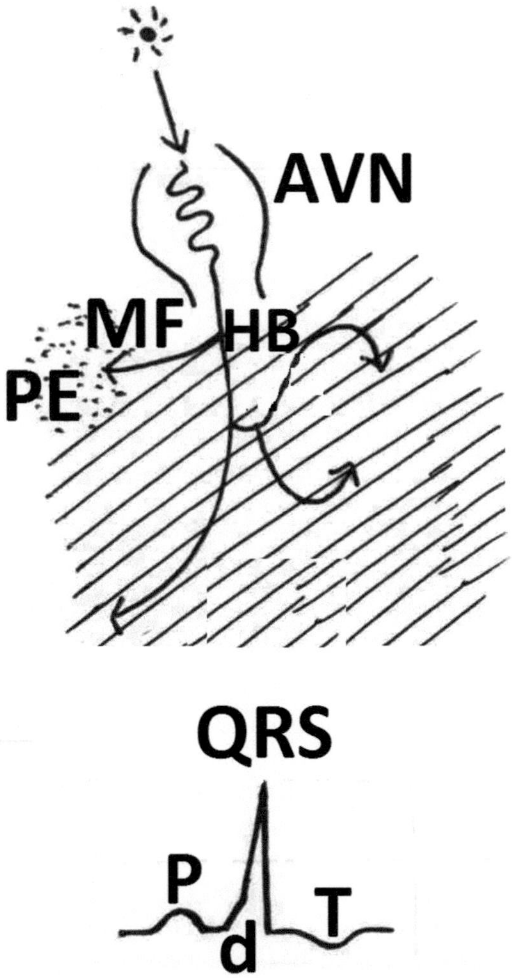

Fig. 7.6 Diagram showing conduction over Mahaim fiber and its effect on the electrocardiogram. Abbreviations: *AVN* atrioventricular node, *HB* His bundle, *PE* preexcited part of the ventricle, *d* delta wave. (Reproduced with permission from: Mittal SR. Short PR interval, ventricular preexcitation. Cardiology Today. 2013;17:43–48)

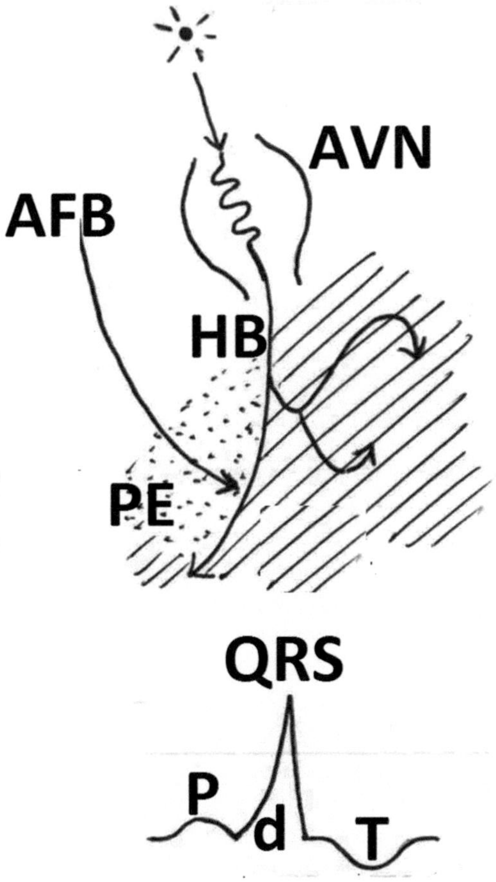

Fig. 7.7 Diagram showing conduction over the atriofascicular bundle and its effects on the electrocardiogram. Abbreviations: *AVN* atrioventricular node, *HB* His bundle, *PE* preexcited part of the ventricle, *d* delta wave. (Reproduced with permission from: Mittal SR. Short PR interval, ventricular preexcitation. Cardiology Today. 2013;17:43–48)

3 Variations in the PR Interval

In ventricular preexcitation, the ventricles are depolarized partly through the accessory pathway and partly through the normal atrioventricular node-His bundle pathway (Fig. 7.8a). QRS configuration depends on the relative velocity of antegrade conduction over the two pathways. When conduction over the accessory pathway is fast, significant part of the ventricular myocardium is preexcited resulting in prominent delta wave and significant broadening of the QRS complex (Fig. 7.8b). On the other end of the spectrum, if conduction over the atrioventricular node-His bundle pathway is faster, it results in near normal depolarization of the ventricles with minimal preexcitation and nearly normal QRS (Fig. 7.8c). In some patients, accessory pathway conducts only retrogradely (ventricle to atria). There is no antegrade conduction from the atria to the ventricles with the result that there is no preexcitation of the ventricles. PR interval and QRS are absolutely normal (Fig. 7.8d). Such accessory pathway is called concealed (hidden) accessory pathway.

The relative velocity of antegrade conduction over the two pathways can vary from time to time depending on the autonomic tone. Exercise and hyperventilation increase sympathetic tone. This results in faster conduction over the atrioventricular node, and preexcitation is less apparent (Figs. 7.9 and 7.10).

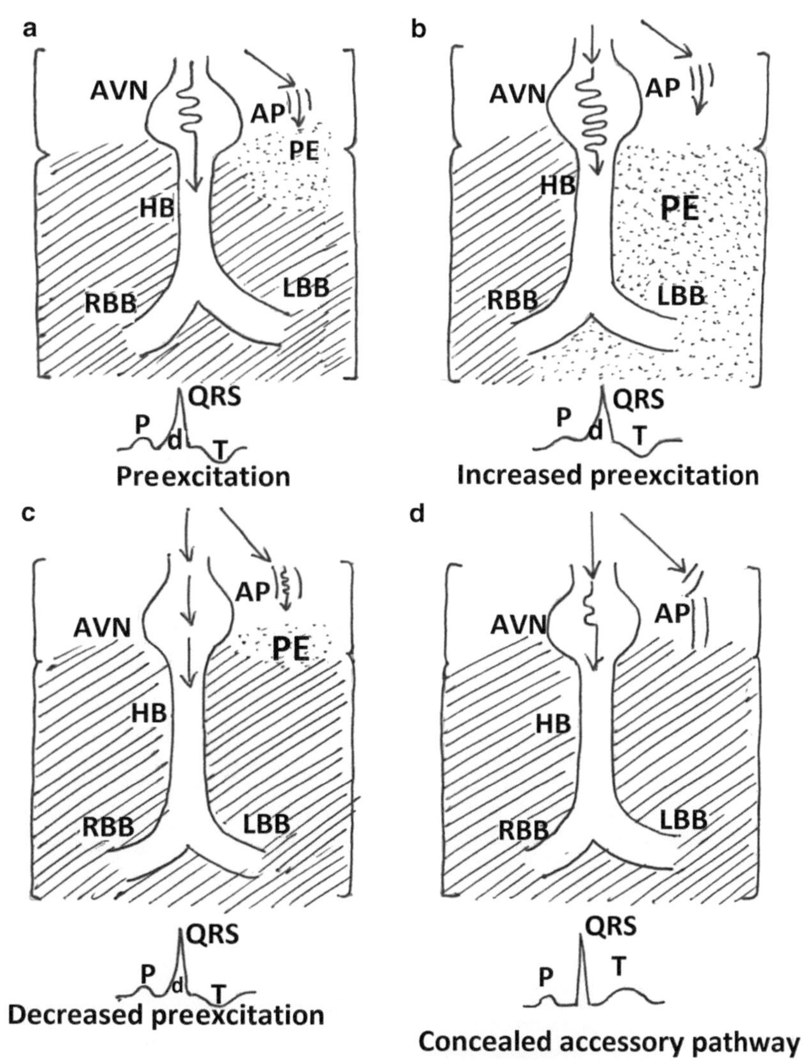

Fig. 7.8 (**a**) Diagram showing ventricular depolarization during preexcitation. (**b**) Showing the effect of fast conduction over the accessory pathway on the ECG. (**c**) Showing the effect of fast conduction over the atrioventricular node on the ECG. (**d**) Showing the effect of concealed accessory pathway. Abbreviations: *AVN* atrioventricular node, *HB* His bundle, *RBB* right bundle branch, *LBB* left bundle branch, *AP* accessory path, *PE* preexcited part of the ventricle, *d* delta wave on the ECG. (Reproduced with permission from: Mittal SR. Short PR interval, ventricular preexcitation. Cardiology Today. 2013;17:43–48)

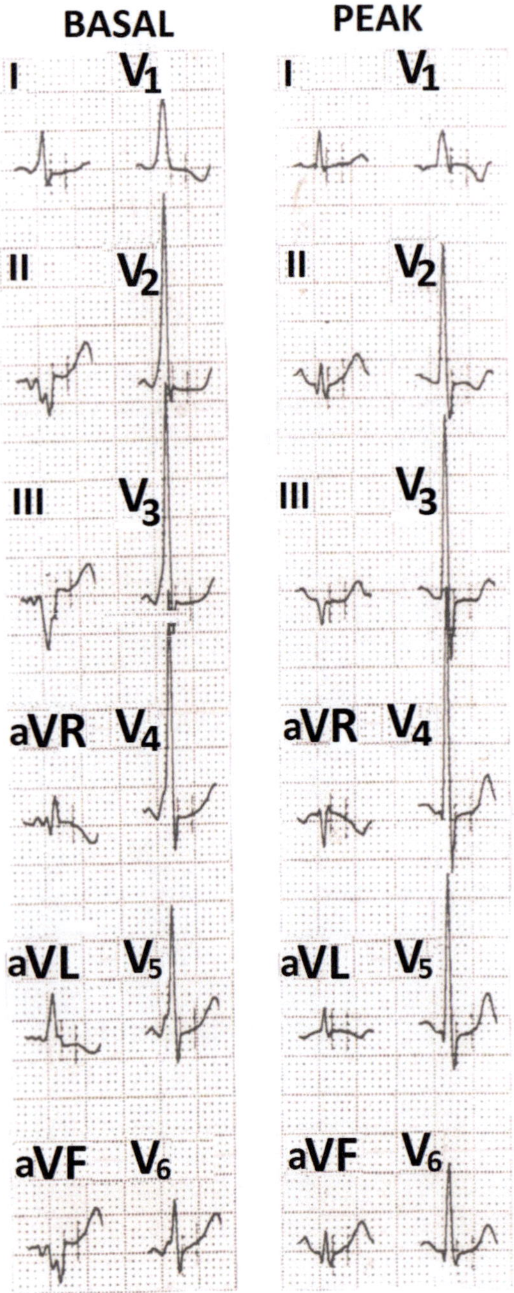

Fig. 7.9 Electrocardiogram showing the disappearance of preexcitation with normalization of the PR interval and the QRS during exercise. (Reproduced with permission from: Mittal SR. Short PR interval, ventricular preexcitation. Cardiology Today. 2013;17:43–48)

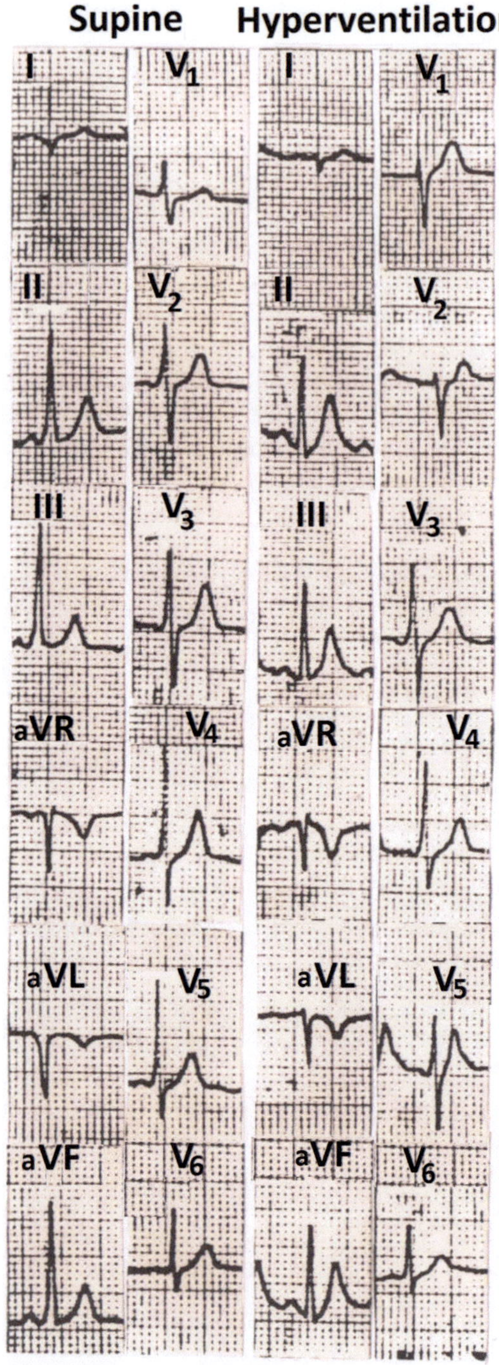

Fig. 7.10 Electrocardiograms are showing disappearance of the delta wave with normalization of the PR interval and the QRS in leads II, III, aVL, and aVF during hyperventilation. (Reproduced with permission from: Mittal SR. Short PR interval, ventricular preexcitation. Cardiology Today. 2013;17:43–48)

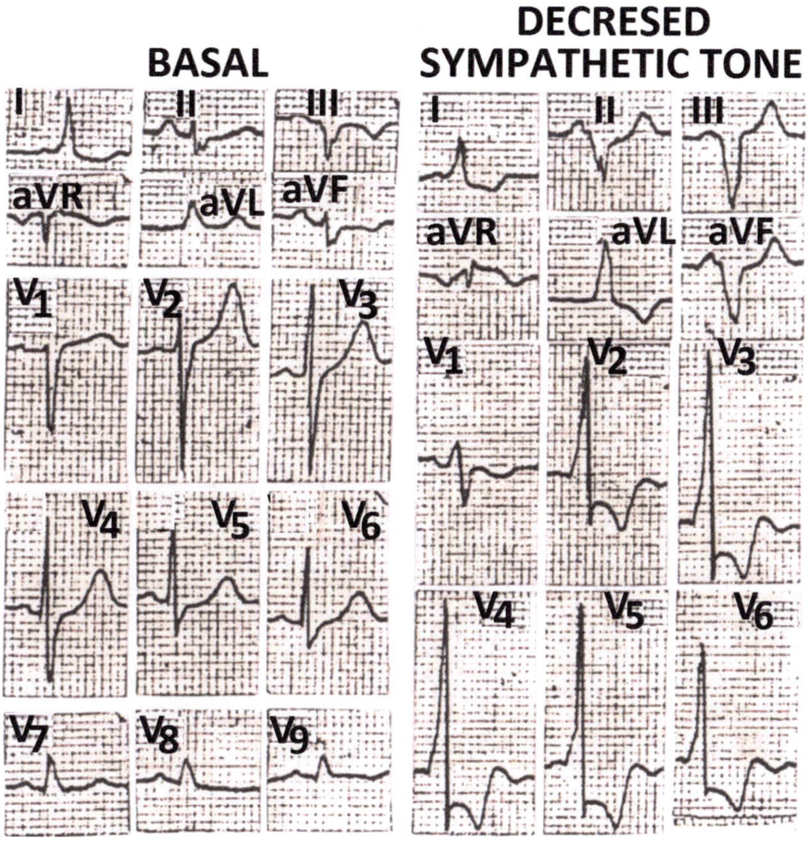

Fig. 7.11 Electrocardiograms showing the effect of decreased sympathetic tone on the PR interval and preexcitation. (Reproduced with permission from: Mittal SR. Short PR interval, ventricular preexcitation. Cardiology Today. 2013;17:43–48)

On the other end, decreased sympathetic tone slows conduction through the atrioventricular node, and preexcitation becomes more apparent (Fig. 7.11).

This may happen during recovery from the exercise stress test (Fig. 7.12).

Use of drugs that block the atrioventricular node, e.g., diltiazem may unmask preexcitation (Fig. 7.13).

During junctional rhythm impulse takes shorter time to cross the atrioventricular node, and preexcitation becomes less apparent (Fig. 7.14a, b).

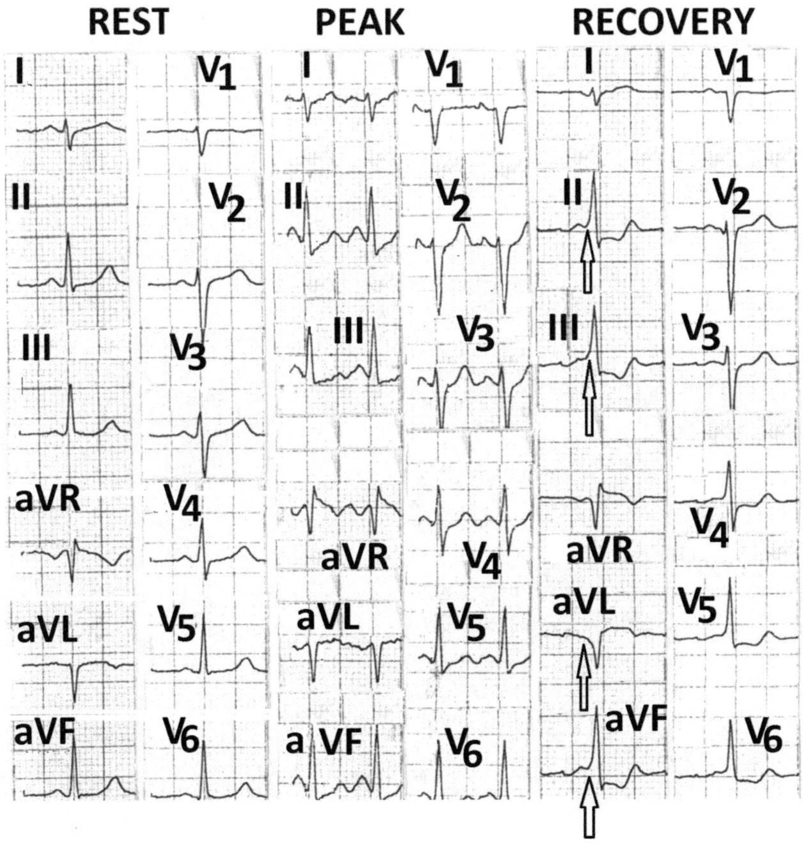

Fig. 7.12 Electrocardiograms showing the appearance of preexcitation with shortening of the PR interval, appearance of the delta wave (marked ⇧), and broadening of the QRS complex during recovery from the exercise stress test. (Reproduced with permission from: Mittal SR. Short PR interval, ventricular preexcitation. Cardiology Today. 2013;17:43–48)

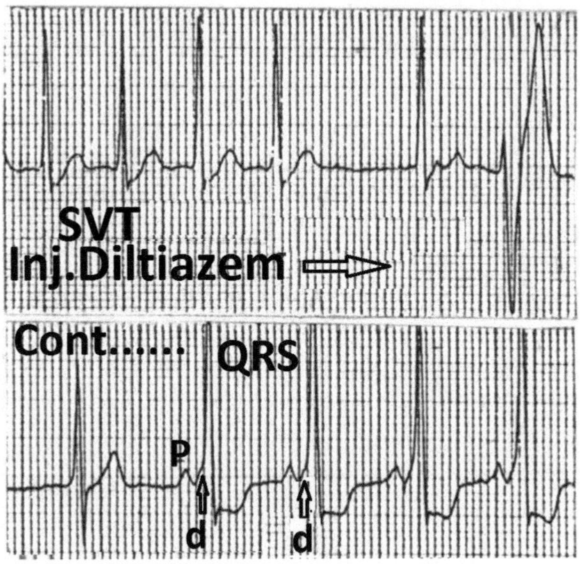

Fig. 7.13 Electrocardiogram is showing the appearance of preexcitation with shortening of the PR interval, appearance of the delta wave (marked d), and broadening of the QRS complex on termination of SVT by I/V diltiazem. *d* delta wave. (Reproduced with permission from: Mittal SR. Short PR interval, ventricular preexcitation. Cardiology Today. 2013;17:43–48)

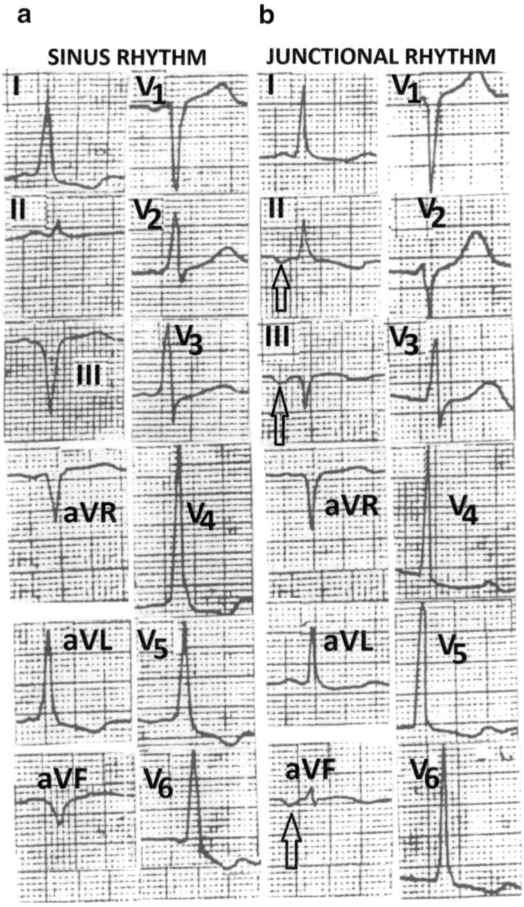

Fig. 7.14 (**a**) Electrocardiogram showing preexcitation. (**b**) Electrocardiogram during junctional rhythm (P wave inverted in leads II, III, and aVF (marked ⇧) with disappearance of preexcitation and normalization of the PR interval and the QRS). (Reproduced with permission from: Mittal SR. Short PR interval, ventricular preexcitation. Cardiology Today. 2013;17:43–48)

4 Intermittent Shortening of the PR Interval (Intermittent Preexcitation)

This term is used when there is intermittent abrupt loss of delta wave with normalization of the PR interval and duration of the QRS complex without apparent change in the autonomic tone (Fig. 7.15a–d).

This condition is different from intermittent change in magnitude of preexcitation observed due to change in the autonomic tone. Intermittent preexcitation suggests a long refractory period of the accessory pathway. Such patients do not have the risk of rapid conduction of impulses over the accessory pathway in event of atrial fibrillation.

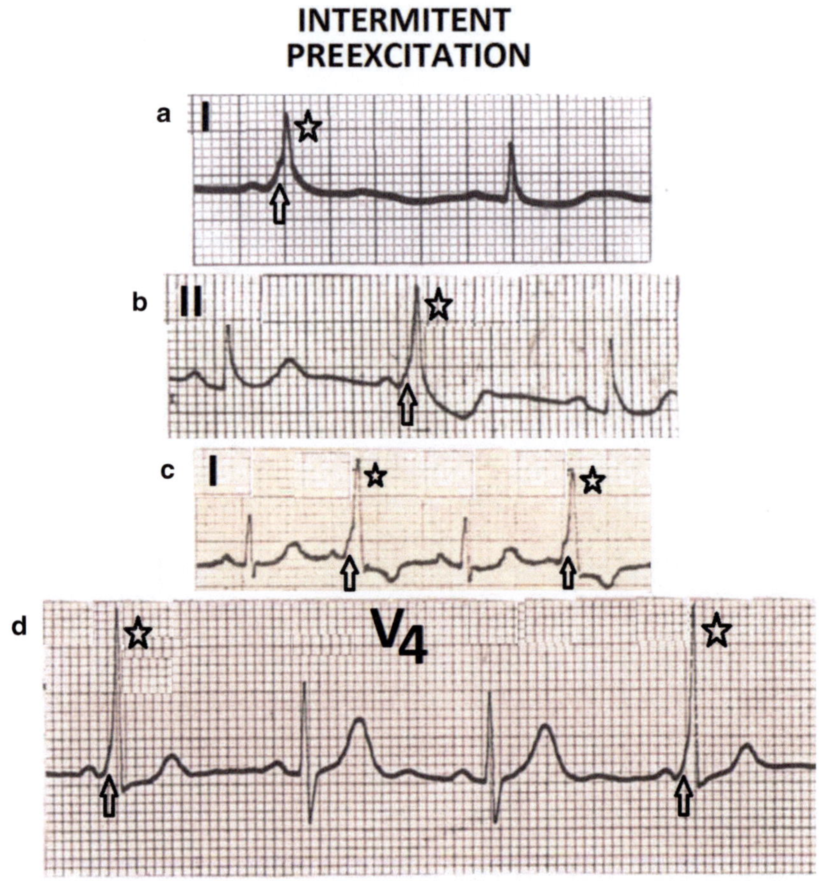

Fig. 7.15 (**a**)–(**d**) Electrocardiograms showing intermittent preexcitation (marked ☆) with intermittent shortening of the PR interval and appearance of the delta wave (marked ⇧). (Reproduced with permission from: Mittal SR. Short PR interval, ventricular preexcitation. Cardiology Today. 2013;17:43–48)

5 Effect of Atrial Ectopic on the PR Interval in Preexcitation

Refractory period of the accessory path is usually longer than that of the atrioventricular node. Atrial ectopic arising in close proximity to the bundle of Kent rapidly reaches the accessory pathway before it has repolarized following the previous sinus beat. Such atrial premature beat, therefore, fails to conduct through the accessory pathway and conducts over the atrioventricular node-His bundle path producing a narrow QRS (Fig. 7.16a, b). This finding also suggests a relatively long refractory period of the accessory pathway. Sometimes atrial ectopics may conduct with preexcitation (Fig. 7.16c). This suggests short refractory period of the accessory pathway.

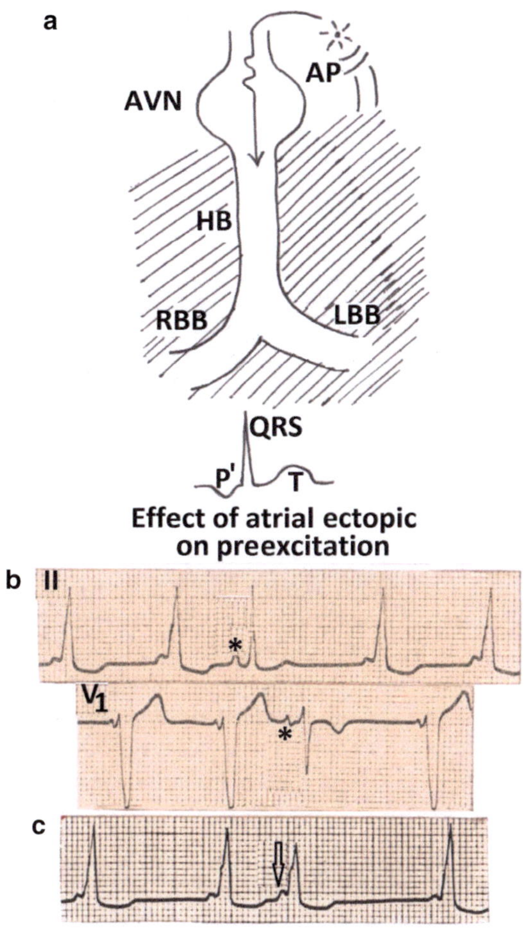

Fig. 7.16 (**a**) Diagram showing the effect of atrial ectopic arising close to the Kent bundle on preexcitation. (**b**) Lead II and V_1 showing atrial ectopics with less of preexcitation and normal PR interval (marked *). (**c**) ECG showing atrial ectopic with persistent preexcitation and short PR interval. (marked ⇩). (Reproduced with permission from: Mittal SR. Short PR interval, ventricular preexcitation. Cardiology Today. 2013;17:43–48)

6 Effect of Preexcitation on Amplitude of the QRS Complex

During normal activation, the two ventricles are depolarized simultaneously, but in opposite directions (Fig. 7.17a). Left ventricular vector dominates and forms the main QRS vector on the ECG. During preexcitation, maximum QRS amplitude usually increases. This could occur because of two factors:

- Preexcitation vector may nullify the QRS vector moving in the opposite direction, thus leaving the main QRS vector unopposed (Fig. 7.17b).
- Preexcitation vector may be in the direction of the main QRS vector, thus increasing its magnitude (Fig. 7.17c).

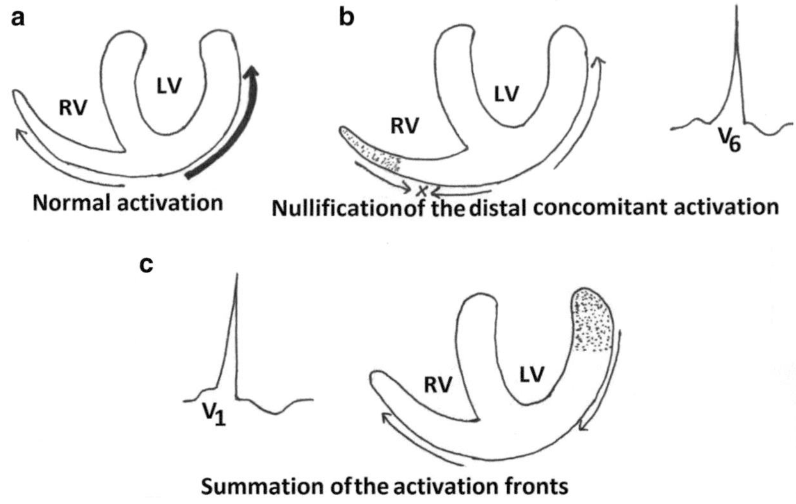

Fig. 7.17 Diagrams showing (**a**) normal depolarization vectors; (**b**) opposition of distal concomitant QRS vector by preexcitation vector; (**c**) summation of the normal QRS vector and the preexcitation vector. (Reproduced with permission from: Mittal SR. Short PR interval, ventricular preexcitation. Cardiology Today. 2013;17:43–48)

7 Repolarization Vector in Preexcitation

The ST segment and the T wave vectors move in the direction opposite to the direction of the delta wave. Magnitude of repolarization abnormalities correlates with the severity of preexcitation. Such ST segment depression and T wave inversion may simulate myocardial ischemia (Fig. 7.11). ST segment depression may increase during stress test giving a false-positive response (Fig. 7.18).

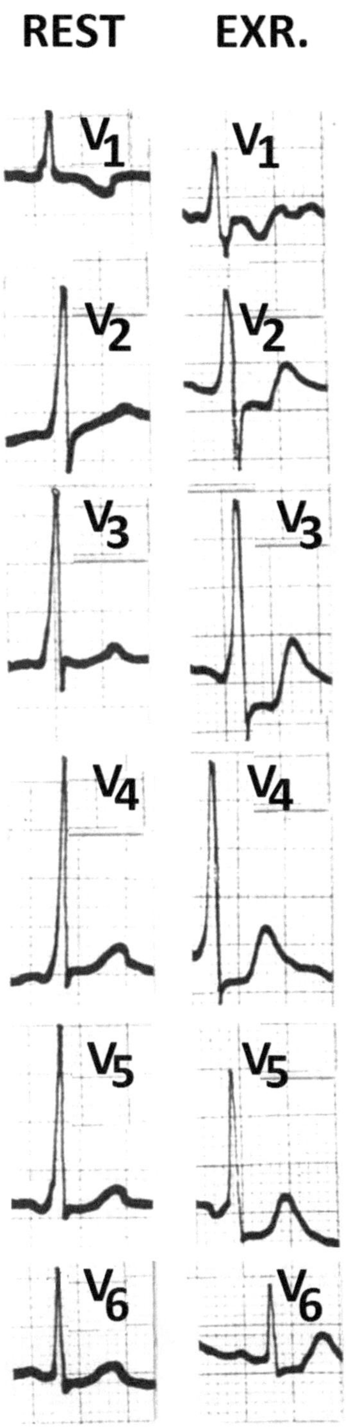

Fig. 7.18 Electrocardiograms showing the effect of exercise on the repolarization abnormalities in preexcitation. (Reproduced with permission from: Mittal SR. Short PR interval, ventricular preexcitation. Cardiology Today. 2013;17:43–48)

Summary

- The PR interval extends from beginning of the P wave to the beginning of the QRS complex. Normal range extends from 0.12 s to 0.20 s. Maximum delay in conduction occurs at the level of the atrioventricular node. PR interval less than 0.12 s is considered a short PR interval. The following conditions can produce short PR interval:
  - Hyperconductive atrioventricular node. The P wave is normal, the PR interval is short, and the QRS complex is normal.
  - Junctional rhythm. The P wave is inverted in leads II, III, and aVF, the PR interval is short, and the QRS complex is normal.
  - Atrionodal pathway connecting lower atrium to the distal atrioventricular node. The P wave is normal, the PR interval is short, and the QRS complex is normal.
  - Atrio-Hisian (James fibers) fibers bypassing the atrioventricular node and connecting lower atrium to the His bundle. The P wave is normal, the PR interval is short, and the QRS complex is normal.
  - Atriofascicular fibers bypassing atrioventricular node and connecting atria to right bundle branch. The P wave is normal, the PR interval is short but may be normal, and the QRS complex is broad and shows delta wave and left bundle branch block pattern.
  - Atrioventricular fibers (Kent bundle) bypassing the atrioventricular node and connecting an atrium to a ventricle. The P wave is normal, the PR interval is short, and the QRS complex is broad and shows delta wave.

MCQs

Q1. PR interval includes conduction across:

(a) The atria
(b) The atrioventricular node
(c) The His-Purkinje system
(d) All

Q2. Junctional rhythm produces:

(a) Inverted P waves in leads II, III, and aVF
(b) Short PR interval
(c) Normal QRS
(d) All

Q3. Exercise in a person with normal resting electrocardiogram produces:

(a) Junctional tachycardia
(b) PR interval less than 0.10 s
(c) Delta wave
(d) None

Q4. Normal P wave, short PR interval, and normal QRS suggest:

(a) Junctional rhythm
(b) Hyperconductive atrioventricular node
(c) Nodofascicular fibers
(d) Atriofascicular bundle

Q5. Normal P wave, short PR interval, and presence of left bundle branch block pattern suggest presence of:

(a) Hyperconductive atrioventricular node
(b) Atrio-Hisian fibers
(c) Atriofascicular bundle
(d) Junctional rhythm

Q6. A beat with inverted P wave in leads II, III, and aVF, short PR interval, and normal QRS suggest:

(a) Junctional ectopic
(b) Atrial ectopic
(c) Ventricular ectopic
(d) Transient preexcitation

Q7. Short PR interval without alteration in QRS can occur if there is:

(a) Sympathetic stimulation
(b) Hyperconductive atrioventricular node
(c) Presence of atrio-Hisian bundle
(d) All

Q8. Accessory pathways are made of:

(a) Connective tissue
(b) Conducting nerve fibers
(c) Muscular tissue
(d) Fibro-fatty tissue

Q9. Which electrocardiographic findings characterize Wolff-Parkinson-White syndrome?

(a) Short PR interval with delta wave and tachyarrhythmias related to preexcitation
(b) Short PR interval with normal QRS with episodes of AVNRT
(c) Normal PR interval with prolonged upstroke of QRS with episodes of tachyarrhythmia.
(d) Short PR interval with delta waves without any tachyarrhythmia

Q10. Concealed accessory pathway produces:

(a) Short PR interval
(b) Delta wave
(c) Broad QRS
(d) None

Q11. Accessory fibers connecting the atria to the His bundle produce:

(a) Short PR interval
(b) Delta waves
(c) Normal QRS
(d) Broad QRS

Q12. Nodofascicular accessory fibers produce:

(a) Long PR interval
(b) Delta wave
(c) Broad QRS
(d) All

Q13. Atriofascicular accessory fibers usually produce:

(a) RBBB pattern
(b) LBBB pattern

(c) Left axis deviation
(d) Right axis deviation

Q14. Which electrocardiographic finding suggests long refractory period of accessory pathway:

(a) Intermittent preexcitation
(b) Normalization of QRS during exercise
(c) Broad QRS tachycardia during atrial fibrillation
(d) All

Answers

(1) d; (2) d; (3) d; (4) b; (5) c; (6) a; (7) d; (8) c; (9) a; (10) d; (11) a, c; (12) b, c; (13) b, c; (14) a, b

References

1. Schamroth L. The Wolff-Parkinson-White and related syndromes. In: Schamroth C, editor. An introduction to electrocardiography. New Delhi: Blackwell Science; 1982. p. 288–306.
2. Wagner GS. Ventricular preexcitation. In: Wagner GS, editor. Marriott's practical electrocardiography. Philadelphia: Wolters Kluwer; 2008. p. 125–38.

Atrioventricular Accessory Path

(The Kent Bundle)

8

1 Introduction

Normally, electrical impulse traverses from the atria to the ventricles through the atrioventricular node, the His bundle, bundle branches, and the Purkinje system to reach the ventricular myocardium. Maximum delay in conduction occurs in the atrioventricular node. If there is an extra path (accessory path) connecting the atria directly to the ventricles (atrioventricular accessory path), it can depolarize a part of the ventricle before its scheduled depolarization through the atrioventricular node—His bundle path. This happens because the accessory path bypasses the delay occurring in the atrioventricular node. This early depolarization of a part of the ventricle is known as preexcitation (early excitation). Atrioventricular accessory pathways are also known as "Kent bundle." These consist of working myocardial cells. The term "preexcitation syndrome" (Wolff-Parkinson-White syndrome) is used only when tachyarrhythmias occur secondary to preexcitation.

2 Localization of the Kent Bundle

Localization of a Kent bundle is important because it is amenable to catheter ablation. The Kent bundle can be present anywhere along the atrioventricular groove or along the junction of the interatrial septum and the interventricular septum (Fig. 8.1). Left lateral pathways are most common and anteroseptal pathways are least common. Site of the Kent bundle produces diagnostic changes in the surface electrocardiogram.

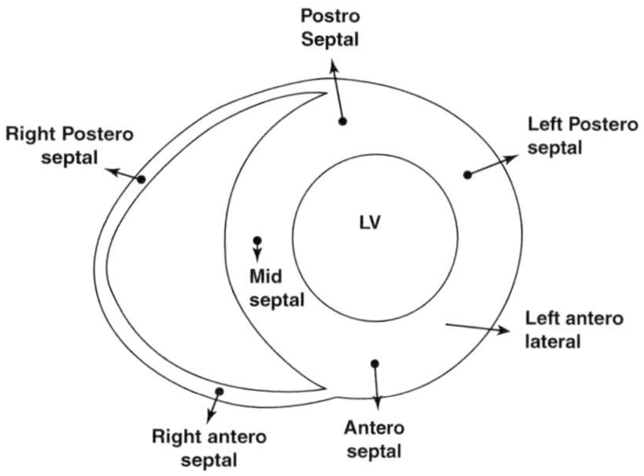

Fig. 8.1 Diagram showing common sites of the Kent bundle. *LV* left ventricle, *RV* right ventricle. (Reproduced with permission from: Mittal SR. Pre-excitation. Cardiology Today. 2013;17:89–95)

2.1 Electrocardiographic Features of Preexcitation from Common Sites of the Kent Bundle [1]

2.1.1 Left Lateral

Atrial end of the bypass tract is far away from the sinoatrial node which is situated in the right upper part of the right atrium. Atrial activation,

S. Mittal, *Insights into Electrocardiograms with MCQs*,
https://doi.org/10.1007/978-981-99-0127-2_8

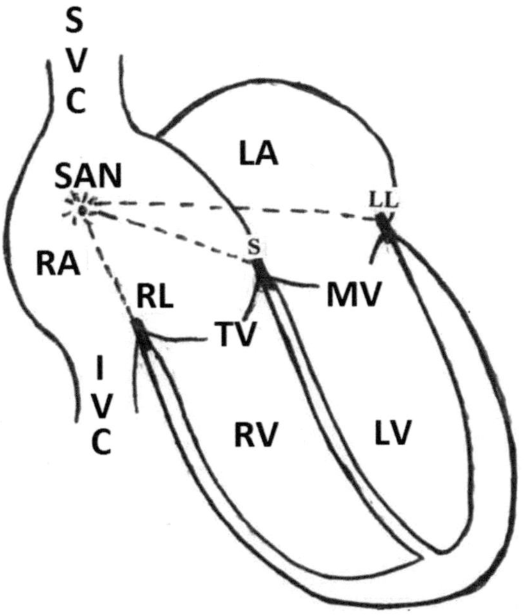

Fig. 8.2 Diagram showing distance of different sites of Kent bundle from the sinoatrial node (SAN). *RA* right atrium, *LA* left atrium, *RV* right ventricle, *LV* left ventricle, *SVC* superior vena cava, *IVC* inferior vena cava, *RL* right lateral pathway, *S* septal pathway, *LL* left lateral pathway. (Reproduced with permission from: Mittal SR. Pre-excitation. Cardiology Today. 2013;17:89–95)

therefore, takes longer time to reach the accessory pathway than the time it takes to reach the atrioventricular node (Fig. 8.2).

The PR interval is, therefore, nearly normal and preexcitation is minimal. Left lateral part of the base of the left ventricle is preexcited through the accessory path, and depolarization proceeds medially, anteriorly, and inferiorly (Fig. 8.3a). Delta wave is, therefore, negative in leads I, aVL, V_5, and V_6 and positive in leads V_1 to V_4, II, III, and aVF (Fig. 8.3b).

2.1.2 Left Posteroseptal

Posteroseptal part of the left ventricle is preexcited, and the initial vector of ventricular depolarization is directed anteriorly and superiorly (Fig. 8.4a). Delta wave is, therefore, negative in leads V_7 to V_9 and leads II, III, and aVF. Delta wave is positive in anterior precordial leads and leads I and aVL. There is prominent broad R wave in right precordial leads (Fig. 8.4b).

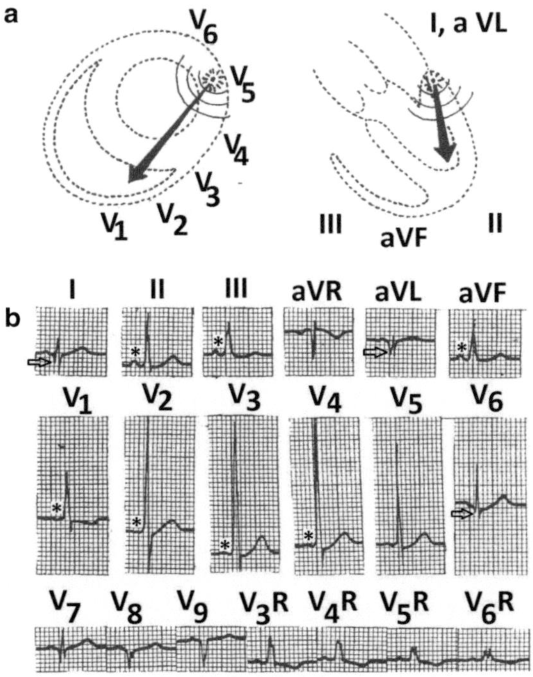

Fig. 8.3 (**a**) Diagram showing direction of the initial ventricular depolarization (the delta wave) vector in the presence of preexcitation via the left lateral Kent bundle. (**b**) Electrocardiogram in the presence of preexcitation via the left lateral Kent bundle. It shows negative delta wave (initial q wave) in leads I, aVL, and V_6 (marked ⇨) and positive delta wave (initial R wave) in leads II, III, aVF, and V_1 to V_4 (marked *). (Reproduced with permission from: Mittal SR. Pre-excitation. Cardiology Today. 2013;17:89–95)

2.1.3 Right Lateral

Lateral wall of the right ventricle is preexcited and the initial vector of ventricular depolarization is directed to the left and superiorly (Fig. 8.5a). Delta wave is, therefore, negative in leads III and aVF and usually in lead II. QRS is negative in leads V_1 to V_3 and positive in leads V_4 to V_6 (Fig. 8.5b).

2.1.4 Right Posteroseptal

Posteroseptal part of the right ventricle is preexcited by the accessory pathway. Initial vector of

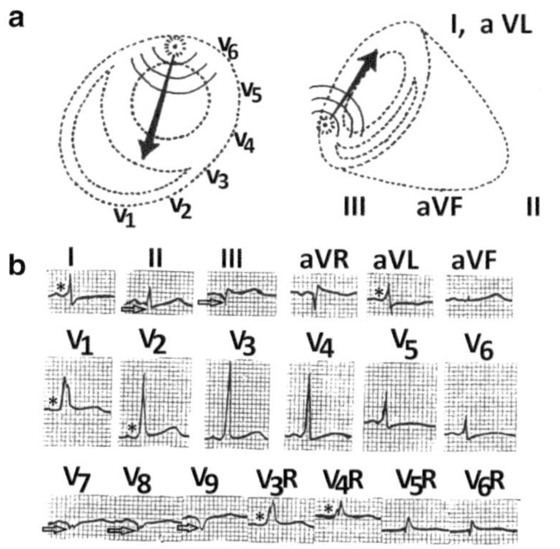

Fig. 8.4 (**a**) Diagram showing direction of the initial ventricular depolarization vector in the presence of preexcitation via left posteroseptal Kent bundle. (**b**) Electrocardiogram in preexcitation via the presence of posteroseptal Kent bundle. It shows negative delta wave (q wave) in leads II, III, and V_7 to V_9 (marked ⇨) and positive delta wave (initial R wave) in leads I, aVL, and V_3R to V_2 (marked *). (Reproduced with permission from: Mittal SR. Pre-excitation. Cardiology Today. 2013;17:89–95)

ventricular depolarization is directed anteriorly and superiorly (Fig. 8.6a). Delta wave is, therefore, negative in leads III and aVF and positive in leads I, aVL, and V_2 to V_6. QRS is negative in lead V_1 and tall and broad in mid-precordial leads V_2 to V_4 (Fig. 8.6b).

2.1.5 Right Anteroseptal

Preexcitation starts in the anteroseptal region and proceeds posteriorly and inferiorly (Fig. 8.7a). It produces broad positive QRS in leads I, aVL, II, III, aVF, and V_4 to V_6. QRS is negative in leads in V_1 to V_3 (Fig. 8.7b).

2.1.6 Midseptal Pathway

Preexcitation starts from mid of the interventricular septum. Depolarization wave front proceeds laterally and toward cardiac apex (Fig. 8.8a). QRS is, therefore, negative in leads V_1 to V_3 and positive in leads V_5, V_6, II, and aVF (Fig. 8.8b). However, as opposed to anteroseptal pathway, QRS is equiphasic or negative in lead III.

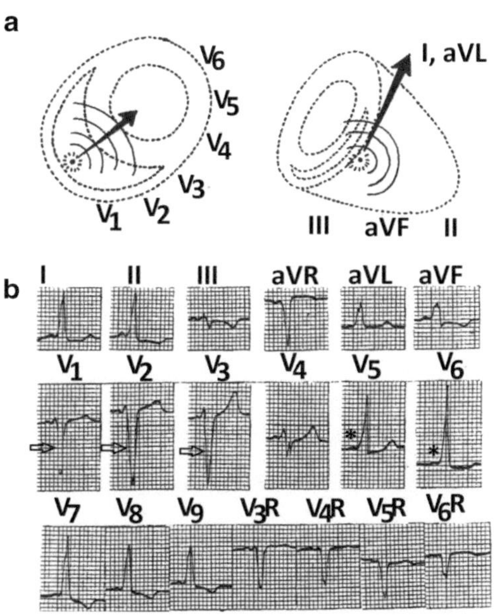

Fig. 8.5 (**a**) Diagram showing direction of the initial ventricular depolarization vector in the presence of preexcitation via right lateral Kent bundle. (**b**) Electrocardiogram in the presence of preexcitation via the right lateral Kent bundle. QRS is negative in leads V_1 to V_3 (marked ⇨) and positive in leads V_5 and V_6 (marked *). (Reproduced with permission from: Mittal SR. Pre-excitation. Cardiology Today. 2013;17:89–95)

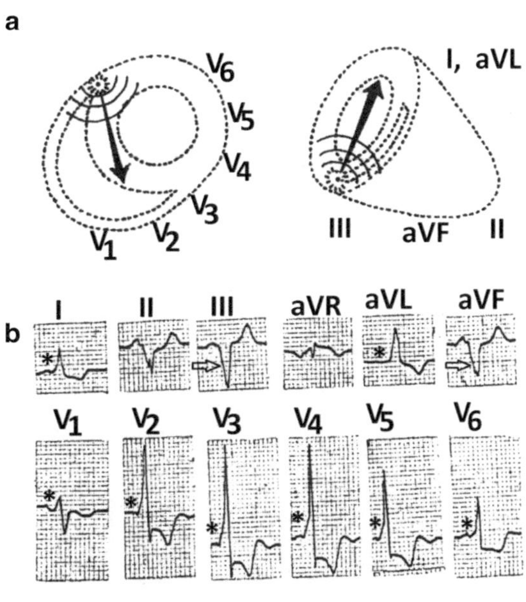

Fig. 8.6 (**a**) Diagram showing direction of the initial ventricular depolarization vector in the presence of preexcitation via the right posteroseptal Kent bundle. (**b**) Electrocardiogram in the presence of preexcitation via the right posteroseptal Kent bundle. Delta wave is negative (initial q wave) in leads III and aVF (marked ⇨) and positive (initial r wave) in leads I, aVL, and V_1 to V_6 (marked *). (Reproduced with permission from: Mittal SR. Pre-excitation. Cardiology Today. 2013;17:89–95)

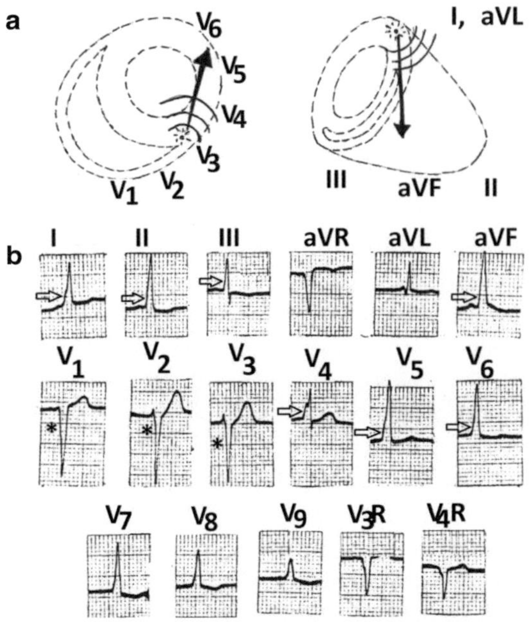

Fig. 8.7 (**a**) Diagram showing direction of the initial ventricular depolarization vector in the presence of preexcitation via the right anteroseptal Kent bundle. (**b**) Electrocardiogram in the presence of preexcitation via the right anteroseptal Kent bundle. It shows positive delta wave (initial r wave) in leads I, II, III, aVF, and V_4 to V_6 (marked ⇨) and negative QRS in leads V_1 to V_3 (marked *). (Reproduced with permission from: Mittal SR. Preexcitation. Cardiology Today. 2013;17:89–95)

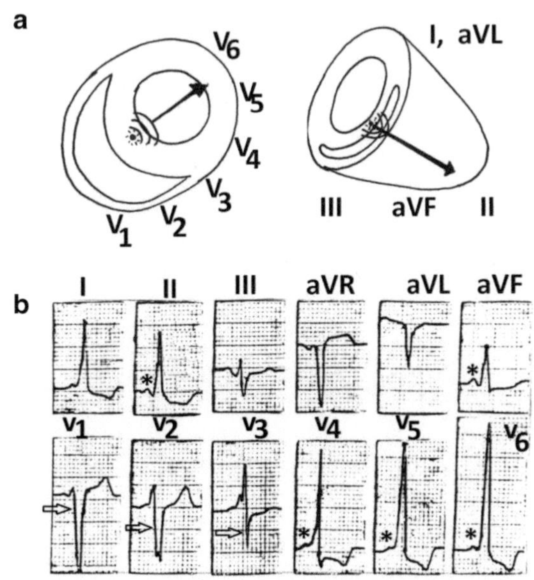

Fig. 8.8 (**a**) Diagram showing direction of the initial ventricular depolarization vector in the presence of preexcitation via the midseptal Kent bundle. (**b**) Electrocardiogram in the presence of preexcitation via the midseptal Kent bundle. It shows negative QRS in leads V_1 to V_3 (marked ⇨). QRS is positive in leads II, aVF, and V_4 to V_6 (marked *). (Reproduced with permission from: Mittal SR. Preexcitation. Cardiology Today. 2013;17:89–95)

2.2 Algorithms for Localization of Kent Bundle

Several algorithms have been suggested for localization of the atrioventricular accessory pathway.

1. *Site and polarity of the delta wave*

   Polarity of the delta wave is helpful in rough localization. Delta wave can be positive, negative, or isoelectric (Fig. 8.9).

Leads and polarity of the delta wave	Site of atrioventricular accessory pathway
Negative delta wave in leads I and aVL	Left side

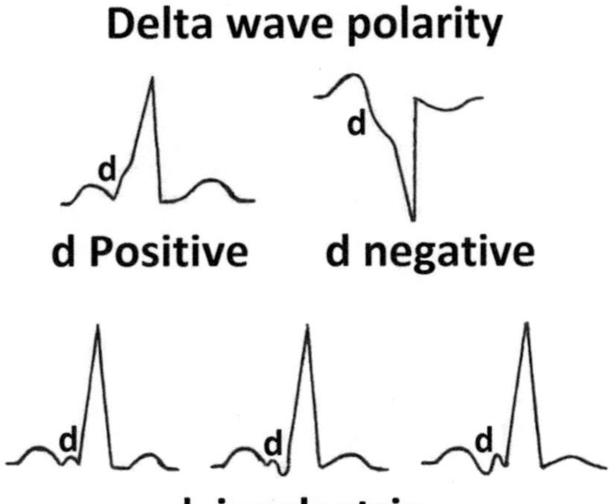

Fig. 8.9 Diagram showing different polarities of the delta wave. (Reproduced with permission from: Mittal SR. Pre-excitation. Cardiology Today. 2013;17:89–95)

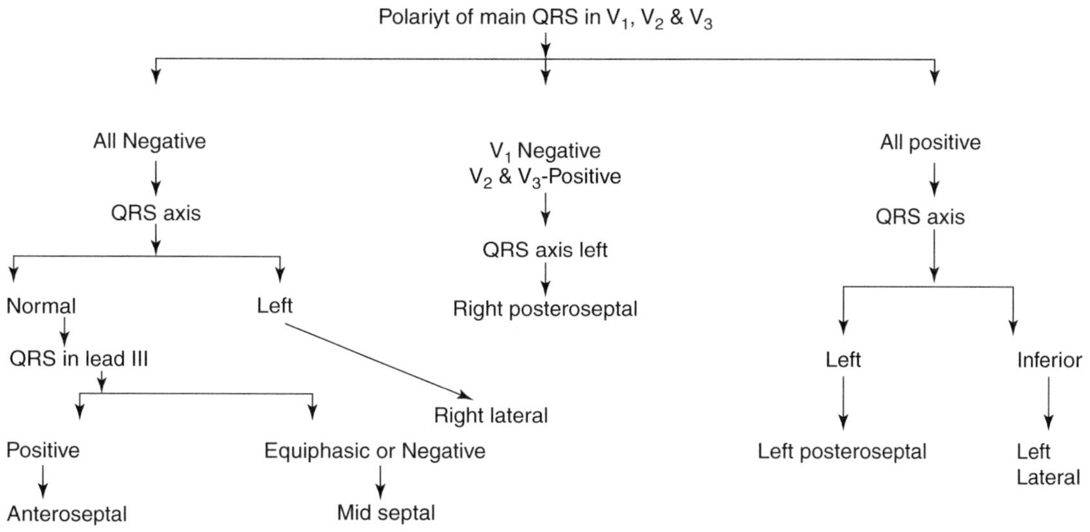

Fig. 8.10 The algorithm using polarity of QRS in leads V_1 to V_3 and mean frontal plane QRS axis [2]

Leads and polarity of the delta wave	Site of atrioventricular accessory pathway
Negative delta wave in lead V_1	Right side
Isoelectric delta wave in lead V_1	Anteroseptal
Negative delta wave in leads II, III, and aVF	Posteroseptal
Positive delta wave in leads II, III, and aVF	Anteroseptal

2. Figure 8.10 depicts *the algorithm using polarity of QRS in leads V_1 to V_3 and mean frontal plane QRS axis* [2]

Surface electrocardiogram provides only gross localization of the area for initial evaluation during the electrophysiologic study. Localization is difficult when electrocardiogram is influenced by other factors, e.g., myocardial infarction, ventricular hypertrophy, or congenital heart disease. Precordial lead placement and chest deformities can also hamper use of algorithm. Minor degrees of preexcitation or presence of more than one accessory pathways also hamper interpretation. Concealed accessory pathways do not manifest on the surface electrocardiogram although they are frequently involved in the atrioventricular reciprocating tachycardias. Absence of preexcitation on surface electrocardiogram, therefore, does not exclude possibility of presence of an accessory pathway.

3 Electrocardiographic Resemblance to Other Cardiac Abnormalities and Their Differentiation

3.1 Bundle Branch Block or Ventricular Hypertrophy

Left posteroseptal accessory pathway produces electrocardiographic findings simulating right bundle branch block or right ventricular hypertrophy. Right anteroseptal accessory pathway simulates left ventricular hypertrophy or left bundle branch block. The presence of short PR interval and a delta wave helps in suspecting preexcitation as the cause for ECG abnormality. Other features can also help differentiation:

1. *Right ventricular hypertrophy*. Preexcitation should be suspected by the absence of other electrocardiographic findings suggestive of right ventricular hypertrophy, e.g., P—pulmonale, right axis deviation, qR configuration with T wave inversion in lead V_1, and clockwise rotation in the precordial leads.
2. *Right bundle branch block*. Absence of triphasic QRS with broad terminal R' in lead V_1 and broad terminal S wave in lead V_6 supports the possibility of preexcitation (Fig. 8.11a) over right bundle branch block (Fig. 8.11b).
3. *Left ventricular hypertrophy*. Presence of q wave in leads V_5 and V_6 and ECG evidence of left atrial enlargement support the diagnosis of left ventricular hypertrophy (Fig. 8.12b) over preexcitation (Fig. 8.12a).

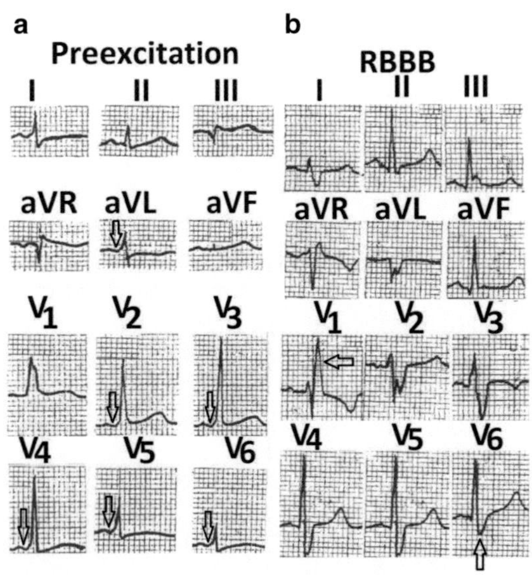

Fig. 8.11 (**a**) Electrocardiogram in preexcitation from left posteroseptal accessory path. PR interval is short and there is slurring of initial part of QRS (delta wave) (marked ⇩). (**b**) Electrocardiogram in right bundle branch block. There is broad terminal R wave in lead V_1 (marked ⇦) and broad terminal S wave in lead V_6 (marked ⇧). (Reproduced with permission from: Mittal SR. Preexcitation. Cardiology Today. 2013;17:89–95)

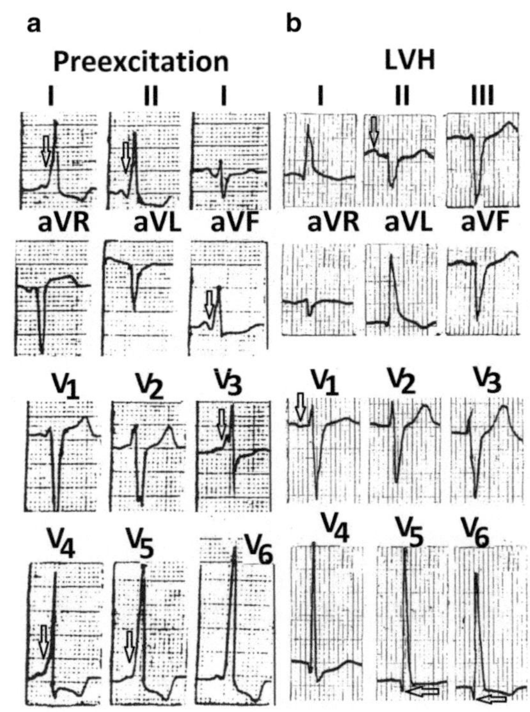

Fig. 8.12 (**a**) Electrocardiogram in preexcitation via the midseptal accessory path. PR interval is short and there is slurring of initial part of QRS delta wave (marked ⇩). (**b**) Electrocardiogram in left ventricular hypertrophy. There is broad P wave in lead II (marked ⇩) and terminal position of the P wave in lead V_1 (marked ⇩). Both these findings suggest left atrial overload. There is left axis deviation in limb leads and initial q wave in leads V_5 and V_6 (marked ⇦)

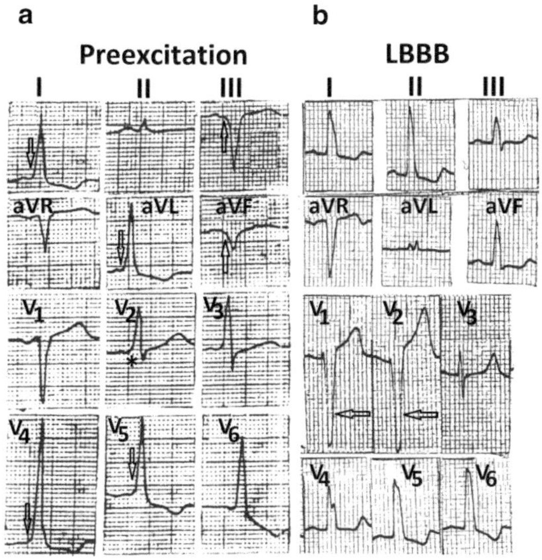

Fig. 8.13 (**a**) Electrocardiogram in preexcitation from the right posteroseptal accessory path. PR interval is short. There is slurring of initial part of the QRS complex (marked ⇩). Delta wave is negative in leads III, aVF, and V_1 (marked ⇧) and positive in lead V_2 (marked *). (**b**) Electrocardiogram in left bundle branch block. There is broad and deep terminal negative deflection (S wave) in leads V_1 and V_2 (marked ⇦). There is no slurring of initial portion of the QRS complex (no delta wave). (Reproduced with permission from: Mittal SR. Pre-excitation. Cardiology Today. 2013;17:89–95)

4. *Left bundle branch block.* Absence of broad terminal positive vector in leads V_5 and V_6 and broad terminal negative vector in leads V_1 and V_2 should raise suspicion of preexcitation (Fig. 8.13a) as compared to left bundle branch block (Fig. 8.13b).

3.2 Myocardial Infarction [3]

Preexcitation can mask, mimic, or exaggerate ECG findings of myocardial infarction.

1. *Masking of myocardial infarction.* An accessory pathway contralateral to the infarction masks the infarction on the surface ECG. For example, an inferoposteriorly located accessory pathway can mask an anterior wall infarction (Fig. 8.14a–c).
2. *Mimicking myocardial infarction.* Left lateral accessory pathway simulates posterior myocardial infarction (Fig. 8.15a). Upright T waves in right-sided chest leads and evidence of concomitant inferior infarction support the possibility of a posterior myocardial infarction (Fig. 8.15b).

   Anteriorly located pathways simulate anterior infarction (Fig. 8.16a). Coved ST segment elevation with deep symmetrical T wave inversion in anterior leads supports the possibility of anterior infarction (Fig. 8.16b).

3. *Exaggeration of ECG findings of myocardial infarction.* An ipsilateral location of accessory path can exaggerate the ECG findings of infarction (Fig. 8.17a, b).

It is also possible that ECG findings of preexcitation disappear if the ventricular end of the accessory pathway is involved in the myocardial infarction.

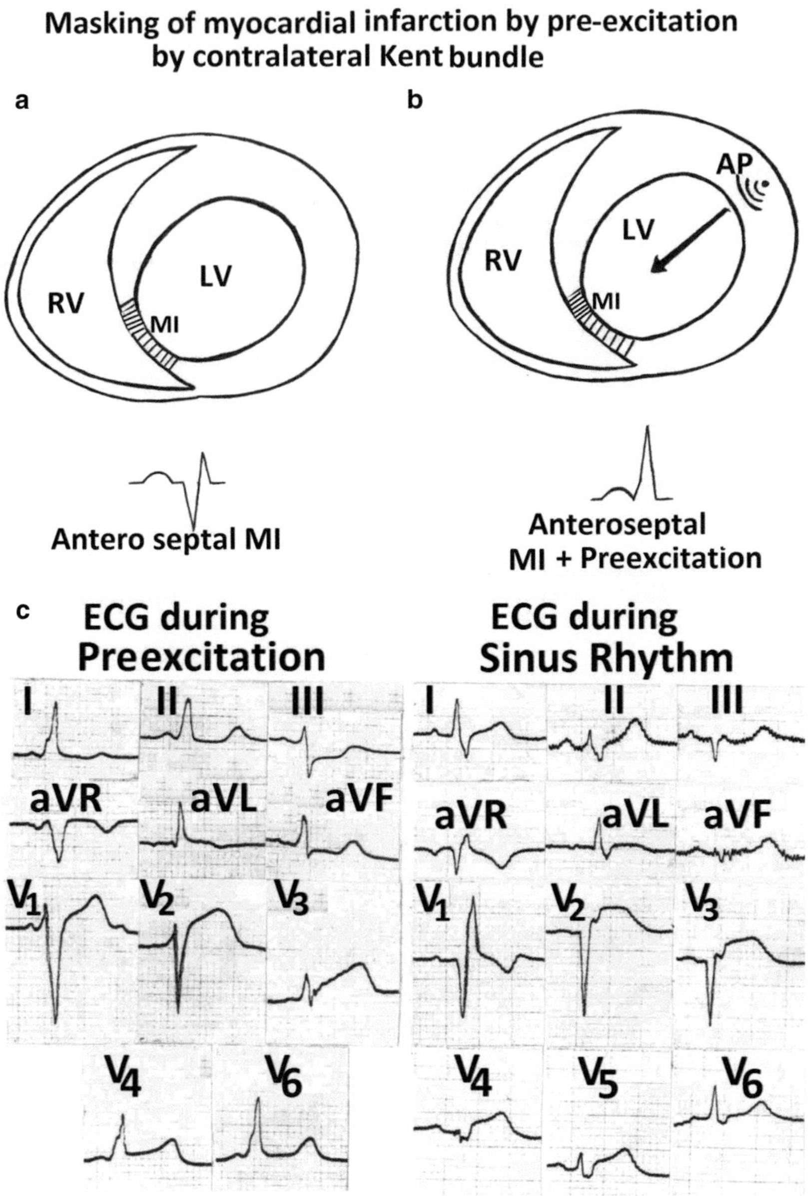

Fig. 8.14 (**a**) QRS in right precordial leads in anterior infarction alone. (**b**) QRS in the presence of anteroseptal MI with contralateral accessory path. *RV* right ventricle, *LV* left ventricle, *AP* accessory pathway, *MI* myocardial infarction. (**c**) Masking of anterior myocardial infarction by preexcitation. (Reproduced with permission from: Mittal SR. Pre-excitation. Cardiology Today. 2013;17:89–95)

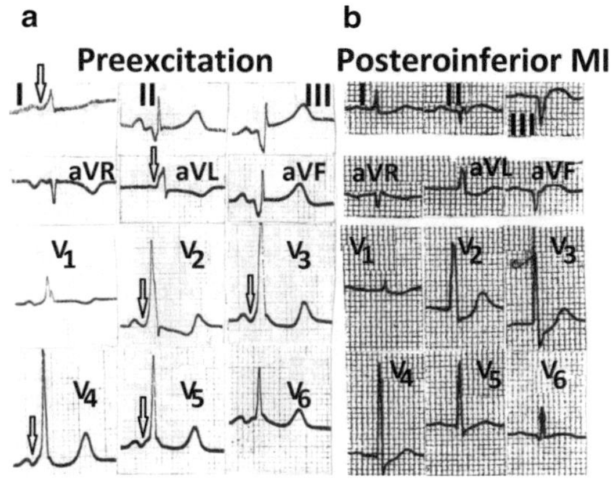

Fig. 8.15 (**a**) Electrocardiogram in preexcitation from the left lateral Kent bundle. PR interval is short. There is slurring of the initial part of QRS—the delta wave (marked ⇩). Leads V_1 to V_6 show positive R wave. (**b**) Electrocardiogram in posteroinferior myocardial infarction. PR interval is normal and there is no slurring of initial part of QRS. (Reproduced with permission from: Mittal SR. Pre-excitation. Cardiology Today. 2013;17:89–95)

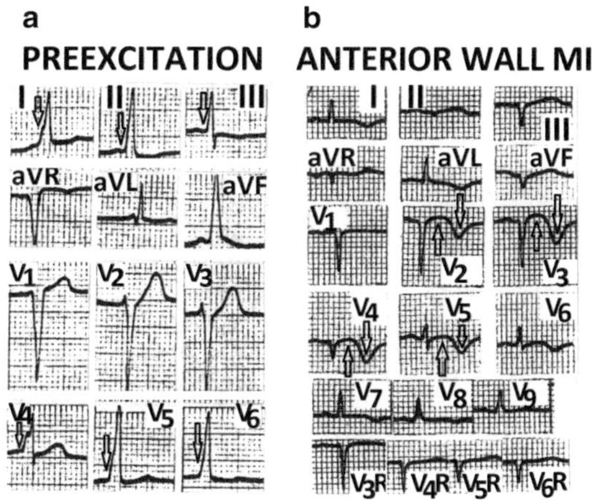

Fig. 8.16 (**a**) Electrocardiogram in the presence of preexcitation via anteriorly located Kent bundle. PR interval is short and there is slurring of initial part of the QRS complex (marked ⇩). (**b**) Electrocardiogram in anterior myocardial infarction. Leads V_2 to V_5 show coved ST segment elevation (marked ⇧) and deep symmetrical T wave inversion (marked ⇩). (Reproduced with permission from: Mittal SR. Pre-excitation. Cardiology Today. 2013;17:89–95)

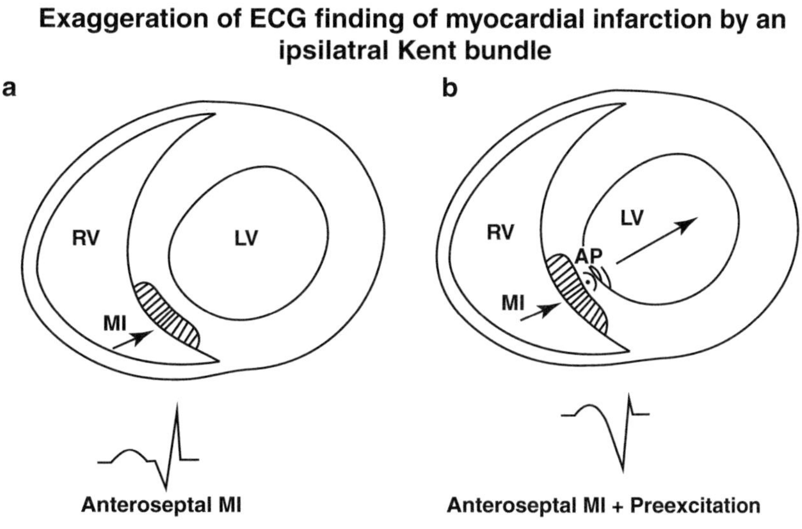

Fig. 8.17 (**a**) QRS in right precordial leads in anterior myocardial infarction. (**b**) QRS in right precordial leads in anterior infarction with ipsilateral Kent bundle. *MI* myocardial infarction, *RV* right ventricle, *LV* left ventricle, *AP* accessory path. (Reproduced with permission from: Mittal SR. Pre-excitation. Cardiology Today. 2013;17:89–95)

4 Other Methods that Can Help in Correct Diagnosis of Underlying Disease

Acute myocardial infarction can be suspected by classical history, raised serum enzymes, sequential changes in the ST segment and the T waves of the electrocardiogram, and regional wall motion abnormality on echocardiography. Diagnosis of old myocardial infarction may be difficult. Past history of classical discomfort and presence of regional wall motion abnormality on echocardiography can be helpful. Holter monitoring can help in the identification of non-preexcited QRS morphology if preexcitation is intermittent (Fig. 8.18a, b).

Intermittent retrograde conduction through a concealed accessory pathway may produce a "P" wave after normal QRS (Fig. 8.19).

Exercise testing can be helpful. In patients with relatively longer refractory period of the accessory pathway, conduction over the accessory pathway may suddenly block at a fast heart rate, thereby unmasking basal QRS configuration (Fig. 8.20a, b).

In other cases increased sympathetic drive during exercise may cause progressive enhancement of conduction over the atrioventricular node-His-Purkinje pathway. This may result in a gradual reduction in preexcitation and normalization of QRS (Fig. 8.21a, b).

Drugs like procainamide and ajmaline can block the accessory pathway without affecting the atrioventricular nodal conduction. These drugs can be helpful in unmasking the basal QRS morphology.

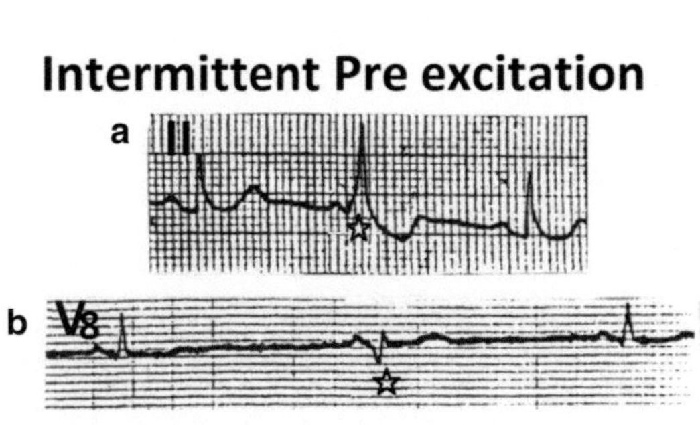

Fig. 8.18 (**a**) Normal and preexcited QRS (☆) in lead II. (**b**) Normal and preexcited QRS (☆) in lead V_8. (Reproduced with permission from: Mittal SR. Pre-excitation. Cardiology Today. 2013;17:89–95)

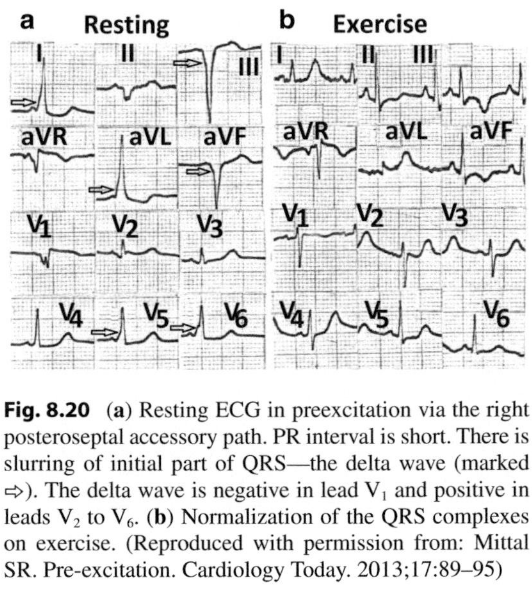

Fig. 8.20 (**a**) Resting ECG in preexcitation via the right posteroseptal accessory path. PR interval is short. There is slurring of initial part of QRS—the delta wave (marked ⇨). The delta wave is negative in lead V_1 and positive in leads V_2 to V_6. (**b**) Normalization of the QRS complexes on exercise. (Reproduced with permission from: Mittal SR. Pre-excitation. Cardiology Today. 2013;17:89–95)

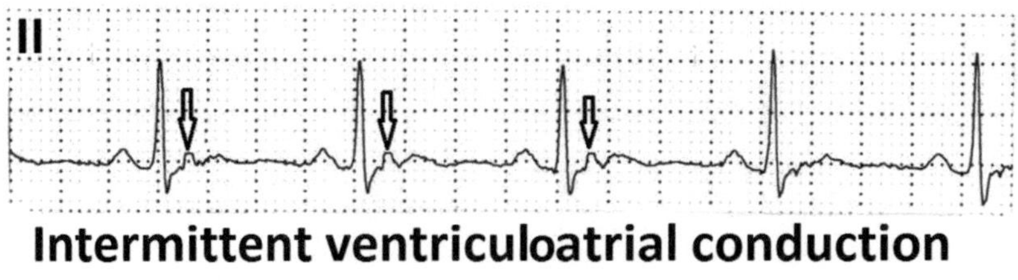

Fig. 8.19 Intermittent retrograde atrial activation probably through a concealed Kent bundle. (Reproduced with permission from: Mittal SR. Pre-excitation. Cardiology Today. 2013;17:89–95)

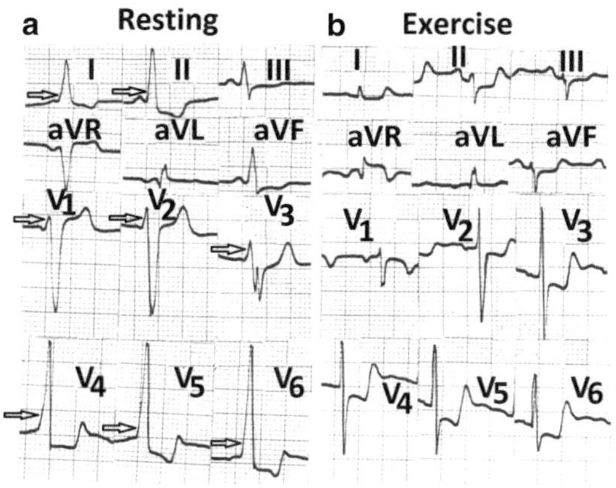

Fig. 8.21 (**a**) Resting ECG in preexcitation via the right anteroseptal accessory path. PR interval is short. There is slurring of initial part of the QRS—delta wave (marked ⇨). The delta wave is positive in leads II, III, and aVF and leads V_1 to V_6. (**b**) Normalization of the QRS complexes and unmasking of anterior ischemia on exercise. (Reproduced with permission from: Mittal SR. Pre-excitation. Cardiology Today. 2013;17:89–95)

Summary

- PR interval includes conduction over the atria, the atrioventricular node, and the His-Purkinje system. The atrioventricular node is the site of maximum delay in the atrioventricular conduction.
- Accessory pathways connecting the atria with the ventricle (atrioventricular accessory pathway—the Kent bundle) and conducting antegradely manifest as short PR interval, a delta wave (slurring on the beginning of QRS), and prolonged duration of the QRS complex.
- Accessory pathways that conduct only retrogradely (i.e., from the ventricle to the atria) do not manifest on the surface electrocardiogram (concealed (hidden) accessory pathway). These are usually along interventricular septum or along left lateral wall.
- Patients of Ebstein anomaly who have WPW syndrome are likely to have more than one accessory pathway.
- Left lateral accessory pathway shows minimal preexcitation on the surface ECG because of the greater distance between the sinoatrial node and the atrial end of the accessory pathway.
- Right-sided accessory path shows maximum preexcitation because the atrial end of the accessory path is close to the sinoatrial node.
- Magnitude of preexcitation on the surface electrocardiogram depends on the relative rates of conduction over the atrioventricular node versus rate of antegrade conduction over the accessory pathway. Increased sympathetic drive accelerates conduction over the atrioventricular node and decreases preexcitation on the ECG. Decreased conduction over the atrioventricular node (vagal stimulation or drugs) increases conduction over the accessory pathway and results in greater preexcitation.
- Intermittent preexcitation without significant change in heart rate suggests that the refractory period of the accessory pathway is long and it is unlikely to conduct at a fast rate in event of atrial tachyarrhythmias.
- Exact localization of accessory pathway needs detailed electrophysiologic study. However, some impression can be drawn from the direction of the delta wave in various leads of the surface electrocardiogram as shown in Fig. 8.22.

Loss of preexcitation during stress testing suggests that during tachycardia, atrioventricular node can conduct at a rate faster than the accessory pathway.

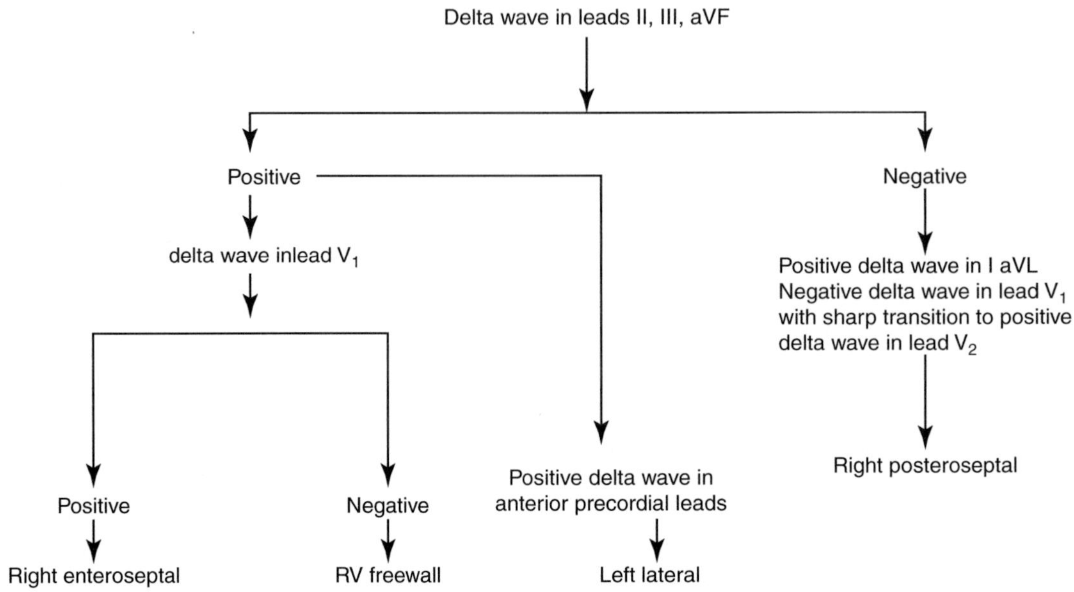

Fig. 8.22 Showing localization of accessory path by direction of the delta wave

MCQs

Q1. Which site of the Kent bundle can produce near normal PR interval during sinus rhythm?

(a) Right ventricular free wall
(b) Left ventricular free wall
(c) Right posteroseptal
(d) Right anteroseptal

Q2. Magnitude of QRS abnormality during preexcitation via a Kent bundle at a given time depends on:

(a) Site of the Kent bundle
(b) Rate of conduction over the Kent bundle
(c) Rate of conduction over the atrioventricular node
(d) All

Q3. Magnitude of preexcitation in a given patient:

(a) Is fixed
(b) Can change gradually
(c) Can change suddenly
(d) Is affected by the balance of sympathetic and parasympathetic system

Q4. WPW syndrome should be diagnosed only when there is:

(a) Short PR interval
(b) Slurring of the initial part of the QRS complex
(c) Preexcitation dependent tachyarrhythmia
(d) Broad QRS tachycardia

Q5. Negative delta wave (initial Q wave) in leads II, III, and aVF with positive delta wave and QRS in leads I and aVL and predominantly negative delta wave in leads V_1 to V_3 with negative QRS in leads V_1 to V_3 suggest:

(a) Left lateral accessory pathway
(b) Right lateral accessory pathway
(c) Right posterior paraseptal accessory pathway
(d) Left posterior paraseptal accessory pathway

Q6. Negative delta wave (initial Q wave) in leads I and aVL with positive delta wave and QRS in leads II, III, and aVF and precordial leads suggest:

(a) Left lateral accessory pathway
(b) Right lateral accessory pathway

(c) Right posterior paraseptal accessory pathway
(d) Left posterior paraseptal accessory pathway

Q7. Negative delta wave (initial Q wave) in leads II, III, aVF, and V_1 and positive delta wave and QRS in leads I, aVL, and V_2 to V_6 suggest:

(a) Right lateral accessory pathway
(b) Left lateral accessory pathway
(c) Right posteroseptal accessory pathway
(d) Left posterior paraseptal accessory pathway

Q8. Negative delta wave (initial Q wave) in leads II, III, and aVF and positive delta wave and QRS in leads V_1 to V_6 suggest:

(a) Right lateral accessory pathway
(b) Left lateral accessory pathway
(c) Right posteroseptal accessory pathway
(d) Left posteroseptal accessory pathway

Q9. Positive delta wave in leads I, II, aVL, and aVF and rS configuration in leads V_1 to V_3 suggest:

(a) Anteroseptal accessory pathway
(b) Right posteroseptal accessory pathway
(c) Left lateral accessory pathway
(d) Right lateral accessory pathway

Q10. Preexcitation can produce:

(a) Abnormal Q wave
(b) Increased QRS voltage
(c) Bundle branch block pattern
(d) All

Q11. Peaked P wave, PR interval of 0.12 s, QRS duration of 0.08 s, QRS axis of +120°, R wave of 6 mm in lead V_1, and RS configuration in lead V_6 suggest:

(a) Left posteroseptal accessory pathway
(b) Left lateral accessory pathway
(c) Right ventricular hypertrophy
(d) Posterior wall infarction

Q12. Normal P wave, PR interval of 0.12 s, QRS axis +60°, QRS duration 0.08 s, and positive R wave with a prominent upright T wave in leads V_1 and V_2 with QR in lead V_6 suggest:

(a) Left lateral accessory pathway
(b) Left posteroseptal accessory pathway
(c) Right ventricular hypertrophy
(d) Posterior infarction

Q13. Normal P wave, PR interval of 0.14 s, QRS axis +75°, QRS duration of 0.08 s, and Qr configuration with T wave inversion in leads I and aVL with rS configuration in lead V_1 suggest:

(a) Left lateral accessory pathway
(b) Left posteroseptal accessory pathway
(c) High lateral infarction
(d) Right posteroseptal accessory pathway

Q14. Normal P wave, PR interval 0.12 s, QRS axis +125°, QRS duration of 0.12 s, rsR' pattern with inverted T wave in lead V_1, and rS configuration in lead V_6 suggest:

(a) Posterior infarction
(b) Right bundle branch block
(c) Left lateral accessory path
(d) Left posteroseptal accessory path

Q15. Normal P wave, PR interval 0.10 s, QRS duration 0.14 s, and QS configuration in leads III, aVF, and V_1 with prominent R in leads V_2 to V_6 suggest:

(a) Inferior infarction
(b) Right posteroseptal accessory pathway
(c) Left posteroseptal accessory pathway
(d) Right lateral accessory pathway

Q16. Normal P wave, PR interval of 0.08 s, QRS axis −60°, QRS duration of 0.12 s, R wave of 20 mm in lead I, R wave of 20 mm in lead aVL, QS in lead V_1, and R wave of 25 mm in lead V_5 suggest:

(a) Left ventricular hypertrophy
(b) Left bundle branch block
(c) Right posteroseptal accessory pathway
(d) Right lateral accessory pathway

Q17. Normal P wave, PR interval of 0.08 s, QRS duration of 0.12 s, QRS axis +30°, Rs configuration in lead V_2, and RR' configuration with T wave inversion in lead V_6 suggest:

(a) Left ventricular hypertrophy
(b) Left bundle branch block
(c) Right posteroseptal accessory pathway
(d) Left lateral accessory pathway

Q18. Concealed accessory pathway is usually on:

(a) Left side
(b) Right side
(c) Posterior paraseptal region
(d) Anterior paraseptal region

Answers

(1) b; (2) d; (3) b, c, d; (4) a, b, c; (5) b; (6) a; (7) c; (8) d; (9) a; (10) d; (11) c; (12) d; (13) c; (14) b; (15) b; (16) a; (17) c; (18) a

References

1. Ellis CR. Wolff-Parkinson-White syndrome workup. Medscape reference. 2012.
2. Schamroth L. The Wolff-Parkinson-White and related syndromes. In: Schamroth C, editor. An introduction to electrocardiography. New Delhi: Blackwell Science; 1982. p. 288–306.
3. Smolders L, Mazidi M, Krucoff MW, et al. Pre-excitation and myocardial infarction: conditions with confusing electrocardiographic manifestations. J Electrocardiol. 2008;41:679.

Tachyarrhythmias Associated with Preexcitation

9

Patients with accessory pathways are susceptible to tachyarrhythmias. These tachyarrhythmias can be grouped as follows [1] (Fig. 9.1):

1. When the accessory pathway is necessary for the initiation and maintenance of tachycardia, e.g., atrioventricular reciprocating tachycardias
2. When the accessory path allows rapid conduction during other supraventricular arrhythmias, e.g., atrial fibrillation/atrial flutter
3. When the accessory pathway is only a bystander

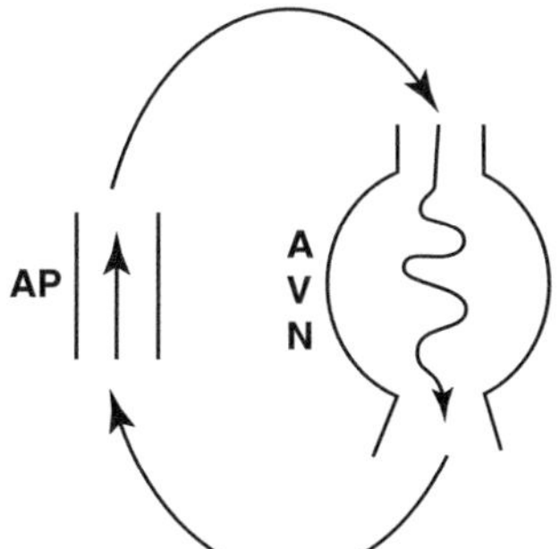

Fig. 9.1 Diagram showing route of electrical impulse in orthodromic AVRT. *AVN* atrioventricular node, *AP* accessory path. (Reproduced with permission from: Mittal SR. Tachyarrhythmias in Pre-excitation. Cardiology Today. 2013;17:177–183)

The term Wolff-Parkinson-White (WPW) syndrome is used only when there is combination of short PR interval and slurring of initial part of the QRS complex (the delta wave) along with tachyarrhythmias that are dependent on preexcitation.

1 Tachyarrhythmias in which the Accessory Pathway Is Necessary for Initiation and Maintenance of the Tachycardia (WPW Syndrome)

Atria, the atrioventricular node, the His bundle, bundle branches, the Purkinje system, and the accessory path are involved in the tachyarrhythmia. These are termed atrioventricular reciprocating tachycardias (AVRT). Atrioventricular accessory pathways (Kent bundle) are frequently involved. Long atriofascicular paths can also produce AVRT. The James fibers, the nodoventricular, and the fasciculoventricular fibers do not participate in reentrant tachyarrhythmias. AVRT could be:

(a) Orthodromic AVRT (about 95%)
(b) Antidromic AVRT (about 5%)

S. Mittal, *Insights into Electrocardiograms with MCQs*,
https://doi.org/10.1007/978-981-99-0127-2_9

1.1 Orthodromic AVRT

A critically timed atrial premature beat, occurring after recovery of the atrioventricular node from the effect of the previous sinus impulse, traverses antegradely over the atrioventricular node and the His-Purkinje system resulting in normal depolarization of the ventricles, thereby producing a narrow QRS. After complete ventricular depolarization, the impulse traverses retrogradely over the accessory pathway resulting in retrograde depolarization of the atria (Fig. 9.1).

P′ wave occurring from the retrograde atrial depolarization, therefore, occurs after completion of the QRS complex. Contour of the P′ wave differs from usual retrograde P′ wave because the atria are usually activated eccentrically rather than from the low right atrial septum (Fig. 9.2a–c).

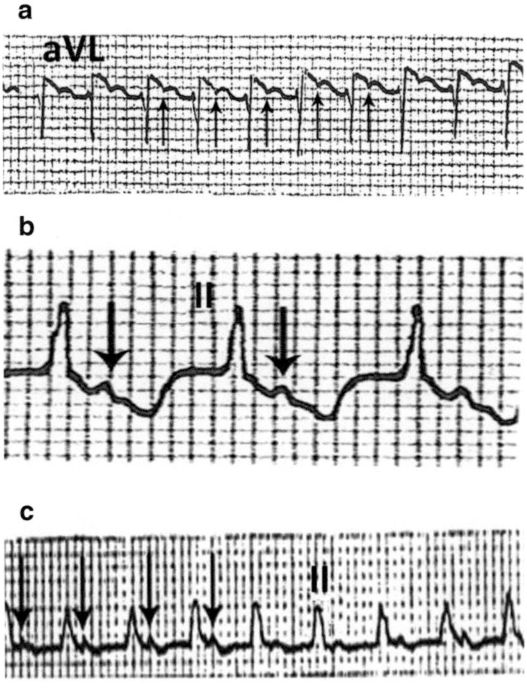

Fig. 9.2 (a–c) Orthodromic AVRT with P′ wave (arrow) occurring after completion of the QRS complex from three different patients. (Reproduced with permission from: Mittal SR. Tachyarrhythmias in Pre-excitation. Cardiology Today. 2013;17:177–183)

A negative P′ wave in lead I suggests a left-sided accessory pathway resulting in atrial depolarization from left to right [2] (Fig. 9.3a, b).

In the case of septal accessory path, retrograde atrial activation occurs from the lower atrial septum. P′ wave, therefore, resembles classical retrograde activation, i.e., P′ waves are inverted in leads II, III, and aVF (Fig. 9.4a, b).

On reaching the atria, the impulse again enters the atrioventricular node and is conducted antegradely to produce another narrow QRS complex and again a retrograde atrial depolarization producing another abnormal P′ wave after the QRS. Activation of the circuit continues producing orthodromic AVRT. During orthodromic AVRT, interval between the preceding QRS and the following P′ wave is less than the interval between the P′ wave and next the QRS. The RP′ interval is less than the P′R interval. If retrograde conduction over the accessory path is slow (Fig. 9.5), distance between the previous QRS and the following P′ wave may be longer than the distance between the P′ wave and following QRS. In such cases the RP′ interval can be longer than the P′R interval. Such AVRTs are also called long RP′ tachycardia (Fig. 9.6a–d).

Such a tachycardia may mimic atrial tachycardia. Sometimes P′ wave is not clearly visible and can result in depression of the ST segment (Fig. 9.7a). When this is seen during orthodromic AVRT, mechanism of arrhythmia is most often reentry involving an accessory pathway [2]. Location of the ST segment depression may suggest location of the accessory path [1]. ST segment depression in leads V_3 to V_6 suggests orthodromic AVRT involving a left lateral accessory path (Fig. 9.7b).

Negative T waves in the inferior leads suggest orthodromic AVRT involving a posteroseptal or posterior path [1]. Negative or notched T waves in leads V_2 or V_3 with retrograde (inverted) p′ waves in at least two inferior leads suggest orthodromic AVRT involving an anteroseptal accessory path [1]. ST segment elevation in lead aVR also suggests an orthodromic AVRT.

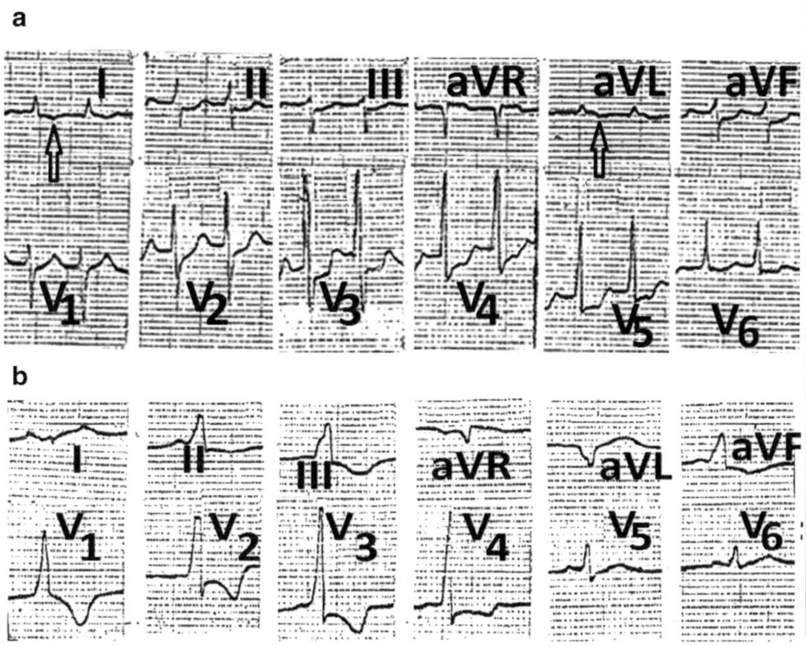

Fig. 9.3 (**a**) Orthodromic AVRT with P′ wave inverted in lead I and aVL (marked arrow). (**b**) ECG during sinus rhythm from the same patient showing preexcitation via left lateral accessory path. (Reproduced with permission from: Mittal SR. Tachyarrhythmias in Pre-excitation. Cardiology Today. 2013;17:177–183)

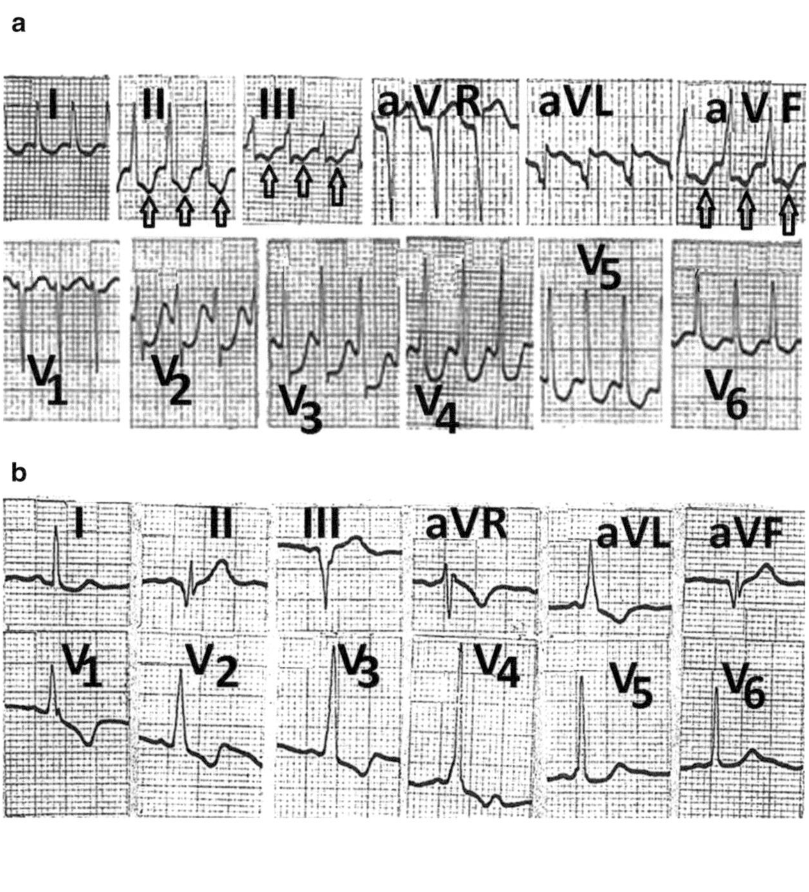

Fig. 9.4 (**a**) Orthodromic AVRT with inverted P′ waves in leads II, III, and aVF (marked ⇧). (**b**) ECG during sinus rhythm from the same patient showing preexcitation via posteroseptal accessory path. (Reproduced with permission from: Mittal SR. Tachyarrhythmias in Pre-excitation. Cardiology Today. 2013;17:177–183)

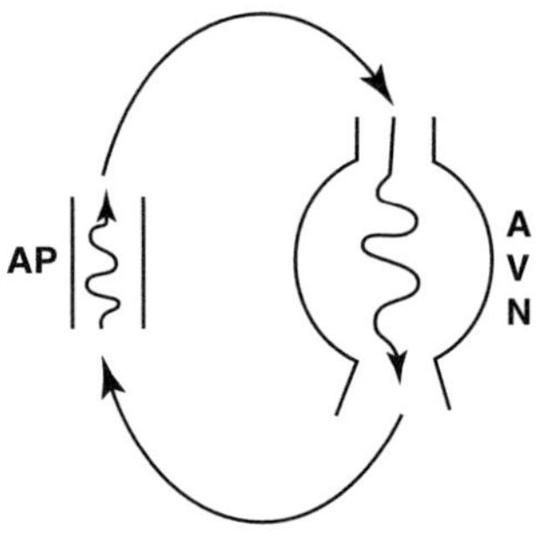

Fig. 9.5 Diagram showing route of electrical impulse in long RP′ AVRT. *AVN* atrioventricular node, *AP* accessory path. (Reproduced with permission from: Mittal SR. Tachyarrhythmias in Pre-excitation. Cardiology Today. 2013;17:177–183)

Frequently retrograde conduction occurs over a concealed accessory path, and preexcitation may not be apparent in sinus rhythm. Premature ventricular contraction can also initiate an orthodromic AVRT. When there are two or more accessory connections, the impulse may conduct antegradely over one accessory path and retrogradely over another accessory path (Fig. 9.8).

At times, an orthodromic AVRT may be difficult to distinguish from usual atrioventricular nodal reentrant tachycardia. Heart rate higher than 200/min with QRS alternans and a clear retrograde p′ wave in the ST segment supports the possibility of an orthodromic AVRT (Fig. 9.9a–c).

Sometimes a posteroseptal accessory pathway conducts only retrogradely. If this retrograde conduction is slow, it can produce permanent junctional reciprocating tachycardia. Such a tachycardia is incessant and responds only to ablation of the accessory pathway [3].

1.2 Antidromic AVRT

1.2.1 Antidromic AVRT Involving a Kent Bundle (Fig. 9.10)

Uncommonly either the refractory period of the accessory path is shorter than that of the atrioventricular node or the atrial ectopic arises in close proximity of the accessory path. In these circumstances, the atrial premature beat conducts rapidly antegradely over the accessory pathway

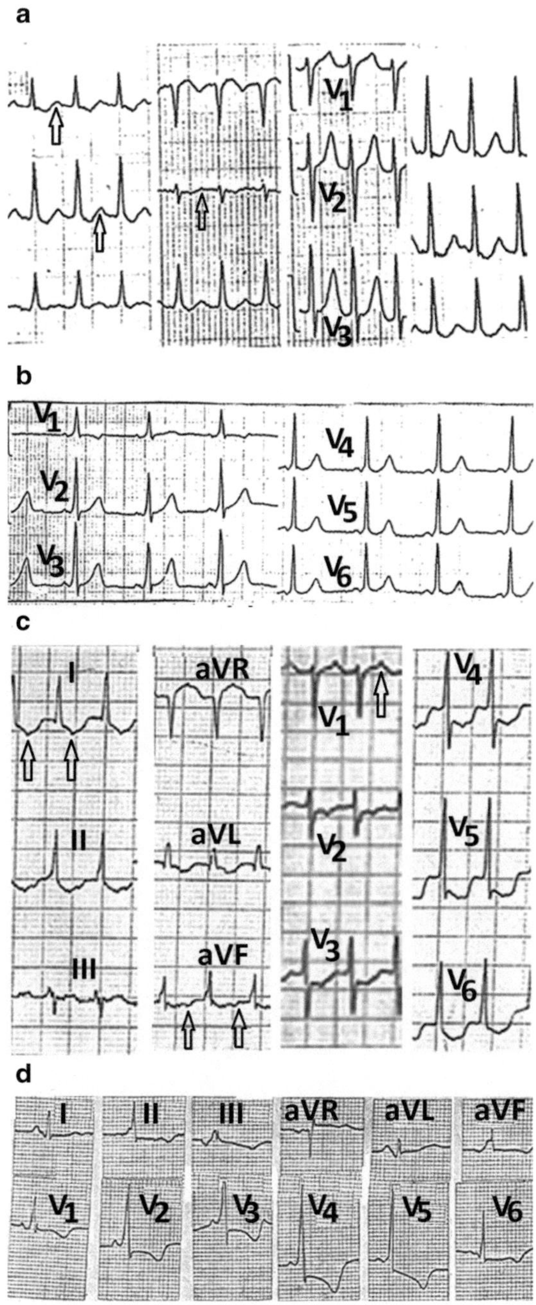

Fig. 9.6 (**a**) Orthodromic AVRT with long RP′ interval (P′ wave marked ⇧). (**b**) ECG during sinus rhythm from the same patient showing preexcitation. (**c**) Orthodromic AVRT with long RP′ from another patient (P′wave marked ⇧). (**d**) ECG during sinus rhythm from the same patient showing preexcitation. (Reproduced with permission from: Mittal SR. Tachyarrhythmias in Pre-excitation. Cardiology Today. 2013;17:177–183)

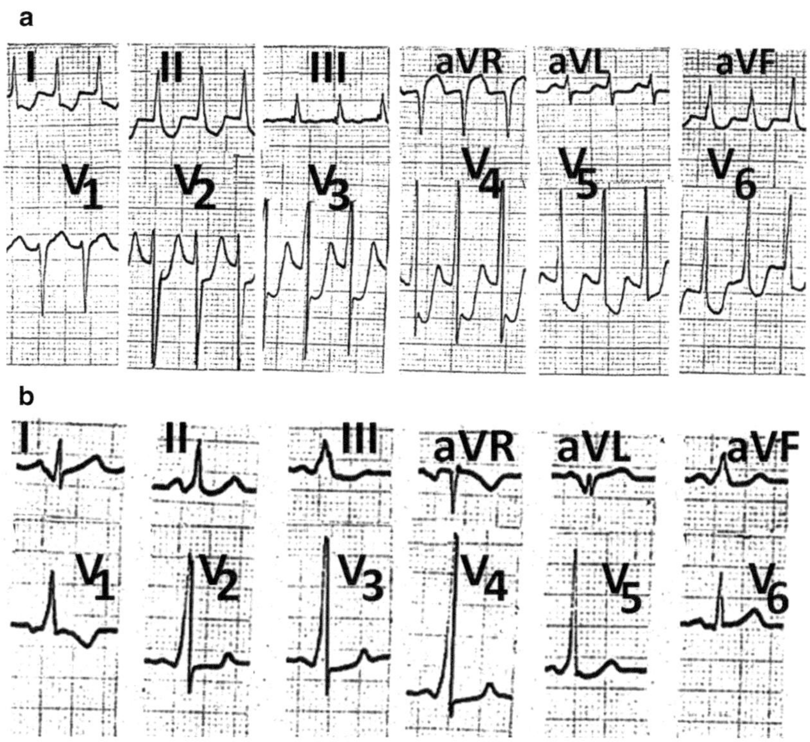

Fig. 9.7 (**a**) Orthodromic AVRT with ST segment depression in leads V_2 to V_6. (**b**) ECG during sinus rhythm from the same patient showing preexcitation from left lateral accessory path. (Reproduced with permission from: Mittal SR. Tachyarrhythmias in Pre-excitation. Cardiology Today. 2013;17:177–183)

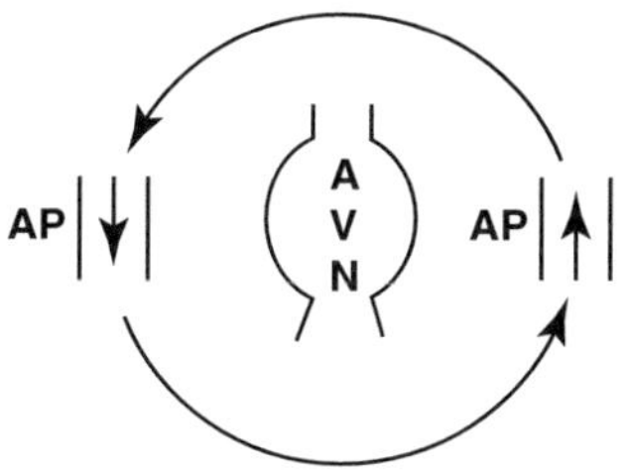

Fig. 9.8 Diagram showing route of electrical impulse in AVRT involving multiple bypass tracts. (Reproduced with permission from: Mittal SR. Tachyarrhythmias in Pre-excitation. Cardiology Today. 2013;17:177–183)

producing preexcitation of a part of either ventricle. This results in a delta wave producing a broad QRS complex. Impulse then enters the normal ventricular conduction system and is conducted retrogradely over the His bundle and atrioventricular node to reach the atria. The impulse faces normal delay in conduction through the atrioventricular node with the result that P′ wave comes after completion of the QRS complex. As the atria are depolarized retrogradely from the atrioventricular node, P′ wave is inverted in leads II, III, and aVF. The impulse then again enters the accessory pathway establishing a circus movement with broad QRS tachycardia (Figs. 9.11a, b and 9.12a, b).

Antidromic AVRT is rare and occurs mostly with the left-sided bypass tracts. It is difficult to differentiate from ventricular tachycardia.

1.2.2 Antidromic AVRT Involving the Atriofascicular Bypass Tract

Atriofascicular bypass tracts can also participate in an antidromic AVRT [4]. These fibers connect the low right atrium (area adjacent to the lateral tricuspid annulus) to the distal part of the right bundle branch or apical region of the right ventricle. This pathway usually represents a duplication of the atrioventricular node and distal conducting system. It typically occupies the right ventricular free wall. Proximal part of this pathway exhibits slow conduction like atrioventricular node. Retrograde conduction is, therefore,

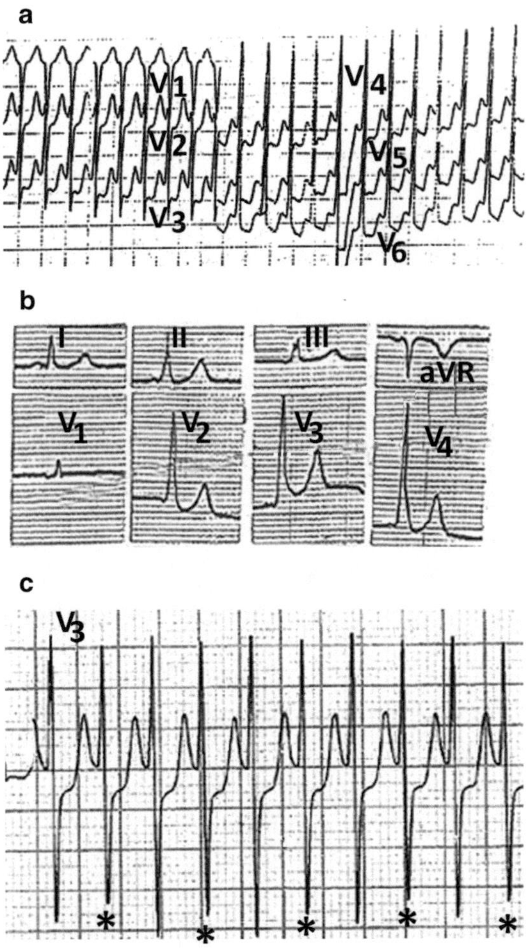

Fig. 9.9 (**a**) Orthodromic AVRT with a rate around 300/min. (**b**) ECG during sinus rhythm from the same patient showing preexcitation. (**c**) Orthodromic AVRT showing electrical alternans (marked *). (Reproduced with permission from: Mittal SR. Tachyarrhythmias in Pre-excitation. Cardiology Today. 2013;17:177–183)

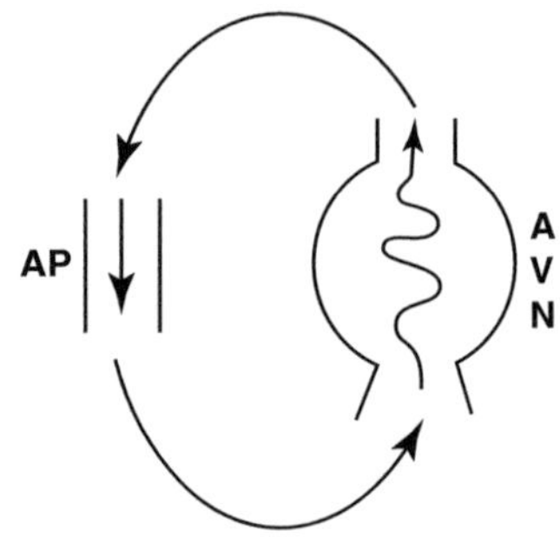

Fig. 9.10 Diagram showing route of electrical impulse in antidromic AVRT. (Reproduced with permission from: Mittal SR. Tachyarrhythmias in Pre-excitation. Cardiology Today. 2013;17:177–183)

absent and only antidromic atrioventricular reentry tachycardia can develop. Tachycardia has left bundle branch block pattern. Development of rate-dependent, right bundle branch block prolongs conduction in the right bundle branch and slows the rate of the tachycardia. Tachycardia rate remains unaffected by aberrancy in the left bundle branch.

1.3 Other Electrocardiographic Findings in AVRT

1. If the rate of tachycardia slows with appearance of the rate-dependent bundle branch block, it suggests that the accessory path is located in the ventricle with bundle branch block.
2. QRS alternans is seen in nearly 1/3 of the patients with AVRT [1] (Fig. 9.9c).

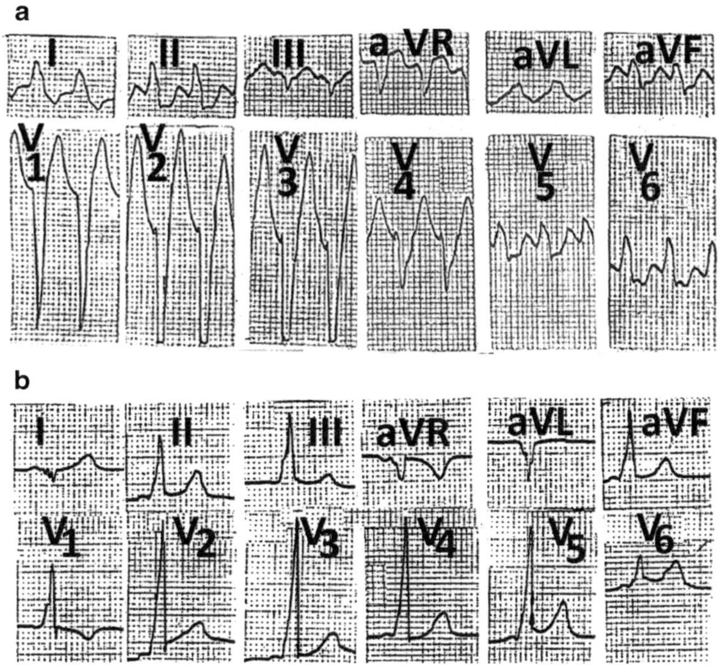

Fig. 9.11 (**a**) Broad QRS tachycardia. (**b**) ECG during sinus rhythm from the same patient showing preexcitation. (Reproduced with permission from: Mittal SR. Tachyarrhythmias in Pre-excitation. Cardiology Today. 2013;17:177–183)

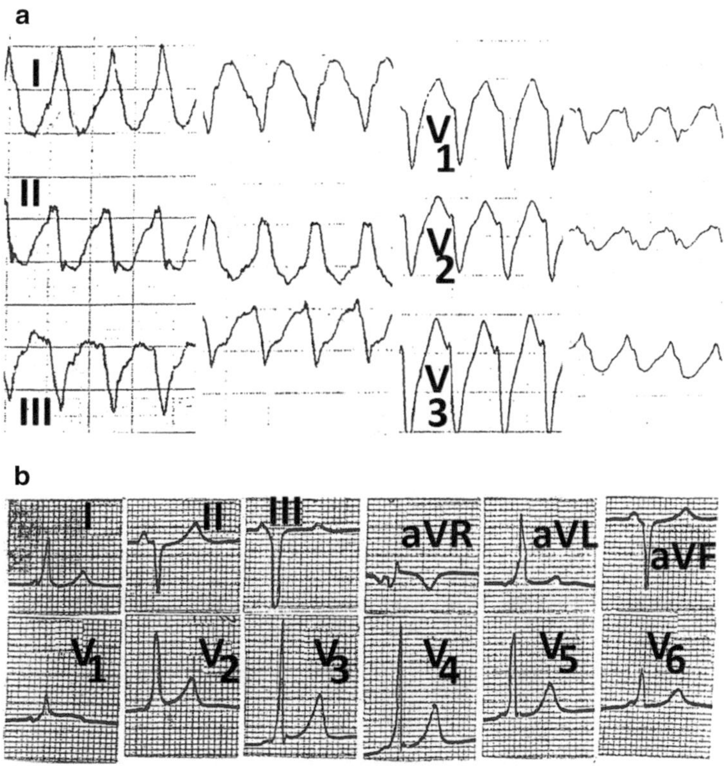

Fig. 9.12 (**a**) Broad QRS tachycardia. (**b**) ECG during sinus rhythm from the same patient showing preexcitation. (Reproduced with permission from: Mittal SR. Tachyarrhythmias in Pre-excitation. Cardiology Today. 2013;17:177–183)

2 Tachycardias where Accessory Pathways Allow Fast Antegrade Conduction

2.1 Atrial Fibrillation [5] (Fig. 9.13)

Atrial fibrillation is relatively more frequent in patients of preexcitation than in general population. It could be initiated by a ventricular ectopic if the retrograde conduction velocity of the accessory path is fast. In this situation, the retrogradely conducted impulse may reach the atria in their vulnerable period and produce atrial fibrillation. Atrial fibrillation could also occur without involvement of the accessory path. Once atrial fibrillation is established, antegrade conduction to the ventricles is usually over the accessory pathway because the refractory period of the accessory path is much shorter than the refractory period of the atrioventricular node. Because of the short refractory period, the accessory path can conduct a large number of atrial impulses to the ventricles. Ventricular rate is, therefore, rapid. It can be in the range of 180–250/min. QRS is irregular and completely bizarre due to irregular conduction over the accessory path and varying degrees of preexcitation (Fig. 9.14). As opposed to atrial fibrillation with aberrancy, widening of the QRS affects the initial part of the QRS unlike widening of the terminal part of the QRS in aberrancy. Further, the magnitude of widening of the QRS is not dependent on long-short cycle sequence.

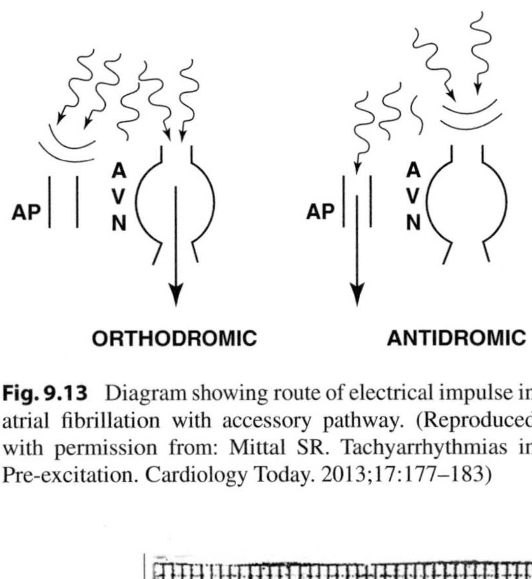

Fig. 9.13 Diagram showing route of electrical impulse in atrial fibrillation with accessory pathway. (Reproduced with permission from: Mittal SR. Tachyarrhythmias in Pre-excitation. Cardiology Today. 2013;17:177–183)

Fast ventricular rate (above 200/min) in a case of atrial fibrillation should raise suspicion of conduction over an accessory path. If the refractory period of the accessory path is very short, large number of atrial impulses may reach the ventricle. In this situation, it is possible that an atrial impulse may reach the ventricle before they are fully repolarized from the effect of the preceding impulse. This situation can cause ventricular fibrillation and sudden death [6]. Intermittent preexcitation or sudden loss of ECG evidence of preexcitation during exercise suggests a relatively long refractory period of the accessory pathway [6]. These patients are not susceptible to have fast ventricular rate during atrial flutter or fibrillation. If refractory period of the accessory path is longer than that

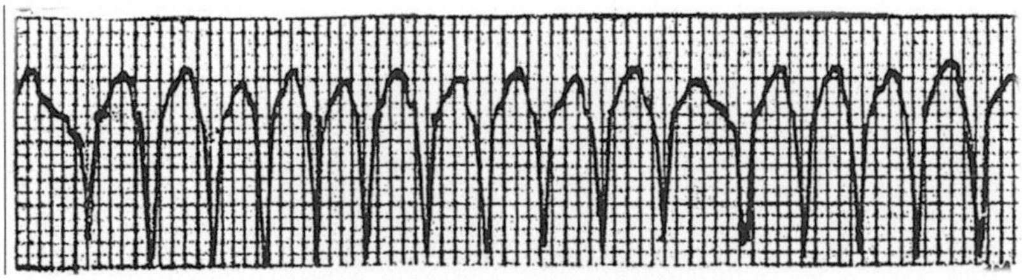

Fig. 9.14 Atrial fibrillation with antegrade conduction over an accessory pathway. It shows irregularly irregular tachycardia with changing degree of widening due to changing degree of preexcitation. (Reproduced with permission from: Mittal SR. Tachyarrhythmias in Pre-excitation. Cardiology Today. 2013;17:177–183)

of the atrioventricular node, antegrade conduction can occur over the atrioventricular node-His bundle path. This occurs infrequently. In this situation ventricular rate is relatively slow (usually less than 180/min). QRS is irregularly irregular but narrow. In patients of preexcitation, atrial fibrillation is usually paroxysmal unless the atria are pathologically enlarged because of other cardiac disease.

2.2 Atrial Flutter

Like atrial fibrillation, atrial flutter can also conduct over an accessory pathway. QRS is rapid, broad, and regular (Fig. 9.15a, b).

Differentiation from ventricular tachycardia is difficult [6]. Conduction over the atrioventricular node produces narrow QRS tachycardia (Fig. 9.16a, b).

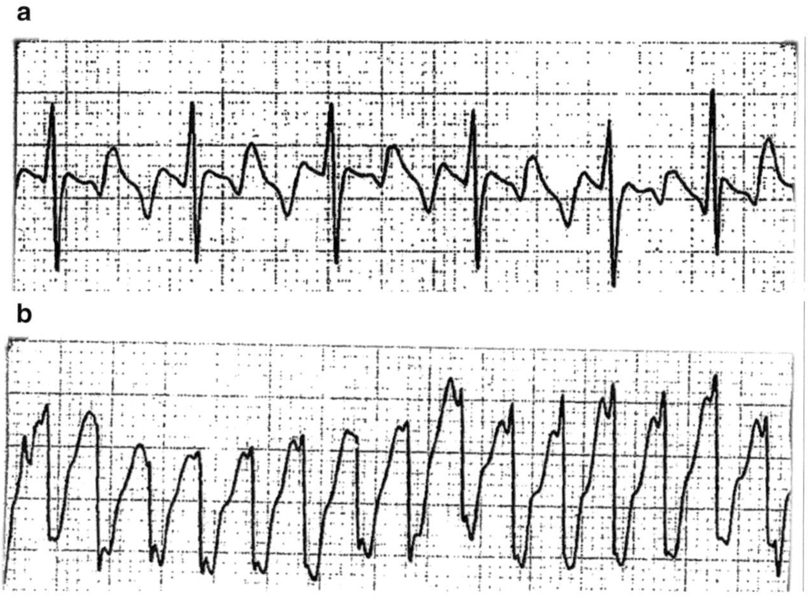

Fig. 9.15 (**a**) Atrial flutter with 2:1 atrioventricular conduction. (**b**) ECG from the same patient showing atrial flutter with 1:1 conduction over an accessory path producing rapid, broad, and regular tachycardia. (Reproduced with permission from: Mittal SR. Tachyarrhythmias in Pre-excitation. Cardiology Today. 2013;17:177–183)

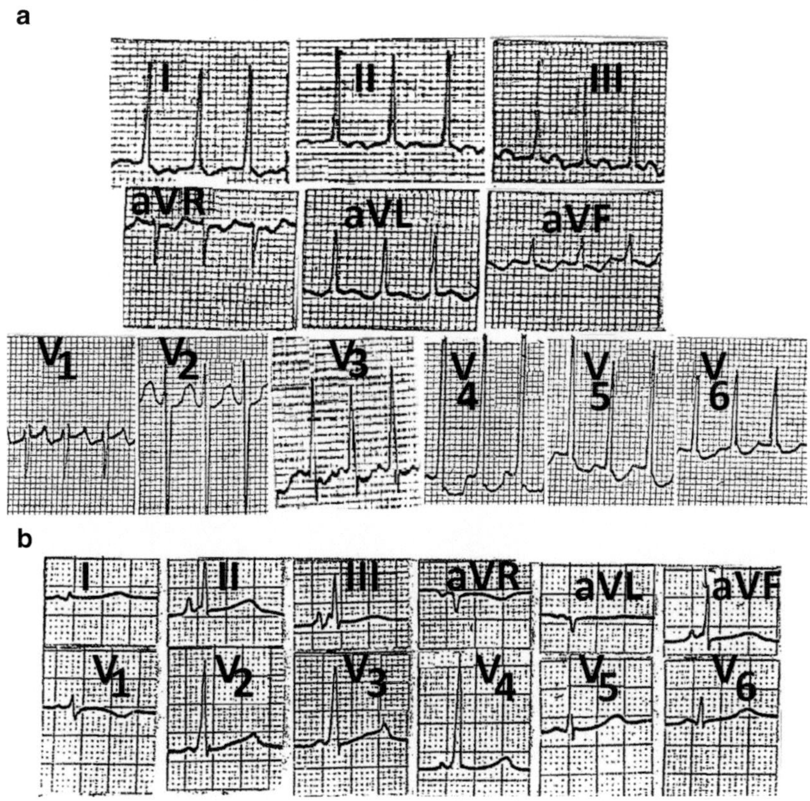

Fig. 9.16 (**a**) Atrial flutter with 2:1 atrioventricular conduction and narrow QRS tachycardia. (**b**) ECG during sinus rhythm from the same patient showing preexcitation. (Reproduced with permission from: Mittal SR. Tachyarrhythmias in Pre-excitation. Cardiology Today. 2013;17:177–183)

3 Tachycardias where the Accessory Path Is Just a Bystander (Fig. 9.17)

Patients with accessory path can have other tachycardias that do not involve the accessory path, e.g., sinus tachycardia, atrial tachycardia, atrioventricular nodal reentry tachycardia, or ventricular tachycardia.

Summary

Patients with preexcitation can have the following:

(A) *Narrow QRS regular tachycardia* (Fig. 9.18a–e)
It can occur in the following situations:

1. Orthodromic atrioventricular reciprocating tachycardia (orthodromic AVRT) with retrograde conduction over the Kent bundle (Fig. 9.18a). Tachycardia rate around 250/min, electrical alternans, and a P′ wave coming clearly after completion of the QRS (in the ST segment) support this possibility.
2. Long RP′ tachycardia (Fig. 9.18b).
3. Atrioventricular nodal reentry tachycardia (AVNRT) (Fig. 9.18c).
4. Atrial flutter with regular 2:1 conduction over the atrioventricular node (Fig. 9.18d).
5. Supraventricular tachycardias in patients with a short PR interval but a normal QRS (no delta wave) in sinus rhythm (Fig. 9.18e).
   (a) Atrioventricular conduction over atrio-Hisian pathway. It is also called Lown-Ganong-Levine (LGL) syndrome.
   (b) Rapid conduction through the atrioventricular node.

(B) *Broad QRS regular tachycardia* (Fig. 9.19)

1. Orthodromic AVRT with preexisting bundle branch block or aberrant ventricular conduction (Fig. 9.19a, b).
2. Antidromic AVRT involving a Kent bundle (Fig. 9.19c).

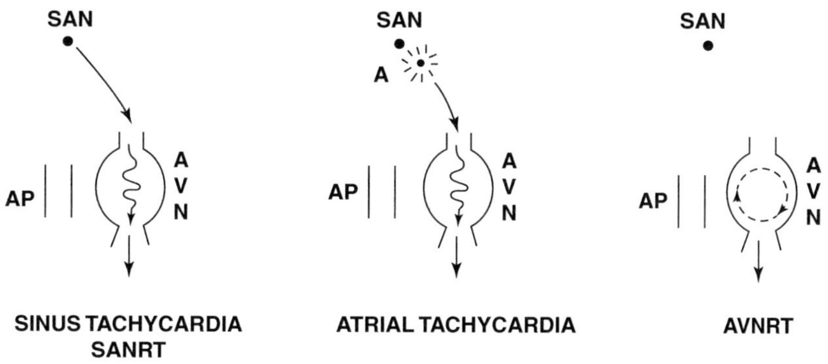

Fig. 9.17 Diagram showing route of electrical impulse in tachycardias when accessory path is only a bystander. (Reproduced with permission from: Mittal SR. Tachyarrhythmias in Pre-excitation. Cardiology Today. 2013;17:177–183)

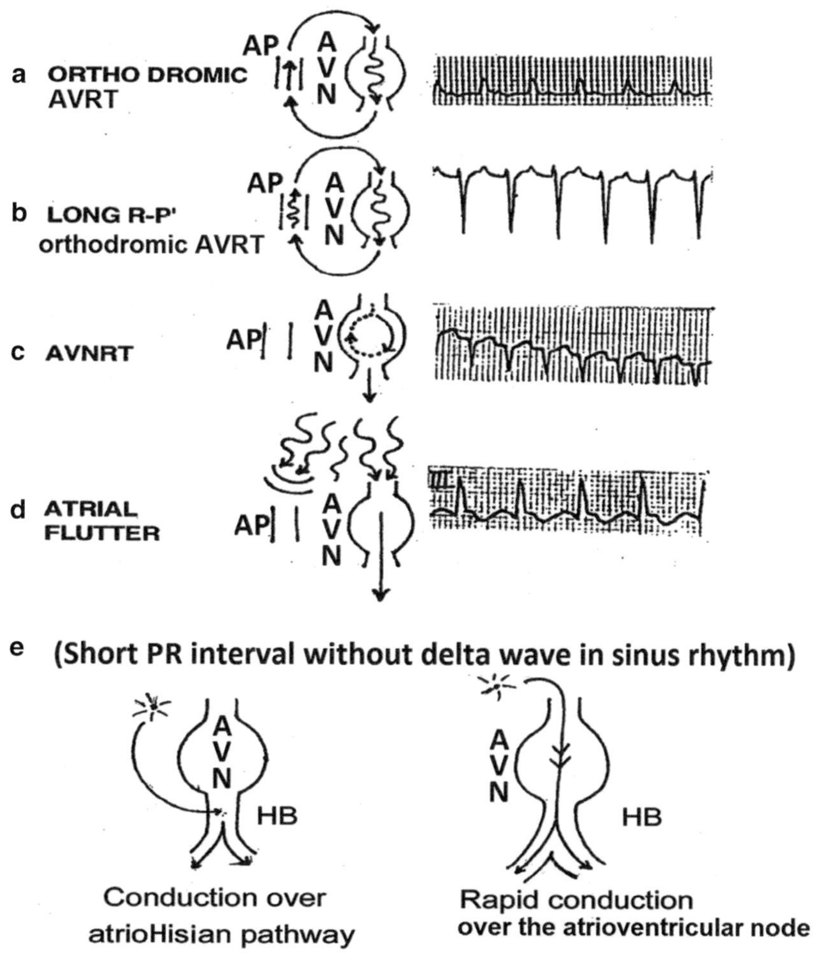

Fig. 9.18 Showing causes of narrow QRS tachycardias. In a case of preexcitation. *AP* accessory pathway, *AVN* atrioventricular node, *HB* His bundle

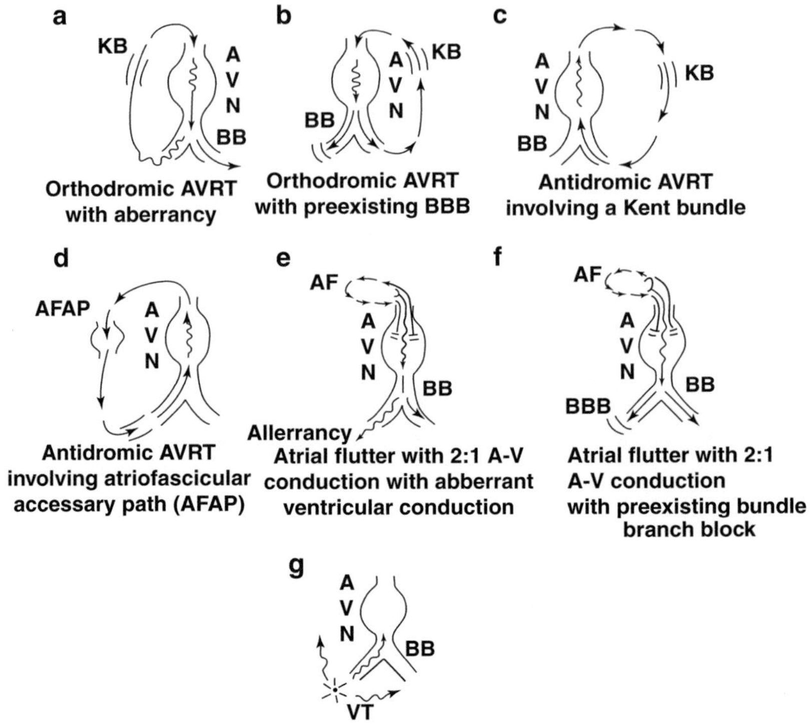

Fig. 9.19 Showing causes of broad QRS regular tachycardia in a case of preexcitation. *KB* Kent bundle, *AVN* atrioventricular node, *BB* bundle branch, *AVRT* atrioventricular reciprocating tachycardia

(a) Left bundle branch block pattern suggests the presence of a right-sided bypass tract.
(b) Right bundle branch block pattern suggests the presence of a left-sided bypass tract.

3. Antidromic AVRT involving atriofascicular bypass tract. It has left bundle branch block pattern (Fig. 9.19d).
4. Atrial flutter with 2:1 atrioventricular conduction with preexisting bundle branch block or aberrant ventricular conduction (Fig. 9.19e, f)
5. Atrial flutter with 2:1 conduction over an accessory pathway.
6. Ventricular tachycardia.

Antidromic AVRT or atrial flutter with conduction over Kent bundle may mimic VT on the surface ECG.

(C) *Irregular tachycardia* (Fig. 9.20)

It can occur in the following situations:

1. Rate more than 200/min
   (a) Atrial fibrillation with antegrade conduction over an accessory path. QRS is broad due to the presence of the delta wave (Fig. 9.20a).
   (b) Atrial fibrillation with rapid conduction over the atrioventricular node. QRS is narrow unless there is aberrant ventricular conduction (Fig. 9.20b).
2. Rate less than 180/min. Atrial fibrillation with antegrade conduction over the atrioventricular node. QRS is usually narrow but can be broad if there is aberrant ventricular conduction or preexisting bundle branch block (Fig. 9.20c).

When ventricular rate is more than 200/min, it may be difficult to differentiate antegrade conduction over the accessory pathway from rapid conduction over the atrioventricular node with aberrant ventricular conduction. In such cases electric cardioversion is safe. If drugs are to be used, the atrioventricular node-blocking drugs (digoxin, diltiazem, verapamil, beta-blockers) should be used only in combination with drugs blocking the accessory pathway (e.g., amioda-

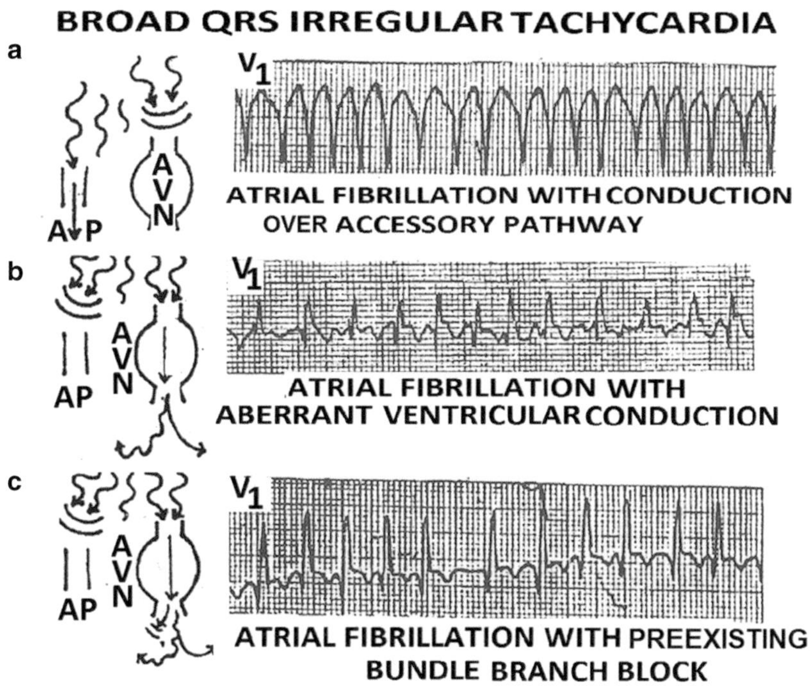

Fig. 9.20 Showing causes of broad QRS irregular tachycardia in the presence of preexcitation

rone). Isolated use of the atrioventricular node-blocking drugs may increase antegrade conduction over the Kent bundle resulting in further increase in ventricular rate. In such a situation, some atrial impulse may reach the ventricle in vulnerable period and precipitate ventricular fibrillation.

MCQs

Q1. In patients with atrioventricular reentry tachycardia, ECG during sinus rhythm must show:

(a) Short PR interval
(b) Delta wave
(c) Broad QRS
(d) None of the above

Q2. A normal electrocardiogram in sinus rhythm excludes any possibility of occurrence of:

(a) AVNRT
(b) Antidromic AVRT
(c) Orthodromic AVRT
(d) None of the above

Q3. Concealed accessory pathway can produce:

(a) Atrioventricular reentry tachycardia
(b) Fast ventricular rate in atrial flutter
(c) Fast ventricular rate in atrial fibrillation
(d) All

Q4. In a case of regular narrow QRS tachycardia, atrioventricular reentry tachycardia can be suspected if there is:

(a) ST segment depression only during tachycardia
(b) P′ wave occurs clearly after the QRS complex
(c) Retrograde P′ wave is negative in lead I
(d) None of the above

Q5. During tachycardia, development of bundle branch block with reduction in the rate of tachycardia suggests:

(a) AVNRT with preexisting bundle branch block
(b) Atrial tachycardia with preexisting bundle branch block
(c) AVRT with accessory pathway on the side of the bundle branch block
(d) AVRT with accessory pathway in the ventricle on the side opposite to that of the bundle branch block

Q6. During tachycardia, which feature suggests retrograde conduction via a septal accessory pathway?

(a) P′ wave occurring before the QRS
(b) P′ waves occurring in the initial or the terminal portion of the QRS
(c) Inverted P′ wave in leads II, III, and aVF following the QRS
(d) Inverted P′ wave in lead I following the QRS

Q7. Regular narrow QRS tachycardia with an inverted P wave falling after the completion of the QRS complex suggests:

(a) Atrial tachycardia
(b) AVNRT
(c) Antidromic atrioventricular reciprocating tachycardia
(d) Orthodromic atrioventricular reciprocating tachycardia

Q8. Regular wide QRS tachycardia with an inverted P wave falling after the completion of the QRS complex suggests:

(a) Atrial tachycardia with preexisting bundle branch block
(b) Ventricular tachycardia
(c) Antidromic atrioventricular reciprocating tachycardia
(d) Orthodromic atrioventricular reciprocating tachycardia

Q9. Irregular wide QRS tachycardia at a rate of 250/min suggests:

(a) Atrial fibrillation with aberrant conduction
(b) Atrial fibrillation with preexisting bundle branch block
(c) Atrial fibrillation with antegrade conduction over an accessory path
(d) Atrial fibrillation with ventricular ectopics

Q10. Accessory pathway can precipitate:

(a) Atrial flutter
(b) Atrial fibrillation
(c) AVNRT
(d) Atrial tachycardia

Q11. A patient with preexcitation during sinus rhythm can have:

(a) Sinus node reentry tachycardia
(b) Atrial tachycardia
(c) AVNRT
(d) None

Answers

(1) d; (2) d; (3) a; (4) a, b, c; (5) c; (6) c; (7) c; (8) d; (9) c; (10) b; (11) a, b, c

References

1. Calkins H. Supraventricular tachycardia: atrioventricular nodal reentry and Wolff-Parkinson-White syndrome. In: Fuster V, Walsh RA, Harrington RA, editors. Hurst's the heart. New York: McGraw Hill; 2011. p. 987–1003.
2. Olgin J, Zipes DP. Specific arrhythmias: diagnosis and treatment. In: Bonow RO, Mann DL, Zipes DP, Liby P, editors. Braunwald's heart disease. Missouri: Elsevier; 2012. p. 771–824.
3. Olshansky B, Chung M, Pogwizd SM, Goldschlager N. Preexcitation syndromes. In: Olshansky B, Chung M, Pogwizd SM, Goldschlager N (eds). Arrhythmia essentials. Jones & Bartlett USA; 2011: 173-184.
4. Ellis CR. Wolff-Parkinson-White Syndrome. Mediscape reference. 2012.
5. Schamroth L. The Wolff-Parkinson-White and related syndromes. In: Schamroth C, editor. An introduction to electrocardiography. New Delhi: Blackwell Science; 1982. p. 288–306.
6. De Luna AB, Goldwasser D, Fiol M, Bayes-Genis A. Surface electrocardiography. In: Fuster V, Walsh RA, Harrington RA, editors. Hurst's the heart. New York: McGraw Hill; 2011. p. 307–70.

10 Atrioventricular Block

Initiation of the P wave represents onset of atrial depolarization, and initiation of the QRS complex denotes onset of ventricular depolarization. PR interval includes time taken for electrical impulse to travel from the sinoatrial node to the ventricles through the atria, the atrioventricular node, the His bundle, bundle branches, and the Purkinje network. Delay or interruption in conduction at any point results in atrioventricular block. On the basis of surface electrocardiogram, atrioventricular blocks are divided into the following categories [1]:

1. First-degree atrioventricular block—There is a delay in conduction of impulse from the atria to the ventricles, but each atrial impulse is conducted to the ventricles. Each P wave is, therefore, followed by a QRS complex.
2. Second-degree atrioventricular block—There is an intermittent block in conduction of an atrial impulse to the ventricles. Some atrial impulses are not conducted to the ventricles with the result that some P waves are not followed by a QRS complex (dropped beats).
3. Third-degree atrioventricular block—There is complete block in atrioventricular conduction. No atrial impulse is conducted to the ventricles. The atria and the ventricles beat independently without any association. Ventricular rate is slower than the atrial rate. There is no relation between the P wave and the QRS complex.

Degree of atrioventricular block does not always correlate with the severity of the disease, prognosis, or the site of block in the conduction system. Degrees of atrioventricular block also do not mean that blocks should progress in this sequence. Any degree of atrioventricular block may be the first presentation (Fig. 10.1a–c).

Depending on the site of the block during electrophysiologic study, atrioventricular blocks are classified as follows:

- Supra-His—Block is above the His bundle, i.e., it is in the atrioventricular node.
- Intra-His—Block is within the His bundle. It is rare [2].
- Infra-His—Block is below the His bundle, i.e., it is at the level of bundle branches and fascicles.

Infra-His blocks are usually associated with organic heart disease and carry relatively poor prognosis. Uncommonly block may be present at more than one site.

S. Mittal, *Insights into Electrocardiograms with MCQs*,
https://doi.org/10.1007/978-981-99-0127-2_10

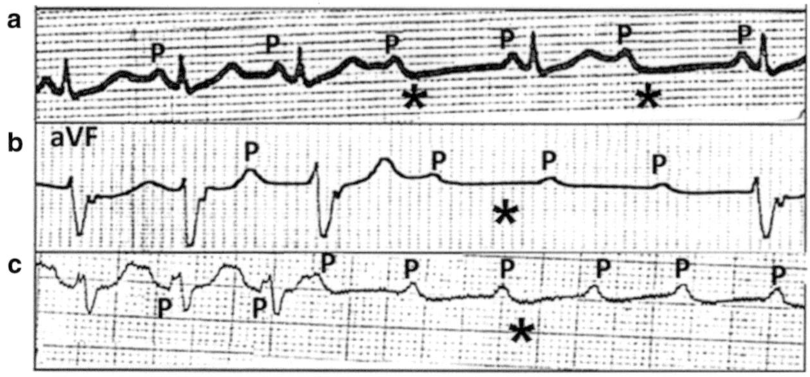

Fig. 10.1 (**a**) Lead II shows sudden appearance of 2:1 atrioventricular block (marked *). (**b**) Lead aVF shows sudden appearance of 3:1 atrioventricular block (marked *). (**c**) Lead II shows sudden onset of high-degree atrioventricular block (marked *). (Reproduced with permission from: Mittal SR. Atrioventricular Block. Cardiology Today. 2013;17:231–235)

1 First-Degree Atrioventricular Block

Delay in conduction results in prolongation of the PR interval (Fig. 10.2).

The PR interval is measured from the beginning of the P wave to the beginning of the QRS complex irrespective of whether the QRS complex begins with a Q wave or a R wave. The normal upper limit is 0.20 s (200 ms). Dual atrioventricular nodal pathways can produce two different PR intervals in the same patient (Fig. 10.3).

If the PR interval is significantly prolonged, P wave may merge with the preceding T wave. The ECG may superficially resemble junctional rhythm. This is frequent with sinus tachycardia (Fig. 10.4).

Deformed or notched T wave gives a clue to merging of the T with the P wave (Fig. 10.5).

Lead V_1 having P wave vector opposite to the T wave vector helps in identifying the P waves merged with the T waves (Fig. 10.6).

Atrial premature beats may conduct with a longer PR interval due to incomplete recovery of the atrioventricular node (partial refractoriness of the atrioventricular node). This is not called atrioventricular block.

First-degree atrioventricular block does not produce symptoms. The patient may have symptoms due to underlying heart disease. Proper timing of the atrial contraction (P wave) before ventricular contraction (QRS) contributes to proper filling of the ventricles just before onset of ventricular systole. This proper synchronization of atrial and ventricular contraction contributes to normal function of the ventricles. Significant prolongation of the PR interval produces loss of the normal atrioventricular synchrony. Atrial contraction is completed much before the onset of the ventricular contraction. This may worsen symptoms if there is underlying heart disease.

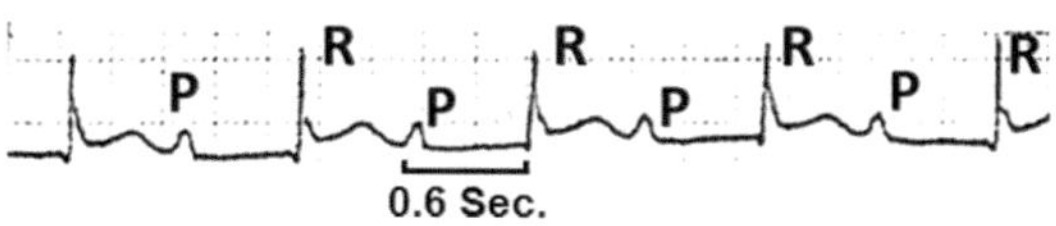

Fig. 10.2 Lead II shows first-degree atrioventricular block. (Reproduced with permission from: Mittal SR. Atrioventricular Block. Cardiology Today. 2013;17:231–235)

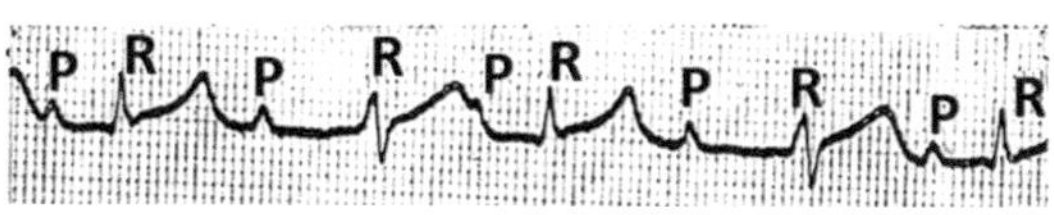

Fig. 10.3 Lead II is showing two different PR intervals. (Reproduced with permission from: Mittal SR. Atrioventricular Block. Cardiology Today. 2013;17:231–235)

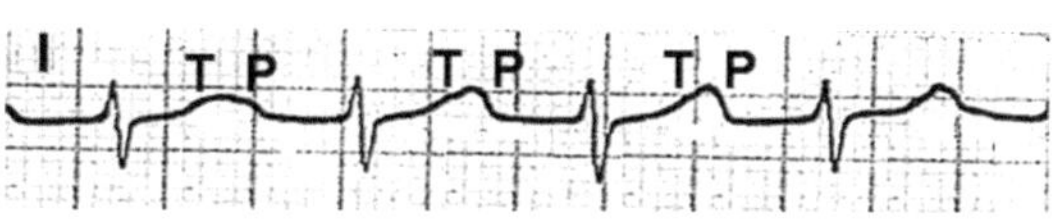

Fig. 10.4 Lead I showing sinus tachycardia with the P wave merged with the T wave. (Reproduced with permission from: Mittal SR. Atrioventricular Block. Cardiology Today. 2013;17:231–235)

Fig. 10.5 Lead II is showing notched T wave due to fusion of the P and the T waves. (Reproduced with permission from: Mittal SR. Atrioventricular Block. Cardiology Today. 2013;17:231–235)

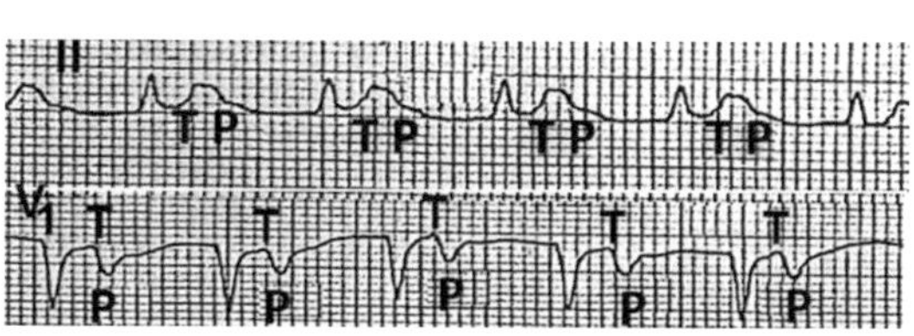

Fig. 10.6 Lead II shows merging of P wave with the preceding T wave. Lead V_1 identifying merging of P and T waves due to opposite vectors. (Reproduced with permission from: Mittal SR. Atrioventricular Block. Cardiology Today. 2013;17:231–235)

2 Second-Degree Atrioventricular Block

Second-degree atrioventricular block may be further categorized into:

1. 2:1 atrioventricular block
2. Mobitz type I atrioventricular block
3. Mobitz type II atrioventricular block
4. High-degree atrioventricular block

2.1 2:1 Atrioventricular Block

When every second P wave is blocked, it produces regular 2:1 atrioventricular block (Fig. 10.7).

In some cases one P wave may be merged with the preceding T wave, and only one P wave may be visible. In such a situation, the ECG may be misinterpreted as sinus bradycardia (Fig. 10.8). Careful observation of the T wave reveals that the T wave is notched and deformed due to merging of the P wave. Review of other leads (usually lead aVR, V_1, and V_3R) may help in the detection of the P wave merging with the T wave (Fig. 10.8).

Merging of the P wave and the T waves can also produce tall T waves (Fig. 10.9).

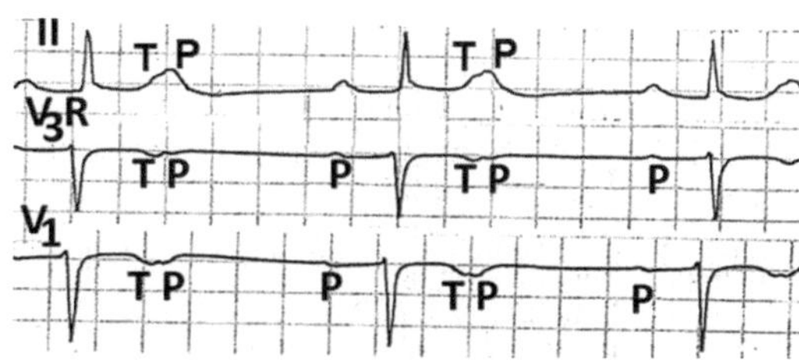

Fig. 10.7 Lead II showing 2:1 atrioventricular block. (Reproduced with permission from: Mittal SR. Atrioventricular Block. Cardiology Today. 2013;17:231–235)

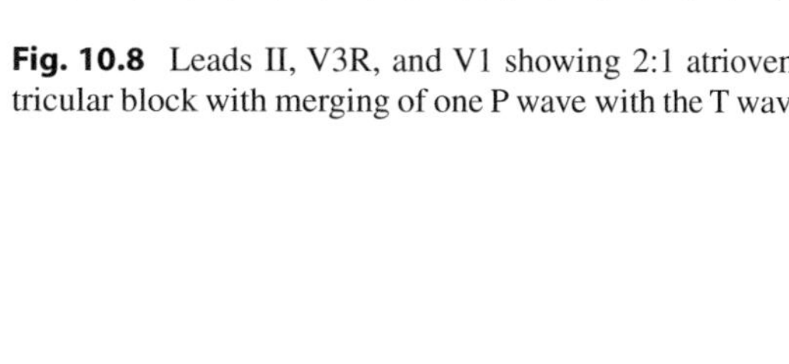

Fig. 10.8 Leads II, V3R, and V1 showing 2:1 atrioventricular block with merging of one P wave with the T wave

Diagnosis may become clear with slight slowing of the heart rate (Fig. 10.10).

Atrial premature beats occurring during or soon after the T wave may also be blocked in the atrioventricular node. This is due to physiological refractoriness of the atrioventricular node. This is not called atrioventricular block. Regularly occurring atrial premature beats may be mistaken for 2:1 atrioventricular block. Configuration of the P' wave produced by an atrial ectopic is different from that of the sinus beat. Further, due to premature occurrence of the atrial ectopic beat, interval between sinus P wave and the P' of the atrial ectopic is less than the interval between P' of the atrial ectopic and the next sinus P wave. A normal atrioventricular node can conduct impulses only up to a rate of 140–170/min. This upper limit declines with advancing age. In cases of atrial tachycardia and atrial flutter, atrial rate is faster (180–300/min) than the number of impulses that can be conducted by a normal atrioventricular node. Therefore, every second or third atrial impulse is conducted to the ventricles even when the atrioventricular node is normal. These situations do not suggest any pathology in atrioventricular conduction. Therefore, these situations are termed 2:1 or 3:1 atrioventricular conduction and not atrioventricular block. The term atrioventricular block should be used only when the atrial rate is 140/min or less [1].

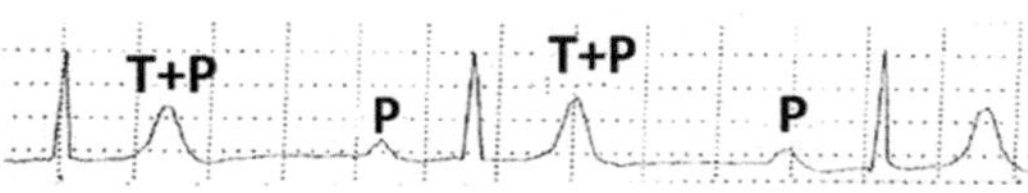

Fig. 10.9 Lead II is showing tall T waves due to merging of one P wave with the T wave. (Reproduced with permission from: Mittal SR. Atrioventricular Block. Cardiology Today. 2013;17:231–235)

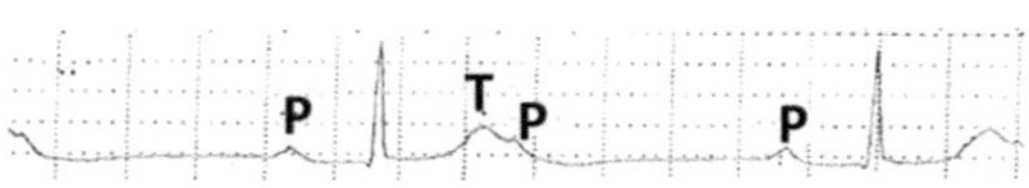

Fig. 10.10 Lead II showing unmasking of fusion of T and P waves with slight slowing of the heart rate. (Reproduced with permission from: Mittal SR. Atrioventricular Block. Cardiology Today. 2013;17:231–235)

2.2 Mobitz Type I Second-Degree Atrioventricular Block

Refractory period of the atrioventricular node is increased. It takes longer time to recover to normal after depolarization by an atrial impulse conducted through it. With each impulse passing through it, the recovery period is progressively prolonged. This results in increasing delay in the atrioventricular conduction till one atrial impulse is not conducted to the ventricles. This is called Wenckebach phenomenon. In the ECG it presents as cycles (grouping of QRS complexes) in which PR interval of the first beat is normal or slightly prolonged. It is followed by a progressive increase in PR interval till one P wave is not conducted to the ventricles, and therefore, it is not followed by a QRS (the dropped beat) (Fig. 10.11).

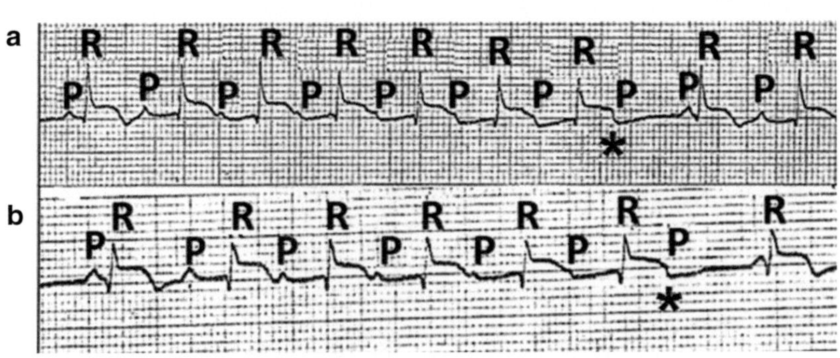

Fig. 10.11 (**a**) Lead II shows Wenckebach phenomenon (* denotes P wave not conducted to the ventricles). (**b**) Lead II shows a small increment in the PR interval after second beat (* denotes P wave not conducted to the ventricles). (Reproduced with permission from: Mittal SR. Atrioventricular Block. Cardiology Today. 2013;17:231–235)

Pause produced by the dropped beat allows recovery of the atrioventricular node. Therefore, after the pause, the PR interval returns back to the initial value, and the cycle is repeated. The pause in the RR interval due to the dropped beat is less than double of the preceding RR interval. Although there is a progressive increase in the PR interval, the increase over the preceding PR interval progressively decreases. The increase in the PR interval is maximum from the first to the second beat of the cycle. In the third beat, the total PR interval is longer than the second beat, but the magnitude of increase in the PR interval is less than the increase from the first to the second beat. It results in a progressive decrease in the RR interval.

The ECG of Wenckebach phenomenon is, therefore, characterized by cycles with the following:

- Progressive increase in the PR interval followed by a dropped beat.
- Progressive decrease in RR interval.
- RR interval, including the blocked atrial impulse (the dropped beat), is shorter than twice the preceding RR interval.
- The PR interval following the dropped beat is smaller than the PR interval preceding the dropped beat.

There can be some variations in the classical sequence [2]:

- The first increment in the PR interval may not be maximum.
- After the first increase in the PR interval, there may not be further increase in PR interval in subsequent beats.
- After the first increase in the PR interval, further increment in the PR interval may be quite small and may be difficult to interpret at the usual speed of ECG recording (25 mm/s) (Fig. 10.11).
- Maximum lengthening of the PR interval may occur in the beat preceding the non-conducted beat (the dropped beat).
- Blocked P wave may be merged with the QRS complex (Fig. 10.12a, b).
- Escape rhythm or ectopics may hamper classical sequence (Fig. 10.13a, b).

  Sometimes blocked P wave may be followed by a 2:1 atrioventricular block (Fig. 10.13c).

  Sometimes, two P waves may be blocked [3] (Fig. 10.13d).
- Classical grouping of QRS may not be seen in all cases [3].

Site of block is usually in the atrioventricular node. Therefore, the QRS is usually normal. However, QRS can be broad if there is accompanying bundle branch block (Fig. 10.14a).

Tachycardia-dependent aberrancy can broaden the second beat of the cycle (Fig. 10.14b—marked ☆). Bradycardia-dependent aberrancy can broaden the QRS following the dropped beat (Fig. 10.14c—marked ☆).

Any variation in the PR interval suggests atrioventricular node as the site of the block [2]. Long

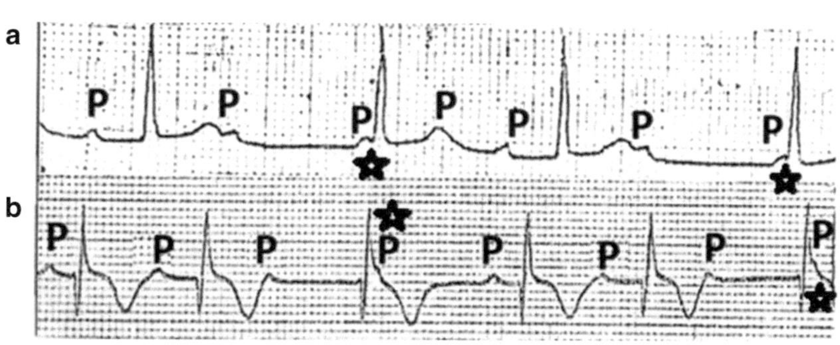

Fig. 10.12 (**a**) Lead II shows partial merging of the blocked P wave with the QRS (marked as ☆). (**b**) Lead aVF shows total merging of the blocked P wave with the QRS (marked as ☆). (Reproduced with permission from: Mittal SR. Atrioventricular Block. Cardiology Today. 2013;17:231–235)

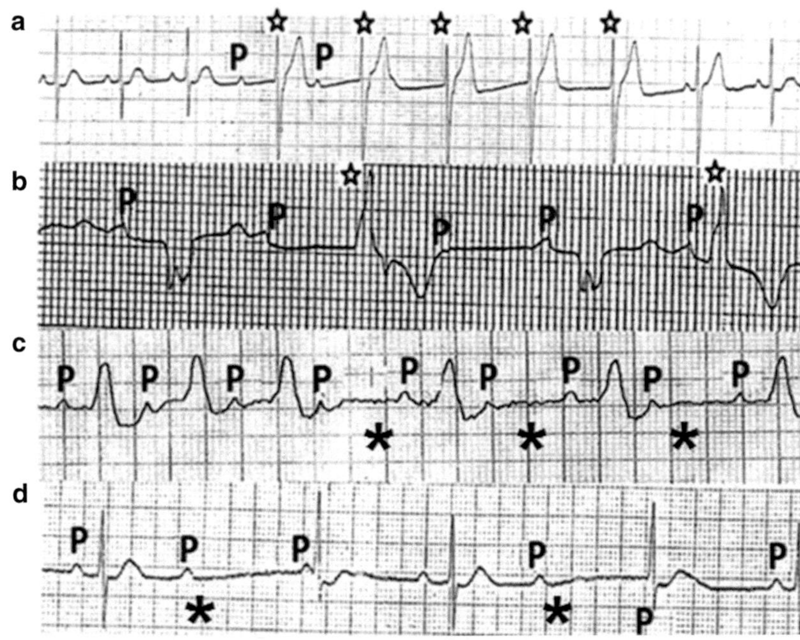

Fig. 10.13 (**a**) Wenckebach phenomenon with intermittent escape rhythm (marked ☆). (**b**) Wenckebach phenomenon with intermittent ventricular ectopics (marked ☆). (**c**) Wenckebach phenomenon with intermittent 2:1 AV block (marked *). (**d**) Wenckebach phenomenon with blocking of more than one P wave (marked *). (Reproduced with permission from: Mittal SR. Atrioventricular Block. Cardiology Today. 2013;17:231–235)

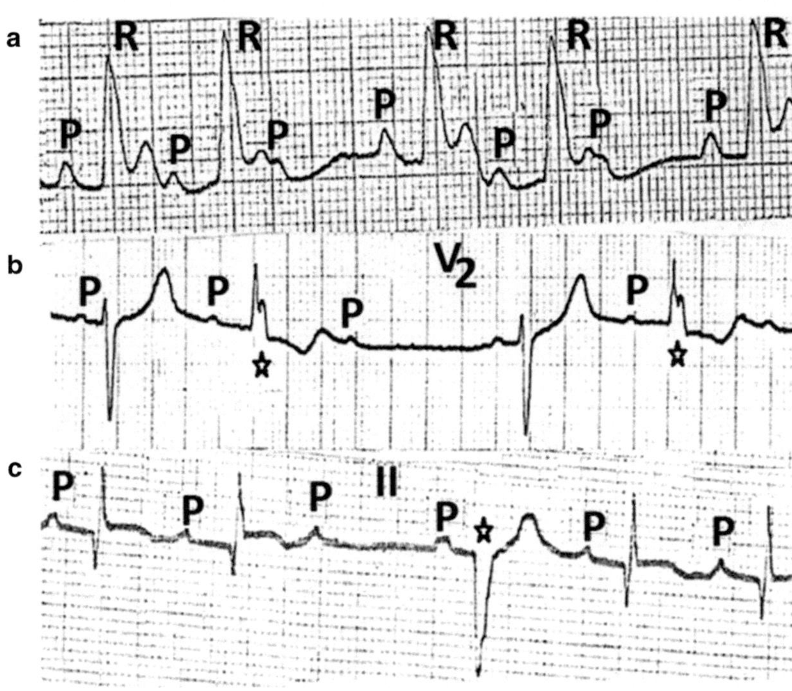

Fig. 10.14 (**a**) Lead II shows Wenckebach phenomenon with wide QRS. (**b**) Lead V_2 showing Wenckebach phenomenon with broad QRS in the second beat of sequence (marked ☆) (aberrancy due to short RR cycle—tachycardia-dependent aberrancy). (**c**) Lead II shows Wenckebach phenomenon with broad QRS following the dropped beat (marked ☆) (aberrancy due to long RR cycle—bradycardia-dependent aberrancy). (Reproduced with permission from: Mittal SR. Atrioventricular Block. Cardiology Today. 2013;17:231–235)

PR interval is usually due to block in the atrioventricular node [4]. Rarely Wenckebach phenomenon can occur in the His-Purkinje system [3]. Mobitz type I atrioventricular block can be present in persons without any other evidence of cardiac disease. Mobitz type I atrioventricular block usually does not progress to complete atrioventricular block and does not carry adverse prognosis as compared to Mobitz type II atrioventricular block.

2.3 Mobitz Type II Second-Degree Atrioventricular Block

Delay in the atrioventricular conduction is constant. There is no progressive delay in the atrioventricular conduction before the "dropped beat" or decrease in the PR interval after the "dropped" beat. ECG is characterized by normal or prolonged but fixed PR interval with intermittent failure of conduction of an atrial impulse to the ventricles (Fig. 10.15).

The lesion is usually at the level of bundle branches. This results in impaired conduction of impulse in the ventricles. The QRS is, therefore, broad (Fig. 10.16).

Rarely Mobitz type II second-degree atrioventricular block may be within the His bundle (intra-His). In this situation ventricular depolarization occurs normally over the bundle branch es, and the QRS is narrow (Fig. 10.17).

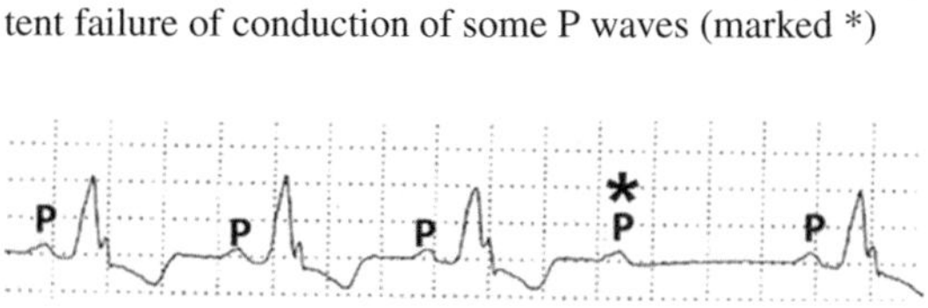

Fig. 10.15 Lead II showing Mobitz type II second-degree atrioventricular block. PR interval is fixed with intermittent failure of conduction of some P waves (marked *)

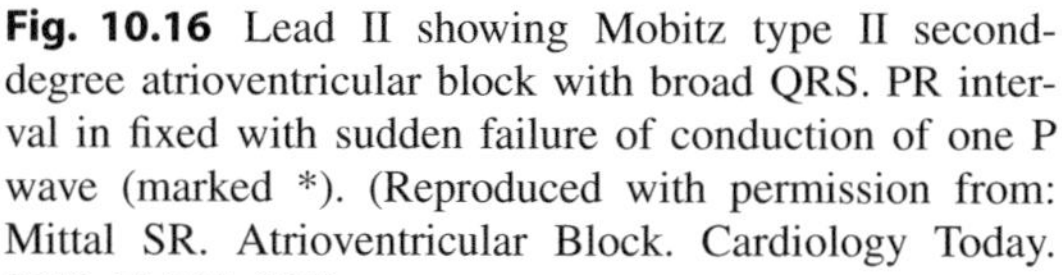

Fig. 10.16 Lead II showing Mobitz type II second-degree atrioventricular block with broad QRS. PR interval in fixed with sudden failure of conduction of one P wave (marked *). (Reproduced with permission from: Mittal SR. Atrioventricular Block. Cardiology Today. 2013;17:231–235)

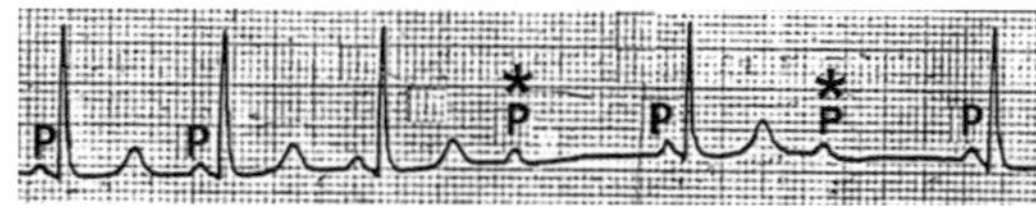

Fig. 10.17 Lead II is showing Mobitz type II second-degree atrioventricular block with narrow QRS (P waves not conducted to the ventricles). (Reproduced with permission from: Mittal SR. Atrioventricular Block. Cardiology Today. 2013;17:231–235)

QRS following the dropped beat may be narrow, unmasking the rate-dependent nature of the broad QRS complex (Fig. 10.18a, b). Mobitz type II second-degree atrioventricular block is mostly due to organic heart disease. It can progress to complete atrioventricular block (Fig. 10.18c, d).

Mobitz type II second-degree atrioventricular block carries an adverse prognosis as opposed to Mobitz type I second-degree atrioventricular block. Permanent pacemaker implantation is needed because progression to complete atrioventricular block may be sudden and unpredictable [5].

Sinoatrial exit block or sinus pause also produces pause in the QRS sequence. However, in these conditions, there is no blocked P wave in the pause among the QRS complexes.

Caveats in the diagnosis of second-degree atrioventricular block [3]:

- Two different degrees of blocks may occur at two different levels at the same time (Fig. 10.19).
- Ectopics arising in the His bundle may not conduct to the ventricles and thus may not manifest on the surface electrocardiogram (concealed His bundle ectopics). Such ectopics depolarize the His bundle. Next atrial impulse is, therefore, blocked in the His bundle. Surface electrocardiogram simulates second-degree atrioventricular block.

2.4 High-Degree Atrioventricular Block

- Atrioventricular block greater than 2:1 (i.e. 3:1 or 4:1) is called a high-degree atrioventricular block. It means that three or more consecutive atrial impulses are not conducted to the ventricles (Fig. 10.20).

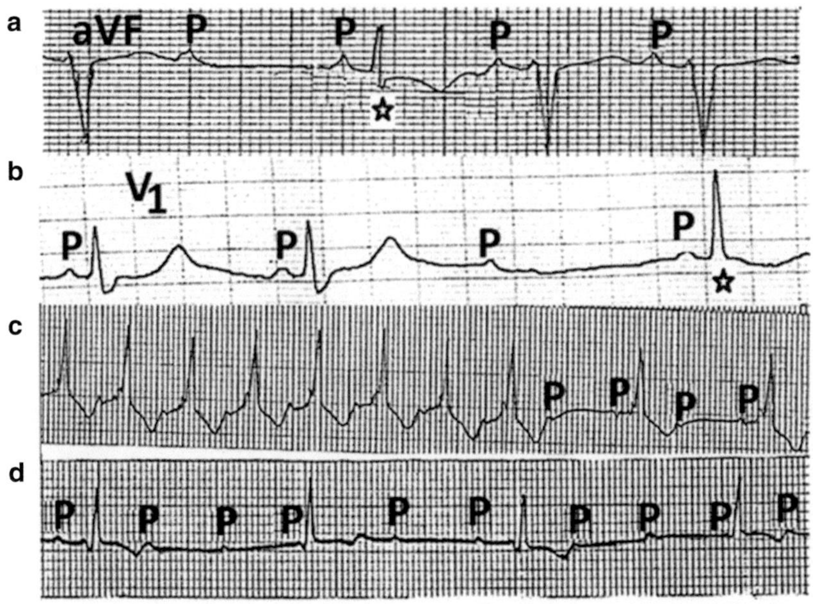

Fig. 10.18 (**a**) Lead aVF shows Mobitz type II second-degree atrioventricular block with normalization of QRS following the dropped beat (marked ☆). (**b**) Lead I showing Mobitz type II second-degree atrioventricular block with normalization of QRS following the dropped beat (marked ☆). (**c**, **d**) Showing advancement of Mobitz type II second-degree atrioventricular block to complete atrioventricular block. (Reproduced with permission from: Mittal SR. Atrioventricular Block. Cardiology Today. 2013;17:231–235)

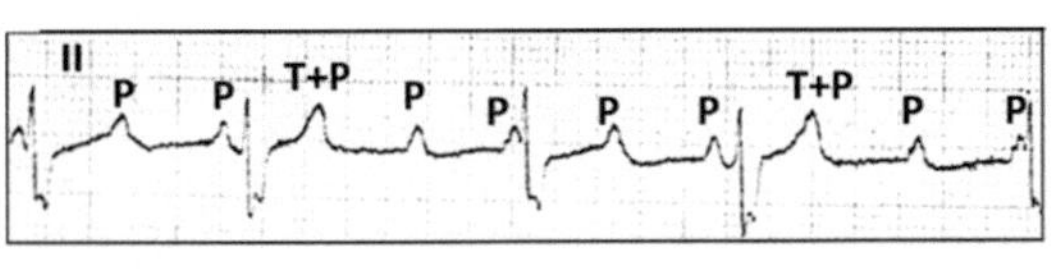

Fig. 10.19 Lead II shows alternate 2:1 and 3:1 atrioventricular block due to block at two levels. (Reproduced with permission from: Mittal SR. Atrioventricular Block. Cardiology Today. 2013;17:231–235)

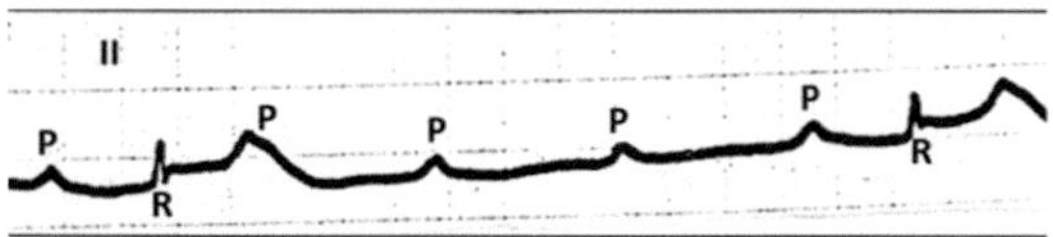

Fig. 10.20 Lead II shows a high-degree atrioventricular block (4:1). (Reproduced with permission from: Mittal SR. Atrioventricular Block. Cardiology Today. 2013;17:231–235)

In cases of atrial tachyarrhythmias, a pause of more than 3 s can be considered as a high-degree atrioventricular block. The QRS complex is narrow or broad depending on the site of the atrioventricular block. Transient acute high-degree atrioventricular block can occur due to acute increase in vagal tone. Cases of chronic high-degree atrioventricular block are at risk of Stokes-Adams' attack and have poor prognosis. Permanent pacemaker implantation is needed if:

- Ventricular rate is less than 40/min.
- QRS are broad, suggesting block at the level of bundle branches.
- The patient has symptoms due to bradyarrhythmia, e.g., presyncope or syncope.

Summary

- When each sinus P wave is followed by a QRS with a PR interval of more than 200 ms, it is labeled as first-degree atrioventricular block. Site of delay in conduction is almost always in the atrioventricular node even when associated with a preexisting bifascicular block.
- When some sinus P waves are not followed by a QRS, it is called a second-degree atrioventricular block.

- Wenckebach-type (Mobitz type I) second-degree atrioventricular block is diagnosed when there is:
  - Progressive prolongation of the PR interval followed by a P wave which is not conducted to the ventricles
  - PR interval following the non-conducted P wave shorter than the PR interval preceding the blocked P wave.
  - Progressive shortening of the RR interval before the blocked P wave
  - Pause having the blocked P wave less than twice of the preceding RR interval
- It can be seen in a condition of increased vagal tone as in well-trained athletes and during inferior myocardial infarction. Conduction delay is almost always in the atrioventricular node. Risk of advancement to complete atrioventricular block is low.
- In Mobitz type II second-degree atrioventricular block:
  - The PR intervals and the RR intervals preceding a non-conducted P wave are constant.
  - Pause having the non-conducted P wave is equal to twice the preceding RR interval.
  - QRS is usually broad.
  - Conduction delay is almost always in the His-Purkinje system. Anterior myocardial infarction usually produces this type of second-degree atrioventricular block. Risk of advancement to complete atrioventricular block is high.
- Failure of conduction of alternate sinus P wave to conduct to the ventricles is called 2:1 atrioventricular block. It can be either at the level of the atrioventricular node or at the level of the His-Purkinje system. Normal QRS duration and very long PR interval suggest block at the level of the atrioventricular node. Exact site of the block may be difficult to decide by surface electrocardiogram in the presence of a wide QRS.
- Failure of two or more consecutive sinus P waves to conduct to the ventricles is defined as a high-degree atrioventricular block.

MCQs

Q1. Normal upper limit of the PR interval is:

(a) 0.12 s
(b) 0.16 s
(c) 0.20 s
(d) 0.24 s

Q2. The PR interval is measured from:

(a) Beginning of the P wave to the beginning of the q wave
(b) End of the P wave to the beginning of the q wave
(c) Beginning of the P wave to the top of the R wave
(d) End of the P wave to the top of the R wave

Q3. If PR interval is significantly prolonged, P wave may fall on:

(a) Preceding QRS
(b) Preceding ST segment
(c) Upstroke of the preceding T wave
(d) Downslope of preceding T wave

Q4. P wave falling on the preceding T wave and not conducted to the ventricles suggests:

(a) Atrial premature beat with normal refractoriness of the atrioventricular node
(b) First-degree atrioventricular block
(c) Second-degree atrioventricular block
(d) Complete atrioventricular block

Q5. PR interval includes:

(a) Intra-atrial conduction
(b) Conduction through the atrioventricular node
(c) Conduction through the His-Purkinje system
(d) All

Q6. Type I second-degree atrioventricular block can occur in the absence of organic disease of the atrioventricular node:

(a) In athletes
(b) During sleep
(c) During carotid massage
(d) All

Q7. In first-degree atrioventricular block, prognosis is poor if:

(a) PR interval is 0.6 s or more
(b) There is associated bundle branch block
(c) There is intermittent type I second-degree atrioventricular block
(d) There is intermittent type II second-degree atrioventricular block

Q8. Regular P waves at a rate of 300/min and regular QRS at a rate of 150/min with constant PR interval suggest:

(a) Normal refractoriness of the atrioventricular node
(b) First-degree atrioventricular block
(c) Second-degree atrioventricular block
(d) Complete atrioventricular block

Q9. Electrocardiographic findings of Mobitz type I second-degree atrioventricular block include:

(a) Progressive prolongation of the PR intervals followed by a P wave which is not conducted to the ventricles
(b) Progressive shortening of the RR interval
(c) Shortening of PR interval after the pause following the dropped beat
(d) All

Q10. Regular P waves with a rate of 60/min, regular broad QRS at a rate of 80/min, and no relation between P and QRS suggest:

(a) Complete atrioventricular block
(b) Second-degree atrioventricular block
(c) Ventriculoatrial block
(d) Accelerated ventricular rhythm

Q11. In Wenckebach sequence, the pause containing the dropped beat is:

(a) Less than twice the preceding RR cycle length
(b) Same as the preceding RR cycle length
(c) Twice the preceding RR cycle length
(d) More than twice of the preceding RR cycle length

Q12. Electrocardiographic features of Mobitz type II second-degree atrioventricular block include:

(a) Regular RR interval
(b) Regular PR interval
(c) Pause including the non-conducted P wave is equal to double of the preceding RR cycle
(d) All

Q13. If a P wave falls on the preceding T wave and conducts with a long PR interval, it suggests:

(a) Sinoatrial exit block
(b) Intra-atrial block
(c) Partial refractoriness of atrioventricular node
(d) First-degree atrioventricular block

Q14. Prolonged PR interval with normal QRS complex suggests delay in conduction in the:

(a) Atrioventricular node
(b) His bundle
(c) Bundle branches
(d) Any of the above

Q15. Sudden block of conduction of a sinus P wave to the ventricles without preceding lengthening of the PR intervals suggests:

(a) First-degree atrioventricular block
(b) Type I second-degree atrioventricular block
(c) Type II second-degree atrioventricular block
(d) Third-degree atrioventricular block

Q16. Which atrioventricular block is more likely to advance to complete atrioventricular block?

(a) Wenckebach block
(b) Wenckebach block with preexisting bundle branch block
(c) Mobitz type II block
(d) All

Answers

(1) c; (2) a; (3) d; (4) a; (5) d; (6) d; (7) d; (8) a; (9) d; (10) d; (11) d; (12) d; (13) c; (14) d; (15) c; (16) c

References

1. Schamroth L. Atrioventricular (atrioventricular) block. In: Schamroth C, editor. An introduction to electrocardiography. New Delhi: Blackwell Science; 1982. p. 375–85.
2. Wagner GS. Atrioventricular block. In: Wagner GS, editor. Marriott's practical electrocardiography. New Delhi: Wolters Kluwer; 2008. p. 401–17.
3. Olgin J, Zipes DP. Specific arrhythmias: diagnosis and treatment. In: Bonow RO, Mann DL, Zipes DP, Libby P, editors. Braunwald's heart disease. Missouri: Elsevier; 2012. p. 771–824.
4. Vijayaraman P, Ellenbogen KA. Bradyarrhythmias and pacemakers. In: Fuster V, Walsh RA, Harrington RA, editors. Hurst's the heart. New York: McGraw Hill; 2011. p. 1025–57.
5. Olshansky B, Chung MK, Pogwizd SM, Goldschlager N. Atrioventricular conduction abnormalities. In: Olshansky B, Chung MK, Pogwizd SM, Goldschlager N, editors. Arrhythmia essentials. Burlington: Jones & Bartlett Learning; 2012. p. 27–55.

11 Third-Degree Atrioventricular Block (Complete Atrioventricular Block), Atrioventricular Dissociation

There is a complete interruption of the atrioventricular conduction. The atria are activated from sinoatrial node or an ectopic focus in the atria. Atrial impulses are not conducted to the ventricles. The ventricles are activated independently from an ectopic focus below the site of the block. The ectopic focus may be in the atrioventricular node, the His bundle, bundle branches, or the Purkinje system. There is no correlation between the atrial and the ventricular depolarizations. The electrocardiogram is characterized by the following:

- Normal P waves at normal rate (60–100/min). P waves may be at a slower rate if there is a concomitant sinoatrial nodal disease. P wave configuration may be abnormal if the atria are activated from an ectopic focus rather than from the sinoatrial node. PP interval including the QRS may be shorter than other PP intervals. It is called ventriculophasic sinus arrhythmia (Fig. 11.1a). PP intervals can also be irregular if there is concomitant sinus node dysfunction. Atrial ectopics can also produce irregular PP cycle (Fig. 11.1b).

Complete atrioventricular block may occur in the setting of atrial tachycardia, atrial flutter (Fig. 11.2a), atrial fibrillation (Fig. 11.2b), or as a part of tachycardia-bradycardia syndrome (Fig. 11.2c).

- The QRS complexes occur at a rate slower than the atrial rate and are regular. QRS complexes may be irregular due to ventricular ectopics (Fig. 11.3a), irregular rate of the

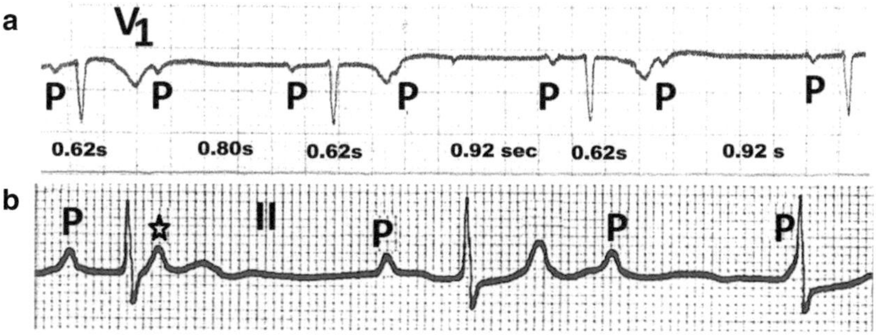

Fig. 11.1 (**a**) Lead V_1 showing complete atrioventricular block with ventriculophasic sinus arrhythmia. (**b**) Lead II showing complete atrioventricular block with intermittent atrial ectopics (marked ☆). (Reproduced with permission from: Mittal SR. Third degree AV Block, AV Dissociation. Cardiology Today. 2013;17:274–277)

S. Mittal, *Insights into Electrocardiograms with MCQs*,
https://doi.org/10.1007/978-981-99-0127-2_11

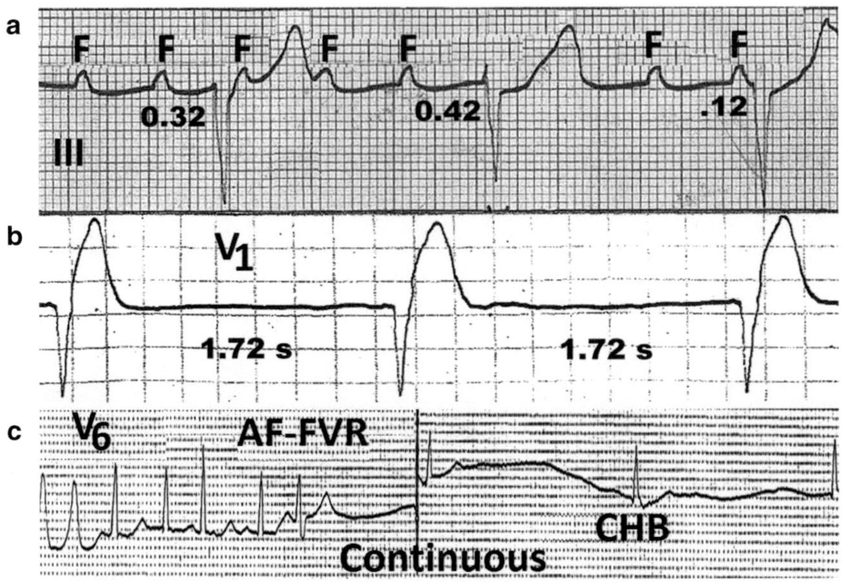

Fig. 11.2 (**a**) Lead III showing atrial flutter (F) with complete atrioventricular block. RR interval is regular with changing FR interval. (**b**) Lead V_1 showing atrial fibrillation (absence of P waves) with complete atrioventricular block. QRS is at a regular slow rate (35/min). (**c**) Lead V_6 (continuous strip) showing atrial fibrillation with a fast ventricular rate (AF-FVR) changing to complete atrioventricular block (CHB) with slow junctional rhythm. (Reproduced with permission from: Mittal SR. Third degree AV Block, AV Dissociation. Cardiology Today. 2013;17:274–277)

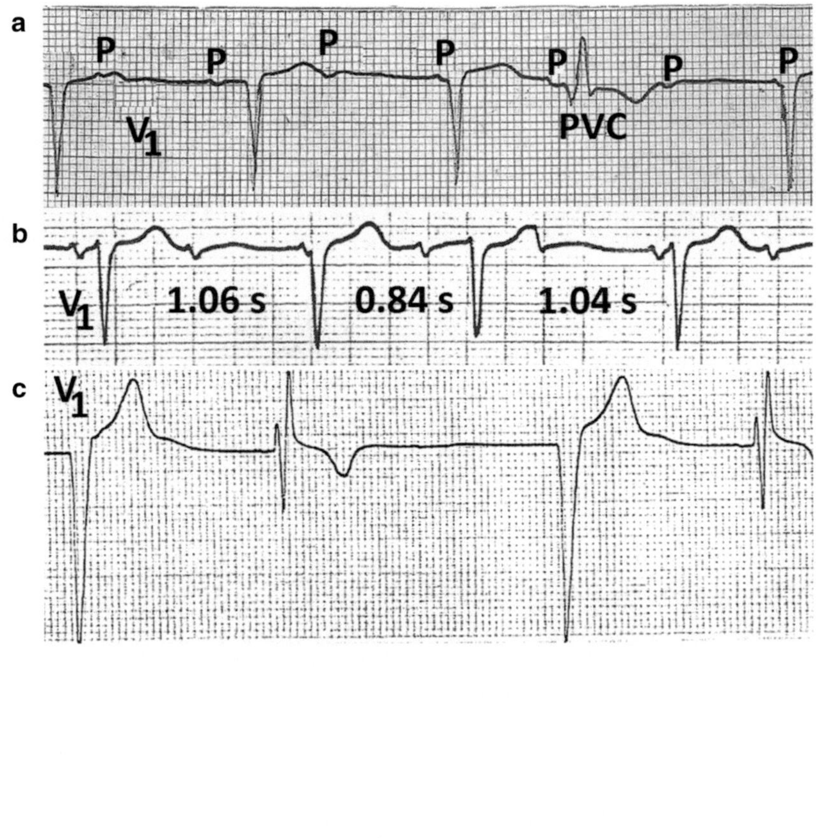

Fig. 11.3 (**a**) Lead V_1 showing complete atrioventricular block with ventricular ectopics (PVC). (**b**) Lead V_1 showing complete atrioventricular block with irregularity of the escape focus. (**c**) Lead V_1 showing complete atrioventricular block with irregularity due to changing site of the escape focus. (Reproduced with permission from: Mittal SR. Third degree AV Block, AV Dissociation. Cardiology Today. 2013;17:274–277)

escape focus (Fig. 11.3b), or change in site of the escape focus [1] (Fig. 11.3c).

QRS configuration is usually uniform. Configuration may change due to capture beats (Fig. 11.4a, b) or ventricular ectopics (Fig. 11.4c).

Changing site of the escape focus or fusion of the P wave with the QRS complex (Fig. 11.5) can also change the QRS configuration.

Group beating suggests a high-degree atrioventricular block with Wenckebach phenomenon rather than complete atrioventricular block (Fig. 11.6).

- There is no relation between the P waves and the QRS complexes. In the setting of atrial flutter, the relation between the flutter waves and the QRS complexes varies (Fig. 11.2a).

Sometimes a critically timed P wave may conduct normally to the ventricles producing a normal QRS (capture beat) (Figs. 11.7 and 11.8).

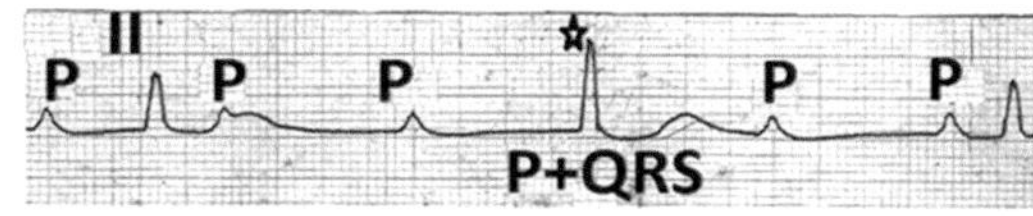

Fig. 11.5 Lead II shows complete atrioventricular block with change in QRS configuration due to fusion of the P with the QRS complex (marked ☆). (Reproduced with permission from: Mittal SR. Third degree AV Block, AV Dissociation. Cardiology Today. 2013;17:274–277)

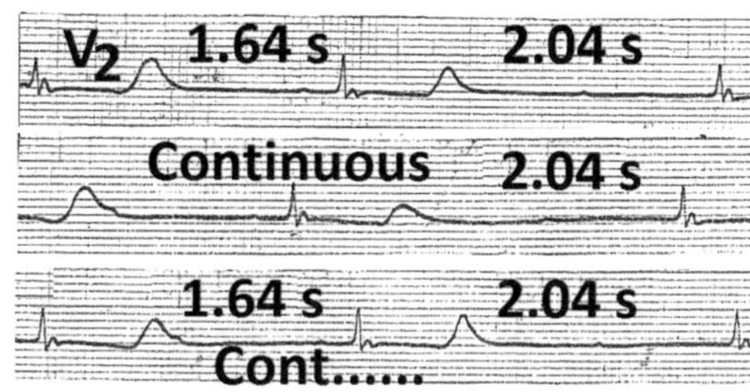

Fig. 11.6 Continuous strip of lead V_2 showing bradycardia with group beating. (Reproduced with permission from: Mittal SR. Third degree AV Block, AV Dissociation. Cardiology Today. 2013;17:274–277)

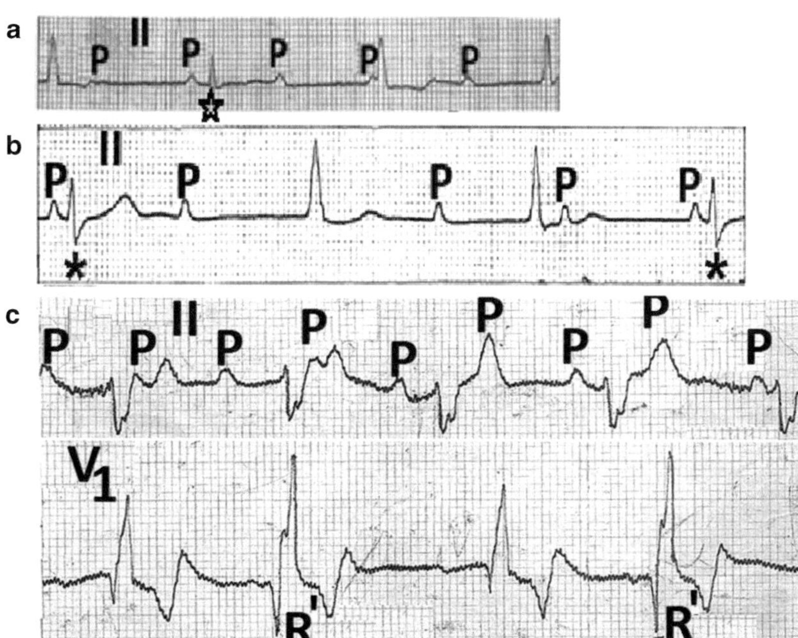

Fig. 11.4 (**a**) Lead II showing complete atrioventricular block with change in the QRS due to intermittent capture beat (marked ☆). (**b**) Lead II shows complete atrioventricular block with change in the configuration of the QRS complex due to intermittent capture beats (marked *). (**c**) Lead II shows complete atrioventricular block. Lead V_1 from the same patient showing change in QRS configuration due to ventricular ectopics (marked R′). (Reproduced with permission from: Mittal SR. Third degree AV Block, AV Dissociation. Cardiology Today. 2013;17:274–277)

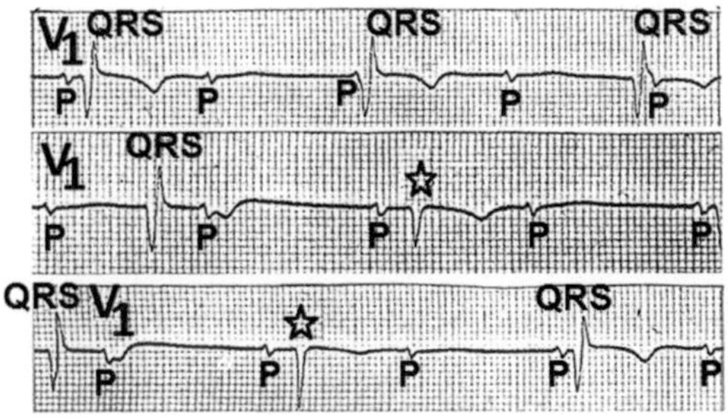

Fig. 11.7 Strips of lead V_1 showing complete atrioventricular block with intermittent capture beats (marked ☆). (Reproduced with permission from: Mittal SR. Third degree AV Block, AV Dissociation. Cardiology Today. 2013;17:274–277)

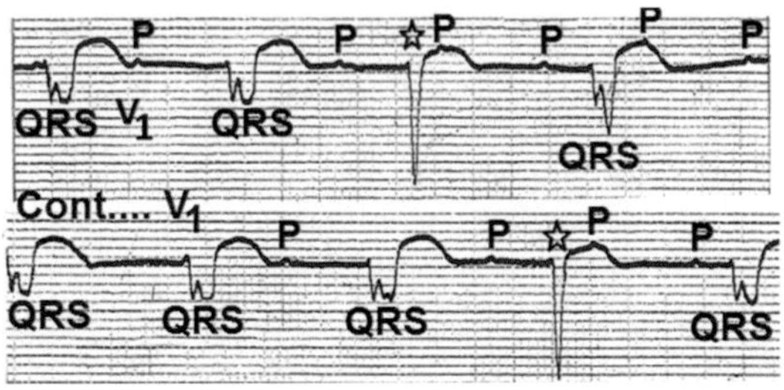

Fig. 11.8 Continuous strips of lead V_1 showing complete atrioventricular block with intermittent capture beats (marked ☆). (Reproduced with permission from: Mittal SR. Third degree AV Block, AV Dissociation. Cardiology Today. 2013;17:274–277)

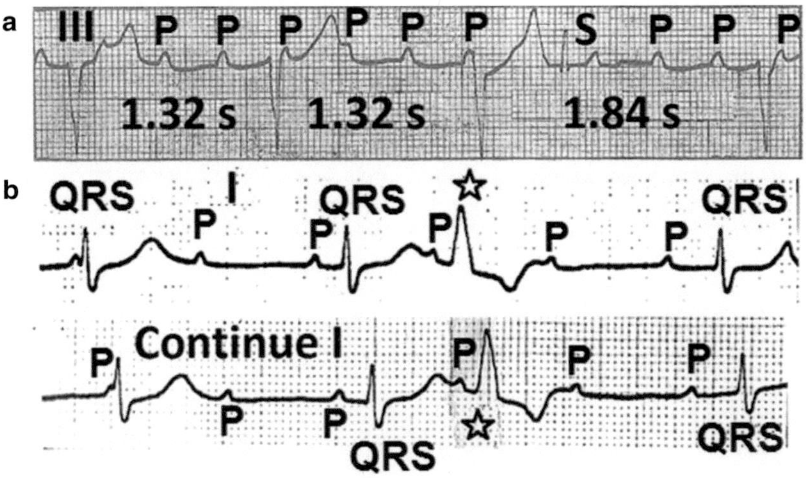

Fig. 11.9 (**a**) Lead III showing atrial tachycardia (marked P) with complete atrioventricular block and resetting of the escape focus by a capture beat (marked S). (**b**) Strips of lead I showing complete atrioventricular block. Early capture beat (marked ☆) showing aberrant ventricular conduction due to long-short cycle sequence. (Reproduced with permission from: Mittal SR. Third degree AV Block, AV Dissociation. Cardiology Today. 2013;17:274–277)

Presence of capture beat suggests that the complete atrioventricular block in these patients is due to impaired conduction rather than an anatomical transaction of the conduction tissue. Such capture beats also depolarize the escape focus and reset its cycle. This results in lengthening of the cycle of the escape focus and irregularity in the rhythm (Fig. 11.9a). Capture beats occurring early after preceding QRS may show tachycardia-dependent aberration (Fig. 11.9b).

If the atrioventricular block is at the level of the atrioventricular node, the escape pacemaker is usually situated in the lower part of the atrioventricular node or the His bundle. The ventricles are, therefore, activated through the normal conduction system, resulting in narrow QRS complexes. Rate of the QRS complexes can also be in the normal range (Fig. 11.10).

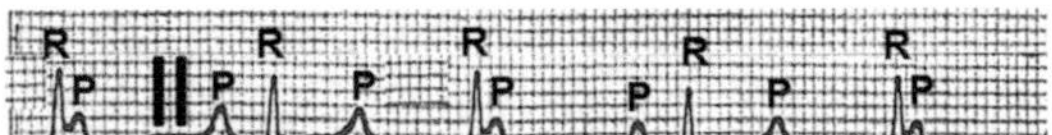

Fig. 11.10 Lead II shows complete atrioventricular block with narrow QRS complexes at a rate around 75/min. (Reproduced with permission from: Mittal SR. Third degree AV Block, AV Dissociation. Cardiology Today. 2013;17:274–277)

Ventricular rate may be slow if atrioventricular node is diseased. Complete atrioventricular block at the level of the atrioventricular node is usually preceded by first-degree and second-degree atrioventricular block. If the block is at the level of the bundle branches or the escape focus arises either from the peripheral part of the bundle branches or the fascicles or more remotely from the Purkinje network. QRS complexes are broad because ventricular activation does not occur through the normal conduction tissue. Ventricular rate depends on the site of the escape focus. More peripheral is the escape focus, the slower is the rate (Fig. 11.11a, b). Such atrioventricular block carries an adverse prognosis and high risk of Stokes-Adams' syncopal attacks.

In some patients one ventricular focus is unable to sustain the ventricular rhythm for a long time. The ventricular focus intermittently shifts to another site. This results in intermittent change in QRS configuration and rate (Fig. 11.12).

Such patients are also at high risk of Stokes-Adams' syncopal attacks. Complete atrioventricular block at the level of bundle branches is usually not preceded by the first-degree or second-degree atrioventricular block [2]. Bundle branch block almost always precedes. Such complete atrioventricular block can appear suddenly specially in the setting of acute anterior myocardial infarction (Fig. 11.13).

Complete atrioventricular block may be intermittent [3] (Fig. 11.14).

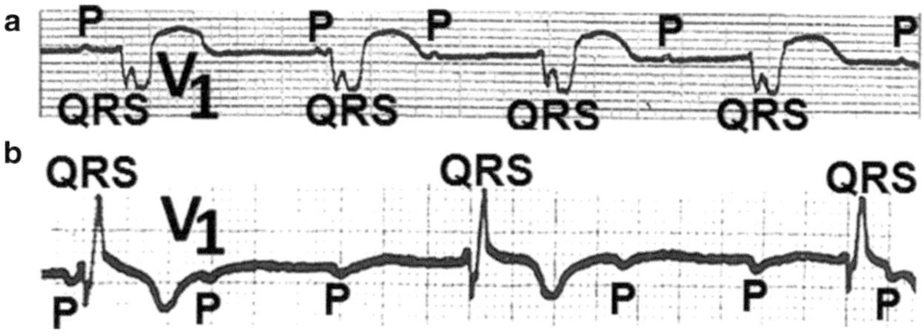

Fig. 11.11 (**a**) Lead V_1 showing complete atrioventricular block with broad QRS at a rate of around 63/min. (**b**) Lead V_1 showing complete atrioventricular block with broad QRS at a rate of around 33/min. (Reproduced with permission from: Mittal SR. Third degree AV Block, AV Dissociation. Cardiology Today. 2013;17:274–277)

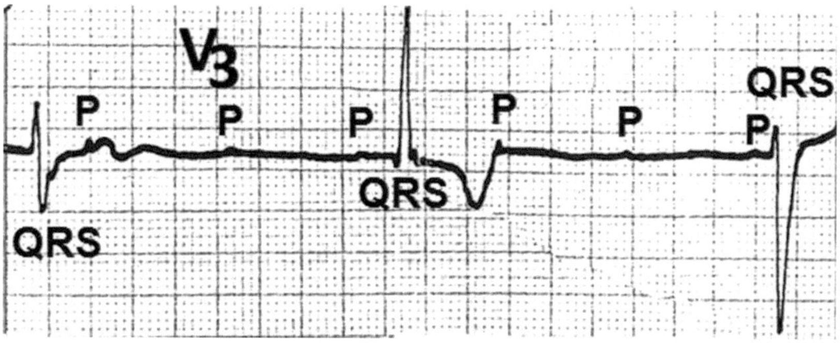

Fig. 11.12 Lead V_3 showing complete atrioventricular block with a changing ventricular focus (changing configurations of the QRS complexes). (Reproduced with permission from: Mittal SR. Third degree AV Block, AV Dissociation. Cardiology Today. 2013;17:274–277)

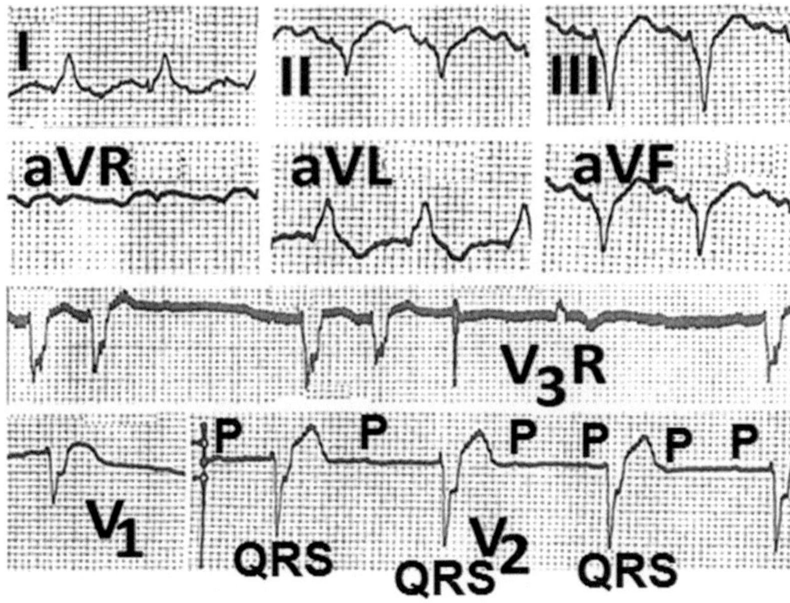

Fig. 11.13 ECG shows sudden onset of complete atrioventricular block in the setting of acute anterior myocardial infarction. (Reproduced with permission from: Mittal SR. Third degree AV Block, AV Dissociation. Cardiology Today. 2013;17:274–277)

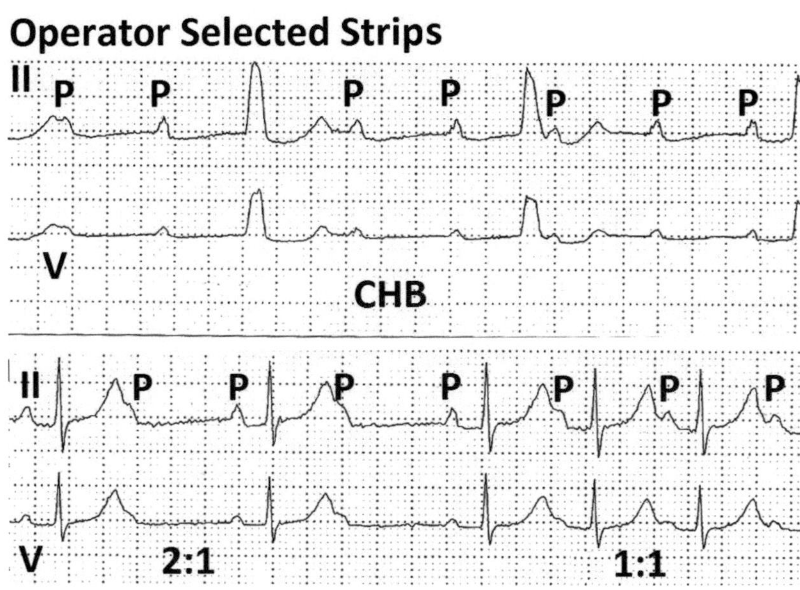

Fig. 11.14 Holter recordings showing complete atrioventricular block with broad QRS reverting to 2:1 and finally to 1:1 conduction. (Reproduced with permission from: Mittal SR. Third degree AV Block, AV Dissociation. Cardiology Today. 2013;17:274–277)

Figure 11.15 shows reversal of complete atrioventricular block in 2:1 atrioventricular block with alternate narrow QRS and, finally, to sinus rhythm with narrow QRS.

Such an ECG also suggests that complete atrioventricular block was due to delay in conduction rather than an anatomical transaction of the conduction tissue. In transient complete atrioventricular block, symptoms may be episodic, and ECG recorded after the episode may be normal. Diagnosis may be missed. Holter monitoring or electrophysiologic studies are indicated if there is strong clinical suspicion of Stokes-Adams' attacks.

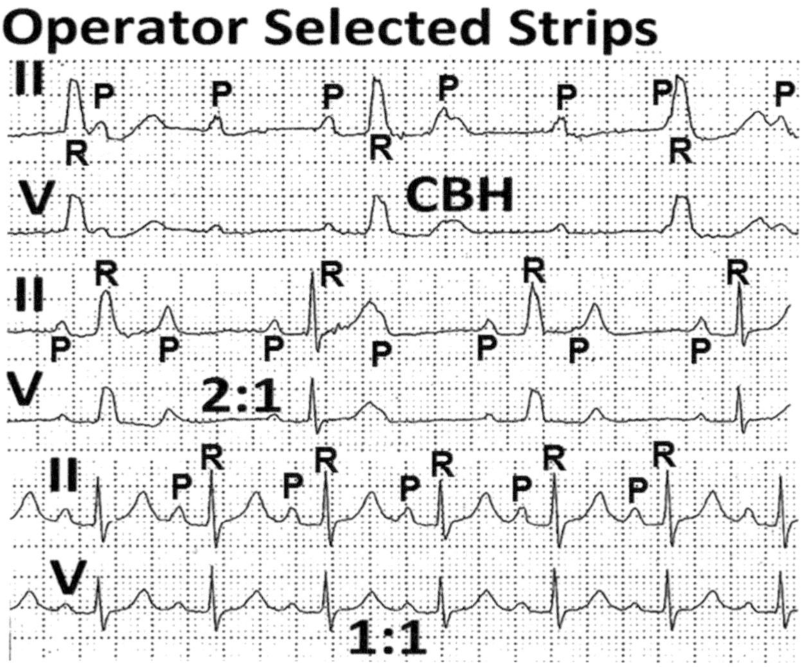

Fig. 11.15 Holter recordings showing reversal of complete atrioventricular block with broad QRS reverting to 2:1 atrioventricular block with alternate narrow QRS and finally to 1:1 conduction. (Reproduced with permission from: Mittal SR. Third degree AV Block, AV Dissociation. Cardiology Today. 2013;17:274–277)

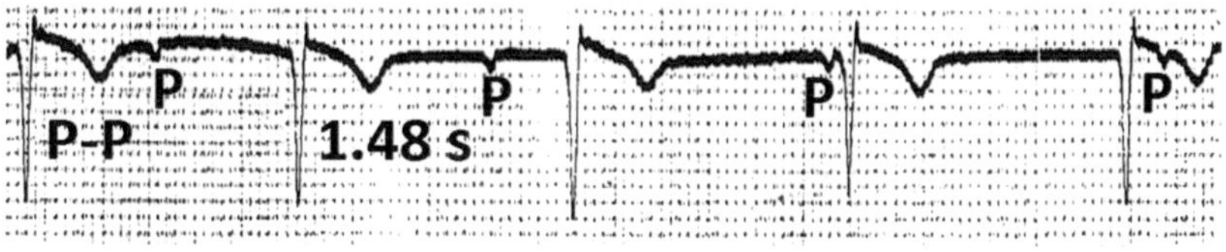

Fig. 11.16 Lead V_1 showing sinus bradycardia with atrioventricular dissociation due to escape junctional focus. QRS is narrow. (Reproduced with permission from: Mittal SR. Third degree AV Block, AV Dissociation. Cardiology Today. 2013;17:274–277)

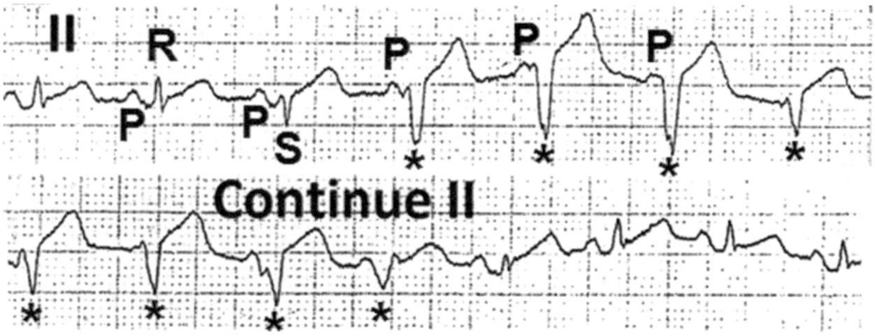

Fig. 11.17 Lead II shows a transient atrioventricular dissociation due to accelerated ventricular rhythm (marked *). (Reproduced with permission from: Mittal SR. Third degree AV Block, AV Dissociation. Cardiology Today. 2013;17:274–277)

Atrioventricular Dissociation

Decreased rate of sinoatrial node (Fig. 11.16) or acceleration of a junctional or ventricular focus (Fig. 11.17) also produce dissociation of the P waves and the QRS complex.

These situations are called atrioventricular dissociation and not atrioventricular block. In atrioventricular dissociation, ventricular rate is faster than the atrial rate, whereas in atrioventricular block, ventricular rate is slower than the

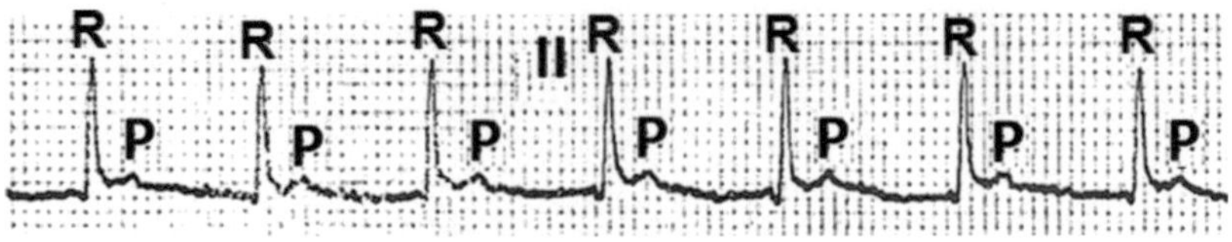

Fig. 11.18 Lead II shows an isorhythmic atrioventricular dissociation. (Reproduced with permission from: Mittal SR. Third degree AV Block, AV Dissociation. Cardiology Today. 2013;17:274–277)

atrial rate. When atrial and ventricular rates are nearly similar but the P wave and the QRS complexes are dissociated, it is called isorhythmic atrioventricular dissociation (Fig. 11.18).

It may be difficult to differentiate such dissociation from complete atrioventricular block. A long strip of ECG or long-term ECG monitoring (Holter monitoring) may be helpful.

Summary

- When no P waves are conducted to the ventricles, ventricular rate is slower than the atrial rate, and there is no relation between the P waves and the QRS complexes, it is called complete or third-degree atrioventricular block. Narrow QRS complexes at a rate of more than 40/min suggest block at the level of the atrioventricular node. These features are seen in congenital complete atrioventricular block, complete atrioventricular block accompanying inferior infarction, or other acquired diseases involving the atrioventricular node. Broad QRS complexes at a rate of less than 40/min suggest block at the level of the His-Purkinje system. This is usual in an acquired complete atrioventricular block.
- Patients with complete atrioventricular block with broad QRS and ventricular rate of less than 40/min are susceptible to syncopal attacks (Stokes-Adams' attacks) and sudden death. ECG following a Stokes-Adams' attack may show diffusely inverted and broad T waves.
- When the ventricular rate is faster than the atrial rate and there is no relation between the P wave and the QRS complexes, it is called atrioventricular dissociation. It is due to acceleration of the junctional or the ventricular focus above the rate of the atrial. There is no delay/block in atrioventricular conduction, and there is no risk of Stokes-Adams' attacks.
- Rarely normal retrograde conduction can occur even in the presence of advanced antegrade atrioventricular block.
- Ectopics arising in the His bundle can also produce surface electrocardiogram simulating atrioventricular block. These can be detected only by electrophysiologic study.

MCQs

Q1. Third-degree atrioventricular block is diagnosed when:

(a) The PR interval is 0.26 s
(b) Some P waves are not conducted to the ventricles
(c) No P waves are conducted to the ventricles
(d) All

Q2. Complete atrioventricular block in always preceded by:

(a) First-degree atrioventricular block
(b) Second-degree atrioventricular block
(c) Both
(d) None

Q3. Regular P waves at a rate of 80/min and regular QRS complexes at a rate of 40/min with changing PR interval suggest:

(a) First-degree atrioventricular block
(b) Second-degree atrioventricular block
(c) Complete atrioventricular block
(d) Accelerated ventricular rhythm

Q4. In complete atrioventricular block at the level of the His-Purkinje system, rate of QRS is:

(a) Between 20 and 40/min
(b) Between 40 and 60/min
(c) Between 60 and 70/min
(d) Between 70 and 80/min

Q5. Complete atrioventricular block may be associated with:

(a) Capture beats
(b) Retrograde ventriculoatrial conduction
(c) Both
(d) None

Q6. Chronic complete atrioventricular block is usually due to:

(a) Sclerodegeneration of the conduction system
(b) Acute myocardial infarction
(c) Myocarditis
(d) Cardiomyopathy

Q7. Regular P waves at a rate of 60/min, regular QRS complexes at a rate of 100/min, and changing PR interval suggest:

(a) First-degree atrioventricular block
(b) Second-degree atrioventricular block
(c) Complete atrioventricular block
(d) Atrioventricular dissociation

Q8. Complete atrioventricular block can be associated with:

(a) Sinus rhythm
(b) Atrial flutter
(c) Atrial fibrillation
(d) All

Q9. Complete atrioventricular block can be associated with:

(a) Ventriculophasic sinus arrhythmia
(b) Ventricular ectopics
(c) Change in the configuration of QRS complexes
(d) None

Q10. Complete atrioventricular block can be associated with:

(a) Irregularity in the PP interval
(b) Irregularity in the QRS complexes
(c) Change in the configuration of the QRS complexes
(d) None

Q11. In complete atrioventricular block, change in the configuration of the QRS complexes can be due to:

(a) Capture beats
(b) Ventricular ectopics
(c) Fusion of the P wave with the QRS complex
(d) All

Q12. Group beating suggests:

(a) First-degree atrioventricular block
(b) Wenckebach phenomenon
(c) Mobitz type II second-degree atrioventricular block
(d) Atrioventricular dissociation

Q13. In the setting of complete atrioventricular block, antegrade conduction of a critically timed P wave to the ventricles suggests:

(a) Impaired conduction across atrioventricular node
(b) Anatomical transaction of the atrioventricular node
(c) Either
(d) Neither

Q14. Complete atrioventricular block at the level of atrioventricular node:

(a) Must be preceded by Mobitz type II second-degree atrioventricular block
(b) Is always associated with broad QRS

(c) Is always associated with a ventricular rate of 30–40/min
(d) None

Q15. Stokes-Adams' attacks are common in:

(a) Complete atrioventricular block at the level of AV node
(b) Complete atrioventricular block at the level of bundle branches
(c) Complete atrioventricular block with changing ventricular focus
(d) All

Q16. Complete atrioventricular block at the level of bundle branches:

(a) Is not preceded by first-degree or second-degree atrioventricular block
(b) Is usually preceded by a bundle branch block
(c) Appears in the setting of inferior infarction
(d) None

Q17. Complete atrioventricular block can be:

(a) Permanent
(b) Transient
(c) Episodic
(d) All

Q18. Atrioventricular dissociation can be due to:

(a) Sinus bradycardia
(b) Accelerated junctional rhythm
(c) Accelerated ventricular rhythm
(d) Atrioventricular block

Q19. In complete atrioventricular block, atrial depolarization can be:

(a) From the sinoatrial node
(b) From the atria
(c) From the atrioventricular junction
(d) From any of the above sites

Q20. Paroxysmal atrioventricular block can be caused by:

(a) Vagotonia
(b) Tachycardia-dependent atrioventricular block
(c) Pause-dependent atrioventricular block
(d) All

Q21. Congenital complete atrioventricular block can occur due to:

(a) Maternal antibodies
(b) Anatomic disruption
(c) Vagotonia
(d) Neither

Q22. In congenital complete atrioventricular block, Stokes-Adams' attacks:

(a) Do not occur
(b) Occur if heart rate is less than 50/min
(c) Occur if there are episodes of paroxysmal tachyarrhythmia
(d) None

Answers

(1) c; (2) d; (3) c; (4) a; (5) c; (6) a; (7) d; (8) d; (9) a, b, c; (10) a, b, c; (11) d; (12) b; (13) a; (14) d; (15) b, c; (16) a, b; (17) d; (18) a, b, c; (19) d; (20) d; (21) a, b; (22) b, c

References

1. Olgin J, Zipes DP. Specific arrhythmias: diagnosis and treatment. In: Bonow RO, Mann DL, Zipes DP, Libby P, editors. Braunwald's heart disease. Philadelphia: Elsevier; 2012. p. 771–824.
2. Wagner GS. Atrioventricular block. In: Wagner GS, editor. Marriot's practical electrocardiography. New Delhi: Wolters Kluwer; 2008. p. 401–17.
3. Vijayaraman P, Ellenbagen KA. Bradyarrhythmias and pacemakers. In: Fuster V, Walsh RA, Harrington RA, editors. Hurst's the heart. New York: McGraw Hill; 2011. p. 1025–57.

12 Site and Etiology of Atrioventricular Block

1 Site of Atrioventricular Block

1.1 Following ECG Features Suggest Block at the Level of the Atrioventricular Node (Supra-His)

- Very long PR interval [1, 2] (Fig. 12.1).
- Variations in the PR interval—it may be spontaneous (Fig. 12.2a, b).

  The variation may be rate dependent (Fig. 12.3a), or it may be a part of Wenckebach phenomenon (Fig. 12.3b).
- Narrow QRS complex irrespective of the degree of the atrioventricular block (Fig. 12.4a–d).

  QRS complex may be broad if there is pre-existing bundle branch block (Fig. 12.5a) or aberrant ventricular conduction. Narrow QRS capture beats in the setting of complete atrioventricular block also support the possibility of block at the level of the atrioventricular node (Fig. 12.5b, c).
- Transient atrioventricular block (Fig. 12.6).

1.2 Block at More than One Level

It is suggested by the presence of:

- Group beating with two different conducting PR intervals (Fig. 12.7a, b)
- Atrioventricular block with changing QRS configuration (Fig. 12.7c)

  This occurs due to atrioventricular block with a rate-dependent aberrancy in a fascicle or bundle branch.

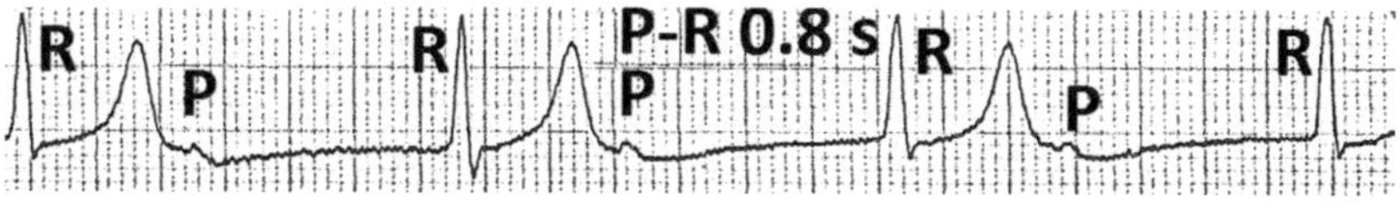

Fig. 12.1 Lead II shows first-degree atrioventricular block with a very long PR interval. (Reproduced with permission from: Mittal SR. Site and Etiology of AV Block. Cardiology Today. 2014;18:75–80)

S. Mittal, *Insights into Electrocardiograms with MCQs*,
https://doi.org/10.1007/978-981-99-0127-2_12

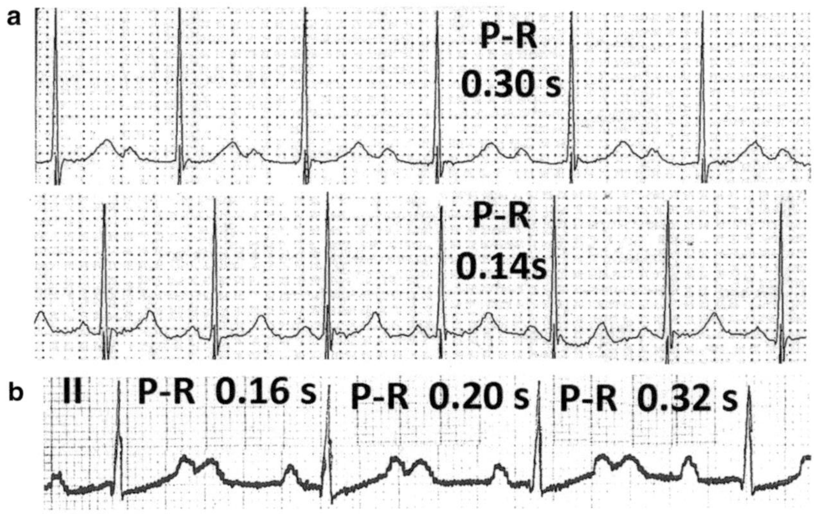

Fig. 12.2 (a) Holter strips of lead II showing spontaneous variation in the PR interval. (b) Lead II showing 2:1 atrioventricular block with spontaneous variation in the PR interval. (Reproduced with permission from: Mittal SR. Site and Etiology of AV Block. Cardiology Today. 2014;18:75–80)

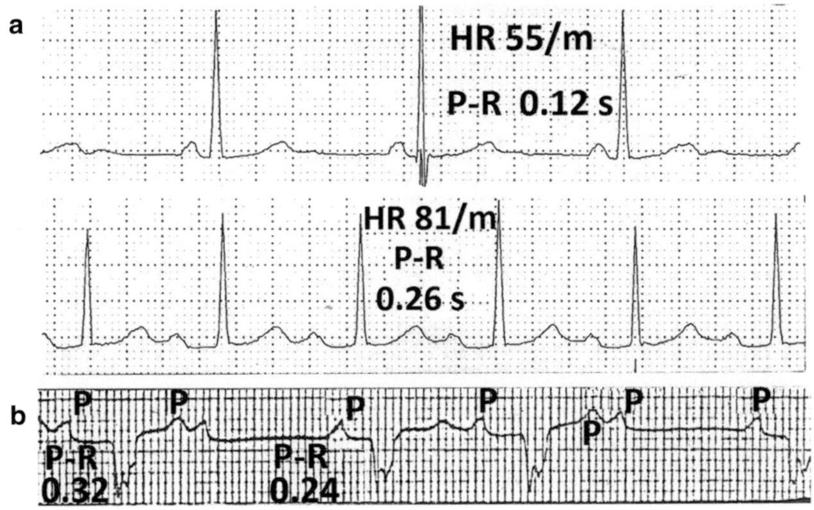

Fig. 12.3 (a) Holter strips of lead II showing rate-dependent first-degree atrioventricular block. (b) The ECG shows variation in the PR interval as a part of Wenckebach phenomenon. (Reproduced with permission from: Mittal SR. Site and Etiology of AV Block. Cardiology Today. 2014;18:75–80)

1.3 Other Investigations

Other investigations that can help in deciding the site of atrioventricular block [3].

1.3.1 Modulation of Autonomic Tone

1. Treadmill stress testing

   Increased sympathetic drive during exercise increases the rate of sinoatrial node and conduction through the atrioventricular node. It may decrease the block if it is at the level of the atrioventricular node. Intra-His and infra-His blocks may increase due to increased number of atrial impulses reaching the diseased bundle branches.

2. Intravenous atropine

   Parasympatholytic effect also increases the rate of sinoatrial node and conduction through the atrioventricular node. Like treadmill stress test, it may improve block at the level of the atrioventricular node (Fig. 12.8) but worsen block at the level of bundle branches.

3. Carotid sinus compression

   It decreases the sinus rate as well as conduction in the atrioventricular node. The response is opposite to that of treadmill stress test and intravenous atropine. Block at the level of the atrioventricular node is increased (Fig. 12.9). Infra-His block may improve due to decreased number of atrial impulses reaching the bundle branches.

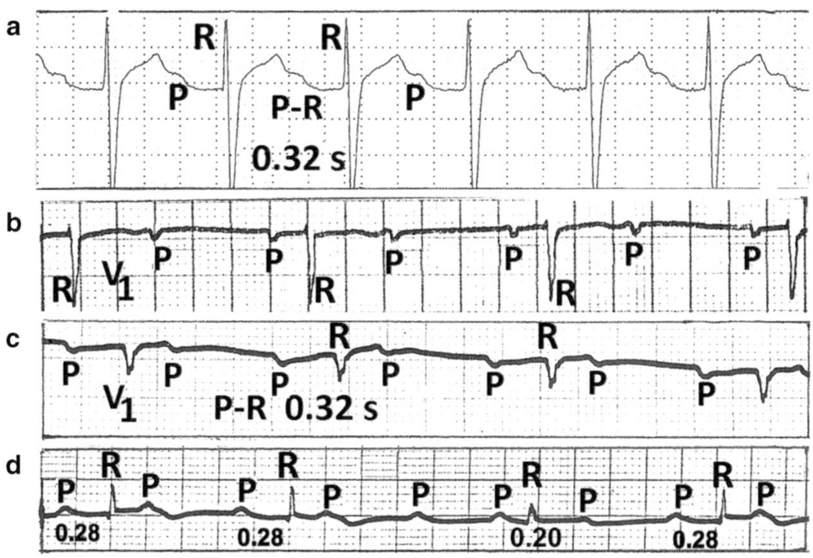

Fig. 12.4 (**a**) Holter strips of lead V_5 showing first-degree atrioventricular block with narrow QRS. (**b**) Lead V_1 showing 2:1 atrioventricular block with narrow QRS. (**c**) Lead V_1 showing 2:1 atrioventricular block with narrow QRS and prolonged PR interval of the conducted beats. (**d**) Lead aVF showing changing atrioventricular block with narrow QRS and shorter PR interval after longer pause. (Reproduced with permission from: Mittal SR. Site and Etiology of AV Block. Cardiology Today. 2014;18:75–80)

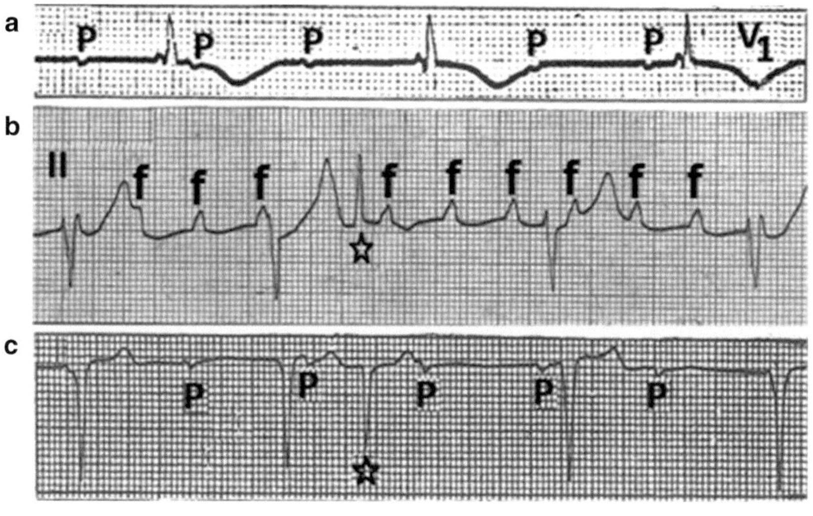

Fig. 12.5 (**a**) Lead V_1 showing complete atrioventricular block with bundle branch block. (**b**) Lead II showing atrial flutter (f) with complete atrioventricular block, broad QRS, and narrow QRS capture beat (marked ☆). (**c**) Lead V_1 showing complete atrioventricular block with broad QRS and narrow QRS capture beat (marked ☆). (Reproduced with permission from: Mittal SR. Site and Etiology of AV Block. Cardiology Today. 2014;18:75–80)

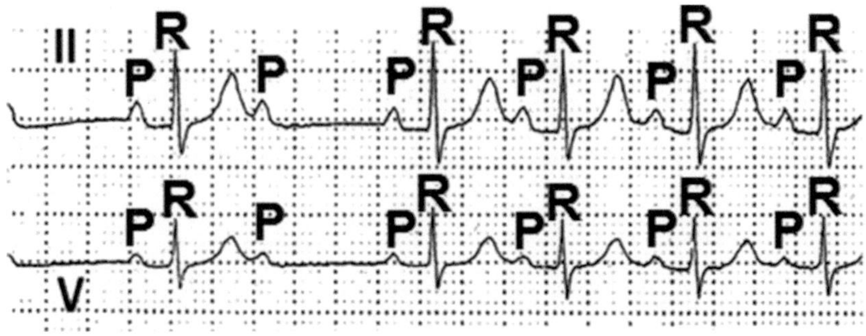

Fig. 12.6 Holter strips showing transient second-degree atrioventricular block with narrow QRS. (Reproduced with permission from: Mittal SR. Site and Etiology of AV Block. Cardiology Today. 2014;18:75–80)

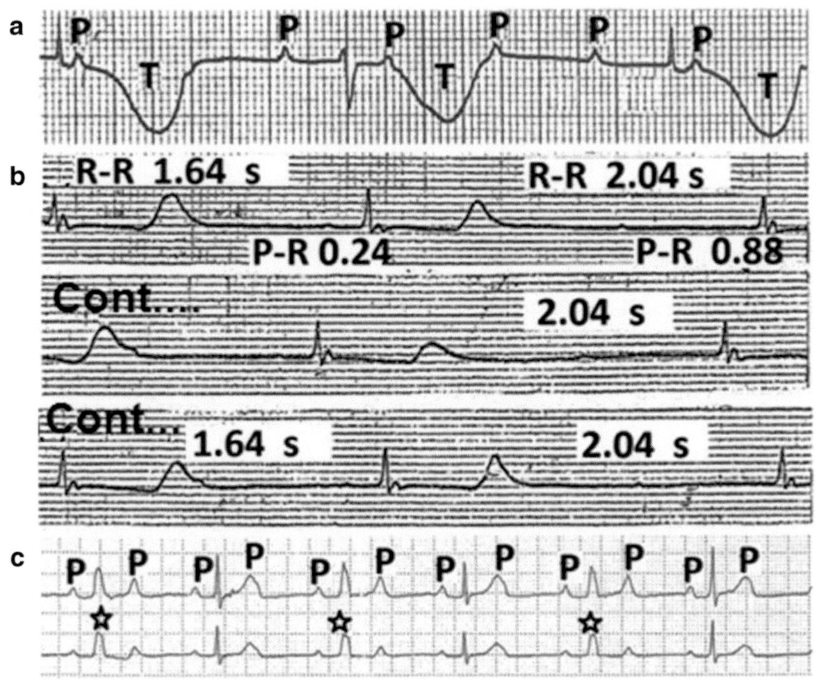

Fig. 12.7 (**a**) Lead II shows a high-degree atrioventricular block with group beating and two different PR intervals with different QRS complexes. (**b**) Continuous strip showing group beating with two different RR cycles and two different PR intervals. (**c**) Holter strip showing 2:1 atrioventricular block with alternate broad QRS complex (marked ☆). (Reproduced with permission from: Mittal SR. Site and Etiology of AV Block. Cardiology Today. 2014;18:75–80)

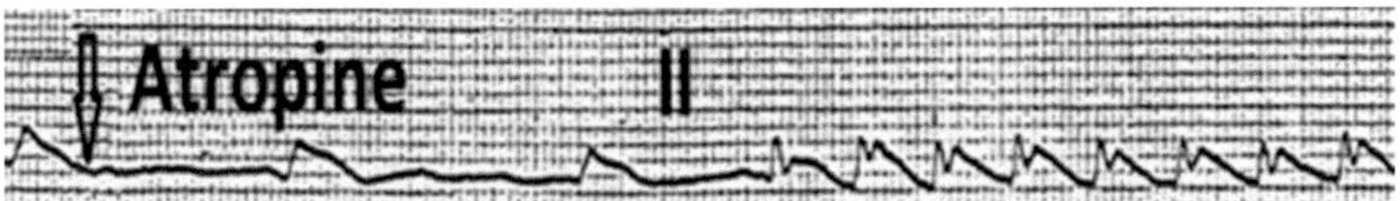

Fig. 12.8 Lead II is showing improvement in the atrioventricular nodal block by I/V atropine. (Reproduced with permission from: Mittal SR. Site and Etiology of AV Block. Cardiology Today. 2014;18:75–80)

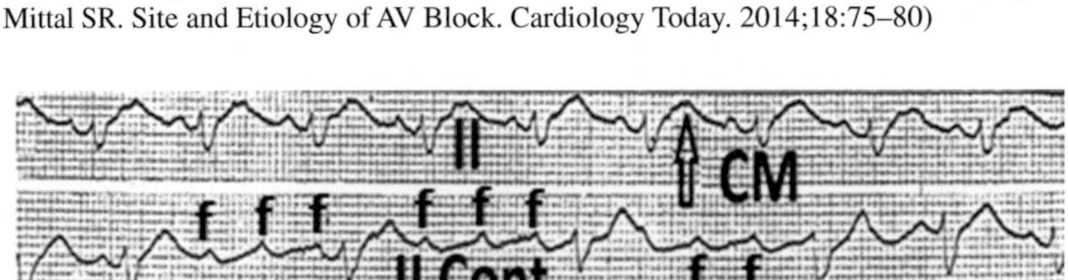

Fig. 12.9 Lead II showing atrial flutter (f) with an increase in the atrioventricular nodal block by carotid massage (CM). (Reproduced with permission from: Mittal SR. Site and Etiology of AV Block. Cardiology Today. 2014;18:75–80)

1.3.2 Holter Monitoring

It can demonstrate the following:

- Transient aggravation of the atrioventricular block
- Transient appearance of a bundle branch block
- Relation of symptoms to the atrioventricular block

1.3.3 External Loop Recorders or Implantable Loop Recorder

These may be needed to evaluate patients with unexplained syncope with a strong suspicion of transient atrioventricular block.

1.3.4 Electrophysiologic Study

A definite conclusion about the site of block needs electrophysiologic study. His bundle electrogram can be recorded by placing a tripolar catheter across the tricuspid valve in such a position that the middle electrode is close to the His bundle and records His bundle activity (H). The distal electrode records ventricular activity (V) and the proximal electrode records atrial activity (A). A prolongation of the interval from the onset of the atrial signal to the His bundle signal (AH interval >135 ms) indicates block at the level of the atrioventricular node. Prolongation of the interval from the His bundle signal to the onset of the ventricular signal (HV interval >55 ms) indicates a block below the His bundle. Permanent pacemaker implantation is recommended if the site of block is infra-His irrespective of findings on the surface ECG.

2 Cause of Atrioventricular Block

2.1 Coronary Artery Disease

Acute inferior infarction may be complicated by first-degree or Mobitz type I second-degree atrioventricular block. It is more frequent if there is associated right ventricular infarction [2]. It may advance to complete atrioventricular block. Site of block is usually in the atrioventricular node. QRS is narrow and ventricular rate is around 50/min (Fig. 12.10a). Stokes-Adams' attacks are uncommon. Mostly, the atrioventricular block

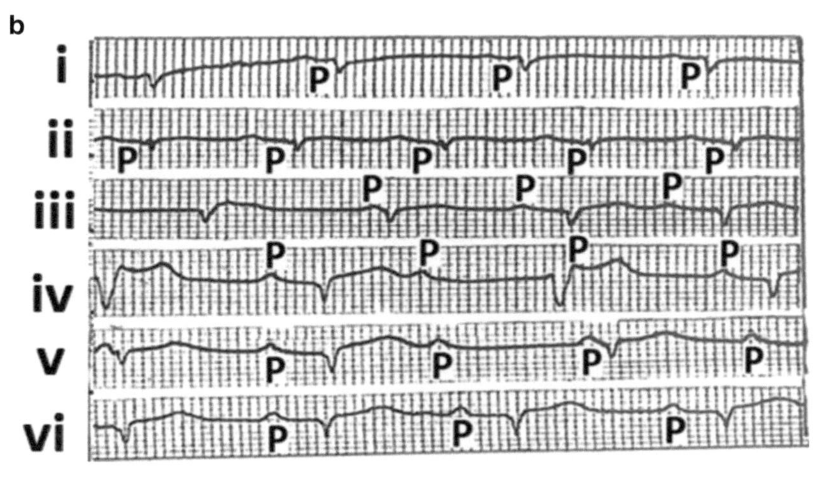

Fig. 12.10 (**a**) Showing inferior myocardial infarction with a 2:1 atrioventricular block with narrow QRS at a rate of 50/min and prolonged PR interval. (**b**) Showing sequential changes in a case of inferior myocardial infarction. (i) Normal PR interval. (ii) First-degree atrioventricular block. (iii)–(v) Wenckebach phenomenon. (vi) Reversal to first-degree atrioventricular block. (Reproduced with permission from: Mittal SR. Site and Etiology of AV Block. Cardiology Today. 2014;18:75–80)

gradually reverts to the normal conduction (Fig. 12.10b) suggesting that pathology in the conduction across the atrioventricular is transient.

Atrioventricular block may not recover if the atrioventricular node artery is blocked resulting in permanent damage to the atrioventricular node. Permanent pacemaker implantation is indicated if, even after 7 days, the ventricular rate is below 40/min and patient is symptomatic because of the atrioventricular block [3].

Acute anterior wall infarction may be complicated by complete atrioventricular block. It is due to direct involvement of the bundle branches, fascicles, and the Purkinje network in the infarction. QRS complexes are broad and at a slow rate (30–40/min). Complete atrioventricular block is usually preceded by a bundle branch block. Complete atrioventricular block usually appears suddenly without any intervening period of first-degree or second-degree atrioventricular block (Fig. 12.11a). Stokes-Adams' attacks occur at the time of transition from normal conduction to atrioventricular block. Myocardial damage is usually extensive. Diagnosis of infarction is usually clear when the escape rhythm has a right bundle branch block pattern (Fig. 12.11b). Infarct may be missed if the escape focus has a left bundle branch block pattern (Fig. 12.11c). Transient change in configuration from a left bundle branch block pattern to right bundle branch block pattern can unmask anterior myocardial infarction (Fig. 12.11d).

Prognosis is usually poor. Usually the atrioventricular block is permanent in the setting of anterior myocardial infarction. Uncommonly it may be transient if the atrioventricular block is secondary to increased vagal tone.

2.2 Idiopathic Progressive Fibrosis of the Conduction Tissue (Lenegre Disease)

It is progressive and may simultaneously involve sinoatrial node and atrioventricular node. Etiology is not clear. These patients usually present as chronic atrioventricular block with sinus node dysfunction (Fig. 12.12a, b).

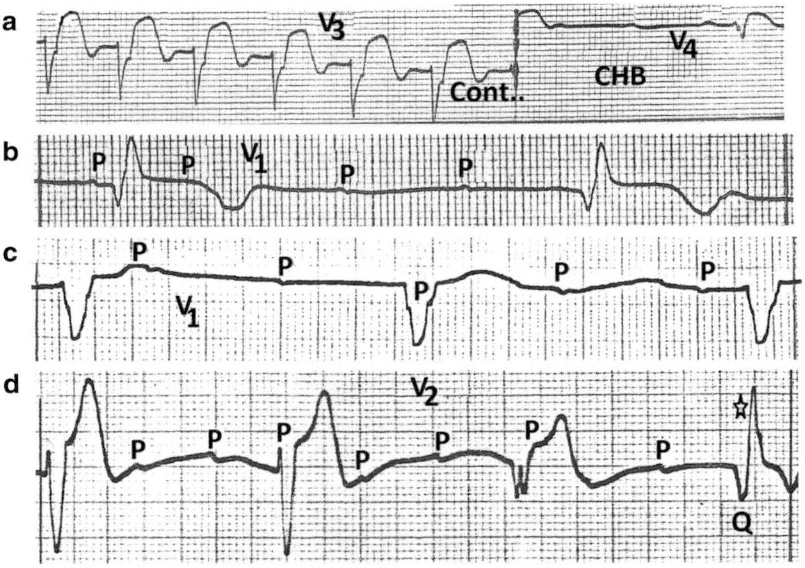

Fig. 12.11 (**a**) Lead V_3 and V_4 showing anterior myocardial infarction with broad QRS and sudden onset of complete atrioventricular block. (**b**) Lead V_1 showing complete atrioventricular block with right bundle branch block configuration of escape focus showing anterior myocardial infarction. (**c**) Lead V_1 showing complete atrioventricular block with left bundle branch block configuration. (**d**) Lead V_2 showing change in the escape focus with unmasking of anterior myocardial infarction (marked ☆). (Reproduced with permission from: Mittal SR. Site and Etiology of AV Block. Cardiology Today. 2014;18:75–80)

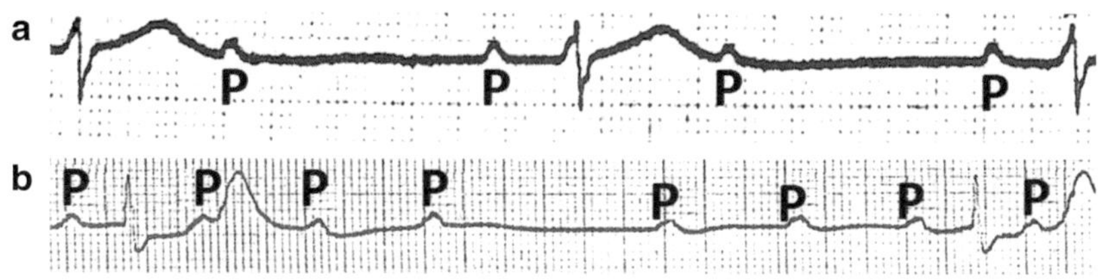

Fig. 12.12 (**a**) Showing sinus bradycardia with chronic 2:1 atrioventricular block. (**b**) Showing sinus node dysfunction with chronic high-degree atrioventricular block. (Reproduced with permission from: Mittal SR. Site and Etiology of AV Block. Cardiology Today. 2014;18:75–80)

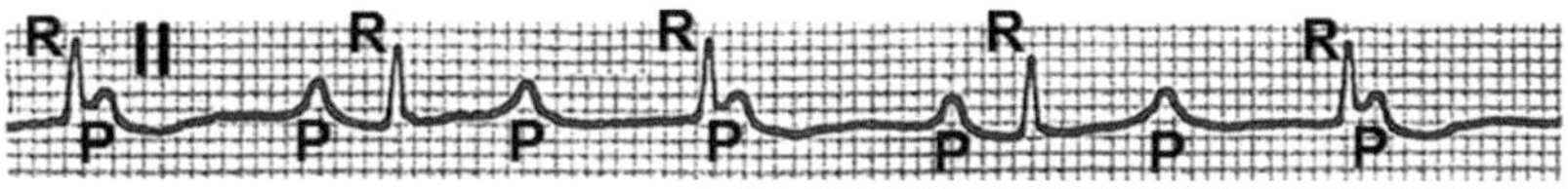

Fig. 12.13 Showing congenital atrioventricular block with narrow QRS and near normal ventricular rate. (Reproduced with permission from: Mittal SR. Site and Etiology of AV Block. Cardiology Today. 2014;18:75–80)

2.3 Involvement in Fibrocalcific Pathology of the Aortic or the Mitral Annulus (Lev's Disease)

Fibrocalcific pathology of the aortic or the mitral annulus can extend to involve the atrioventricular node. It usually occurs in the elderly.

2.4 Congenital Atrioventricular Block

Corrected transposition of the great vessels and endocardial cushion defects can be associated with congenital complete atrioventricular block. Congenital atrioventricular block can also occur as an isolated anomaly. It may be caused by maternal lupus and other maternal antibodies. Block is in the atrioventricular node. The escape rhythm arises from the distal part of the atrioventricular node. QRS is, therefore, narrow with a rate between 50 and 60/min (Fig. 12.13).

Patients do not have any cardiac disability. If the ventricular rate is less than 50/min, patients may feel fatigue, weakness, or lightheadedness on effort. Stokes-Adams' attacks are uncommon but can occur with extreme bradycardia.

2.5 Drugs

Usually there is an underlying conduction disorder. The block is usually of first degree or second degree. Complete atrioventricular block is rare. The block is reversible on discontinuation of the drug.

1. Digitalis toxicity:

   Atrioventricular block is usually due to concomitant hypokalemia. Atrial tachycardia with atrioventricular block is suggestive of digitalis toxicity (Fig. 12.14a). Complete atrioventricular block with frequent ectopics (Fig. 12.14b) or accelerated junctional rhythm (Fig. 12.14c) also support the possibility of digitalis toxicity. The block is reversible on discontinuation of the drug and management of hypokalemia.
2. Beta adrenergic blockers
3. Calcium channel blockers—diltiazem and verapamil
4. Antiarrhythmics:
   - Amiodarone.
   - Propafenone, flecainide, procainamide, quinidine, and disopyramide.
   - These drugs can slow or block conduction in the His-Purkinje system and can pro-

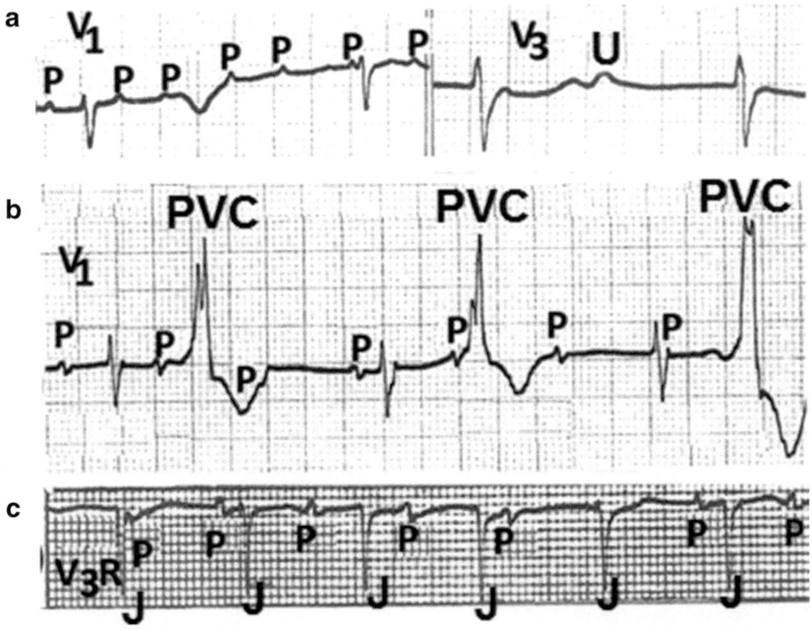

Fig. 12.14 (**a**) Lead V_1 showing atrial tachycardia with complete atrioventricular block. Lead V_3 is showing prolonged QT interval with prominent U wave suggestive of hypokalemia. (**b**) Lead V_1 showing atrial tachycardia with complete atrioventricular block and frequent ventricular ectopics (PVC). (**c**) Lead V_3R showing atrial tachycardia with accelerated junctional rhythm (marked J). (Reproduced with permission from: Mittal SR. Site and Etiology of AV Block. Cardiology Today. 2014;18:75–80)

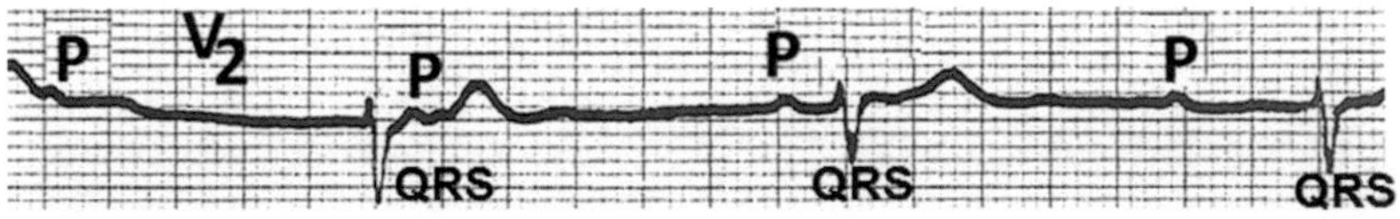

Fig. 12.15 Showing sinus bradycardia with complete atrioventricular block and narrow QRS. (Reproduced with permission from: Mittal SR. Site and Etiology of AV Block. Cardiology Today. 2014;18:75–80)

duce Mobitz type II second-degree atrioventricular block.
- Lidocaine can also worsen atrioventricular block.

5. Lithium

2.6 Vagotonia

Increased vagal tone without any organic cardiac disease can also produce first-degree or Mobitz type I second-degree atrioventricular block. This can occur in athletes, during sleep in healthy persons, and during sleep apnea [3]. Acute increase in vagal tone can produce transient high-degree atrioventricular block [1]. This can occur during vomiting, intubation, or suction. Block is transient. Vagotonia affects the sinoatrial node as well as the atrioventricular node. Such atrioventricular blocks are, therefore, associated with sinus bradycardia (Fig. 12.15).

2.7 Acute Rheumatic Carditis

First-degree atrioventricular block is common. It can advance to second-degree atrioventricular block. Block reverses with control of rheumatic activity.

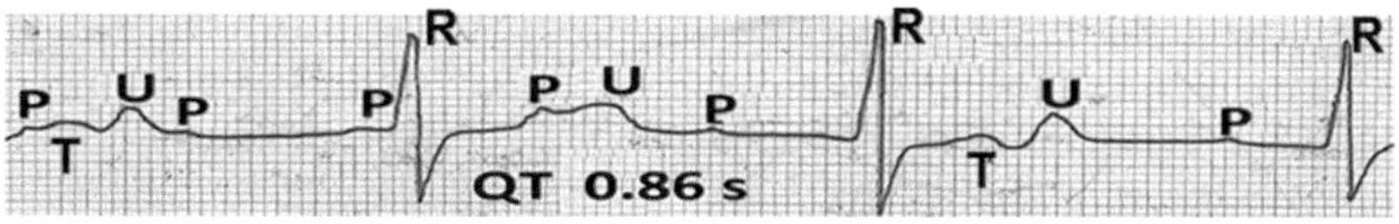

Fig. 12.16 Lead V_3 showing complete atrioventricular block with QT prolongation and prominent U wave suggestive of hypokalemia. (Reproduced with permission from: Mittal SR. Site and Etiology of AV Block. Cardiology Today. 2014;18:75–80)

2.8 Uncommon Causes

1. Tachycardia-dependent atrioventricular block (phase 3 block) [2]. The transient atrioventricular block may follow a run of tachycardia.
2. Bradycardia-dependent atrioventricular block (phase 4 block) [2]. Paroxysmal atrioventricular block can occur during relative bradycardia.
3. Intracardiac surgery in the areas close to the atrioventricular node or the His bundle may produce atrioventricular block, e.g., aortic valve replacement, mitral valve replacement, closure of ventricular septal defect, correction of tetralogy of Fallot, atrioventricular canal defects, and myectomy for hypertrophic obstructive cardiomyopathy. Block is not reversible.
4. Infective endocarditis may be associated with extension of the infective process in the perivalvular region (valve ring abscess) or apex of the interventricular septum. Block may not revert with treatment [4].
5. Infiltrative diseases—amyloidosis, sarcoidosis, and hemochromatosis.
6. Reiter syndrome. It can produce chronic first-degree or second-degree atrioventricular block.
7. Chagas' disease. It can produce chronic first-degree or second-degree atrioventricular block.
8. Neuromuscular diseases—myotonic muscular dystrophy, Erb limb girdle muscular dystrophy, *X*-linked muscular dystrophy, and peroneal muscular dystrophy.
9. Collagen vascular diseases—systemic lupus erythematosus, rheumatoid arthritis, and scleroderma.
10. Radio frequency ablation of the atrioventricular node for control of ventricular rate in atrial tachyarrhythmia.
11. Lyme disease.
12. Electrolyte imbalance—hyper- or hypokalemia (Fig. 12.16).

Summary

- Very long PR interval, variation in PR intervals, narrow QRS complex, group beating, and transient atrioventricular block suggest block at the level of the atrioventricular node.
- Increased sympathetic drive (treadmill stress test or I/V atropine) decreases block at the level of atrioventricular node but increases infra-His block. Vagal stimulation (e.g., carotid sinus compression) has opposite effect.
- During electrophysiologic study, prolongation of the AH interval suggests block of the level of the atrioventricular node. Prolongation of the HV interval suggests infra-His block.
- In the setting of inferior myocardial infarction, site of atrioventricular block is usually at the level of the atrioventricular node. Usually, the block gradually advances and gradually recovers.
- In the setting of anterior myocardial infarction, atrioventricular block is usually at the level of the bundle branches. Atrioventricular block usually appears suddenly and is usually permanent.
- Congenital atrioventricular block is usually at the level of the atrioventricular node.
- Beta-blockers, diltiazem, and verapamil delay conduction at the level of the atrioventricular node. Amiodarone, propafenone, and fle-

cainide delay conduction at the level of the His bundle-Purkinje system.

- Vagotonia delays conduction at the level of the atrioventricular node and is usually accompanied by sinus bradycardia.
- Chronic atrioventricular block in elderly is usually due to sclerodegeneration of the conduction tissue.

MCQs

Q1. Type I second-degree atrioventricular block with a normal QRS usually occurs at the level of:

(a) The atrioventricular node
(b) The His bundle
(c) The Bundle branches
(d) The Purkinje system

Q2. Prolonged PR interval with QRS showing bundle branch pattern suggests delay in conduction at:

(a) The atrioventricular node
(b) The His-Purkinje system
(c) Either atrioventricular node or the His-Purkinje system
(d) Atrioventricular node with preexisting bundle branch block

Q3. Type II second-degree atrioventricular block occurs more commonly in the setting of:

(a) Inferior myocardial infarction
(b) Right ventricular infarction
(c) Posterior infarction
(d) Anterior infarction

Q4. Type II second-degree atrioventricular block with bundle branch block pattern usually occurs at the level of:

(a) The atrioventricular node
(b) The His bundle
(c) The Bundle branches
(d) The Purkinje system

Q5. In 2:1 atrioventricular block with normal QRS, block is usually at the level of:

(a) The atrioventricular node
(b) The His bundle
(c) The bundle branches
(d) Any of the above

Q6. In 2:1 atrioventricular block with broad QRS, block is usually at the level of:

(a) Junction of intra-atrial conduction tissue with the atrioventricular node
(b) The atrioventricular node
(c) The His bundle
(d) The bundle branches

Q7. Complete atrioventricular block with QRS duration of more than 0.12 s suggests:

(a) Block at the level of the atrioventricular node with preexisting bundle branch block
(b) Block at the level of the atrioventricular node with bradycardia dependent aberrant ventricular conduction
(c) Block at the level of bundle branches
(d) All

Q8. Chronic complete atrioventricular block is usually at the level of:

(a) The atrioventricular node
(b) The His bundle
(c) The bundle branches
(d) The Purkinje system

Q9. Complete atrioventricular block in the setting of acute anterior myocardial infarction in usually at the level of:

(a) The atrioventricular node
(b) The His bundle
(c) The bundle branches
(d) The Purkinje system

Q10. Complete atrioventricular block in the setting of acute inferior infarction is usually at the level of:

(a) The atrioventricular node
(b) The His bundle
(c) The bundle branches
(d) The Purkinje system

Q11. In congenital complete atrioventricular block:

(a) Ventricular rate is usually above 50/min
(b) QRS duration is usually normal
(c) QRS configuration is usually normal
(d) All

Q12. In the setting of complete atrioventricular block, ventricular rate around 50/min with narrow QRS suggests that the site of atrioventricular block is at the level of:

(a) The atrioventricular node
(b) The His bundle
(c) The bundle branches
(d) The Purkinje system

Answers

(1) a; (2) c, d; (3) d; (4) c, d; (5) a; (6) d; (7) d; (8) c; (9) c; (10) a; (11) d; (12) a

References

1. Olgin J, Zipes DP. Specific arrhythmias: diagnosis and treatment. In: Banow RO, Mann DL, Zipes DP, Libby P, editors. Braunwald's heart disease. Philadelphia: Elsevier; 2012. p. 771–824.
2. Olshansky B, Chung MK, Pogwizd SM, Goldschlager N. Atrioventricular conduction abnormalities. In: Olshansky B, Chung MK, Pogwizd SM, Goldschlager N, editors. Arrhythmia essentials. Burlington: Jones & Bartlett Learning; 2012. p. 27–55.
3. Schamroth LS. Atrioventricular (AV) blocks. In: Schamroth C, editor. An introduction to electrocardiography. New Delhi: Blackwell Science; 1982. p. 375–85.
4. Vijayaraman P, Ellenbogen KA. Bradyarrhythmias and pacemakers. In: Fuster V, Walsh RA, Harrington RA, editors. Hurst's the heart. New York: McGraw Hill; 2011. p. 1025–57.

13 Stokes-Adams' Attacks

Patients with high-degree atrioventricular block or complete atrioventricular block are susceptible to have episodes of syncope (transient loss of consciousness). These are also called Stokes-Adams' attack. Such episodes occur because of transient cerebral ischemia due to sudden onset/aggravation of bradycardia resulting in a period of asystole (absence of the QRS complex on the ECG). When the circulation becomes stable at a slower rate, the brain also gets adjusted to the slow heart rate, and further episodes of syncope may not occur even at a slow heart rate. Such episodes are likely to occur in following situations:

- Second-degree atrioventricular blocks suddenly worsening into complete atrioventricular block with slow ventricular rate.
- Sudden onset of complete atrioventricular block with slow ventricular rate.
- Sudden onset of high-degree atrioventricular block (Fig. 13.1a, b, c).
- Unstable ventricular escape focus—when one ventricular focus cannot maintain an adequate rate, the focus transiently shifts to another site (escape focus). Stokes-Adams' attacks occur during the pause before another pacemaker takes up (Fig. 13.2).
- Paroxysm of fast ventricular tachycardia.
- Transient ventricular fibrillation.

Patients with high-degree atrioventricular block or complete atrioventricular block can have polymorphic ventricular tachycardia (Figs. 13.3 and 13.4), torsades de pointes (Figs. 13.5a, b, c

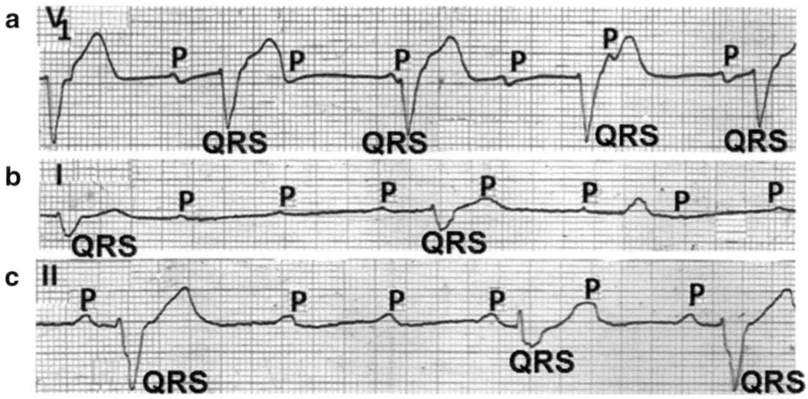

Fig. 13.1 (**a**) Lead V_1 showing complete atrioventricular block with broad QRS. (**b** and **c**). Lead I and II showing sudden aggravation of atrioventricular block. (Reproduced with permission from: Mittal SR. Stokes Adams Attacks. Cardiology Today. 2014;18:34–36)

S. Mittal, *Insights into Electrocardiograms with MCQs*,
https://doi.org/10.1007/978-981-99-0127-2_13

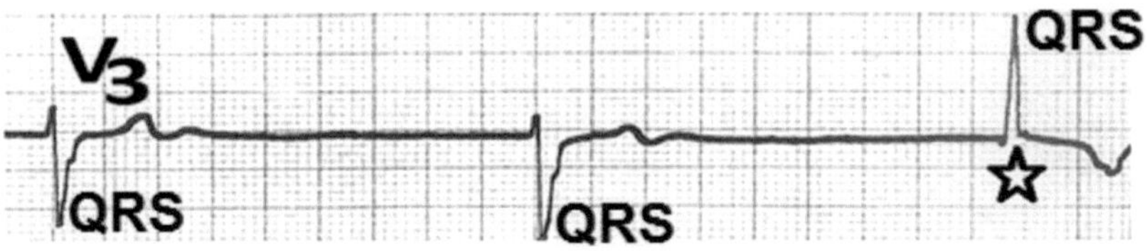

Fig. 13.2 Lead V_3 showing complete atrioventricular block with change in ventricular focus (marked☆). (Reproduced with permission from: Mittal SR. Stokes Adams Attacks. Cardiology Today. 2014;18:34–36)

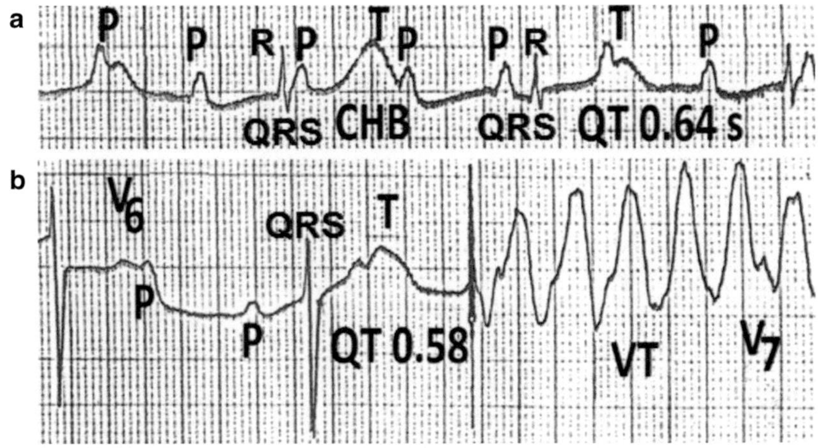

Fig. 13.3 (**a**) Lead II showing complete atrioventricular block with prolonged QT interval. (**b**) Lead V_6 showing complete atrioventricular block without any ectopics. Lead V_7 showing sudden onset of VT. (Reproduced with permission from: Mittal SR. Stokes Adams Attacks. Cardiology Today.2014;18:34–36)

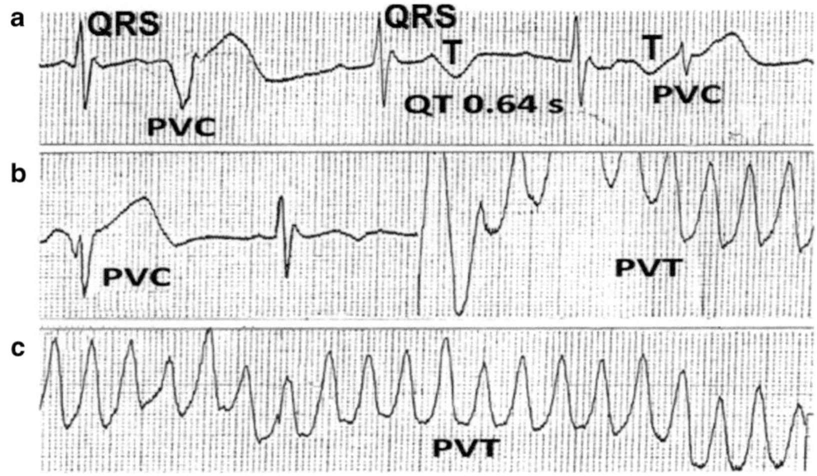

Fig. 13.4 (**a**) ECG showing complete atrioventricular block with prolonged QT interval and ventricular ectopics (PVC). (**b**) ECG showing onset of polymorphic VT (PVT). (**c**) ECG showing polymorphic VT (PVT). (Reproduced with permission from: Mittal SR. Stokes Adams Attacks. Cardiology Today. 2014;18:34–36)

and 13.6), or ventricular fibrillation (Fig. 13.7). These ventricular tachyarrhythmias can also be responsible for Stokes-Adams' attacks.

These tachyarrhythmias may or may not be preceded by ventricular ectopics and may be the cause of sudden cardiac death. Width or rate of the QRS complex also does not predict the risk of tachyarrhythmias. Prolongation of the QT interval is a predictor of ventricular tachyarrhythmias. Patients with persistent ischemia or electrolyte imbalance are also at high risk. Tachyarrhythmias may be bradycardia dependent. Such tachyar-

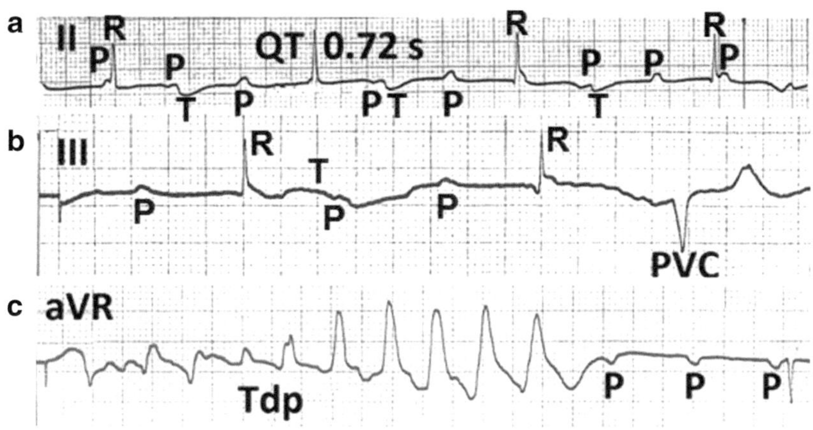

Fig. 13.5 (**a**) Lead II showing complete atrioventricular block with prolonged QT interval. (**b**) Lead III showing complete atrioventricular block with one ventricular ectopic (PVC). (**c**) Lead aVR showing transient torsades de pointes (Tdp). (Reproduced with permission from: Mittal SR. Stokes Adams Attacks. Cardiology Today. 2014;18:34–36)

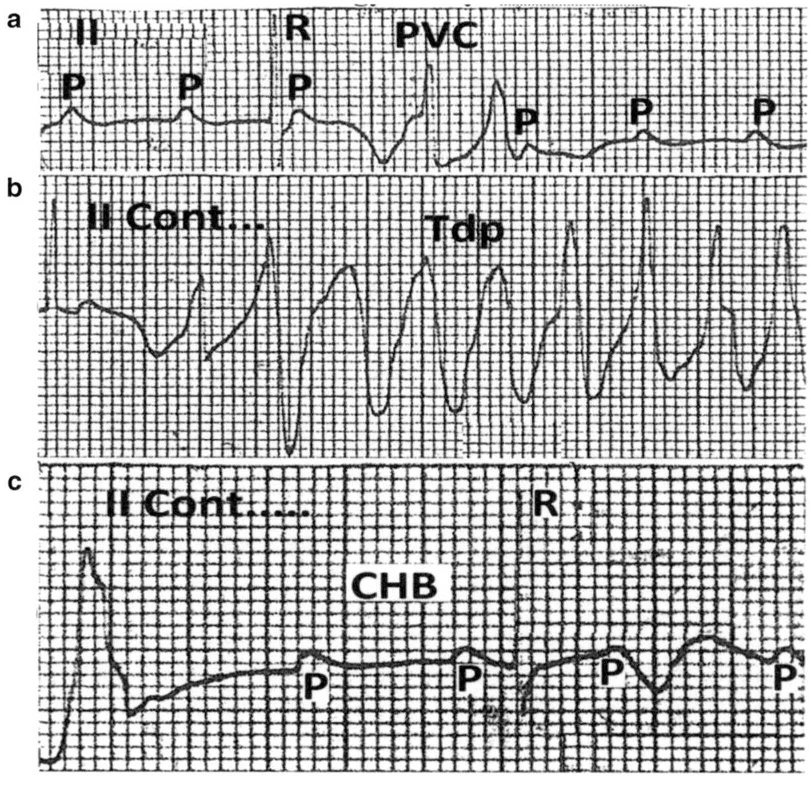

Fig. 13.6 (**a**) Lead II showing complete atrioventricular block with ventricular ectopics (PVC). (**b**) Continuous strips showing sudden onset of torsades de pointes. (**c**) Showing spontaneous termination of torsades de pointes. (Reproduced with permission from: Mittal SR. Stokes Adams Attacks. Cardiology Today. 2014;18:34–36)

rhythmias are likely to be controlled by pacing. In some cases tachyarrhythmias may be because of underlying cardiac pathology and may not respond unless the primary cause is managed.

ECG of patients with Stokes-Adams' attacks may show large, broad, and inverted T waves with marked prolongation of the QT interval (Figs. 13.8 and 13.9).

Such "giant" T wave inversion in the setting of complete atrioventricular block is almost pathognomonic of recent syncopal attack [1].

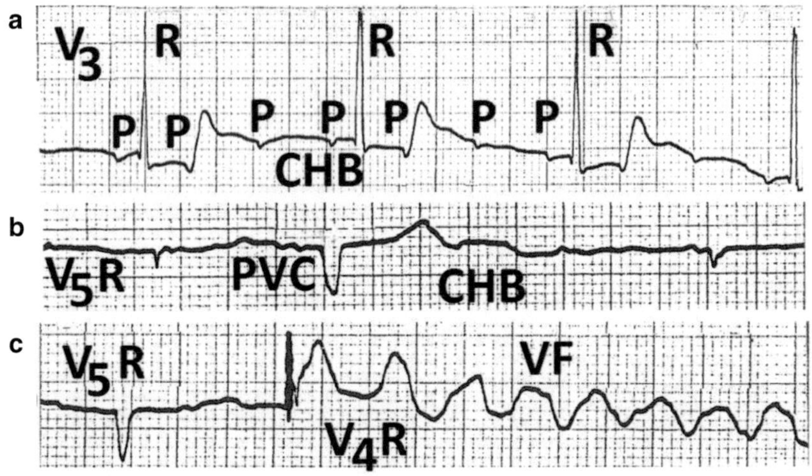

Fig. 13.7 (**a**) Lead V_3 showing complete atrioventricular block. (**b**) Lead V_5R showing complete atrioventricular block with ventricular ectopic (PVC). (**c**) Lead V_5R and V_4R showing sudden onset of VF. (Reproduced with permission from: Mittal SR. Stokes Adams Attacks. Cardiology Today. 2014;18:34–36)

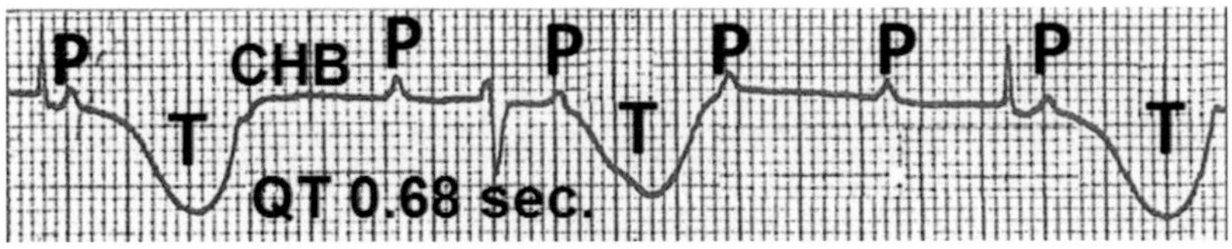

Fig. 13.8 Showing complete atrioventricular block with large, broad, and inverted T waves and prolonged QT interval. (Reproduced with permission from: Mittal SR. Stokes Adams Attacks. Cardiology Today. 2014;18:34–36)

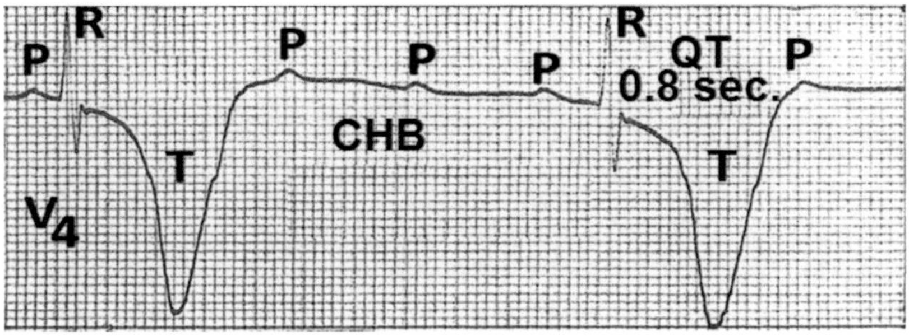

Fig. 13.9 Lead V_4 showing complete atrioventricular block with large, giant, and inverted T waves with prolonged QT interval. (Reproduced with permission from: Mittal SR. Stokes Adams Attacks. Cardiology Today. 2014;18:34–36)

Summary

- Stokes-Adams' attacks are episodes of syncope due to transient ventricular asystole, ventricular flutter, or fibrillation.
- In the setting of atrioventricular block, Stokes-Adams' attacks usually occur during transition from incomplete to complete atrioventricular block. Attacks are uncommon once the complete atrioventricular block has been stable for a month or more.
- Following Stokes-Adams' attacks due to transient atrioventricular block, transient ventricular flutter, or fibrillation, an electrocardiogram may be deceptively normal. In the setting of recent onset of complete atrioventricular block, ECG may show large, broad, and inverted T waves with prolonged QT interval.
- In the setting of complete atrioventricular block, Stokes-Adams' attacks may be due to tachyarrhythmias like polymorphic VT, torsades de pointes, or ventricular fibrillation.

These tachyarrhythmias may not be preceded by ventricular ectopics.

- Long sudden sinus pause without an escape focus can also produce Stokes-Adams attack.
- In the setting of acute anterior infarction, Stokes-Adams' attacks may be sudden and may not be preceded by first-degree or second-degree atrioventricular block.
- In the setting of complete atrioventricular block, the presence of intermittent capture beat or retrograde ventriculoatrial conduction suggests a lower risk of Stokes-Adams' attack.
- In the setting of complete atrioventricular block, risk of Stokes-Adams' attacks depends on the site of the escape rhythm. Risk progressively increases as the escape focus shifts distally from the atrioventricular node to the Purkinje network. More peripheral is the pacemaker, the slower is the rate and higher is the risk of Stokes-Adams' attack.
- If the escape focus is unstable and intermittently shifts the site, it also increases the risk of Stokes-Adams' attack.
- Atrioventricular dissociation does not produce Stokes-Adams' attacks.

MCQs

Q1. Electrocardiographic findings suggestive of a preceding Stokes-Adams attack include:

(a) Complete atrioventricular block
(b) Prolonged QTc interval
(c) Large, broad and inverted T waves
(d) All

Q2. During atrioventricular block, Stokes-Adams' attacks can occur:

(a) When second-degree atrioventricular block suddenly changes to complete atrioventricular block
(b) When there is paroxysm of ventricular flutter
(c) When there is slowing of ventricular escape rhythm
(d) All

Q3. Stokes-Adams' attacks must always be preceded by:

(a) First-degree atrioventricular block
(b) Second-degree atrioventricular block
(c) Third-degree atrioventricular block
(d) None

Q4. Which block in most susceptible for Stokes-Adams' attacks?

(a) Interatrial block
(b) Block at the level of the atrioventricular node
(c) Intra-His block
(d) Infra-His block

Q5. Stokes-Adams' attack can occur when:

(a) Sinus rate is less than 50/minute
(b) There is atrial fibrillation with a ventricular rate of 180/minute
(c) There is paroxysm of ventricular flutter
(d) There is transient ventricular fibrillation

Q6. Stokes-Adams' attacks are more likely to occur in:

(a) Recent onset of complete atrioventricular block
(b) Chronic complete atrioventricular block
(c) Mobitz type I, second-degree atrioventricular block
(d) All

Q7. Broad, inverted T waves with prolongation of QT interval are seen:

(a) In subarachnoid hemorrhage
(b) In cerebral hemorrhage
(c) Following Stokes-Adams' attacks
(d) All

Q8. Stokes-Adams' attacks are likely to occur in:

(a) Mobitz type II second-degree atrioventricular block
(b) Presence of Wenckebach phenomenon

(c) Atrial fibrillation with slow ventricular rate
(d) Atrial fibrillation with fast ventricular rate

Q9. Stokes-Adams' attacks are likely to occur in:

(a) Complete atrioventricular block with narrow QRS at a rate of 50/minute
(b) Complete atrioventricular block with broad QRS at a rate of 35/minute
(c) Complete atrioventricular block with unstable ventricular focus
(d) All

Q10. Stokes-Adams' attacks are more likely to occur in:

(a) Congenital complete atrioventricular block
(b) Complete atrioventricular block in the setting of acute inferior myocardial infarction
(c) Complete atrioventricular block in the setting of acute anterior myocardial infarction
(d) Ventricular bigeminy

Q11. Stokes-Adams' attack is more likely to occur in:

(a) Junctional tachycardia
(b) Ventricular tachycardia
(c) Atrial fibrillation with antegrade conduction over atrioventricular node
(d) None

Q12. Stokes-Adams' attacks can occur in:

(a) Sick sinus syndrome
(b) Concealed accessory pathway
(c) Ventricular trigeminy
(d) All

Q13. Stokes-Adams' attacks are more likely to occur in the setting of:

(a) Atrioventricular V dissociation
(b) Long sinus pause with slow escape rhythm
(c) Complete atrioventricular block with intermittent narrow QRS (capture beats)
(d) Complete atrioventricular block with intermittent retrograde (ventriculoatrial) conduction to atria

Q14. Stokes-Adams' attacks are most likely to occur in a complete atrioventricular block with the escape rhythm arising from:

(a) The atrioventricular node
(b) The His bundle
(c) The bundle branches
(d) The Purkinje network

Q15. Stokes-Adams' attacks in the setting of acute anterior infarction are always preceded by:

(a) First-degree atrioventricular block
(b) Wenckebach phenomenon
(c) Mobitz type II second-degree atrioventricular block
(d) None

Q16. Following an episode of Stokes-Adams' attack, the ECG may show:

(a) No abnormality
(b) Broad inverted T wave
(c) Myocardial infarction
(d) Any of above

Answers

(1) d; (2) d; (3) d; (4) d; (5) c, d; (6) a; (7) d; (8) a; (9) b, c; (10) c; (11) d; (12) a; (13) b; (14) d; (15) d; (16) d

Reference

1. Schamroth L. Atrioventricular (AV) block. In: Schamroth C, editor. An introduction to electrocardiography. Blackwell Science: New Delhi; 1988. p. 375–85.

Cardiac Pacing Systems and Their Normal Functioning

14

1 Nomenclature [1]

Cardiac pacing systems are described by a four-letter code. The first letter indicates the paced chamber. The second letter indicates the sensed chamber. The letter A is used for the atrium. The letter V is used for the ventricle. Letter D denotes dual chamber, i.e., atrium as well as ventricle. The word "biventricular" is used when both ventricles are paced. Right ventricle is paced from the right ventricular apex, and the left ventricle is paced through the lateral cardiac vein. The third letter indicates the response of the device to a sensed stimulus. Letter I represents inhibition of the pacing stimulus. Letter T indicates triggering of the pacing stimulus. Letter D represents dual (both). The fourth letter R represents rate responsiveness. This means the capacity of the system to increase or decrease pacing rate depending on metabolic demand. Commonly only three initial letters are used. For example:

AAI indicates atrial pacing, atrial sensing, and inhibitory mode.

VVI denotes ventricular pacing, ventricular sensing, and inhibitory mode.

DDD denotes dual chamber (atrium as well as ventricular) pacing.

- Dual chamber sensing
- Dual response (inhibitory as well as triggering) in response to a sensed stimulus

Applying a magnet over the pulse generator changes it to automatic pacing mode with inhibition of sensing function. Most of the currently available pacemakers are "programmable," i.e., pacing rate, sensing threshold, pacing current, and atrioventricular delay can be changed as needed.

In "unipolar" systems, one electrode (pole) is at the tip of the pacing lead and covering of the pulse generator acts as another pole. In "bipolar" systems, the second pole is close to the tip of the lead.

In the AAI, pacing lead is positioned in the right atrial appendage (Fig. 14.1a). In VVI pacing, lead is positioned in the apex of the right ventricle (Fig. 14.1b). In DDD pacemaker, atrial lead is positioned in the right atrial appendage, and ventricular lead is positioned in the right ventricular apex (Fig. 14.1c). In biventricular pacing, right ventricular lead is positioned in the right ventricular apex, and the left ventricular lead is positioned in the lateral cardiac vein through the coronary sinus (Fig. 14.1d). Left ventricular pacing can also be done from epicardial surface. This is rarely needed if transvenous access to right ventricle is not feasible.

S. Mittal, *Insights into Electrocardiograms with MCQs*,
https://doi.org/10.1007/978-981-99-0127-2_14

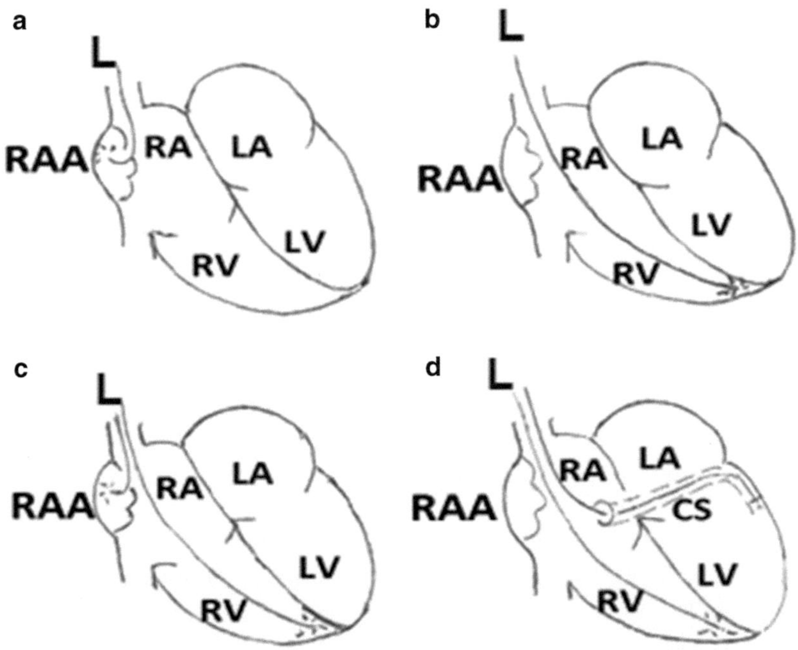

Fig. 14.1 Pacing lead positions in (**a**) AAI pacing, (**b**) VVI pacing, (**c**) DDD pacing, and (**d**) biventricular pacing. *RAA* right atrial appendage, *RA* right atrium, *RV* right ventricle, *LA* left atrium, *LV* left ventricle, *CS* coronary sinus, *L* pacing leads. (Reproduced with permission from: Mittal SR. Cardiac Pacing Systems. Cardiology Today. 2015; 19:148–154)

2 Indications for Various Pacemakers

AAI can be used in patients with sick sinus syndrome but normal atrioventricular conduction and normal ventricular functions. VVI pacemakers can be used in patients with chronic atrial fibrillation with complete atrioventricular block. VVI is not ideal for patients of sinus rhythm with complete atrioventricular block. VVI pacing results in atrioventricular dissociation with loss of atrial booster to the ventricular contraction. Synchronized contraction of the atria and the ventricles is necessary for optimal cardiac output. DDD pacemakers are ideal for patients of complete atrioventricular block specially so when the ventricular functions are impaired. Such patients need synchronized contractions of the atria and the ventricles for adequate cardiac output. DDD pacemakers are also ideal for patients with sick sinus syndrome with complete atrioventricular block. Biventricular pacing is beneficial in some patients of left bundle branch block with QRS duration of more than 120 ms and echocardiographic evidence of delay in depolarization of the left ventricle.

3 Surface ECG in a Patient with Implanted Pacemaker

Pacemaker discharge produces a sharp, narrow deflection (pacemaker spike). This spike is large in unipolar pacemakers (Fig. 14.2) because of large circuit between two poles (Fig. 14.3a).

On the other hand, spike is small in a bipolar pacemaker [2] due to small electric circuit between two poles (Figs. 14.3b and 14.4).

Careful inspection of all leads is needed to detect these small spikes (Fig. 14.5) which may be isoelectric in some leads.

The broad and changing amplitude of the spike should suggest varying degree of fusion of the spike with the intrinsic QRS complex of the patient (Fig. 14.6).

Regular, uniformly broad spikes can occur due to leakage of current in the myocardium (Fig. 14.7).

In pacing from the right ventricular apex, mean frontal plane axis of the spike is directed upward, i.e., the spike is negative in leads II, III, and aVF (Fig. 14.2).

Pacing from the right ventricular outflow tract (RVOT) causes inferior axis of the pacing spike,

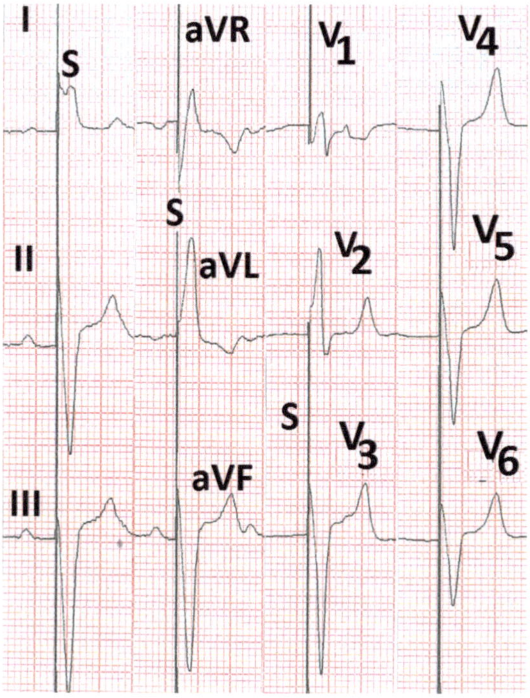

Fig. 14.2 Electrocardiogram from a patient with unipolar VVI pacing, showing large spikes (S) and left axis of the QRS complex. (Reproduced with permission from: Mittal SR. Cardiac Pacing Systems. Cardiology Today. 2015;19:148–154)

i.e., the pacing spike is positive in leads II, III, and aVF (Fig. 14.8).

In AAI pacemakers, the pacing spike is followed by an atrial depolarization (P wave). This is followed by a normal PR interval and a normal QRS complex (Fig. 14.9).

In VVI pacing there is atrioventricular dissociation with changing PR interval. Ventricular pacing has no relation with the P wave. Pacemaker spike is followed by a broad and bizarre QRS complex. When pacing is done from the right ventricular apex, QRS vector is directed upward and to the left with the result that the mean frontal plane QRS axis is in the left upper quadrant (Fig. 14.10a). In pacing from the right ventricular outflow, mean frontal plane QRS axis is directed inferiorly (Fig. 14.10b).

Pacing from the right ventricle causes slow depolarization of the left ventricle, resulting in left bundle branch block configuration. In DDD pacemakers (sensing the right atrial depolarization and pacing the right ventricle), ECG looks like VVI pacing except that each ventricular depolarization is preceded by a normal P wave and a fixed PR interval (Fig. 14.11).

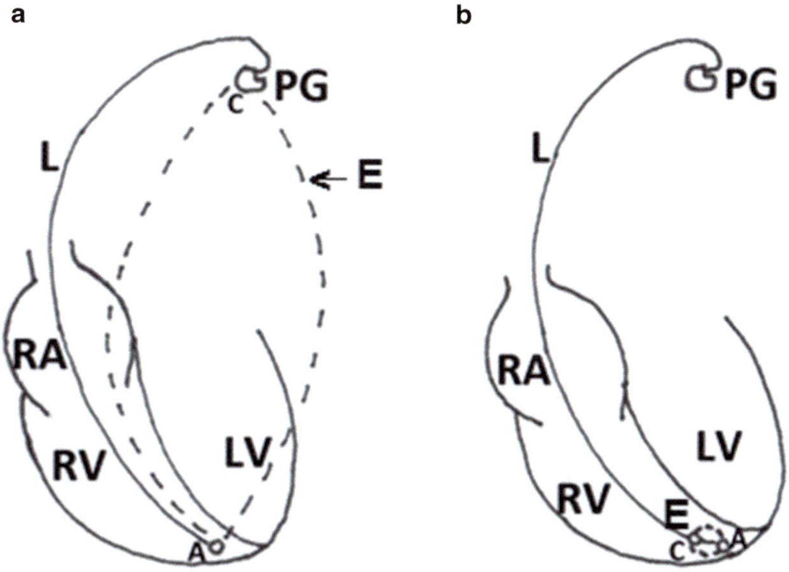

Fig. 14.3 Diagram showing electric circuit (E) between anode (A) and cathode (C) in unipolar (**a**) and bipolar (**b**) VVI pacing. *PG* pulse generator, *L* lead, *RA* right atrium, *RV* right ventricle, *LV* left ventricle. (Reproduced with permission from: Mittal SR. Cardiac Pacing Systems. Cardiology Today. 2015;19:148–154)

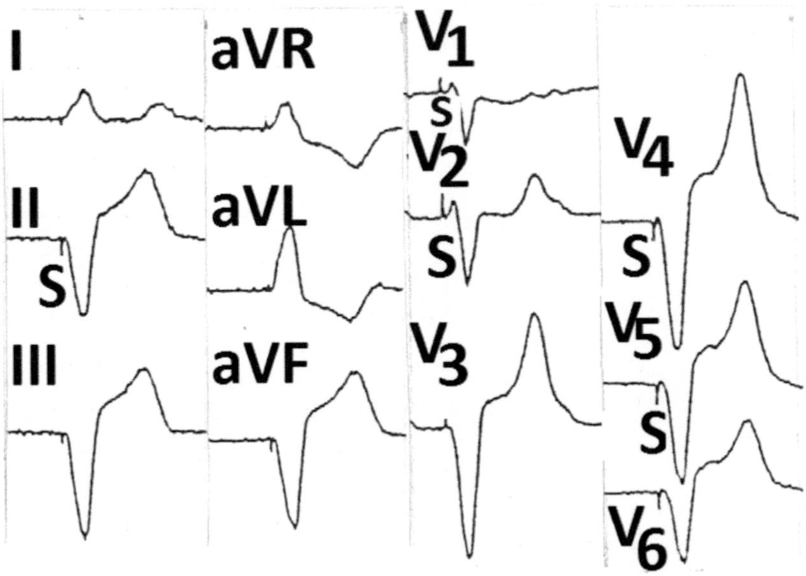

Fig. 14.4 Electrocardiogram from a patient with bipolar pacing, showing relatively small spike (marked -S). (Reproduced with permission from: Mittal SR. Cardiac Pacing Systems. Cardiology Today. 2015;19:148–154)

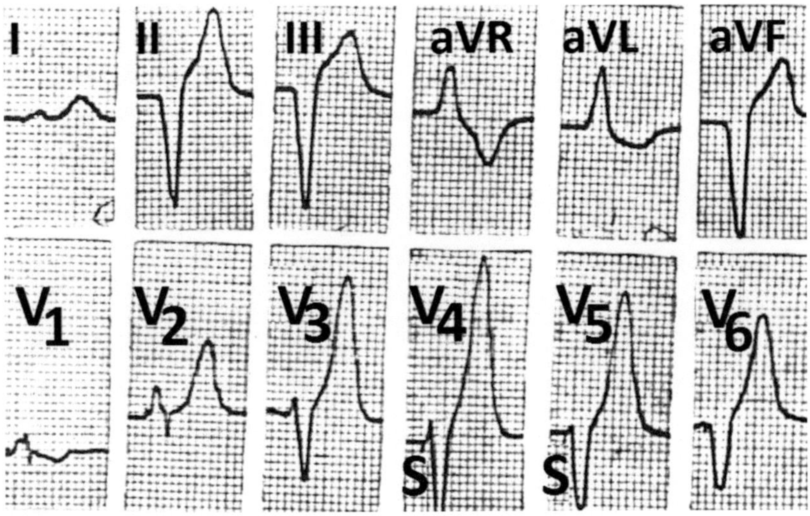

Fig. 14.5 Electrocardiogram from a patient with bipolar pacing, showing very small spikes clearly visible only in leads V4 and V5 (marked -S). (Reproduced with permission from: Mittal SR. Cardiac Pacing Systems. Cardiology Today. 2015;19:148–154)

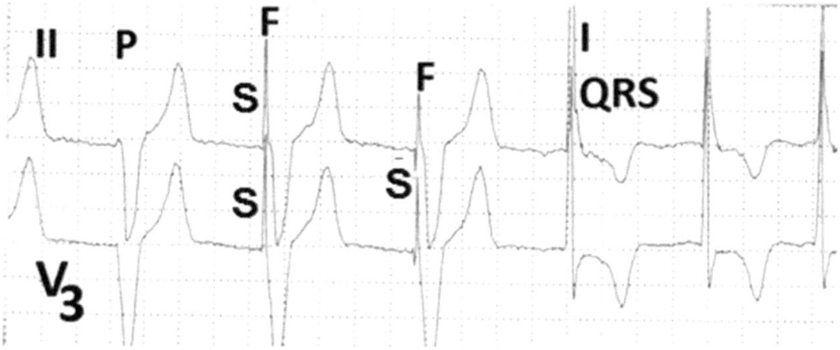

Fig. 14.6 Holter recording from a patient of atrial fibrillation showing paced beats (P) and fusion beats (F) with wide spike (marked -S) with variable amplitude and intrinsic QRS (I). (Reproduced with permission from: Mittal SR. Cardiac Pacing Systems. Cardiology Today. 2015;19:148–154)

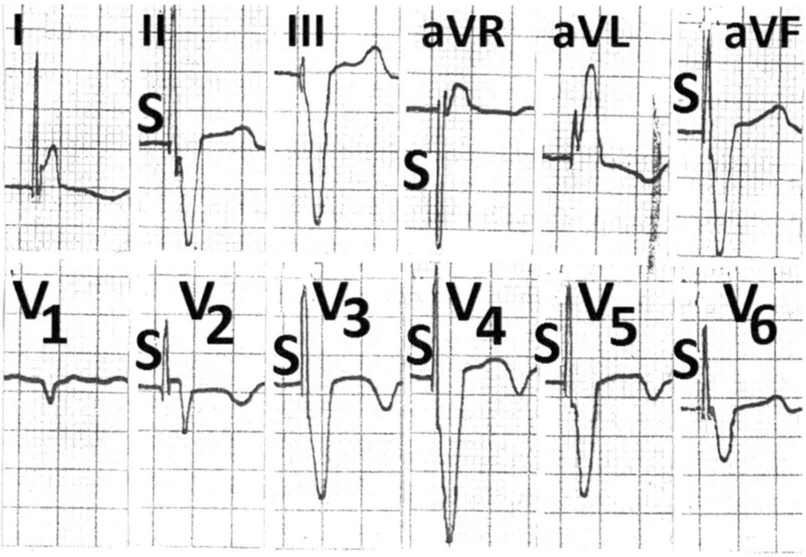

Fig. 14.7 Electrocardiogram from a patient with permanent pacemaker with regular wide spikes (S). (Reproduced with permission from: Mittal SR. Cardiac Pacing Systems. Cardiology Today. 2015;19:148–154)

The QRS complex is broad and bizarre. If DDD pacemaker paces the atrium as well as the ventricle, there is an atrial pacing spike which is followed by a P wave, a fixed PR interval, second pacing spike from the ventricular pacing, and then a broad QRS (Fig. 14.12).

Biventricular pacing produces a single spike from simultaneous depolarization of the two ventricles. QRS is relatively less broad due to simultaneous depolarization of both the ventricles (Fig. 14.13).

Mean frontal plane QRS axis is to the right, and the QRS has right bundle branch block morphology due to dominance of the left ventricular vector. Left ventricular epicardial pacing also produces right bundle branch block morphology (Figs. 14.14 and 14.15).

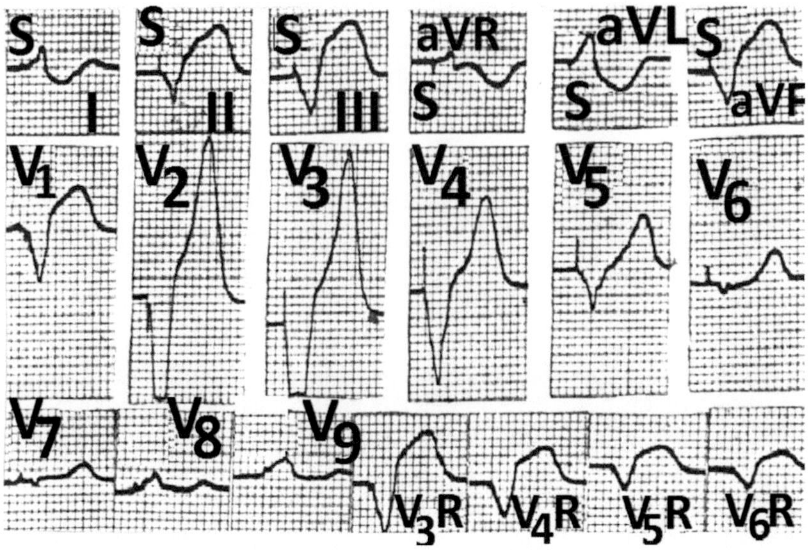

Fig. 14.8 Electrocardiogram from a patient with permanent pacemaker with inferior axis of the pacing spike (marked -S). (Reproduced with permission from: Mittal SR. Cardiac Pacing Systems. Cardiology Today. 2015;19:148–154)

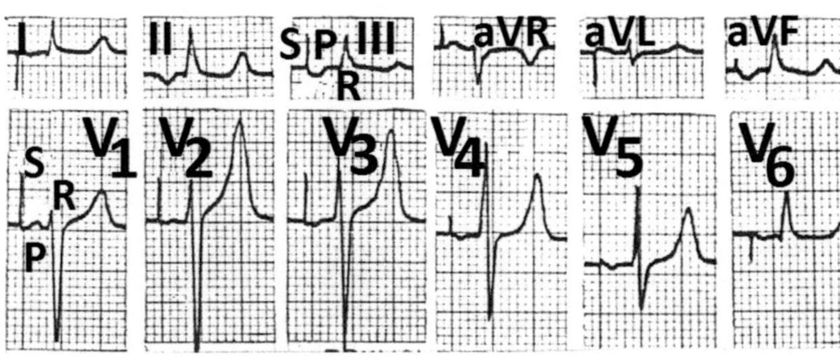

Fig. 14.9 Electrocardiogram from a patient with AAI pacemaker. Spike (S) is followed by a P wave, normal PR interval, and a narrow QRS(R). (Reproduced with permission from: Mittal SR. Cardiac Pacing Systems. Cardiology Today. 2015;19:148–154)

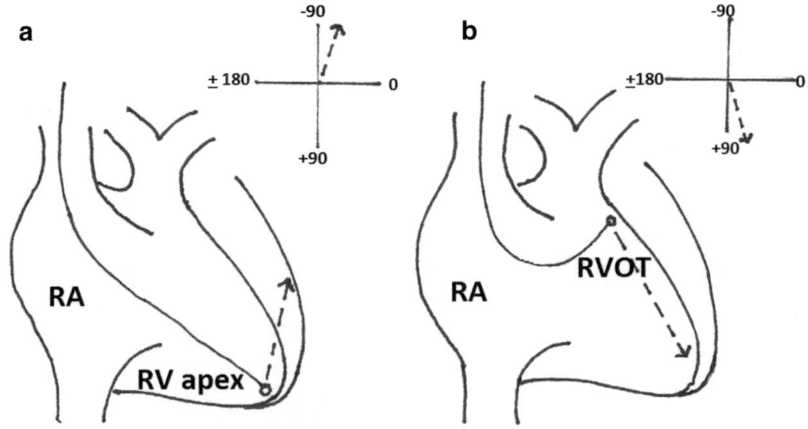

Fig. 14.10 Diagram to illustrate the mean frontal plane QRS axis in VVI pacing. (**a**) Pacing from the right ventricular apex causes the superior axis. (**b**) Pacing from the right ventricular outflow tract (RVOT) causes inferior axis. (Reproduced with permission from: Mittal SR. Cardiac Pacing Systems. Cardiology Today. 2015;19:148–154)

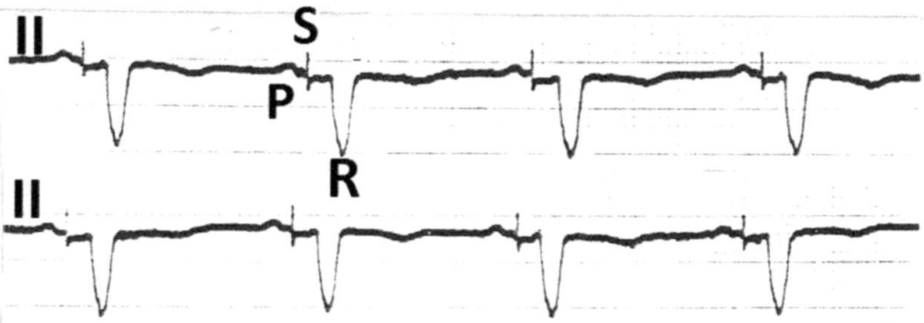

Fig. 14.11 Electrocardiogram from a patient with DDD pacing, showing sensing of each P wave, followed by a pacing spike (S) and then a QRS(R). (Reproduced with permission from: Mittal SR. Cardiac Pacing Systems. Cardiology Today. 2015;19:148–154)

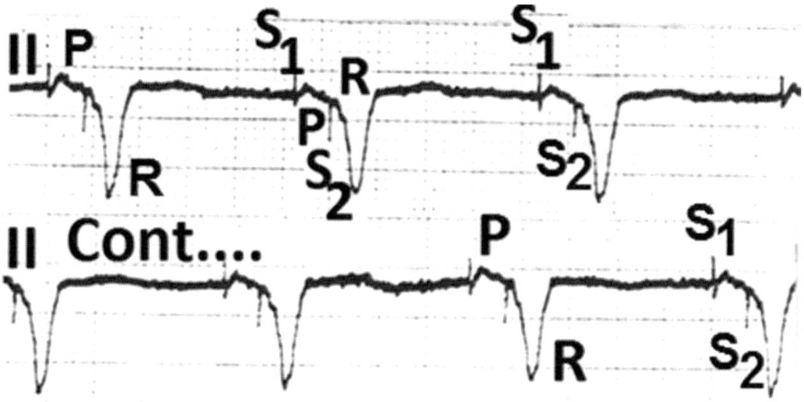

Fig. 14.12 Electrocardiogram from a patient with DDD pacemaker showing an atrial pacing spike (S1) followed by a P wave, a spike from ventricular pacing spike (S2), and then a QRS (marked R). (Reproduced with permission from: Mittal SR. Cardiac Pacing Systems. Cardiology Today. 2015;19:148–154)

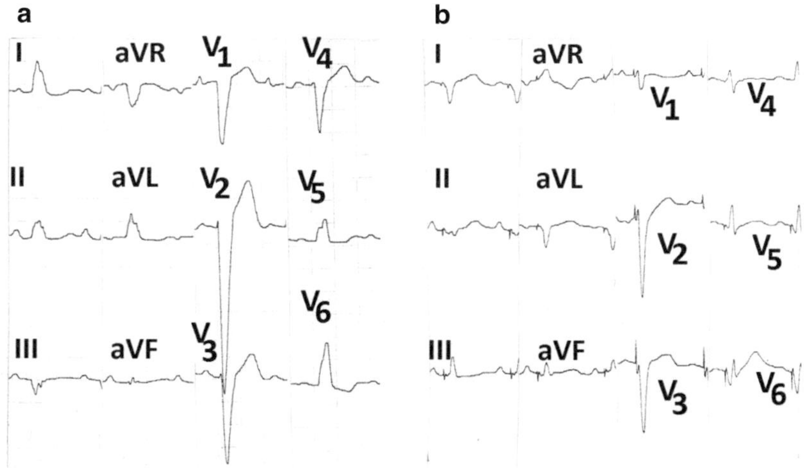

Fig. 14.13 (**a**) Electrocardiogram showing left bundle branch block, (**b**) electrocardiogram after biventricular pacing. (Reproduced with permission from: Mittal SR. Cardiac Pacing Systems. Cardiology Today. 2015;19:148–154)

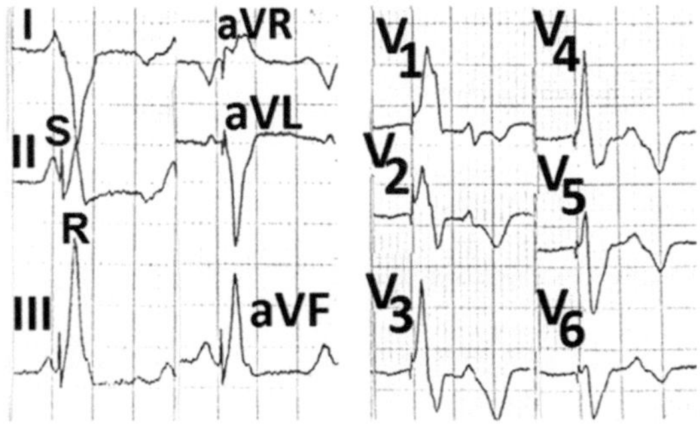

Fig. 14.14 Electrocardiogram from a patient with left ventricular epicardial pacing. S, ventricular pacing spike; R, QRS complex. (Reproduced with permission from: Mittal SR. Cardiac Pacing Systems. Cardiology Today. 2015;19:148–154)

4 Normal Functioning of the Pacemaker

4.1 Normal Sensing

Each QRS (whether produced by the intrinsic rhythm of the patient or by an ectopic beat or produced by the pacemaker) is sensed, and it resets the pacemaker cycle for the preset RR interval. This shows that the sensing function of the pacemaker is normal. Electrocardiogram (Fig. 14.16a) shows that the pacemaker senses the QRS of the sinus beat (S) and postpones the next pacing stimulus (marked *). If the pacemaker did not

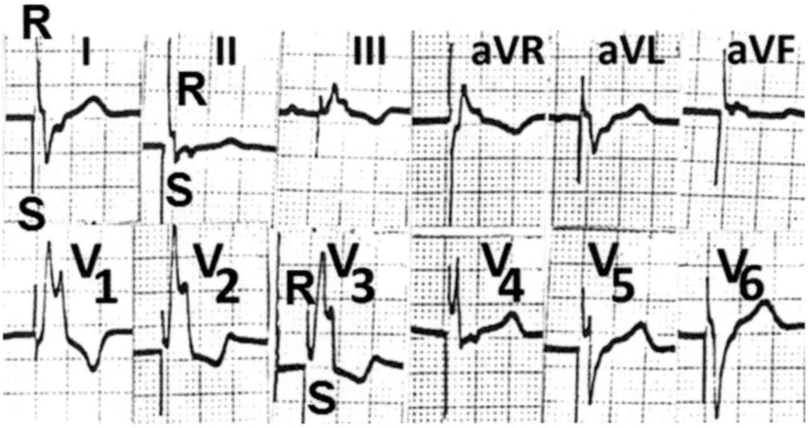

Fig. 14.15 Electrocardiogram from a patient with unipolar left ventricular epicardial pacing. S, ventricular pacing spike; R, QRS complex. (Reproduced with permission from: Mittal SR. Cardiac Pacing Systems. Cardiology Today. 2015;19:148–154)

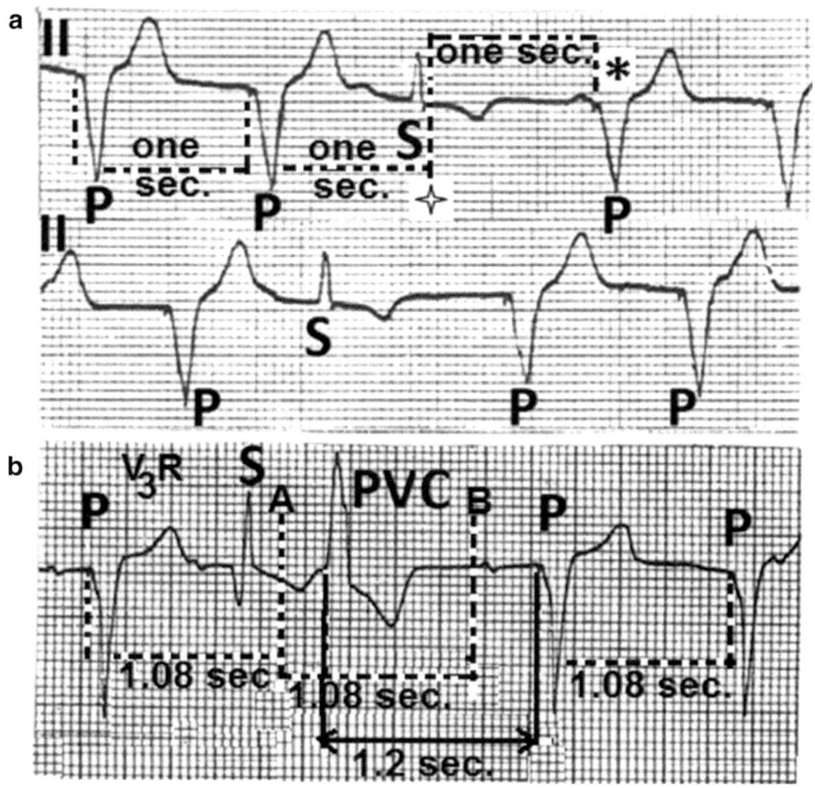

Fig. 14.16 Electrocardiogram (**a**) showing normal sensing (S) and pacing (P). (**b**) Electrocardiogram showing normal sensing of sinus beats (S) as well as premature ventricular contraction (PVC) and normal pacing. (Reproduced with permission from: Mittal SR. Cardiac Pacing Systems. Cardiology Today. 2015;19:148–154)

sense the intrinsic QRS (S), next pacemaker stimulus should have come at the fixed interval marked ✧). Figure 14.16b shows that the pacemaker senses the sinus beat (marked S) and postpones the next pacing stimulus. During this period, there is a ventricular ectopic (marked PVC). Pacemaker again senses the QRS complex of this ectopic and again postpones the pacing stimulus. If the pacemaker did not sense the QRS of the sinus beat, next pacing stimulus should have come at the point marked A. If the pacemaker did not sense the QRS of the ventricular ectopic, next pacing stimulus should have come at the point marked B.

Electrocardiogram in Fig. 14.17 also shows postponement of pacing stimulus by sensing of the QRS of a ventricular ectopic (marked PVC) and again sensing of QRS of a sinus beat (marked S). If the sensing function of the pacemaker was not normal, pacing spikes should have come at points marked A and B.

Pause after a sensed sinus beat is slightly longer than the interval between two paced beats. It is called hysteresis (Fig. 14.18).

It is a normal function. This is because the pacing system waits for a little more time than the preset RR interval, for the normal heart beat to come and maintain the normal sinus rhythm. Another important point to remember is that a period of >0.04 second is required for the intrinsic ventricular activation to reach the pacemaker lead positioned in the right ven-

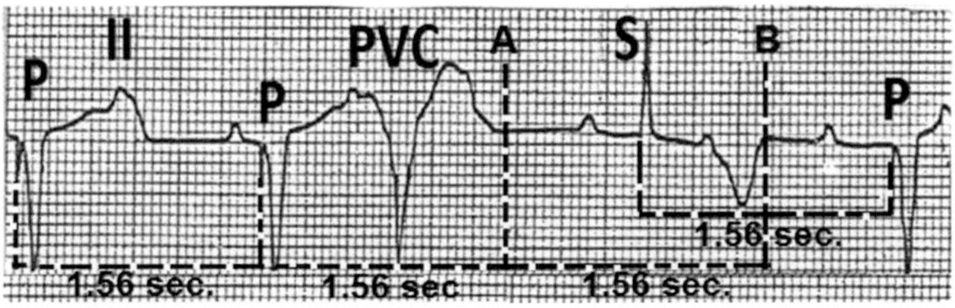

Fig. 14.17 Electrocardiogram showing normal sensing of a premature ventricular contraction (marked PVC) followed by sensing of the sinus beat (marked S) and then normal pacing (marked P). (Reproduced with permission from: Mittal SR. Cardiac Pacing Systems. Cardiology Today. 2015;19:148–154)

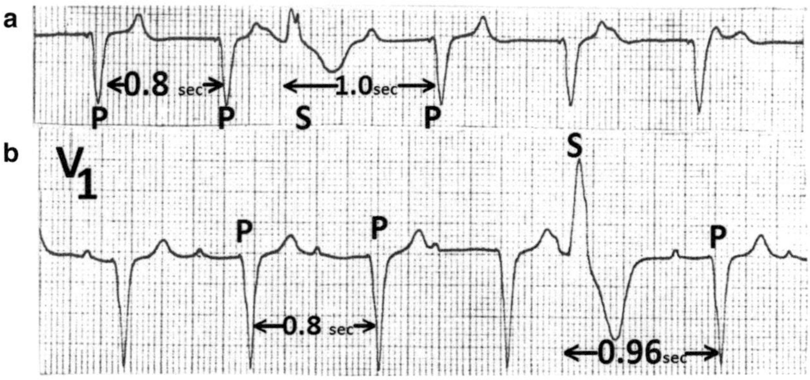

Fig. 14.18 (**a** and **b**) Electrocardiograms from two patients (**a** and **b**) showing longer pause after a sensed beat (S) (hysteresis). P—paced beats. (Reproduced with permission from: Mittal SR. Cardiac Pacing Systems. Cardiology Today. 2015;19:148–154)

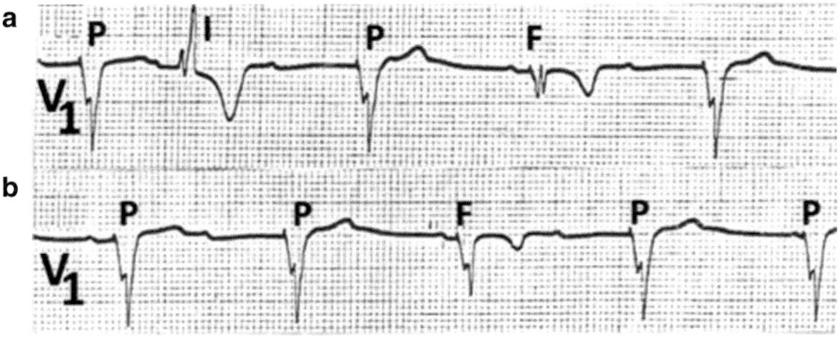

Fig. 14.19 Electrocardiograms (**a** and **b**) showing paced beats (P), intrinsic QRS (I), and fusion beats (F). (Reproduced with permission from: Mittal SR. Cardiac Pacing Systems. Cardiology Today. 2015;19:148–154)

tricular apex and to be sensed by the pacemaker [3]. If the duration between a sinus beat and next scheduled paced beat is less than 0.04 second, both may occur simultaneously producing a fusion beat [4]. If the intrinsic QRS comes first, the initial part of QRS is produced by the intrinsic beat and the later part by the paced QRS. (Fusion beat) On the other end, if paced beat comes first, initial part of the fusion beat is formed by the paced beat, and the later part is formed by the intrinsic QRS (Fig. 14.19a, b).

If the pacemaker spike comes slightly later, it only deforms intrinsic QRS (pseudo fusion beat) (Fig. 14.20a, b). Such fusion beats occur with normal sensing of the pacemaker and do not suggest any error in the sensing function of the pacemaker.

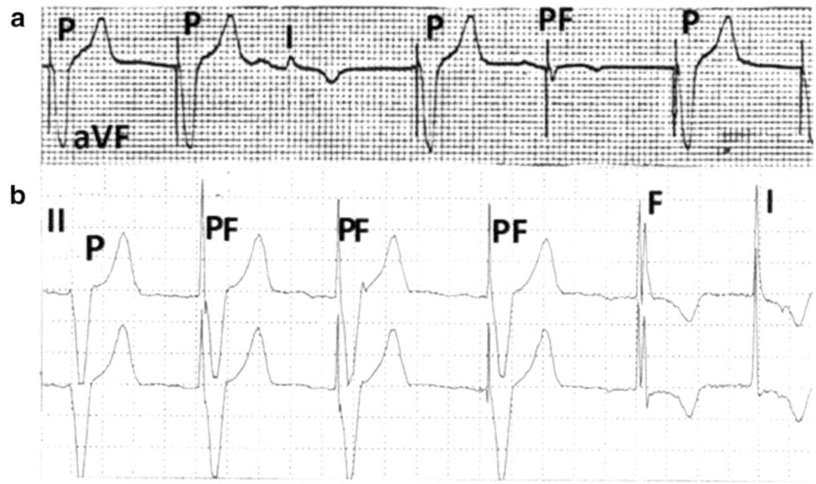

Fig. 14.20 (**a** and **b**) Showing paced beats (P), fusion beats (F), pseudo fusion beats (PF), and intrinsic QRS (I). (Reproduced with permission from: Mittal SR. Cardiac Pacing Systems. Cardiology Today. 2015;19:148–154)

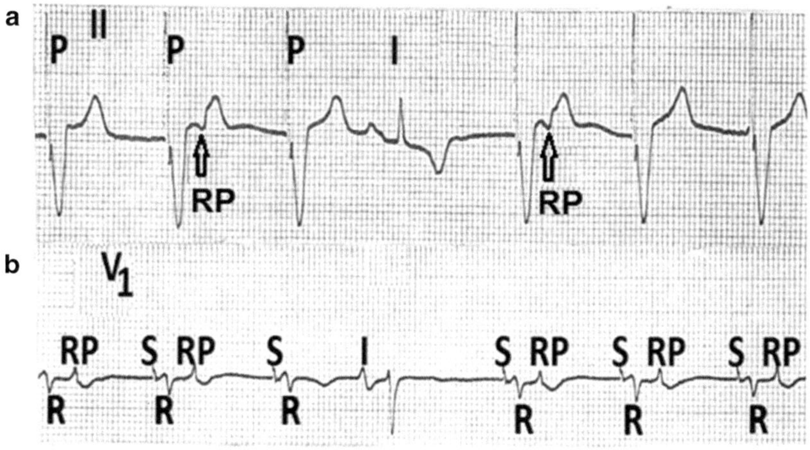

Fig. 14.21 Electrocardiograms. (**a**) Lead II showing intermittent retrogradely conducted P waves (marked RP) after the paced beats (marked P). I—intrinsic QRS. (**b**) Lead V_1 showing pacing spike (S), paced QRS (R), retrogradely conducted P waves (RP), and intrinsic heart beat (I). (Reproduced with permission from: Mittal SR. Cardiac Pacing Systems. Cardiology Today. 2015;19:148–154)

4.2 Normal Pacing

If normal sinus beat does not come in the preset RR interval, the pacemaker system triggers a stimulus and causes ventricular depolarization. RR interval between two paced beats is constant, depending on the minimal rate fixed at the time of programming. Figures 14.16 and 14.17 show normal pacing function. If the pacemaker stimulus does not occur after the preset RR interval in absence of any intrinsic QRS, it suggests pacing dysfunction.

4.3 Retrograde Conduction of Paced QRS

Paced QRS may conduct retrogradely over the atrioventricular node producing an abnormal P wave falling on the ST segment or the T wave following the paced QRS (Fig. 14.21).

This occurs intermittently whenever the atrioventricular node is not refractory due to depolarization by sinus P wave. Such retrograde P wave may again conduct antegradely producing a narrow QRS (echo beat). Regular sensing of such

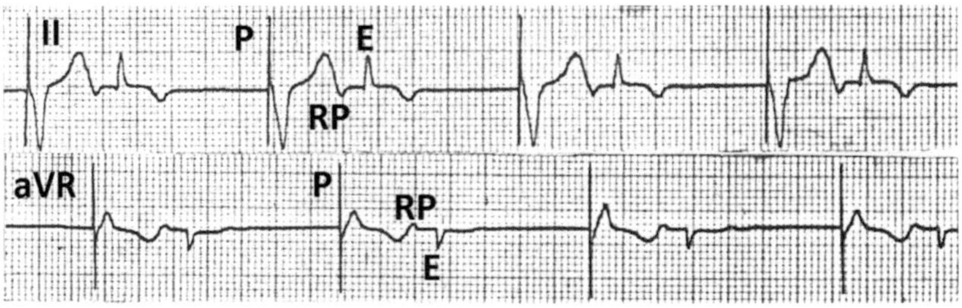

Fig. 14.22 Electrocardiogram showing regular retrograde P wave (RP) followed by antegrade conduction producing a narrow QRS (echo beat—E). (Reproduced with permission from: Mittal SR. Cardiac Pacing Systems. Cardiology Today. 2015;19:148–154)

beat may result in reduced rate of pacing (Fig. 14.22).

Retrograde conduction of the paced QRS complex and "echo beats" suggests that there is no anatomical pathology in the atrioventricular node.

Summary

- Nomenclature of permanent pacemakers has four letters. The first letter indicates the paced chamber (A, atrium; V, ventricle; D, dual). The second letter indicates the sensed chamber. The third letter indicates response to sensing (I, inhibited; T, triggered). The fourth letter indicates the presence or absence of rate adaptive pacing (0, none, R, rate modulation). Modes commonly used are VVI or DDD.
- In a VVI pacemaker, there is a single lead positioned in the right ventricle. It senses spontaneous ventricular depolarization and gives an impulse if there is no spontaneous ventricular depolarization for the preset interval. Atria and ventricle contract independently. There is no atrioventricular synchronization. On electrocardiogram each pacing spike is followed by a paced QRS. There is no relation between the P waves and the QRS complexes. However, paced QRS complexes may be followed by a retrogradely conducted P wave falling on the ST segment or on the T wave following the paced QRS. If there is a spontaneous QRS, it is sensed by the pacemaker, and the subsequent paced beat is postponed for the preset minimum pacing RR interval.
- In DDD → (dual chamber) pacing, there are two leads. One is in the right atrium and the other is in the right ventricle. Normally, the right atrial P wave is sensed. It is followed by a prefixed atrioventricular delay, a ventricular pacing spike, and a paced QRS. In case of sinus bradycardia, the sequence of events on the surface ECG is atrial pacing spike →, paced P wave → preset, atrioventricular delay →, ventricular pacing spike →, and paced QRS. Atrioventricular synchrony is maintained. There is also an upper limit for atrial sensing to avoid fast ventricular rate during atrial fibrillation. DDD pacing is preferred in most circumstances as it maintains atrioventricular synchrony. VVI pacing can be useful in the presence of atrial fibrillation to avoid fast ventricular rate.
- In biventricular pacing one lead is positioned in the right ventricle, and the other is positioned in a tributary of the coronary sinus overlying the left ventricle. This is performed in cases of left bundle branch block with a wide QRS. The aim is to synchronize the contractions of the two ventricles.
- In unipolar pacing systems, pacing spikes are large. In bipolar pacing systems, pacing spikes are very small and may not be visible in some leads.
- In a case of pacing from right ventricular apex, QRS shows left bundle branch block configuration with left axis deviation. In a case of pac-

ing from right ventricular outflow tract, QRS shows left bundle branch block configuration with inferior axis.

- Biventricular pacing in a case of left bundle branch block produces a QRS which is narrower than that before pacing.

MCQs

Q1. Which of the following is correct for the VOO mode of pacing?

(a) Ventricular pacing
(b) Ventricular sensing
(c) Inhibition from ventricular sensing
(d) All

Q2. Applying magnet over the pulse generator of a VVI pacemaker changes it to:

(a) VOO mode
(b) VVO mode
(c) VOI mode
(d) None

Q3. In VVI pacemaker, which parameter is not programmable?

(a) Atrioventricular delay
(b) Sensing threshold
(c) Pulse width
(d) Pacing current

Q4. Large pacing spike suggests:

(a) Unipolar pacing
(b) Bipolar pacing
(c) DDD pacing
(d) Biventricular pacing

Q5. Normal QRS in a patient with pacemaker suggests:

(a) AAI pacing
(b) DDD pacing
(c) Biventricular pacing
(d) VVI pacing

Q6. Right bundle branch block pattern in a patient with pacemaker suggests:

(a) Biventricular pacing
(b) Pacing from the RVOT
(c) Left ventricular epicardial pacing
(d) DDD pacing

Q7. AAI pacemaker is useful in:

(a) Sinus pause
(b) Sick sinus syndrome with complete atrioventricular block
(c) Atrial flutter with complete Atrioventricular block
(d) Atrial fibrillation

Q8. Pacemaker of choice for a patient of complete atrioventricular block with left ventricular diastolic dysfunction is:

(a) AAI
(b) VVI
(c) DDD
(d) Biventricular

Q9. Pacemaker of choice in chronic atrial fibrillation with complete atrioventricular block in:

(a) AAI
(b) AAI(R)
(c) VVI
(d) DDD

Q10. Double pacing spikes suggest:

(a) Unipolar pacing
(b) Bipolar pacing
(c) DDD pacing
(d) Biventricular pacing

Q11. Left bundle branch block configuration with inferior axis suggests:

(a) Pacing from the right ventricular apex
(b) Pacing from the right ventricular outflow tract

(c) Left ventricular epicardial pacing
(d) Biventricular pacing

Q12. Time taken for the normal activation of the heart to reach the right ventricular apex is:

(a) 0.01 second
(b) 0.02 second
(c) 0.03 second
(d) 0.04 second

Q13. Slightly longer interval after a sensed beat suggests:

(a) Sensing failure
(b) Pacing failure
(c) Hysteresis
(d) None

Q14. Fusion beat suggests:

(a) Sensing failure
(b) Pacing failure
(c) Battery depletion
(d) Normal function

Q15. In VVI pacing, retrograde activation of the atria can produce:

(a) Atrial fibrillation
(b) Atrial tachycardia
(c) Atrial flutter
(d) Echo beat

Q16. Pacing spike is small in:

(a) Unipolar pacing
(b) Bipolar pacing
(c) Dual chamber pacing
(d) Biventricular pacing

Q17. DDD pacing means:

(a) Atrial as well as ventricular sensing and pacing
(b) Ventricular sensing and pacing
(c) Atrial sensing and ventricular pacing
(d) Ventricular sensing and atrial pacing

Q18. VVI pacing means:

(a) Ventricular pacing
(b) Ventricular sensing
(c) Inhibition by ventricular sensing
(d) All

Q19. DDD pacing:

(a) Preserves atrioventricular synchrony
(b) Causes atrioventricular dissociation
(c) Can cause 1:1 ventriculoatrial association
(d) All

Q20. In biventricular pacing, one lead is placed in the right ventricular apex and the; the other lead is placed:

(a) In the left ventricular apex
(b) On the left ventricular epicardium
(c) In tributary of the coronary sinus
(d) In the right atrium

Q21. Pacing from the right ventricular apex produces:

(a) Left bundle branch block configuration of QRS with left axis
(b) Left bundle branch block configuration of QRS with inferior axis
(c) Left bundle branch block configuration of QRS with right axis
(d) Right bundle branch block configuration of QRS with left axis

Q22. Biventricular pacing in a case of left bundle branch block can produce:

(a) Widening of QRS
(b) Narrowing of QRS
(c) Right bundle branch block
(d) None

Answers

(1) A, (2) A, (3) A, (4) A, (5) A, (6) C, (7) A, (8) C, (9) C, (10) C, (11) B, (12) D, (13) C, (14) D, (15) D, (16) B, (17) A, (18) D, (19) A, (20) C, (21) A, (22) B

References

1. Olshansky B, Chang MK, Pogowizd SM, Goldschlager N. Cardiac pacing and pacemaker rhythms. In: Olshansky B, Chung MK, Pogowizd SM, Goldschlager N, editors. Arrhythmia Essentials. Burlington: Jones Bartlett Learning; 2012. p. 240–60.
2. Vijayraman P, Ellenbogen KA. Bradyarrhythmias and pacemakers. In: Fuster V, Walsh RA, Harrington RA, editors. Hurst's the heart. New York: Mc Graw Hill; 2011. p. 1025–57.
3. Haisty WK, Lim TH, Wagner GS. Artificial cardiac pacemakers. In: Wagner GS, editor. Marriot's practical electrocardiography. Philadelphia: Wolters Kluwer; 2001. p. 423–44.
4. Swerdlow CD, Wang PJ, Zipes DP. Pacemakers and implantable cardioverter-defibrillators. In: Mann DP, Zipes DP, Libby P, Bonow RO, editors. Braunwald's heart disease. Philadelphia: Elsevier; 2015. p. 721–42.

15 Pacemaker Malfunctions

1 Failure of Output [1] (Absence of the Pacing Spike)

In bipolar systems, the pacing artefact is small, especially if it is isoelectric in some surface leads. Simultaneous recording of multiple leads may help in the detection of pacing spike in such cases. True absence of pacing spike (output) can occur in the following situations:

(i). Intermittent delay of the pacing spike
  – Oversensing of a tall T wave (Figs. 15.1 and 15.2) or a P wave can reset the pacemaker cycle and delay the next pacing spike.

(ii). Transient loss of the pacing spike

This occurs when the pulse generator senses electrical potentials generated by sources other than those generated in the heart.
  – Oversensing of myopotentials (electric potentials generated by muscles in the vicinity of the pulse generator)—it is common in unipolar systems.
  – A loose set screw may cause electrical potentials that are sensed by the pulse generator.
  – Electromagnetic signals from welding and electrocautery.

(iii). Total absence of pacing spike
  – Circuitry failure—there is no delivery of the pacing output of the device. Magnet application changes pacemaker to asynchronous mode and restores pacing in cases of oversensing but does not restore pacing spike in circuitry failure.

S. Mittal, *Insights into Electrocardiograms with MCQs*,
https://doi.org/10.1007/978-981-99-0127-2_15

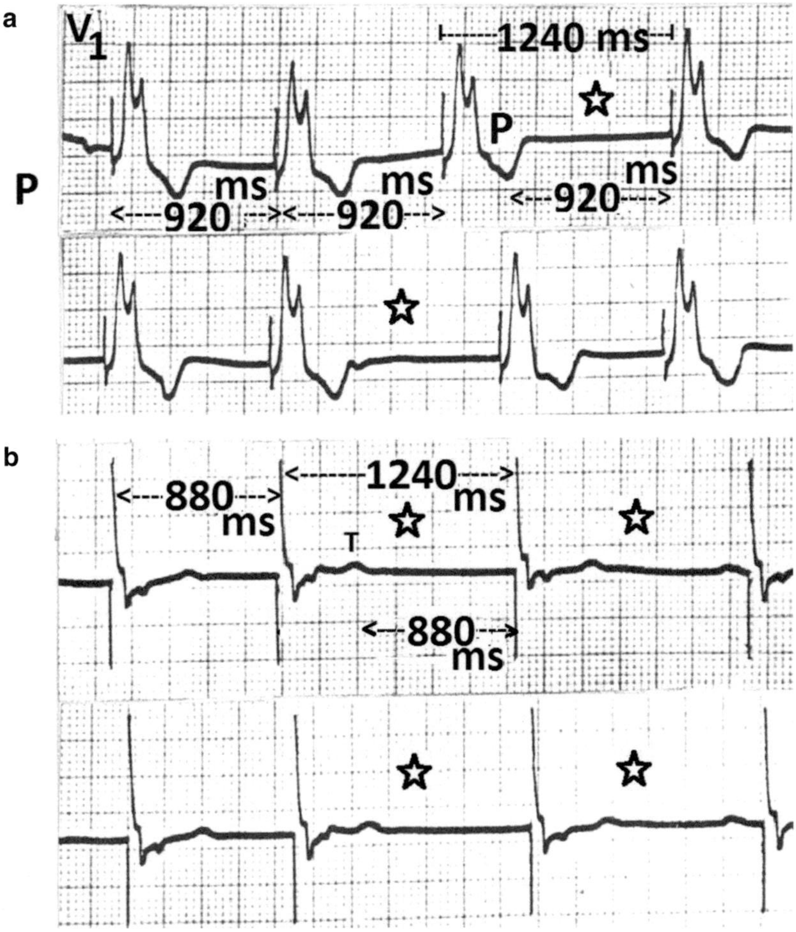

Fig. 15.1 (**a**) Electrocardiograms from a patient with left ventricular pacing and (**b**) right ventricular pacing showing pause due to oversensing of the T wave (marked as ☆). In (**a**) regular pacing interval (RR interval) is 920 ms. Intermittent oversensing of the T wave resets the pacemaker cycle so that the next spike comes 920 msec after the sensed T wave (RR interval 1240 ms). In (**b**) regular pacing interval is 880 msec. Oversensing of the T wave results in an RR interval of 1240 sec. (Reproduced with permission from: Mittal SR. Pacemaker Malfunction. Cardiology Today. 2015;19:180–186)

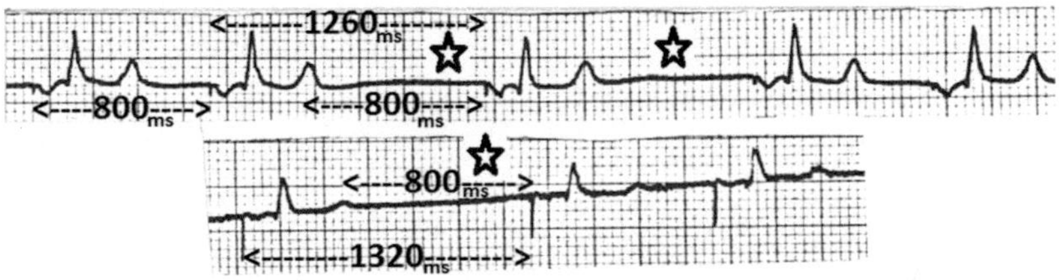

Fig. 15.2 Electrocardiogram from a patient of AAI pacemaker showing pause due to oversensing of the T wave (marked as ☆). The regular pacing interval was 880 msec. Sensing of the T wave resulted in resetting of the pacing cycle with the result that the next pacing spike comes 800 msec after the T wave. Next pacing spike occurs after 1320 ms. (Reproduced with permission from: Mittal SR. Pacemaker Malfunction. Cardiology Today. 2015;19:180–186)

2 Failure to Capture (Pacing Failure)

Pacing stimulus is present but it fails to depolarize the myocardium (Fig. 15.3).

Sometimes artefacts can resemble pacemaker spikes. These are irregularly irregular and are not followed by a QRS complex [2] (Fig. 15.4).

Other causes of pacing failure include the following:

(i) Too low programmed voltage output. Normally the myocardial threshold increases for a few weeks after pacing due to fibrosis around the tip of the electrode. If the initial output is low, it may become inadequate after a few weeks.

(ii) Breaking of pacing lead insulation.

(iii) Fracture of lead conductor—pacing failure can be intermittent.

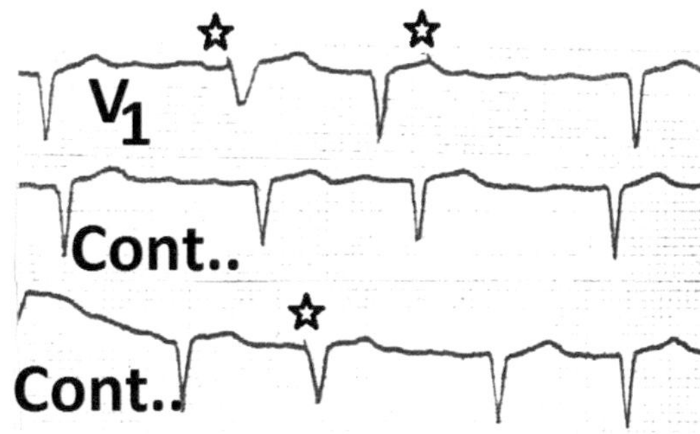

Fig. 15.4 Electrocardiogram shows artefacts simulating pacing spike (marked as ☆). (Reproduced with permission from: Mittal SR. Pacemaker Malfunction. Cardiology Today. 2015;19:180–186)

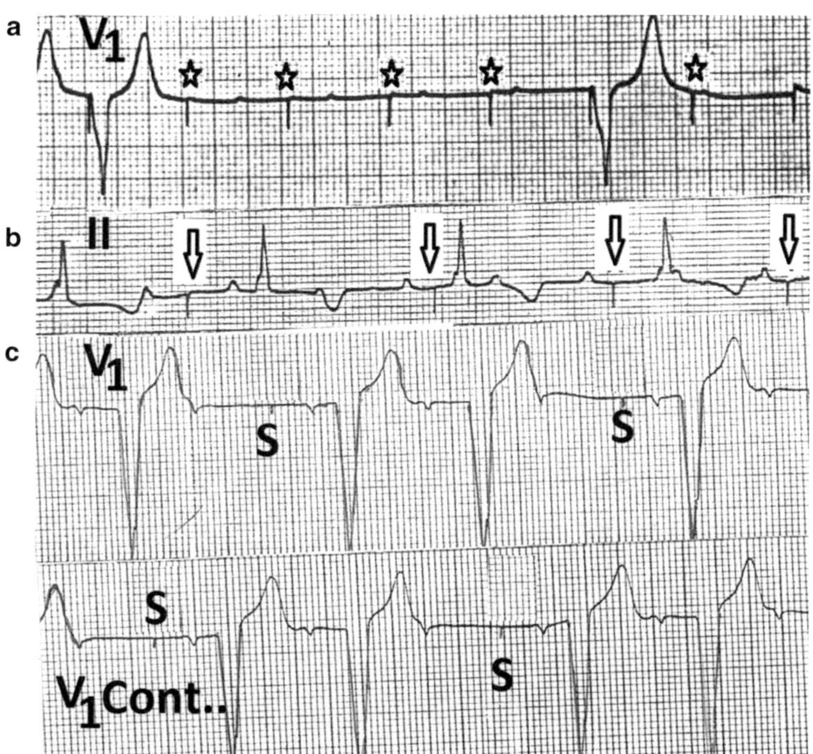

Fig. 15.3 Electrocardiograms (**a**) showing intermittent pacing failure (marked as ☆). First and sixth pacing spikes are followed by a QRS complex. Second, third, fourth, and fifth pacing spikes (marked ☆) are not followed by a QRS complex. (**b**) showing total pacing failure (marked as ⇩), no pacing spike is followed by a QRS; (**c**) total pacing failure in a case with left bundle branch block with Wenckebach phenomenon (pacing spikes, marked S). There is no QRS after any pacing spike. (Reproduced with permission from: Mittal SR. Pacemaker Malfunction. Cardiology Today. 2015;19:180–186)

(iv) Increased myocardial stimulation threshold—it can occur with hyperkalemia or flecainide treatment.
(v) Refractoriness of the myocardial tissue—it can be due to myocardial ischemia or infarction.
(vi) Battery depletion.

Use of manufacturer's specific pacing system analyzer is often necessary to find the exact cause [1]. Commonly used parameters and their interpretation are shown below.

Pacing threshold	Impedance	Most possible causes
Increased	Normal	Lead dislodgment
Increased	Decreased	Lead insulation break
Increased	Increased	Lead conductor failure
Increased	Increased	Loose set screw
Normal	Normal	Battery depletion
Increased	Normal	Drugs, electrolytes, ischemia, myocardial infarction

3 Undersensing

Failure to sense an intracardiac signal falling outside the refractory period of the paced beat suggests sensing failure (Figs. 15.5, 15.6 and 15.7).

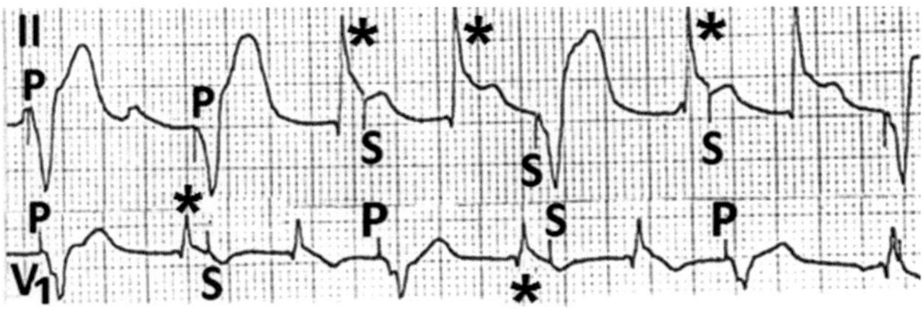

Fig. 15.5 Electrocardiogram shows sensing failure. Normally the pacemaker should sense an intrinsic QRS and should not give a pacing stimulus. In this case the pacemaker is giving pacing stimulus (spike marked S) even after an intrinsic QRS. This shows failure of pacemaker system to sense the intrinsic QRS (sensing failure, marked *). Pacing spikes are followed by a paced QRS whenever the ventricle is repolarized from the previous QRS (marked P). Therefore, pacing function is normal. (Reproduced with permission from: Mittal SR. Pacemaker Malfunction. Cardiology Today. 2015;19:180–186)

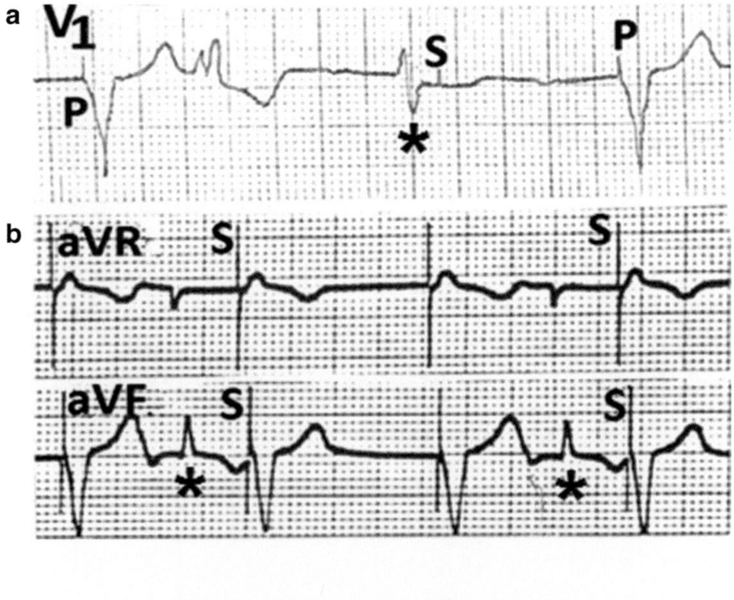

Fig. 15.6 (**a**, **b**) Electrocardiograms from two patients showing sensing failure (S). Pacing spikes are present even after an intrinsic QRS (marked *). This shows that the pacemaker is failing to sense intrinsic QRS. (Reproduced with permission from: Mittal SR. Pacemaker Malfunction. Cardiology Today. 2015;19:180–186)

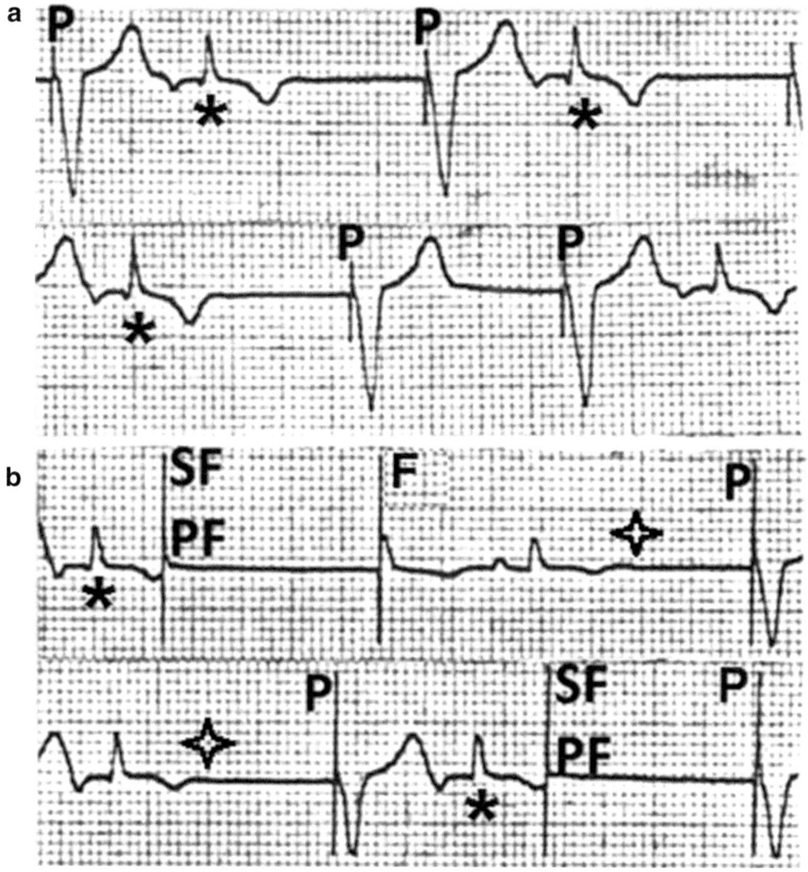

Fig. 15.7 Electrocardiograms. (**a**) Pacemaker senses intrinsic QRS (marked *). And next pacing stimulus is postponed. This shows that the sensing function of the pacemaker is normal. Each pacing spike (marked P) is followed by a QRS. This shows normal pacing function. (**b**) There is intermittent occurrence of pacing spike even after an intrinsic QRS (marked *). At other times, the pacemaker senses the intrinsic QRS and postpones next pacing spike (marked ✧). Thus there is intermittent sensing failure (marked SF). Some pacing spikes are followed by a paced QRS (marked P). At other times the pacing spikes are not followed by a paced QRS. This shows intermittent pacing failure (marked PF). At times there is fusion of pacing spike with intrinsic QRS (marked F). (Reproduced with permission from: Mittal SR. Pacemaker Malfunction. Cardiology Today. 2015;19:180–186)

Ectopics falling within the programmed refractory period of the pacemakers are not sensed [1]. It does not mean sensing failure.

Other causes of sensing failures:

(i) Premature complexes have different local electrograms than a sinus beat and may not be sensed even if the system normally senses sinus beats [3].
(ii) Intracardiac electrogram can deteriorate [1] because of inflammation, infarction, or scar formation at the tissue lead interface.
(iii) Poor signals may not be sensed even if the system is normal [2]. Appropriate programming of the system rectifies the problem.
(iv) Cardioversion or defibrillation can also attenuate intracardiac electrogram [1] resulting in transient undersensing.
(v) Lead fracture or insulation break cause sensing as well as pacing failure.
(vi) Lead displacement (even microdisplacement) can cause sensing failure.
(vii) Failure to sense may be incorrectly suspected when the pacing system did not have sufficient time to sense an intrinsic beat. A

period of >0.04 sec is required for intrinsic activation to reach pacemaker lead positioned in the right ventricular apex [4].

4 Oversensing

(i) Sensing of electrical signals not generated in the heart, e.g., myopotentials, electrocautery, and electronic devices. This results in failure of output from the generator. It results in absence of any pacing spike or paced QRS. This results in asystole. This problem is more common in unipolar systems. Other conditions that can produce electromagnetic interference include therapeutic diathermy, radio-frequency ablation, transcutaneous electric nerve stimulation, spinal cord stimulation, extracorporeal lithotripsy, electric razors, cellular phones, and high-voltage power lines.[1]

(ii) Lead insulation break or fracture results in sensing of electric signals due to potential difference within the lead. With early lead problems, malfunction is typically intermittent and may be exacerbated by certain body position or motion.

(iii) Cross talk is a specific form of oversensing in which the atrial pacing stimulus is oversensed by the ventricular lead resulting in inhibition of ventricular pacing [3]. High atrial lead output, low sensing parameter of the ventricular lead, and short duration of the ventricular blanking after atrial pacing predispose to this problem. Ventricular blanking is the preset period during which the ventricular lead cannot sense an atrial pacing stimulus. This problem is seen in dual chamber pacing.

(iv) Failure to insert the lead fully into the pulse generation or failure to tighten the screw can also predispose to oversensing. [3]

(v) In VVI pacing, oversensing by the ventricular lead causes inhibition of pacing resulting in inappropriate pause. In DDD pacing, oversensing of the atrial lead can cause inappropriate triggering of the ventricular stimulus, whereas oversensing by the ventricular lead causes inhibition of pacing.

5 Erratic Prolongation of Spike to Spike Interval

- Hysteresis. Following an intrinsic QRS, the pacemaker waits for a little longer period than the preset RR interval to allow occurrence of a sinus beat. Therefore, interval between an intrinsic QRS and next pacing spike is slightly longer than the interval between other pacing spikes. This is normal and not a malfunction.
- Oversensing of the T wave. T wave activity in the region of the lead may not correspond to the surface ECG. Surface ECG may not show a prominent T wave.
- Oversensing of the P wave may be caused by insulation break in atrial portion of the lead. Electrocardiographic diagnosis may be difficult.
- Oversensing of the T wave or the P wave can be suspected by surface ECG if the magnitude of delay in the next pacing spike corresponds with the interval between the previous pacing spike and the suspected P or T wave.

6 Constant Pacing at a Rate Slower than the Programmed Rate

It can be due to battery depletion [3].

7 Constant Pacing at a Rate Faster than the Programmed Rate

(i) It can occur due to sensing of supraventricular tachycardia, atrial flutter, or atrial fibrillation by DDD device (Fig. 15.8). In DDD device, atrial depolarization is sensed by atrial lead. This is followed by a preset PR delay and then a ventricular stimulation. In atrial tachycardia or atrial flutter, each atrial depolarization is sensed and is followed by a ventricular depolarization. Thus sensing of fast rate of atrial tachycardia or atrial flutter can result in fast ventricular rate.

(ii) Pacemaker-mediated tachycardia (PMT)

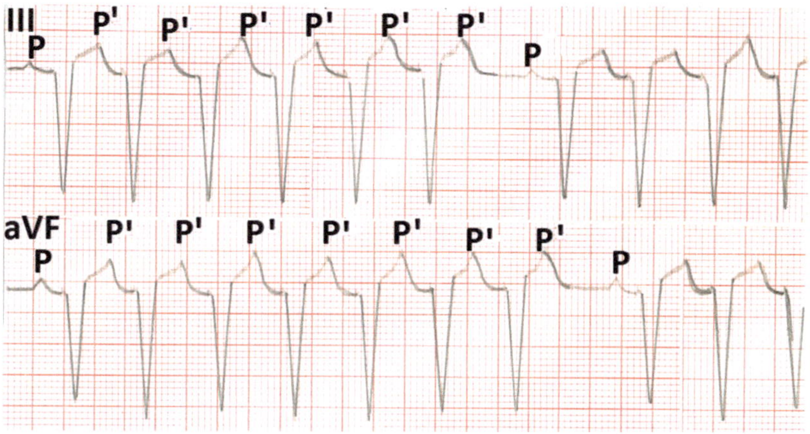

Fig. 15.8 Electrocardiogram is showing sensing of atrial tachycardia (P′) by DDD pacemaker, producing fast ventricular rate. (Reproduced with permission from: Mittal SR. Pacemaker Malfunction. Cardiology Today. 2015;19:180–186)

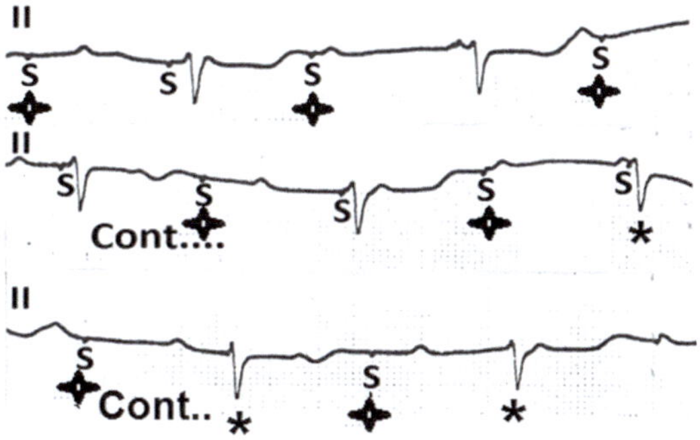

Fig. 15.9 Electrocardiogram showing sensing failure (pacing spikes coming in spite of intrinsic QRS—marked *) as well as pacing failure (absence of paced QRS after pacing spikes (marked ✧) (S—pacing spikes). (Reproduced with permission from: Mittal SR. Pacemaker Malfunction. Cardiology Today. 2015;19:180–186)

This occurs due to regular retrograde conduction of the ventricular impulse to the atria (ventriculoatrial conduction) resulting in regular atrial sensing and subsequent ventricular stimulation. Use of magnet can change the pacemaker to asynchronous mode (VOO) and terminate PMT.

Combination of sensing as well as pacing failure (Fig. 15.9) and oversensing is almost diagnostic of lead-related problems.

8 Complications

(a) Pacemaker Syndrome [4]

This is seen in VVI pacing.

- 1:1 retrograde conduction to the atria can cause atrial contraction against closed atrioventricular valve (i.e., during ventricular contraction). This results in back pressure in the pulmonary and the systemic veins and symptoms and signs of systemic and pulmonary venous congestion.
- Normally atrial contraction just precedes onset of ventricular contraction. A properly timed P wave increases ventricular filling. This transiently increases end-diastolic length of the myocardial fibers. This increases myocardial contractility. Dissociation of normal atrial and ventricular synchrony can result in impairment of ventricular function, fall in cardiac output, fatigue, fall in systolic blood pressure, dizziness or near syncope, and precipitation of heart failure in patients with preexisting ventricular dysfunction.

(b) Displacement of Pacing Leads

(i). Atrial lead is more likely to be displaced if it is not actively fixed. In such situation, there is loss of atrial sensing and pacing. Pacemaker works as VVI pacemaker.

(ii). Even microdisplacement of the ventricular lead results in failure of sensing and pacing. Such small displacement may not be detectable on skiagram of chest. Displacement from right ventricular apex to right ventricular outflow shifts the mean frontal plane axis of the pacing spike (Fig. 15.10) and the paced QRS complex inferiorly (Fig. 15.11).

(c) Spontaneous change of QRS configuration from left bundle branch block pattern to right bundle branch block pattern (Fig. 15.12) should raise suspicion of perforation of the interventricular septum with passage of the pacing lead into the left ventricle.

(d) Diaphragmatic pacing (presenting as hiccup with each paced heartbeat) or pericardial rub suggests perforation of the myocardium. Diaphragmatic pacing can also occur if pacing output is more than normal. In this situation diaphragmatic pacing may be abolished by reducing the output of the pacemaker.

a

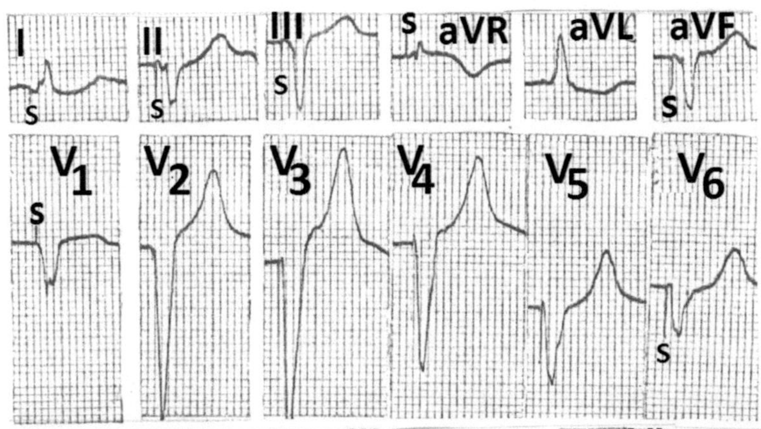

b

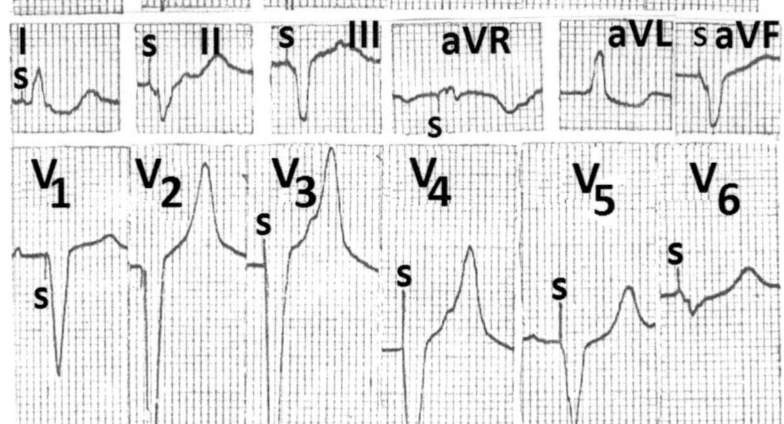

Fig. 15.10 Electrocardiograms showing changes in the axis of the pacing spike. (**a**) Pacing spike is negative in leads I, II, III, aVF, V_5, and V_6 and positive in leads aVR and V_1; (**b**) pacing spike is positive in leads I, II, III, aVF, and V_3 to V_6 and negative in leads aVR, V_1, and V_2. (Reproduced with permission from: Mittal SR. Pacemaker Malfunction. Cardiology Today. 2015;19:180–186)

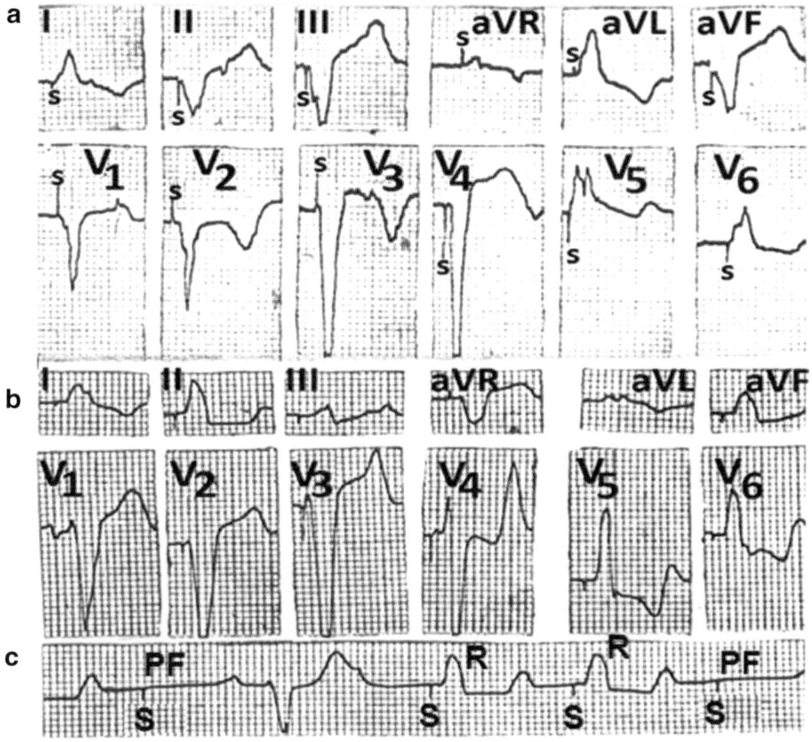

Fig. 15.11 Electrocardiograms showing (**a**) mean frontal plane QRS axis directed superiorly (QRS is negative in leads II, III, and aVF), (**b**) mean frontal plane QRS axis directed inferiorly (positive QRS in leads II, III, and aVF), and (**c**) intermittent pacing failure (no QRS after pacing spike marked S(PF)). (Reproduced with permission from: Mittal SR. Pacemaker Malfunction. Cardiology Today. 2015;19:180–186)

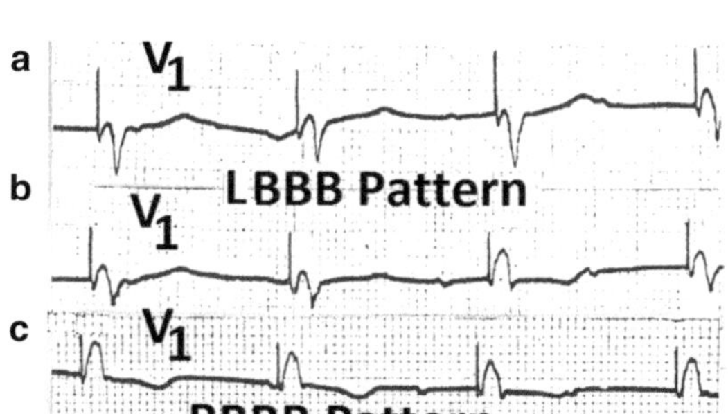

Fig. 15.12 Electrocardiogram showing progressive change in the polarity of the QRS from LBBB pattern (**a**) to intermediate pattern (**b**) and finally to RBBB pattern (**c**). (Reproduced with permission from: Mittal SR. Pacemaker Malfunction. Cardiology Today. 2015;19:180–186)

9 ECG Diagnosis of Myocardial Infarction in Patients with Implanted Pacemaker [5]

Diagnosis of old infarction is difficult due to left bundle branch block morphology of the paced QRS. As in a case of left bundle branch block, Q wave in leads I, aVL, V_5, and V_6 or in leads II, III, and aVF may be suggestive of myocardial infarction. Deep symmetrical inversion of the T waves in precordial leads also suggests myocardial infarction (Fig. 15.13).

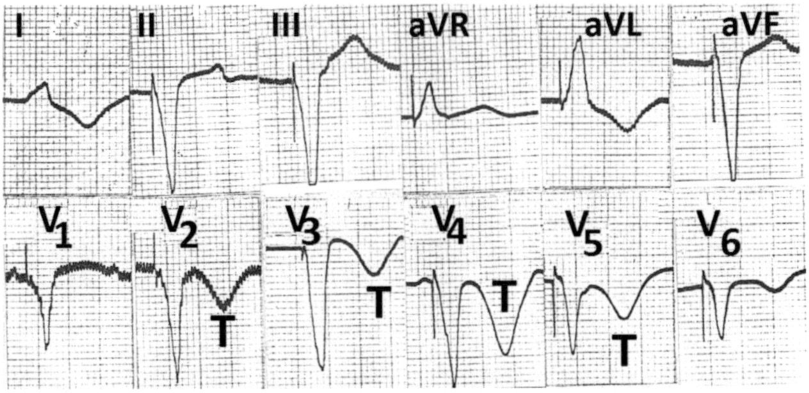

Fig. 15.13 Electrocardiogram from a patient with permanent pacemaker showing deep symmetrical inversion of the T waves (marked T) in leads V_2 to V_6 suggestive of underlying anterior infarction. (Reproduced with permission from: Mittal SR. Pacemaker Malfunction. Cardiology Today. 2015;19:180–186)

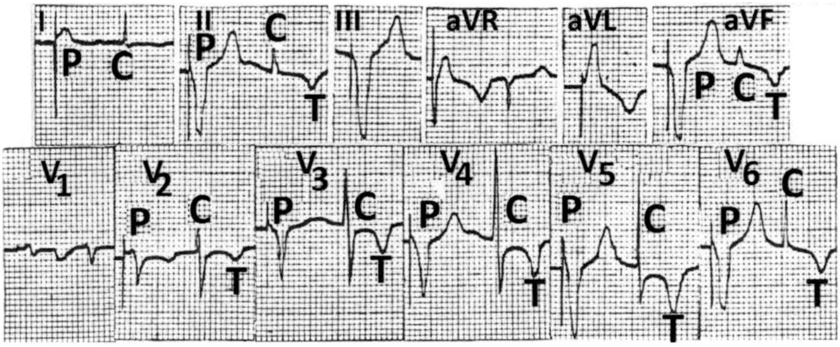

Fig. 15.14 Electrocardiogram showing alternating paced (P) and capture (C) beats. Capture beats show symmetrical inversion of the T waves (marked T) in leads II and aVF and leads V_3 to V_6 suggestive of underlying myocardial infarction. (Reproduced with permission from: Mittal SR. Pacemaker Malfunction. Cardiology Today. 2015;19:180–186)

Capture beats can also reveal underlying myocardial infarction (Fig. 15.14).

Acute infarction can be diagnosed by sequential changes in the ST segment and the T wave (Figs. 15.15 and 15.16).

Clinical profile, serum enzyme estimation, and echocardiography are useful. Spontaneous emergence of faster intrinsic heart rate can also reveal underlying myocardial infarction (Fig. 15.17).

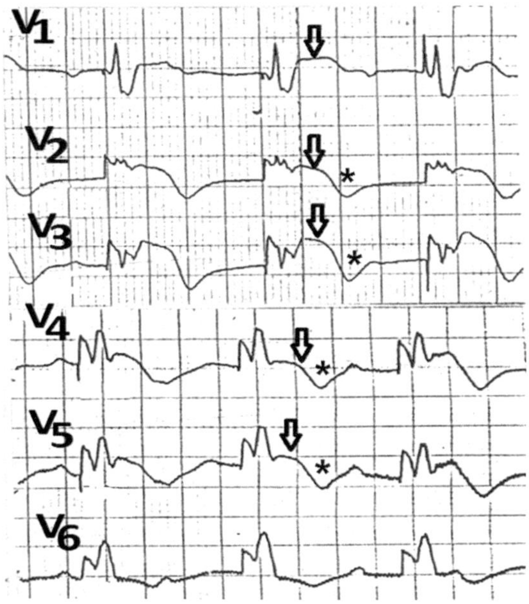

Fig. 15.15 Electrocardiogram from a patient with permanent pacemaker showing ST segment elevation (marked ⇩) and inversion of the T waves (marked *) in leads V_2 to V_5 suggestive of acute anterior infarction. (Reproduced with permission from: Mittal SR. Pacemaker Malfunction. Cardiology Today. 2015;19:180–186)

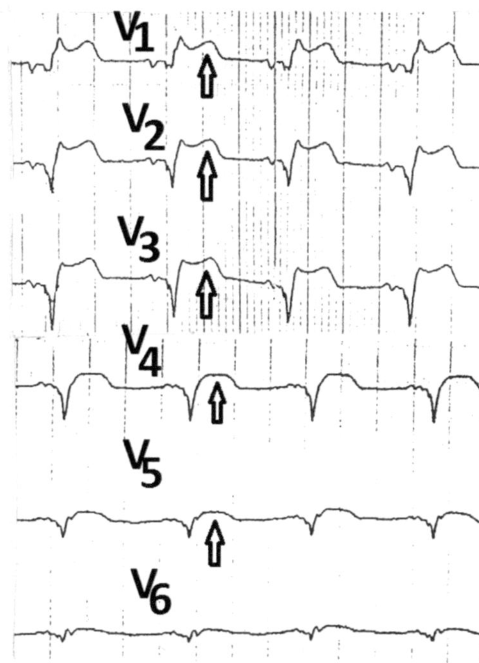

Fig. 15.17 Electrocardiogram from a patient with permanent pacemaker showing accelerated intrinsic heart rate with elevation of the ST segment in leads V_1 to V_6 (marked as ⇧) suggestive of anterior myocardial infarction. (Reproduced with permission from: Mittal SR. Pacemaker Malfunction. Cardiology Today. 2015;19:180–186)

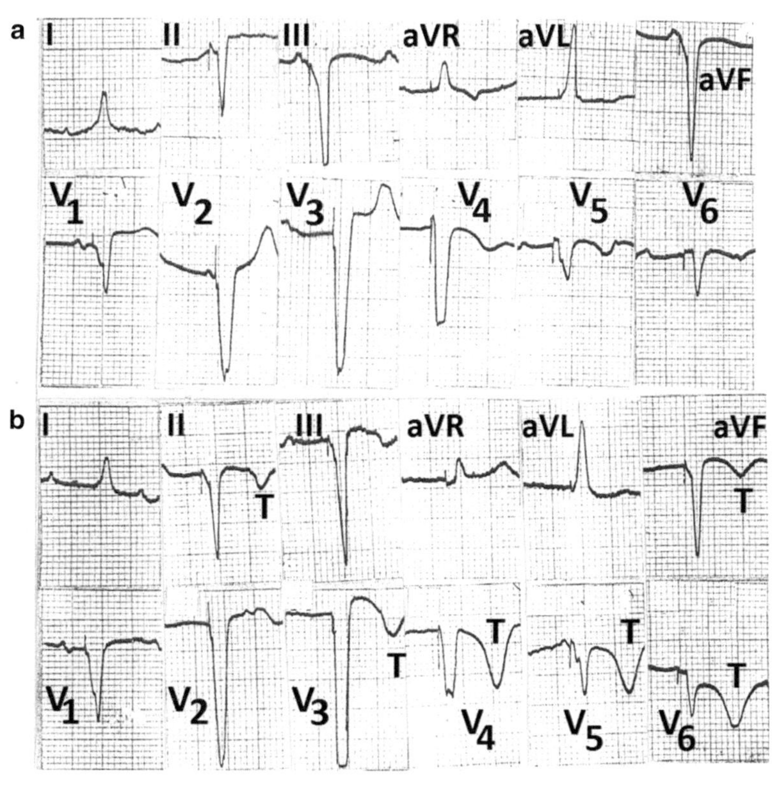

Fig. 15.16 (**a**) Electrocardiogram from a patient with permanent pacemaker. (**b**) Sequential symmetrical inversion of the T waves in leads II, III, aVF, and V_3 to V_6 (marked T) is suggestive of recent infarction. (Reproduced with permission from: Mittal SR. Pacemaker Malfunction. Cardiology Today. 2015;19:180–186)

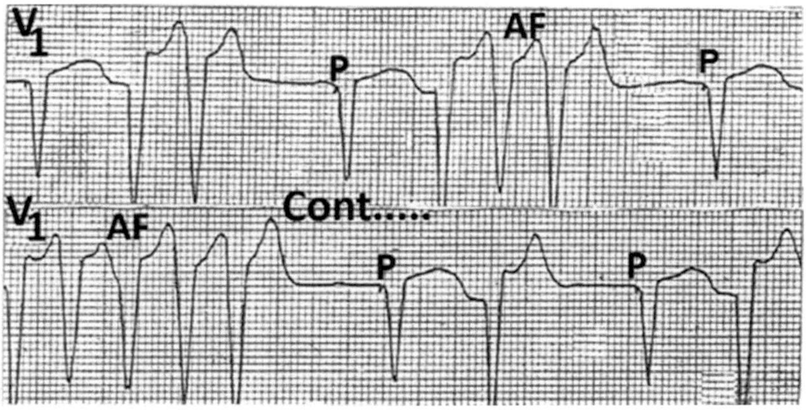

Fig. 15.18 Electrocardiogram from a patient with permanent pacemaker showing atrial fibrillation (AF) with intermittent fast ventricular rate and intermittent pacing (P— paced beats). (Reproduced with permission from: Mittal SR. Pacemaker Malfunction. Cardiology Today. 2015;19:180–186)

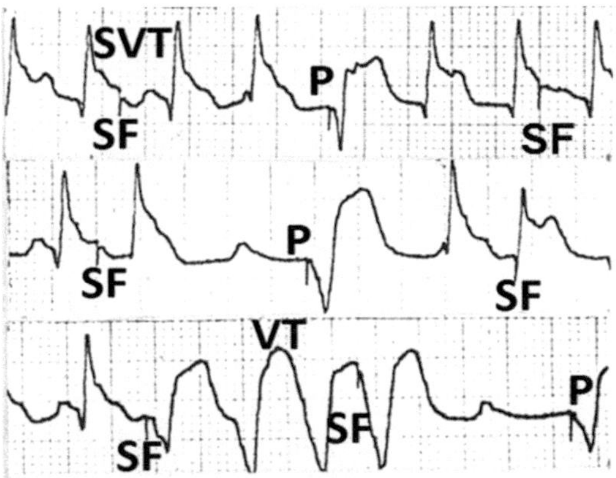

Fig. 15.19 Electrocardiogram from a patient with permanent pacemaker showing intermittent supraventricular tachycardia (SVT) and ventricular tachycardia (VT) with sensing failure (occurrence of pacing spike after intrinsic QRS marked SF). P—paced beats. (Reproduced with permission from: Mittal SR. Pacemaker Malfunction. Cardiology Today. 2015;19:180–186)

10 Limitation

Pacing devices only manage bradyarrhythmias. They do not manage concomitant supraventricular or ventricular tachyarrhythmias (Figs. 15.18 and 15.19).

Summary

- *Hysteresis*. This phenomenon is seen in on-demand intermittent pacing. Following an on-demand paced beat, the next paced beat is delayed by few milliseconds, while the pacemaker waits for the next spontaneous cardiac impulse to come. This interval is thus slightly longer than the interval set for regular pacing. This is a normal function and not a pacemaker malfunction.
- *Electromagnetic interference*. Any electrical stimulation in the body can change the pacemaker function to VVO mode. This can happen during electrocautery, lithotripsy, diathermy, radio-frequency catheter ablation, transcutaneous electric nerve stimulation, or spinal cord stimulation [5]. This is a normal protection against failure of pacing due to oversensing.
- *Pacing failure*. When there is no pacing stimulus in spite of the absence of an expected spontaneous P or QRS, it is called pacing failure. This can be due to oversensing of a noncardiac electric potential, tall T or P waves, lead connection failure, or lead fracture.

Magnet application changes pacing to VOO mode with appearance of regular pacing spikes. Magnet application does not restore pacing spikes in the setting of device malfunction.

- *Failure to capture.* The presence of a pacemaker stimulus (pacing spike) which is not followed by a paced P wave or QRS complex is known as failure to capture. It is usually due to lead displacement or elevated pacing threshold at the electrode-myocardial interface. Elevation of threshold can be due to fibrosis, ischemia, or infarction around the tip of the pacing lead. Loose screw, insulation break, lead fracture, or battery depletion can also result in failure to capture. Combination of failure to pace and capture usually indicates a pacing system failure.
- *Cross talk.* Oversensing of the atrial pacing stimulus by the ventricular lead is known as "cross talk." It results in inhibition of ventricular pacing.
- *Battery depletion.* Automatic decrease in the pacing rate or change of pacemaker function to VOO mode suggests battery depletion.

Differentiation of Pacemaker Malfunction

(A) **Pacemaker spike**

(1). **Absent pacemaker spikes**

(a) **QRS faster than the preset pacing rate:**

It happens if patient's intrinsic heart rate is faster than the pacemaker rate (Figs. 15.20 and 15.21). Applying magnet over the pulse generator converts the pacemaker to asynchronous mode. Appearance of regular spike with intermittent capture suggests normal function (Figs. 15.20 and 15.21).

(b) **QRS slower than the preset pacing rate:**

- Dry pocket (fibrosis around the pulse generator), battery exhaustion, or lead fracture can reduce the pacing rate.

(2). **Intermittent absence of pacemaker spikes**

- Lead fracture—fractured ends make intermittent contact. It is not corrected by a magnet.
- Oversensing of myopotentials or extraneous electric signals.
- It is corrected by use of magnet.

(3). **Regular pacing spike**

(a) Rate

- Rate slower than the preset rate
  Intermittent—hysteresis
  Regular—battery depletion, sensing of the T wave, sensing of an echo beat
- Rate faster than the preset rate—pacemaker-mediated tachycardia (PMT)

(b) Amplitude – Large—unipolar unit

- Small—bipolar unit
- Smaller than at the time of implantation—lead fracture resulting in high lead impedance

(c) Duration – Broad fusions with intrinsic QRS and direct current leakage into the heart due to a short circuit in output circuit

(d) Axis – Negative in leads II, III, and aVF—pacing from the right ventricular apex

- Positive in leads II, III, and aVF—pacing from the right ventricular outflow
- Change from previous record—displacement

(e) Changing direction and amplitude (Figs. 15.22 and 15.23). Insulation fracture

(4). Delayed spike after a sensed QRS—hysteresis

(B) **Distance between the pacing spike and the paced QRS**

- Normally there is no appreciable distance due to rapid entry of the electric impulse in the branches of the conducting system (Fig. 15.24).

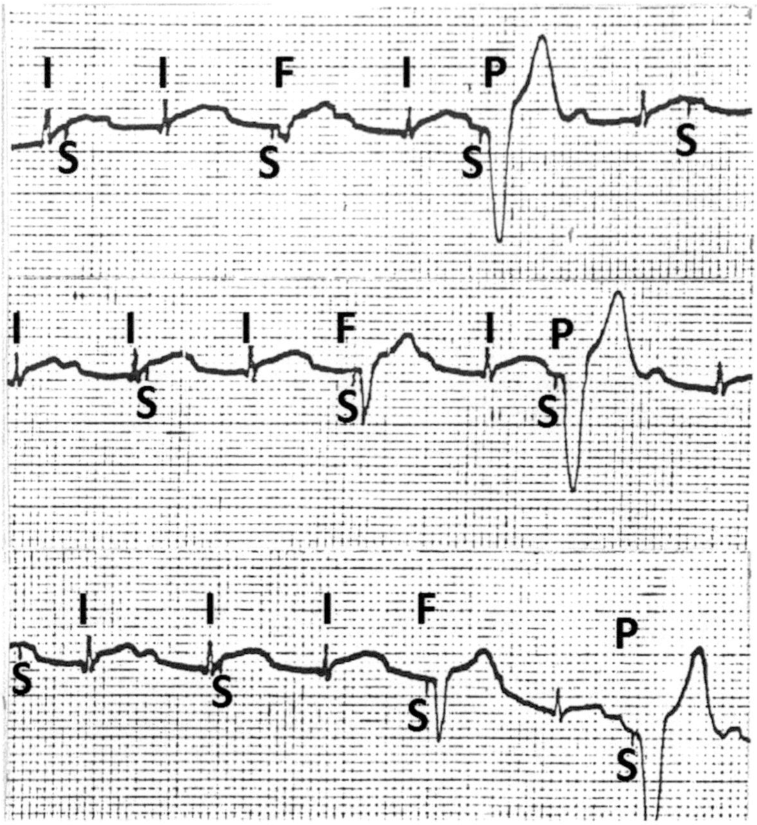

Fig. 15.20 Electrocardiogram from a patient with permanent pacemaker showing fast intrinsic heart rate (I). Application of magnet on the pulse generator results in regular spikes (S) with intermittent fusion beats (F) and paced beats (P). (Reproduced with permission from: Mittal SR. Pacemaker Malfunction. Cardiology Today. 2015;19:180–186) S—pacing spike

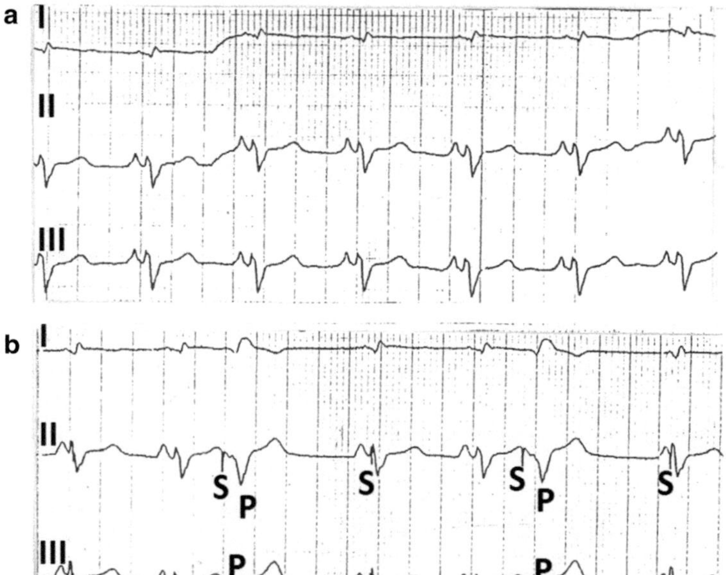

Fig. 15.21 (**a**) Electrocardiogram from a patient with permanent pacemaker with accelerated intrinsic heart rate. (**b**) Application of magnet over the pulse generator results in appearance of regular spikes (S) with intermittent paced beats (P). (Reproduced with permission from: Mittal SR. Pacemaker Malfunction. Cardiology Today. 2015;19:180–186)

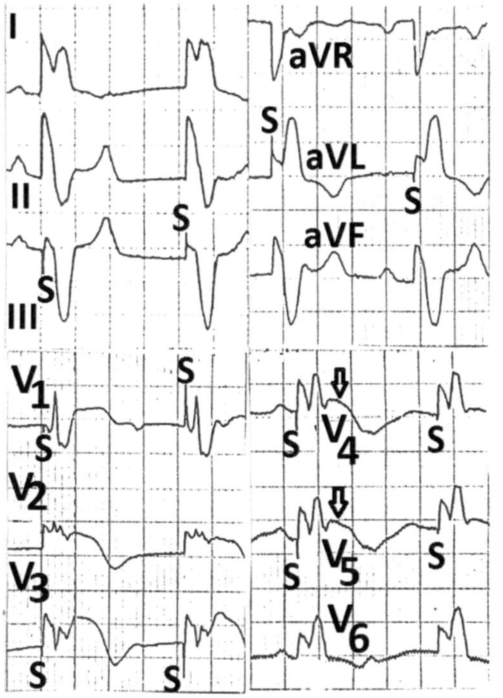

Fig. 15.22 Electrocardiogram is showing changing direction and amplitude of the pacing spike (marked S). ST segment is elevated in leads V_2–V_3 (marked as ⇩) suggestive of anterior myocardial infarction. (Reproduced with permission from: Mittal SR. Pacemaker Malfunction. Cardiology Today. 2015;19:180–186)

– Increased duration (Fig. 15.24) suggests decreased conduction around the electrode tip. This could be due to ischemia, infarction, or fibrosis.

(C) **QRS complex**

(a) **Configuration.** Left bundle branch block configuration—right ventricular pacing

Right bundle branch block configuration—left ventricular pacing, biventricular pacing

Recent change in QRS configuration—displacement

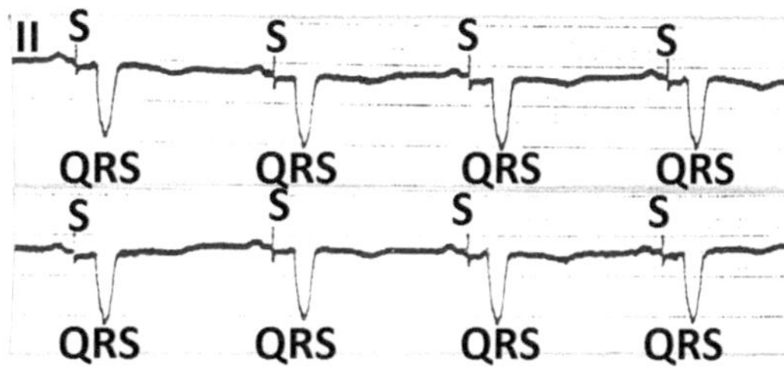

Fig. 15.24 Electrocardiogram from a patient of DDD pacing, showing increased distance between the spike and the paced QRS. (Reproduced with permission from: Mittal SR. Pacemaker Malfunction. Cardiology Today. 2015;19:180–186)

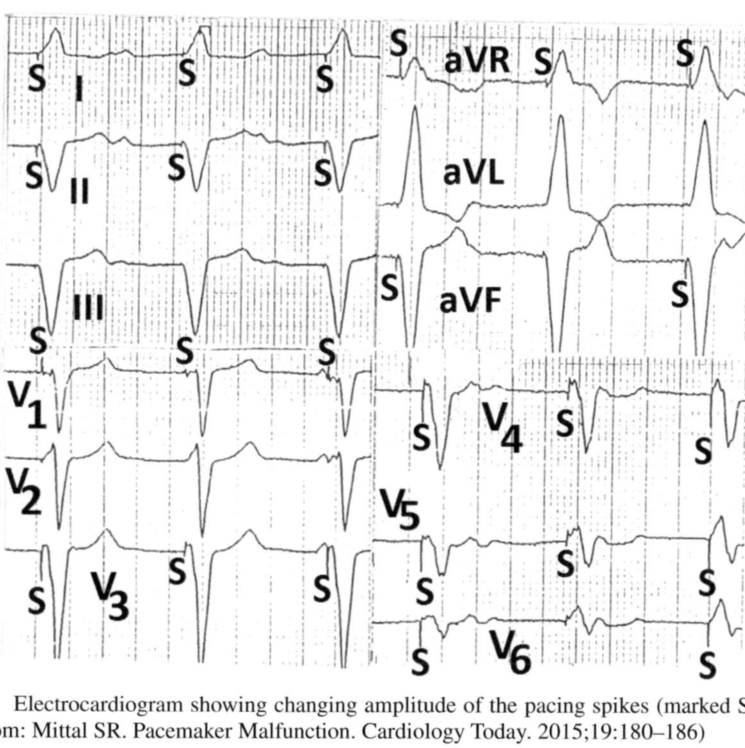

Fig. 15.23 Electrocardiogram showing changing amplitude of the pacing spikes (marked S). (Reproduced with permission from: Mittal SR. Pacemaker Malfunction. Cardiology Today. 2015;19:180–186)

(b) **Axis.** Left upper quadrant—pacing from right ventricular apex
Inferior—pacing from right ventricular outflow
(c) **Fusion or pseudo fusion beats** do not indicate sensing failure.

(D) **Sensing failure** [6]
(a) Ventricular ectopics arising at a distance from the tip of the lead
(b) Low amplitude of QRS of intracardiac electrogram due to infarction, inflammation, or fibrosis
(c) Lead displacement, even microdisplacement
(d) Myocardial perforation
(e) Lead fracture
(f) Fault in pulse generator

(E) **Failure to capture (pacemaker spike without subsequent QRS)**
(a) Raised threshold—Hypokalemia, hypocalcemia, flecainide
Ischemia, infarction, fibrosis, inflammation
(b) Lead displacement
(c) Myocardial perforation

MCQs

Q1. Change in the axis of the pacemaker spike as compared to previous ECG suggests:

(a) Insulation break
(b) Lead fracture
(c) Lead displacement
(d) All

Q2. Increase in the distance between the pacing spike and onset of the paced QRS as compared to previous ECG could be due to:

(a) Ischemia
(b) Infarction
(c) Fibrosis
(d) All

Q3. Change in the QRS configuration from LBBB to RBBB pattern suggests:

(a) Lead displacement into right ventricular outflow tract
(b) Lead fracture
(c) Perforation of interventricular septum
(d) Insulation break

Q4. Failure to sense ventricular ectopics with normal sensing of intrinsic cardiac impulse needs:

(a) Replacement of pulse generator
(b) Change of lead
(c) Programming of pulse generator
(d) None

Q5. Presence of a pacemaker spike with failure of ventricular depolarization can be due to:

(a) Hypokalemia
(b) Myocardial perforation
(c) Lead fracture
(d) Battery exhaustion

Q6. Lead impedance less than 300 ohms suggests:

(a) Insulation break
(b) Lead fracture
(c) Microdisplacement
(d) Myocardial perforation

Q7. Lead impedance more than 1200 ohms suggests:

(a) Insulation break
(b) Lead fracture
(c) Poor electrode location
(d) Macro displacement

Q8. Lead impedance is not increased by:

(a) Lead fracture
(b) Poor connection of the lead to the pulse generator

(c) Loose set screws
(d) Lead displacement

Q9. Diaphragmatic stimulation suggests:

(a) Oversensing
(b) Insulation rupture
(c) Gross change in lead position
(d) Myocardial perforation

Q10. Intermittent loss of pacing/sensing suggests:

(a) Lead displacement
(b) Myocardial perforation
(c) Lead fracture
(d) Pseudomalfunction

Q11. Delayed spike after sensing of a normal QRS suggests:

(a) Hysteresis
(b) Sensing malfunction
(c) Pacing malfunction
(d) All

Q12. Asystole produced by exercising muscles close to the implant site suggests:

(a) Sensing failure
(b) Pacing failure
(c) Oversensing
(d) All

Q13. Which cause of low-voltage QRS causes undersensing?

(a) Myocardial infarction
(b) Pericardial effusion
(c) Emphysema
(d) Obesity

Q14. Amplitude of intracardiac R wave signal is not reduced in:

(a) Lead fracture
(b) Microdisplacement
(c) Myocardial perforation
(d) Emphysema

Q15. Fusion beat suggests:

(a) Sensing failure
(b) Pacing failure
(c) Oversensing
(d) Nil

Q16. Regular bradycardia in a patient with permanent pacemaker is not due to:

(a) T wave sensing
(b) Battery depletion
(c) Pacing failure
(d) Myopotential inhibition

Q17. Pacemaker-mediated tachycardia can occur in:

(a) VVI pacing
(b) DDD pacing
(c) Biventricular pacing
(d) All

Q18. Marked variation in spike amplitude can occur in:

(a) Lead fracture
(b) Insulation break
(c) Oversensing of extracorporeal signals
(d) Battery depletion

Q19. Tachycardia in a patient with a DDD pacemaker can be due to:

(a) Atrial tachycardia
(b) Ventricular tachycardia
(c) Pacemaker-mediated tachycardia
(d) All

Q20. Deterioration in cardiac status after VVI pacing suggests:

(a) Pacing failure
(b) Pacemaker syndrome

(c) Myocardial ischemia
(d) Unmasking of cardiac failure

Q21. Cannon waves in a patient with implanted pacemaker suggest which mode of pacing:

(a) AAI
(b) VVI
(c) DDD
(d) Biventricular

Q22. Pacemaker stimulus falling on the T wave of an intrinsic QRS suggests:

(a) Sensing failure
(b) Pacing failure
(c) Failure to capture
(d) None

Q23. Failure of appearance of a pacing stimulus in spite of a long pause suggests:

(a) Sensing failure
(b) Pacing Failure
(c) Failure of capture
(d) Hysteresis

Q24. Pacemaker stimulus not followed by a paced QRS suggests:

(a) Sensing failure
(b) Pacing failure
(c) Failure of capture
(d) Electromagnetic interference

Q25. Atrial pacing stimulus followed by a P wave but not followed by ventricular pacing spike and paced QRS suggests:

(a) Hysteresis
(b) Failure of capture
(c) Electromagnetic interference
(d) Cross talk

Q26. Hysteresis suggests:

(a) Failure of capture
(b) Failure to pace
(c) Normal function
(d) Failure to sense

Q27. Automatic decrease in the pacing rate of a pacemaker suggests:

(a) Hysteresis
(b) Electromagnetic interference
(c) Failure to sense
(d) Battery depletion

Q28. Battery depletion can produce:

(a) Failure to sense
(b) Failure to pace
(c) Change of VVI to VOO mode
(d) Hysteresis

Q29. Pacemaker syndrome can occur in:

(a) VVI pacing
(b) DDD pacing
(c) DDDR pacing
(d) All

Q30. Pathology at the tissue-lead interface can produce:

(a) Failure to sense
(b) Failure to capture
(c) Failure to pace
(d) All

Q31. Lead fracture can cause:

(a) Hysteresis
(b) Cross talk
(c) Electromagnetic interference
(d) Sensing failure

Q32. Insulation break can cause:

(a) Sensing failure
(b) Pacing Failure
(c) Failure of capture
(d) Hysteresis

Q33. Magnet application over pulse generator changes:

(a) VVI to VOO mode
(b) VOO mode to VVI mode
(c) VVI to DDD mode
(d) DDD to VVI mode

Q34. Hysteresis is suggested when a paced beat occurs:

(a) Before its scheduled time
(b) At its scheduled time
(c) Slightly later than the scheduled time
(d) Much after the scheduled time

Q35. Electromagnetic interference changes:

(a) VOOO to VVI mode
(b) VVI to VOO mode
(c) DDD to VVI mode
(d) None

Answers

(1) c; (2) d; (3) c; (4) d; (5) a, b; (6) a; (7) b; (8) d; (9) d; (10) c; (11) a; (12) c; (13) a; (14) d; (15) d; (16) d; (17) a; (18) a; (19) d; (20) b; (21) b; (22) a; (23) b; (24) c; (25) d; (26) c; (27) d; (28) c; (29) a; (30) a, b; (31) d; (32) a, c; (33) a; (34) c; (35) b

References

1. Vijayraman P, Ellenbogen KA. Bradyarrhythmias and pacemakers. In: Fuster V, Walsh RA, Harrington RA, editors. Hurst's The Heart. New York: Mc Graw Hill; 2011. p. 1025–57.
2. Olshansky B, Chang MK, Pogowizd SM, Goldschlager N. Cardiac pacing and pacemaker rhythms. In: Olshansky B, Chung MK, Pogowizd SM, Goldschlager N, editors. Arrhythmia Essentials. Burlington: Jones Bartlett Learning; 2012. p. 240–60.
3. Swerdlow CD, Wang PJ, Zipes DP. Pacemakers and implantable cardioverter-defibrillators. In: Mann DP, Zipes DP, Libby P, Bonow RO, editors. Braunwald's Heart Disease. Philadelphia: Elsevier; 2015. p. 721–42.
4. Haisty WK, Lim TH, Wagner GS. Artificial cardiac pacemakers. In: Wagner GS, editor. Marriot's Practical Electrocardiography. Philadelphia: Wolters Kluwer; 2001. p. 423–44.
5. Singh R, Mittal SR, Mathur D. Status of surface electrocardiography in evaluating pacemaker malfunctions. Indian Journal of Electrocardiology. 1996;2:7–10.
6. Sethi KK, Rajagopal S. Diagnosis of pacemaker malfunction. Indian Journal of Electrocardiology. 1989;1:4–8.

16 Ventricular Ectopic Beats

1 Definition

Normally the heart is stimulated by the impulse arising in the sinoatrial node. If the focus of stimulation is other than the sinoatrial node, it is called an ectopic beat, i.e., heart beat arising from an abnormal focus. When the abnormal focus is in the atria, the atrioventricular node, or the His bundle (i.e., from structures above the ventricles), the ectopic is called a supraventricular ectopic. When the abnormal focus is in the ventricles, it is called ventricular ectopic beat.

2 Differentiation

Site of origin of an ectopic beat has diagnostic, therapeutic, and prognostic implication. Electrocardiographic differentiation of supraventricular and ventricular ectopics is given below.

	Supraventricular (Fig. 16.1a) ectopic beat	*Ventricular* (Fig. 16.1b) ectopic beat
Preceding P wave	Abnormal or absent	Normal (like a P wave in other sinus beats) or absent
ORS duration	Narrow (0.08 sec) exception-presence of preexisting bundle branch block in sinus rhythm or aberrant ventricular conduction	Wide (0.12 sec) exception-Ectopic arising from proximal part of a bundle branch or fascicle
Initial vector of QRS in lead V_1	Identical to other sinus beats (Fig. 16.1c)	Different (Fig. 16.1d)

Sometimes the differentiation may be difficult from surface ECG alone. Sometimes the same patient may have supraventricular as well as ventricular ectopics (Fig. 16.1e).

S. Mittal, *Insights into Electrocardiograms with MCQs*,
https://doi.org/10.1007/978-981-99-0127-2_16

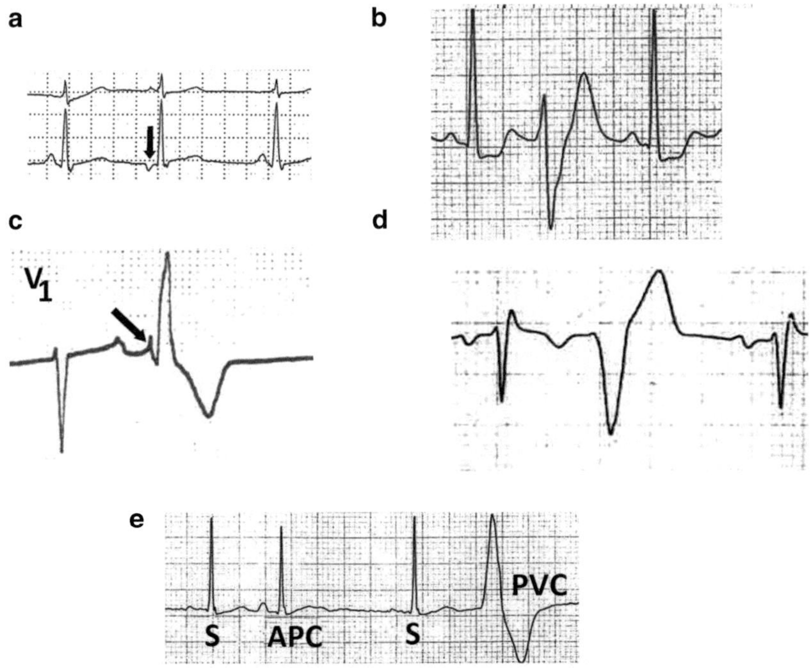

Fig. 16.1 Electrocardiograms showing (**a**) supraventricular (atrial) ectopic (marked with ↓), (**b**) ventricular ectopic, (**c**) supraventricular ectopic showing initial vector of QRS in lead V_1 identical to other sinus beats (marked with arrow), (**d**) ventricular ectopic with initial QRS vector in lead V_1 different from sinus beats, and (**e**) supraventricular (marked APC) as well as ventricular ectopic (marked PVC) in the same patient. (Reproduced with permission from: Mittal SR. Ventricular Ectopics Beats. Cardiology Today. 2011;15:39–43)

3 Premature Beat vs Escape Beat

When the ectopic beat occurs prematurely (earlier than the scheduled next sinus beat), it is called a premature beat (Fig. 16.2a). It is not useful for circulation and can be treated if indicated. If the ectopic beat occurs later than the scheduled next sinus beat, it is called an escape beat (Fig. 16.2b). Escape beats are useful for maintaining circulation whenever there is a longer sinus pause. Escape beats do not need to be suppressed.

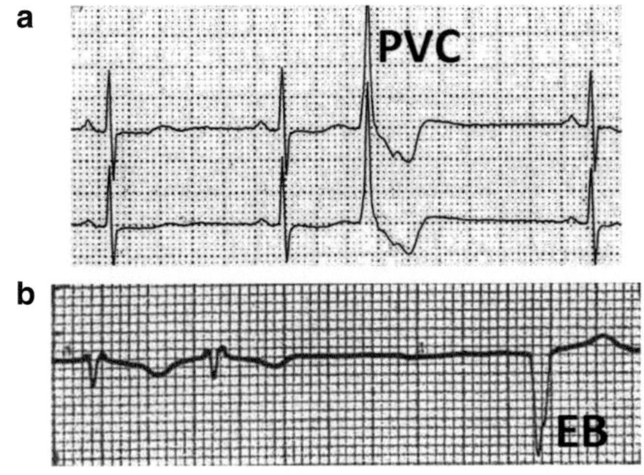

Fig. 16.2 Electrocardiograms (**a**) showing premature ventricular beat (PVC) and (**b**) showing escape beat (EB) arising from the ventricle. (Reproduced with permission from: Mittal SR. Ventricular Ectopics Beats. Cardiology Today. 2011;15:39–43)

4 Site of Origin of the Ventricular Ectopic Beat

Ventricular ectopic beats can arise from any part of either ventricle. Those arising from the left ventricle or left side of the interventricular septum produce a right bundle branch block configuration (rsR' configuration in lead V_1 and the qRS configuration in lead V_6) (Fig. 16.3a). In the presence of anterior myocardial infarction, left ventricular ectopics produce QR configuration in leads V_1 to V_4 (Fig. 16.3b). Ectopics arising from the right ventricle or right side of the interventricular septum produce left bundle branch block configuration (rs configuration in lead V_1 and RsR' configuration in lead V_6) (Fig. 16.3c). Ectopics arising from the right ventricular outflow tract (RVOT) have a left bundle branch block morphology with inferior axis (left bundle branch block morphology in lead V_1 with tall monophasic R in leads II, III, and aVF) (Fig, 16.3d). Identification can be useful because ectopics arising from the right ventricular outflow tract respond to beta-blockers and calcium channel blockers (verapamil, diltiazem) and are amenable to radio-frequency ablation. Ectopics arising from the right ventricle can be a manifestation of arrhythmogenic right ventricular cardiomyopathy, which can be excluded only by detailed echocardiographic evaluation and MRI. Ectopics arising from other areas of the ventricles may have variable shapes.

5 Configuration

Ventricular ectopics with identical configuration usually arise from the same focus (unifocal or uniform) (Fig. 16.4a). If configuration is changing, they may be bifocal (two foci) (Fig. 16.4b) or multifocal (multiple points of origin) (Fig. 16.4c). Same focus can also produce different configuration (unifocal multiform) due to varying intraventricular conduction.

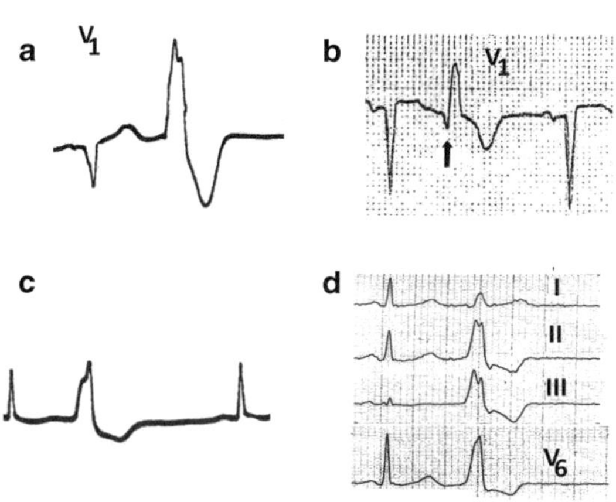

Fig. 16.3 Electrocardiograms showing (**a**) ventricular ectopic with right bundle branch block configuration, (**b**) ventricular ectopic with right bundle branch block configuration from a patient with anterior myocardial infarction showing QR configuration (marked with ↑), (**c**) ventricular ectopic showing left bundle branch block configuration, and (**d**) ventricular ectopic arising from the right ventricular outflow tract showing left bundle branch block morphology (monophasic R in lead V_6) with inferior axis (positive QRS in leads I, II, and III). (Reproduced with permission from: Mittal SR. Ventricular Ectopics Beats. Cardiology Today. 2011;15:39–43)

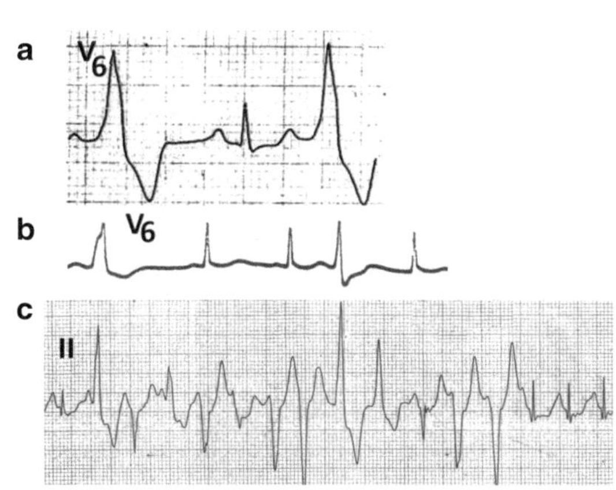

Fig. 16.4 Electrocardiograms (**a**) showing uniform ventricular ectopics, (**b**) showing bifocal ventricular ectopics, and (**c**) showing multiform ventricular ectopics. (Reproduced with permission from: Mittal SR. Ventricular Ectopics Beats. Cardiology Today. 2011;15:39–43)

6 Coupling Interval

The interval between the ectopic beat and the previous sinus beat is known as the coupling interval. Ectopics arising from the same focus usually have a fixed coupling interval (Fig. 16.5a). Reentry is the usual underlying mechanism in such cases. Those arising from different foci can have different coupling intervals (Fig. 16.5b). Ectopics due to increased automaticity of a single focus can also have different coupling intervals. However, the interval between such ectopics is either fixed or multiple of a fixed interval. Such ectopics are also called parasystolic ectopic beats. Coupling interval can be very short—R wave of the ectopic beat falling on the T wave of the previous sinus beat (R on T phenomenon). On the other end of the spectrum, the coupling interval may be long, and ectopic may appear after initiation of the P wave of the next sinus beat (late diastolic ectopics) (Fig. 16.5c). Late diastolic ventricular ectopics can fuse with the QRS of next sinus beat producing fusion beat (QRS configuration of the fusion beat is formed partly by the QRS of the sinus beat and partly by the QRS of the ectopic beat).

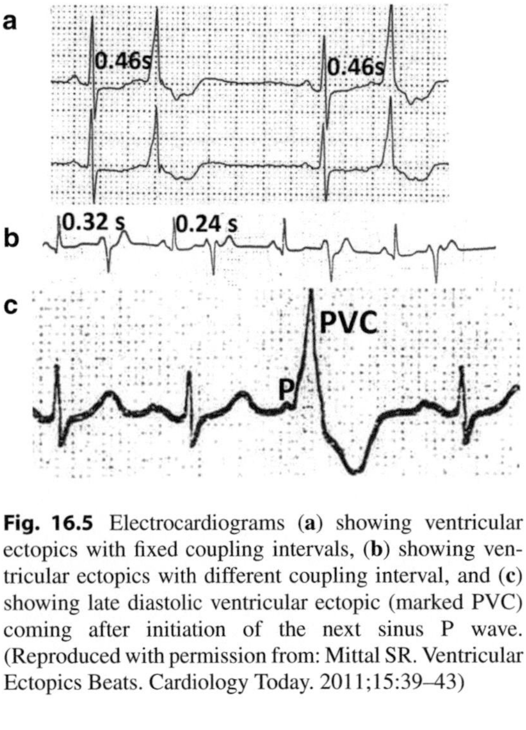

Fig. 16.5 Electrocardiograms (**a**) showing ventricular ectopics with fixed coupling intervals, (**b**) showing ventricular ectopics with different coupling interval, and (**c**) showing late diastolic ventricular ectopic (marked PVC) coming after initiation of the next sinus P wave. (Reproduced with permission from: Mittal SR. Ventricular Ectopics Beats. Cardiology Today. 2011;15:39–43)

7 Frequency (Fig. 16.6)

When ectopics alternate with sinus beat, it is called bigeminy.

When each ectopics occurs after two sinus beats, it is called trigeminy.

When ectopics occur after three sinus beats, it is called quadrigeminy.

There is some paradox in nomenclature. Although, by nomenclature, quadrigeminy appears more serious, the frequency of ectopics is less than in bigeminy or trigeminy. Frequency of ectopics is highly variable and varies significantly from time to time and day to day. Average load can be determined only by prolonged ambulatory monitoring.

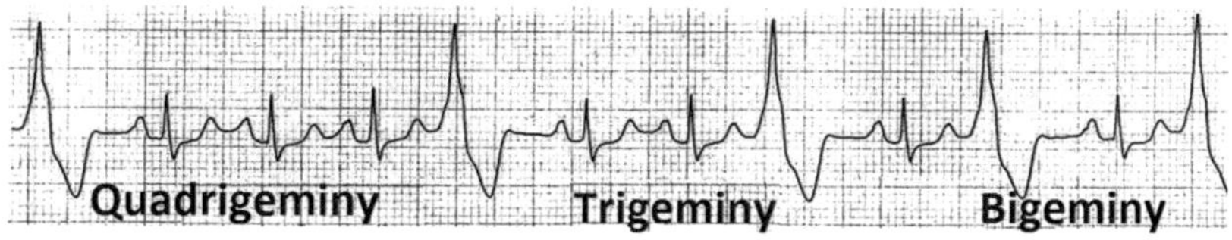

Fig. 16.6 Electrocardiograms showing quadrigeminy, trigeminy, and bigeminy in the same strip. (Reproduced with permission from: Mittal SR. Ventricular Ectopics Beats. Cardiology Today. 2011;15:39–43)

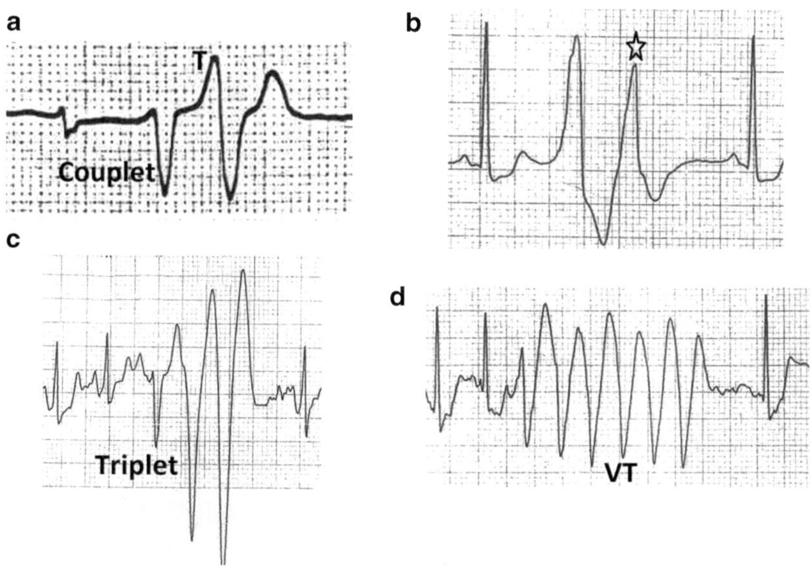

Fig. 16.7 Electrocardiograms (**a**) showing ventricular ectopics occurring in group of two with second ventricular ectopic falling on the T wave of the first ventricular ectopic, (**b**) showing group of two ventricular ectopics with slightly different configuration of the second ectopic (marked as ☆), (**c**) showing three ventricular ectopics in succession (triplet), and (**d**) showing ill-sustained ventricular tachycardia. (Reproduced with permission from: Mittal SR. Ventricular Ectopics Beats. Cardiology Today. 2011;15:39–43)

8 Repetitiveness

Ventricular ectopics can recur one after the other. When they occur in group of two, they are called couplets (Fig. 16.7a). In case of couplet, R wave of the second ectopic may fall on the T wave of the first ectopic (Fig. 16.7a). In case of couplets, QRS configuration of second ectopic may be slightly different from the first (Fig. 16.7b (marked ☆)) probably due to partial aberration in ventricular conduction because of incomplete repolarization following the first ectopic. "Triplet" suggests the occurrence in the group of three (Fig. 16.7c). By definition, occurrence of 3 to 30 ectopics in succession is called nonsustained ventricular tachycardia (Fig. 16.7d).

9 The ST Segment and the T Wave

These are directed away from the QRS of the ectopic. ST segment of the ventricular ectopic beat may be deformed by the next normal sinus P wave (having the configuration of sinus P wave) (Fig. 16.8a). Sinus P wave following a ventricu-

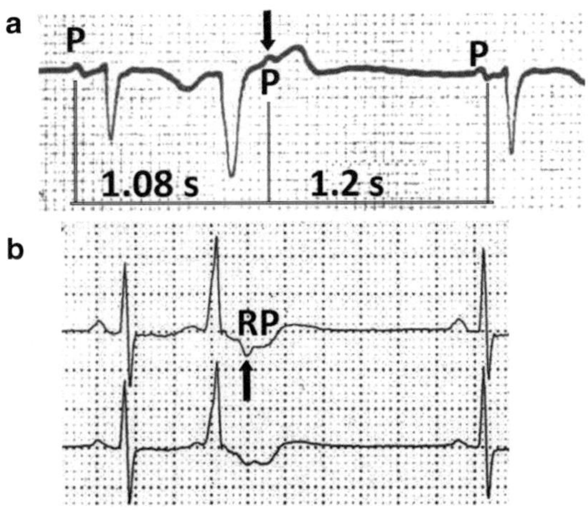

Fig. 16.8 Electrocardiograms (**a**) showing ventriculophasic sinus arrhythmia with sinus P wave falling on the ST segment of the preceding ventricular ectopic beat (marked with ↓) and (**b**) showing ventricular ectopic with retrogradely conducted P wave (RP marked with ↑). (Reproduced with permission from: Mittal SR. Ventricular Ectopics Beats. Cardiology Today. 2011;15:39–43)

lar ectopic may occur earlier than its scheduled time (marked ⇩). This is known as ventriculophasic sinus arrhythmia (Fig. 16.8a). Retrogradely conducted P waves are inverted in leads II, III, and aVF (Fig. 16.8b) (marked ⇧).

Retrograde conduction may occur regularly or occasionally.

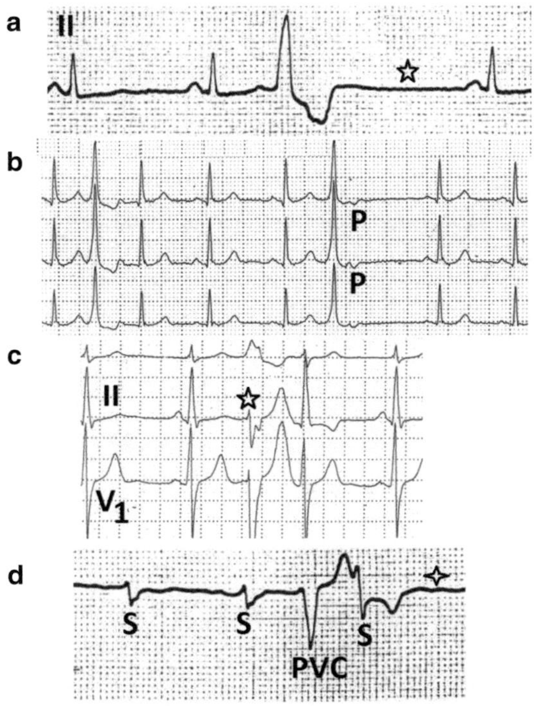

Fig. 16.9 Electrocardiograms (**a**) showing ventricular ectopic with compensatory pause (marked ☆), (**b**) showing ventricular ectopic blocking next sinus P wave (marked P), (**c**) showing interpolated ventricular ectopic (marked as ☆), and (**d**) showing compensatory pause occurring in the sinus beat following the ventricular ectopic beat (marked ✧). (Reproduced with permission from: Mittal SR. Ventricular Ectopics Beats. Cardiology Today. 2011;15:39–43)

10 Changes in the ECG Following a Ventricular Ectopic Beat

(a) *Compensatory pause*. Pause following an ectopic is usually longer than the sinus cycle length (Fig. 16.9a). It is called compensatory pause. It is highly variable even with unifocal ectopics and depends on several factors including previous cycle length, coupling interval, retrograde conduction of the ventricular ectopic to the atria and sinus node or blocking of the next sinus P wave, etc. (Fig. 16.9b marked P). Sometimes, there may be no significant compensatory pause after the ventricular ectopic. Such ectopics are called interpolated ventricular ectopics (Fig. 16.9c) (marked ☆). In patients with sick sinus syndrome, there may be abnormally long compensatory pause. In some cases of interpolated ectopics, compensatory pause may follow the sinus beat after the ectopic (postponed compensatory pause) (Fig. 16.9d marked ✧). This is usually due to blocking of the second P wave following the ectopic.

(b) *P wave*. Ventricular ectopic may be followed by an atrial or junctional ectopic beat with abnormal P wave configuration (Fig. 16.10 marked ✧). This may suggest sinus node dysfunction. In patients with sinus tachycardia producing fusion of the T and the P waves, post-ectopic pause can unmask sinus P wave, thus clarifying the nature of the tachycardia.

(c) *PR interval*. PR interval of the next sinus beat may be prolonged due to partial retrograde depolarization of the atrioventricular node by the ventricular ectopic (Fig. 16.11a marked ⇩). As the retrograde conduction does not reach the atria, there is no manifestation of retrograde conduction on the electrocardiogram, i.e., there is no abnormal P wave following the ventricular ectopic. This phenomenon is known as concealed (hidden) retrograde conduction. Sometimes the next P wave may be blocked even when occurring after the T wave (Fig. 16.11b marked ⇧). This suggests impaired atrioventricular conduction.

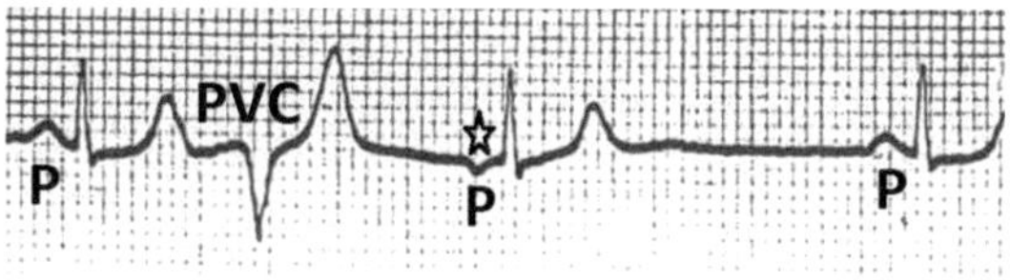

Fig. 16.10 Electrocardiograms showing ventricular ectopic followed by a P wave with abnormal configuration (marked as ☆). (Reproduced with permission from: Mittal SR. Ventricular Ectopics Beats. Cardiology Today. 2011;15:39–43)

(d) *QRS of the next sinus beat*. It may show broadening due to aberrant intraventricular conduction (Fig. 16.12a marked ↑) or fusion of the sinus beat with the second

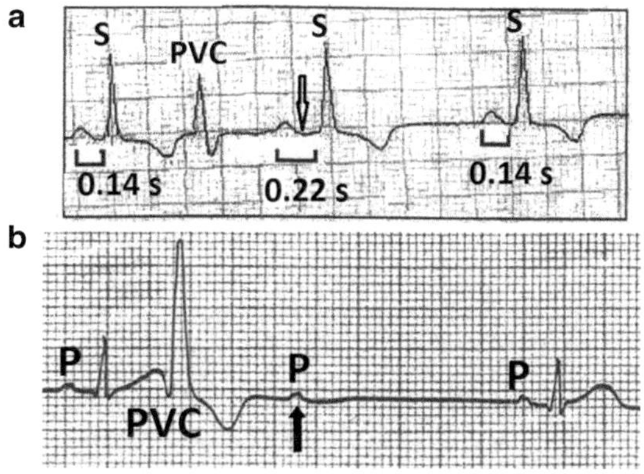

Fig. 16.11 Electrocardiograms (**a**) showing prolongation of the PR interval (marked ⇩) in the sinus beat following the ventricular ectopic beat due to concealed retrograde conduction of the ventricular ectopic producing partial refractoriness of the atrioventricular node and (**b**) showing blocked P wave (marked with ↑) following ventricular ectopic beat. (Reproduced with permission from: Mittal SR. Ventricular Ectopics Beats. Cardiology Today. 2011;15:39–43)

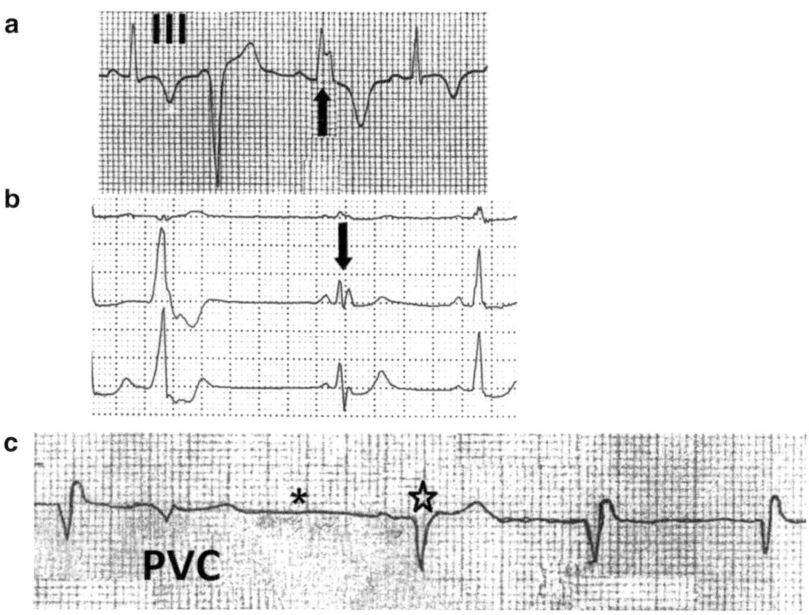

Fig. 16.12 Electrocardiograms (**a**) showing aberrant conduction of the sinus beat following ventricular ectopic beat (marked ↑), (**b**) showing fusion of the next sinus beat with ectopic beat (marked with ↓), and (**c**) showing normalization of the sinus beat (marked as ☆) after the compensatory pause (marked *). (Reproduced with permission from: Mittal SR. Ventricular Ectopics Beats. Cardiology Today. 2011;15:39–43)

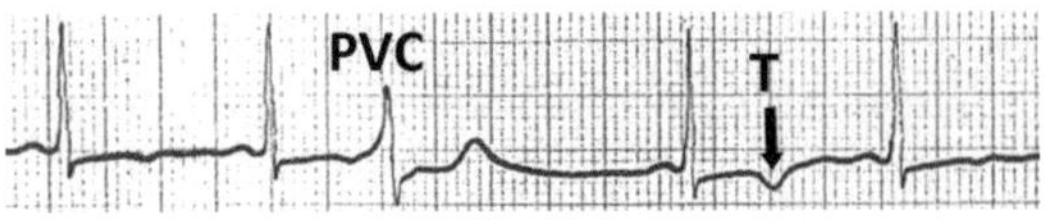

Fig. 16.13 Electrocardiograms showing change in the T wave of the sinus beat (marked ↓) following a ventricular ectopic (marked PVC). (Reproduced with permission from: Mittal SR. Ventricular Ectopics Beats. Cardiology Today. 2011;15:39–43)

ectopic (Fig. 16.12b marked ↓). In patients with preexisting bundle branch block in the sinus rhythm, QRS complex following the compensatory pause may sometimes become normal (Fig. 16.12c marked ☆). This phenomenon suggests that the bundle branch block is rate dependent and disappears due to the longer compensatory pause following the ventricular ectopic (relative bradycardia).

(e) *T wave polarity*. T wave of the sinus beat following a ventricular ectopic beat may change due to abnormal ventricular repolarizations (Fig. 16.13 marked ↓). It is considered a sign of myocardial ischemia [4].

(a) *U wave*. The "U" wave may invert in the sinus beat following the ventricular ectopic beat. It is also considered as suggestive of underlying myocardial ischemia.

11 Clinical Significance [1, 2, 3]

- Incidence of ventricular ectopics increases with advancing age.
- Clinical significance does not depend on morphology, prematurity, or repetitiveness.
- Patients with underlying heart disease, systolic failure, and active ischemia are more susceptible to development of ventricular fibrillation.
- In more than 50% of cases, ventricular tachycardia/ventricular fibrillation starts without any warning arrhythmia.
- Polymorphic nonsustained VT is more likely to degenerate into VF.
- Ectopics appearing during exercise do not adversely affect prognosis. Those appearing during recovery are more significant from the point of view of adverse prognosis [5].

Summary

- Premature ventricular complexes (PVC) are broad complexes occurring earlier than the scheduled sinus beat. The T wave is broad and is in the opposite direction.
- Premature ventricular contractions are not preceded by an abnormal P wave. However, late occurring PVC may follow a sinus P wave with no or short PR segment.
- Retrograde conduction to atria can produce a negative P wave deforming the ST segment or the T wave of the premature ventricular conduction.
- Coupling interval from previous sinus beat may be fixed or variable. Uniform premature ventricular contraction with variable coupling interval but having mathematical distance from each other is known as parasystole and suggests increased automaticity as the underlying mechanism.
- Premature ventricular contraction may be interpolated or may have incomplete or complete compensatory pause. Compensatory pause does not help in differentiating premature ventricular contraction from supraventricular ectopics.
- In interpolated premature ventricular contraction, next sinus P wave is conducted with mild prolongation of the PR interval due to partial refractoriness of the atrioventricular node by retrograde conduction of premature ventricular contraction.
- When a premature ventricular contraction is followed by a retrograde P wave which returns back to excite the ventricle after mildly increased PR interval, it is known as an "echo beat."
- Premature ventricular contraction usually has a duration of 120 msec or more. However, it may not be so wide if:
  - There is fusion of a late occurring premature ventricular contraction with the next sinus beat.
  - Premature ventricular contraction arises from the ventricle showing bundle branch block in resting ECG.
  - Premature ventricular contraction arises at a higher point equidistant from both bundle branches.

- Alternating sinus beat and premature ventricular contraction is called "bigeminy." Two sinus beats followed by a premature ventricular contraction is called "trigeminy." Premature ventricular contraction following three sinus beats is called "quadrigeminy." In contrast to their nomenclature, these patterns show decreasing frequency of premature ventricular contraction.
- Bigeminal sequence (rule of bigeminy) does not differentiate ventricular from a supraventricular premature complex.
- Two successive premature ventricular contractions are called "couplet." Three successive premature ventricular contractions are called "triplet." More than three successive premature ventricular contractions are called ventricular tachycardia. This term denotes repetitiveness of the ectopic and is more sinister than isolated ectopics.
- Configuration of the premature ventricular contraction on the surface ECG can give a reasonable impression about the site of their origin. Left bundle branch block pattern in lead V_1 with inferior axis suggests an origin in the right ventricular outflow tract. Right bundle branch block pattern suggests an origin in the left ventricle. Superior axis suggests origin in the inferior wall, whereas inferior axis suggest origin in the anterior wall. Dominant S wave in leads V_3 and V_4 suggests apical origin, and dominant R wave in leadsV_3 and V_4 suggests basal origin. Site of origin alone does not have clinical significance. Significance depends on the magnitude of structural disease in the culprit ventricle. "Multiform PVCs" need not always be "multifocal."
- Unifocal premature ventricular contraction (having a fixed coupling interval) can have different forms due to the different paths of conduction. Premature ventricular contraction with changing coupling interval and changing configuration which clearly define different sites of origin can be considered multifocal. Definite diagnosis may need an electrophysiologic study. Definitely multifocal premature ventricular contraction is likely to occur in more diffuse disease.
- In an irregularly irregular rhythm, the following features suggest the possibility of a premature ventricular contraction over a supraventricular beat with aberration—short coupling interval, fixed coupling, long pause after abnormal beat, and runs of bigeminy.
- Isolated premature ventricular contraction frequently occurs in persons without any evidence of structural heart disease. In these cases premature ventricular contraction alone does not suggest adverse prognosis and does not require antiarrhythmic therapy. Initially there was some impression that very early premature ventricular contraction (R wave of the ectopic falling on the T wave of previous beat—R on T) or very late (end-diastolic) premature ventricular contraction falling near next sinus P wave is vulnerable to trigger ventricular flutter. However, coupling interval alone may not be very relevant in the absence of any underlying risk factor for ventricular flutter. Seriousness also does not depend only on the coupling interval.
- Premature ventricular contraction is clinically significant when they occur in the presence of structural heart disease, heart failure, or ongoing ischemia. Significance is increased in patients with cardiac diseases with increased risk of ventricular fibrillation, e.g., Brugada syndrome or long QT syndrome and short QT syndrome or idiopathic catecholaminergic polymorphic ventricular tachycardia. It is important to remember that early stages of arrhythmogenic right ventricular dysplasia may be missed on clinical examination.
- High frequency of premature ventricular contraction occurring frequently throughout the day and premature ventricular contraction with very wide QRS complexes may be associated with increased risk of development of dilated cardiomyopathy. Regularly occurring interpolated premature ventricular contraction actually doubles the heart rate and thus causes more stress on the myocardium than premature ventricular contraction occurring as regular bigeminy.
- Increased frequency of premature ventricular contraction during exercise stress test or during recovery may also be associated with increased risk of catecholaminergic polymorphic ventricular tachycardia.

– New appearance or increased frequency during recovery following an exercise stress test may suggest underlying myocardial ischemia.

MCQs

Q1. A broad QRS with normal PR interval, rsR' pattern in lead V_1, and qRS pattern in lead V_6 suggests:

(a) Premature ventricular complex
(b) Supraventricular premature beat with aberration
(c) Supraventricular premature beat with preexcitation
(d) All

Q2. Which electrocardiographic factures of a wide QRS complex support the diagnosis of a premature ventricular contraction?

(a) Initial deflection identical to that of the sinus beat
(b) P wave with normal PR interval precedes the wide QRS complex
(c) Rr' complex in lead V_1
(d) Interval between the P wave preceding the wide QRS complex and the P wave following the broad QRS complex is equal to the interval between three preceding sinus beats

Q3. Which electrocardiographic finding in an irregular rhythm with one broad QRS complex supports the diagnosis of a premature ventricular contraction?

(a) RR interval of the sinus cycle preceding the broad QRS complex is shorter than the interval between the preceding narrow QRS and the broad QRS complex (short cycle-long cycle sequence)
(b) There is a long-short cycle sequence
(c) Other similar long-short cycle sequence in the rhythm strip is not followed by a broad QRS
(d) There is no relatively longer pause after the wide QRS

Q4. PR interval of the sinus beat following an interpolated premature ventricular contraction is:

(a) Normal
(b) Reduced
(c) Increased
(d) Any one of above

Q5. Which of the following findings in lead V_1 suggest left ventricular origin of a ventricular ectopic beat?

(a) Rr' configuration
(b) rSR' configuration
(c) rS configuration
(d) QS configuration

Q6. Which electrocardiographic findings suggest right ventricular origin of a premature ventricular contraction?

(a) Wide r followed by S in lead V_1
(b) R or rR' in lead V_6
(c) rS or QS deeper in lead V_4 than in lead V_1
(d) rsR' in lead V_1

Q7. Premature ventricular contraction with left bundle branch block configuration and inferior axis suggests origin in:

(a) Right ventricular outflow tract
(b) Left ventricular outflow tract
(c) left ventricular apex
(d) Left anterior fascicle

Q8. Premature ventricular contraction with right bundle branch block configuration with left axis suggests origin in:

(a) Right ventricular outflow tract
(b) Left posteroinferior fascicle
(c) Left anterosuperior fascicle
(d) Left ventricular outflow tract

Q9. Premature ventricular contraction with right bundle branch block configuration and right axis suggests origin in:

(a) Right ventricular outflow tract
(b) Left ventricular apex
(c) Left posteroinferior fascicle
(d) Left anterosuperior fascicle

Q10. A premature ventricular beat can have:

(a) Preceding sinus beat
(b) QRS duration less than 120 milliseconds
(c) Incomplete compensatory pause
(d) T wave concordant to major deflection of QRS

Q11. A premature ventricular contraction can have incomplete compensatory pause if there is:

(a) Retrograde conduction to the atria and the sinus node
(b) No retrograde conduction to the atria and the sinus node
(c) Ventricular echo beat
(d) Interpolated premature ventricular contraction

Q12. In the presence of a bundle branch block in sinus rhythm, premature ventricular contraction can be narrow if premature ventricular contraction arises:

(a) From the ventricle having bundle branch block
(b) From the ventricle opposite to the side of bundle branch block
(c) From higher point in the interventricular septum
(d) None

Q13. A combination of sinus P wave, short PR interval, and broad QRS can occur if:

(a) Sinus beat conducts through as accessory pathway
(b) Premature ventricular beat follows onset of a sinus P wave
(c) There is atrial ectopic with aberrant conduction
(d) There is retrograde conduction of a premature ventricular beat

Q14. Trigeminy means:

(a) Two sinus beats followed by one premature ventricular beat
(b) Two premature ventricular beat following one sinus beat
(c) One premature ventricular beat following three sinus beats
(d) Three premature ventricular beat following one sinus beat

Q15. A premature ventricular beat is suspected to be parasystolic if:

(a) Premature ventricular beat is multiformic
(b) Premature ventricular beat occurs in couplet
(c) Coupling interval between preceding sinus beat and the premature ventricular beat is variable
(d) Interval between consecutive premature ventricular beat is multiple of some interval

Q16. Premature ventricular beat may be serious if they:

(a) Have variable coupling
(b) Are multiformic
(c) Occur as "triplet"
(d) Fall on top of the preceding T wave

Q17. Premature ventricular beat can be serious if they occur:

(a) As trigeminy
(b) With long coupling interval
(c) In the presence of ongoing myocardial ischemia
(d) In the presence of heart failure

Q18. Which electrocardiographic feature of a premature ventricular beat may be associated with development of cardiomyopathy?

(a) Duration of premature ventricular beat >0.16 second
(b) Very frequent premature ventricular beat
(c) Interpolated premature ventricular beat
(d) None

Q19. Very wide QRS of a premature ventricular beat may be seen in:

(a) Myocarditis
(b) Cardiomyopathy
(c) Extensive myocardial infarction
(d) Hypokalemia

Q20. New appearance of premature ventricular beats in a patient with heart failure receiving diuretic beta-blocker, ACE inhibitors, and digitalis should raise suspicion of:

(a) Digitalis overdose
(b) Inadequate dose of beta-blocker
(c) Underdose of ACE inhibitors
(d) Hypokalemia

Answers

(1) b; (2) c, d; (3) a, c; (4) c; (5) a; (6) a, b, c; (7) a; (8) b; (9) d; (10) a, b, c; (11) a, c, d; (12) a; (13) a, b; (14) a; (15) c, d; (16) a, c, d; (17) c, d; (18) a, b; (19) a, b, c; (20) a, d

References

1. Olgin JE, Zipes DP. Specific arrhythmias and treatment. In: Libby P, Bonow RO, Mann DL, Zipes DP, editors. Braunwald's heart disease. Philadelphia: Saunders; 2008. p. 863–922.
2. Rho RW, Page RL. Ventricular arrhythmias. In: Fuster V, O'Rourke RA, Walsh RA, Wilson P, editors. Hurst's the heart. New York: Mc Graw Hill; 2008. p. 1003–19.
3. Wagner GS, Wang TY. Premature beats. In: Wagner GS, editor. Marriott's practical electrocardiography. Philadelphia: Lippincott Williams & Wilkins; 2008. p. 261–86.
4. Schamroth L. Coronary insufficiency. In: Schamroth C, editor. An introduction to electrocardiography. Blackwell Science: New Delhi; 1982. p. 157–65.
5. Thomas GS, Balouch M. Rhythm and conduction disturbances in stress testing. In: Thomas GS, Wann LS, Ellestad MH, editors. Ellestad's stress testing. New Delhi: Oxford; 2018. p. 199–228.

17 Broad QRS Tachycardias

When the QRS duration of a tachycardia is 0.12 second or more, it is called "broad QRS tachycardia" [1]. It can be ventricular or supraventricular in origin. Differentiation is important from etiological, therapeutic, and prognostic point of view.

1 Causes of Broad QRS Tachycardia

(a) Originating from the ventricles—ventricular tachycardia (Fig. 17.1)
(b) Originating from above the ventricles—supraventricular tachycardias with either of the following conditions:
   (i). Tachycardia-dependent aberrant (slow) intraventricular conduction (Fig. 17.2). It commonly involves the right bundle branch.
   (ii). Preexisting bundle branch block (Fig. 17.3).
   (iii). Antegrade conduction over an accessory pathway (Fig. 17.4). It results in preexcitation of a part of a ventricle resulting in early onset of QRS. Fusion of this early depolarization (delta wave) with the ventricular depolarization occurring through atrioventricular node results in broadening of the QRS complex.
   (iv). Hyperkalemia, amiodarone, group 1A and 1C antiarrhythmics.

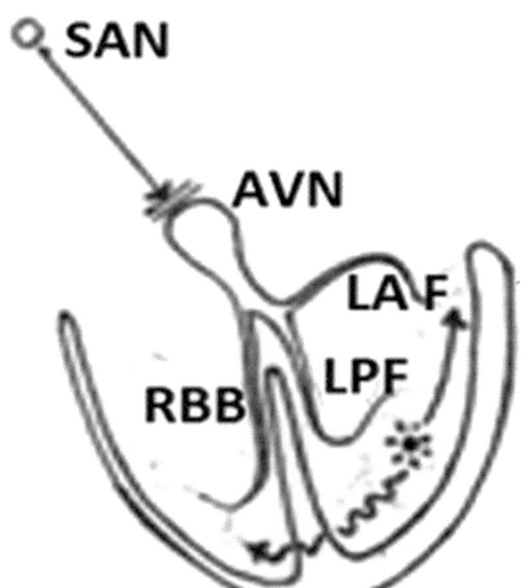

Fig. 17.1 Diagram showing that in tachycardia arising from one ventricle, QRS is broad due to slow conduction of impulse in the other ventricle. *SAN* sinoatrial node, *AVN* atrioventricular node, *LAF* left anterior fascicle, *LPF* left posterior fascicle, *RBB* right bundle branch. (Reproduced with permission from: Mittal SR. Broad QRS Tachycardia. Cardiology Today. 2012;16:265–270)

S. Mittal, *Insights into Electrocardiograms with MCQs*,
https://doi.org/10.1007/978-981-99-0127-2_17

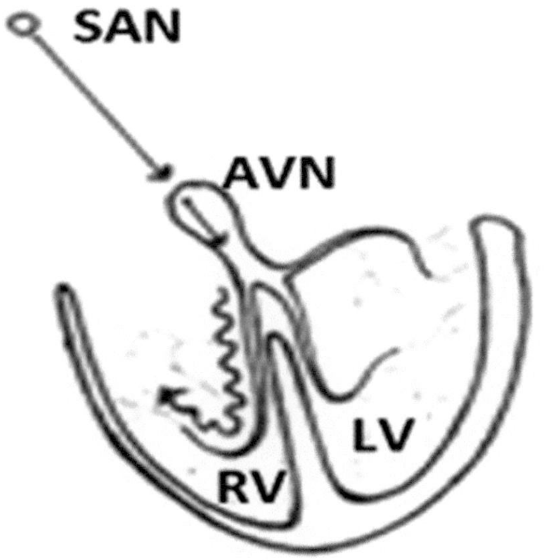

Fig. 17.2 Diagram showing that tachycardia-dependent slow conduction in a bundle branch or fascicle can cause broad QRS in a case of supraventricular tachycardia. *SAN* sinoatrial node, *AVN* atrioventricular node, *LV* left ventricle, *RV* right ventricle. (Reproduced with permission from: Mittal SR. Broad QRS Tachycardia. Cardiology Today. 2012;16:265–270)

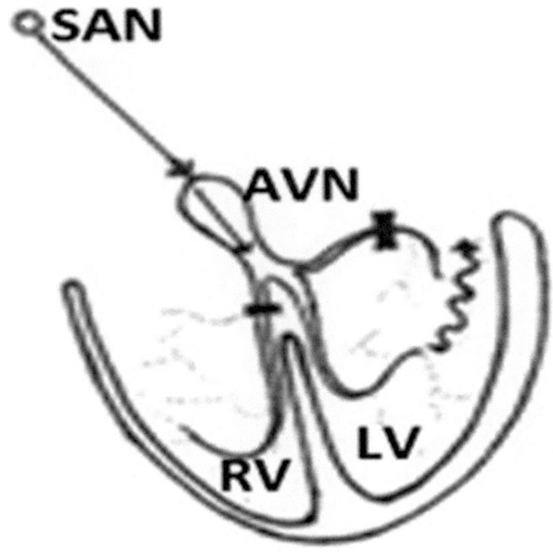

Fig. 17.3 Diagram showing that a preexisting bundle branch block or fascicular block can produce broad QRS in a case of supraventricular tachycardia. *SAN* sinoatrial node, *AVN* atrioventricular node, *LV* left ventricle, *RV* right ventricle. (Reproduced with permission from: Mittal SR. Broad QRS Tachycardia. Cardiology Today. 2012;16:265–270)

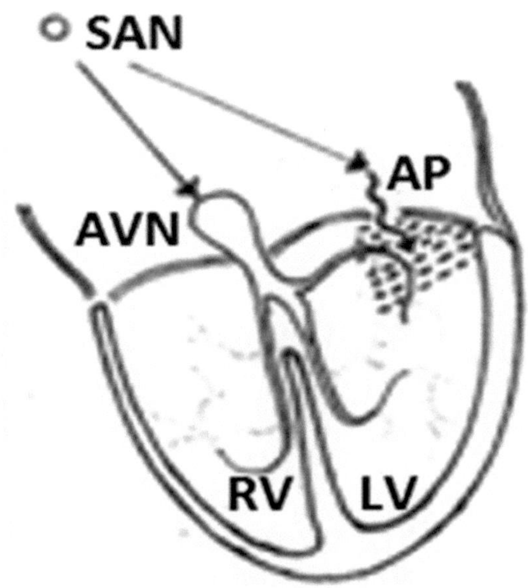

Fig. 17.4 Diagram showing that antegrade conduction through an accessory path (AP) can produce broad QRS in a case of supraventricular tachycardia. *SAN* sinoatrial node, *AVN* atrioventricular node, *LV* left ventricle, *RV* right ventricle. (Reproduced with permission from: Mittal SR. Broad QRS Tachycardia. Cardiology Today. 2012;16:265–270)

2 Differentiating Electrocardiographic Findings

(a) **Atrioventricular dissociation**. If atria are depolarized from the sinoatrial node (rate around 70 to 80/min.) and ventricles are depolarized at a fast rate (>120/min.) from the focus of tachycardia in the ventricle, the P waves and the QRS complexes occur independently. There is no association between the P waves and the QRS complexes (atrioventricular dissociation) (Fig. 17.5).

P waves are best seen in leads II and V_1 as intermittent deformation of the QRS complex, the ST segment, or the T wave (Fig. 17.6a, b).

All leads should be examined to detect atrioventricular dissociation as the P waves may be clear in a particular lead only. This finding strongly suggests that the fast rate is arising from a site below the atria (either the atrioventricular node or the ventricle). However, atrioventricular dissociation is not seen in all patients of ventricular tachycardia. Usually the ventricular impulse conducts retrogradely to atria resulting in "ventriculoatrial association" (Fig. 17.7).

Usually retrogradely conducted P wave merges with the QRS complex and is not identified. At times the retrogradely conducted P waves may follow the QRS complex. Because of retrograde depolarization of the atria, the P waves are inverted in leads II, III, and aVF [2]. In such a situation, it may be difficult to differentiate ventricular tachycardia with retrograde conduction from atrial tachycardia with aberrant ventricular conduction. If the beginning of the tachycardia has been recorded, it may help [3]. Abnormal P wave following the first broad QRS complex supports the diagnosis of ventricular tachycardia (Fig. 17.8a). On the other hand, an abnormal P wave preceding the first broad

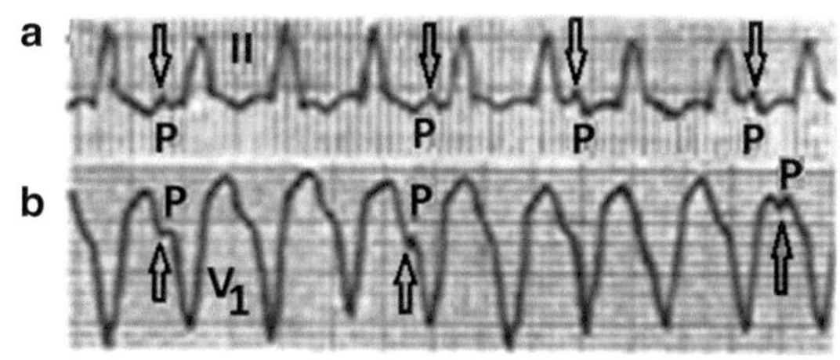

Fig. 17.6 (**a**) Lead II showing dissociation between the P waves and the QRS complexes. (**b**) Lead V_1 showing dissociation between the P waves and the QRS complexes. (Reproduced with permission from: Mittal SR. Broad QRS Tachycardia. Cardiology Today. 2012;16:265–270)

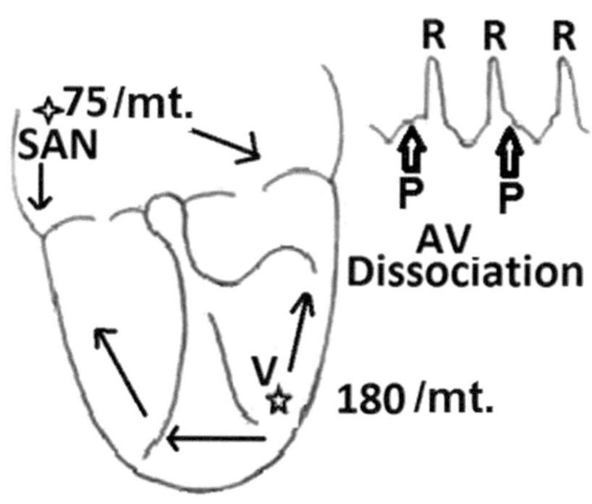

Fig. 17.5 Diagram showing the mechanism of atrioventricular dissociation in ventricular tachycardia. *SAN* sinoatrial node, *V* ventricular focus. (Reproduced with permission from: Mittal SR. Broad QRS Tachycardia. Cardiology Today. 2012;16:265–270)

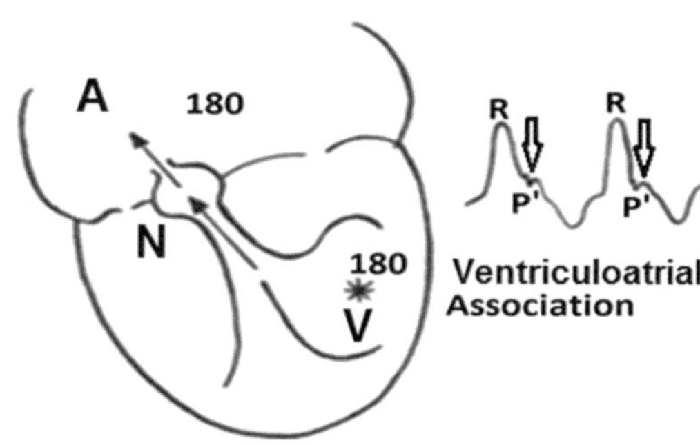

Fig. 17.7 Diagram showing the mechanism of ventriculoatrial association. Abbreviations: *A* atria, *V* ventricle, *P′* retrogradely conducted P waves, *N* atrioventricular node. (Reproduced with permission from: Mittal SR. Broad QRS Tachycardia. Cardiology Today. 2012;16:265–270)

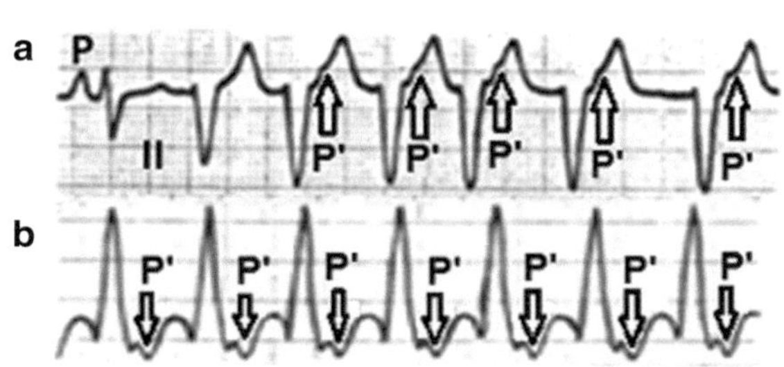

Fig. 17.8 (**a**) Lead II showing the beginning of a broad QRS tachycardia. First abnormal P′ follows first broad QRS. (**b**) Lead II showing 1:1 retrograde conduction to atria (p′—retrogradely conducted P waves) (ventriculoatrial association). (Reproduced with permission from: Mittal SR. Broad QRS Tachycardia. Cardiology Today. 2012;16:265–270)

QRS complex suggests atrial tachycardia with aberrancy. Retrograde conduction to atria may be regular (1:1) (Fig. 17.8b).

There may be intermittent ventriculoatrial block (2:1 or 3:2, etc.) (Fig. 17.9a, b).

(b) **Capture and fusion beats**. In the presence of atrioventricular dissociation, some of the sinus impulses may cross the atrioventricular node to reach the ventricles when they are in repolarized state. Such a sinus impulse depolarizes the ventricles in a normal fashion resulting in a narrow QRS. Such a beat is called "capture beat" (Fig. 17.10).

Such beats produce intermittent narrow QRS complex in a run of broad QRS tachycardia (Fig. 17.11).

Sinus P wave precedes a capture beat [3]. If the sinus impulse reaches the ventricles just before the next broad QRS complex of ventricular tachycardia, the two impulses may fuse in the ventricles (fusion beat) (Fig. 17.12).

In a fusion beat, initial part of the QRS resembles the supraventricular narrow QRS complex and the later part resembles the broad QRS complex of the ventricular tachycardia (Fig. 17.13a, b).

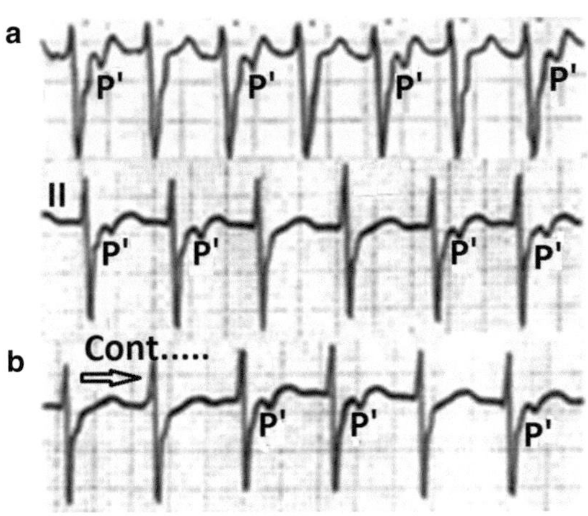

Fig. 17.9 (**a**) Lead II showing 2:1 retrograde conduction to the atria (P′). (**b**) Lead II shows irregular retrograde conduction to the atria (P′). (Reproduced with permission from: Mittal SR. Broad QRS Tachycardia. Cardiology Today. 2012;16:265–270)

The presence of fusion beats and capture beats strongly supports the diagnosis of ventricular tachycardia as a cause of broad QRS tachycardia. Such beats are, however, seen only infrequently (low sensitivity) especially so if the ventricular rate is faster than 160/min [2].

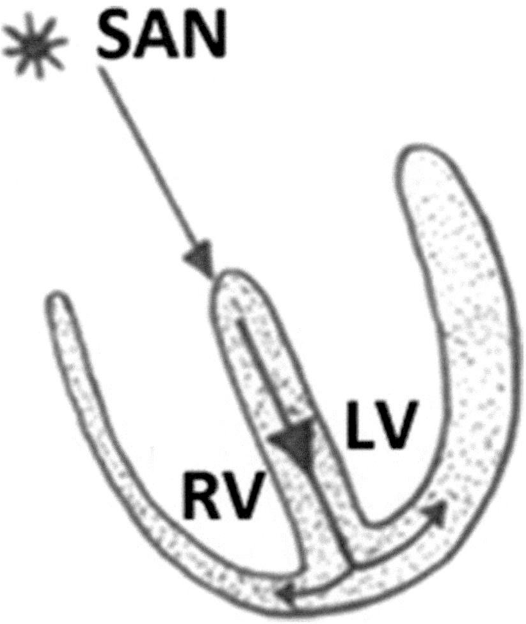

Fig. 17.10 Diagram showing the mechanism of capture beat. *SAN* sinoatrial, *RV* right ventricle, *LV* left ventricle. (Reproduced with permission from: Mittal SR. Broad QRS Tachycardia. Cardiology Today. 2012;16:265–270)

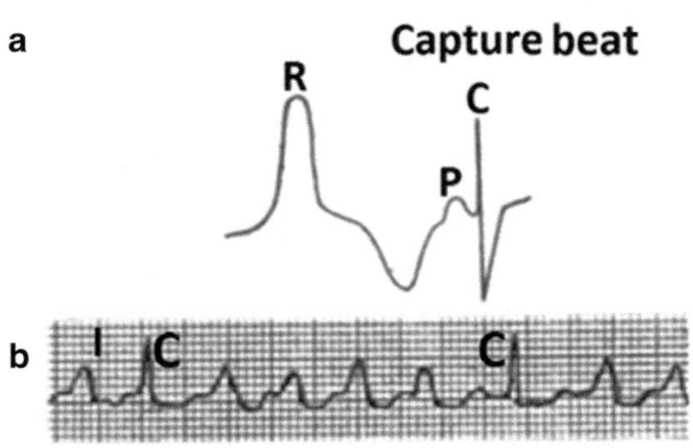

Fig. 17.11 (**a**) Diagram showing capture beat (C). (**b**) Lead I showing capture beats (marked C) in a run of broad QRS tachycardia. (Reproduced with permission from: Mittal SR. Broad QRS Tachycardia. Cardiology Today. 2012;16:265–270)

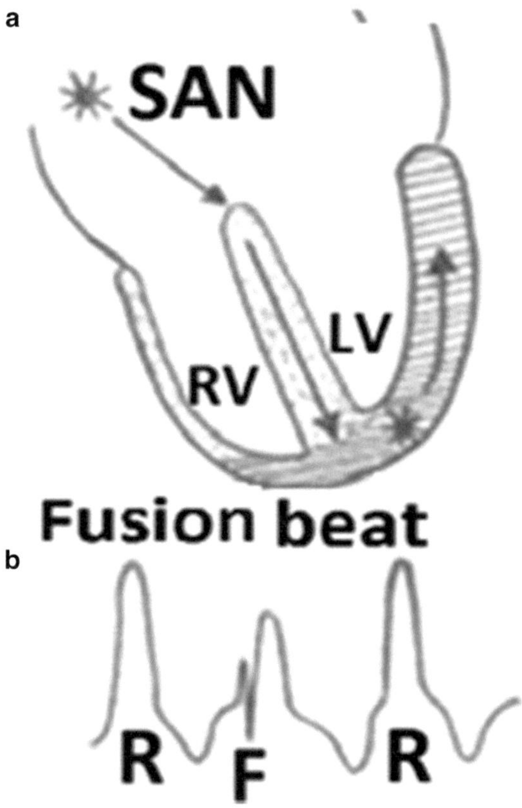

Fig. 17.12 (**a**) Diagram showing the mechanism of fusion beat. *SAN* sinoatrial node, *RV* right ventricle, *LV* left ventricle. (**b**) Diagram showing fusion beat (F). R—broad QRS tachycardia. (Reproduced with permission from: Mittal SR. Broad QRS Tachycardia. Cardiology Today. 2012;16:265–270)

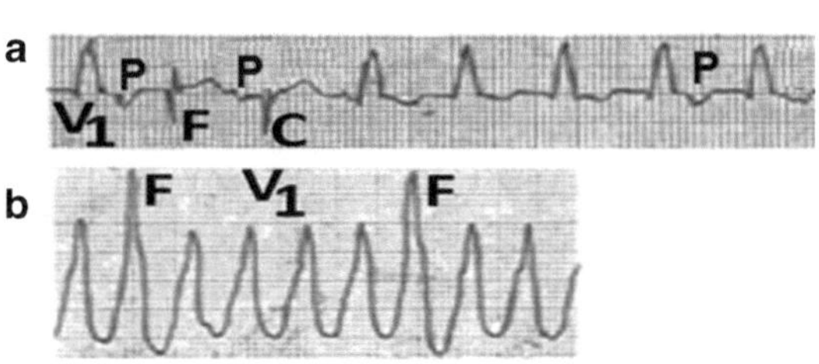

Fig. 17.13 (**a**) Lead V_1 showing broad QRS tachycardia with capture beat (marked C) and fusion beat (marked F), P—sinus P waves. (**b**) Lead V_1 showing broad QRS tachycardia with fusion beats (marked F). (Reproduced with permission from: Mittal SR. Broad QRS Tachycardia. Cardiology Today. 2012;16:265–270)

(c) **Evidence of old myocardial infarction** [1]. Evidence of old myocardial infarction in a previous ECG or during a broad QRS tachy-

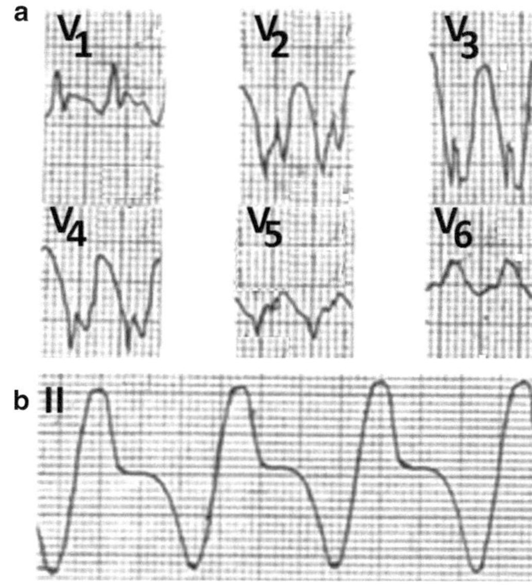

Fig. 17.14 (**a**) Broad QRS tachycardia with QRS duration of >0.16 second. (**b**) Broad QRS due to hyperkalemia. (Reproduced with permission from: Mittal SR. Broad QRS Tachycardia. Cardiology Today. 2012;16:265–270)

cardia supports the possibility of ventricular tachycardia. This finding is, however, nonspecific. Patients with old myocardial infarction can also have supraventricular tachycardia. Also, there may be no unequivocal electrocardiographic signs of myocardial infarction in a broad QRS tachycardia with left bundle branch block configuration.

(d) **Extremely prolonged QRS duration** [2, 4]. QRS duration usually does not exceed 0.12 second in a supraventricular tachycardias with broad QRS. QRS duration of ≥0.14 second usually suggests involvement of the peripheral part of the conduction system of the ventricles. Therefore, an extremely prolonged QRS duration, in a case of broad QRS tachycardia, indirectly suggests ventricular myocardium as the site of origin of the arrhythmia (Fig. 17.14a). QRS duration of >0.14 second in a wide QRS tachycardia of right bundle branch block morphology and >0.16 second in a broad QRS tachycardia with left bundle branch block morphology suggests ventricular tachycardia. This finding, however, has low sensitivity as large numbers of ventricular tachycardias have QRS duration of 0.14 second or less. It is also important to remember that hyperkalemia itself can produce abnormally wide QRS complex (Fig. 17.14b).

(e) **Extremely deviated frontal plane QRS axis** [2, 4]. In supraventricular tachycardia with broad QRS, mean frontal plane QRS axis usually ranges between −90° and +180°. A frontal plane QRS axis between −90° and ±180° (right upper quadrant axis or northwest axis) (Fig. 17.15) supports the possibility of ventricular tachycardia.

It manifests as dominantly negative QRS in leads I, II, and III and positive QRS in lead aVR (Fig. 17.16).

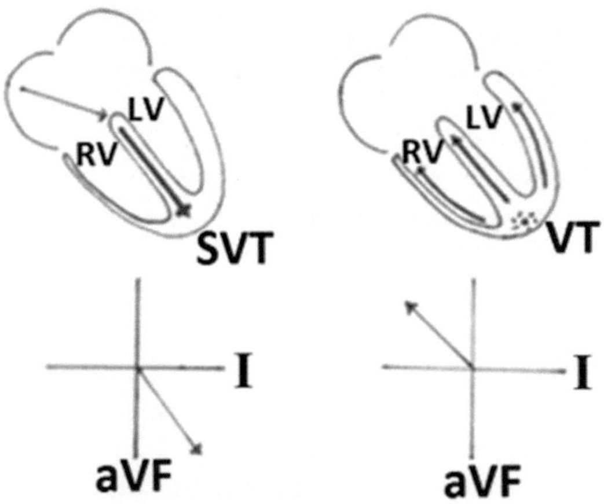

Fig. 17.15 Showing mechanism of right upper quadrant frontal plane QRS axis in ventricular tachycardia. (Reproduced with permission from: Mittal SR. Broad QRS Tachycardia. Cardiology Today. 2012;16:265–270)

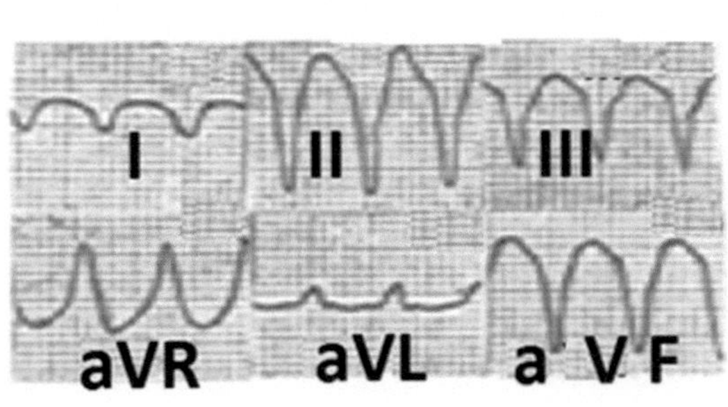

Fig. 17.16 ECG shows right upper quadrant mean frontal plane QRS axis in a case of ventricular tachycardia. (Reproduced with permission from: Mittal SR. Broad QRS Tachycardia. Cardiology Today. 2012;16:265–270)

(f) **Monomorphic QRS complexes in precordial leads**. In a supraventricular impulse, lead V_1 and lead V_6 show opposite configuration of QRS because lead V_1 faces the right ventricle and lead V_6 faces the left ventricle (Fig. 17.17a). When an impulse originates from the ventricles, it is possible that the mean horizontal plane QRS vector moves towards or away from all precordial leads producing identical monomorphic QRS complexes in leads V_1 to V_6 (either all positive or all negative) [1] (Fig. 17.17b).

Representative electrocardiograms are shown in Fig. 17.18a, b.

This finding is strongly suggestive of ventricular tachycardia but has low sensitivity. Atrial fibrillation with preexcitation (antegrade conduction over the left lateral accessory pathway) can also produce dominantly positive QRS in all precordial leads [4]. Irregular rhythm can help in differentiation from ventricular tachycardia.

(g) **Abnormal QRS configuration:**

(i). **Broad QRS tachycardia with dominant positive deflection in lead V_1**

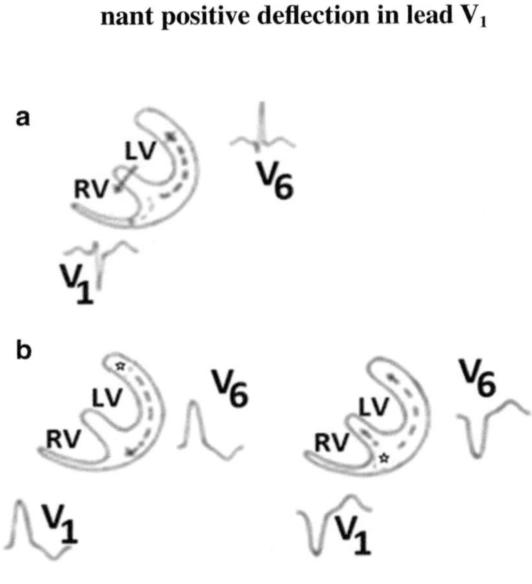

Fig. 17.17 (**a**) Showing mechanism of difference in QRS configuration in lead V_1 and V_6 in a supraventricular impulse. (**b**) Showing mechanism of monomorphic QRS complexes in precordial leads in a case of ventricular tachycardia. Abbreviations: *LV* left ventricle, *RV* right ventricle. (Reproduced with permission from: Mittal SR. Broad QRS Tachycardia. Cardiology Today. 2012;16:265–270)

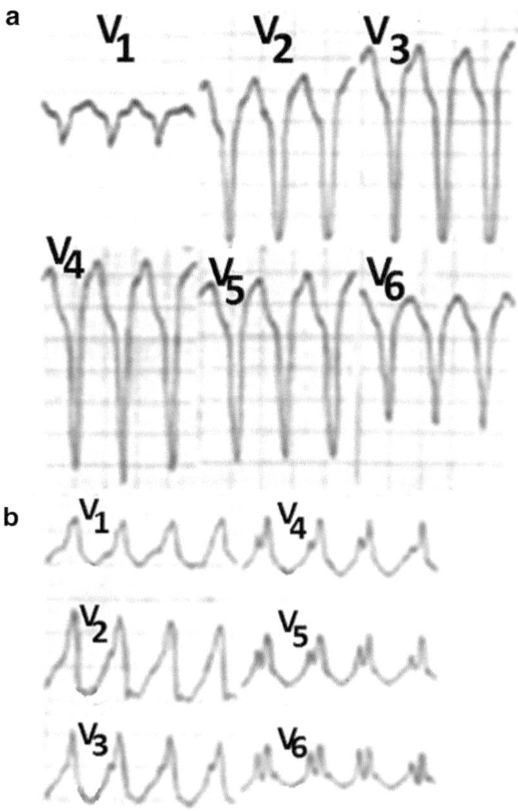

Fig. 17.18 (**a**) Electrocardiogram showing QS complexes in leads V_1 to V_6. (**b**) Electrocardiogram showing positive QRS complexes in leads V_1 to V_6. (Reproduced with permission from: Mittal SR. Broad QRS Tachycardia. Cardiology Today. 2012;16:265–270)

In a supraventricular tachycardia with dominant positive deflection in lead V_1, QRS complexes show classical electrocardiographic features of right bundle branch block. It means that lead V_1 shows rsR'configuration and lead V_6 shows qRS configuration (Fig. 17.19). On the other hand, R, Rsr', or qR configuration in lead V_1 and rs or QS configuration in lead V_6 support the possibility of a ventricular tachycardia [1, 2] (Fig. 17.20).

It is, however, important to remember that right ventricular hypertrophy and posterior myocardial infarction can also produce monophasic R wave in lead V_1.

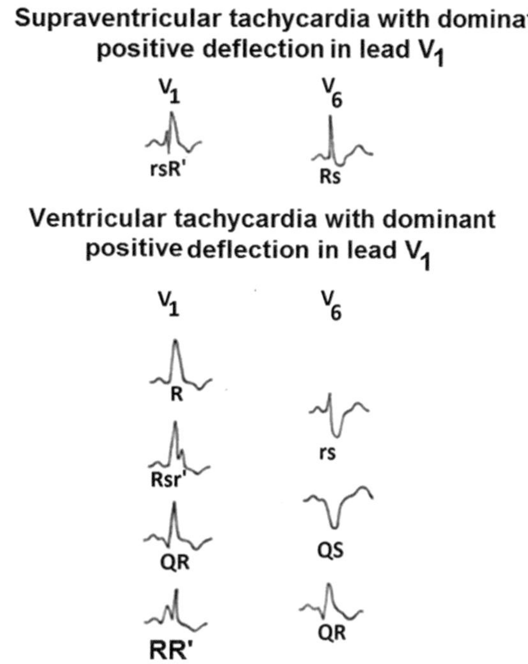

Fig. 17.19 Showing QRS configurations in leads V_1 and V_6 in cases of supraventricular tachycardia and ventricular tachycardia with dominant positive deflection in lead V_1. (Reproduced with permission from: Mittal SR. Broad QRS Tachycardia. Cardiology Today. 2012;16:265–270)

(ii). **Broad QRS tachycardia with dominant positive deflection in lead V_6** Supraventricular tachycardias have classical left bundle branch block pattern, i.e., they produce QS or rS configuration in leads V_1 to V_3 and rsR' configuration in lead V_6 (Fig. 17.21). Broad R ($\geq$ 0.04 second), notch on the downslope of the S wave, or onset of R to nadir of S > 100 milliseconds in lead V_1 or V_2 supports the possibility of ventricular tachycardia [1, 2]. qR or QS configuration in lead V_6 also supports the possibility of ventricular tachycardia [1, 4] (Fig. 17.22).

(h) **Termination by ventricular ectopic**. Supraventricular tachycardia is not affected by a ventricular premature beat. Termination of tachycardia by a ventricular ectopic supports the possibility of ventricular tachycardia.

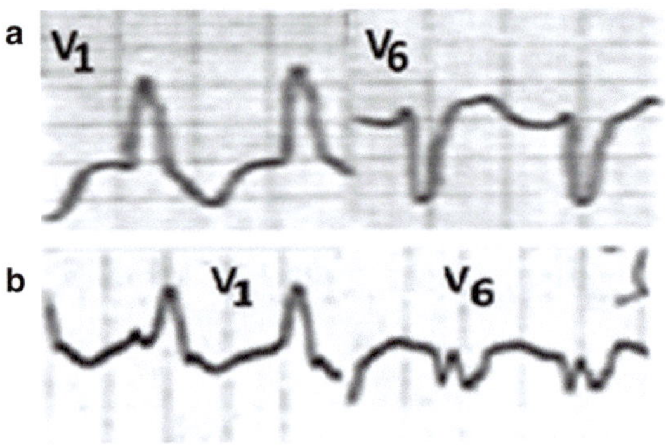

Fig. 17.20 Electrocardiograms (**a**) showing abnormal QRS configurations in leads V_1 and V_6 in two cases of ventricular tachycardia. (**b**) Electrocardiogram showing leads V_1 and V_6 from another case of ventricular tachycardia. (Reproduced with permission from: Mittal SR. Broad QRS Tachycardia. Cardiology Today. 2012;16:265–270)

(i) **Initial QRS vector in lead V_1 different from the initial QRS vector of sinus rhythm**. In sinus rhythm, initial part of QRS is formed by left to right depolarization of the interventricular septum. In a case of supraventricular tachycardia, the initial vector remains identical to that of the sinus beat. On the other hand, if the impulse arises from the ventricles, the initial QRS vector is different from that of the sinus beat (Fig. 17.23).

(j) **Initiation of broad QRS tachycardia by a fusion beat**. In supraventricular tachycardia with aberrancy, the first and the subsequent QRS complexes of broad QRS tachycardia are identical. If the first beat of a broad QRS tachycardia shows a fusion complex, it suggests that the fast rate is arising from the ventricles (Fig. 17.24).

(k) **Other findings supporting supraventricular origin of tachycardia:**

   (i). Irregular rhythm without change in QRS configuration.

   (ii). Electrocardiogram showing irregularly irregular rhythm without any change in QRS configuration is suggestive of atrial fibrillation with preexisting bundle branch block (Fig. 17.25).

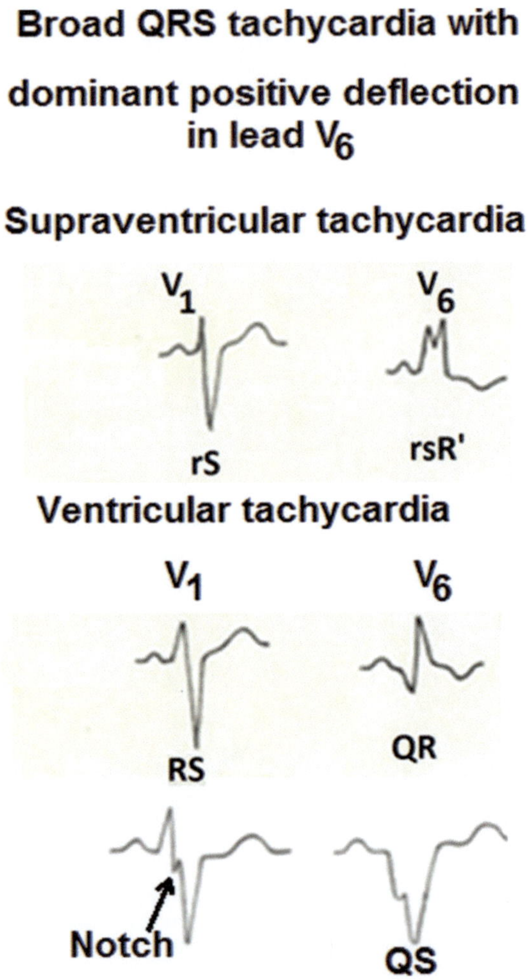

Fig. 17.21 Showing QRS configuration in cases of supraventricular tachycardia and in cases of ventricular tachycardia with dominant positive deflection in lead V_6. (Reproduced with permission from: Mittal SR. Broad QRS Tachycardia. Cardiology Today. 2012;16:265–270)

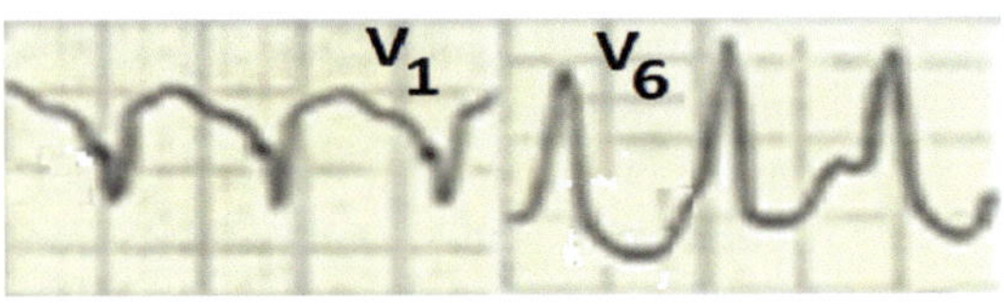

Fig. 17.22 Electrocardiogram showing leads V_1 and V_6 from a case of ventricular tachycardia. (Reproduced with permission from: Mittal SR. Broad QRS Tachycardia. Cardiology Today. 2012;16:265–270)

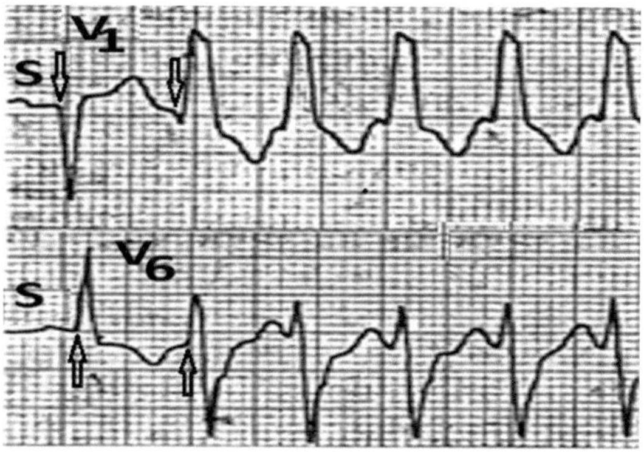

Fig. 17.23 Showing onset of supraventricular tachycardia with aberrancy. Initial QRS vector of the broad QRS tachycardia is identical to that of the sinus beat -S (marked arrows). First and subsequent QRS complexes of the broad QRS tachycardia are also identical. (Reproduced with permission from: Mittal SR. Broad QRS Tachycardia. Cardiology Today.2012;16:265–270)

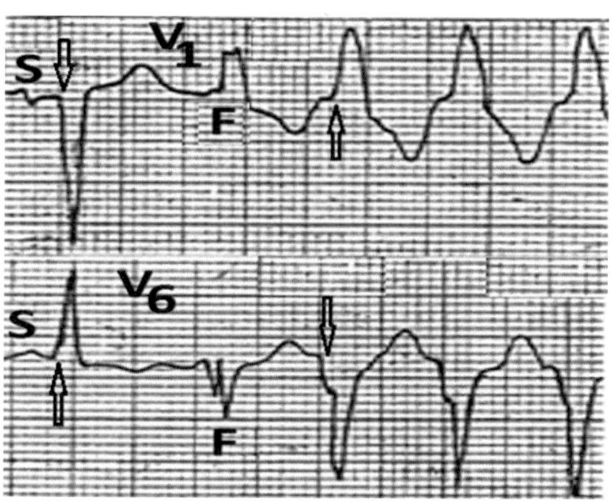

Fig. 17.24 Showing onset of ventricular tachycardia. The first beat of the broad QRS tachycardia is a fusion beat (F). Initial QRS vector of broad QRS tachycardia is also different from that of the sinus beat (marked as arrows). (Reproduced with permission from: Mittal SR. Broad QRS Tachycardia. Cardiology Today. 2012;16:265–270)

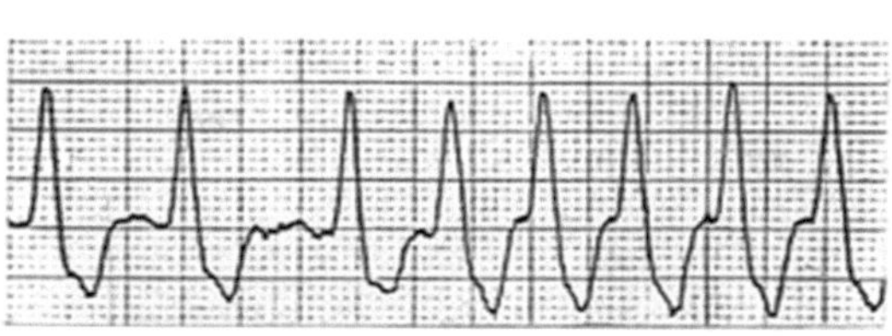

Fig. 17.25 Electrocardiogram showing atrial fibrillation with preexisting bundle branch block. (Reproduced with permission from: Mittal SR. Broad QRS Tachycardia. Cardiology Today. 2012;16:265–270)

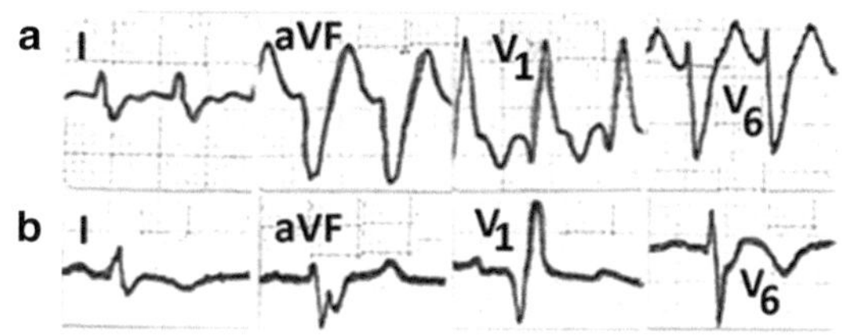

Fig. 17.26 Showing broad QRS tachycardia (**a**) with QRS complexes identical to those seen during the sinus rhythm (**b**). (Reproduced with permission from Mittal SR. Broad QRS Tachycardia. Cardiology Today. 2012;16:265–270)

(iii). Irregular rhythm with rate-dependent change in QRS configuration suggests atrial fibrillation with aberrant ventricular conduction.
(iv). Atrioventricular block.
(v). Slowing of tachycardia rate by vagal maneuver.
(vi). Grossly irregular wide QRS tachycardia with a rate exceeding 200/min. suggests atrial fibrillation with conduction over an accessory pathway.
(vii). Wide QRS tachycardia with QRS configuration similar to that in the sinus rhythm suggests preexisting bundle branch block (Fig. 17.26).

Summary

- Broad QRS tachycardia is usually a ventricular tachycardia. However supraventricular tachycardias can also have broad QRS in the following situations:
  - Preexisting bundle branch block
  - Aberrant ventricular conduction
  - Antegrade conduction over an accessory pathway
- Capture beats, fusion beats, and atrioventricular dissociation strongly support the diagnosis of ventricular tachycardia but occur infrequently.
- The beginning of a broad QRS tachycardia with an atrial ectopic beat supports the possibility of supraventricular tachycardia with aberrant conduction.

- Some features of QRS support possibility of ventricular tachycardia over a supraventricular tachycardia:
  - Mean frontal plane QRS axis in the right upper quadrant.
  - QRS duration of more than 140 milliseconds.
  - Symmetrical positive or negative deflection of the QRS in leads V_1 to V_6. Antegrade conduction of supraventricular tachycardia over left lateral accessory pathway can also produce positive QRS complexes in leads V_1 to V_6.
  - In ventricular tachycardia with dominant positive deflection in lead V_1:
  - Lead V_1 showing monophasic (R) or biphasic (RR') configuration
    - QR configuration
    - Rr' configuration
    - Initial deflection different from that of the sinus beat

    Lead V_6 showing rs or QS configuration

    rsR' pattern in lead V_1 and qRS configuration in lead V_6 support possibility of supraventricular tachycardia with aberrancy.
  - In ventricular tachycardia with dominantly positive deflection in lead V_6:
    - Lead V_1 showing broad R (>40 milliseconds)
    - Deepest S in lead V_4 rather than in lead V_2
    - Lead V_6 showing qR or QS configuration

    Narrow r with deep S in lead V_1 and RR' configuration in lead V_6 support possibility of supraventricular tachycardia with aberrancy.
  - QRS morphology may be affected by previous myocardial infarction.
- In a case of preexisting bundle branch block, ventricular tachycardia can be suspected if the QRS configuration during tachycardia is different from the QRS in the sinus rhythm.
- In irregular rhythm:
  - Identical QRS complexes suggest preexisting bundle branch block.
  - Increasing abnormality in QRS with increasing rate suggests aberrancy.
- Most of the abovementioned criteria have low sensitivity (seen infrequently). Also, these criteria have been developed from patients with coronary artery disease. Their specificity and sensitivity in diagnosing ventricular tachycardia from other causes is not clear. In a hemodynamically compromised patient, a broad QRS tachycardia should be treated as ventricular tachycardia without investing time in detailed analysis of the ECG.

MCQs

Q1. Broad QRS tachycardia with normal P wave and normal PR interval can be due to:

(a) Sinus tachycardia with preexisting bundle branch block
(b) Sinus tachycardia with rate-dependent bundle branch block
(c) Sinus tachycardia with antegrade conduction over an accessory pathway
(d) All

Q2. Abnormal P wave (not inverted in leads II, III, and aVF) preceding each QRS in a case of regular wide QRS tachycardia suggests:

(a) Sinus node reentry tachycardia with preexisting bundle branch block
(b) Atrial tachycardia with aberrant ventricular conduction
(c) Atrioventricular nodal reentry tachycardia with aberrant ventricular conduction
(d) All

Q3. In a case of regular wide QRS tachycardia, P wave following each QRS and inverted in leads II, III, and aVF suggests:

(a) Atrioventricular reentry tachycardia with antegrade conductions over accessory pathway
(b) Atrioventricular reentry tachycardia with orthodromic conduction

(c) Atrioventricular nodal reentry tachycardia with aberrant ventricular conduction
(d) Ventricular tachycardia with retrograde conduction

Q4. In a case of irregular broad QRS tachycardia, fixed morphology of QRS suggests:

(a) Aberrant ventricular conduction
(b) Preexisting bundle branch block
(c) Antegrade conduction over an accessory pathway
(d) None

Q5. In a case of irregular broad QRS tachycardia, changing configuration of QRS suggests:

(a) Aberrant ventricular conduction
(b) Preexisting bundle branch block
(c) Conduction over an accessory pathway
(d) Ventricular ectopics

Q6. Irregularly irregular broad QRS tachycardia with fixed QRS morphology suggests:

(a) Atrial fibrillation with aberrant ventricular conduction
(b) Atrial tachycardia with changing atrioventricular conduction and preexisting bundle branch block
(c) Atrial flutter with aberrant intraventricular conduction
(d) Ventricular tachycardia

Q7. Regular monomorphic wide QRS tachycardia with atrioventricular dissociation suggests:

(a) Sinus node reentry tachycardia with preexisting bundle branch block
(b) Atrial tachycardia with preexisting bundle branch block
(c) Accelerated junctional rhythm
(d) Ventricular tachycardia

Q8. Which of the following is unlikely to produce wide QRS tachycardia?

(a) Atrial fibrillation with antegrade conduction over an accessory pathway
(b) Orthodromic atrioventricular reentry tachycardia with tachycardia-dependent aberrant conduction
(c) Junctional tachycardia with retrograde conduction
(d) Ventricular tachycardia with 1:1 retrograde conduction

Q9. Which of the following is unlikely to produce broad QRS tachycardia?

(a) Atrioventricular reentry (reciprocating) tachycardia with antegrade conduction over accessory pathway
(b) Atypical atrioventricular nodal reentry tachycardia
(c) Atrioventricular reentry (reciprocating) tachycardia with antegrade conduction over one accessory pathway and retrograde conduction over another accessory pathway
(d) Atrioventricular reciprocating tachycardia using nodofascicular fibers

Q10. A narrow QRS complex during a broad QRS tachycardia suggests diagnosis of:

(a) Supraventricular tachycardia with aberrant ventricular conduction
(b) Supraventricular tachycardia with fixed bundle branch block
(c) Ventricular tachycardia
(d) Antidromic atrioventricular reentry (reciprocating) tachycardia

Q11. A regular broad QRS tachycardia with positive deflection in all precordial leads suggests:

(a) Ventricular tachycardia
(b) Antidromic atrioventricular reentry (reciprocating) tachycardia with antegrade conduction over left lateral accessory pathway
(c) Supraventricular tachycardia with aberrant conduction over the right bundle branch
(d) Supraventricular tachycardia with aberrant conduction over the left bundle branch

Q12. A regular broad QRS tachycardia with QS configuration in leads V_1 to V_6 with deepest QS in leads V_4 and V_5 suggests:

(a) Ventricular tachycardia
(b) Supraventricular tachycardia with old extensive anterior infarction
(c) Supraventricular tachycardia with left anterosuperior fascicular block
(d) Supraventricular tachycardia with left posteroinferior fascicular block

Q13. Irregular broad QRS tachycardia at a rate of 120/minute with changing QRS configuration can be due to:

(a) Atrial flutter with 2:1 atrioventricular conduction and preexisting bundle branch block
(b) Atrial fibrillation with aberrant ventricular conduction
(c) Atrial tachycardia with aberrant ventricular conduction
(d) All

Q14. A regular broad QRS tachycardia with QRS duration of 0.12 seconds, rsR' pattern in lead V_1, and qRS configuration in lead V_6 suggests:

(a) Supraventricular tachycardia with aberrant conduction over the right bundle branch
(b) Supraventricular tachycardia with preexisting left bundle branch block
(c) Ventricular tachycardia arising from the left ventricle
(d) All

Q15. A regular broad QRS tachycardia with QRS duration of 0.12 second, rS configuration in lead V_1, and RR' configuration in V_6 suggests:

(a) Supraventricular tachycardia with preexisting left bundle branch block
(b) Supraventricular tachycardia with aberrant conduction over the left bundle branch
(c) Ventricular tachycardia arising from the right ventricle
(d) Supraventricular tachycardia with antegrade conduction over the right posteroseptal accessory pathway

Q16. Regular broad QRS tachycardia can occur in:

(a) Atrial tachycardia with preexisting bundle branch block
(b) Atrioventricular nodal reentry tachycardia with tachycardia-dependent aberrant conduction
(c) Sinus tachycardia with antegrade conduction over an accessory path
(d) All

Q17. Regular broad QRS tachycardia with fusion beats is diagnostic of:

(a) Atrioventricular reentry (reciprocating) tachycardia with preexisting bundle branch block
(b) Atrial tachycardia with rate-dependent aberrant conduction
(c) Sinus tachycardia with antegrade conduction over an accessory pathway
(d) Ventricular tachycardia

Q18. Absence of which electrocardiographic finding excludes any possibility of ventricular tachycardia:

(a) Capture beats
(b) Fusion beats
(c) Atrioventricular dissociation
(d) None

Q19. Beginning of a broad QRS tachycardia with an abnormal P wave suggests possibility of:

(a) Sinus tachycardia with preexisting bundle branch block
(b) Atrial tachycardia with aberrant ventricular conduction
(c) Ventricular tachycardia with retrograde conduction
(d) Atrioventricular reciprocating tachycardia with antidromic conduction

Q20. Broad QRS tachycardia with normal P wave, PR interval of 0.12 second, and mean frontal plane QRS axis of $+120^0$ supports the possibility of:

(a) Sinus tachycardia with aberrant conduction over the right bundle branch
(b) Atrioventricular nodal reentry tachycardia with preexisting right bundle branch
(c) Ventricular tachycardia arising from the left ventricle
(d) Sinus tachycardia with conduction along right lateral accessory pathway

Q21. Broad QRS tachycardia with left bundle branch configuration, QRS duration of 0.16 seconds, R wave of more than 40 millisecond in lead V_1, and deepest S in lead V_4 suggests possibility of:

(a) Sinus tachycardia with preexisting left bundle branch block
(b) Sinus tachycardia with aberrant conduction over left bundle branch
(c) Ventricular tachycardia arising from the right ventricle
(d) Atrioventricular reciprocating tachycardia with antegrade conduction over the left lateral accessory path

Q22. Broad QRS tachycardia with normal P wave, PR interval of 0.16 second with monophasic R in lead V_1, and QS configuration in lead V_6 suggests possibility of:

(a) Sinus tachycardia with preexisting right bundle branch block
(b) Atrial tachycardia with aberrant conduction over the right bundle branch
(c) Ventricular tachycardia arising from left ventricle
(d) All

Q23. Irregularly irregular broad QRS tachycardia with fixed configuration of QRS complex suggests:

(a) Atrial fibrillation with aberrant conduction
(b) Atrial fibrillation with preexisting bundle branch block
(c) Atrial fibrillation with antegrade conduction over an accessory path
(d) All

Q24. Irregularly irregular broad QRS tachycardia with changing QRS configuration could be due to:

(a) Atrial fibrillation with aberrant conduction
(b) Atrial fibrillation with antegrade conduction over an accessory path
(c) Atrial fibrillation with preexisting bundle branch block
(d) Atrial tachycardia with preexisting bundle branch block

Q25. Grossly irregular wide QRS tachycardia with a rate of 250/minute suggests:

(a) Atrial fibrillation with aberrant ventricular conduction
(b) Atrial fibrillation with antegrade conduction over an accessory path
(c) Orthodromic atrioventricular reciprocating tachycardia
(d) Ventricular tachycardia

Q26. In a case of regular broad QRS tachycardia, inverted P wave on the ST segment suggests:

(a) Atrial tachycardia with fixed bundle branch block
(b) Atrial tachycardia with changing atrioventricular block
(c) Atrioventricular nodal reentry tachycardia with preexisting bundle branch block
(d) Ventricular tachycardia with 1:1 retrograde conduction

Q27. In a case of broad QRS tachycardia, which of the following findings suggest a supraventricular origin?

(a) Atrioventricular block
(b) Slowing by carotid massage
(c) Termination by a ventricular ectopic
(d) Onset with an atrial ectopic beat

Q28. Monomorphic QRS complexes in leads V_1 to V_6 in case of regular broad QRS tachycardia could occur in:

(a) Ventricular tachycardia
(b) Orthodromic atrioventricular reciprocating tachycardia with antegrade conduction over left lateral accessory path
(c) Atrial tachycardia with aberrant ventricular conduction
(d) Sinus tachycardia with preexisting bundle branch block

Answers

(1) a, b; (2) b; (3) a, d; (4) b; (5) a; (6) b; (7) d; (8) c; (9) b; (10) c; (11) a, b; (12) a; (13) b; (14) a; (15) a; (16) d; (17) d; (18) d; (19) b; (20) a; (21) c; (22) c; (23) b; (24) a, b; (25) b; (26) c, d; (27) a, b, d; (28) a, b

References

1. Olgin JE, Zipes DP. Specific arrhythmias: diagnosis & treatment. In: Libby P, Bonow RO, Mann DL, Zipes DP, editors. Braunwald's heart disease. Philadelphia: Saunders; 2008. p. 863–922.
2. Gilbert M, Wagner GS. Re-entrant ventricular tachyarrhythmias. In: Wagner GS, editor. Marriot's practical electrocardiography. Philadelphia: Wolters Klawer; 2008. p. 347–64.
3. Schamroth L. Ventricular rhythm. In: Schamroth C, editor. An introduction to electrocardiography. France: Blackwell Science; 1982. p. 349–66.
4. Schamroth L. Aberrant ventricular conduction. In: Schamroth C, editor. An introduction to electrocardiography. France: Blackwell Science; 1982. p. 411–28.

18 Ventricular Tachyarrhythmias

1 Definition

Three or more consecutive QRS complexes originating from the ventricle with a rate of more than 120/min are termed ventricular tachycardia [1]. When the rate is less than 120/min, it is termed accelerated idioventricular rhythm. QRS duration is 120 milliseconds or more with ST segment and T wave vector pointing opposite to the major QRS deflection [2].

2 Accelerated Idioventricular Rhythm (Fig. 18.1)

It is commonly observed following thrombolytic therapy for acute myocardial infarction or during percutaneous coronary intervention. It can occur in other cardiac diseases but at times can occur without any detectable heart disease. There is acceleration of ventricular rhythm above the sinus rate with atrioventricular dissociation.

It is generally a transient rhythm and does not cause hemodynamic instability. It does not need any active management.

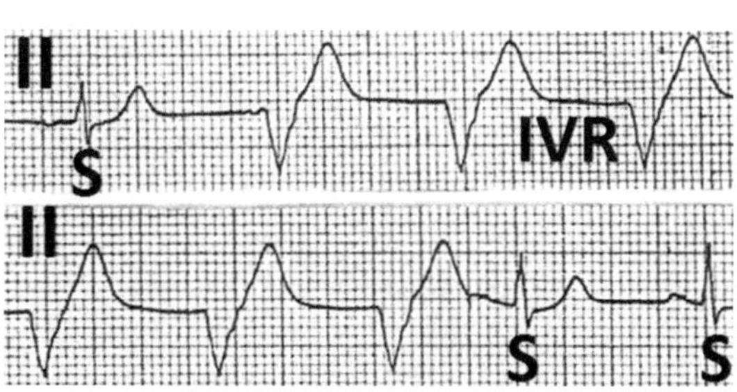

Fig. 18.1 Electrocardiogram showing accelerated idioventricular rhythm (IVR). *S* sinus beat. (Reproduced with permission from: Mittal SR. Ventricular Tachyarrhythmias. Cardiology Today. 2012;16:219–226)

3 Ventricular Tachycardia (VT)

3.1 Nomenclature [1, 2]

- Nonsustained. Ventricular tachycardia lasting for less than 30 seconds is termed nonsustained or ill-sustained ventricular tachycardia.
- Paroxysmal. Ventricular tachycardias which occur in episodes are called paroxysmal ventricular tachycardias.
- Repetitive. When sinus beats are interposed between the runs of nonsustained ventricular tachycardia, it is termed repetitive ventricular tachycardia (Fig. 18.2).
- Monomorphic. When the QRS morphology is uniform from beat to beat, it is called monomorphic ventricular tachycardia (Fig. 18.2).
- Polymorphic. When the QRS morphology varies during the same episode of ventricular tachycardia, it is known as polymorphic ventricular tachycardia (Fig. 18.3).
- Right bundle branch block pattern. When the ECG during ventricular tachycardia resembles right bundle branch block, it is termed right

S. Mittal, *Insights into Electrocardiograms with MCQs*,
https://doi.org/10.1007/978-981-99-0127-2_18

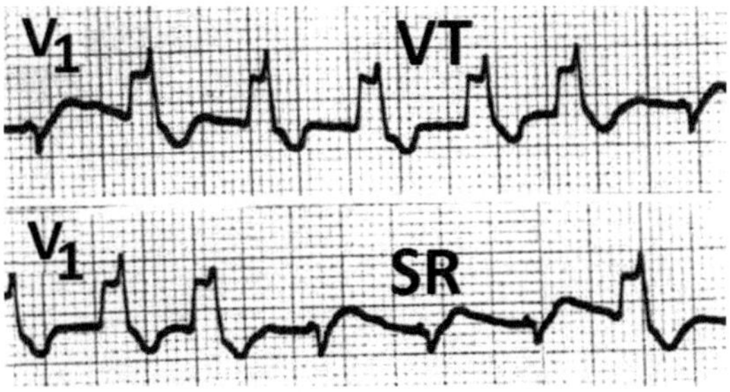

Fig. 18.2 Electrocardiogram showing monomorphic nonsustained ventricular tachycardia. *SR* sinus rhythm. (Reproduced with permission from: Mittal SR. Ventricular Tachyarrhythmias. Cardiology Today. 2012;16:219–226)

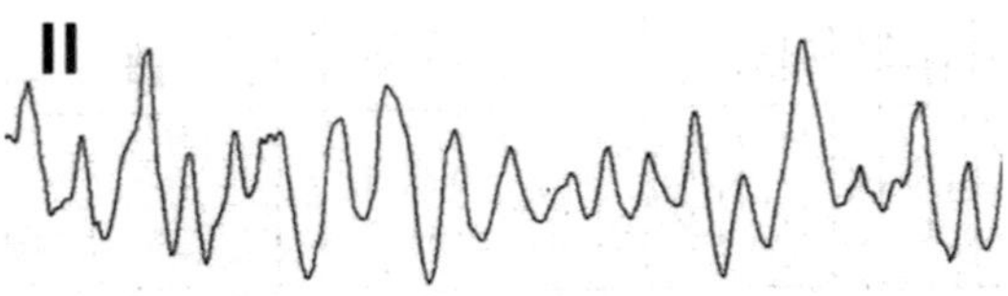

Fig. 18.3 Electrocardiogram is showing polymorphic ventricular tachycardia. (Reproduced with permission from: Mittal SR. Ventricular Tachyarrhythmias. Cardiology Today. 2012;16:219–226)

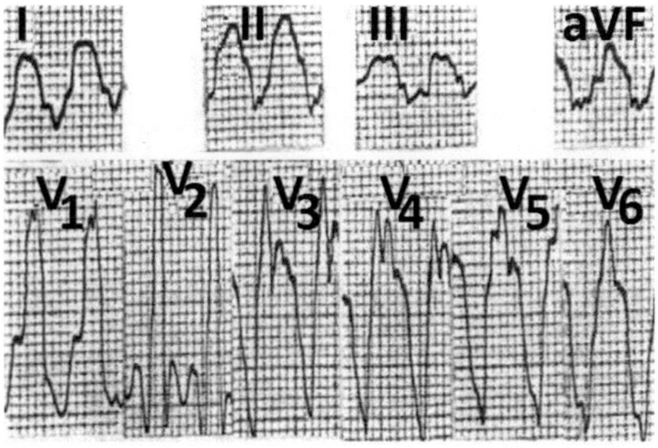

Fig. 18.4 Electrocardiogram showing ventricular tachycardia with right bundle branch block configuration. (Reproduced with permission from: Mittal SR. Ventricular Tachyarrhythmias. Cardiology Today. 2012;16:219–226)

bundle branch block pattern ventricular tachycardia (Fig. 18.4).

– Terminal portion of the QRS complex is positive in lead V_1. These are usually left ventricular in origin.
– Left bundle branch block pattern. When the ECG during ventricular tachycardia resembles left bundle branch block, it is termed left bundle branch block pattern ventricular tachycardia (Fig. 18.5).

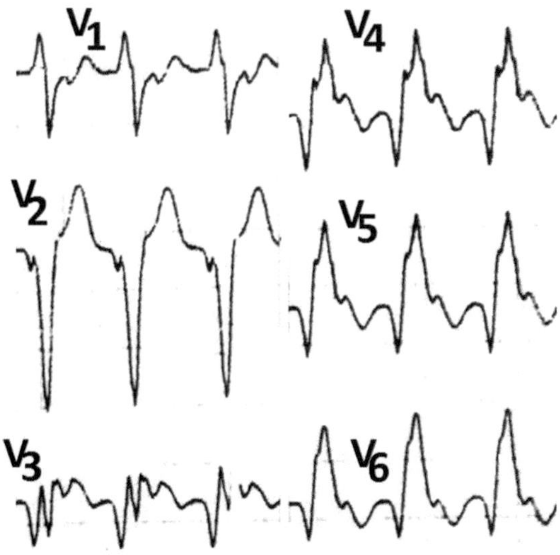

Fig. 18.5 Electrocardiogram showing ventricular tachycardia with left bundle branch block configuration. (Reproduced with permission from: Mittal SR. Ventricular Tachyarrhythmias. Cardiology Today. 2012;16:219–226)

Terminal portion of the QRS complex is negative in lead V_1. These are usually right ventricular in origin. Ventricular tachycardias arising from right side of the interventricular septum and right ventricular outflow tract also exhibit left bundle branch block pattern.

3.2 Specific Types of Ventricular Tachycardia [3–6]

(a) **Monomorphic ventricular tachycardia**. ECG features depend on the site of origin of the ventricular tachycardia and its etiology.
  (i). **Sites of origin:**
    - **Septal versus free wall**. Narrow QRS suggests septal origin of the ventricular tachycardia. Short duration of the QRS complex in septal origin is due to rapid spread of excitation down the septal Purkinje network and bundle branches (Fig. 18.6a). QRS duration of greater

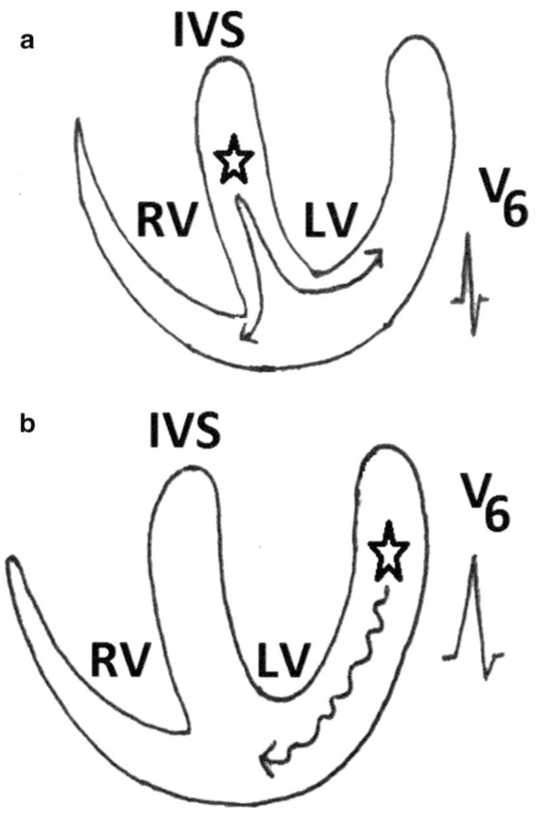

Fig. 18.6 (**a**) Diagram showing rapid spread of impulse in septal origin of ventricular tachycardia (reproduced with permission from: Mittal SR. Ventricular Tachyarrhythmias. Cardiology Today. 2012;16:219–226). (**b**) Diagram showing slow conduction of impulse in left ventricular free wall origin of the ventricular tachycardia. (Reproduced with permission from: Mittal SR. Ventricular Tachyarrhythmias. Cardiology Today. 2012;16:219–226)

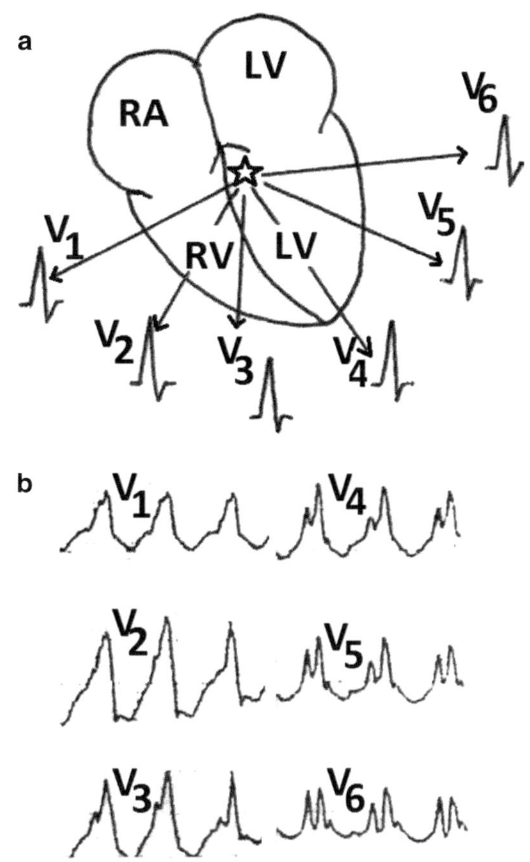

Fig. 18.7 (**a**) Diagram showing the basal origin of ventricular tachycardia and its effect on the precordial leads. (**b**) Electrocardiogram showing the positive QRS complexes in all precordial leads in a ventricular tachycardia arising from base of the heart. (Reproduced with permission from: Mittal SR. Ventricular Tachyarrhythmias. Cardiology Today. 2012;16:219–226)

than 140 ms predicts slow conduction due to free wall origin of the ventricular tachycardia (Fig. 18.6b).

- **Basal versus apical origin of ventricular tachycardia.** Differentiating ECG features include the following:
  - QRS axis. In basal origin of the ventricular tachycardia, QRS axis is inferior due to spread of the impulse toward the inferior leads. On the other hand, in apical origin of the ventricular tachycardia, QRS axis is superior due to spread of excitation away from the inferior leads.
  - Concordance of QRS complexes in the precordial leads. In basal origin of ventricular tachycardia, QRS is positive in all precordial leads due to spread of excitation toward precordial leads (Fig. 18.7a, b).

    In the apical origin of the ventricular tachycardia, QRS is negative in all precordial leads due to spread of excitation away from precordial leads (Fig. 18.8a, b).
- **Epicardial versus endocardial origin**. Endocardial origin of ventricular tachycardia (Fig. 18.9a) results in the rapid depolarization of the ventricles

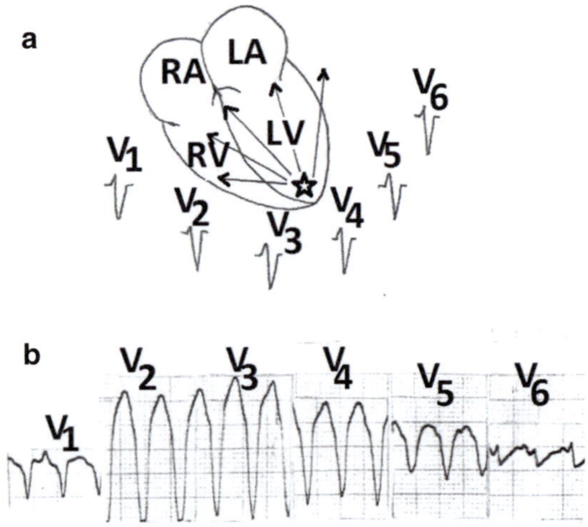

Fig. 18.8 (**a**) Diagram showing apical origin of the ventricular tachycardia and its effect on the precordial leads. (**b**) Electrocardiogram showing negative QRS complex in all precordial leads in a ventricular tachycardia originating from apical portion of the heart. (Reproduced with permission from: Mittal SR. Ventricular Tachyarrhythmias. Cardiology Today. 2012;16:219–226)

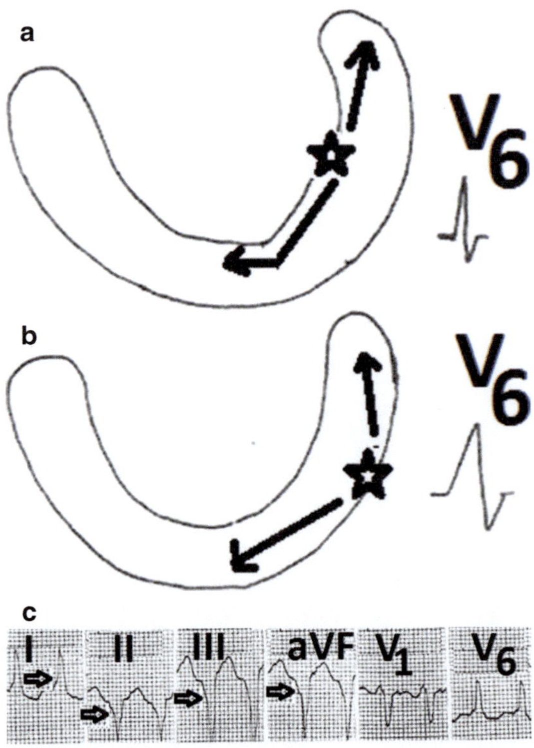

Fig.18.9 (**a**) Diagram showing endocardial origin of ventricular tachycardia (reproduced with permission from: Mittal SR. Ventricular Tachyarrhythmias. Cardiology Today. 2012;16:219–226). (**b**) Diagram showing epicardial origin of the ventricular tachycardia (reproduced with permission from: Mittal SR. Ventricular Tachyarrhythmias. Cardiology Today. 2012;16:219–226). (**c**) ECG showing pseudo delta wave (arrows) in ventricular tachycardia of epicardial origin. (Reproduced with permission from: Mittal SR. Ventricular Tachyarrhythmias. Cardiology Today. 2012;16:219–226)

along the specialized conduction system resulting in a narrow QRS complex.

In contrast, in epicardial origin of ventricular tachycardia (Fig. 18.9b), there is intramyocardial delay in conduction. This produces slurred initial portion of QRS (pseudo delta wave) (Fig. 18.9c) resulting in a wide QRS.

(ii). **Etiology:**

- **Ischemic cardiomyopathy**. Ventricular tachycardia usually has left bundle branch block morphology. Prognosis is usually poor.
- **Dilated cardiomyopathy**. Ventricular tachycardia usually has left bundle branch morphology. Prognosis is usually poor.
- **Bundle Branch reentry ventricular tachycardia**. Usually there is detectable structural heart disease. Commonly antegrade conduction is over the right bundle branch and retrograde conduction is over the left bundle branch (Fig. 18.10a). During ventricular tachycardia, there is left bundle branch block pattern in lead V_1 with left superior axis (Fig. 18.10b).

  Uncommonly antegrade conduction over left bundle branch block produces right bundle branch block configuration. Tachycardia is usually rapid and sustained. It is usually amenable to catheter ablation.
- **Idiopathic**. These are defined as ventricular tachycardias in patients without structural heart disease or coronary artery disease. Some patients may have early developing

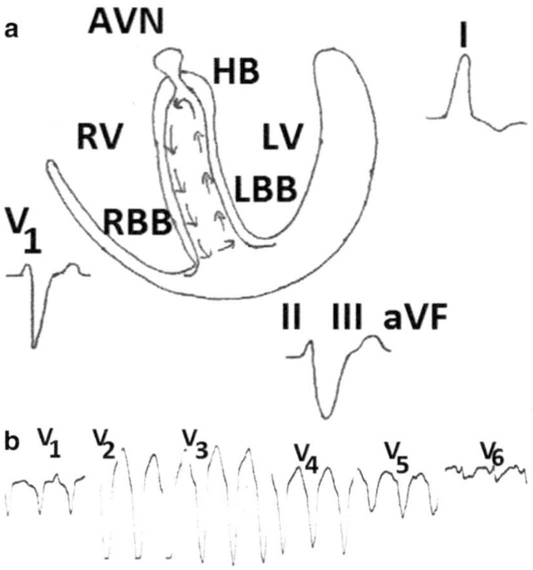

Fig. 18.10 (**a**) Diagram showing bundle branch reentry ventricular tachycardia. *AVN* atrioventricular node, *RBB* right bundle branch, *LBB* left bundle branch (reproduced with permission from: Mittal SR. Ventricular Tachyarrhythmias. Cardiology Today. 2012;16:219–226). (**b**) ECG shows bundle branch reentry ventricular tachycardia. (Reproduced with permission from: Mittal SR. Ventricular Tachyarrhythmias. Cardiology Today. 2012;16:219–226)

cardiomyopathy, which may become apparent only later. These are monomorphic and are amenable to drug therapy and ablation. Prognosis is usually good. These are classified according to the site of origin.

- *Right ventricular outflow tract ventricular tachycardia*. During ventricular tachycardia, there is left bundle branch block pattern in V_1 with inferior axis in the frontal plane (tall monophasic R wave in leads II, III, and aVF) (Fig. 18.11). Representative tracing is shown in (Fig. 18.12).

  Precordial transition is seen in leads V_4 to V_6. QS complex in lead aVR is equal to or larger than that in lead aVL. Lead I shows R wave. Such tachycardias may be provoked by exercise and emotional stress. These may be relieved by vagal maneuvers, beta-blockers, verapamil, diltiazem, adenosine, and radio-frequency ablation. Prognosis is usually good.

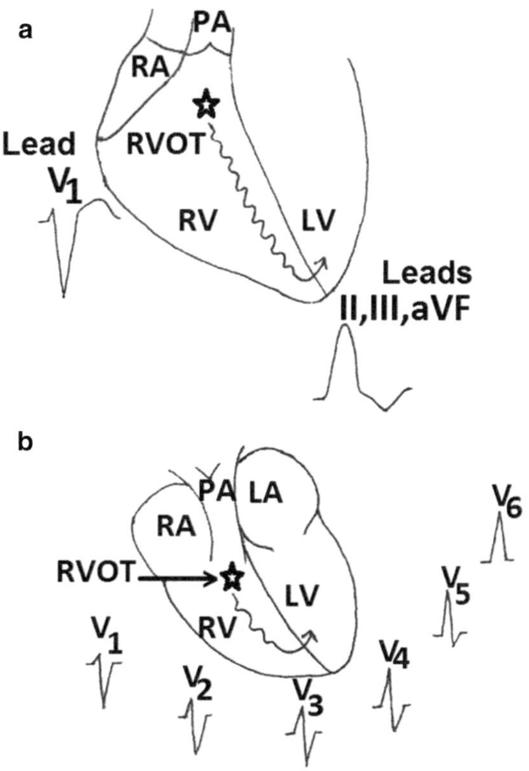

Fig. 18.11 (**a** and **b**) Diagram showing direction of ventricular depolarization and its effect on ECG in ventricular tachycardia arising in right ventricular outflow tract. (Reproduced with permission from: Mittal SR. Ventricular Tachyarrhythmias. Cardiology Today. 2012;16:219–226)

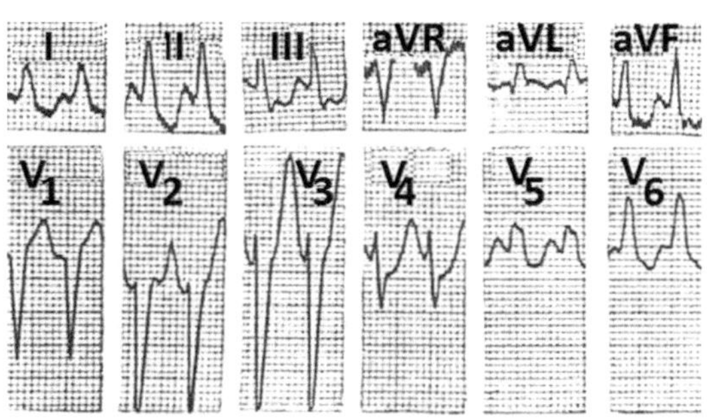

Fig. 18.12 ECG showing right ventricular outflow tract ventricular tachycardia. (Reproduced with permission from: Mittal SR. Ventricular Tachyarrhythmias. Cardiology Today. 2012;16:219–226)

- *Left ventricular outflow tract ventricular tachycardia*. During ventricular

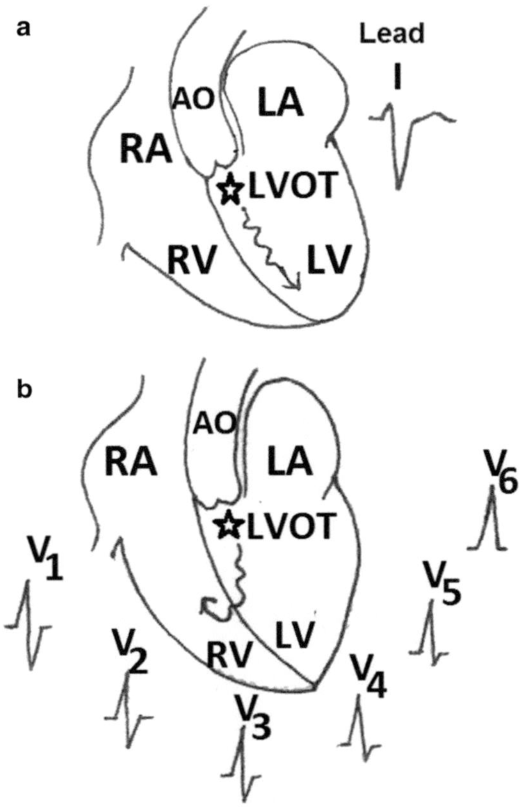

Fig. 18.13 (**a** and **b**) Diagram showing direction of ventricular depolarization and its effect on ECG in ventricular tachycardia originating in left ventricular outflow tract. (Reproduced with permission from: Mittal SR. Ventricular Tachyarrhythmias. Cardiology Today. 2012;16:219–226)

tachycardia there is S wave in lead I (Fig. 18.13a). Early precordial R wave transition is present in lead V_1 (Fig. 18.13b). QS wave depth in lead aVL is larger than that in lead aVR.

- *Mitral annular ventricular tachycardia.* These are characterized by right bundle branch block pattern, transition in lead V_1 or V_2, S wave in lead V_6, and R or RS pattern in leads V_2 to V_6 (Fig. 18.14a, b).
- *Tricuspid annular ventricular tachycardia.* These are characterized by narrow QRS complex and left bundle branch block pattern (Fig. 18.15a, b).
- *Fascicular ventricular tachycardia (left septal VT).* Origin is usually

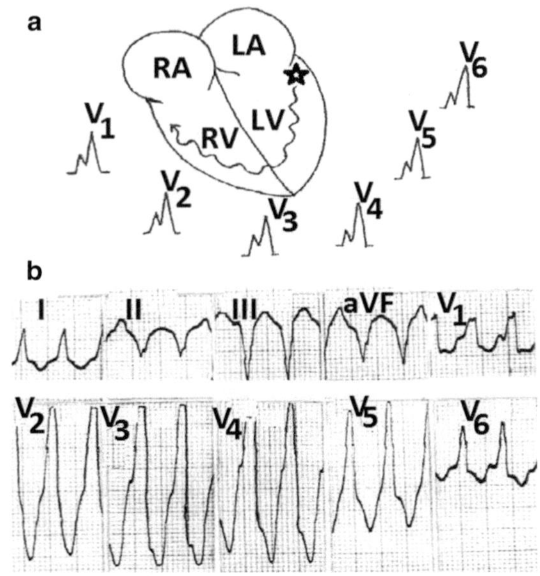

Fig. 18.14 (**a**) Diagram showing direction of ventricular depolarization and its effect on precordial leads in ventricular tachycardia originating around mitral annulus. (**b**) Electrocardiogram showing mitral annular ventricular tachycardia. (Reproduced with permission from: Mittal SR. Ventricular Tachyarrhythmias. Cardiology Today. 2012;16:219–226)

from left posterior fascicle in the inferior portion of the left side of the interventricular septum (Fig. 18.16a). Usually the heart is structurally normal. During ventricular tachycardia, QRS shows right bundle branch block configuration with left superior axis (Fig. 18.16b).

QRS is relatively narrow (< 140 msec). These ventricular tachycardias are sensitive to verapamil or diltiazem. Adenosine is usually not effective.

(b) Polymorphic ventricular tachycardia. These can be due to:

(i). **Arrhythmogenic right ventricular dysplasia (ARVD).**

In sinus rhythm, QRS duration is >110 ms in lead V_1. There is a small high-frequency deflection in the terminal portion of the QRS complex in leads V_1 to V_3 (Epsilon wave)

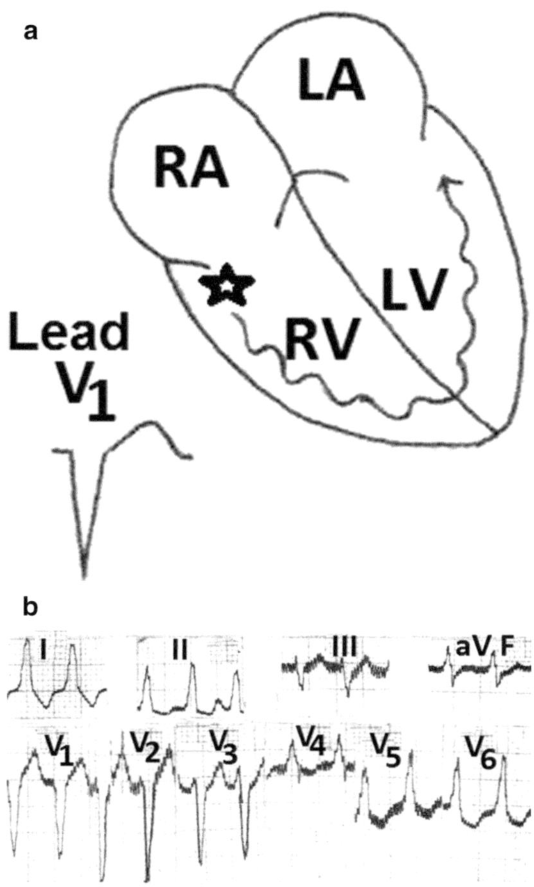

Fig. 18.15 (**a**) Diagram showing direction of ventricular depolarization and its effect on ECG in tricuspid annular ventricular tachycardia (reproduced with permission from: Mittal SR. Ventricular Tachyarrhythmias. Cardiology Today. 2012;16:219–226). (**b**) Electrocardiogram showing tricuspid annular ventricular tachycardia. (Reproduced with permission from: Mittal SR. Ventricular Tachyarrhythmias. Cardiology Today. 2012;16:219–226)

(Fig. 18.17a). The T wave is inverted in leads V_1 toV_3.

During ventricular tachycardia, there is left bundle branch block morphology in V_1 with superior axis in frontal plane (negative deflection in leads II, III, and aVF) (Fig. 18.17b). Multiple morphologies of ventricular tachycardia with left bundle branch block morphology and superior axis should raise suspicion of ARVD.

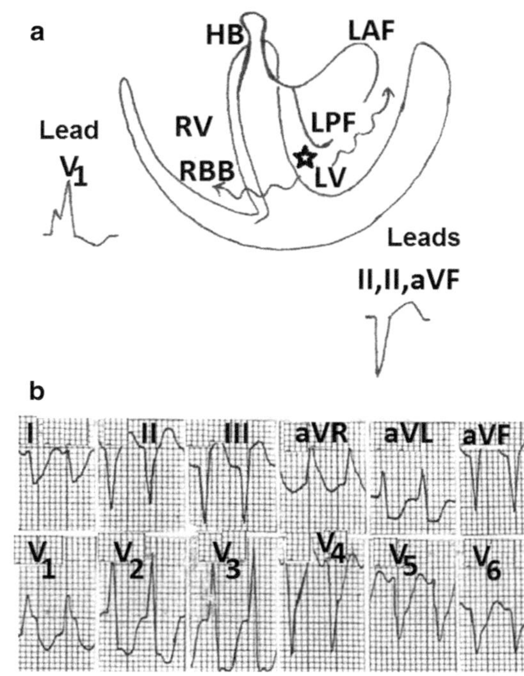

Fig. 18.16 (**a**) Diagram showing direction of ventricular depolarization and its effect on electrocardiogram in fascicular ventricular tachycardia. *HB* His bundle, *LAF* left anterior fascicle, *LPF* left posterior fascicle, *RBB* right bundle branch. (**b**) Electrocardiogram showing fascicular ventricular tachycardia. (Reproduced with permission from: Mittal SR. Ventricular Tachyarrhythmias. Cardiology Today. 2012;16:219–226)

(ii). **Catecholaminergic polymorphic ventricular tachycardia** [4].

It is seen in children and adolescents without any overt structural heart disease and normal QT interval. It is usually induced during physical stress. With increasing exercise isolated premature ventricular contractions develop into ill-sustained ventricular tachycardia. During VT, QRS is either bidirectional (alternating polarity of QRS in frontal plane leads and frontal plane axis of alternate QRS changing between −90° to +130°) or is of irregular polymorphic configuration. It is sensitive to beta-blockers.

(iii). Digitalis excess can also produce bidirectional VT (Fig. 18.18).

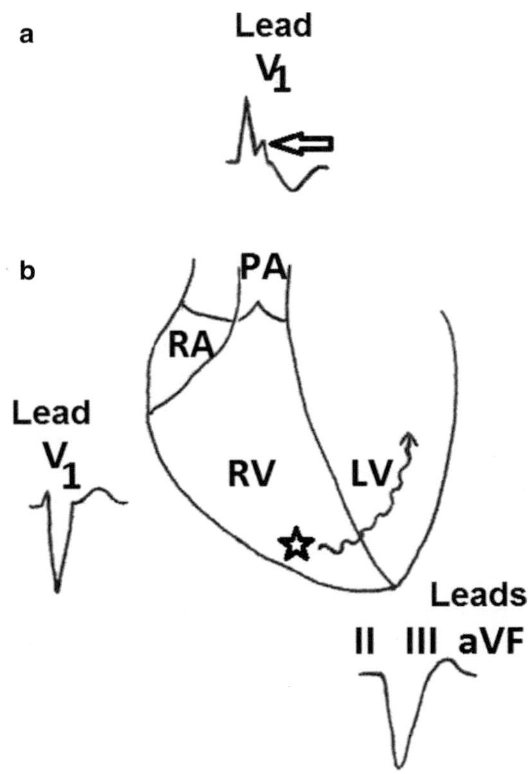

Fig. 18.17 (**a**) Diagram showing Epsilon wave in lead V_1 (marked arrow). (**b**) Diagram showing direction of ventricular depolarization in ventricular tachycardia due to arrhythmogenic right ventricular dysplasia and its effect on ECG. (Reproduced with permission from: Mittal SR. Ventricular Tachyarrhythmias. Cardiology Today. 2012;16:219–226)

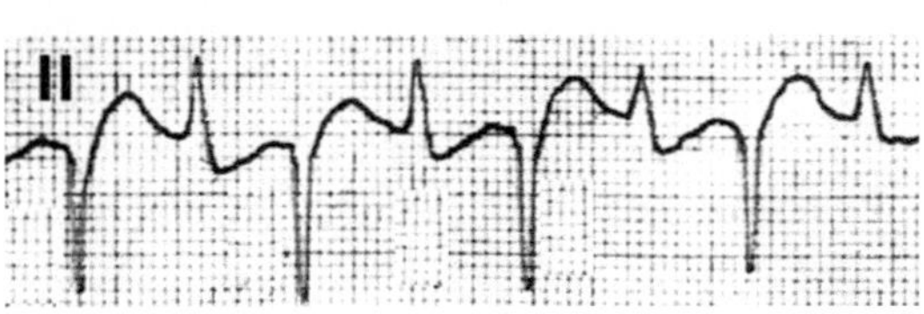

Fig. 18.18 Electrocardiogram showing bidirectional ventricular tachycardia. (Reproduced with permission from: Mittal SR. Ventricular Tachyarrhythmias. Cardiology Today. 2012;16:219–226)

(iv). **Brugada syndrome**.

Characteristic ECG findings in sinus rhythm are coved ST segment elevation of 2 mm or more in at least two of the three right precordial leads (V_1 toV_3) with incomplete or complete right bundle branch block (Fig. 18.19a, b) in the absence of signs of acute myocardial ischemia. ECG pattern may not be stable and can vary over time.

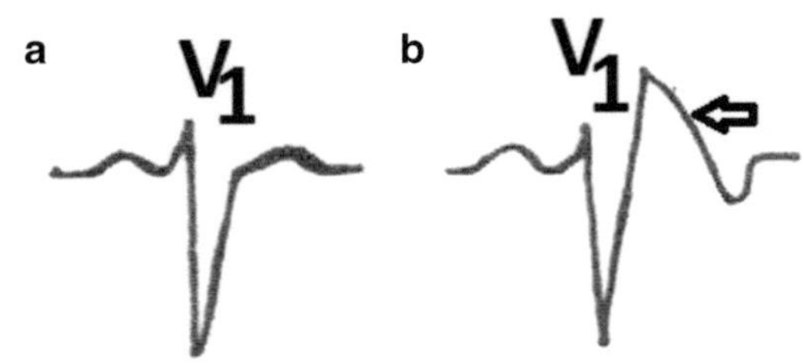

Fig. 18.19 (**a**) Showing lead V_1 in a normal person. (**b**) Showing coved ST segment elevation (arrow) in lead V_1 in Brugada syndrome. (Reproduced with permission from: Mittal SR. Ventricular Tachyarrhythmias. Cardiology Today. 2012;16:219–226)

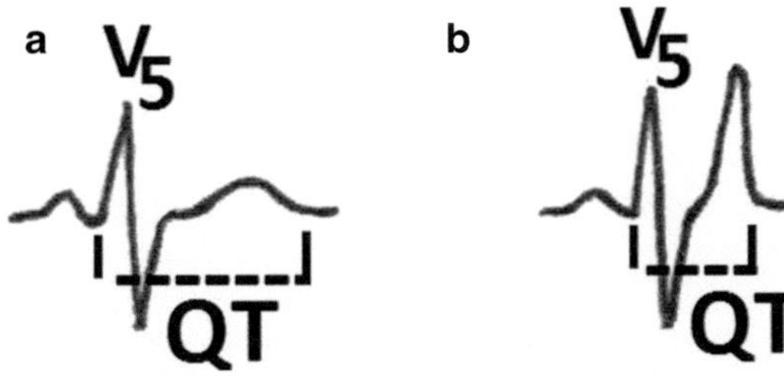

Fig. 18.20 (**a**) Diagram showing normal QT interval in lead V_5. (**b**) Diagram showing short QT interval with tall and peaked T wave in lead V_5. (Reproduced with permission from: Mittal SR. Ventricular Tachyarrhythmias. Cardiology Today. 2012;16:219–226)

(v). **Short QT syndrome**.

ECG pattern in normal QT interval is shown in Fig. 18.20 (a). Characteristic finding in sinus rhythm in short QT syndrome is QTc interval of less than 350 ms. T waves are usually tall and peaked resembling hyperkalemia (Fig. 18.20b).

(vi). **Torsades de pointes**.

Cycle length, axis, and morphology of QRS changes constantly in a pattern that simulates twisting of points around a central axis (Fig. 18.21).

It may be repetitive, nonsustained, or sustained and may degenerate into

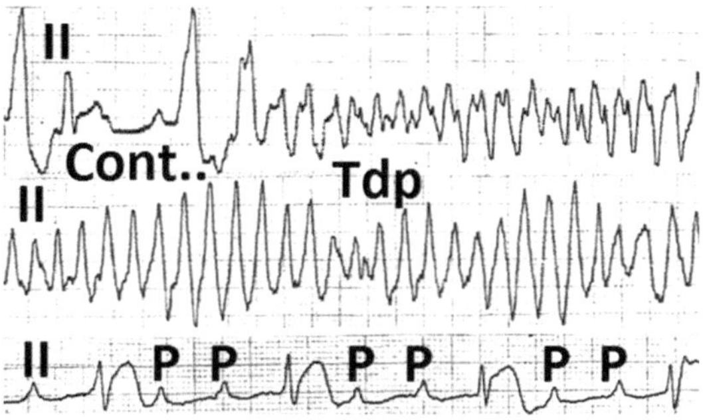

Fig. 18.21 Electrocardiogram is showing "torsades de pointes (Tdp)" in a case of inferior infarction with complete atrioventricular block and normal QTc interval. (Reproduced with permission from: Mittal SR. Ventricular Tachyarrhythmias. Cardiology Today. 2012;16:219–226)

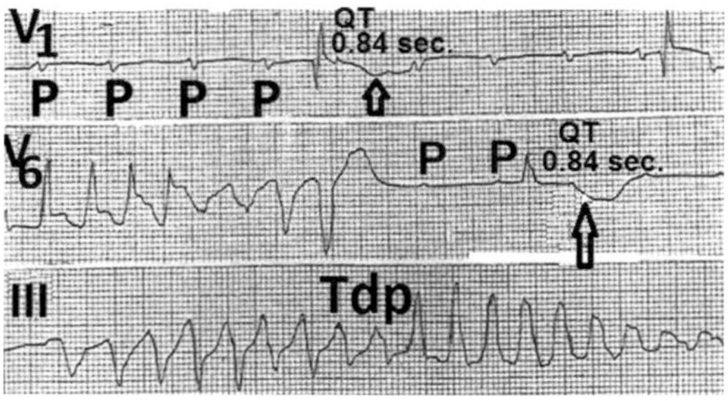

Fig. 18.22 Electrocardiogram is showing "torsades de pointes" (Tdp) in a case of complete atrioventricular block and prolonged QT interval (marked arrow). (Reproduced with permission from: Mittal SR. Ventricular Tachyarrhythmias. Cardiology Today. 2012;16:219–226)

ventricular fibrillation. Rate of ventricular tachycardia is usually fast (200–250/min.). Bradycardia and QT prolongation (QTc >500 ms.) usually precede "torsades de pointes" (Fig. 18.22).

QT prolongation may be apparent only in the beat prior to the onset of torsades, i.e., following the initiating premature ventricular contraction. Long-short RR cycle sequence may precede the onset of "torsades de points" from acquired causes. Premature ventricular beats superimposed on the terminal portion of long T wave may also precipitate "torsades de pointes." Some patients may not have preceding bradycardia or QT prolongation. Some patients may have biphasic or notched T wave.

(vii). **Proarrhythmic effect of antiarrhythmic drugs.**

Class 1A and 1C drugs and amiodarone have the potential to produce proarrhythmias. It should be suspected if:

- Morphology of ventricular tachycardia is different from ventricular tachycardia recorded before starting the drug.
- Frequency of ventricular tachycardia increases after addition or change of antiarrhythmic drug.
- There is prolongation of QT interval after addition or change of antiarrhythmic drug.
- There is fresh appearance of polymorphic or "torsades de pointes" configuration.

4 Ventricular Flutter [2]

It is characterized by very rapid (150–300/min.), uniform, and regular ventricular rhythm without clear definition of QRS complex, ST segment, and T wave (Fig. 18.23).

There is prompt deterioration to ventricular fibrillation. Coarse flutter is more likely to respond to electric cardioversion. Beta-adrenergic agents and calcium may help cardioversion.

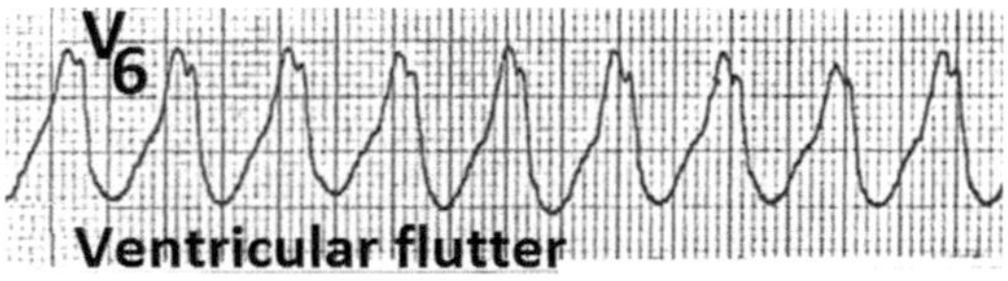

Fig. 18.23 Electrocardiogram showing ventricular flutter. (Reproduced with permission from: Mittal SR. Ventricular Tachyarrhythmias. Cardiology Today. 2012;16:219–226)

5 Ventricular Fibrillation [2, 5]

It is characterized by the following:

- Irregular undulations of varying contour and amplitude (Fig. 18.24).
- There is preceding very rapid ventricular tachycardia, ventricular flutter, or premature ventricular beat falling on the T wave of the preceding QRS (R on T).
- Primary VF—when VF occurs in patients without preexisting cardiac failure, e.g., in acute myocardial infarction. It responds relatively well to electrical defibrillation.
- Secondary VF—when VF occurs as a terminal event in patients with cardiac failure, resuscitation is usually unsuccessful.
- Fine fibrillation has a poor prognosis as compared to coarse fibrillation.

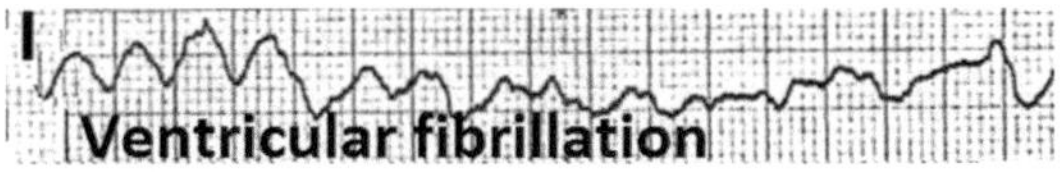

Fig. 18.24 Electrocardiogram showing ventricular fibrillation. (Reproduced with permission from: Mittal SR. Ventricular Tachyarrhythmias. Cardiology Today. 2012;16:219–226)

Electrocardiographic Differential Diagnosis of Broad QRS Ventricular Rhythm

1. Rate < 120/min
   - Accelerated idioventricular rhythm
2. Rate > 120/min
   - Ventricular tachycardia
     - Monomorphic:
       Left bundle branch block pattern
       - Inferior axis
         - Precordial transition in leads V_4 to V_6—RVOT VT
         - Precordial transition in V_1—LVOT VT
       - Superior axis
         - Bundle branch reentry VT
         - Arrhythmogenic right ventricular cardiomyopathy
         - With narrow QRS—tricuspid annular VT

       Right bundle branch block pattern
       - Superior axis
         - Narrow QRS—fascicular VT
         - Transition in leads V_1 or V_2 and R or Rs configuration in leads V_2–V_6 mitral annular VT
     - Bidirectional:
       Catecholaminergic VT
       Digitalis excess
       Severe myocardial disease
     - Polymorphic:
       With QT interval prolongation (QTc >550 msec)
       - Torsades de pointes

       Without QT prolongation
       - Catecholaminergic VT
       - VT in Brugada syndrome

       Short QT syndrome (QTc < 350 msec)

Summary

- Gradual acceleration of ventricular rate beyond the sinus rate is known as accelerated idioventricular rhythm. Fusion beats or isorhythmic dissociation may be seen at the onset or termination. Digitalis overdose and successful reperfusion following thrombolytic therapy are common causes. There is no risk of deterioration into VT or VF.
- Occurrence of three or more premature ventricular contractions in a run and terminating spontaneously is known as nonsustained (ill-sustained) ventricular tachycardia. In absence of structural heart disease, these do not require antiarrhythmic therapy.
- Origin of ventricular tachycardia can be decided by the configuration of the QRS complex on the surface electrocardiogram.
  - Left bundle branch block configuration suggests an origin from the right ventricular free wall or the interventricular septum.
  - Right bundle branch block configuration suggests an origin from the left ventricular free wall.
  - Positive QRS in all precordial leads V_1 to V_6 suggests origin from left ventricular base.

- Negative QRS in all precordial leads V_1 to V_6 suggests origin from left ventricular apex.
- Negative QRS in leads II, III, and aVF suggests an inferior wall origin.
- Positive QRS in leads II, III, and aVF suggest an origin in the outflow tract.
- Epicardial origin is suggested if intrinsic deflection exceeds 55% of the QRS duration.

Thus:

- Right bundle branch block pattern with left superior axis suggests an origin in left posterior fascicle (fascicular VT). These are relatively narrow and sensitive to verapamil.
- Left bundle branch block morphology with inferior axis suggests origin of VT from right ventricular outflow tract. These VTs are also sensitive to verapamil.
- Left bundle branch block configuration in lead V_1 is also seen in bundle branch reentry VT because of antegrade conduction over the right bundle branch and retrograde conduction over the left bundle branch.
- Arrhythmogenic right ventricular dysplasia produces VT with left bundle branch block configuration in lead V_1 with superior QRS axis (negative QRS in leads II, III, and aVF). Left bundle branch block pattern with multiple changing morphologies of VT supports the possibility of arrhythmogenic right ventricular dysplasia. This is because of diffuse involvement of the right ventricle. Resting ECG in sinus rhythm shows incomplete or complete right bundle branch block pattern with small positive deflection in the terminal part of the QRS in leads V_1 to V_3 (Epsilon wave).
- Changing configuration of QRS in alternate beats suggests bidirectional VT. VT shows a right bundle branch block pattern with QRS axis alternating between −60° to +120° because of alternating conduction over the left anterior fascicle and the left posterior fascicle. Digitalis excess used to be a common cause. Catecholaminergic polymorphic VT can be bidirectional to start with.
  - VTs with well-defined QRS complexes with constantly changing morphology are called polymorphic VT.
  - Rapid polymorphic VT with QRS complexes that appear to twist around a central axis is termed "torsades de pointes." It usually occurs in the setting of prolonged QTc interval.
  - Bidirectional/polymorphic VT induced by exercise is called catecholaminergic polymorphic VT.
  - Grossly irregular, fast, broad QRS tachycardia is seen in atrial fibrillation with antegrade conduction over an accessory pathway.
  - Repolarization abnormalities without any structural heart disease but abnormal ECG can also produce VTs. These include J wave syndrome, long QT syndromes, short QT syndrome, and Brugada syndrome.

MCQs

Q1. Ventricular tachycardia is defined as occurrence of:

(a) Three or more ectopics in succession
(b) Run of ectopics lasting more than 1 minute
(c) Ventricular rhythm at a rate of 80/minute
(d) Irregularly irregular broad QRS tachycardia

Q2. Which of the following suggests ventricular tachycardia:

(a) P wave slower than QRS and not related to QRS
(b) Abnormal P wave preceding each QRS
(c) Inverted P wave following each QRS
(d) Sinus P wave preceding each QRS

Q3. In the beginning of a broad QRS tachycardia, abnormal P wave follows the first broad QRS complex. It suggests a diagnosis of:

(a) Atrial tachycardia with aberrant conduction
(b) AVNRT with preexisting bundle branch block

(c) Ventricular tachycardia with retrograde conduction
(d) Antidromic atrioventricular reentry (reciprocating) tachycardia

Q4. Bidirectional VT usually suggests:

(a) Digitalis overdose
(b) Amiodarone overdose
(c) Presence of myocardial scar
(d) Catecholaminergic polymorphic VT

Q5. LBBB morphology of VT suggests origin of VT in:

(a) Right ventricular outflow tract
(b) Right ventricle
(c) Left ventricle
(d) Interventricular septum

Q6. RBBB morphology of VT suggests origin of VT in:

(a) Left anterior fascicle
(b) Left posterior fascicle
(c) Interventricular septum
(d) Right ventricle

Q7. VT arising in right ventricular outflow tract has:

(a) LBBB configuration
(b) Inferior axis
(c) Precordial transition in lead V_3 or later
(d) All

Q8. Narrow (<140 milliseconds) QRS VT with RBBB configuration with left axis is likely to be arising from:

(a) Left posterior fascicle
(b) Left anterior fascicle
(c) RVOT
(d) LVOT

Q9. VT with positive deflection in leads V_1 to V_6 suggests an origin in:

(a) RVOT
(b) LVOT
(c) Posterior RV
(d) Posterior LV

Q10. VT with negative deflection in lead V_1 to V_6 suggests origin from:

(a) LV apex
(b) RVOT
(c) LVOT
(d) Posterior LV

Q11. Termination of a broad QRS tachycardia by carotid massage supports the possibility of:

(a) AVNRT
(b) VT arising from the RVOT
(c) Sinus tachycardia with aberrant conduction
(d) Atrial tachycardia with preexisting bundle branch block

Q12. Termination of a broad QRS tachycardia by intravenous injection of diltiazem suggests the possibility of:

(a) Atrial tachycardia with aberrant ventricular conduction
(b) AVNRT with preexisting bundle branch block
(c) VT arising from the right ventricular outflow tract
(d) Fascicular VT

Q13. Which VTs are likely to respond to verapamil:

(a) LBBB pattern with inferior axis
(b) RBBB pattern with left axis
(c) Scar VT
(d) LBBB pattern with left axis

Q14. Wide QRS tachycardia appearing alternately above and below the baseline at a rate of 250/minute suggests:

(a) Bidirectional VT
(b) Torsades de pointes
(c) Ventricular fibrillation
(d) SVT with alternating bundle branch block

Q15. Torsades de pointes is preceded by:

(a) Long QT interval
(b) Prominent U wave
(c) R wave of premature ventricular contraction falling on the preceding T wave
(d) None

Q16. Very wide, uniform bizarre regular QRS complexes at a rate of 150/minute with merging of the QRS, the ST segment, and the T wave suggests:

(a) Ventricular tachycardia
(b) Ventricular flutter
(c) Ventricular fibrillation
(d) Supraventricular tachycardia with aberrant conduction

Answers

(1) a; (2) a, c; (3) c; (4) a, d; (5) a, b, d; (6) a, b; (7) d; (8) a; (9) c, d; (10) a; (11) a, b; (12) b, c, d; (13) a, b; (14) b; (15) a, b, c; (16) b

References

1. Gilbert M, Wagner GS. Re-entrant ventricular tachyarrhythmias. In: Wagner GS, editor. Marriott's practical electrocardiography. Philadelphia: Wolters Kluwer; 2005. p. 347–64.
2. Schamroth L. Ventricular rhythms. In: Schamroth C, editor. An introduction to electrocardiography. France: Blackwell Science; 1982. p. 349–66.
3. Rho RW, Page RL. Ventricular arrhythmias. In: Fuster V, Walsh RA, Harrington RA, editors. Hurst's the heart. New York: Mc Graw Hill; 2011. p. 1006–24.
4. Priori SG, Napolitano C. Genetics of channelopathies and clinical implications. In: Fuster V, Walsh RA, Harrington RA, editors. Hurst's the heart. New York: Mc Graw Hill; 2011. p. 897–910.
5. Olgin JE, Zipes DP. Specific arrhythmias: diagnosis and treatment. In: Libby P, Bonow RO, Mann DL, Zipes DP, editors. Braunwald's heart disease. Philadelphia: Saunders; 2008. p. 863–922.
6. Darrat Y. Predicting the origin of PVC based on EKG characteristics. Indian Journal of Electrocardiology. 2012;1:9–11.

The Q Wave

19

1 Nomenclature

An initial negative deflection of the QRS complex is called the "Q" wave. In the absence of an R wave, a purely negative deflection is called "QS" complex. A narrow and small Q wave is labeled "q" wave, whereas a broad and deep Q wave is called "Q" wave.

2 Genesis of the Normal "Q" Wave

Normal q wave represents depolarization of the interventricular septum. Normal septal depolarization occurs from left to right and anteriorly (Fig. 19.1). with the result that right-sided leads (aVR, V_3R, and V_4R) and anterior lead (V_1) show an initial positive deflection (r wave). On the other hand, left-sided leads (I, aVL) and posterolateral leads (V_5 to V_9) show an initial negative deflection (q wave). These are also called septal q waves.

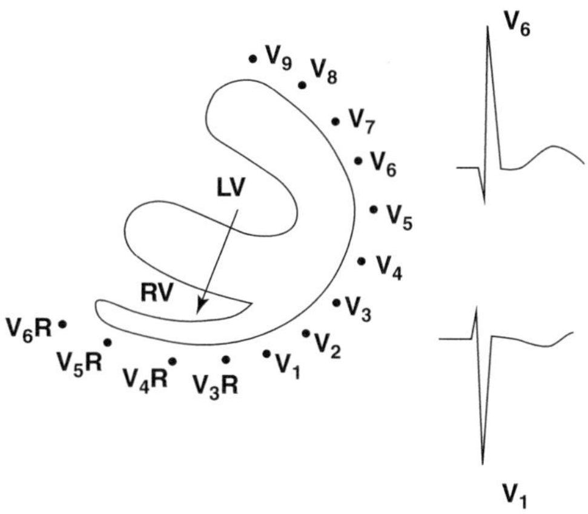

Fig. 19.1 Diagram showing vector of septal depolarization (arrow) and its effect on the QRS complex of chest leads. (Reproduced with permission from Mittal SR. Q wave. Cardiology Today. 2016;20:28–38)

Lead V_5R and V_6R face the right atrium and thus pick up cavity potential. These right-sided chest leads may show qR configuration (Fig. 19.2).

S. Mittal, *Insights into Electrocardiograms with MCQs*,
https://doi.org/10.1007/978-981-99-0127-2_19

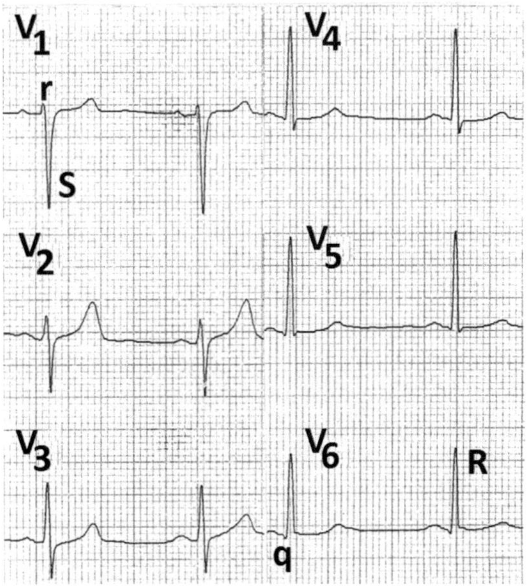

Fig. 19.2 Electrocardiogram showing the effect of septal depolarization on precordial leads. Lead V_1 shows rS configuration and lead V_6 shows a qR configuration. (Reproduced with permission from Mittal SR. Q wave. Cardiology Today. 2016;20:28–38)

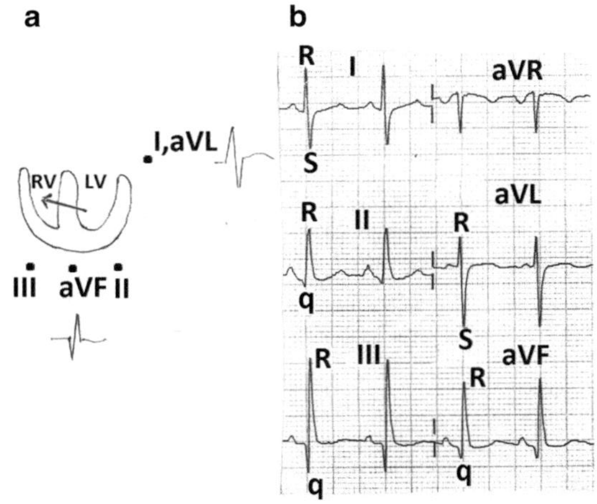

Fig. 19.3 (a) Diagram showing septal depolarization vector (arrow) in vertical position of heart and its effect on the QRS of limb leads. (b) Electrocardiogram from a person with vertical heart. Leads I and aVL show RS configuration. Leads II, III, and aVF show qR configuration. (Reproduced with permission from Mittal SR. Q wave. Cardiology Today. 2016;20:28–38)

3 Normal Q Wave

It is usually less than 30 msec wide and < 25% of the ensuing R wave in voltage [1]. In lead III, even in normal persons, Q wave can be broad as well as deep because the major part of the QRS and the T wave vector is directed away from the positive pole of lead III. For the same reason, the T wave can also be inverted in lead III.

3.1 Normal Variations in the "q" Wave

- Vector of septal depolarization can vary depending on the position of the heart in thorax [2]. In vertical position of the heart (mean frontal plane QRS axis directed toward +90°), septal vector is directed away from the inferior leads (II, III, and aVF) with the result that these leads show qR configuration (Fig. 19.3a, b). Lead I shows an RS pattern.

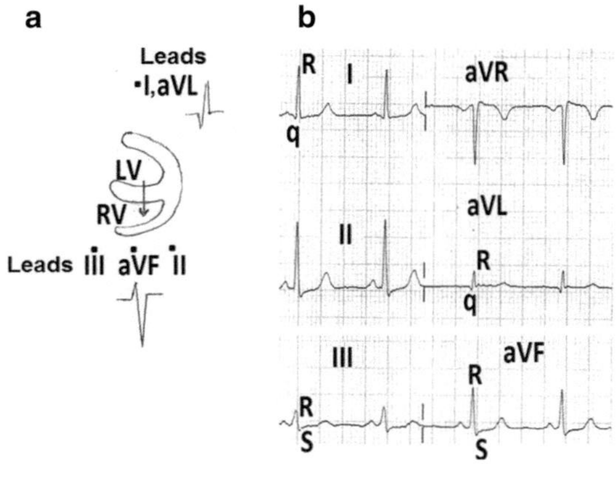

Fig. 19.4 (a) Diagram showing septal depolarization vector (arrow) in horizontal position of the heart and its effect on the QRS complex of the limb leads. (b) Electrocardiogram from a person with horizontal heart. Leads I and aVL show qR pattern and leads II, III, and aVF show Rs configuration. (Reproduced with permission from Mittal SR. Q wave. Cardiology Today. 2016;20:28–38)

- In horizontal position of the heart (mean frontal plane QRS axis around 0 degree), septal vector is directed inferiorly. This results in rS configuration in leads II, III, aVF, and a qR configuration in leads I and aVL (Fig. 19.4a, b).

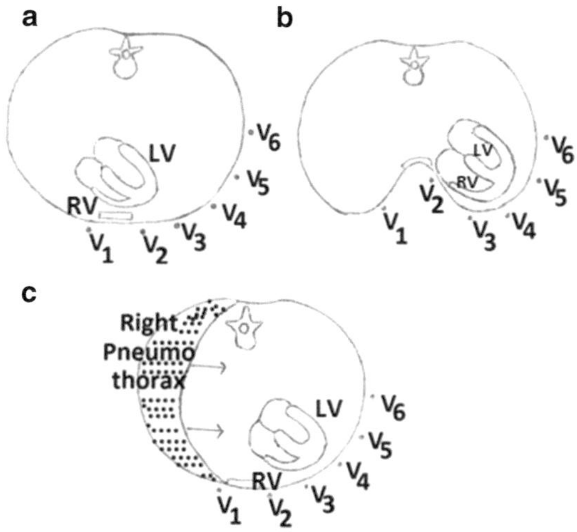

Fig. 19.5 Diagram showing (**a**) normal relation of the chest leads to the heart. (**b**) Relation of the chest leads to the heart in pectus excavatum. (**c**) Relation of the chest leads to the heart in right-sided pneumothorax or pleural effusion. *LV* left ventricle; *RV* right ventricle. (Reproduced with permission from Mittal SR. Q wave. Cardiology Today. 2016;20:28–38)

- Severe pectus excavatum and severe scoliosis with convexity to the right can displace the heart to the left hemithorax. This may result in lead V_1 facing the right atrium rather than surface of the right ventricle. Lead V_1 and even V_2 may show initial q wave (Fig 19.5a, b). Pushing of heart to left by massive right pleural effusion or pneumothorax (Fig. 19.5c) can also produce similar finding.
- In mirror image dextrocardia, left-sided chest leads face the atria and may show a q wave (Fig. 19.6a, b).
- In technical dextrocardia (right arm electrode connected to the left arm), leads I and aVL show Q wave and negative P wave. Wrong connection of right arm electrode to the left foot produces Q wave in leads II, III, and aVF. Higher placement of leads V_1 and V_2 can produce q wave in these leads.
- On deep inspiration, heart assumes a vertical position. As the heart becomes more vertical, the mean frontal plane QRS axis shifts in the positive half of lead III. Q wave in lead III diminishes or may even disappear. However,

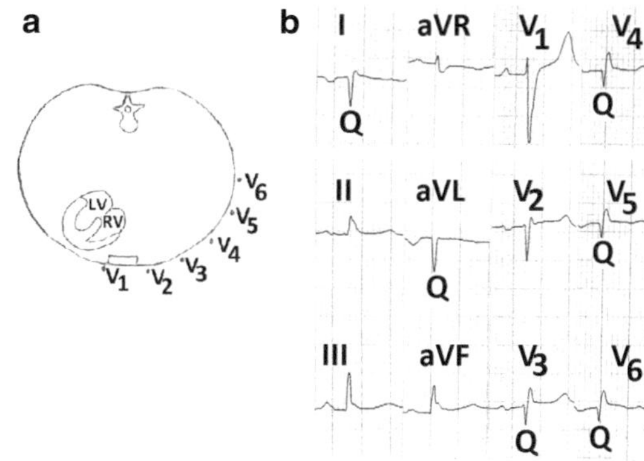

Fig. 19.6 (**a**) Diagram showing the relation of chest leads to the heart in dextrocardia. (**b**) Electrocardiogram from a case of dextrocardia showing q wave in leads I, aVL, and V_4 to V_6. (Reproduced with permission from Mittal SR. Q wave. Cardiology Today. 2016;20:28–38)

this finding does not necessarily prove that such "q" wave is a normal variation. Q waves in inferior leads from organic cause may also be reduced in depth on deep inspiration.

4 Other Causes of Noninfarct Q Waves [2, 3]

(a) **Fascicular Blocks**
   - Left anterosuperior fascicular block produces a qR configuration in leads I and aVL (Fig. 19.7).
   - Left posteroinferior fascicular block produces qR configuration in leads II, III, and aVF (Fig. 19.8). Amplitude of the R wave is increased.
   - In these conditions, q waves are neither broad nor slurred. Absence of sequential ST-segment or T wave changes also helps in excluding myocardial infarction.

(b) **Left Bundle Branch Block.**
It can produce QS complexes in leads V_1 to V_3 (Fig. 19.9). The ST-segment is elevated in these leads. Absence of sequential T wave changes helps in excluding concomitant myocardial infarction.

(c) **Preexcitation:** Negative delta wave produces a "q" wave.

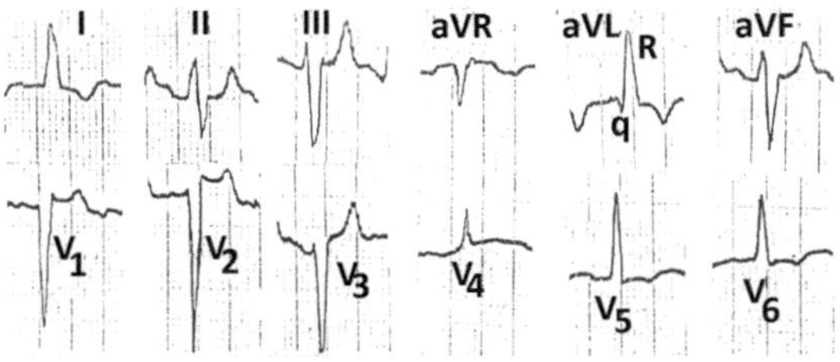

Fig. 19.7 Electrocardiogram from a case of left anterosuperior fascicular block showing q wave in lead aVL. (Reproduced with permission from Mittal SR. Q wave. Cardiology Today. 2016;20:28–38)

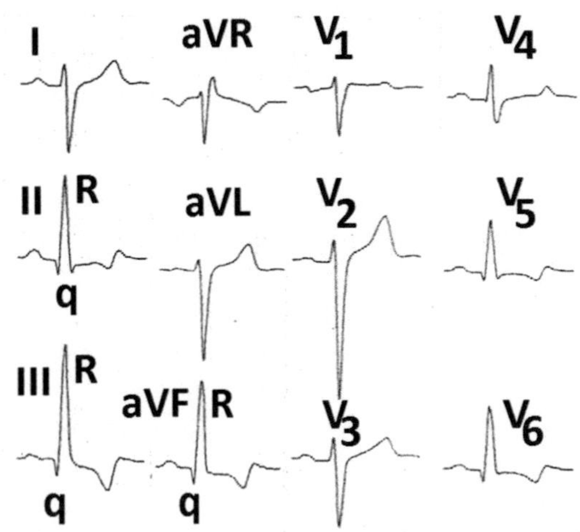

Fig. 19.8 Electrocardiogram from a case of left postero-inferior fascicular block showing q waves in leads II, III, and aVF. (Reproduced with permission from Mittal SR. Q wave. Cardiology Today. 2016;20:28–38)

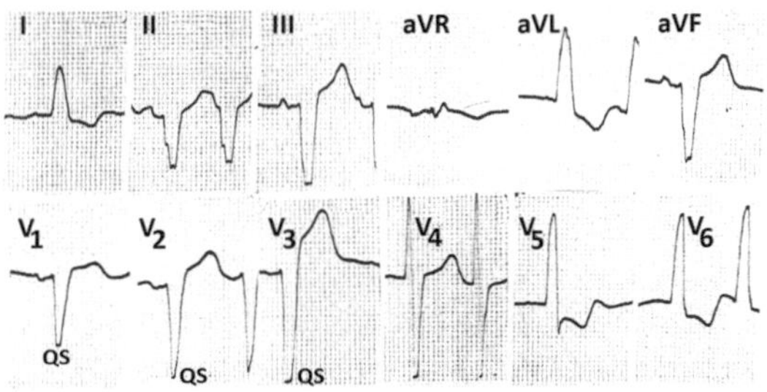

Fig. 19.9 Electrocardiogram from a case of LBBB showing QS pattern in leads V_1 to V_3. (Reproduced with permission from Mittal SR. Q wave. Cardiology Today. 2016;20:28–38)

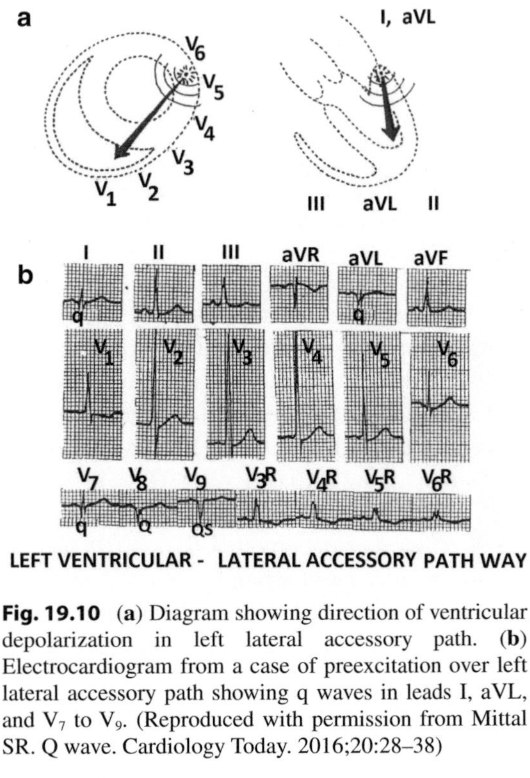

Fig. 19.10 (**a**) Diagram showing direction of ventricular depolarization in left lateral accessory path. (**b**) Electrocardiogram from a case of preexcitation over left lateral accessory path showing q waves in leads I, aVL, and V_7 to V_9. (Reproduced with permission from Mittal SR. Q wave. Cardiology Today. 2016;20:28–38)

– In the left lateral accessory path, delta wave is directed inferiorly. Thus, leads I, aVL, V_5, and V_6 show a q wave (Fig. 19.10).
– In preexcitation via right posteroseptal accessory path, delta wave is directed superiorly. This produces "q" waves in leads II, III, and aVF (Fig. 19.11).
– The PR interval is short. Absence of ST-segment elevation and sequential changes in the T waves helps in excluding concomitant myocardial infarction.

(d) **Ventricular Enlargement**

- Left ventricular volume overload (e.g., chronic moderately severe aortic regurgitation or large left to right shunt across patent ductus arteriosus) produces prominent (but not wide) q waves in leads V_5 andV_6 (Fig. 19.12). R wave voltage is also increased in these leads.

  In the advanced stage, the ST-segment may be depressed and the T wave may be inverted. Left ventricular pressure overload can produce QS configuration in leads V_1 and V_2.

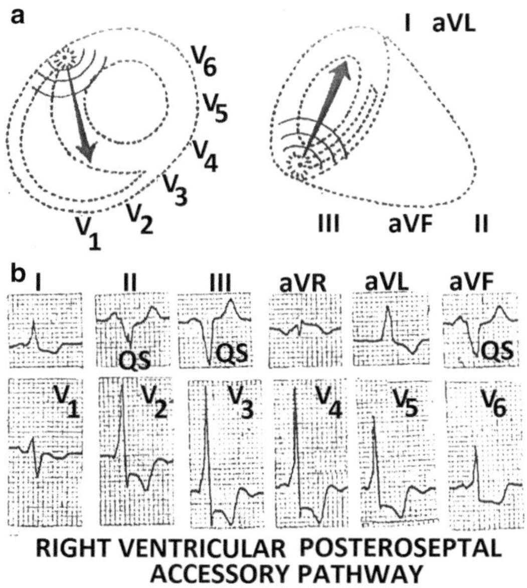

Fig. 19.11 (**a**) Diagram showing direction of ventricular depolarization in a case of preexcitation due to posteroseptal accessory pathway. (**b**) Electrocardiogram from a case of preexcitation due to posteroseptal accessory path showing q waves in leads II, III, and aVF. (Reproduced with permission from Mittal SR. Q wave. Cardiology Today. 2016;20:28–38)

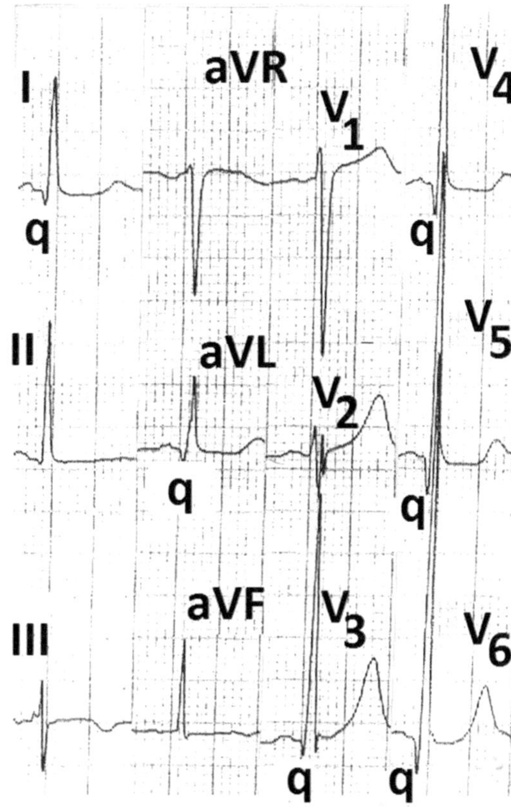

Fig. 19.12 Electrocardiogram from a case of severe aortic regurgitation showing increased R wave voltage with q waves in leads I, aVL, and V_3 to V_6. (Reproduced with permission from Mittal SR. Q wave. Cardiology Today. 2016;20:28–38)

- Right ventricular enlargement with right atrial enlargement (especially in cases of cor pulmonale) can produce qr. configuration in lead V_1 (Fig. 19.13).

(e) **Nonischemic Myocardial Necrosis**

- Acute myocarditis can produce abnormal q waves in leads V_5 and V_6 due to myocardial necrosis. Sinus tachycardia, ectopics, nonspecific intraventricular conduction defects, ST-segment elevation or depression, T wave inversion in leads V_5,V_6, and QT prolongation can also be present.
- Dilated cardiomyopathy can produce abnormal (even broad and slurred) Q waves in leads V_5 and V_6 due to replacement of myocardium by fibrous time. Increased left ventricular voltage and left anterosuperior fascicular block are other findings. Some patients can develop left bundle branch block.
- Duchenne muscular dystrophy selectively involves posterior wall of the left ventricle, resulting in q wave in leads V_7 to V_9. Reciprocal prominent R wave in leads V_1 and V_2 is a common finding.

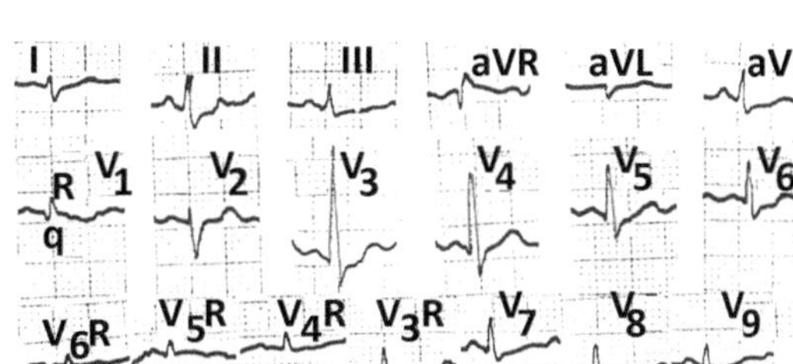

Fig. 19.13 Electrocardiogram from a case of cor pulmonale showing q wave in lead V_1. (Reproduced with permission from Mittal SR. Q wave. Cardiology Today. 2016;20:28–38)

(f) **Hypertrophic Cardiomyopathy**

Asymmetrical septal hypertrophy produces prominent and relatively deep but narrow Q waves in leads I, aVL, and V_4 to V_6 (Fig. 19.14). Tall R wave, deep symmetrical T wave inversion, and "shelf-like" ST-segment are other findings. Absence of sequential changes differentiate it from myocardial infarction.

(g) **Emphysema**

Downward displacement of the diaphragm results in vertical heart position. The chest leads V_1 and V_2, even when placed correctly, are higher for the vertical heart. This can result in QS configuration in leads V_1 and V_2 (Fig. 19.15). Low-voltage QRS, small QRS in lead I, and low voltage of R wave in lead V_6 are other supportive findings.

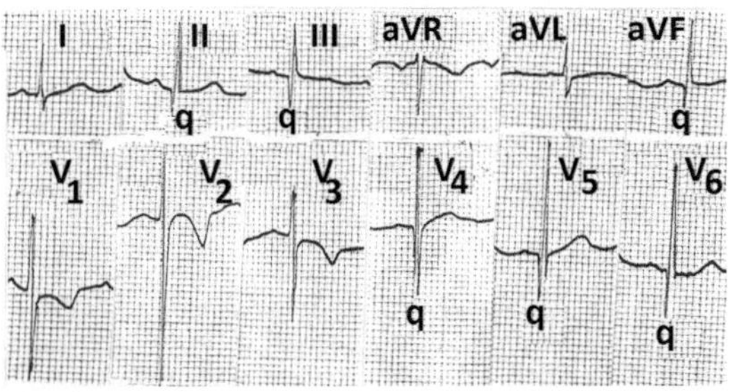

Fig. 19.14 Electrocardiogram from a case of hypertrophic cardiomyopathy showing prominent q waves in leads II, III, aVF, and V_3 to V_6. (Reproduced with permission from Mittal SR. Q wave. Cardiology Today. 2016;20:28–38)

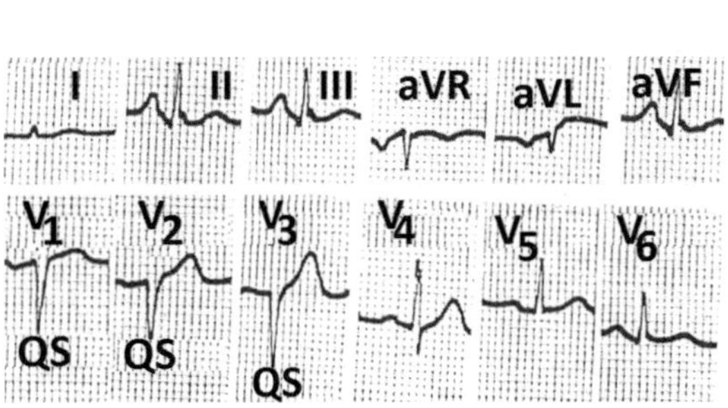

Fig. 19.15 Electrocardiogram from a case of emphysema showing QS pattern in leads V_1 to V_3. (Reproduced with permission from Mittal SR. Q wave. Cardiology Today. 2016;20:28–38)

(h) **Myocardial Infiltration**

Amyloidosis, sarcoidosis, and tumors can also produce abnormal Q waves due to loss of excitable myocardium.

(i) **Acute Pulmonary Thromboembolism**

Lead III shows a qR pattern. The q wave is neither deep nor broad. S wave in lead I and T wave inversion in lead III are other findings (S_I Q_{III} inverted T_{III} pattern). Sinus tachycardia, right axis deviation, clockwise rotation, "P- pulmonale," and transient incomplete right bundle branch block are other common findings. T waves are inverted in leads V_1 to V_3. There is no coving of ST-segment. ECG changes usually regress within days to weeks.

5 Abnormal Q Waves [1, 4]

- Any Q wave in leads V_2 to V_4 should be considered abnormal (Fig. 19.16).
- Q waves in leads V_3 and V_4 more than those in leads V_5 and V_6 are abnormal (Fig. 19.17).
- Slurring on the downslope of a QS complex suggests the presence of a q wave (Fig. 19.18).
- Absence of small q waves in leads V_5 and V_6 should be considered abnormal.
- Q wave of any size can be normal in leads III and aVF.
- Abnormality of a Q wave should be decided primarily by the width ($\leq$0.04 second) and

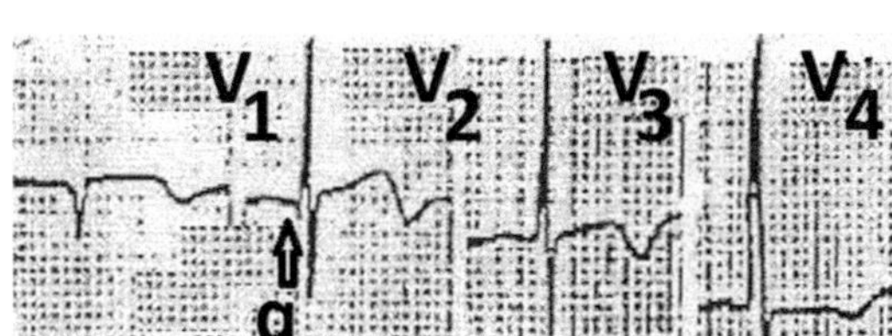

Fig. 19.16 Electrocardiogram showing small q wave in lead V_2 (marked) and terminal T wave inversion in leads V_1 to V_4 suggestive of anterior infarction. (Reproduced with permission from Mittal SR. Q wave. Cardiology Today. 2016;20:28–38)

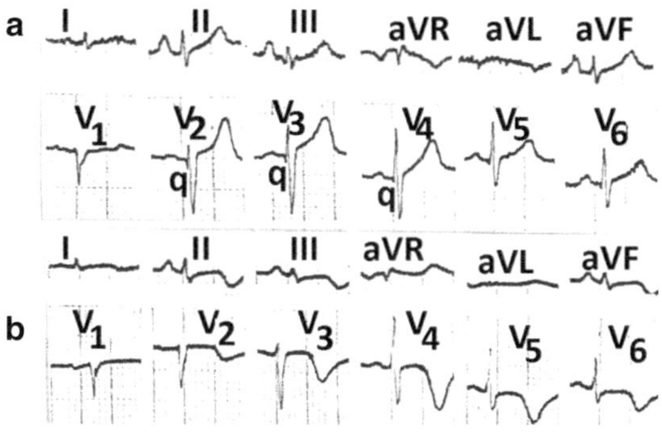

Fig. 19.17 (**a**) Electrocardiograms showing small q waves in leads V_2 to V_4 more than those in leads V_5 and V_6. (**b**) Sequential ECG shows anterior myocardial infarction. (Reproduced with permission from Mittal SR. Q wave. Cardiology Today. 2016;20:28–38)

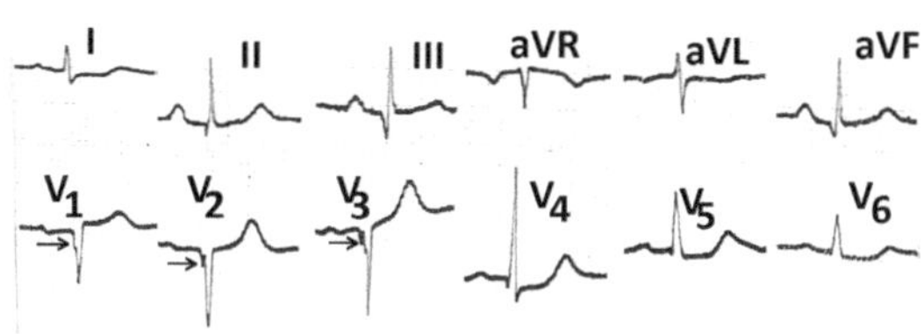

Fig. 19.18 Electrocardiogram showing slurring on the descending limb of the QS complex (marked) in leads V_1 to V_3. (Reproduced with permission from Mittal SR. Q wave. Cardiology Today. 2016;20:28–38)

slurring. Amplitude can vary with overall amplitude of the QRS complex.

- Abnormal Q waves can be normally present in leads III, aVR, V_1, and sometimes in lead aVL.

6 Q Wave Myocardial Infarction

Necrosis of sufficient myocardium results in loss of depolarization forces and produces an abnormal Q wave. Nontransmural infarcts can also produce Q waves. On the other hand, transmural infarcts may not produce a Q wave [2].

6.1 Localization of Myocardial Infarction by Q Wave [1, 2]

The relation of precordial leads and limb leads to cardiac surfaces is shown in Fig. 19.19a, b, c.

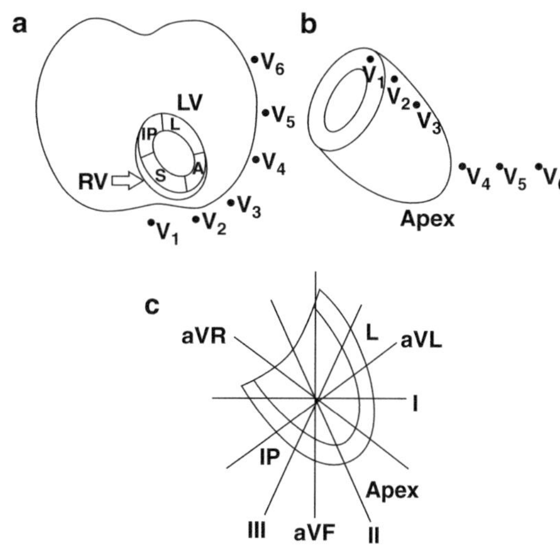

Fig. 19.19 (**a**) Diagram showing the relation of precordial leads to cardiac surfaces in the horizontal plane. (**b**) Diagram showing the relation of precordial leads to cardiac surfaces in the frontal plane. (**c**) Diagram showing the relation of limb leads to cardiac surfaces in the frontal plane. *A* anterior, *L* lateral, *S* septum, *IP* inferoposterior. (Reproduced with permission from Mittal SR. Q wave. Cardiology Today. 2016;20:28–38)

Leads showing Q wave	Site of infarction	Most probable vessel occluded
I, aVL	High lateral	First diagonal
I, aVL, V_6	Lateral	Dominant diagonal
I, aVL, V_2–V_3	Mid anterior	Dominant diagonal
V_1, V_2 (Fig. 19.20)	Septal	Septal branch of LAD
V_2 to V_5	Anterior and apical	Left anterior descending coronary artery after origin of D_1
V_4–V_6	Apicolateral	Dominant diagonal / OM
Leads I, aVL, V_2 to V_6 (Fig. 19.21)	Extensive anterior	Proximal LAD
Leads II, III, aVF	Inferior	Distal RCA/LCX
Leads II, III, aVF		
+		
I, aVL, V_5, V_6 (Fig. 19.22)	Inferolateral	Dominant RCA/ proximal LCX
Leads II, III, aVF + V_7 to V_9 (Fig. 19.23)	Inferior + posterolateral	LCX
Leads II, III, aVF, V_3R to V_6R (Fig. 19.24)	Inferior and RV	Proximal RCA

Leads showing Q wave	Site of infarction	Most probable vessel occluded
Leads II, III, aVF + V_1 to V_3 (Fig. 19.25)	Inferior and septal	LAD + PDA
Leads V_3R to V_6R (Fig. 19.26)	Right ventricle	Proximal nondominant RCA

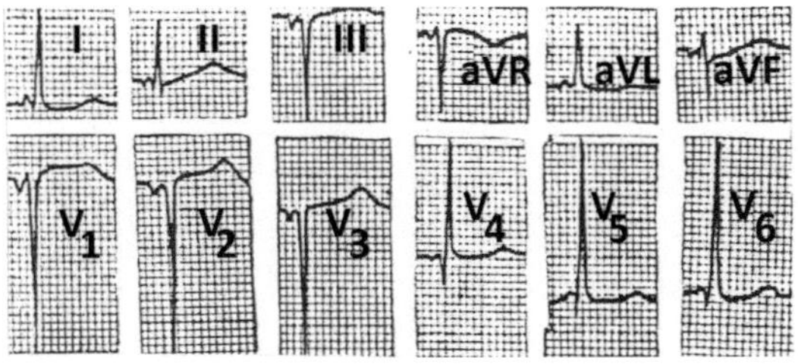

Fig. 19.20 Electrocardiogram showing QS in leads V_1 to V_3 suggestive of septal infarction. (Reproduced with permission from Mittal SR. Q wave. Cardiology Today. 2016;20:28–38)

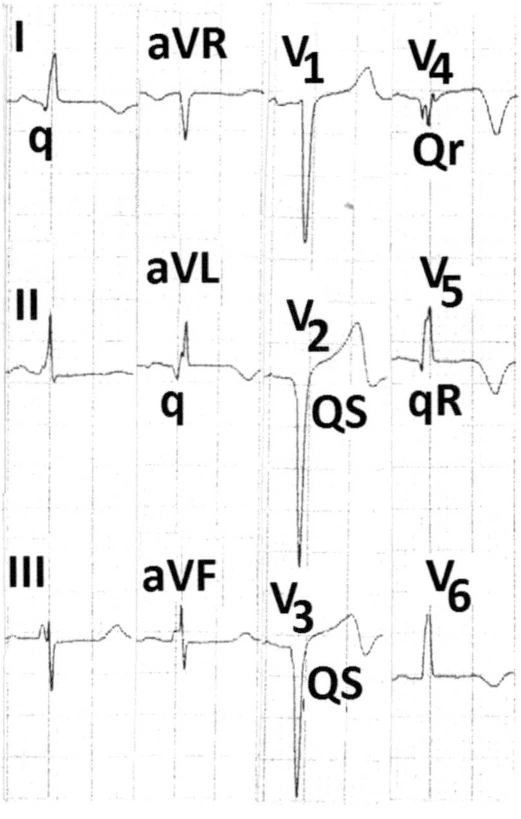

Fig. 19.21 Electrocardiogram showing Q waves in leads I, aVL, and V_1 to V_6 suggestive of extensive anterior infarction. (Reproduced with permission from Mittal SR. Q wave. Cardiology Today. 2016;20:28–38)

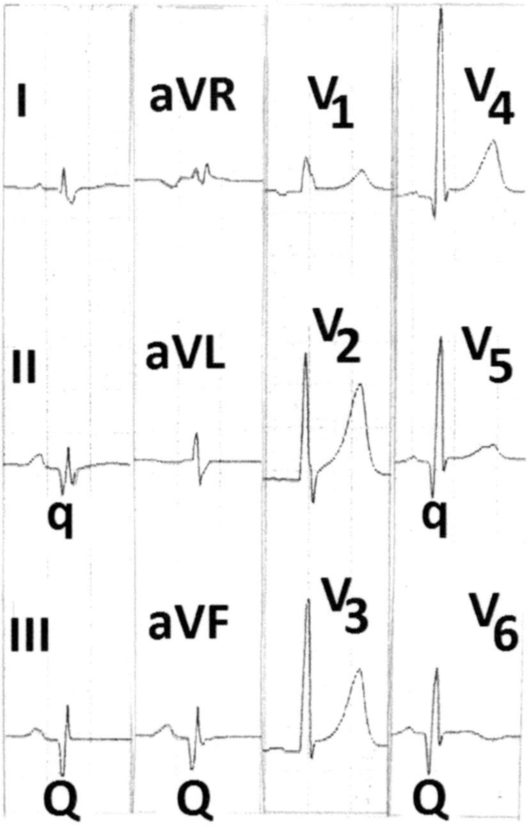

Fig. 19.22 Electrocardiogram showing Q waves in leads II, III, aVF, V_5, and V_6 suggestive of inferolateral infarction. Lead V_1 shows reciprocal changes as prominent R wave and upright T wave. (Reproduced with permission from Mittal SR. Q wave. Cardiology Today. 2016;20:28–38)

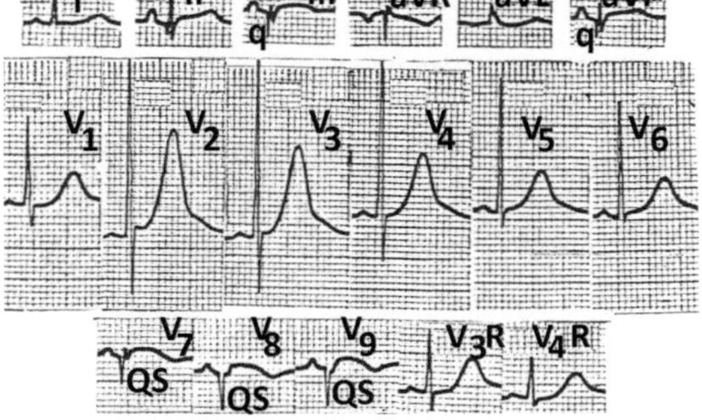

Fig. 19.23 Electrocardiogram showing Q waves in leads II, III, aVF, and V_7 to V_9 suggestive of inferoposterior infarction. Lead V_1 shows reciprocal changes in the form of prominent R wave and upright T wave. (Reproduced with permission from Mittal SR. Q wave. Cardiology Today. 2016;20:28–38)

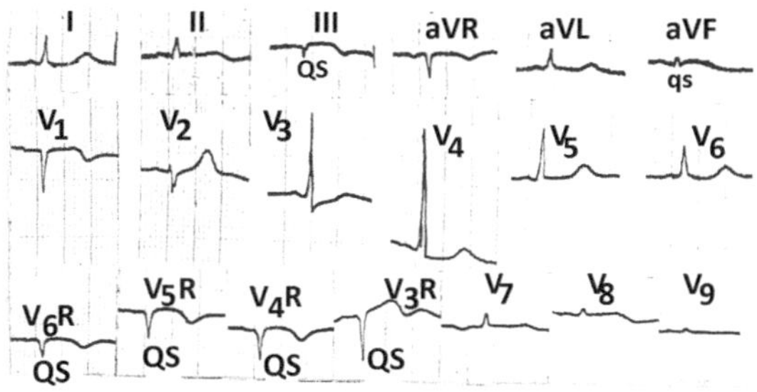

Fig. 19.24 Electrocardiogram showing Q waves in leads III, aVF, and V_6R to V_3R suggestive of inferior and right ventricular infarction. (Reproduced with permission from Mittal SR. Q wave. Cardiology Today. 2016;20:28–38)

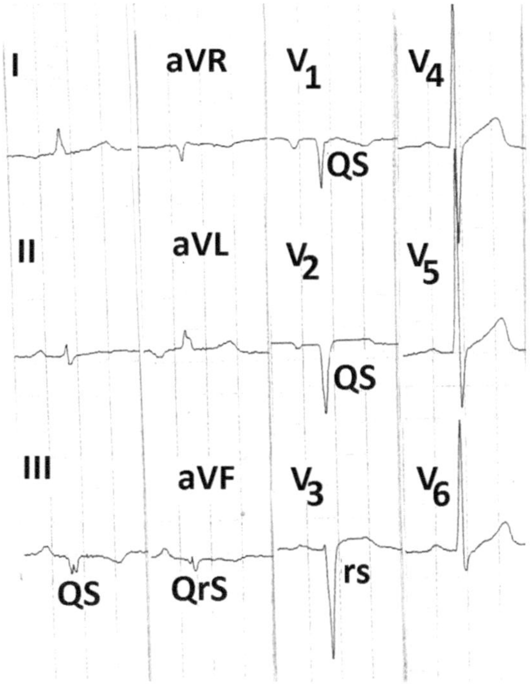

Fig. 19.25 Electrocardiogram showing Q waves in leads III, aVF, and V_1 and V_2 suggestive of inferior and septal infarction. (Reproduced with permission from Mittal SR. Q wave. Cardiology Today. 2016;20:28–38)

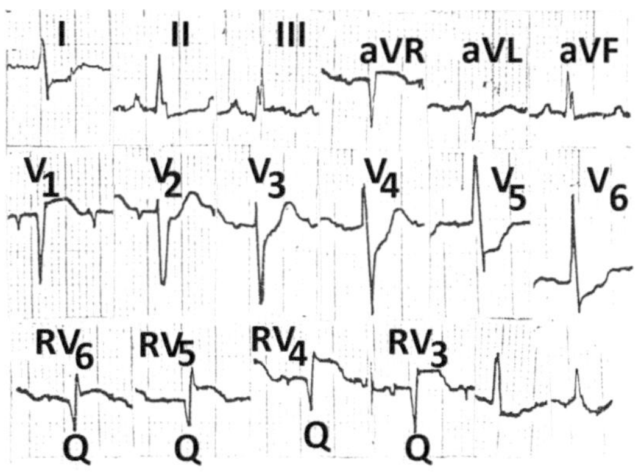

Fig. 19.26 Electrocardiogram showing Q waves in leads V_6R to V_3R suggestive of right ventricular infarction. (Reproduced with permission from Mittal SR. Q wave. Cardiology Today. 2016;20:28–38)

7 Q Wave Equivalent R Wave in Lead V_1

R wave ≥40 milliseconds.

> 3 mm in amplitude

R/S ≥ 0.5

These findings suggest lateral infarction. Initially, these findings were supposed to suggest “direct posterior“ infarction. Now it is felt that there is no true posterior wall of the left ventricle, and it is the vector of the lateral wall infarction that is directed toward lead V_1. These findings are usually associated with Q wave in leads I, aVL, and V_6, diminished voltage of R wave in lead V_6, and positive T wave in lead V_1.

7.1 Other Causes of Prominent R Wave in Lead V_1

- Early precordial transition in normal persons. Lead V_1 is normal. Lead V_2 shows prominent R wave with upright T wave (Fig. 19.27).
- Dextroversion: Left ventricle is displaced more anteriorly. In this condition, there are no q waves in leads I, aVL, and V_6. It can be congenital or acquired, e.g., in fibrosis or collapse of the right lung, pneumothorax, or massive effusion on the left side or eventration of the left dome of diaphragm.

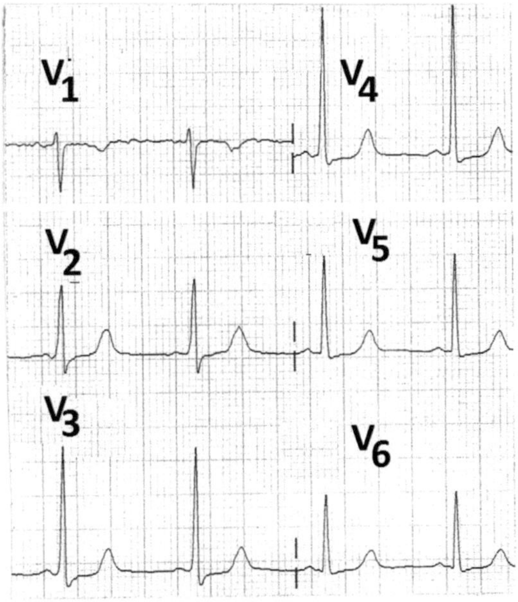

Fig. 19.27 Electrocardiogram shows the early precordial transition. Lead V_1 shows rS configuration and lead V_2 shows Rs configuration. (Reproduced with permission from Mittal SR. Q wave. Cardiology Today. 2016;20:28–38)

- Duchenne muscular dystrophy: Pathological q waves are present in leads I, aVL, and V_5 to $V_{9.}$
- Preexcitation from left lateral and left posteroseptal accessory path: Short PR interval and presence of delta wave help in diagnosis.
- Right ventricular hypertrophy: "P- pulmonale," T wave inversion in lead V_1, right axis deviation, and clockwise rotation help differentiation.
- Rarely patients with hypertrophic cardiomyopathy, especially those with right ventricular involvement, can have prominent R wave in lead V_1.

8 Differential Diagnosis of Causes of Q Waves in Different Leads

(a) **In Limb Leads**

- Leads I and aVL
  - Normal in vertical electrical axis.
  - Dextrocardia, technical dextrocardia, high lateral myocardial infarction, left lateral accessory pathway, and left anterosuperior hemiblock.
- III
  - Common in normal persons.
  - Horizontal electric axis of heart: Q wave reduces in depth during deep inspiration.
  - Inferior MI: Q wave is broad and slurred. Usually, q wave is also present in leads II and aVF. Usually, there is coved ST-segment with inversion of T wave.
  - Left posteroinferior fascicular block: q wave is narrow. R wave voltage is increased. There are no sequential ST-segment and T wave changes.
  - Left posteroseptal and right posteroseptal accessory paths.
  - Acute pulmonary thromboembolism.
- QS in leads II, III, and aVF.
  - Extensive inferior infarction.
  - Inferior infarction with left anterior hemiblock.
  - Left bundle branch block with left anterior hemiblock.
- AVR- Q wave in lead aVR is normal.

(b) **Precordial Leads**

- Leads V_1, V_2.
  - If septal forces are directed precisely at right angles to the recording lead, initial r wave may not be present and qRS complex may appear as QS in lead V_1. Rarely such QS complex may also appear in lead V_2.
  - Septal infarction: Notch or slurring, especially on the initial down stroke, supports the presence of a q wave. Marked ST-segment elevation and inversion of the T wave also support possibility of infarction.
  - Higher placement of leads V_1 and V_2.
  - Emphysema.
  - Left anterosuperior fascicular block especially when electrodes of leads V_1 and V_2 are placed higher.

- Marked right atrial enlargement can produce qR configuration in lead V_1. It can be present in tricuspid incompetence, cor pulmonale, or Ebstein anomaly.
- Preexcitation from right anteroseptal and midseptal accessory path.
- Left ventricular hypertrophy (Pressure overload).
- Left bundle branch block.
- Corrected transposition of great arteries.
- Acute pulmonary embolism.

• Leads V_3,V_4.
  - Apical infarction.

• Leads V_5,V_6.
  - Left ventricular volume overload: q waves are prominent but not wide.
  - Vagotonia: q waves are prominent but not abnormal in width.
  - Myocardial diseases: Q wave may be abnormal and may also affect limb leads.
    Myocarditis.
    Dilated cardiomyopathy.
    Hypertrophic cardiomyopathy.
    Duchenne muscular dystrophy: also affects leads V_7-V_9.
    Friedreich's ataxia.
  - Apical and anterolateral infarction.

(c) **Transient Q Waves:**
- Aborted infarction: electric stunning but not infarction.
- Severe hypotension: e.g., Acute pancreatitis.
  Allergic shock.
- Tachyarrhythmia.
- Transient fascicular blocks or bundle branch block.
- Wrong placement of leads.
- Protracted hypoglycemia.
- Phosphorus poisoning.
- Hyperkalemia.

(d) **Cyclic Change in Depth of Q Wave**
- Cardiac tamponade.
- Bronchial asthma.

Summary

- First negative wave of the QRS complex is called a Q wave.
- Normally deep (but not broad) Q wave can be present in lead III and to some extent in the lead aVF in horizontal heart. Lead aVR usually shows rs, QS, or Qr pattern. Lead V_1 normally shows rS complex, but may show a QS complex. q wave of any magnitude in leads V_2 or V_3 is always abnormal. Small narrow q waves can be present normally in leads I, aVL, V_5, and V_6 due to normal depolarization of the interventricular septum from left to right.
- Abnormality of the Q wave should be decided by duration (>0.03 second). Amplitude should be considered abnormal only if it exceeds 25% of the amplitude of the ensuing R wave except in leads III and aVR.
- Abnormal Q wave suggests loss of excitable myocardium. This can be due to myocardial necrosis or replacement of normal myocardium by amyloid, sarcoid, or tumor. Myocardial necrosis in myocarditis or replacement by fibrous tissue in dilated cardiomyopathy can also produce abnormal Q wave.
- Q waves in leads I and aVL suggest high lateral myocardial infarction caused by occlusion of the first diagonal or obtuse marginal coronary arteries. Q waves in leads V_2 to V_4 suggest infarction involving left ventricular apex. Q waves in leads II, III, and aVF suggest inferior infarction due to occlusion of the posterior descending coronary artery.
- Posterior descending coronary artery can be a branch of right coronary artery or left circumflex coronary artery depending on the dominance of the coronary circulation.
- Q waves of old myocardial infarction (especially inferior infarction) may disappear from electrocardiogram over a period of time. A transient Q wave may be seen in cases of stunned (but not necrosed) myocardium as in prinzmetal's angina or recanalized coronary artery.

- Noninfarct Q waves (usually deep but not broad) can be present in leads I, aVL or II, III, aVF, and V_5, V_6 in left ventricular hypertrophy from any cause including hypertrophic cardiomyopathy.
- Right ventricular hypertrophy with right atrial enlargement can produce QR complex in lead V_1 and lead aVR.
- QS pattern can be present in leads V_1 to V_3 in the presence of left bundle branch block. Leads V_1 to V_3 recorded in a higher intercostal space than the standard positions can also show QS pattern especially in the presence of left anterior fascicular block or emphysema.
- Chest deformities like pectus excavatum and kyphoscoliosis can produce QS configuration in lead V_1 (not in lead V_2).
- Negative delta wave can produce q wave depending on location of the accessory pathway.
- Pulmonary thromboembolism is known to produce Q wave with T wave inversion in lead III. However, T wave inversion in leads V_1 to V_4 is more frequent.

MCQs

Q1. In a normally positioned heart, depolarization of the interventricular septum is directed

(a) Anteriorly
(b) Superiorly
(c) Laterally
(d) Posteriorly

Q2. Duration of a normal q wave is

(a) <30 milliseconds
(b) 30–40 milliseconds
(c) 40–60 milliseconds
(d) >60 milliseconds

Q3. Isolated presence of QS complex in lead V_1 suggests

(a) Normal variation
(b) Septal infarction
(c) Anterior infarction
(d) Right ventricular infarction

Q4. rS pattern in lead I suggests

(a) Normal variation
(b) Left bundle branch block
(c) Right bundle branch block
(d) Left posteroinferior hemiblock

Q5. QS complex in V_1 and V_2 does not occur in

(a) Higher placement of chest leads
(b) Vertical heart
(c) Dextrocardia
(d) Isolated left anterior fascicular block

Q6. Q wave in leads V_3,V_4 does not suggest

(a) High lateral infarction
(b) Apical infarction
(c) Anterior transmural infarction
(d) Anterior subendocardial infarction

Q7. Slow progression of r wave can occur in

(a) Pectus excavatum
(b) Dextrocardia
(c) Congenitally corrected transposition of the great arteries
(d) All

Q8. Which ECG finding is abnormal in lead V_1?

(a) R duration ≥40 milliseconds
(b) R amplitude ≥3 mm
(c) R/S ration >0.5
(d) All

Q9. RS configuration in V_1 suggests

(a) Apical infarction
(b) Lateral infarction
(c) Posterior infarction
(d) Inferior infarction

Q10. Q wave in leads V_1 and V_2 suggest

(a) Septal infarction
(b) Anterior infarction
(c) Right ventricular infarction
(d) High lateral infarction

Q11. Proximal occlusion of dominant left circumflex coronary artery produces Q waves in leads

(a) I, aVL, V_1 to V_4
(b) I, aVL, V_5 and V_6
(c) I, aVL, V_5,V_6 and II, III, aVF
(d) II, III, aVF

Q12. Distal occlusion of long type III left anterior descending coronary artery produces Q wave in leads

(a) I and aVL
(b) I, aVL, V_5, V_6
(c) V_5, V_6
(d) V_5, V_6, II, III, and aVF

Q13. Occlusion of the first diagonal coronary artery produces Q wave in

(a) I and aVL
(b) V_3,V_4
(c) V_5,V_6
(d) V_1,V_2

Q14. Occlusion of the left main coronary artery is suggested by

(a) ST-segment elevation in lead aVR > V_1
(b) Q wave in leads V_1 to V_4
(c) Q wave in leads I, aVL
(d) ST-segment elevation in leads V_1 to V_6

Q15. Occlusion of nondominant left circumflex coronary artery does not produce

(a) $RV_1 \geq SV_1$
(b) $RV_2 \geq SV_2$
(c) ST-segment elevation in leads II, III, and aVF
(d) q wave in leads I and aVL

Q16. Left-sided pneumothorax with mediastinal shift to right can produce

(a) R wave in V_2
(b) T wave inversion in leads V_1 to V_3
(c) Q wave in lead V_6
(d) QS complex in lead I

Q17. Negative P wave, QS complex, and T wave inversion in lead I suggest

(a) Technical dextrocardia
(b) Dextrocardia
(c) Dextroversion
(d) Levocardia

Q18. Left anterosuperior fascicular block shows

(a) qR complex in lead I
(b) Qr complex in lead aVL
(c) RS complex in lead III
(d) QS complex in leads V_1 to V_3

Q19. Left bundle branch block can produce QS complex in

(a) Leads I and aVL
(b) Leads V_1 to V_3
(c) Leads V_4 to V_6
(d) Leads II, III, and aVF

Q20. Right posteroseptal accessory pathway produces q wave in

(a) Leads I and aVL
(b) Leads V_5, V_6
(c) Leads II, III, and aVF
(d) Leads V_1, V_2

Q21. qr. configuration in lead V_1 suggests

(a) Right atrial enlargement
(b) Left atrial enlargement
(c) Incomplete right bundle branch block
(d) Right ventricular hypertrophy

Q22. qR configuration with increased R wave voltage in lead V_5 and V_6 can occur in

(a) Gross aortic regurgitation
(b) Large left to right shunt across PDA
(c) Myocarditis
(d) All

Q23. Prominent R wave in leads V_1 and V_2 cannot be due to

(a) Duchenne muscular dystrophy
(b) Lateral infarction
(c) Emphysema
(d) Severe pulmonary valve stenosis

Q24. Asymmetrical septal hypertrophy does not produce

(a) Deep symmetrical T wave inversion in leads V_2 to V_4
(b) Tall R waves in leads V5,V6
(c) Wide Q in leads V5,V6
(d) Shelf-like ST-segment

Q25. The second negative wave of the QRS complex is called

(a) Q wave
(b) Q' wave
(c) S wave
(d) s' wave

Q26. Q wave can be present in lead III in

(a) Horizontal heart
(b) Left anterior hemiblock
(c) Left posterior hemiblock
(d) Vertical heart

Q27. Q wave can be present in lead I and aVL in

(a) Horizontal heart
(b) Vertical heart
(c) Uncomplicated left bundle branch block
(d) Left posterior hemiblock

Q28. QS complex can be present in normal persons in leads

(a) I
(b) aVR
(c) V_1
(d) V_6

Q29. Q wave in lead V_3 is considered normal if

(a) Duration is less than 0.04 seconds
(b) Depth is less than 0.1 millivolt
(c) Depth is less than 25% of the ensuing R wave
(d) None of the above

Q30. Which finding is always abnormal?

(a) Presence of q wave in lead III
(b) Presence of q wave in lead V_5
(c) Presence of q wave in lead V_2
(d) Presence of Q wave in lead aVR

Q31. Q wave in leads other than lead III and aVR are considered abnormal if

(a) Duration is less than 0.02 second
(b) Duration is more than 0.03 second
(c) Depth is more than 0.1 millivolt
(d) Depth is more than 25% of the amplitude of the ensuing R wave

Q32. Q wave without myocardial necrosis can be present in

(a) Cardiac amyloidosis
(b) Cardiac tumors
(c) Sarcoid disease of the heart
(d) All

Q33. Abnormal Q waves can be present in

(a) Myocarditis
(b) Dilated cardiomyopathy
(c) Hypertrophic cardiomyopathy
(d) None

Q34. Q wave with ST-segment elevation and T wave inversion in leads I and aVL suggest

(a) High lateral infarction
(b) Mid lateral infarction
(c) Infarction of left ventricular apex
(d) Infarction of the interventricular septum

Q35. Q wave in leads V_2 to V_4 suggests infarction of

(a) Basal part of the interventricular septum
(b) Lateral wall of the left ventricle
(c) Left ventricular apex
(d) Basal inferior part of the left ventricle

Q36. Well-developed Q waves can disappear if

(a) There is stunning of myocardium without necrosis
(b) There is a small infarction
(c) There is infarction in electrically opposite wall of the myocardium
(d) None

Q37. Right ventricular hypertrophy with dilatation of the right atrium can produce QR complex in

(a) Lead aVR
(b) Lead aVL
(c) Lead V_1
(d) Lead V_4

Q38. QS pattern can be present in leads V_1 to V_3 in cases of

(a) Left bundle branch block
(b) Leads placed in higher than standard position
(c) Left anterior fascicular block
(d) Emphysema

Q39. Right posteroseptal accessory pathway produces abnormal Q wave in

(a) Leads II, III, aVF
(b) Leads I and aVL
(c) Lead V_1
(d) Lead V_5

Q40. Left lateral accessory pathway produces abnormal Q wave in

(a) Leads II, III, and aVF
(b) Leads I and aVL
(c) Lead V_1
(d) Lead V_6

Q41. Q wave can be present in lead III in

(a) Pulmonary thromboembolism
(b) Right posteroseptal accessory pathway
(c) Left lateral accessory pathway
(d) Left posterior fascicular block.

Answers

(1) a (2) a (3) a (4) a (5) c, d (6) a (7) d (8) d (9) b, c (10) a (11) c (12) d (13) a (14) a (15) c (16) a, b (17) a, b (18) a (19) b (20) c (21)a (22) a, b (23) c (24) c (25) c (26) d (27) a (28) b, c (29) d (30) c (31) b, d (32) d (33) a, b, c (34) b (35) c (36) a, b, c (37) a, c (38) a, b, c, d (39) a (40) b (41) a, b, d.

References

1. De Luna AB, Goldwasser D, Fiol M, Bayes-Genis A. Surface electrocardiography. In: Fuster V, Walsh RA, Harrington RA, editors. Hurst's the heart. New York: Mc Graw Hill; 2011. p. 307–70.
2. Mirvis DM, Goldberger AL. Electrocardiography. In: Banow RO, Mann DL, Zipes DP, Libby P, editors. Braunwald's heart disease. Missouri: Elsevier; 2012. p. 126–63.
3. Schamroth L. Myocardial infarction. In: Schamroth C, editor. An introduction to electrocardiography. France: Blackwell Science; 1982. p. 131–56.
4. Wagner GS, Lim TH. Interpretation of the normal electrocardiogram. In: Wagner GS, editor. Marriott's practical electrocardiography. Philadelphia: Wolters Kluwer; 2008. p. 43–70.

Subtle ECG Changes in Early Phase of Anterior and Lateral Myocardial Infarction

20

Classical ECG changes of anterior myocardial infarction are easily picked up (Fig. 20.1).

However, more than 50% of patients with acute anterior myocardial infarction may have only subtle ECG changes during the early hours of the infarction. These changes are transient and appear in the hyperacute phase. Knowledge of these changes is necessary so that the diagnosis is not missed during early hours following infarction and the patient gets timely proper treatment. Such subtle ECG changes include.

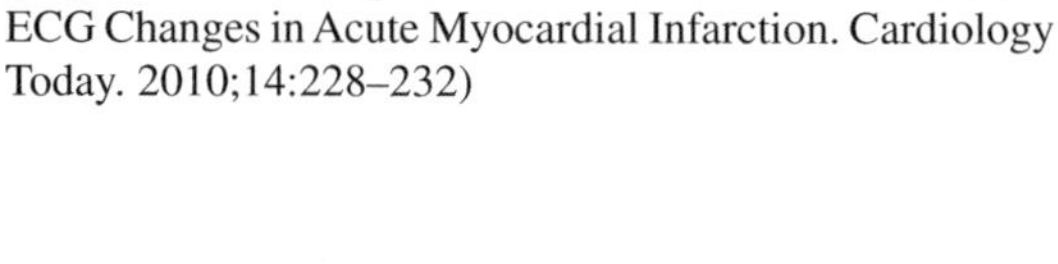

Fig. 20.1 Electrocardiogram showing classical ST-segment elevation in anterior myocardial infarction. (Reproduced with permission from Mittal SR. Subtle ECG Changes in Acute Myocardial Infarction. Cardiology Today. 2010;14:228–232)

1 Early uptake of the ST-segment

Normally the ST-segment gradually merges with the T wave (Fig. 20.2, marked *). If the ST-segment rises from the j point, it is known as early uptake of the ST-segment (Fig. 20.2 marked as ↓). This is usually the earliest change in the ST-segment during acute myocardial infarction [1]. It is considered to be related to epicardial injury [2]. Frequently, the ST-segment depression

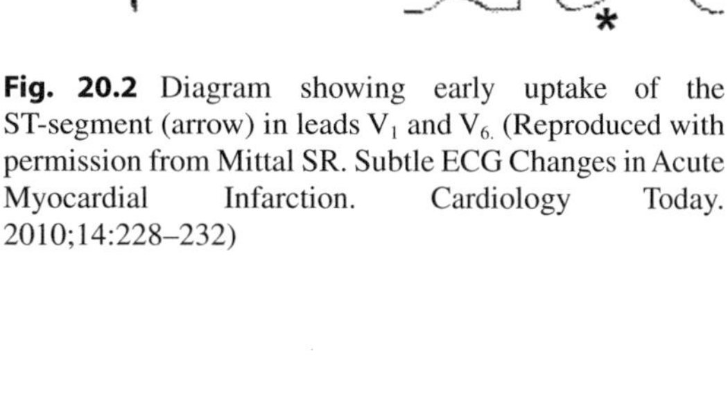

Fig. 20.2 Diagram showing early uptake of the ST-segment (arrow) in leads V_1 and V_6. (Reproduced with permission from Mittal SR. Subtle ECG Changes in Acute Myocardial Infarction. Cardiology Today. 2010;14:228–232)

S. Mittal, *Insights into Electrocardiograms with MCQs*,
https://doi.org/10.1007/978-981-99-0127-2_20

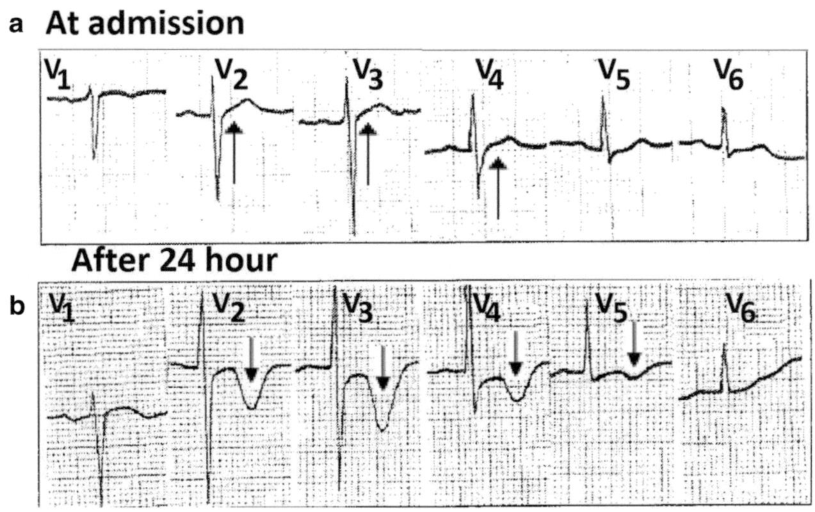

Fig. 20.3 (**a**) Electrocardiogram at admission showing early uptake of the ST-segment in leads V_2 to V_4 (marked ↑). (**b**) ECG recorded after 24 hours showing classical deep symmetrical T wave inversions in leads V_2 to V_6 (marked ↓) diagnostic of anterior myocardial infarction. (Reproduced with permission from Mittal SR. Subtle ECG Changes in Acute Myocardial Infarction. Cardiology Today. 2010;14:228–232)

in electrically opposite leads may be more impressive.

Representative electrocardiograms recorded at admission (Fig. 20.3a) showing early uptake of ST-segment and subsequent electrocardiogram showing classical ECG changes (Fig. 20.3b) are shown below.

2 Tall T Waves in Leads V_2 and V_3

Normally amplitude of the T waves is less than that of the preceding r wave in leads V_1 to V_3. If the amplitude of the T wave is more than the amplitude of the preceding r wave in these leads,

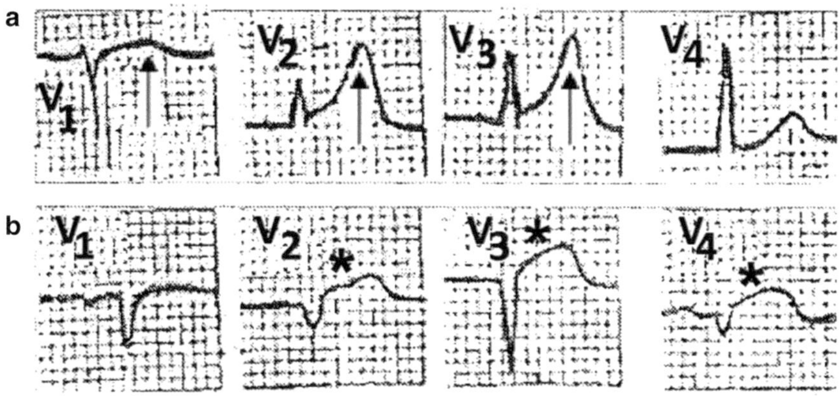

Fig. 20.4 (**a**) Electrocardiogram at admission showing T waves taller than R waves in leads V_1 to V_3 (marked ↑). (**b**) ECG recorded after few hours showing loss of R with classical coved ST-segment elevation in leads V_2 to V_4 (marked *) with terminal T wave inversion in lead V_4 diagnostic of anterior myocardial infarction. (Reproduced with permission from Mittal SR. Subtle ECG Changes in Acute Myocardial Infarction. Cardiology Today. 2010;14:228–232)

it is usually abnormal (Fig. 20.4a, b). It is seen during hyperacute phase of myocardial infarction [1, 2]. This change is seen in the earliest stage of myocardial infarction [3]. Such T wave change can precede the change in the ST-segment [4, 5].

3 ST-Segment Depression

Normally the ST-segment rises slowly after the J point and merges with the T wave. In some cases of anterior myocardial infarction, early ECG may show only ST-segment depression (Fig. 20.5a). ECG taken later may reveal classical ECG changes of anterior infarction (Fig. 20.5b). It is likely that initial ST-segment depression is due to subendocardial ischemia, which finally presents as subendocardial infarction [4, 6–8].

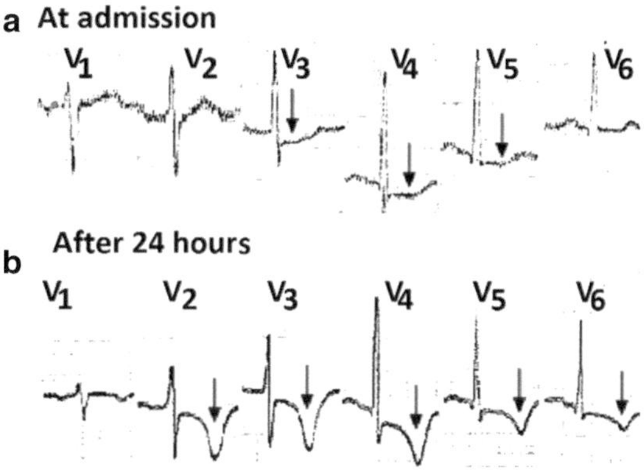

Fig. 20.5 (**a**) Electrocardiogram at admission showing ST-segment depression in leads V_3 to V_6 (marked ↓). (**b**) Electrocardiogram recorded after 24 hours showing classical deep symmetrical inversion of the T waves in leads V_2 to V_6. (marked ↓). (Reproduced with permission from Mittal SR. Subtle ECG Changes in Acute Myocardial Infarction. Cardiology Today. 2010;14:228–232)

4 ST-Segment Elevation in Leads aVR and V_1

Normally in leads aVR and V_1, the ST-segment shows gradual downslope and merges with the negative T wave. ST-segment elevation in these leads (Fig. 20.6a) could be an early marker of myocardial infarction (Fig. 20.6b) especially when there are concomitant ST-segment depressions in inferior and lateral precordial leads (Fig. 20.6a). ST-segment elevation in lead aVR more than in lead V_1 with ST-segment depression in other leads suggests left main coronary artery disease or severe multivessel disease [4, 8, 9].

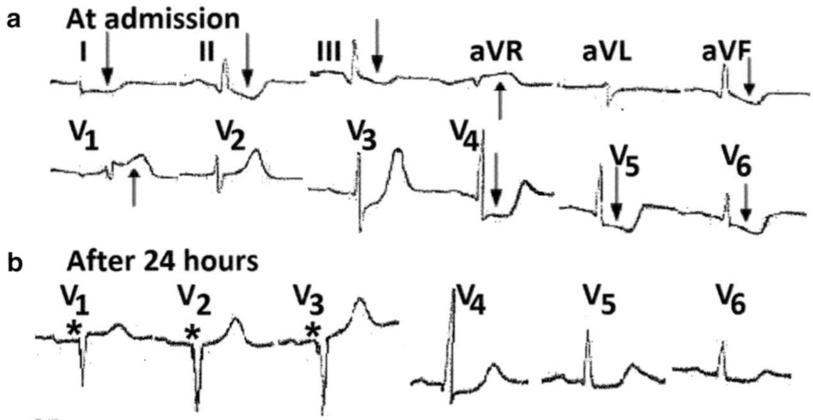

Fig. 20.6 (**a**) Electrocardiogram at admission showing ST-segment elevation in leads aVR and V_1 (marked ↑) with ST-segment depression in leads I, II, III, aVF, and V_3 to V_6 (marked ↓). (**b**) Electrocardiogram recorded after 24 hours showing classical QS pattern in leads V_1 to V_3 (marked *) with reduction in reciprocal ST-segment depressions confirming anterior myocardial infarction. (Reproduced with permission from Mittal SR. Subtle ECG Changes in Acute Myocardial Infarction. Cardiology Today. 2010;14:228-232)

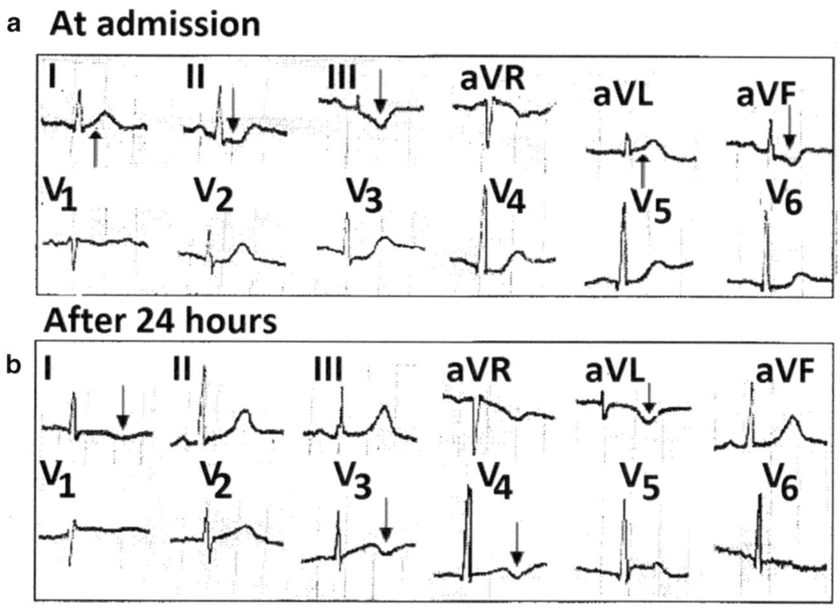

Fig. 20.7 (**a**) Electrocardiogram at admission showing early uptake of the ST-segment in leads I and aVL (marked ↑) with reciprocal ST-segment depression in leads II, III, and aVF (marked ↓). (**b**) Electrocardiogram recorded after 24 hours showing T wave inversion in leads I, aVL, V_3, and V_4 (marked ↓) suggestive of anterior infarction. (Reproduced with permission from Mittal SR. Subtle ECG Changes in Acute Myocardial Infarction. Cardiology Today. 2010;14:228–232)

5 Isolated ST-Segment Elevation in Leads I and aVL (High Lateral Myocardial Infarction)

Leads I and aVL represent high lateral surface of the left ventricle. ST-segment elevation localized to these leads (Fig. 20.7a) suggests high lateral infarction (Fig. 20.7b) especially if there reciprocal ST-segment depressions in leads II, III, and aVF [4, 10].

6 ECG Changes Seen Only in Higher Intercostal Space (High Lateral Myocardial Infarction)

Sometimes, changes of high lateral infarction may be clear only in precordial leads recorded one intercostal space above the conventional sites [11, 12] (Fig. 20.8a, b).

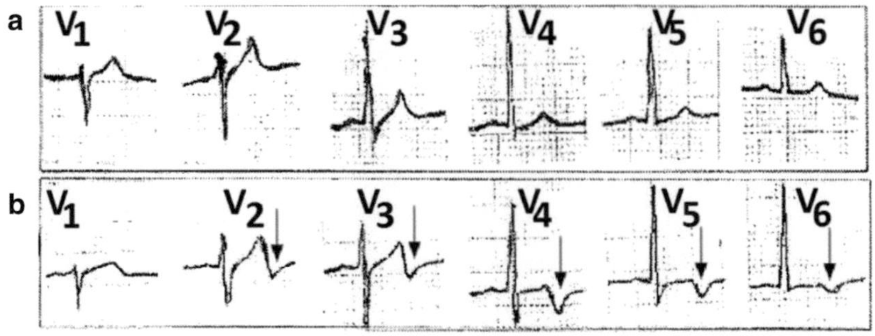

Fig. 20.8 (**a**) Conventional precordial leads showing normal ECG. (**b**) Precordial leads recorded in higher intercostal space showing T wave inversion in leads V_2 to V_6 (marked ↓) diagnostic of anterior infarction. (Reproduced with permission from Mittal SR. Subtle ECG Changes in Acute Myocardial Infarction. Cardiology Today. 2010;14:228-232)

7 Small "q" Wave in Leads V_2 and V_3

Normally the QRS complex in leads V_2 and V_3 shows rS or RS configuration. In normal persons, these leads never show a q wave. Therefore, any q wave in their leads (even if tiny) is suggestive of myocardial infarction [12, 13] (Fig. 20.9a, b).

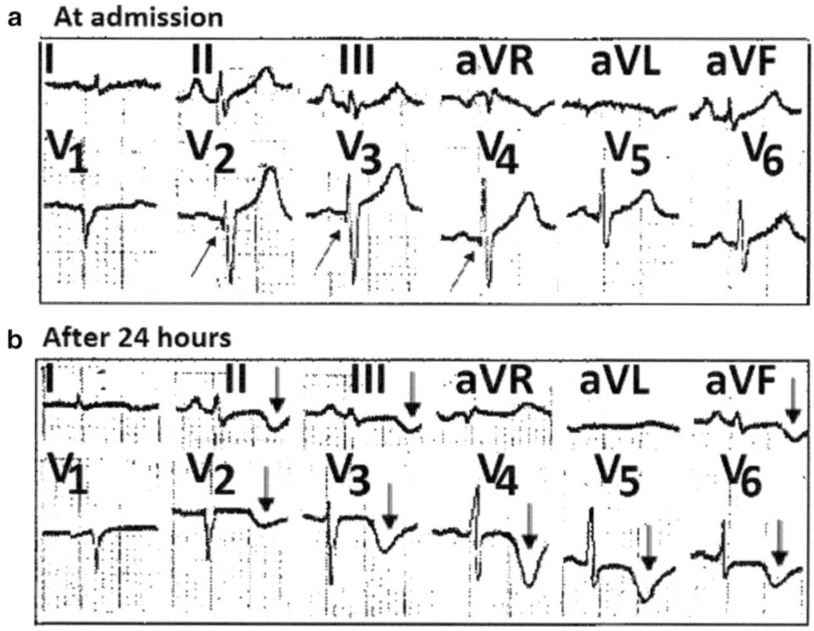

Fig. 20.9 (**a**) Admission electrocardiogram showing small q waves in leads V_2 to V_4 (arrows). (**b**) ECG recorded after 24 hours showing sequential T wave inversions in leads II, III, aVF, and V_2 to V_6 (marked ↓) diagnostic of myocardial infarction. (Reproduced with permission from Mittal SR. Subtle ECG Changes in Acute Myocardial Infarction. Cardiology Today. 2010;14:228–232)

8 Nonprogression of the "R" Wave

Amplitude of the R wave represents amount of excitable (viable) myocardium. Normally the height of the "R" wave increases progressively from lead V_1 to V_5 (Fig. 20.10).

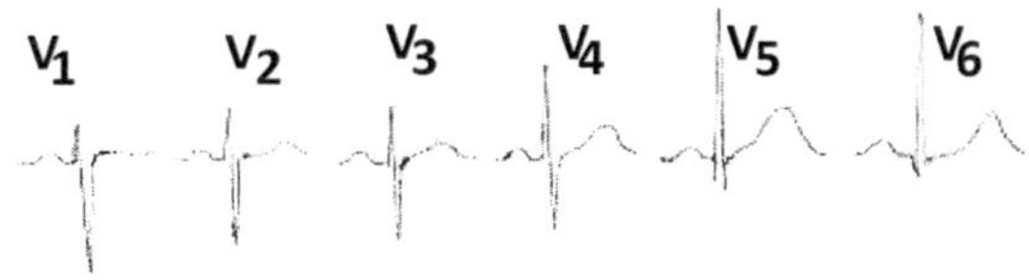

Fig. 20.10 Normal progression of the R wave in the precordial leads. (Reproduced with permission from Mittal SR. Subtle ECG Changes in Acute Myocardial Infarction. Cardiology Today. 2010;14:228–232)

Decrease in the amplitude or R wave suggests loss of viable myocardium. Therefore, nonprogression of the R wave in precordial leads is a marker of infarction. However, this is a nonspecific finding. Left bundle branch block, left ventricular hypertrophy, emphysema, left pneumothorax, dextrocardia, pectus excavatum, and congenital absence of the pericardium can also produce this finding. This finding should, therefore, be interpreted in a given clinical context. Notch on the descending limb of the QS complex (Fig. 20.11) suggests that a Q wave is present as a part of the QS complex. This supports the possibility of a septal infarction over nonspecific QS patterns seen in leads V_1 to V_3 [10].

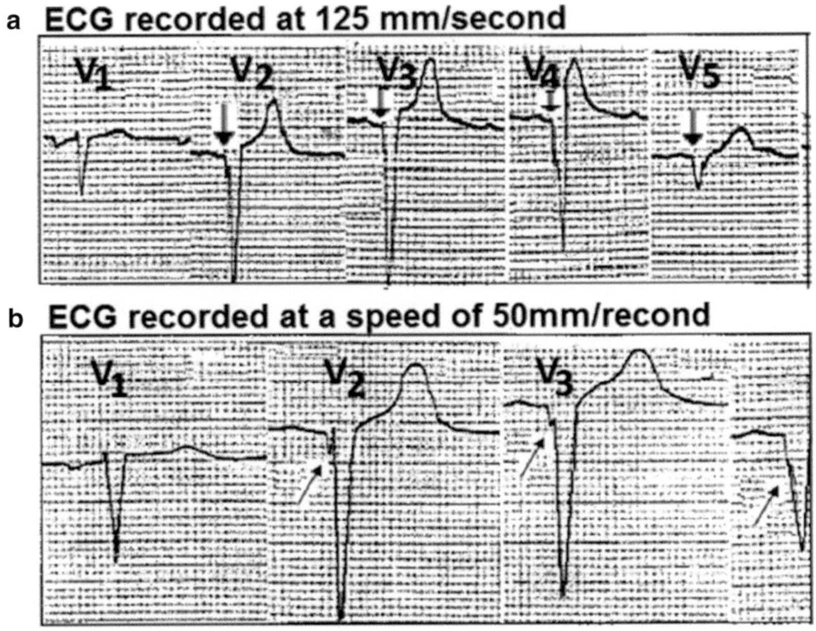

Fig. 20.11 (**a**) Electrocardiogram showing nonprogression of R wave (marked ↓) in precordial leads. (**b**) Electrocardiogram recorded at 50 mm/second speed clearly showing slurring on the downstroke of the QRS complex in V_2 to V_4 (marked ↑) suggestive of presence of Q wave. (Reproduced with permission from Mittal SR. Subtle ECG Changes in Acute Myocardial Infarction. Cardiology Today. 2010;14:228-232)

9 Reverse Progression of the R Wave

Normally height of the R wave increases from lead V_1 to V_5 (Fig. 20.10). Paradoxically, if height of the R wave suddenly decreases in a precordial lead, it is a marker of anterior myocardial infarction [10] (Fig. 20.12).

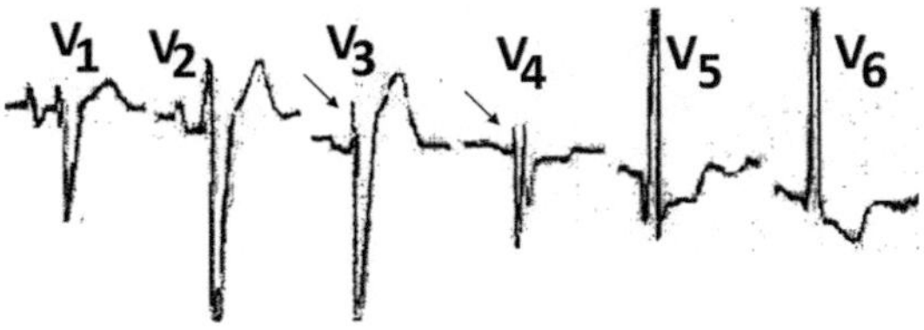

Fig. 20.12 Electrocardiogram showing reverse progression of the R wave from lead V_2 to V_4. (marked ↓). (Reproduced with permission from Mittal SR. Subtle ECG Changes in Acute Myocardial Infarction. Cardiology Today. 2010;14:228–232)

10 Ventricular Ectopics with Right Bundle Branch Block Configuration Showing "q" Wave

Some persons show a QS or rS configuration. Diagram of myocardial infarction in these cases is not clear as several condition (mentioned above in heading (h)) can produce nonprogression of the r wave in leads V_1 to V_3 (Fig. 20.13a). Sudden appearance of R wave (rather than a rs wave) in lead V_4 (Fig. 20.13a) strongly supports the possibility that the QS configuration in leads V_1 to V_3 is because of infarction of the interventricular septum. Notch on the upstroke of the R wave in lead V_4 (Fig. 20.13a) further supports the possibility of a Q wave in leads V_1 to V_3. If such a patient develops ventricular premature beats with right bundle branch block configuration, diagnosis become clear. Presence of a Q wave in the ventricular ectopics with right bundle branch block pattern (Fig. 20.13b) confirms diagnosis of anterior infarction [12, 14].

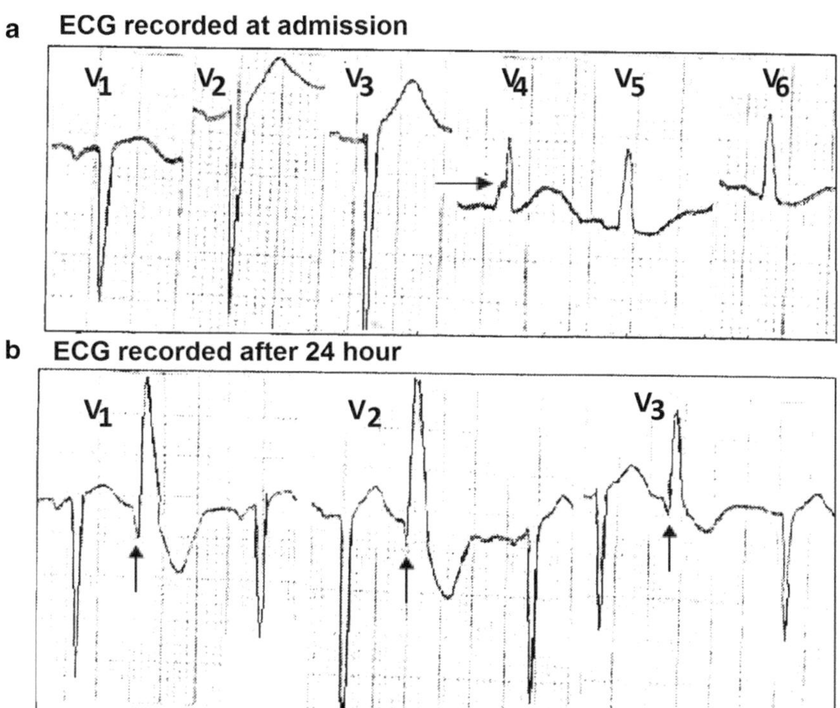

Fig. 20.13 (**a**) Electrocardiogram showing nonprogression of r wave in leads V_1 to V_3 with a notch on the upstroke of the QRS (marked →) in lead V_4. This finding per se also suggests that the nonprogression of the r wave in leads V_1 to V_3 is because of true loss of viable myocardium. (**b**) ECG recorded from the same patient at a different time. Ectopic beats with right bundle branch block configuration showing clear Q waves (marked ↑) confirming the diagnosis of anterior myocardial infarction. (Reproduced with permission from Mittal SR. Subtle ECG Changes in Acute Myocardial Infarction. Cardiology Today. 2010;14:228–232)

11 Mid QRS Change

Q wave forms the initial part and the S wave forms the terminal part of the QRS complex. R wave forms the middle part of the QRS complex. Changes in the R wave are, therefore, called mid QRS changes. Conventionally abnormalities of the Q wave are considered suggestive of myocardial necrosis. However, significance of changes in the R wave as markers of myocardial necrosis is not widely appreciated. R wave represents excitable (viable) myocardium. Normally the upstrokes and downstrokes of the R wave are equal and smooth (Fig. 20.14).

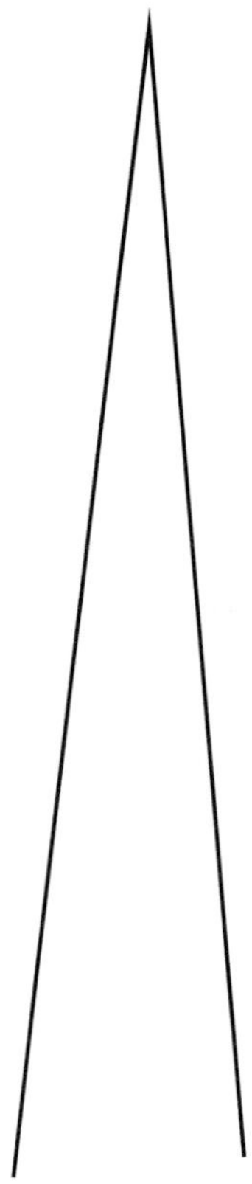

Fig. 20.14 Diagram showing the configuration of a normal R wave. (Reproduced with permission from Mittal SR. Subtle ECG Changes in Acute Myocardial Infarction. Cardiology Today. 2010;14:228–232)

Loss of the R wave voltage with slurring of ascending limb and top or descending limb in the absence of any finding suggestive of a bundle branch block suggests necrosis of some part of the myocardium (Fig. 20.15).

Representative electrocardiographic tracings are shown in Fig. 20.16a, b.

QRS complexes showing changes in the middle or late part of QRS are also called fragmented QRS complex. Fragmented QRS also strongly suggests the diagnosis of infarction in the involved leads [15, 16]. In addition to necrosis of part of the myocardium, peri-infarction block can also contribute to fragmented QRS [12].

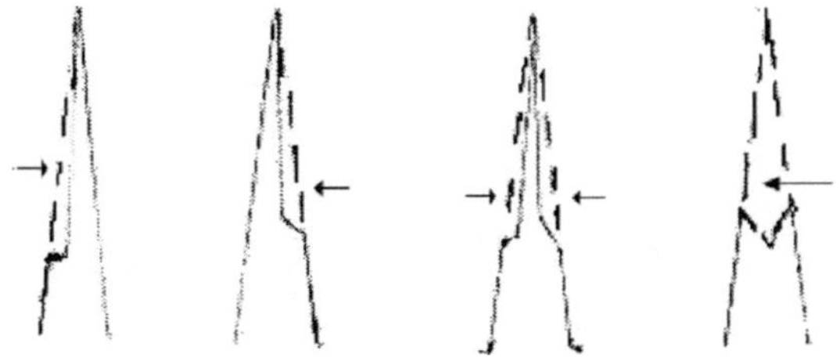

Fig. 20.15 Diagram showing mid QRS changes (changes in the configuration of the R wave) due to loss of different parts of the electrical potential of the R wave (marked → and ←). (Reproduced with permission from Mittal SR. Subtle ECG Changes in Acute Myocardial Infarction. Cardiology Today. 2010;14:228–232)

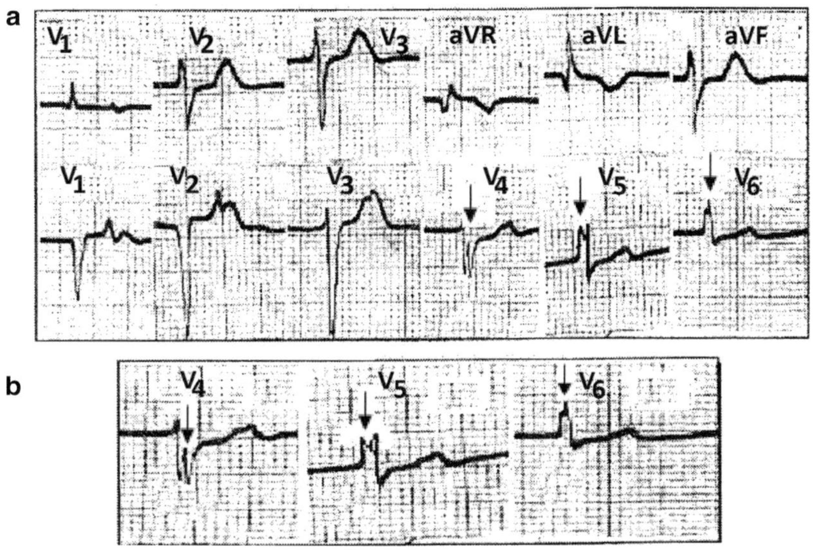

Fig. 20.16 (**a**) Electrocardiogram showing q wave with T wave inversion in leads aVL. Leads V_4 to V_6 show slurring (notching) of the QRS (marked ↓). (**b**) Magnified view of leads V_4 to V_6 showing notching in the QRS. (Reproduced with permission from Mittal SR. Subtle ECG Changes in Acute Myocardial Infarction. Cardiology Today. 2010;14:228–232)

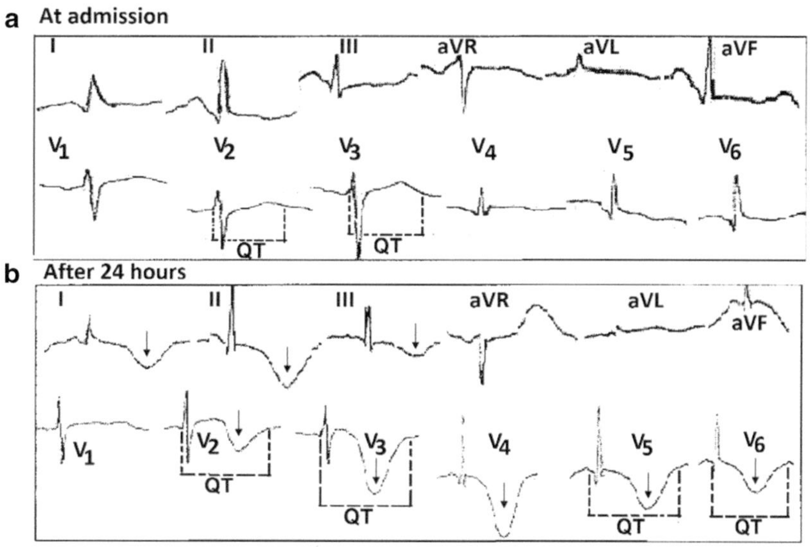

Fig. 20.17 (**a**) ECG recorded at admission showing mild prolongation of the QT interval. (**b**) ECG recorded after 24 hours showing deep inversion of the T waves in leads I, II, III, aVF, and V_2 to V_6 with clear prolongation of the QT interval. (Reproduced with permission from Mittal SR. Subtle ECG Changes in Acute Myocardial Infarction. Cardiology Today. 2010;14:228–232)

12 Prolongation of the QT Interval

Ischemic myocardium takes longer period to repolarize [4]. This results in prolongation to the QT interval in the electrocardiogram recorded early after onset of chest discomfort [5, 12, 17] (Fig. 20.17a). ECG recorded after the completion of myocardial infarction may show classical changes (Fig. 20.17b).

13 Inversion of the U Wave

Normally the U wave has the same polarity as that of the preceding T wave. It is, therefore, positive in leads V_2 to V_6. Any inversion of the U wave is abnormal (Fig. 20.18a) and suggests myocardial ischemia. In the clinical context of acute coronary syndrome, it can be the earliest electrocardiographic marker of myocardial infarction [4, 8, 9, 12] (Fig. 20.18b).

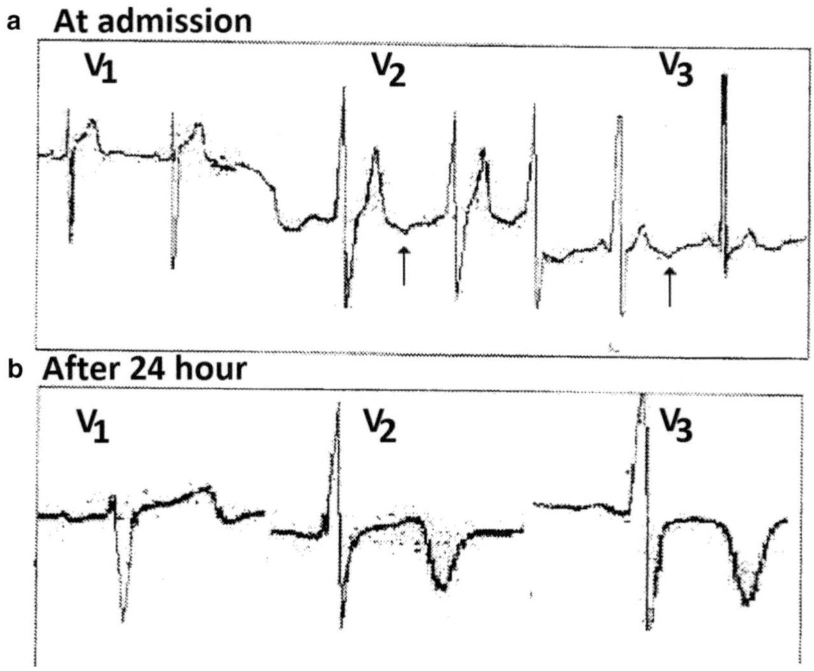

Fig. 20.18 (**a**) Admission ECG shows only U wave inversion (marked ↑). (**b**) ECG recorded after 24 hours showing classical deep symmetrical inversion of the T waves (marked ↓) in leads V_2 and V_3 diagnostic of myocardial infarction. (Reproduced with permission from Mittal SR. Subtle ECG Changes in Acute Myocardial Infarction. Cardiology Today. 2010;14:228–232)

14 Pseudonormalization of the ST-Segment During Evolution of the ECG Changes

During the acute stage of myocardial infarction, the ST-segment is elevated (Fig. 20.19a, marked ↑). With evolution, the ST-segment settles (Fig. 20.19b). If the first ECG is recorded at this time, the diagnosis of myocardial infarction may be missed (pseudonormalization) before the T waves finally become inverted [1, 8, 12] (Fig. 20.19c).

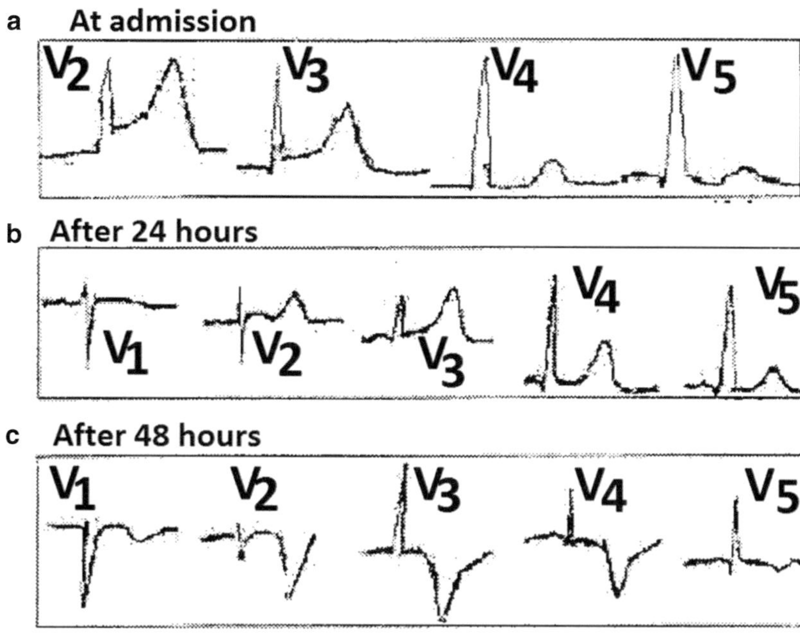

Fig. 20.19 (**a**) ECG recorded at admission showing classical ST-segment elevation in leads V_2 and V_3 (marked ↑). (**b**) Electrocardiogram recorded after 24 hours showing apparently normal ECG (pseudonormalization). (**c**) ECG recorded after 48 hours showing classical sequential T wave inversion in leads V_1 to V_6 (marked ↓) confirming anterior myocardial infarction. (Reproduced with permission from Mittal SR. Subtle ECG Changes in Acute Myocardial Infarction. Cardiology Today. 2010;14:228–232)

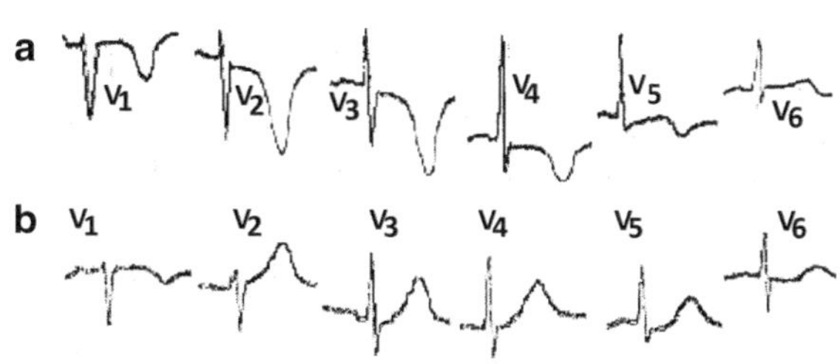

Fig. 20.20 (**a**) ECG shows symmetrical T wave inversion in leads V_2 to V_4 diagnostic of anterior myocardial infarction. (**b**) ECG showing "pseudonormalization" of T wave during recurrence of chest pain. (Reproduced with permission from Mittal SR. Subtle ECG Changes in Acute Myocardial Infarction. Cardiology Today. 2010;14:228-232)

15 Pseudonormalization with Recurrence of Ischemia in the Same Area

An old myocardial infarction shows symmetrical and deep inversion of the T waves (Fig. 20.20a). Recurrence of ischemia in the same territory results in change in polarity of the T waves with the result that the ECG appears apparently normal [4] (Fig. 20.20b) and diagnosis of myocardial ischemia is missed.

16 Late Appearance of the ECG Changes

In a case of myocardial infarction, initial ECG may be apparently normal [7, 18] (Fig. 20.21a). Diagnostic ECG changer may appear only after a few days (Fig. 20.21b marked ↓).

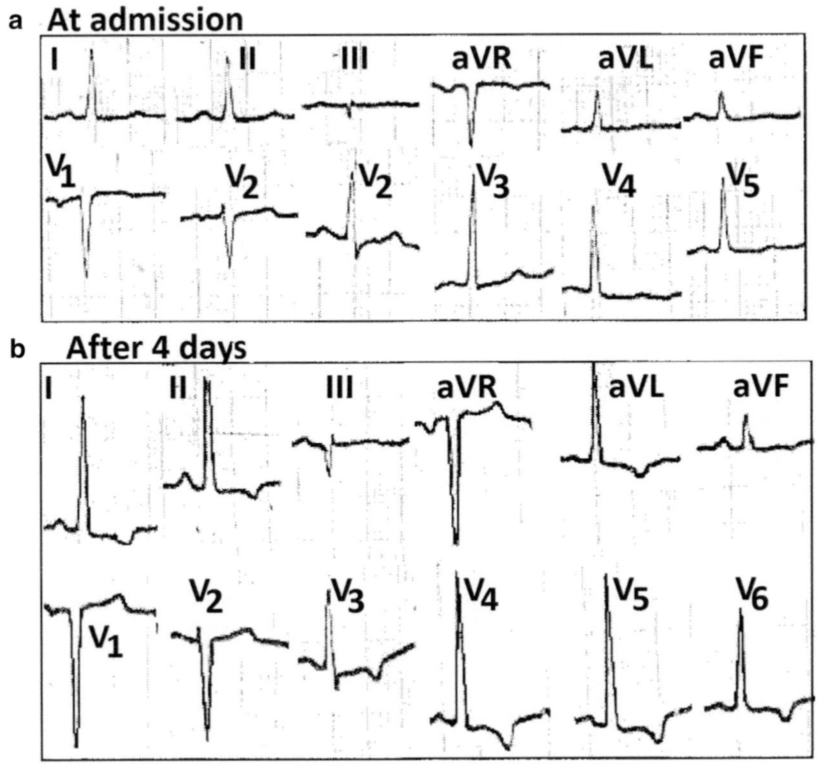

Fig. 20.21 (**a**) ECG recorded at admission is apparently normal. (**b**) ECG recorded after four days showing T wave inversion in leads I, II, aVL, aVF, and V_3 to V_6 diagnostic of anterior subendocardial infarction (marked ↓). (Reproduced with permission from Mittal SR. Subtle ECG Changes in Acute Myocardial Infarction. Cardiology Today. 2010;14:228–232)

17 Presentation as Atrial Fibrillation

Acute myocardial infarction may be accompanied by atrial fibrillation. It can be due to acute atrial overload, concomitant atrial infarction, or pericarditis. At times, it may be the dominant abnormality in the presenting ECG (Fig. 20.22a, b).

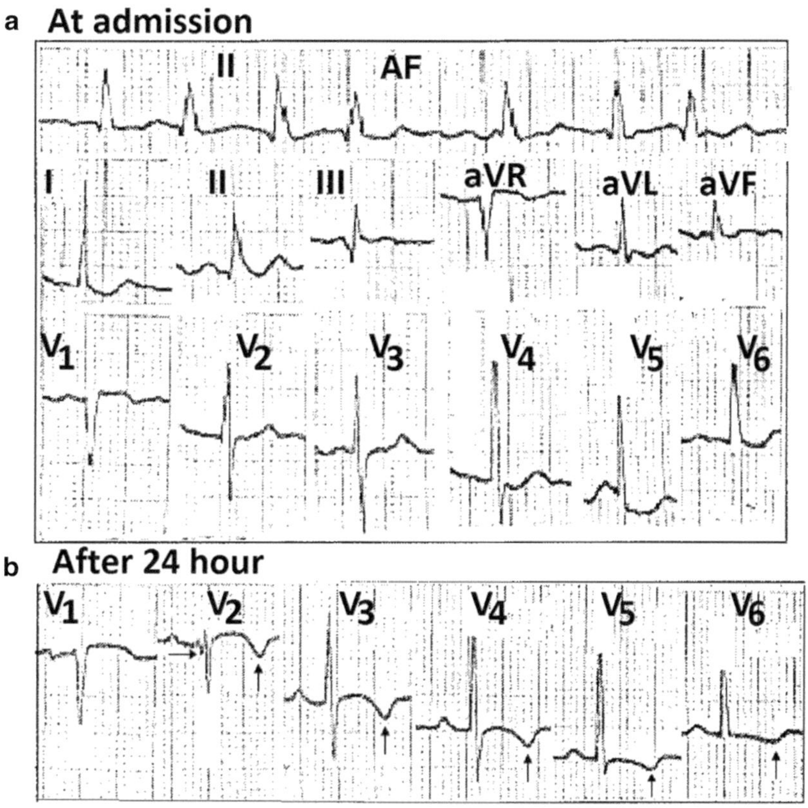

Fig. 20.22 (**a**) ECG recorded at admission showing atrial fibrillation (lead II, marked AF). (**b**) ECG recorded after 24 hours showing small q wave (marked →) with loss of R wave in lead V_2 and T wave inversion in leads V_2 to V_6 (marked ↑) diagnostic of myocardial infarction. (Reproduced with permission from Mittal SR. Subtle ECG Changes in Acute Myocardial Infarction. Cardiology Today. 2010;14:228–232)

18 Masking of Old Anterior Myocardial Infarction by Fresh Inferior Myocardial Infarction

Elevation of the ST-segment in leads II, III, and aVF due to acute inferior infarction produces reciprocal ST-segment depression in precordial leads. Such reciprocal ST-segment depression can mark the ST-segment elevation of previous anterior infarction (Fig. 20.23a, b). Fresh posterior infarction can also mask old anterior infarction by producing reciprocal prominent R waves in leads V_1 to V_3 [12].

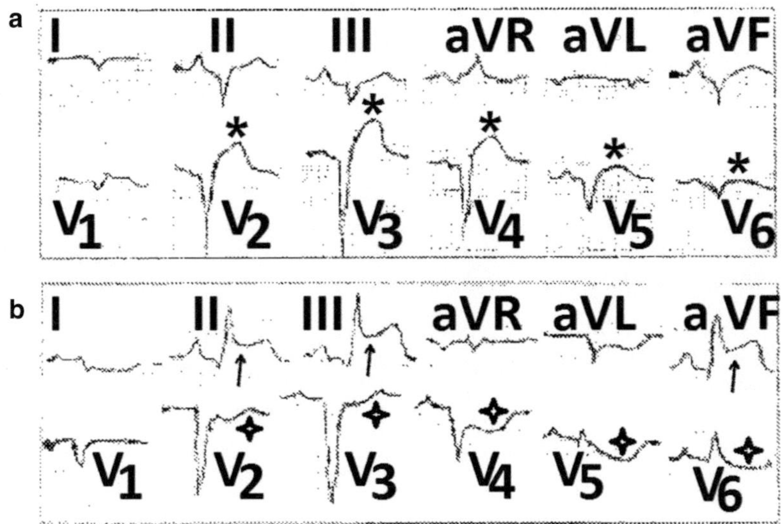

Fig. 20.23 (**a**) ECG showing classical anterior myocardial infarction (marked *). (**b**) ECG showing masking of ST-segment elevation in leads V_1 to V_6 (marked ✧) by fresh inferior myocardial infarction (marked ↑). (Reproduced with permission from Mittal SR. Subtle ECG Changes in Acute Myocardial Infarction. Cardiology Today. 2010;14:228–232)

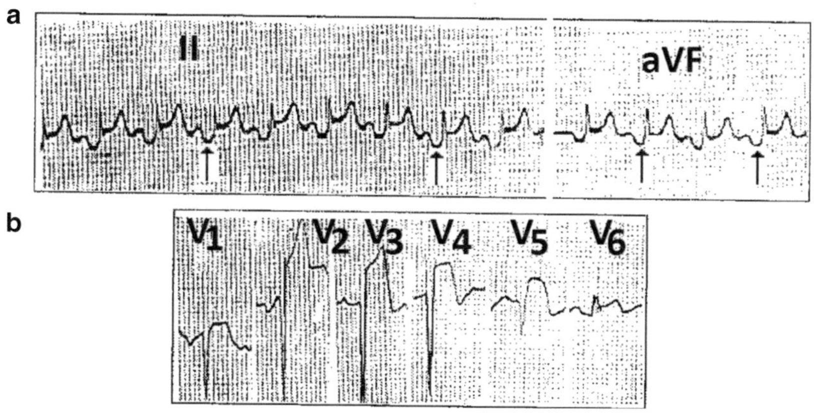

Fig. 20.24 (**a**) ECG (lead II and aVF) showing depression of the PR segment (marked ↑) suggestive of concomitant atrial infarction. (**b**) ECG from same patient showing classical anterior myocardial infarction. (Reproduced with permission from Mittal SR. Subtle ECG Changes in Acute Myocardial Infarction. Cardiology Today. 2010;14:228–232)

19 Atrial Infarction

Atrial infarction can produce elevation or depression of the PR segment [1, 4] (Fig. 20.24a, b). Acute ventricular myocardial infarction may be accompanied by atrial infarction [19].

Summary

- In hyperacute stage, there is early uptake of the ST-segment at the J point, which merges with upslope of the T wave. The ST-segment depression in electrically opposite leads may be more impressive.
- In the setting of acute coronary syndrome, elevation of the ST-segment suggests transmural ischemia. ST-segment depression in the electrically opposite wall could be due to reciprocal change or due to subendocardial ischemia in the area supplied by another coronary artery. Transmural ischemia without infarction can occur in Prinzmetal's angina and some cases of Takotsubo cardiomyopathy.
- Presence or absence of Q wave does not help in differentiating subendocardial from transmural myocardial infarction.
- Necrosis of myocardial tissue can produce abnormal Q wave (representing dead tissue) or only decrease in amplitude of R wave (representing loss of viable myocardium). R wave amplitude of <0.6 mv suggests myocardial infarction.
- Slow progression of the R wave is a nonspecific finding. In addition to non-ST elevation anterolateral myocardial infarction, it can also be present in left ventricular hypertrophy, left anterior fascicular block, left pneumothorax, pectus excavatum, and congenitally corrected transposition of the great arteries. Precordial leads placed above the standard position can also result in slow progression of the R wave. This ECG finding can also be present in patients with emphysema. Heart is pulled down by the low-placed diaphragm. Placing the precordial leads in a space lower than the standard position normalizes the ECG.
- Nonprogression of the r wave in precordial leads can be suggestive of myocardial infarction. However, left bundle branch block, left ventricular hypertrophy, emphysema, left pneumothorax with mediastinal shift, uncomplicated dextrocardia, pectus excavatum, and congenital absence of pericardium can also produce this finding.

- Reverse progression of the R wave in precordial leads suggests loss of the R wave. Indirectly it shows loss of viable myocardium. This finding is suggestive of underlying myocardial infarction. Dextrocardia also produces reverse progression of the R wave.
- Slurring of ascending limb, and top or descending limb of QRS in the absence of any QRS widening suggestive of bundle branch block.
- Fragmentation of QRS in at least two contiguous leads can be suggestive of myocardial infarction in the setting of acute coronary syndrome.
- Localized widening of QRS (periinfarction block), which is not typical of any bundle branch block can also be a marker of myocardial infarction.
- ST-segment elevation in lead aVR more than in lead V_1 with ST-segment depression in other leads suggests left main coronary artery disease or severe multivessel disease. ST-segment elevation in leads I, aVL, and V_2 to V_6 suggests proximal LAD occlusion.
- ST-segment elevation in leads I and aVL without ST-segment depression in leads V_1 and V_2 suggests mid-lateral myocardial infarction due to occlusion of the first diagonal artery.
- In the clinical context of acute coronary syndrome, deep and symmetrical inversion of the T wave in leads V_1 to V_4 suggests high-grade stenosis of the left anterior descending artery.
- Normalization of previously deep, symmetrically inverted T waves (pseudonormalization) in the clinical context of acute coronary syndrome suggests reocclusion of previously occluded blood vessel.
- In some cases, U wave inversion can be the earliest electrocardiographic sign of myocardial infarction.
- The above findings are nonspecific. They are important only when there is new appearance in classical clinical setting of acute coronary syndrome.

MCQs

Q1. In hyperacute mid-anterior myocardial infarction, ST-segment depression is seen in leads

(a) I, aVL
(b) II, III, and aVF
(c) V_1 to V_3
(d) V_4 to V_6

Q2. In the context of acute coronary syndrome, ST-segment depression suggests

(a) Transmural ischemia
(b) Subendocardial ischemia
(c) Reciprocal change
(d) All

Q3. Presence of Q wave suggests

(a) Transmural infarction
(b) Subendocardial ischemia
(c) Either
(d) Neither

Q4. In ST-segment elevation myocardial infarction, follow-up ECGs can show

(a) Pseudonormalization of the ST-segment elevation
(b) Abnormal Q waves
(c) Inversion of the T waves
(d) All

Q5. In the context of ongoing acute coronary syndrome, deep symmetrical T wave inversion in leads V_1 to V_4 suggests critical stenosis of

(a) Left main coronary artery
(b) Left anterior descending coronary artery
(c) Posterior descending coronary artery
(d) First septal coronary artery

Q6. In the context of acute coronary syndrome, which of the following can be an early electrocardiographic sign of myocardial infarction?

(a) Early uptake of the ST-segment
(b) T waves taller than R waves
(c) Inversion of U wave
(d) Inversion of T wave

Q7. Which of the following ECG findings suggests left main stem disease?

(a) ST elevation in lead aVR more than the ST elevation in lead V_1
(b) ST elevation in lead V_1 more than ST elevation in lead aVR
(c) ST elevation in leads I, aVL, V_2 to V_6
(d) ST depression in leads II, III, aVF, and V_2 to V6

Q8. ST-segment elevation in leads I and aVL without ST-segment depression in leads V_1 and V_2 suggests occlusion of

(a) Proximal D_1
(b) Proximal LAD
(c) Proximal LCX
(d) All

Q9. ST-segment elevation only in leads V_1 and V_2 suggests occlusion of

(a) Proximal LAD
(b) First septal
(c) First diagonal
(d) Proximal LCX

Q10. ST-segment elevation in leads V_1 to V_6 without the involvement of I and aVL suggests occlusion of

(a) LAD after the origin of D_1
(b) LAD before the origin of D_1
(c) LCX after the origin of OM_1
(d) Proximal LCX

Q11. Slow progression of the R wave is seen in

(a) Emphysema
(b) Right pneumothorax
(c) Left inferoposterior hemiblock
(d) Pectus excavation

Q12. Reverse progression of the R wave is seen in

(a) Anterolateral myocardial infarction
(b) Dextrocardia
(c) Congenitally corrected transposition of the great arteries
(d) Left anterosuperior hemiblock

Q13. Which is the earliest electrocardiographic change in acute myocardial infarction?

(a) M-pattern of the ST-segment elevation
(b) Coved ST-segment
(c) Deep symmetrical T wave inversion
(d) Early uptake of ST-segment

Q14. Which conditions can produce ST-segment elevation?

(a) Transmural infarction
(b) Subendocardial ischemia
(c) Prinzmetal's angina
(d) Takotsubo cardiomyopathy

Q15. Occlusion of which coronary artery may not produce diagnostic findings on ECG?

(a) Distal occlusion of long LAD wrapping around apex
(b) Nondominant circumflex
(c) Nondominant RCA
(d) First septal

Q16. ST-segment elevation in leads I and aVL suggests occlusion of

(a) D_1

(b) OM_1
(c) Distal LAD
(d) Posterior descending coronary artery

Q17. Normalization of previously deep, symmetrically inverted T waves in leads V_1 to V_4 suggests the possibility of

(a) Reperfusion
(b) Recurrence of infarction in the same zone
(c) Hyperacute posterior infarction
(d) All

Q18. ST-segment elevation localized to leads V_1 and V_2 suggests the occlusion of

(a) LAD before S_1
(b) LAD after D_1
(c) Septal branch of LAD
(d) Distal LAD

Answers

(1) b (2) b, c (3) c (4) d (5) b (6) a, b, c (7) a, d (8) a (9) b (10) a (11) a, d (12) a, b (13) d (14) a, c, d (15) b (16) a, b (17) a, b (18) c

References

1. Schamroth L. Myocardial infarction. In: Schamroth C, editor. An introduction to Electrocardiography. New Delhi: Willey India; 1982. p. 131–56.
2. Wagner GS, Lim TH. Insufficient blood supply. In: Wagner GS, editor. Marriott's Practical electrocardiography. New Delhi: Wolters Kluwer; 2008. p. 163–78.
3. Dressler W, Roester H. High T waves in the earliest stage of myocardial infarction. Am Heart J. 1947;34:627–45.
4. Mirvis DM, Goldberger AL. Electrocardiography. In: Zipes DP, Libby P, Bonow RO, Mann DL, Tomaselli GF, editors. Braunwald's Heart Disease. Philadelphia: Elsevier; 2019. p. 117–51.
5. de Luna AB, Goldwasser D, Fiol M, Bayes-Genis A. Surface electrocardiography. In: Fuster V, Harrington RA, Narula J, Eapen ZJ, editors. Hurst's the Heart. New York: Mc Graw Hill; 2017. p. 252–317.
6. Schamroth L. Coronary insufficiency. In: Schamroth C, editor. An introduction to electrocardiography. New Delhi: Willey India; 1982. p. 131–56.
7. Goldberger AL, Goldberger ZD, Shvilkin A, editors. Clinical electrocardiography. Philadelphia: Elsevier; 2018. p. 92–100.
8. de Luna AB. Electrocardiographic pattern of ischemia, injury and necrosis. In: deLuna AB, editor. Basic electrocardiography. Futura, UK: Blackwell; 2007. p. 68–116.
9. Surawicz B, Knilans TK. Acute ischemic electrocardiographic patterns. In: Surawicz B, Knilans TK (eds). Chou's. Electrocardiography in clinical practice. Saunders, Philadelphia ; 2008:124-161.
10. Goldberger AL, Goldberger ZD, Shvilkin A. Myocardial ischemia and infarction. Part I-ST-segment elevation and Q wave syndromes. In: Goldberger AL, Goldberger ZD, Shvilkin A, editors. Clinical electrocardiography. Philadelphia: Elsevier; 2018. p. 73–91.
11. Pearce MI, Kossowsky W, Levine R. Isolated abnormalities in higher precordial leads, an infrequent sign of myocardial infarction. Am Heart J. 1966;72:442–5.
12. Surawicz B, Knilans TK. Myocardial infarction and electrocardiographic pattern simulating myocardial infarction. In: Surawicz B, Knilans TK, editors. Electrocardiography in clinical practice. Philadelphia: Saunders; 2018. p. 162–204.
13. Pahlm O, Haisty WK Jr, Wagner NB, et al. Specificity and sensitivity of QRS criteria for diagnosis of single and multiple myocardial infarcts. Am J Cardiol. 1991;68:1300–4.
14. Lichtenberg SB, Schwartz MJ, Case RB. Value of premature contraction morphology in the detection of myocardial infarction. J Electrocardiol. 1980;13:167–71.
15. Das MK, Khan B, Jacob S, Kumar A, Mahenthiran J. Significance of a fragmented QRS complex versus a Q wave in patients with coronary artery disease. Circulation. 2006;113:2459–501.
16. Bozbeyoglu E, Yildirimurk O, Yazici S, et al. Fragmented QRS on admission electrocardiography predicts long term mortality in patients with non ST-segment elevation myocardial infarction. Ann Noninvasive Electrocardiol. 2016;21:352–7.
17. Kenigsberg DN, Khanal S, Kowalski M, Krishnan SC. Prologation of the QTc interval is seen uniformly during early transmural ischemia. J Am Coll Cardiol. 2007;49:1299–305.
18. Welch RD, Zalenski RJ, Frederick PD, et al. Prognostic valve of a normal or nonspecific initial electrocardiogram in acute myocardial infarction. JAMA. 2001;286:1977–84.
19. Scirica BM, Libby P, Marrow DA. ST-elevation myocardial infarction: Pathophysiology and clinical evaluation. In: Zipes DP, Libby P, Bonow RO, Mann DL, Tomaselli GF, editors. Brounwald's Heart Disease. Philadelphia: Elsevier; 2019. p. 1095–122.

21 Inferior or Posterior Myocardial Infarction

Classical ST-segment elevation in inferior leads is easily diagnosed (Fig. 21.1). Right ventricular (lead V3R to V6R) or posterior (leads V_7–V_9) infarction is frequently associated (Fig. 21.1) but may be missed if these leads are not recorded.

Some patients may have only minor changes at the time of presentation. These may be missed if not interpreted carefully. These are mentioned below.

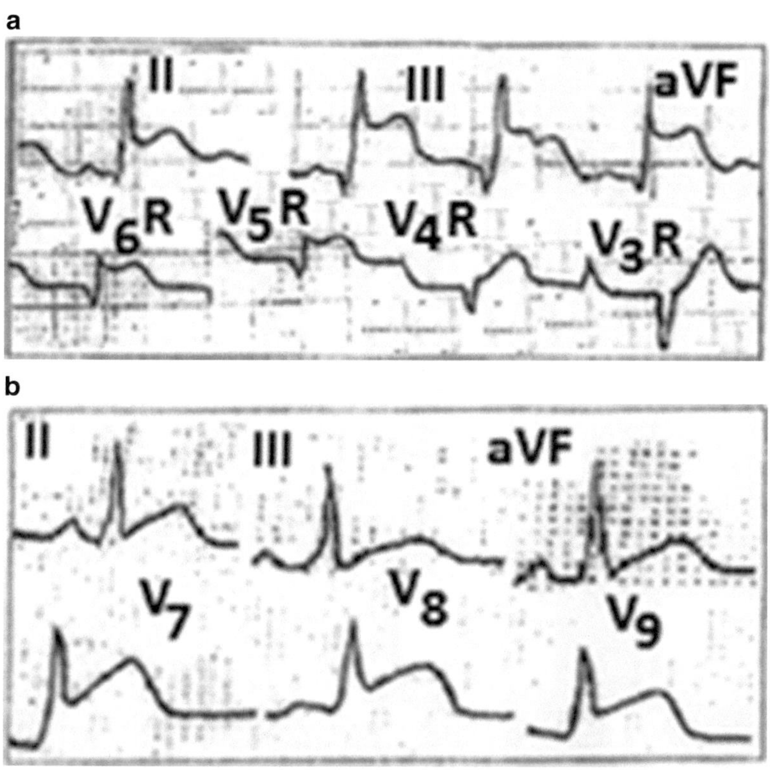

Fig. 21.1 Electrocardiogram showing (**a**) classical inferior and right ventricular infarction. (**b**) Classical inferior and posterior infarction. (Reproduced with permission from Mittal SR. ECG Changes in Acute Myocardial Infarction. Cardiology Today. 2010;14:267–270)

S. Mittal, *Insights into Electrocardiograms with MCQs*,
https://doi.org/10.1007/978-981-99-0127-2_21

1 Notching of QRS (Mid QRS Change) in Leads II, III, and aVF

Such a change in the R wave (in the absence of abnormal Q wave) suggests loss of part of the myocardium on the inferior surface (nontransmural infarction) [1, 2]. It is suggestive of myocardial infarction especially when there are reciprocal ST-segment depressions in leads I and aVL (Fig. 21.2a, b).

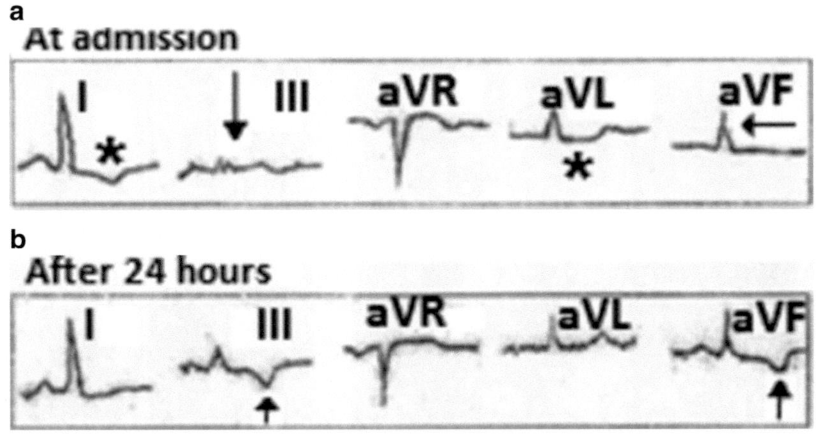

Fig. 21.2 (**a**) Electrocardiogram on admission showing notching of the R wave in leads III (marked ↓) and aVF (marked ←) with reciprocal ST-segment depression in leads I and aVL (marked *). (**b**) Electrocardiogram recorded after 24 hours showing sequential T wave inversion in leads III and aVF (marked ↑) with normalization of the reciprocal ST-segment changes in leads I and aVL. (Reproduced with permission from Mittal SR. ECG Changes in Acute Myocardial Infarction. Cardiology Today. 2010;14:267–270)

2 Prominent "R" Wave in Lead V_1

Normally lead V_1 shows a QS or an rS complex. Prominent and slightly widened R wave in lead V_1 in the absence of other ECG signs of right ventricular hypertrophy, right bundle branch block, or dextrocardia suggests a posterior infarction (Fig. 21.3). Such an R wave in lead V_1 is, in fact, a reciprocal change of Q wave in the posterior leads. Concomitant ST-segment depression in leads V_1 to V_3 is also reciprocal of the ST-segment elevation in leads V_7 to V_9 [3, 4].

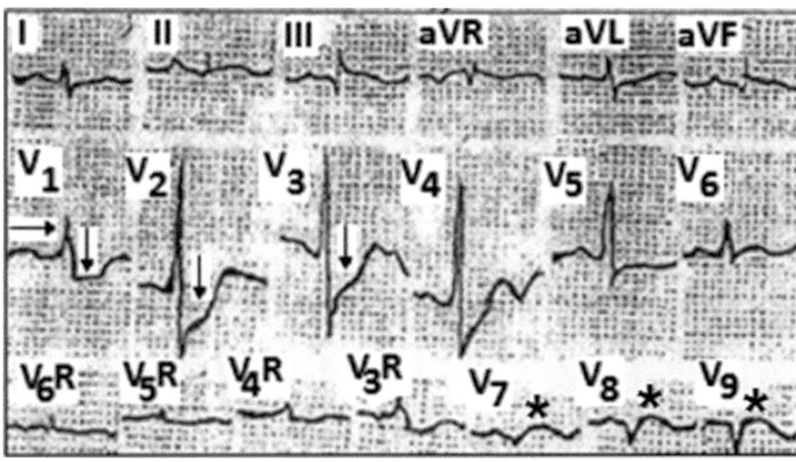

Fig. 21.3 Electrocardiogram showing prominent R wave in lead V_1 (marked →) with ST-segment depression in leads V_1 to V_3 (marked ↓) suggestive of direct posterior infarction (coved ST-segment elevation with T wave inversion in leads V_7, V_8, and V_9 marked *). (Reproduced with permission from Mittal SR. ECG Changes in Acute Myocardial Infarction. Cardiology Today. 2010;14:267–270)

3 Slight Elevation of the ST-Segment in Leads II, III, and aVF

Soon after onset of infarction, there might be only slight elevation in inferior leads without any loss of the R wave [3] (Fig. 21.4a). Sequential ECG may show loss of the R wave with appearance of abnormal Q wave (Fig. 21.4b).

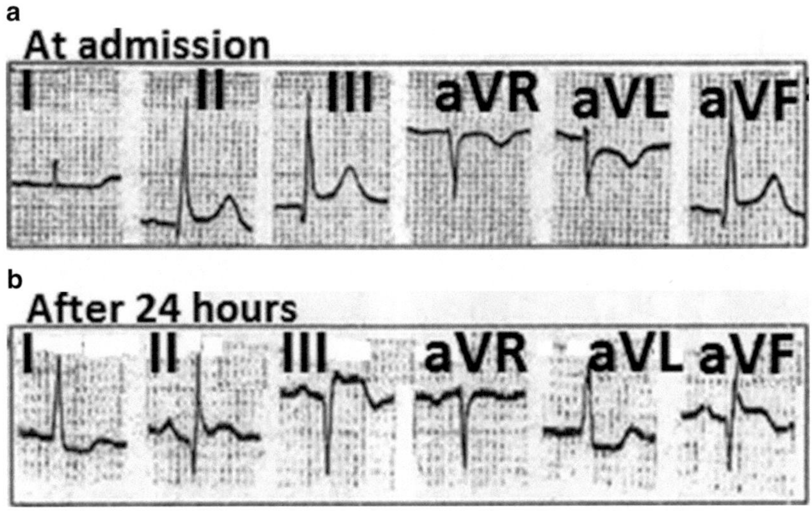

Fig. 21.4 (**a**) Electrocardiogram at admission showing slight elevation of the ST-segment in leads II, III, and aVF. (**b**) Electrocardiogram recorded after 24 hours showing abnormal Q waves, elevation of the ST-segments, and inversion of the T waves in leads II, III, and aVF. These findings are diagnostic of acute inferior infarction. (Reproduced with permission from Mittal SR. ECG Changes in Acute Myocardial Infarction. Cardiology Today. 2010;14:267–270)

4 Shallow T Wave Inversion in Leads II, III, and aVF

Shallow T wave inversion can be present in leads III and aVF in some normal persons. Persistent T wave inversion can be seen in ventricular hypertrophy or prolongation of the QT interval. Some patients with inferior infarction may show only symmetrical but not deep inversion of the T waves in leads II, III, and aVF in early stage [5] (Fig. 21.5a). Sequential deep and symmetrical T wave inversion suggests inferior infarction. This becomes more relevant in the presence of left anterior hemiblock when normally the T wave is upright in these leads (Fig. 21.5b).

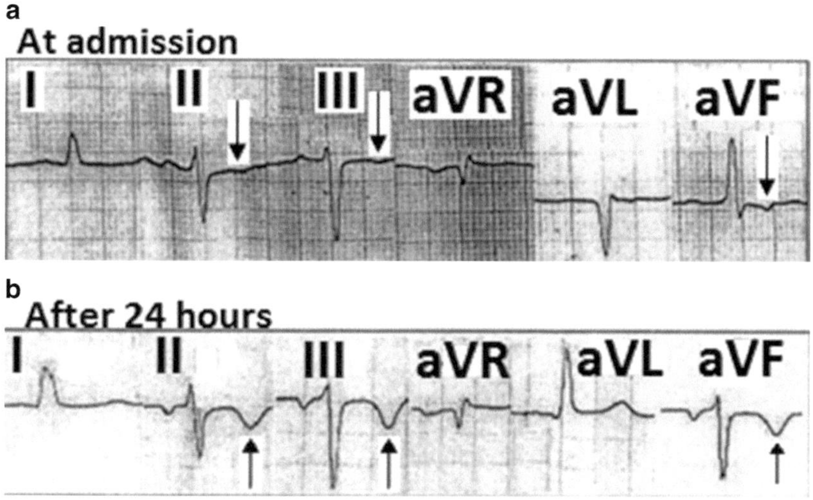

Fig. 21.5 (**a**) Electrocardiogram recorded at admission showing left anterior hemiblock with shallow inversion of the T waves in leads III and aVF (marked ↓). (**b**) Electrocardiogram recorded after 24 hours showing deep and symmetrical inversion of T waves in leads III and aVF (marked ↑) in the presence of left anterior hemiblock suggestive of inferior infarction. (Reproduced with permission from Mittal SR. ECG Changes in Acute Myocardial Infarction. Cardiology Today. 2010;14:267–270)

5 ST-Segment Depression in Leads I, aVL, and V_2 to V_6

In some patients, reciprocal ST-segment depression in leads I, aVL, and V_2-V_6 may be more prominent than ST-segment elevation in inferior leads [3, 6, 7] (Fig. 21.6a) (marked ↓). Sequential electrocardiogram may show classical ST-segment elevation in inferior leads with normalization of the reciprocal ST-segment depressions (Fig. 21.6b).

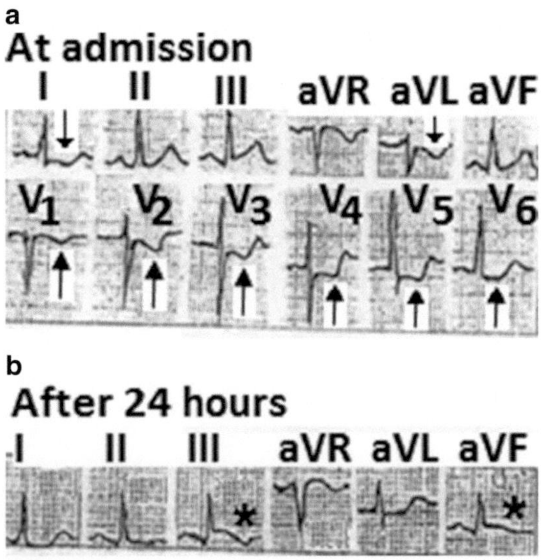

Fig. 21.6 (**a**) Electrocardiogram recorded at admission showing prominent ST-segment depression in leads I, aVL (marked ↓), and V_2 to V_6 (marked ↑). (**b**) Electrocardiogram recorded after 24 hours showing inferior infarction (marked *) with normalization of the reciprocal ST-segment depressions. (Reproduced with permission from Mittal SR. ECG Changes in Acute Myocardial Infarction. Cardiology Today. 2010;14:267–270)

6 Pseudonormalization of ST-Segment Elevation and Prominent T Waves

During the evolution of infarction from ST-segment elevation to inversion of the T waves, a stage occurs when the ST-segment elevation settle but inversion of the T waves or appearance of the Q waves has not yet occurred. During this stage, the ECG looks apparently normal (pseudonormalization) [8]. Diagnosis may be missed if the ECG is taken during this period of pseudonormalization (Fig. 21.7).

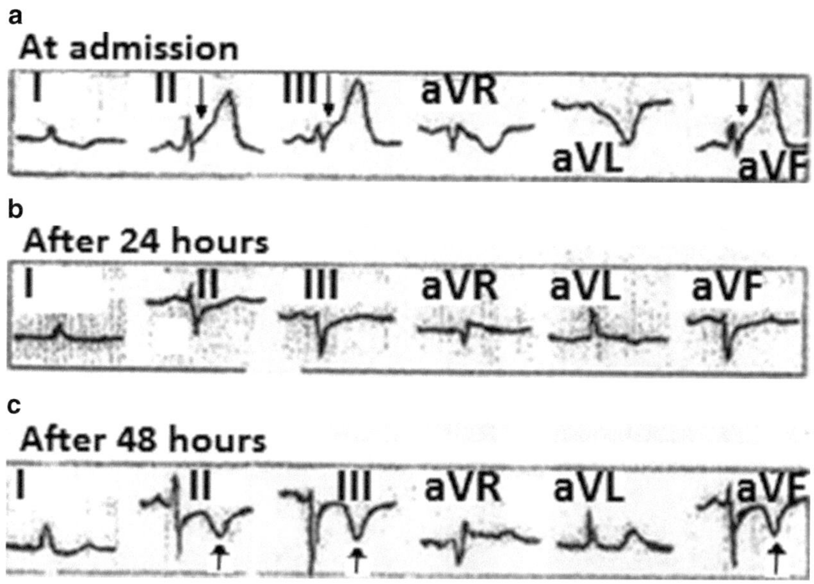

Fig. 21.7 (**a**) Electrocardiogram recorded at admission showing ST-segment elevation in leads II, III, and aVF (marked ↓) suggestive of inferior infarction with reciprocal ST-segment depression in leads I and aVL. (**b**) Electrocardiogram recorded after 24 hours showing transient pseudonormalization. (**c**) Electrocardiogram recorded after 48 hours showing classical T wave inversion in leads II, III, and aVF (marked ↑) with normalization of the reciprocal ST-segment depressions in leads I and aVL. (Reproduced with permission from Mittal SR. ECG Changes in Acute Myocardial Infarction. Cardiology Today. 2010;14:267–270)

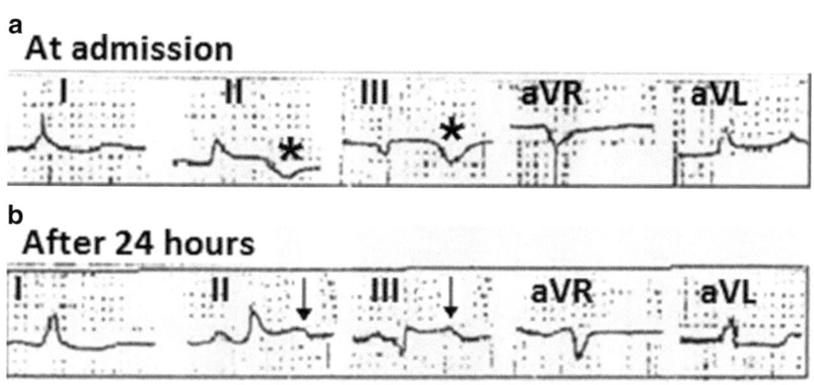

Fig. 21.8 (**a**) Electrocardiogram recorded at admission showing ST-segment elevation and T wave inversion in leads II and III (marked *) diagnostic of inferior infarction. (**b**) Electrocardiogram recorded during recurrence of pain showing normalization of the previously inverted T wave (marked ↓) due to recurrence of infarction in leads II and III. (Reproduced with permission from Mittal SR. ECG Changes in Acute Myocardial Infarction. Cardiology Today. 2010;14:267–270)

7 Normalization of the T Wave in a Patient with T Wave Inversion due to Previous Inferior Infarction

Recurrence of infarction in the same territory can result in transient normalization of inverted T waves [9] (Fig. 21.8). The electrocardiogram appears normal and diagnosis of myocardial infarction may be missed if old electrocardiogram is not available for comparison or sequential electrocardiograms are not recorded.

8 Prominent Upright T Wave in Leads V_1 and V_2

In the setting of inferior myocardial infarction, this finding suggests reciprocal changes secondary to concomitant posterior myocardial infarction [3, 5, 8] (Fig. 21.9).

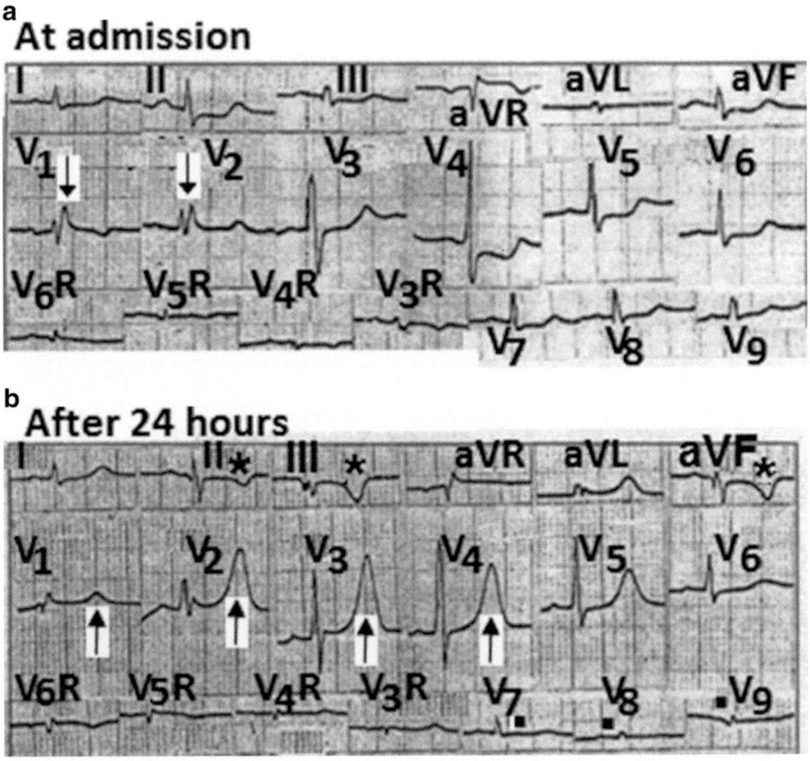

Fig. 21.9 (**a**) Electrocardiogram recorded at admission showing prominent R wave in leads V_1 and V_2 (marked ↓). (**b**) Electrocardiogram recorded after 24 hours showing loss of R wave and inversion of T waves in leads II, III, and aVF (marked *), an increase in the amplitude of T waves in leads V_1 to V_4 (marked ↑), and decrease in amplitude of the R wave in leads V_7, V_8, and V_9 (marked ■) suggestive of direct posterior infarction. (Reproduced with permission from Mittal SR. ECG Changes in Acute Myocardial Infarction. Cardiology Today. 2010;14:267–270)

9 Inferior–Posterior–Lateral Myocardial Infarction

Involvement of inferior, posterior, and lateral surface (Fig. 21.10) suggests occlusion of the left circumflex coronary artery in a left dominant system. In a case where the ST-segment elevation in inferior leads is not very clear and posterior leads have not been recorded, ST-segment depression in leads aVL and V_1 to V_3 (reciprocal changes) [3, 6, 7, 10, 11] should raise suspicion of acute inferoposterior infarction.

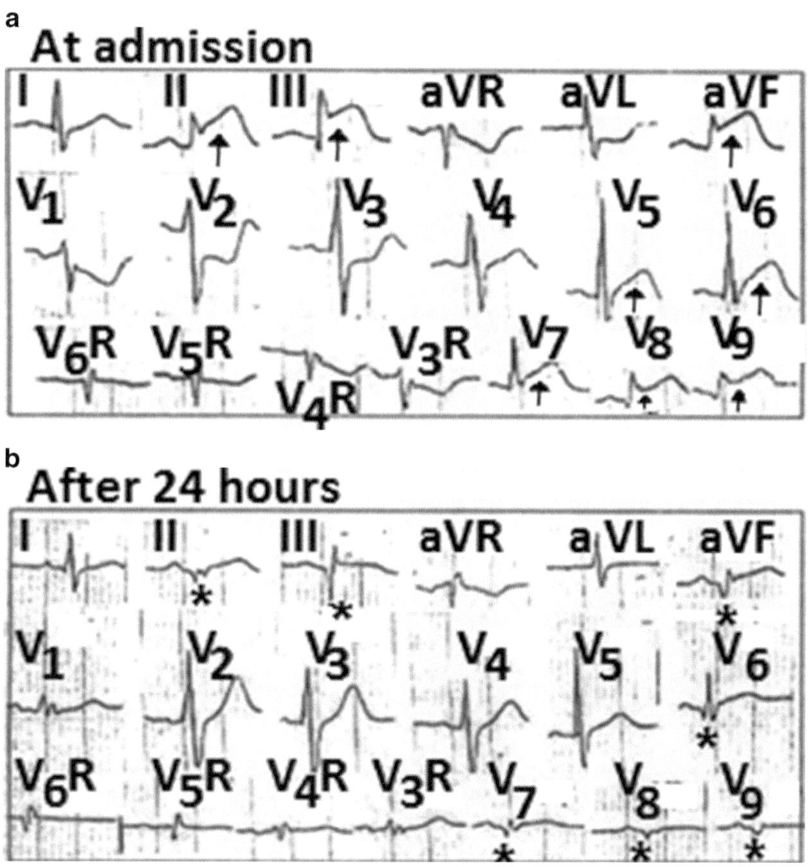

Fig. 21.10 (**a**) Electrocardiogram recorded at admission showing ST-segment elevation in leads II, III, aVF, and V_5 to V_9 (marked ↑) suggestive of inferior, posterior, and lateral infarction. ST-segment is depressed in leads aVL and V_1 to V_3 (reciprocal changes). (**b**) Electrocardiogram recorded after 24 hours showing loss of R wave and development of Q wave in leads II, III, aVF, and V_6 to V_9 (marked *). (Reproduced with permission from Mittal SR. ECG Changes in Acute Myocardial Infarction. Cardiology Today. 2010;14:267–270)

10 Isolated Right Ventricular Infarction

Changes confined to right-sided chest leads (V_3R to V_6R) without involvement of the inferior and the posterior leads (Fig. 21.11a) suggest isolated right ventricular infarction [12]. It is usually due to occlusion of the nondominant right coronary artery or right ventricular branch of the right coronary artery. Diagnosis of isolated right ventricular infarction is likely to be missed if right-sided chest leads are not recorded at admission because ST-segment elevation in right-sided chest leads normalize within 24 hours in most of the cases [13] (Fig. 21.11b). In such a situation, elevation of the ST-segment in lead V_1 should raise suspicion of right ventricular infarction.

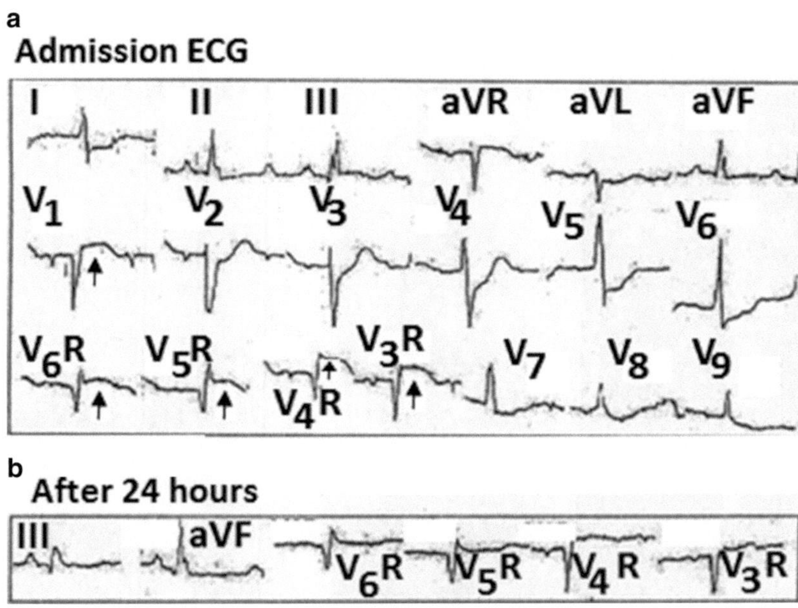

Fig. 21.11 (**a**) Electrocardiogram recorded at admission showing ST-segment elevation in leads V_1 to V_6 R (marked ↑). This finding is diagnostic of acute right ventricular infarction. (**b**) Electrocardiogram recorded after 24 hours showing normalization of the ST-segment elevation in leads V_3R to V_6R. (Reproduced with permission from Mittal SR. ECG Changes in Acute Myocardial Infarction. Cardiology Today. 2010;14:267–270)

11 First-Degree Atrioventricular Block

First-degree atrioventricular block is frequent in cases of inferior infarction [3, 5, 8, 10]. This occurs because the atrioventricular node artery arises from the right coronary artery or the left circumflex artery [14]. Some patients with inferior infarction may present with atrioventricular block (Fig. 21.12a) as the dominant finding in the first ECG. Subsequent ECG may reveal classical electrocardiographic findings of inferior infarction (Fig. 21.12b). Atrioventricular block is at the level of atrioventricular node resulting in narrow QRS complexes.

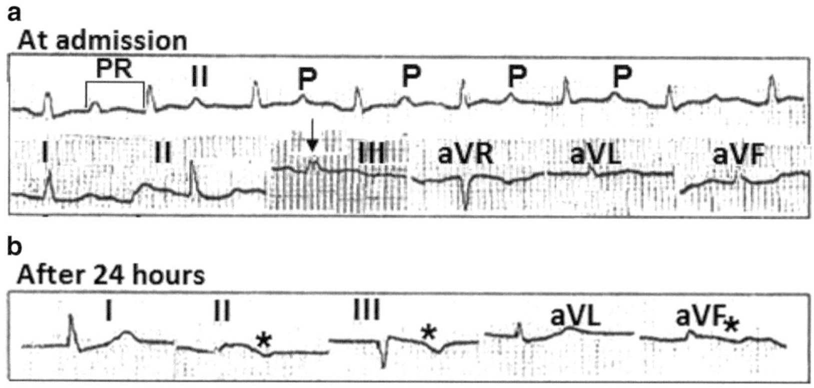

Fig. 21.12 (**a**) Electrocardiogram recorded at admission showing first-degree atrioventricular block with the notch in the R wave of lead III (mid QRS change, marked ↓). (**b**) Electrocardiogram recorded after 24 hours showing classical inferior infarction (marked *). (Reproduced with permission from Mittal SR.ECG Changes in Acute Myocardial Infarction. Cardiology Today. 2010;14:267–270)

12 Complete Atrioventricular Block

In some cases of infarction, complete atrioventricular block may be the prominent finding in the initial electrocardiogram (Fig. 21.13a). Inferior infarction may be clear only in the subsequent electrocardiograms (Fig. 21.13b). In isolated inferior myocardial infarction, atrioventricular block is usually at the level of the atrioventricular node, resulting in narrow QRS at a rate of around 50/minute. Block is more common in patients with inferior myocardial infarction caused by occlusion of the right coronary artery proximal to the branch to the atrioventricular node. First-degree or type 1 second-degree atrioventricular block usually precedes complete atrioventricular block [15]. Broad QRS and slow ventricular rate should suggest concomitant or old anterior myocardial infarction with bundle branch block.

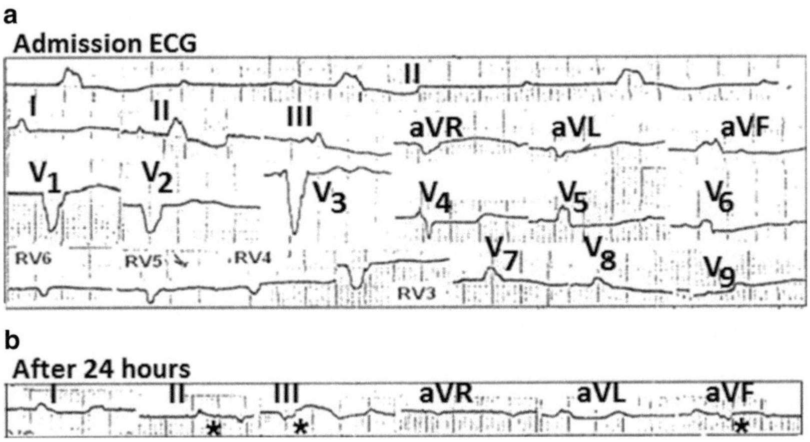

Fig. 21.13 (**a**) Electrocardiogram recorded at admission showing complete atrioventricular block with broad QRS at a slow rate. (**b**) Electrocardiogram recorded after 24 hours showing ST-segment elevation and T wave inversion in leads II, III, and aVF (marked *) diagnostic of inferior infarction. (Reproduced with permission from Mittal SR. ECG Changes in Acute Myocardial Infarction. Cardiology Today. 2010;14:267–270)

13 Atrial Fibrillation

At times, the first ECG may show only atrial fibrillation (Fig. 21.14a, lead V_1 marked AF) and reciprocal ST-segment depression in leads V_2 to V_6 (marked ↓). There may be no unequivocal evidence of myocardial infarction. Myocardial infarction may be clear only in subsequent ECG (Fig. 21.14b). Atrial fibrillation may be a manifestation of concomitant atrial infarction, pericarditis, or acute atrial overload. Occlusion of the right coronary artery with the left circumflex coronary artery proximal to its left atrial branch can result in left atrial ischemia and resultant atrial fibrillation [16].

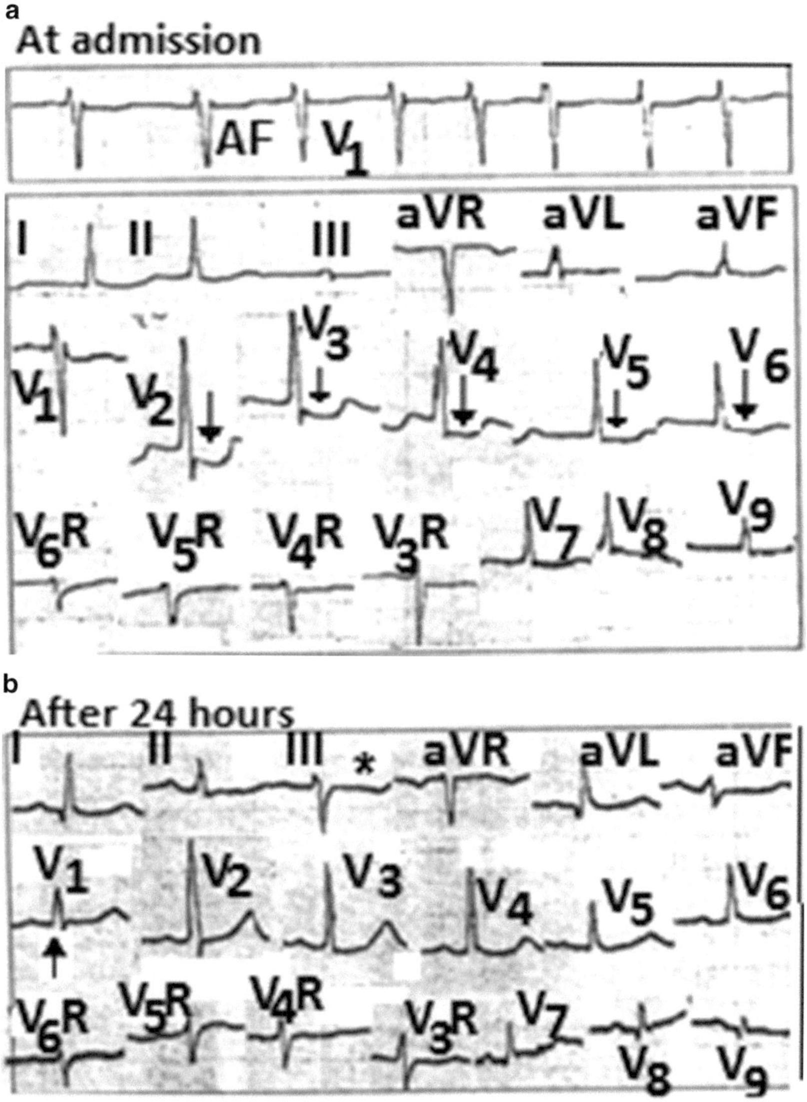

Fig. 21.14 (**a**) Electrocardiogram recorded at admission showing atrial fibrillation (marked AF in lead V_1) and ST-segment depression in leads V_2 to V_6 (marked ↓). (**b**) Electrocardiogram recorded after 24 hours showing T wave inversion in lead III (marked *), prominent R wave in lead V_1 (marked ↑), and loss of R wave voltage in leads V_7 to V_9 suggestive of inferoposterior infarction. (Reproduced with permission from Mittal SR. ECG Changes in Acute Myocardial Infarction. Cardiology Today. 2010;14:267–270)

14 Accelerated Ventricular Rhythm

Some patients with inferior infarction may present with accelerated ventricular rhythm with atrioventricular dissociation (Fig. 21.15). This could be due to autothrombolysis and reperfusion. Accelerated ventricular rhythm is a marker of reperfusion [17].

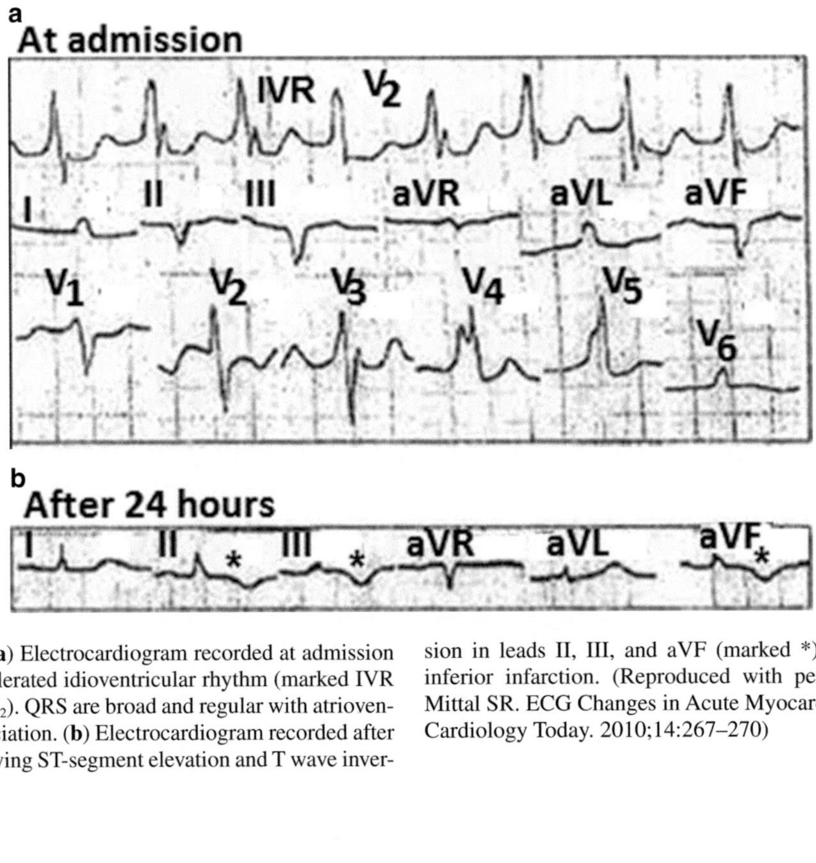

Fig. 21.15 (**a**) Electrocardiogram recorded at admission showing accelerated idioventricular rhythm (marked IVR in long lead V_2). QRS are broad and regular with atrioventricular dissociation. (**b**) Electrocardiogram recorded after 24 hours showing ST-segment elevation and T wave inversion in leads II, III, and aVF (marked *) diagnostic of inferior infarction. (Reproduced with permission from Mittal SR. ECG Changes in Acute Myocardial Infarction. Cardiology Today. 2010;14:267–270)

Summary

- Occlusion of nondominant circumflex coronary artery may not produce diagnostic Q wave or ST-segment elevation in conventional ECG leads.
- Direct posterior (now also termed lateral) infarction is diagnosed by indirect signs in leads V_1 and V_2. R wave duration of 0.04 second, and R/S ratio $\geq$ 0.5 with positive T waves in leads V_1 and V_2 suggest the diagnosis.
- Right ventricular hypertrophy, right bundle branch block, posterior or left lateral accessory pathway, and Duchenne muscular dystrophy also produce prominent R wave in leads V_1 and V_2. Other electrocardiographic findings help differentiation.
- ST-segment depression in leads V_1 to V_3 can be due to anterior subendocardial ischemia or reciprocal to posterior infarction secondary to occlusion of the right coronary artery or left circumflex coronary artery.
- Combination of anterior and inferior infarctions suggests either occlusion of left anterior descending coronary artery wrapping around the left ventricular apex or a multivessel disease.
- Inferior infarction can occur due to occlusion of the posterior descending artery arising from either left circumflex coronary artery or right coronary artery or occlusion of distal left anterior descending coronary artery wrapping around left ventricular apex.
- Some cases of inferior infarction may present with notching of QRS, early uptake of ST-segment, or shallow T wave inversion in leads II, III, and aVF.
- In some cases, reciprocal ST-segment depression in leads I, aVL, and V_2 to V_6 may be more prominent than ST-segment elevation in leads II, III, and aVF in the presenting ECG.
- Some cases may present with atrioventricular block, atrial fibrillation, or accelerated idioventricular rhythm.

MCQs

Q1. In the context of acute coronary syndrome, prominent R wave with positive T wave in leads V_1 and V_2 suggests

(a) Anteroseptal infarction
(b) Anterolateral infarction
(c) Posterior infarction
(d) Apical infarction

Q2. Prominent R wave in lead V_1 is seen i:

(a) Duchenne muscular dystrophy
(b) Myocarditis
(c) Hypertrophic cardiomyopathy
(d) All

Q3. Prominent R wave in lead V_1 is seen in

(a) Left lateral accessory pathway
(b) Left posteroseptal accessory pathway
(c) Right posteroseptal accessory pathway
(d) Right lateral accessory pathway

Q4. ST-segment depression in leads V_1 to V_3 suggests

(a) Anterior subendocardial ischemia
(b) Anterior transmural infarction
(c) Posterior transmural infarction
(d) Apical transmural infarction

Q5. Occlusion of the distal left anterior descending coronary artery wrapping around the left ventricular apex produces

(a) Anteroseptal infarction
(b) Antero lateral infarction
(c) Mid anterior infarction
(d) Inferior infarction

Q6. Inferior infarction is always associated with

(a) ST-segment elevation in leads V_7 to V_9
(b) ST-segment elevation in leads V_3R to V_6R

(c) ST-segment depression in leads I and aVL
(d) All

Q7. Early uptake of the ST-segment in leads III and aVF with ST-segment depression in leads I and AVL suggests

(a) Inferior infarction
(b) Mid lateral infarction
(c) Direct posterior infarction
(d) All

Q8. Deep symmetrical inversion of the T waves in leads III and aVF can occur in

(a) Horizontal heart
(b) Inferior infarction
(c) Left anterosuperior fascicular block
(d) Ventricular hypertrophy

Q9. Inferior infarction produces ST-segment depression in leads

(a) II, III, aVF
(b) I, aVL
(c) V_2 to V_6
(d) None

Q10. In the context of an inferior infarction, T wave can be normal

(a) In an old infarction
(b) Recurrence of ischemia in inferior region
(c) Infarction involving leads I and aVL
(d) Lateral infarction

Q11. Prominent positive T waves in leads V_1 and V_2 in a case of inferior infarction suggest

(a) Direct posterior infarction
(b) Right ventricular infarction
(c) Mid lateral infarction
(d) Septal infarction

Q12. Combination of inferior, posterior, and lateral infarction suggests occlusion of

(a) Proximal dominant left circumflex coronary artery
(b) Posterior left ventricular branch of dominant right coronary artery
(c) Posterior descending coronary artery
(d) First obtuse marginal artery

Q13. ST-segment elevation localized to leads V_3R to V_6R without any involvement of inferior leads suggests occlusion of

(a) Proximal dominant right coronary artery
(b) Nondominant right coronary artery
(c) Right ventricular branch of right coronary artery
(d) All

Q14. In the setting of an inferior infarction, atrioventricular block can be

(a) First degree
(b) Mobitz type 1
(c) Complete atrioventricular block
(d) Any of above

Q15. In the setting of an inferior infarction, concomitant atrial infarction can be suspected if there is

(a) Atrial fibrillation
(b) Elevation of the PR segment
(c) Depression of the PR segment
(d) All

Q16. Atrioventricular block is more frequent when inferior infarction in associated with

(a) Posterior infarction
(b) Lateral infarction
(c) Right ventricular infarction
(d) Any one of above

Answers

(1) c (2) a (3) a, b (4) a, c (5) d (6) c (7) a (8) b, d (9) b, c (10) a, b, c (11) a (12) a (13) b, c (14) d (15) d (16) c

References

1. Warner RA, Battaghia J, Hill NE. Terminal portion of the QRS in the electrocardiographic diagnosis of inferior myocardial infarction. Am J Cardiol. 1985;55:896–9.
2. Babitt DG, Binkley PF, Schaat SF. Clinical significance of terminal QRS abnormalities in the setting of inferior myocardial infarction. J Electrocardiol. 1991;24:85–90.
3. Schamroth L. Myocardial infarction. In: Schamroth C, editor. An introduction to electrocardiography. New Delhi: Willey India; 1982. p. 131–56.
4. Wagner GS, Lim TH. In: Wagner GS, editor. Marriott's Practical Electrocardiography. New Delhi: Wolters Kluwer; 2008. p. 179–207.
5. deLuna AB, Goldwasser D, Fiol M, Bayes-Genis A. Surface electrocardiography. In: Fuster V, Harrington RA, Narula J, Eapen ZJ, editors. Hurst's The Heart. New York: Mc Graw Hill; 2017. p. 252–317.
6. Wagner GS, Lim TH. Insufficient blood supply. In: Wagner GS, editor. Marriott's Practical Electrocardiography. New Delhi: Wolters Kluwer; 2008. p. 163–78.
7. de Luna AB. Electrocardiographic patterns of ischemia, injury and necrosis. In: de Luna AB, editor. Basic electrocardiography. Futura, UK: Blackwell; 2007. p. 68–116.
8. Surawicz B, Knilons TK. Myocardial infarction and electrocardiographic patterns simulating myocardial infarction. In: Surawicz B, Knilons TK, editors. Chou's Electrocardiography in clinical practice. Philadelphia: Saunders; 2019. p. 117–51.
9. Mirvis DM, Goldberger AL. Electrocardiography. In: Zipes DP, Libby P, Bonow RO, Mann DL, Tomaselli GF, editors. Braunwald's Heart Disease. Philadelphia: Elsevier; 2019. p. 117–81.
10. Goldbeger AL, Goldberger ZD, Shvilkin A. Myocardial ischemia and infarction, Part I: ST-segment elevation and Q wave syndromes. In: Goldberger AL, Goldberger ZD, Shvilkin A, editors. Goldberger's clinical Electrocardiography. Philadelphia: Elsevier; 2018. p. 73–91.
11. Goldberger AL, Goldberger ZD, Shvilkin A. Myocardial ischemia and infarction, Part II: Non-ST-segment elevation and non Q wave syndromes. In: Goldberger AL, Goldberger ZD, Shvilkin A, editors. Goldberger's clinical Electrocardiography. Philadelphia: Elsevier; 2018. p. 92–103.
12. Forman MB, Goodin J, Phelan B, et al. Electrocardiographic changes associated with isolated right ventricular infarction. J Am Coll Cardiol. 1984;4:640–3.
13. Braat S, Brugada P, De Zwana C, et al. Value of electrocardiogram in diagnosing right ventricular infarction in patients with an acute inferior wall myocardial infarction. Br Heart J. 1983;49:368–72.
14. Goldbeger AL, Goldberger ZD, Shvilkin A. Atrioventricular conduction abnormalities, Part I: Delays, blocks and dissociation syndromes. In: Goldberger AL, Goldberger ZD, Shvilkin A, editors. Goldberger's clinical Electrocardiography. Philadelphia: Elsevier; 2018. p. 172–82.
15. Surawicz B, Knilons TK. Atriventricular block; concealed conduction; Gap phenomenon. In: Surawicz B, Knilons TK, editors. Chou's Electrocardiography in clinical practice. Philadelphia: Saunders; 2008. p. 456–80.
16. Hod H, Lew AS, Keltal M, et al. Early atrial fibrillation in the setting of acute myocardial infarction: a consequence of impaired left atrial perfusion. Circulation. 1987;75:146–50.
17. Gorgles APM, Vos MA, Letsch IS, et al. Usefulness of the accelerated idioventricular rhythm as a marker for myocardial necrosis and reperfusion during thrombolytic therapy in acute myocardial infarction. Am J Cardiol. 1988;61:231–5.

22 Mid–Late QRS Changes Suggestive of Myocardial Necrosis

Such changes occur when there is alteration of ventricular activation after initial 40 ms [1]. Sequence of activation of various parts of the ventricles is shown in Fig. 22.1.

Such changes may occur as an isolated abnormality or with the Q wave (Fig. 22.2).

Such changes are usually caused by nontransmural necrosis, usually involving basal segments. These changes can present as

- Loss of voltage of the R wave.
- Notching or slurring of the R wave.
- Peri-infarction block.
- Rsr' pattern in the absence of bundle branch block.
- Fragmented QRS.
- Slow (poor) or nonprogression of the R wave in the precordial leads.
- Reverse progression of the R wave in the precordial leads.

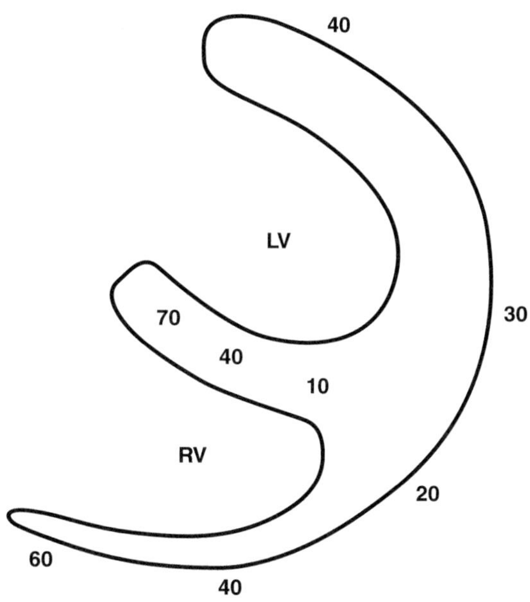

Fig. 22.1 Diagram showing the sequence of activation of various parts of ventricular myocardium in msec. (Reproduced with permission from Mittal SR. Mid-late QRS Changes Suggestive of Myocardial Necrosis. Cardiology Today. 2015;19:211–216)

S. Mittal, *Insights into Electrocardiograms with MCQs*,
https://doi.org/10.1007/978-981-99-0127-2_22

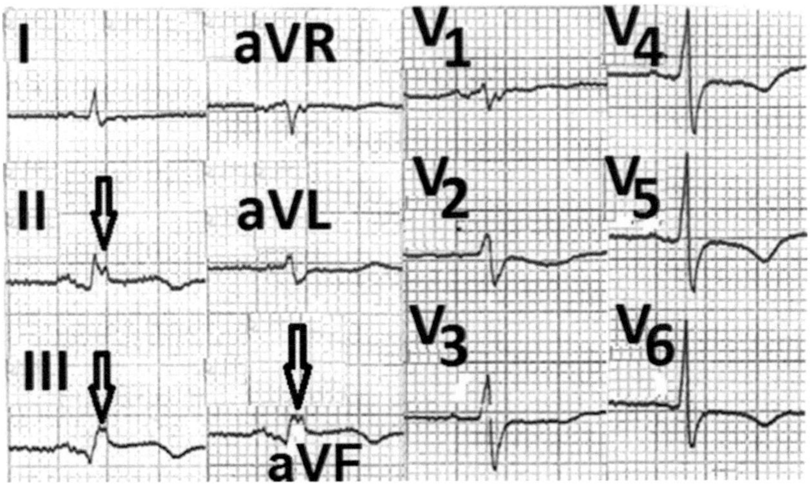

Fig. 22.2 Electrocardiogram showing mid-late QRS change (notching of QRS, marked as ⇩) along with Q wave in leads II, III, and aVF. These leads also show elevation of the ST-segment and inversion of the T waves. (Reproduced with permission from Mittal SR. Mid-late QRS Changes Suggestive of Myocardial Necrosis. Cardiology Today. 2015;19:211–216)

1 Marked Loss of R Wave Voltage [2]

It is important if it is a sequential change associated with a clinical event

suggestive of infarction. This is due to loss of the electric potential because of necrosed myocardium underlying the affected lead (Fig. 22.3).

Technical error (low standard setting; e.g., 5 mm/mV) should be excluded. Pericardial effusion can also cause generalized reduction in the QRS voltage. Long-standing low-voltage QRS can also be due to obesity, emphysema, or pericardial constriction. In the context of acute coronary syndrome, low voltage of the QRS complex in the admission electrocardiogram is associated with previous myocardial infarction and higher in-hospital mortality [3]. At times only the width of the top of the QRS complex may be diminished and the top of the QRS complex may look like a spike (Fig. 22.4).

At times there might be a notch on the top of the QRS complex (Fig. 22.5).

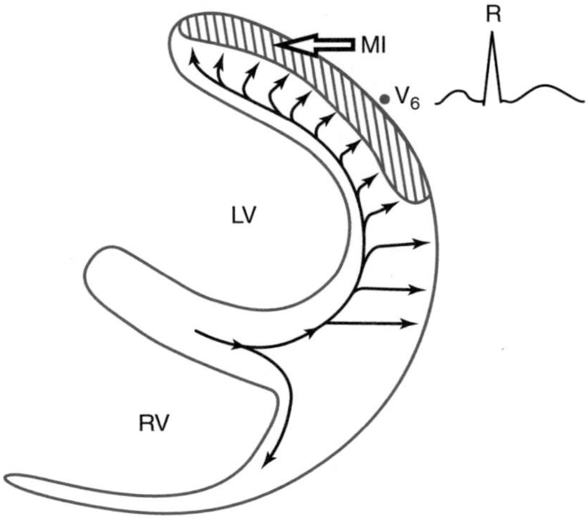

Fig. 22.3 Diagram showing the genesis of loss of R wave amplitude due to nontransmural necrosis of the underlying myocardium. *LV* left ventricle, *RV* right ventricle, *MI* myocardial infarction. (Reproduced with permission from Mittal SR. Mid- late QRS Changes Suggestive of Myocardial Necrosis. Cardiology Today. 2015;19:211–216)

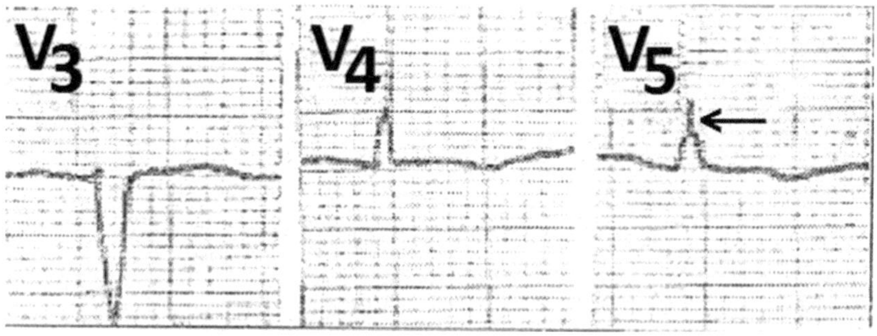

Fig. 22.4 Electrocardiogram showing the reduced width of the top of the QRS complex (looking like a spike-marked as an ←). (Reproduced with permission from Mittal SR. Mid-late QRS Changes Suggestive of Myocardial Necrosis. Cardiology Today. 2015;19:211–216)

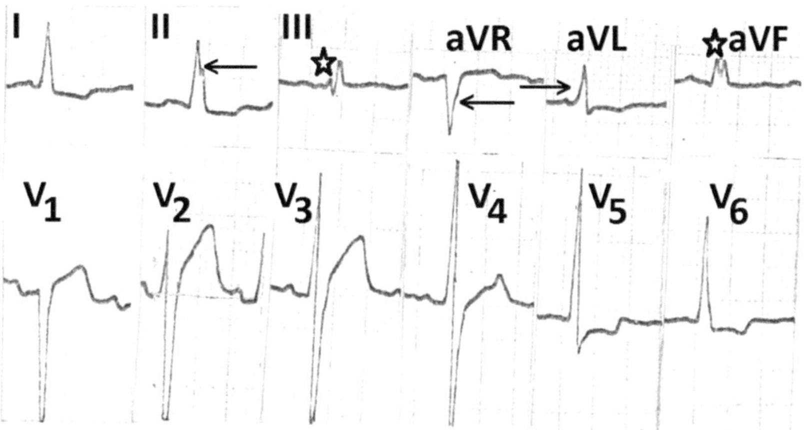

Fig. 22.5 Electrocardiogram showing notch on the top of the QRS complex in leads III and aVF (marked as ☆). Lead II, aVR, and aVL show slurring (marked as arrows)

2 Slow (Poor) Progression of the R Wave [2]

This finding is produced due to nontransmural infarction of decreasing thickness underlying leads V_1 to V_3 (Fig. 22.6).

Representative electrocardiogram is shown in Fig. 22.7.

This is, however, a nonspecific finding. Left ventricular hypertrophy can also produce slow progression of the R wave. Increased voltage and ST-segment and T wave changes in lateral leads help in differentiation. Pulmonary thromboembolism can be another cause. It is associated with sinus tachycardia and T wave inversion in right to mid precordial leads (V_1 to V_4). S wave in lead I with Q wave and T wave inversion in lead III (S_1, Q_3, inverted T_3) can support the diagnosis but are neither sensitive nor specific. Chronic obstructive airway disease with emphysema can also produce slow progression due to the downward displacement of the heart. "P- pulmonale," low voltage, especially in lead I and V_6, supports the diagnosis. In case of chronic obstructive airway disease, recording precordial leads in lower intercostal space may show normal progression of the R wave.

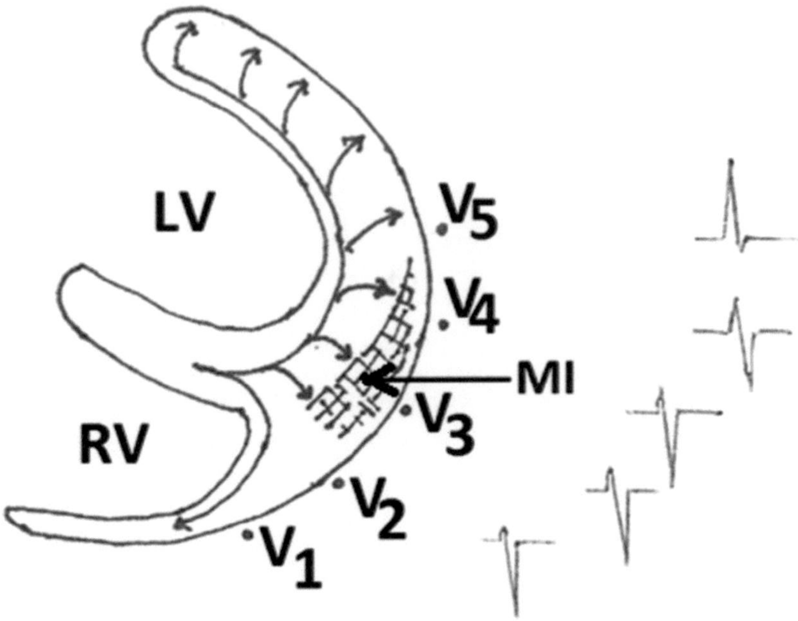

Fig. 22.6 Diagram showing the genesis of slow (poor) progression of the R wave due to nontransmural myocardial necrosis of decreasing thickness (marked as MI). (Reproduced with permission from Mittal SR. Mid-late QRS Changes Suggestive of Myocardial Necrosis. Cardiology Today. 2015;19:211–216)

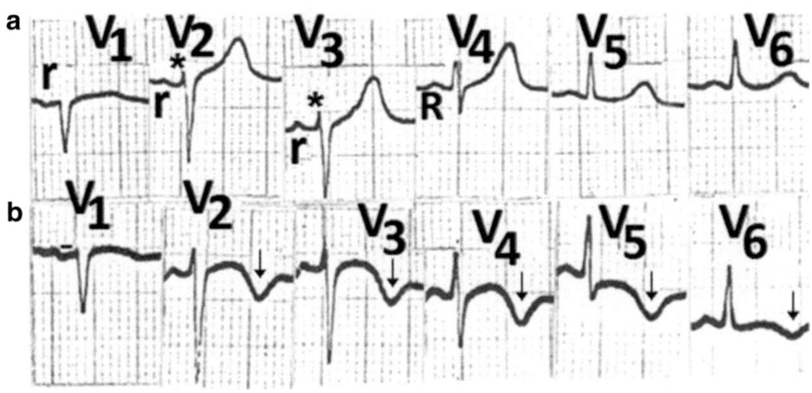

Fig. 22.7 (**a**) Electrocardiogram recorded at the time of admission showing slow progression of the r wave in leads V_2 to V_3 (marked *). (**b**) Subsequent electrocardiogram showing sequential changes of anterior infarction (marked ↓). (Reproduced with permission from Mittal SR. Mid-late QRS Changes Suggestive of Myocardial Necrosis. Cardiology Today. 2015;19:211–216)

3 Nonprogression of the R Wave

It is produced by nontransmural infarction of nearly similar thickness
underlying leads V_1 to V_3 (Fig. 22.8).

Representative electrocardiogram is shown in Fig. 22.9.

Sudden appearance of prominent R wave in lead V_4 strongly supports the conclusion that nonprogression of the r wave in leads V_1 to V_3 is due to myocardial necrosis. It is relevant in a given clinical setting. Several other conditions can produce nonprogression of the R wave. These include left bundle branch block, left ventricular hypertrophy, emphysema, left pneumothorax with mediastinal shift, uncomplicated dextrocardia, pectus excavatum, and congenital absence of the pericardium.

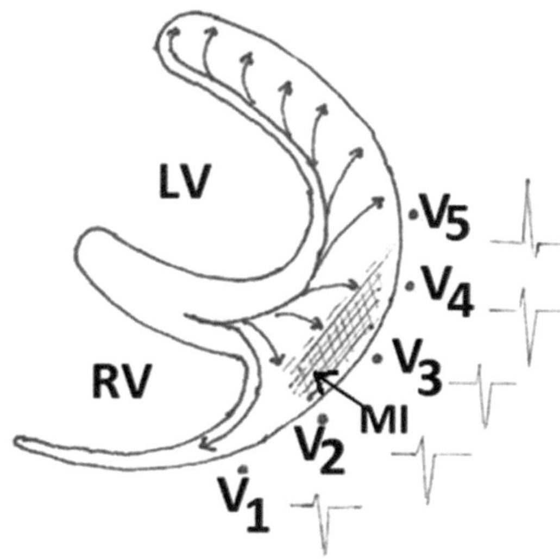

Fig. 22.8 Diagram showing the genesis of nonprogression of the R wave due to nontransmural myocardial infarction (marked as MI). *LV* left ventricle, *RV* right ventricle. (Reproduced with permission from Mittal SR. Mid-late QRS Changes Suggestive of Myocardial Necrosis. Cardiology Today. 2015;19:211–216)

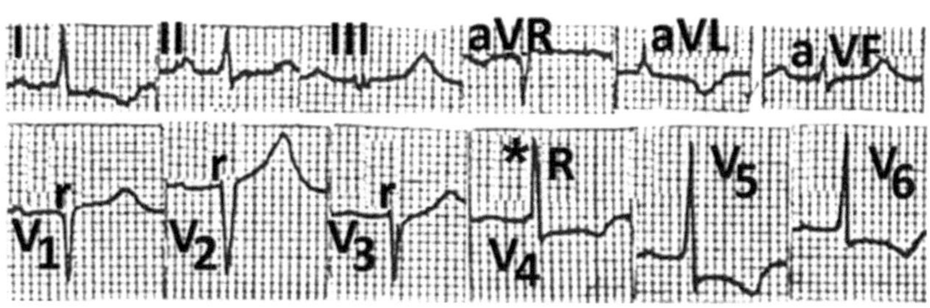

Fig. 22.9 Electrocardiogram showing nonprogression of the R wave (marked r) in leads V_1 to V_3. There is sudden appearance of prominent R wave in lead V_4 (marked *). (Reproduced with permission from Mittal SR. Mid-late QRS Changes Suggestive of Myocardial Necrosis. Cardiology Today. 2015;19:211–216)

4 Reverse Progression of the R Wave

Sudden reduction in the height of the R wave in precordial leads suggests necrosis of the underlying myocardium (Fig. 22.10).

It is more specific than "slow progression" or "nonprogression" of the R wave. Representative electrocardiograms are shown in Figs. 22.11 and 22.12.

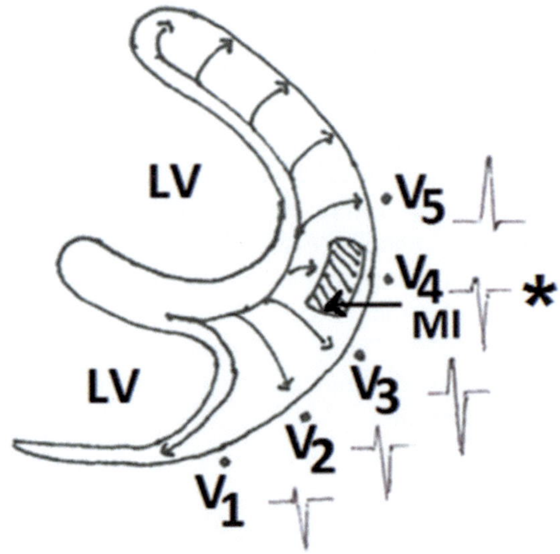

Fig. 22.10 Diagram showing the genesis of reverse progression of the R wave from lead V_3 to V_4 (marked *) by nontransmural infarction (marked as MI). (Reproduced with permission from Mittal SR. Mid-late QRS Changes Suggestive of Myocardial Necrosis. Cardiology Today. 2015;19:211–216)

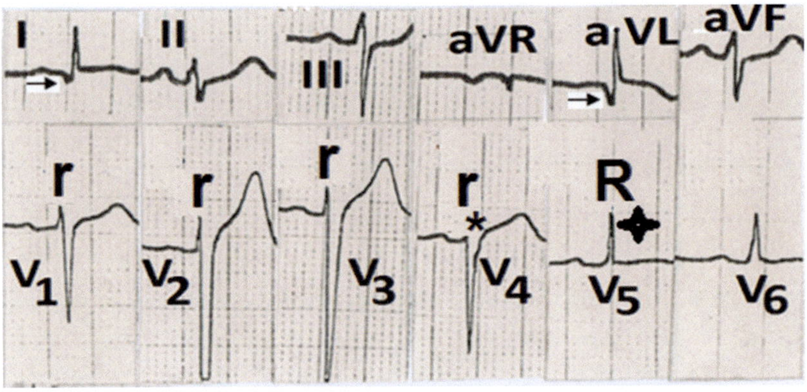

Fig. 22.11 Electrocardiogram showing reverse progression of r wave from lead V_3 to V_4 (marked *). There is sudden appearance of prominent R wave in lead V_5 (marked ✧). Lead I and aVL also show prominent Q wave (marked →). (Reproduced with permission from Mittal SR. Mid-late QRS Changes Suggestive of Myocardial Necrosis. Cardiology Today. 2015;19:211–216)

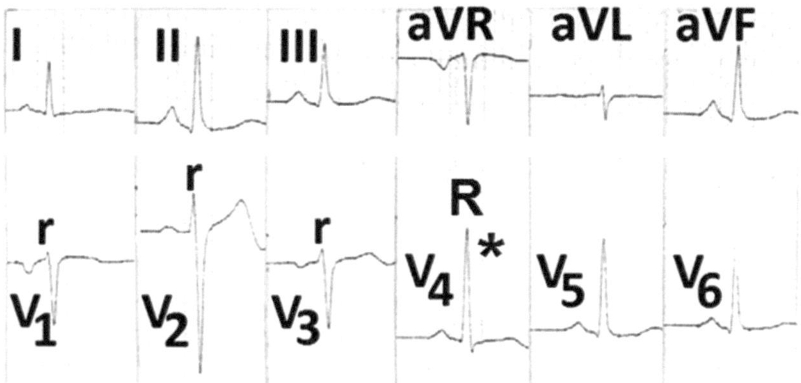

Fig. 22.12 Electrocardiogram showing reverse progression of the r wave from lead V_2 to V_3. There is sudden appearance of prominent R wave in lead V_4 (marked *). (Reproduced with permission from Mittal SR. Mid-late QRS Changes Suggestive of Myocardial Necrosis. Cardiology Today. 2015;19:211–216)

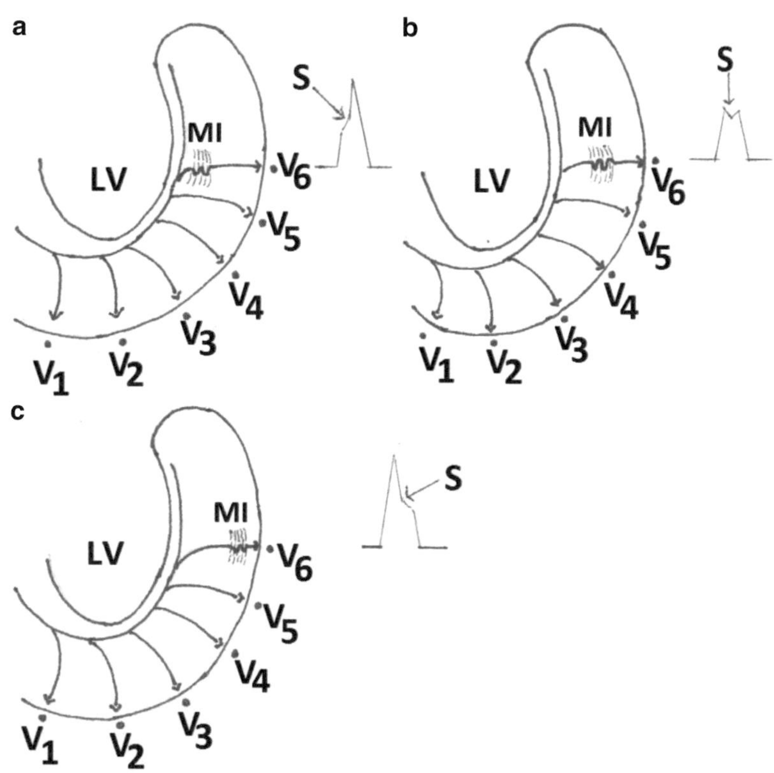

Fig. 22.13 Diagram showing the genesis of (**a**) slurring on the upstroke of the R wave (marked as S and arrow), (**b**) notch on the top of the R wave (marked as S and arrow), and (**c**) slurring on the downslope of the R wave (marked as S and arrows). *MI* myocardial infarction. (Reproduced with permission from Mittal SR. Mid-late QRS Changes Suggestive of Myocardial Necrosis. Cardiology Today. 2015;19:211–216)

5 Slurring of the QRS Complex

Slurring can be present on the upstroke of the R wave (Fig. 22.13a), on the top of the R wave (Fig. 22.13b) or on the downstroke of the R wave (Fig. 22.13c).

Representative electrocardiogram is shown in Fig. 22.14.

Nontransmural subendocardial infarction produces slurring in the initial half of the QRS. Slurring on the downstroke is produced by subepicardial infarction. This finding is relevant only in the absence of QRS widening due to bundle branch block. Fresh appearance in a classical clinical setting supports diagnosis of myocardial necrosis.

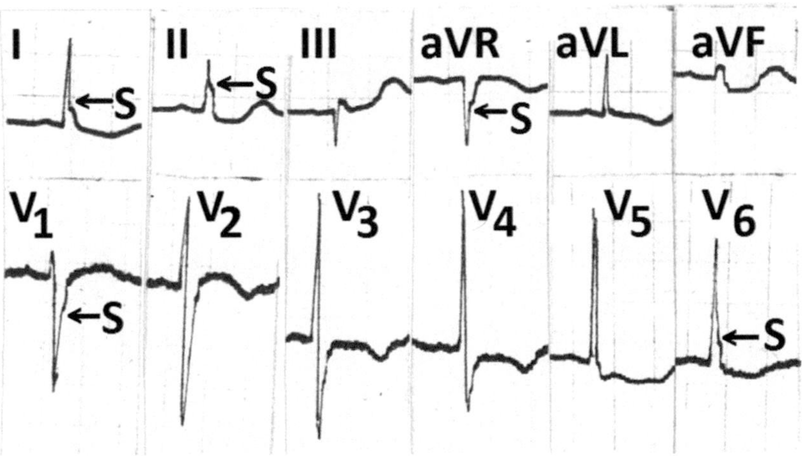

Fig. 22.14 Electrocardiogram showing slurring (marked as ← S) of the QRS complex in leads I, II, aVR, V_1, and V_6. (Reproduced with permission from Mittal SR. Mid-late QRS Changes Suggestive of Myocardial Necrosis. Cardiology Today. 2015;19:211–216)

6 Fragmented QRS

It is defined as an additional R' wave or notching of the R or S wave or presence of fragmentation (more than one R' wave) in two contiguous leads corresponding to the territory of a major coronary artery [4]. This definition is applicable only when the QRS is narrow. This finding is produced by patchy necrosis of the myocardium (Fig. 22.15).

Representative electrocardiograms are shown in Figs. 22.16, 22.17, and 22.18.

The fragmented QRS complex is more sensitive and specific in detecting myocardial scar than a Q wave [4]. This finding is important, especially if it is a new finding in the classical clinical setting. In the context of coronary artery disease, this electrocardiographic finding also correlates with recurrent cardiac events and mortality [5]. Fragmented QRS suggests inhomogeneous intraventricular conduction and is present in several cardiac disorders other than coronary artery disease [6].

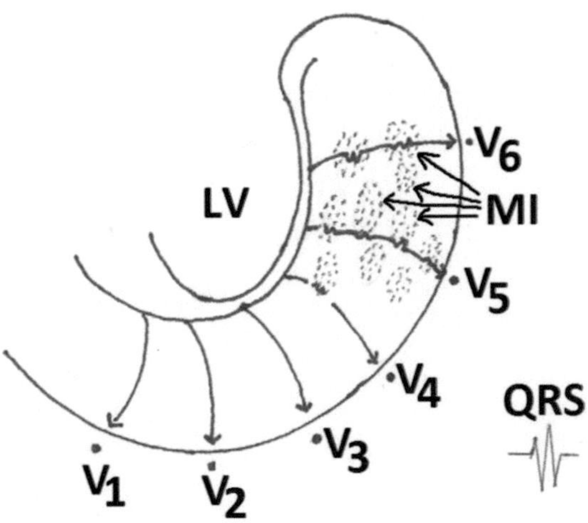

Fig. 22.15 Diagram showing multiple patchy myocardial infarcts producing fragmented QRS. *MI* myocardial infarction. (Reproduced with permission from Mittal SR. Mid-late QRS Changes Suggestive of Myocardial Necrosis. Cardiology Today. 2015;19:211–216)

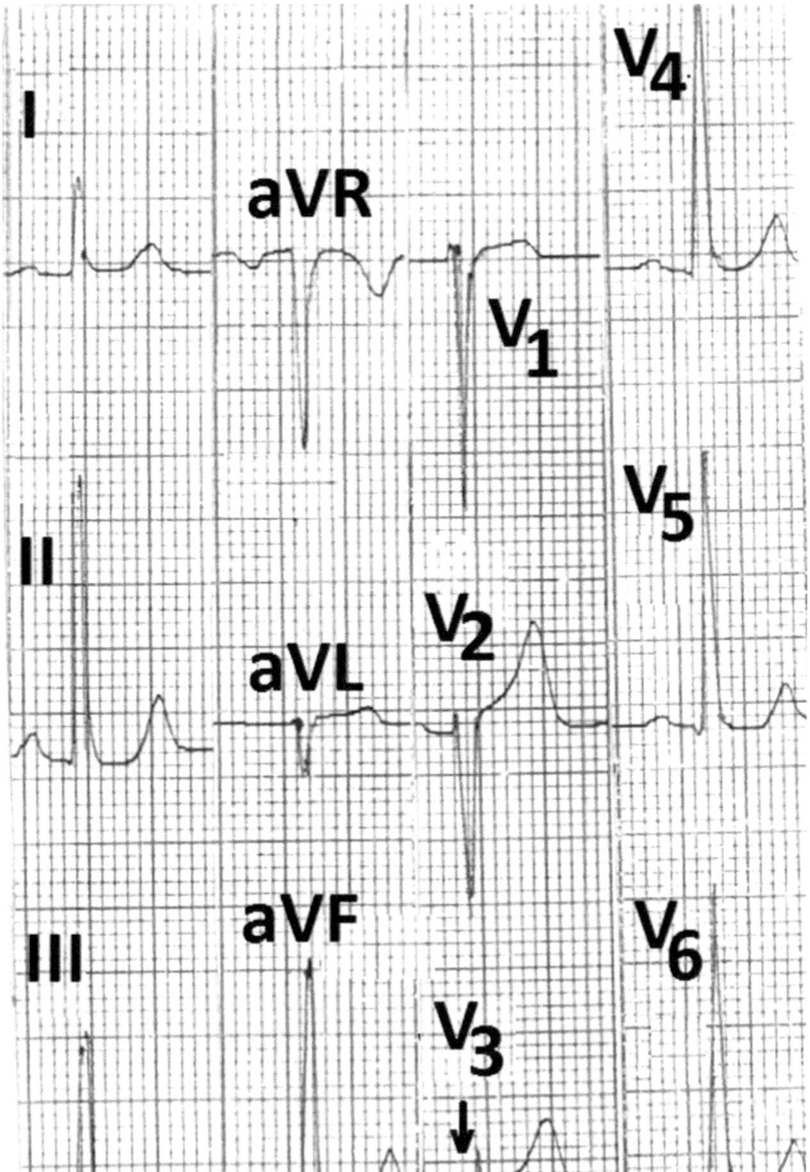

Fig. 22.16 Electrocardiogram showing fragmented QRS in lead V_3 (marked ↓). (Reproduced with permission from Mittal SR. Mid-late QRS Changes Suggestive of Myocardial Necrosis. Cardiology Today. 2015;19:211–216)

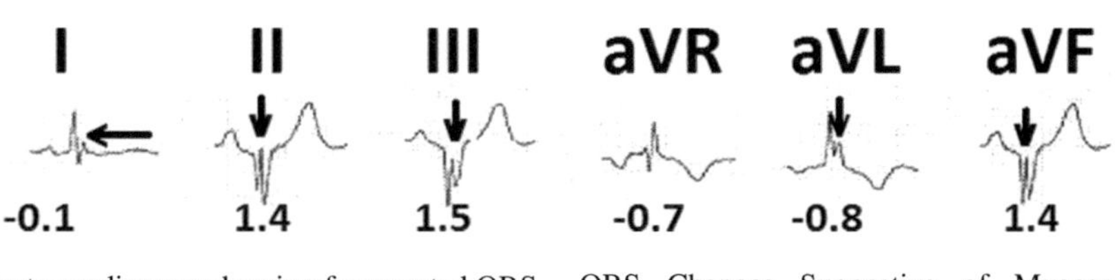

Fig. 22.17 Electrocardiogram showing fragmented QRS in leads I, II, III, aVL, and aVF (marked as arrows). (Reproduced with permission from Mittal SR. Mid-late QRS Changes Suggestive of Myocardial Necrosis. Cardiology Today. 2015;19:211–216)

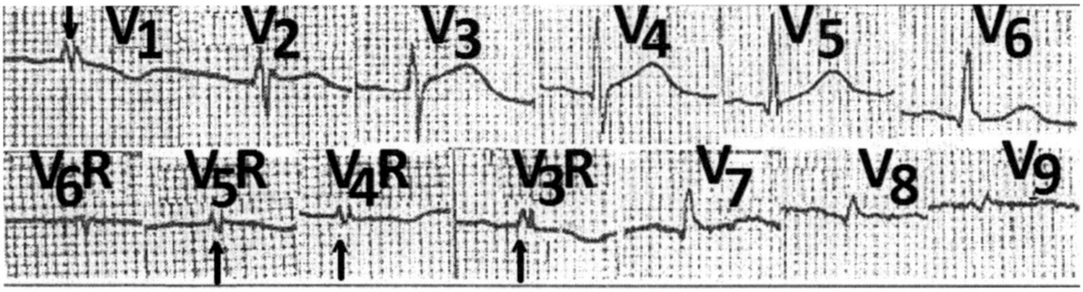

Fig. 22.18 Electrocardiogram showing fragmented QRS in leads V_1, V_3R, V_4 R, and V_5 R (marked as arrows). (Reproduced with permission from Mittal SR. Mid-late QRS Changes Suggestive of Myocardial Necrosis. Cardiology Today. 2015;19:211–216)

7 Peri-infarction Block

Focal block around an area of infarction produces localized widening of the QRS complex, which is not typical of any bundle branch block (Fig. 22.19).

Representative electrocardiogram is shown in Fig. 22.20.

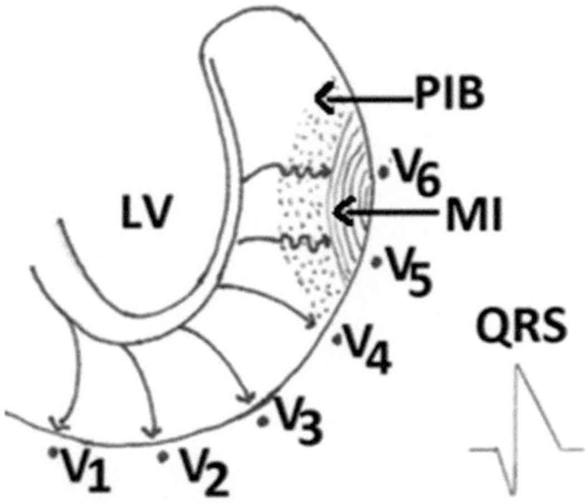

Fig. 22.19 Diagram showing peri-infarction block and its effect on the QRS complex. *MI* myocardial infarction, *PIB* peri-infarction block. (Reproduced with permission from Mittal SR. Mid-late QRS Change Suggestive of Myocardial Necrosis. Cardiology Today. 2015;19:211–216)

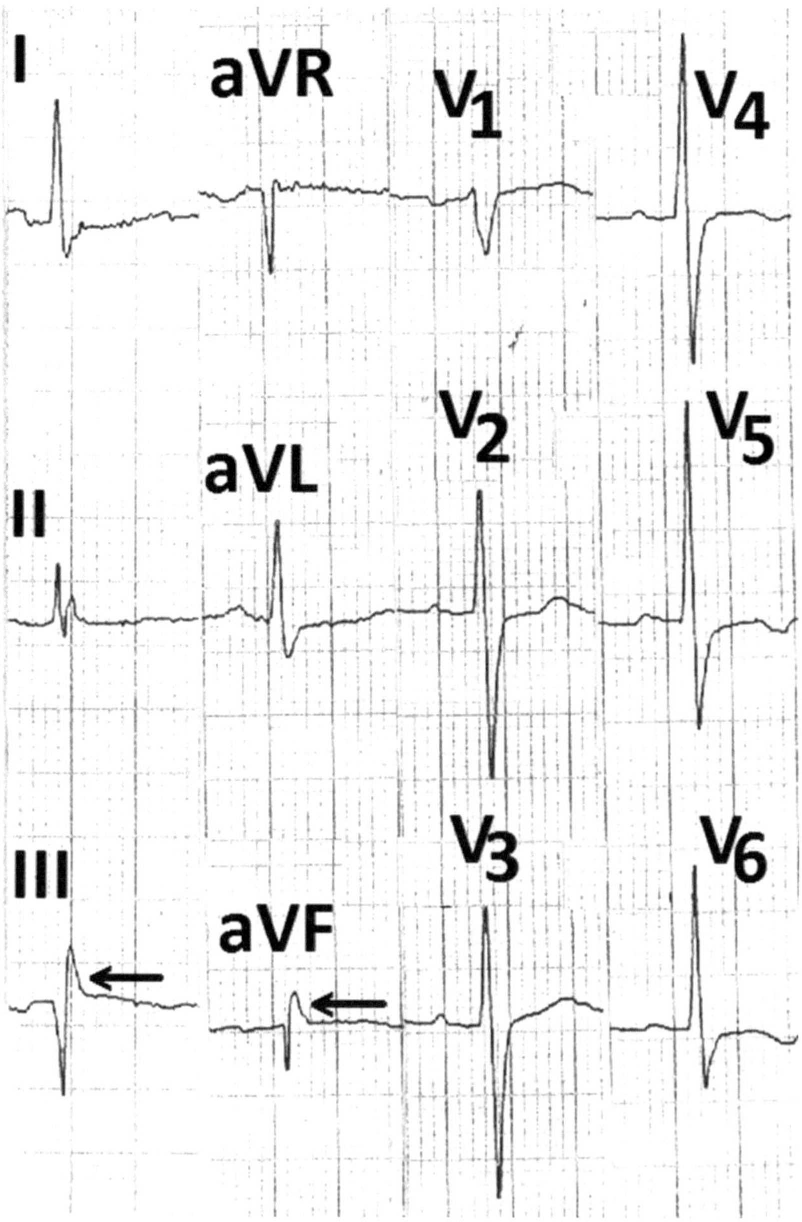

Fig. 22.20 Electrocardiogram showing peri-infarction block in leads III and aVF (marked ←). (Reproduced with permission from Mittal SR. Mid-late QRS Changes Suggestive of Myocardial Necrosis. Cardiology Today. 2015;19:211–216)

Summary

- Nontransmural myocardial necrosis may not produce a Q wave. QRS complex may show any of the following changes.
- Significant loss of the R wave voltage in correctly placed leads. At times, there may be only partial loss of the R wave voltage, which may present as a notch on the top of the R wave or the top of the R wave may look like a spike.
- Slow (poor) progression of the R wave in the precordial leads. Left ventricular hypertrophy, pulmonary thromboembolism, and emphysema can also produce this finding.
- Nonprogression of the R wave in the precordial leads. Left bundle branch block, left ventricular hypertrophy, emphysema, left pneumothorax with mediastinal shift, uncomplicated dextrocardia, pectus excavatum, and congenital absence of the pericardium can also produce this finding.
- Reverse progression of the R wave (sudden reduction of R wave amplitude) in a precordial lead. This finding is specific of myocardial necrosis.
- Slurring of the ascending limb, top, or the descending limb of the R wave in absence of any QRS widening suggestive of bundle branch block.
- Fragmentation of the QRS complex in at least two contiguous leads.
- Localized widening of QRS, which is not typical of any bundle branch block (peri-infarction block).
- Above findings are nonspecific. They are important only when there is a new appearance in the classical clinical setting of acute coronary syndrome.

MCQs

Q1. Which part of the ventricles is depolarized last?

(a) Base of the right ventricle
(b) Base of the left ventricle
(c) Base of the interventricular septum
(d) Apex of the heart

Q2. Loss of R wave voltage is not caused by

(a) Multiple infarcts
(b) Technical error
(c) Obesity
(d) Last trimester of pregnancy

Q3. Slow progression of the R wave is not seen in

(a) Subendocardial infarction
(b) Pulmonary embolism
(c) Emphysema
(d) Dextroversion

Q4. Nonprogression of the R wave is not seen in

(a) Left bundle branch block
(b) Nontransmural myocardial infarction
(c) Pectus excavatum
(d) Technical dextrocardia

Q5. Reverse progression of the R wave is seen in

(a) Nontransmural myocardial infarction
(b) Emphysema
(c) Dextrocardia
(d) Left pneumothorax

Q6. Isolated slurring of a narrow QRS is seen in

(a) Nontransmural infarction
(b) Bundle branch block
(c) Peri-infarction block
(d) Transmural myocardial infarction

Q7. What is not correct for fragmented QRS?

(a) It suggests inhomogeneous depolarization
(b) QRS is narrow
(c) It is associated with recurrent cardiac events
(d) It is specific of myocardial infarction

Q8. Fractioned QRS is not seen in

(a) Systemic hypertension
(b) Hypertrophic cardiomyopathy
(c) Aortic stenosis
(d) None

Q9. Fractioned QRS is not seen in

(a) Sarcoidosis
(b) Brugada syndrome
(c) Arrhythmogenic right ventricular dysplasia
(d) Large VSD

Q10. Fractioned QRS is not seen in

(a) Repaired TOF
(b) Large primum ASD
(c) Ebstein anomaly
(d) Corrected transposition

Answers

(1) c, (2) d, (3) d, (4) d, (5) a, (6) a, (7) d, (8) d, (9) d, (10) b, d.

References

1. De Luna AS, Goldwasser D, Fiol M, Bayes-Genis A. Surface electrocardiography. In: Fuster V, Walsh RA, Harrington RA, editors. Hurst's The Heart. New York: Mc Graw Hill; 2017. p. 302–70.
2. Mirvis DM, Goldberger AL. Electrocardiography. In: Bonow RO, Mann DL, Zipes DP, Libby P, editors. Braunwald's Heart Disease. Elsevier: Missouri; 2012. p. 126–65.
3. Tan NS, Goodman SG, Yan RT, et al. Prognostic significance of low QRS voltage on the admission electrocardiogram in acute coronary syndromes. Int J Cardiol. 2015;190:34–9.
4. Das MK, Khan B, Jacob S, Kumar A, Mahenthiran J. Significance of a fragmented QRS complex versus a Q wave in patients with coronary artery disease. Circulation. 2006;113:2495–501.
5. Das MK, Saha C, El Masry H, et al. Fragmented QRS on a 12 lead ECG: a prediction of mortality and cardiac events in patients with coronary artery disease. Heart Rhythm. 2007;4:1385–92.
6. Jain R, Singh R, Sundermurthy Y, Das MK. Fragmented ECG as a risk marker in cardiovascular diseases. Curr Cardiol Rev. 2014;10:277–86.

Right Ventricular Infarction

23

1 Arterial Supply of Right Ventricle and Its Correlation with Electrocardiographic Findings

We should recapitulate the arterial supply of the right ventricle (Fig. 23.1a and b) for better understanding of the location of the electrocardiographic findings in cases of right ventricular infarction.

The first branch is the conus artery. It supplies right ventricular outflow tract and anastomoses with the conus branch of the left anterior descending artery. Occlusion of this branch can cause infarction in the region of the right ventricular outflow tract except in cases where supply from the conus branch of the left anterior descending coronary artery is adequate. Right ventricular outflow tract may also escape infarction if the conus artery arises independently from the aorta. Infarcts of this area can be picked up only if leads V_3R and V_4R are placed higher than their standard position.

Occlusion of the sinus node artery (Fig. 23.2a) results in ischemic damage to the sinus node, resulting in sinus bradycardia (Fig. 23.2b), which may persist even after recovery of acute myocardial infarction.

The artery supplying the infundibular septum arises either from the proximal right coronary artery or from the conus artery. Occlusion of these vessels can produce changes in lead V_1 and even in lead V_2.

Marginal branches, including the acute marginal branch, supply the anterior surface of the right ventricle (Fig. 23.1). Isolated occlusion of these vessels results in isolated infarction of the anterior surface of the right ventricle without any involvement of its inferior surface. Changes of infarction are seen only in leads V_3 R and V_4 R without any change in inferior leads.

After the acute marginal branch, the right coronary artery supplies the inferior wall of the right ventricle through the posterior descending coronary artery (Fig. 23.1). Occlusion distal to the marginal branch, therefore, produces changes in the inferior leads and leads V_5 R and V_6 R. No changes are produced in leads V_3 R and V_4R as there is no damage to the right ventricular anterior wall. In nearly 70% of the human hearts, posterior descending artery arises from the right coronary artery. In these cases, the right coronary artery also supplies the inferior part of the interventricular septum (Fig. 23.1b), atrioventricular node (Fig. 23.3a), the His bundle, and the right bundle branch. In these persons, occlusion of the right coronary artery can produce infarction of the inferior surface of the left ventricle, atrioventricular block (Fig. 23.3b), and right bundle branch block.

Atrioventricular node may escape ischemic injury if there is a good collateral flow to the pos-

S. Mittal, *Insights into Electrocardiograms with MCQs*,
https://doi.org/10.1007/978-981-99-0127-2_23

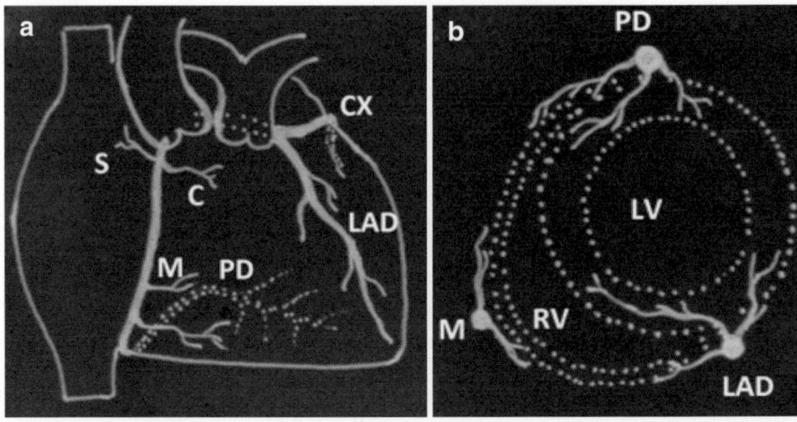

Fig. 23.1 Diagram showing the arterial supply of the right ventricle (**a**) in frontal plane and (**b**) horizontal plane. *S* artery to SA node; *C* conus artery; *M* marginal branches; *PD* posterior descending coronary artery; *LAD* left anterior descending coronary artery; *CX* left circumflex coronary artery; *LV* left ventricle; *RV* right ventricle.

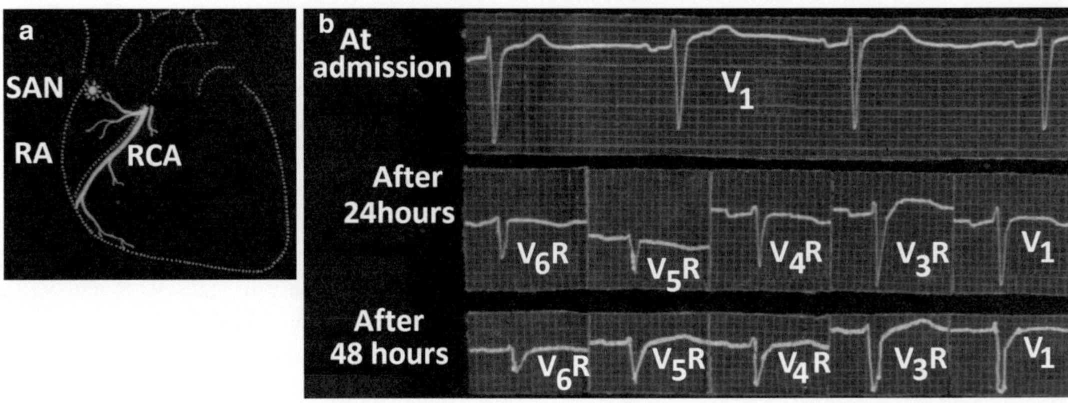

Fig. 23.2 (**a**) Diagram showing the involvement of sinoatrial node (SA). *RA* right atrium; *RCA* right coronary artery. (**b**) Electrocardiogram from a case of right ventricular infarction showing sinus bradycardia and ST segment elevation in leads V_1,V_3 R, and V_4R.

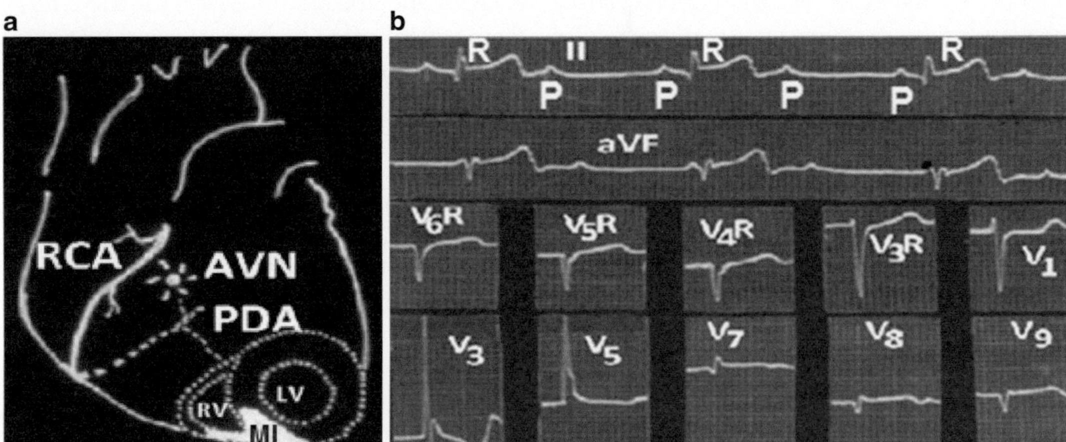

Fig. 23.3 (**a**) Diagram showing the arterial supply of a dominant right coronary artery (RCA). *AVN* atrioventricular node; *PDA* posterior descending coronary artery; *MI* infarction of the inferior surface of the left ventricle (LV) and the right ventricle (RV). (**b**) Electrocardiogram showing inferior, posterior, and right ventricular infarction with complete atrioventricular block (leads II and aVF).

terior descending coronary artery or if the atrioventricular nodal artery arises from the circumflex artery. Dominant right coronary artery also supplies branches to the posterior surface of the left ventricle (Fig. 23.1). In such instances, right ventricular infarction may be accompanied by left ventricular posterior infarction (Fig. 23.3b). In the remaining 30% persons who have dominant left circulation, the posterior descending coronary artery arises from the left circumflex coronary artery. In these cases, the posterior descending coronary artery also supplies the inferior surface of the right ventricle. In these cases, inferoposterior infarction of the right ventricle can occur without significant disease of the right coronary artery.

Juxtaseptal portion of the anterior wall of the right ventricle is supplied by branches of the left anterior descending coronary artery (Fig. 23.4a, b) (arrow).

Occlusion of the left anterior descending coronary artery before these branches leads to infarction of the juxtaseptal part of the anterior surface of the right ventricle in addition to anteroseptal infarction of the left ventricle. In such an instance, ST-segment elevation of leads V_2 to V_4 may extend to leads V_1, V_3 R, and even V_4 R.

Right ventricle has rich collateral circulation (Fig. 23.5 a, b).

Septal branches of the left anterior descending coronary artery anastomose with the septal branches of the posterior descending coronary artery arising from the right coronary artery [1]. Collateral flow through the moderator band artery also plays an important role in salvaging the right ventricular myocardium in the event of occlusion of the right coronary artery [2]. Electrocardiographic findings may, therefore, differ in patients with similar occlusion of the right coronary artery.

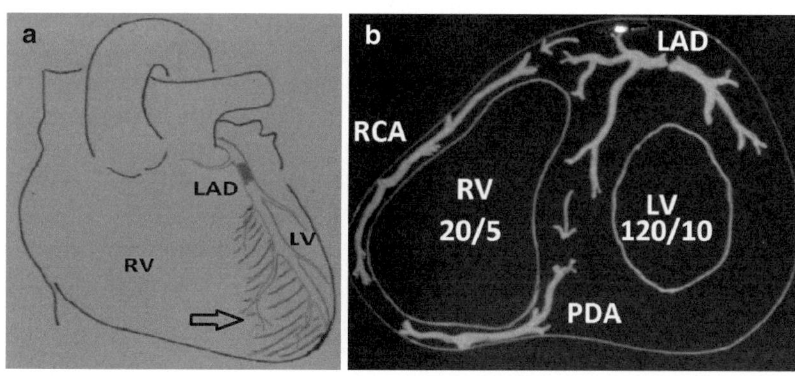

Fig. 23.4 (**a**, **b**) Diagram showing the arterial supply of the juxtaseptal portion of the anterior wall of the right ventricle (arrow). *RV* right ventricle; *LV* left ventricle; *LAD* left anterior descending coronary; *RCA* right coronary artery; *PDA* posterior descending coronary artery.

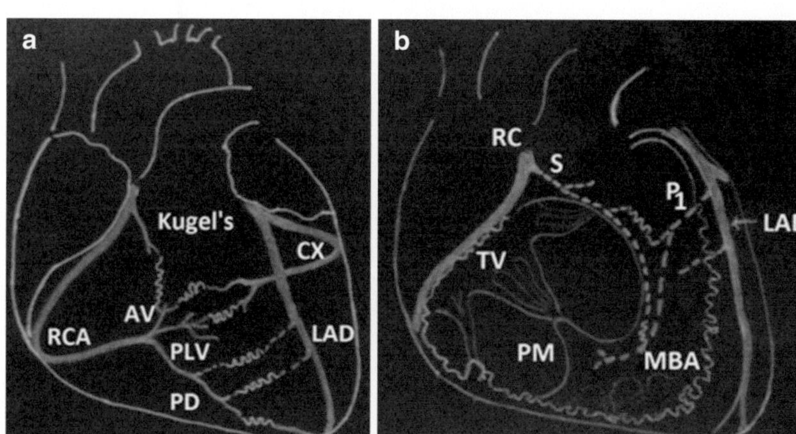

Fig. 23.5 Diagram showing collaterals to the right coronary artery. *RCA* right coronary artery, Kugel's artery; *AV* atrioventricular nodal artery; *PLV* posterior left ventricular artery; *PD* posterior descending artery; *LAD* left anterior descending artery; *CX* left circumflex artery; *S* septal branch; *MBA* moderator band artery; *PM* papillary muscle; *TV* tricuspid valve.

2 Correlation of Various Right Precordial Electrocardiographic Leads to Various Parts of the Right Ventricle (Fig. 23.6)

Lead V_1 overlies the juxtaseptal portion of the anterior wall of the right ventricle. Therefore, involvement of the anterior part of the interventricular septum and adjacent right ventricular anterior wall can produce ST-segment elevation in lead V_1 in addition to leads V_2 toV_4. Leads V_3 R and V_4 R overlie anterior wall of the right ventricle. ST- segment elevation in these leads suggests infarction of the right ventricular anterior wall. Leads V_5 R and V_6 R look at the posteroinferior surface of the right ventricle [3, 4, 5]. ST-segment elevation in these leads suggests infarction of the posteroinferior surface of the right ventricle (Fig. 23.6).

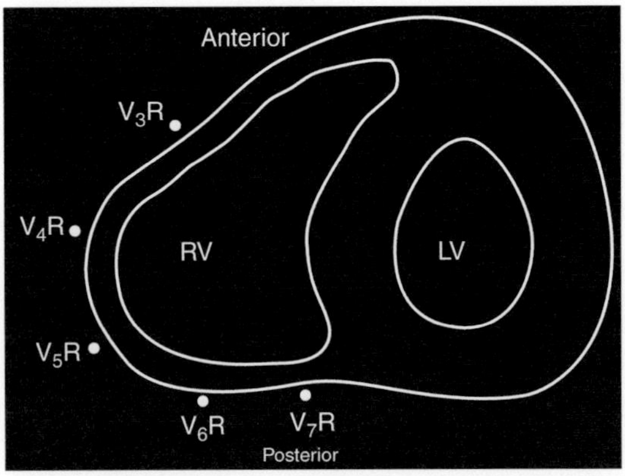

Fig. 23.6 Diagram showing relation of right-sided chest leads to different parts of the right ventricle. *RV* right ventricle; *LV* left ventricle

It is, therefore, important to record all right precordial leads, so that diagnosis of right ventricular infarction is not missed. Leads V_3 R and V_4 R are diagonally opposite to leads V_5 R and V_6 R. Therefore, ST-segment elevation localized to leads V_3 R, V_4 R or leads V_5 R and V_6 R may produce reciprocal ST-segment depression in opposite leads.

3 Correlation of Left-Sided Electrocardiographic Leads to Right Precordial Leads

Limb leads I and aVL are diagonally opposite to the right precordial leads in the frontal plane (Fig. 23.7a). ST-segment elevation in leads I and aVL, therefore, produces ST-segment depression in leads V_3 R and V_4 R masking right ventricular infarction. On the other hand, ST-segment depression in leads I and aVL may produce ST-segment elevation in leads V_3 R and V_4 R, giving a false positive impression of right ventricular infarction.

Leads V_7 ,$V_{8,}$ and V_9 are diagonally opposite to leads V_{1} ,V_3 R, and V_4 R in the horizontal plane (Fig. 23.7b).

ST-segment elevation in leads V_3 R and V_4 R can, therefore, produce reciprocal ST-segment depression in leads V_6 to V_9 (Fig. 23.7c). Similarly, ST-segment elevation in leads V_7, V_8, V_9 can produces ST-segment depression in leads V_1, V_3 R, and V_4 R. This can mask concomitant right ventricular infarction.

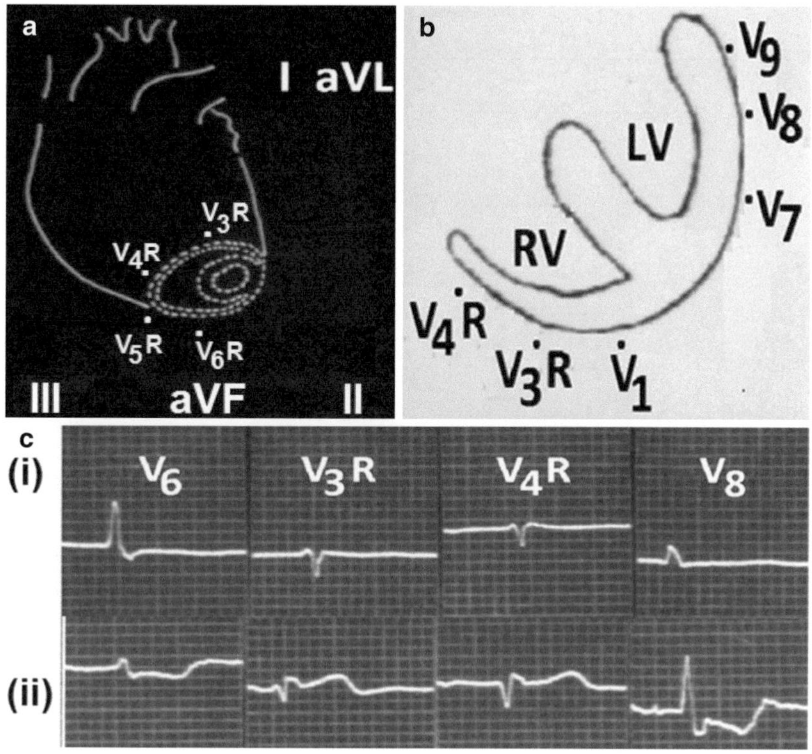

Fig. 23.7 (**a**) Diagram showing the relation of leads I and aVL with the right-sided chest leads and inferior leads in frontal plane. (**b**) Diagram showing the relation of posterior chest leads (V_7 to V_9) with right-sided chest leads (V_3 R toV_4 R) in the horizontal plane. (**c**) Electrocardiogram, from a patient before (i) and after (ii) isolated right ventricular anterior myocardial infarction. It shows ST-segment elevation in leads V_3 R and V_4 R and reciprocal ST-segment depression in leads V_6 and V_8.

4 Normal Configuration of the Electrocardiographic Wave Forms in the Right-Sided Chest Leads

Normally right-sided chest leads show rs configuration with isoelectric ST-segment and shallow T wave inversion. Leads V_5R and V_6R can show qr pattern (Figs. 23.8 and 23.9). R and Rs patterns are seen in right ventricular hypertrophy, right bundle branch block, or infarction of the left ventricular posterior wall. QS pattern in leads V_3 R and V_4 R has been considered as a marker of right ventricular infarction [6]. This electrocardiographic finding, however, has lower diagnostic value than elevation of the ST-segment in these leads.

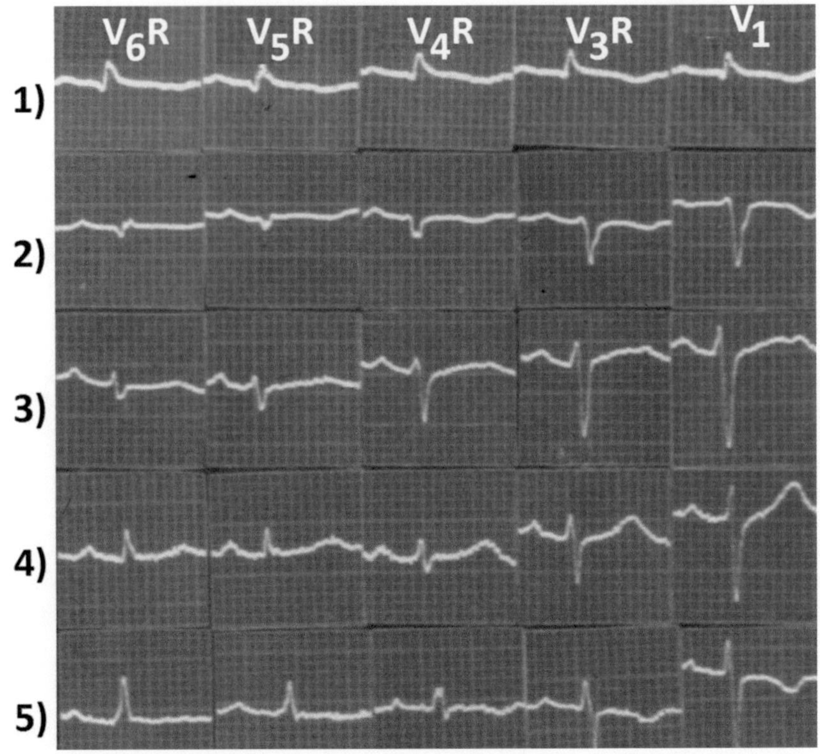

Fig. 23.8 Electrocardiogram in right-sided chest leads in different normal persons

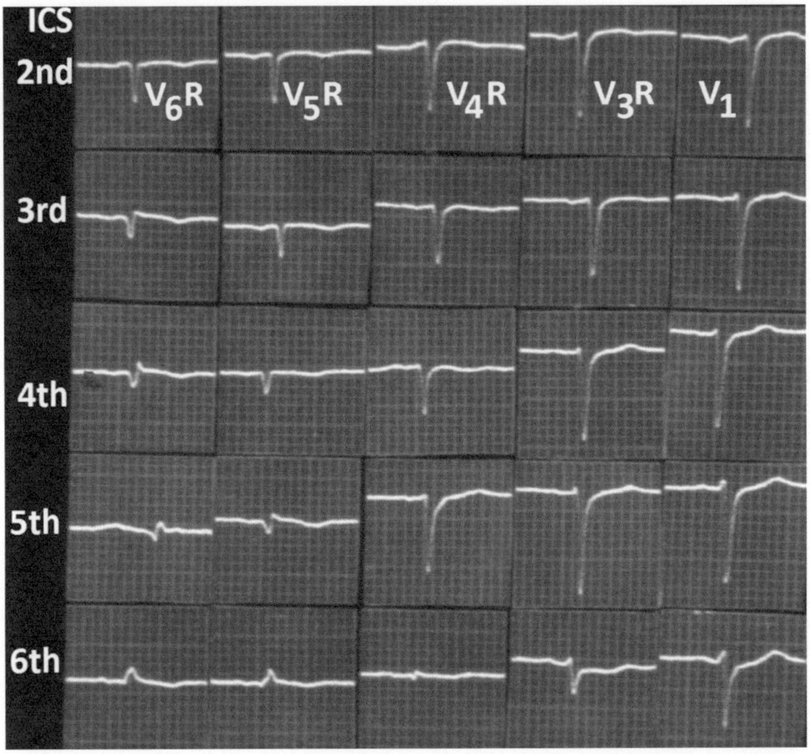

Fig. 23.9 Electrocardiogram in different intercostal spaces (ICS) in right-sided chest leads in a normal person

5 Electrocardiographic Diagnosis of Right Ventricular Anterior Infarction

ST-segment elevation in the right-sided chest leads is considered suggestive of right ventricular anterior infarction (Fig. 23.10).

Elevation is usually very small probably due to small muscle mass of the right ventricle. Elevation of 0.5 mm in lead V_4 R has been shown to have a sensitivity of 76% and specificity of 86%. Specificity increases at the cost of sensitivity if criteria of 1 mm ST-segment elevation are taken as diagnostic [6]. Different right-sided chest leads have different sensitivity and specificity. Different patients may have ST-segment elevation in different right precordial leads depending on the site of the right ventricular infarction (Fig. 23.11).

Combination of leads V_4R, V_5R, and V_6 R has been shown to have maximum sensitivity (90%) and specificity (93%) [5]. The number of leads showing ST-segment elevation may correlate with the extent of right ventricular infarction [3]. At times, recording of all conventional right-sided chest leads (V_3R to V_7R) may not show clear elevation of the ST-segment [3]. In such situations, recording right-sided chest leads in a higher (Fig. 23.12) or lower intercostal space can reveal diagnostic changes [7].

Initially, there was a feeling that right ventricular infarction was always associated with inferior infarction of the left ventricle. This is not correct. Infarction of the right ventricular anterior wall can occur as an isolated entity if there is isolated occlusion of a marginal branch of the right coronary artery. Right ventricular anterior infarction can also occur along with left ventricular anteroseptal infarction due to occlusion of the left anterior descending coronary artery, which also supplies anterior juxtaseptal part of right ventricle.

5.1 Effect of Size of Right Ventricular Infarction

Autopsy studies have shown that infarcts involving less than 25% of the right ventricular myocardium are not picked on surface electrocardiogram [8].

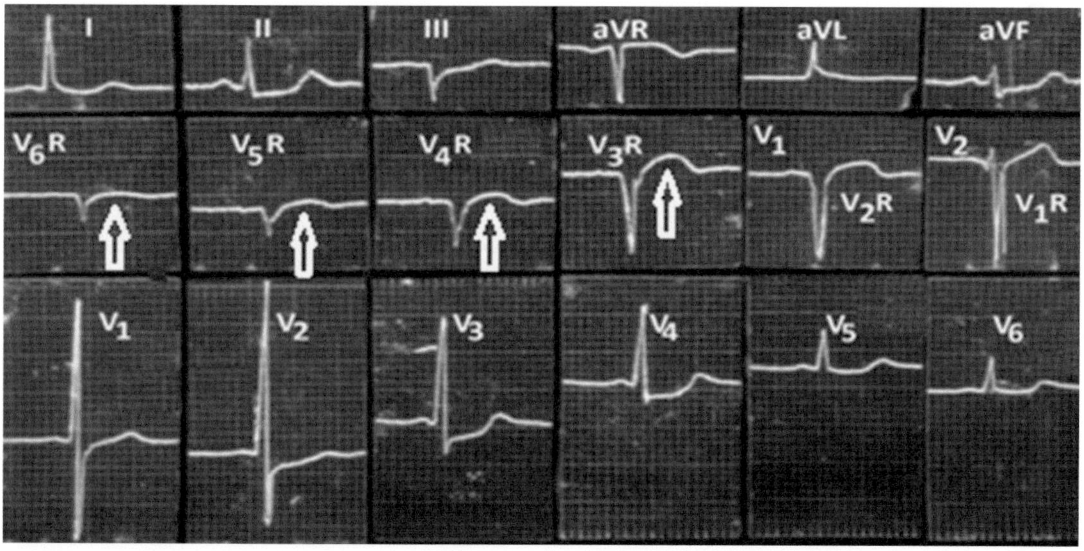

Fig. 23.10 Electrocardiogram from a case of right ventricular anterior infarction showing ST-segment elevation in leads V_3 R to V_6 R (marked ⇧)

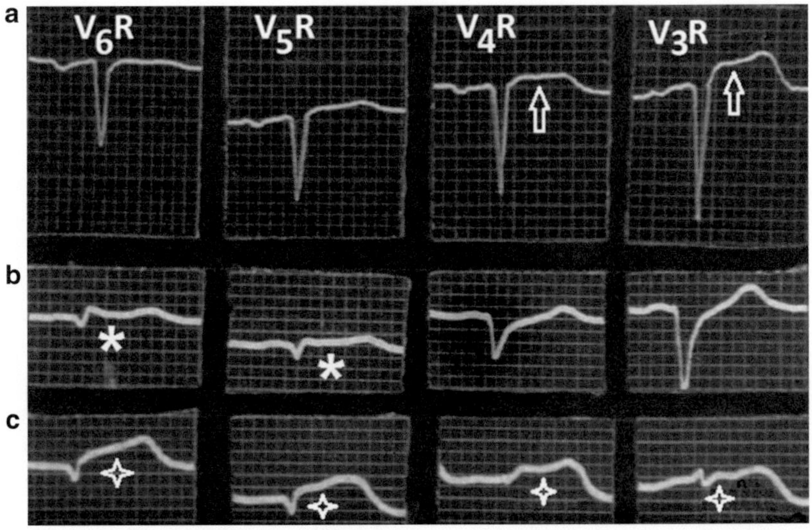

Fig. 23.11 Right-sided chest leads from different patients with right ventricular infarction. (**a**) ST-segment elevation in leads V_3 R, V_4 R (marked ⇧). (**b**) ST-segment elevation in leads V_5 R, V_6 R (marked *). (**c**) ST-segment elevation in leads V_3 R to V_6 R (marked ✧).

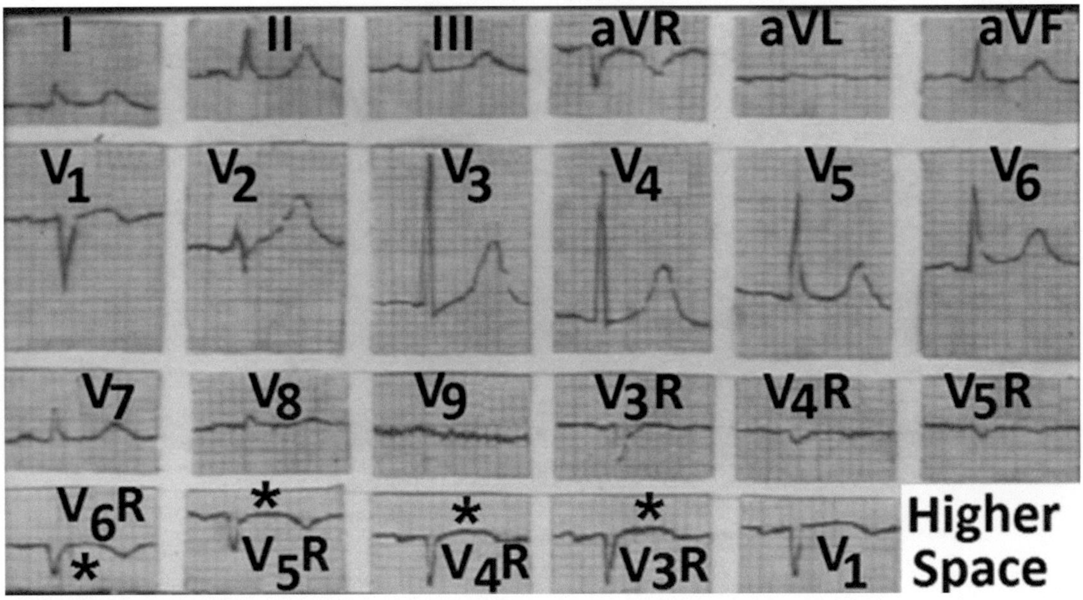

Fig. 23.12 Electrocardiograms from a case of right ventricular infarction. ST-segment elevation is better seen in right-sided chest leads recorded in higher intercostal space (marked *).

5.2 Effect of Time on the ST-Segment Elevation

In a case of left ventricular infarction, it takes a few days for the ST-segment to settle down. In a case of right ventricular infarction, the ST-segment settles down quickly. In nearly 40% of cases, elevation of the ST-segment in the right-sided chest leads can settle down within 20–24 hours of onset of pain [9] (Figs. 23.13 and 23.14) and in most of the cases it settles down by the third or the fourth day [8].

The diagnosis may be missed in some cases if the right-sided chest leads have not been recorded in the first 24–48 h. This rapid healing occurs because of profuse collateral circulation to the right ventricle (Fig. 23.5). Right-sided chest leads should, therefore, be recorded in all cases presenting with symptoms suggestive of myocardial ischemia. It is important to remember that the ST-segment elevation represents transmural ischemia, which may not progress to necrosis in all cases. This may also contribute to rapid settling down of the ST-segment.

5.3 Differentiation from Left Ventricular Anteroseptal Infarction

Right ventricular infarction can lead to right ventricular dilatation. In these cases, electrocardiographic findings can extend up to lead V_2 or even up to lead V_3 mimicking left ventricular anteroseptal infarction [10, 11] (Figs. 23.15 and 23.16).

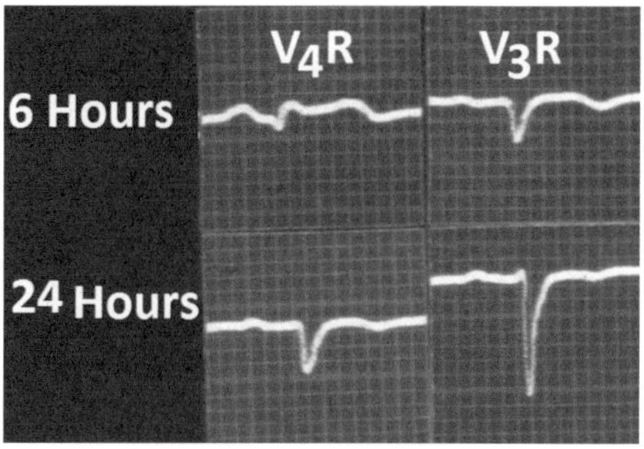

Fig. 23.13 Electrocardiograms (leads V_3 R and V_4 R) from a case of right ventricular infarction recorded six hours and 24 h after onset of chest pain. The ST-segment elevation settled within 24 h.

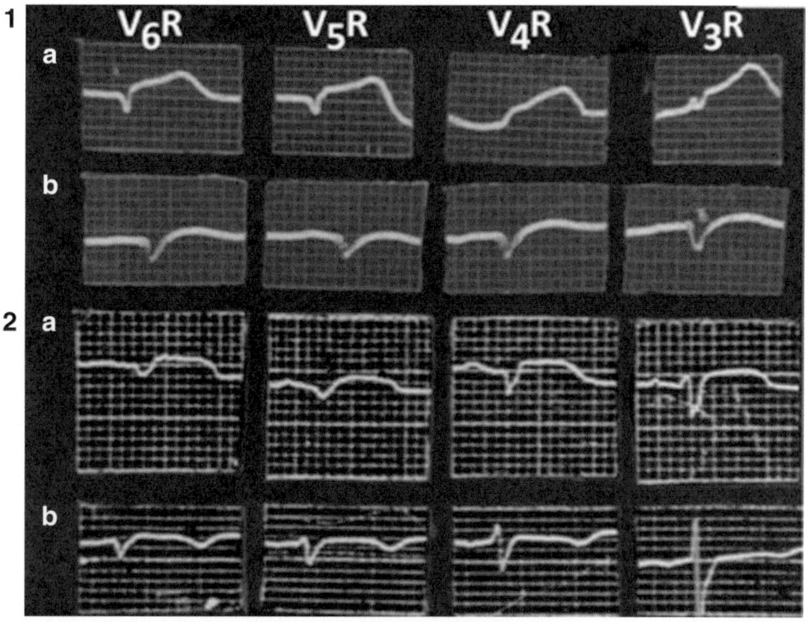

Fig. 23.14 Leads V_3 R to V_6 R recorded from two patients (1 and 2) of right ventricular infarction at the time of admission (**a**) and 24 h after admission (**b**) showing rapid settling of the ST-segment

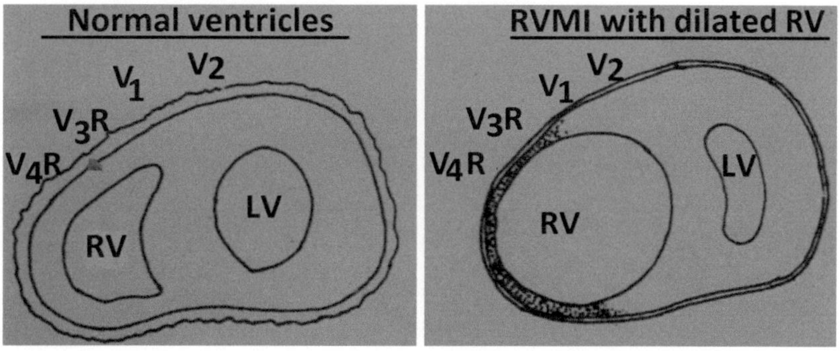

Fig. 23.15 Diagram showing that in the case of dilatation of the right ventricle, leads V_1 and V_2 may overlie the right ventricle and may show ST-segment elevation. *RVMI* right ventricular myocardial infarction.

Similarly dilatation of the left ventricle can bring the interventricular septum and anterior wall of the left ventricle under leads V_3R and V_4R (Fig. 23.17a). In such cases, changes of left ventricular anteroseptal infarction can extend to leads V_1,V_3R, and at times up to lead V_4R (Fig. 23.17b) due to dilatation of left ventricle bringing right-sided chest leads close to the interventricular septum.

Leads V_4R to V_2, therefore, show a common zone. It may be difficult to dogmatically identify the culprit vessel. Some electrocardiographic

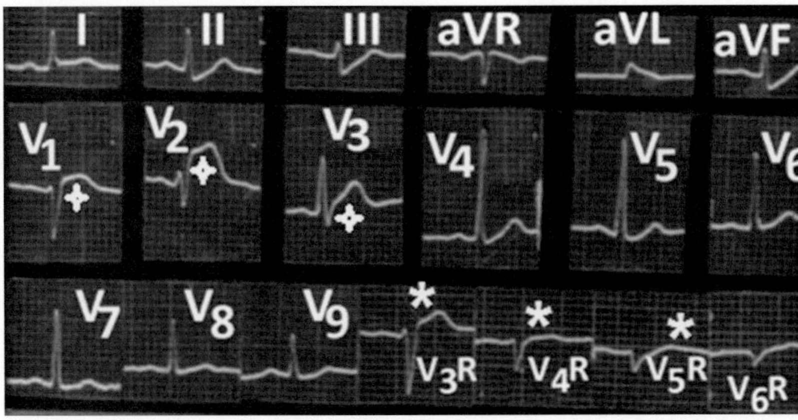

Fig. 23.16 Electrocardiogram from a case of isolated right ventricular anterior myocardial infarction (marked *) showing ST segment elevation in leads V_1 to V_3 (marked ✧)

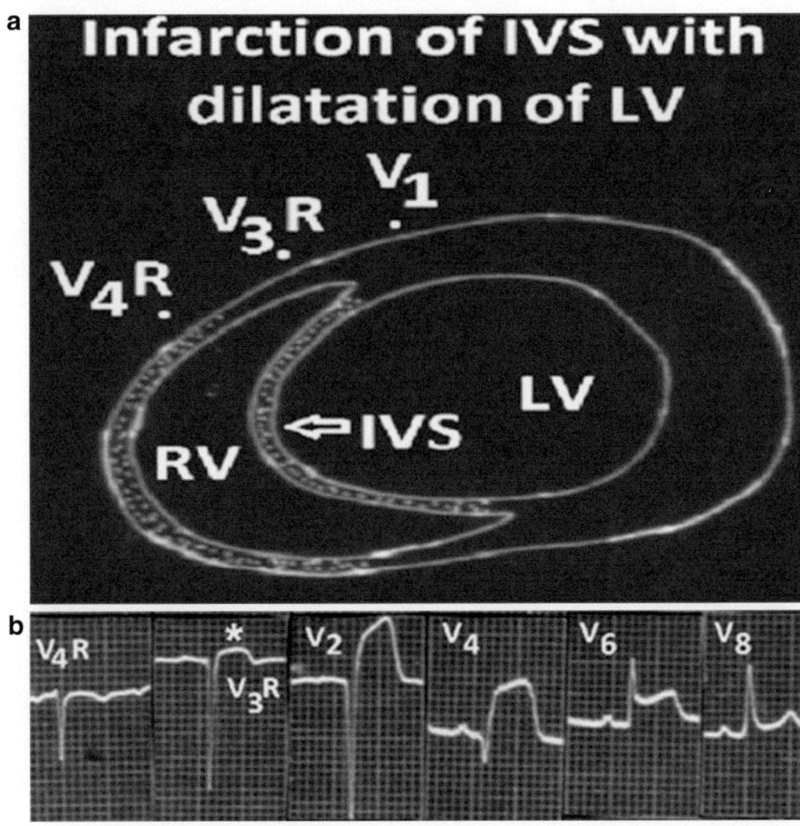

Fig. 23.17 (**a**) Diagram showing that dilatation of the left ventricle can bring infarcted interventricular septum under leads V_3 R and V_4 R. (**b**) Electrocardiogram from a case of left ventricular anterior infarction showing ST-segment elevation in lead V_3 R (marked *).

features may help in differentiation. In the case of left ventricular anterior infarction, the magnitude of ST-segment elevation increases from lead V_1 to lead V_5 with minimal elevation in lead V_1. On the other hand, in a case of infarction of the anterior wall of the right ventricle, ST-segment elevation is maximal in lead V_1 and decreases toward lead V_5 [10] (Fig. 23.18).

5.4 Effects of Infarction of Other Walls of the Left Ventricle

(i) Left ventricular inferior infarction can also mask anterior right ventricular infarction by producing reciprocal ST-segment depression in leads V_3 R and V_4 R [12] (Fig. 23.19).

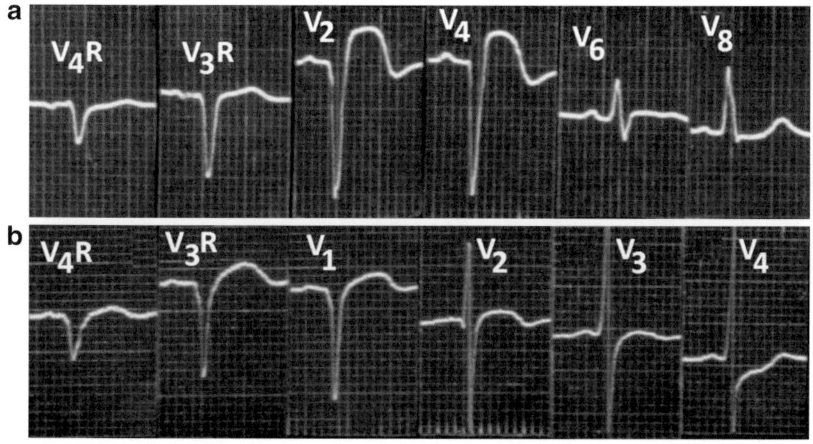

Fig. 23.18 Electrocardiograms (**a**) from a case of anterior left ventricular infarction showing ST-segment elevation increasing from leads V_1 to V_4. (**b**) Electrocardiogram from a case of anterior right ventricular infarction showing ST-segment elevation decreasing from leads V_1 to V_4.

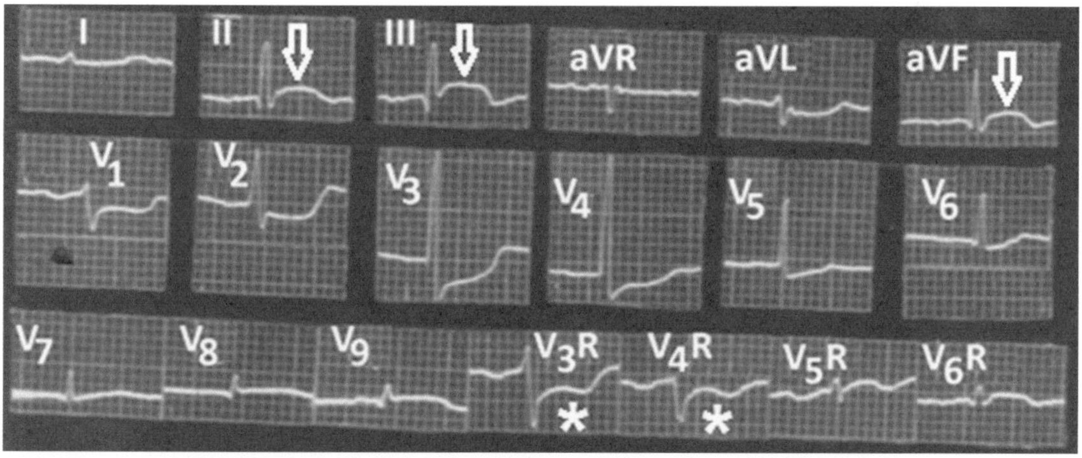

Fig. 23.19 Electrocardiogram in a case of inferior infarction (marked ⇩ in leads II, III, and aVF) showing reciprocal ST-segment depression in leads V_3 R and V_4 R (marked *)

Experimental studies have also shown that ST-segment elevation produced in the right precordial leads due to isolated occlusion of the right coronary artery is abolished by subsequent occlusion of left circumflex coronary artery and ischemia of inferior wall [13].

(ii) As discussed previously, leads I, aVL and leads V_7 to V_9 are diagonally opposite to leads V_3 R to V_6 R in frontal and horizontal plane, respectively. Right ventricular leads, therefore, record reciprocal changes from these left ventricular leads (Figs. 23.20 and 23.21).

As left ventricular wall thickness is much greater than the thickness of the right ventricular wall, ST-segment elevation in the left-sided leads masks concomitant right ventricular infarction on the surface electrocardiogram [14].

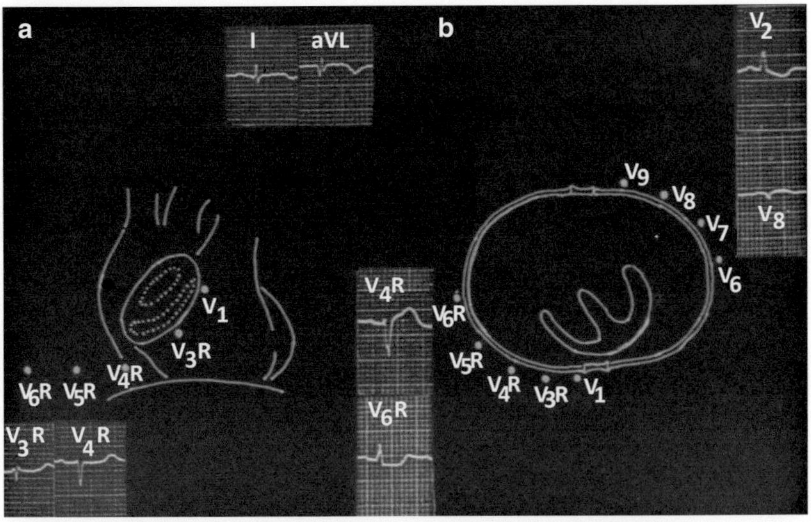

Fig. 23.20 Diagram and electrocardiogram showing reciprocal changes in right-sided chest leadsV_3 R and V_4 R from primary changes (**a**) in leads I and aVL; (**b**) leads V_7 and V_8

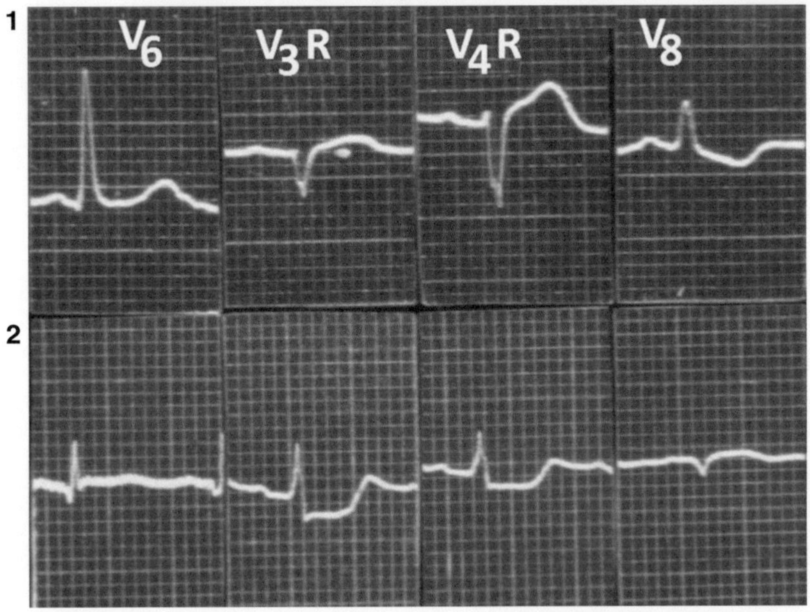

Fig. 23.21 Electrocardiogram from patient 1 showing ST-segment elevation in leads V_3 R and V_4 R secondary to ST-segment depression in lead V_8. Patient 2 showing ST-segment depression in leads V_3 R and V_4 R secondary to ST-segment elevation in lead V_8.

5.5 Causes of ST Segment Elevation in Leads V_1 to V_4 R Other Than Right Ventricular Infarction

- Infarction of the interventricular septum.
- Left bundle branch block.
- ST-segment depression in leads V_7 , V_8, V_9.
- ST-segment depression in leads I and aVL.
- Brugada syndrome.

5.6 Effect of Ventricular Hypertrophy on Diagnosis of Right Ventricular Myocardial Infarction

Left ventricular hypertrophy with ST-segment depression and T wave inversion in leads V_5 and V_6 produces reciprocal ST-segment elevation in leads V_1 to V_4 R (Fig. 23.22).

This may give a false impression of right ventricular infarction. In patients with additional ischemia of the left ventricle, the ST-segment vector may shift earlier than the T wave vector. This results in ST-segment depression even when the T wave is still isoelectric or slightly positive in leads V_7 toV_9. ST-segment elevation with T wave inversion is then recorded in the right sided chest leads exactly mimicking right ventricular infarction. Subendocardial ischemia of the left ventricular anterolateral wall also affects the right-sided chest leads in a similar fashion [15].

Right ventricular hypertrophy produces prominent R wave with ST- segment depression in leads V_4 R to V_2 (Fig. 23.23).

This may mask ST-segment elevation caused by right ventricular infarction. It is important because patients with right ventricular hypertrophy are more susceptible to have right ventricular infarction [16].

5.7 Effect to Acute Pulmonary Embolism

Patients with acute pulmonary embolism can have ST-segment elevation in right-sided chest leads. Acute myocardial ischemia due to acute pulmonary artery hypertension could be responsible. Small q waves and T wave inversion may be present in lead III. Absence of ST-segment

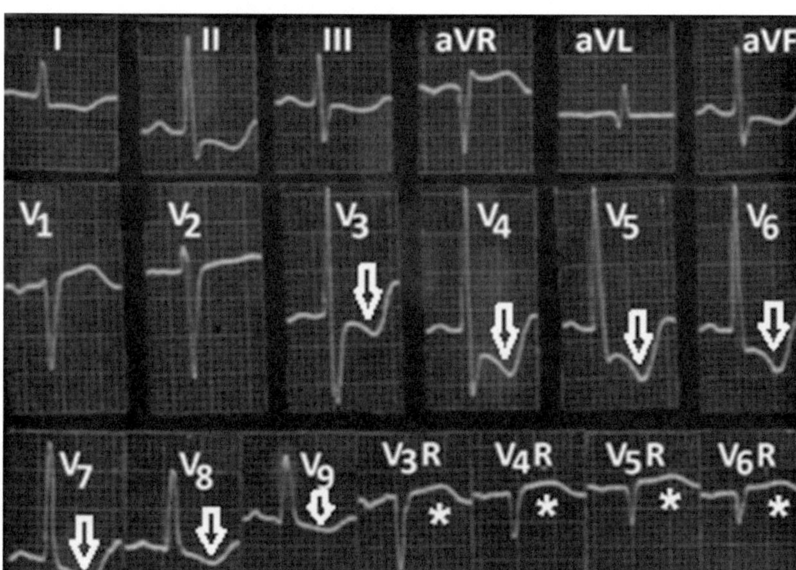

Fig. 23.22 Electrocardiogram from a case of left ventricular hypertrophy with ST-segment depression in the left ventricular leads (V_3 to V_9 marked ⇩)with reciprocal ST-segment elevation in right-sided chest leads (V_3R to V_6R marked *)

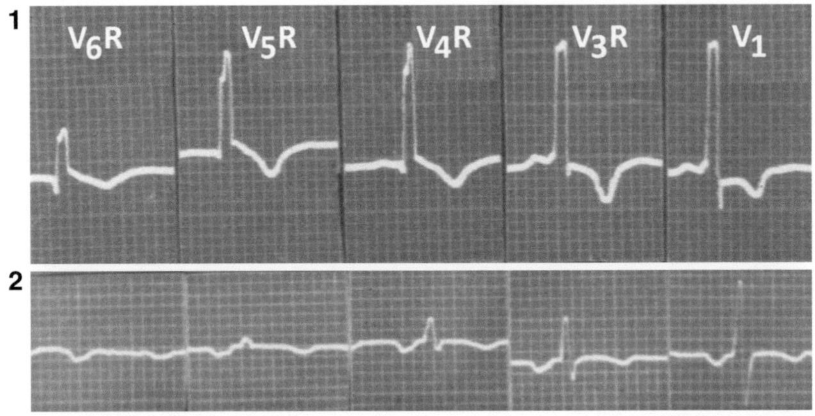

Fig. 23.23 Electrocardiogram from two different cases (1 and 2) of severe right ventricular hypertrophy showing prominent R waves with inverted T waves in leadsV_1 to V_6 R

elevation in inferior leads, right axis deviation, and T wave inversion in leads V_1 to V_3 support possibility of pulmonary embolism [17, 18].

5.8 Effects of Bundle Branch Block on the Diagnosis of Right Ventricular Infarction

(i) Left bundle branch block produces rS or QS pattern with upsloping ST-segment elevation and upright T waves in right-sided chest leads (Fig. 23.24a, b). This interferes with electrocardiographic diagnosis of right ventricular infarction. Coved ST-segment elevation with T wave inversion in leads V_3R and V_4 R is suggestive of concomitant right ventricular infarction (Fig. 23.24c). It is, however, important to remember that infarction of the interventricular septum may also produce similar changes [19].

(ii) Extensive and severe injury to the right ventricle can per se produce transient incomplete right bundle branch block pattern probably due to conduction delay in the peripheral part of the right bundle branch [20]. Such incomplete right bundle branch block is not accompanied by abnormal Q wave or ST-segment elevation in right-sided chest leads.

(iii) Right bundle branch block produces prominent rsR' with downsloping ST-segment and inverted T waves in right precordial leads (Fig. 23.25a). This cancels ST-segment elevation produced by right ventricular infarction. ST-segment elevation with concordant T wave in right-sided leads suggests right ventricular infarction (Fig. 23.25b).

It is important to remember that Brugada syndrome also produces a right bundle branch pattern with ST-segment elevation in lead V_1. Loss of initial r wave alone does not suggest right ventricular infarction. It can occur in any condition with dilatation of the right atrium, e.g., in pulmonary embolism. This occurs because with right atrial dilatation, lead V_1 lies over the right atrium and records cavity potential. r wave in right bundle branch block occurs from depolarization of the interventricular septum. Therefore, initial r wave may persist even in the presence of right ventricular infarction if the interventricular septum is not involved in the infarction. Infarction of interventricular septum results in loss of initial r wave [21]. Septal focal block is likely to have similar effect. Dilatation of the left ventricle with displacement of the interventricular septum anteriorly and to right can also distort the septal vector in such a way that it no longer remains perpendicular to the right-sided chest leads and r wave is, therefore, not recorded even if interventricular septum is not involved in infarction.

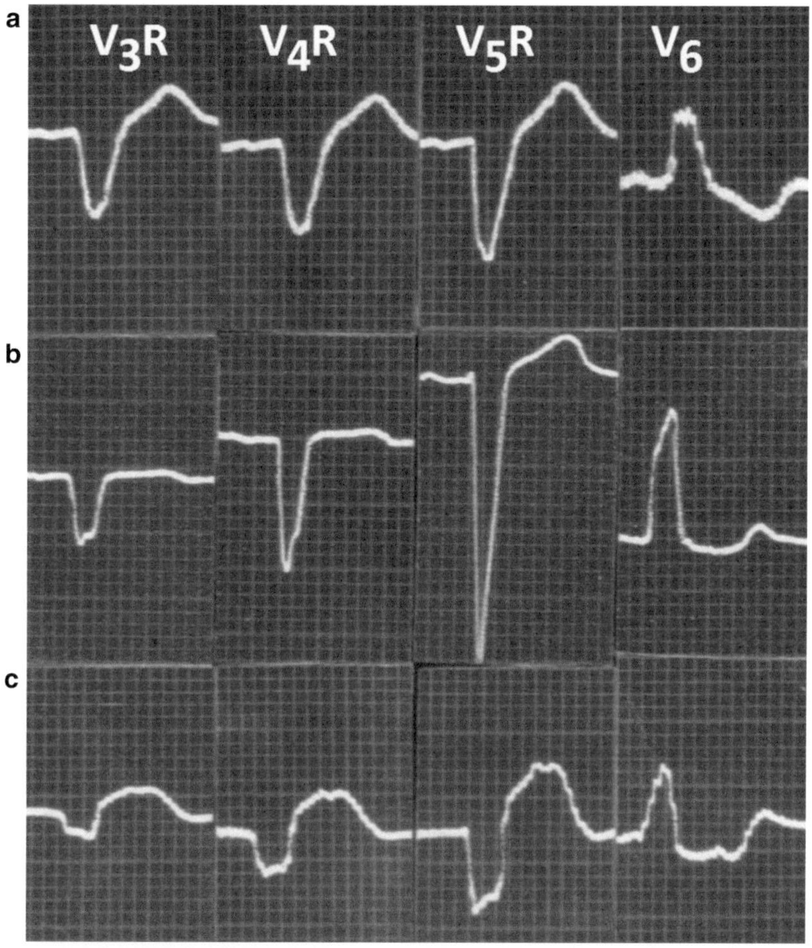

Fig. 23.24 (**a** and **b**) Electrocardiogram showing right-sided chest leads from two cases of left bundle branch block. (**c**) Left bundle branch block with coved ST-segment elevation in right-sided chest leads is suggestive of concomitant right ventricular infarction.

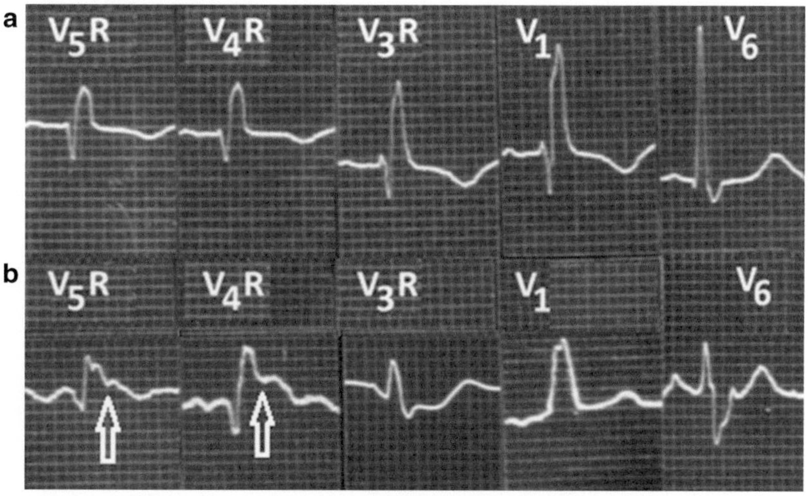

Fig. 23.25 (**a**) Configuration of right-sided chest leads in classical right bundle branch block. (**b**) Electrocardiogram showing RBBB with ST-segment elevation in right sided chest leads (marked ⇧).

5.9 Effects of Fascicular Blocks

5.9.1 Left Anterior Fascicular Block

Left anterosuperior fascicular block, per se, can produce "q" wave in right-sided chest leads. In the presence of block in the anterosuperior fascicle, septal vector arises from posteroinferior fascicle. The septal vector is, therefore, directed inferiorly and posteriorly, resulting in an initial q wave in the right-sided chest leads. Recording electrocardiogram in the intercostal space lower than the conventional points results in the disappearance of these "q" waves.

5.9.2 Left Posteroinferior Fascicular Block

Block in the left posteroinferior fascicle results in activation of the left ventricle through the anterior fascicle. Vector of left ventricular depolarization is, therefore, directed anteriorly and significantly reduces depth of the q wave in right-sided chest leads.

5.10 Effect of Left-Sided Accessory Pathway

Left-sided accessory pathway produces prominent R wave in right-sided chest leads (Fig. 23.26). This may mask concomitant right ventricular infarction.

5.11 Effect of Cardiac Malpositions

In "dextroposition," heart is shifted to the right because of other intrathoracic abnormalities. In "dextroversion," heart rotates to right along its supero inferior axis but the relation of cardiac chambers and great vessels remains normal. In both these conditions, leads V_3 R and V_4 R lie over the interventricular septum and part of the left ventricular anterior wall (Fig. 23.27).

Infarction of the interventricular septum in these cases produces ST-segment elevation in leads V_3 R and V_4 R [22] (Fig. 23.28).

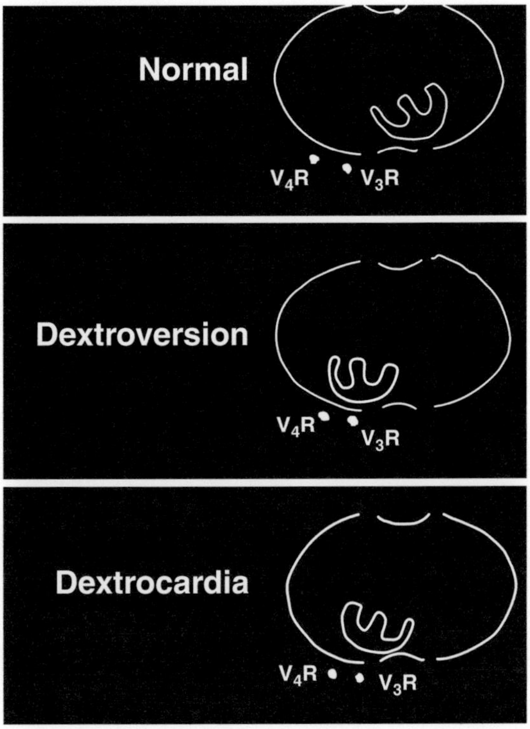

Fig. 23.27 Diagram showing the effect of cardiac malpositions on the relation of ventricles and interventricular septum to the right-sided chest leads

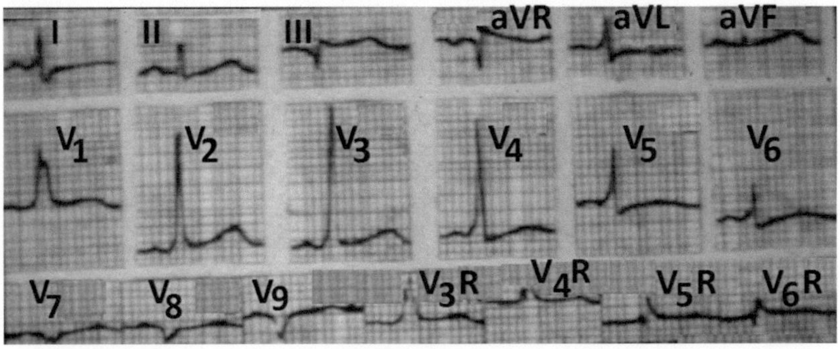

Fig. 23.26 Electrocardiographic pattern in right-sided chest leads from a case of the left lateral accessory pathway

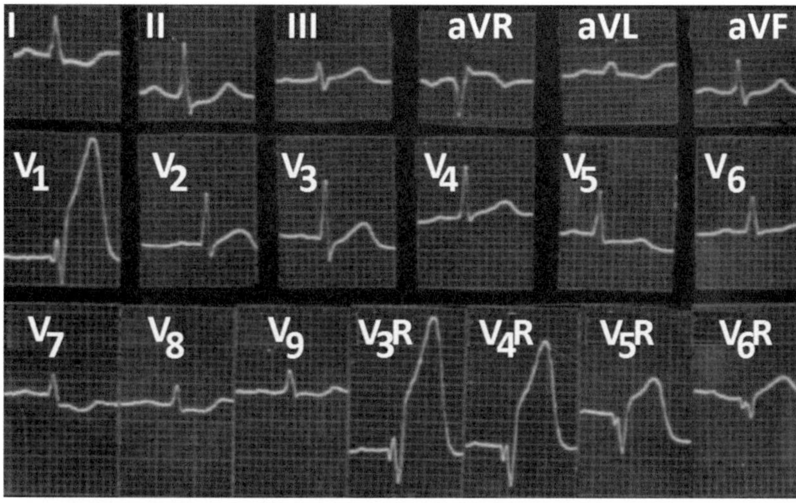

Fig. 23.28 Electrocardiogram from a case of dextroposition with infarction of the interventricular septum showing ST-segment elevation in the right-sided chest leads

In "dextrocardia," the anatomical left ventricle faces the right-sided chest leads. Infarction of the interventricular septum and the left ventricular anterior wall is, therefore, reflected in right-sided chest leads.

6 Electrocardiographic Diagnosis of Infarction of the Right Ventricular Inferior Wall

Inferior surface of the heart comprises of the inferior surface of both the ventricles. This surface is supplied by the posterior descending coronary artery, which can arise from either the right coronary artery or from the left circumflex coronary artery depending on which artery is dominant. Occlusion of the dominant vessel results in inferior infarction. Inferior leads II, III, and aVF look at the inferior surface of the heart from a distance (left foot). These leads look at the whole inferior surface of the heart. These leads, therefore, show abnormal Q wave and ST-segment elevation irrespective of whether the inferior surface of the left or the right ventricle is involved in infarction. Electrocardiographic findings that can help identification of the culprit artery are as follows.

6.1 Leads I and aVL

- Isoelectric or elevated ST-segment in lead I and aVL suggest occlusion of the left circumflex coronary artery (Fig. 23.29a). This happens because occlusion of the proximal part of the left circumflex coronary artery causes high lateral infarction involving leads I and aVL.

ST-segment depressions in leads I and aVL suggests occlusion of the right coronary artery. Leads I and aVL are diagonally opposite to leads V_3R and V_4R in frontal plane. ST-segment elevation in leads V_3R and V_4R, therefore, produces ST-segment depression in leads I and aVL (Fig. 23.29b).

6.2 The Magnitude of ST-Segment Elevation in Lead II Versus That in Lead III

ST-segment elevation in lead III exceeding that in lead II suggests right ventricular inferior infarction [4] (Fig. 23.30).

On the contrary, ST segment elevation greater in lead II supports the diagnosis of left ventricular inferior infarction. This happens because the positive pole of lead II lies to the left and that of lead III lies to the right on the hexaxial reference system in frontal plane (Fig. 23.31).

a

b

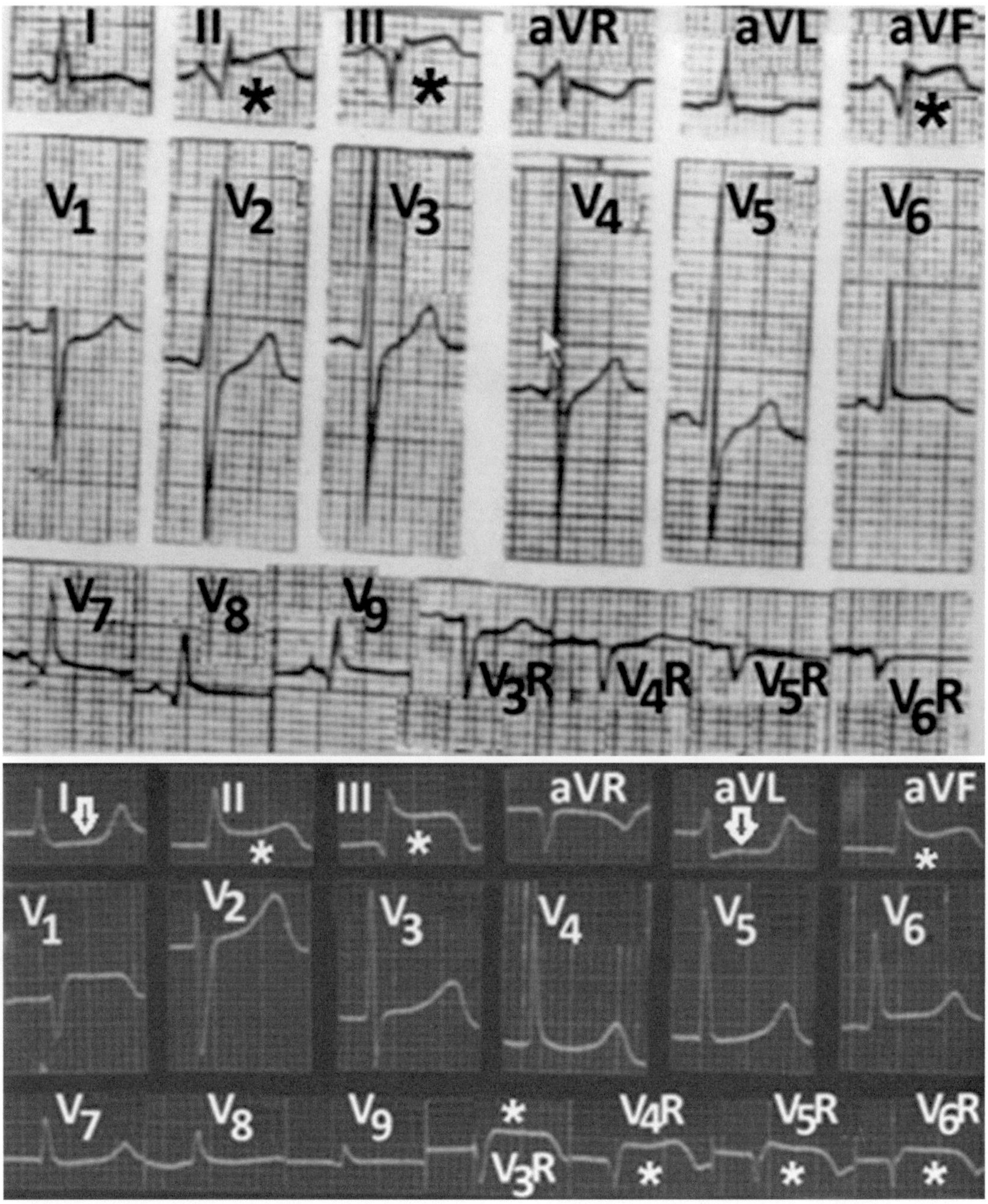

Fig. 23.29 (**a**) Electrocardiogram of inferior infarction (marked *) from occlusion of the left circumflex coronary artery. Leads I and aVL show isoelectric ST-segment. (**b**) Electrocardiogram showing inferior infarction with ST-segment depression in leads I and aVL (marked ⇩) suggestive of occlusion of right coronary artery with inferior and right ventricular infarction (marked *).

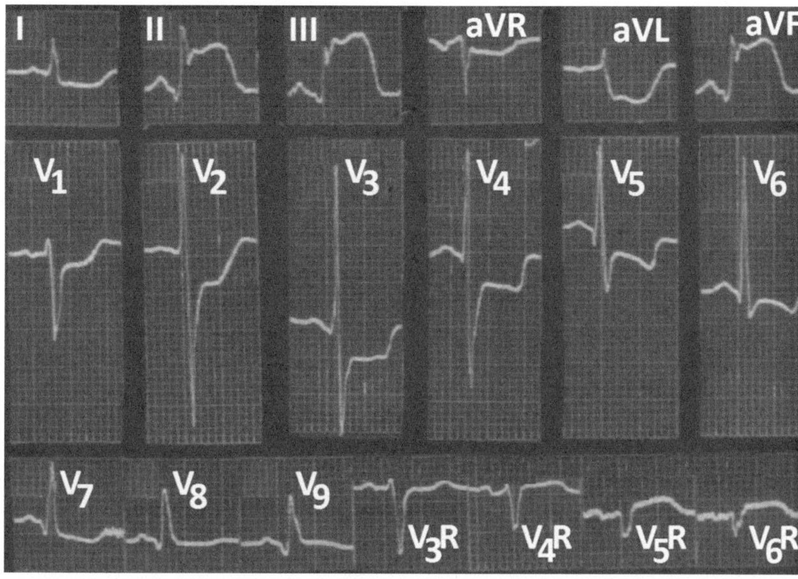

Fig. 23.30 Electrocardiogram from a case of right ventricular inferior infarction due to occlusion of the right coronary artery. ST-segment elevation in lead III is more than that in lead II. There is ST-segment depression in leads I and aVL and ST-segment elevation in leads V_5 R and V_6R.

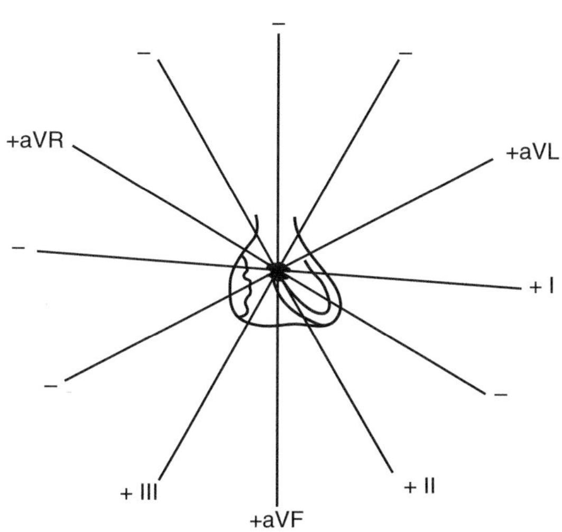

Fig. 23.31 Diagram showing that the left ventricular inferior wall faces positive pole of lead II and inferior wall of the right ventricle faces positive pole of lead III

6.3 ST-Segment in Leads V_1 to V_3

ST-segment elevation in lead V_1 supports the possibility of right ventricular infarction. This happens because of infarction of the anterior juxtaseptal portion of the right ventricle. ST-segment depression in leads V_2 and V_3 supports the diagnosis of right ventricular infarction [14] (Fig. 23.32). This happens because right ventricular posterior leads V_5 R and V_6 R are diagonally opposite to leads V_2 and V_3.

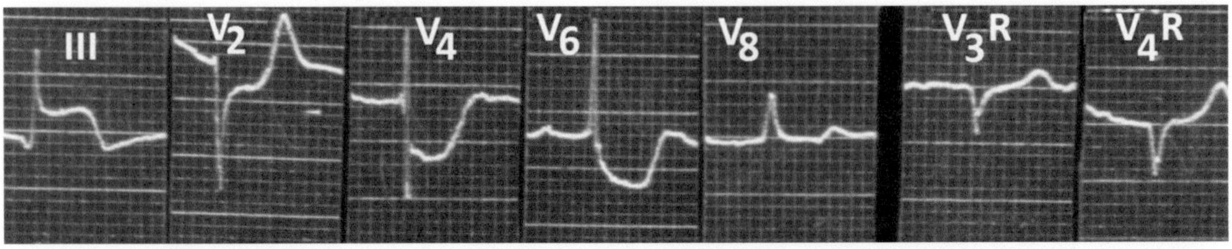

Fig. 23.32 Electrocardiogram from a case of right ventricular infarction showing ST-segment elevation in lead III with ST-segment depression in lead V_2

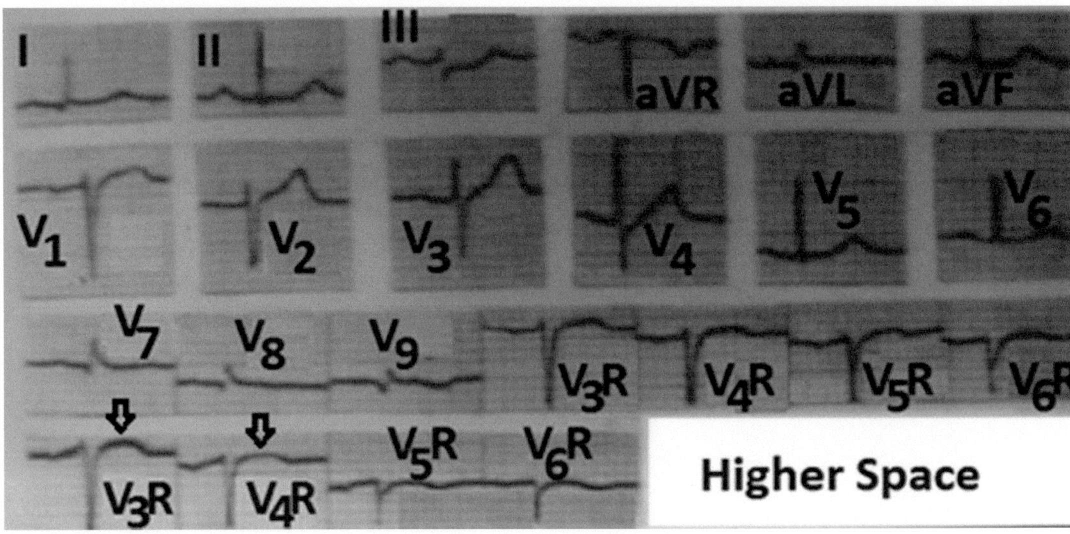

Fig. 23.33 Electrocardiogram showing ST-segment elevation in right-sided chest leads recorded from a higher intercostal space (marked ⇩)

7 [G] Isolated (Pure) Right Ventricular Infarction

This can occur in the following situations:

(1) Isolated occlusion of the conus branch of right coronary artery. ST-segment elevation is likely to be seen in lead V_3 R and V_4 R recorded in a higher intercostal space (Figs. 23.12 and 23.33).
(2) Isolated occlusion of the marginal branches of the right coronary artery (Fig. 23.34).

ST-segment elevation is seen only in leads V_3 R and V_4 R (Fig. 23.35).

At times, changes may be more prominent in leads V_1 to V_3 simulating infarction of the interventricular septum (Fig. 23.16).

(3) Distal occlusion of a nondominant right coronary artery. No ST-segment elevation is seen in leads V_3R, V_4R (Fig. 23.36).
(4) Distal occlusion of the dominant right coronary artery with good collateral flow to the posterior descending coronary artery (Fig. 23.37).
(5) Distal occlusion of the right coronary artery in a codominant circulation when the inferior surface of the left ventricle is supplied by posterior descending coronary artery arising from the left circumflex coronary artery. This produces isolated infarction of the inferior surface of the right ventricle (Fig. 23.19).

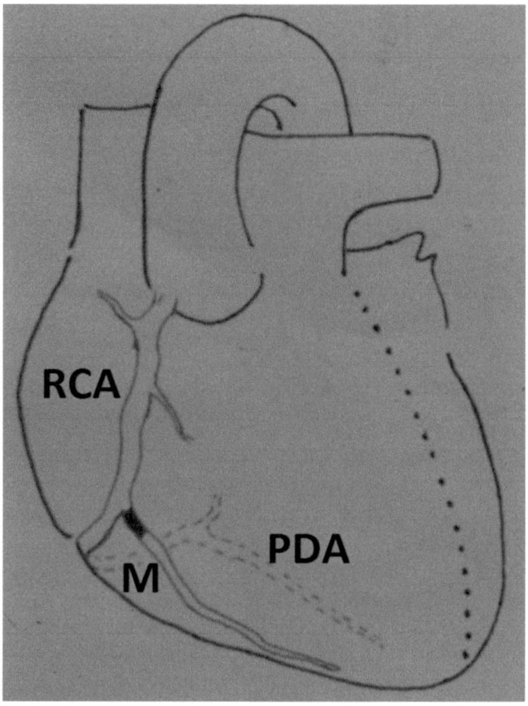

Fig. 23.34 Diagram showing the mechanism of isolated infarction of the right ventricular anterior wall due to occlusion of a marginal branch of RCA. *RCA* right coronary artery; *M* marginal branch; *PDA* posterior descending artery

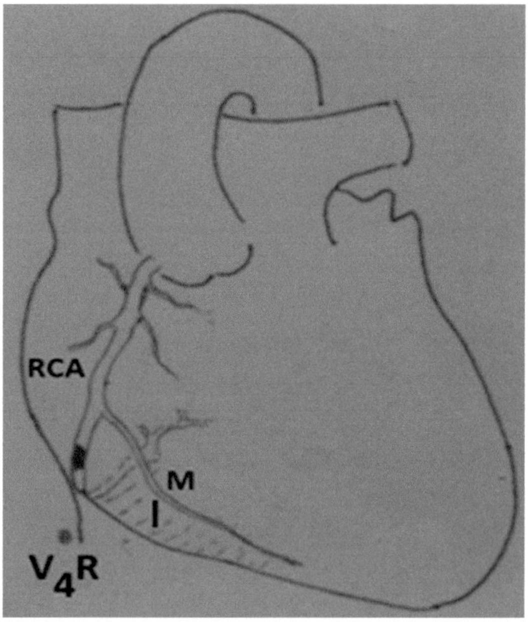

Fig. 23.36 Diagram showing distal occlusion of a nondominant right coronary artery. *M* marginal branch; *I* area of infarction

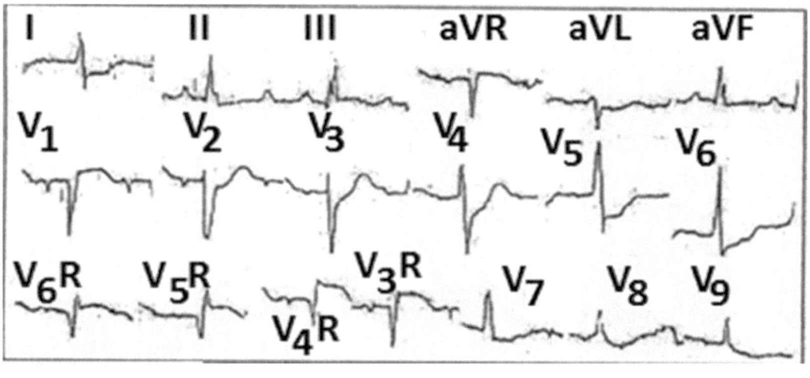

Fig. 23.35 Electrocardiogram showing ST-segment elevation in leads V_3R to V_6R only

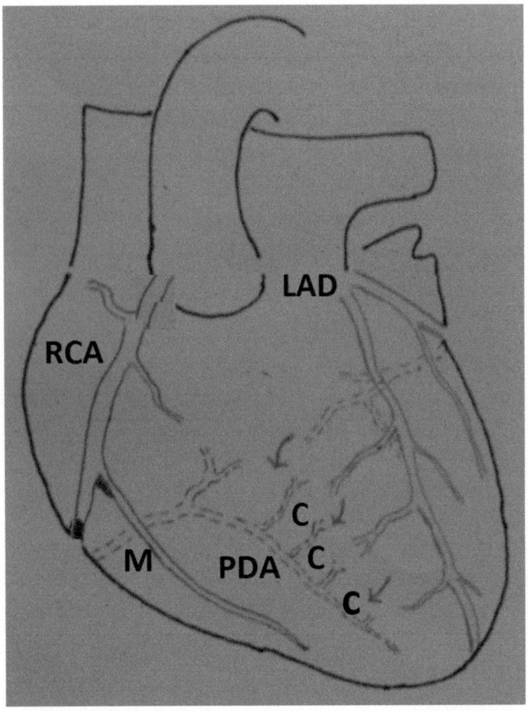

Fig. 23.37 Diagram showing distal occlusion of the dominant right coronary artery with good collateral flow to the posterior descending coronary artery. *RCA* right coronary artery; *M* marginal branch; *PDA* posterior descending coronary artery; *C* collaterals; *LAD* left anterior descending coronary artery.

Summary

- Occlusion of the conus branch of the right coronary artery can produce infarction of the right ventricular outflow tract. Infarction is picked up only if leads V_3R and V_4R are placed higher than their standard position.
- Occlusion of the artery to the sinoatrial node can produce sinus bradycardia.
- Occlusion of the marginal branches of the right coronary artery results in isolated infarction of the anterior surface of the right ventricle without any involvement of the inferior surface. Electrocardiographic changes are seen only in leads V_3R and V_4R without any change in the inferior leads.
- Occlusion of the right coronary artery distal to the marginal branches produces changes in the inferior leads and leads V_5R and V_6R.
- In dominant right circulation, occlusion of the right coronary artery can produce atrioventricular nodal block, right bundle branch block, and left ventricular posterior infarction in addition to inferior infarction.
- Occlusion of the proximal portion of the left anterior descending coronary artery can also produce changes in leads V_1, V_3R and even V_4R due to involvement of branches supplying juxtaseptal portion of the right ventricular anterior wall.
- Right precordial leads are diagonally opposite to leads I and aVL in the frontal plane and leads V_7 to V_9 in the horizontal plane. Leads V_3R to V_6R can, therefore, show changes if leads I, aVL and V_7 to V_9 are affected.
- In right ventricular infarction, ST-segment elevation is very small and may disappear within 24–48 h. Diagnosis of the right ventricular infarction is, therefore, likely to be missed if the ECG is not recorded within a few hours of onset of pain.
- In left ventricular anterior infarction, the magnitude of ST-segment elevation increases from lead V_1 to V_5 with minimal elevation in lead V_1. In infarction of the right ventricular anterior wall, ST-segment elevation is maximal in lead V_1 and decreases toward lead V_5.
- Infarction of the interventricular septum, left bundle branch block, ST-segment depression in leads I, aVL, V_7 to V_9, and Brugada syndrome can also produce an ST-segment elevation in leads V_1 to V_4R.
- Right ventricular hypertrophy and right bundle branch block can mask electrocardiographic findings of right ventricular infarction.
- Absence of ST-segment elevation in inferior leads, right axis deviation, and T wave inversion in leads V_1 to V_3 support the possibility of acute pulmonary embolism.

In the setting of inferior infarction:

- ST-segment elevation in leads I and aVL suggest occlusion of the left circumflex coronary artery. ST-segment depression in leads I and aVL support occlusion of the right coronary artery.

- ST-segment elevation in lead III exceeding that in lead II suggests occlusion of the right coronary artery. ST-segment elevation greater in lead II than in lead III supports occlusion of the left circumflex coronary artery.
- ST-segment elevation in lead V_1 with ST-segment depression in leads V_2 and V_3 support the diagnosis of right ventricular infarction.

– Isolated infarction of the right ventricle can occur due to isolated occlusion of the conus branch or the marginal branches of the right coronary artery; occlusion of the distal part of a nondominant right coronary artery can also produce isolated infarction of the right ventricle.

MCQs

Q1. Which electrocardiographic findings in a case of inferior infarction suggest occlusion of the right coronary artery?

(a) ST-segment elevation in lead I
(b) ST-segment elevation in lead II > in lead III
(c) ST-segment elevation in lead III > in lead II
(d) ST-segment depression in lead V_2

Q2. Which electrocardiographic findings in a case of inferior infarction suggest occlusion of the left circumflex coronary artery:

(a) ST-segment elevation in lead I
(b) ST-segment elevation in lead II > in lead III
(c) ST-segment elevation in lead V_1
(d) ST-segment elevation in lead aVL

Q3. Which electrocardiographic findings suggest occlusion of the right coronary artery?

(a) ST-segment elevation in lead V_1
(b) ST-segment elevation in leads V_2 and V_3
(c) ST-segment depression in leads V_2 ,V_3
(d) ST-segment elevation in lead V_4R

Q4. Normal configuration of QRS in lead V_4R is

(a) rs
(b) R
(c) Qr
(d) QS

Q5. In right ventricular infarction, ST-segment

(a) Remains persistently elevated
(b) Settles within few hours
(c) Settles within few days
(d) Settles within few months

Q6. Infarction of which wall of the left ventricle can affect electrocardiographic findings of right ventricular infarction in leads V_3 R and V $_4$R?

(a) Anterior wall
(b) Lateral wall
(c) Posterior wall
(d) None

Q7. Which electrocardiographic finding can suggest right ventricular infarction in the presence of right bundle branch block?

(a) Coved ST segment elevation in leads V_1 to V_4R
(b) Coved ST segment elevation in leads V_2 to V_4
(c) ST-segment elevation in lead III
(d) ST-segment depression in leads V_5 ,V_6

Q8. Which conditions can produce ST-segment elevation in lead V_1?

(a) Left bundle branch block
(b) Brugada syndrome

(c) Left ventricular hypertrophy with ST segment depression in leads V_4 to V_6
(d) All

Q9. Which finding in lead V_4R suggests occlusion of distal RCA?

(a) ST-segment elevation
(b) Upright T wave
(c) Inverted T wave
(d) Qr configuration

Q10. Which finding in lead V_4R suggests occlusion of proximal RCA?

(a) ST-segment elevation
(b) Upright T wave
(c) Inverted T wave
(d) rs configuration

Q11. Which finding in lead V_4R suggests occlusion of the left circumflex coronary artery?

(a) ST-segment elevation
(b) Upright T wave
(c) Inverted T wave
(d) rs configuration

Q12. In the setting of inferior infarction, which electrocardiographic findings suggest occlusion of the left circumflex coronary artery?

(a) ST-segment elevation in lead I
(b) ST-segment elevation in lead aVL
(c) ST-segment depression in lead V_4R
(d) All

Q13. Inferior infarction with ST-segment depression in leads V_1, V_2, and V_3 suggests infarction of

(a) Left ventricular anterior wall
(b) Interventricular septum
(c) Posterior wall of the left ventricle
(d) Left ventricular apex

Q14. Inferior infarction with ST-segment elevation in leads V_1 to V_6 suggests occlusion of

(a) Proximal left anterior descending coronary artery curving around the apex
(b) Dominant left circumflex artery
(c) Dominant right coronary artery
(d) Nondominant right coronary artery

Q15. ST-segment elevation localized to leads II, III, and aVF can occur with occlusion of

(a) Distal left anterior descending coronary artery wrapping around the apex
(b) Posterior descending coronary artery
(c) Proximal left circumflex coronary artery
(d) Proximal right coronary artery

Q16. ST-segment elevation in leads II, III, aVF, V_4 R, V_3R, and V_1 suggests occlusion of

(a) Proximal right coronary artery
(b) Mid right coronary artery
(c) Proximal left circumflex coronary artery
(d) Distal left circumflex coronary artery

Q17. ST-segment depression in lead I suggests occlusion of :\

(a) Proximal right coronary artery
(b) Proximal left circumflex coronary artery
(c) Proximal left anterior descending coronary artery
(d) Posterior descending coronary artery

Q18. Inferior infarction with ST-segment elevation in lead V_1 and ST segment depression in lead V_2 suggests

(a) Right ventricular infarction
(b) Left ventricular posterolateral infarction
(c) Left ventricular anteroseptal infarction
(d) Infarction of the left ventricular apex

Q19. In right ventricular infarction ST-segment elevation can extend up to

(a) Lead V_1
(b) Lead V_2
(c) Lead V_3
(d) Lead V_6

Answers:

(1) c,d (2) a,b,d (3) a,c,d (4) a (5) b (6) c (7) a,c (8) d (9) b (10) a (11) c (12) d (13) c (14) a (15) b (16) a (17) a (18) a (19) c

References

1. Levine DC. Pathway and functional significance of the coronary collateral circulation. Circulation. 1974;50:831–7.
2. Haupt HM, Mutchins GM, Moore GW. Right ventricular infarction- role of the moderator band artery in determining infarct size. Circulation. 1983;67:1268–72.
3. Anderson HR, Folk E, Niclson D. Right ventricular infarction: diagnostic accuracy of electrocardiographic right chest leads V_3R to V_7R investigated prospectively in 43 consecutive fatal cases from a coronary care unit. Br Heart J. 1989;61:514–20.
4. Anderson HR, Nielsen D, Falk E. Right ventricular infarction: diagnostic value of ST elevation in lead III exceeding that of lead II during inferior/posterior infarction and comparison with right chest leads V_3R to V_7 R. Am Heart J. 1989;117:82–5.
5. Croft CH, Nicod P, Corbett JR, et al. Detection of acute right ventricular infarction by right precordial electrocardiography. Am J Cardiol. 1982;50:421–7.
6. Morgera T, Albert E, Silvestri F, et al. Right precordial ST and QRS changes in the diagnosis of right ventricular infarction. Am Heart J. 1984;108:13–8.
7. Mittal SR, Sexena R. Letter to Editor. Int J Cardiol. 1991;32:115.
8. Erhardt LR, Sjogren A, Wahlberg I. Single right sided precordial lead in the diagnosis of right ventricular involvement in inferior myocardial infarction. Am Heart J. 1976;91:558–65.
9. Klein HO, Tordjman T, Ninio R, et al. The early recognition of right ventricular infarction. Diagnostic accuracy of the electrocardiographic V_4R. Circulation. 1983;67:558–65.
10. Geft IL, Shah PK, Rodriguez L, et al. ST elevations in leads V_1 to V_5 may be caused by right coronary artery occlusion and acute right ventricular infarction. Am J Cardiol. 1984;53:991–6.
11. Halkett JA, Commerford PJ, Millar RS. Right ventricular infarction mimicking extensive anterior infarction. Chest. 1986;90:617–8.
12. Bratt SH, Gorgels AP, Bar FW, et al. Value of ST- T segment in led V_4 R in inferior wall acute myocardial infarction to predict the site of coronary artery occlusion. Am J Cardiol. 1988;62:140–2.
13. Lew AS, Laramee P, Shah PK, et al. Ratio of ST segment depression in lead V_2 to ST segment elevation in lead aVF in evolving inferior acute myocardial infarction: an aid to the early recognition of right ventricular ischemia. Am J Cardiol. 1986;57:1047–51.
14. Mittal SR, Mantri V, Gokhroo RK. Masking of electrocardiographic features of right ventricular infarction by true posterior left ventricular infarction. Int J Cardiol. 1991;31:112–3.
15. Lew AS, Maddahi J, Shah PK, et al. Factors that determine the direction and magnitude of precordial ST segment deviation during inferior wall myocardial infarction. Am J Cardiol. 1985;55:883–8.
16. Kopelman HA, Forman MB, Wilson BH, et al. Right ventricular myocardial infarction in patients with chronic lung disease. Possible role of right ventricular hypertrophy. J Am Coll Cardiol. 1985;5:1302–7.
17. Goldhaber SZ. Pulmonary embolism. In: Zipes DP, Libby P, Bonow RO, Mann DL, Tomaselli GF, editors. Braunwald's heart disease. Elsevier: Philadelphia; 2019. p. 1681–98.
18. Rodrigo Fernandez J, Tapson V, Fuster V, Ibanez B. Pulmonary embolism. In: Fuster V, Harrington RA, Narula J, Eapen ZJ, editors. Hurst's the heart. Mc Graw Hill: New York; 2017. p. 1809–34.
19. Mittal SR, Garg R, Nayak RR. False localization of septal infarction to right sided chest leads in presence of left bundle branch block. Asian J Clin Cardiol. 2000;3:17–8.
20. Kataoka H, Tamura A, Yano S, et al. Intraventricular conduction delay in acute right ventricular ischemia. Am J Cardiol. 1989;64:94–5.
21. Rosenbaum MB, Girotti LA, Lazzari JO, et al. Abnormal Q wave in right sided chest leads provoked by onset of right bundle branch block in patients with anteroseptal infarction. Br Heart J. 1982;47:227–32.
22. Mittal SR. Right ventricular infarction: limitation of surface electrocardiography in diagnosis. Medicine Update API;1995:170–176.

Left Bundle Branch Block (Delay in Left Ventricular Depolarization Delay)

24

1 Nomenclature

Delay in left ventricular depolarization is a more appropriate term because

(1) Site of delay in conduction need not always be in the left bundle branch. It can be at any of the following sites [1]:- (Fig. 24.1)
   (a) His bundle fibers extending into the left bundle branch.
   (b) Left bundle branch per se.
   (c) Left anterosuperior fascicle + left posteroinferior fascicle.
   (d) Purkinje network of the left ventricle.
   (e) Any combination of the above.
(2) Underlying pathophysiology is usually slow conduction rather than any anatomical "block" in conduction.
(3) Electrocardiographic findings may be transient (Fig. 24.2) and/or rate dependent (Figs. 24.3, 24.4, 24.5, 24.6 and 24.7).
(4) QRS duration may progress (Fig. 24.8) from <0.12 s to 0.14 s or even more. This suggests progressively increasing delay in depolarization rather than a block in depolarization.

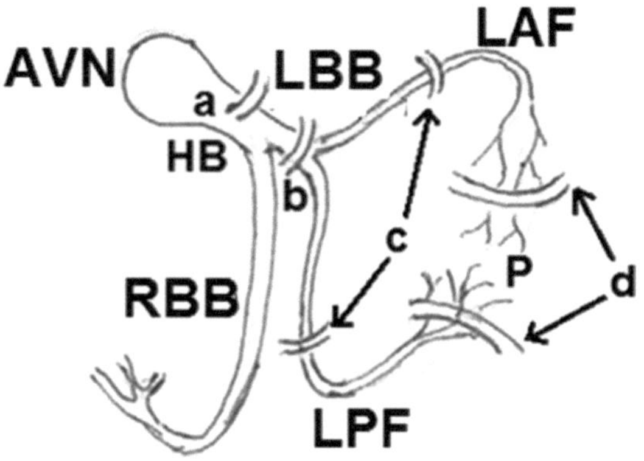

Fig. 24.1 Diagram showing sites for delay in depolarization of the left ventricle. AVN—Atrioventricular node, HB—His bundle, LAF—Left anterior fascicle, LPF—Left posterior fascicle, P—Purkinje network of the left ventricle, RBB—Right bundle branch. (Reproduced with permission from: Mittal SR. Left Ventricular Depolarization Delay. Cardiology Today- 2015;19:31–39)

S. Mittal, *Insights into Electrocardiograms with MCQs*,
https://doi.org/10.1007/978-981-99-0127-2_24

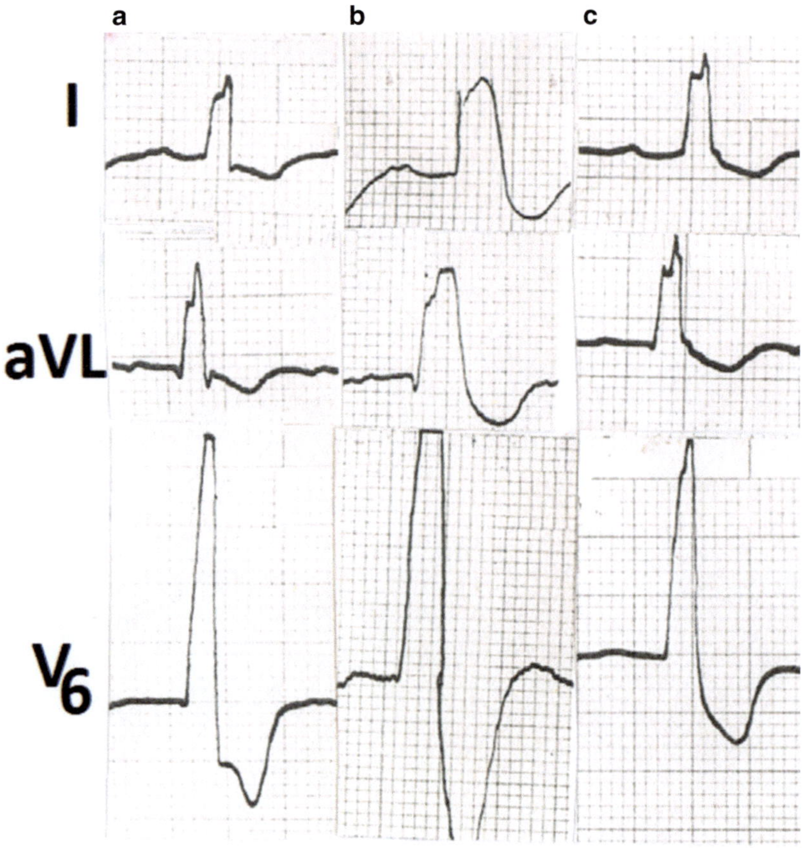

Fig. 24.2 Electrocardiograms showing transient increase in delay in left ventricular depolarization (**b**). (Reproduced with permission from: Mittal SR. Left Ventricular Depolarization Delay. Cardiology Today- 2015;19:31–39)

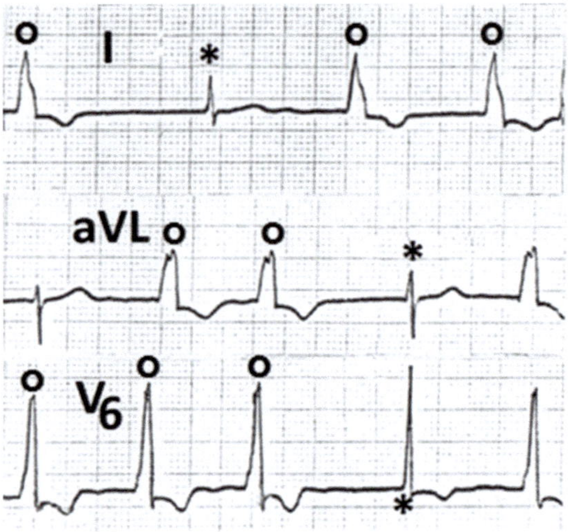

Fig. 24.3 Electrocardiogram showing atrial fibrillation with left ventricular depolarization delay in beats at fast rate (marked O) and normal QRS in beats after pause (marked *) (tachycardia-dependent LBBB). (Reproduced with permission from Mittal SR. Left Ventricular Depolarization Delay. Cardiology Today- 2015;19:31–39)

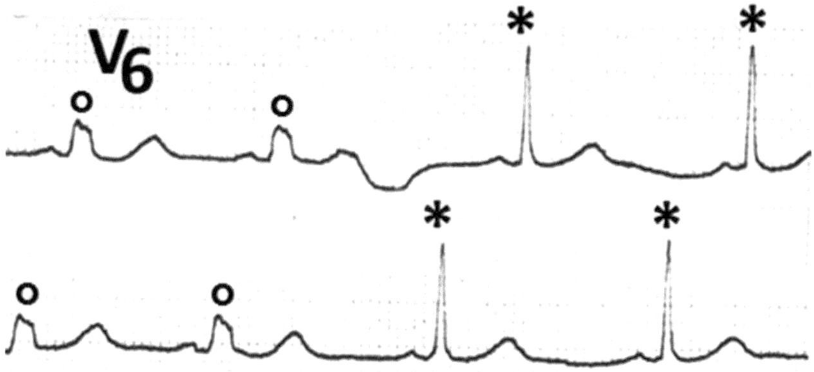

Fig. 24.4 Electrocardiogram (lead V6) showing tachycardia-dependent LBBB (marked O). QRS becomes normal on slight prolongation of the R-R interval (marked *). (Reproduced with permission from: Mittal SR. Left Ventricular Depolarization Delay. Cardiology Today- 2015;19:31–39)

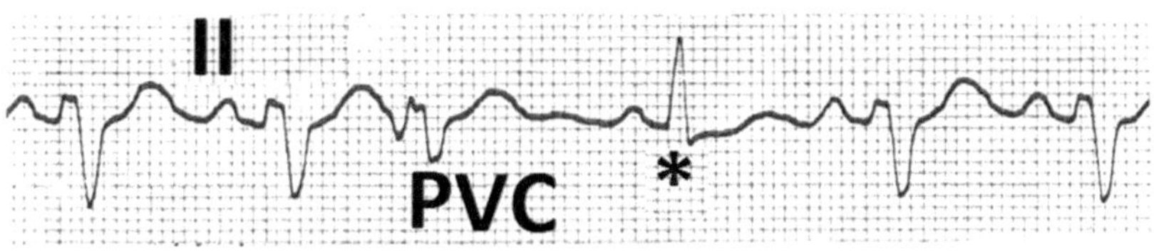

Fig. 24.5 Electrocardiogram (lead II) showing normalization of the QRS complex (marked *) after the pause following a premature ventricular contraction (PVC). It suggests that the underlying LBBB is tachycardia dependent. (Reproduced with permission from: Mittal SR. Left Ventricular Depolarization Delay. Cardiology Today- 2015;19:31–39)

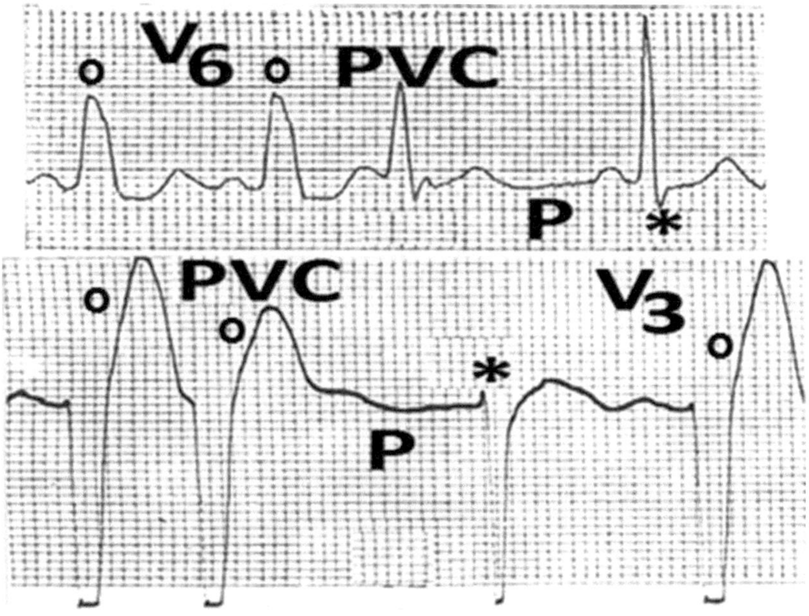

Fig. 24.6 Electrocardiogram (lead V6 and lead V3) showing LBBB (marked O) and normalization of the QRS (marked *) in the beat following post-ectopic pause (marked P). PVC—Premature ventricular contraction. (Reproduced with permission from: Mittal SR. Left Ventricular Depolarization Delay. Cardiology Today- 2015;19:31–39)

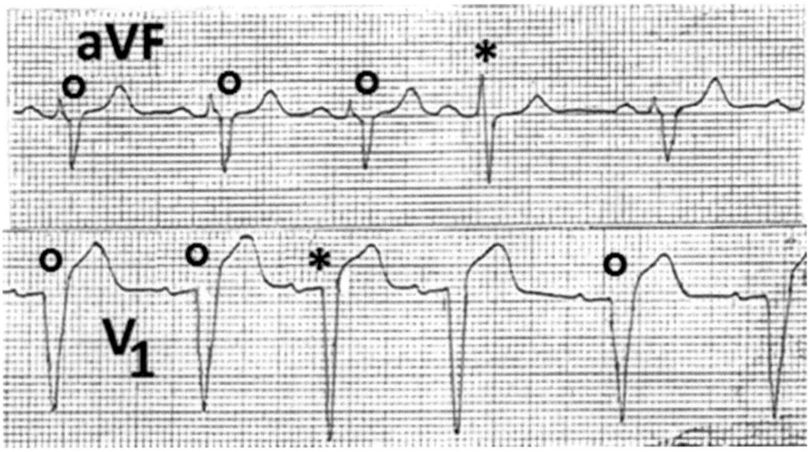

Fig. 24.7 Electrocardiogram showing bradycardia-dependent LBBB (marked O). QRS following short cycle (marked *) is narrow. (Reproduced with permission from: Mittal SR. Left Ventricular Depolarization Delay. Cardiology Today- 2015;19:31–39)

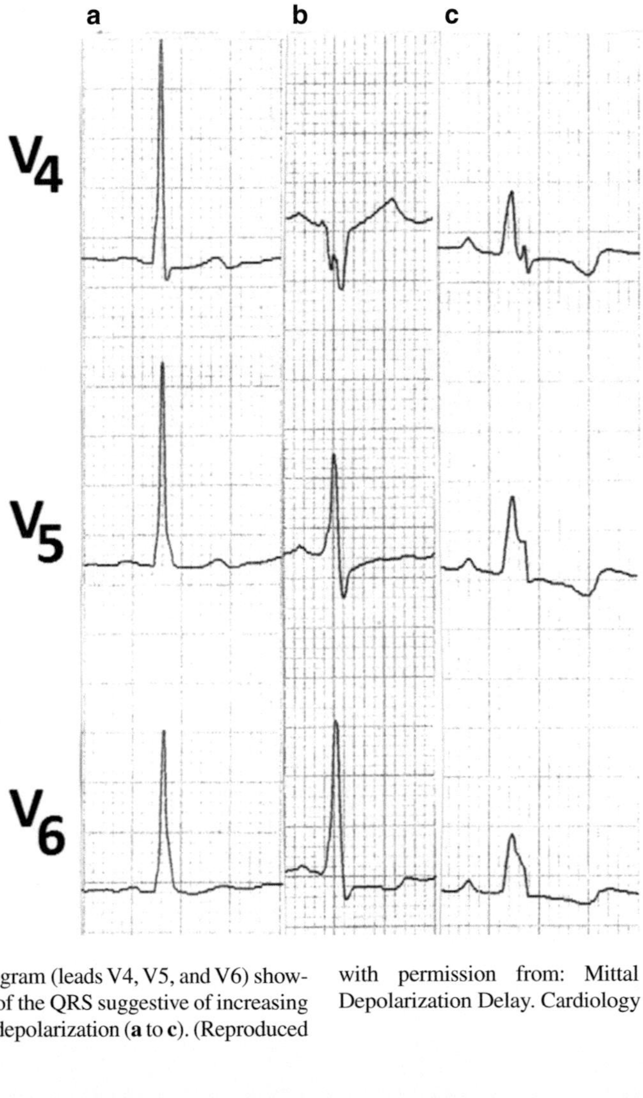

Fig. 24.8 Electrocardiogram (leads V4, V5, and V6) showing increasing duration of the QRS suggestive of increasing delay in left ventricular depolarization (**a** to **c**). (Reproduced with permission from: Mittal SR. Left Ventricular Depolarization Delay. Cardiology Today- 2015;19:31–39)

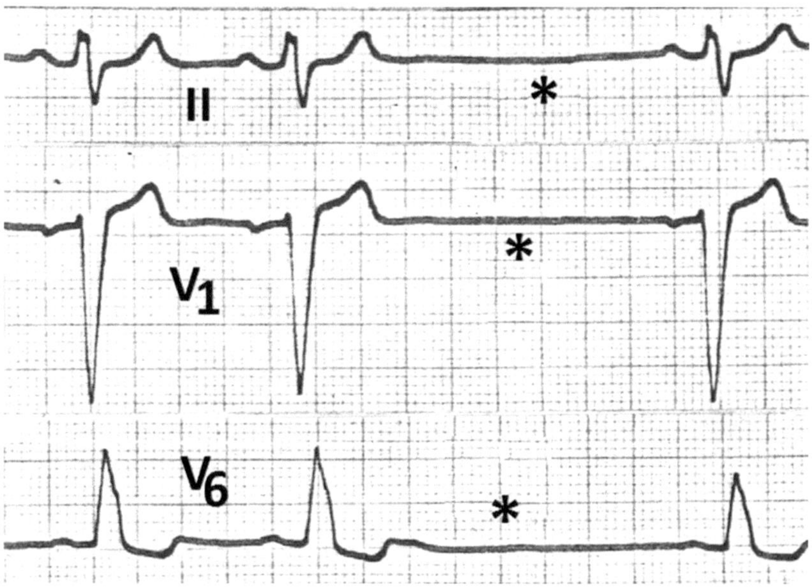

Fig. 24.9 Electrocardiogram showing delayed depolarization of the left ventricle with intermittent sinoatrial exit block (marked *) suggestive of diffuse conduction tissue disease. (Reproduced with permission from: Mittal SR. Left Ventricular Depolarization Delay. Cardiology Today- 2015;19:31–39)

2 Etiology

It is almost always in the presence of structural heart disease. Common causes are [2]

(1) Coronary artery disease.
(2) Cardiomyopathy.
(3) Left ventricular hypertrophy.
(4) Idiopathic degeneration of the conduction tissue (Fig. 24.9).
(5) Encroachment form calcific aortic valve stenosis.
(6) Infiltrative myocardial diseases.
(7) Hyperkalemia.
(8) Antiarrhythmic drugs—e.g., Flecainide.
(9) Rarely, this electrocardiographic finding may be seen in the presence of atriofascicular accessory pathway.

3 Electrocardiographic Diagnosis

Delay in left ventricular depolarization may appear acutely (e.g., in the setting of acute myocardial infarction) or may develop gradually. In gradual development of delay in left ventricular depolarization, following changes appear in sequence (Fig. 24.10):

(1) Loss of septal q wave. Due to loss of septal q, leads V_5 and V_6 show only R wave. Lead V1 may have a QS configuration or may have a small initial r (rs configuration) due to depolarization of the right ventricular free wall. Loss of septal q may also be seen in septal fibrosis, small septal infarction, or left ventricular pressure overload [3].
(2) Subsequently, there is slurring or notching on the upstroke of the R wave.
(3) It is followed by progressive widening of the QRS complex with delay in time to peak of the R wave in left precordial leads (V_5 and V_6).
(4) Left ventricular depolarization delay is considered as advanced when:
    (a) QRS duration is ≥120 ms. Some authorities feel that duration of >140 ms for males and > 120 ms for females should be considered a prerequisite for definition especially when considering resynchronization therapy [4].
    (b) Time to peak of R wave is >60 ms in leads V_5 and V_6.

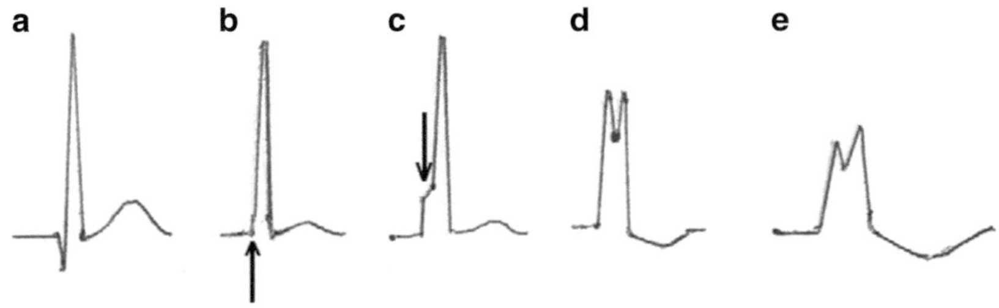

Fig. 24.10 Diagram showing the effect of progressively increasing delay in left ventricular depolarization. (**a**) Normal (**b**) Loss of septal q (marked ↑) (**c**) Mild broadening of the QRS complex with slurring on the upstroke (marked ↓). (**d**) & (**e**) Progressive widening of the QRS and inversion of the T wave. (Reproduced with permission from: Mittal SR. Left Ventricular Depolarization Delay. Cardiology Today- 2015;19:31–39)

4 Genesis of Electrocardiographic Findings

(1) Normally the interventricular septum is depolarized from left to right through the septal fibers of the left bundle branch. Initial QRS vector is, therefore, directed from left to right. This produces initial q wave in leads I, aVL, V5, and V6 and initial r wave in leads V1 and V2. In delayed conduction in the left bundle branch, depolarization of the interventricular septum occurs through the right bundle branch. Interventricular septum is, therefore, depolarized from right to left. Initial QRS vector is, therefore, directed from right to left. This results in appearance of an initial r wave (loss of normal septal q wave) in leads I, V5, and V6 and appearance of an initial q wave in leads V1 and V2. A narrow q wave may be present in lead aVL in the absence of myocardial pathology. [5] An initial small r wave may persist in leads V1 and V2 in some patients due to depolarization of the right ventricle (Fig. 24.11a).

Subsequently the unopposed depolarization of the right ventricle transiently moves the QRS complex vector towards right. This may produce a notch in the QRS complex (Fig. 24.11b).

This notch may not be apparent in all cases. Finally, the left ventricle is depolarized slowly through the myocardium (not via the conduction tissue) which conducts slowly. This produces a broad terminal R wave in leads I, aVL, V5, V6, and a broad S wave in leads V1 and V2 (Fig. 24.11c). Thus leads I, aVL, V5, and V6 show rR' configuration and leads V1 and V2 show a QS (or rS) configuration.

Representative electrocardiogram is shown in Fig. 24.12.

Frontal plane and horizontal plane QRS vectors are shown in Fig. 24.13a, b.

The notch in the QRS complex may not be apparent in all cases and QRS complex may appear monophasic with prolonged time to peak of R wave. ST-segment and T wave vectors are directed opposite to the terminal QRS vector with the result that ST-segment and T wave are upright in leads V1 and V2 and inverted in leads V5 and V6. However, positive T wave in leads with upright QRS (I, aVL, and V_6) (positive concordance) may be normal [5] (Fig. 24.14).

ST-segment may be straight or may show minimal curve (Fig. 24.15).

4.1 Leads V3 and V4 (The Transition Zone)

Initial QRS vector is parallel to these leads. Initial r wave may, therefore, be smaller than in leads V1 and V2. Apparently it may look like "reverse progression." However, total amplitude of the QRS complex is also diminished. Ratio of the r wave to the total amplitude of the QRS complex is, therefore, not significantly diminished (Fig 24.14).

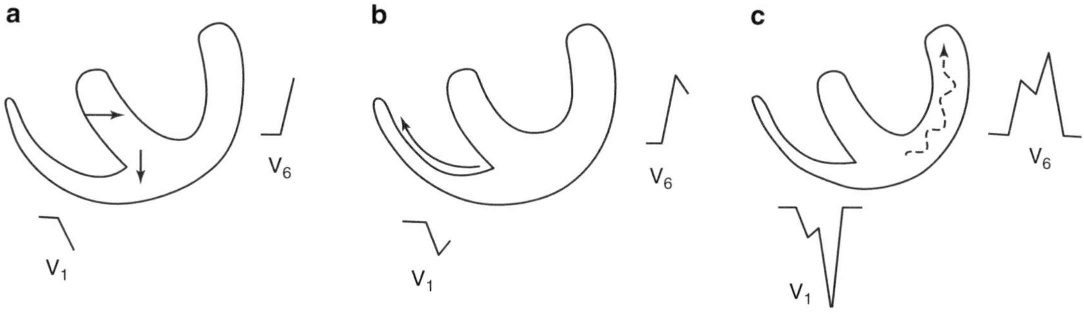

Fig. 24.11 (**a**) Diagram showing septal depolarization and its effect on the leads V_1 and V_6 in left ventricular depolarization delay. (Reproduced with permission from: Mittal SR. Left Ventricular Depolarization Delay. Cardiology Today- 2015;19:31–39). (**b**) Diagram showing right ventricular depolarization and its effect on the leads V_1 and V_6 in left ventricular depolarization delay. (Reproduced with permission from : Mittal SR. Left Ventricular Depolarization Delay. Cardiology Today- 2015;19:31-39). (**c**) Diagram showing slow left ventricular depolarization and its effect on the leads V_1 and V_6 in left ventricular depolarization delay. (Reproduced with permission from: Mittal SR. Left Ventricular Depolarization Delay. Cardiology Today- 2015;19:31–39)

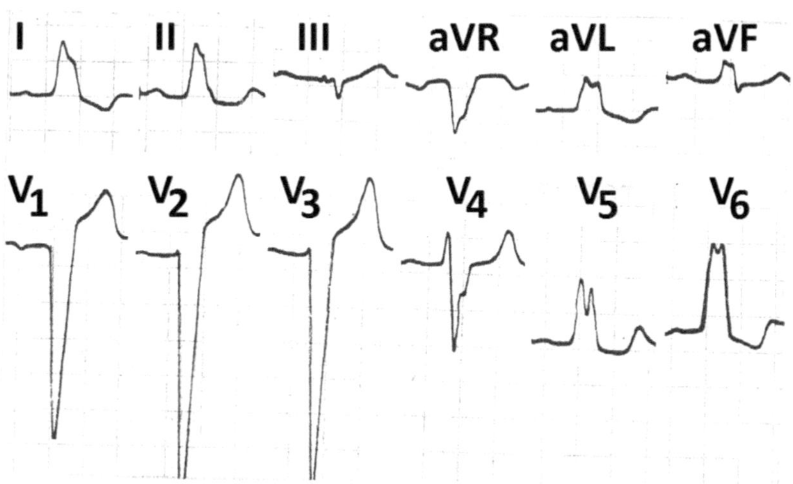

Fig. 24.12 Electrocardiogram showing findings of uncomplicated left ventricular depolarization delay. (Reproduced with permission from: Mittal SR. Left Ventricular Depolarization Delay. Cardiology Today- 2015;19:31–39)

Fig. 24.13 Diagram showing QRS vector and its effect on the QRS complex in different leads (**a**) in frontal plane and (**b**) in horizontal plane. (Reproduced with permission from: Mittal SR. Left Ventricular Depolarization Delay. Cardiology Today-2015;19:31–39)

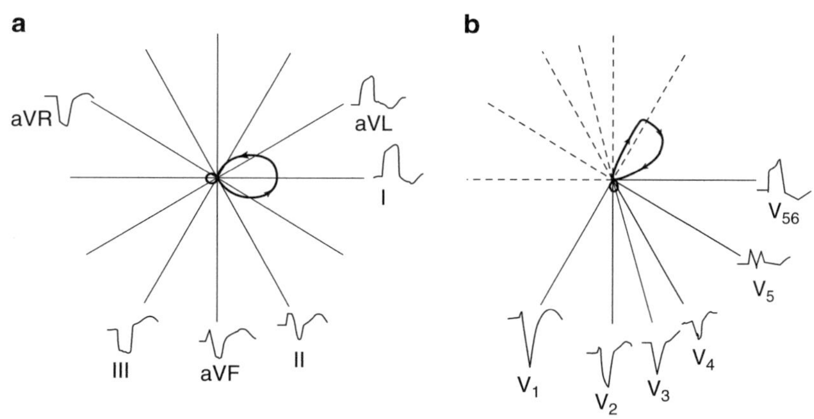

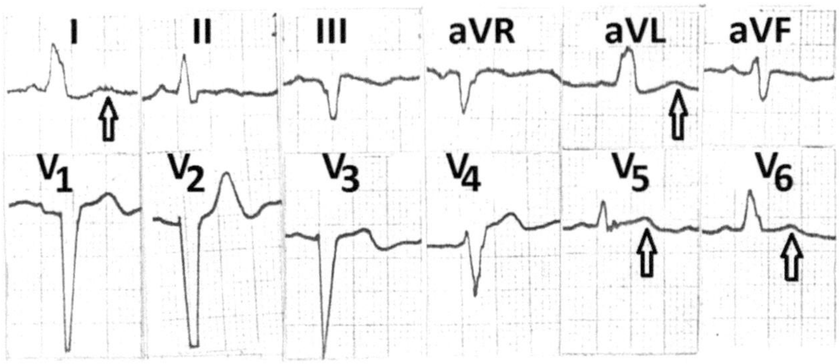

Fig. 24.14 Electrocardiogram showing left ventricular depolarization delay with positive T waves (marked ⇧) in leads I, aVL, V5, and V6 (Positive concordance). (Reproduced with permission from: Mittal SR. Left Ventricular Depolarization Delay. Cardiology Today- 2015;19:31–39)

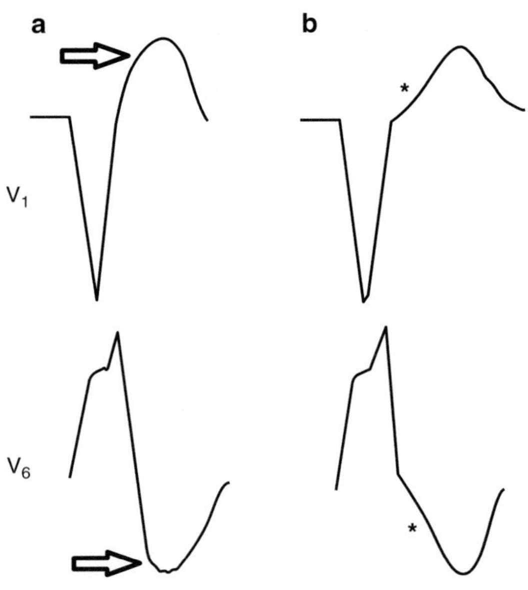

Fig. 24.15 Diagram showing different configurations of the ST-segment in leads V1 andV6 (**a**) straight (marked ⇨) (**b**) slightly curved (marked *). (Reproduced with permission from: Mittal SR. Left Ventricular Depolarization Delay. Cardiology Today- 2015;19:31–39)

4.2 Leads V7 to V9

These leads show rR' configuration or a monophasic slurred R. The ST-segment is downsloping and the T waves are inverted (Fig. 24.16a).

4.3 Leads V3 to V6R

These leads show rS (or QS) configuration with elevated ST-segment and upright T waves (Fig. 24.16b).

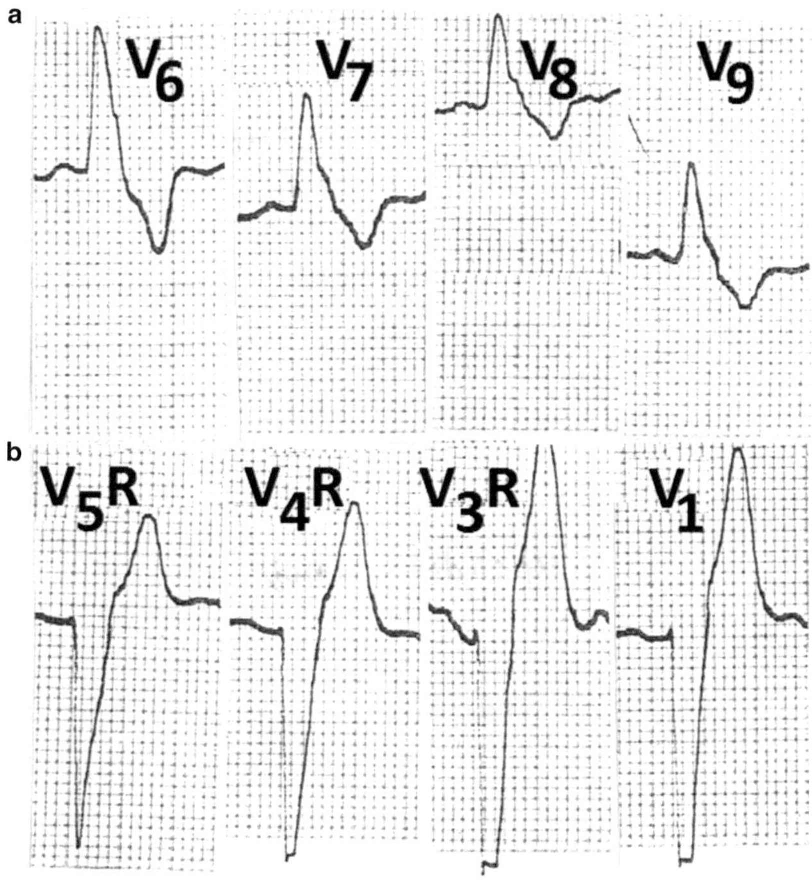

Fig. 24.16 Electrocardiograms from a case of left ventricular depolarization delay showing QRS complex , ST-segment, and T wave configuration in (**a**) posterior chest leads (V7 to V9) and (**b**) in the right-sided chest leads (V3R to V5R). (Reproduced with permission from: Mittal SR. Left Ventricular Depolarization Delay. Cardiology Today- 2015;19:31–39)

5 Other Electrocardiographic Findings

5.1 S Wave in Lead V6

Properly placed V6 does not show a prominent S wave in uncomplicated left bundle branch block. S wave may be present in

(a) Lateral infarction (Fig. 24.17)
(b) Dominant left anterosuperior depolarization delay.
(c) Hyperkalemia.
(d) Clockwise rotation (Fig. 24.18).

5.2 Left Axis Deviation

It may be present due to any of the following combinations:

(a) Delay in the main left bundle branch + additional delay in the left anterosuperior fascicle (Fig. 24.19a).

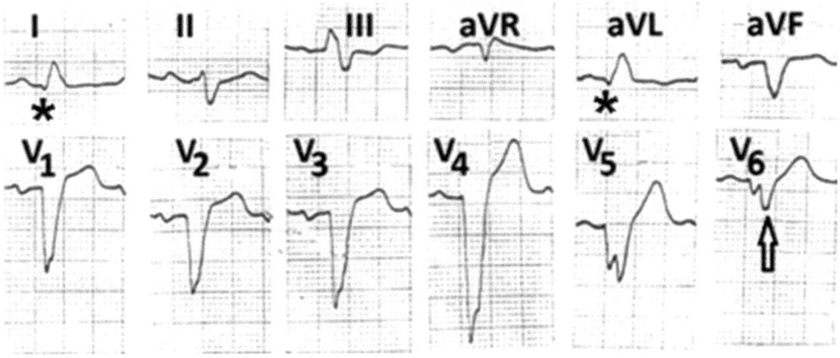

Fig. 24.17 Electrocardiogram from a case of left ventricular depolarization delay showing S wave in lead V6 (marked ⇧) due to anterolateral infarction. QRS duration is 0.18 seconds. Leads I and aVL show q wave (marked *). (Reproduced with permission from: Mittal SR. Left Ventricular Depolarization Delay. Cardiology Today- 2015;19:31–39)

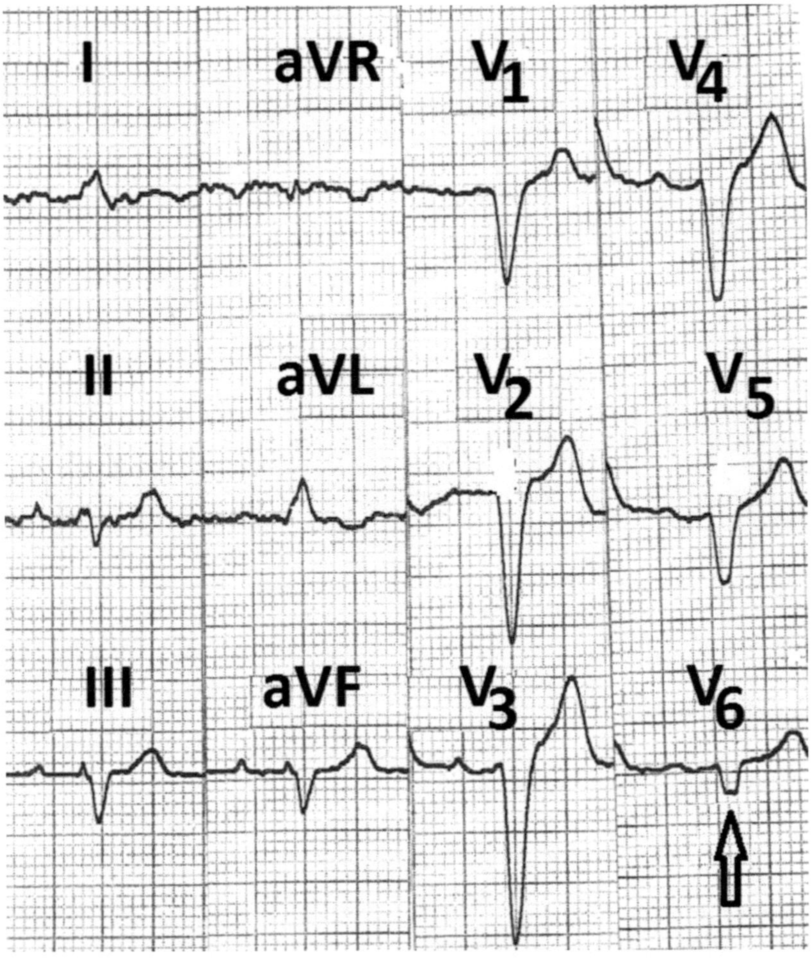

Fig. 24.18 Electrocardiogram from a case of left ventricular depolarization delay showing S wave in lead V6 (marked ⇧) due to clockwise rotation. Limb leads and leads V5 and V6 show low voltage due to emphysema. (Reproduced with permission from: Mittal SR. Left Ventricular Depolarization Delay. Cardiology Today- 2015;19:31–39)

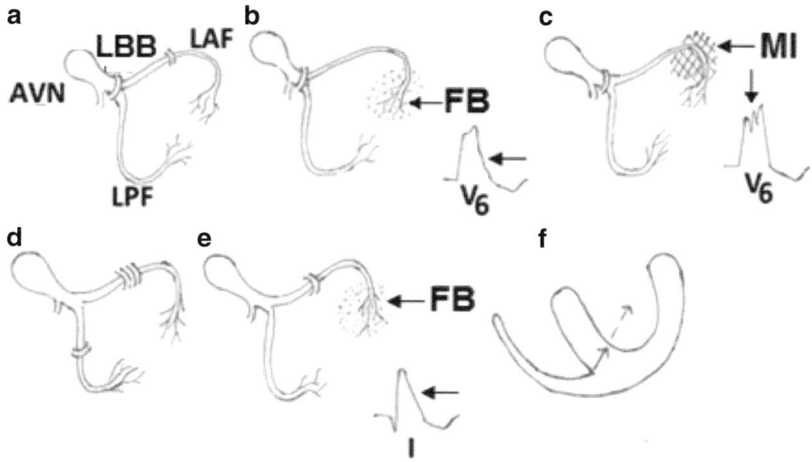

Fig. 24.19 Diagram showing causes of left axis deviation in the presence of LBBB. (**a**) LBBB + left anterior fascicular block (**b**) LBBB + focal block (marked FB) in anterosuperior region. There is slurring of the terminal part of the QRS complex (marked as ←)(**c**) LBBB + infarction in anterosuperior region (marked MI). There is notching on the top of the QRS complex (marked ↓). (**d**) LPFB + dominant left anterior fascicular block (**e**) left anterior fascicular block + anterosuperior focal block (marked FB). Lead I shows qR configuration with terminal slurring of the QRS complex (marked ←). (**f**) Depolarization of the left ventricle form the right ventricular apex rather than form mid septum. (Reproduced with permission from: Mittal SR. Left Ventricular Depolarization Delay. Cardiology Today- 2015;19:31–39)

(b) Delay in the main left bundle branch + additional left anterosuperior focal block (Fig. 24.19b). It produces notching and slurring of the R wave towards its end in leads I, aVL, V5, and V6.
(c) Delay in the main left bundle branch + additional anterolateral myocardial infarction (Fig. 24.19c). It produces notching and slurring of the R wave on its top.
(d) Delay in both left-sided fascicles with greater delay in the anterosuperior region (Fig. 24.19d).
(e) Delay in the anterosuperior fascicle with additional focal block in the anterolateral region (Fig. 24.19e). This produces a qR in lead I. R wave shows terminal slurring.
(f) Exit of impulse from right ventricular apex rather than from mid septum (Fig. 24.19f).

The representative electrocardiograms of left ventricular depolarization delay with left axis deviation are shown in Figs. 24.20 and 24.21.

In the setting of acute myocardial infarction, delay in anterosuperior fascicle may precede development of LBBB (Fig. 24.22).

5.3 Right Axis Deviation

It can be due to

(a) Biventricular enlargement. LBBB + delayed conduction over the right bundle branch (Fig. 24.23a).
(b) Delay in the left bundle branch + additional delay in the left posteroinferior fascicle (Fig. 24.23b).
(c) Delay in both left-sided fascicles with greater delay in the posteroinferior fascicle (Fig. 24.23c).

Representative electrocardiograms are shown in Figs. 24.24 and 24.25.

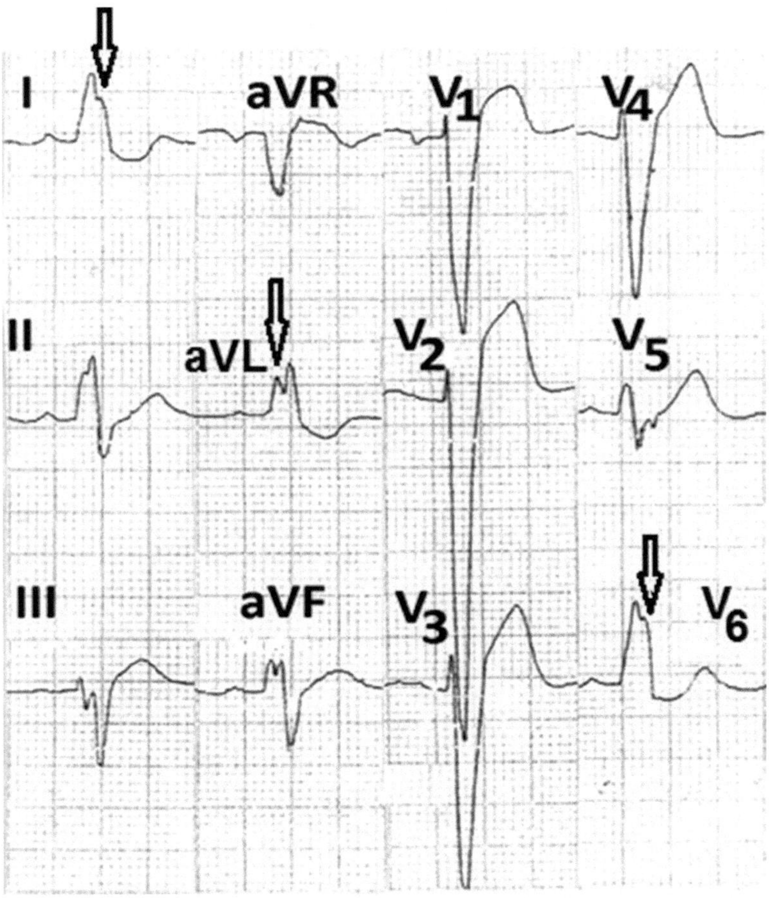

Fig. 24.20 Electrocardiogram showing left ventricular depolarization delay with left axis deviation. There is notch on the top of the QRS complex (marked ⇩) (Reproduced with permission from: Mittal SR. Left Ventricular Depolarization Delay. Cardiology Today- 2015;19:31–39)

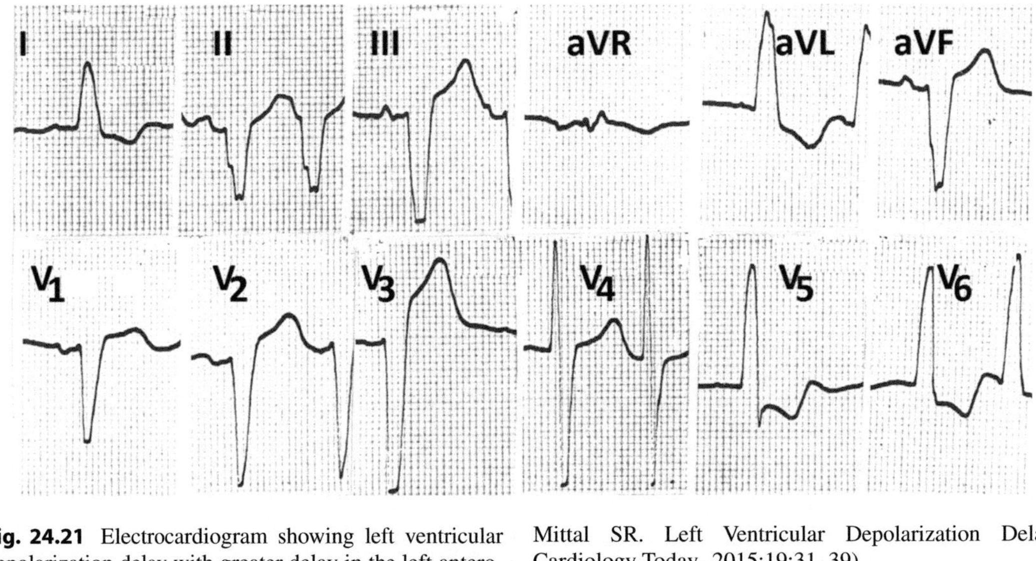

Fig. 24.21 Electrocardiogram showing left ventricular depolarization delay with greater delay in the left antero-superior fascicle. (Reproduced with permission from: Mittal SR. Left Ventricular Depolarization Delay. Cardiology Today- 2015;19:31–39)

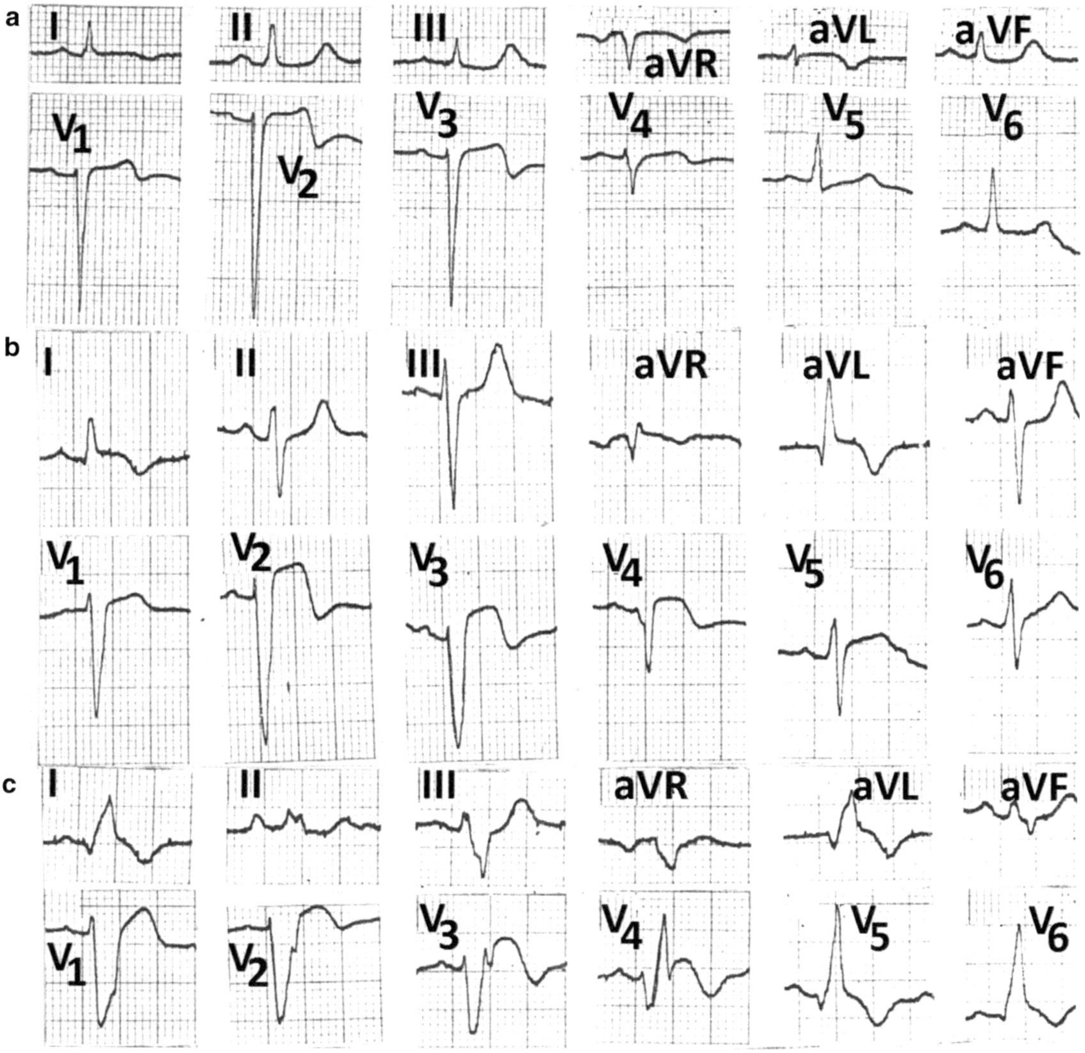

Fig. 24.22 Electrocardiograms showing (**a**) anterior infarction (**b**) fresh LAFB (**c**) fresh LBBB. (Reproduced with permission from: Mittal SR. Left Ventricular Depolarization Delay. Cardiology Today- 2015;19:31–39)

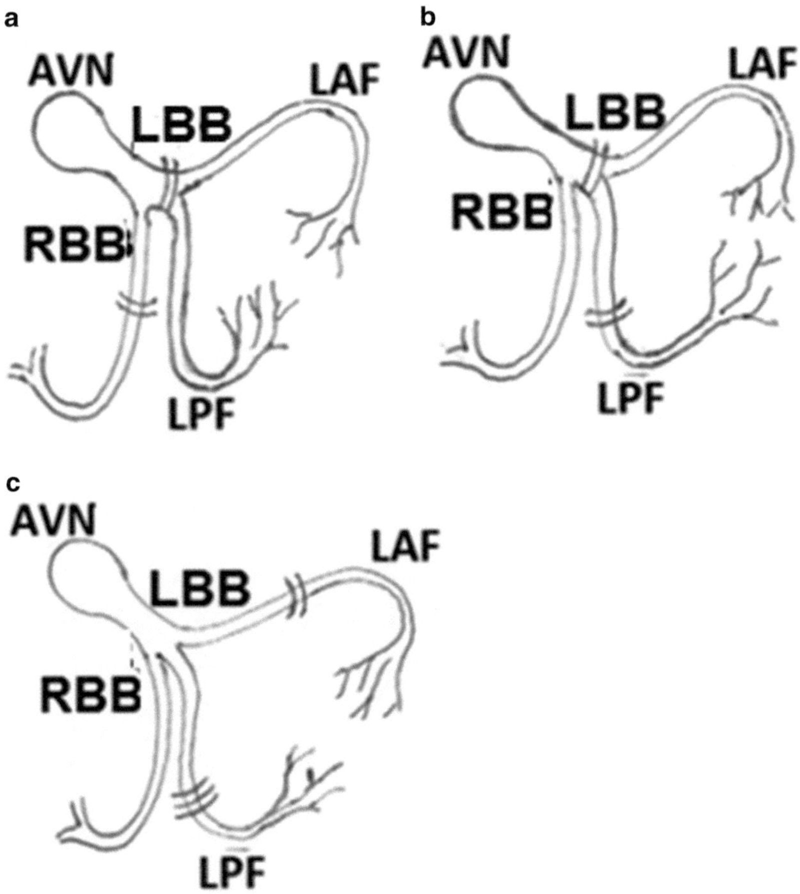

Fig. 24.23 Diagram showing causes of delayed left ventricular depolarization with right axis deviation (**a**) LBBB+ RV depolarization delay (**b**) LBBB + additional delay in left inferoposterior fascicle (LPF) (**c**) Delay in left anterosuperior fascicle (LAF) with greater delay in left inferoposterior fascicle (LPF). (Reproduced with permission from: Mittal SR. Left Ventricular Depolarization Delay. Cardiology Today- 2015;19:31–39)

5.4 Prominent Negative Deflection of P Wave in Lead V_1 and/or Broad and Notched P Wave in Lead II or Precordial Leads (Fig. 24.24)

It could be due to:

(a) Left atrial enlargement due to:
   (i) Left ventricular hypertrophy.
   (ii) Left ventricular failure.
   (iii) Mitral regurgitation.
(b) Interatrial block.

5.5 Intermittent Narrow QRS

It can be due to:

(a) Left ventricular ectopy partly neutralizing the delay in left ventricular depolarization (Figs. 24.27 and 24.28).
(b) Rate-dependent LBBB with bradycardia-dependent normalization of QRS.

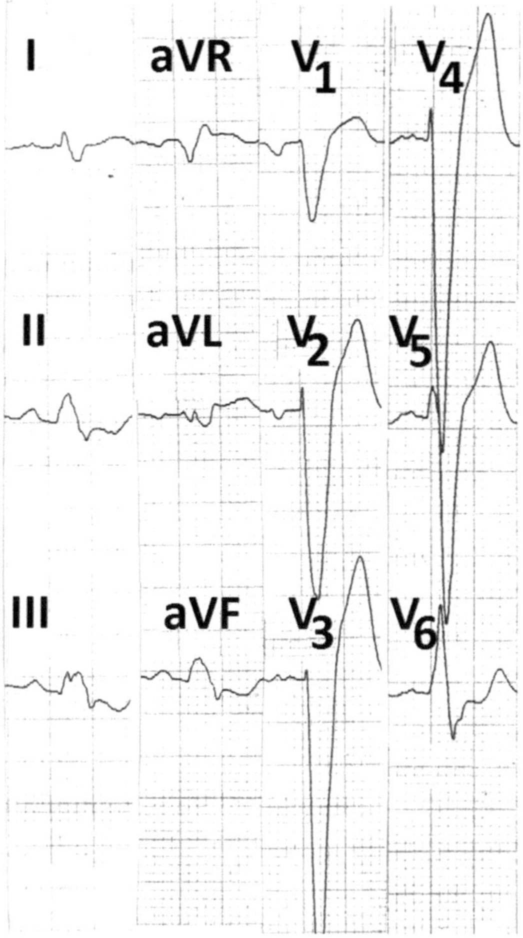

Fig. 24.24 Electrocardiogram showing LBBB + right axis deviation. (Reproduced with permission from: Mittal SR. Left Ventricular Depolarization Delay. Cardiology Today- 2015;19:31–39)

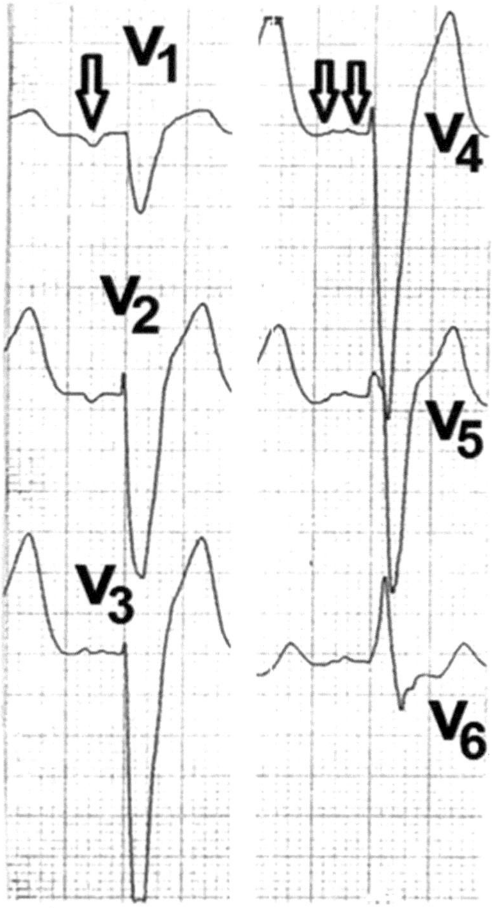

Fig. 24.26 Electrocardiogram showing LBBB with evidence of left atrial enlargement. Lead V1 shows prominent negative P wave (marked ⇩). Lead V4 shows broad and notched P wave (marked ⇩⇩). (Reproduced with permission from: Mittal SR. Left Ventricular Depolarization Delay. Cardiology Today- 2015;19:31–39)

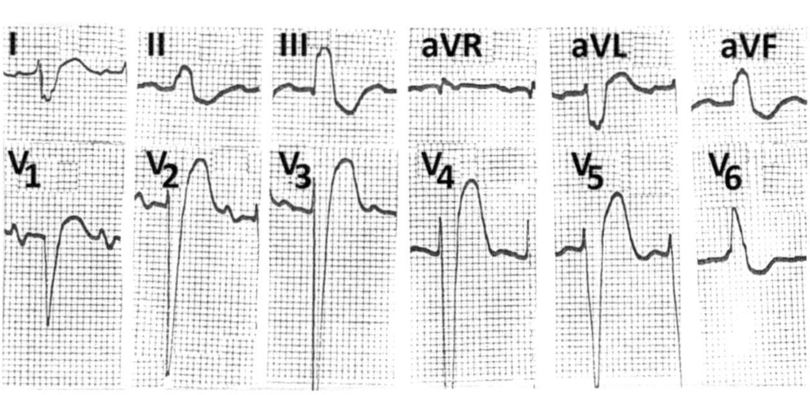

Fig. 24.25 Electrocardiogram showing LBBB + additional delays in left inferoposterior fascicle. (Reproduced with permission from: Mittal SR. Left Ventricular Depolarization Delay. Cardiology Today- 2015;19:31–39)

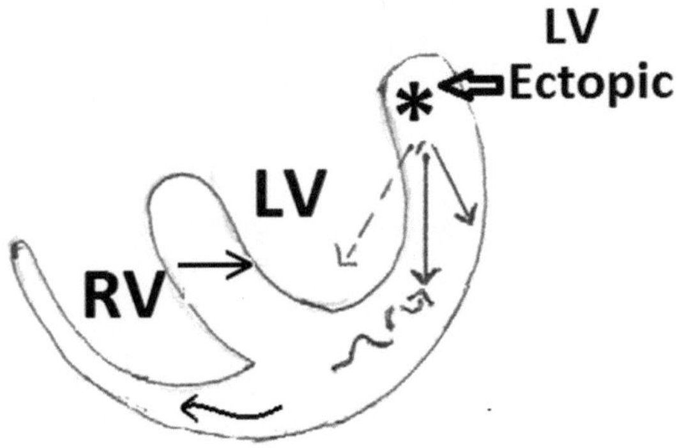

Fig. 24.27 Diagram showing that an ectopy arising in the left ventricle (marked *) can partly neutralize the delay in left ventricular depolarization. (Reproduced with permission from: Mittal SR. Left Ventricular Depolarization Delay. Cardiology Today- 2015;19:31–39)

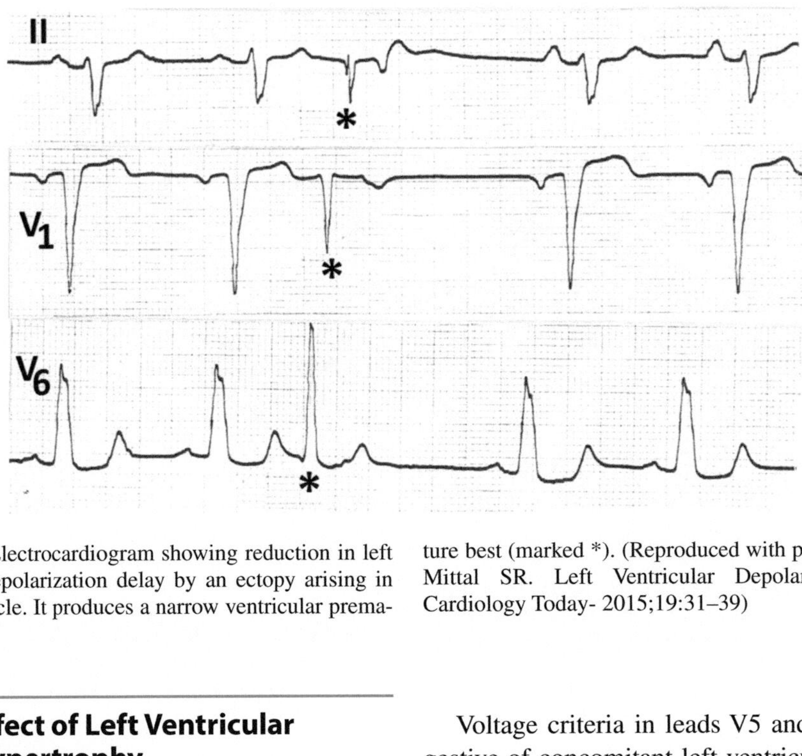

Fig. 24.28 Electrocardiogram showing reduction in left ventricular depolarization delay by an ectopy arising in the left ventricle. It produces a narrow ventricular premature best (marked *). (Reproduced with permission from: Mittal SR. Left Ventricular Depolarization Delay. Cardiology Today- 2015;19:31–39)

6 Effect of Left Ventricular Hypertrophy on Electrocardiographic Findings

Left ventricular depolarization delay masks left ventricular hypertrophy (Figs. 24.29 and 24.30).

Voltage criteria in leads V5 and V6 are suggestive of concomitant left ventricular hypertrophy but have low sensitivity. Concomitant dominant left anterosuperior fascicular block increases QRS voltages in limb leads and may mimic LVH (Fig. 24.31).

Deep S wave in leads V1,V2, and left axis deviation may mimic LVH [1] (Fig. 24.32).

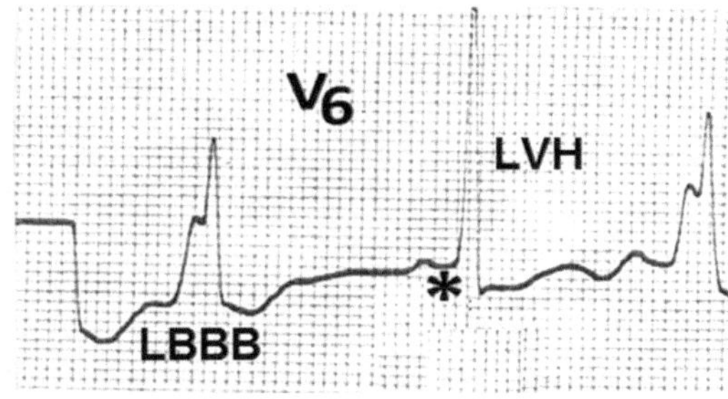

Fig. 24.29 Electrocardiogram showing intermittent LBBB with masking of left ventricular hypertrophy (marked *) during LBBB. (Reproduced with permission from: Mittal SR. Left Ventricular Depolarization Delay. Cardiology Today- 2015;19:31–39)

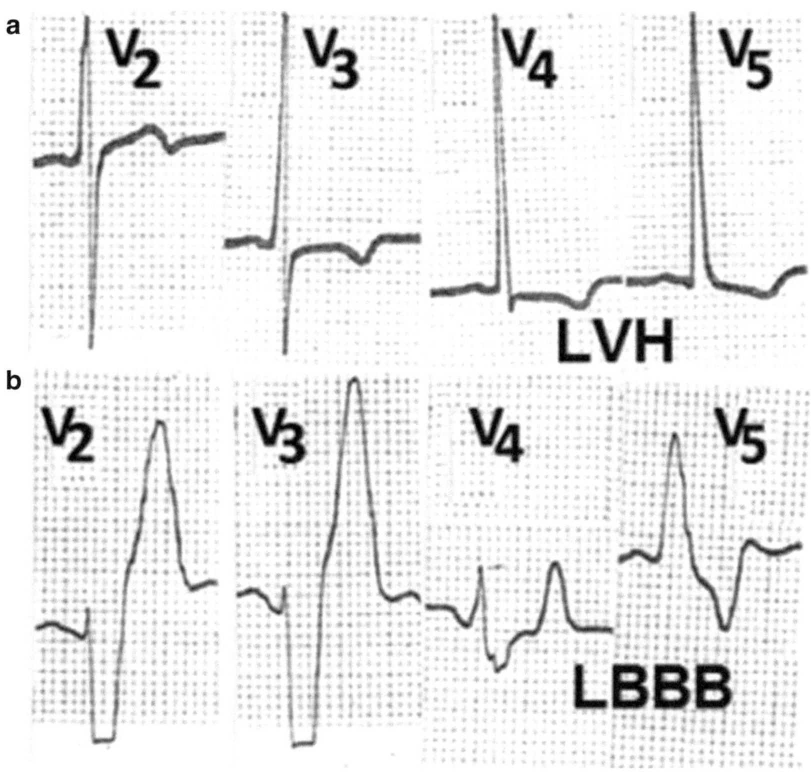

Fig. 24.30 Electrocardiograms showing (**a**) Left ventricular hypertrophy during normal conduction (**b**) Masking of left ventricular hypertrophy during LBBB. (Reproduced with permission from: Mittal SR. Left Ventricular Depolarization Delay. Cardiology Today- 2015;19:31–39)

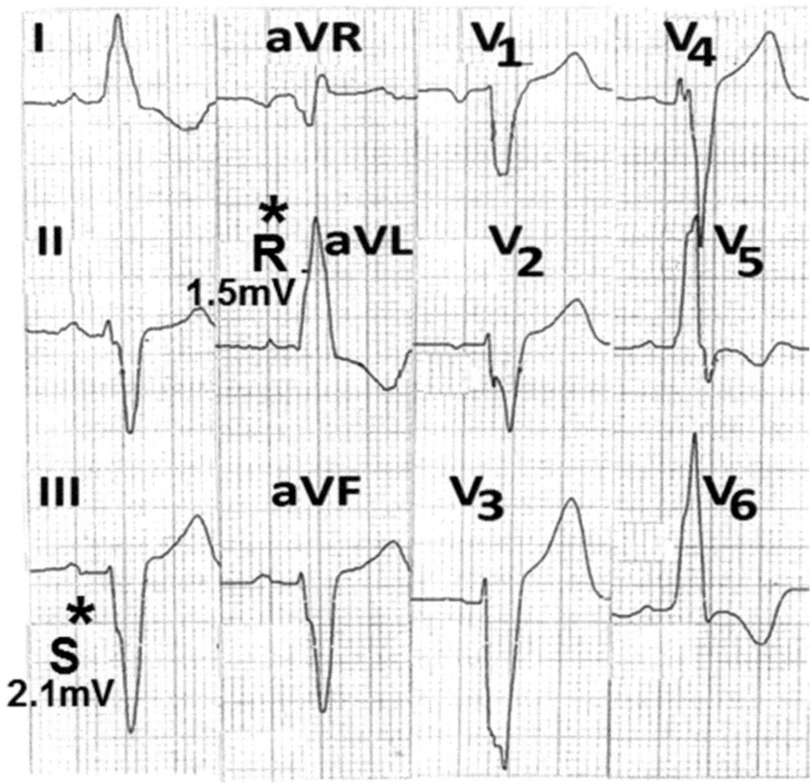

Fig. 24.31 Electrocardiogram showing LBBB with left anterosuperior fascicular block mimicking voltage criteria of LVH in limb leads (marked *). (Reproduced with permission from: Mittal SR. Left Ventricular Depolarization Delay. Cardiology Today- 2015;19:31–39)

7 Clinical Significance of Delayed Left Ventricular Depolarization

(1) Even in persons without overt heart disease, left ventricular depolarization delay is associated with a higher than normal risk of cardiovascular mortality from infarction or heart failure [1].
(2) In the context of coronary artery disease, left ventricular depolarization delay is associated with more extensive disease, more severe left ventricular dysfunction, and reduced survival rates.
(3) Due to delayed depolarization, contraction of the posterior and lateral walls of the left ventricle is delayed as compared to the interventricular septum. Interventricular septum is already relaxed by the time posterolateral walls of the left ventricle contract. Contraction of the left ventricular free wall, therefore, pushes blood towards interventricular septum rather than out into aorta (Fig. 24.32).

This asynchrony between different walls (Fig. 24.33) results in reduction in left ventricular ejection fraction [6].

Patients of LBBB presenting with left ventricular failure, QRS duration >130 ms and a delay of more than 60 ms between septal and lateral walls may improve with resynchronization device implantation (biventricular pacing) (Fig. 24.34).

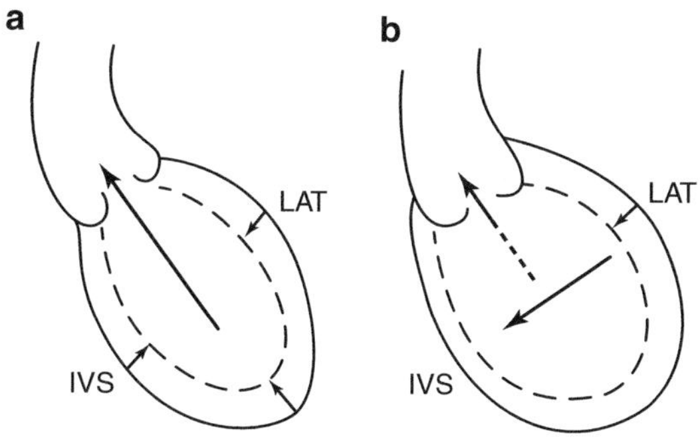

Fig. 24.32 Diagram (**a**) showing synchronous contraction of all wall of the left ventricular pushing blood into the aorta (**b**) Delayed depolarization of the left ventricular posterolateral wall resulting in pushing of blood towards the already relaxed interventricular septum (IVS) with the result that the amount of blood ejected out into the aorta is reduced (reduced ejection fraction). IVS—Interventricular septum. LAT—Lateral wall of LV. (Reproduced with permission from: Mittal SR. Left Ventricular Depolarization Delay. Cardiology Today- 2015;19:31–39)

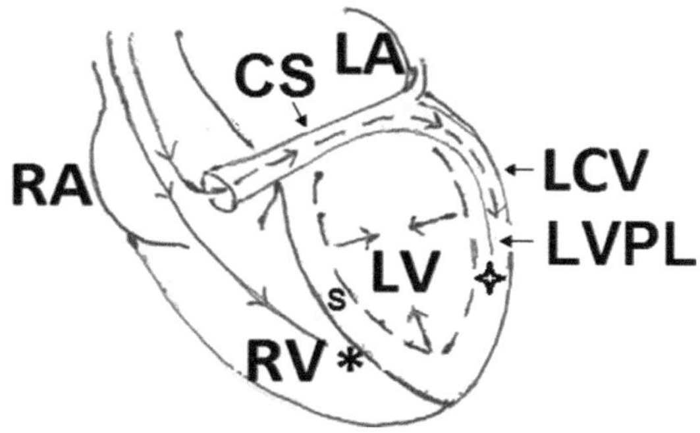

Fig. 24.34 Diagram showing biventricular pacing to resynchronize contraction of the interventricular septum (marked as S) and left ventricular posterolateral wall. RA— Right atrium, RV—Right ventricle, LV—Left ventricle, CS—Coronary sinus, LCV Lateral cardiac vein. (* -Right ventricular pacing, ✦—Left ventricular pacing, S—Interventricular septum, LVPL—Left ventricular posterolateral wall. (Reproduced with permission from: Mittal SR. Left Ventricular Depolarization Delay. Cardiology Today- 2015;19:31–39)

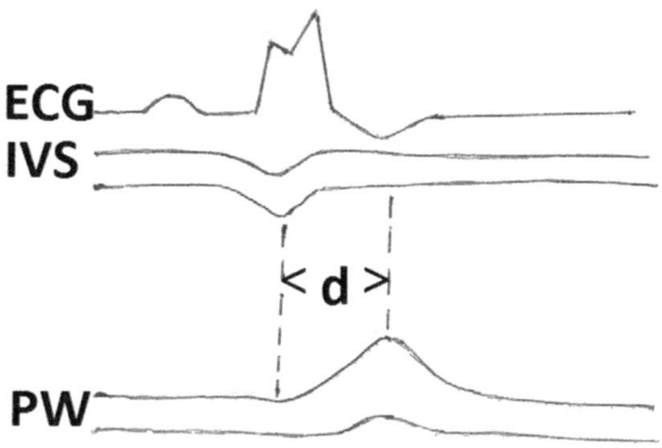

Fig. 24.33 Diagram of M—mode echocardiogram shows delay in contraction between interventricular septum (IVS) and left ventricular posterior wall (PW) (marked as ←d→). This results in a reduced ejection fraction. (Reproduced with permission from: Mittal SR. Left Ventricular Depolarization Delay. Cardiology Today- 2015;19:31–39)

(4) Delayed activation of the left ventricle results in delayed left ventricular ejection. This results in delayed closure of the aortic valve. Aortic component of the second sound is, therefore, delayed and comes after the pulmonary component of the second sound. Second sound is, therefore, split in expiration and single in inspiration (Paradoxical split of second sound).
(5) Left axis deviation is associated with more severe conduction system disease [1]. Right axis deviation suggests biventricular enlargement. Increasing duration of QRS (Fig. 24.35) correlates with greater asynchrony and increasing severity of LV dysfunction.
(5) Left atrial enlargement may result in atrial fibrillation (Fig. 24.36).
(6) Exercise-induced left bundle branch block [7] is associated with
   - Higher risk of development of LBBB at rest on follow-up.
   - Higher risk of major cardiovascular events in future.

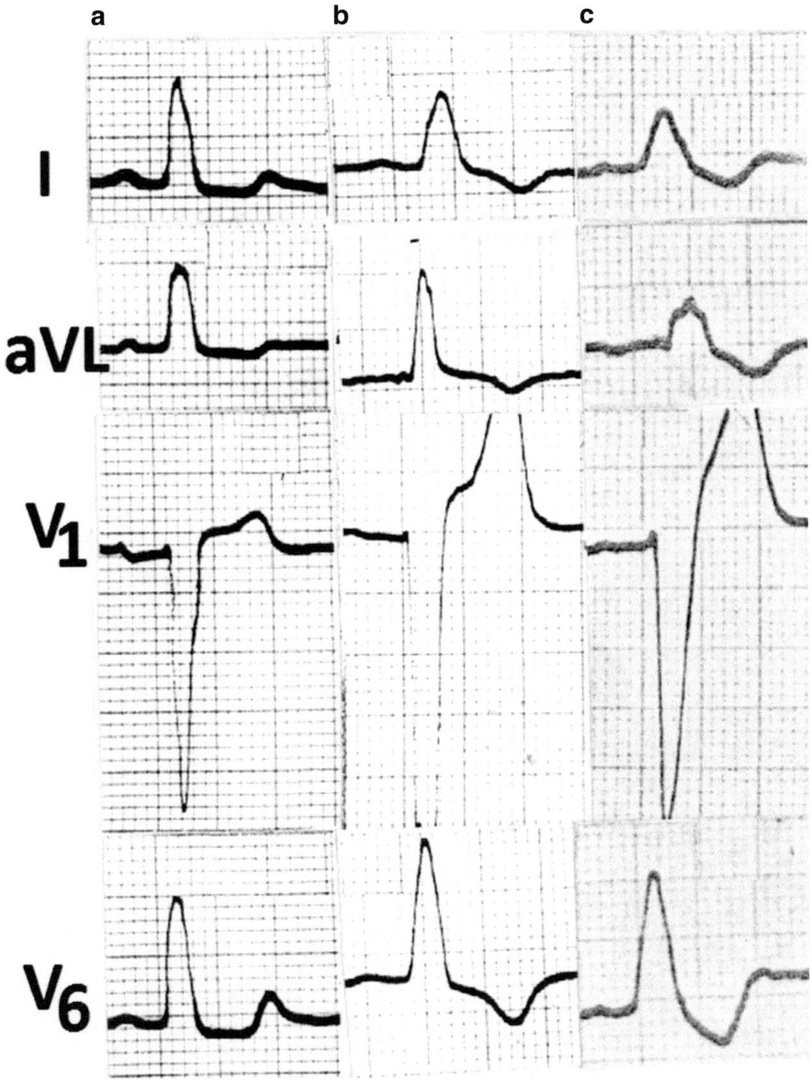

Fig. 24.35 Electrocardiogram showing increasing duration of QRS from (**a**) to (**c**). (Reproduced with permission from: Mittal SR. Left Ventricular Depolarization Delay. Cardiology Today- 2015;19:31–39)

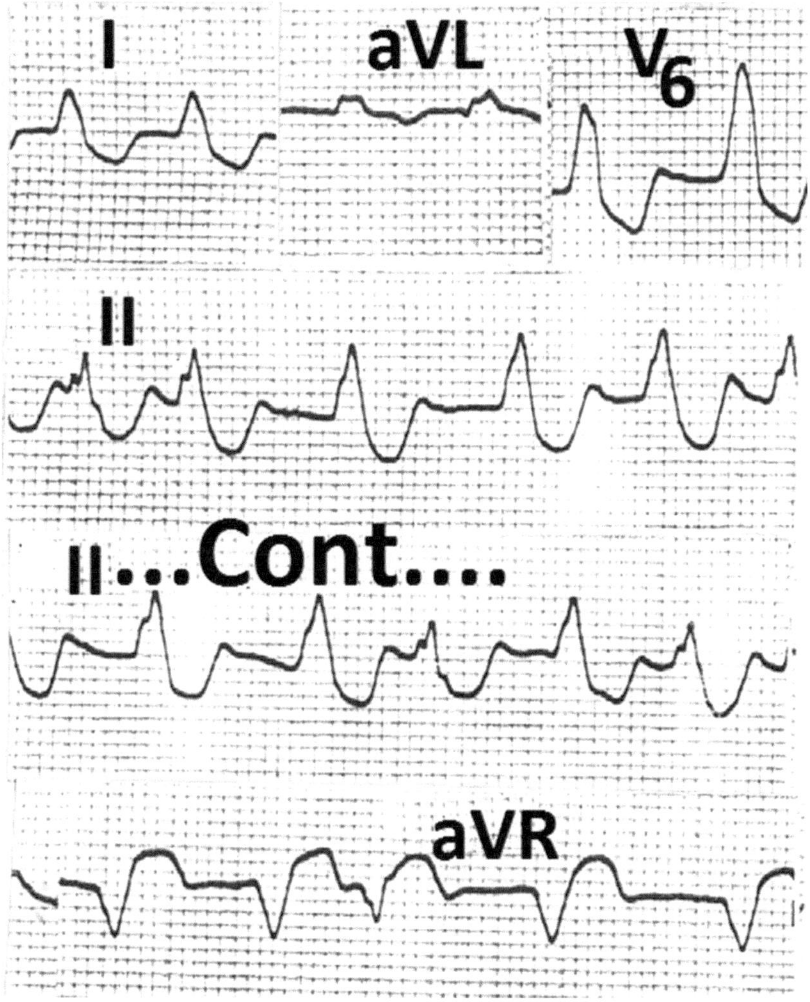

Fig. 24.36 Electrocardiogram showing LBBB with atrial fibrillation. (Reproduced with permission from: Mittal SR. Left Ventricular Depolarization Delay. Cardiology Today- 2015;19:31–39)

Summary

- Partial LBBB is diagnosed when the duration of the QRS complex is less than 0.12 s, but lead V_1 shows QS or rS pattern with positive T wave and leads I and V_6 show single R wave with loss of q in lead V_6.
- Progression to advanced LBBB results in progressive decrease in the height of R wave and increase in the duration of the R wave in leads I, aVL, V_5, and V_6.
- Advanced LBBB is diagnosed when in addition to findings of partial LBBB, QRS duration is more than 0.12 s, and time from onset to peak of the R wave in leads I and V_6 is 60 ms or more. ST-segment and T waves are opposite to the direction of the terminal part of QRS.
- Mean frontal plane QRS axis is usually normal. rS configuration in leads II, III, and aVF suggests left axis deviation. Leads V_5 and V_6 may show RS configuration. It suggests

greater delay in depolarization over the left anterior fascicle and more severe conduction system disease. Right axis deviation suggests greater delay in depolarization over the left posterior fascicle or additional delay in conduction over the right bundle branch.
- Electrocardiographic evidence of left atrial enlargement, QRS duration of more than 155 ms, and positive voltage criteria for left ventricular hypertrophy ($SV_2 + RV_6$ >45 mm.) suggest underlying left ventricular hypertrophy. Severe conduction delay alone can prolong the QRS duration to more than 155 ms. Mitral regurgitation and left ventricular failure can contribute to left atrial enlargement without meaning left ventricular hypertrophy.
- Presence of LBBB alone is also associated with higher than normal morbidity and mortality from cardiovascular events in the future.
- Some persons can develop tachycardia-dependent LBBB during an exercise stress test. These persons are likely to develop LBBB at rest in future.

MCQs

Q1. Which is not a cause of LBBB?

(a) Arrhythmogenic dysplasia
(b) Coronary artery disease
(c) Lenegre disease
(d) Mahaim fibers

Q2. LBBB is usually not a manifestation of :

(a) LVH
(b) Calcific AS
(c) Rheumatic heart disease
(d) Chagas disease

Q3. Chagas disease does not produce:

(a) Myocarditis
(b) Atrioventricular block
(c) Intraventricular conduction defects
(d) Endocarditis.

Q4. Incomplete LBBB is characterized by:

(a) Loss of septal q wave
(b) QS configuration in leads V1 to V3
(c) QRS duration > 0.12 s
(d) Notch on the top of QRS complex in lead V6

Q5. Loss of the septal q wave is not seen in:

(a) Septal infarction
(b) LV pressure overload
(c) Uncomplicated LBBB
(d) LV volume overload

Q6. Resynchronization therapy is most effective when QRS duration is:

(a) 60 ms
(b) 100 ms
(c) 120 ms
(d) >120 ms

Q7. In advanced LBBB, QS configuration in lead V1 suggests :

(a) RV infarction
(b) Septal infarction
(c) Anterior infarction
(d) None

Q8. In uncomplicated LBBB, reverse progression of r from lead V1 to V4 suggests :

(a) Septal infarction
(b) Anterior infarction
(c) Advanced block
(d) None

Q9. In uncomplicated LBBB, lead V4R shows :

(a) RS configuration
(b) Prominent R wave

(c) Inverted T wave
(d) QS configuration

Q10. In uncomplicated LBBB, lead V8 shows :

(a) RS configuration
(b) QR configuration
(c) Prominent R wave
(d) Upright T wave

Q11. In LBBB, S wave in lead V6 cannot be due to :

(a) Medial placement of lead V6
(b) Lateral infarction
(c) Hyperkalemia
(d) Septal infarction

Q12. LBBB with a frontal plane QRS axis of -60^0 can be due to additional:

(a) Septal infarction
(b) Lateral infarction
(c) LAFB
(d) LVH

Q13. LBBB with a frontal plane QRS axis of + 130^0 can be due to additional :

(a) RV infarction
(b) LPFB
(c) RVH
(d) None

Q14. In the presence of LBBB, premature narrow QRS complex suggests:

(a) Tachycardia-dependent LBBB
(b) Supraventricular ectopy
(c) Left ventricular ectopy
(d) Right ventricular ectopy

Q15. What is correct for LBBB?

(a) Mimics LVH in leads V1,V2
(b) Mimics LVH in leads V5,V6
(c) Masks LVH in leads V1,V2
(d) Mimics LVH in limb leads

Q16. In the presence of LBBB, presence of left ventricular hypertrophy correlates with:

(a) QRS Voltage in V6
(b) QRS duration in V6
(c) Mean frontal plane QRS axis
(d) Intrinsicoid deflection in lead V6

Q17. Exercise-induced LBBB is associated with :

(a) Short-term risk of development of atrioventricular block
(b) Short-term risk of acute myocardial infarction
(c) Risk of future development of LBBB at rest
(d) Short-term risk of LVF

Q18. In LBBB, QRS duration does not correlate with:

(a) LV dysfunction
(b) LV asynchrony
(c) LVH
(d) Long-term prognosis

Q19. Echocardiographic findings in LBBB include:

(a) Early systolic posterior motion of IVS
(b) Absence of systolic thickening of IVS
(c) Absence of systolic thickening of posterolateral wall of LV
(d) Diastolic dysfunction of LV

Q20. In LBBB, response to resynchronization therapy correlates best with :

(a) QRS voltage
(b) QRS duration
(c) LV dysfunction
(d) LV dyssynchrony

Q21. LBBB produces:

(a) Lateral displacement of the left ventricular apex
(b) S3 in mitral area

(c) Soft aortic component of the second sound
(d) Paradoxical split of the second sound

Q22. Which noninvasive test can be useful in a patient with LBBB complaining of effort angina?

(a) Treadmill stress test
(b) Echocardiography
(c) Stress SPECT
(d) Cardiac MRI

Q23. QRS duration less than 0.12 s, rS configuration with positive T wave in leads V_1, V_2, and R wave in lead I, V_5, V_6 suggests:

(a) Left anterior hemiblock
(b) Incomplete LBBB
(c) Complete LBBB
(d) Left posterior hemiblock

Q24. Absence of q wave in leads $V_{5,}$ V_6 can suggest:

(a) Septal infarction
(b) Incomplete LBBB
(c) Normal variant
(d) None

Q25. Progression of incomplete to advanced LBBB results in progressive:

(a) Increase in the height of R wave in lead V_5
(b) Decrease in the height of R wave in lead V_5
(c) Appearance of S wave in lead V_6
(d) Appearance of q wave in leads I and aVL

Q26. QRS duration of 0.14 s, QS pattern with positive T wave in leads V_1 to V_3, RR' pattern with downsloping ST segment, and negative T wave in leads I, aVL, and $V_{5,}$ V_6 suggests:

(a) Incomplete LBBB
(b) Advanced LBBB with septal infarction
(c) Uncomplicated advanced LBBB
(d) Advanced LBBB with lateral infarction

Q27. In uncomplicated advanced LBBB, mean frontal plane QRS axis shows:

(a) Normal axis
(b) Left axis deviation
(c) Right axis deviation
(d) Right upper quadrant axis

Q28. Advanced LBBB with rS configuration in leads II, III, aVF suggests greater delay in conduction over

(a) Left anterior fascicle
(b) Left posterior fascicle
(c) Right bundle branch
(d) All

Q29. Advanced LBBB with rS configuration in lead I and prominent qR in leads II, III, aVF suggest:

(a) Greater delay in conduction over the left anterior fascicle
(b) Greater delay in conduction over the left posterior fascicle
(c) Additional right bundle branch block
(d) Additional RVH

Q30. Which ECG findings is specific of diagnosis of LVH in the presence of advanced LBBB ?

(a) ECG evidence of left atrial enlargement
(b) QRS duration of 0.16 s or more
(c) Left axis deviation
(d) None

Q31. Which of the following is associated with higher than normal cardiovascular mortality and morbidity in the future?

(a) Advanced RBBB without structural heart disease
(b) Advanced LBBB without structural heart disease
(c) Left anterosuperior hemiblock
(d) All

Answers

(1) a, (2) c, (3) d, (4) a, (5) d, (6) d, (7) d, (8) d, (9) d, (10) c, (11) d, (12) c, (13) b, (14) c, (15) a, (16) a, (17) c, (18) c, (19) a, (20) d, (21) d, (22) c, (23) b, (24) a, b, c, (25) b, (26) c, (27) a, (28) a, (29) b, d, (30) d, (31) b.

References

1. Mirvis DM, Goldberger AL. Electrocardiography. In: Bonow RO, Mann DL, Zipes DP, Libby P, editors. Braunwald's Heart Disease. Missouri: Elsevier; 2012. p. 126–65.
2. Olshansky B, Chung MK, Pogwizd SM, Goldschlager N. Bradyarrhythmias- conduction system abnormalities. In: Olshansky B, Chung MK, Pogwizd SM, Goldschlager N, editors. Arrhythmia Essentials. Burlington: Jones & Bartlett; 2012. p. 26–78.
3. DeLuna AB, Goldwasser D, Fiol M, Bayes- Genis A. Surface electrocardiography. In: Fuster V, Walsh RA, Harrington RA, editors. Hurst's the heart. Mc Graw Hill: New York; 2011. p. 307–70.
4. Nikoo MH, Aslani A, Jorat MV. LBBB : State of the art criteria. Int Cardiovasc Res J. 2013;7:39–40.
5. Surawicz B, Childers R, Deal BJ, Gettes LS. AHA/ ACCF/ HRS recommendations for the standardization and interpretation of the electrocardiogram. Part III: intraventricular conduction disturbances. J Am Coll Cardiol. 2009;53:976–81.
6. Grines CL, Bashore TM, Boudoulas H, et al. Functional abnormalities in isolated left bundle branch block. The effect of interventricular asynchrony. Circulation. 1989;79:845–53.
7. Said SAM, Bultje- Peters M, Nijhuls RLG. Exercise induced left bundle branch block: an infrequent phenomenon: report of two cases. World J Cardiol. 2013;5:359–63.

25 Left Bundle Branch Block (Delay in Left Ventricular Depolarization) and Myocardial Necrosis

Myocardial necrosis can be ischemic or nonischemic (e.g., myocarditis or dilated cardiomyopathy). Electrocardiogram cannot identify the cause of necrosis. Definite electrocardiographic diagnosis of myocardial necrosis is difficult in the presence of delay in left ventricular depolarization. Several criteria have been suggested in the literature.

1 Diagnosis of Acute Myocardial Infarction

ST-segment elevation with sequential changes in leads with positive QRS complex suggests acute myocardial infarction (Figs. 25.1, 25.2, 25.3, and 25.4).

ST-segment concordance alone has low specificity [1]. Leads V1 and V2 may show significant (> 0.5 mV) ST-segment elevation even without an infarction [2] (Figs. 25.5 and 25.6).

ST-segment depression of >1 mm in leads V1 and V2 (Fig. 25.7) (negative concordance) also suggests myocardial necrosis.

S. Mittal, *Insights into Electrocardiograms with MCQs*,
https://doi.org/10.1007/978-981-99-0127-2_25

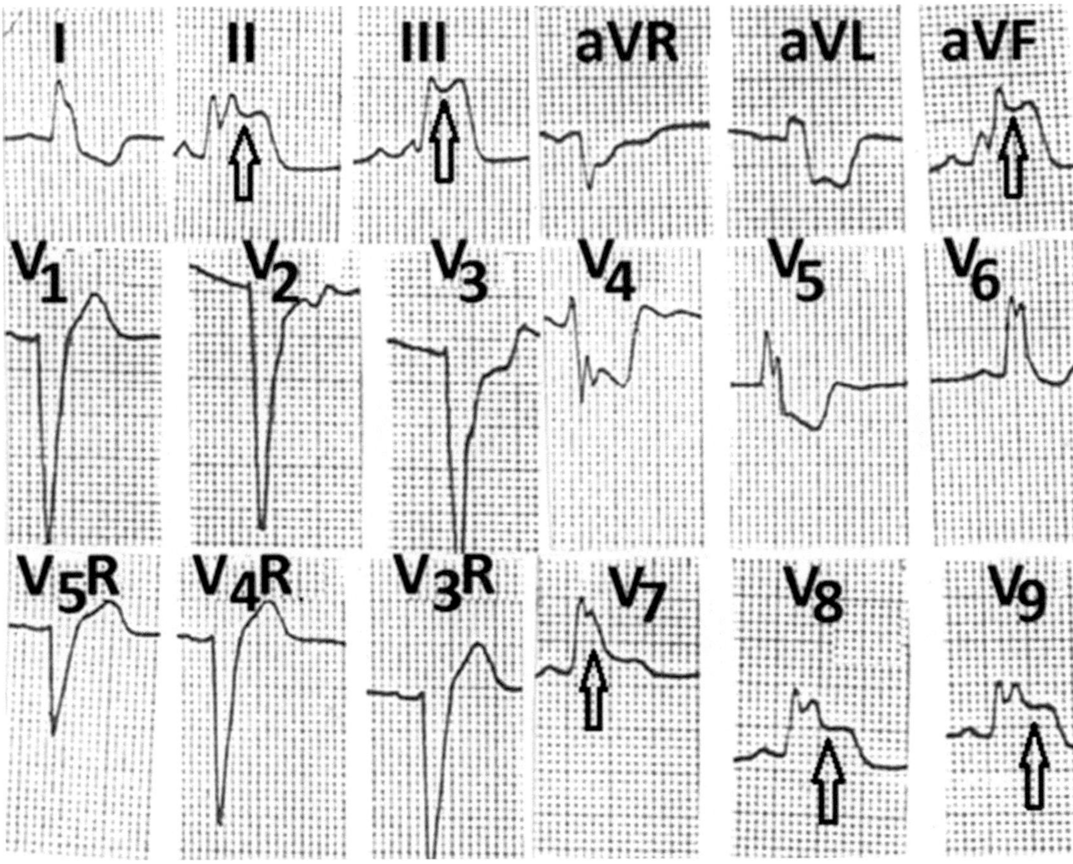

Fig. 25.1 Electrocardiogram showing delay in left ventricular depolarization with acute inferior and posterior infarction. ST-segment is elevated in leads II, III, aVF, and V_7 to V_9 (marked ⇧). (Reproduced with permission from Mittal SR. Left Ventricular Depolarization Delay and Myocardial Necrosis. Cardiology Today-2015;19:68–107)

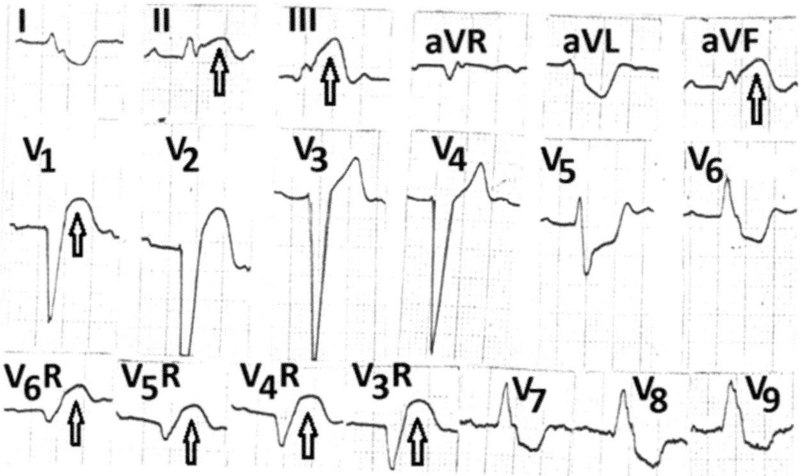

Fig. 25.2 Electrocardiogram showing delay in left ventricular depolarization with acute inferior and right ventricular infarction. ST-segment is elevated in leads II, III, aVF, and V_6R to V_1 (marked ⇧). (Reproduced with permission from Mittal SR. Left Ventricular Depolarization Delay and Myocardial Necrosis. Cardiology Today- 2015;19:68–107)

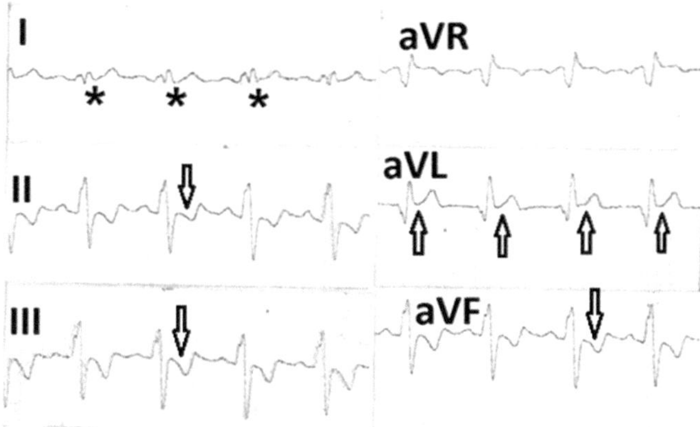

Fig. 25.3 Electrocardiogram showing delay in left ventricular depolarization with acute high lateral infarction. There is early uptake of ST-segment in lead I (marked ⇧) and Q wave with ST-segment elevation in lead aVL (marked *). There is reciprocal ST-segment depression in leads II, III, and aVF (marked ⇩). (Reproduced with permission from Mittal SR. Left Ventricular Depolarization Delay and Myocardial Necrosis. Cardiology Today- 2015;19:68–107)

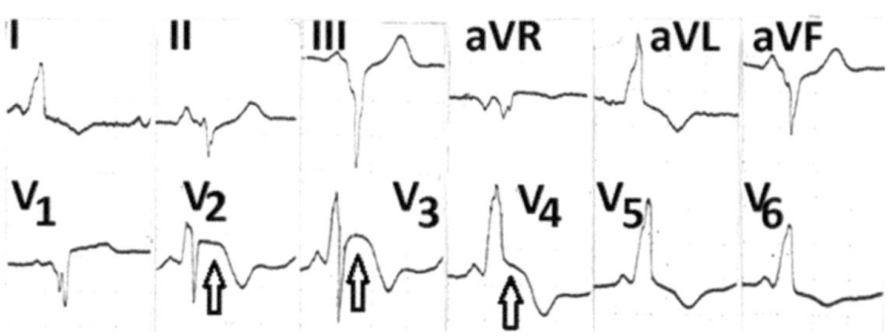

Fig. 25.4 Electrocardiogram showing left ventricular depolarization delay with acute anterior infarction. ST-segment is elevated in leads V_2 to V_4 (marked ⇧). (Reproduced with permission from: Mittal SR. Left Ventricular Depolarization Delay and Myocardial Necrosis. Cardiology Today- 2015;19:68–107)

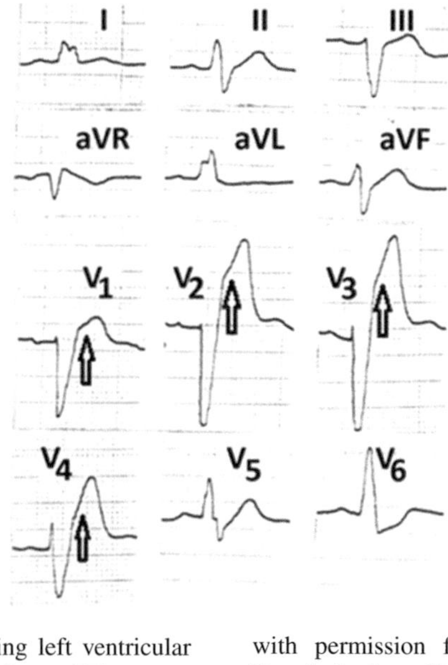

Fig. 25.5 Electrocardiogram showing left ventricular depolarization delay with noninfarct ST-segment elevation in leads V1 to V4. (marked ⇧). (Reproduced with permission from Mittal SR. Left Ventricular Depolarization Delay and Myocardial Necrosis. Cardiology Today- 2015;19:68–107)

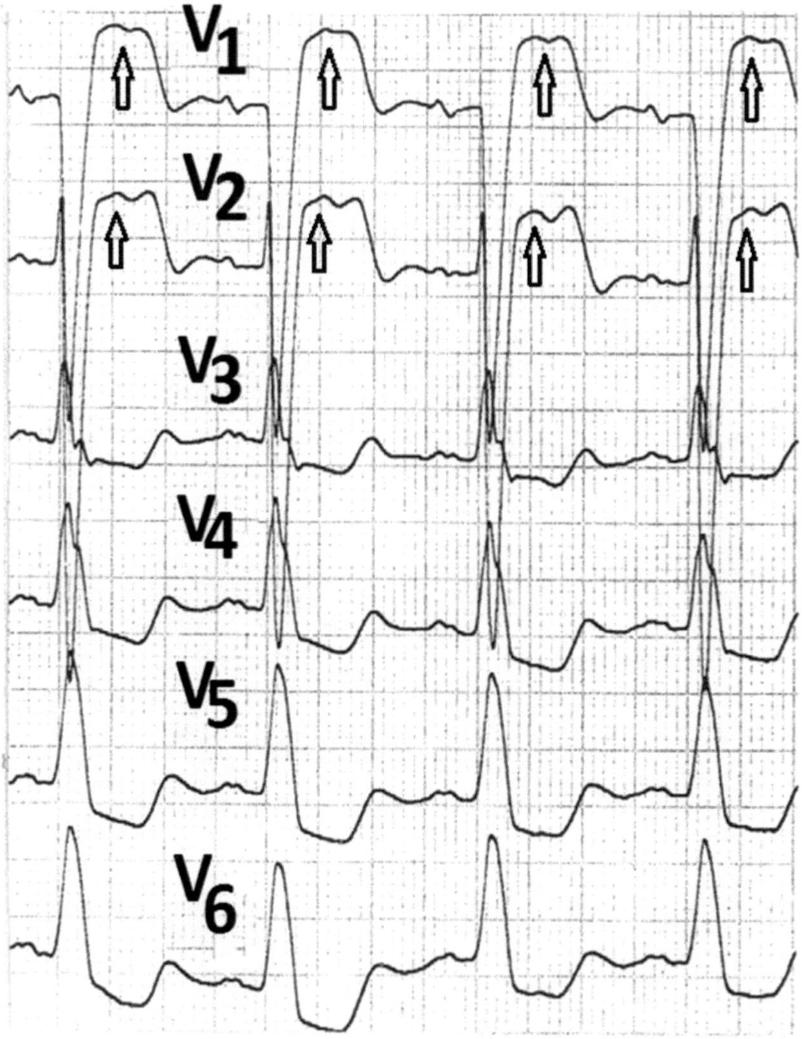

Fig. 25.6 Electrocardiogram showing left ventricular depolarization delay with noninfarct ST-segment elevation in leads V1,V2. (marked ⇧). (Reproduced with permission from Mittal SR. Left Ventricular Depolarization Delay and Myocardial Necrosis. Cardiology Today- 2015;19:68–107)

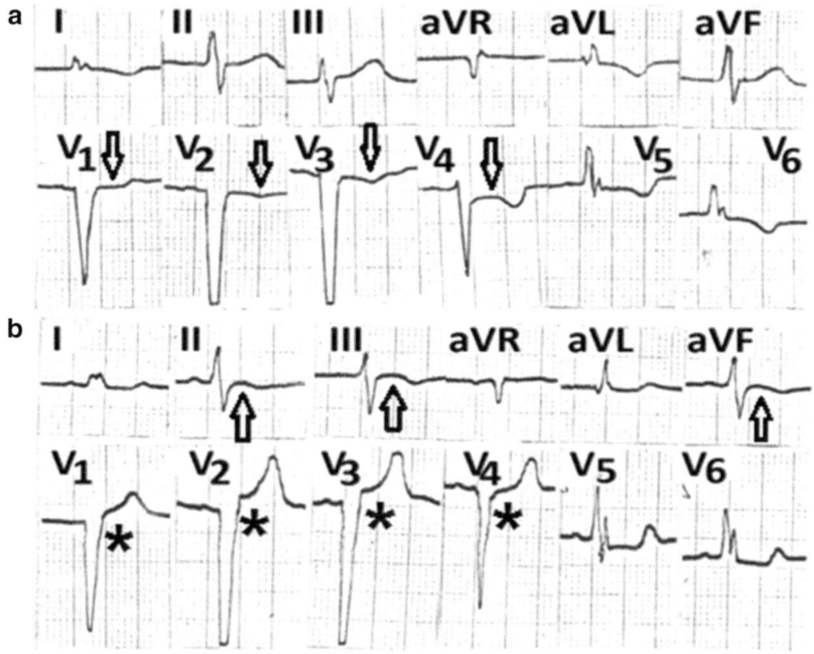

Fig. 25.7 Electrocardiogram showing left ventricular depolarization delay with (**a**) transient ST-segment depression in leads V1 to V4 (marked ⇩) and (**b**) normalization of the ST-segment in leads V1 to V4 (marked *) with ST-segment elevation in leads II, III, and aVF (marked ⇧) suggestive of inferior infarction. (Reproduced with permission from Mittal SR. Left Ventricular Depolarization Delay and Myocardial Necrosis. Cardiology Today- 2015;19:68–107)

2 Old Myocardial Infarction

2.1 Changes in QRS Configuration

2.1.1 Septal Infarction

Presence of Q wave in leads I, aVL, V5 and V6 suggests septal infarction (Figs. 25.8a, b, 25.9, and 25.10). These Q waves are probably produced from posterior and superior forces from the spared basal portion of the interventricular septum [3].

Initial part of the R wave in leads I, aVL, and V_6 may be small, slurred, rounded, or dome shaped (Fig. 25.11).

Left anterosuperior depolarization delay (left anterosuperior fascicular block) with focal block also produces broad QRS with Q wave in leads I and aVL

(Fig. 25.12). This condition, however, produces slurring of the terminal portion of the R wave without Q wave in leads V_5 and V_6.

Q wave localized only to lead aVL can be due to the electrical position of the heart and does not necessarily suggest myocardial infarction. QS configuration in leads V1 and V2 is part of electrocardiographic findings of LBBB and does not suggest myocardial infarction. Abnormal Q waves in leads V_3, V_4 suggest infarction (Figs. 25.13 and 25.14).

Reverse progression of the r wave from leads V_1 to V_3 (Fig. 25.15) is not suggestive of infarction if the depth of the S wave is also progressively decreasing.

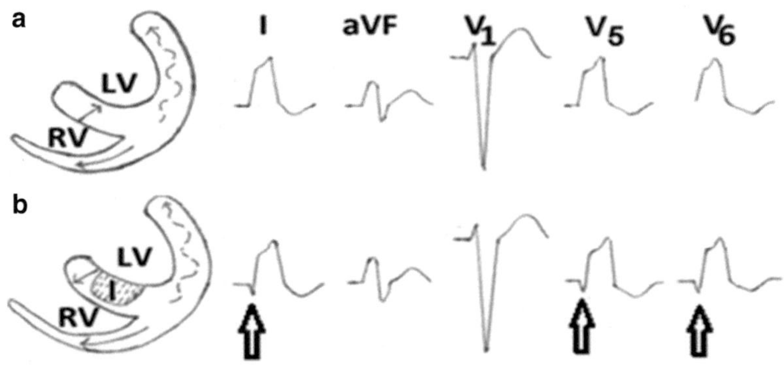

Fig. 25.8 Diagram showing depolarization of the ventricles (**a**) in left ventricular depolarization delay (LBBB) and (**b**) in LBBB + septal infarction. Q wave is present in leads I, V_5, and V_6 (marked ⇧). *I* infarction; *LV* left ventricle, *RV* right ventricle. (Reproduced with permission from Mittal SR. Left Ventricular Depolarization Delay and Myocardial Necrosis. Cardiology Today- 2015;19:68–107)

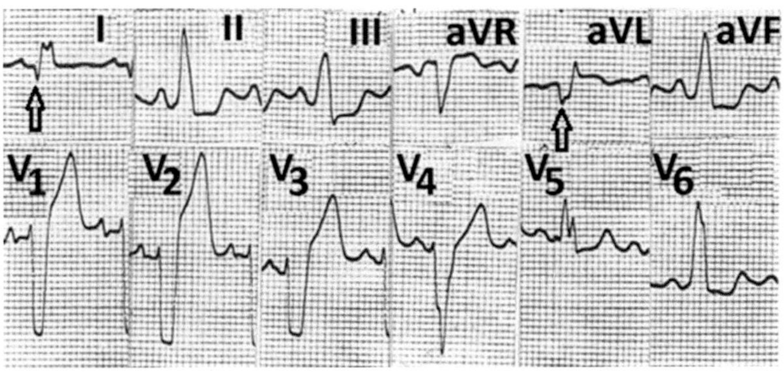

Fig. 25.9 Electrocardiogram showing left ventricular depolarization delay with Q wave in leads I and aVL (marked ⇧) suggestive of high lateral infarction. (Reproduced with permission from Mittal SR. Left Ventricular Depolarization Delay and Myocardial Necrosis. Cardiology Today-2015;19:68–107)

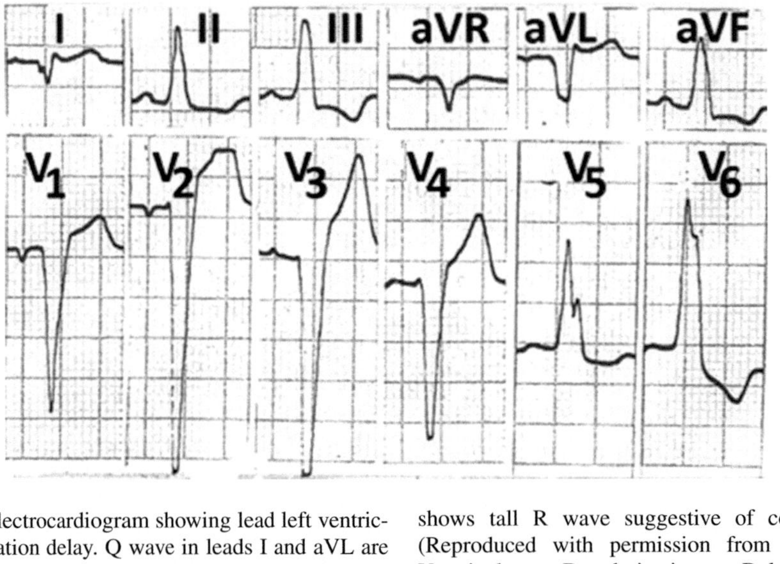

Fig. 25.10 Electrocardiogram showing lead left ventricular depolarization delay. Q wave in leads I and aVL are suggestive of septal infarction. P wave in lead V_1 is inverted, suggesting left atrial enlargement. Lead V_6 shows tall R wave suggestive of concomitant LVH. (Reproduced with permission from Mittal SR. Left Ventricular Depolarization Delay. Cardiology Today-2015;19:31–39)

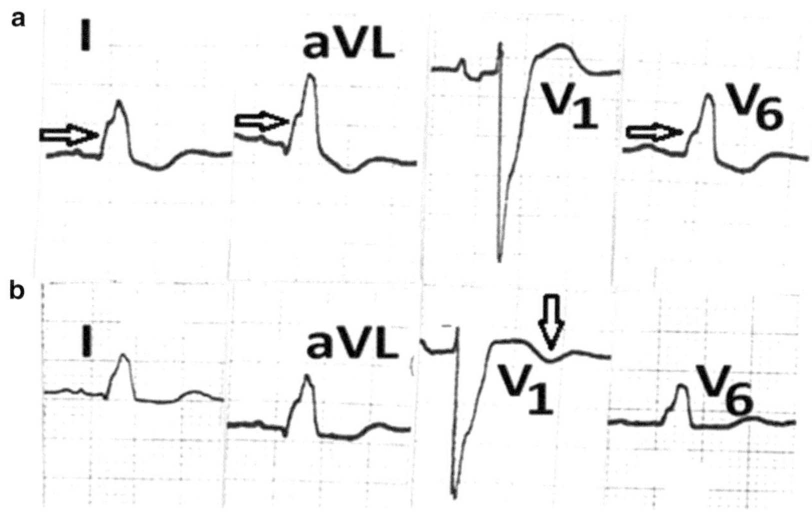

Fig. 25.11 Electrocardiogram showing left ventricular depolarization delay. (**a**) The initial part of the R wave is small in leads I, aVL, and V6 (marked ⇨). (**b**) Sequential T wave inversion in lead V1 (marked ⇩). (Reproduced with permission from Mittal SR. Left Ventricular Depolarization Delay and Myocardial Necrosis. Cardiology Today- 2015;19:68–107)

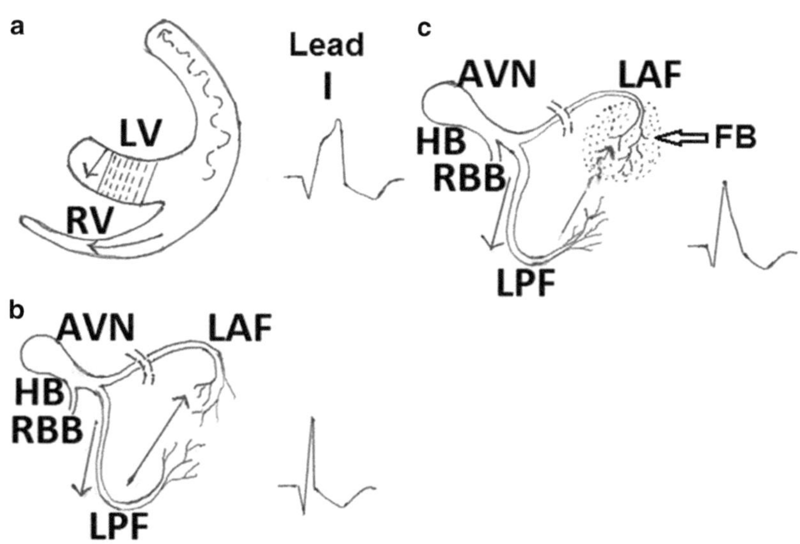

Fig. 25.12 Diagram showing causes of broad QRS with Q wave in lead I. (**a**) Left ventricular depolarization delay + septal infarction. (**b**) Isolated left anterior fascicular block. (**c**) Left anterior fascicular block + focal block in anterosuperior region. *LV* left ventricle; *RV* right ventricle; *AVN* AV node; *HB* His bundle; *RBB* right bundle branch; *LAF* left anterior fascicle; *LPF* left posterior fascicle; *FB* focal block; *I* lead I. (Reproduced with permission from Mittal SR. Left Ventricular Depolarization Delay and Myocardial Necrosis. Cardiology Today- 2015;19:68–107)

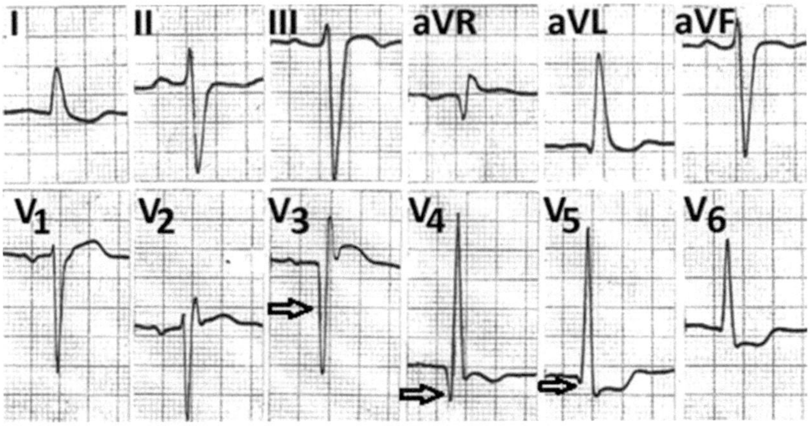

Fig. 25.13 Electrocardiogram showing left ventricular depolarization delay with Q wave in leads V3, V4, and V5 (marked ⇨) suggestive of anterior infarction. (Reproduced with permission from Mittal SR. Left Ventricular Depolarization Delay and Myocardial Necrosis. Cardiology Today- 2015;19:68–107)

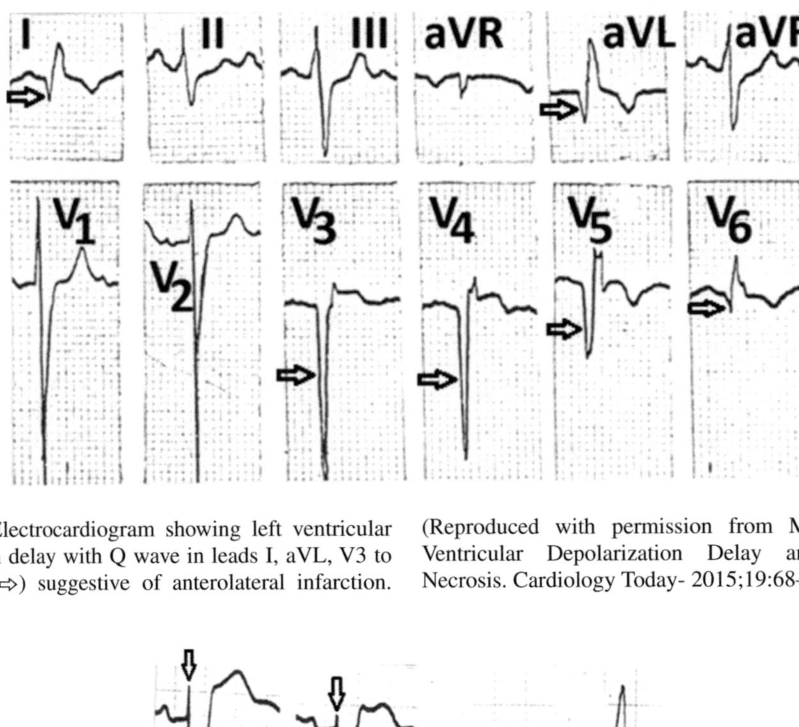

Fig. 25.14 Electrocardiogram showing left ventricular depolarization delay with Q wave in leads I, aVL, V3 to V6 (marked ⇨) suggestive of anterolateral infarction. (Reproduced with permission from Mittal SR. Left Ventricular Depolarization Delay and Myocardial Necrosis. Cardiology Today- 2015;19:68–107)

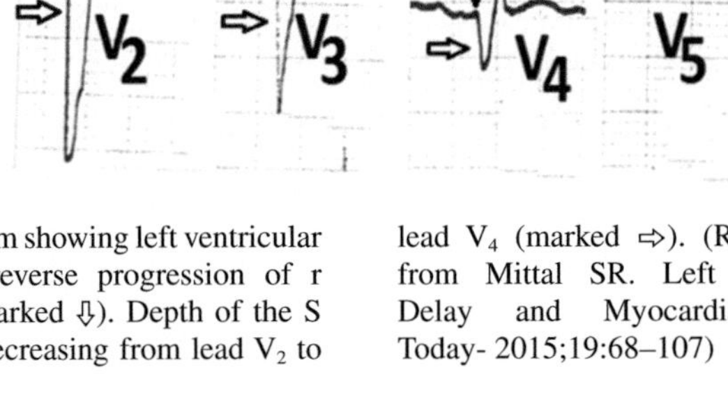

Fig. 25.15 Electrocardiogram showing left ventricular depolarization delay with reverse progression of r from lead V2 to lead V4 (marked ⇩). Depth of the S wave is also progressively decreasing from lead V_2 to lead V_4 (marked ⇨). (Reproduced with permission from Mittal SR. Left Ventricular Depolarization Delay and Myocardial Necrosis. Cardiology Today- 2015;19:68–107)

2.1.2 Inferior Infarction

Isolated inferior infarction does not change QRS vectors in left ventricular depolarization delay (Fig. 25.16).

QSR configuration, therefore, does not change. Abnormal Q wave may be present in inferior leads even in the absence of infarction. Concomitant left axis deviation due to dominant left anterosuperior fascicular block can also mask inferior infarction (Fig. 25.17).

2.1.3 Lateral Infarction

It also does not change the initial septal vector (Fig. 25.18) and, therefore, does not produce Q

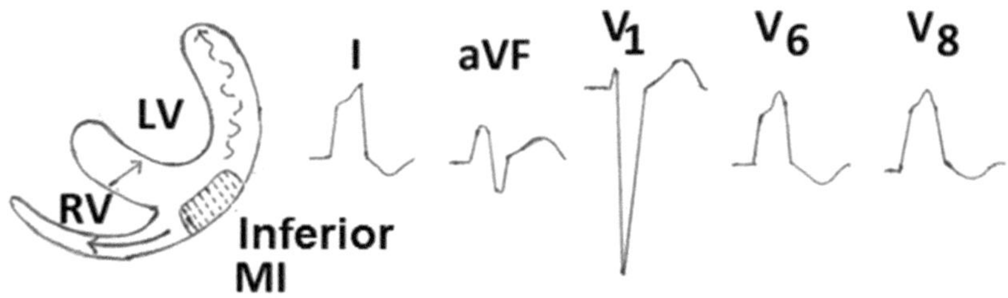

Fig. 25.16 Diagram showing depolarization of ventricles in LBBB+ inferior infarction. (Reproduced with permission from: Mittal SR. Left Ventricular Depolarization Delay and Myocardial Necrosis. Cardiology Today- 2015;19:68–107)

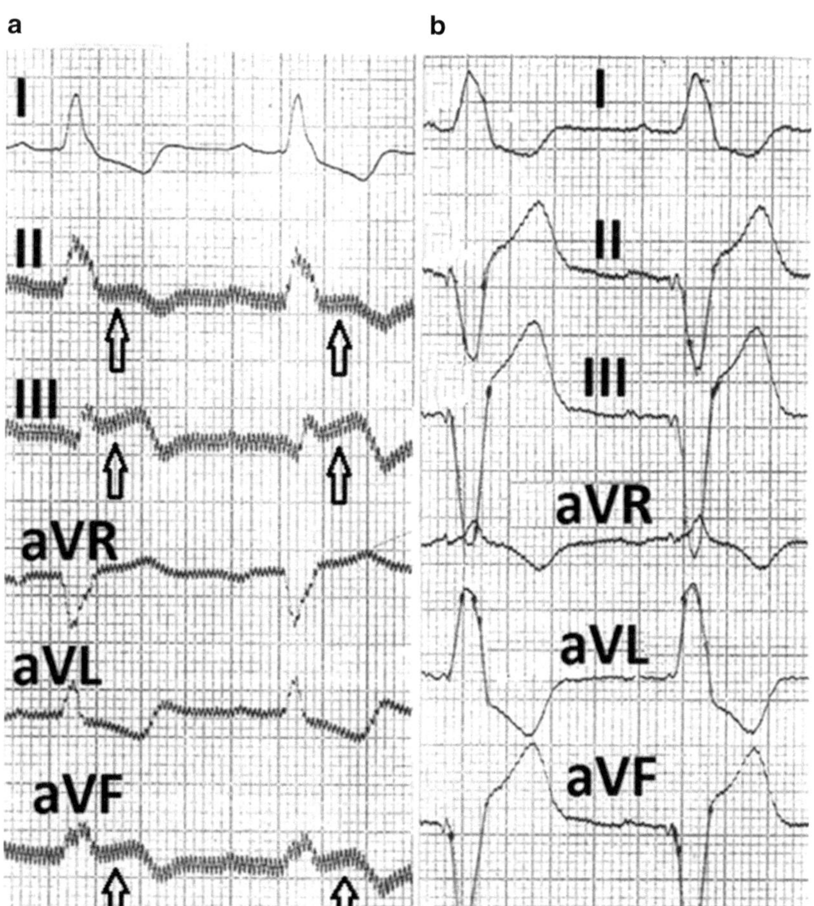

Fig. 25.17 Electrocardiogram showing left ventricular depolarization delay with (**a**) acute inferior infarction [ST-segment elevation in leads II, III, and aVF (marked ⇧)]. (**b**) Masking of inferior infarction by left anterosuperior fascicular block. (Reproduced with permission from Mittal SR. Left Ventricular Depolarization Delay and Myocardial Necrosis. Cardiology Today- 2015;19:68–107)

wave. Depolarization of the posterobasal portion of the left ventricular free wall produces an abnormal S wave in properly placed lead V6.

2.1.4 Septal+ Lateral Infarction

It produces abnormal Q wave and S wave in leads I, aVL, V5, and V6 (Fig. 25.19). With extensive damage, these leads may show QS configuration.

2.1.5 Cabrera Sign

Notch on the ascending part of a wide S wave in mid precordial leads (V_3 and V_4) suggests myocardial necrosis (Fig. 25.20 a, b).

It has low specificity. Representative electrocardiogram is shown in Fig. 25.21.

2.1.6 Chapman Sign

Notch on the ascending limb of a wide R in lead I, aVL, V5, and V6 suggests myocardial necrosis

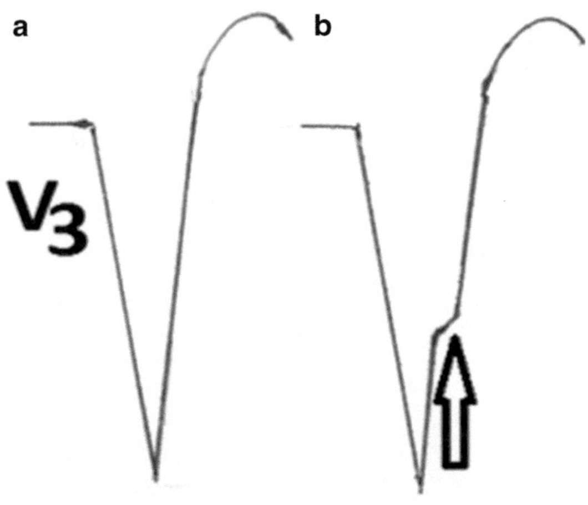

Fig. 25.20 Diagram showing (**a**) lead V3 in uncomplicated left ventricular depolarization delay. (**b**) Notch on the ascending part of a wide S wave in lead V_3 (marked ⇧). (Reproduced with permission from Mittal SR. Left Ventricular Depolarization Delay and Myocardial Necrosis. Cardiology Today- 2015;19:68–107)

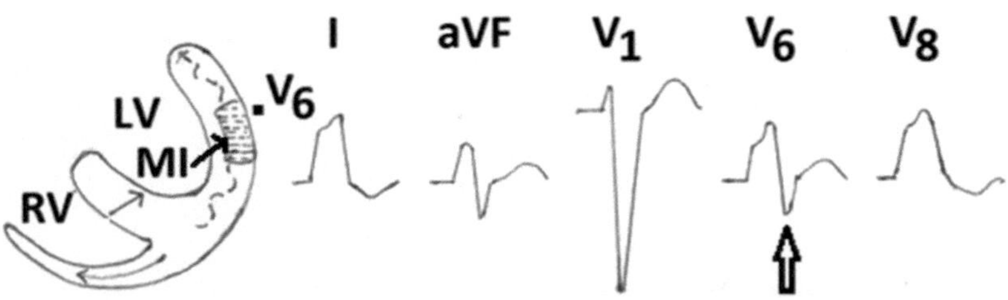

Fig. 25.18 Diagram showing depolarization of the ventricles in LBBB + lateral infarction. Lead V_6 shows S wave (marked ⇧). (Reproduced with permission from Mittal SR. Left Ventricular Depolarization Delay and Myocardial Necrosis. Cardiology Today- 2015;19:68–107)

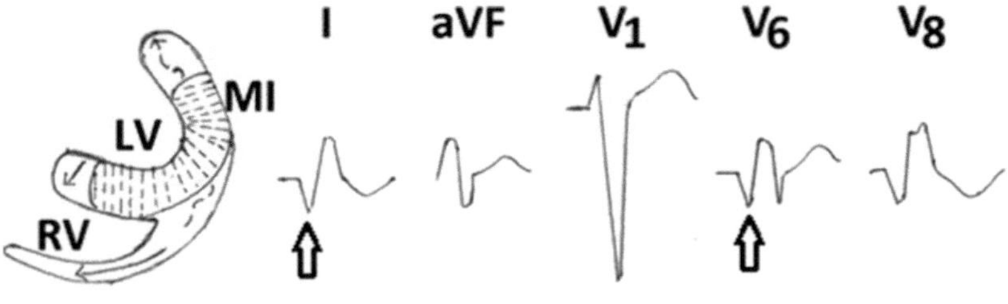

Fig. 25.19 Diagram showing depolarization of ventricles in LBBB+ anteroseptal and lateral infarction. Leads I and V_6 show abnormal Q waves (marked ⇧). (Reproduced with permission from: Mittal SR. Left Ventricular Depolarization Delay and Myocardial Necrosis. Cardiology Today- 2015;19:68–107)

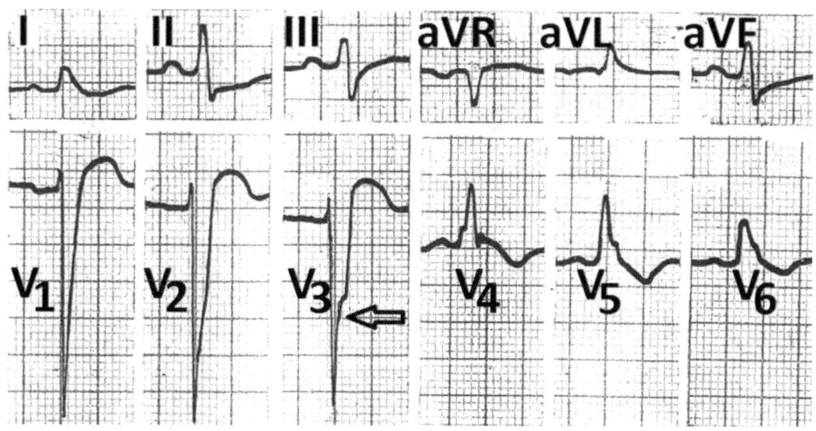

Fig. 25.21 Electrocardiogram showing Cabrera sign-notch in the ascending part of a wide S in mid precordial lead-V3 (marked ⇦). (Reproduced with permission from Mittal SR. Left Ventricular Depolarization Delay and Myocardial Necrosis. Cardiology Today- 2015;19:68–107)

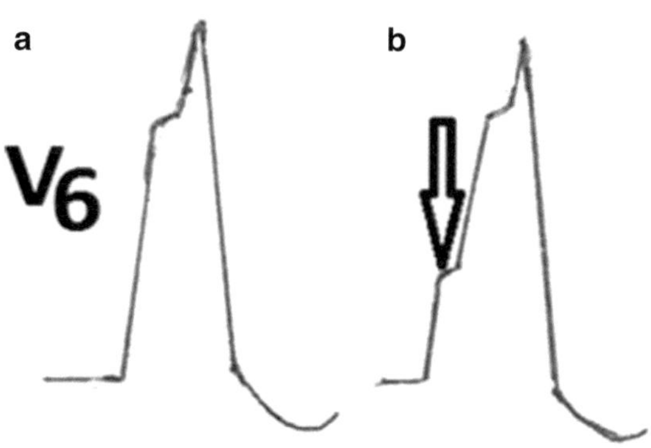

Fig. 25.22 Diagram showing (**a**) lead V6 in uncomplicated left depolarization delay. (**b**) Notch on the ascending limb of R wave in lead V_6 (marked ⇩); Chapman sign. (Reproduced with permission from Mittal SR. Left Ventricular Depolarization Delay and Myocardial Necrosis. Cardiology today- 2015;19:68–107)

(Figs. 25.22 and 25.23). This sign has low sensitivity.

Representative electrocardiogram is shown in Fig. 25.23.

2.1.7 Multiple Notches

Multiple notches on a broad QRS (Fig. 25.24) suggest fragmented conduction in the left ventricular Purkinje system and indirectly suggest myocardial necrosis.

2.1.8 Generalized Diminution in the Voltage of the QRS Complex

In the absence of any other possibility (e.g., obesity, pericardial effusion, emphysema), generalized diminution in the voltage of the QRS complex suggests diminished amplitude of various QRS vectors. Indirectly it may suggest loss of viable myocardium (myocardial necrosis) representative electrocardiogram is shown in Fig. 25.25.

2.2 Changes in QRS Duration

QRS duration depends on functional status of the Purkinje system. QRS duration of >0.14 second strongly (Fig. 25.26) suggests myocardial damage.

2.3 Ventricular Ectopics

Abnormal Q wave in ventricular ectopics with right bundle branch block configuration (Figs. 25.27, 25.28, 25.29) strongly suggests myocardial necrosis.

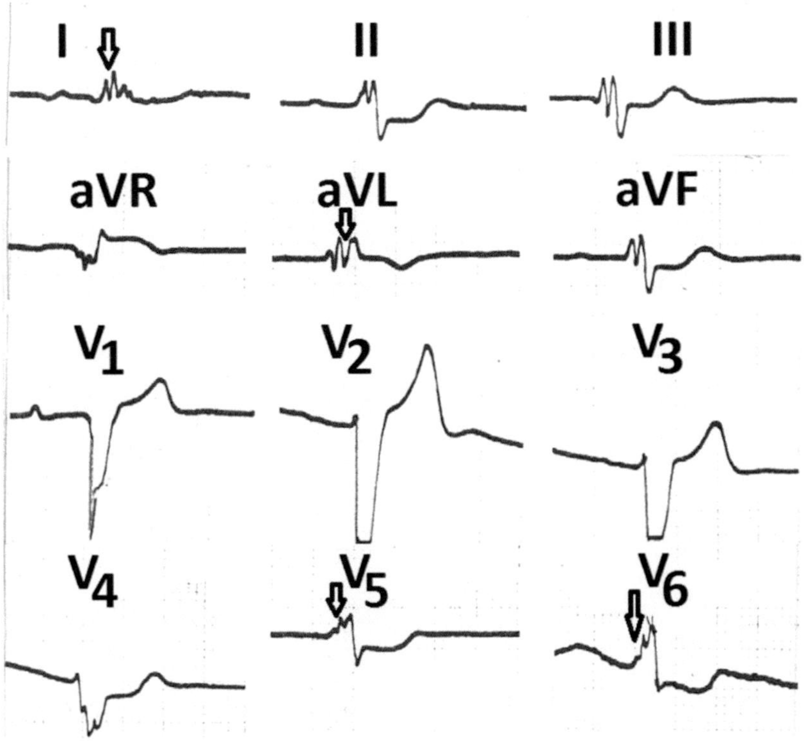

Fig. 25.23 Electrocardiogram showing Chapman sign – notch in the ascending limb of a wide R in leads I, aVL V5, and V6 (marked ⇩). (Reproduced with permission from Mittal SR. Left Ventricular Depolarization Delay and Myocardial Necrosis. Cardiology Today- 2015;19:68–107)

2.4 Changes in T Waves

Deep symmetrical T wave inversion in leads V1 to V3 (Fig. 25.30, 25.31 and 25.32) strongly suggests myocardial damage.

Similar T waves in other leads are also important (Figs. 25.32 and 25.33).

Positive T wave in leads with upright QRS (positive concordance) may be normal [4].

Most of the criteria mentioned above have low sensitivity. It is important to remember that LBBB can mask any myocardial infarction (Figs. 25.34 and 25.35).

Absence of any of the criteria mentioned above does not exclude myocardial infarction [5]. If there is strong clinical suspicion, the clinician should depend on other modalities of investigation, e.g., echocardiography [6] or biomarkers.

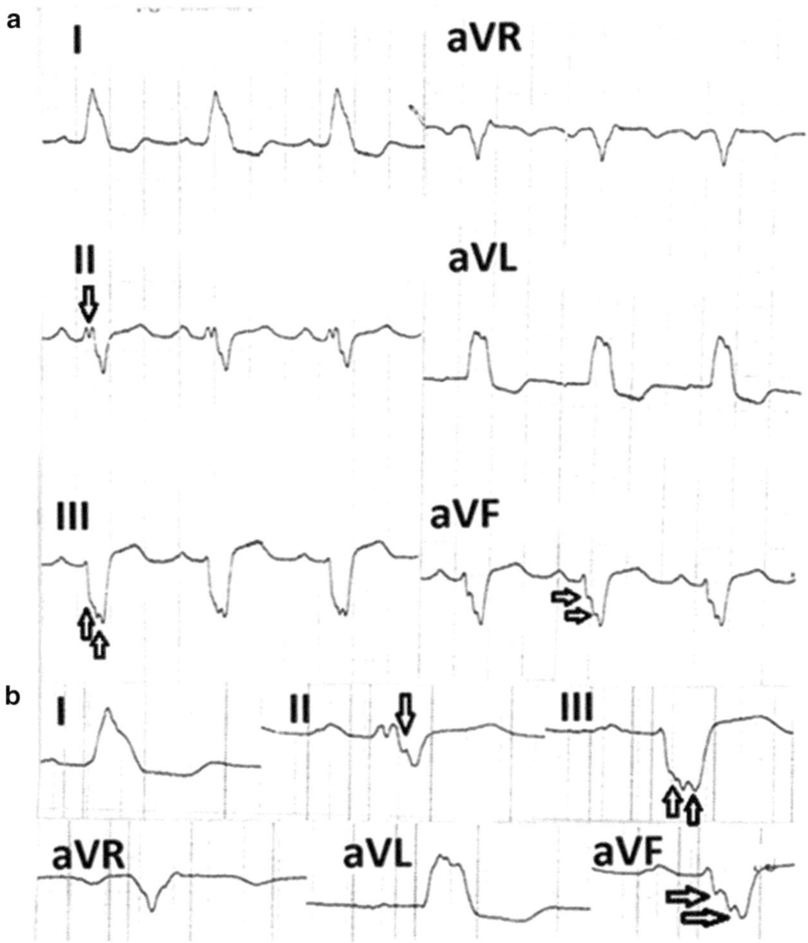

Fig. 25.24 Electrocardiogram (**a**) showing multiple notches in the QRS complex (marked arrow). (**b**) Electrocardiogram recorded at speed of 50 mm/sec. Notches in the QRS complex are better seen (marked arrows). (Reproduced with permission from Mittal SR. Left Ventricular Depolarization Delay and Myocardial Necrosis. Cardiology Today- 2015;19:68–107)

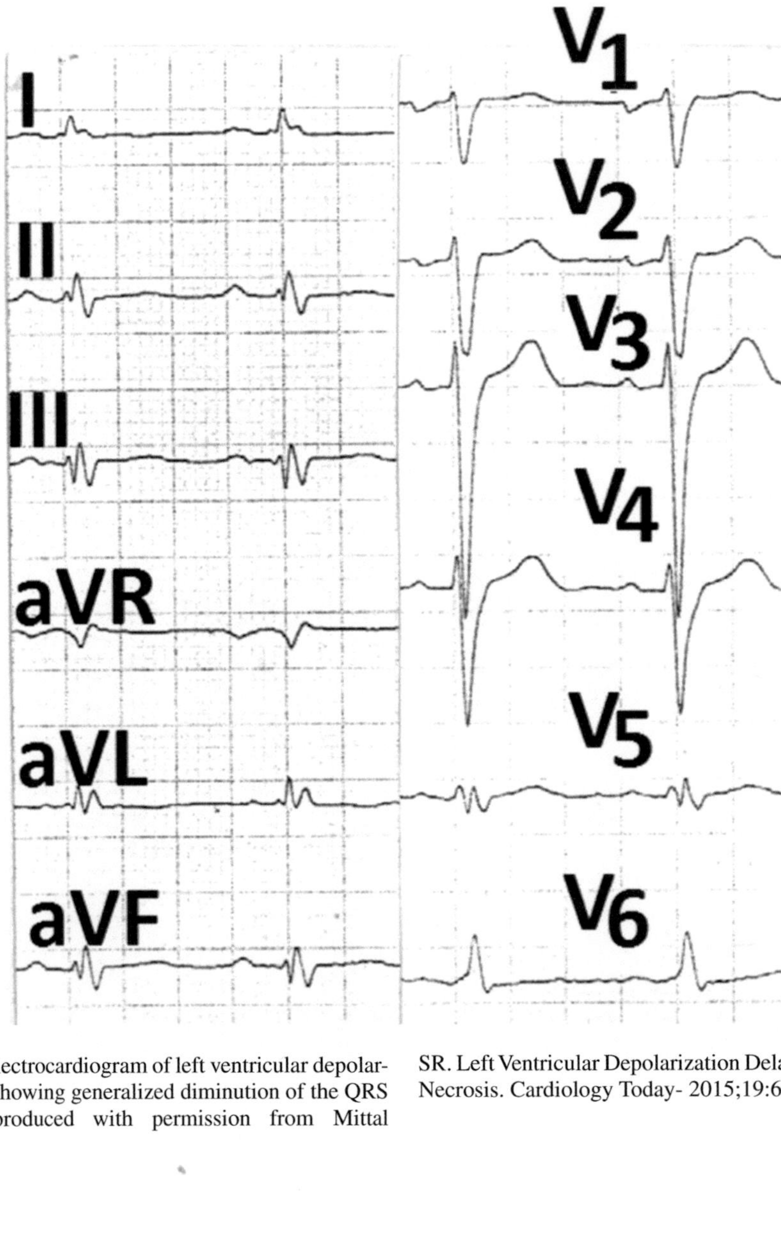

Fig. 25.25 Electrocardiogram of left ventricular depolarization delay showing generalized diminution of the QRS voltage. (Reproduced with permission from Mittal SR. Left Ventricular Depolarization Delay and Myocardial Necrosis. Cardiology Today- 2015;19:68–107)

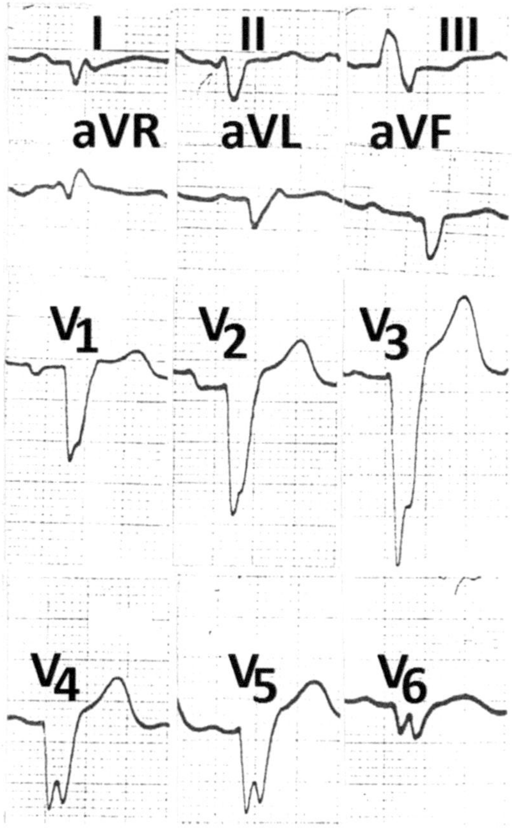

Fig. 25.26 Electrocardiogram of left ventricular depolarization delay showing QRS duration of 0.16 second. (Reproduced with permission from Mittal SR. Left Ventricular Depolarization Delay and Myocardial Necrosis. Cardiology Today- 2015;19:68–107)

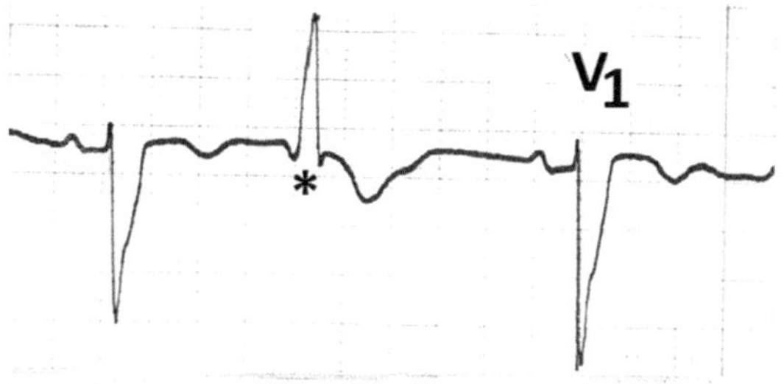

Fig. 25.27 Electrocardiogram (lead V1) showing left ventricular depolarization delay and ventricular ectopic with right bundle branch block configuration showing Q wave (marked *). (Reproduced with permission from Mittal SR. Left Ventricular Depolarization Delay and Myocardial Necrosis. Cardiology Today- 2015;19:68–107)

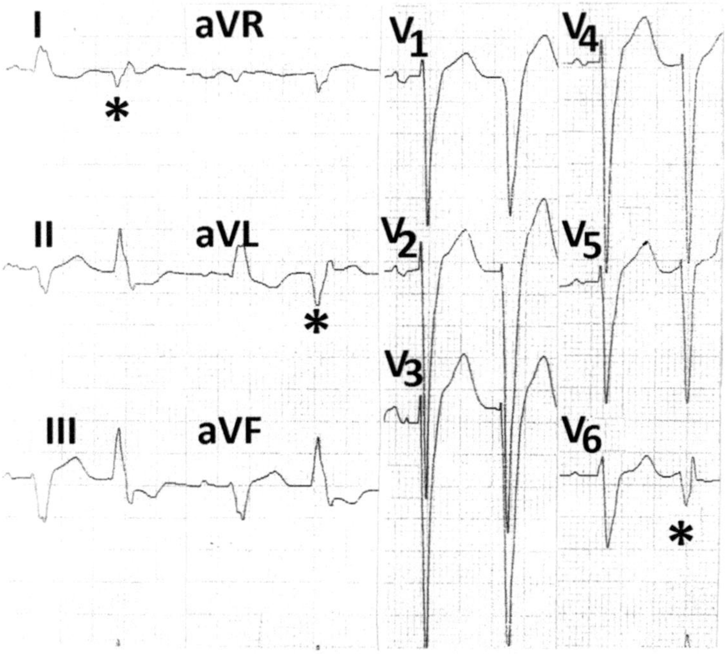

Fig. 25.28 Electrocardiogram showing left ventricular depolarization delay with ventricular ectopics showing Q wave in leads I, aVL, and V6 (marked *). (Reproduced with permission from Mittal SR. Left Ventricular Depolarization Delay and Myocardial Necrosis. Cardiology Today- 2015;19:68–107)

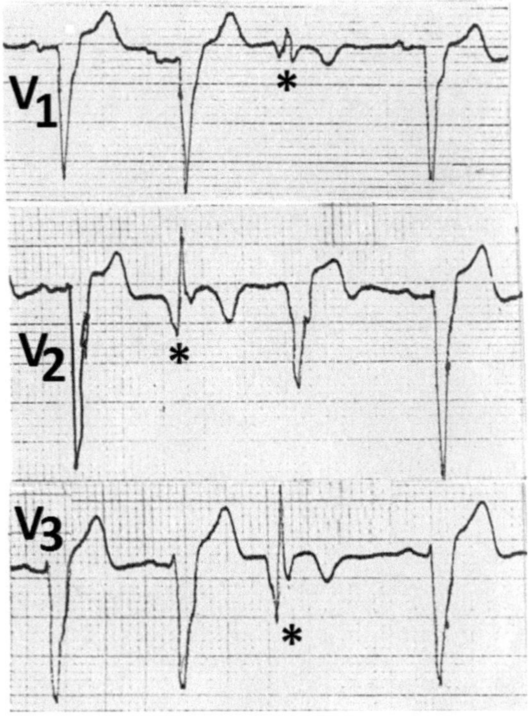

Fig. 25.29 Electrocardiogram showing left ventricular depolarization delay with ventricular ectopics having right bundle branch block pattern showing Q wave (marked *). (Reproduced with permission from Mittal SR. Left Ventricular Depolarization Delay and Myocardial Necrosis. Cardiology Today- 2015;19:68–107)

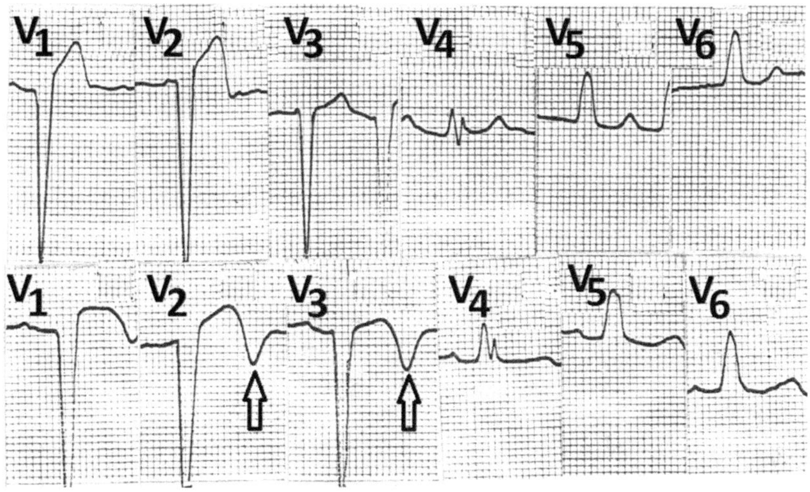

Fig. 25.30 Electrocardiogram showing left ventricular depolarization delay with T wave inversion in leads V1 to V3 (marked ⇧). (Reproduced with permission from Mittal SR. Left Ventricular Depolarization Delay and Myocardial Necrosis. Cardiology Today- 2015;19:68–107)

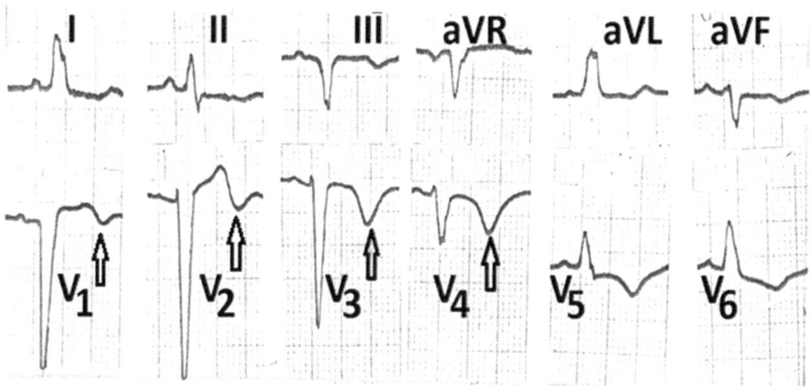

Fig. 25.31 Electrocardiogram showing left ventricular depolarization delay with T wave inversion in leads V1 to V4 (marked ⇧). (Reproduced with permission from Mittal SR. Left Ventricular Depolarization Delay and Myocardial Necrosis. Cardiology Today- 2015;19:68–107)

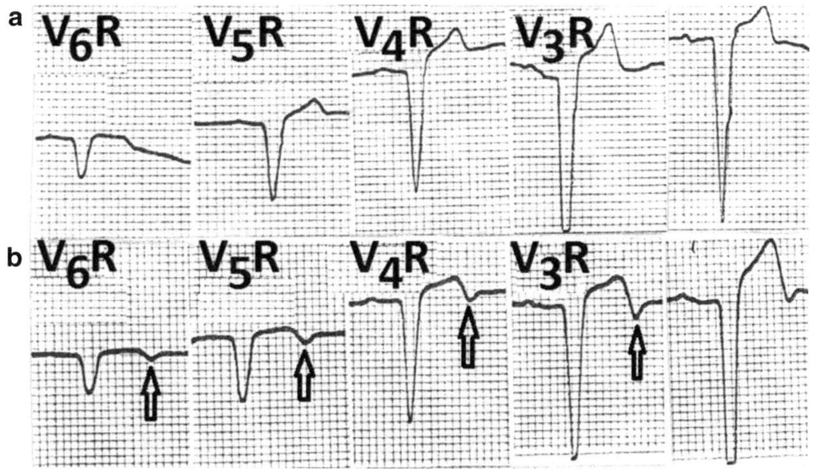

Fig. 25.32 (**a**) Electrocardiogram showing left ventricular depolarization delay. (**b**) Sequential T wave inversion in leads V6R to V1(marked ⇧) suggestive of myocardial necrosis. (Reproduced with permission from Mittal SR. Left Ventricular Depolarization Delay and Myocardial Necrosis. Cardiology Today- 2015;19:68–107)

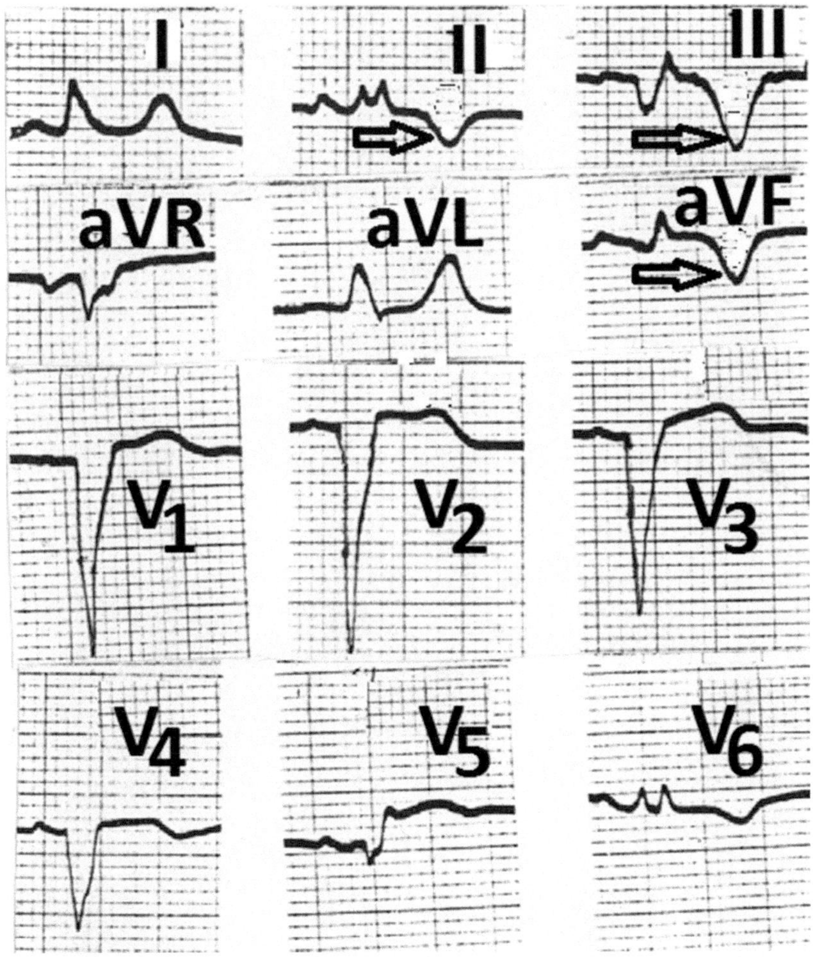

Fig. 25.33 Electrocardiogram showing left ventricular depolarization delay with symmetrical (arrowhead) T wave inversion in leads II, III, and aVF (marked ⇨) suggestive of old inferior infarction. (Reproduced with permission from Mittal SR. Left Ventricular Depolarization Delay and Myocardial Necrosis. Cardiology Today- 2015;19:68–107)

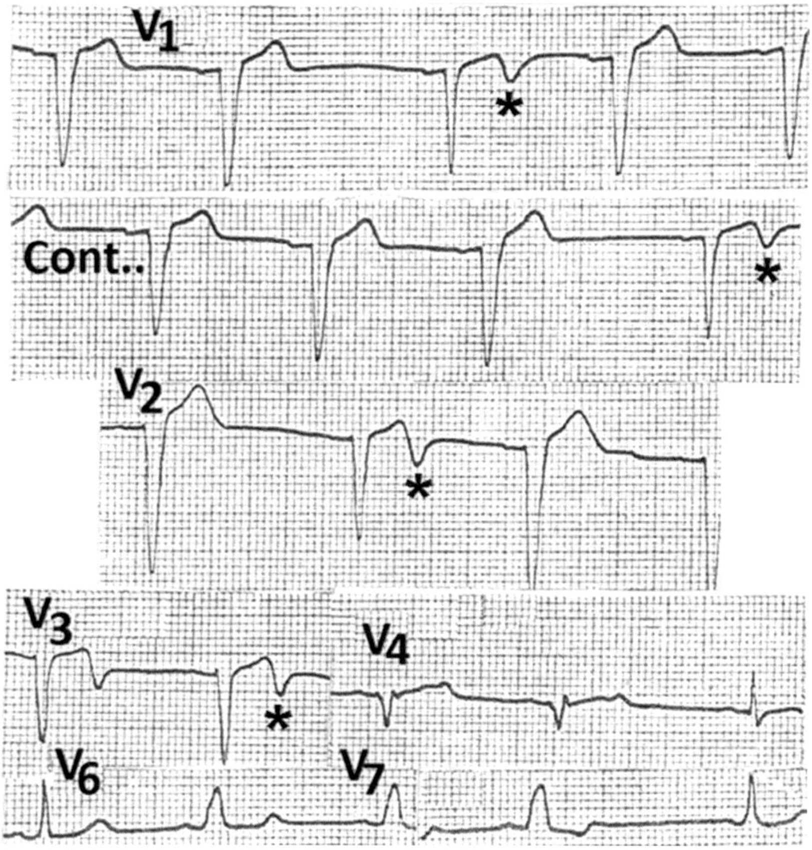

Fig. 25.34 Electrocardiogram showing intermittent tachycardia-dependent left ventricular depolarization delay masking T wave inversion seen in bradycardia-dependent narrow QRS (marked *) in leads V1 to V3. (Reproduced with permission from Mittal SR. Left Ventricular Depolarization Delay and Myocardial Necrosis. Cardiology Today- 2015;19:68–107)

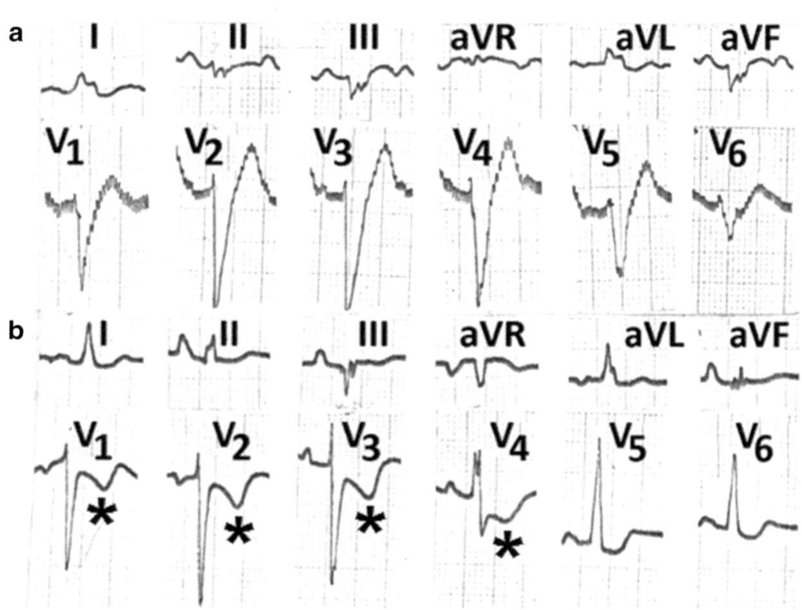

Fig. 25.35 (**a**) Electrocardiogram showing left ventricular depolarization delay. (**b**) Disappearance of left ventricular depolarization delay with unmasking of T wave inversions in leads V1 to V4 (marked *). (Reproduced with permission from Mittal SR. Left Ventricular Depolarization Delay and Myocardial Necrosis. Cardiology Today- 2015;19:68–107)

Summary

Definite electrocardiographic diagnosis of myocardial infarction is difficult in the presence of left bundle branch block (delay in left ventricular depolarization). The following findings suggest the possibility of underlying myocardial necrosis:

- Q wave of 40 ms or more in leads I, aVL/II, III, aVF/V_5,V_6.
- ST-segment and T waves concordant to the direction of the terminal part of the QRS complex, i.e.,
- ST-segment elevation with positive T wave in leads V_5 and V_6.
- ST-segment depression with inverted T wave in leads V_1 to V_3.
- Notch on the ascending limb of a wide S wave in mid precordial leads (V_3 and V_4) or ascending limb of R wave in leads I, aVL, and V_5,V_6.
- The following electrocardiographic findings also suggest the possibility of myocardial necrosis in the presence of left bundle branch block.
  - QRS duration of more than 0.14 s.
  - Premature ventricular contractions with broad Q waves.
  - Generalized diminution in the voltage of the QRS complex.
  - Multiple notches in the QRS complex.

If there is strong clinical suspicion, one should depend on non-electrocardiographic methods, e.g., echocardiography or cardiac biomarkers.

MCQs

Q1. Which electrocardiographic criteria can differentiate ischemic from nonischemic myocardial necrosis?

(a) QRS axis
(b) QRS voltage
(c) QRS duration
(d) None

Q2. Which electrocardiographic finding is suggestive of myocardial necrosis in the presence of left bundle branch block?

(a) ST-segment depression in lead V1
(b) QS configuration in lead V1
(c) Positive T wave in lead V6
(d) Q wave in lead aVL

Q3. Which electrocardiographic finding is suggestive of myocardial necrosis in the presence of left bundle branch block?

(a) Q with terminal sturring of the R wave in lead I
(b) QR in lead aVL
(c) QR in lead aVF
(d) QR in lead V6

Q4. In the presence of left bundle branch block, Q wave in lead V6 suggests

(a) Normal variation
(b) Lateral placement of electrode
(c) Left anterosuperior fascicular block
(d) Septal infarction

Q5. In the setting of left ventricular depolarization delay, reverse progression of the r wave with decreasing depth of S wave in leads V1 to V3 suggests

(a) Normal variation
(b) Septal infarction
(c) Anterior infarction
(d) Left ventricular hypertrophy

Q6. In the setting of left bundle branch block, additional left anterosuperior depolarization delay can

(a) Mask inferior infarction
(b) Mimic inferior infarction
(c) Mask anterior infarction
(d) Mimic high lateral infarction

Q7. In the setting of left bundle branch block, lateral infarction produces

(a) S wave in lead V6
(b) Q wave in lead V6
(c) Q wave in lead I
(d) S wave in lead aVL

Q8. In the setting of left bundle branch block, QrS configuration in lead V6 suggests

(a) Anterolateral infarction
(b) Posterolateral infarction
(c) High lateral infarction
(d) Anteroseptal and lateral infarction

Q9. In the setting of left bundle branch block, myocardial necrosis is most reliably suggested by

(a) Notch on the ascending limb of S wave in lead V3
(b) Monophasic QRS in lead V6
(c) QRS duration of > 0.16 second
(d) Notch on the ascending limb of the R wave in lead V6

Q10. Which feature of a ventricular ectopic suggests myocardial necrosis in context of left bundle branch block?

(a) Presence of Q wave
(b) Voltage of the R wave
(c) Depression of the ST-segment
(d) Inversion of the T wave

Q11. Which finding suggests myocardial necrosis in the presence of left bundle branch block?

(a) Inversion of the T wave in leads V1 andV2
(b) Inversion of the T wave in lead V5 and V6
(c) Positive T wave in leads I and aVL
(d) Positive T wave in leads V5 and V6

Q12. Left bundle branch block can mask

(a) Inferior infarction
(b) Lateral infarction
(c) Right ventricular infarction
(d) All

Q13. Absence of which electrocardiographic finding excludes myocardial necrosis in the presence of left bundle branch block?

(a) Q wave in lead V6
(b) Cabrera sign
(c) Chapman sign
(d) None

Q14. What is not correct for left bundle branch block in the setting of acute myocardial infarction?

(a) Reflects occlusion of proximal LAD
(b) Carries risk of sudden progression to atrioventricular block
(c) Reflects extensive myocardial damage
(d) Is usually transient

Q15. Which of the following is diagnostic of myocardial infarction in the presence of advanced LBBB?

(a) Q wave of 40 ms in leads I and aVL
(b) ST-segment elevation with tall T waves in leads V_1 to V_3
(c) ST-segment depression with T wave inversion in leads V_5, V_6
(d) All

Q16. Which electrocardiographic findings suggest the diagnosis of myocardial infarction in the presence of advanced LBBB?

(a) Absence of q wave in leads I and aVL
(b) Q wave of 40 ms or more in leads V_5 and V_6
(c) ST-segment depression with T wave inversion in leads V_1 and V_2
(d) ST-segment elevation with tall T waves in leads V_1 and V_2

Q17. Which of the following suggest the possibility of myocardial infarction in the presence of advanced LBBB?

(a) Q wave of 40 ms or more in leads II, III, and aVF
(b) ST-segment elevation with positive T wave in leads V_5 and V_6

(c) Notch on ascending limb of R wave in leads I and aVL
(d) All

Answers

(1) d, (2) a, (3) d, (4) d, (5) a, (6) a, (7) a, (8) d, (9) c, (10) a, (11) a, (12) d, (13) d, (14) d. (15) a, (16) b, c, (17) d

References

1. Liakopoulos V, Kellerth T, Christensen K. Left bundle branch block and suspected myocardial infarction: does chronicity of the branch block matter? Eur Heart J Acute Cardiovasc Care. 2013;2:182–9.
2. Gregg RE, Helfenbein ED, Zhou SH. Combining Sgarbossa and Selvester ECG criteria to improve STEMI detection in the presence of LBBB. Comput Cardiol. 2010;37:277–80.
3. Mirvis DM, Goldberger AL. Electrocardiography. In: Bonow RO, Mann DL, Zipes DP, Libby P, editors. Braunwald's Heart Disease. Missouri: Elsevier; 2012. p. 126–65.
4. Surawicz B, Childers R, Deal BJ, Gettes LS. AHA/ ACCF/ HRS recommendations for the standardization and interpretation of the electrocardiogram. Part III: Intraventricular conduction disturbances. J Am Coll Cardiol. 2009;55:976–81.
5. Tabas JA, Rodriguez RM, Seligman HK, Goldschlager NF. Electrocardiographic criteria for detecting acute myocardial infarction in patients with left bundle branch block: a meta-analysis. Ann Emerg Med. 2008;52:329–36.
6. Parekh P, Agrawal N, Vasavada A, Vinchurkar M. LBBB masking the ECG changes of inferior wall infarction: a caution to be vigilant. BMJ Case Rep. https://doi.org/10.1136/bcr-2014-205095.

26 Left Anterior Fascicular Block (LAFB) (Delay in Left Anterosuperior Depolarization) (Left Anterior Hemiblock LAHB)

1 Nomenclature

(a) Left anterosuperior depolarization delay is a preferred term because
   (i) It more completely describes the area of depolarization by the fascicle and helps in better understanding of the electrocardiographic findings.
   (ii) Left bundle branch fans into a number of interconnecting fibers (Fig. 26.1) with significant interindividual variability. It is difficult to define two independent and anatomically separate branches of the left bundle.
   (iii) Chronic diseases produce progressive delay in the rate of conduction rather than sudden "block."
   (iv) Acute appearance of left anterosuperior conduction delay in the setting of acute myocardial infarction may be transient (Fig. 26.2a,b).
   (v) Myocardial fibrosis in the anterosuperior portion of the left ventricle without isolated disease of the fascicle also produces similar electrocardiographic findings.
   (vi) The term "fascicular blocks" is used to describe two diagonally opposite electrocardiographic findings. These terms do not represent pathological findings strictly confined to a part of the left ventricular conduction system. There must be several intermediate and overlapping variants that are not yet described.

(b) By definition, mean frontal plane QRS axis between -30^0 and -90^0 is called left anterior fascicular block. It is, however, important to remember that a change in the QRS axis may not be acute. QRS axis may change gradu-

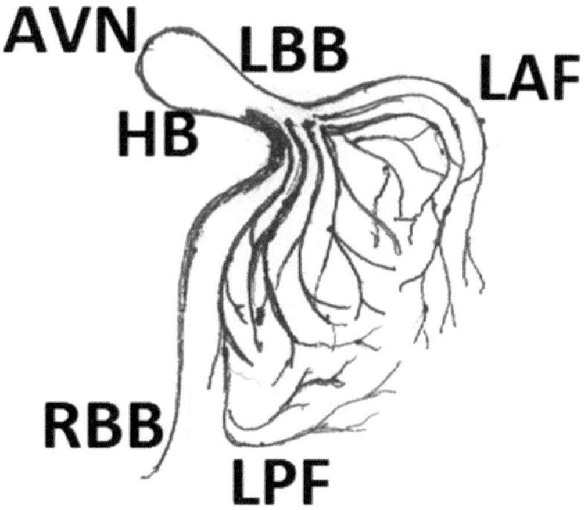

Fig. 26.1 Diagram showing that left bundle branch fans into a number of interconnecting fibers soon after its origin. *AVN* atrioventricular node, *HB* His bundle, *LBB* left bundle branch, *LAF* left anterosuperior fascicle, *LPF* left posteroinferior fascicle, *RBB* right bundle branch. (Reproduced with permission from Mittal SR. Left Anterosuperior Depolarization Delay. Cardiology Today- 2014;18:175–182)

S. Mittal, *Insights into Electrocardiograms with MCQs*,
https://doi.org/10.1007/978-981-99-0127-2_26

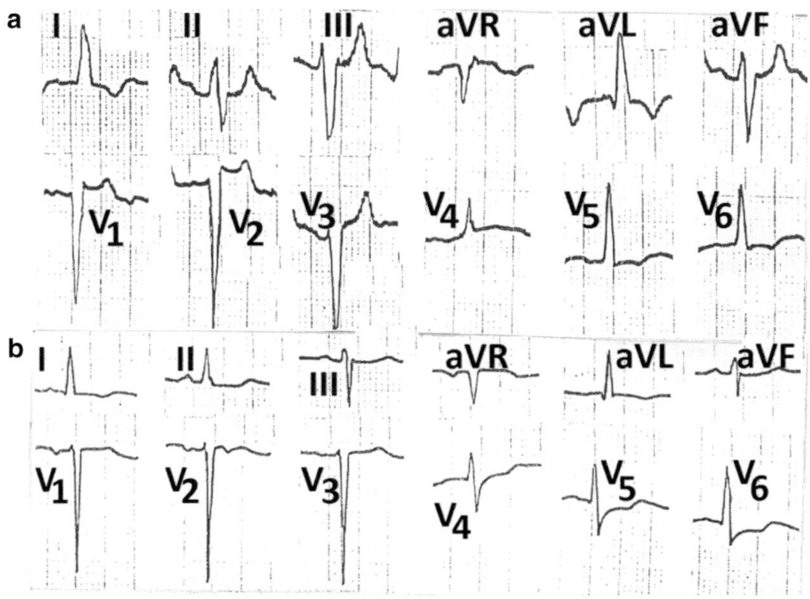

Fig. 26.2 Electrocardiogram (**a**) showing transient left anterosuperior conduction delay in the setting of anterior myocardial infarction. (**b**) Recovery in conduction delay in subsequent ECG. (Reproduced with permission from Mittal SR. Left Anterosuperior Depolarization Delay. Cardiology Today- 2014;18:175–182)

ally from less than -30^0 to more than -30^0. A progressive change from $+30^0$ to 0^0 also suggests progressively delayed depolarization of the anterosuperior part of the left ventricle even when it does not strictly qualify to be called LAFB. Therefore, exact frontal plane QRS axis should be recorded for better follow-up of patients with intraventricular conduction delay.

2 Etiology

(a) Left anterosuperior fascicle is more vulnerable to damage [1] because
  (i) These fibers are relatively thin and long.
  (ii) These fibers have single arterial supply from the septal branches of the left anterior descending coronary artery.
  (iii) These fibers are close to the aortic valve and can be involved in calcific process involving the aortic valve.

(b) This electrocardiographic finding may be present in persons without any other evidence of cardiac disease, especially in the elderly.

(c) Other conditions that are associated with variable degrees of left-axis deviation [1, 2] include
  (i) Acute myocardial infarction.
  (ii) Chronic coronary artery disease.
  (iii) Long-standing systemic hypertension with left ventricular hypertrophy. Usually, there is fibrosis in anterosuperior region.
  (iv) Cardiomyopathy: dilated or hypertrophic.
  (v) Severe aortic regurgitation. Stretching of the anterosuperior fibers is considered responsible.
  (vi) Encroachment from the calcified aortic valve.
  (vii) Pacing from the apex of the right ventricle.
  (viii) Congenital heart diseases
    – Some cases of ventricular septal defects, especially posterior inlet VSD.
    – Tricuspid atresia.
    – Endocardial cushion defects.
    – Hypoplastic right ventricle.
    – Single ventricle.

3 Electrocardiographic Diagnostic Criteria

(a) Frontal plane mean QRS axis between -45^0 and -90^0.
   Minor degree of LAFB can have axis between 0^0 and -30^0.
(b) rS configuration in leads II, III, and aVF. S wave in lead III is larger than the S wave in lead II.
(c) qR configuration in lead aVL with intrinsicoid deflection of > 45 ms.
(d) qR configuration in aVR.
(e) Normal to high QRS voltage in limb leads.
(f) No significant widening of the QRS complex in uncomplicated LAFB.

4 Genesis of the ECG Changes

4.1 Effects on Limb Leads

Normally, the anterosuperior fascicle depolarizes the anterosuperior part of the left ventricle, and the posteroinferior fascicle simultaneously depolarizes the posteroinferior part of the left ventricle. Subsequently, the two depolarization vectors move toward each other. Simultaneous depolarization of the whole left ventricle results in a narrow QRS with mean frontal plane QRS axis between 0^0 and $+90^0$ (Fig. 26.3a). In left anterosuperior fascicular block, depolarization of the anterosuperior part of the left ventricle does not occur through the anterosuperior fascicle. Normal conduction through the posteroinferior fascicle results in initial depolarization of the posteroinferior part of left ventricle. Subsequently, there is depolarization of the anterosuperior region of the left ventricle through the myocardium. The major QRS vector, therefore, moves toward the anterosuperior part of the left ventricle with the result that the mean frontal plane QRS axis moves between -45^0 and -90^0 (Fig. 26.3 b).

Initial depolarization through the posteroinferior fascicle moves inferiorly, posteriorly, and slightly to the right. This produces an initial r wave in inferior leads (II, III, and aVF) and an initial q wave in superiorly oriented leads (I, aVL, and aVR) (Fig. 26.4a). Subsequently, the major QRS vector moves laterally and superiorly producing terminal S wave in leads II, III, and aVF and terminal R wave in leads I, aVL, and aVR (Fig. 26.4b).

As the terminal vector is unopposed, it increases the amplitude of the terminal part of the QRS complex in limb leads. As most part of the terminal vector moves maximally away from the positive pole of lead III, S wave is deeper in lead

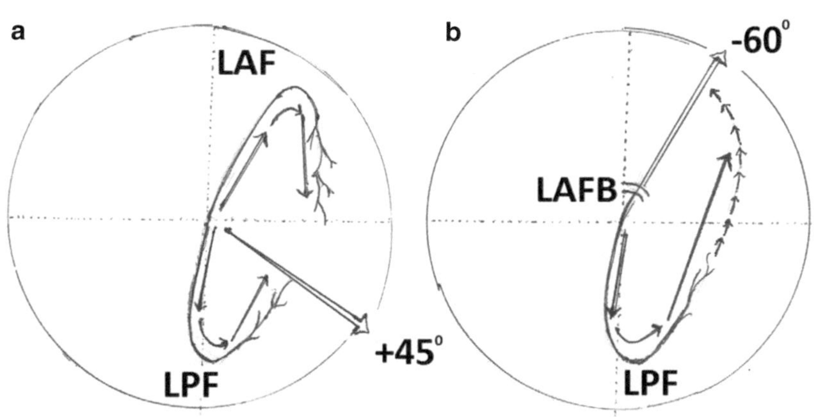

Fig. 26.3 (**a**) Diagram showing normal simultaneous depolarization through the left anterior fascicle (LAF) and the left posterior fascicle (LPF) with normal frontal plane QRS axis. (**b**) Delayed depolarization of the left ventricular anterosuperior region in the presence of LAFB with resultant abnormal left axis. *LAF* left anterior fascicle, *LPF* left posterior fascicle, *LAFB* left anterior fascicular block. (Reproduced with permission from Mittal SR. Left Anterosuperior Depolarization Delay. Cardiology Today- 2014;18:175–182)

III than in leads aVF and II. LAFB is unlikely if the S wave in lead II is deeper than the S wave in lead III.

Terminal depolarization of the anterosuperior portion of the left ventricle occurs through muscle and not through the specialized conduction tissue. Terminal vector, therefore, moves slightly slowly, resulting in a mild increase in the intrinsicoid deflection in leads aVL and aVR (>45 ms). However, the total duration of the QRS complex does not increase by more than 0.20 s. The total duration of QRS complex, therefore, remains <100 ms in uncomplicated LAFB.

4.2 Effect on Chest Leads [1, 3]

4.2.1 Leads V_1 and V_2

Normally, the initial r wave in leads V_1 and V_2 is produced by normal septal depolarization from left to right (Fig. 26.5a). In LAFB, normal septal activation is replaced by initial depolarization of the posteroinferior part of the left ventricle. This initial vector, which moves inferiorly and posteriorly, is also directed slightly to the right. Therefore, it also produces initial r wave in leads V_1 and V_2 (Fig. 26.5b).

In LAFB, QRS vector is more vertically oriented. r waves of lead V_1 and V_2 are, therefore,

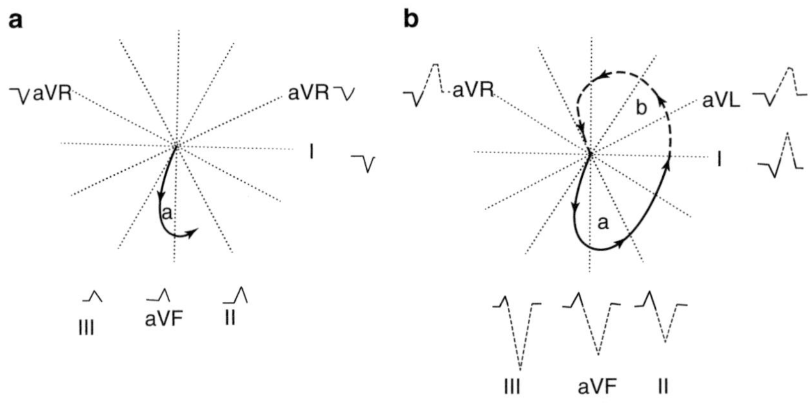

Fig. 26.4 Diagram showing the effect of the left anterosuperior fascicular block on limb leads. (**a**) Initial depolarization through the left posterior fascicle. (**b**) Subsequent slow depolarization of the anterosuperior part of the left ventricle (marked with broken line). (Reproduced with permission from Mittal SR. Left Anterosuperior Depolarization Delay. Cardiology Today- 2014;18:175–182)

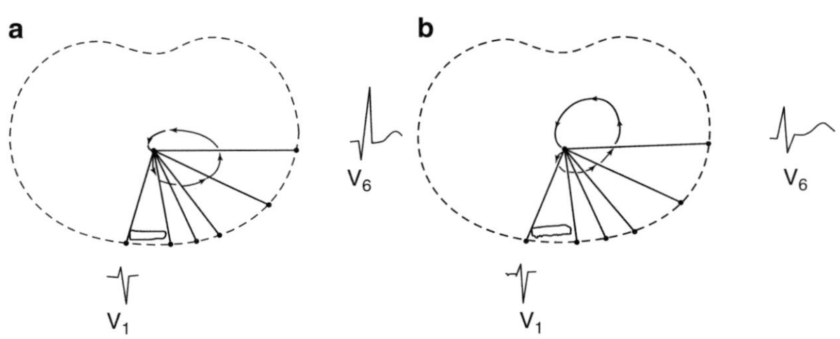

Fig. 26.5 Diagram showing the effect of left anterosuperior fascicular block on the chest leads. (**a**) Normal QRS vector in the horizontal plane. (**b**) QRS vector in horizontal plane in the presence of left anterosuperior fascicular block. (Reproduced with permission from Mittal SR. Left Anterosuperior Depolarization Delay. Cardiology Today- 2014;18:175–182)

very sensitive to lead a placement. Slightly higher placement of electrodes results in loss of the initial r wave in leads V_1 and V_2 (Fig. 26.6a,b).

In vertical heart as seen in long, slender physique or emphysema, even normally placed electrodes of leads V_1 and V_2 are relatively higher. Such patients can, therefore, also show loss of initial r wave even in normally placed electrodes of leads V_1 and V_2 in the presence of LAFB.

4.2.2 Leads V_5 and V_6

Normally, initial septal depolarization from the left to the right produces initial q wave in leads V_5 and V_6. In LAFB, this normal septal depolarization is replaced by initial depolarization of the inferoposterior part of the left ventricle. This results in loss of the initial q wave in leads V_5 and V_6 (Fig. 26.5b). In normal persons, mean QRS vector is directed laterally toward leads V_5 and V_6. This produces prominent R wave in these leads. In LAFB, the mean QRS vector is directed vertically away from the left precardial leads. This results in attenuation of the R and persistence of prominent S wave in leads V_5 and V_6 (Fig. 26.5b).

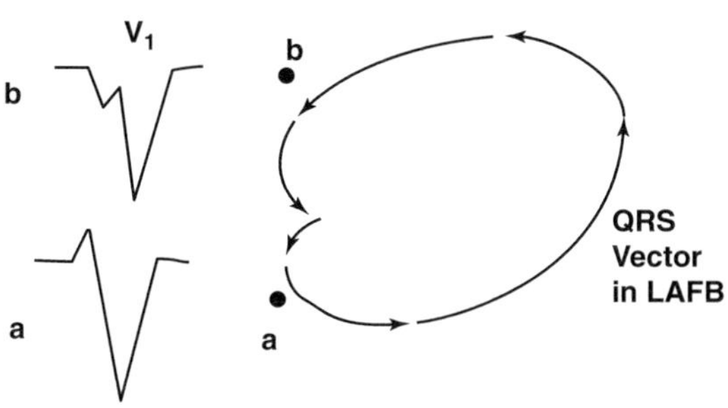

Fig. 26.6 Diagram showing the effect of QRS vector in LAFB on the vertical site of placement of leads V_1 and V_2. (**a**) Conventional site. (**b**) Slightly higher location. (Reproduced with permission from Mittal SR. Left Anterosuperior Depolarization Delay. Cardiology Today- 2014;18:175–182)

4.2.3 Right-Sided Chest Leads

As the initial vector moves slightly to the right, right-sided chest leads show rs configuration (Fig. 26.7).

4.2.4 Posterior Chest Leads

As the initial vector moves inferiorly and posteriorly, there is no q wave in leads V_7 to V_9 (Fig. 26.7).

4.3 Effect on T Waves

T waves are directed opposite to the terminal vector of the QRS complex. T waves are, therefore, inverted in leads I, aVL and upright in leads II, III, aVF, and the chest leads. In cases with QRS axis between 0^0 and -30^0, T waves may be inverted in leads II, III, and aVF.

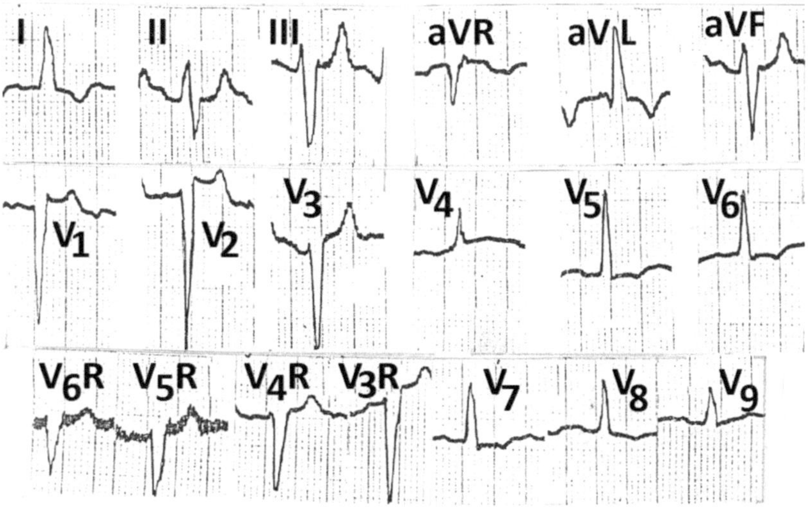

Fig. 26.7 ECG shows the effect of LAFB on right-sided chest leads (V_3R to V_6R) and posterior chest leads (V_7 to V_9). (Reproduced with permission from Mittal SR. Left Anterosuperior Depolarization Delay. Cardiology Today- 2014; 18:175–182)

5 Significance of Alterations in the QRS Complex

5.1 QRS Duration of 0.12 S or More [1]

(i) Widening of the terminal part of the QRS complex suggests additional abnormalities.
  - Figure 26.8a shows classical LAFB. In cases with additional infarction in the anterosuperior region of the left ventricle, there is loss of R zzzwave in leads I and aVL (Fig. 26.8b).
  - Intraventricular conduction delay, e.g., right bundle branch block. There is slurred S wave in leads I and V_6 with slurred R'wave in lead V_1. It produces $S_1S_2S_3$ complex (Figs. 26.8c and 26.9).
  - Fibrosis in the anterosuperior region of the left ventricle produces ECG findings simulating LAFB+anterosuperior myocardial infarction.

(ii) Widening of the whole QRS
  - It should suggest the possibility of hyperkalemia.

5.2 Absence of q Wave in Leads I and aVL

In classical LAFB, initial QRS vector is directed inferiorly, posteriorly, and slightly to the right . This produces initial q wave in lead aVL and initial r wave in lead V_1 (Fig. 26.10a). Absence of q wave in leads I and aVL can occur if LAFB is associated with other conduction abnormalities.

(i) Infarction of the interventricular septum. In this situation, normal depolarization of the interventricular septum (from left to right and anteriorly) is replaced by depolarization of the basal interventricular septum from right to left , posteriorly and superiorly (Fig. 26.10b). This results in loss of initial r wave in lead V_1 and initial q wave in lead aVL.

(ii) Incomplete left bundle branch block: r wave of lead V_1 is preserved. Conduction delay in the left bundle branch results in earlier activation of the right bundle branch. Interventricular septum is, therefore, depolarized from right to left and anteriorly.

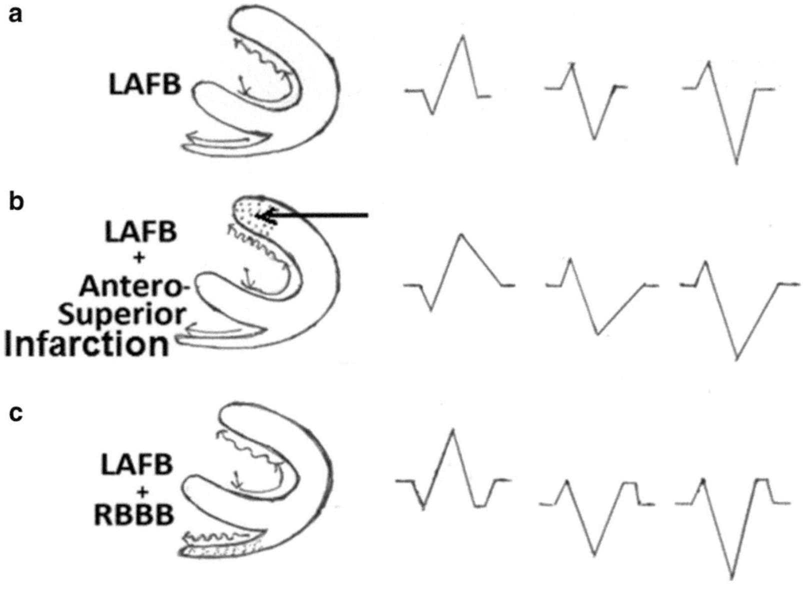

Fig. 26.8 Diagram showing the causes of wide QRS complex in the presence of LAFB. (**a**) Classical LAFB, (**b**) LAFB + infarction/ fibrosis in the anterosuperior region of the left ventricle (marked as ←), and (**c**) LAFB+ right bundle branch block. (Reproduced with permission from Mittal SR. Left Anterosuperior Depolarization Delay. Cardiology Today- 2014;18:175–182)

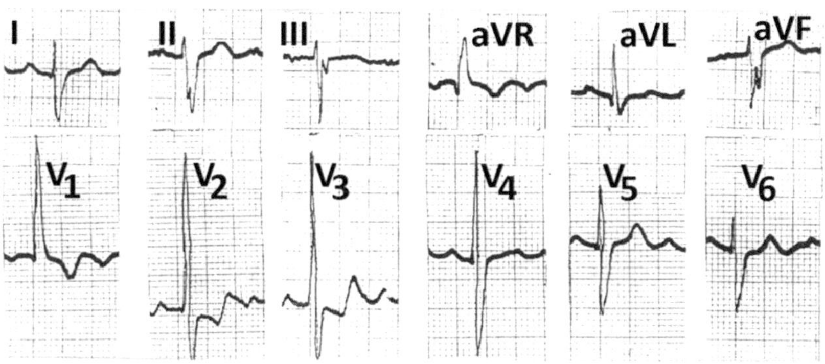

Fig. 26.9 Electrocardiogram showing LAFB with right bundle branch block. (Reproduced with permission from Mittal SR. Left Anterosuperior Depolarization Delay. Cardiology Today- 2014;18:175–182)

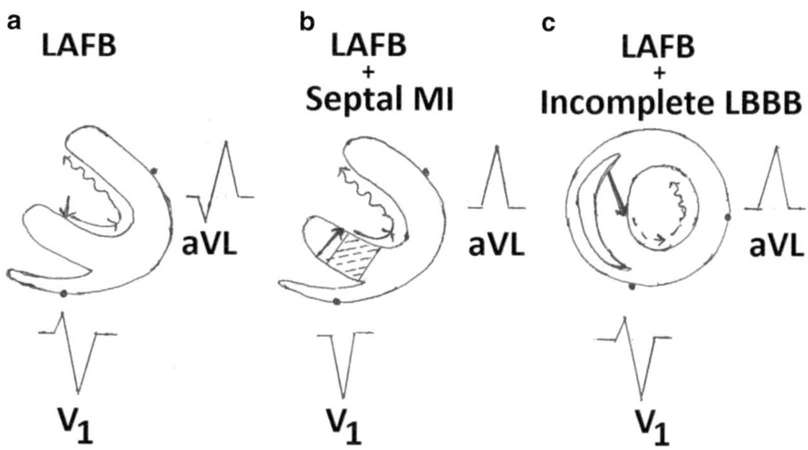

Fig. 26.10 Diagrams showing the direction of initial QRS vector in the causes of absence of q wave in leads I and aVL in the presence of LAFB. (**a**) Classical LAFB, (**b**) LAFB + septal infarction, and (**c**) LAFB+ incomplete left bundle branch block. (Reproduced with permission from Mittal SR. Left Anterosuperior Depolarization Delay. Cardiology Today- 2014;18:175–182)

Initial QRS vector is, therefore, directed to the left and anteriorly. This results in loss of q waves in leads I and aVL and preservation of r in lead V_1 (Fig. 26.10c).

5.3 QS or Qr in Lead I

Diminution or loss of R wave in lead I suggests additional lateral myocardial infarction. Initial depolarization of the interventricular septum in isolated LAFB (from left to right and inferiorly) in maintained. This results in the preservation of initial q wave in leads I and aVL (Fig. 26.11a). The magnitude of the later QRS vector produced by depolarization of the lateral wall of the left ventricle (moving upwards) is reduced due to loss of viable myocardium in the lateral wall (Fig. 26.11b). This results in diminished amplitude or loss of R wave in leads I and aVL. Q wave of more than 0.04 s also supports the possibility of lateral myocardial infarction. This happens because of the summation of q wave produced by initial QRS vector moving inferiorly with Q wave produced by infarction of the lateral wall of the left ventricle.

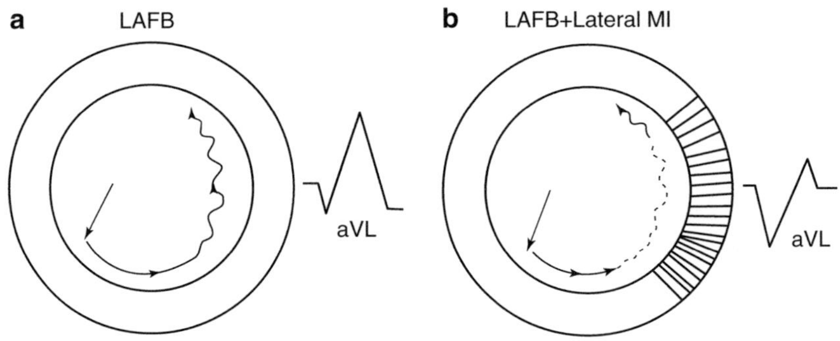

Fig. 26.11 Diagrams showing the effect of LAFB and lateral infarction on leads I and aVL. (**a**) Classical LAFB; (**b**) LAFB + lateral infarction. (Reproduced with permission from Mittal SR. Left Anterosuperior Depolarization Delay. Cardiology Today- 2014;18:175–182)

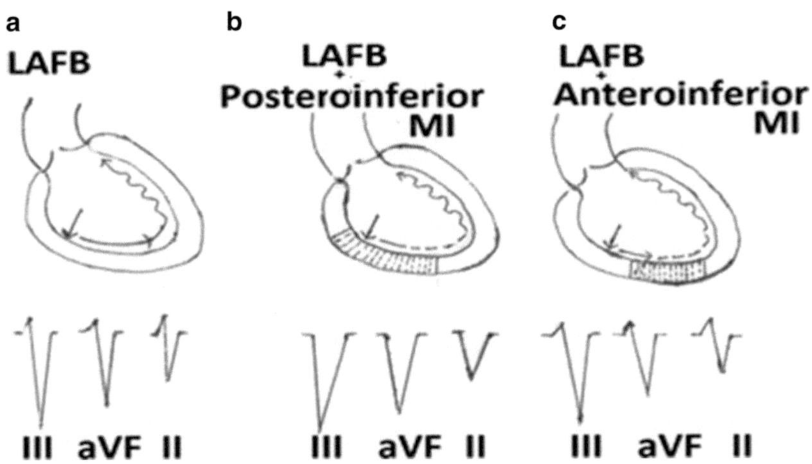

Fig. 26.12 (**a**) Diagram showing the effect of classical LAFB on ECG. (**b**) Effect of LAFB + posteroinferior infarction. (**c**) Effect of LAFB + anteroinferior infarction sparing the posterior part of the left ventricle, which is the site of initial depolarization in LAFB. (Reproduced with permission from Mittal SR. Left Anterosuperior Depolarization Delay. Cardiology Today- 2014;18:175–182)

5.4 QS Complexes in Leads II, III, and aVF

Initial depolarization of the inferoposterior part of the left ventricle produces initial r wave in leads II, III, and aVF in uncomplicated LAFB (Fig. 26.12a). Inferior infarction involving this region produces QS complexes in leads II, III, and aVF (Fig. 26.12b). Nontransmural inferior infarction sparing the posterior portion may spare initial r wave (Fig. 26.12c).

Notching (slurring) in the initial r wave in inferior leads suggests additional inferior focal block that can be due to old healed inferior myocardial infarction (Fig. 26.13).

5.5 q Waves in Leads V_2 to V_6

It suggests concomitant anterior infarction. q wave of lead aVL also becomes broad (Fig. 26.14).

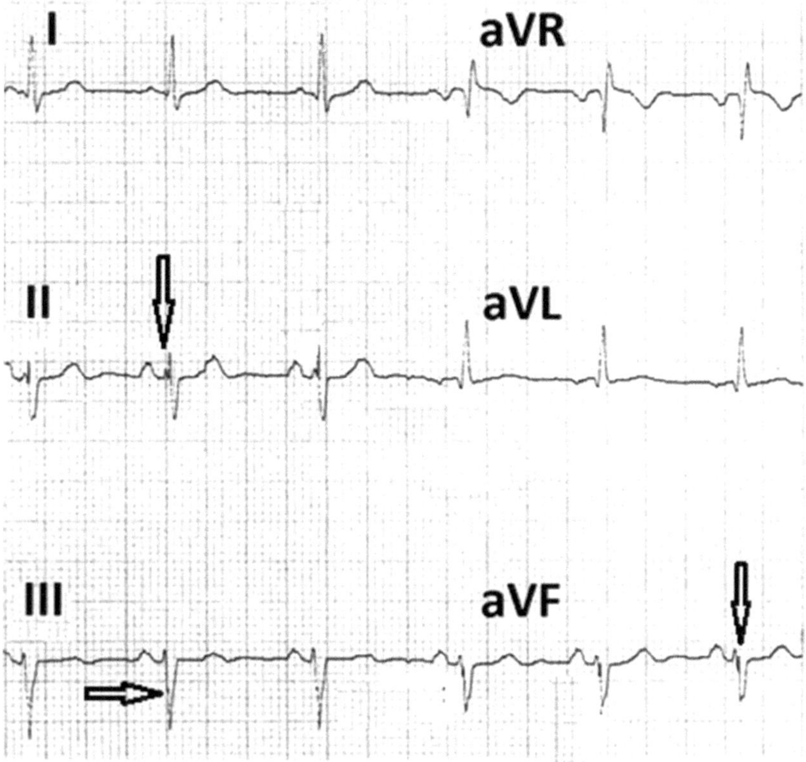

Fig. 26.13 ECG shows LAFB with notching (slurring) (marked arrows) of the initial part of the QRS complex in leads II, III, and aVF suggestive of old inferior infarction. (Reproduced with permission from Mittal SR. Left Anterosuperior Depolarization Delay. Cardiology Today- 2014;18:175–182)

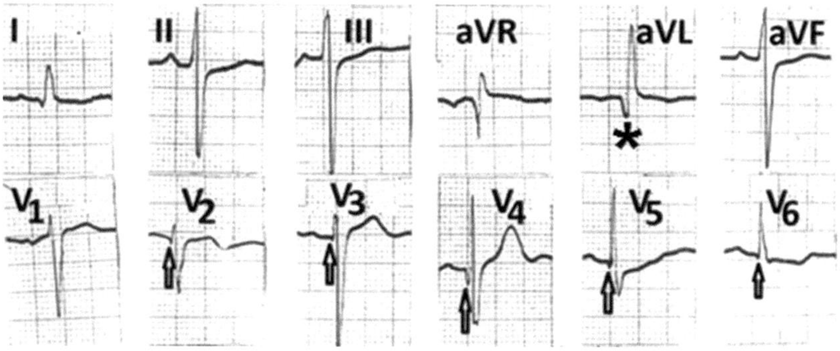

Fig. 26.14 ECG shows LAFB with the q waves in leads V_2 to V_6 (marked arrows) and broad q wave in lead aVL (marked *). (Reproduced with permission from Mittal SR. Left Anterosuperior Depolarization Delay. Cardiology Today- 2014;18:175–182)

6 Significance of Alteration in T Waves

In uncomplicated LAFB, polarity of the T wave is opposite to that of the terminal part of the QRS. Therefore, T waves are positive in leads II, III, aVF, V_5, and V_6.T wave inversion in leads II, III, aVF, and V_5, V_6 is abnormal (Fig. 26.15). Deep, symmetrical (arrowhead) T wave inversions suggest the underlying myocardial necrosis or fibrosis.

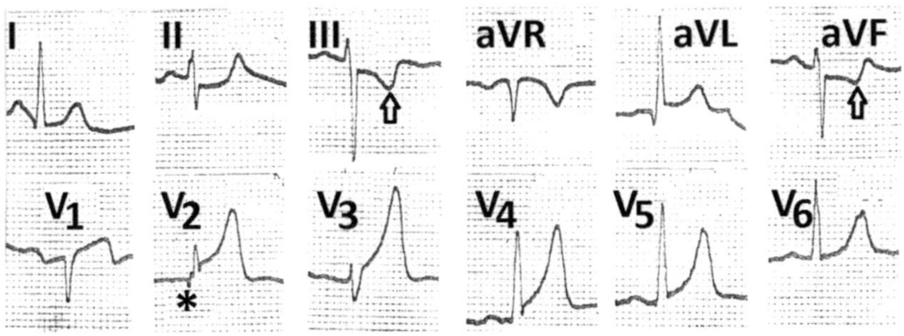

Fig. 26.15 ECG shows LAFB with T wave inversion in leads III and aVF (marked as ⇧) suggestive of concomitant inferior infarction. QS configuration in lead V_1 and qR configuration lead in V_2 (marked *) is suggestive of concomitant septal infarction. (Reproduced with permission from Mittal SR. Left Anterosuperior Depolarization Delay. Cardiology Today- 2014;18:175–182)

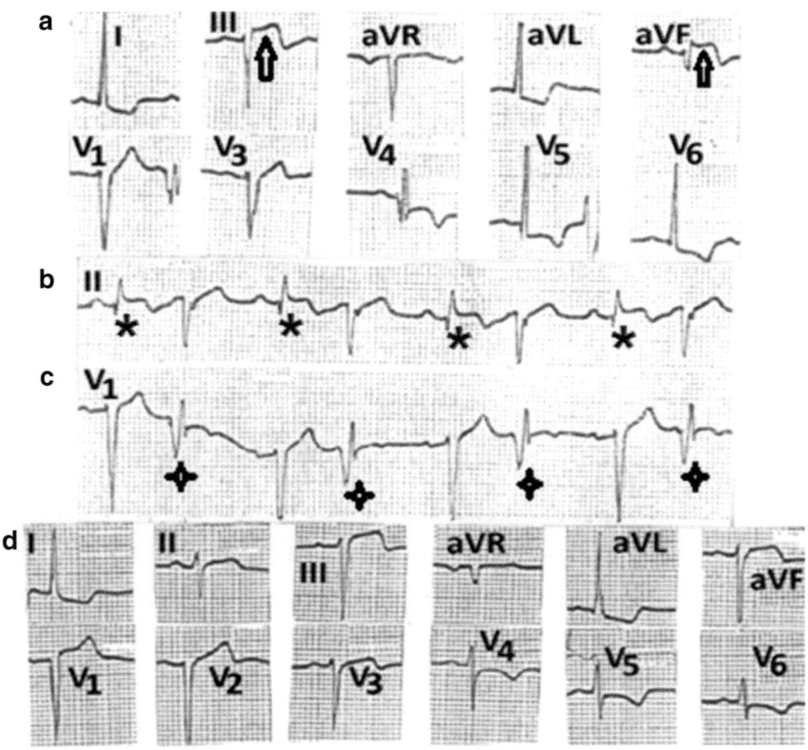

Fig. 26.16 ECG showing (**a**) inferior myocardial infarction (marked ⇧). (**b**) Rhythm strips of lead II showing q with ST-segment elevation (marked *). (**c**) Rhythm strip of lead V_1 showing ventricular ectopics with RBBB pattern showing Q wave suggestive of anterior infarction (marked ✧). (**d**) Appearance of LAFB masking inferior infarction and appearance of S wave in leads V_4 to V_6. (Reproduced with permission from Mittal SR. Left Anterosuperior Depolarization Delay. Cardiology Today- 2014;18:175–182)

7 Masking and Mimicking of Other Cardiac Disorders

7.1 Ischemic Heart Disease

(i) Initial r wave and upright T waves in leads II, III, and aVF can mask old inferior myocardial infarction (Fig. 26.16a,b,c,d).

(ii) Initial r wave in leads V_1 and V_2 can mask old anteroseptal myocardial infarction (Fig. 26.17).

Figure 26.18a,b,c also shows the effect of LAFB on anterior myocardial infarction.

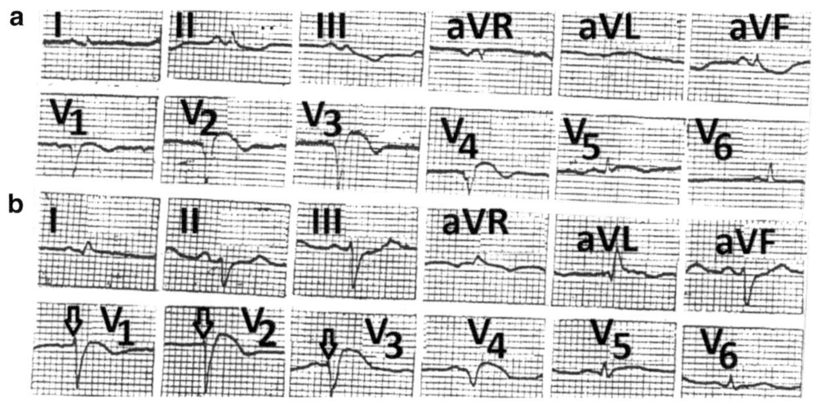

Fig. 26.17 ECG shows the effect of LAFB on anterior myocardial infarction. (**a**) Anterior myocardial infarction. (**b**) Development of LAFB results in the appearance of initial r wave in lead V_1 to V_3 (marked ⇩). (Reproduced with permission from Mittal SR. Left Anterosuperior Depolarization Delay. Cardiology Today- 2014;18:175–182)

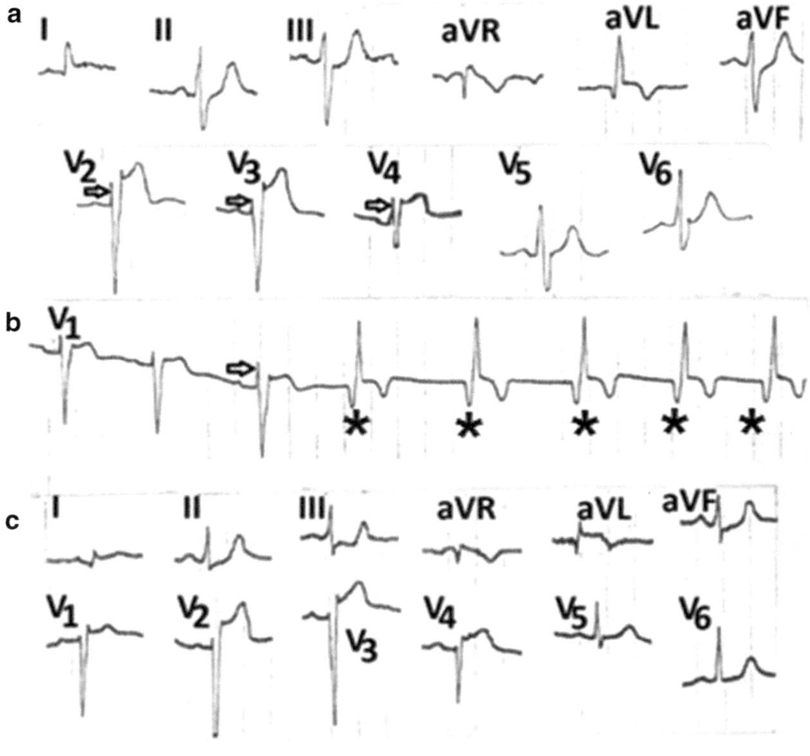

Fig. 26.18 ECG shows the effect of LAFB on anterior myocardial infarction. (**a**) Anterior myocardial infarction with LAFB. There is initial r wave in leads V_1 to V_4 (marked ⇨) and prominent S wave in V_5 and V_6. (**b**) Rhythm strip of lead V_1 shows accelerated ventricular rhythm with right bundle branch block configuration that unmasks anterior infarction (marked *). (**c**) Disappearance of LAFB is accompanied by diminished r wave in leads V_1 to V_4 and disappearance of S from leads V_5 and V_6. (Reproduced with permission from Mittal SR. Left Anterosuperior Depolarization Delay. Cardiology Today- 2014;18:175–182)

(iii) Upright T waves in leads V_5 and V_6 can mask T wave inversions of old lateral myocardial infarction.
(iv) qR configuration in leads I and aVL can mimic high lateral myocardial infarction.

7.2 Right Ventricular Hypertrophy

LAFB obscures all conventional signs of right ventricular hypertrophy from limb leads, e.g., right-axis deviation and terminal slurred S wave in lead I. Presence of S waves in leads V_5 and V_6 gives wrong impression of clockwise rotation and right ventricular dilatation. "P-pulmonale" can give a clue to the underlying right ventricular hypertrophy.

7.3 Left Ventricular Hypertrophy

LAFB masks left ventricular hypertrophy in chest leads (due to diminished R wave amplitude in leads V_5 and V_6) and overdiagnoses it in the limb leads (due to tall R wave in leads I and aVL) [1]. S_{III}>1.5 mV is arbitrarily considered suggestive of concomitant hypertrophy of the anterolateral wall of the left ventricle [1]. Severe left ventricular hypertrophy may produce prominent R wave with T wave inversion in leads V_5 ,V_6.

8 Differential Diagnosis: Other Causes of Abnormal Left-Axis Deviation Include

8.1 Left Ventricular Hypertrophy [3]

QRS axis is between 0^0 and -45^0. Increased QRS voltage in chest leads, T wave inversion in leads V_5 ,V_6 and "P–mitrale" support the diagnosis of left ventricular hypertrophy (Fig. 26.19).

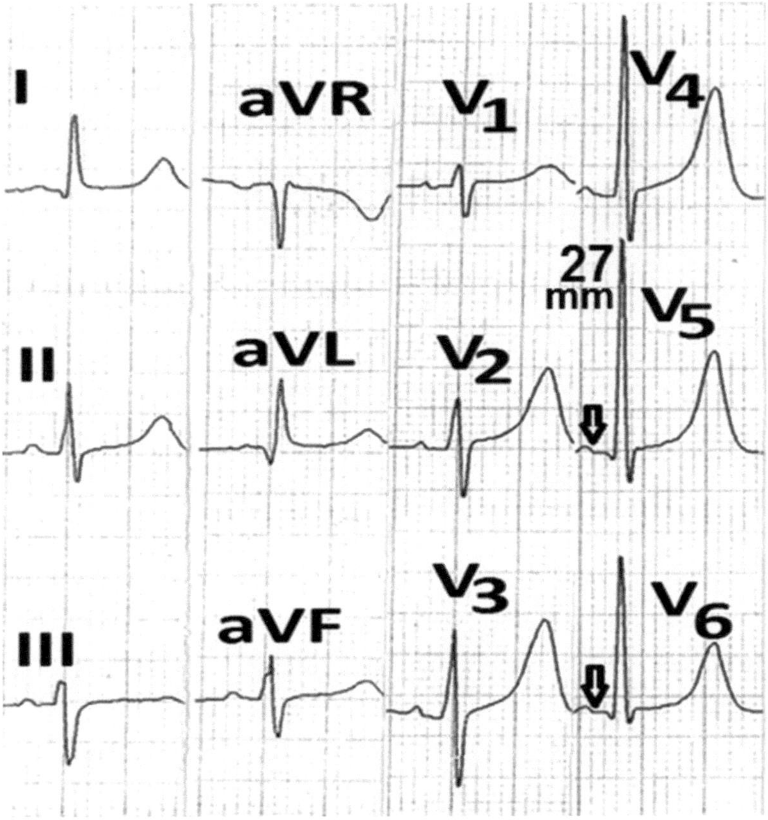

Fig. 26.19 ECG shows left-axis deviation in left ventricular hypertrophy. R wave amplitude in lead V5 is 27 mm and there is "p–mitrale" in leads V5 and V6 (marked ⇩). (Reproduced with permission from Mittal SR. Left Anterosuperior Depolarization Delay. Cardiology Today-2014;18:175–182)

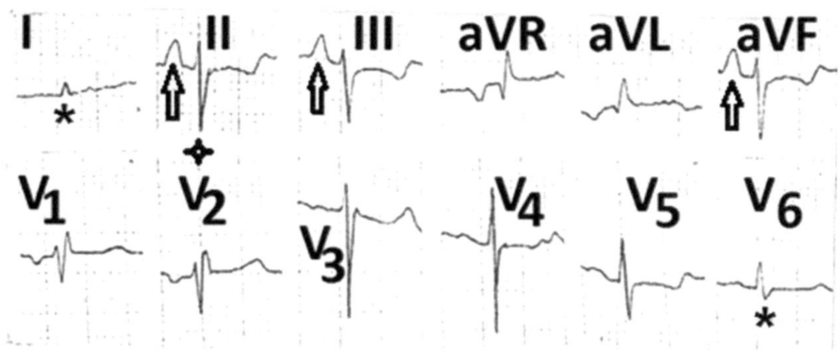

Fig. 26.20 ECG shows left-axis deviation in emphysema. It shows low voltage in leads I and V_6 (marked *). "P-pulmonale" (marked ⇧) and clockwise rotation. S wave in lead II (marked ⇧) is deeper than S wave in lead III. In LAFB, S wave in lead III is deeper than S wave in II. (Reproduced with permission from Mittal SR. Left Anterosuperior Depolarization Delay. Cardiology Today- 2014;18:175–182)

8.2 Emphysema

Low voltage in leads I and V_6 differentiates it from LAFB. "P-pulmonale" and $S_1S_2S_3$ support the diagnosis of emphysema (Fig. 26.20).

8.3 Inferior Myocardial Infarction

Leads II, III, and aVF show QS configuration (marked as *) with inverted T waves (Fig. 26.21).

8.4 Posteroseptal Preexcitation [2]

Short PR interval and slurring of the initial part of the QRS complex (the delta wave) suggest this possibility (Fig. 26.22a,b).

8.5 Left Coronary Injection

Lead I and aVL show inversion of the T waves, slight depression of the ST-segment, and prolongation of QT interval. These changes are due to acute, transient anterolateral left ventricular focal block. The absence of changes in the early part of the QRS differentiates these changes from LAFB.

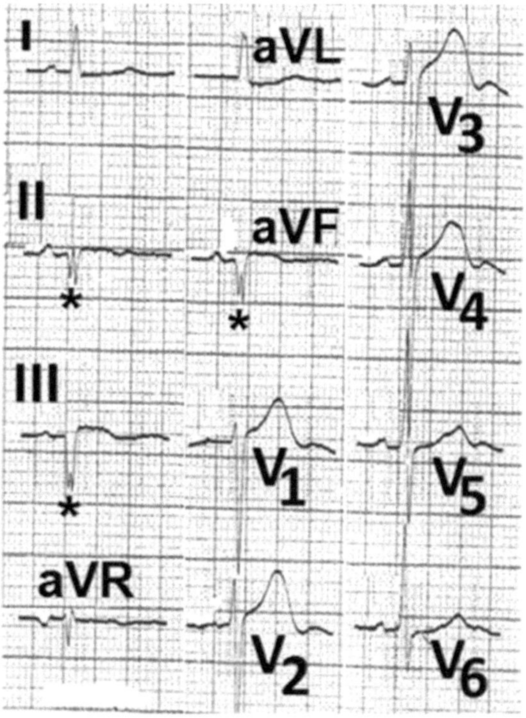

Fig. 26.21 ECG shows left-axis deviation due to inferior myocardial infarction. Lead III and aVF show QS configuration rather than rS configuration as seen in LAFB. (Reproduced with permission from Mittal SR. Left Anterosuperior Depolarization Delay. Cardiology Today- 2014;18:175–182)

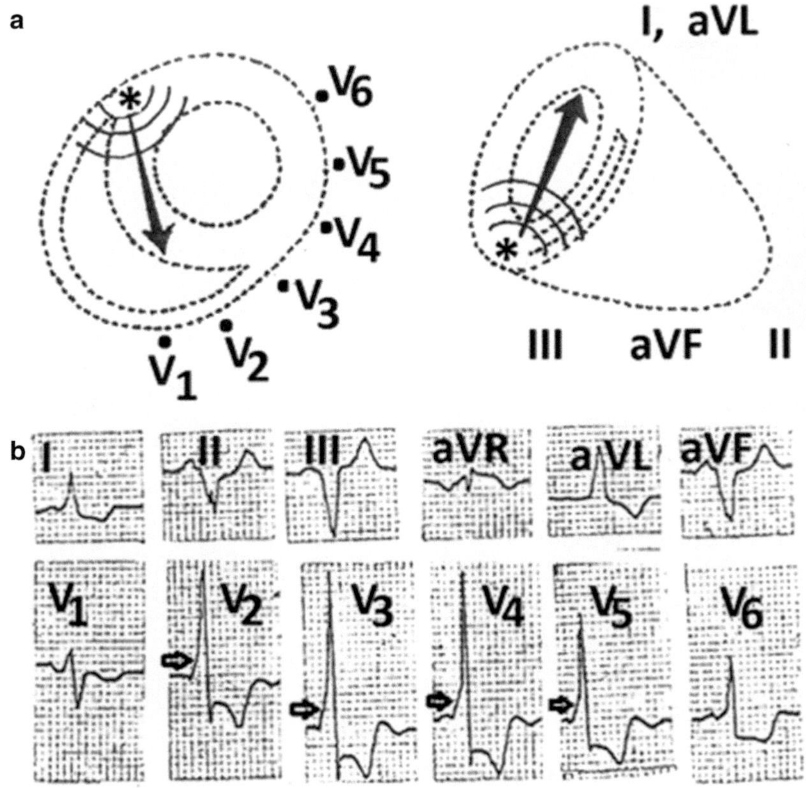

Fig. 26.22 (**a**) Diagram showing the direction of ventricular depolarization in a case of posteroseptal accessory pathway. (**b**) ECG shows left-axis deviation due to preexcitation via posteroseptal accessory pathway. PR interval is short (clear in leads II, III, and aVF). There is slurring of the initial part of the QRS (the delta wave, marked as ⇨). (Reproduced with permission from Mittal SR. Left Anterosuperior Depolarization Delay. Cardiology Today-2014;18:175–182)

Summary

- Left anterosuperior fascicular block (hemiblock) is diagnosed by QRS duration of less than 120 ms, mean frontal plane QRS axis of -45^0 to -90^0, qR pattern in leads I and aVL, rS configuration in leads II, III, aVF, and increased QRS voltages in limb leads. S wave in lead III is larger than in lead II and r is more prominent in lead II than in lead III. S wave is present up to lead V_6 (delayed transition zone pattern). Time to peak of R wave in lead aVL is more than 45 ms.
- Other conditions that can produce a left-axis deviation and need differentiation include inferior infarction, pacing from the apex of right ventricle, S_1 S_2 S_3 syndrome.
- QS pattern in place of the rs pattern in leads II, III, and aVF suggests additional inferior infarction.
- Left anterosuperior fascicular block can mask old inferior infarction by producing rs configuration in leads II, III, and aVF.

MCQs

Left Anterosuperior Fascicular Block

Q1. In the setting of acute inferior infarction, new appearance of left anterosuperior fascicular block suggests occlusion of

(a) Proximal right coronary artery
(b) Proximal left circumflex coronary artery
(c) First septal perforator
(d) First diagonal

Q2. Proximal part of the left anterosuperior fascicle is supplied by

(a) Septal perforator from left anterior descending coronary artery
(b) Septal perforator from posterior descending coronary artery
(c) Atrioventricular nodal artery
(d) First diagonal

Q3. Left-axis deviation is not seen in

(a) Subaortic VSD
(b) Coronary artery disease
(c) Systemic hypertension
(d) Severe aortic regurgitation

Q4. New-onset left anterosuperior fascicular block is not a manifestation of

(a) Acute myocardial ischemia
(b) Rate-dependent aberrancy
(c) Hyperkalemia
(d) Digitalis overdose

Q5. In left anterosuperior fascicular block, mean frontal plane QRS axis is between

(a) 0^0 and -30^0
(b) -30^0 and -45^0
(c) -45^0 and -90^0
(d) -90^0 and -120^0

Q6. Mean frontal plane QRS axis between 0^0 and -30^0 is not present in

(a) Left anterior fascicular block
(b) Left ventricular hypertrophy
(c) Systemic hypertension
(d) Cardiomyopathy

Q7. Which of the following does not produce a left-axis deviation?

(a) Pacing from the apex of the right ventricle
(b) Inferior myocardial infarction
(c) Left lateral accessory path
(d) Left coronary injection

Q8. In left anterosuperior fascicular block:

(a) $S_I > S_{II}$
(b) $S_{II} > S_{III}$
(c) $S_{III} > S_{II}$
(d) $SV_6 > S_{III}$

Q9. What is correct for left anterosuperior fascicular block?

(a) $R_{II} > R_{III}$
(b) $R_{III} > R_{II}$
(c) $R_I < R_{III}$
(d) $RV_6 > R_{III}$

Q10. Intrinsicoid deflection (time to peak of R wave) is > 40 ms in lead AVL is seen in

(a) Left anterior fascicular block
(b) Left ventricular hypertrophy
(c) Right ventricular hypertrophy
(d) Systemic hypertension

Q11. What is not seen in uncomplicated left anterosuperior fascicular block?

(a) Poor progression of R wave in precordial leads
(b) Small q wave in leads V_2 to V_6
(c) R wave in aVR follows R wave in aVL
(d) QRS duration is > 120 ms

Q12. What is not seen in uncomplicated left anterosuperior fascicular block?

(a) T wave inversion in lead aVL
(b) T wave upright in lead V_5
(c) T wave inversion in lead III
(d) q wave in lead aVL

Q13. Left anterosuperior fascicular block with wide QRS does not occur with additional

(a) Right bundle branch block
(b) High lateral infarction

(c) Septal infarction
(d) Hyperkalemia

Q14. Left anterosuperior fascicular block does not mask

(a) Left ventricular hypertrophy in limb leads
(b) Left ventricular hypertrophy in chest leads
(c) Old inferior infarction
(d) High lateral infarction

Q15. In the presence of the left anterosuperior fascicular block, left ventricular hypertrophy can be suspected by

(a) Mean frontal plane QRS axis leftward than -30^0
(b) QRS duration > 120 ms
(c) $S_{III} > 1.5$ mv
(d) $SV_6 > 1$ mv

Q16. Advanced left anterosuperior fascicular block produces

(a) QRS duration >120 ms
(b) Slurring on the ascending limb of R wave in lead I
(c) Slurring on the descending limb of R wave in lead aVL
(d) Slurring on the ascending limb of R wave in lead V_6

Q17. What is correct about isolated left anterosuperior fascicular block?

(a) It is often benign
(b) Advances to LBBB
(c) Advances to atrioventricular block
(d) Results in left ventricular failure

Q18. Left anterosuperior fascicle is

(a) Thinner than left posteroinferior fascicle
(b) Is present in left ventricular outflow
(c) Is less frequently damaged than left posteroinferior fascicle
(d) None

Q19. In left anterosuperior fascicular block, QRS duration is

(a) Less than 80 ms
(b) Less than 120 ms
(c) More than 120 ms
(d) More than 140 ms

Q20. In left anterosuperior fascicular block, qR pattern is present in leads

(a) I, aVL
(b) II, III, aVF
(c) V_5, V_6
(d) V_7 to V_9

Q21. In left anterosuperior fascicular block, rS pattern is present in leads

(a) I, aVL
(b) II, III, aVF
(c) V_5, V_6
(d) V_7 to V_9

Q22. In left anterosuperior fascicular block, QRS voltage is increased in

(a) Limb leads
(b) Leads V_1 to V_3
(c) Leads V_7 to V_9
(d) Leads V_3R to V_6R

Q23. In left anterosuperior fascicular block

(a) S wave in lead II is larger than that in lead III
(b) S wave in lead III is larger than that in lead II
(c) r wave in lead II is more prominent than that in lead III
(d) r wave in lead III is more prominent than that in lead II

Q24. In left anterosuperior fascicular block, lead V_6 shows

(a) rR' pattern
(b) qR pattern
(c) RS pattern
(d) rS pattern

Q25. Which myocardial infarction can produce left axis?

(a) Inferior infarction
(b) Lateral infarction
(c) Posterior infarction
(d) Right ventricular infarction

Q26. Left anterosuperior fascicular block can mask

(a) Septal infarction
(b) Lateral infarction
(c) Inferior infarction
(d) Right ventricular infarction

Q27. In left anterosuperior fascicular block with inferior infarction, leads II, III, and aVF show

(a) qR pattern
(b) rs pattern
(c) QS pattern
(d) rR' pattern

Answers

(1)c, (2)a, (3)a, (4)d, (5)c, (6)a, (7)c, (8)c, (9)a, (10)a, (11)d, (12)c, (13)c, (14)a, (15)c, (16)c, (17)a. (18) a, b (19) b, (20) a, (21)b, (22)a, (23)b, c, (24)c,, (25)a, (26)c, (27)c

References

1. Schamroth L. The hemiblocks (fascicular blocks). In: Schamroth C, editor. An introduction to electrocardiography. Blackwell Science: France; 1982. p. 105–15.
2. De luna AB, Goldwasser B, Fiol M, Bayes-Genis A. Surface electrocardiography. In: Fuster V, Walsh RA, Harrington RA, editors. Hurst's The Heart. New York: Mc Graw Hill; 2011. p. 302–70.
3. Mirvis DM, Goldberger AL. Electrocardiography. In: Bonow RO, Mann DL, Zipes DP, Libby P, editors. Braunwald's Heart Disease. Elsevier: Missouri; 2012. p. 126–65.

27 Left Posteroinferior Fascicular Block (Delay in Depolarization of the Posteroinferior Fascicle)

1 Nomenclature

Delay in depolarization of the left posteroinferior fascicle is a better term because

(a) It better describes the area of depolarization by the fascicle and helps in better understanding of the electrocardiographic findings.
(b) Left bundle branch fans into a number of interconnecting fibers with significant inter-individual variability. It is difficult to define two independent and separate anatomical halves of the left bundle.
(c) Posteroinferior depolarization delay may occur due to diseases other than those directly involving the fascicles.
(d) Chronic diseases may produce progressive delay in conduction rather than a sudden block. Only patients with sudden appearance of ECG findings during acute myocardial infarction can be considered to have a fascicular "block."
(e) Many persons have isolated electrocardiographic findings without the clinical criteria considered necessary for the diagnosis of left posterior fascicular block (LPFB). These patients could be having delayed depolarization of the posteroinferior left ventricle for reasons not yet clear.

2 Etiology

(a) Classical left posterior fascicular block is less common than left anterior fascicular block because
   – Left posterior fascicle is thicker than the left anterior fascicle.
   – Left posterior fascicle has dual arterial supply from circumflex artery and right coronary artery.
   – It is located near the left ventricular inflow tract [1] (Fig. 27.1) and is, therefore, not involved in diseases of the aortic valve.
(b) It is unusual in otherwise healthy persons [1].
(c) Coronary artery disease is the commonest cause.

S. Mittal, *Insights into Electrocardiograms with MCQs*,
https://doi.org/10.1007/978-981-99-0127-2_27

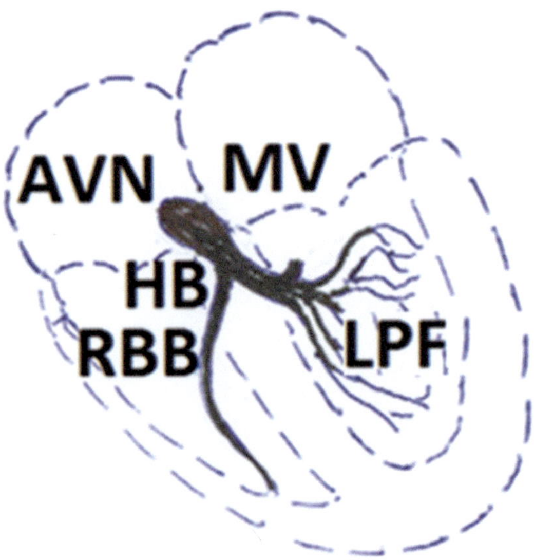

Fig. 27.1 Diagram showing the position of the left posterior fascicle in the left ventricular inflow tract. *AVN* atrioventricular node; *HB* His bundle; *LPF* left posterior fascicle; *MV* mitral valve; *RBB* right bundle branch. (Reproduced with permission from Mittal SR Left Postero-inferior Depolarization Delay. Cardiology Today- 2014;18:214–221)

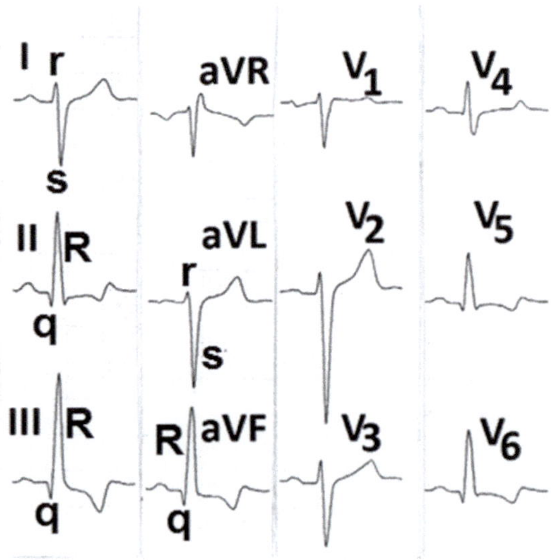

Fig. 27.2 Electrocardiogram showing left posterior fascicular block. Leads I and aVL show rS pattern and leads II, III, and aVF show qR pattern. R wave in lead III is taller than R wave in lead II. (Reproduced with permission from Mittal SR. Left Postero-inferior Depolarization Delay. Cardiology Today- 2014;18:214–221)

3 Diagnostic Criteria

3.1 Electrocardiographic Criteria

(i) Duration of the QRS complex is 100 ms or less.

(ii) Mean frontal plane QRS axis is between + 90^0 and $+120.^0$

Usually, QRS axis does not deviate beyond + 120^0 in uncomplicated left posterior fascicular block [2].

(iii) Leads I and aVL show RS or rs configuration [3].

(iv) Leads II, III, and aVF have qR configuration. R wave is usually tallest in lead III [2].

(v) QRS voltage is increased in limb leads [2].

(vi) S wave is present up to lead V_6 [3].

(vii) ST-segment and T waves are opposite to the terminal part of the QRS complex.

Representative electrocardiograms are shown in Figs. 27.2 and 27.3.

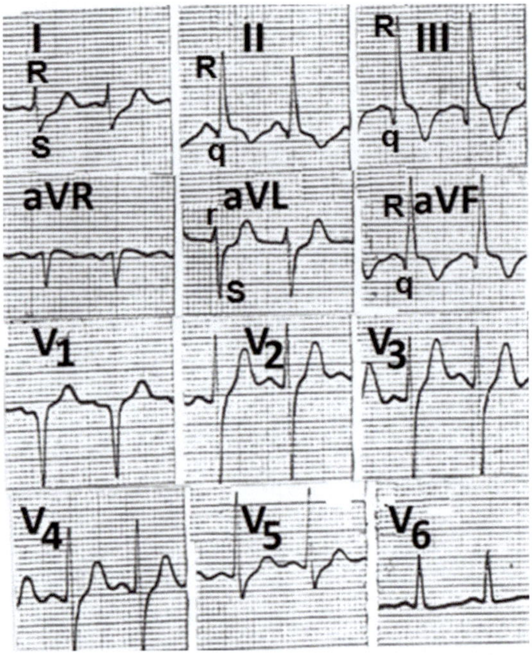

Fig. 27.3 Electrocardiogram showing left posterior fascicular block. Leads I and aVL show rs configuration. Leads II, III, and aVF show qR configuration R wave is taller in lead III than in lead II. (Reproduced with permission from Mittal SR. Left Postero-inferior Depolarization Delay. Cardiology Today-2014;18:214–221)

3.2 Necessary Criteria

Absence of other conditions producing right-axis deviation is considered necessary for the diagnosis of left posterior fascicular block. These include

- Asthenic habitus, chest deformities.
- Right ventricular overload.
- Lateral infarction.

Some authorities feel that evidence of left ventricular abnormalities must exist for the diagnosis of left posterior fascicular block [3].

3.3 Caveats

- Some of the patients mentioned above could be having a posteroinferior depolarization delay without left posterior fascicular block.
- Considering left ventricular abnormalities as necessary for the diagnosis of left posterior fascicular block is likely to increase specificity at the coast of sensitivity.
- Initial stages of posteroinferior depolarization delay may not qualify the rigid criteria mentioned above and may contribute to wrong impression about the incidence and etiology of left posterior fascicular block.

4 Genesis of Electrocardiographic Findings

Major QRS vector is directed inferiorly, posteriorly, and to the right. Frontal plane QRS axis is, therefore, shifted to right (Fig. 27.4).

Left ventricular depolarization starts through the unblocked anterosuperior fibers. Initial mean frontal plane QRS vector is, therefore, directed anteriorly, superiorly, and to the left. This produces initial r wave in leads I, aVL, and initial q wave in leads II, III, and aVF (Fig. 27.5a). Subsequently, the impulse spreads to the posteroinferior region of the left ventricle. The later part of the QRS vector is, therefore, directed inferiorly, posteriorly, and to the right. This results in terminal S wave in leads I, aVL, and terminal R wave in leads II, III, and aVF (Fig. 27.5b).

Mean QRS voltages are increased due to lack of opposition from the blocked posteroinferior fascicle. As major QRS vector is toward positive pole of lead III, R wave is usually tallest in lead III [2]. In horizontal plane, the late QRS vector is directed posteriorly (Fig. 27.6a,b) with the result that the S wave persists up to lead V6.

Like other intraventricular conduction disorders, repolarization vector is directed opposite to the terminal vector of the QRS complex. This results in low to inverted T waves in leads II, III, aVF, and positive T waves in leads I and aVL. It may give a wrong impression of inferior ischemia.

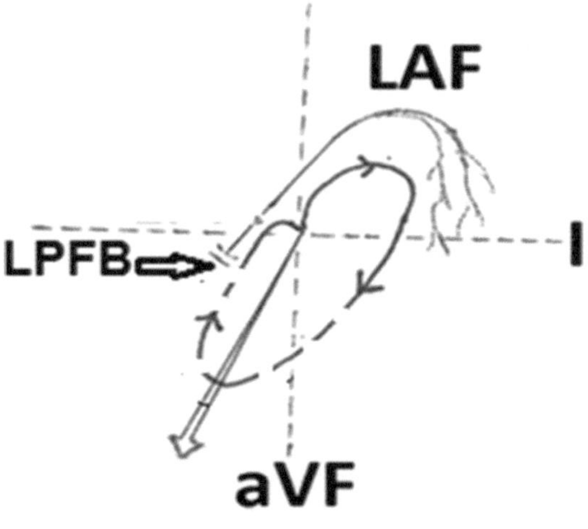

Fig. 27.4 Diagram showing frontal plane QRS vector and mean frontal plane QRS axis (marked arrow) in the left posterior fascicular block. *LAF* left anterior fascicle; *LPF* left posterior fascicle. (Reproduced with permission from Mittal SR. Left Postero- inferior Depolarization Delay. Cardiology Today-2014;18:214–221)

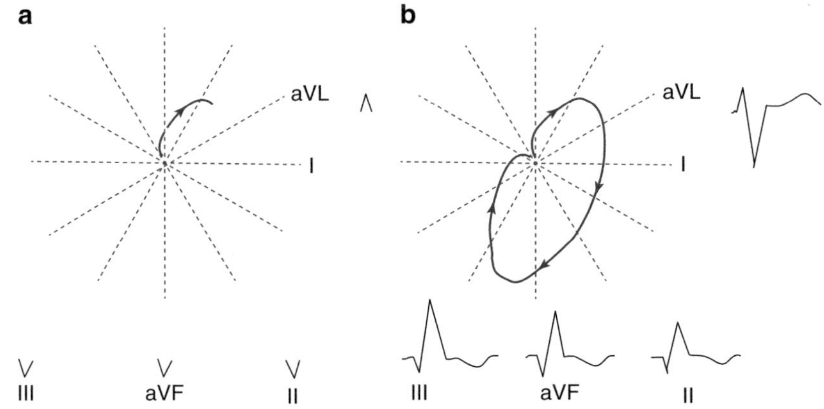

Fig. 27.5 (**a**, **b**) Diagram showing the QRS vector and its effect on genesis of changes in the QRS complexes in the limb leads in left posterior fascicular block. (Reproduced with permission from Mittal SR Left Postero-inferior Depolarization Delay. Cardiology Today-2014;18:214–221)

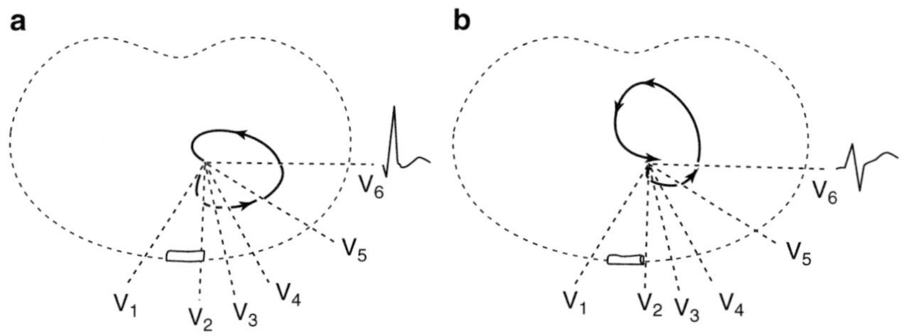

Fig. 27.6 (**a**) Diagram showing QRS vector in the horizontal plane. (**b**) Late QRS vector is directed posteriorly (marked ⇦) in the presence of left posterior fascicular block, resulting in persistence of S wave in lead V_6. (Reproduced with permission from Mittal SR Left Postero-inferior Depolarization Delay. Cardiology Today-2014;18:214–221)

5 Effect of Other Cardiac Disorders

5.1 Acute Myocardial Infarction

ST-segment elevation and sequential changes in the ST-segment and T waves help in the diagnosis of acute myocardial infarction.

5.2 Old Anterolateral Myocardial Infarction

In uncomplicated anterolateral myocardial infarction, initial QRS vector is directed inferiorly and to the right. This produces a Q wave in leads I, aVL, and an initial R wave in leads II, III, and aVF (Fig. 27.7a). Terminal vector is directed superiorly and to the left producing terminal R wave in leads I, aVL, and terminal S wave in leads II, III, and aVF (Fig. 27.7a). T wave is inverted in leads I, aVL, and positive in leads II, III, and aVF. In left posterior fascicular block, the QRS vector is in opposite direction (Fig. 27.7b). When anterolateral myocardial infarction is complicated by left posterior fascicular block, there is superimposition of the two vectors.

The initial vector is dominantly affected by the infarction, whereas the terminal vector is dominantly affected by the left posterior fascicular block. Initial as well as the terminal QRS vectors are, therefore, directed away from leads I and aVL. This produces a QS complex in leads I, aVL, and an R wave in leads II, III, and aVF [3] (Fig. 27.7c). Total amplitude of the QRS complex is also reduced.

5.3 Infarction or Fibrosis in the Area Depolarized by the Left Anterior Fascicle

In the setting of left posterior fascicular block, initial depolarization via the anterosuperior fascicle produces initial r wave in leads I, aVL, and initial q wave in leads II, III, and aVF (Fig. 27.8a). Infarction or fibrosis in the anterosuperior region reduces the magnitude of this vector. Amplitude of r wave in leads I, aVL, and q in leads II, III, and aVF is reduced (Fig. 27.8b).

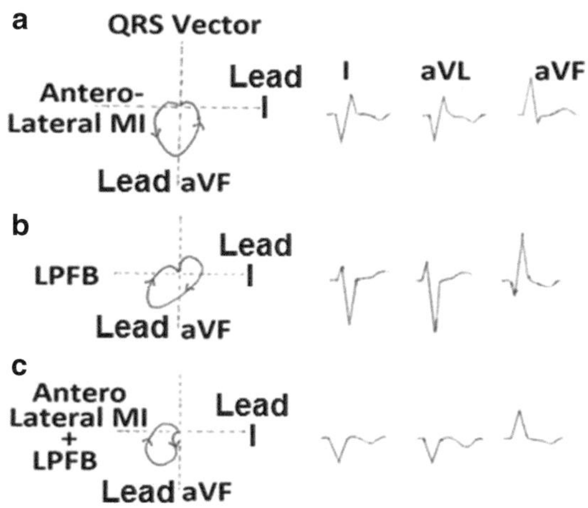

Fig. 27.7 Diagram showing frontal plane QRS vector and its effect on the QRS complexes in limb leads in (**a**) anterolateral infarction, (**b**) left posterior fascicular block, and (**c**) anterolateral infarction with left posterior fascicular block. (Reproduced with permission from Mittal SR Left Postero-inferior Depolarization Delay. Cardiology Today-2014;18:214–221)

Representative electrocardiogram is shown in Fig. 27.9.

5.4 Old Inferior MI

Left posterior fascicular block produces tall R waves in leads II, III, and aVF. The q wave of old inferior infarction may appear relatively small compared with the accompanying R wave (Fig. 27.10).

However, criteria of wide Q wave are still applicable. Amplitude of R wave is also reduced due to the loss of inferior vector (Fig. 27.11).

Symmetrical arrowhead inversion of T waves in inferior leads also supports the diagnosis of old infarction (Fig. 27.12).

5.5 Right Bundle Branch Block

Left posterior fascicular block in the presence of right bundle branch block results in

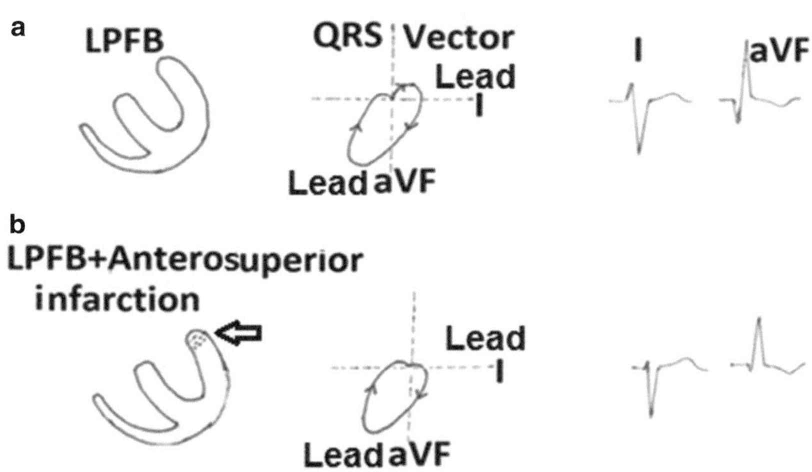

Fig. 27.8 Diagram showing frontal plane QRS vector and its effect on the QRS complex in limb leads (**a**) in left posterior fascicular block and (**b**) in left posterior fascicular block with infarction in the anterosuperior region of the left ventricle (marked ⇦). (Reproduced with permission from Mittal SR Left Postero-inferior Depolarization Delay. Cardiology Today-2014;18:214–221)

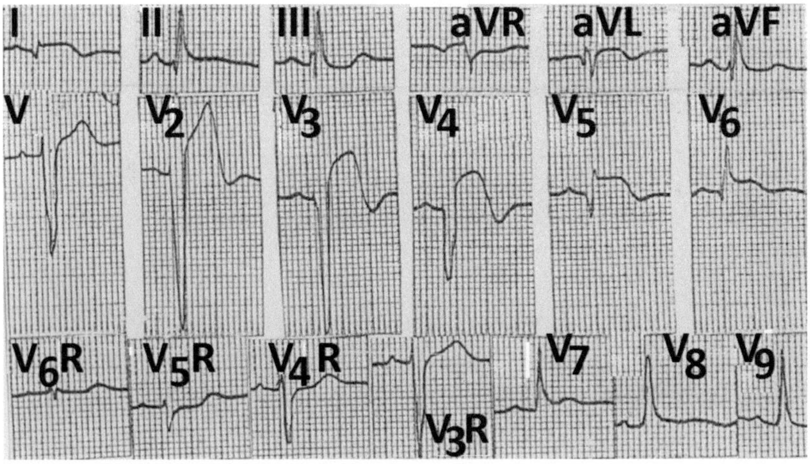

Fig. 27.9 Electrocardiogram showing anterior myocardial infarction with left posterior fascicular block. (Reproduced with permission from Mittal SR. Left Postero-inferior Depolarization Delay. Cardiology Today-2014;18:214–221)

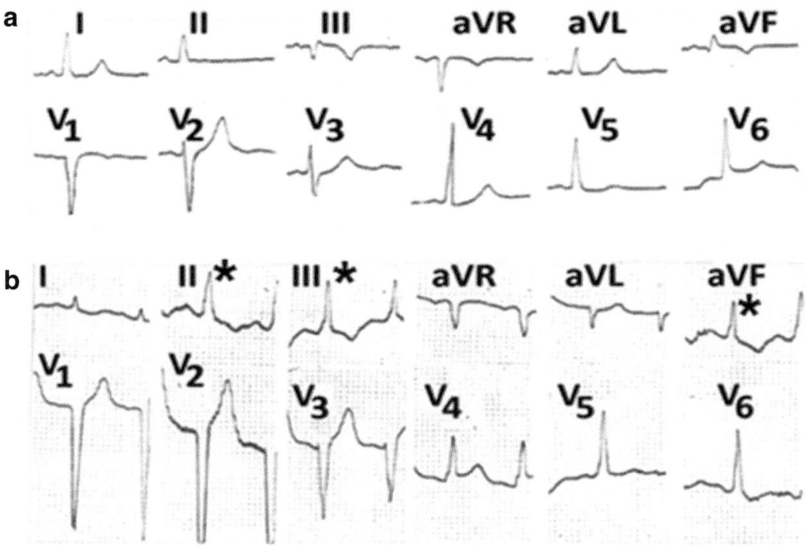

Fig. 27.10 Electrocardiograms showing (**a**) inferior myocardial infarction and (**b**) development of left posterior fascicular block produces tall R waves in leads II, III, and aVF (marked *). (Reproduced with permission from Mittal SR Left Postero- inferior Depolarization Delay. Cardiology Today-2014;18:214–221)

- Mean frontal plane QRS axis of more than + 120^0. QRS axis should be calculated from the amplitude (and not the area) of the initial 0.04 s.
- Extremely tall R wave in II, III, and aVF supports this possibility.

This combination is frequently associated with diffuse involvement of the atrioventricular and intraventricular conduction tissue.

Other causes of right bundle branch block with mean frontal plane QRS axis of more than + 120^0

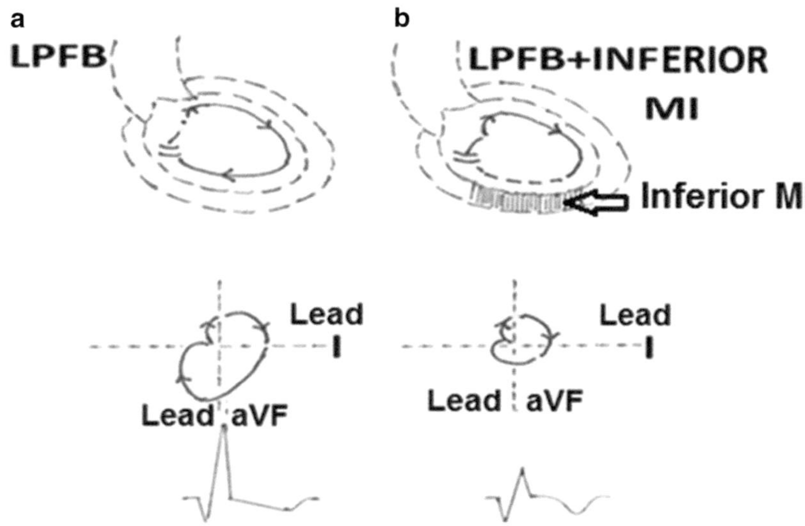

Fig. 27.11 Diagram showing frontal plane QRS vector and its effect on the inferior leads (**a**) in left posterior fascicular block and (**b**) in left posterior fascicular block with inferior myocardial infarction. (Reproduced with permission from Mittal SR Left Postero-inferior Depolarization Delay. Cardiology Today-2014;18:214–221)

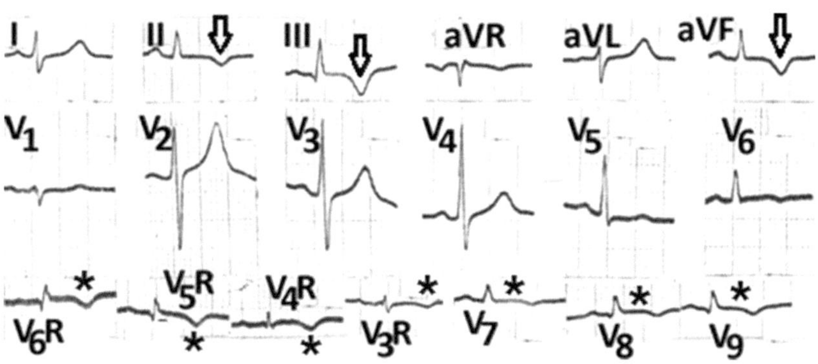

Fig. 27.12 Electrocardiogram showing left posterior fascicular block with inferior infarction (symmetrical T wave inversion in leads II, III, and aVF marked ⇩). Posterior chest leads (V_7 to V_9) and right-sided chest leads (V_3R to V_6R) also show infarction pattern (marked *). (Reproduced with permission from Mittal SR Left Postero-inferior Depolarization Delay. Cardiology Today- 2014;18:214–221)

- Right bundle branch block + right ventricular hypertrophy.
- Right bundle branch block + vertical heart.
- Right bundle branch block + extensive lateral infarction.

However, unlike right bundle branch block + left posterior fascicular block, the abovementioned conditions are not associated with tall R waves in leads II, III, and aVF.

5.6 QRS Duration of More Than 0.10 Second

It suggests additional intraventricular conduction delay (Fig. 27.13).

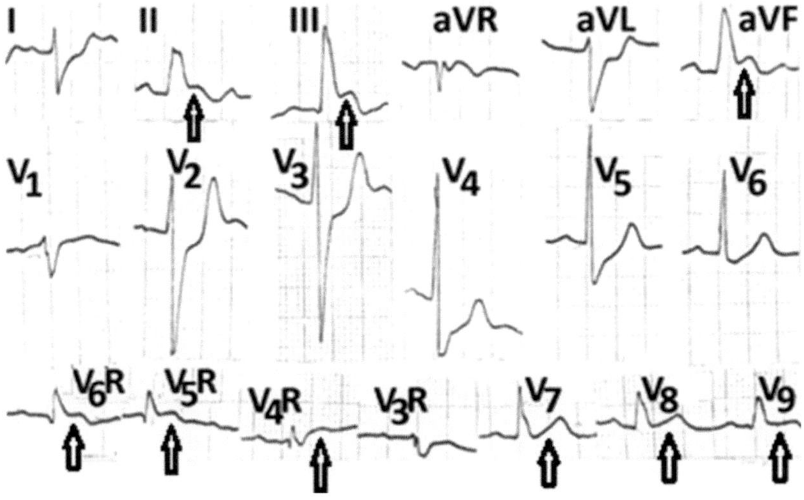

Fig. 27.13 Electrocardiogram showing left posterior fascicular block with broad QRS. There is ST-segment elevation in leads II, III, aVF, V_7 to V_9, and V_3R to V_6R (marked ⇧) suggestive of additional inferior, posterior, and right ventricular infarction. (Reproduced with permission from Mittal SR Left Postero-inferior Depolarization Delay. Cardiology Today-2014;18:214–221)

5.7 Changes in the T Wave

Deep symmetrical inversion of the T wave is abnormal. Polarity of the T wave in the direction of the terminal part of the QRS complex (Fig. 27.14) is also abnormal.

If there is terminal S wave in leads II, III, and aVF, a positive T wave may be normal (Fig. 27.15).

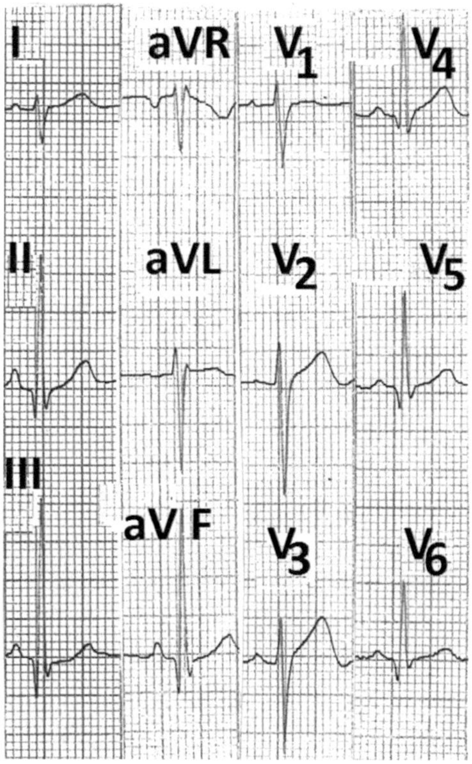

Fig. 27.15 Electrocardiogram showing left posterior fascicular block with terminal S wave in leads II, III, and aVF with positive T waves in these leads. T waves are opposite to the terminal part of the QRS complex. (Reproduced with permission from Mittal SR. Left Postero-inferior Depolarization Delay. Cardiology Today-2014;18:214–221)

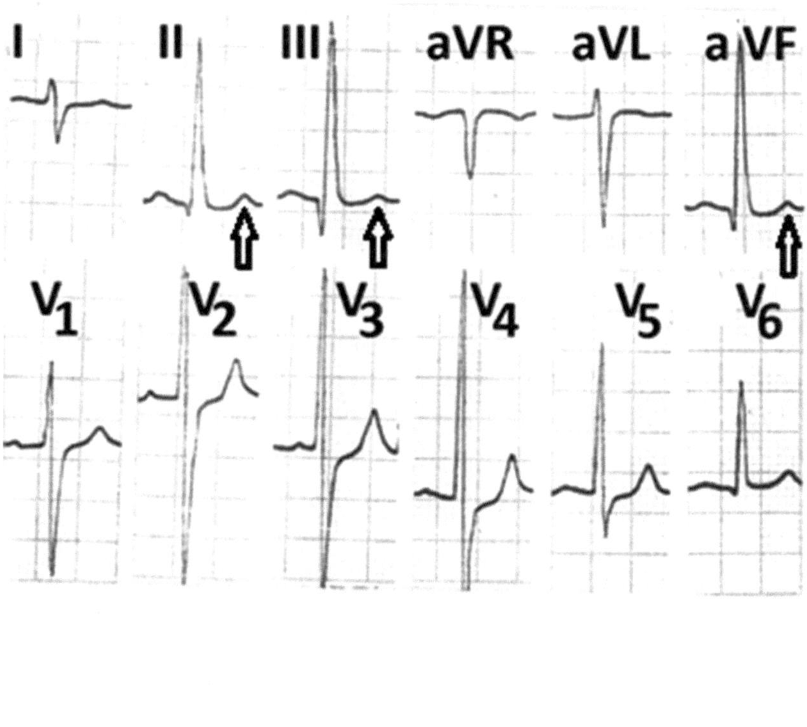

Fig. 27.14 Electrocardiogram showing left posterior fascicular block with positive T wave in leads II, III, and aVF (marked ⇧). T wave is in the direction of the terminal part of QRS complex. (Reproduced with permission from Mittal SR Left Postero- inferior Depolarization Delay. Cardiology Today-2014;18:214–221)

6 Differential Diagnosis

Causes of right axis deviation other than left posterior fascicular block.

(a) Right ventricular hypertrophy

The following electrocardiographic findings help in differentiation:

– "P- pulmonale" and prominent R wave in lead V_1 support the diagnosis of right ventricular hypertrophy (Fig 27.16).
– There is no increase in the size of the R waves in leads II, III, and aVF in cases of the right ventricular hypertrophy.
– Left posterior fascicular block also produces persistent S wave in leads V_5 and V_6. This finding alone should not be taken as a sign of right ventricular hypertrophy.

(b) Vertical heart (Fig 27.17)

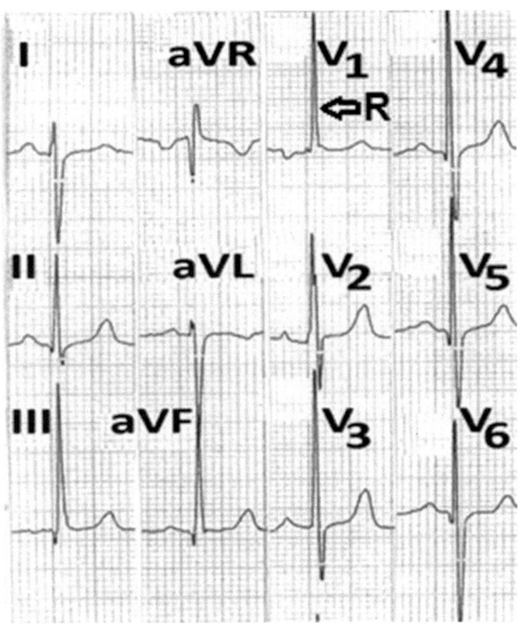

Fig. 27.16 Electrocardiogram showing right ventricular hypertrophy mimicking left posterior fascicular block. (Reproduced with permission from Mittal SR Left Postero-inferior Depolarization Delay. Cardiology Today-2014;18:214–221)

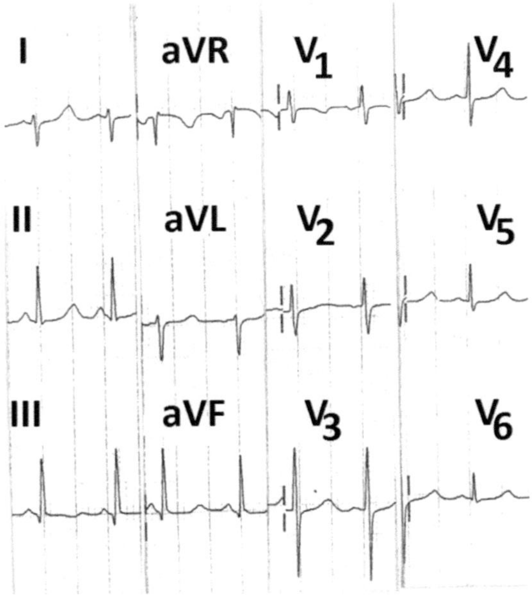

Fig. 27.17 Electrocardiogram from a patient of emphysema mimicking left posterior fascicular block. QRS shows low voltage especially in leads I, V_5, and V_6. (Reproduced with permission from Mittal SR Left Postero-inferior Depolarization Delay. Cardiology Today-2014;18:214–221)

For example, slender body

– Emphysema: QRS voltage is low. In left posterior fascicular block, QRS voltages are increased in limb leads.

(c) Chest deformities.

(d) Extensive lateral infarction.

Abnormal Q waves, elevation of ST-segment, and symmetrical inversion of the T wave in leads I, aVL and V5, V6 support the diagnosis of lateral infarction (Fig. 27.18). QRS voltage in limb leads is reduced as opposed to increased voltage in left posterior fascicular block.

(e) Preexcitation from the left-lateral accessory path.

Short PR interval and presence of the delta wave suggest the diagnosis of preexci-

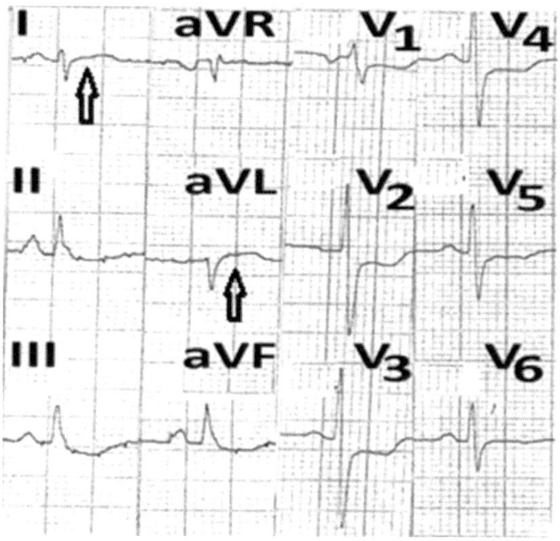

Fig. 27.18 Electrocardiogram from a patient of lateral infarction mimicking left posterior fascicular block. QRS in limb leads shows low voltage. ST segment is elevated in leads I and aVL (marked ⇧). (Reproduced with permission from Mittal SR Left Postero-inferior Depolarization Delay. Cardiology Today-2014;18:214–221)

tation. In preexcitation via the left-lateral accessory path lead, V_1 shows a prominent R wave. This finding is not seen in left posterior fascicular block.

The direction of QRS vector and representative electrocardiogram is shown in Fig. 27.19.

(f) Some cases of hyperkalemia may also show right-axis deviation. Tall tented T waves support the possibility of hyperkalemia.

(g) Technical dextrocardia.

Inverted P wave and T wave in lead I, positive P wave, and T wave in lead aVR help suspicion. QRS amplitude is normal in limb leads as opposed to increased amplitude in left posterior fascicular block.

The most important feature of left posterior fascicular block is not the degree of right-axis deviation but the increase in the size of R wave in leads II, III, and aVF.[2]

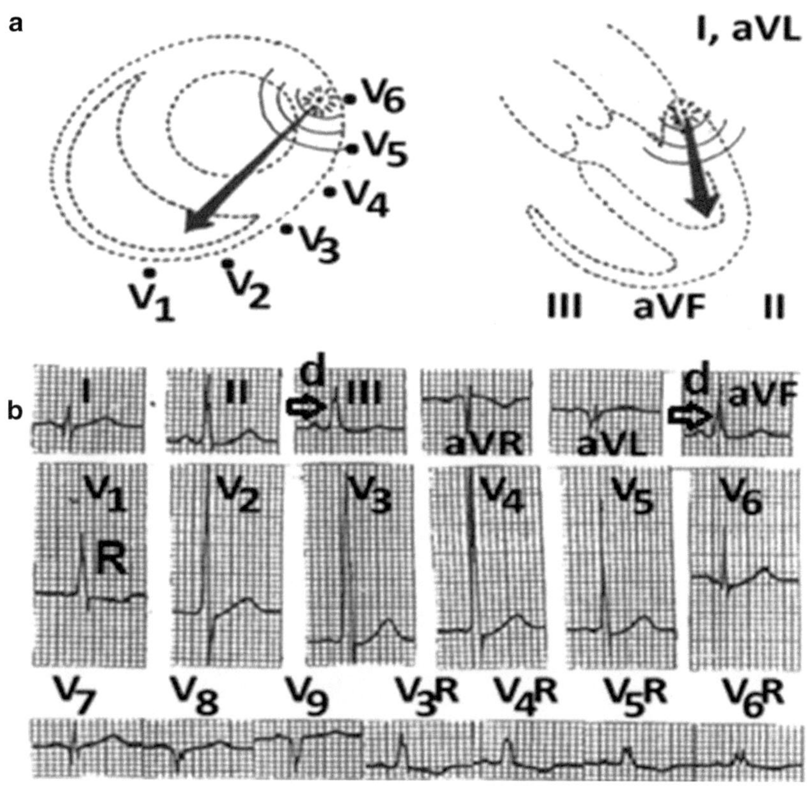

Fig. 27.19 (**a**) Major QRS vector in preexcitation via the left-lateral accessory path. (**b**) Electrocardiogram from a patient of preexcitation via the left-lateral accessory path. There is slurring of initial part of the QRS (delta wave marked d). There is prominent R wave in lead V_1. (Reproduced with permission from Mittal SR. Left Postero-inferior Depolarization Delay. Cardiology Today-2014;18:214–221)

Summary

Left posteroinferior fascicular block (hemi block) is diagnosed by QRS duration of less than 120 ms, mean frontal plane QRS axis of $+90^0$ to $+120^0$, rS configuration in leads I and aVL, qR configuration in leads II, III, and aVF, and increased QRS voltages in limb leads. The time to peak of R wave in lead aVF is more than 45 ms.

Right ventricular enlargement, asthenic built, and lateral infarction should be excluded before diagnosing left posteroinferior fascicular block.

MCQs

Q1. Left posteroinferior fascicular block is not characterized by

(a) QRS duration < 120 ms
(b) rS configuration in lead I
(c) qR configuration in lead aVF
(d) Mean frontal plane QRS axis between 0^0 and $+90^0$

Q2. Left posteroinferior fascicle is supplied by

(a) Left anterior descending + left circumflex coronary artery
(b) Left anterior descending + right coronary artery
(c) Left circumflex + right coronary artery
(d) First septal + first diagonal

Q3. What is the least frequent in persons without structural heart disease?

(a) Left anterior fascicular block
(b) Left posterior fascicular block
(c) Right bundle branch block
(d) Ectopic beats

Q4. Left posteroinferior fascicle supplies

(a) Interventricular septum
(b) Left ventricular outflow
(c) Left ventricular inflow
(d) Left ventricular lateral wall

Q5. Uncomplicated left posteroinferior fascicular block produces broad Q wave in

(a) Lead I
(b) Lead aVF
(c) Lead V6
(d) None

Q6. What is correct for the left posteroinferior fascicular block?

(a) S wave in lead I > S wave in aVL
(b) S wave in lead aVL < S wave in lead V6
(c) R wave in lead II > R wave in lead III
(d) R wave in lead III > R wave in lead aVF

Q7. Left ventricular lateral wall accessory path produces prominent R wave in

(a) Lead I
(b) Lead aVL
(c) Lead II
(d) Lead V9

Q8. Diagnosis of left posteroinferior fascicular block needs exclusion of

(a) Vertical heart
(b) Left ventricular hypertrophy
(c) Left ventricular failure
(d) None

Q9. Left posteroinferior fascicular block can mask

(a) Septal infarction
(b) Inferior infarction
(c) High lateral infarction
(d) Right ventricular infarction

Q10. The left posteroinferior fascicular block can mimic

(a) Left ventricular hypertrophy in limb leads
(b) Right ventricular hypertrophy in precardial leads
(c) Inferior ischemia
(d) Lateral ischemia

Q11. Which condition does not mimic left posteroinferior fascicular block?

(a) Technical dextrocardia
(b) Right ventricular overload
(c) Lateral infarction
(d) Left ventricular hypertrophy

Q12. Which finding of inferior infarction is affected by left posteroinferior fascicular block?

(a) Q/R ratio
(b) Width of the Q wave
(c) T wave
(d) QRS duration

Q13. Advanced left posteroinferior fascicular block produces

(a) QRS duration > 120 ms.
(b) Decrease in the height of the R wave in lead III
(c) Slurring of the r wave in lead I
(d) Slurring of the descending part of the R wave in lead aVF.

Q14. qR configuration in leads I, aVL, and rS configuration in leads II, III, and aVF with increased QRS voltages in limb leads suggest

(a) Left anterosuperior fascicular block
(b) Left posteroinferior fascicular block
(c) Advanced right bundle branch block
(d) None

Q15. rS configuration in leads I, aVL, and qR configuration in leads II, III, and aVF, QRS duration less than 120 ms with increased voltages in limb leads suggest

(a) Left anterosuperior fascicular block
(b) Left posteroinferior fascicular block
(c) Advanced left bundle branch block
(d) Advanced right bundle branch block

Q16. QRS duration of less than 120 ms, mean frontal plane QRS axis of $+120^0$, qR configuration in leads II, III, and aVF with increased QRS voltages in limb leads suggest

(a) Right ventricular hypertrophy
(b) Left ventricular hypertrophy
(c) Left posteroinferior fascicular block
(d) Advanced right bundle branch block

Q17. Which of the following can produce right-axis deviation?

(a) Right ventricular hypertrophy
(b) Asthenic built
(c) Lateral infarction
(d) All

Q18. In left posteroinferior fascicular block, QRS duration is

(a) More than 120 ms
(b) More than 140 ms
(c) Less than 120 ms
(d) Less than 80 ms

Q19. In left posteroinferior fascicular block, mean frontal plane QRS axis is between

(a) 0^0 to -30^0
(b) 0^0 to $+45^0$
(c) $+45^0$ to $+90^0$
(d) $+90^0$ to $+120^0$

Q20. In left posteroinferior fascicular block, q wave with tall R wave is present in leads

(a) I, aVL
(b) II, III, aVF
(c) V_1 to V_3
(d) V_4 to V_6

Q21. In left posteroinferior fascicular block, rS configuration is seen in leads

(a) I, aVL
(b) II, III, and aVF
(c) V_1 to V_3
(d) V_5, V_6

Q22. Left posteroinferior fascicle is

(a) Thicker than left anterior fascicle
(b) Located in left ventricular outflow
(c) Less frequently injured
(d) All

Answers

(1) d (2) c (3) b (4) c (5) d (6) d (7) c (8) a (9) b (10) c (11) d (12) a (13) d (14)a (15)b (16) c (17) d (18) c (19) d (20) b (21) a (22) a, c

References

1. Mirvis DM, Goldberger AL. Electrocardiography. In: Bonow RO, Mann DL, Zipes DP, Libby P, editors. Braunwald's Heart Disease. Missouri: Elsevier; 2012. p. 126–65.
2. Schamroth L. The hemiblocks (fascicular block). In: Schamroth C, editor. An introduction to electrocardiography. Blackwell Science: France; 1982. p. 105–15.
3. DeLuna AB, Goldwasser D, Fiol M, Bayes-Genis A. Surface electrocardiography. In: Fuster V, Walsh RA, Harrington RA, editors. Hurst's The Heart. New York: Mc Graw Hill; 2011. p. 307–70.

28 Right Bundle Branch Block (RBBB) (Delay in Right Ventricular Depolarization)

1 Nomenclature

The term "right bundle branch block" has some limitations.

(i) Site of delay in conduction need not always be in the right bundle branch. It may be in the His bundle fibers connecting to the tight bundle branch or in the Purkinje network in the right ventricular free wall (Fig. 28.1).
(ii) Right ventricular outflow tract (RVOT) is the last part of the right ventricle to be depolarized (Fig. 28.2).

Incomplete right bundle branch block (Fig. 28.3) as seen in patients with atrial septal defects is usually due to delayed depolarization of the right ventricular outflow tract rather than due to block in the right bundle branch.

(iii) Electrocardiographic findings of RBBB are mostly due to physiological delay in depolarization of the right ventricle rather than an anatomical block in the right bundle branch except in cases of new-onset RBBB in the setting of acute anterior myocardial infarction.
(iv) Rate-dependent RBBB (Fig. 28.4) is due to rate-dependent aberrant conduction and not due to "block" in conduction.

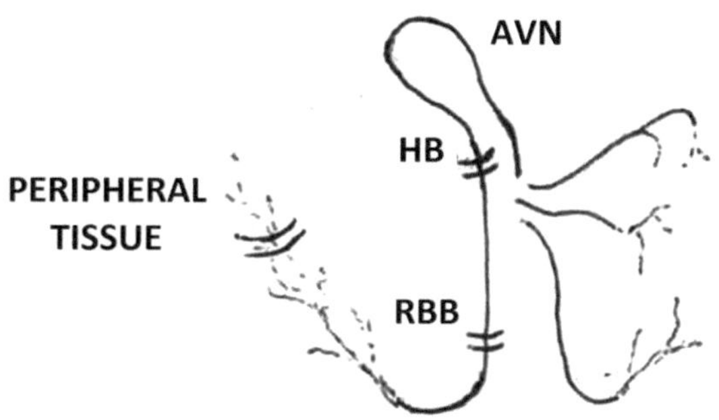

Fig. 28.1 Diagram showing sites of delay in the depolarization of the right ventricle. *AVN* atrioventricular node; *HB* His bundle; *RBB* right bundle branch. (Reproduced with permission from Mittal SR. Right Bundle Branch Block (RBBB). Cardiology Today- 2014;18:126–134)

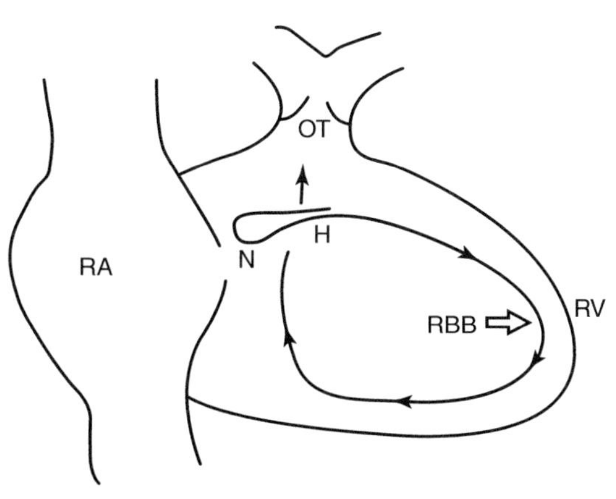

Fig. 28.2 Diagram showing last depolarization of the right ventricular outflow tract (OT). *RA* right atrium; *RV* right ventricle; *N* atrioventricular node; *H* His bundle; RBB right bundle branch. (Reproduced with permission from Mittal SR. Right Bundle Branch Block (RBBB). Cardiology Today-2014;18:126–134)

S. Mittal, *Insights into Electrocardiograms with MCQs*,
https://doi.org/10.1007/978-981-99-0127-2_28

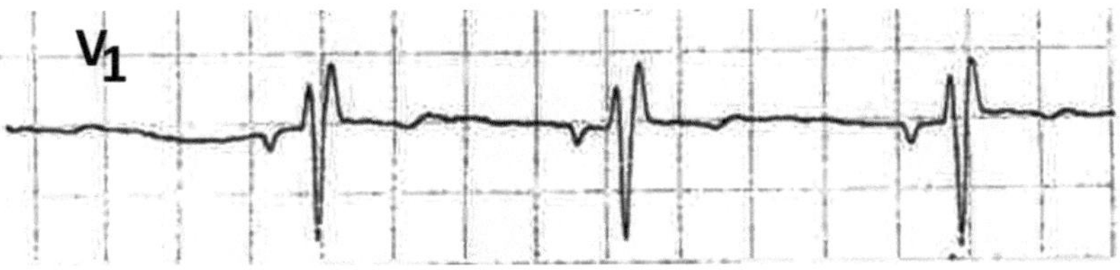

Fig. 28.3 Lead V_1 showing incomplete right bundle branch block. (Reproduced with permission from Mittal SR. Right Bundle Branch Block (RBBB). Cardiology Today-2014;18:126–134)

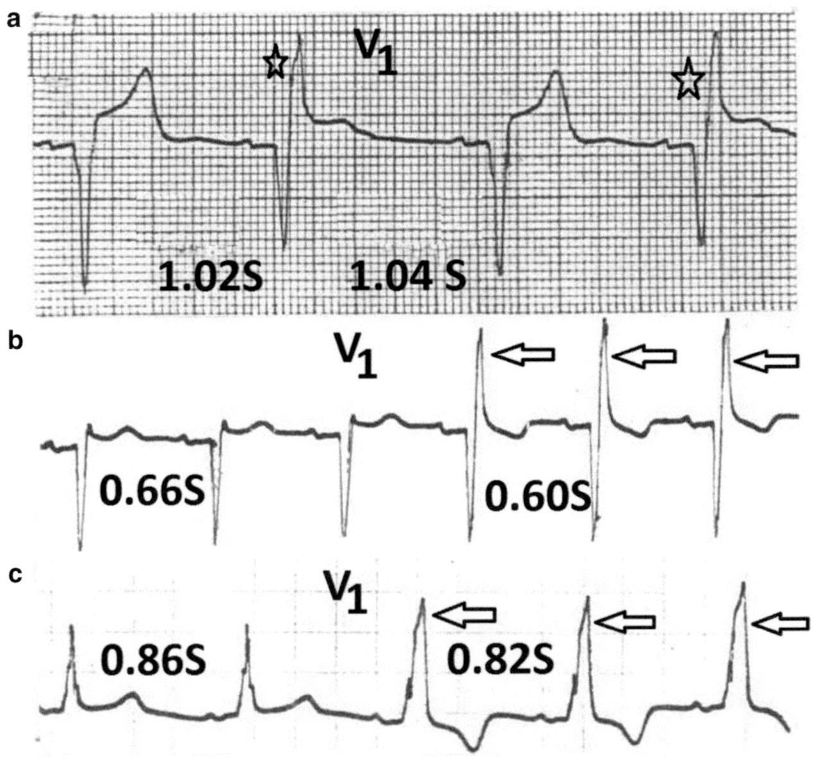

Fig. 28.4 (**a**) Electrocardiogram showing rate-dependent appearance of RBBB (marked ☆). (**b, c**) Electrocardiogram showing rate-dependent increase in RBBB (marked ⇦). (Reproduced with permission from Mittal SR. Right Bundle Branch Block (RBBB). Cardiology Today-2014;18:126–134)

(v) QRS configuration in lead V_1 may progressively change from rsr' to rR' without significant change in QRS duration (Fig. 28.5).

(vi) In slowly developing RBBB, QRS duration progressively increases (Fig. 28.6) and at a given time it may not strictly qualify as "incomplete" or "complete." In advanced stages, QRS duration may exceed 0.12 s (Fig. 28.6).

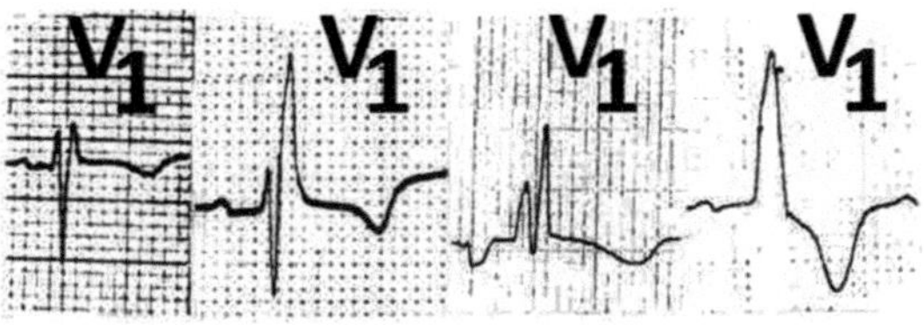

Fig. 28.5 Electrocardiograms of lead V_1 showing different configurations of the QRS complex in lead V_1 in delayed depolarization of the right ventricle. (Reproduced with permission from Mittal SR. Right Bundle Branch Block (RBBB). Cardiology Today-2014;18:126–134)

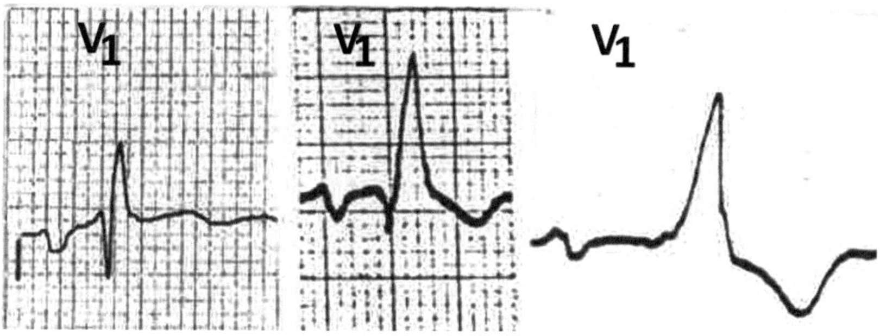

Fig. 28.6 Electrocardiograms of lead V_1 showing RBBB with increasing duration of QRS in lead V_1. (Reproduced with permission from Mittal SR. Right Bundle Branch Block (RBBB). Cardiology Today- 2014;18:126–134)

The term "delayed right ventricular depolarization delay" better covers several variations. Other findings should be added as suffix to give a full description of the ECG for better follow-up. One example could be

"Delayed right ventricular depolarization, QRS duration 0.12 s, rR' in lead V_1."

2 Etiology

(1) QRS duration < 100 ms
  (a) Physiological variants [1]
    (i) Post-term infants
    (ii) Scanty Purkinje fibers in anteroseptal zone
  (b) Positional variant:
    Wrong placement of electrode of lead V_1 in higher intercostal space.
(2) QRS duration between 100 and 120 ms (incomplete RBBB)
  (a) Right ventricular enlargement
    (i) Atrial septal defect.
    (ii) Partial anomalous pulmonary venous drainage
    (iii) Cor pulmonale [2]
    (iv) Tricuspid regurgitation
  (b) Skeletal deformities
    (i) Pectus excavatum
    (ii) Straight back
  (c) Left anterior hemiblock
    r' ray be recorded if electrode of lead V_1 is placed in higher intercostal space.
(3) QRS duration > 120 ms
  (a) Transient
    (i) Following right ventricular catheterization. It can last up to 24 h [3].
    (ii) Rate dependent
      – Usually tachycardia dependent (Fig. 28.4a,b,c)
      – During—sinus tachycardia
      – Stress testing
      – Atrial tachyarrhythmias with fast ventricular rate

      It may be present in patients without any other evidence of organic heart disease. On follow-up, some patients may develop RBBB at normal heart rates. Rate-dependent aberrant conduction frequently involves the right bundle branch. It is due to physiologically longer refractory period of the right bundle branch.
      – Uncommonly bradycardia dependent
      – It occurs usually in the presence of organic heart disease
    (iii) Hyperkalemia
  (b) Sometimes in the absence of other structural heart diseases [4]
    Sclerodegeneration or fibrosis of the conduction tissue may be responsible in some cases.
  (c) Systemic hypertension [5]
  (d) Ischemic heart disease
    Right bundle branch is supplied by the septal perforators arising from the

left anterior descending coronary artery. Occlusion of these vessels can damage right bundle branch.
(e) Acute pulmonary embolism
(f) Cardiomyopathy
(g) Surgery involving right ventriculotomy
For example, repair of tetralogy of Fallot, or large ventricular septal defect.

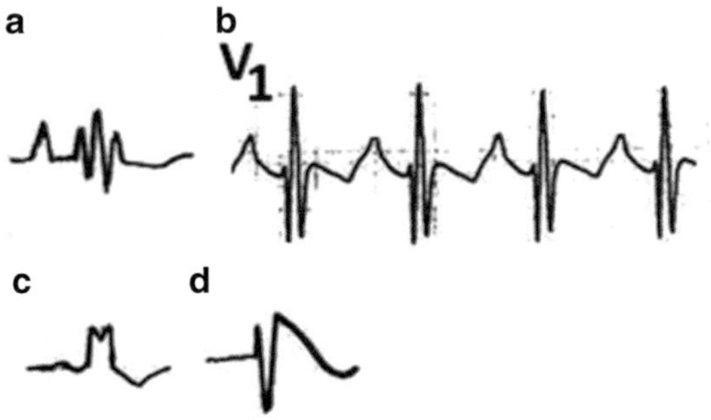

Fig. 28.7 Electrocardiogram patterns in lead V_1 in atypical RBBB. (**a**) Ebstein anomaly. (**b**) Electrocardiogram (lead V_1) from a case of Ebstein anomaly. (**c**) Arrhythmogenic right ventricular dysplasia. (**d**) Brugada syndrome. (Reproduced with permission from Mittal SR. Right Bundle Branch Block (RBBB). Cardiology Today-2014;18:126–134)

3 Electrocardiographic Features

(1) Typical RBBB
(a) Terminal slurred r' to R' in leads V_1 and V_2 producing rsr', rsR', or rR' configuration in these leads (Fig. 28.5). QRS duration increases with increasing delay in depolarization of the right ventricle (Fig. 28.6).
(b) Terminal slurred s wave in leads I and V_6.
(c) Mean frontal plane QRS axis is usually normal unless there is concomitant left posterior fascicular block.
(d) ST-segment and T wave are opposite to the terminal part of the QRS complex.
(2) Atypical RBBB [1]
(a) Ebstein anomaly—There are many deflections in the QRS complex producing rsr's' pattern. There are tall and peaked P waves due to right atrial enlargement (Fig. 28.7a,b).
(b) Arrhythmogenic right ventricular dysplasia:
Lead V_1 shows a single R wave of low voltage with notches and/or slurring on the plateau (Fig. 28.7c).
(c) Brugada syndrome:
QRS is usually < 0.12 s. There is persistent ST segment elevation with T wave inversion in leads V_1 to V_3 (Fig. 28.7d).

4 Mechanism of ECG Changes

4.1 Ventricular Depolarization (QRS)

Normally activation of the interventricular septum occurs from left to right and anteriorly (Fig. 28.8a, vector 1). It produces initial r wave in leads V_1 and V_2 and initial q waves in leads I, V_5, and V_6. Septal depolarization is followed by simultaneous depolarization of the free walls of left and right ventricles (Fig. 28.8b, **vectors 2a and 2b**).

Left ventricular depolarization (**vector 2a**) dominates as left ventricle is thicker than the right ventricle. Net vector, therefore, moves toward the left ventricle producing terminal S wave in leads V_1,V_2 and terminal R wave in leads I,V_5,V_6.

In RBBB, activation of the interventricular septum occurs normally (Fig. 28.9a, **vector 1**) producing initial r wave in leads V_1, V_2 and initial q wave in leads I, V_5, V_6. This is followed by activation of free wall of the left ventricle (Fig. 28.9b**, vector 2**) producing S wave in leads V_1, V_2

and R wave in leads I, V_5, V_6. Subsequently, the right ventricle depolarizes slowly through myocardial tissue and not through the specialized conduction tissue (Fig. 28.9c, **vector 3**). It produces a terminal r' in leads V_1,V_2 and terminal s wave in leads I, V_5,V_6. Figure 28.9d shows repolarization of the ventricles.

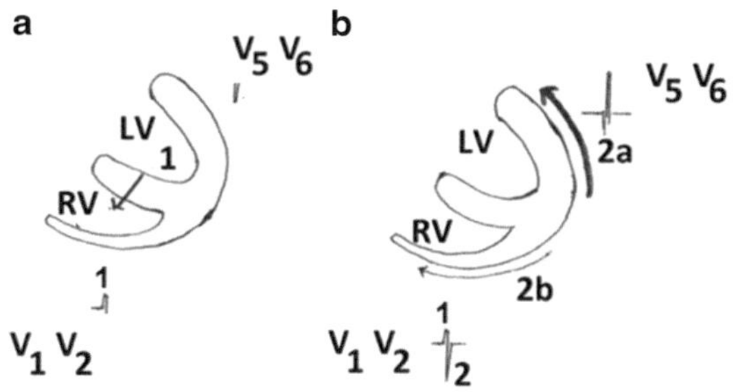

Fig. 28.8 Diagram (**a**) showing normal depolarization of the interventricular septum. (**b**) Depolarization of the ventricular free walls. (Reproduced with permission from Mittal SR. Right Bundle Branch Block (RBBB). Cardiology Today-2014;18:126–134)

Figure 28.10a shows normal depolarization of the two ventricles. In initial stages of RBBB, depolarization of the right ventricle occurs slowly but simultaneously with depolarization of the left ventricle (Fig. 28.10b, **vector 2b).** This partly opposes left ventricular depolarization (Fig. 28.10b, vector 2a). This reduces the depth of the S wave and produces a notch on the upstroke of the S wave in lead V_1 (Fig. 28.10 b). This is the earliest stage of incomplete RBBB (Fig. 28.11). With progressive delay in depolarization of the right ventricle, vector 2b follows vector 2a rather than occurring with vector 2a. Right ventricular depolarization (vector 2b) is, therefore, not opposed by the left ventricular depolarization forces (vector 2a). Late and unopposed depolarization of the right ventricle produces terminal r' in leads V_1,V_2 (Fig. 28.10c) and

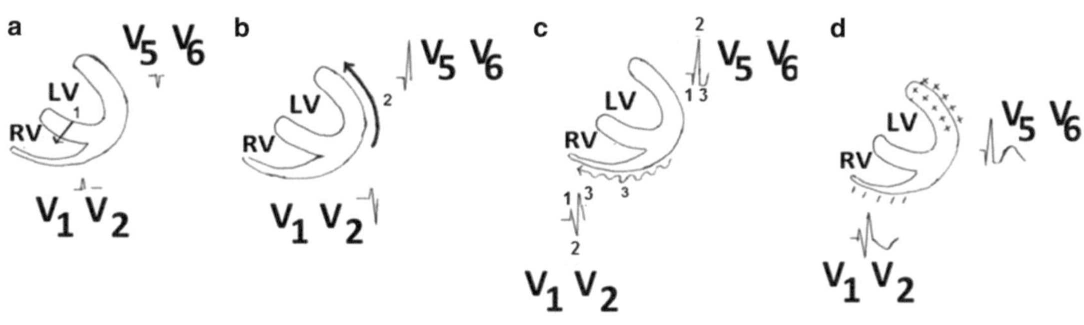

Fig. 28.9 Diagram showing ventricular depolarization and repolarization in the presence of RBBB. (**a**) Septal depolarization. (**b**) Depolarization of the left ventricular free wall. (**c**) Depolarization of the right ventricular free wall. (**d**) Repolarization. (Reproduced with permission from Mittal SR. Right Bundle Branch Block (RBBB). Cardiology Today-2014;18:126–134)

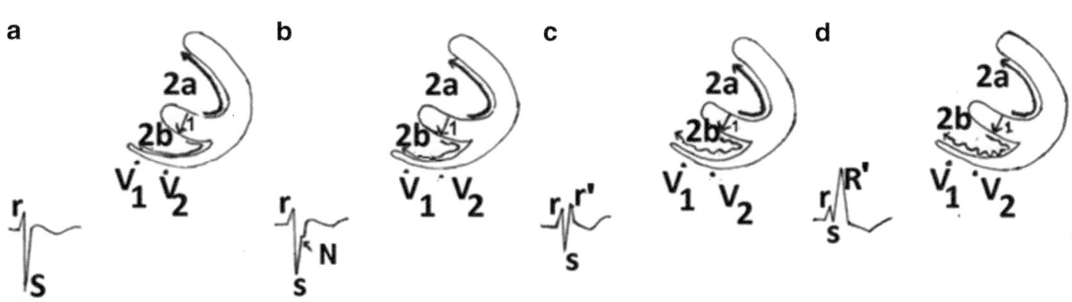

Fig. 28.10 Diagram showing the impact of increasing delay in depolarization of the right ventricle on configuration of the QRS complex in lead V_1. *N* notch on the upstroke of the S wave. (Reproduced with permission from Mittal SR. Right Bundle Branch Block (RBBB). Cardiology Today- 2014;18:126–134)

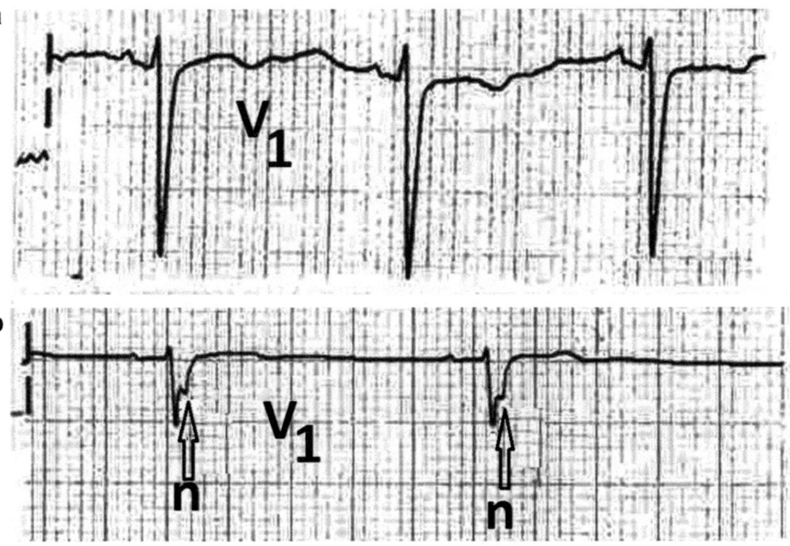

Fig. 28.11 ECG (**a**) showing QRS in lead V_1 in normal person (**b**) showing notch (n-arrow) in the S wave in early RBBB. (Reproduced with permission from Mittal SR. Right Bundle Branch Block (RBBB). Cardiology Today-2014;18:126–134)

terminal s wave in leads I, V_5, and V_6. Slow depolarization of the right ventricle results in slurring (broadening) of the terminal vector (Fig. 28.10c). With increasing delay in the depolarization of the right ventricle, there is progressive broadening of the terminal vector. rsr' in lead V_1 progressively changes to rsR' (Fig. 28.10d).

Figure 28.11 shows electrocardiogram from a case of early RBBB.

(1) Ventricular repolarization

Repolarization forces are directed toward the earlier activated left ventricle and away from the right ventricle. ST-segment and T wave are, therefore, directed opposite to the terminal QRS vector, resulting in downsloping ST-segment and inverted T waves in leads V_1, V_2 and positive T waves in leads V_5, V_6 (Fig. 28.9d).

4.2 Posterior Leads (V_7 to V_9)

These leads are directionally opposite to leads V_1 and V_2. These leads, therefore, show qrs configuration with upright T wave (Fig. 28.12).

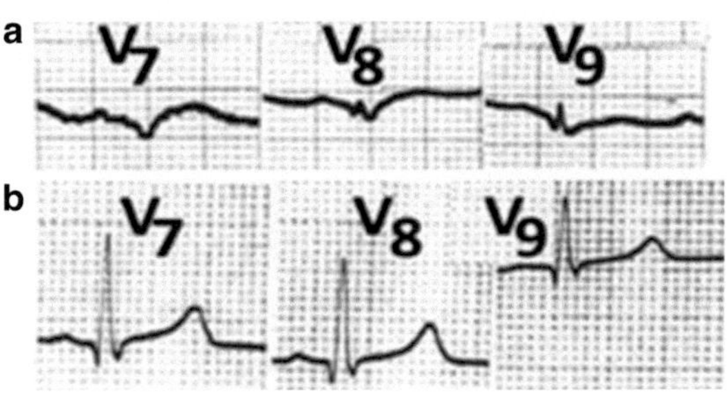

Fig. 28.12 (**a**, **b**) Electrocardiograms showing posterior chest leads from two cases of RBBB. (Reproduced with permission from Mittal SR. Right Bundle Branch Block (RBBB). Cardiology Today-2014;18:126–134)

4.3 Right-Sided Chest Leads (V_3R to V_6R)

Leads V_3R and V_4R resemble lead V_1 (rsR' configuration with inverted T wave) (Fig. 28.13b). Leads V_5R and V_6R may show loss of initial r wave as the initial septal depolarization is directed anteriorly and is not picked up by the leads V_5R and V_6R (Fig. 28.13a). These leads may, therefore, show a qr configuration. ST-segment and T wave, however, remain inverted (Fig. 28.13b).

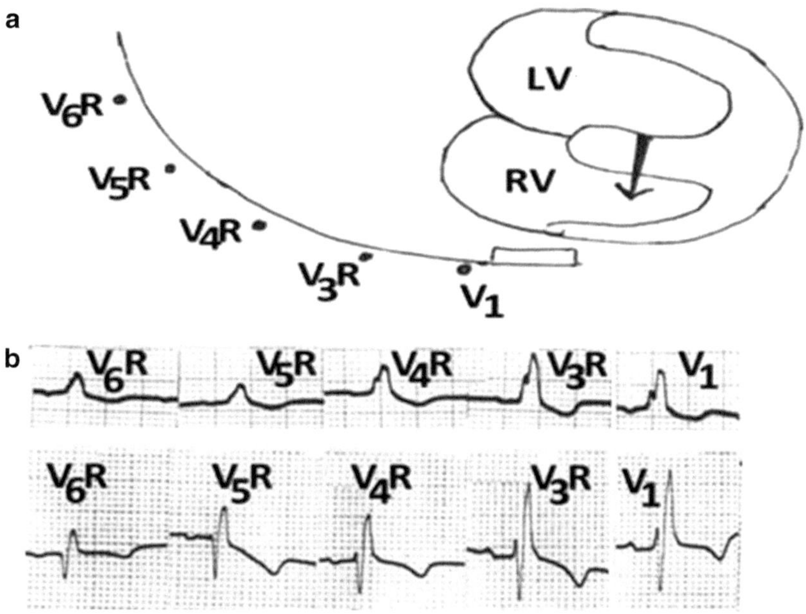

Fig. 28.13 (**a**) Diagram showing relation of right-sided chest leads to the vector of septal depolarization. (**b**) Configuration of QRS complex, ST-segment, and T wave in right-sided chest leads in the presence of RBBB from two patients. (Reproduced with permission from Mittal SR. Right Bundle Branch Block (RBBB). Cardiology Today-2014;18:126–134)

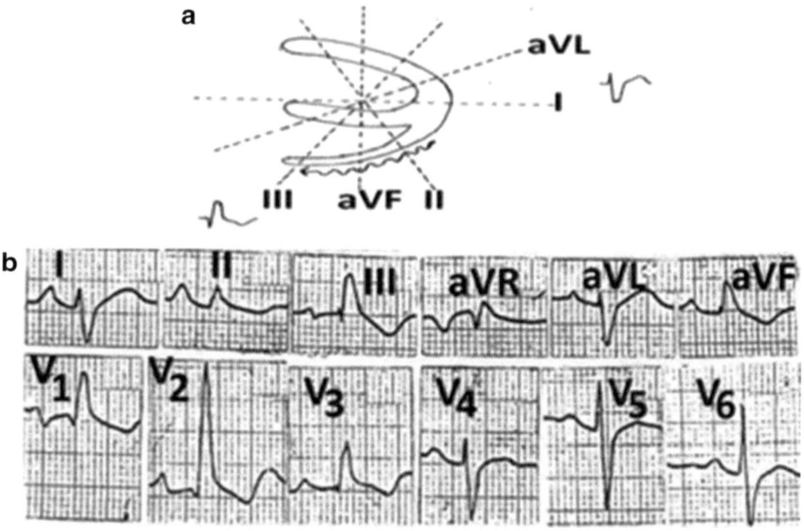

Fig. 28.14 (**a**, **b**) Diagram showing the relation of right ventricular depolarization vector to inferior leads (**a**) in a horizontal heart and (**b**) electrocardiogram from a case of right bundle branch block with horizontal heart. (Reproduced with permission from Mittal SR. Right Bundle Branch Block (RBBB). Cardiology Today-2014;18:126–134)

4.4 Limb Leads

Configuration in the limb leads depends on the cardiac position. In horizontal position, right ventricular depolarization vector moves toward the inferior leads (Fig. 28.14a). These leads, therefore, show qR configuration (Fig. 28.14b).

In vertical position, right ventricular depolarization vector moves away from inferior leads (Fig. 28.15a). These leads, therefore, show rs configuration (Fig. 28.15b).

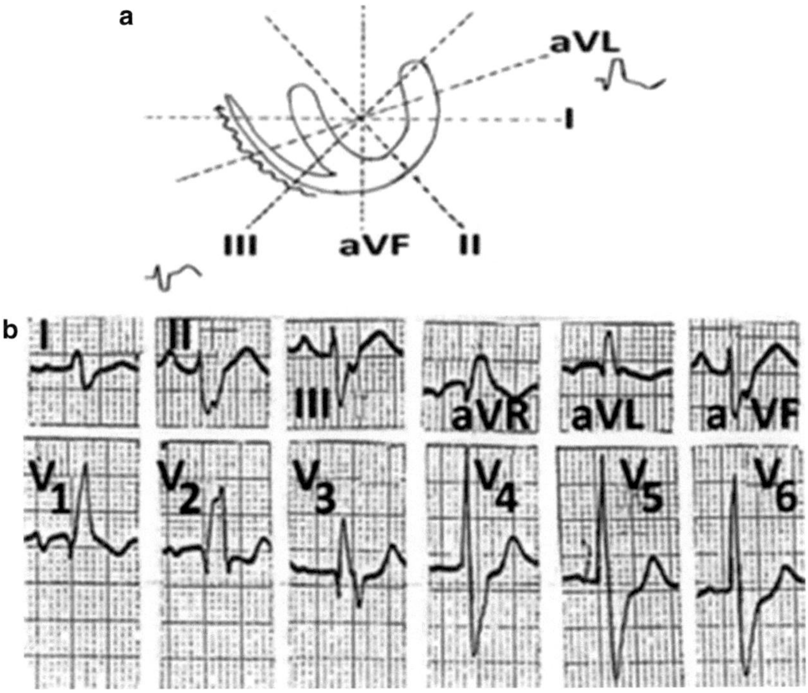

Fig. 28.15 (**a**) Diagram showing relation of the vector of right ventricular depolarization to inferior leads in a vertical heart. (**b**) ECG from a case of RBBB with vertical position of the heart. (Reproduced with permission from Mittal SR. Right Bundle Branch Block (RBBB). Cardiology Today-2014;18:126–134)

5 Effect of Associated Cardiac Disorders

5.1 Atrial Enlargement

RBBB par se does not affect the P wave. Atrial enlargement can, therefore, be diagnosed by conventional criteria. Tall P waves in leads II, III, aVF and prominent upright P wave in lead V_1 suggest right atrial enlargement (Fig. 28.16a). "P mitral" and/or prominent terminal negative component of the P wave in lead V_1 suggest concomitant left atrial enlargement (Fig. 28.16b).

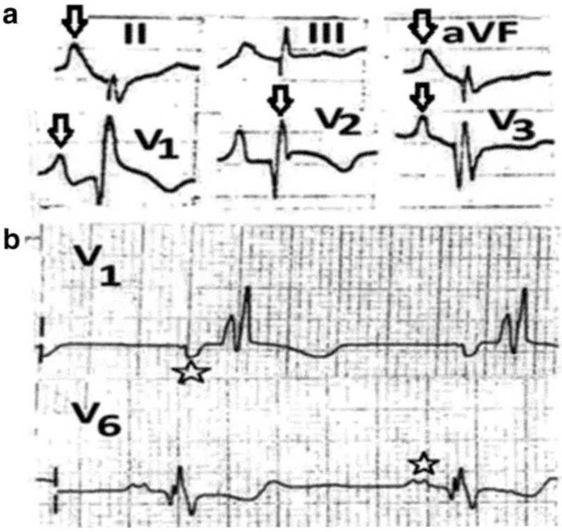

Fig. 28.16 (**a**) ECG showing RBBB with tall P wave in leads II, III, aVF and V_1 to V_3 (marked ⇩) suggestive of right atrial enlargement. (**b**) ECG showing RBBB with negative P wave in lead V_1 (marked ⇩) suggestive of left atrial enlargement. (Reproduced with permission from Mittal SR. Right Bundle Branch Block (RBBB). Cardiology Today-2014;18:126–134)

5.2 Ventricular Hypertrophy

Right ventricular hypertrophy increases the vector of depolarization of the right ventricular free wall (vector 3) (Fig. 28.17a vs. b). This increases the amplitude of the terminal R' wave in lead V_1 (Fig. 28.17c).

Left ventricular hypertrophy increases the magnitude of the vector of left ventricular depolarization (vector 2) (Fig. 28.18a vs. b). This produces a left-axis deviation in frontal plane leads and increases amplitude of the R wave in leads V_5,V_6 (Fig. 28.18c). Greater magnitude of left ventricular hypertrophy can produce T wave inversion in leads V_4 to V_6 (Fig. 28.18d).

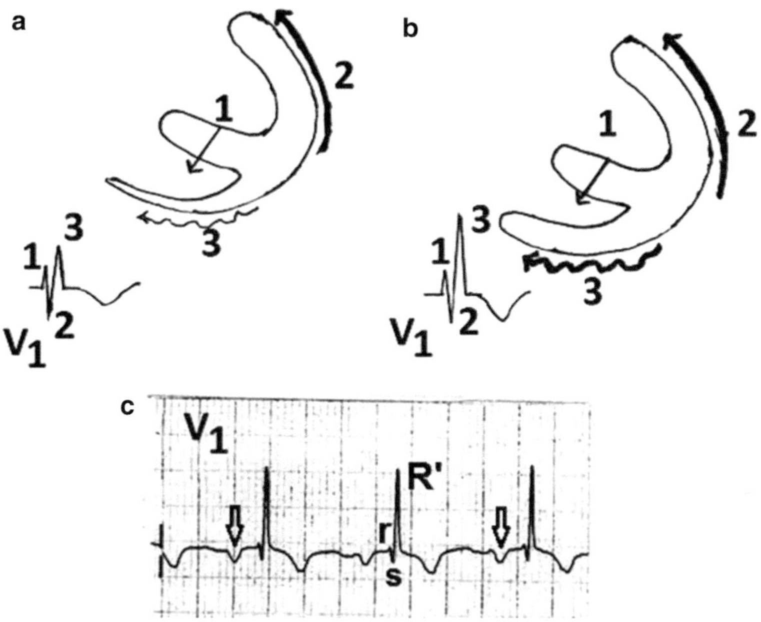

Fig. 28.17 (**a**) Diagram showing vectors in uncomplicated RBBB. (**b**) Diagram showing vectors in RBBB + right ventricular hypertrophy. (**c**) Electrocardiogram showing RBBB with tall terminal R' wave in lead V_1 suggestive of right ventricular hypertrophy. P wave is also inverted (marked ⇩) suggestive of concomitant left atrial enlargement. (Reproduced with permission from Mittal SR. Right Bundle Branch Block (RBBB). Cardiology Today- 2014;18:126–134)

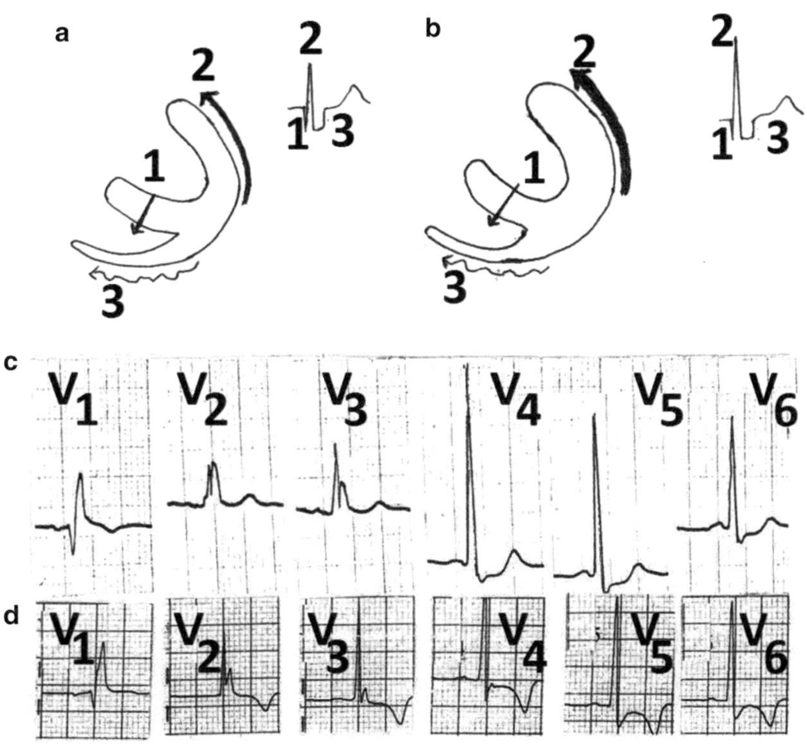

Fig. 28.18 (**a**) Diagram showing vectors in uncomplicated RBBB. (**b**) Diagram showing vectors in RBBB + left ventricular hypertrophy. (**c**) Electrocardiogram showing RBBB with left ventricular hypertrophy (increased R wave voltage in leads V_4 and V_5). (**d**) Electrocardiogram showing RBBB with LVH and inverted T waves in leads V_3 to V_6. (Reproduced with permission from Mittal SR. Right Bundle Branch Block (RBBB). Cardiology Today-2014;18:126–134)

5.3 Ventricular Dilatation

In uncomplicated RBBB, right ventricular leads (V_1 ,V_2) show rsR' pattern and left ventricular leads (V_5, V_6) show qRS pattern. Transition zone (transition from right ventricular pattern to the left ventricular pattern) (rR's with flat or shallow inversion of T wave) is in leads V_3 or V_4 (Fig. 28.19a).

Presence of left ventricular pattern (qRs with upright T wave) in lead V_2 or V_3 (Figs. 28.19b and 28.20a) suggests left ventricular dilatation with counterclockwise rotation (CCWR). On the other hand, persistence of the transition zone pattern in leads V_5,V_6 (Figs. 28.19c and 28.20b) suggests right ventricular dilatation with clockwise rotation (CWR).

5.4 Myocardial Infarction

RBBB does not affect initial septal depolarization with the result that initial 0.04 s of the QRS complex in unaffected. Q wave occurs during the initial 0.04 s of the QRS complex. Q waves of myocardial infarction are, therefore, not masked. In uncomplicated RBBB, ST-segment and T waves are directed opposite to the terminal QRS. ST-segment and T wave in the same direction as terminal portion of the QRS suggest myocardial damage (Fig. 28.21).

Transient ST-segment elevation without any sequential appearance of abnormal Q wave or T wave changes can occur in vasospastic angina. In uncomplicated RBBB, T wave inversion is asymmetrical. Symmetrical inversion (resembling an arrowhead) suggests myocardial damage.

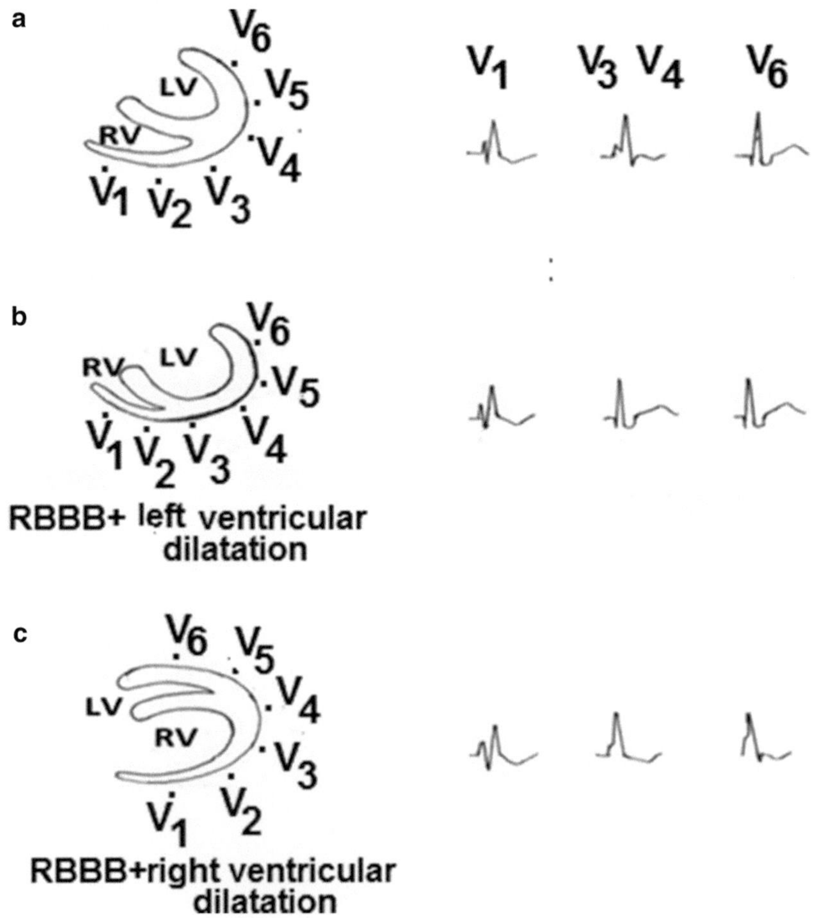

Fig. 28.19 (**a**) Diagram showing transition zone QRS complex in leads V_3 and V_4 in uncomplicated RBBB. (**b**) Diagram showing QRS complexes in RBBB+ left ventricular dilatation. Transition zone is shifted between leads V_1 to V_3 (counterclockwise rotation). (**c**) Diagram showing QRS complexes in RBBB + right ventricular dilation. Transition zone is shifted to lead V_6 (clockwise rotation). (Reproduced with permission from Mittal SR. Right Bundle Branch Block (RBBB). Cardiology Today-2014;18:126–134)

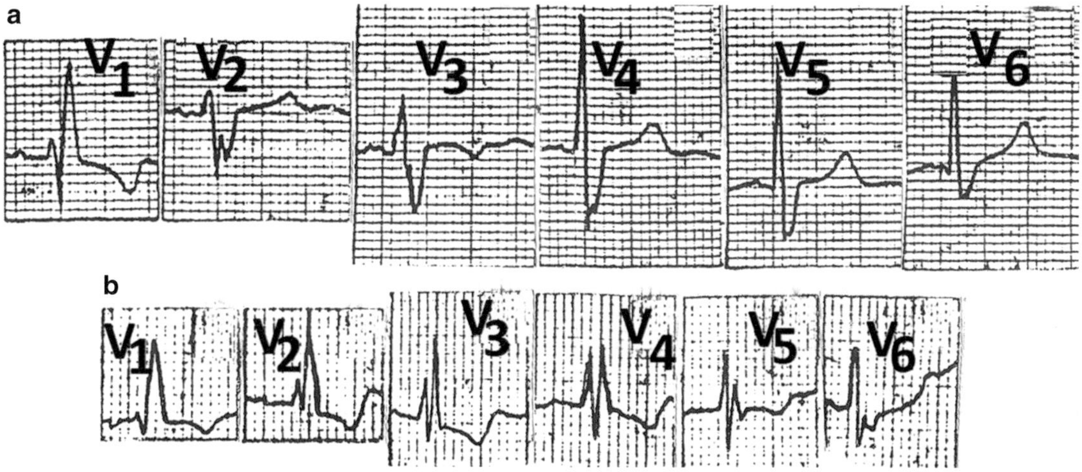

Fig. 28.20 (**a**) Electrocardiogram showing RBBB with left ventricular dilatation. Transition zone is between leads V_1 and V_2. Lead V_1 shows right ventricular pattern (rsR' with inverted T wave). Lead V_2 shows left ventricular pattern (rs configuration with positive T wave). (**b**) Electrocardiogram showing RBBB with right ventricular dilatation. Transition zone is between leads V_5 and V_6. Lead V_5 shows right ventricular pattern (rsr' with inverted T wave). Lead V_6 shows left ventricular pattern (Rs with positive T wave). (Reproduced with permission from Mittal SR. Right Bundle Branch Block (RBBB). Cardiology Today-2014;18:126–134)

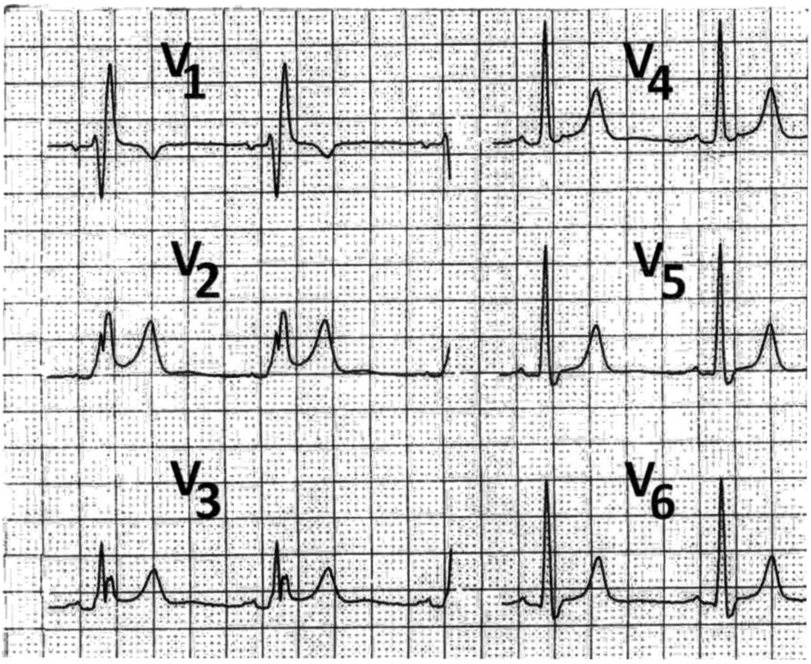

Fig. 28.21 Electrocardiogram showing RBBB with T waves in the direction of the terminal part of the QRS in leads V_2 and V_3. (Reproduced with permission from Mittal SR. Right Bundle Branch Block (RBBB). Cardiology Today-2014;18:126–134)

5.4.1 Septal Infarction

Nontransmural septal infarction produces coved ST-segment with symmetrical (arrowhead) T wave inversion in leads V_1 to V_3 (Fig. 28.22).

In uncomplicated RBBB, these leads show down sloping ST segment with asymmetrical T wave inversion (Fig. 28.5). Transmural septal infarction produces loss of initial anterior vector of septal depolarization, resulting in loss of initial r wave in leads V_1,V_2 and loss of initial q wave in leads V_5,V_6 (Fig. 28.23a vs. b).

Leads V_1,V_2, therefore, show a QR pattern (Fig. 28.24) rather than an rs'R pattern.

5.4.2 Anterior Myocardial Infarction

ST-segment elevation (Fig. 28.25a), symmetrical T wave inversion (Fig. 28.25b), reverse progression of r wave (Fig. 28.25c), and abnormal Q waves in leads V_2 to V_4 suggest anterior MI.

5.4.3 Other Causes of qR Pattern in Lead V_1

(i) Gross right atrial enlargement, e.g., in chronic cor pulmonale, acute pulmonary embolism, and Ebstein anomaly (Fig. 28.26a & b). Due to gross dilatation of the right atrium, lead V_1 overlies the right atrium and records cavity potential. Vector of septal depolarization (marked 1 in Fig. 28.26) moves away from lead V_1, resulting in initial q wave in lead V_1.

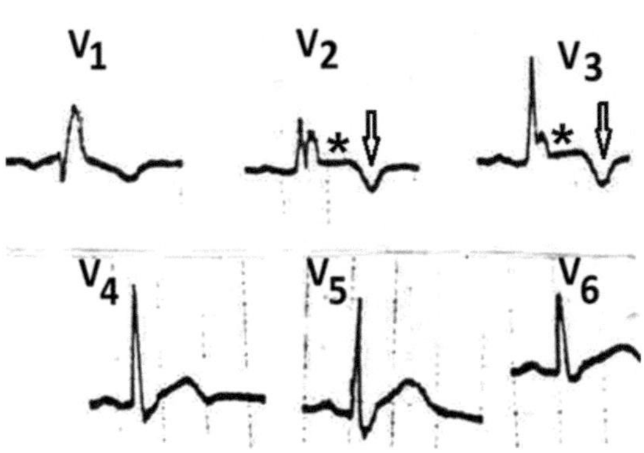

Fig. 28.22 Electrocardiogram showing RBBB with coved ST-segment (marked *) and symmetrical (arrowhead) T wave inversion (marked ⇩) in leads V_2 and V_3. (Reproduced with permission from Mittal SR. Right Bundle Branch Block (RBBB). Cardiology Today- 2014;18:126–134)

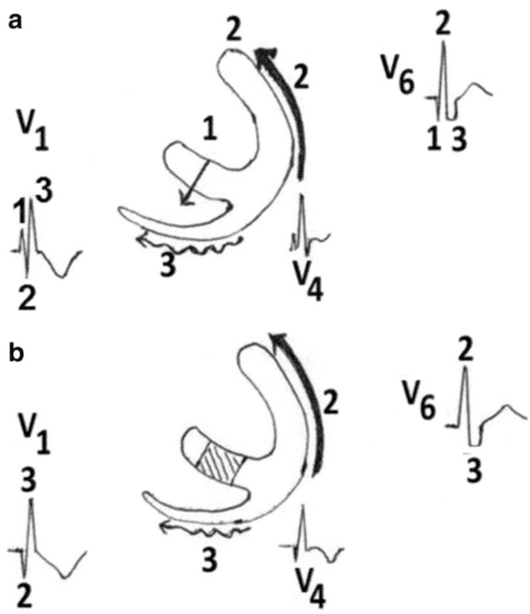

Fig. 28.23 (**a**) Diagram showing QRS in uncomplicated RBBB. (**b**) Diagram showing effect of transmural septal infarction on QRS complex in the presence of RBBB. (Reproduced with permission from Mittal SR. Right Bundle Branch Block (RBBB). Cardiology Today- 2014;18: 126–134)

Other signs of right atrial enlargement, e.g., "P-pulmonale," are present (Fig. 28.27). Q wave does not extend beyond lead V_2 (Fig. 28.27).

Other differentiating features include

(i) q wave of leads V_5,V_6, is preserved. (ii) R wave is not broad in leads V_1,V_2. (iii) There is no slurred s wave in leads I,V_5,V_6. (iv) There are no ST segments or T wave changes suggestive of infarction.

(ii) Focal block in the left bundle branch fibers depolarizing the septum. Septal depolarization occurs from right to left, resulting in q wave in leads V_1 and V_2 and loss of initial q wave in leads V_5 and V_6 (Fig. 28.28).

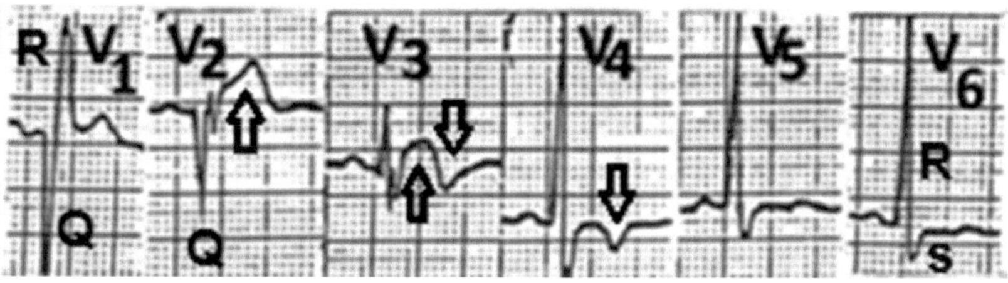

Fig. 28.24 ECG showing RBBB with transmural septal infarction. Lead V_1 shows QR configuration. Lead V_6 shows Rs configuration. Lead V_2 shows QRS with ST-segment elevation (marked ⇧). Lead V_3 shows qRs with ST-segment elevation (marked ⇧) and inversion of T wave (marked ⇩). Lead V_4 shows inversion of T wave (marked ⇩) (Reproduced with permission from Mittal SR. Right Bundle Branch Block (RBBB). Cardiology Today- 2014;18:126–134)

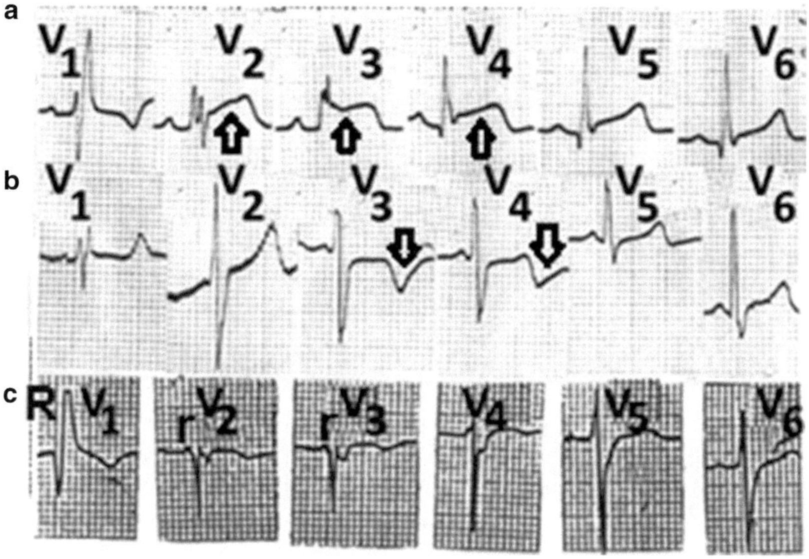

Fig. 28.25 (**a**) Electrocardiogram showing RBBB with ST-segment elevation in leads V_2 to V_4 (marked ⇧) suggestive of acute anterior myocardial infarction. (**b**) Electrocardiogram showing RBBB with T wave inversion in leads V_3 and V_4 (marked ⇩) suggestive of non-ST-segment elevation anterior myocardial infarction. (**c**) Electrocardiogram showing RBBB with reverse progression of R wave from leads V_1 to V_2 and V_3 suggestive of anterior myocardial infarction. (Reproduced with permission from Mittal SR. Right Bundle Branch Block (RBBB). Cardiology Today- 2014;18:126–134)

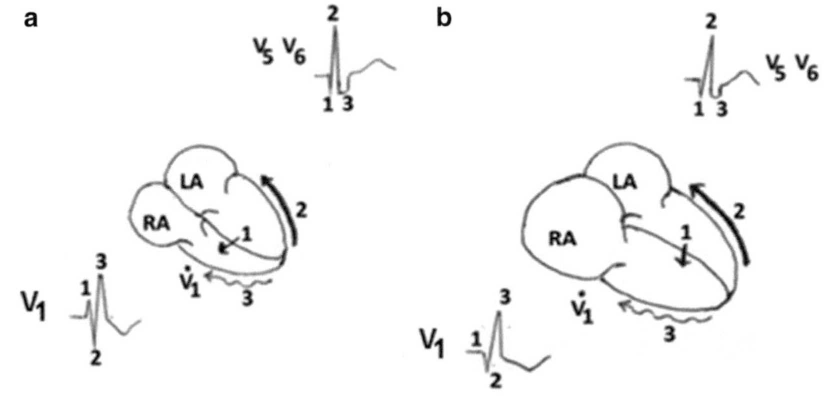

Fig. 28.26 (**a**) Diagram showing QRS configuration in lead V_1 and leads V_5, V_6 in uncomplicated RBBB. (**b**) Diagram showing the effect of right atrial enlargement on QRS configuration in lead V_1 and leads V_5, V_6. 1 vector of septal depolarization; 2 vector of depolarization of left ventricular free wall; 3, vector of depolarization of right ventricular free wall; *RA* right atrium; *LA* left atrium. (Reproduced with permission from Mittal SR. Right Bundle Branch Block (RBBB). Cardiology Today-2014;18:126–134)

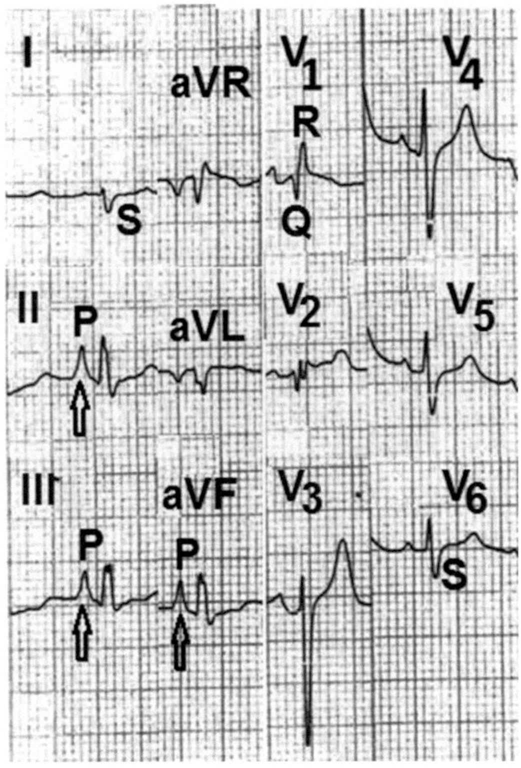

Fig. 28.27 Electrocardiogram from a patient of cor pulmonale showing QR pattern in lead V_1 with low voltage in limb leads, right-axis deviation (prominent S wave in lead I), P- pulmonale (tall and peaked P waves in leads II, III, and a⇧F, marked ⇧) clockwise rotation (persistence of S wave up to lead V_6). (Reproduced with permission from Mittal SR. Right Bundle Branch Block (RBBB). Cardiology Today-2014;18:126–134)

R wave is not slurred in leads V_1 and V_2. There is no slurring of S wave in leads I, aVL, V_5, and V_6. There are no ST-segments or T wave changes of myocardial infarction.

5.4.4 Lateral Myocardial Infarction

There is an abnormal Q wave and loss of R wave in lead V_6 (Fig. 28.29).

ST-segment elevation (Fig. 28.30a) or symmetrical (arrow head) T wave inversion (Fig.28.30b) in leads V_5,V_6 suggests lateral myocardial infarction. Changes confined to leads I and aVL suggest high lateral infarction (Fig. 28.30c).

(d) Inferior infarction

Transient ST-segment elevation with sequential changes in leads II, III, and aVF suggest acute inferior infarction. Abnormal Q waves support the diagnosis of old infarction (Fig. 28.31).

5.4.5 Posterior Infarction

In uncomplicated RBBB, posterior leads (V_7 to V_9) show qrs configuration with upright T wave (Fig. 28.12). Isolated posterior infarction can produce a prominent R wave in lead V_1. Diagnosis of posterior infarction in the presence of RBBB is,

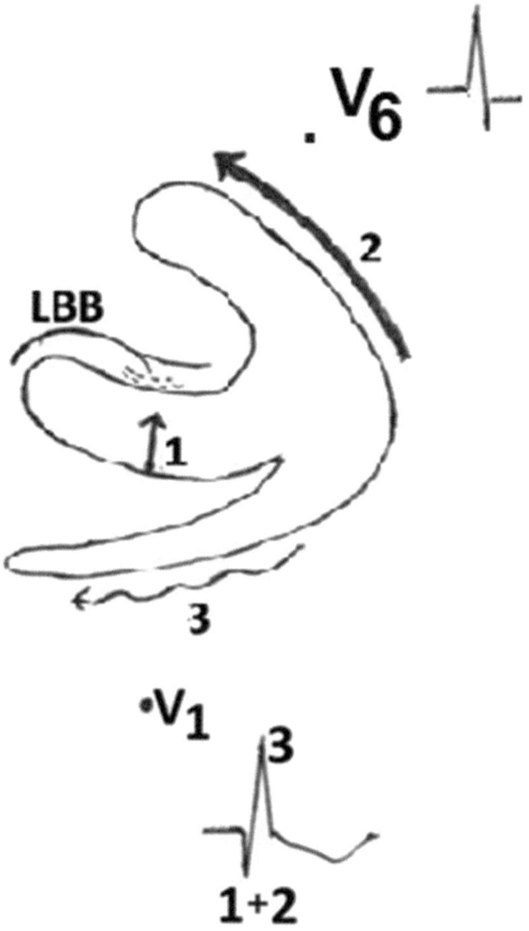

Fig. 28.28 Diagram showing the effect of block in the septal fibers of the left bundle branch on the QRS configuration in the presence of RBBB. *LBB* left bundle branch; 1, septal depolarization; 2, depolarization of left ventricular free wall; 3, depolarization of right ventricular free wall. (Reproduced with permission from Mittal SR. Right Bundle Branch Block (RBBB). Cardiology Today- 2014;18:126–134)

therefore, difficult. The following ECG findings can suggest concomitant posterior infarction:

(i) Symmetrical (arrowhead) T wave inversion in posterior leads (Fig. 28.32).
(ii) Upright T wave in lead V_1 (Figs. 28.32 and 28.33).
(iii) Evidence of concomitant lateral myocardial infarction, e.g., abnormal Q waves and/or T wave inversion in leads V_5 and V_6 (Fig. 28.33).

5.4.6 Right Ventricular Infarction

In uncomplicated RBBB, leads V_1 to V_4R show rsR' pattern (Fig. 28.12). QR pattern in these leads suggests the possibility of right ventricular infarction. Concomitant changes of inferior infarction further support this possibility (Fig. 28.34).

In septal infarction, changes are present in leads V_1 to V_3 and not in V_3R and V_4R. Posterior infarction can produce reciprocal changes (rs pattern with positive T waves) in leads V_1 to V_4R. These findings can mask concomitant right ventricular infarction. Leads V_5R and V_6R can show qR pattern even in uncomplicated RBBB. New appearance of RBBB can mask q wave and ST-segment elevation of right ventricular infarction (Fig. 28.35a,b).

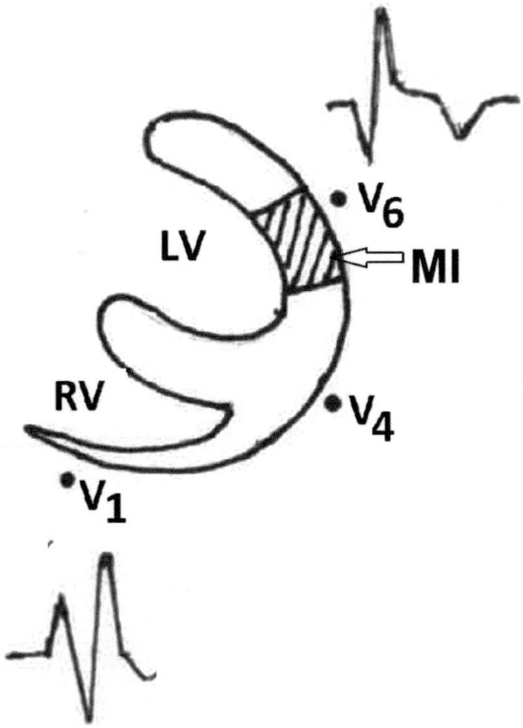

Fig. 28.29 Diagram showing the effect of lateral MI on ECG of RBBB. (Reproduced with permission from Mittal SR. Right Bundle Branch Block (RBBB). Cardiology Today-2014;18:126–134)

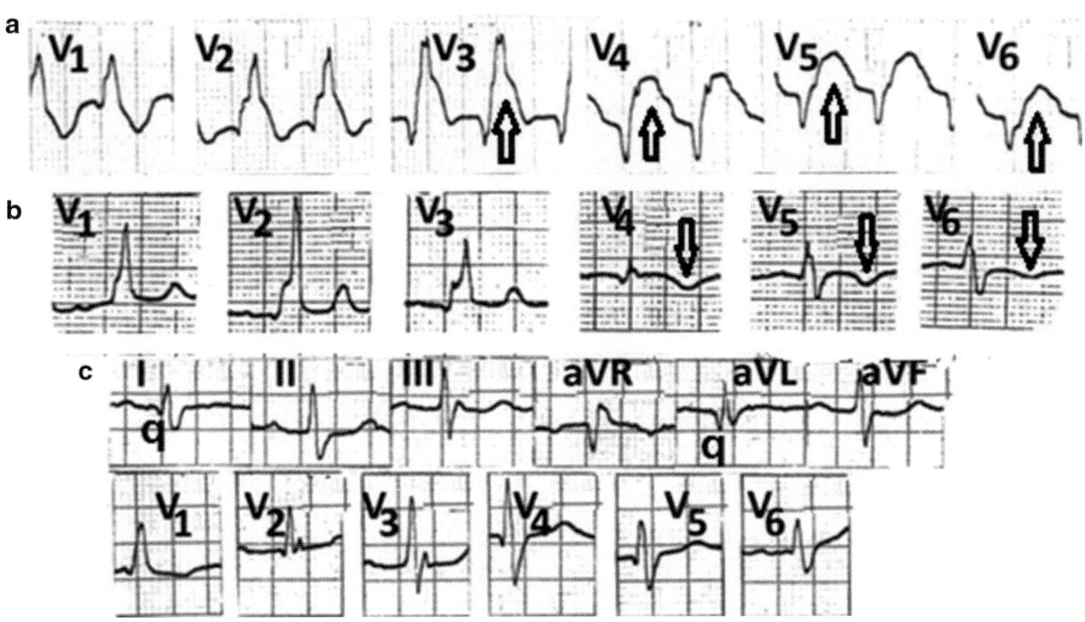

Fig. 28.30 (**a**) Electrocardiogram showing RBBB with ST-segment elevation in leads V_3 to V_6 (marked ⇧). (**b**) Electrocardiogram showing RBBB with symmetrical T wave inversion in leads V_4 to V_6 (marked ⇩). (**c**) Electrocardiogram showing RBBB with Q waves in leads I and aVL. (Reproduced with permission from Mittal SR. Right Bundle Branch Block (RBBB). Cardiology Today-2014;18:126–134)

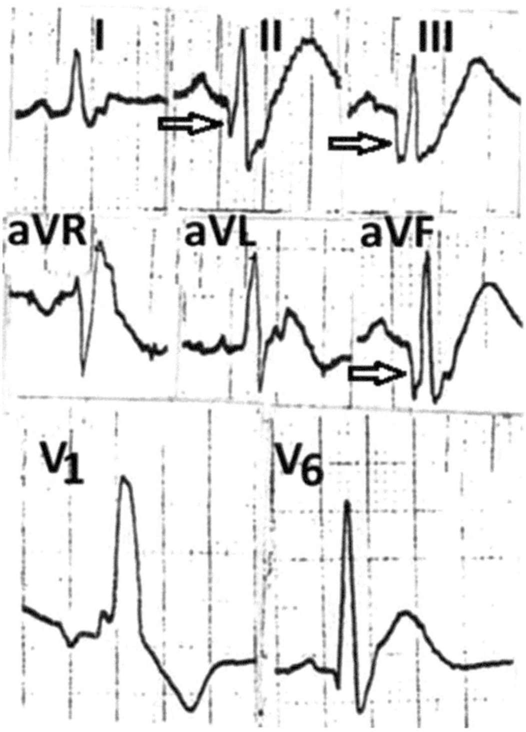

Fig. 28.31 Electrocardiogram showing RBBB with inferior myocardial infarction. Leads II, III, and aVF show abnormal Q waves (marked ⇨). (Reproduced with permission from Mittal SR. Right Bundle Branch Block (RBBB). Cardiology Today- 2014;18:126–134)

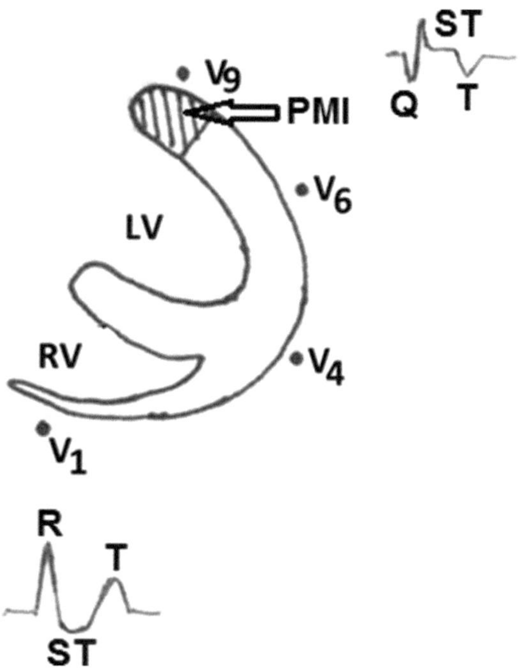

Fig. 28.32 Diagram showing the effect of transmural posterior infarction (marked PMI) on the electrocardiogram of RBBB. Lead V_9 shows abnormal Q wave, ST-segment elevation, and deep symmetrical T wave inversion. Lead V_1 shows dominant R wave, ST-segment depression, and positive T wave. (Reproduced with permission from Mittal SR. Right Bundle Branch Block (RBBB). Cardiology Today- 2014;18:126–134)

Fig. 28.33 Electrocardiogram showing RBBB with upright T wave in leads V_1 to V_3 (marked ⇧) and inverted T waves in leads V_4 to V_7 (marked ⇩) suggestive of non-ST-segment elevation posterolateral infarction. (Reproduced with permission from Mittal SR. Right Bundle Branch Block (RBBB). Cardiology Today- 2014;18:126–134)

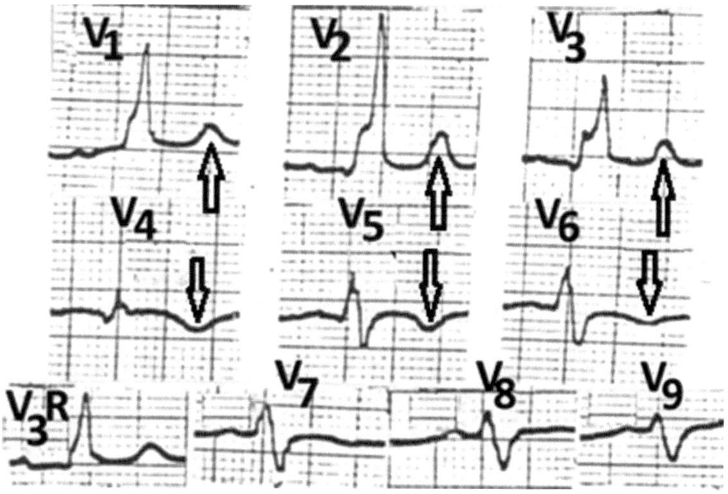

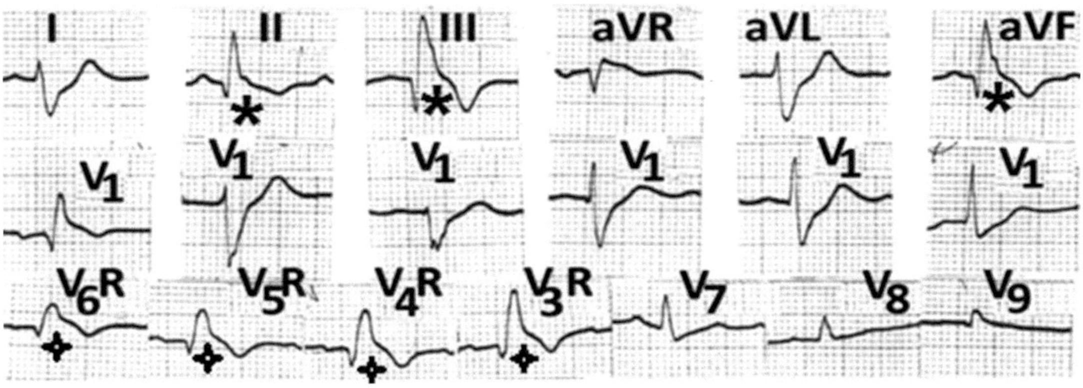

Fig. 28.34 Electrocardiogram showing RBBB with inferior infarction (marked *) with right ventricular infarction (marked ✧). (Reproduced with permission from Mittal SR. Right Bundle Branch Block (RBBB). Cardiology Today- 2014;18:126–134)

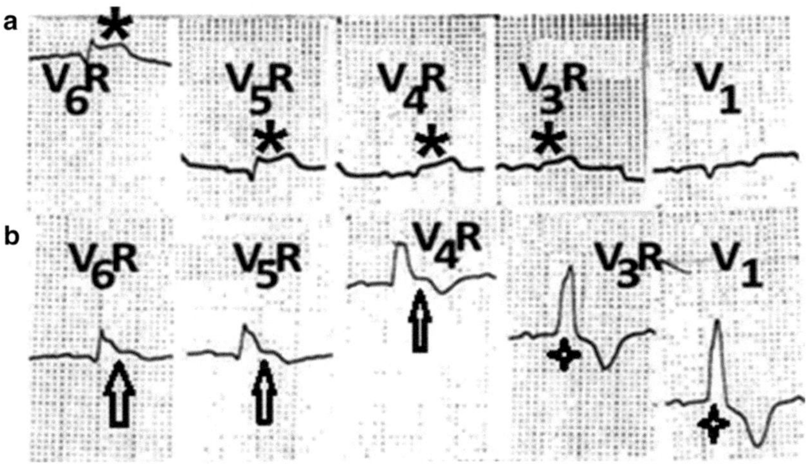

Fig. 28.35 (**a**) Electrocardiogram showing right ventricular infarction. Q wave with ST-segment elevation in leads V_3R to V_6R (marked *). (**b**) Electrocardiogram from same patient. Appearance of RBBB masks Q wave and ST-segment elevation in leads V_1 and V_3R (marked ✧). ST-segment is elevated in leads V_4R to V_6R (marked ⇧). (Reproduced with permission from Mittal SR. Right Bundle Branch Block (RBBB). Cardiology Today- 2014;18: 126–134)

ST-segment elevation in right-sided chest leads in the presence of RBBB supports the diagnosis of right ventricular infarction.

6 Differential Diagnosis

Causes of prominent R wave in lead V_1

1. RBBB
2. Right ventricular hypertrophy: Diagnosis is supported by the presence of P- pulmonale, right-axis deviation in frontal plane, qR or RS morphology in lead $V_{1,}$ and absence of terminal slurring of the QRS in lead V_6 (Fig. 28.36).
3. Posterior myocardial infarction (Fig. 28.37): Isolated posterior myocardial infarction produces R or Rs configuration in lead $V_{1.}$ Initial R wave is prominent.

   There is no slurred terminal R' wave in lead V_1 or slurred s wave in lead V_6. The T wave is upright in lead V_1. Leads V_7 to V_9 show a Qr pattern with coved ST-segment and inverted T wave. Changes in lead V_1 are recip-

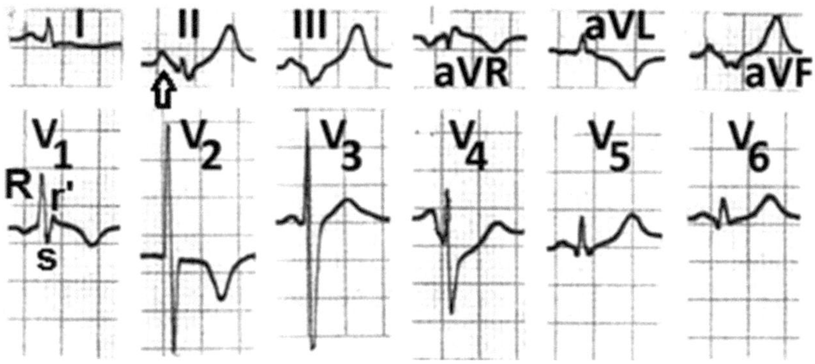

Fig. 28.36 Electrocardiogram showing low voltage in limb leads and lead V_5 and V_6 (suggestive of emphysema), P- pulmonale (peaked P wave in lead II; marked ⇧), and Rsr' in lead V_1. In lead V_1, initial R is more prominent. There is no terminal slurred R' in lead V_1 and no terminal slurred S wave in lead V_6. (Reproduced with permission from Mittal SR. Right Bundle Branch Block (RBBB). Cardiology Today-2014;18:126–134)

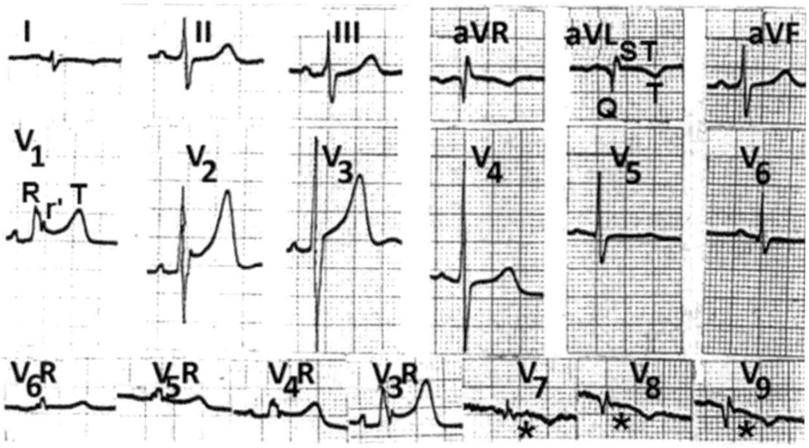

Fig. 28.37 Electrocardiogram showing posterior infarction. Lead V_1 shows positive T wave. Initial R is prominent and slurred in lead V_1. Leads V_7 to V_9 show Qr pattern with ST-segment elevation and inverted T waves (marked *), and lead aVL also shows Q wave with coved ST segment and inverted T wave. Findings suggest posterolateral infarction. (Reproduced with permission from Mittal SR. Right Bundle Branch Block (RBBB). Cardiology Today-2014;18:126–134)

rocal changes secondary to abnormal Q wave and inverted T waves in posterior leads.

4. Duchenne muscular dystrophy: Posterior myocardium is selectively involved in the dystrophic process. ECG resembles posterior myocardial infarction. Clinical profile helps differentiation.
5. Preexcitation via left lateral accessory pathway (Figs. 28.38 and 28.39). Short PR interval and initial rather than terminal slurring of

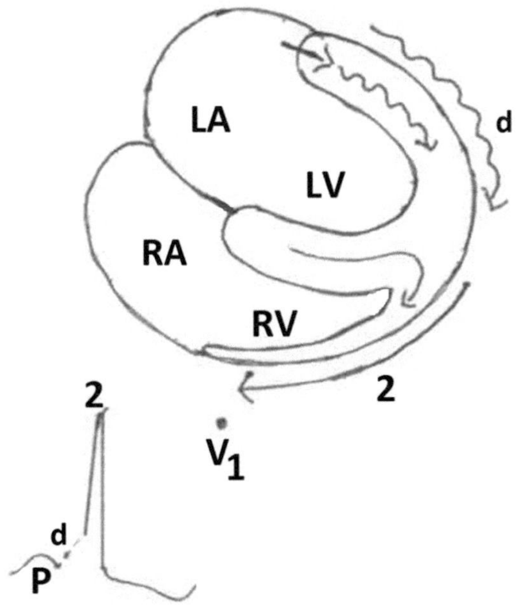

Fig. 28.38 Diagram showing the effect of preexcitation via the left lateral accessory path on lead V_1. The PR interval is short, and there is initial slurring of the QRS complex due to presence of the delta wave. (d) (Reproduced with permission from Mittal SR. Right Bundle Branch Block (RBBB). Cardiology Today- 2014;18:126–134)

the QRS complex identifies preexcitation. There is no slurred s wave in lead V_6. Posterior leads (V_7 to V_9) show Q wave. but ST segment and T waves are normal.

6. Dextrocardia: P wave inversion in lead I, prominent R in right-sided chest leads, deep S in left precordial leads, and absence of terminal slurring of QRS complex in lead V_1 and absence of slurred s wave in lead V_6 differentiates it from RBBB (Fig. 28.40).
7. Hyperkalemia: Diffuse and uniform depression of intraventricular conduction produces uniform prolongation of the whole QRS (Fig. 28.41).

RBBB prolongs only terminal QRS. Other ECG findings that help differentiation include tall peaked T waves with a narrow base, merging of the ST-segment and the T waves with QRS, and decreased amplitude or disappearance of the P wave.

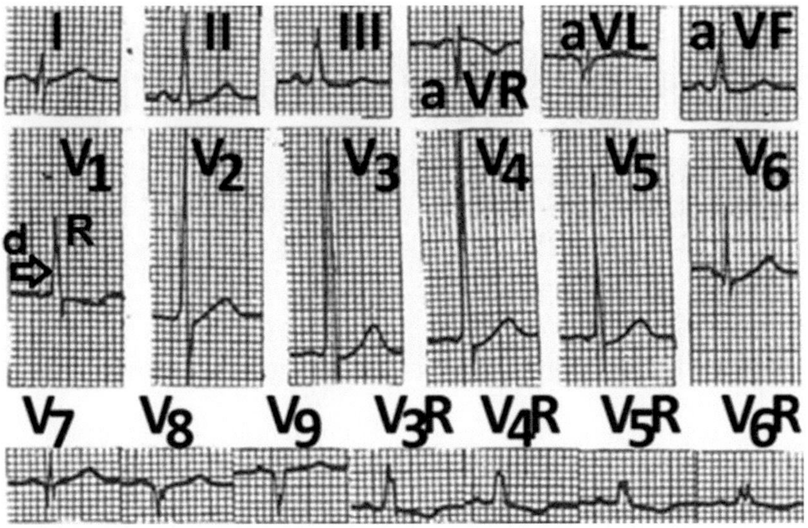

Fig. 28.39 Electrocardiogram from a patient of preexcitation via left lateral accessory pathway. Lead V_1 shows short PR interval, initial slurring of the R wave due to delta wave (marked ⇨ d). Posterior leads (V_7 to V_9) do not show any elevation of ST-segment or inversion of T waves. These findings exclude any possibility of posterior infarction. (Reproduced with permission from Mittal SR. Right Bundle Branch Block (RBBB). Cardiology Today- 2014;18:126–134)

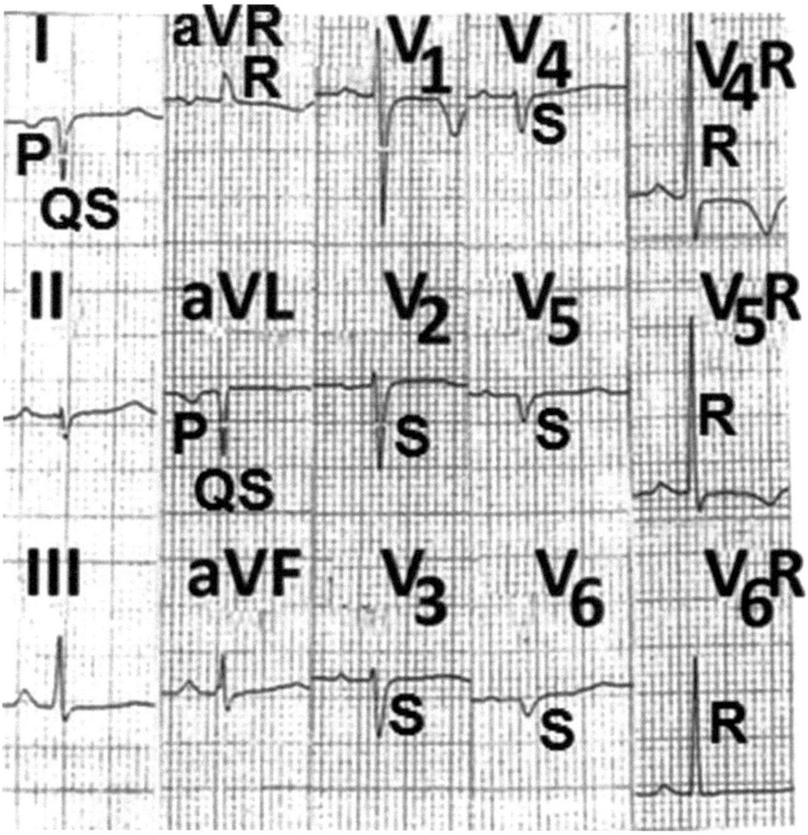

Fig. 28.40 Electrocardiogram from a patient of dextrocardia. Leads I and aVL show negative P wave and QS configuration. Lead aVR shows R wave. Leads V_1 to V_6R show prominent R wave. Precordial leads V_2 to V_6 show dominant S wave. (Reproduced with permission from Mittal SR. Right Bundle Branch Block (RBBB). Cardiology Today- 2014;18:126–134)

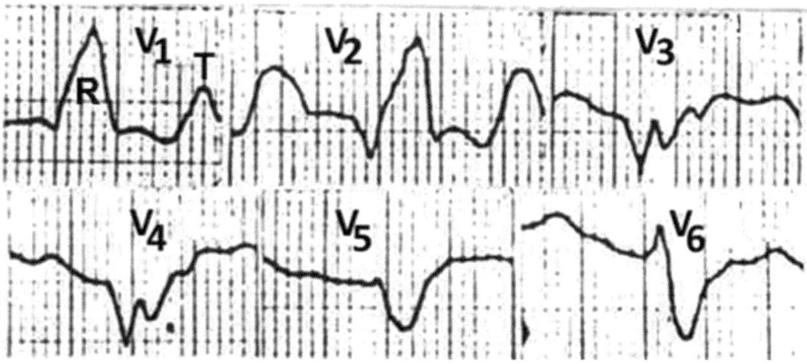

Fig. 28.41 Electrocardiogram from a patient of hyperkalemia showing prominent R wave in lead V_1. There is diffuse broadening of the QRS complex and the T wave is positive. (Reproduced with permission from Mittal SR. Right Bundle Branch Block (RBBB). Cardiology Today-2014;18:126–134)

7 Clinical Significance

1. Many persons with RBBB may not have evidence of other structural heart diseases [4]. Hypertension is frequently associated. Patients without structural heart disease usually do not advance to complete atrioventricular block.
2. New-onset RBBB without other evidence of structural heart disease is associated with higher risk of cardiovascular morbidity and mortality in the future.
3. RBBB associated with any cardiac disease suggests more extensive pathology compared to patients without any conduction disorder.
4. Appearance during an acute coronary syndrome suggests occlusion of left anterior descending coronary artery proximal to the first septal perforator branch that gives blood supply to the right bundle branch [1].
5. In the setting of acute anterior myocardial infarction, fresh appearance of RBBB is associated with higher risk of development of atrioventricular block [6].

Summary

- Partial RBBB is diagnosed when QRS duration is less than 120 milliseconds, and lead V_1 shows rsr' or rsR' pattern with a terminal r in lead aVR and terminal s wave in leads aVL and V_6. These ECG findings are, however, nonspecific. This pattern can also be seen in some normal persons, athletes, pectus excavatum, right ventricular enlargement, and atrial septal defect. In Brugada syndrome, this pattern is associated with ST-segment elevation in leads V_1 and V_2.
- Progression to advanced RBBB results in progressive decrease in the depth of s wave and increase in amplitude and duration of the R' in lead V_1.
- Advanced RBBB is diagnosed when QRS duration is 120 milliseconds or more, lead V_1 shows rsR' pattern with slurring of R' wave , leads I, aVL, and V_6 show qRS configuration with slurring of the S wave, and lead aVR shows terminal slurred R wave. Time to peak of R in lead V_1 is more than 50 milliseconds. ST-segment and T wave are opposite to the direction of the terminal part of the QRS complex. Thus, the T waves are inverted in right precordial leads and are positive in leads I, aVL, V_5, and V_6.
- Mean frontal plane QRS axis is normal in uncomplicated RBBB.
- Right-axis deviation, right atrial abnormality, and amplitude of R' wave more than 1.5 mV in lead V_1 suggest underlying right ventricular hypertrophy.
- Left-axis deviation and left atrial abnormality suggest underlying left ventricular hypertrophy. Voltage criteria are not sensitive in diagnosing left ventricular hypertrophy in the presence of RBBB.
- Diagnosis of Q wave myocardial infarction is not affected by the presence of RBBB.
- ST-segment and T wave concordant to the direction of the terminal part of the QRS complex are abnormal. Thus, ST-segment elevation and positive T wave in right precordial leads suggest underlying myocardial necrosis. Similarly, ST-segment depression and deep symmetrical T wave inversion in leads I, aVL, V_5 and V_6 are abnormal.
- In the absence of any structural heart disease or additional involvement of any other fascicles of the left bundle branch, RBBB alone does not increase the risk of cardiac morbidity or mortality.

MCQs

Q1. RBBB distorts

(a) Early 1/3rd of the QRS complex
(b) Middle 1/3rd of the QRS complex
(c) Last 1/3rd of the QRS complex
(d) Complete the QRS complex

Q2 .In RBBB, interventricular septum is activated from

(a) Left to right
(b) Right to left
(c) Anterior to posterior
(d) Posterior to anterior

Q3. In RBBB, which part of the heart is activated last?

(a) Interventricular septum
(b) Left ventricular free wall
(c) Right ventricular free wall
(d) Left ventricular apex

Q4. The terminal r' wave in lead V_1 , terminal "r" in lead aVR, and terminal narrow "s" in leads aVL and V_6 with a QRS duration of 0.10 s and frontal plane QRS axis of $+60^0$ suggest

(a) Incomplete RBBB
(b) Complete RBBB
(c) Complete RBBB + left anterosuperior fascicular block
(d) Complete RBBB + left posteroinferior fascicular block

Q5. Mean frontal plane QRS axis of + 90^0, QRS duration of 0.14 s, terminal slurred S in lead I and $V_{6,}$ and terminal slurred R in lead aVR suggest

(a) Incomplete RBBB
(b) Complete RBBB
(c) Incomplete LBBB
(d) None

Q6. Q wave of 0.04 s, followed by a slurred R wave in leads V_1 and V_2 , frontal plane QRS axis of + 75^0 , terminal slurred S in leads I and aVL, and QRS duration of 0.14 s suggest

(a) Incomplete RBBB with septal infarction
(b) Complete RBBB with septal infarction
(c) Complete RBBB with direct posterior infarction
(d) Incomplete RBBB with left ventricular hypertrophy

Q7. QRS duration of 0.12 s, rsR' in leads V_1 and $V_{2,}$ mean frontal plane QRS axis of $+75^0$, qRS with symmetrical, and pointed T wave inversion in leads I, aVL,$V_{5,}$ and V_6 suggest

(a) Incomplete RBBB with septal infarction
(b) Complete RBBB with anterior infarction
(c) Complete RBBB with RVH
(d) Complete RBBB with lateral subendocardial infarction

Q8. QRS duration of 0.10 s with rsr' pattern in lead V_1 can be seen in

(a) Incomplete RBBB
(b) Atrial septal defect
(c) Pectus excavatum
(d) None

Q9. Advancement of incomplete RBBB toward complete RBBB is suggested by

(a) Decrease in the height of initial r wave in lead V_1
(b) Decrease in the depth of the s wave in lead V_1
(c) Decrease in the height of r' wave in lead V_1
(d) Increase in the height of r' wave in lead V_1

Q10. rsR' in leads V_1 ,V_2 with R' of 2 millivolts, QRS duration of 0.12 s, qR with slurred S in V_5 suggest

(a) Complete RBBB with left ventricular hypertrophy
(b) Complete RBBB with right ventricular hypertrophy
(c) Incomplete RBBB with right ventricle volume overload
(d) Complete RBBB with left ventricle volume overload

Q11. In the presence of complete RBBB, left ventricular hypertrophy can be suspected by the presence of

(a) Left-axis deviation
(b) Prominent negative terminal force of P wave in lead V_1
(c) Voltage criteria
(d) All

Q12. In the presence of complete RBBB, ST-segment elevation with positive T wave in leads V_1 and V_2 suggest

(a) Right ventricular hypertrophy
(b) Left ventricular hypertrophy
(c) Biventricular enlargement
(d) None

Q13. In the presence of complete RBBB, ST-segment elevation with symmetrical T wave inversion in leads V_5 and V_6 suggests

(a) Lateral infarction
(b) Left ventricular hypertrophy
(c) Direct posterior infarction
(d) None

Answers

(1) c (2) a (3) c (4) a (5) b (6) b (7) d (8) a, b, c (9) b, d (10) b (11) a, b (12) d (13)a

References

1. De Luna AB, Goldwasser D, Fiol M, Bayes - Genis A. Surface electrocardiography. In: Fuster V, Walsh RA, Harrington RA, editors. Hurst's The Heart. New York: Mc Graw Hill; 2011. p. 307–70.
2. Schamroth L. Right bundle branch block. In: Schamroth C, editor. An introduction to electrocardiography. Blackwell Science: France; 1982. p. 89–98.
3. Olshansky B, Chung MK, Pogwizd SM, Goldschlager N. Bradyarrhythmias - conduction system abnormalities. In: Olshansky B, Chung MK, Pogwizd SM, Goldschlager N, editors. Arrhythmia essentials. Burlington: Jones & Bartlett; 2012. p. 26–78.
4. Wagner GS. Intraventricular conduction abnormalities. In: Wagner GS, editor. Marriott's practical electrocardiography. Philadelphia: Wolters Kluwer; 2008. p. 97–124.
5. Mirvis DM, Goldberger AL. Electrocardiography. In: Bonow RO, Mann DL, Zipes DP, Libby P, editors. Braunwald's Heart disease. Missouri: Elsevier; 2012. p. 126–67.
6. Antman EM, Marrow DA. ST segment elevation myocardial infarction. In: Bonow RO, Mann DL, Zipes DP, Libby P, editors. Braunwald's Heart Disease. Elsevier: Missouri; 2012. p. 1111–70.

29 Multifascicular Blocks (Delayed Conduction in More Than One Fascicle)

It denotes delayed conduction in more than one fascicle. Right bundle branch is considered as one fascicle.

1 Bifascicular Block

It denotes delayed conduction in any two fascicles. The following combinations suggest bifascicular block (Fig. 29.1a,b,c).

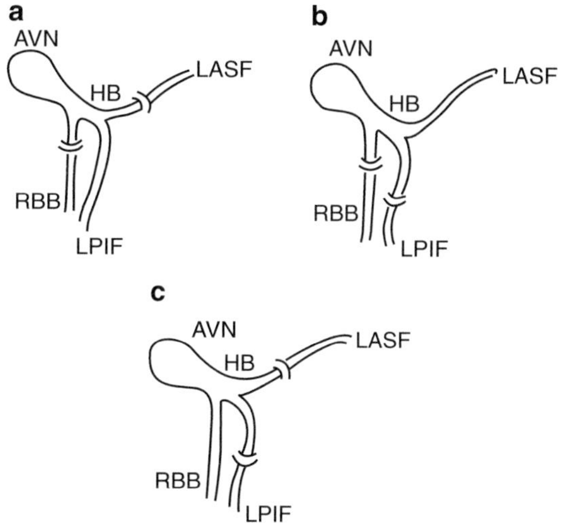

Fig. 29.1 Diagram showing sites of bifascicular block. (**a**) RBBB+ left anterosuperior fascicular block. (**b**) RBBB+ left posteroinferior fascicular block. (**c**) Left anterosuperior fascicular block + left posteroinferior fascicular block. (Reproduced with permission from Mittal SR. Multifascicular blocks. Cardiology Today-2015;19:100-106)

1.1 Electrocardiographic Diagnosis

1. Right bundle branch block + left anterosuperior fascicular block.

   Limb leads are influenced by the anterosuperior fascicular block. Mean frontal plane QRS axis is to the left of -45^0. Leads I and aVL show qR configuration and leads II, III,

S. Mittal, *Insights into Electrocardiograms with MCQs*,
https://doi.org/10.1007/978-981-99-0127-2_29

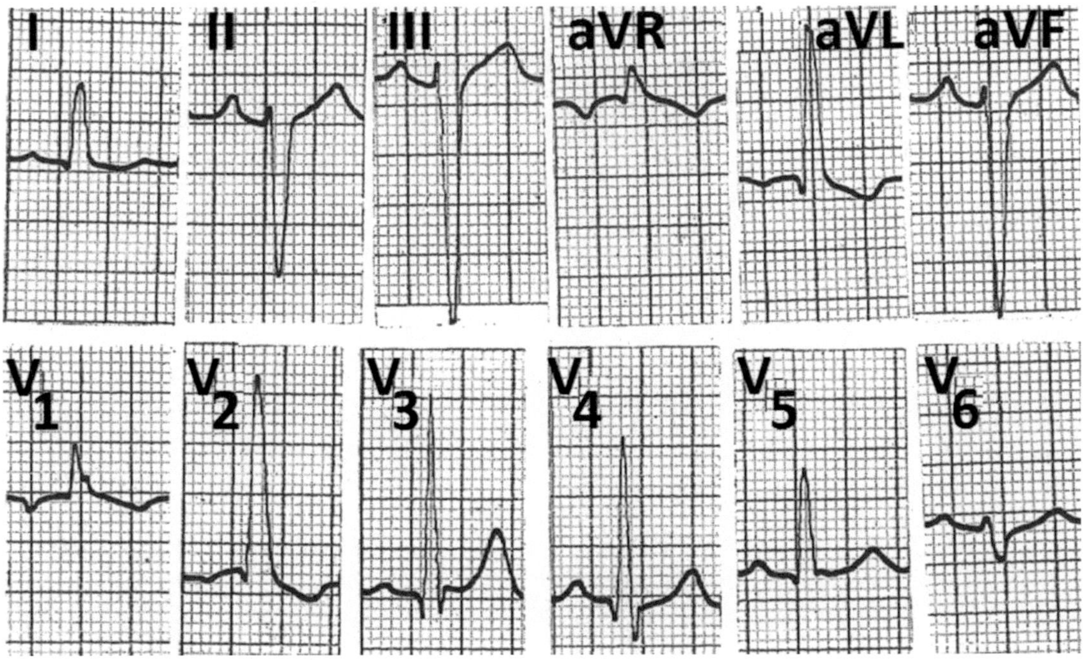

Fig. 29.2 Electrocardiogram showing right bundle branch block with left anterosuperior fascicular block. (Reproduced with permission from Mittal SR. Multifascicular blocks. Cardiology Today-2015;19:100-106)

and aVF show rS configuration due to left anterosuperior fascicular block. Precordial leads show right bundle branch block pattern (Fig. 29.2).

Additional left anterior focal block or high degree of left anterosuperior fascicular block can produce dominance of left anterosuperior forces with increased duration of R wave in leads I and aVL. Additional loss of initial QRS vector can result in loss of initial q wave in leads I and aVL. In this situation, leads I and aVL have only a broad R wave simulating left bundle branch block. Precordial leads, however, maintain right bundle branch block configuration. Electrocardiographic findings of right bundle branch block are not seen in the limb leads. This situation is also called "masked" (hidden) bundle branch block [1] (Figs. 29.3, 29.4, and 29.5).

2. Right bundle branch block + left posteroinferior fascicular block.

   Limb leads are influenced by the posteroinferior fascicular block. Mean frontal plane QRS axis is, therefore, to the right of $+120^0$. Leads I and aVL show rS configuration. Leads II, III, and aVF show qR configuration (Fig. 29.6).

3. Delayed conduction in left anterosuperior fascicle and left posteroinferior fascicle.

   This combination presents as LBBB on surface electrocardiogram (Fig. 29.7a). Surface electrocardiogram cannot differentiate this possibility from other causes of LBBB. Electrophysiological study is necessary for correct diagnosis. Development of atrioventricular block during exercise (Fig. 29.7b) supports the possibility of infra- His pathology, i.e., the disease is at the level of the two fascicle of the left bundle branch rather than at the level of main left bundle branch.

1.2 Clinical Significance

1. Right bundle branch block with left anterosuperior fascicular block is more common. In the context of coronary artery disease, this combi-

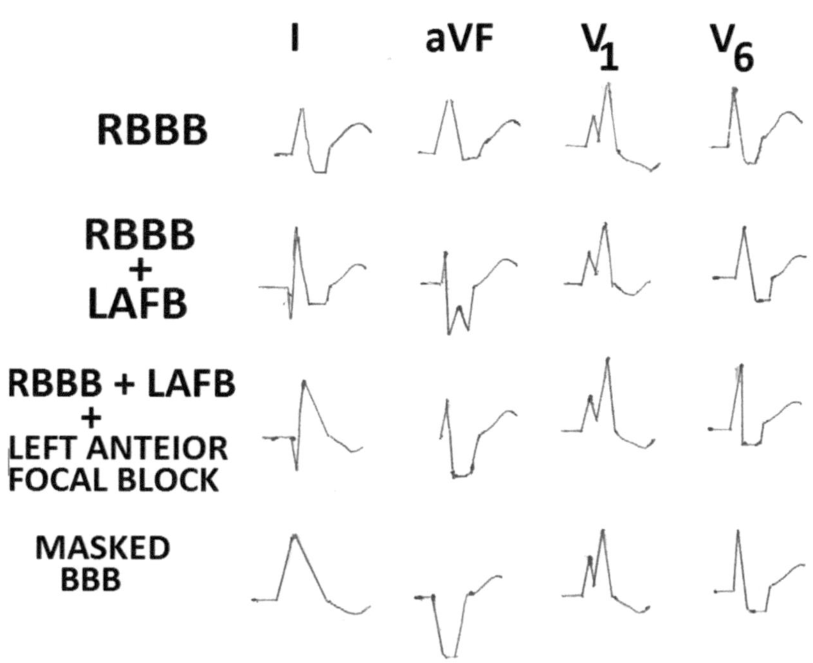

Fig. 29.3 Diagram showing the development of "masked" bundle branch block. *RBBB* right bundle branch block; *LAFB* left anterior fascicular block; *BBB* bundle branch block. (Reproduced with permission from Mittal SR. Multifascicular blocks. Cardiology Today-2015;19:100-106)

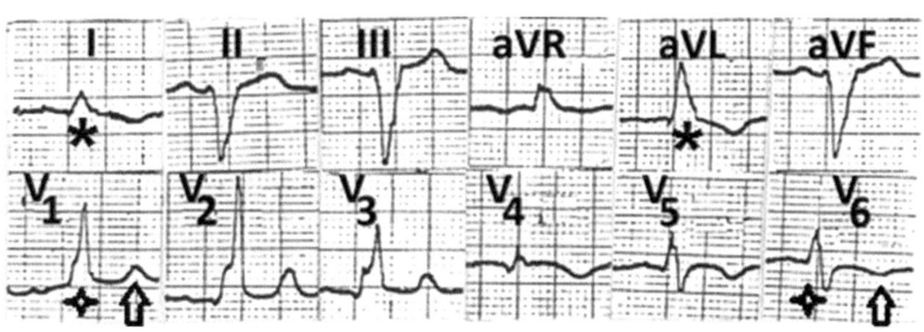

Fig. 29.4 Electrocardiogram showing "masked" bundle branch block. Leads I and aVL look like left bundle branch block (marked *). Leads V_1 and V_6 show right bundle branch block pattern (marked ✧). Polarity of T waves is in the direction of the terminal QRS vector (positive in leads V_1 to V_3 and negative in leads V_4 to V_6 (marked ⇧).This finding suggests concomitant myocardial necrosis. (Reproduced with permission from Mittal SR. Multifascicular blocks. Cardiology Today-2015;19:100-106)

nation is due to involvement of the left anterior descending coronary artery. It is not prone to advance into complete atrioventricular block and per se does not constitute an indication for permanent pacemaker implantation.

2. Right bundle branch block with left posteroinferior fascicular block is less common due to dual arterial supply of posteroinferior fascicle (right coronary artery + left circumflex coronary artery). This combination is associated with more extensive myocardial damage and is prone to sudden development of complete atrioventricular block (Fig. 29.8). Therefore, this combination is an indication for permanent pacemaker implantation.
3. Risk of sudden development of Stokes–Adams' attack in a given patient is also affected by the autonomic tone and coronary flow at a given time. Electrophysiological study, therefore, may not be able to document the risk of cardiogenic syncope in every patient in the laboratory environment. If there

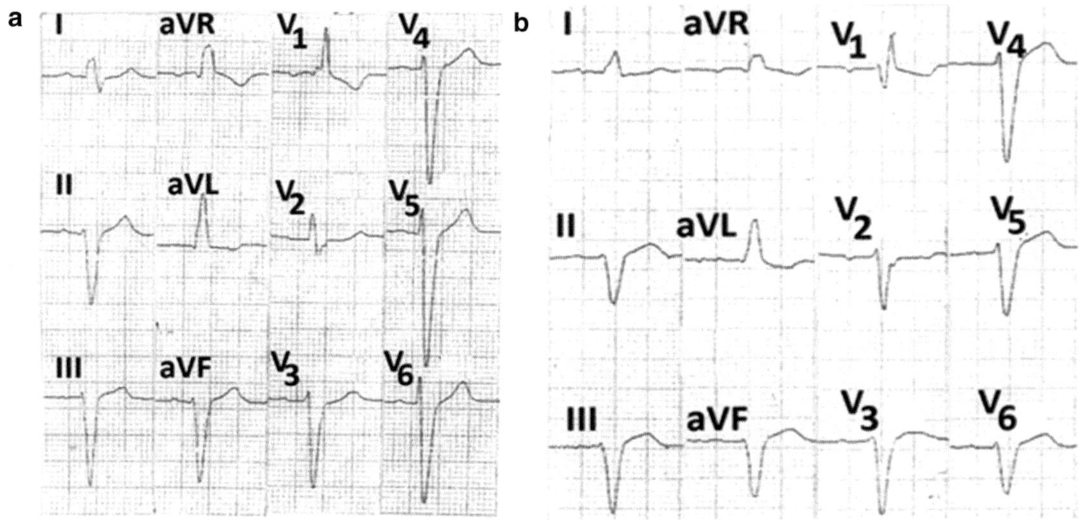

Fig. 29.5 (**a**) Electrocardiogram shows RBBB + left anterosuperior fascicular block. (**b**) Electrocardiogram from the same patient showing the development of "masked" bundle branch block. (Reproduced with permission from Mittal SR. Multifascicular blocks. Cardiology Today-2015;19:100- 106)

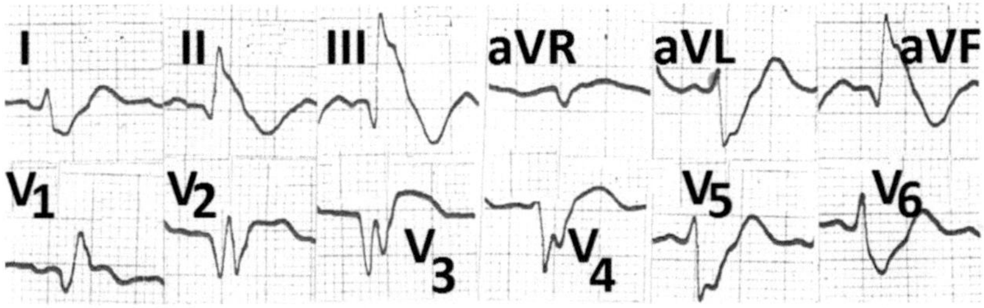

Fig. 29.6 Electrocardiogram showing right bundle branch block + left posteroinferior fascicular block. Leads V_1 to V_4 also show broad and deep Q waves with elevation of the ST-segment. These findings suggest concomitant anteroseptal infarction. (Reproduced with permission from Mittal SR. Multifascicular blocks. Cardiology Today-2015;19:100-106)

is strong clinical suspicion of cardiac cause of syncope, implantable loop recorder may be more information. Implantable loop recorder is more sensitive in detecting the cause of syncope even in patients with negative electrophysiological study [2].

4. Dual-chamber pacing has been shown to significantly reduce episodes of presyncope/syncope in patients with bifascicular block and syncope of unexplained origin [3].
5. In the presence of history of syncope or presyncope, a QRS width >140 ms, renal failure, and H-V interval > 64 ms are independent predictors of the risk of future development of atrioventricular block in patients with bifascicular block [4].

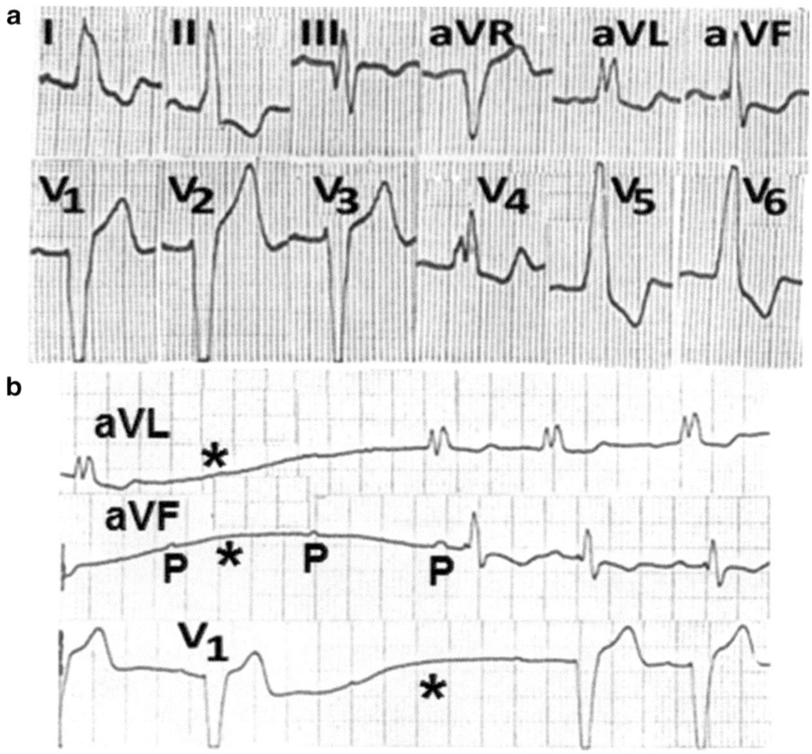

Fig. 29.7 (**a**) Electrocardiogram shows left bundle branch block. (**b**) Electrocardiogram from the same patient showing the development of atrioventricular block (marked *) during exercise. (Reproduced with permission from Mittal SR. Multifascicular blocks. Cardiology Today-2015;19:100–106)

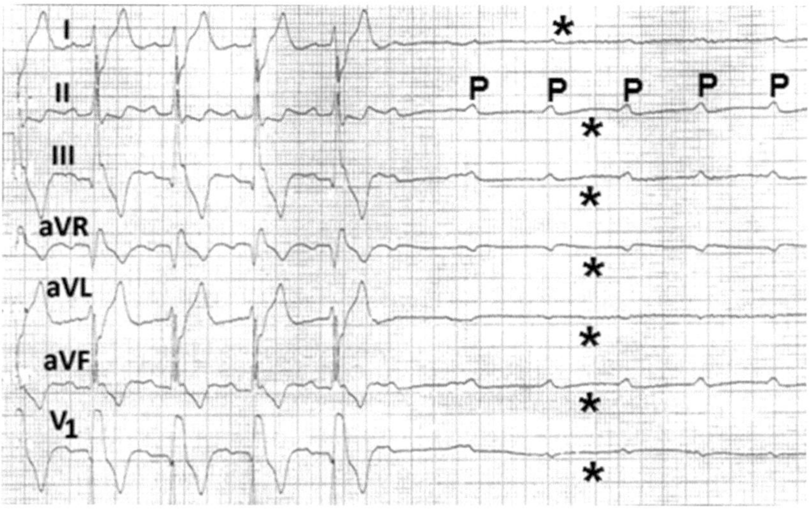

Fig. 29.8 Electrocardiogram showing sudden development of complete atrioventricular block (marked *) in a patient with RBBB + left posteroinferior fascicular block. (Reproduced with permission from Mittal SR. Multifascicular blocks. Cardiology Today-2015;19:100-106)

2 Trifascicular Block

2.1 Nomenclature

In fact, the block is not present simultaneously in all three fascicles as this will present as complete atrioventricular block on the surface electrocardiogram [5]. Trifascicular block can be diagnosed on the surface electrocardiogram only when conduction is present (although at a slower rate) in at least one fascicle.

2.1.1 Electrocardiographic Diagnosis

The following electrocardiographic findings present at different times or in the same electrocardiogram suggest trifascicular disease:

1. Right bundle branch block with alternating left anterosuperior fascicular block or left posteroinferior fascicular block in the same ECG or at different times (Fig. 29.9a).
2. Alternating complete right bundle branch block and left bundle branch block at different times or in same ECG (Figs. 29.9b, 29.10, and 29.11).

Intermittent concealed retrograde conduction of an ectopic arising in the distal conduction tissue (a fascicle or bundle branch) can also produce alternating fascicular or bundle branch block on surface electrocardiogram. Differentiation is important because patients with intermittent concealed retrograde conduction do not have the risk of developing complete atrioventricular block and, therefore, do not need permanent pacemaker implantation.

3. Bifascicular block with prolonged PR interval (Figs. 29.9c, 29.12, and 29.13).

Prolonged PR interval can also be due to slow conduction in the atrioventricular node rather than a disease of the third fascicle [6] (Fig. 29.9c). Differentiation is important because patients with trifascicular block are more prone to develop sudden complete atrioventricular block. Significant prolongation of the QRS complex supports the possibility of trifascicular block (Fig. 29.14).

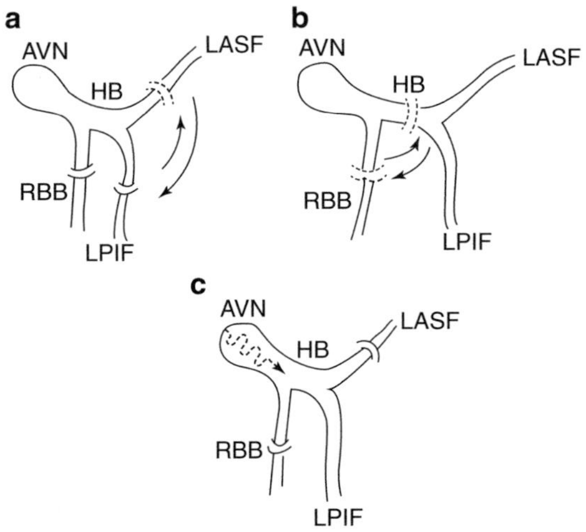

Fig. 29.9 Diagram showing combination of findings suggestive of trifascicular block. (**a**) RBBB+ alternating left anterosuperior fascicular block or left posteroinferior fascicular block or (**b**) alternating left and right bundle branch block. (**c**) RBBB + left anterosuperior fascicular block or left posteroinferior fascicular block + prolonged PR interval. *AVN* atrioventricular node; *HB* His bundle; *RBB* right bundle branch; *LASF* left anterosuperior fascicle; *LPIF* left posteroinferior fascicle. (Reproduced with permission from Mittal SR. Multifascicular blocks. Cardiology Today- 2015;19:100-106)

Definite differentiation is possible only by electrophysiological study. Prolongation of the H-V interval suggests trifascicular block.

4. The following electrocardiographic findings can also suggest trifascicular block.
(a) LBBB advancing to complete atrioventricular block with RBBB configuration of the QRS complex (Fig. 29.15a,b).
(b) RBBB advancing to complete atrioventricular block with LBBB configuration of the QRS complex [7] (Fig. 29.16a,b).
(c) RBBB+ left anterosuperior fascicular block advancing to complete atrioventricular block with RBBB+ left posteroinferior fascicular block configuration of the QRS complex (Fig. 29.17a,b).
(d) Complete atrioventricular block with the QRS complex configuration alternating between RBBB pattern and LBBB pattern (Fig. 29.18a, b).

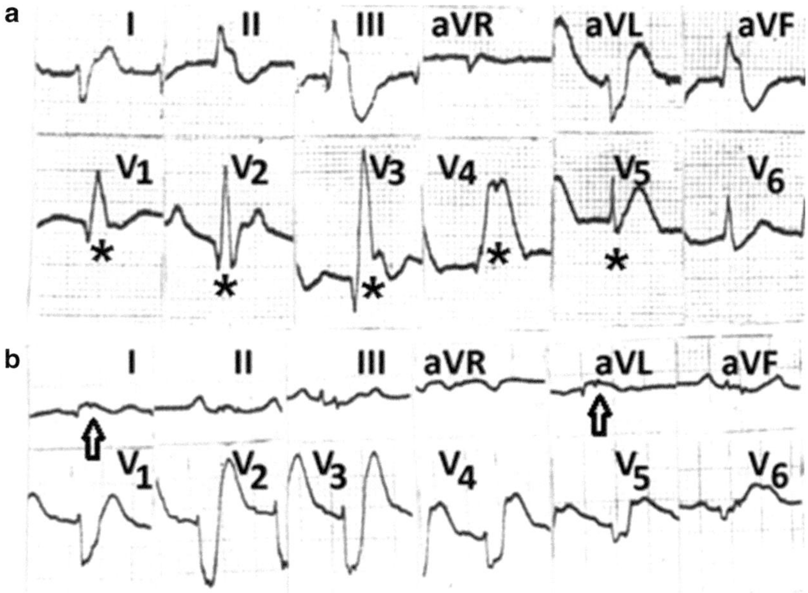

Fig. 29.10 (**a**) Electrocardiogram showing RBBB+ left posteroinferior fascicular block + anterior myocardial infarction (marked *). (**b**) Electrocardiogram recorded from the same patient showing LBBB. Infarction pattern in leads I and aVL (marked ⇧) was not clear during left posteroinferior fascicular block (**a**). Changes of infarction in leads $V_1$to V_5 seen during left posteroinferior fascicular block (**a**) are masked during LBBB (**b**). (Reproduced with permission from Mittal SR. Multifascicular blocks. Cardiology Today-2015;19:100-106)

Fig. 29.11 (**a**) Electrocardiogram showing RBBB+ symmetrical inversion of T waves in leads V_3 and V_4 (marked ⇩) suggestive of non-STE elevation myocardial infarction. (**b**) Electrocardiogram recorded from the same patient showing LBBB. (Reproduced with permission from Mittal SR. Multifascicular blocks. Cardiology Today-2015;19:100-106)

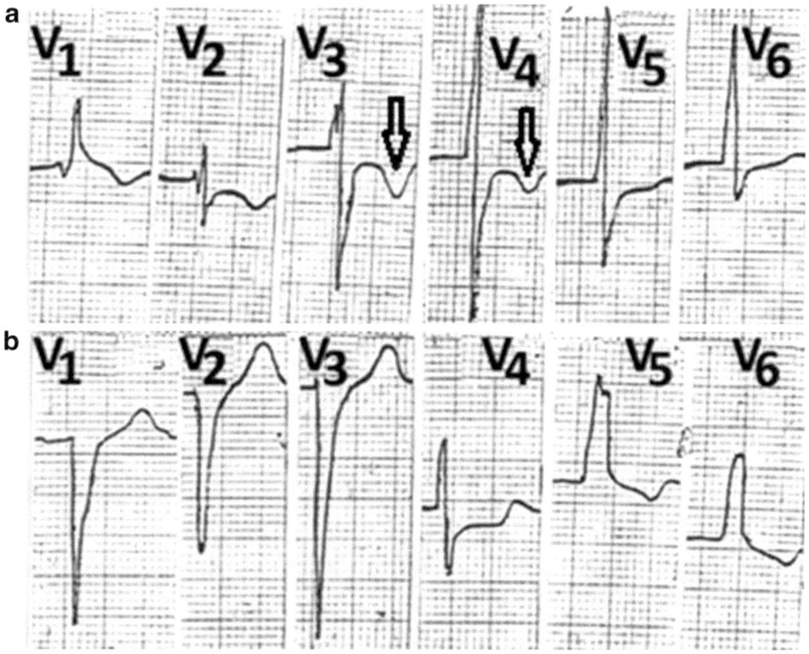

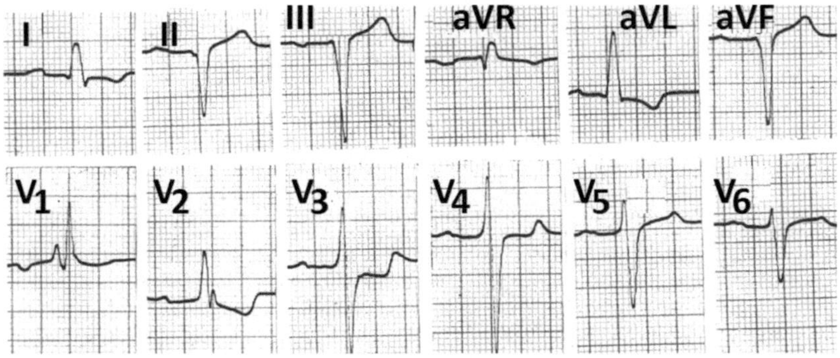

Fig. 29.12 Electrocardiogram showing RBBB+ left anterosuperior fascicular block+ prolonged PR interval. (Reproduced with permission from Mittal SR. Multifascicular blocks. Cardiology Today-2015;19:100-106)

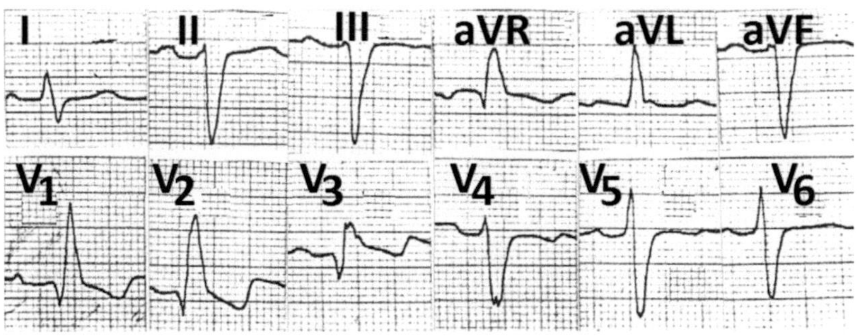

Fig. 29.13 Electrocardiogram showing RBBB+ left anterosuperior fascicular block + prolonged PR interval. Abnormal Q waves in leads V_1 to V_3 are suggestive of old anterior myocardial infarction. (Reproduced with permission from Mittal SR. Multifascicular blocks. Cardiology Today- 2015;19:100-106)

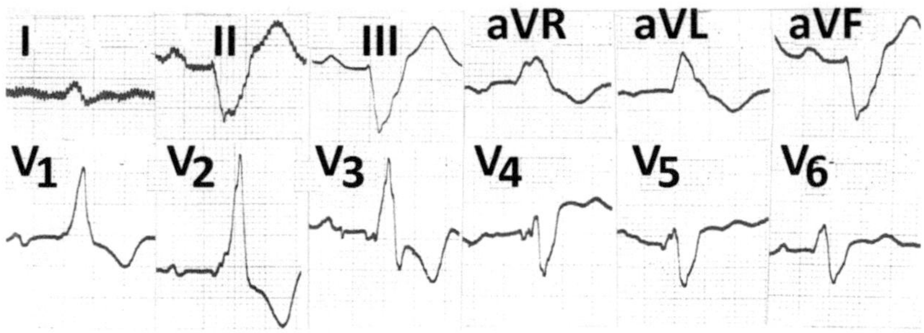

Fig. 29.14 Masked bundle branch block + prolonged P-R interval. QRS duration is 0.24 s. (Reproduced with permission from Mittal SR. Multifascicular blocks. Cardiology Today-2015;19:100-106)

2.2 Clinical Significance

1. Trifascicular block denotes advanced conduction system disease.
2. Usually, there is significant myocardial disease.
3. There is a high risk of sudden, complete atrioventricular block. Permanent pacemaker implantation is indicated if H-V internal is ≥ 100 ms.
4. Alternating bundle branch block can also be a manifestation of digitalis toxicity.

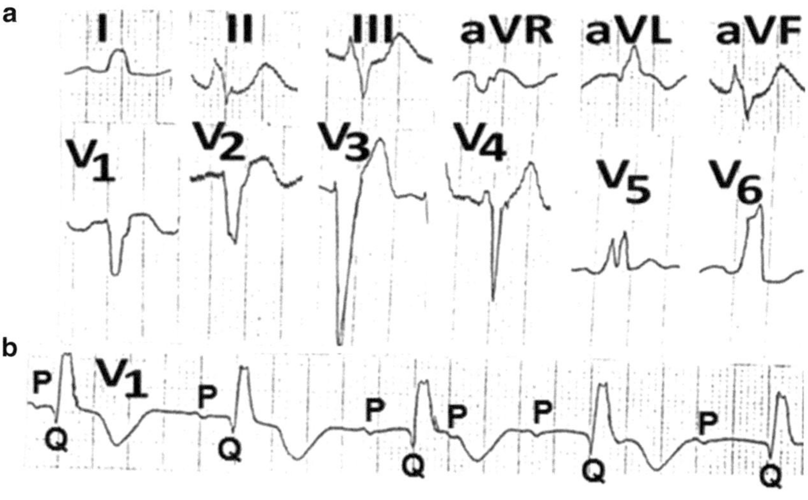

Fig. 29.15 (**a**) Electrocardiogram shows LBBB+ left-axis deviation. (**b**) Electrocardiogram (lead V_1) from the same patient showing the development of complete atrioventricular block with RBBB configuration of the QRS complex Q waves suggest anterior infarction. (Reproduced with permission from Mittal SR. Multifascicular blocks. Cardiology Today-2015;19:100-106)

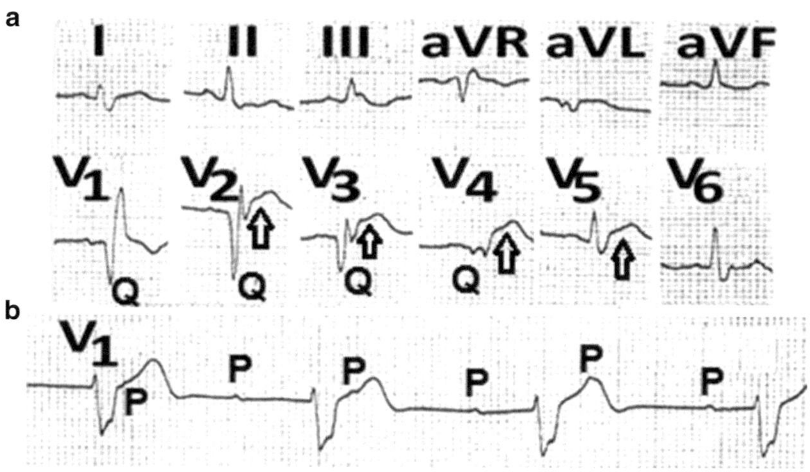

Fig. 29.16 (**a**) Electrocardiogram shows right bundle branch block with anterior infarction. (Q wave in leads V_1 to V_4 and elevation of the ST-segment in leads V_2 to V_5 -marked ⇧.) (**b**) Electrocardiogram (lead V_1) from the same patient showing the development of complete atrioventricular block with LBBB configuration. (Reproduced with permission from Mittal SR. Multifascicular blocks. Cardiology Today-2015;19:100-106)

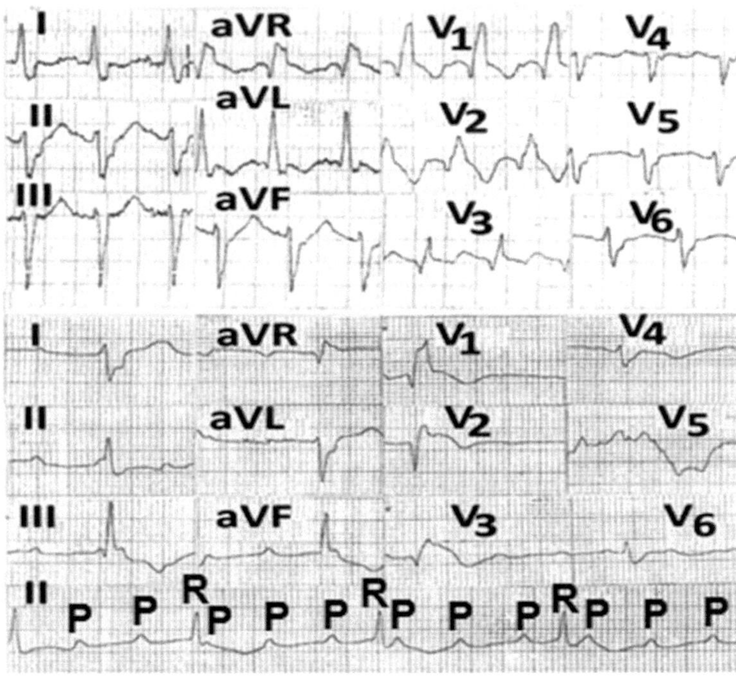

Fig. 29.17 (**a**) Electrocardiogram shows RBBB+ left anterosuperior fascicular block + anterior infarction (Q waves in leads V_1 to V_3). (**b**) Electrocardiogram from the same patient showing the development of complete atrioventricular block with RBBB+ left posteroinferior fascicular block configuration of QRS.

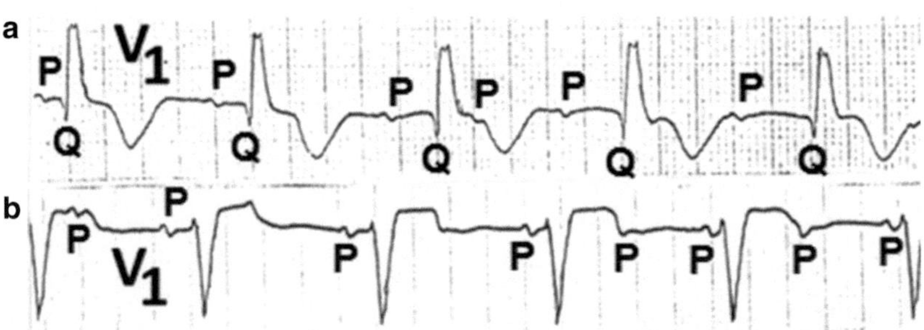

Fig. 29.18 Electrocardiogram (**a**) lead V_1 showing complete atrioventricular block with RBBB configuration of the QRS complex and anterior infarction (**b**) lead V_1 from the same patient showing complete atrioventricular block with LBBB pattern of the QRS complexes. (Reproduced with permission from Mittal SR. Multifascicular blocks. Cardiology Today- 2015;19:100-106)

Summary

RBBB with left anterosuperior fascicular block is diagnosed when precordial leads show features of RBBB but limb leads show some features of left anterosuperior fascicular block like frontal plane QRS axis of -45^0 to -90^0 and rs configuration in leads II, III, and aVF. However, unlike pure RBBB, lead I shows tall R without s wave. Limb leads, therefore, resemble LBBB.

RBBB with left posteroinferior fascicular block should be suspected when precordial leads show RBBB but limb leads show features of left posteroinferior fascicular block like frontal plane QRS axis more than $+90^0$ and rs configuration in leads I and aVL.

Bilateral bundle branch block should be suspected when alternate beats show RBBB and LBBB pattern in the presence of regular sinus rhythm.

Trifascicular block should be suspected when

- Precordial leads show RBBB pattern and limb leads show left anterior fascicular block and left posterior fascicular block pattern in alternate beats in regular sinus rhythm.
- Presence of alternating RBBB and LBBB at different times or in the same ECG.
- Bifascicular block with prolonged PR interval and significant prolongation of the QRS interval.
- Define diagnosis in possible only by electrophysiological study.

MCQs

Q1. Addition of left anterosuperior fascicular block over right bundle branch block affects

(a) Limb leads
(b) Right precordial leads
(c) Mid precordial leads
(d) Posterior chest leads

Q2. In RBBB+ left anterosuperior fascicular block, mean frontal plane QRS axis is between

(a) 0 to -45^0
(b) 0 to $+90^0$
(c) $+90^0$ to $+120^0$
(d) Right of + 120^0

Q3. "Masked" bundle branch block suggests masking of

(a) RBBB in limb leads
(b) RBBB in chest leads
(c) LBBB in limb leads
(d) LBBB in chest leads

Q4. "Masked" bundle branch block occurs in setting of

(a) RBBB+ left anterosuperior fascicular block
(b) RBBB + left posteroinferior fascicular block
(c) LBBB + left ventricular hypertrophy
(d) LBBB + right ventricular hypertrophy

Q5. Which conditions do not produce masked bundle branch block?

(a) Left anterior focal block
(b) High degree of left anterior hemi block
(c) Left ventricular hypertrophy
(d) Right ventricular hypertrophy

Q6. In right bundle branch block + left posteroinferior fascicular block, mean frontal plane QRS axis is between

(a) 0 to -45^0
(b) 0 to $+90^0$
(c) $+90^0$ to $+120^0$
(d) Right of + 120^0

Q7. Which combination can present as LBBB on surface electrocardiogram?

(a) RBBB +left anterior fascicular block
(b) RBBB + left posterior fascicular block
(c) Left anterior fascicular block + left posterior fascicular block
(d) Left anterior fascicular block + left mid fibers block

Q8. LBBB + prolonged P-R interval increasing on exercise suggest

(a) Unifascicular block
(b) Bifascicular block
(c) Trifascicular block
(d) Bifascicular block +delayed conduction in atrioventricular node

Q9. LBBB + HV interval >120 ms suggest

(a) Unifascicular block
(b) Bifascicular block
(c) Trifascicular block
(d) Unifascicular block +delayed conduction in atrioventricular node

Q10. Prophylactic pacing is indicated in

(a) LBBB + PR interval of 0.24 s
(b) RBBB + left anterosuperior fascicular block + anterior MI

(c) RBBB + left posteroinferior fascicular block + anterior MI
(d) None

Q11. Which combination suggests trifascicular block?

(a) RBBB + left anterosuperior fascicular block + prolonged PR interval (AH interval > 120 ms)
(b) RBBB + left posteroinferior fascicular block + prolonged PR interval (AH interval > 120 ms)
(c) LBBB + prolonged PR interval (HV interval > 120 ms)
(d) LBBB + prolonged PR interval (HV interval <60 ms)

Q12. Which combination does not suggest trifascicular block?

(a) RBBB + alternating left anterosuperior fascicular block/left posteroinferior fascicular block
(b) Alternating complete RBBB and LBBB
(c) Bifascicular block + AH interval > 120 ms
(d) Bifascicular block + HV interval > 120 ms

Q13. Prophylactic pacing is indicated in

(a) RBBB + LASFB + AH interval > 120 ms
(b) RBBB + LPIFB + AH interval > 120 ms
(c) RBBB + LPIFB + HV interval > 100 ms
(d) RBBB + AH interval > 120 ms

Q14. Concealed retrograde conduction of on ectopic arising in the fascicular or the bundle branch cannot produce

(a) Alternating BBB pattern on the surface ECG
(b) Alternating fascicular block pattern on the surface ECG
(c) Sudden increase in PR interval
(d) Atrial ectopics in surface ECG

Q15. rsR' in lead V_1, RS in lead V_6, frontal plane QRS axis of -60^0, rS configuration in leads II, III, aVF, and R wave in lead I suggest the possibility of

(a) RBBB with anterosuperior fascicular block
(b) RBBB with posteroinferior fascicular block
(c) RBBB with inferior infarction
(d) RBBB with lateral infarction

Q16. rsR' in lead V_1, frontal plane QRS axis of $+120^0$, and rS configuration in lead I and aVL suggest the possibility of

(a) RBBB with anterosuperior fascicular block
(b) RBBB with posteroinferior fascicular block
(c) RBBB with lateral myocardial infarction
(d) RBBB with anterior infarction

Q17. LBBB pattern in limb leads and RBBB pattern in precordial leads should suggest the possibility of

(a) RBBB with left anterosuperior fascicular block
(b) RBBB with left posteroinferior fascicular block
(c) Bilateral bundle branch block
(d) Trifascicular block

Q18. Alternate beat showing RBBB and alternate premature beat showing LBBB pattern suggest

(a) Bilateral bundle branch block
(b) RBBB with alternate premature contractions arising from the right ventricle
(c) RBBB with alternate premature contraction arising from the left ventricle
(d) Trifascicular block

Q19. rsR' pattern in lead V_1 and alternate beat showing qR and rS configuration in lead I in regular sinus rhythm suggest

(a) Bilateral bundle branch block
(b) RBBB with alternate premature ventricular contraction
(c) Trifascicular block
(d) RBBB with aberrant conduction in alternate beats

Answers

(1) a (2) a (3) a (4) a (5) c,d (6) d (7) c (8) d (9) c (10) c (11) c (12) c (13) c (14) d, (15) a (16) b (17) a (18) b (19) c.

References

1. De Luna AB, Goldwasser D, Fiol M, Bayes-GA. Surface Electrocardiography. In: Fuster V, Walsh RA, Harrington RA, editors. Hurst's The Heart. Mc Graw Hill: New York; 2011. p. 307–70.
2. Da Costa A, Defaye P, Romeyer- Bouchard C, et al. The clinical impact of the implantable loop recorder in patients with isolated syncope, bundle branch block and negative workup: a randomized multicentre prospective study. Archiv Cardiovasc Dis. 2013;106:146–54.
3. Santini M, Castro A, Giada F, et al. Prevention of syncope through permanent cardiac pacing in patients with bifascicular block and syncope of unexplained origin. The PRESS study. Circ Arrhythm Electrophysiol. 2013;6:101–7.
4. Marti-Almor J, Cladellas M, Bazan V, et al. Novel predictors of progression of atrioventricular block in patients with chronic bifascicular block. Rev Esp Cardiol. 2010;63:400–8.
5. Olshansky B, Chung MK, Pogwizd SM, Goldschlager N. Bradyarrhythmias- conduction system abnormalities. In: Olshansky B, Chung MK, Pogwizd SM, Goldschlager N, editors. Arrhythmia Essentials. Burlington: Jones and Bartlett Learning; 2012. p. 26–105.
6. Mirvis DM, Goldberger AL. Electrocardiography. In: Mann DL, Zipes DP, Libby P, Bonow RO, editors. Braunwald's Heart disease. Philadelphia: Saunders; 2015. p. 114–54.
7. Femenia F, Cuesta A, Mauricio A. A rare form of trifascicular block with intermittent complete atrioventricular block in a patient with Chagas disease. Cardiol J. 2009;16:582–4.

30 Other Localized Intraventricular Conduction Defects

1 Left Septal Fascicular Block (Left Middle Fibers Block)

Normally, interventricular septum is supplied by fibers from the left bundle branch. These fibers are known as septal fibers or middle fibers. These fibers are responsible for normal depolarization of the interventricular septum from left to right and anteriorly, resulting in an initial r wave in lead V_1 and an initial q wave in leads V_5,V_6 (Fig. 30.1a). Slow conduction in these fibers (incomplete block) increases the initial rightward vector. This may produce prominent R wave in lead V_1 and RS morphology in lead V_2 [1] (Fig. 30.1b). q waves in leads V_5,V_6 are preserved. Definite electrocardiographic diagnosis is difficult because of several differential diagnosis. Transient ECG changes associated with acute clinical event support the diagnosis.

Total block of conduction in these fibers (complete block) may result in loss of initial left to right vector. Interventricular septum is depolar-

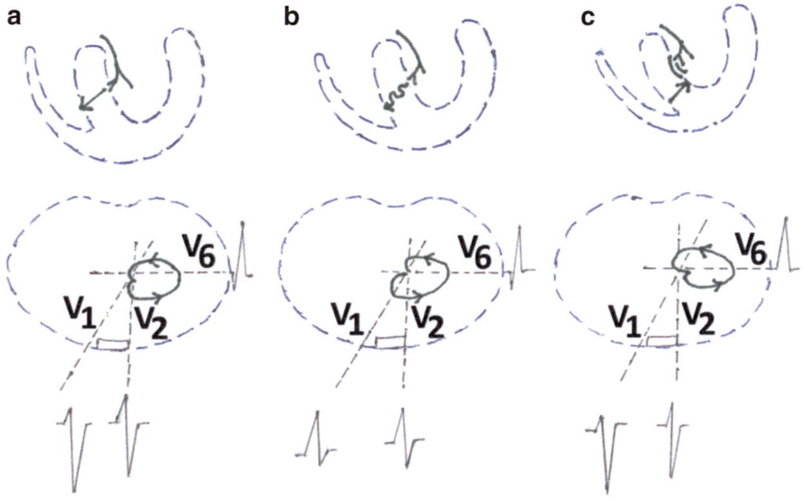

Fig. 30.1 Diagram showing the effect of septal depolarization on the QRS vector in the horizontal plane and configuration of QRS in leads V_1, V_2, and V_6. (**a**) Normal depolarization. (**b**) Incomplete left septal block. (**c**) Complete left septal block. (Reproduced with permission from Mittal SR. Other Localized Intraventricular Conduction Defects. Cardiology Today-2014;18:259–263)

S. Mittal, *Insights into Electrocardiograms with MCQs*,
https://doi.org/10.1007/978-981-99-0127-2_30

ized from right bundle branch with the result that the initial QRS vector moves from right to left. This results in loss of septal q waves in leads V_5,V_6 (Fig. 30.1c). Small r may be present in lead V_1 due to activation of free wall of the right ventricle through the right bundle branch.

2 Transient Intraventricular Conduction Defects

2.1 Rate-Dependent Conduction Defects (Aberration) [2]

2.1.1 Tachycardia Dependent

Such aberration can be seen even in the absence of other structural heart diseases. Such aberration can appear during sinus tachycardia, e.g., during stress testing (Fig. 30.2a) or during supraventricular tachyarrhythmias (Fig. 30.2b). Right bundle branch is frequently involved. With time, the aberration appears at relatively lesser increase in heart rate.

2.1.2 Bradycardia Dependent

Such aberration is seen less frequently. Aberrant conduction occurs after a longer pause (Fig. 30.3). Usually, it is seen only in the presence of significant disease of the conduction system. It is frequently associated with myocardial dysfunction.

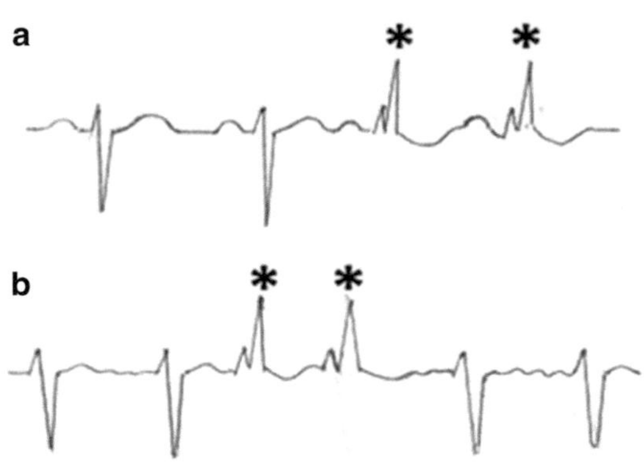

Fig. 30.2 Diagram showing rate-dependent conduction defects. (**a**) Tachycardia-dependent aberrant conduction in sinus rhythm. (marked *) (**b**) Tachycardia-dependent aberrant conduction in atrial fibrillation. (marked *). (Reproduced with permission from Mittal SR. Other Localized Intraventricular Conduction Defects. Cardiology Today-2014;18:259–263)

2.2 Concealed Conduction in the Bundle Branches [2]

This is due to an ectopic beat not manifesting on the surface ECG (concealed or hidden) but depolarizing part of the conduction tissue. This can be antegrade or retrograde.

2.2.1 Antegrade

This occurs from an ectopic arising in the His bundle or proximal bundle branches. Ectopic depolarizes part of the distal conduction tissue (Fig. 30.4) but is not strong enough to conduct up to the ventricles with the result that there is no premature ventricular ectopic beat (QRS) on the

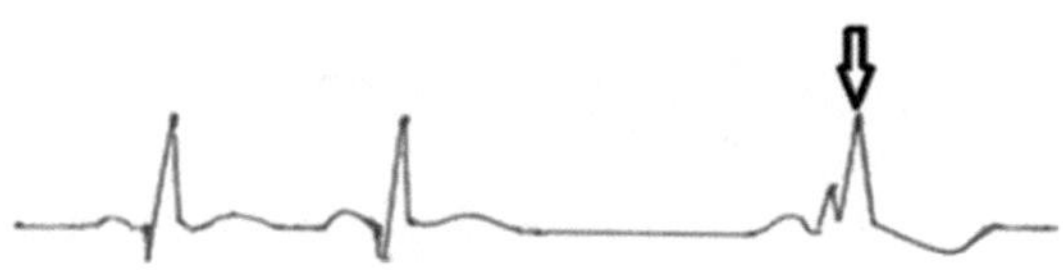

Fig. 30.3 Bradycardia-dependent aberrant conduction (marked with arrow). (Reproduced with permission from Mittal SR. Other Localized Intraventricular Conduction Defects. Cardiology Today-2014;18:259–263)

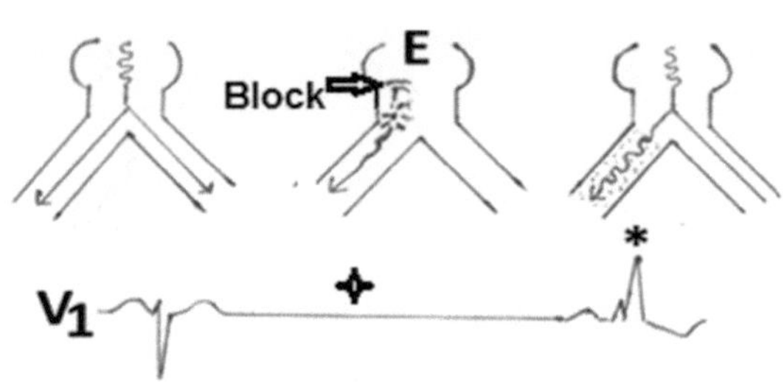

Fig. 30.4 Diagram showing the effect of concealed antegrade conduction on the electrocardiogram. of the next sinus beat. His bundle ectopic (E) is conducted antegradely over the right bundle branch and partly depolarizes it but does not reach the ventricles with the result that there is no QRS complex on the ECG. Ectopic in blocked retrogradely (marked ⇨) so that there is also no P wave on the ECG during the period marked ✧. Next sinus beat finds the right bundle in partly depolarized state with the result that there is slow conduction over the right bundle branch (marked *). (Reproduced with permission from: Mittal SR. Other Localized Intraventricular Conduction Defects. Cardiology Today-2014;18:259–263)

surface ECG. When the next sinus impulse comes through the atrioventricular node, it finds part of the conduction tissue in partial or absolute refractory period. This results in slow or no conduction through that part of the intraventricular conduction tissue with the result that there is widening of the QRS complex.

2.2.2 Retrograde

Ectopic impulse arising in the distal conduction tissue (bundle branches or peripheral network of conduction tissue) conducts retrogradely without antegrade conduction to the ventricles. This produces depolarization of proximal part of the conduction tissue. Next, sinus impulse finds the conduction tissue in a depolarized state, resulting in slow or no conduction through that part of the conduction tissue with the broadening of QRS (Fig. 30.5).

2.3 Effect of Changing Cycle Length on Refractoriness of the Conduction Tissue (Ashman Phenomenon) [2]

Refractoriness of the conduction tissue is dependent on the length of the preceding cycle. Longer the preceding cycle length, longer is the subsequent refractory period. Supraventricular impulse following such a long cycle finds the conduction tissue in partial refractory period and conducts slowly. On surface ECG, it looks like a broad QRS in the short cycle that follows a long cycle (long–short cycle sequence). Right bundle branch is commonly involved. This phenomenon is frequent in supraventricular tachyarrhythmias with irregular atrioventricular conduction, e.g., atrial fibrillation (Fig 30.6).

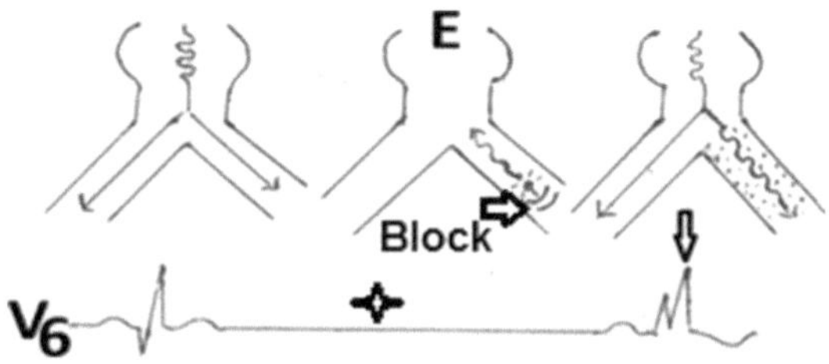

Fig. 30.5 Diagram showing that on ectopic from the left bundle branch conducts retrogradely toward the atrioventricular node. It partly depolarizes the left bundle branch but does not reach the atria with the result that there is no P wave on the ECG. There is an antegrade block (marked ⇨) so that the ectopic also does not depolarize the ventricles. Thus, there is neither a P wave nor a QRS complex on the ECG during the period marked (marked ✧). Next sinus beat finds the left bundle branch in partly depolarized state and conduct slowly over the left bundle branch, resulting in a broad QRS (marked ⇩). (Reproduced with permission from Mittal SR. Other Localized Intraventricular Conduction Defects. Cardiology Today-2014;18:259–263)

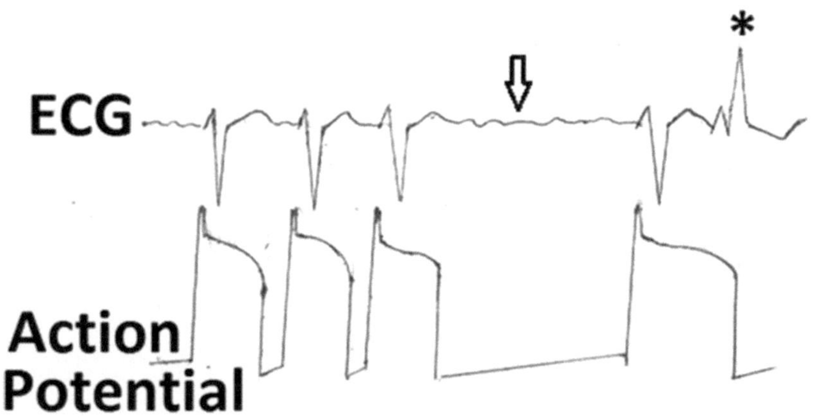

Fig. 30.6 Diagram showing Ashman phenomenon in atrial fibrillation. There is prolongation of the action potential (refractory period) after a long pause (marked ⇩). Subsequent supraventricular impulse finds the distal conduction tissue in partly depolarized state and conducts slowly. It results in a wide QRS complex (marked *). (Reproduced with permission from Mittal SR. Other Localized Intraventricular Conduction Defects. Cardiology Today-2014;18:259–263)

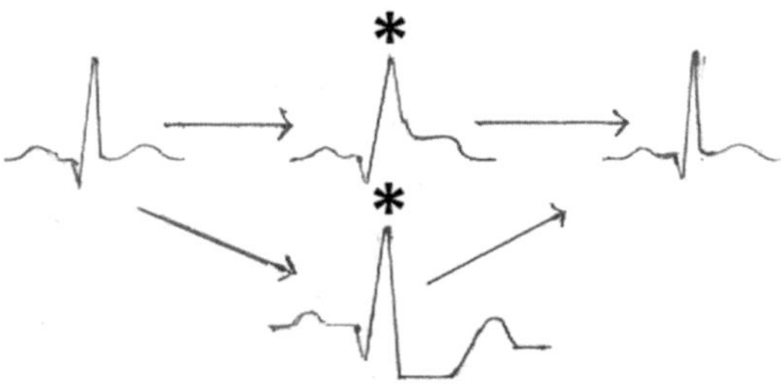

Fig. 30.7 Diagram showing transient intraventricular conduction defect (broadening of QRS) during ischemia (marked *). (Reproduced with permission from Mittal SR. Other Localized Intraventricular Conduction Defects. Cardiology Today-2014;18:259–263)

2.4 Peri-Ischemic Block [2]

It manifests as reversible widening of the QRS complex in leads with transient ST segment elevation or depression (Fig. 30.7). It is due to transient focal delay in conduction in the area of ischemia.

3 Other Focal Blocks

3.1 Localized Notching in QRS [2]

Such blocks present as multiple deflections or high-frequency notches within the QRS without prolongation of the QRS complex (Figs. 30.8 and 30.9).

Such high-frequency notches are usually as a result of localized fibrosis due to old small infarcts. Such changes can also be seen in long-standing systemic hypertension with left ventricular hypertrophy (Figs. 30.10, 30.11, and 30.12).

3.2 Peri-Infarction Block [2]

There is localized delay in conduction within the tissue surrounding the infarcted region. This results in widening of the QRS complex localized to the leads showing myocardial infarction. Figure 30.13(a) shows QRS vector and lead aVF

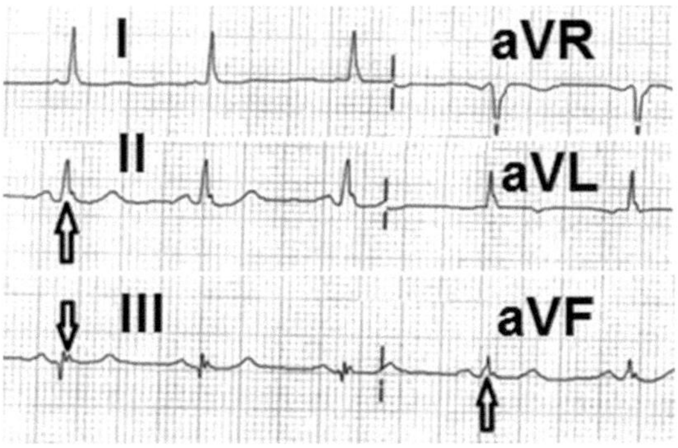

Fig. 30.8 Electrocardiogram showing focal block (notching of the QRS) in leads II, III, and aVF (marked with arrows). (Reproduced with permission from Mittal SR. Other Localized Intraventricular Conduction Defects. Cardiology Today-2014;18:259–263)

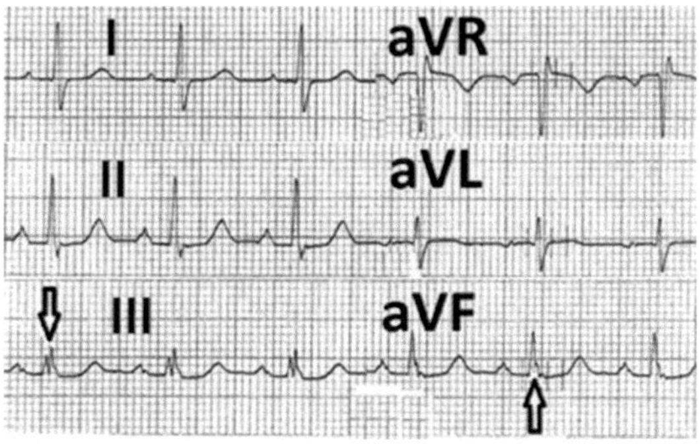

Fig. 30.9 Electrocardiogram showing focal block (notching of the QRS complex marked arrows) in leads III and aVF. (Reproduced with permission from Mittal SR. Other Localized Intraventricular Conduction Defects. Cardiology Today-2014;18:259–263)

in a case of inferior infarction. Figure 30.13(b) shows broadening of the QRS with the appearance of peri-infarction block (marked PIB) (Fig. 30.13).

Initial QRS vector is directed away from the infarction, resulting in Q wave. Terminal forces are directed toward infarction producing delayed terminal and slurred r or R wave. Such blocks can be focal or may involve a fascicle [3]. Representative electrocardiograms are shown in Figs. 30.14, 30.15, 30.16, 30.17, and 30.18.

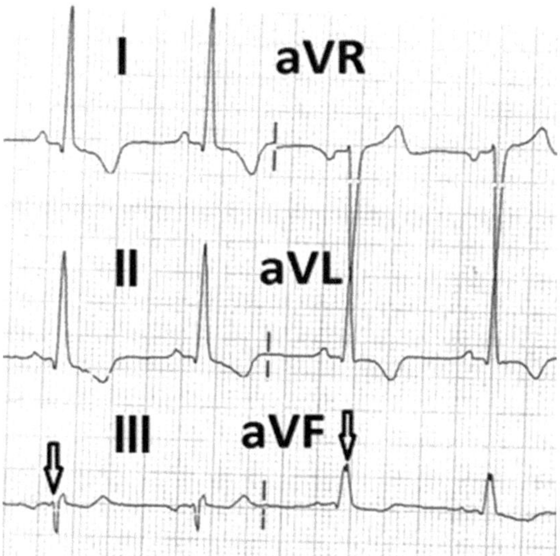

Fig. 30.10 Electrocardiogram showing focal block (notching of the QRS marked ⇩) in leads III and aVF in the setting of left ventricular hypertrophy. (Reproduced with permission from Mittal SR. Other Localized Intraventricular Conduction Defects. Cardiology Today-2014;18:259–263)

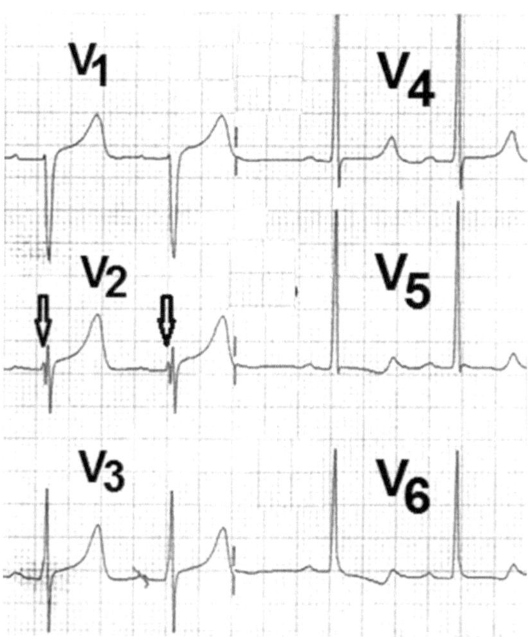

Fig. 30.11 Electrocardiogram showing focal block (high-frequency notching in the QRS complex marked ⇩) in lead V_2 in the setting of left ventricular hypertrophy. (Reproduced with permission from Mittal SR. Other Localized Intraventricular Conduction Defects. Cardiology Today- 2014;18:259–263)

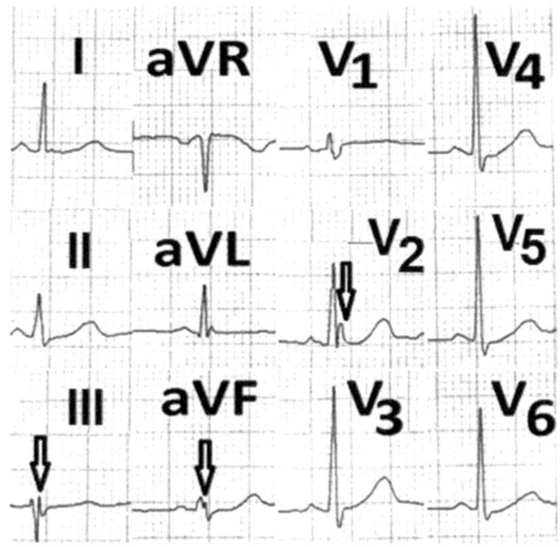

Fig. 30.12 Electrocardiogram showing focal block (notching of the QRS complex (marked ⇩) in leads III, aVF, and V_2. (Reproduced with permission from Mittal SR. Other Localized Intraventricular Conduction Defects. Cardiology Today-2014;18:259–263)

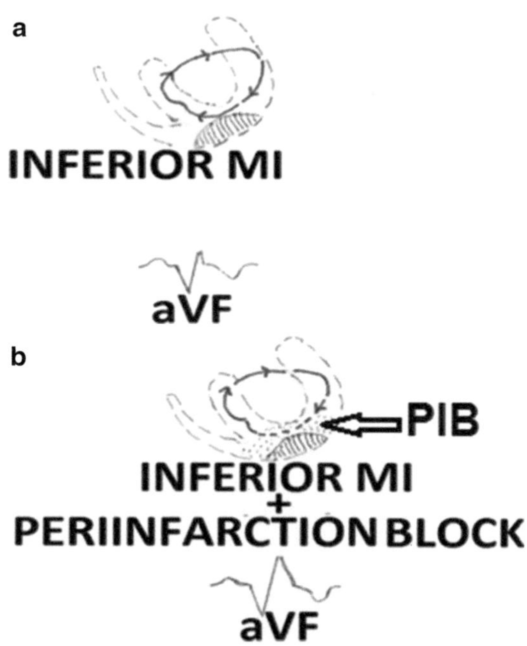

Fig. 30.13 (**a**) Diagram showing QRS vector and lead aVF in isolated inferior infarction. (**b**) Diagram showing the effect of peri-infarction block on the QRS complex. (Reproduced with permission from Mittal SR. Other Localized Intraventricular Conduction Defects. Cardiology Today- 2014;18:259–263)

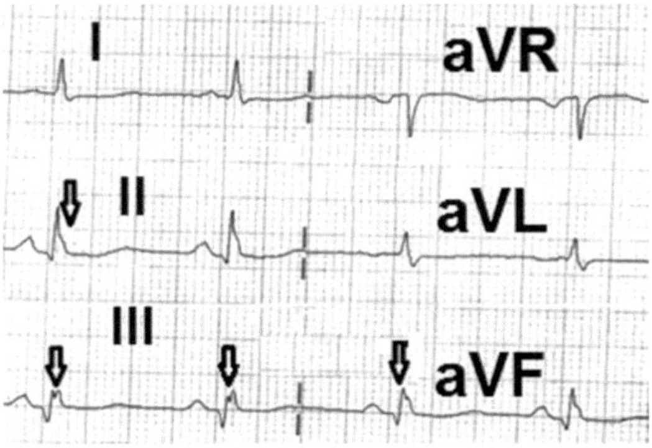

Fig. 30.14 Electrocardiogram showing peri-infarction block (marked ⇩) in leads II, III, and aVF. (Reproduced with permission from Mittal SR. Other Localized Intraventricular Conduction Defects. Cardiology Today-2014;18:259–263)

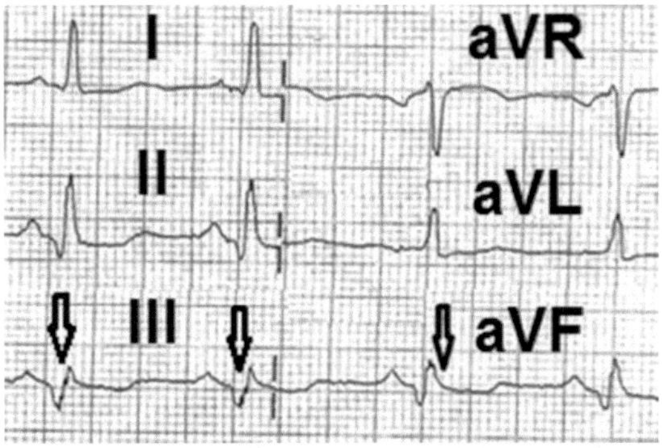

Fig. 30.15 Electrocardiogram showing peri-infarction block (marked ⇩) in leads III and aVF. (Reproduced with permission from Mittal SR. Other Localized Intraventricular Conduction Defects. Cardiology Today-2014;18:259–263)

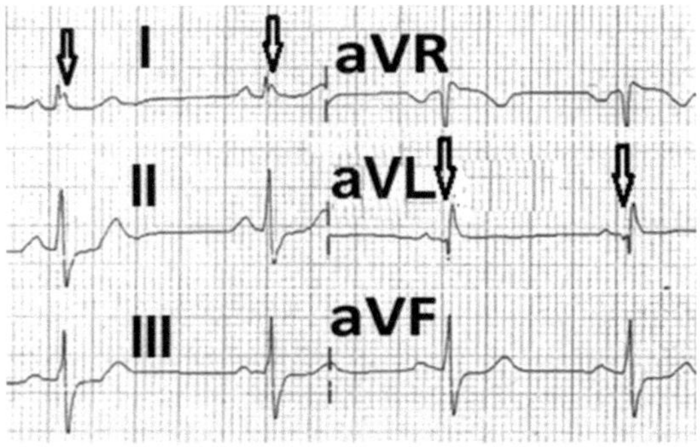

Fig. 30.16 Electrocardiogram showing peri-infarction block (marked ⇩) in leads I and aVL. (Reproduced with permission from Mittal SR. Other Localized Intraventricular Conduction Defects. Cardiology Today- 2014;18:259–263)

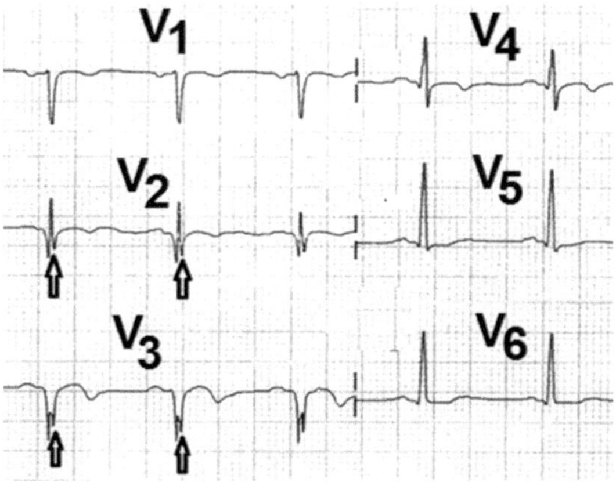

Fig. 30.17 Electrocardiogram showing peri-infarction block (marked ⇧) in leads V_2 and V_3. (Reproduced with permission from Mittal SR. Other Localized Intraventricular Conduction Defects. Cardiology Today- 2014;18:259–263)

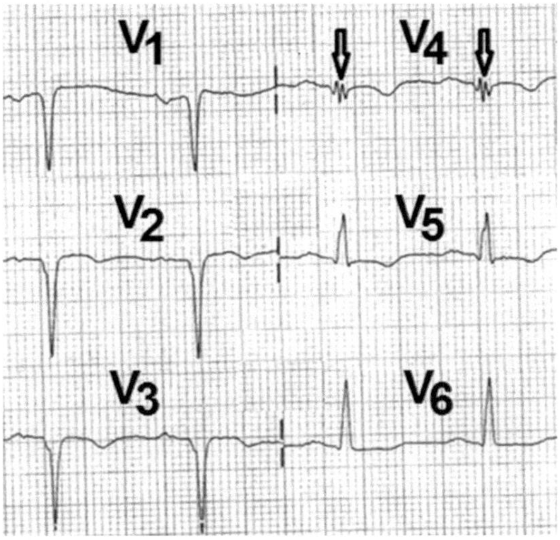

Fig. 30.18 Electrocardiogram showing peri-infarction block (marked ⇩) in leads V_4. (Reproduced with permission from Mittal SR. Other Localized Intraventricular Conduction Defects. Cardiology Today-2014;18:259–263)

Summary

- Incomplete block of the left middle fibers produces prominent R wave in lead V_1 and RS morphology in V_2. Complete block results in loss of septal q wave in leads V_5, V_6.
- Tachycardia-dependent aberrant conduction usually affects the right bundle branch. It can be seen in the absence of other structural heart diseases.
- Bradycardia-dependent aberration is seen in the presence of significant disease of conduction system and is frequently associated with myocardial dysfunction.
- Ectopics arising in the His bundle or bundle branches produce widening of the next QRS complex due to concealed depolarization of the conduction tissue.
- Ashman phenomenon produces a broad QRS in the short cycle following a long cycle.
- Peri-ischemic block manifests as reversible widening of the QRS complex in leads showing transient ST segment elevation or depression due to ischemia.
- Old infarcts can produce multiple high-frequency notches in the QRS complexes without prolongation of QRS duration.
- Localized widening of the QRS complex in leads involved in myocardial infarction is due to peri-infarction block.

MCQs

Q1. Normally which part of the myocardium in depolarized last?

(a) Apex of LV
(b) Base of LV
(c) Base of IVS
(d) Base of RV

Q2. Normally interventricular septum is depolarized from

(a) Left to right
(b) Right to left
(c) Base to apex
(d) Anterior to posterior

Q3. Loss of septal q wave is not seen in

(a) LBBB
(b) Septal infarction
(c) Left septal fascicular block
(d) RBBB

Q4. Prominent R wave in lead V_1 is not seen in

(a) Right ventricular hypertrophy
(b) Dextrocardia
(c) Incomplete left septal fascicular block
(d) Emphysema

Q5. What is not correct for tachycardia-dependent aberrant conduction?

(a) It is usually benign
(b) It can occur in the absence of structural heart disease
(c) It frequently involves the left bundle branch
(d) It is not a marker of significant coronary artery disease

Q6. What is not correct for bradycardia-dependent LBBB?

(a) It is uncommon
(b) It suggests significant conduction tissue disease
(c) It is usually associated with myocardial dysfunction
(d) It is usually benign

Q7. Concealed retrograde conduction in the left bundle branch presents as

(a) Inverted P wave
(b) Ventricular ectopic with RBBB configuration
(c) Prolonged PR interval in the next sinus beat
(d) LBBB configuration in the next sinus beat

Q8. Concealed antegrade conduction of a His bundle ectopic in right bundle branch presents as

(a) Atrial ectopic

(b) Ventricular ectopic with LBBB configuration
(c) Prolonged PR interval in the next sinus beat
(d) RBBB configuration in the next sinus beat

Q9. Ashman phenomenon is

(a) Abnormal phenomenon
(b) It is seen in the short pause following a long pause
(c) It is seen in the long pause following a short pause
(d) It is seen in sinus tachycardia

Q10. Peri-ischemic block is seen

(a) Before ischemia
(b) After recovery from ischemia
(c) Produces widening of QRS during ischemia
(d) Prolongs PR interval

Q11. Peri-infarction block produces

(a) High-frequency notching in the QRS complex
(b) QRS duration >0.12 s in leads involved in the infarction
(c) Abnormal P wave configuration
(d) Abnormal T waves

Q12. High-frequency notching within the QRS without widening of QRS can be seen in

(a) Old small infarcts
(b) Long-standing systemic hypertension
(c) Left ventricular hypertrophy
(d) All

Answers

(1) c (2) a (3) d (4) d (5) c (6) d (7) d (8) d (9) b (10) c (11) b (12) d

References

1. De Luna AB, Goldwasser D, Fiol M, Bayes GA. Surface electrocardiography. In: Fuster V, Walsh RA, Harrington RA, editors. Hurst's The Heart. Mc Graw Hill: New York; 2011. p. 307–70.
2. Mirvis DM, Goldberger AL. Electrocardiography. In: Bonow RO, Mann DL, Zipes DP, Libby P, editors. Braunwald's heart disease. Missouri: Elsevier; 2012. p. 126–65.
3. Schamroth L. The hemiblocks (fascicular block). In: Schamroth C, editor. An introduction to electrocardiography. France: Blackwell Science; 1982. p. 105–18.

31 Chronic Left Ventricular Volume Overload

1 Etiology

Chronic moderately severe aortic regurgitation (AR), chronic moderate mitral regurgitation (MR), moderate left to right shunt across a VSD, atrioventricular septal defect, or patient ductus arteriosus.

Can produce chronic left ventricular volume overload. Mild lesions do not cause significant volume overload and, therefore, do not produce appreciable changes in the electrocardiogram. Acute mitral regurgitation acute aortic regurgitation, acute rupture of interventricular septum, or acute rupture of aneurysm of sinus of Valsalva produce acute left ventricular volume overload. However, these conditions do not produce electrocardiographic signs of left ventricular overload because of short duration of illness and small increase in the left ventricular volume.

2 Classical Electrocardiographic Findings of Chronic Left Ventricular Volume Overload and Their Genesis

(a) The q Wave

Narrow (<25 ms) but deep (0.2 mV or greater) q waves are present in leads facing LV free wall [1]. These reflect hypertrophy of the basal part of the interventricular septum. Q wave vector is, therefore, prominent and directed superiorly, posteriorly, and to the right, away from leads II, $V_{5,}$ and V_6 [2]. With passage of time, these initial forces diminish but total QRS amplitude increases [3].

(b) The R Wave

Normally, the interventricular septum is depolarized from left to right and anteriorly. This produces small q wave in leads $V_{5,}$ V_6. Subsequently, there is depolarization of the left ventricular free wall that produces positive QRS and T waves in leads $V_{5,}$ V_6 (Fig. 31.1a). QRS and T wave vectors in the horizontal plane in a normal person are shown in Fig. 31.1b. In chronic left ventricular volume overload, hypertrophy of the left ventricle increases amplitude of the QRS complex in leads V_5 and V_6 (Fig. 31.1c). This is because the leftward and posteriorly directed QRS vector becomes prominent (Fig. 31.1d).

Figure 31.2 shows electrocardiogram from a case of chronic left ventricular volume overload.

Several factors contribute to this increase in QRS vector

- Due to dilatation of the left ventricle, myocardium is closer to the precordial electrodes [4].
- Brody effect [1]. Ventricular dilatation increases the size of the highly conductive intraventricular blood pool. It increases the

S. Mittal, *Insights into Electrocardiograms with MCQs*,
https://doi.org/10.1007/978-981-99-0127-2_31

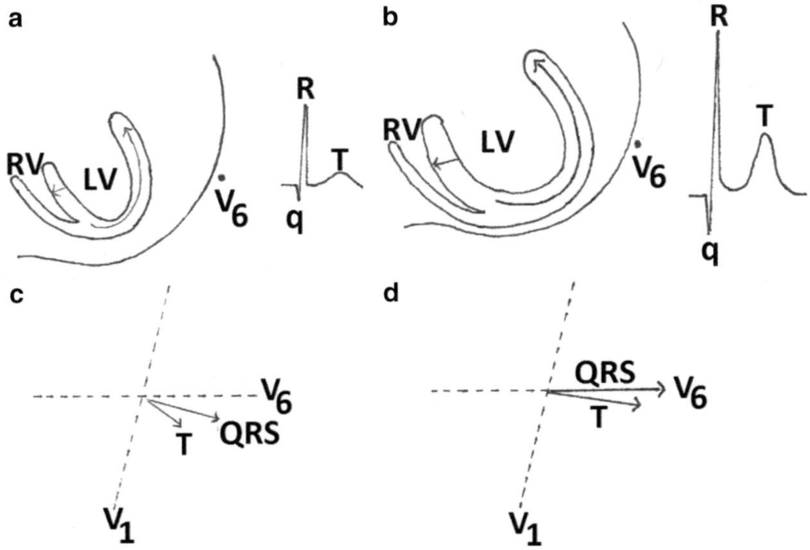

Fig. 31.1 (**a**) Diagram showing depolarization of the interventricular septum and left ventricular free wall in a normal person and corresponding QRS complex and T wave in lead V_6. (**b**) Diagram showing QRS and T wave vectors in the horizontal plane in a normal person. (**c**) Diagram showing depolarization of the interventricular septum and left ventricular free wall in chronic left ventricular volume overload and corresponding configuration of QRS complex and T wave in lead V_6. (**d**) Diagram showing QRS and T wave vectors in the horizontal plane in chronic left ventricular volume overload. (Reproduced with permission from Mittal SR. Chronic Left Ventricular Volume Overload. Cardiology Today-2016;20:61–65)

potentials produced by transmural activation fronts.

– In chronic severe mitral regurgitation, large left atrium displaces the heart forward and closer to the anterior chest wall [5].

(c) Unlike systolic overload, usually there are no deep S waves in right precordial leads (V_1 to V_3).

(d) ST-segment is minimally elevated and concave upward in lead V_6.

(e) T wave amplitude is also increased in leads showing prominent R wave. These T waves tend to be symmetrical. With increasing hypertrophy, the T waves may be directed opposite to the QRS. This is due to widening of the angle between the vectors of the QRS complex and the T wave (Fig. 31.3).

Representative electrocardiogram is shown in Fig. 31.4.

Tall R wave in V_6 occurs with chronic moderately severe aortic regurgitation and chronic moderate mitral regurgitation [5]. These are less prominent in patent ductus arteriosus and least prominent in ventricular septal defect. S wave in lead V_1 is largest in chronic moderately severe aortic regurgitation and patent ductus arteriosus. They are least prominent in chronic moderate mitral regurgitation [5].

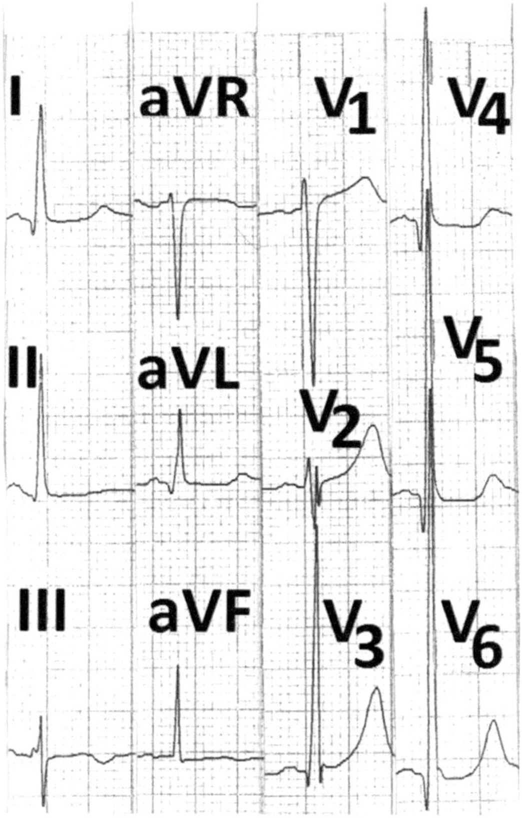

Fig. 31.2 Electrocardiogram from a patient with chronic left ventricular volume overload showing relatively deep but narrow q wave, tall R wave, and prominent, symmetrical, positive T wave in lead V_6 and deep S wave in lead V_1. (Reproduced with permission from Mittal SR. Chronic Left Ventricular Volume Overload. Cardiology Today-2016;20:61–65)

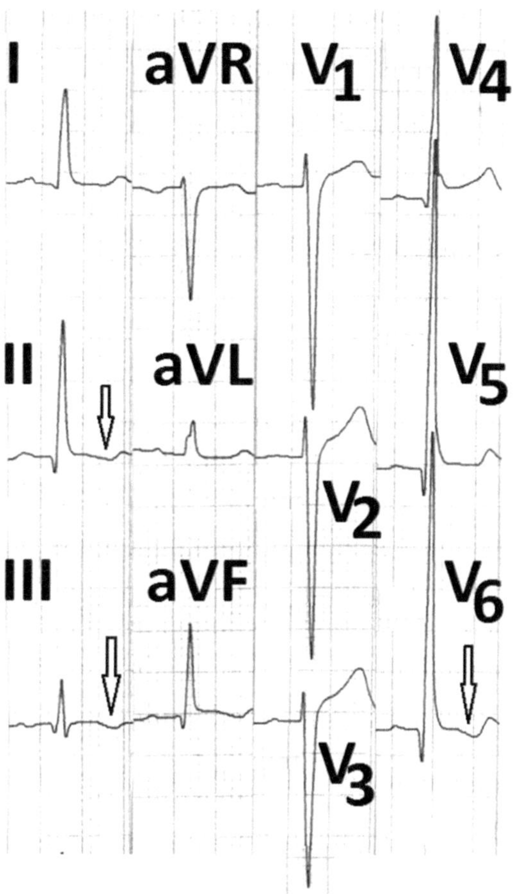

Fig. 31.4 Electrocardiogram from a patient of advanced chronic left ventricular volume overload showing inversion of the T wave in leads II, III, and V_6 (marked ⇩). (Reproduced with permission from Mittal SR. Chronic Left Ventricular Volume Overload. Cardiology Today-2016;20:61–65)

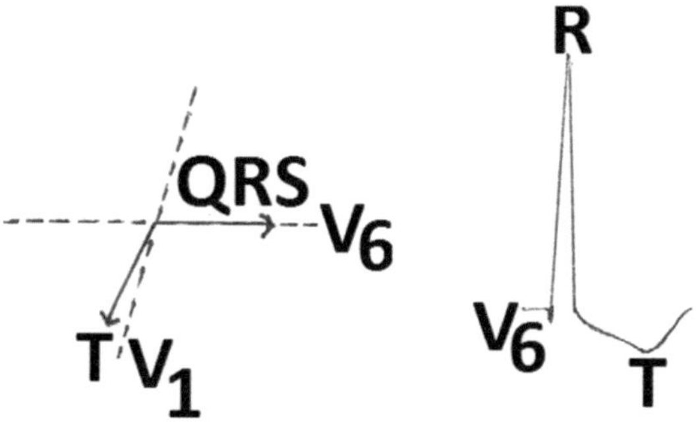

Fig. 31.3 Diagram showing QRS and T wave vectors and corresponding QRS complex and the T wave configuration in lead V_6 in advanced chronic left ventricular volume overload. (Reproduced with permission from Mittal SR. Chronic Left Ventricular Volume Overload. Cardiology Today-2016;20:61–65)

3 Electrocardiographic Differences Amongst Various Causes of Chronic Moderate Left Ventricular Volume Overload [5]

3.1 Chronic Moderate Mitral Regurgitation (MR)

Electrocardiographic evidence of left ventricular enlargement is not seen in all the cases. It occurs only in approximately 15% of all the patients with mitral regurgitation and one-third of the patients with severe mitral regurgitation [3]. Symptomatic patients with mitral valve prolapse can have inverted or biphasic T waves and nonspecific ST-segment changes in leads II, III, and aVF and occasionally in anterolateral leads V_4 to V_6 [3]. These may normalize spontaneously or with effort [6].

Figure 31.5a shows normal cardiac chambers. Normal QRS vector in horizontal plane and normal QRS configuration in leads V_1 and V_6 are shown in Fig. 31.5b. Cardiac chambers, QRS vector, and QRS configuration in chronic, moderate mitral regurgitation are shown in Fig. 31.5c and d. QRS vector is directed posteriorly in the horizontal plane. This results in increased amplitude of R wave in lead V_6 and increased depth of S wave in lead V_1. In severe chronic MR, giant left atrium displaces the heart forward. It results in anterior displacement of the horizontal plane QRS axis [2] with the result that it is less aligned to and more obliquely oriented to lead V_1 axis (Fig. 31.5e and f).

S wave in lead V_1 is consequently attenuated (Fig. 31.6). Electrocardiographic signs of left atrial enlargement are common with significant mitral regurgitation (Fig. 31.6).

Frontal plane QRS axis is usually normal. Left-axis deviation suggests left anterior hemiblock, which is a pointer to congestive cardiomyopathy as a cause of mitral regurgitation. Atrial fibrillation supports the possibility of mitral regurgitation. It is common with rheumatic mitral regurgitation. Giant left atrium facilitates development and sustenance of atrial fibrillation [7]. Normally R wave amplitude is more in lead V_5 than in lead V_6. In patients with chronic severe mitral regurgitation, R wave in lead V_6 may be taller than an R wave in lead V_5.

3.2 Ventricular Septal Defect with Large Left to Right Shunt

In infants with large left to right shunts, mean frontal plane QRS axis tends to remain to the right with little or no regression of the right ventricular voltage. LV forces gradually increase, resulting in a pattern of biventricular hypertrophy [8] (Fig. 31.7).

However, R wave in lead V_6 is relatively less tall compared to patients with chronic moderately severe aortic regurgitation or chronic moderate mitral regurgitation. S in lead V_1 is also not deep. Electrocardiographic signs of left atrial enlargement are common with large shunt (Fig. 31.8).

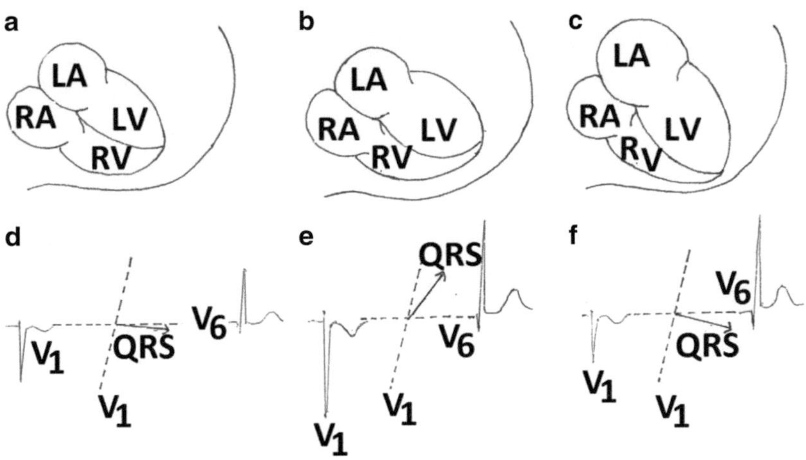

Fig. 31.5 Diagram showing (**a**) normal cardiac chambers, (**b**) normal QRS vector in horizontal plane and QRS configuration in leads V_1 and V_6, (**c**) cardiac chambers in chronic left ventricular volume overload due to moderate mitral regurgitation, (**d**) QRS vector and QRS configuration in leads V_1 and V_6 in chronic, moderate mitral regurgitation, (**e**) cardiac chambers in chronic severe mitral regurgitation, and (**f**) QRS vector and QRS configuration in leads V_1 and V_6 in chronic severe mitral regurgitation. (Reproduced with permission from Mittal SR. Chronic Left Ventricular Volume Overload. Cardiology Today-2016;20:61–65)

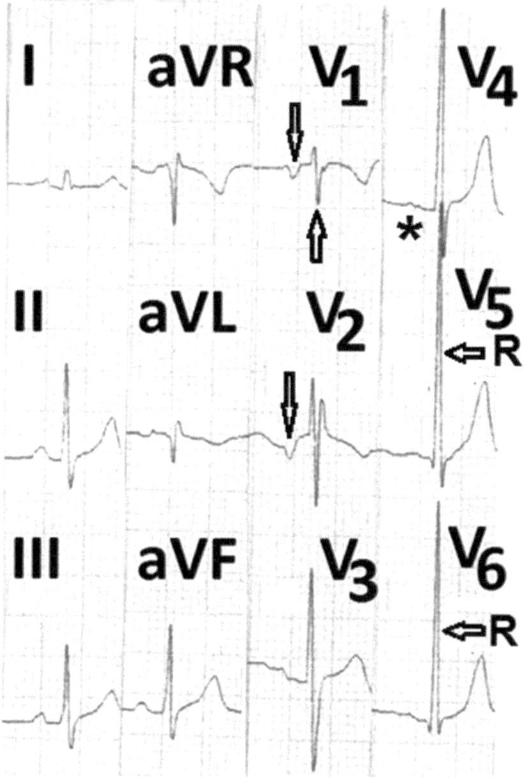

Fig. 31.6 Electrocardiogram in chronic, severe mitral regurgitation showing left atrial enlargement (negative P waves in leads V_1 and V_2 marked ⇩). Broad and notched P wave is present in lead V_4 (marked *). Tall QRS complexes are seen in leads V_5 and V_6 (marked ⇦ R). S wave is small in lead V_1 (marked ⇧). (Reproduced with permission from Mittal SR. Chronic Left Ventricular Volume Overload. Cardiology Today-2016;20:61–65)

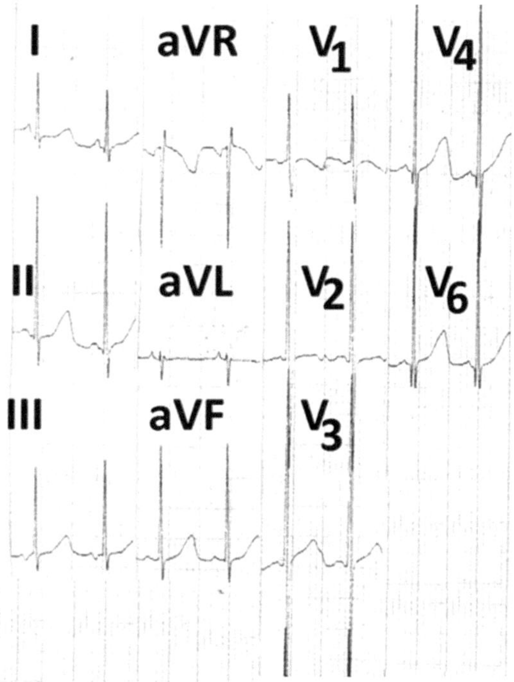

Fig. 31.7 Electrocardiogram from a child with ventricular septal defect and large left to right shunt showing tall R waves in lead V_4 and V_6 as well as in lead V_1. Large equiphasic QRS complexes are seen in leads V_2 and V_3. These findings suggest biventricular enlargement. (Reproduced with permission from Mittal SR. Chronic Left Ventricular Volume Overload. Cardiology Today-2016;20:61–65)

3.3 Atrioventricular Septal Defect [8]

Mean frontal plane QRS axis is oriented superiorly (left-axis deviation) with right bundle branch block pattern in precordial leads.

3.4 Chronic Moderate Aortic Regurgitation [3]

PR interval may be prolonged if aortic regurgitation is caused by an inflammatory pathology, e.g., ankylosing spondylitis. Mean frontal plane QRS axis is usually normal. It may be shifted to left in around 30% of patients with chronic severe aortic regurgitation [7]. This is probably due to stretching of the left anterosuperior fascicle. Electrocardiographic signs of left atrial enlargement are not frequent unless patient develops left ventricular failure. Atrial fibrillation is rare in isolated chronic moderately severe aortic regurgitation. In chronic severe aortic regurgitation, ST-segment may be depressed in leads V_5 and V_6. Electrocardiographic differences between different causes of LV volume overload are shown in Table 31.1.

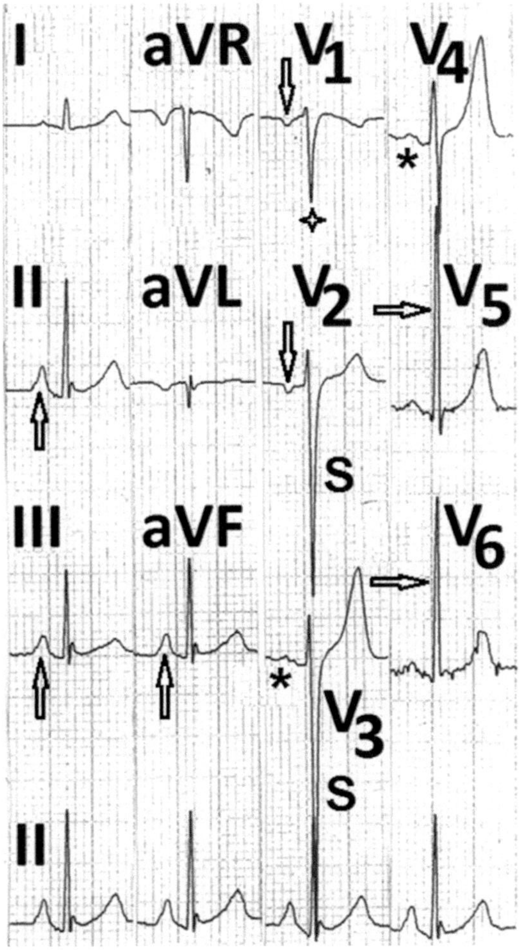

Fig. 31.8 Electrocardiogram from an adult with VSD and large left to right shunt. P waves are negative in leads V_1 and V_2 (marked ⇩). P wave is broad and notched in lead V_3 and V_4 (marked *). These findings suggest left atrial overload. P wave in leads II, III, and aVF is tall and peaked (marked ⇧). These suggest concomitant right atrial overload. Tall R waves are present in leads V_5 and V_6 (marked ⇨). These suggest left ventricular overload. S wave in lead V_1 is small (marked ✧). Deep S waves in leads V_2 and V_3 also suggest left ventricular volume overload. (Reproduced with permission from Mittal SR. Chronic Left Ventricular Volume Overload. Cardiology Today-2016;20:61–65)

Table 31.1 Differences in ECG between different causes of LV volume overload [5]

	AR (Fig. 31.2)	MR (Fig. 31.6)	PDA	VSD (Fig. 31.8)
Rhythm	Sinus, ventricular ectopics may be present	Atrial fibrillation supports the diagnosis of chronic severe mitral regurgitation, supraventricular ectopics can be present	Sinus	Sinus
P wave	Normal	Left atrial enlargement, biatrial enlargement if there is pulmonary artery hypertension	Left atrial enlargement	Biatrial enlargement
Frontal plane QRS axis	Normal or left-axis deviation	Normal	Normal	Normal or right axis deviation if pulmonary artery hypertension, left axis deviation if atrioventricular septal defect
Tall R inV_6	+++	+++	++	++
S wave in lead V_1	+++	Normal or reduced	+++ Reduced with pulmonary artery hypertension	++ Reduced with pulmonary artery hypertension
RV_6/SV_1	Around 1	Around 2	Around 1	Around 1

4 Clinical Correlation

- ECG does not accurately predict the severity of aortic regurgitation or cardiac weight [3].
- "Strain" pattern (ST depression and T wave inversion) correlates with dilatation and hypertrophy of the left ventricle [3].
- Inversion of the U wave suggests compromised LV functions [7].
- Left atrial enlargement in case of chronic severe aortic regurgitation suggests left ventricular failure.
- Biatrial enlargement suggests concomitant pulmonary artery hypertension.

Summary

Aortic Regurgitation

- In mild left ventricular volume overload, an electrocardiogram is apparently normal.
- In moderate left ventricular volume overload, q wave in lead V_5 and V_6 becomes prominent and amplitude of R wave in lead V_6 becomes more than the amplitude of R wave in lead V_5. ST-segment is isoelectric or minimally elevated.
- In gross left ventricular volume overload,

  Depth of q wave in leads V_5 and V_6 increases.

  Amplitude of R wave in leads V_5 and V_6 increases.

  Depth of S wave in lead V_1 increases.

  ST-segment in lead V_5 and V_6 is minimally elevated.

  T wave in leads V_5 and V_6 are positive and may be tall and peaked.
- In the later stage, leads V_5 and V_6 show the following changes:

  q wave depth decreases.

  R wave amplitude remains unchanged.

  ST-segment becomes downsloping and T wave becomes inverted.

Mitral Regurgitation

- There are no diagnostic changes in mild and even moderate mitral regurgitation.
- Only one-third of the patients with severe mitral regurgitation show electrocardiographic signs of left ventricular volume overload.
- Electrocardiographic signs of left atrial enlargement are common. Magnitude of left atrial enlargement correlates with the magnitude of mitral regurgitation and indirectly with the magnitude of left ventricular volume overload.
- In contrast to aortic regurgitation, S wave depth in lead V_1 does not increase.
- Mean frontal plane QRS axis remains normal.

Mitral Valve Prolapse

- These patients may show shallow T wave inversion in leads II, III, aVF, V_5, and V_6. These changes do not correlate with left ventricular volume overload.

Myocarditis/Dilated Cardiomyopathy

- In these conditions, voltage of the QRS complex is increased. ST-segment and T wave changes are, however, due to damage to myocardium and do not correlate with severity of volume overload.

Acute Mitral or Aortic Regurgitation

- Produce sinus tachycardia. There is no chamber enlargement because of short duration of illness.

MCQs

Q1. Which condition produces significant left ventricular volume overload?

(a) Atrioventricular septal defect with severe pulmonic stenosis
(b) Aortopulmonary window with severe pulmonary hypertension

(c) Patient ductus arteriosus with large left to right shunt
(d) All

Q2. Which condition produces significant left ventricular volume overload?

(a) Large ventricular septal defect with normal pulmonary vascular resistance
(b) Secundum atrial septal defect with large left to right shunt
(c) Ventricular septal defect with severe pulmonic stenosis
(d) Coronary arteriovenous fistula

Q3. Significant pulmonary artery hypertension is common in

(a) Chronic aortic regurgitation
(b) Nonrheumatic mitral regurgitation
(c) Muscular ventricular septal defect
(d) Aortopulmonary window

Q4. Prominent R wave in lead V_6 is common with

(a) Rheumatic mitral regurgitation
(b) Chordal rupture
(c) Rupture of sinus of Valsalva in left ventricle
(d) Rupture of the interventricular septum

Q5. Chronic left ventricular volume overload is characterized by

(a) Absence of septal q wave in lead V6
(b) Broad and deep q wave in lead V6
(c) Prominent but narrow q wave in lead V6
(d) Prominent q in leads V3 and V4

Q6. Chronic moderately severe aortic regurgitation is characterized by

(a) Tall R wave in lead V6
(b) Tall R wave in leads V3 and V4
(c) Deep S wave in lead V1
(d) Deep S wave in lead V6

Q7. Chronic mitral regurgitation does not produce

(a) Tall R wave in lead V6
(b) Deep S wave in lead V1
(c) Small S wave in lead V1
(d) Large equiphasic QRS in leads V3 and V4

Q8. Ventricular septal defect with large left to right shunt does not produce

(a) Deep S wave in lead V1
(b) Large equiphasic QRS in leads V3 and V4
(c) "P- mitral"
(d) Tall R wave in lead V6

Q9. Atrial fibrillation supports the possibility of

(a) Chronic rheumatic mitral regurgitation
(b) Chronic severe aortic regurgitation
(c) Ventricular septal defect with large left to right shunt
(d) Patent ductus arteriosus with large left to right shunt

Q10. Left-axis deviation can be seen in

(a) Atrioventricular septal defect
(b) Chronic severe aortic regurgitation
(c) Perimembranous ventricular septal defect
(d) Chronic rheumatic mitral regurgitation

Q11. Biatrial enlargement is not seen in

(a) Chronic aortic regurgitation
(b) Chronic rheumatic mitral regurgitation
(c) Ventricular septal defect with pulmonary artery hypertension
(d) Patent ductus arteriosus with large left to right shunt

Q12. In chronic left ventricular volume overload, which finding closely correlates with dilatation and hypertrophy of the left ventricle?

(a) R wave amplitude in lead V6
(b) S wave depth in lead V1
(c) ST-segment and T wave inversion in leads V5 to V6
(d) U wave inversion

Q13. Normally R wave amplitude is maximal in

(a) Lead I
(b) Lead aVL
(c) Lead V5
(d) Lead V6

Q14. In moderately severe chronic aortic regurgitation

(a) q wave becomes prominent in lead V5
(b) R wave amplitude increase in lead V6
(c) T wave is inverted in leads V5 and V6
(d) ST-segment is depressed in leads I and aVL

Q15. In long-standing severe aortic regurgitation

(a) q wave depth decreases in lead V5
(b) ST segment is depressed in lead V5
(c) T wave is tall and peaked in V5
(d) Mean frontal plane QRS axis of −90°

Q16. In moderately severe chronic mitral regurgitation, there is

(a) Left-axis deviation
(b) Deep S wave in lead V1
(c) Left atrial enlargement
(d) T wave inversion is inverted in lead V6

Q17. In mild mitral regurgitation due to mitral valve prolapse, there may be

(a) Shallow T wave inversion in leads II, III, aVF
(b) Shallow T wave inversion in leads V5, V6
(c) Deep symmetrical T wave inversion in leads V5, V6
(d) Deep q wave in leads V5, V6

Q18. In case of myocarditis, magnitude of volume overload correlates with

(a) Amplitude of R wave in lead V5
(b) Width of Q wave in lead V5
(c) Inversion of T wave in lead V5
(d) None

Q19. In acute severe mitral regurgitation, there is

(a) Gross left atrial enlargement
(b) Deep prominent q wave in lead V6
(c) Tall R wave in lead V6
(d) None

Q20. Acute aortic regurgitation produces

(a) Sinus tachycardia
(b) Deep q in leads V5 and V6
(c) Tall R in leads V5 and V6
(d) All

Answers (1) c (2) a (3) d (4) a (5) c (6) a, c (7) b, d (8) a (9) a (10) a, b (11) a, d (12) c, d (13) c (14) a, b (15) a, b (16) c (17) a, b (18) d (19) d (20) a

References

1. Mirvis DM, Goldberger AL. Electrocardiography. In: Bonow RO, Mann DL, Zipes DP, Libby P, editors. Braunwald's heart disease. St. Louis: Elsevier; 2012. p. 126–65.
2. Schamroth L. Left ventricular hypertrophy. In: Schamroth C, editor. An introduction to electrocardiography. France: Blackwell Science; 1982. p. 67–77.
3. Otto CM, Bonow RO. Valvular heart disease. In: Mann DL, Zipes DP, Libby P, Bonow RO, editors. Braunwald's heart disease. Philadelphia: Elsevier; 2015. p. 1446–514.
4. Wagner GS, Lim TH. Chamber enlargement. In: Wagner GS, editor. Marriott's practical electrocardiography. Philadelphia: Wolters Kluwer; 2008. p. 71–96.
5. Schamroth L, Schamroth C, Sareli P, Hummel D. Electrocardiographic differentiation of the causes of left ventricular diastolic overload. Chest. 1986;89:95–9.
6. Schamroth L. The billowing mitral leaflet syndrome. In: Schamroth C, editor. An introduction to electrocardiography. France: Blackwell Science; 1982. p. 221–4.
7. Schamroth L. Rheumatic heart disease. In: Schamroth C, editor. An introduction to electrocardiography. France: Blackwell Science; 1982. p. 206–20.
8. Brown DW, Fulton DR. Congenital heart disease in children and adolescents. In: Fuster V, Walsh RA, Harrington RA, editors. Hurst's the heart. New York: McGraw Hill; 2011. p. 1827–83.

Left Ventricular Systolic Overload (Pressure Overload)

32

Although imaging modalities provide more direct assessment of left ventricular hypertrophy, the electrocardiogram is a very cheap and widely available facility. Further, it gives additional important information regarding the electrical activity of the heart that has a potential impact on prognosis [1].

1 Terminology

Left ventricular hypertrophy is not an appropriate term because the thickness of the left ventricle may be within the normal range in initial stages. Further, the conventional criteria described for the diagnosis of left ventricular hypertrophy are not present during early stage. "Systolic overload" or "pressure overload" give correct impression of pathophysiology.

Terms such as "probable" or "possible" are significantly dependent on individual interpretation [2] and are not reproducible. For the sake of follow-up, it is better to write the electrocardiographic findings in detail. ST-segment and T wave changes do not always correlate with hemodynamic load [2]. Using the term "strain" is not correct in all cases. It is, however, better to describe the ST-segment and the T wave changes in detail for better follow-up. The ST-segment can be normal, horizontal, or downsloping. The T wave can be positive, flat, or inverted. Report of electrocardiogram should include details of abnormalities present in any lead.

2 Etiology

- Systemic hypertension
- Left ventricular outflow tract obstructions
- Coarctation of the aorta

3 Electrocardiographic Changes [3]

These can be classified as follows:

1. Abnormalities in the P wave
2. Abnormalities in the QRS complex
3. Abnormalities in the ST-segment and the T waves
4. Inversion of U wave
5. Prolongation of the QT interval [4]

3.1 Abnormalities of the P Wave: Left Atrial Enlargement

Electrocardiographic signs of left atrial enlargement provide supportive evidence to the diagno-

S. Mittal, *Insights into Electrocardiograms with MCQs*,
https://doi.org/10.1007/978-981-99-0127-2_32

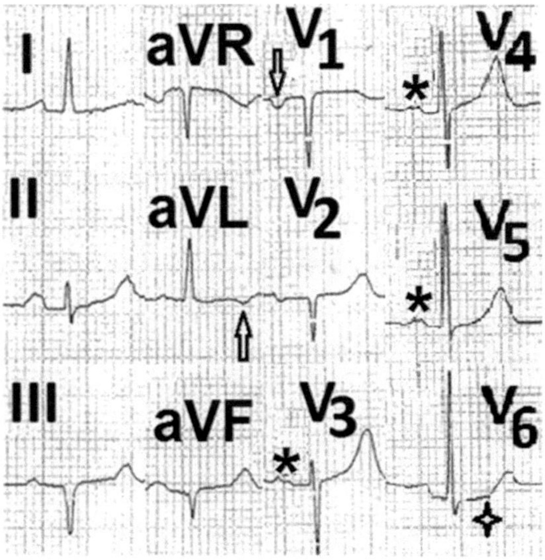

Fig. 32.1 Electrocardiogram from a case of left ventricular pressure overload showing prominent negative P wave in lead V_1 (marked ⇩). Broad and notched P waves are present in leads V_3 to V_5 (marked *). Additional features of left ventricular pressure overload are leftward deviation of the mean frontal plane QRS axis (QRS is positive in lead I and negative in lead aVF), inversion of the T wave in lead aVL (marked ⇧), and mild depression of the ST-segment in lead V_6 (marked ✧). (Reproduced with permission from Mittal SR. Left Ventricular Systolic Overload. Cardiology Today-2016;20:101–110)

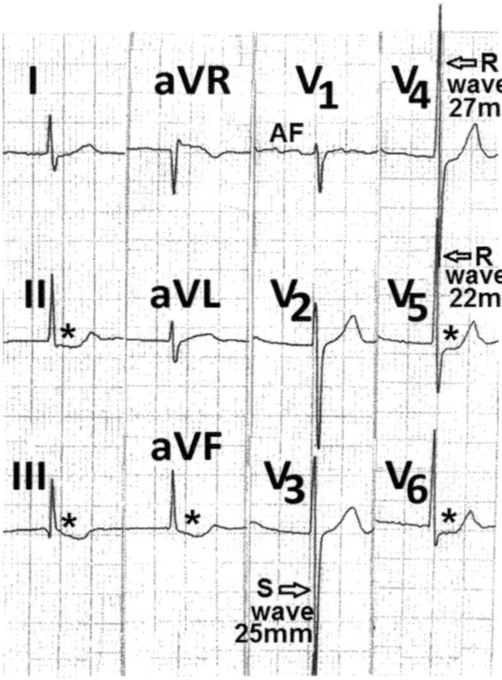

Fig. 32.2 Electrocardiogram showing atrial fibrillation [absence of P wave and wavy baseline in lead V_1 (marked AF)] in a case of left ventricular pressure overload. Additional features of left ventricular pressure overload are deep S wave (25 mm) in lead V_3 (marked ⇨), tall R wave in leads V_4 and V_5 (marked ⇦), and ST-segment depression in leads II, III, aVF, and V_5, V_6 (marked *). (Reproduced with permission from Mittal SR. Left Ventricular Systolic Overload. Cardiology Today-2016;20:101–110)

sis of left ventricular systolic overload provided there is no mitral valve disease. It is an indirect sign. Left ventricular hypertrophy results in progressive reduction in compliance of the left ventricle. Left atrium has to contract forcefully to fill the stiff left ventricle. This results in progressive enlargement of the left atrium. P wave in lead V_1 becomes progressively more negative (Fig. 32.1).

Wide and notched P waves appear in lead II and leads V_4 to V_6. Electrocardiographic signs of left atrial overload are useful in suspecting left ventricular systolic overload even in the presence of left bundle branch block. In advanced stage, patients can have atrial fibrillation (Fig. 32.2). However, this is uncommon in isolated pressure overload of the left ventricle.

3.2 Abnormalities of the QRS Complex

These can be further subclassified as follows [3]:

(a) Increased voltage of the QRS deflection
(b) Decrease in amplitude of the initial q wave in left precordial leads
(c) Increase in the left ventricular activation time
(d) Counterclockwise electrical rotation in the horizontal plane
(e) Left-axis deviation in the frontal plane
(f) Fragmentation of the QRS

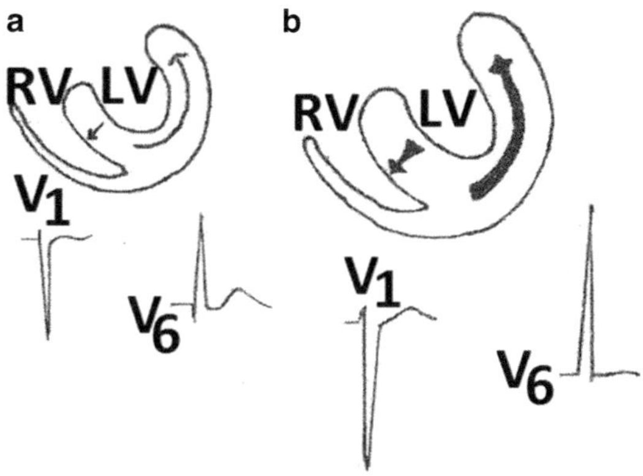

Fig. 32.3 Diagram showing (**a**) normal QRS vector (arrow) with normal left ventricular muscle mass and (**b**) increased QRS vector (arrow) due to increased muscle mass of the left ventricle. *LV* left ventricle, *RV* right ventricle. (Reproduced with permission from Mittal SR. Left Ventricular Systolic Overload. Cardiology Today-2016;20:101–110)

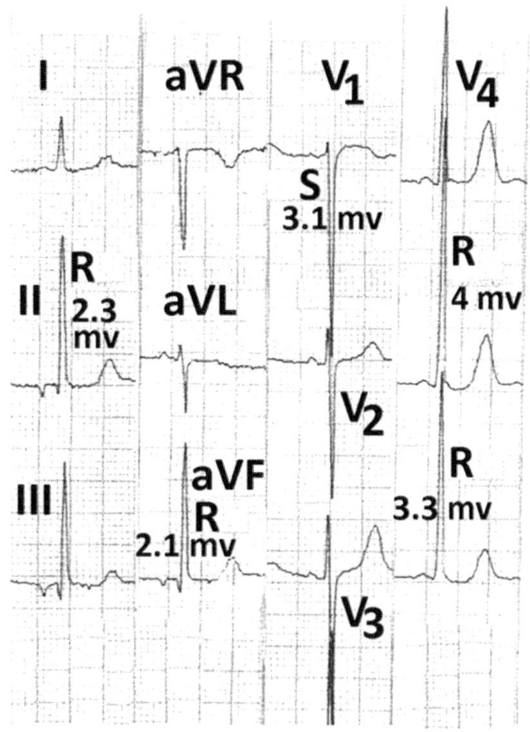

Fig. 32.4 Electrocardiogram showing amplitude of the R wave in lead II = 2.3 mV, R wave in lead aVF = 2.1 mV, S wave in lead V_1 = 3.1 mV, and R wave in lead V_5 = 4 mV. (Reproduced with permission from Mittal SR. Left Ventricular Systolic Overload. Cardiology Today-2016;20:101–110)

3.2.1 Increased Voltage of the QRS Deflection

It is because of an increase in the magnitude of QRS vector due to increased muscle mass of the left ventricle (Fig. 32.3).

QRS vector is also directed more posteriorly than normal with the result that the amplitude of the negative component of the QRS (the S wave) increases in right precordial leads (V_1, V_2, and V_3). Several voltage criteria have been proposed for the diagnosis of left ventricular hypertrophy in the past.

In Limb Leads

- Amplitude of the R wave or the S wave in any limb lead ≥2 mV [5], i.e., 20 mm or 20 small squares (Fig. 32.4).
- Amplitude of the R wave in lead aVL > 1.1 mV [6] (Fig. 32.5).

In Precordial Leads

- Amplitude of the S wave in leads V_1 or V_2 or R wave in leads V_5 or V_6 ≥ 3 mV [5] (Figs. 32.4 and 32.6).
- S wave amplitude in lead V_1 + R wave amplitude in leads V_5 or V_6 ≥ 3.5 mV [6] (Fig. 32.4).
- Or
- R wave amplitude in lead V_5 or R wave amplitude in lead V_6 > 2.6 mV (Figs. 32.4, 32.5, and 32.6).
- R wave amplitude in lead V_6 ≥ R wave amplitude in lead V_5 [3] (Fig. 32.5).

Combination of Limb and Precordial Leads

- Females: R wave amplitude in lead aVL + S wave amplitude in lead V_3 > 2 mV [7] (Fig. 32.7).

Males: R wave in lead aVL + S wave in lead V_3 > 2.8 mV [7].

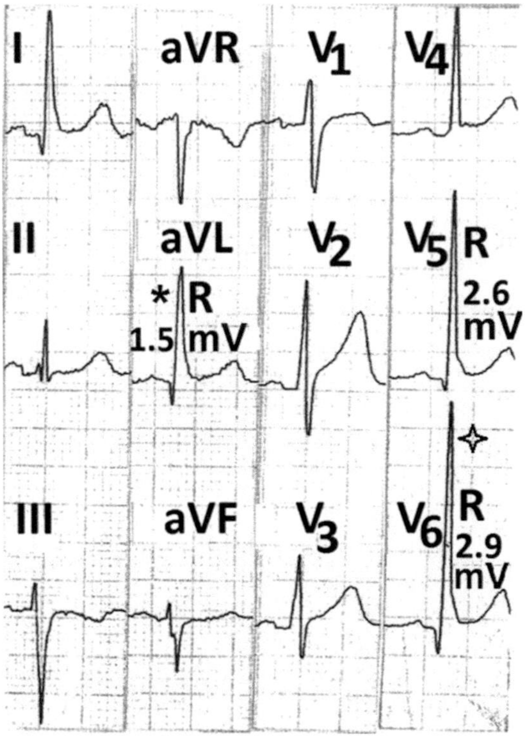

Fig. 32.5 Electrocardiogram showing R wave amplitude in lead aVL = 1.5 mV (marked *). Additional features of left ventricular pressure overload are leftward deviation of the mean frontal plane axis of the QRS complex (prominent R wave in lead I and dominant S wave in lead III), counterclockwise rotation in horizontal plane (transition zone between lead V_1 and V_2—prominent S wave in lead V_1 and prominent R wave in lead V_2), R wave in lead V_6 = 2.9 mV(marked ✧), and R wave amplitude in lead V_6 more than R wave amplitude in lead V_5. (Reproduced with permission from Mittal SR. Left Ventricular Systolic Overload. Cardiology Today-2016;20:101–110)

– Sum of the QRS voltage of 12 leads >175 mm [8], sum of 12 leads voltage >120 mm [9] in hypertensive (Figs. 32.4, 32.5, 32.6, and 32.7).
– Total 12 lead QRS voltage minus total voltage of leads aVR, aVL, and aVF >110 mm [10].

Caveats

– Proper standardization of machine (10 mm/mV) is necessary for correct interpretation of voltage criteria. Recording at 5 mm/mV leads to underestimation of voltages (Fig. 32.8) and recording at 15 mm/ mV leads to overestimation of voltages.
– Correct placement of chest leads is also necessary as various criteria are specific for particular leads.

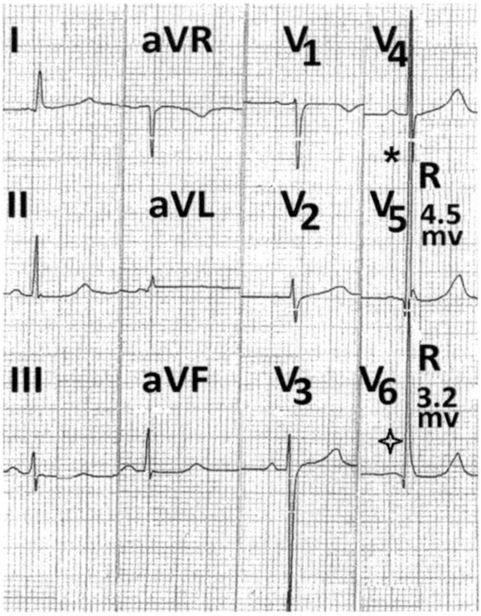

Fig. 32.6 Electrocardiogram showing R wave amplitude in lead V_5 = 4.5 mV and R wave amplitude in lead V_6 = 3.2 mV (marked ✧). (Reproduced with permission from Mittal SR. Left Ventricular Systolic Overload. Cardiology Today-2016;20:101–110)

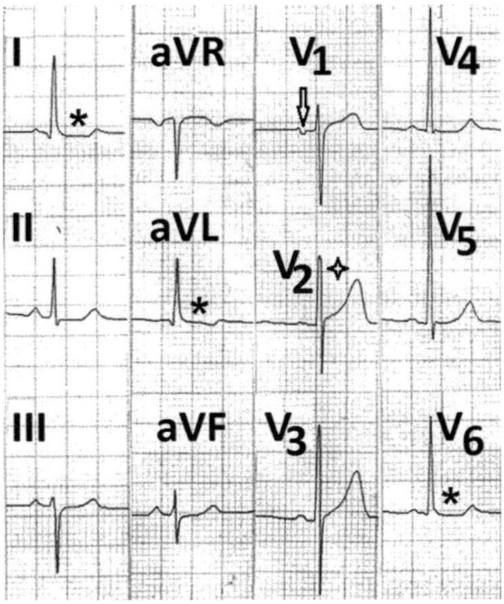

Fig. 32.7 Electrocardiogram showing R wave amplitude in lead aVL + S wave amplitude in lead V_3 = 2.5 mV. Additional features of left ventricular pressure overload are counterclockwise rotation in the horizontal plane (RS configuration in lead V_2 (marked ✧), negative P wave in lead V_1 (marked ⇩), left-axis deviation in the frontal plane, SV_1 + RV_5 = 4.1 mV, ST-segment depression in leads I, aVL and V_6 (marked *). (Reproduced with permission from Mittal SR. Left Ventricular Systolic Overload. Cardiology Today-2016;20:101–110)

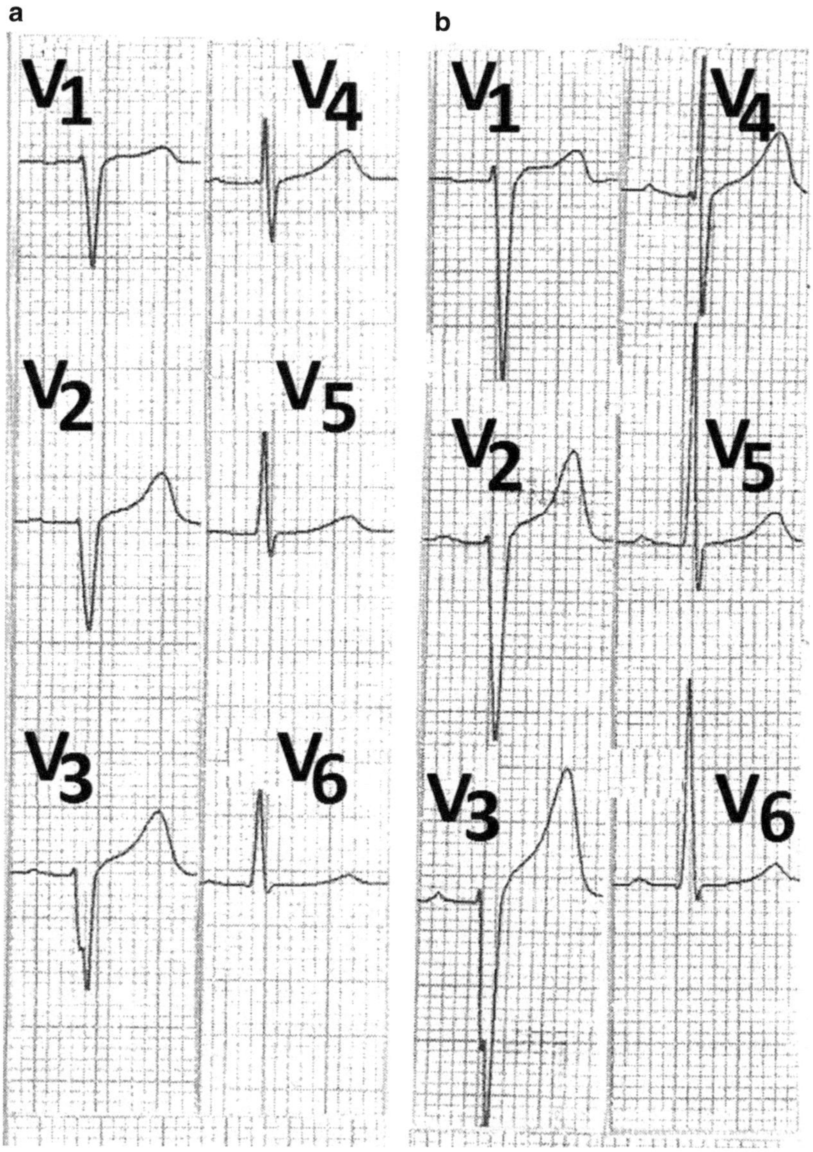

Fig. 32.8 Electrocardiograms (**a**) showing underestimation of voltages at standardization of 5 mm/mV and (**b**) voltage criteria of LVH at the standardization of 10 mm/mV. (Reproduced with permission from Mittal SR. Left Ventricular Systolic Overload. Cardiology Today-2016;20:101–110)

- Most of the ECGs do not strictly fulfill all the abovementioned voltage criteria. An ECG may fulfill only one or two criteria. A large number of ECGs may not fulfill any voltage criteria even when voltages are apparently increased or meet non voltage criteria. Some patients may have increased voltages in leads V_3 and/or V_4 and may not meet voltage criteria in leads V_1, V_2, V_5, and V_6. Variations are likely to be due to variation in the site of hypertrophy. All patients may not have uniform concentric left ventricular hypertrophy. Relatively greater hypertrophy of a particular wall may result in variation in the site of abnormal voltage (Figs. 32.9, 32.10, and 32.11).
- Various voltage criteria listed above have high specificity but low sensitivity [11]. This is because most of them have been devel-

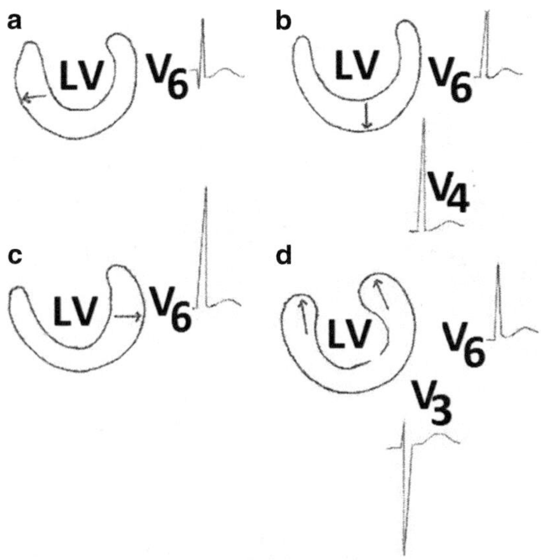

Fig. 32.9 Diagram showing the effect of variation in site of left ventricular hypertrophy on the QRS voltage in different chest leads. (**a**) Hypertrophy of mid septum is likely to produce prominent q wave in lead V_6. (**b**) Relatively greater hypertrophy of left ventricular apex is likely to produce increased R wave voltage in leads V_3 and V_4. (**c**) Hypertrophy of lateral wall of the left ventricle is likely to produce increased R wave voltage in leads V_5 and V_6. (**d**) Hypertrophy of base of the heart is likely to produce increased posteriorly directed voltages with reciprocal deep S waves in leads V_3 and V_4. (Reproduced with permission from Mittal SR. Left Ventricular Systolic Overload. Cardiology Today-2016;20:101–110)

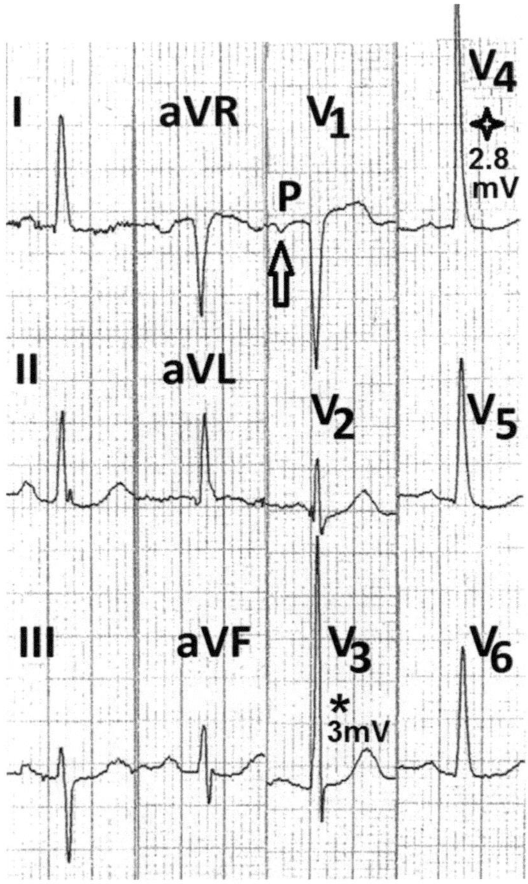

Fig. 32.10 Electrocardiogram showing prominent R wave in leads V_3 (3 mV) (marked ⁎) and V_4 (marked ✧) (2.8 mV). Other features of LVH are negative P wave in lead V_1 (marked ⇧), leftward deviation of the QRS axis in the frontal plane, and shifting of the transition zone toward right-sided chest leads (lead V_1 shows QS pattern and lead V_2 shows Rs pattern). (Reproduced with permission from Mittal SR. Left Ventricular Systolic Overload. Cardiology Today-2016;20:101–110)

oped from autopsy studies that considered the presence or absence of left ventricular hypertrophy. Left ventricular hypertrophy does not occur overnight. It occurs gradually [11]. Changes in sequential ECGs are, therefore, more useful in detecting early changes. Limited accuracy of voltage criteria also reflects a limited relationship between the structural changes of left ventricular hypertrophy and the associated electrophysiological changes at cellular level [4]. Diagnostic accuracy also varies with the population studied [12]. Specificity is more among hypertensives than in the general population. Women and obese persons have low voltage compared to asthenic subjects. Flat chest (Fig. 32.12) and early repolarization (Fig. 32.13) can have false positive voltage criteria.

– Standard criteria for "normal" are also not available for children. This is important because children of hypertensive parents can have early LVH even when their blood pressure readings are within "normal" range.
– Effect of other ECG abnormalities on voltage criteria.

Left anterosuperior fascicular block

It per se produces large R wave in leads I, aVL, and small R wave with deep S wave in leads V_5,V_6. Thus, it mimics LVH in limb leads and masks LVH in leads V_5 and V_6.

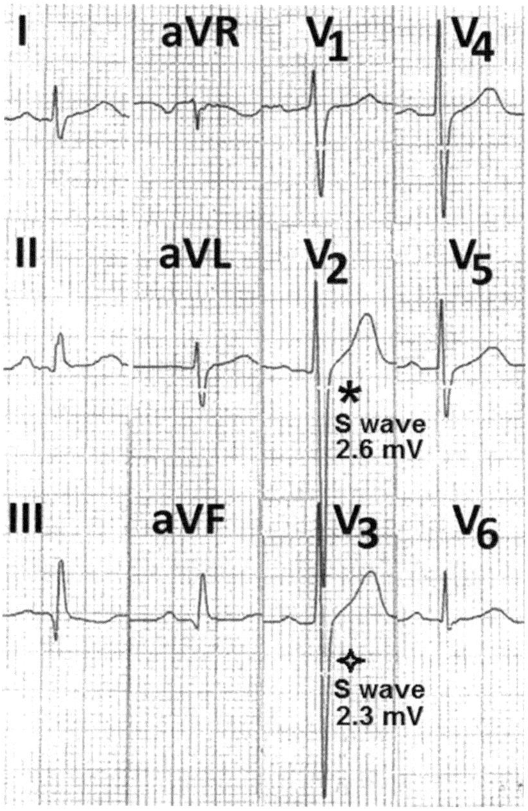

Fig. 32.11 Electrocardiogram showing prominent S wave in leads V_2 (2.6 mV—marked *) and V_3 (2.3 mV—marked ✧). (Reproduced with permission from Mittal SR. Left Ventricular Systolic Overload. Cardiology Today-2016;20:101–110)

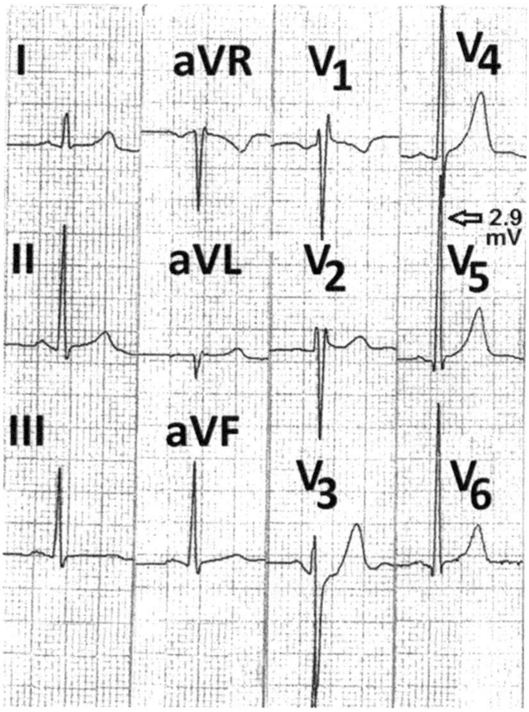

Fig. 32.12 Electrocardiogram from a patient with flat chest without left ventricular hypertrophy showing increased QRS voltages (RV_5 = 2.9 my—marked ⇦ and SV_1 + RV_5 = 4 mV). (Reproduced with permission from Mittal SR. Left Ventricular Systolic Overload. Cardiology Today-2016;20:101–110)

Left bundle branch block

It per se produces deep S waves in leads V_1 to V_3, thus mimicking LVH. Diagnosis is further compromised by the fact that LVH is a common comorbidity with LBBB and LVH per se can produce a QRS duration of 120 ms or more. P wave abnormalities suggestive of left atrial enlargement, QRS duration of ≥155 ms, and presence of voltage criteria in leads V_5, V_6 support diagnosis of LVH with high specificity but low sensitivity [2].

Right bundle branch block

Delayed activation of the right ventricle cancels the left ventricular forces. This reduces the amplitude of the S wave in right precordial leads and the R wave in left precordial leads. Left atrial abnormality, leftward deviation of the frontal plane QRS axis, and presence of voltage criteria in leads V_5 and V_6 support the diagnosis of concomitant LVH [2].

Myocardial infarction

Loss of viable myocardium can reduce R wave voltage and mask LVH.

Preexcitation

It can mask or mimic LVH on ECG (Fig. 32.14).

3.2.2 Decrease in Amplitude of Initial q Wave in Left Precordial Leads (Fig. 32.15)

It is probably related to interstitial fibrosis in the interventricular septum. It reduces voltage of initial QRS vector produced by depolarization of the interventricular septum. It may suggest early (partial) left bundle branch block [13].

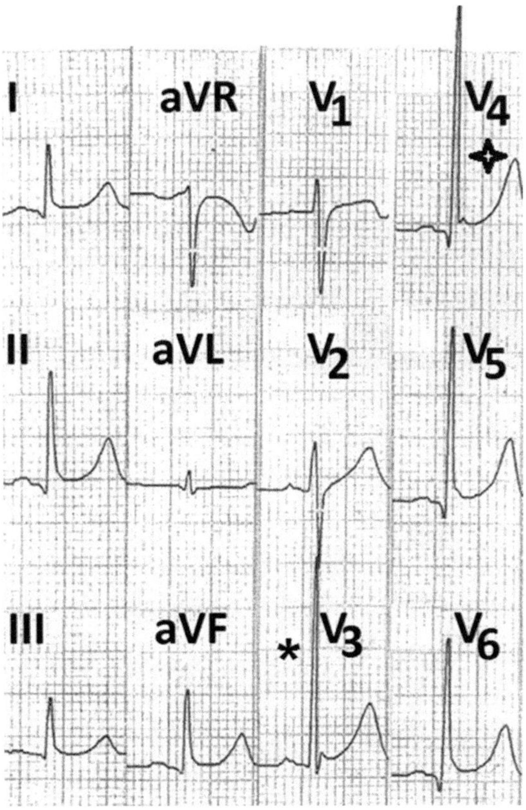

Fig. 32.13 Electrocardiogram from a case of early repolarization without LVH showing increased QRS voltages (RV_3 = 2.4 mV—marked * and RV_4 = 2.6 my—marked ✧). (Reproduced with permission from Mittal SR. Left Ventricular Systolic Overload. Cardiology Today-2016;20:101–110)

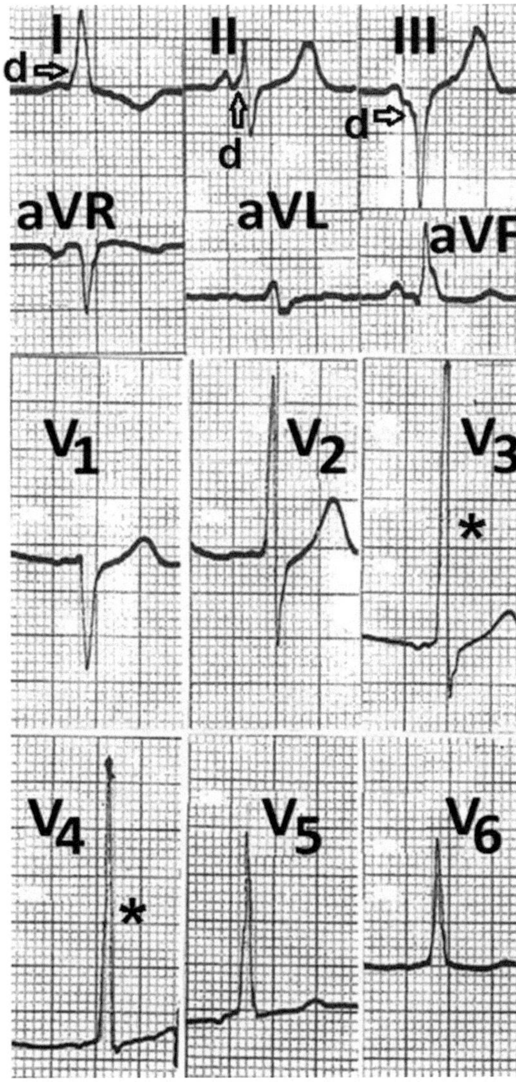

Fig. 32.14 Electrocardiogram from a case of preexcitation. PR interval is short (clear in lead II and lead III). There is slurring of initial part of QRS—delta wave (marked d) with widening of the QRS complex. R wave voltage is increased in leads V_3 (3 mV) and V_4 (3 mV) giving false impression of LVH. (Reproduced with permission from Mittal SR. Left Ventricular Systolic Overload. Cardiology Today-2016;20:101–110)

3.2.3 Increase in Left Ventricular Activation Time (VAT)

Left ventricular activation time is the time from the beginning of the QRS to the peak of the R wave in leads V_5 and V_6. Normal ventricular activation time in left-oriented leads is 0.04 s [3]. It is an indirect measure of the time taken for the impulse to travel the thickness of the left ventricle from endocardium to epicardium [3] (Fig. 32.16). It may be accompanied by indentation or notching on upstroke of the QRS complex.

3.2.4 Leftward Deviation of the Mean Frontal Plane QRS Axis

With increasing LVH, mean frontal plane QRS axis shifts to the left, i.e., toward lead aVL. Lead II becomes biphasic and lead III is inverted (Fig. 32.17).

Left-axis deviation increases with increasing left ventricular hypertrophy As the angle between the vectors of the QRS complex and T wave widens, T wave becomes progressively flat and finally inverted in leads I and aVL (Fig. 32.18).

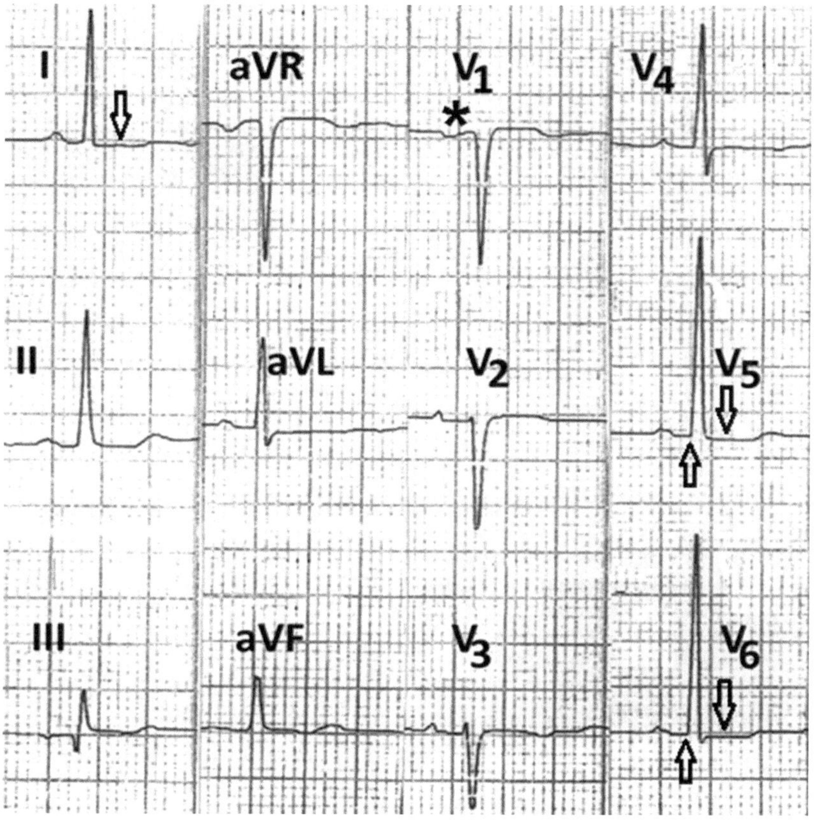

Fig. 32.15 Electrocardiogram showing loss of septal q wave in leads V_5, V_6 (marked ⇧). Other features of LVH include $SV_1 + RV_6 = 3.5$ my, slight ST-segment depression in leads I, V_4 to V_6 (marked ⇩), flattened T wave in I, aVL, V_5, V_6, and negative P wave in lead V_1 (marked *). (Reproduced with permission from Mittal SR. Left Ventricular Systolic Overload. Cardiology Today-2016;20:101–110)

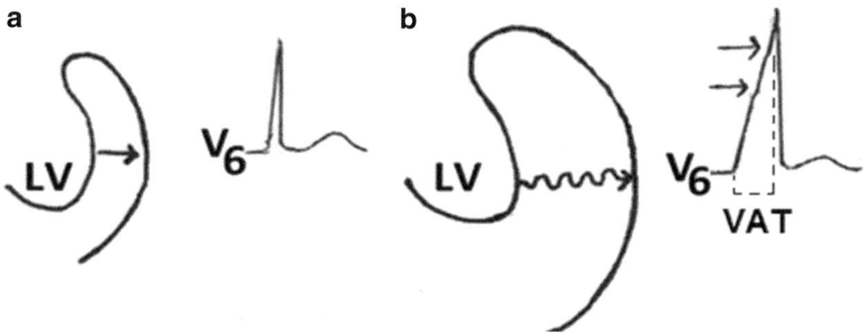

Fig. 32.16 Diagram showing (**a**) normal thickness of the left ventricle and normal ventricular activation time. (**b**) Increased thickness of left ventricular wall resulting in increased ventricular activation line (VAT) and notches (arrows) on the upstroke of the QRS complex. (Reproduced with permission from Mittal SR. Left Ventricular Systolic Overload. Cardiology Today-2016;20:101–110)

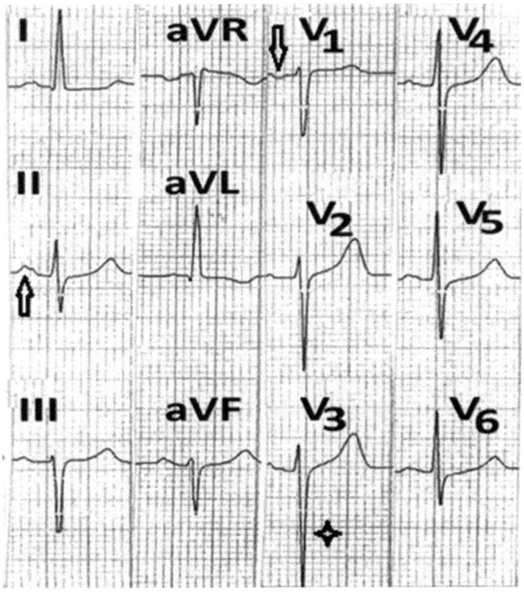

Fig. 32.17 Electrocardiogram from a case of LVH showing leftward deviation of the mean frontal plane QRS axis. Other supportive evidence is negative P wave in lead V_1 (marked ⇩), and broad and notched P wave in lead II (marked ⇧ and deep S wave in lead V_3 (marked ✧). (Reproduced with permission from Mittal SR. Left Ventricular Systolic Overload. Cardiology Today-2016;20:101–110)

Finally, the patient may develop classical left anterosuperior fascicular block. Sequential ECGs are helpful in detecting early changes. Leftward deviation of mean frontal plane QRS axis also results in RS configuration with decreased R wave voltage in leads V_5 and V_6. This can mask voltage criteria in leads V_5 and V_6. Significant left-axis deviation is an adverse prognostic sign [3]. When present, it indicates that LVH is long-standing with some degree of myocardial fibrosis [14]. Conditions other than LVH can also produce left-axis deviation. Left-axis deviation should, therefore, be used only as a supportive evidence [2].

3.2.5 Counterclockwise Rotation in the Horizontal Plane

Normally, leads V_1 and V_2 lie over the right ventricle and show rS configuration. Leads V_5 and V_6 lie over the left ventricle and show Rs configuration. Transition from rS configuration to Rs configuration (the transition zone) occurs in leads V_3 and V_4. In left ventricular hypertrophy, transition zone is shifted toward right-sided chest leads (V_3

Fig. 32.18 Diagram showing progressive (**a**) change in QRS vector, (**b**) increase in QRS—T angle, and (**c**) their effect on frontal plane leads. (Reproduced with permission from Mittal SR. Left Ventricular Systolic Overload. Cardiology Today-2016;20:101–110)

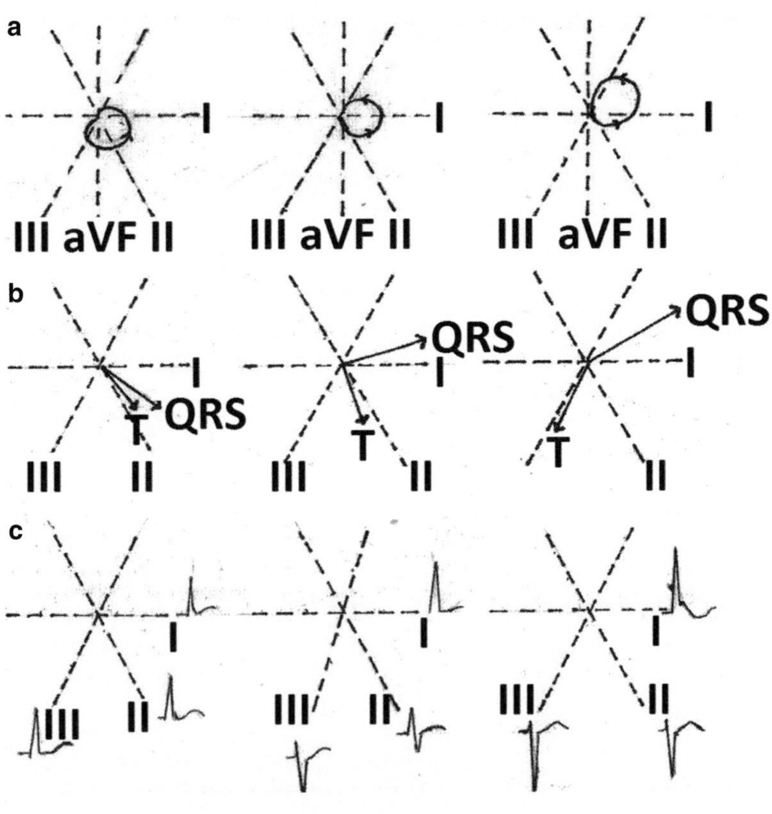

or even V_2). This is known as counterclockwise rotation in horizontal plane. It suggests dominance of left ventricular depolarization vector (Figs. 32.5, 32.7, and 32.10).

3.2.6 Fragmentation of the QRS Complex

Fragmentation of the QRS complex is defined as the presence of notching of the R wave or the S wave or the presence of more than one R' wave in two contiguous leads corresponding to a major coronary artery territory [15] (Figs. 32.19 and 32.20).

Fragmented QRS suggests myocardial fibrosis [16]. In hypertensives, it is associated with worse LVH [17] and more severe diastolic dysfunction [16]. Presence of fragmented QRS independently correlates with severity of aortic stenosis [18] and is a better indicator of myocardial fibrosis than traditional ECG criteria of LVH (except depression of the ST-segment and inversion of the T waves) [19].

3.3 Secondary Abnormalities of the ST-Segment and T Waves

Initially, the ST-segment becomes flat in leads I, aVL, V_5, and V_6 (Fig. 32.21).

Subsequently, it becomes depressed (Fig. 32.22) and finally inverted (Fig. 32.23).

Right-sided leads V_1, V_2, and aVR show reciprocal elevation of the ST-segment. As the angle between the vectors of the QRS complex and the T wave widens, the T wave vector is directed progressively away from the QRS vector. In the horizontal plane, T wave is directed toward lead V_1 (Fig. 32.24) with the result that T wave becomes positive in lead V_1 (Fig. 32.25). At times, this

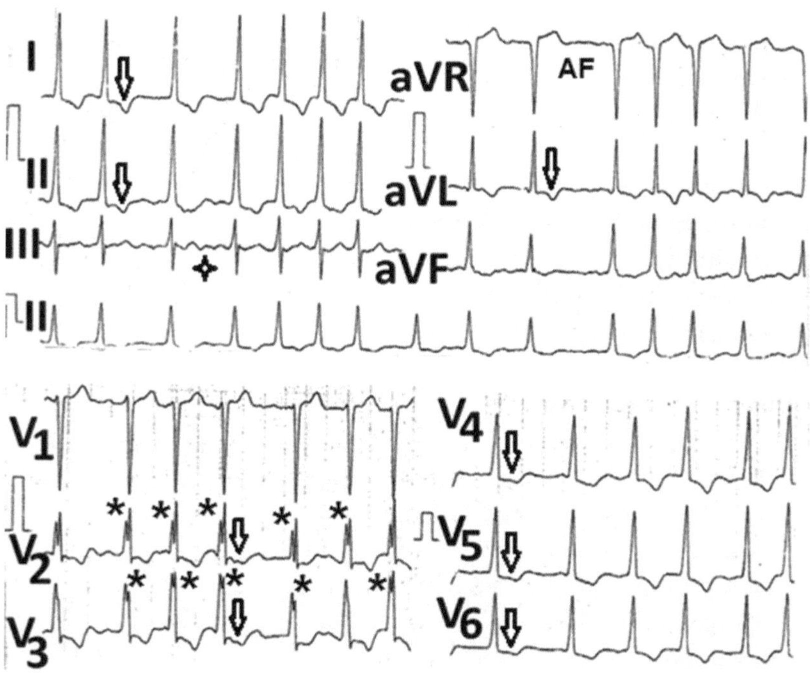

Fig. 32.19 Electrocardiogram showing fragmentation of the QRS complex in leads V_2 and V_3 (marked *). Additional findings are atrial fibrillation (irregularly irregular rhythm with absence of P wave; marked AF in lead aVR) and wavy baseline in lead III (marked ✧), depression of the ST-segment, and inversion of the T wave in leads I, II, aVL, and V_2 to V_6 (marked ⇩). (Reproduced with permission from Mittal SR. Left Ventricular Systolic Overload. Cardiology Today-2016;20:101–110)

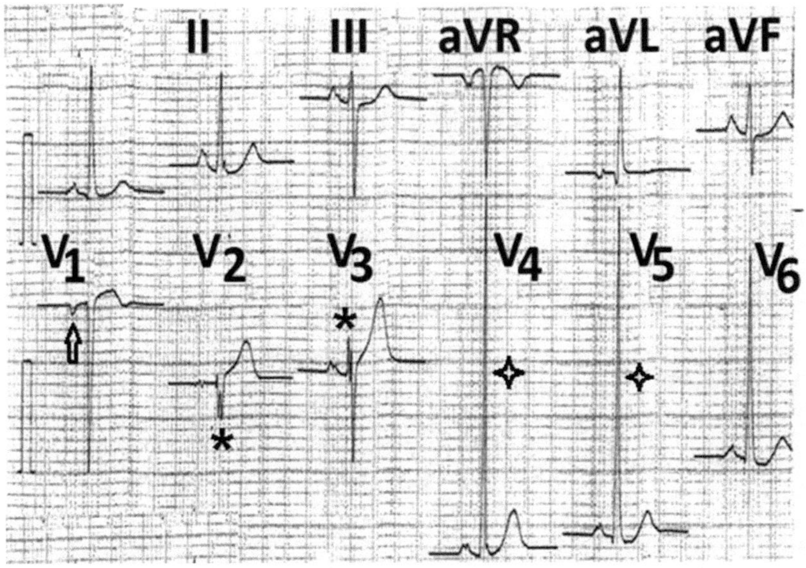

Fig. 32.20 Electrocardiogram showing fragmentation of the QRS complex in leads V_2 and V_3 (marked *). Additional findings include tall R wave in lead V_4 (3.3 mV) and V_5 (3 mV—marked ✧), leftward deviation of the mean frontal plane QRS axis, and negative P wave in lead V_1 (marked ⇧). (Reproduced with permission from Mittal SR. Left Ventricular Systolic Overload. Cardiology Today-2016;20:101–110)

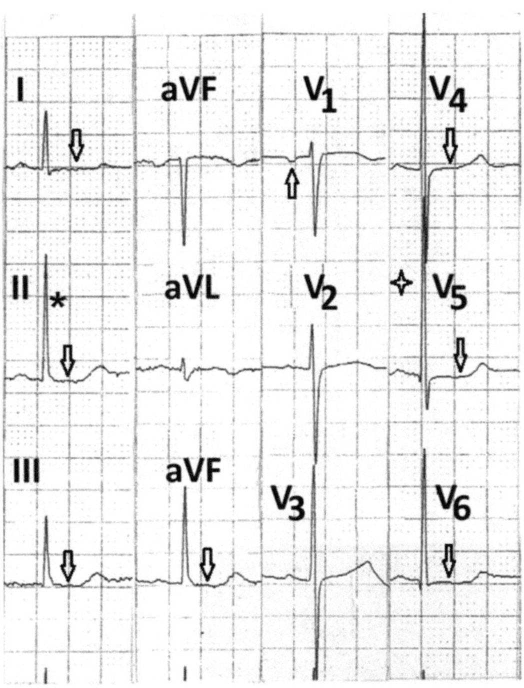

Fig. 32.21 Electrocardiogram shows horizontal ST-segment in leads I, II, III, aVF, and V_4 to V_6 (marked ⇩). Additional features of LVH include tall R wave in lead II (2 mV—marked *) and lead V_5 (marked ✧) (2.8 mV) and negative P wave in lead V_1 (marked ⇧). (Reproduced with permission from Mittal SR. Left Ventricular Systolic Overload. Cardiology Today-2016;20:101–110)

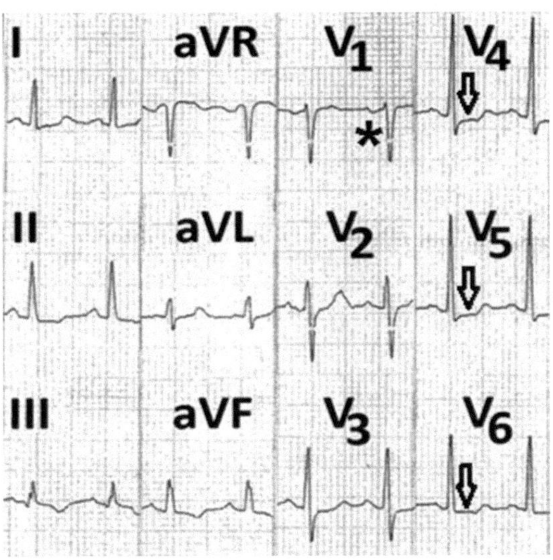

Fig. 32.22 Electrocardiogram shows horizontal depression of the ST-segment in leads V_4 to V_6 (marked ⇩). P wave in lead V_1 is negative (marked *), suggesting left atrial overload. (Reproduced with permission from Mittal SR. Left Ventricular Systolic Overload. Cardiology Today-2016;20:101–110)

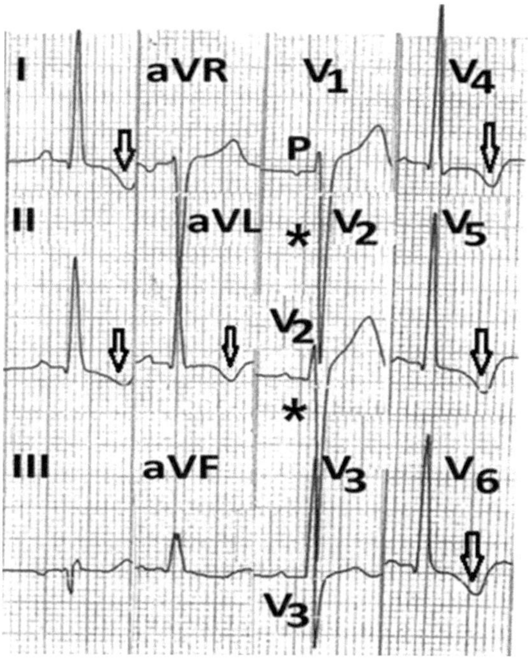

Fig. 32.23 Electrocardiogram showing downsloping depression of the ST-segment with inverted T waves in leads I, II, aVL, and V_4 to V_6 (marked ⇩). Other findings of LVH include tall R wave in lead I (2 mV), deep S wave in leads V_1 (3 mV), and V_2 (2.8 mV) (marked *) and $SV_1 + RV_5$ (5.4 mV). (Reproduced with permission from Mittal SR. Left Ventricular Systolic Overload. Cardiology Today-2016;20:101–110)

change in the polarity of the T wave may be the earliest change.

Progressively height of the T wave increases in lead V_1 and decreases in lead V_6 with the result that T wave in lead V_1 becomes taller than T wave in lead V_6 (Fig. 32.25). With the progressive widening of the QRS- T angle (Fig. 32.24), the T wave becomes inverted in V_6. In frontal plane, initially the T wave in lead III becomes taller than the T wave in lead I. Subsequently, T wave becomes inverted in lead I. ST-segment and T wave changes are due to altered repolarization. However, subendocardial ischemia secondary to high wall tension [19] and limited blood flow to the subendocardium [4] can also contribute to the ST-segment and T wave changes. Patients with repolarization abnormality have relatively severe LVH [2].

3.4 Inversion of the U Wave in Left Precordial Leads

U wave inversion in leads V_4 to V_6 (Fig. 32.26) suggests compromised left ventricular function in case of LVH.

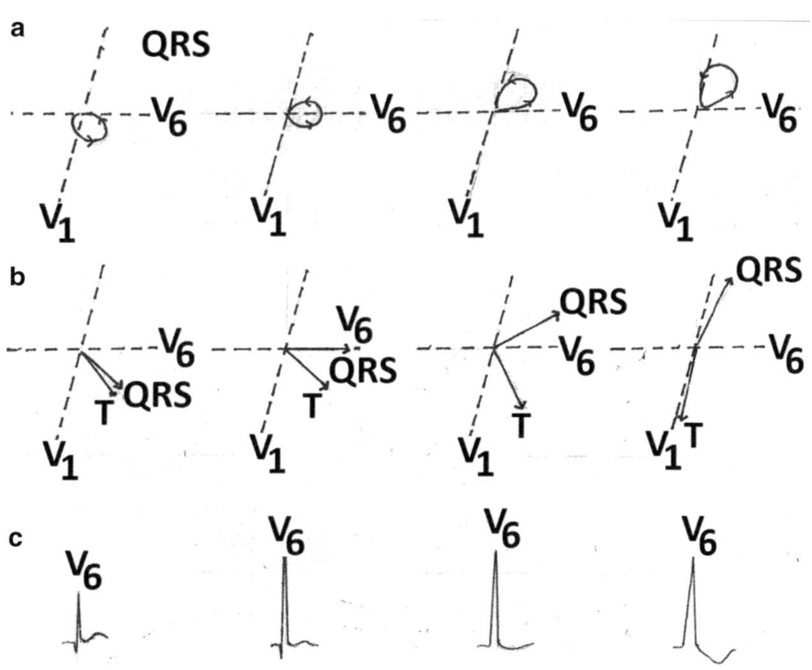

Fig. 32.24 Diagram showing progressive change in (**a**) the QRS loop; (**b**) the QRS complex and T wave vectors in the horizontal plane; and (**c**) their effect on lead V_6. (Reproduced with permission from Mittal SR. Left Ventricular Systolic Overload. Cardiology Today-2016;20:101–110)

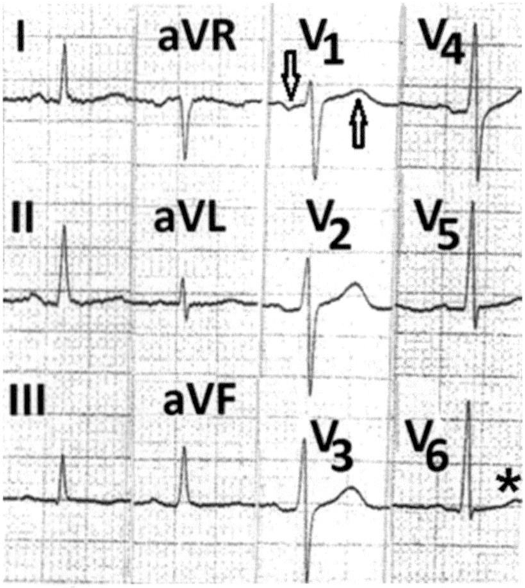

Fig. 32.25 Electrocardiogram showing positive T wave in lead V_1 (marked ⇧). P wave in lead V_1 is inverted (marked ⇩). T wave in lead V_1 is taller than the T wave in lead V_6 (marked *). (Reproduced with permission from Mittal SR. Left Ventricular Systolic Overload. Cardiology Today-2016;20:101–110)

It is a sensitive marker of LVH but is not specific as it can also be present in acute coronary syndrome.

3.5 Prolongation of the QT Interval

Slight prolongation of the QT interval can occur (Fig. 32.27).

Diagnostic significance of such prolongation is not clear [2]. Abnormal prolongation should have a secondary cause [2].

3.6 Clinical Correlation

3.6.1 Systemic Hypertension

- Signs of left atrial enlargement are sensitive and may appear early [2].
- ST-segment and T wave changes usually occur in patients with QRS changes, but may appear alone [4].

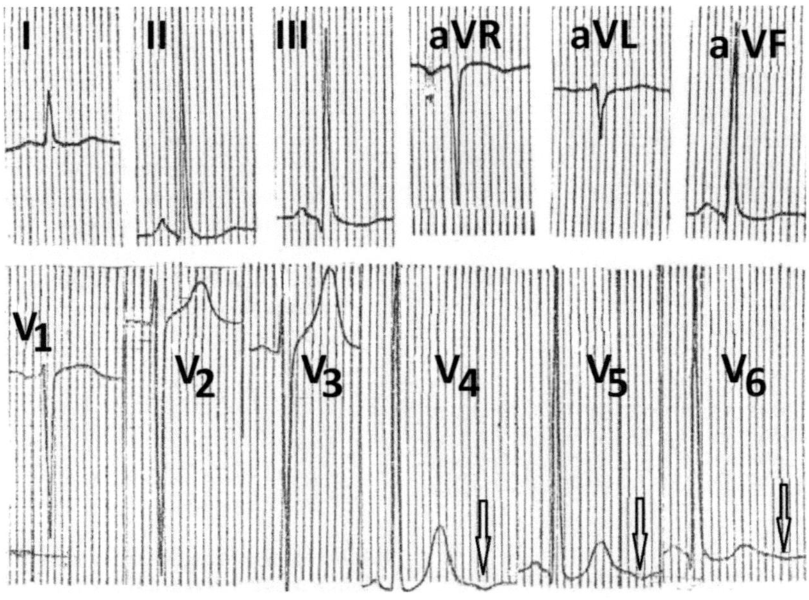

Fig. 32.26 Electrocardiogram shows inversion of the U wave in leads V_4 to V_6 (marked ⇩). Other findings include increased voltages of the QRS complexes in leads II, III, aVF, and V_2 to V_6. (Reproduced with permission from Mittal SR. Left Ventricular Systolic Overload. Cardiology Today-2016;20:101–110)

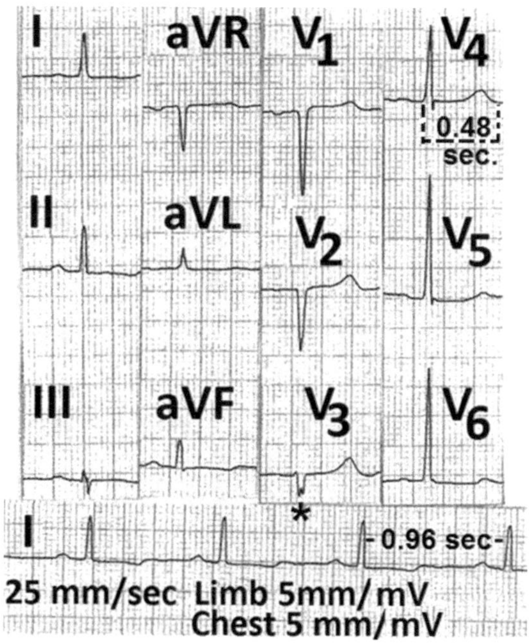

Fig. 32.27 Electrocardiogram recorded at 5 mm/mV showing prolonged QTc interval (QT interval in lead $V_4 = 0.48$ s. R-R interval in rhythm ship of lead II = 0.96 s. (QTc = 0.494 s). Other findings of LVH include increased voltages of the QRS complex and fragmentation of QRS in lead V_3 (marked *). (Reproduced with permission from Mittal SR. Left Ventricular Systolic Overload. Cardiology Today-2016;20:101–110)

- Presence of LVH identifies patients with increased risk of cardiovascular morbidity and mortality [20].
- Regression of ECG signs of LVH reduces risk of cardiovascular events [21].

3.6.2 Aortic Stenosis (AS)

- Although LVH is present in around 85% of the patients of severe aortic stenosis, absence of LVH does not exclude the presence of critical AS [4].
- Correlation between the left ventricle voltages in precordial leads and severity of obstruction is poor in the adults but good in children with congenital AS [4]. In children and adolescents, voltage criteria in left precordial leads are not useful in distinguishing severe from mild to moderate obstruction. Diminished anterior forces and increased posterior forces in right precordial leads ($SV_1 > 30$ mm) and absence of q in lead V_6 suggest severe stenosis [22]. Flat, biphasic, or inverted T wave in lead V_6 also supports the possibility of severe stenosis [22].
- Although left atrial enlargement is frequent in patients with severe isolated AS, atrial fibrillation is uncommon [22].
- In calcific AS, calcific infiltrate into the conduction system can cause various degrees of atrioventricular block [21]. It is more common with concomitant mitral annulus calcification [22].
- ECG "strain" pattern (depression of the ST-segment and/or inversion of the T wave) is a specific marker of mid wall myocardial fibrosis and predicts adverse clinical outcome in aortic stenosis [23].

4 Conclusion

Electrocardiogram is more sensitive than other modalities of investigation in detecting early impact of pressure overload on the heart. Early changes occur at cellular level. These changes affect relaxation of the left ventricle. Subsequently, there is slow progressive change in the vector of the QRS complex and the T wave. Thickness of the left ventricular wall may still be within the normal range. All findings are not present in all cases. As the electrocardiographic changes occur only very slowly, initial electrocardiogram may not fulfill any of the classical electrocardiography criteria. This does not exclude the effect of pressure overload on the left ventricle. Any changes in the sequential electrocardiograms are important. These can manifest as any of the following.

1. Any increase in negativity of the terminal deflection of the P wave in lead V_1.
2. Any decrease in depth of the q wave in leads V_5, V_6.
3. Any increase in amplitude of the R wave.
4. Any increase in depth of the S wave in lead V_1 to V_3.
5. Any deviation of the mean frontal plane QRS axis toward left.

6. Any increase in positivity of the T wave in lead V_1 and any decrease in positivity of the T wave in lead V_6.

   Any of these changes can be suggestive of early left ventricular pressure overload.

Summary

- Leads I, aVL, V_5, and V_6 face the left ventricle. Leads V_1 and V_2 face the right ventricle and are opposite to the left ventricle.
- Hypertrophy of the left ventricle is a slow process.
- In early stages of left ventricular pressure overload, an electrocardiogram may be apparently within normal limit. It may not fulfill any of the classical criteria for the diagnosis of left ventricular hypertrophy. Absence of the conventionally described criteria does not exclude the effect of pressure overload on the left ventricle. Evaluation of sequential electrocardiograms can give a clue to early changes.
- With increasing left ventricular hypertrophy.

  T wave amplitude in lead V_6 starts diminishing and T wave in lead V_1 becomes taller than the T wave in lead V_6.

  R wave amplitude may be increased in leads I, aVL, V_5, and V_6. Deep S waves in leads V_1 and V_2 are, however, more frequent.

  Repolarization changes are usually more apparent than increase in the amplitude of the R wave. Initially, there is flattening of the ST-segment with decrease in amplitude of the T wave. Subsequently, the ST-segment becomes downsloping and the T wave becomes biphasic (−ve/+ve). Finally, the T waves become inverted. Repolarization changes have low sensitivity but high specificity.

  Mean frontal plane QRS axis is normal or can be around 0°. Clear left-axis deviation (more than −30°) is unusual in isolated left ventricular hypertrophy.

  Chronic pressure overload hampers left ventricular relaxation with resultant left atrial enlargement. At this stage, an electrocardiogram may show left atrial overload.

  Severe hypertrophy can produce an inversion of the U wave in leads V_4 to V_6.
- Effect of additional conduction abnormalities.

  Left anterosuperior fascicular block

  Increases amplitude of the R wave in leads I and aVL

  Decreases amplitude of the R wave in leads V_5,V_6

  Left posteroinferior fascicular block can mask voltage criteria in leads I and aVL.

  Left bundle branch block

  Decreases voltage in leads V_5 and V_6.

  Per se can produce deep S wave in leads V_1 and V_2.

  Right bundle branch block

  Decreases amplitude of the R wave in leads V_5 and V_6.

  Produces absence of deep S wave in leads V_1,V_2.

  Produces S wave in lead V_6.

MCQs

Q1. Which is the least sensitive in detecting LVH?

(a) Skiagram chest
(b) Electrocardiogram
(c) Echocardiography
(d) Cardiac MRI

Q2. Left atrial enlargement is most common in

(a) Systemic hypertension
(b) Aortic stenosis
(c) LBBB
(d) Preexcitation

Q3. Atrial fibrillation is uncommon in

(a) Mitral valve disease
(b) Long-standing hypertension
(c) Isolated aortic stenosis
(d) Atrial infarction

Q4. LVH is suggested by

(a) R wave amplitude in lead II > 2 mV
(b) S wave depth in lead $V_1 \leq 2$ mV

(c) R wave amplitude in lead $V_5 > 2$ mV
(d) R wave amplitude in lead $V_5 \geq$ R wave amplitude in lead V_6

Q5. LVH is suggested by

(a) R wave amplitude in lead aVL > 1.5 mV
(b) R wave amplitude in lead aVL + S wave depth in lead $V_3 > 3.0$ mV
(c) S wave depth in lead V_2 + R wave amplitude in lead $V_5 \geq 3$ mV
(d) All

Q6. Various voltage criteria have

(a) High sensitivity
(b) High specificity
(c) High accuracy
(d) None

Q7. Voltage criteria may be falsely positive in

(a) Obese
(b) Flat chest
(c) Women
(d) Emphysema

Q8. Voltage criteria may be falsely negative in

(a) Asthenia
(b) Early repolarization
(c) Emphysema
(d) Standard of 1 mV = 15 mm

Q9. Left anterosuperior fascicular block

(a) Mimics LVH in limb leads
(b) Mimics LVH in leads V_1 to V_3
(c) Masks LVH in limb leads
(d) Masks LVH in leads V_5 and V_6

Q10. LBBB per se produces

(a) Deep Q wave in leads V_1 to V_3
(b) Tall R wave in leads I and aVL
(c) Tall R wave in leads V_5 and V_6
(d) Deep S wave in lead aVR

Q11. Which findings support the possibility of LVH in the presence of LBBB?

(a) Left atrial enlargement
(b) QRS duration of 130 ms
(c) Deep S wave in leads V_1 to V_3
(d) Voltage criteria in leads V_5 and V_6

Q12. Which conditions can mask voltage criteria of LVH in leads V_5, V_6?

(a) RBBB
(b) Lateral infarction
(c) Right lateral accessory pathway
(d) None

Q13. Loss of septal q is not seen in

(a) Septal infarction
(b) Hypertension
(c) Aortic regurgitation
(d) Mitral regurgitation

Q14. Normal ventricular activation time in lead V_5 is

(a) < 0.02 s
(b) < 0.03 s
(c) < 0.04 s
(d) > 0.06 s

Q15. Mean frontal plane QRS axis of 0–15° in a case of hypertension suggests

(a) Long-standing hypertension
(b) Accelerated hypertension
(c) Isolated diastolic hypertension
(d) None

Q16. Fragmentation of the QRS complex in a case of hypertension suggests

(a) Myocardial fibrosis
(b) Worse LVH

(c) Severe diastolic dysfunction
(d) All

Q17. An early sign of LVH in hypertension is

(a) R wave amplitude in lead V_6 more than that in lead V_5
(b) Left-axis deviation
(c) T wave in lead V_1 taller than T wave in lead V_6
(d) ST-segment depression in leads I, V_5, and V_6

Q18. Which ECG signs independently correlate with severity of aortic stenosis?

(a) Repolarization abnormalities
(b) Fragmentation of QRS in leads V_4 to V_6
(c) Voltage criteria
(d) All

Q19. Abnormalities of the ST-segment and the T wave are not due to

(a) Repolarization abnormality
(b) Subendocardial ischemia
(c) Low wall tension
(d) Digitatis

Q20. Inversion of the U wave is not seen in

(a) Severe LVH
(b) Acute coronary syndrome
(c) Hyperkalemia
(d) Hypocalcemia

Q21. What is correct about ST-segment and T wave changes?

(a) Can occur without increase in QRS voltage
(b) Suggest increased cardiovascular risk
(c) Suggest early LVH
(d) Usually regress with treatment

Q22. Which ECG signs correlate with severity of aortic stenosis?

(a) Increased posterior forces
(b) Absence of q wave in lead V_6
(c) Inverted T wave in lead V_6
(d) All

Q23. Which leads are opposite to the left ventricle?

(a) I and aVL
(b) II, III, and aVF
(c) V_1 and V_2
(d) V_5 and V_6

Q24. What are the early electrocardiographic changes in systemic hypertension?

(a) Leftward deviation of the mean frontal plane QRS axis
(b) Increased negativity of the P wave in lead V_1
(c) R wave amplitude in lead V_6 more than that in lead V_5
(d) Positive T wave in lead V_1

Q25. What is the common electrocardiographic finding in left ventricular systolic overload?

(a) Mean frontal plane QRS axis around 0°
(b) Deep S wave in leads V_1 and V_2
(c) Inversion of the U wave in leads V_1 and V_2
(d) Inversion of the T waves in leads II, III, aVF

Q26. In left ventricular pressure overload, mean frontal plane QRS axis is usually

(a) Around +60°
(b) Around 0°
(c) Around −60°
(d) Around +120°

Q27. What is the effect of the left anterosuperior fascicular block on electrocardiographic findings of left ventricular pressure overload?

(a) Increase in amplitude of the R wave in leads I and aVL
(b) Increase in amplitude of the R wave in leads II, III, and aVF
(c) Increase in amplitude of R wave in leads V_5 and V_6
(d) Appearance of S wave in leads V_5 and V_6

Q28. In left ventricular pressure overload, T wave amplitude

(a) Decreases in lead V_6
(b) Increases in lead V_6
(c) Decreases in lead V_1
(d) Increases in lead V_1

Q29. In left ventricular pressure overload

(a) q wave in leads V_5 and V_6 decreases
(b) q wave in leads V_5 and V_6 increases
(c) r wave amplitude in lead V_1 increases
(d) r wave amplitude in lead V_1 decreases

Q30. What is the effect of left posteroinferior fascicular block on electrocardiographic findings of left ventricular pressure overload?

(a) Increase in voltage of the R wave in leads I and aVL
(b) Decrease in voltage of the R wave in leads I and aVL
(c) Decrease in amplitude of the R wave in leads V_5 and V_6
(d) Increase in depth of the S wave in leads II, III, and aVF

Q31. In left ventricular pressure overload, ST-segment is

(a) Depressed in lead aVL
(b) Elevated in lead I
(c) Depressed in lead V_1
(d) Elevated in lead V_5

Q32. Development of left bundle branch block in case of left ventricular pressure overload

(a) Reduces height of the R wave in lead V_6
(b) Increases height of the R wave in lead V_1
(c) Increases depth of the S wave in lead V_1
(d) Produces S wave in lead V_6

Q33. Development of Right bundle branch block in case of left ventricular pressure overload produces

(a) Increase in height of the R wave in leads V_5 and V_6
(b) Decrease in depth of the S wave in leads V_1 and V_2
(c) Increase in height of the r wave in lead V_1
(d) Appearance of the S wave in lead V_6

Answers

(1) a (2)a (3) c (4) a (5) d (6) b (7) b (8) c (9) a,d (10) a (11)a, d (12) a,b (13) c,d (14) c (15) a (16) d (17) c (18) d (19) c (20) c,d (21) a, b (22) d (23) c (24) a,b,d (25) a,b (26) b (27) a,d (28) a,d (29) a (30) b,c,d (31) a (32) a,c (33) b,c,d.

References

1. Bacharova L. Electrical and structural remodeling in left ventricular hypertrophy. Ann Noninvasive Electrocardiol. 2007;12:260–73.
2. Hancock EW, Deal BJ, Mirvis DM, Okin P, Kligfield P, Gettes LS. AHA/ACCF/HRS recommendations for the standardization and interpretation of the electrocardiogram. Part V: electrocardiogram changes associated with cardiac chamber hypertrophy. J Am Coll Cardiol. 2009;53:992–1002.
3. Schamroth L. Left ventricular hypertrophy. In: Schamroth C, editor. An introduction to electrocardiography. France: Blackwell Science; 1982. p. 67–77.
4. Mirvis DM, Goldberger AL. Electrocardiography. In: Mann DL, Zipes DP, Libby P, Bonow RO, editors. Braunwald's heart disease. Philadelphia: Elsevier; 2015. p. 114–52.
5. Romhilt DW, Estes EH Jr. A point score system for the ECG diagnosis of left ventricular hypertrophy. Am Heart J. 1968;75:752–8.
6. Sokolow M, Lyon T. The ventricular complex in left ventricular hypertrophy as obtained by unipolar precordial and limb leads. Am Heart J. 1949;37:161–86.
7. Casale PN, Devereux RB, Alonso DR, et al. Improved sex specific criteria of left ventricular hypertrophy for clinical and computer interpretation of electrocardiograms: validation with autopsy findings. Circulation. 1987;75:565–72.
8. Odom H II, Davis JL, Dinh H, Baker BJ, Roberts WC, Murphy ML. QRS voltage measurements in autopsied men free of cardiopulmonary disease: a basis for evaluating the total QRS voltage as an index of left ventricular hypertrophy. Am J Cardiol. 1986;58:801–4.
9. Kumar D, Bajaz R, Chhabra L, Spodick DH. Refinement of total 12 leads QRS voltage criteria for diagnosing left ventricular hypertrophy. W J Cardiovasc Dis. 2013;3:210–4.

10. Rodriguez PL. Usefulness of total 12 lead QRS voltage for determining the presence of left ventricular hypertrophy in systemic hypertension. Am J Cardiol. 1991;68:261–2.
11. Pewsner D, Juni P, Egger M, et al. Accuracy of electrocardiography in diagnosis of left ventricular hypertrophy in arterial hypertension: systematic review. BMJ. 2007;335:711–20.
12. Mirvis DM, Goldberger AL. Electrocardiography. In: Bonow RO, Mann DL, Zipes DP, Libby P, editors. Braunwald's heart disease. St. Louis: Elsevier; 2012. p. 126–65.
13. DeLuna AB, Goldwasser D, Fiol M, Bayes-Genis A. Surface electrocardiography. In: Fuster V, Walsh RA, Harrington RA, editors. Hurst's the heart. New York: McGraw Hill; 2011. p. 307–70.
14. Schamroth L. Systemic hpyertension. In: Schamroth C, editor. An introduction to electrocardiography. France: Blackwell Science; 1982. p. 205–7.
15. Jain R, Singh R, Yamini S, Das MK. Fragmented ECG as a risk marker in cardiovascular diseases. Curr Cardiol Rev. 2014;10:277–86.
16. Steger A, Sinnecker D, Berkefeld A, et al. Fragmented QRS: relevance in clinical practice. Herzschrittmacherther Electrophysiol. 2015;26:234–41.
17. Zhang B, Zhen Y, Shen D, Zhang G. Significance of fragmented QRS complexes for identifying left ventricular hypertrophy in patients with hypertension. Ann Noninvasive Electrocardiol. 2015;20:175–80.
18. Kadi H, Demir AK, Ceyhan K, Damar IH, Karman K, Zorla C. Association of fragmented QRS complexes on ECG with left ventricular diastolic function in hypertensive patients. Arch Turk Soc Cardiol. 2015;43:149–56.
19. Nakamura N, Hirata K, Imanishi T, et al. Electrocardiographic strain and endomyocardial radial strain in hypertensive patients. Int J Cardiol. 2011;150:319–24.
20. Okin PM, Oikarinen L, Viitasalo M, et al. The prognostic value of changes in the electrocardiographic strain pattern during antihypertensive therapy. Circulation. 2009;49:1883–91.
21. Brown DW, Fullon DR. Congenital heart disease in children and adolescents. In: Fuster V, Walsh RA, Harrington RA, editors. Hurst's the heart. New York: McGraw Hill; 2011. p. 1827–83.
22. Otto CM, Bonow RO. Valvular heart disease. In: Mann DL, Zipes DP, Libby P, Bonow RO, editors. Braunwald's heart disease. Philadelphia: Elsevier; 2015. p. 1446–514.
23. Shah ASV, Chin CWL, Vassiliou V, et al. Left ventricular hypertrophy with strain and aortic stenosis. Circulation. 2014;130:1607–16.

Chronic Right Ventricular Volume Overload

33

1 Introduction

Slight or even moderate degree of right ventricular volume overload may not produce abnormalities in the electrocardiogram (ECG). Sensitivity of echocardiography is much higher than electrocardiogram (ECG).

2 Etiology

(a) *Atrial septal defect (ASD)*
- Ostium secundum defect
- Ostium primum defect
- Sinus venosus defect

(b) *Atrial septal defect + rheumatic mitral stenosis (Lutembacher's syndrome)*

(c) *Atrial septal defect + mitral regurgitation*
- Rheumatic
- Cleft in mitral leaflet in primum ASD
- Mitral valve prolapse

(d) *Anomalous pulmonary venous drainage*
- Partial
- Total

(e) *Nonpulmonary artery hypertensive pulmonary regurgitation*
- Congenital pulmonary valve regurgitation
- Idiopathic dilatation of the pulmonary artery

Pulmonary artery hypertensive pulmonary regurgitation produces pressure overload (systolic overload) plus volume overload (diastolic overload).

(f) *Nonpulmonary artery hypertensive tricuspid regurgitation*
- Tricuspid valve prolapse
- Isolated right ventricular dilated cardiomyopathy
- Tricuspid valve endocarditis
- Ebstein anomaly
- Isolated rheumatic tricuspid value disease

3 Electrocardiogram

3.1 Small ASD with Small Left to Right Shunt

ECG is usually normal. Some patients may have an incomplete RBBB pattern (rsr′ pattern in lead V_1 without prolongation of QRS duration). T waves are inverted in leads V_1 and V_2.

S. Mittal, *Insights into Electrocardiograms with MCQs*,
https://doi.org/10.1007/978-981-99-0127-2_33

3.2 Large Secundum ASD with Large Left to Right Shunt

- Mean frontal plane QRS axis is +90° or greater [1]. Lead V_1 shows rsR′ configuration (Fig. 33.1). R′ is usually less than 7 mm [2].

Terminal R′ wave in lead V_1 is because the terminal QRS vector is directed to the right in horizontal plane (Fig. 33.2).

Two factors contribute to this rightward terminal vector.

(a) Delayed conduction in the right bundle branch (RBB) [2]. It is due to lengthening of the RBB due to dilatation of the right ventricle (Fig. 33.3).
(b) Hypertrophy of crista supraventricularis (Fig. 33.4).

In chronic right ventricular volume overload, hypertrophy involves crista supraventricularis. In systolic overload, hypertrophy involves the right ventricular free wall. Crista supraventricularis is among the last regions of the right ventricle to depolarize.

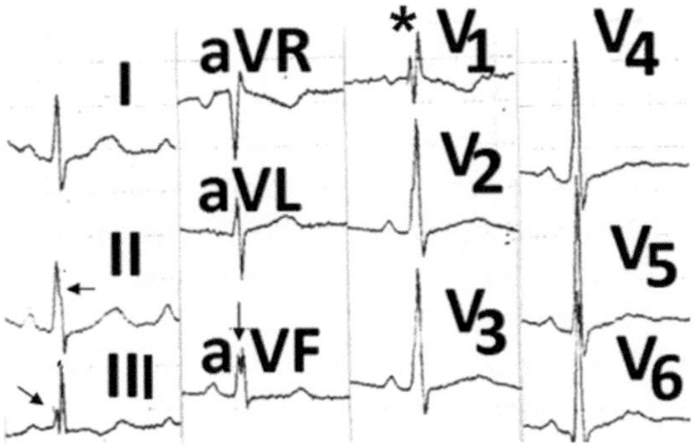

Fig. 33.1 ECG from a case of secundum ASD showing rsr′ pattern in lead V_1 (marked *). Notch near the apex of the R wave in leads II, III, aVF (marked arrows—crochetage) supports the diagnosis of secundum ASD. (Reproduced with permission from Mittal SR. Chronic Right Ventricular Volume Overload. Cardiology Today-2016;20:183–186)

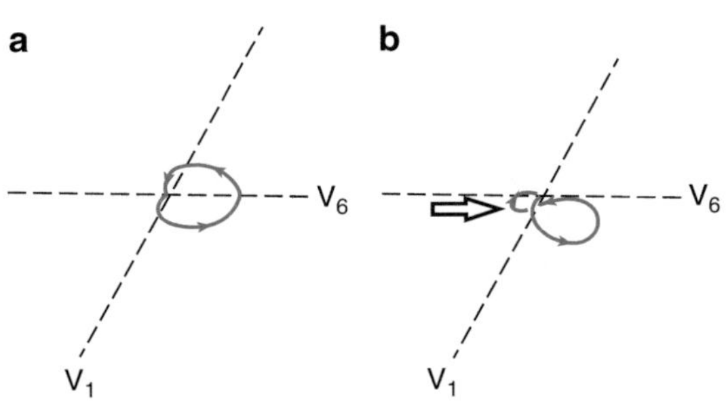

Fig. 33.2 Diagram showing QRS vector in the horizontal plane is (**a**) normal (**b**) in case of secundum ASD showing terminal vector directed to the right (marked⇨). (Reproduced with permission from Mittal SR. Chronic Right Ventricular Volume Overload. Cardiology Today-2016;20:183–186)

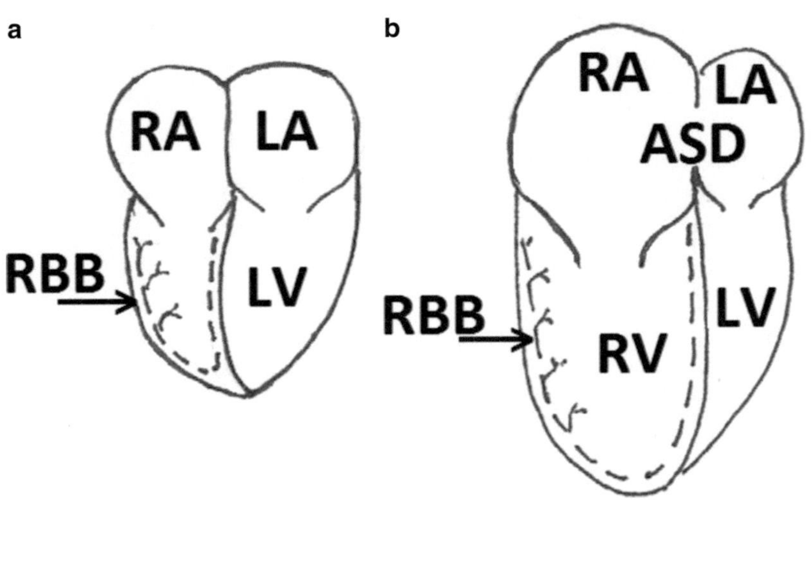

Fig. 33.3 Diagram showing (**a**) RBB in normal person and (**b**) lengthening of RBB in ASD. *RA* right atrium; *LA* left atrium; *RV* right ventricle; *LV* left ventricle; *ASD* atrial septal defect. (Reproduced with permission from Mittal SR. Chronic Right Ventricular Volume Overload. Cardiology Today-2016;20:183–186)

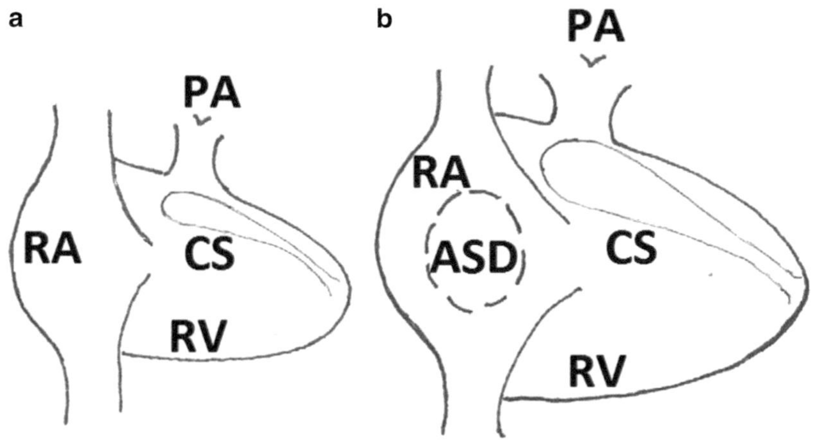

Fig. 33.4 Diagram showing (**a**) crista supraventricularis (CS) in normal person and (**b**) hypertrophy of crista supraventricularis in ASD. *RA* right atrium; *RV* right ventricle; *PA* pulmonary artery; *ASD* atrial septal defect. (Reproduced with permission from Mittal SR. Chronic Right Ventricular Volume Overload. Cardiology Today-2016;20:183–186)

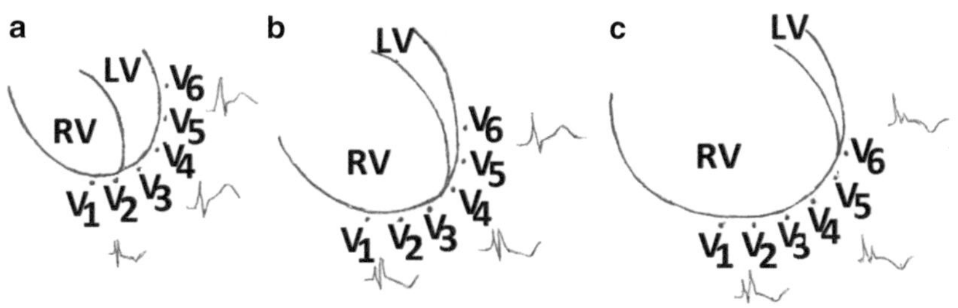

Fig. 33.5 Diagram showing the effect of volume of left to right shunt on the precordial leads: (**a**) small shunt showing T wave inversion in leads V_1 and V_2; (**b**) moderate shunt showing T wave inversion up to lead V_4; (**c**) large shunt showing T wave inversion up to lead V_6. (Reproduced with permission from Mittal SR. Chronic Right Ventricular Volume Overload. Cardiology Today-2016;20:183–186)

- The right ventricular volume depends on the magnitude of the left to right shunt. Larger the right ventricular volume, greater the clockwise rotation in the horizontal plane. In moderate shunts with moderate right ventricular dilatation, transition zone is shifted to left and T wave inversion is seen up to lead V_4. In very large shunts with gross right ventricular dilatation, transition zone is shifted further to left and T wave inversion is seen up to leads V_5 and V_6 [3] (Fig. 33.5).
- Beyond fourth decade, some patients can present with atrial flutter or fibrillation [4]. These arrhythmias arise from the dilated right atrium.

3.3 Secundum ASD with Pulmonary Artery Hypertension (PAH)

Development of PAH results in

- Rightward deviation of the frontal plane P wave axis (P- pulmonale) [5].
- Increasing rightward deviation of the mean frontal plane QRS axis.
- Increasingly tall R wave in lead V_1 [2]. S wave shows progressive diminution. Some patients may show qR configuration in lead V_1 due to gross dilatation of the right atrium.

Representative ECG tracings are shown in Figs. 33.6 and 33.7.

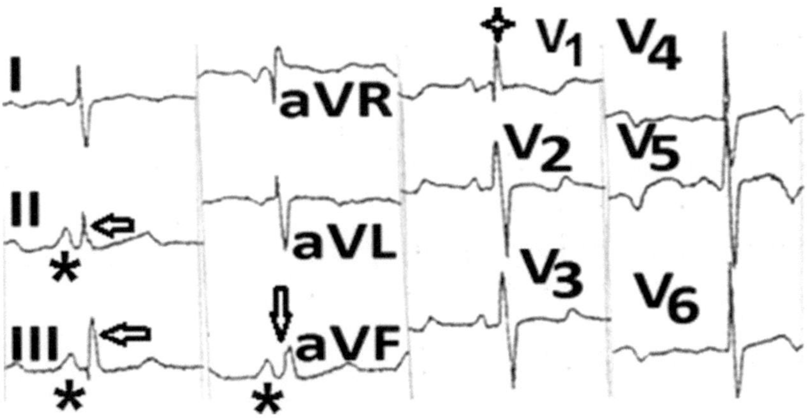

Fig. 33.6 Electrocardiogram from a patient with secundum ASD with early pulmonary artery hypertension showing mild right-axis deviation in the frontal plane (lead I showing dominant S wave), P- pulmonale (tall and peaked P waves in leads II, III and aVF (marked *), and prominent R′ wave in lead V_1 (marked✧). Notch near the top of the R wave in leads II, III, and aVF (marked arrows—crochetage) supports the diagnosis of secundum ASD. (Reproduced with permission from Mittal SR. Chronic Right Ventricular Volume Overload. Cardiology Today-2016;20:183–186)

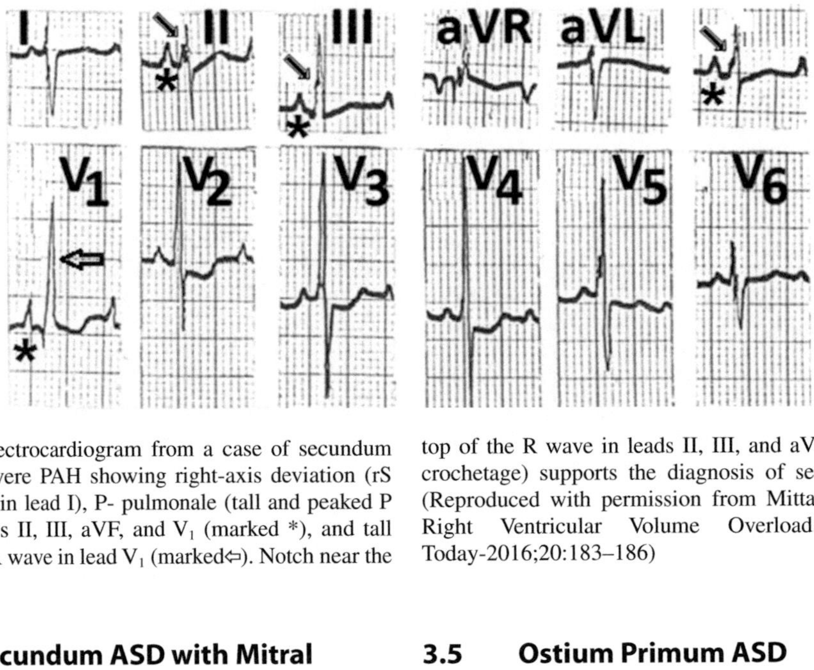

Fig. 33.7 Electrocardiogram from a case of secundum ASD with severe PAH showing right-axis deviation (rS configuration in lead I), P- pulmonale (tall and peaked P waves in leads II, III, aVF, and V_1 (marked *), and tall monophasic R wave in lead V_1 (marked⇦). Notch near the top of the R wave in leads II, III, and aVF (marked—⇘ crochetage) supports the diagnosis of secundum ASD. (Reproduced with permission from Mittal SR. Chronic Right Ventricular Volume Overload. Cardiology Today-2016;20:183–186)

3.4 Secundum ASD with Mitral Valve Disease

Biatrial enlargement (tall and peaked P waves in leads II and aVF with prominent negative component of P wave in lead V_1) suggests the possibility of concomitant mitral valve disease (Fig. 33.8). ASD with mitral regurgitation presents with biventricular enlargement.

3.5 Ostium Primum ASD

Maldevelopment of the left anterosuperior fascicle results in leftward deviation of the mean frontal plane QRS axis (Fig. 33.9) [1].

PR interval may be prolonged [6].

If there is mitral regurgitation due to cleft in mitral leaflet, there may be additional left atrial enlargement (prominent negative P wave in lead V_1) (Fig. 33.10).

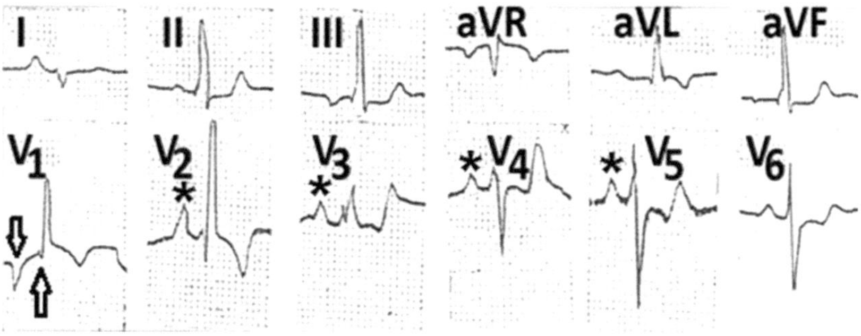

Fig. 33.8 Electrocardiogram from a case of secundum ASD with rheumatic mitral stenosis with pulmonary artery hypertension. It shows right-axis deviation (dominant S wave in lead I) with qR configuration in lead V_1 (marked ⇧). Lead V_1 shows prominent negative P wave (marked ⇩) suggestive of left atrial enlargement. Leads V_2 to V_5 show tall and peaked P waves (marked *) suggestive of right atrial enlargement. (Reproduced with permission from Mittal SR. Chronic Right Ventricular Volume Overload. Cardiology Today-2016;20:183–186)

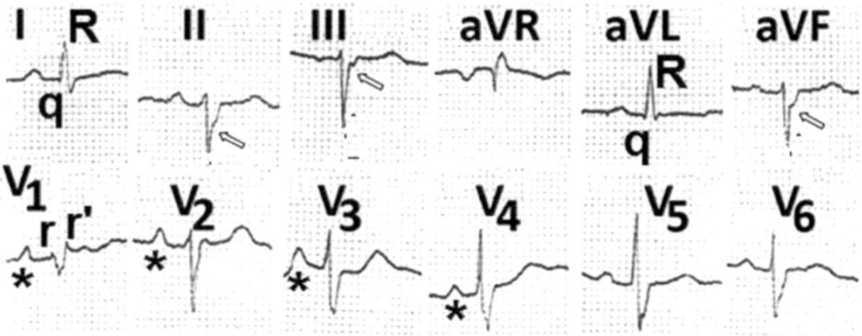

Fig. 33.9 Electrocardiogram from a case of primum ASD showing left anterosuperior fascicular block in limb leads and rsr′ configuration in lead V_1. Leads V_1 to V_6 show peaked P waves (marked *) suggestive of right atrial enlargement. Initial q waves in leads I and aVL suggest counterclockwise depolarization in frontal plane. Notch on the upstroke of the S wave in leads II, III, and aVF (marked ⇘) supports the diagnosis of primum ASD. (Reproduced with permission from Mittal SR. Chronic Right Ventricular Volume Overload. Cardiology Today-2016;20:183–186)

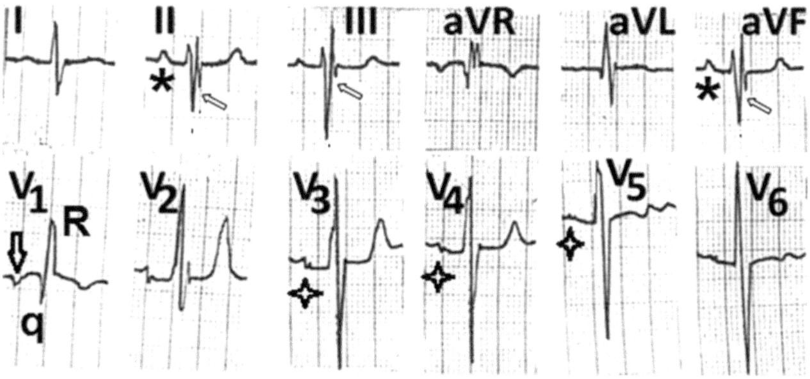

Fig. 33.10 Electrocardiogram from a case of primum ASD with mitral regurgitation and pulmonary artery hypertension. It shows left-axis deviation and left atrial enlargement (prominent negative P wave in lead V_1 marked ⇩ and notched P waves in leads V_3 to V_5 marked✧). Lead V_1 shows qR configuration suggestive of pulmonary artery hypertension and right atrial enlargement. Peaked P waves in leads II and aVF (marked *) also suggest right atrial enlargement. Notch on the upstroke of S wave in leads II, III, and aVF (marked ⇘) supports the diagnosis of primum ASD. (Reproduced with permission from Mittal SR. Chronic Right Ventricular Volume Overload. Cardiology Today-2016;20:183–186)

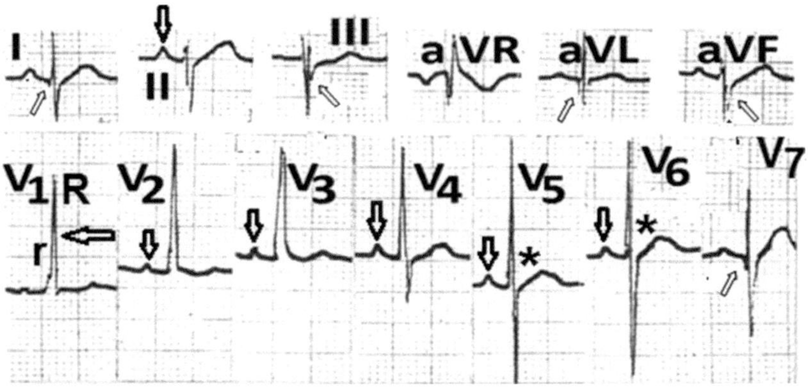

Fig. 33.11 Electrocardiogram from a case of primum ASD with pulmonary artery hypertension. Peaked P waves in leads II and V_2 to V_6 (marked ⇩) suggest right atrial overload. Limb leads show left-axis deviation. Lead V_1 shows rR′ (marked ⇦) pattern with loss of s wave suggestive of right ventricular hypertrophy. Leads V_5 and V_6 show large equidiphasic QRS complex (marked *) suggestive of biventricular enlargement. Lead V_7 shows q wave (marked ⇗) suggestive of left ventricular overload in the presence of right ventricular hypertrophy. Initial q wave in leads I and aVL (marked ⇗) suggests counterclockwise depolarization in frontal plane. Leads III and aVF show notch on the upstroke of S wave (marked ⇖) suggestive of primum ASD. (Reproduced with permission from Mittal SR. Chronic Right Ventricular Volume Overload. Cardiology Today-2016;20:183–186)

Mitral regurgitation may also produce concomitant left ventricular volume overload. Left to right shunt through concomitant VSD can also produce additional left ventricular volume overload. Pulmonary artery hypertension is common, resulting in additional right ventricular systolic overload (tall R wave with T wave inversion in leads V_1 and V_2) with volume overload of the left ventricle (q wave with prominent R wave in leads V_5 and V_6) (Fig. 33.11).

3.6 Sinus Venosus (Superior Vena Cava Type) Defect

Such defects produce negative P waves in inferior leads (leads II, III, and aVF) due to low atrial pacemaker. It may be due to abnormalities of the sinoatrial node.

3.7 Anomalous Pulmonary Venous Drainage

ECG is normal in anomalous connection of a single pulmonary vein. Anomalous connection of more pulmonary veins results in right ventricular volume overload and incomplete right bundle branch block pattern (rsr′) in lead V_1.

3.8 Nonpulmonary Artery Hypertensive Pulmonary Regurgitation [7]

– *Congenital pulmonary valve regurgitation*

Electrocardiogram shows incomplete right bundle branch block pattern with positive T wave in right precordial leads.

– *Idiopathic dilatation of the pulmonary artery*

ECG is usually normal. Sometimes, rsr′ pattern may be present. T waves are not positive in leads V_1 to V_3.

4 Differential Diagnosis

Other causes of right ventricular dilatation on echocardiography include the following.

4.1 Arrhythmogenic Right Ventricular Dysplasia [8]

It is characterized by monomorphic ventricular tachycardia of left bundle branch block morphology with superior axis. During sinus rhythm, there is T wave inversion in precordial leads V_1 to V_3. Occasionally, an "Epsolin wave" can be present at the end of the QRS complex and at the beginning of the ST-segment. It is most obvious in leads V_1 and V_2 in the absence of right bundle branch block. It is a specific finding but is not sensitive.

4.2 Isolated or Dominant Right Ventricular Dilated Cardiomyopathy

It is rare. ECG may show right atrial enlargement, right bundle branch block, nonspecific ST-segment and T wave changes, and ventricular arrhythmias.

4.3 Isolated Tricuspid Regurgitation Without Pulmonary Artery Hypertension

Isolated mild tricuspid regurgitation is not uncommon. It does not produce significant right ventricular volume overload and the electrocardiogram is normal. Moderately severe tricuspid regurgitation can produce significant right ventricular volume overload. Electrocardiogram shows evidence of right atrial enlargement and rsr′ pattern in lead V_1. Ebstein anomaly shows right atrial enlargement with low-voltage fragmented QRS in lead V_1.

Summary

- The most common cause of right ventricular volume overload is secundum atrial septal defect.
- Rhythm is usually sinus. Sinus arrhythmia is usually absent due to similar filling of the right ventricle during inspiration and expiration. Patients aged more than 40 years may show atrial fibrillation, atrial flutter, or supraventricular tachycardia arising from the dilated right atrium.
- P waves are usually normal and positive in leads II, III, and aVF. They may be peaked but are usually not tall. P wave may be broad if terminal portion is written by enlarged right atrium.
- PR interval is normal but may be prolonged in some familial cases and some cases of Holt–Oram syndrome.
- Delayed depolarization of the hypertrophied right ventricular outflow tract results in a terminal r′ in lead V_1 and a terminal s wave in leads V_5 V_6. However, a similar rsr′ pattern in lead V_1 can be seen in some normal persons when lead V_1 is placed higher than normal or in some cases of chest deformities. Mild slurring of the terminal R′ wave and notch near the apex of the R wave in inferior leads (crochetage) helps differentiating changes due to atrial septal defect from normal variants. QRS duration is normal (usually less than 0.10 s) except in the elderly where the pattern may resemble a complete right bundle branch block. The mean frontal plane QRS axis is normal (usually between +60° and + 110°). Depolarization is clockwise with initial q wave in leads II, III, and aVF. There is no q wave in leads V_5 and V_6 as the shunt does not involve the left ventricle. With gross right ventricular dilatation, the transition zone in precordial leads may shift to the left. With increasing pulmonary vascular resistance, P waves become peaked, mean frontal plane QRS axis shift to the right, terminal R wave in lead V_1 becomes taller, and prominence of S wave in leads V_5 and V_6 increases. Configuration in lead V_1 progressively changes from rsr′ to rR′, R, and, finally, to qR (q wave is because of right atrial enlargement).
- Uncommon causes of chronic right ventricular volume overload include primum atrial septal defect, atrioventricular septal defect, or common atrium. Rarely isolated tricuspid

regurgitation can cause right ventricular volume overload.

- Electrocardiogram is normal in patent foramen ovale or small ASD with less than 2:1 left to right shunt.
- In partial anomalous pulmonary venous drainage, electrocardiographic findings resemble secundum ASD.
- In superior vena cava sinus venosus defect, P waves are inverted in leads II, III, and aVF and upright in lead aVL. Other electrocardiographic findings are like secundum atrial septal defect.
- Left-axis deviation, counterclockwise depolarization (initial q in leads I and aVL), and notch on the upstroke of the S wave in leads II, III, and aVF suggest the diagnosis of ostium primum atrial septal defect or common atrium. QRS configuration in lead V_1 is similar to that in ostium secundum atrial septal defect.
- Isolated tricuspid regurgitation causes right atrial enlargement.

MCQs

Q1. Which method is the most sensitive in detecting mild right ventricular volume overload?

(a) ECG
(b) Skiagram of chest
(c) Echocardiography
(d) Vectorcardiography

Q2. Small secundum ASD with the small left to right shunt does not produce

(a) rsr′ in lead V_1
(b) Normal frontal plane QRS axis
(c) rsR′ in lead V_1
(d) qR in lead V_1

Q3. In large secundum ASD with large left to right shunt, mean frontal plane QRS axis is between

(a) 0° and − 30°
(b) 0° and + 60°
(c) +90° and + 120°
(d) +150° and 180°

Q4. Large secundum ASD with the large left to right shunt without pulmonary artery hypertension can produce

(a) qR in lead V_1
(b) rR′ in lead V_1
(c) rsR′ in lead V_1
(d) T wave inversion in leads V_1 to V_4

Q5. The onset of pulmonary vascular occlusive diseases in a case of atrial septal defect is suggested by

(a) Rightward deviation of frontal plane P wave axis
(b) Increasing leftward deviation of mean frontal plane QRS axis
(c) Increasingly tall R wave in lead V_1
(d) Progressive increase in S wave depth in lead V_1

Q6. Which ECG finding is suggestive of Lutembacher's syndrome?

(a) Right atrial enlargement
(b) Biatrial enlargement
(c) rsR′ in lead V_1
(d) qR in lead V_1

Q7. Ostium primum ASD is characterized by

(a) Left anterosuperior fascicular block
(b) Left posteroinferior fascicular block
(c) Right bundle branch block
(d) Left bundle branch block

Q8. Ostium primum ASD with mitral regurgitation produces

(a) Biatrial enlargement
(b) Biventricular volume overload
(c) Marked clockwise rotation in horizontal plane
(d) Atrioventricular block

Q9. Ostium primum ASD with VSD without pulmonary hypertension produces

(a) Biatrial enlargement
(b) Q with prominent R in leads V_5 and V_6
(c) Tall R with T wave inversion in leads V_1 and V_2
(d) Atrioventricular block

Q10. In the case of right ventricular volume overload, the superior P wave axis suggests

(a) Primum ASD
(b) Lutembacher's syndrome
(c) Superior vena cava-type sinus venosus defect
(d) Inferior vena cava-type sinus venosus defect

Q11. Anomalous drainage of the left upper pulmonary vein into the coronary sinus produces

(a) Normal ECG
(b) rsR′ in V_1
(c) Complete RBBB
(d) Left axis deviation

Q12. Idiopathic dilatation of the pulmonary artery does not produce

(a) rsr′ complex in V_1
(b) Upright T wave in leads V_1 and V_2
(c) Inverted T waves in leads V_1 and V_2
(d) qR in V_1

Q13. Which of the following conditions does not cause significant right ventricular volume overload?

(a) Partial anomalous pulmonary venous drainage
(b) Superior vena cava-type sinus venosus atrial septal defect
(c) Coronary sinus atrial septal defect
(d) None

Q14. Secundum atrial septal defect with 1.8:1 left to right shunt produces

(a) Counterclockwise depolarization
(b) Tall peaked P waves
(c) rsR′ pattern in lead V_1
(d) None

Q15. Normal electrocardiogram excludes any possibility of

(a) Patent foramen ovale
(b) Small secundum ASD
(c) Ostium primum ASD
(d) Common atrium

Q16. rsr′ pattern in lead V_1 can be seen

(a) In some normal persons
(b) In secundum ASD with less than 2:1 left to right shunt
(c) If lead V1 is placed in higher intercostal space
(d) All

Q17. In uncomplicated secundum ASD with small left to right shunt, P wave is usually

(a) Normal
(b) Tall and peaked
(c) Broad
(d) Biphasic

Q18. In case of right ventricular volume overload, P waves inverted in leads II, III, and aVF suggest

(a) Secundum ASD
(b) Primum ASD
(c) Superior vena cava-type sinus venosus defect
(d) Atrioventricular septal defect

Q19. In case of right ventricular volume overload, counterclockwise depolarization in frontal plane suggests

(a) Secundum ASD
(b) Superior vena cava sinus venosus type ASD
(c) Ostium primum ASD
(d) Partial anomalous pulmonary venous connection

Q20. In case of right ventricular volume overload, left-axis deviation suggest possibility of

(a) Superior vena cava-type sinus venous defect
(b) Ostium primum type ASD

(c) Common atrium
(d) Total anomalous pulmonary venous drainage

Q21. "Crochetage" in case of right ventricular volume overload suggests the possibility of

(a) Primum ASD
(b) Secundum ASD
(c) Total anomalous pulmonary venous drainage
(d) Superior vena cava-type sinus venosus ASD

Q22. Classical secundum ASD with left to right shunt shows

(a) Normal P wave
(b) Normal PR interval
(c) Inferior QRS axis
(d) qR in lead V_1

Q23. Right ventricular volume overload with prolonged PR interval can be seen in

(a) Familial ASD
(b) Holt–Oram syndrome
(c) Ostium primum ASD
(d) All

Q24. In secundum ASD, lead V_5 shows

(a) QRS
(b) qRS
(c) QS
(d) RS

Q25. In gross right ventricular volume overload, transition zone in precordial leads is seen between leads

(a) V_1 and V_2
(b) V_2 and V_3
(c) V_3 and V_4
(d) V_4 and V_5

Q26. In case of right ventricular volume overload, notch on the upstroke of the S wave in leads II, III, and aVF suggests diagnosis of

(a) Ostium secundum ASD
(b) Ostium primum ASD
(c) Common atrium
(d) Total anomalous pulmonary venous connection

Q27. Large secundum ASD in a child shows

(a) Respiratory sinus arrhythmia
(b) Atrial fibrillation
(c) Both
(d) None

Q28. In case of right ventricular volume overload, increasing pulmonary vascular resistance is suggested by

(a) Increasing height of P wave
(b) Rightward shift of mean frontal plane QRS axis
(c) Decreasing depth of s wave and increasing height of terminal R in lead V_1
(d) Prolongation of PR interval

Q29. Which electrocardiographic finding can be seen in lead V_1 in severe low-pressure isolated tricuspid regurgitation?

(a) Peaked P wave
(b) q wave
(c) rsr′ pattern
(d) All

Q30. In right ventricular volume overload due to secundum ASD, lead aVR shows

(a) QS
(b) Qr
(c) qR
(d) Q with slurred R

Q31. In superior vena cava-type sinus venosus ASD, P wave is

(a) Inverted in leads II, III, and aVF
(b) Positive in lead aVL
(c) Positive in lead aVR
(d) Positive in leads II, III, and aVF

Q32. In case of right ventricular volume overload with clockwise depolarization in frontal plane, positive broad and notched P wave in lead II and broad deep terminal negative P wave in lead V_1 should suggest the possibility of

(a) Lutembacher's syndrome
(b) Superior vena cava-type sinus venosus ASD with cleft mitral leaflet
(c) Common atrium
(d) Primum ASD with congenital mitral regurgitation

Answers

(1) c (2)c,d (3) c (4) c, d (5) a,c (6) b (7) a (8) a,b,c (9) a,b (10) c (11)a (12)b, d (13) a (14) d (15) c, d (16) d (17) a (18) c (19) c (20) b,c (21) b (22) a,b,c (23) d (24) d (25) d (26) b,c (27) d (28) a,b,c (29) d (30) d (31) a,b,c (32) a

References

1. Brown DW, Fulton DR. Congenital heart disease in children and adolescents. In: Fuster V, Walsh RA, Harrington RA, editors. Hurst's the heart. New York: McGraw Hill; 2011. p. 1827–83.
2. Webb GD, Smalthorn JF, Therrien J, Redington AN. Congenital heart disease. In: Mann DL, Zipes DP, Libby P, Bonow RO, editors. Braunwald's heart disease. Philadelphia: Elsevier; 2015. p. 1391–445.
3. Choudhri M. ECG interpretation in congenital heart disease. In: Gua S, editor. Cardiology update. New Delhi: Jaypee; 2015. p. 942–56.
4. Dash PK, Satpathy M. Atrial septal defect (secundum type). In: Satpathy M, editor. Clinical diagnosis of congenital heart disease. New Delhi: Jaypee; 2008. p. 74–82.
5. Schamroth L. Congenital heart disease. In: Schamroth C, editor. An introduction to electrocardiography. France: Blackwell Science; 1982. p. 67–77.
6. Maheshwari S. Atrial septal defect (primum type). In: Satpathy M, editor. Clinical diagnosis of congenital heart disease. New Delhi: Jaypee; 2008. p. 83–6.
7. Desai N, Kumar R, Mahadevan C, Prakash VS. Congenital pulmonary regurgitation. In: Satpathy M, editor. Clinical diagnosis of congenital heart disease. New Delhi: Jaypee; 2008. p. 208–12.
8. Falk RH, Hershberger RE. The dilated, restrictive and infiltrative cardiomyopathies. In: Mann DL, Zipes DP, Libby P, Bonow RO, editors. Braunwald's heart disease. Philadelphia: Elsevier; 2015. p. 1551–73.

34 Chronic Right Ventricular Pressure Overload (Systolic Overload)

ECG is not sensitive in detecting mild to moderate right ventricular (RV) pressure overload. This is because, for the right ventricular hypertrophy (RVH) to be manifested on ECG, it must be severe enough to overcome the masking effect of the dominant left ventricular (LV) forces [1]. RV mass never overcomes mass of LV [2]. ECG changes produced by right ventricular hypertrophy (RVH) are more as a result of delay in activation of the right ventricle rather than an increase in RV mass [2]. Echocardiographic diagnosis of RVH is also difficult because of its triangular geometry and difficulty in evaluation of thickness of right ventricular free wall.

1 Etiology

1.1 Pulmonary Artery Hypertension (PAH)

(a) Primary PAH
(b) PAH secondary to
    - Lung diseases (cor pulmonale).
        - Chronic obstructive airway disease (COAD).
        - Nonobstructive pulmonary diseases.
    - Pulmonary venoocclusive disease
    - Diseases of left side of the heart
    - Left to right shunts
    - Chronic/recurrent pulmonary thromboembolism.

1.2 Right Ventricular Outflow Tract Obstruction

Infundibular stenosis/pulmonary valvular stenosis/supravalvular stenosis/pulmonary artery branch stenosis or any combination of the above with or without other congenital defects.

1.3 Congenital Heart Disease with Right Ventricle Attached to the Aorta

Right ventricle is exposed to systemic pressure in

(a) D-Transposition of great arteries
(b) Double outlet right ventricle
(c) Normal neonates have right ventricular dominance due to normally higher pulmonary vascular resistance (PVR) of intrauterine life. ECG gradually normalizes with normalization of PVR.

S. Mittal, *Insights into Electrocardiograms with MCQs*,
https://doi.org/10.1007/978-981-99-0127-2_34

2 Electrocardiographic Changes

Several ECG findings have been described in association with RVH depending on the underlying cause [2]. General findings are discussed.

(a) Initially, there is rightward shift of the mean frontal plane QRS axis [3] (Fig. 34.1).

It manifests as prominent S wave in lead I and prominent R wave in lead aVR.

Other causes of right-axis deviation (RAD) include

- Inferoposterior fascicular block
- Vertical heart
- Lateral infarction [2]

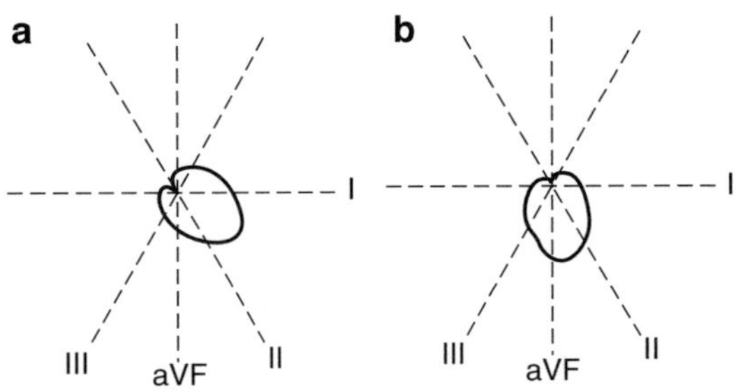

Fig. 34.1 Diagram showing QRS loops in frontal plane (**a**) in a normal person and (**b**) rightward shift in chronic right ventricular pressure overload. (Reproduced with permission from Mittal SR. Chronic Right Ventricular Pressure Overload. Cardiology Today-2016;20:215–221)

Patients with Noonan syndrome, atrioventricular septal defect, and some cases of COAD can have left-axis deviation.

(b) Subsequently, there is
- Anterior shift of the QRS axis in horizontal plane [3] (Fig. 34.2) producing
  - Increasing R wave amplitude in lead V_1
  - Increasing depth of the S wave in left-sided leads, i.e., leads I, aVL, V_5, and V_6.
- At times, the QRS loop is directed toward right and posteriorly (Fig. 34.3) producing negative QRS deflection in anterior chest leads.
- Rightward deviation of the P wave axis [3] producing tall and peaked P waves in leads II, III, and aVF. At times, the P waves may not be tall but are usually peaked. This finding is consistent with right atrial overload. It is seen even when there is over inflation of lungs as in cases of COAD [4].
- Secondary ST-segment and T wave abnormalities [3]. ST-segment depression and T wave inversion are present in leads V_1 and V_2.

(c) Lastly, there is an S_1, Q_3 pattern or S_I, S_{II}, S_{III} pattern—It reflects rightward and superiorly directed QRS loop in frontal plane (Fig. 34.4).

S_I, S_{II}, S_{III} pattern can also occur secondary to positional changes or it may simply be

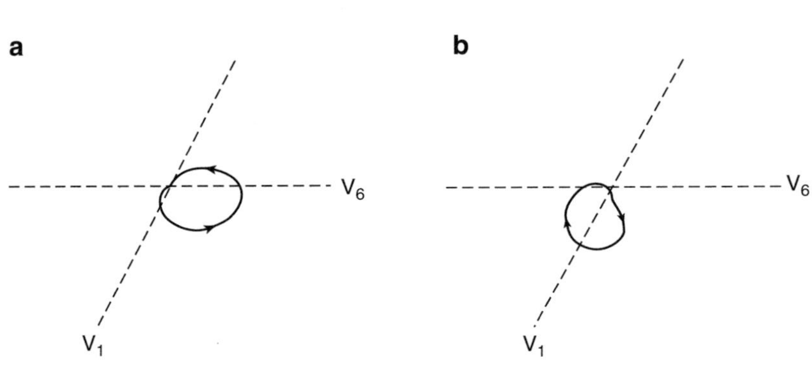

Fig. 34.2 Diagram showing QRS loops in the horizontal plane (**a**) in a normal person and (**b**) rightward and anterior shift in chronic right ventricular pressure overload. (Reproduced with permission from Mittal SR. Chronic Right Ventricular Pressure Overload. Cardiology Today-2016;20:215–221)

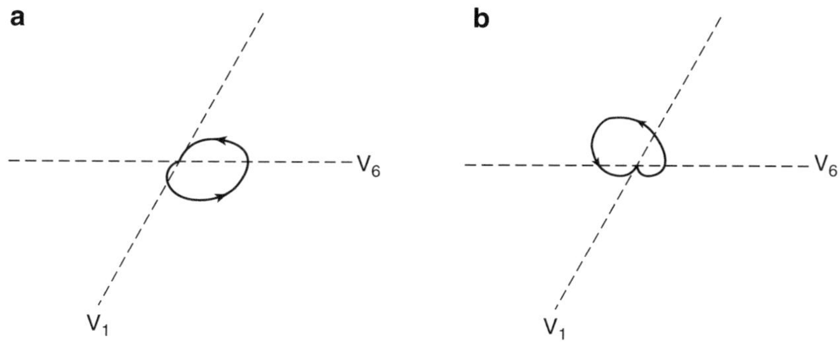

Fig. 34.3 Diagram showing QRS loops in horizontal plane (**a**) in normal person and (**b**) rightward and posterior shift in some cases of chronic right ventricular pressure overload. (Reproduced with permission from Mittal SR. Chronic Right Ventricular Pressure Overload. Cardiology Today-2016;20:215–221)

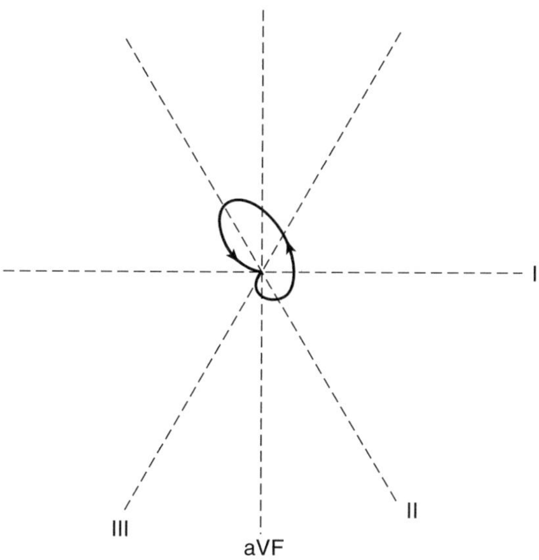

Fig. 34.4 Diagram showing QRS loop-directed rightward and superiorly in frontal plane. (Reproduced with permission from Mittal SR. Chronic Right Ventricular Pressure Overload. Cardiology Today-2016;20:215–221)

a normal variant. Abnormal P wave favors diagnosis of RVH [2].

(d) QRS duration is not prolonged [3]. However, right bundle branch block (RBBB) may be present in some cases as a part of aging [5].

Classical criteria have high specificity but low sensitivity in general population because

(a) Criteria have been developed from patients showing the presence or absence of RVH on autopsy. However, in clinical setting, RVH does not develop overnight. Hypertrophy increase gradually [6, 7]. Various criteria mentioned above may not defect early stages of RVH.

(b) Other studies are based on a clinical diagnosis of conditions known to have RVH [7]

(c) Several etiologies have their own direct influence on the electrocardiogram that is independent of RVH. Therefore, diagnosis of early right ventricular hypertrophy can be made only by comparison with a previously recorded electrocardiogram. Any of the following electrocardiographic findings should suggest diagnosis of early right ventricular hypertrophy:
   - Any increase in amplitude or peaking of P waves in leads II, III, aVF, and/or leads V_1 to V_4.
   - Any magnitude of rightward shift of the mean frontal plane QRS axis in the absence of any other etiology, e.g., left inferoposterior fascicular block or lateral infarction.
   - Any magnitude of increase in clockwise rotation in horizontal plane (any shift of transition zone toward left-sided chest leads).
   - Any increase in amplitude of the R wave in lead V_1 and lead aVR.
   - Any decrease in depth of the S wave in lead V_1
   - Any increase in depth of the S wave in lead V_5 and V_6

It is important to remember that these criteria can be used only in electrocardiogram recorded with proper standardization, correctly placed chest leads, and absence of development of any new intraventricular conduction defects, myocardial infarction, or preexcitation.

3 ECG Findings Depending on the Underlying Disease

3.1 Chronic Obstructive Airway Disease (COAD)

It affects the electrocardiogram in the following ways:

(a) Hyperinflated lungs cover the heart, diminishing the transmission of electric potentials to the surface electrodes (Fig. 34.5).

This results in low voltage of QRS deflections. Other causes of low-voltage QRS include incorrect standardization, obesity, myxoedema, pericardial effusion, and pericardial constriction. These conditions should be excluded. Hyperinflation with interposition of left lung and chest leads also results in slow progression or nonpregression of r wave in leads V_1 to V_3.

(b) Hyperinflation of the lungs (emphysema) also results in downward displacement of the diaphragm and the heart. This results in
- Vertical position of the heart
- Low voltage of the QRS complexes in the limb leads, especially in leads I and aVL.
- Rightward shift of the mean frontal plane QRS axis (Fig. 34.6).

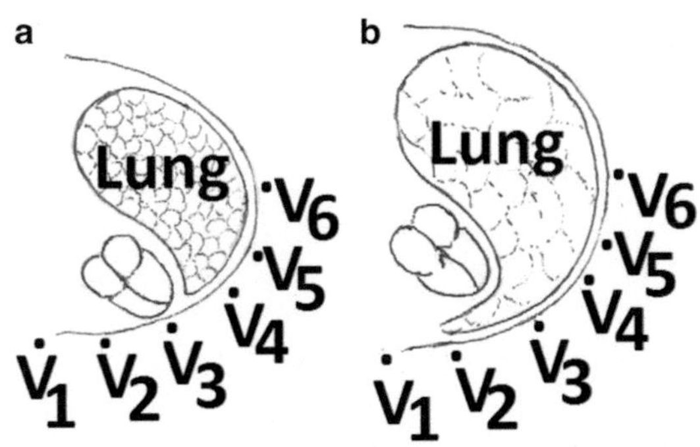

Fig. 34.5 Diagram showing the position of the left lung, heart, and chest leads (**a**) in normal person and (**b**) in hyperinflation of lungs. (Reproduced with permission from Mittal SR. Chronic Right Ventricular Pressure Overload. Cardiology Today-2016;20:215–221)

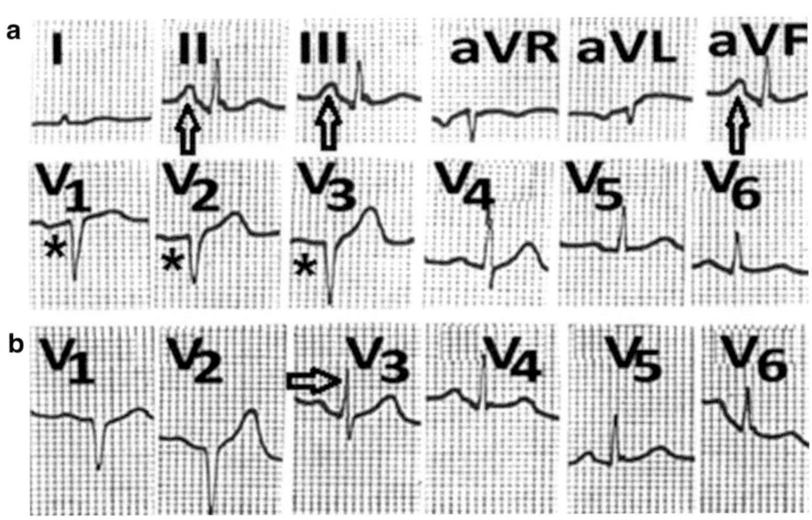

Fig. 34.6 Electrocardiogram from a patient with chronic obstructive airway disease showing (**a**) low voltage in limb leads, rightward deviation of mean frontal plane QRS axis, prominent P waves in leads II, III, aVF (marked ⇧), and nonprogression of the r wave in leads V_1 to V_3 (marked *); (**b**) precordial leads recorded two intercostal space lower showing the appearance of the r wave in lead V_3 (marked ⇨). (Reproduced with permission from Mittal SR. Chronic Right Ventricular Pressure Overload. Cardiology Today-2016;20:215–221)

As the heart becomes more vertical, recording leads V_1 to V_3 in one or two intercostal space lower than their conventional position shows normal progression of r wave in leads V_1 to V_3.

- Uncommonly the mean frontal plane QRS axis is shifted toward left (Fig. 34.7).

The mechanism of such left axis is not clear. It is possible that there is additional anterosuperior fascicular block.

- Altered position of cardiac chambers in relation to conventional position of precordial electrodes. Conventionally placed precordial leads become relatively higher due to downward displacement of the heart (Fig. 34.8).

It results in slow or nonprogression of the r wave in precordial leads [8]. Recording precordial leads two intercostal spaces lower can reveal r wave (Fig. 34.6). S wave persists in all precordial leads due to right ventricular dilatation and clockwise rotation. At times, lead V_1 may show qR pattern due to right atrial enlargement (Fig. 34.9).

- Rightward shift of the P wave axis (> + 60°) (Fig. 34.10).

This results in prominent P wave in leads II, III, and aVF. P wave in lead V_1 may be normal or biphasic with initial component peaked and slightly increased in amplitude. P wave in lead V_1 may become entirely negative due to downward displacement of heart (Fig. 34.11).

Conventionally positioned leads V_1 and V_2 become close to the sinoatrial node. Atrial activation is, therefore, directed away from these leads (Fig. 34.12).

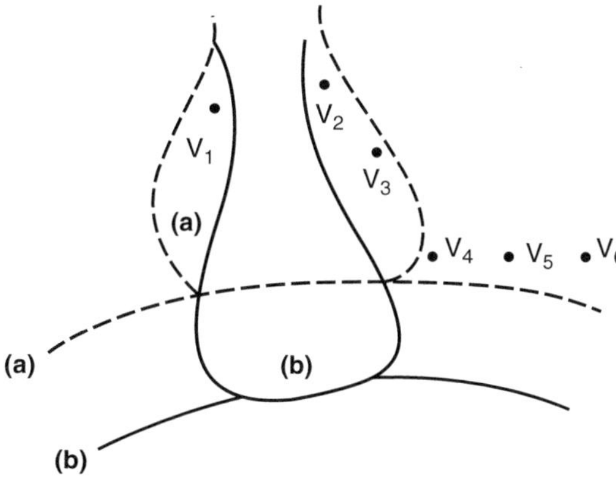

Fig. 34.8 Diagram showing altered relation of conventional position of precordial leads to the heart (**a**) in normal person (**b**) in case of vertical heart due for COAD. (Reproduced with permission from Mittal SR. Chronic Right Ventricular Pressure Overload. Cardiology Today-2016;20:215–221)

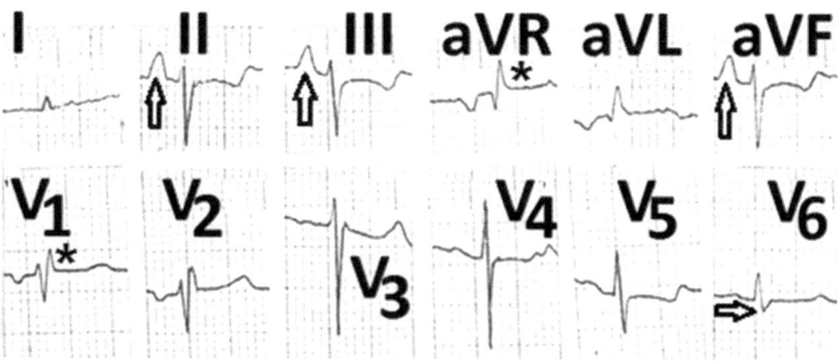

Fig. 34.7 Electrocardiogram from a patient with chronic obstructive airway disease showing leftward deviation of the mean frontal plane QRS axis (QRS in lead I is positive and is dominantly negative in leads II, III, and aVF). Tall and peaked P wave in leads II, III, and aVF (marked ⇧), prominent terminal R wave in leads aVR, and V_1 (marked *), clockwise rotation in horizontal plane (persistence of S wave up to lead V_6 (marked ⇨) support the diagnosis of RVH. Low voltage of QRS in leads I and V_6 supports the possibility of emphysema. (Reproduced with permission from Mittal SR. Chronic Right Ventricular Pressure Overload. Cardiology Today-2016;20:215–221)

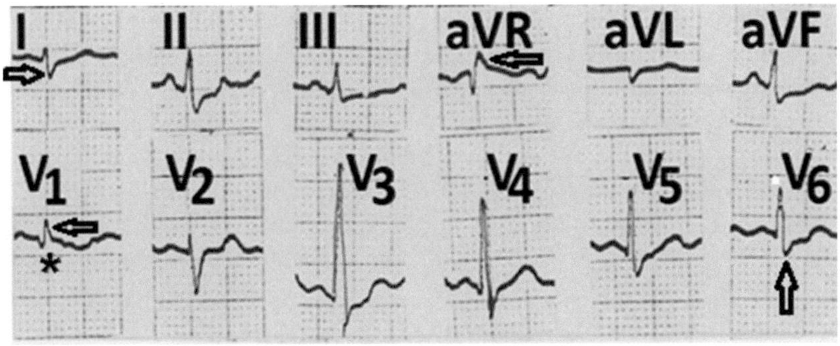

Fig. 34.9 Electrocardiogram from a patient with COAD showing qR pattern in lead V_1 (marked *). Other findings supporting the diagnosis of RVH are rightward deviation of the mean frontal plane QRS axis (dominant S wave in lead I; marked ⇨), prominent terminal R wave in lead aVR and V_1 (marked ⇦), and clockwise rotation in horizontal plane (persistence of s wave in lead V_6) (marked ⇧). (Reproduced with permission from Mittal SR. Chronic Right Ventricular Pressure Overload. Cardiology Today-2016;20:215–221)

Fig. 34.10 Diagram showing the P wave axis in the frontal plane (**a**) in normal person and (**b**) in COAD

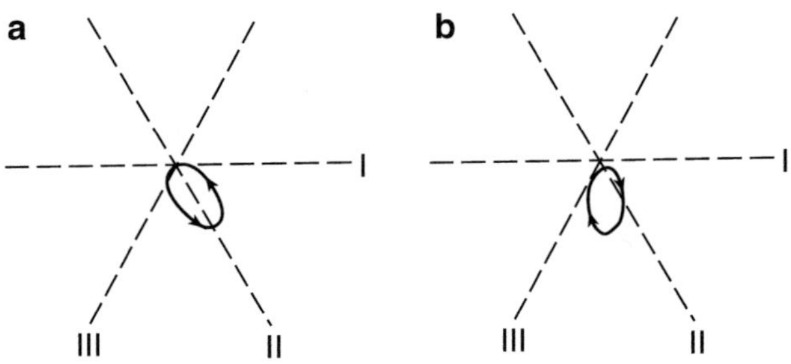

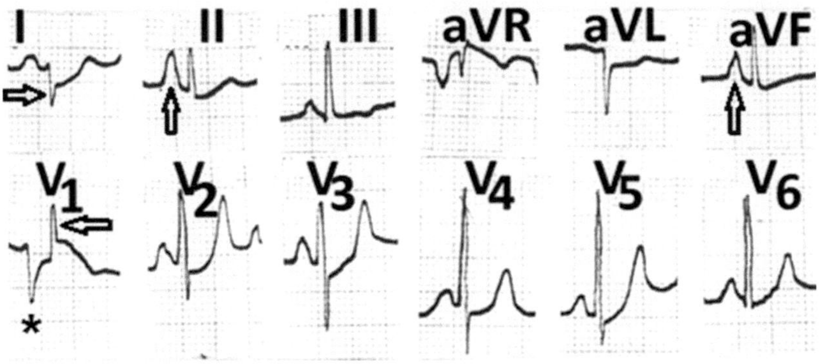

Fig. 34.11 Electrocardiogram from a case of COAD showing right-axis deviation (dominant S wave in lead I, marked ⇨) with P- pulmonale (tall P waves in leads II and aVF, marked ⇧). Lead V_1 shows negative P wave (marked *) and prominent R wave (marked ⇦). (Reproduced with permission from Mittal SR. Chronic Right Ventricular Pressure Overload. Cardiology Today-2016;20:215–221)

- Mean frontal plane axis of the P wave, the QRS complex, and the T wave are directed toward +90°, i.e., perpendicular to the standard lead I. This lead, therefore, shows very small, equiphasic or flat deflection [8]. In the horizontal plane, QRS vector is directed away from lead V_6. This results in low R wave amplitude in lead V_6.

(c) Incomplete or complete right bundle branch block may appear transiently during acute exacerbation [8]

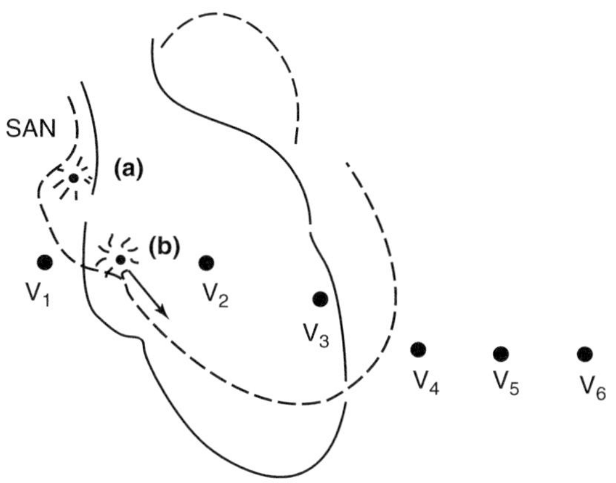

Fig. 34.12 Diagram showing the relation of the sinoatrial node to leads V_1 and V_2 (**a**) in normal position of the heart (marked with the broken line) (**b**) in vertical heart due to COAD (marked with continuous line). In COAD, sinoatrial node becomes close to lead V_1 and V_2 with the result that the vector of atrial depolarization moves away from these leads. (Reproduced with permission from Mittal SR. Chronic Right Ventricular Pressure Overload. Cardiology Today-2016;20:215–221)

The following findings suggest concomitant right ventricular hypertrophy [8]:

- Frontal plane QRS axis ≥ +110° (Fig. 34.13)
- Evidence of right atrial overload—P-pulmonale (Fig. 34.13). P waves are peaked in leads II, III, and aVF. It may not be tall [8].

Presence of ECG abnormalities is helpful in diagnosing cor pulmonale, but absence of such findings does not exclude the diagnosis [9]. Either S_I, S_{II}, S_{III}, or right atrial overload or both are negative prognostic markers [5]. QRS axis between +90° and +180° is also a negative prognostic marker [10].

3.2 Chronic Nonobstructive Lung Diseases [7]

Right ventricular hypertrophy produces

- P-pulmonale
- Rightward deviation of the mean frontal plane QRS axis
- Deep S waves in precordial leads

There are no ECG signs of lung overinflation, e.g., low voltage QRS in leads I and V_6 or QS configuration in leads V_1 to V_3.

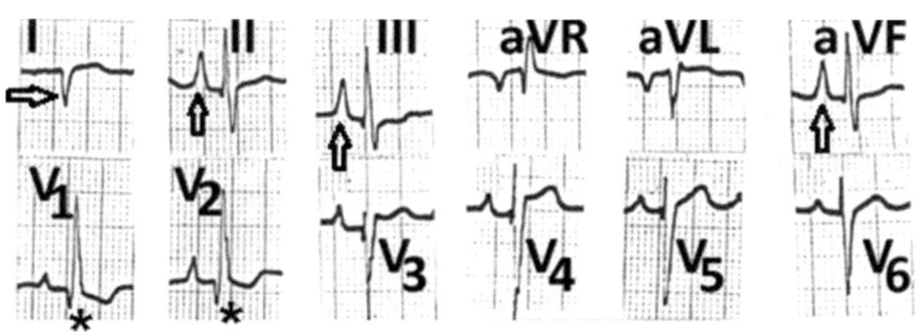

Fig. 34.13 Electrocardiogram from a patient with cor pulmonale due to chronic nonobstructive lung disease showing P-pulmonale (marked as ⇧), rightward deviation of mean frontal plane QRS axis (dominant S wave in lead I, marked as ⇨), qR in lead V_1 and V_2 (marked *) suggestive of right atrial enlargement, and deep S wave in leads V_3 to V_6 (clockwise rotation in horizontal plane). There is normal voltage of the QRS as opposed to patients with emphysema. (Reproduced with permission from Mittal SR. Chronic Right Ventricular Pressure Overload. Cardiology Today-2016;20:215–221)

3.3 Congenital Heart Diseases

3.3.1 Pulmonic Stenosis (PS)

Mild Obstruction

Mean frontal plane QRS axis is around +90° [11]. Anterior precordial leads (V_1, V_2) show rsR' pattern [12] with T wave inversion.

Moderate Obstruction

Mean frontal plane QRS axis is around +120° [11]. Lead V_1 shows dominant or exclusive R wave. ST-segment depression with asymmetrically inverted T waves is seen in leads V_1 to V_3 (Fig. 34.14).

Severe Obstruction

Mean frontal plane QRS axis is around +150° [11] (Figs. 34.15 and 34.16).

Patients with a dysplastic pulmonary valve frequently have a superior axis [12] with counter-clockwise depolarization in frontal plane (q wave in leads I and aVL; Fig. 34.17).

There is evidence of right atrial overload (P ≥ 2.5 mm in lead II and/or ≥ 1.5 mm in lead V_1). Prominent P wave in the context of congenital heart disease is also termed P congenitale. In P-congenitale, P wave axis is slightly to left as compared to P-pulmonale, where P wave axis is to right [2] (Fig. 34.18).

Leads V_1 and V_2 show tall pure R wave [12]. Lead V_3 may also show pure R wave [2]. qR configuration in lead V_1 is almost always a sign of severe obstruction [12]. With increasing severity, T wave inversion extends progressively further to the left [11]. However, T

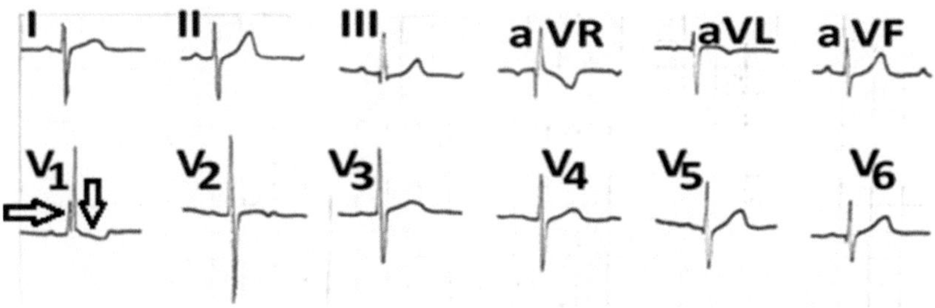

Fig. 34.14 Electrocardiogram from a patient with moderate PS showing mean frontal plane QRS axis around +120° and prominent R wave (marked as ⇨) with T wave inversion (marked as ⇩) in lead V_1. (Reproduced with permission from Mittal SR. Chronic Right Ventricular Pressure Overload. Cardiology Today-2016;20:215–221)

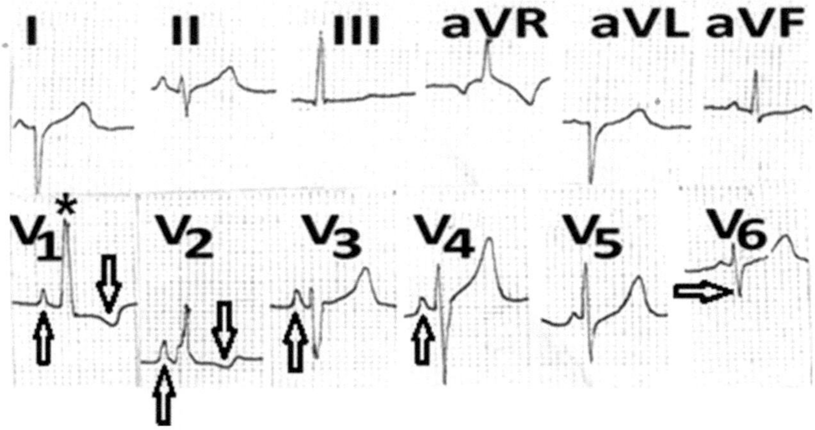

Fig. 34.15 Electrocardiogram from a case of severe PS showing extreme right-axis deviation, prominent and peaked P wave in leads V_1 to V_4 (marked as ⇧), prominent R wave in lead V_1 (marked *), clockwise rotation (persistence of S wave up to lead V_6 (marked ⇨), and T wave inversion in leads V_1 and V_2 (marked ⇩). (Reproduced with permission from Mittal SR. Chronic Right Ventricular Pressure Overload. Cardiology Today-2016;20:215–221)

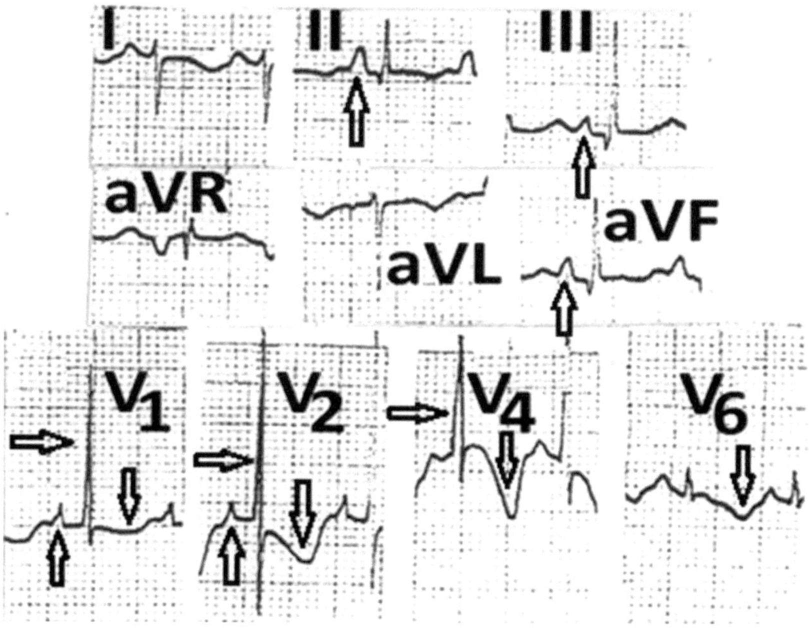

Fig. 34.16 Electrocardiogram from another case of severe PS. It shows right-axis deviation (prominent S wave in lead I and prominent R wave in lead aVF). Peaked P waves in leads II, III, aVF, and V_1, V_2 (marked as ⇧) prominent R wave in leads V_1 to V_4 (marked as ⇨) with T inversion in leads V_1 to V_6 (marked as ⇩). (Reproduced with permission from Mittal SR. Chronic Right Ventricular Pressure Overload. Cardiology Today-2016;20:215–221)

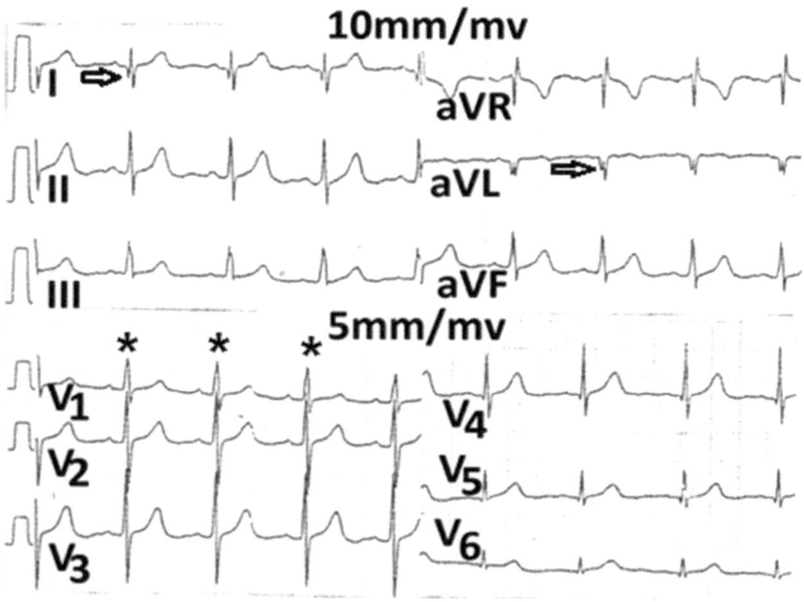

Fig. 34.17 Electrocardiogram from a case of PS due to dysplastic valve showing counter clockwise loop (q wave in leads I and aVL, marked as ⇨). Lead V_1 shows prominent R wave (marked *). (Reproduced with permission from Mittal SR. Chronic Right Ventricular Pressure Overload. Cardiology Today-2016;20:215–221)

waves tend to be deepest in leads V_1 and V_2, depth progressively diminishing toward lead V_6 [11].

Other Causes of Prominent R Wave in Lead V_1

Preexcitation, lateral myocardial infarction, RBBB, arrhythmogenic right ventricular dysplasia, post-term infant, and scanty Purkinje fibers in anteroseptal zone are other causes of prominent R wave in lead V_1.

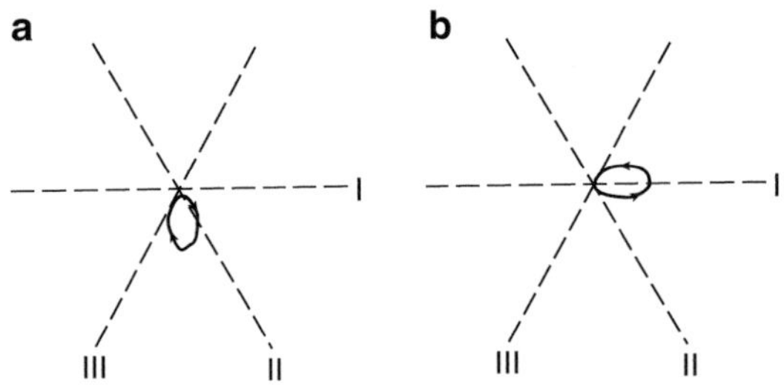

Fig. 34.18 Diagram showing P wave vector in frontal plane (**a**) in P-pulmonale and (**b**) P- congenitale. (Reproduced with permission from Mittal SR. Chronic Right Ventricular Pressure Overload. Cardiology Today-2016;20:215–221)

3.3.2 Tetralogy of Fallot (TOF)

Lead V_1 is similar to that in isolated pulmonic stenosis, i.e., it shows prominent R wave with T wave inversion. Classically, there is a rapid change from lead V_1 to lead V_2 (Fig. 34.19). Lead V_2 shows RS or rS configuration with positive T wave [11, 12]. Further, compared to isolated pulmonic stenosis, right-axis deviation is more marked and T wave inversion is not marked.

In young patients, P waves are often normal because right atrium is not enlarged and its contraction is not increased. In patients with restrictive VSD, there may be evidence of right atrial enlargement (Fig. 34.20).

Adult TOF with right-sided failure also produces evidence of right atrial enlargement (Fig. 34.21).

Extreme right-axis deviation (> + 150°) should suggest the possibility of double outlet right ventricle (DORV) (Fig. 34.22) or D- transposition of great arteries (D-TGA) [12] (Fig. 34.23).

Left-axis deviation suggests the possibility of underlying

- Atrioventricular septal defect or
- Single ventricle

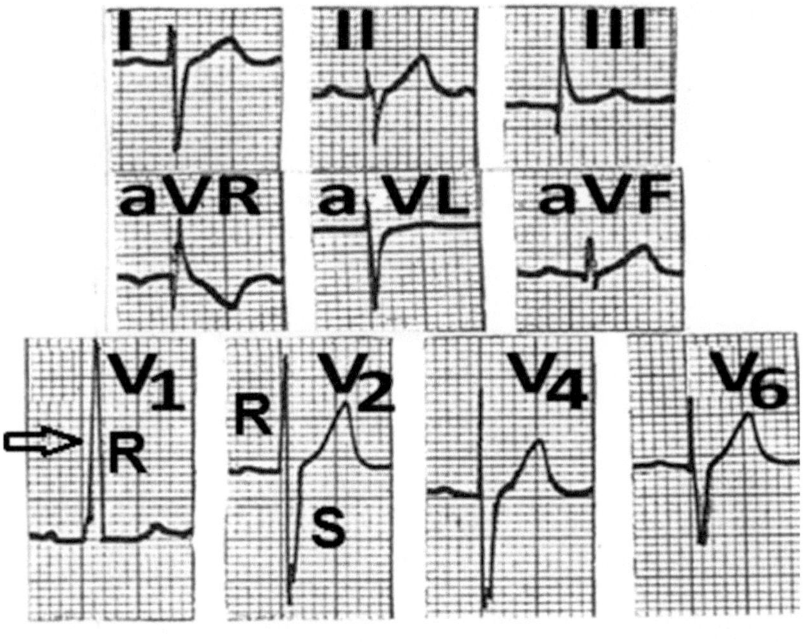

Fig. 34.19 Electrocardiogram from a case of TOF. Lead V_1 shows tall R wave (marked as ⇨) with positive T wave. Lead V_2 shows sudden change to RS configuration with positive T wave. P waves are normal. (Reproduced with permission from Mittal SR. Chronic Right Ventricular Pressure Overload. Cardiology Today-2016;20:215–221)

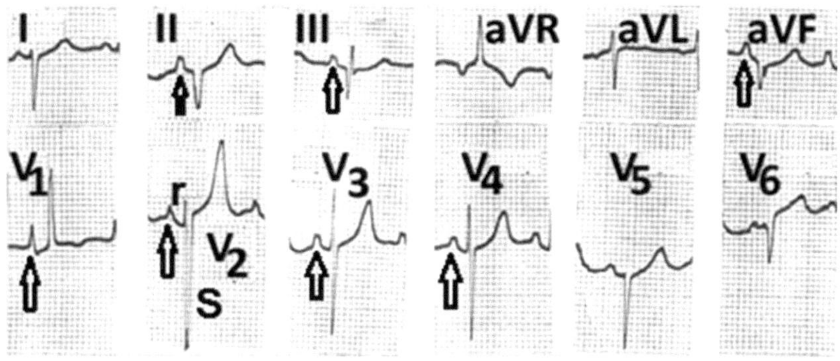

Fig. 34.20 Electrocardiogram from a case of TOF with restrictive VSD. It shows right-axis deviation (dominant S wave in lead I) and sudden change from prominent R wave in lead V_1 to rS in lead V_2. P wave are peaked in leads II, III, aVF, and V_1 to V_4 (marked ⇧). These suggest right atrial enlargement. (Reproduced with permission from Mittal SR. Chronic Right Ventricular Pressure Overload. Cardiology Today-2016;20:215–221)

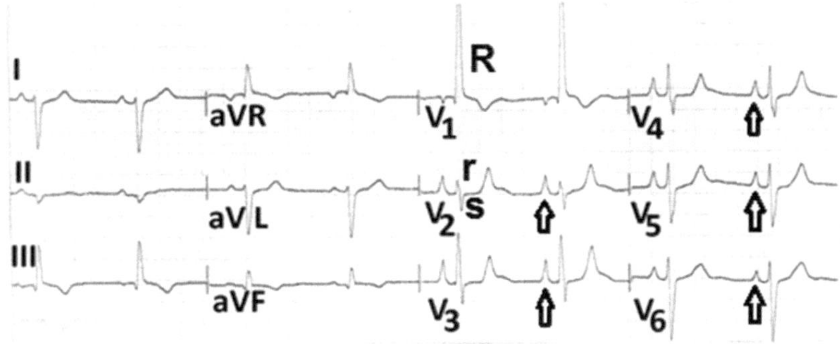

Fig. 34.21 Electrocardiogram from an adult patient with TOF with right-sided failure. It shows right-axis deviation and sudden change from R wave in lead V_1 to rs in lead V_2. P wave is tall and peaked in leads V_2 to V_6 (marked ⇧). (Reproduced with permission from Mittal SR. Chronic Right Ventricular Pressure Overload. Cardiology Today-2016;20:215–221)

3.3.3 Dextroposition of the Great Arteries (DTGA) with Intact Interventricular Septum

- ECG findings include
  - Tall or peaked P waves because of increased right atrial pressure.
  - Abnormal right-axis deviation.
  - Positive T waves in leads V_1 and V_3R beyond 4 days of age.

Representative electrocardiogram is shown in Fig. 34.24.

3.3.4 Double Outlet Right Ventricle [13]

Right-axis deviation and right ventricular hypertrophy resemble TOF. Differentiating features include right atrial enlargement, prolonged PR interval (due to an unusually long course of common atrioventricular bundle), counterclockwise loop in frontal plane (q wave in leads I and aVL), qR pattern in leads V_1, V_2 ,and broad and slurred S wave in leads V_5 and V_6.

Representative electrocardiogram is shown in Fig. 34.25.

3.3.5 General Criteria for the Diagnosis of Right Ventricular Hypertrophy in Neonates and Infants [14]

- Mean frontal plane QRS axis > +135°.
- Tall R wave in lead V_1 (more than 10 mm in newborn).
- R/s > 1 in lead V_1 in the absence of right bundle branch block.

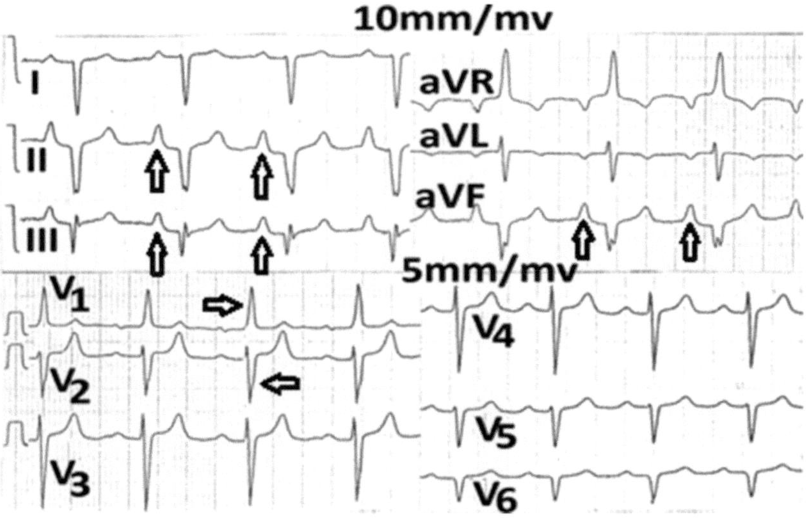

Fig. 34.22 Electrocardiogram from a case of tetralogy-like physiology due to DORV, VSD, and PS. The ECG shows extreme right-axis deviation with sudden change from pure R wave in lead V_1 (marked as ⇨) and lead V_1 to rs configuration in lead V_2 (marked as ⇦). P waves in leads II, III, and aVF are tall and peaked (marked as ⇧). These P waves suggest right atrial overload. (Reproduced with permission from Mittal SR. Chronic Right Ventricular Pressure Overload. Cardiology Today-2016;20:215–221)

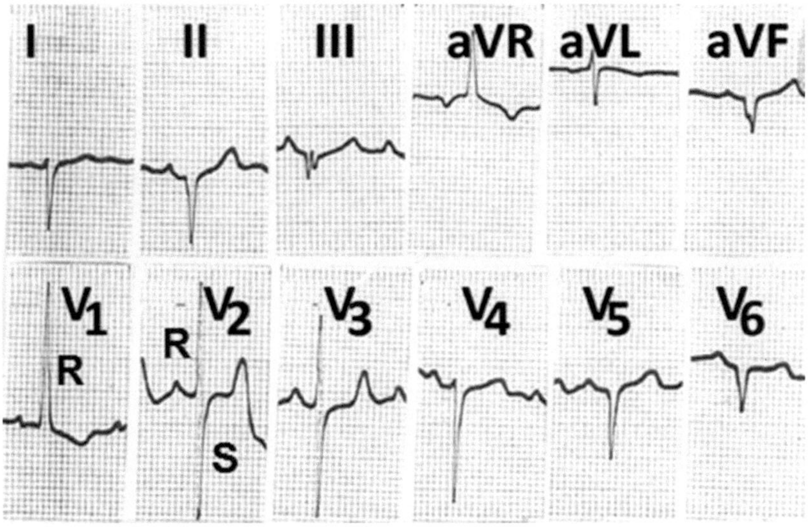

Fig. 34.23 Electrocardiogram from a case of tetralogy-like physiology due to DTGA +VSD+ PS. There is extreme right-axis deviation with sudden change from R wave in lead V_1 to RS configuration in lead V_2. (Reproduced with permission from Mittal SR. Chronic Right Ventricular Pressure Overload. Cardiology Today-2016;20:215–221)

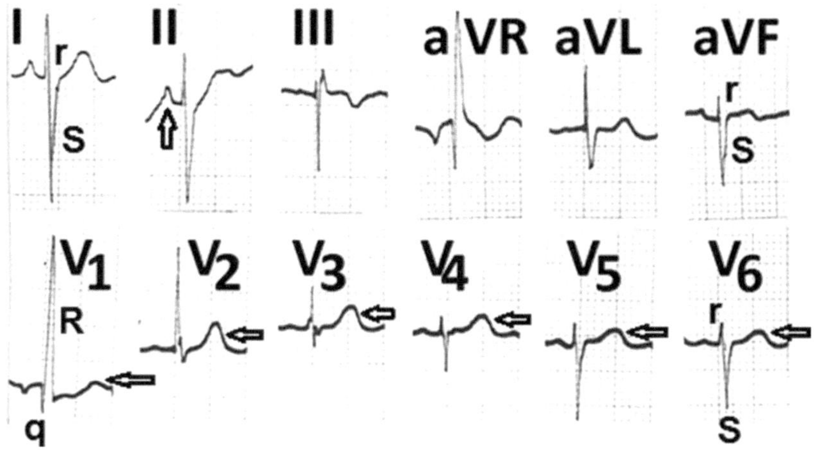

Fig. 34.24 Electrocardiogram from a case of DTGA with intact interventricular septum showing extreme right-axis deviation (rS configuration in leads I and aVF), peaked P wave in lead II (marked ⇧), and right ventricular hypertrophy (qR in lead V_1 and rs in lead V_6) with positive T waves in precordial leads (marked ⇦). (Reproduced with permission from Mittal SR. Chronic Right Ventricular Pressure Overload. Cardiology Today-2016;20:215–221)

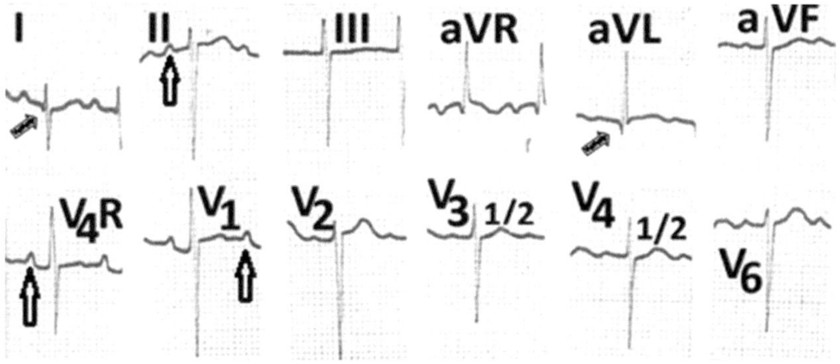

Fig. 34.25 Electrocardiogram from a case of DORV showing extreme right-axis deviation (prominent S wave in leads I and aVF) with counterclockwise loop (q in leads I and aVL, marked as ⇗). P waves are peaked in leads II, V_4R, and V_1 (marked ⇧). This finding suggests right atrial overload. (Reproduced with permission from Mittal SR. Chronic Right Ventricular Pressure Overload. Cardiology Today-2016;20:215–221)

- qR configuration in lead V_1.
- Positive T waves in lead V_1 and V_2 after 5 days of birth.

3.4 Chronic Pulmonary Thromboembolism [15]

Most frequently observed changes are

- Right-axis deviation (> + 90°).
- P-pulmonale.
- Negative T waves in precordial leads V_1 to V_5 and leads II, III, and aVF.

3.5 Other Causes of Pulmonary Artery Hypertension

Right-axis deviation, right atrial overload, and prominent R wave in lead V_1 are the common findings. However, it is important to remember that in some cases ECG may be "normal" in spite of significant PAH [16].

Summary

- Mild or even moderate right ventricular pressure overload may not produce any diagnostic change in the electrocardiogram.
- Right-axis deviation more than +90° is the earliest change, but is a nonspecific finding. Mean frontal plane QRS axis of more than +120° is important. However, left inferoposterior fascicular block and lateral infarction can also produce right-axis deviation in the absence of right ventricular pressure overload.
- Severe right ventricular pressure overload produces
- Increasing prominence of the R wave in leads aVR and V_1.
- Increasing prominence of the S wave in leads I, aVL, V_5, and V_6.
- Increasing depression of the ST-segment and increasing depth of the T wave inversion in leads V_1,V_2.
- ECG evidence of right atrial overload.

These findings have high specificity but low sensitivity. Concomitant left anterosuperior fascicular block can mask prominence of S wave in leads I and aVL and increase S wave prominence in leads V_5,V_6. Left posteroinferior fascicular block can aggravate S wave changes in leads I and aVL.

- In congenital valvular pulmonary stenosis, dominant R waves extending from lead V_1 to V_3, suggest severe right ventricular hypertrophy. Right atrial overload (prominent peaked P waves in leads II, III, aVF, and V_1,V_2) also supports the diagnosis of gross RVH. On the contrary, in neonates, positive T waves in leads V_1 and V_2 suggest severe right ventricular hypertrophy. Posterior myocardial infarction and left lateral accessory pathway also produce prominent R wave in lead V_1 but do not suggest right ventricular pressure overload. In these conditions, T waves are usually positive in precordial leads.
- In contrast, in tetralogy of Fallot, prominent R wave in lead V_1 changes to rS configuration in lead V_2 in spite of significant right ventricular pressure overload. Peaked P wave is not seen in classical TOF.
- In patients with chronic cor pulmonale, R wave is usually not prominent in lead V_1. Peaked (not tall) P waves are common. S_I, S_{II}, S_{III} pattern also suggests right ventricular pressure overload, but it is a nonspecific finding. Depth of the S wave in lead II is more than in lead III as opposed to left anterosuperior fascicular block.

MCQs

Q1. In chronic right ventricular pressure overload, right ventricular mass is usually

(a) Less than LV mass
(b) Equal to LV mass
(c) Greater than LV mass
(d) Significantly greater than LV mass

Q2. ECG changes produced in chronic right ventricular pressure overload are due to

(a) Right ventricular hypertrophy
(b) Delay in activation of the right ventricle
(c) Right ventricular ischemia
(d) Right ventricular failure

Q3. Right ventricular dominance is not seen in

(a) Right ventricular infundibular stenosis
(b) L-TGA
(c) DORV
(d) Normal neonates

Q4. ECG signs of chronic right ventricular pressure overload may not be present in some cases of

(a) Severe pulmonic stenosis
(b) D-TGA
(c) DORV
(d) Pulmonary artery hypertension

Q5. $S_1S_2S_3$ pattern suggests

(a) Rightward shift of mean frontal plane QRS axis

(b) Anterior shift of the horizontal plane QRS axis
(c) Posterior shift of the horizontal plane QRS axis
(d) Rightward and superior shift of the mean frontal plane QRS axis

Q6. Right-axis deviation is not seen in

(a) Anterosuperior fascicular block
(b) Right ventricular hypertrophy
(c) Vertical heart
(d) Lateral infarction

Q7. Left-axis deviation in the presence of other electrocardiographic signs of RVH can occur in

(a) Severe valvular pulmonary stenosis
(b) Noonan syndrome
(c) Atrioventricular septal defect
(d) Chronic airway disease

Q8. Emphysema alone does not produce

(a) Low voltage in lead I
(b) Low voltage in lead V_6
(c) Prominent R in lead V_1
(d) P-pulmonale

Q9. Emphysema alone can produce

(a) Right bundle branch block
(b) Negative P wave in lead V_1
(c) QS configuration in leads V_1 to V_3
(d) P-pulmonale

Q10. In the presence of emphysema, right ventricular hypertrophy is suggested by

(a) Frontal plane QRS axis > +110°
(b) Peaked P waves in leads II, III, and aVF
(c) Negative P waves in lead V_1
(d) QS configuration in leads V_1 to V_3

Q11. Cor pulmonale secondary to chronic obstructive airway disease usually does not produce

(a) Right-axis deviation
(b) Peaked P waves in leads II, III, and aVF
(c) Tall R wave in right-sided chest leads
(d) Prominent R in lead aVR

Q12. QS configuration in leads V_1 to V_3 does not occur in

(a) Chronic obstructive airway disease
(b) Old anterior myocardial infarction
(c) Severe PS
(d) PAH due to severe interstitial lung disease

Q13. Severe pulmonic stenosis does not produce

(a) Mean frontal plane QRS axis > +180°
(b) R wave in leads V_1 to V_3
(c) P-congenitale
(d) T wave inversion in leads V_1 to V_3

Q14. Prominent R wave in lead V_1 is not seen in

(a) Right-sided accessory pathway
(b) Lateral infarction
(c) Arrhythmogenic right ventricular dysplasia
(d) Neonates

Q15. Tetralogy of Fallot with extreme RAD suggests

(a) DORV+VSD + PS
(b) LTGA+VSD + PS
(c) Atrioventricular septal defect + PS
(d) Single ventricle +PS

Q16. Tetralogy of Fallot with LAD suggests

(a) DORV+ VSD + PS
(b) DTGA+VSD + PS
(c) Atrioventricular septal defect +PS
(d) L-TGA + VSD + PS

Q17. Tetralogy of Fallot with extreme RAD and counterclockwise loop should suggest

(a) DORV+VSD + PS
(b) DTGA+VSD + PS

(c) Atrioventricular septal defect +PS
(d) Single ventricle +PS

Q18. Which ECG finding suggests RVH in a neonate?

(a) Mean frontal plane QRS axis around +90°
(b) rS configuration in lead V_1
(c) R wave of 6 mm in lead V_1
(d) Positive T waves in lead V_1 and V_2 after 5 days of birth

Q19. Mild right ventricular hypertrophy produces

(a) Mean frontal plane QRS axis deviation of more than +120°
(b) Tall and peaked P wave
(c) Both
(d) None

Q20. Moderate right ventricular hypertrophy produces

(a) Frontal plane QRS axis of more than +140°
(b) Tall and peaked P waves
(c) R/S more than 1 in lead V1
(d) None

Q21. Severe right ventricular pressure overload produces

(a) Prominent Q wave in lead aVR
(b) Deep S wave in lead V1
(c) R/S more than 1 in lead V1
(d) All

Q22. Severe right ventricular hypertrophy produces

(a) Prominent R wave in lead aVR
(b) Prominent S wave in leads I, aVL
(c) Prominent S wave in leads V_5, V_6
(d) None

Q23. Left inferoposterior fascicular block can produce

(a) Right-axis deviation
(b) Prominent R wave in lead V_1
(c) Prominent S wave in leads I and aVL
(d) All

Q24. Left anterosuperior fascicular block can produce

(a) Prominent S wave in lead I
(b) Prominent S wave in lead aVL
(c) Prominent S wave in leads V_5, V_6
(d) None

Q25. In mild congenital pulmonary valve stenosis, electrocardiogram shows

(a) Peaked P waves
(b) QRS axis more than +120°
(c) R/S in more than I in lead V_1
(d) None

Q26. In severe congenital pulmonary valve stenosis, electrocardiogram shows

(a) Prominent R wave extending from lead V_1 to V_3
(b) Peaked P wave in leads II, III, aVF, V_1,V_2
(c) Mean frontal plane QRS axis more than +120°
(d) None

Q27. In a neonate, right ventricular pressure overload is suggested by

(a) R wave more than 10 mm in lead V_1
(b) Positive T wave in leads V_1 and V_2 after 5 days of birth
(c) Deep T wave inversion in leads V_1 and V_2
(d) All

Q28. Prominent R wave in lead V_1 can be due to

(a) Severe valvular pulmonary stenosis
(b) Posterior myocardial infarction
(c) Left lateral accessory pathway
(d) All

Q29. Right ventricular pressure overload in a case of Fallot's tetralogy shows

(a) Right atrial enlargement
(b) Prominent R wave extending from leads V_1 to V_3
(c) Deep T wave inversion from leads V_1 to V_4
(d) None

Q30. Common electrocardiographic findings in chronic cor pulmonale include

(a) Extreme right-axis deviation
(b) Peaked (but not tall) P waves
(c) S_1, S_2, S_3 pattern
(d) Nil

Q31. S_1, S_2, S_3 pattern is seen in

(a) Cor pulmonale
(b) Some normal persons
(c) Chest deformities
(d) None

Q32. Common electrocardiographic findings in cor pulmonale include

(a) Low amplitude of QRS
(b) Leftward shift of transition zone in precordial lead
(c) S in lead III deeper than S in lead II
(d) All

Q33. In left anterosuperior fascicular block

(a) QRS amplitude is increased in limb lead
(b) S wave in lead III is deeper than in lead II
(c) There is no S wave in lead I
(d) None

Answers

(1) a (2) b (3) b (4) d (5) d (6) a (7) b,c (8) c,d (9) b,c (10) a,b (11) c,d (12) c,d (13) a (14) a (15) a (16) c (17) a (18) d (19) d (20) d (21) c (22) a,b,c (23) a,c (24) c (25) d (26) a,b,c (27) a,b (28) d (29) d (30) b,c (31) a,b,c (32) a,b (33) a,b,c

References

1. Mirvis DM, Goldberger AL. Electrocardiography. In: Bonow RO, Mann DL, Zipes DP, Libby P, editors. Braunwald's heart disease. St. Louis: Elsevier; 2012. p. 126–65.
2. De Luna AB, Goldwasser D, Fiol M, Bayes-Genis A. Surface electrocardiography. In: Fuster V, Walsh RA, Harrington RA, editors. Hurst's the heart. New York: Mc Graw Hill; 2011. p. 307–70.
3. Wagner GS, Lim TH. Chamber enlargement. In: Wagner GS, editor. Marriott's practical electrocardiography. Philadelphia: Wolters Kluwer; 2008. p. 71–96.
4. Ikeda K, Kubota I, Takahashi K, Yasui S. P wave changes in obstructive and restrictive lung disease. J Electrocardiol. 1985;18:233–8.
5. Lncalzi RA, Fuso L, Rosa MD, Napoli AD, Basso S, Pagliari G, et al. Electrocardiographic signs of chronic cor pulmonale. Circulation. 1999;99:1600–5.
6. Mirvis DM, Goldberger AL. Electrocardiography. In: Mann DL, Zipes DP, Libby P, Bonow RO, editors. Braunwald's heart disease. Philadelphia: Elsevier; 2015. p. 114–52.
7. Hancock EW, Deal BJ, Mirvis DM, Okin P, Kligfield P, Gettes LS. AHA/ ACCF/ HRS recommendations for the standardization and interpretation of the electrocardiogram. Part V. Electrocardiogram changes associated with cardiac chamber hypertrophy. J Am Coll Cardiol. 2009;53:992–1002.
8. Schamroth L. Emphysema: chronic obstructive airway disease. In: Schamroth C, editor. An introduction to electrocardiography. France: Blackwell Science; 1982. p. 233–8.
9. Miller MA, Sweeng JM, Lawrance EC, Brigham KL. Chronic cor pulmonale. In: Fuster V, Walsh RA, Harrington RA, editors. Hurst's the heart. New York: McGraw Hill; 2011. p. 1655–67.
10. Kok JA. Simple electrocardiographic features for prognosis in severe chronic bronchial obstruction. Scand J Respir Dis. 1975;56:273–84.
11. Schamroth L. Congenital heart disease. In: Schamroth C, editor. An introduction to electrocardiography. France: Blackwell Science; 1982. p. 261–87.
12. Choudhri M. ECG interpretation in congenital heart disease. In: Guha S, editor. Cardiology update 2015. New Delhi: Jaypee; 2016. p. 942–56.
13. Patnaik AN. Double outlet right ventricle. In: Satpathy M, editor. Clinical diagnosis of congenital heart disease. New Delhi: Jaypee; 2008. p. 280–9.
14. Kulkarni ML, Satpathy M. Electrocardiogram – clinically relevant to congenital heart disease. In: Satpathy M, editor. Clinical diagnosis of congenital heart disease. Jaypee: New Delhi; 2008. p. 27–32.

15. Lewczuk J, Ajlan AW, Piszko P, Jagas J, Mikulewicj M, Wrabec K. Electrocardiographic signs of right ventricular overload in patients who underwent pulmonary embolism event(s). Are they useful in diagnosis of chronic thromboembolic pulmonary hypertension? J Electrocardial. 2004;37:219–25.
16. Baist RJ, Rubin LJ. Pulmonary hypertension. In: Fuster V, Walsh RA, Harrington RA, editors. Hurst's the heart. New York: McGraw Hill; 2011. p. 1609–33.

Acute Right Ventricular Pressure Overload: Acute Pulmonary Embolism

35

Electrocardiographic (ECG) changes depend on the extent of occlusion of the pulmonary vasculature. The greater the extent of occlusion, more likely are the "classic" ECG alterations [1].

1 Small Pulmonary Embolism

ECG may be entirely normal [2]. Some patients may show mild sinus tachycardia (Figs. 35.1 and 35.2) and mild deviation of the mean frontal plane QRS axis to the right as compared to any ECG recorded before the onset of illness.

Changes occur very early and are transient. These changes are, therefore, likely to be missed if ECG is not recorded in the early hours of the illness. Clinical setting and comparison with a previous ECG is important because such changes are also present in other causes of pulmonary artery hypertension (PAH) [3]. A high index of clinical suspicion is necessary to diagnose small pulmonary embolism. ECG cannot be relied upon to diagnose or rule out small embolism. An ECG is helpful in excluding myocardial infarction.

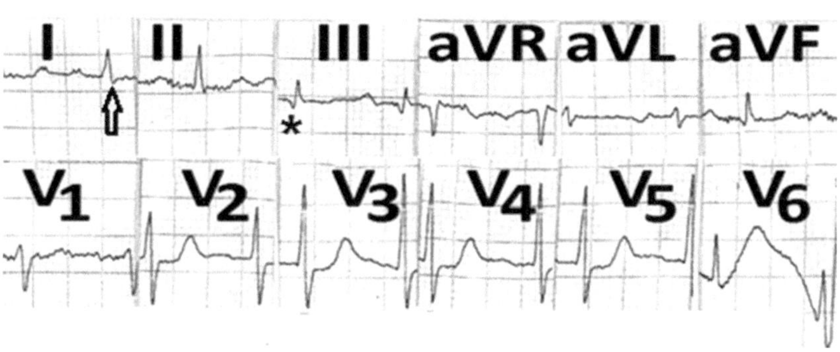

Fig. 35.1 Electrocardiogram is showing mild sinus tachycardia, small s wave in lead I (marked ⇧), and small q wave in lead III (marked *). (Reproduced with permission from: Mittal SR. Acute Right Ventricular Pressure Overload—Acute Pulmonary Embolism. Cardiology Today-2017;21:41–46)

S. Mittal, *Insights into Electrocardiograms with MCQs*,
https://doi.org/10.1007/978-981-99-0127-2_35

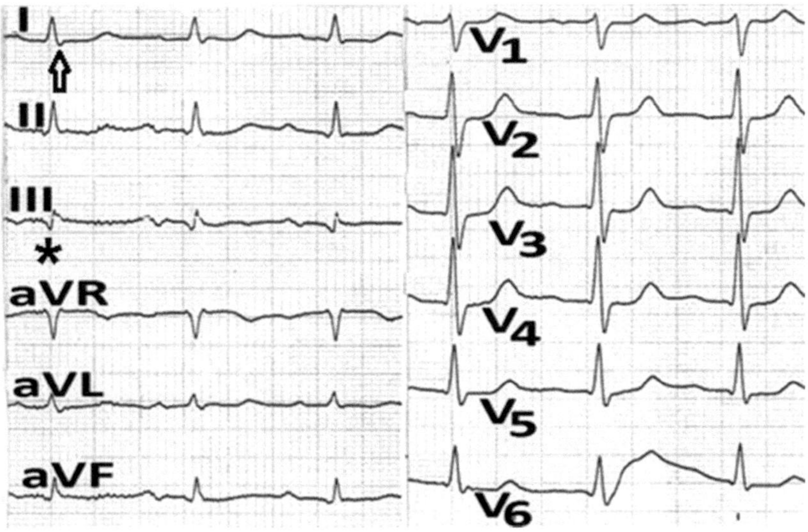

Fig. 35.2 Electrocardiogram is showing mild sinus tachycardia, small s wave in lead I (marked as ⇧), and small q wave in lead III (marked *). (Reproduced with permission from: Mittal SR. Acute Right Ventricular Pressure Overload—Acute Pulmonary Embolism. Cardiology Today-2017;21:41–46)

2 Submassive Pulmonary Embolism [4]

1. Sinus tachycardia is the commonest finding (Fig. 35.3).

Sinus tachycardia is a compensatory response to decreased right ventricular stroke volume. Some patients can present with atrial flutter/atrial fibrillation or atrial tachycardia [5].

2. Other findings with decreasing frequency are as follows:
   (a) S wave of any magnitude in lead I (Fig. 35.3). It is due to the rightward shift of the mean frontal plane QRS axis. Even small S wave in lead I is important (Figs. 35.1 and 35.2), specially so if it is a new transient change.
   (b) S_1, q_3 inverted T_3/S_1 inverted T_3 (Fig. 35.3)/S_1, q_3 patterns. Q wave is neither broad nor deep.
   (c) ST-segment elevation in inferior and/or anterior leads (V_1 to V_3) [6]. These

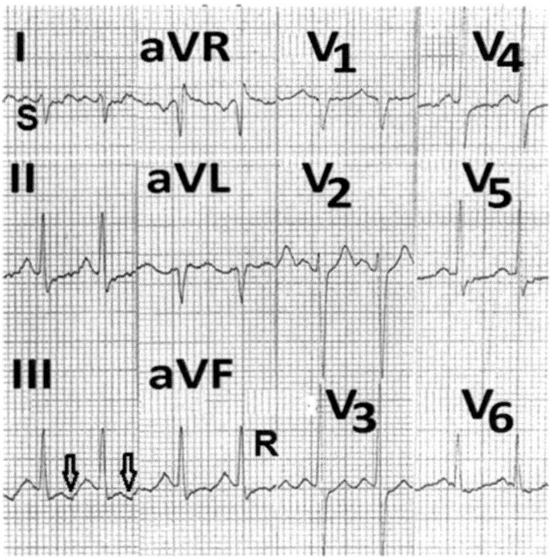

Fig. 35.3 Electrocardiogram is showing mild sinus tachycardia, right axis deviation (dominant S wave in lead I and dominant R wave is and aVF) and T wave inversion in lead III (marked ⇩). (Reproduced with permission from: Mittal SR. Acute Right Ventricular Pressure Overload—Acute Pulmonary Embolism. Cardiology Today-2017;21:41–46)

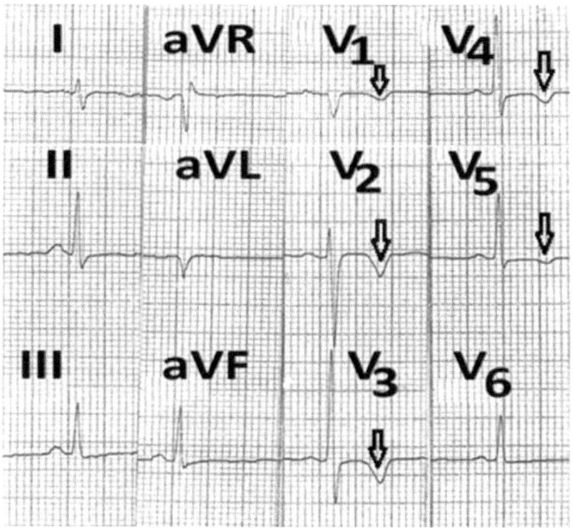

Fig. 35.4 Electrocardiogram from the same patient whose initial ECG is shown in Fig. 35.3. Electrocardiogram recorded on second day of admission showing right axis deviation, normalization of T wave in lead III and sequential T wave inversion in leads V_1 to V_5 (marked as ⇧). Maximum T wave inversion is seen in lead V_2 and V_3. (Reproduced with permission from: Mittal SR. Acute Right Ventricular Pressure Overload—Acute Pulmonary Embolism. Cardiology Today-2017;21:41–46)

changes are due to right ventricular ischemia secondary to acute increase in pulmonary artery pressure.

3. Sequential changes include:
   (a) T wave inversion in leads V_1 to V_3 [7] (Fig. 35.4).

These T wave inversions appear over the next few days. The mechanism of this ECG change and reason for the delay after onset of symptoms is not clear. Progressive right ventricular (RV) dilatation and secondary right ventricular ischemia could be responsible. Ciurzynski et al. [8] found a direct correlation between right ventricular end—diastolic diameter and sum of negative T waves in leads V_1 to V_4. Raised serum levels of troponin have been observed with submassive and massive pulmonary embolism [1, 9].

(b) P-pulmonale may appear after a few days [8]. It suggests chronic pulmonary artery hypertension with right atrial overload. It has very low sensitivity in diagnosing acute pulmonary embolism. However, in a case with strong clinical suspicion, sequential development of P-pulmonale supports the possibility of pulmonary embolism.

3 Massive Pulmonary Embolism

Differentiation between submassive and massive pulmonary embolism by ECG alone is difficult. Some findings are, however, more frequent with massive embolism.

1. Transient right bundle branch block [10]. This finding has high specificity but low sensitivity. Transient RBBB can also be rate dependent. In this situation, it does not necessarily denote massive embolism.
2. ST-segment elevation in leads V_1 to V_3 [11]. It is probably because of acute transmural ischemia of the right ventricle because of acute increase right ventricular afterload.
3. ST-segment elevation with Q wave in any of right precordial leads (V_3R to V_6R). This is a very early but transient change [12, 13].
4. Prominent R wave with ST-segment elevation in lead aVR [10, 14]. This is again, due to acute transmural ischemic or infarction of the right ventricle. ST-segment elevation due to right ventricular infarction normalize very quickly due to significant collateral circulation.
5. ST-segment depression in leads I, V_4 to V_6 [14, 15]. The reason for this change is not clear. Hypoxia could be responsible. Concomitant LVH can be another possibility.
6. T wave inversion in leads III, aVF, and V_1 to V_6 [1, 16, 17]. With increasing severity and dilatation of the right ventricle, the interventricular septum shift toward the left ventricle. Inversion of the T waves moves toward the left, that is, from lead III to lead aVF and II and from lead V_1 to V_6.
7. QS in leads V_1 to V_3 [1]. These are also probably due to dilatation of the right ventricle.

8. Low amplitude of the QRS deflection [5]. The mechanism of this change is not clear. Global ischemia secondary to hypoxia and/or hypotension could contribute to this electrocardiographic finding.
9. QT prolongation [17] with global T wave inversion. This change could be due to global ischemia.

4 ECG and Right Ventricular Dysfunction in Acute Pulmonary Embolism

T wave inversion in leads V_1 through V_3 is the strongest independent predictor of right ventricular dysfunction [18]. Normalization of the negative T waves is associated with improvement in right ventricular dysfunction [19]. S_1, Q_3 inverted T_3, complete RBBB, P-pulmonale, and right axis deviation also have high specificity for diagnosis of right ventricular enlargement [20]. None of the ECG abnormalities is independently sensitive for RV enlargement [20]. There is significant correlation with total number of abnormalities in the ECG [21].

4.1 ECG and Prognosis in Acute Pulmonary Embolism

1. *Daniel's 21-point ECG scale* [22]

Electrocardiographic Finding	Points
Tachycardia >100/min	2
Incomplete right bundle branch block	2
Complete right bundle branch block	3
Negative T waves in leads V_1 to V_4	4
Negative T wave in lead V_1 (mm)	
<1	0
1–2	1
>2	2
Negative T wave in lead V_2	
<1	1
1–2	2
>2	3
Negative T wave in lead V_3	
<1	1
1–2	2
>2	5
Q wave in lead III >1.5 mm	1
Negative T wave in lead III	1
S_1 Q_3 negative T_3 complex	2

Score ≥ 10 points correlated with a pulmonary artery pressure exceeding 50 mmHg and mortality over 50%. Score < 3 denotes better prognosis [23]. The main limitation of the score is that it gives more importance to the amplitude of a negative T wave. Negativity of the T waves in leads V_1 to V_3 may be influenced by RBBB or other cardiovascular diseases. Negative T waves in these leads can be present even in some healthy persons.

2. *Twi ST score* [18]

ECG Finding	Points
T wave inversion in leads V_1 through V_3	3
S wave in lead I	2
Tachycardia	3

Score ≤ 2 was highly sensitive for absence of adverse clinical events. Score ≥ 5 was highly specific for adverse clinical events.

3. *Other ECG findings associated with adverse prognosis*
   (a) Atrial arrhythmias [24].
   (b) Low voltage of QRS in limb lead [24].
   (c) Q wave in lead aVF [24].
   (d) ST-segment elevation in lead III [14].
   (e) ST-segment elevation in lead V_1 [14].
   (f) ST-segment elevation or depression in left precordial leads [14].
   (g) QR in lead V_1 [14, 25].
   (h) Longer the time taken for normalization of T wave inversion following institution of therapy implies a worse prognosis [1].

5 ECG Findings Difficult to Explain by Right Ventricular Overload

1. "P-mitrale" or prominent negative deflection of the P wave in lead V_1.
2. Positive voltage criteria for left ventricular hypertrophy and/or ST-segment depression in left-sided leads.

These changes could be due to preexisting hypertension or ischemic heart disease.

3. Left axis deviation—Lynch et al. [26] observed that leftward shift of the mean frontal plane QRS axis was common in pulmonary embolism. They observed that an axis of <30° at the time of onset of symptoms was twice as frequent as right axis deviation.

6 Differential Diagnosis

6.1 Inferior Myocardial Infarction

ECG in patients with acute pulmonary embolism often mimics inferior infarction [14].

The following points can help with differentiation [5]:

(a) Inferior infarction can present with bradycardia or atrioventricular block. On the contrary, sinus tachycardia is invariable with acute pulmonary embolism.
(b) In inferior myocardial infarction, Q wave is wide and more than 25% of the ensuing R wave (Fig. 35.5). Q of pulmonary embolism does not have these features.
(c) Inferior myocardial infarction is usually not associated with changes in leads V_1 to V_3. This combination of ECG changes is common in pulmonary embolism. This results from the fact that leads V_1 and V_2 reflect changes occurring in the anterior wall of the right ventricle and lead III overlies the inferior wall of the right ventricle [14] (Figs. 35.6 and 35.7).
(d) In inferior myocardial infarction, there is a sequential increase in ECG changes, whereas in pulmonary embolism, changes are transient and normalize over days. Only rarely acute pulmonary embolism may present with ST-segment elevation and sequential changes in the inferior leads [6]. Such changes could be secondary to acute trans-

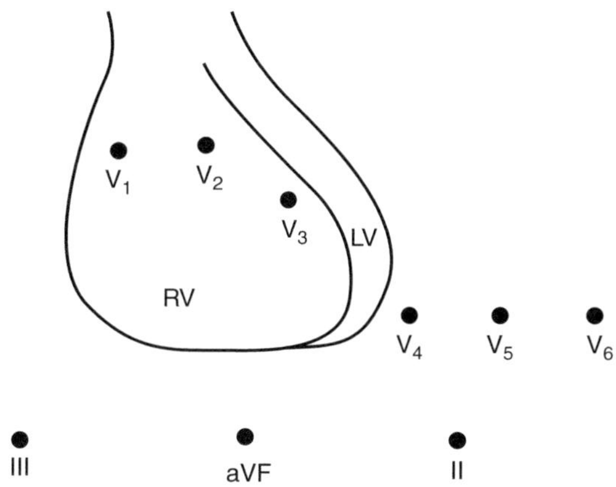

Fig. 35.6 Diagram showing relation of the electrocardiographic leads to the ventricles in frontal plane. *LV* left ventricle, *RV* right ventricle. (Reproduced with permission from: Mittal SR. Acute Right Ventricular Pressure Overload—Acute Pulmonary Embolism. Cardiology Today-2017;21:41–46)

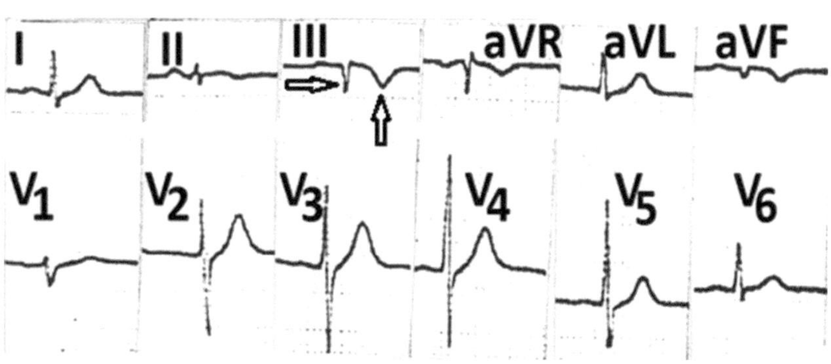

Fig. 35.5 Electrocardiogram from a patient with inferior infarction showing deep and broad Q wave (marked as ⇧) with sequential T wave inversion in lead III (marked as ⇩). (Reproduced with permission from: Mittal SR. Acute Right Ventricular Pressure Overload—Acute Pulmonary Embolism. Cardiology Today-2017;21:41–46)

mural ischemia or infarction of the right ventricle due to acute increase in afterload and concomitant hypoxia.

6.2 Left Ventricular Anterior Myocardial Infarction

Electrocardiographic findings that help in differentiation are:

(a) ST-segment elevation preceding T wave inversion in precordial leads supports the possibility of acute coronary syndrome. ST-segment elevation in precordial leads is rarely seen in acute pulmonary embolism [11].

(b) Pulmonary embolism is strongly associated with simultaneous inversion of T waves in leads III and V_1, V_2.

(c) In pulmonary embolism, maximum magnitude of T wave inversion is seen in leads V_1, V_2 [27] unlike acute anterior myocardial infarction where maximum inversion is seen in leads V_3 to V_6 (Fig. 35.8).

(d) Right axis deviation, "P-pulmonale" and qR configuration in lead V_1 support the possibility of pulmonary embolism.

(e) Q wave and/or T wave inversion in leads I and aVL support diagnosis of acute lateral myocardial infarction.

(f) ECG evidence of left atrial overload supports possibility of acute left ventricular infarction.

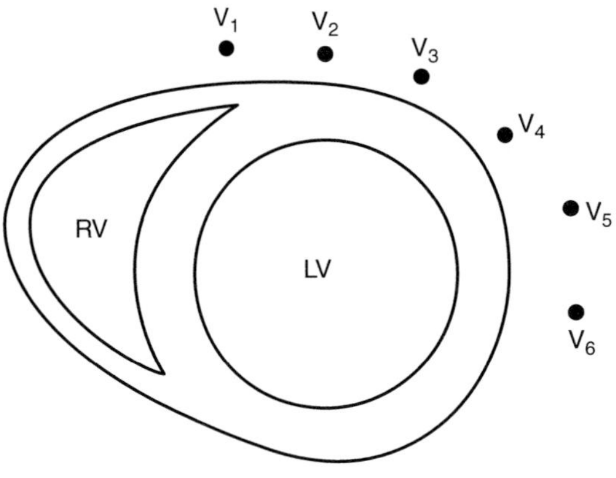

Fig. 35.7 Diagram showing relation of electrocardiographic leads to the ventricles in horizontal plane. *LV* left ventricle, *RV* right ventricle. (Reproduced with permission from: Mittal SR. Acute Right Ventricular Pressure Overload—Acute Pulmonary Embolism. Cardiology Today-2017;21:41–46)

6.3 Right Ventricular Infarction (Fig. 35.9)

Electrocardiographic differentiation from pulmonary embolism is difficult.

(a) Classical findings of inferior infarction with q wave and ST-segment elevation in right sided chest leads support the diagnosis of right ventricular infarction (Fig. 35.9).

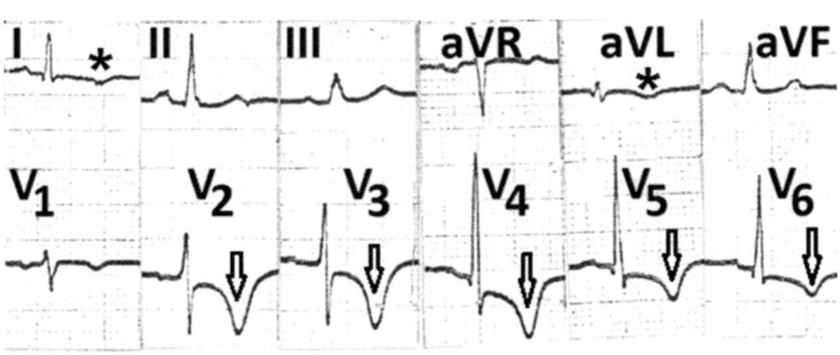

Fig. 35.8 Electrocardiogram from a patient with non ST-segment elevation anterior infarction showing deep T wave inversions from leads V_2 to V_6 (marked ⇧). The T wave is also inverted in leads I and aVL (marked *). Leads II, III, and aVF are normal. (Reproduced with permission from: Mittal SR. Acute Right Ventricular Pressure Overload—Acute Pulmonary Embolism. Cardiology Today-2017;21:41–46)

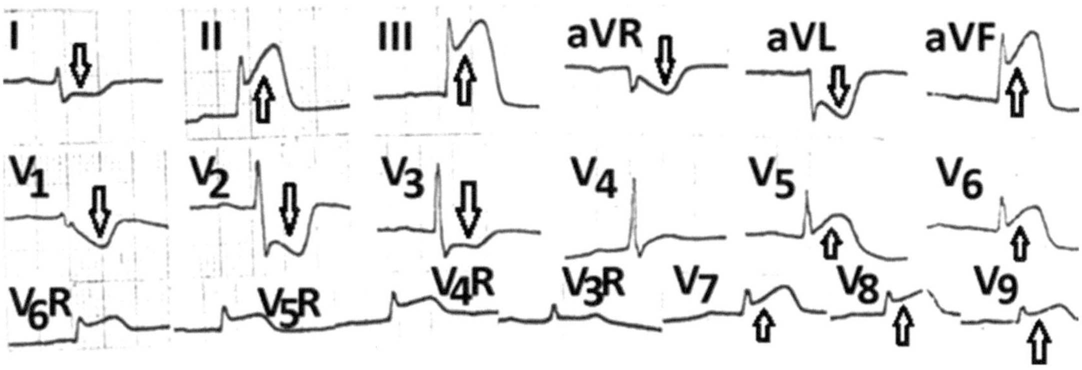

Fig. 35.9 Electrocardiogram from a patient with right ventricular infarction. Leads II, III, aVF, and V_5 to V_9 also show ST-segment elevation (marked ⇧) suggestive of concomitant inferior-posterior and lateral wall infarction. Leads I, aVR, aVL, and V_1 to V_3 show reciprocal ST-segment depression (marked ⇧). (Reproduced with permission from: Mittal SR. Acute Right Ventricular Pressure Overload—Acute Pulmonary Embolism. Cardiology Today-2017;21:41–46)

(b) Deep T wave inversion in leads V_1 & V_2 favor diagnosis of pulmonary embolism.
(c) Bradycardia or atrioventricular block support the diagnosis of right ventricular infarction.
(d) Right axis deviation and P-pulmonale support the possibility of pulmonary embolism.

Summary

- Small or submassive pulmonary embolism may not produce any diagnostic electrocardiographic findings other than sinus tachycardia. There may be a mild rightward shift in mean frontal plane QRS axis. These findings alone are not diagnostic but may be helpful in a given clinical setting specially if a previous ECG is available for comparison. Electrocardiogram is also useful because it helps in excluding other causes of sudden onset of breathlessness.
- Even massive pulmonary embolism may not produce any diagnostic electrocardiographic finding other than sinus tachycardia and nonspecific ST-segment or T wave abnormalities. Electrocardiographic exclusion of any cardiac cause for acute onset breathlessness gives an indirect clue to the diagnosis of acute massive pulmonary embolism in a given clinical context.
- Some electrocardiographic findings suggestive of acute pulmonary embolism are:

  New and sudden change in QRS axis by +30° or more.

  New and sudden appearance of prominent R wave in leads aVR and V_1.

  New and sudden appearance of S_1, Q_{III}, and inverted T_{III} pattern. Q wave and an inverted T wave in lead III are nonspecific findings frequently seen in the normal population. New appearance of S wave in lead I is more specific specially when associated with prominent S wave in leads V_5 and V_6.

  New and sudden appearance of ST-segment depression and T wave inversion in leads V_1 to V_3.

  New and sudden appearance of incomplete or complete right bundle branch block. This can also happen in acute occlusion of left anterior descending coronary artery or its major septal branch.

  Above-mentioned electrocardiographic changes are neither sensitive nor specific. These changes can be present in acute asthma or acute decompensation of chronic cor pulmonale. Clinical correlation and comparison with electrocardiogram recorded before onset of symptoms is very important.

- Preexisting left ventricular abnormalities, for example, left ventricular hypertrophy or old myocardial infarction, may dominate the electrocardiogram even in the presence of massive pulmonary embolism.
- Hypotension and hypoxia secondary to massive pulmonary embolism may produce electrocardiographic signs of left ventricular global ischemia.

MCQs

Q1. In small pulmonary embolism heart rate is:

(a) Around 90/min
(b) >150/min
(c) <60/min
(d) <45/min

Q2. In pulmonary embolism, ECG changes are:

(a) Early
(b) Transient
(c) Delayed
(d) Permanent

Q3. Commonest finding in submassive pulmonary embolism is:

(a) Sinus tachycardia
(b) Atrial flutter
(c) Atrial fibrillation
(d) Atrial tachycardia

Q4. Commonest finding in submassive pulmonary embolism is:

(a) S wave in lead I
(b) Q wave in lead III
(c) Inverted T wave in lead III
(d) P-pulmonale

Q5. In submassive pulmonary embolism, Q wave in lead III is:

(a) >0.04 s
(b) Narrow
(c) > 25% of ensuring R wave
(d) >50% of ensuring R wave

Q6. In submassive pulmonary embolism, commonest sequential change is:

(a) T wave inversion in leads V_1 to V_3
(b) T wave inversion in leads V_5 and V_6
(c) T wave inversion in leads I and aVL
(d) ST-segment depression in leads V_5 and V_6

Q7. In submassive pulmonary embolism, common ECG finding is:

(a) S_1, q_{III}, inverted T_{III}
(b) ST elevation in leads II, III, aVF
(c) ST elevation in leads V_1 to V_3
(d) ST depression in lead V_5 and V_6

Q8. In a case of pulmonary embolism transient right bundle branch block suggests:

(a) Submassive pulmonary embolism
(b) Massive pulmonary embolism
(c) Rate-related bundle branch block
(d) Age-related bundle branch block

Q9. Which ECG finding is sensitive in diagnosing massive pulmonary embolism?

(a) ST-segment elevation in leads V_1 to V_3
(b) ST-segment elevation in leads V_3R to V_6R
(c) ST-segment elevation in lead aVR
(d) None

Q10. In the context of pulmonary embolism, which ECG findings suggest right ventricular: dysfunction?

(a) T inversion in leads V_1 to V_3
(b) Complete RBBB
(c) P-Pulmonale
(d) None

Q11. In the context of pulmonary embolism, which ECG findings correlates with prognosis?

(a) Tachycardia
(b) T wave inversion in leads V_1 to V_3
(c) S_I, q_{III}, inverted T_{III} pattern
(d) Incomplete right bundle branch block

Q12. In the context of pulmonary embolism, which ECG finding does not correlate with adverse prognosis?

(a) Atrial arrhythmias
(b) Low voltage in limb leads
(c) q wave in lead aVF
(d) T wave inversion in leads V_1 to V_3

Q13. Which ECG finding is seen in pulmonary embolism?

(a) Atrioventricular block
(b) ST-segment elevation in lead III
(c) Progressive increase in T wave inversion in lead III
(d) q in lead III

Q14. In pulmonary embolism, maximum magnitude of T wave inversion is seen in leads:

(a) V_1 to V_3
(b) V_4 to V_6
(c) V_7 to V_9
(d) V_3R to V_6R

Q15. Pulmonary embolism is associated with simultaneous T wave inversion in leads:

(a) III and V_1–V_2
(b) II and V_4–V_5
(c) I, aVL, V_5–V_6
(d) III and V_3R–V_6R

Q16. Commonest electrocardiographic finding in small pulmonary embolism is:

(a) S_I, q_{III}, inverted T_{III} pattern
(b) Peaked P waves
(c) Right axis deviation more than +120°
(d) None

Q17. Commonest electrocardiographic findings in submassive pulmonary embolism are:

(a) Sinus tachycardia
(b) Right atrial enlargement
(c) New right axis deviation
(d) All

Q18. Commonest electrocardiographic finding in massive pulmonary embolism is:

(a) Sinus tachycardia
(b) P-pulmonale
(c) S_1, q_{III}, inverted T_{III} pattern
(d) Prominent R wave in lead V_1

Q19. Which is the most specific electrocardiographic finding in submassive pulmonary embolism?

(a) New appearance of complete right bundle branch block
(b) New T wave inversion in leads V_1 to V_3
(c) S_I, Q_{III}, inverted T_{III} pattern
(d) None

Q20. New right bundle branch block can appear in:

(a) Massive pulmonary embolism
(b) Sudden occlusion of major septal coronary artery
(c) Sudden occlusion of right coronary artery
(d) All

Q21. Acute change in right axis deviation by +30° or more can occur in:

(a) Submassive pulmonary embolism
(b) Acute asthma
(c) Development of left posteroinferior fascicular block
(d) All

Q22. New appearance of right axis deviation can occur in:

(a) Small pulmonary embolism
(b) Acute decompensation of chronic cor pulmonale
(c) Lateral infarction
(d) All

Q23. Electrocardiogram within normal limits can exclude:

(a) Mild pulmonary embolism
(b) Massive pulmonary embolism
(c) Left ventricular failure
(d) None

Q24. Electrocardiographic findings of left ventricular hypertrophy or old anterior myocardial infarction can exclude the possibility of:

(a) Mild pulmonary embolism
(b) Submissive pulmonary embolism
(c) Massive pulmonary embolism
(d) None

Q25. Old inferior myocardial infarction produces:

(a) Q_{III} and inverted T_{III} pattern
(b) Q with upright T in lead III
(c) S_I, Q_{III}, inverted T_{III} pattern
(d) R with upright T in lead III

Q26. Acute, right ventricular pressure overload can be suggested by new and sudden appearance of:

(a) Prominent R wave in leads aVR and V_1
(b) Prominent S wave in leads I, aVL, V_6
(c) ST-segment depression extending from V_1 to V_3
(d) All

Answers

(1) a (2) a,b (3) a (4) a (5) b (6) a (7) a (8) b,c (9) d (10) a,b,c (11)a,b (12) c (13) d (14) a (15) a (16) d (17) a,c (18) a (19) b (20) a,b (21) d (22) b,c (23) b,c (24) d (25) a (26) b.

References

1. Abecasis J, Monge J, Alberca D, Grenho MF, Arroja I, Aleixo AM. Electrocardiographic presentation of massive and sub-massive pulmonary embolism. Rev Port Cardiol. 2008;27:591–610.
2. Goldhaber SZ. Pulmonary embolism. In: Mann DL, Zipes DP, Libby P, Bonow RO, editors. Braunwald's heart disease. Philedelphia: Elsevier; 2015. p. 1664–81.
3. De Luna AB, Goldwasser D, Fiol M, Bayes-Genis A. Surface electrocardiography. In: Fuster V, Walsh RA, Harrington RA, editors. Hurst's the heart. New York: McGraw Hill; 2011. p. 307–70.
4. Mittal SR, Maheshwari M. Electrocardiographic changes in submassive pulmonary embolism. Indian Heart J. 2005;57:80–1.
5. Schamroth L. Acute pulmonary embolism. In: Schamroth C, editor. An introduction to electrocardiography. Blackwell Science: France; 1982. p. 227–32.
6. Emren SV, Arikan ME, Senoz O, Varis E, Akan E. Acute pulmonary embolism mimicking inferior myocardial infarction. Arch Turk Soc Cardiol. 2014;42:290–3.
7. Ferrari E, Imbert A, Chevalier T, Mihoubi A, Morand P, Baudouy M. ECG in pulmonary embolism. Predictive valve of negative T waves in precordial leads. 80 case reports. Chest. 1997;111:537–43.
8. Ciurzynski M, Kurzyna M, Ficygowski W, Wocial AK, Szule M, Chlebus M, et al. Electrocardiographic changes in patients with hemodynamically significant pulmonary embolism. Kardiol Pol. 2000;52:454–9.
9. Kostrubiec M, Jankowski K, Pedowska- Wloszek J, Puncewicz B, Pacho S, Lichodziejewska B, et al. Signs of myocardial ischemia on electrocardiogram correlate with elevated plasma cardiac troponin and right ventricular systolic dysfunction in acute pulmonary embolism. Cardiol J. 2010;17:157–62.
10. Petrov DB. Appearance of right bundle branch block in electrocardiograms of patients with pulmonary embolism as a marker for obstruction of the main pulmonary trunk. J Electrocardiol. 2001;34:185–8.
11. Ozer N, Yorgun H, Canpolet U, Ates AH, Aksoyck S. Pulmonary embolism presenting with evolving electrocardiographic abnormalities mimicking anteroseptal myocardial infarction: a case report. Med Princ Pract. 2011;20:577–80.
12. Chia BL, Tan HC, Lim YT. Right sided chest lead electrocardiographic abnormalities in acute pulmonary embolism. Int J Cardiol. 1997;61:43–6.
13. Grand A, Taine B, Huret JF, Pernot F, Lagabrielle D. An overlooked electrocardiographic sign of acute embolic corpulmonale: elevation of the ST segment in right precordial leads. Ann Cardiol Angiol. 1985;34:361–2.
14. Kukla P, Dlugopolski R, Krupa E, Furtake R, Mirek-Bryniarska E, Szeleme R, et al. How pulmonary embolism mimics acute coronary syndrome. Kardislogia Polska. 2011;69:235–40.

15. Ahonen A. Electrocardiographic changes in massive pulmonary embolism. Analysis of the changes in the ST-segment and T wave. Acta Med Scand. 1977;201:543–5.
16. Wang F, Kang JQ, Chen SY, Wang GF. Acute pulmonary embolism with electrocardiographic changes mimicking acute coronary syndrome: a case report and literature review. Int J Clin Exp Med. 2015;8:1038–42.
17. Panukollu G, Gowde RM, Khan IA, Wilbur SL, Vasavada BC, Sacchi TJ. QT interval prolongation with global T wave inversion: a novel ECG finding in acute pulmonary embolism. Ann Noninvasive Electrocardiol. 2004;9:94–8.
18. Hariharan P, Dudzinski DM, Okechukwu I, Takayesu JK, Chang Y, Kabrtiel C. Association between electrocardiographic findings, right heart strain and short term adverse clinical events in patients with acute pulmonary embolism. Clin Cardiol. 2015;38:236–42.
19. Choi BY, Park DG. Normalization of negative T wave on electrocardiography and right ventricular dysfunction in patients with an acute pulmonary embolism. Korean J Intern Med. 2012;27:53–9.
20. Stein PD, Matta F, Sabra MJ, Treadaway B, Vijapura C, Warren R, et al. Relation of electrocardiographic changes in pulmonary embolism to right ventricular enlargement. Am J Cardiol. 2013;112:1958–61.
21. Kostrubiec M, Hrynkiewiez A, Pedowska-Wtoszek J, Pacho S, Ciurzynski M, Jankowski K, et al. Is it possible to use standard electrocardiography for risk assessment of patients with pulmonary embolism? Kardiol Pol. 2009;67:744–50.
22. Daniel KR, Courtney DM, Kline JA. Assessment of cardiac stress from massive pulmonary embolism with 12 leads ECG. Chest. 2001;120:474–81.
23. Tosi MS, Merlino JD, Leeper KU. Electrocardiographic score and short term outcomes of acute pulmonary embolism. Am J Cardiol. 2007;100:1172–6.
24. Geibel A, Zehender M, Kasper W, Olschawski M, Klima C, Konstantinides SV. Prognostic valve of the ECG on admission in patients with acute major pulmonary embolism. Eur Respir J. 2005;25:843–8.
25. Fedullo PF. Pulmonary embolism. In: Fuster V, Walsh RA, Harrington RA, editors. Hurst's the heart. New York: McGraw Hill; 2011. p. 1634–54.
26. Lynch RE, Stem PD, Bruee TA. Leftward shift of frontal plane QRS axis as a frequent manifestation of acute pulmonary embolism. Chest. 1972;61:443–6.
27. Kosuge M, Ebina T, Hibi K, Tsukahara K, Iwahashi N, Umemura S, et al. Differences in negative T waves between acute pulmonary embolism and acute coronary syndrome. Circ J. 2014;78:483–9.

36 Biventricular Enlargement/Hypertrophy/Overload

1 Etiology

Common causes include

- Mitral regurgitation with pulmonary artery hypertension. This is common with rheumatic etiology.
- Right ventricular hypertrophy from any cause with concomitant systemic hypertension/gross aortic regurgitation/moderately severe mitral regurgitation.
- Ventricular septal defect or atrioventricular septal defect with increase in pulmonary vascular resistance.
- Cardiomyopathy.

2 Electrocardiographic Diagnosis

Electrocardiographic diagnosis of early biventricular hypertrophy (BVH) is difficult because normally greater left ventricular forces require a much greater degree of right ventricular hypertrophy, before it can manifest on ECG [1]. This is because the opposing forces of hypertrophy of the two ventricles cancel each other [2]. Further, intraventricular conduction defects or accompanying hypertrophy of the left ventricle also affect electrocardiographic diagnosis of biventricular hypertrophy. Several criteria are used for electrocardiographic diagnosis of biventricular hypertrophy.

2.1 Specific Criteria

(a) In the presence of voltage criteria of left ventricular hypertrophy (LVH), one or more of the following findings suggest biventricular hypertrophy [3, 4].

- Rightward deviation of the mean frontal plane QRS axis.
- Clockwise rotation in horizontal plane, that is, shift of precordial transition zone toward left lateral chest leads (Fig. 36.1).
- Deep S wave in leads V_5, V_6.
- Electrocardiographic signs of right atrial overload (Figs. 36.2 and 36.3).
- Tall R wave in right precordial leads—R/s ratio greater than 1 [5] (Figs. 36.1, 36.2, and 36.3).

Caution

- Asthenic individuals may have voltage criteria of left ventricular hypertrophy even in the absence of true LVH
- Emphysema and obesity may mask voltage criteria of left ventricular hypertrophy.

(b) *Katz-Wachtel sign* [5]

Tall R wave with deep S wave in leads V_2 to V_4 with combined amplitude greater than 60 mm is called Katz-Wachtel sign [3]. It is more likely to occur in children with ventricular septal defect or atrioventricular septal defect with pulmonary

S. Mittal, *Insights into Electrocardiograms with MCQs*,
https://doi.org/10.1007/978-981-99-0127-2_36

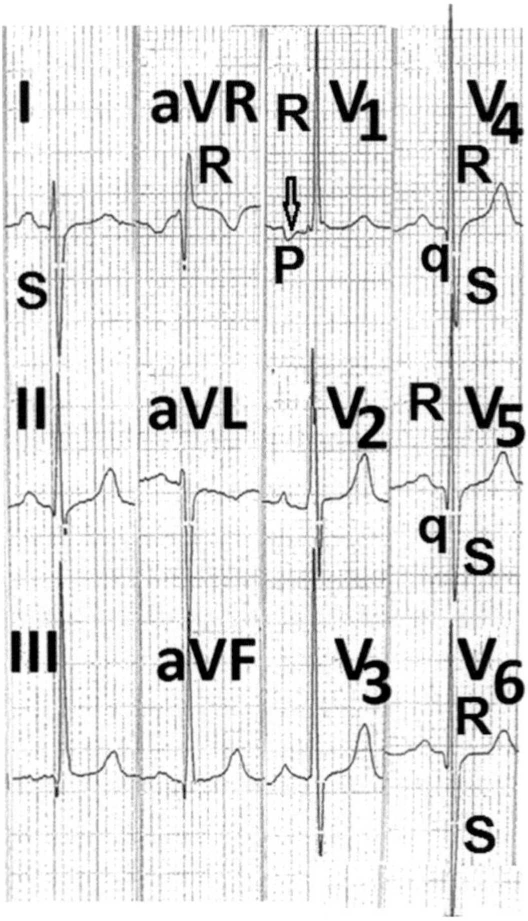

Fig. 36.1 Electrocardiogram from a patient of ventricular septal defect with pulmonary artery hypertension. Rightward deviation of mean frontal plane QRS axis (dominant S wave in lead I), dominant R wave in lead V_1, prominent R wave in lead aVR, and clockwise rotation (persistence of S wave up to lead V_6) suggest right ventricular hypertrophy. Dominant negative deflection of P wave in lead V_1 (marked as ⇩) and prominent q wave and tall R wave in lead V_4 and V_5 suggest concomitant left ventricular overload. (Reproduced with permission from: Mittal SR. Biventricular Enlargement/Hypertrophy. Cardiology Today-2017;21:85–89)

artery hypertension. In these conditions, left ventricular volume overload is associated with right ventricular pressure overload (Figs. 36.4, 36.5, and 36.6).

Specific criteria have high specificity but low sensitivity [2].

2.2 Nonspecific Criteria

Most of the patients do not fulfil standard criteria because effects of enlargement of one ventricle cancel the effects of enlargement of another ventricle. Several subtle ECG findings may also suggest early biventricular hypertrophy.

1. Right axis deviation with any of the following ECG findings.
   (a) qR configuration in leads V_5, V_6. This suggests concomitant left ventricular volume overload (Fig. 36.7).
   (b) Predominant R wave in leads $V_{5,}$ V_6 [2] but not fulfilling classical voltage criteria for LVH.
   (c) RS in lead V_1 and very deep S wave in lead V_2 [2, 6] (Figs. 36.8, and 36.9).
   (d) Prominent R wave with ST-segment depression in lead $V_{5,}$ V_6 suggestive of LVH (Fig. 36.10).
   (e) Left atrial enlargement [2] (Figs. 36.11 and 36.12) (except in mitral stenosis).

Caution Left inferoposterior hemiblock should not be wrongly interpreted as right axis deviation.

2. $S_1S_2S_3$ pattern with prominent R wave or qR in leads $V_{5,}$ $V_{6.}$

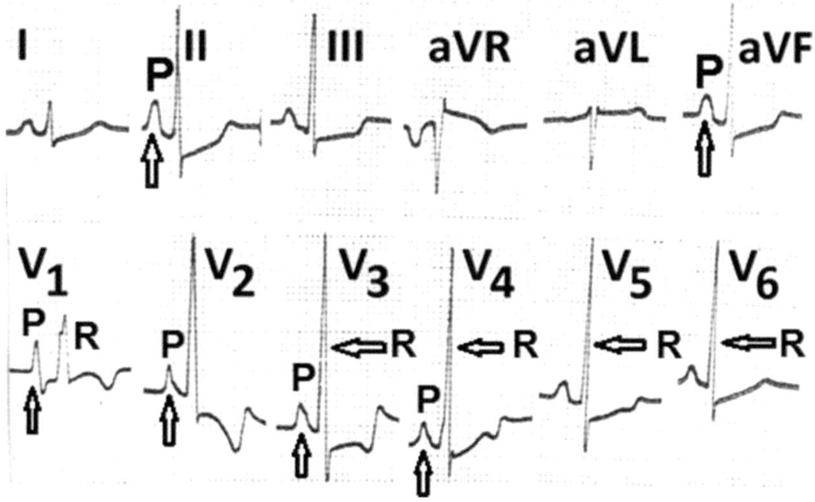

Fig. 36.2 Electrocardiogram shows dominant R wave in lead V_1 and tall peaked P waves in leads II, aVF, and V_1 to V_4 (marked as ⇧). These findings are suggestive of right atrial overload with right ventricular hypertrophy. Leads V_3 to V_6 show dominant R wave (marked ⇦) suggestive of concomitant left ventricular hypertrophy. (Reproduced with permission from: Mittal SR. Biventricular Enlargement/Hypertrophy. Cardiology Today-2017;21:85–89)

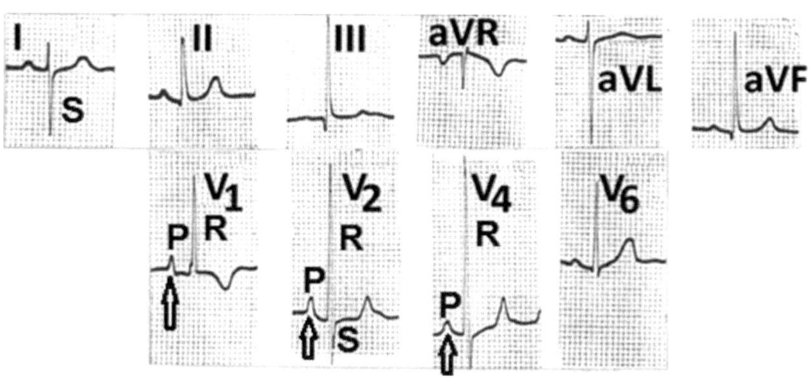

Fig. 36.3 Electrocardiogram from a patient of tetralogy of Fallot with aortic regurgitation. It shows right axis deviation (dominant S wave in lead I, peaked P waves in leads V_1 to V_4 (marked ⇧), and dominant R wave in lead V_1). These findings suggest right ventricular hypertrophy. Leads V_2 to V_4 show tall R wave and lead V_6 shows qR configuration even in the presence of right ventricular hypertrophy. These findings suggest concomitant left ventricular overload. (Reproduced with permission from: Mittal SR. Biventricular Enlargement/Hypertrophy. Cardiology Today-2017;21:85–89)

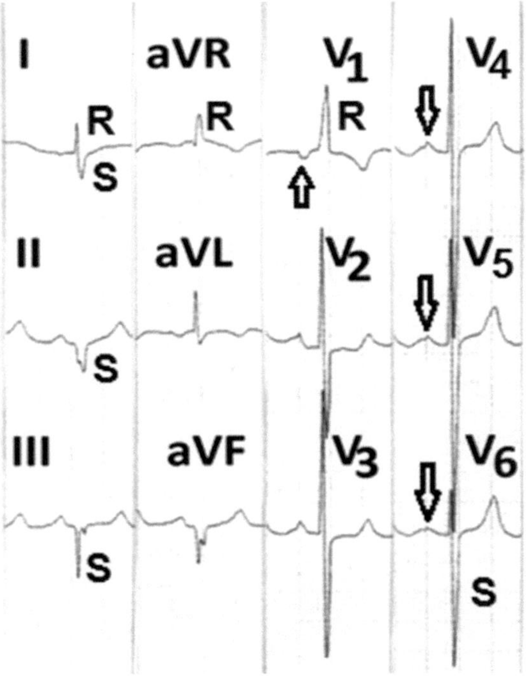

Fig. 36.4 Electrocardiogram from a case of Eisenmenger syndrome. It shows right axis deviation (dominant S wave in leads I, II, and III), dominant R waves in leads aVR and V_1, and clockwise rotation (dominant S wave in lead V_6). These findings are suggestive of right ventricular hypertrophy. P wave in lead V_1 is negative (marked as ⇧) and is broad and notched in leads V_3 to V_6 (marked ⇩) These findings suggest left atrial overload. Leads V_3 to V_5 show tall, equidiphasic (equal R wave and S wave) QRS (Katz-Wachtel sign) suggestive of biventricular overload. (Reproduced with permission from: Mittal SR. Biventricular Enlargement/Hypertrophy. Cardiology Today-2017;21:85–89)

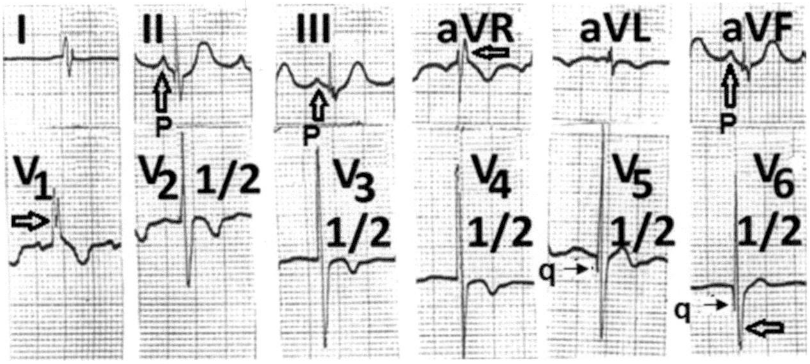

Fig. 36.5 Electrocardiogram from a case of VSD with pulmonary artery hypertension. Leads V_2 to V_3 are recorded at ½ standard (5 mm = 1 mV). Lead V_1 shows RsR' pattern (marked ⇨). Lead aVR shows terminal, slurred R wave (marked ⇦) and lead V_6 shows terminal, slurred S wave (marked as ⇦). QRS duration is 0.08 s (two small squares). These findings suggest incomplete right bundle branch block. P waves are peaked in leads II, III, and aVF (marked ⇧). This suggests right atrial overload. Leads V_5 and V_6 show prominent q waves (marked →). This suggests left ventricular volume overload. Leads V_3 to V_5 show tall equidiphasic (nearly equal R wave and S wave) QRS completes (>60 mm) (Katz-Wachtel sign), suggestive of biventricular overload. (Reproduced with permission from: Mittal SR. Biventricular Enlargement/Hypertrophy. Cardiology Today-2017;21:85–89)

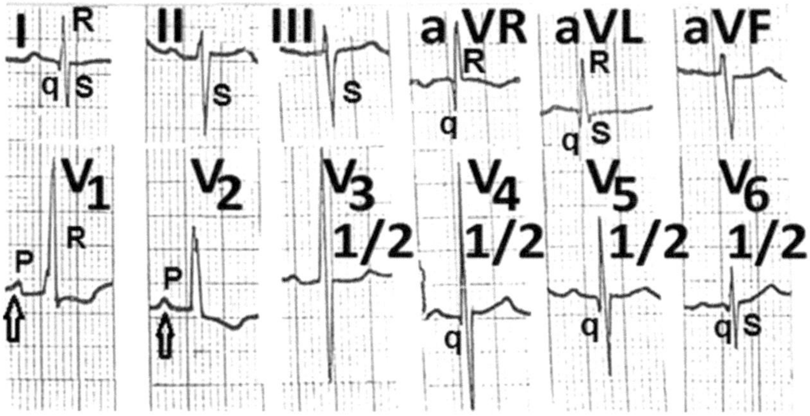

Fig. 36.6 Electrocardiogram from a case of atrioventricular septal defect with pulmonary artery hypertension. Leads V_3 to V_6 are recorded at half standard (5 m = 1 mV). It shows S_I, S_{II}, S_{III} pattern, dominant terminal R wave in lead aVR, prominent R wave in lead V_1, persistence of S wave up to lead V_6, and peaked P waves in lead V_1 and V_2 (marked ⇧). These findings suggest right ventricular hypertrophy. Leads I and aVL show initial q waves. This suggests counterclockwise loop in frontal plane. Leads V_4 to V_6 show persistent q wave in spite of RVH. This suggests concomitant left ventricular volume overload. Leads V_3 and V_4 show tall, equidiphasic QRS (70 and 80 mm, respectively). This suggests positive Katz-Wachtel sign suggestive of biventricular overload. (Reproduced with permission from: Mittal SR. Biventricular Enlargement/Hypertrophy. Cardiology Today-2017;21:85–89)

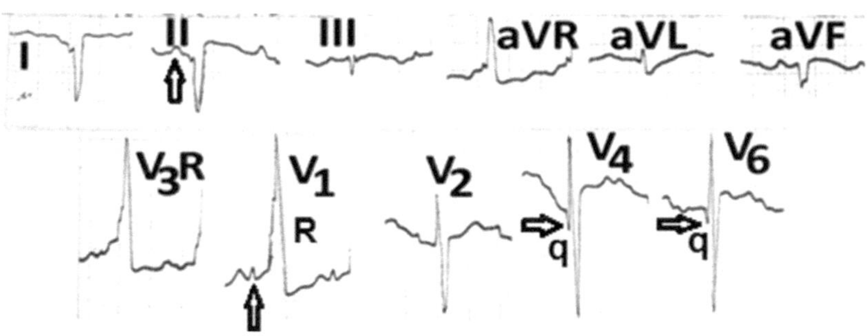

Fig. 36.7 Electrocardiogram from a case of VSD with right ventricular outflow tract obstruction showing right axis deviation (prominent S wave in lead I), prominent R wave in lead V_1, and peaked P wave in leads II and V_1 (marked ⇧). These findings suggest RVH. Leads V_5 and V_6 show prominent q waves (marked ⇨) in spite of RVH. This finding suggests concomitant left ventricular overload. (Reproduced with permission from: Mittal SR. Biventricular Enlargement/Hypertrophy. Cardiology Today-2017;21:85–89)

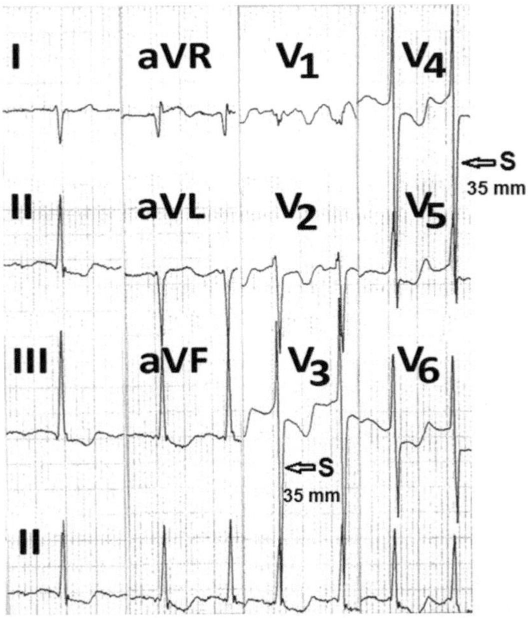

Fig. 36.8 Electrocardiogram is showing atrial fibrillation (absence of P wave and irregularly irregular rhythm of QRS in lead II), right axis deviation (dominant S wave in lead I), deep S wave in leads V_3 (35 mm) and V_4 (35 mm). (Reproduced with permission from: Mittal SR. Biventricular Enlargement/Hypertrophy. Cardiology Today-2017;21:85–89)

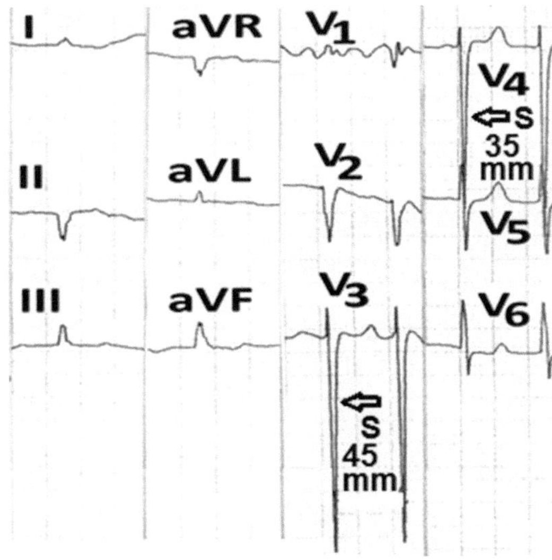

Fig. 36.9 Electrocardiogram is showing atrial fibrillation (absence of P wave and irregularly irregular rhythm of QRS). Lead V_1 shows qR configuration suggestive of right atrial overload. There are deep S waves in leads V_3 (45 mm) and V_4 (35 mm) (marked ⇦). These are suggestive of left ventricular overload. (Reproduced with permission from: Mittal SR. Biventricular Enlargement/Hypertrophy. Cardiology Today-2017;21:85–89)

Fig. 36.10 Electrocardiogram is showing right axis deviation (dominant S wave in lead I), P-Pulmonale (tall peaked P waves in leads V_2, V_3) (marked ⇧), qR configuration in leads V_1 and V_2 (suggestive of right ventricular overload with right atrial enlargement). These findings suggest RVH. ST-segment depression in leads II, III, aVF, and V_4 to V_6 (marked ⇩) suggest concomitant LVH. (Reproduced with permission from: Mittal SR. Biventricular Enlargement/Hypertrophy. Cardiology Today-2017;21:85–89)

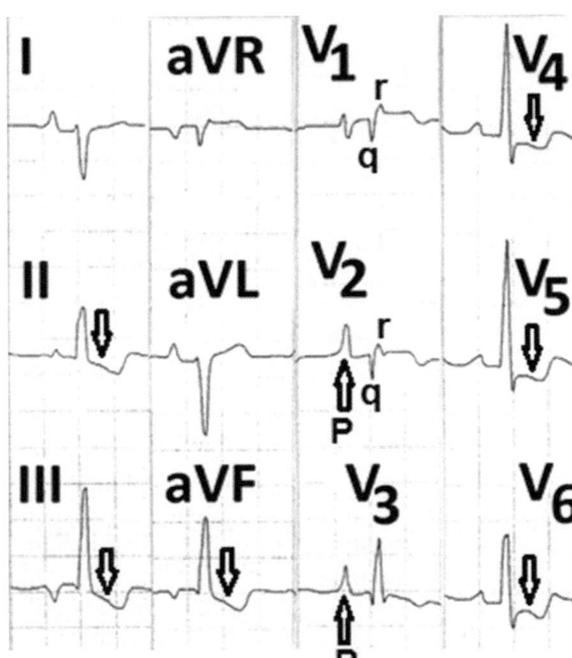

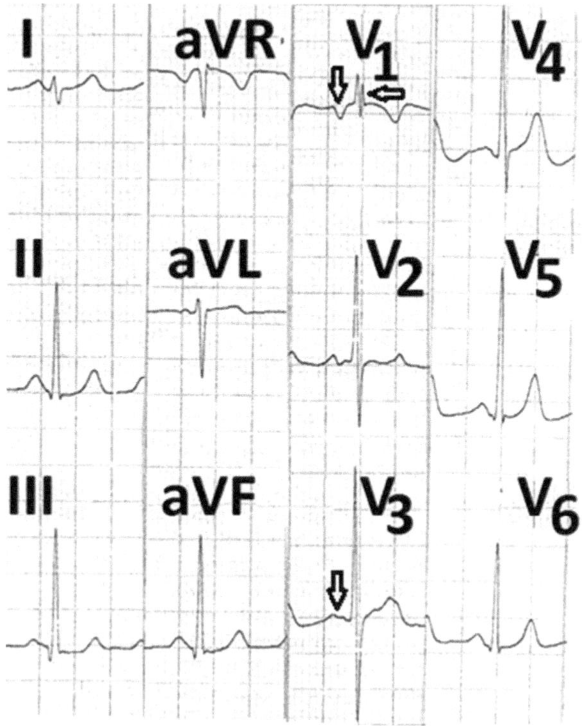

Fig. 36.11 Electrocardiogram is showing right axis deviation (prominent S wave in lead I), incomplete RBBB (RsR' pattern without prolongation of QRS duration in lead V_1 (marked ⇦). These findings are suggestive of RVH. There is evidence of left atrial overload (prominent negative deflection of P wave in lead V_1, marked as ⇩) and broad and notched P wave in lead V_3 (marked ⇩). Other findings that support the presence of concomitant LVH are persistence of small q wave in leads V_5 and V_6, tall R waves in leads V_4 and V_5, and large equidiphasic QRS in lead V_3. (Reproduced with permission from: Mittal SR. Biventricular Enlargement/Hypertrophy. Cardiology Today-2017;21:85–89)

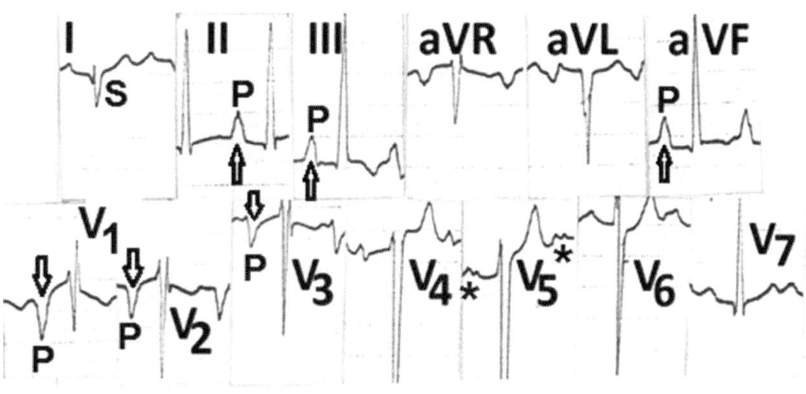

Fig. 36.12 Electrocardiogram shows right axis deviation (dominant S wave in lead I), tall peaked P waves is leads II, III, and aVF—marked ⇧ (suggestive of right atrial overload) and incomplete RBBB in lead V_1 (RsR' pattern without prolongation of the QRS duration). These findings suggest RVH. Prominent negative P waves in leads V_1 to V_3 (marked as ⇩) and wide and notched P wave in lead V_5 (marked as *) suggest left atrial overload, lead V_7 shows persistence of q wave with dominant R wave even in the presence of RVH. This finding suggests concomitant LVH. (Reproduced with permission from: Mittal SR. Biventricular Enlargement/Hypertrophy. Cardiology Today-2017;21:85–89)

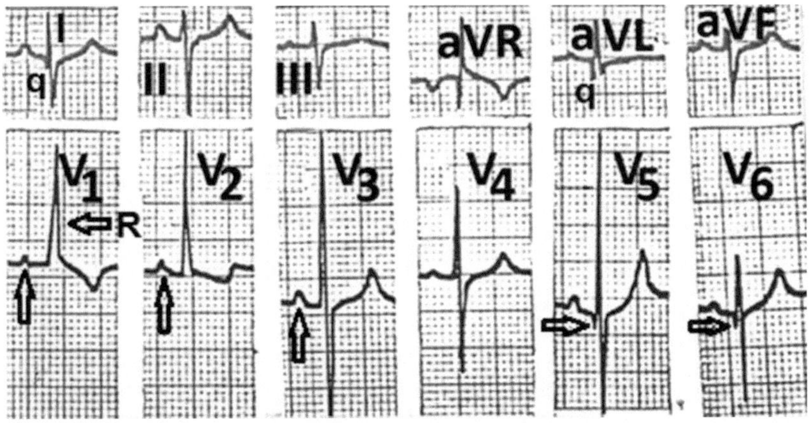

Fig. 36.13 Electrocardiogram showing S_I S_{II} S_{III} pattern. Lead V_1 shows tall R wave (marked ⇦). Lead V_6 shows persistent S wave (clockwise rotation in horizontal plane). Peaked P wave are present in leads V_1, V_2, and V_3 (marked ⇧). All these findings suggest right ventricular overload. Lead V_3 shows tall equidiphasic QRS complex and leads V_5 and V_6 show persistence of q wave (marked ⇨) in spite of RVH. These findings suggest concomitant left ventricular overload. (Reproduced with permission from: Mittal SR. Biventricular Enlargement/Hypertrophy. Cardiology Today-2017;21:85–89)

3. Tall equiphasic RS complexes in leads V_3 and V_4 not fulfilling criteria of 60 mm (Fig. 36.13).
4. Tall R wave in right as well as left precordial leads [4].

These criteria are less specific but more sensitive.

3 Conclusions

Both ventricles may hypertrophy simultaneously as in cases of ventricular septal defect or atrioventricular septal defect. Diagnosis is not difficult in these cases. In other cases, one ventricle hypertrophies in the beginning from a primary pathology and the other ventricle hypertrophies gradually as a consequence of primary pathology, that is, in cases of mitral valve disease. In yet another group of patients, the second ventricle hypertrophies due to an independent pathology, that is, development of pulmonary artery hypertension in a case of left ventricular hypertrophy or development of left ventricular hypertrophy in cases with presenting RVH, that is, development of systemic hypertension or left ventricular outflow tract obstruction. Early diagnosis of biventricular hypertrophy in the latter two groups is difficult. It is too late by the time the electrocardiogram shows diagnostic signs of biventricular hypertrophy. Early diagnosis can be made by comparison of sequentially and correctly recorded electrocardiograms.

In a case with left ventricular hypertrophy, early development of additional right ventricular hypertrophy can be suspected by the new appearance of the following electrocardiographic findings:

- Any magnitude of rightward deviation of mean frontal plane QRS axis.
- Any increase in amplitude or peaking of P waves is leads II, III, and aVF and/or leads V_1 to V_4.
- Any increase in amplitude of R wave in lead V_1.
- Any increase in depth of S wave in correctly placed lead V_6.

In a case of right ventricular hypertrophy, early development of left ventricular hypertrophy can be suspected by new appearance of any of the following findings:

- Any increase or decrease in the amplitude of q wave in correctly placed leads V_5, V_6.
- Any increase in amplitude of R wave in correctly placed leads V_5, V_6.
- Any magnitude of ST-segment depression in leads I, aVL, V_5, V_6.
- Any decrease in amplitude of T wave in leads V_5, V_6.
- Any increase in amplitude of T wave in lead V_1.

However, it is important to remember that while comparing two electrocardiograms:

- Both ECGs should be recorded at same conventional standardization of 10 mm/mV.
- Chest leads should be correctly placed in both ECGs.
- New development of any intraventricular conduction defect, myocardial ischemia, and myocardial infarction should be excluded.
- Significance of any minute change should be decided by subsequently recorded electrocardiograms.

Summary

- The forces produced by the enlargement of the right ventricle are directed anteriorly. Forces produced by enlargement of the left ventricle are directed posteriorly. Forces from enlargement of one ventricle, therefore, cancel the forces generated by enlargement of another ventricle.
- Thicker left ventricle generates greater forces. Hence right ventricle has to develop significant hypertrophy before the appearance of electrocardiographic signs of biventricular enlargement. Mild or even moderate right ventricular enlargement may not produce diagnostic changes of biventricular enlargement on electrocardiogram.
- Electrocardiographic signs of left ventricular hypertrophy with one or more of the following electrocardiographic findings of right ventricular hypertrophy suggest biventricular enlargement:

  Right axis deviation more than +90° (inferoposterior hemiblock and asthenic built should be excluded as the cause of right axis deviation).

  Right atrial enlargement suggested by peaked P waves in leads II, III, aVF, and/or lead V_1, V_2.

  Shift of transition zone (equal R wave and S wave) toward left precordial leads (V_5, V_6).

  Tall R wave in right precordial leads.

  Prominent R wave in lead V_1 with deep S in lead V_2.
- In cases of ventricular septal defect or atrioventricular septal defect, large equidiphasic QRS complexes in mid precordial leads suggest biventricular hypertrophy.
- The presence of electrocardiographic signs of right ventricular enlargement following signs in leads V_5, V_6 suggests biventricular enlargement:

  Prominent q wave.

  Tall R wave.

  Relatively tall T wave.

MCQs

Q1. Which ECG findings suggest biventricular enlargement?

(a) LVH + LAD
(b) LVH + RAD
(c) LVH + Clockwise rotation
(d) LVH + Counter-clockwise rotation

Q2. Which ECG finding does not suggest biventricular enlargement?

(a) LVH + deep S in V_5, V_6
(b) LVH + Tall R in V_1
(c) LVH + left atrial overload
(d) LVH + right atrial overload

Q3. Katz-Wachtel sign is seen in:

(a) Leads V_3R to V_1
(b) Leads V_3 and V_4
(c) Leads V_7 and V_8
(d) Leads I and aVL

Q4. In Katz-Wachtel phenomenon R + S amplitude should be more than:

(a) 30 mm
(b) 40 mm
(c) 50 mm
(d) 60 mm

Q5. Katz-Wachtel phenomenon is commonly seen in:

(a) VSD+ PAH
(b) Atrioventricular septal defect with PAH
(c) LV to RA shunt
(d) TOF + AR

Q6. Which ECG finding is not suggestive of biventricular hypertrophy?

(a) R in V_1 + qR in V_5
(b) R in V_1 + S in V_6
(c) R in V_1 + deep S in V_2
(d) RAD + left atrial overload

Q7. Causes of right axis deviation include:

(a) RVH
(b) Left posteroinferior fascicular block
(c) Left anterosuperior fascicular block
(d) RBBB

Q8. Which of the following conditions can show voltage criteria of LVH?

(a) Asthenic individuals
(b) Emphysema
(c) Obesity
(d) Pericarditis

Q9. Which electrocardiographic combination suggests biventricular enlargement?

(a) RsR' in lead V_1 with rS in lead V_5
(b) RsR' in lead V_1 with qRs in lead V_6
(c) RsR' in lead V_1 with left axis deviation
(d) All

Q10. Combination of which electrocardiographic findings suggests biventricular enlargement?

(a) R wave with inversion of T wave in lead V_6 with left posteroinferior fascicular block
(b) Right bundle branch block with left antero-superior fascicular block
(c) R wave with inversion of T wave in lead V_1 and qR with upright T wave to V_6.
(d) None

Q11. Combination of which electrocardiographic findings suggests biventricular hypertrophy?

(a) RBBB with left anterosuperior fascicular block
(b) RBBB with left posteroinferior fascicular block
(c) Bilateral bundle branch block
(d) None

Q12. Combination of which electrocardiographic findings is suggestive of biventricular hypertrophy?

(a) Wide and notched P wave in lead II with peaked positive P wave in lead V_1
(b) R/S more than 1 in lead V_1 with deep S in V_2
(c) qR in lead V_1 with qRs in V_5
(d) All

Q13. Combination of which electrocardiographic findings suggest biventricular hypertrophy?

(a) Equidiphasic large R/S complexes in V_3, V_4 with mean frontal plane QRS axis of +60°

(b) Right axis deviation with tall R and T wave inversion in lead V_5
(c) Right axis deviation with RsR' in lead V_1
(d) None

Q14. Combination of which electrocardiographic findings do not suggest biventricular hypertrophy?

(a) Wide and notched P wave in lead II with R more than 30 mm in lead V_5
(b) Peaked P wave is lead II with right axis deviation
(c) Normal P wave with RsR' in lead V_1
(d) qR in lead V_1 with deep, narrow q, tall R, and prominent S in lead V_6

Q15. In the presence of electrocardiographic signs of left ventricular hypertrophy, mean frontal plane QRS axis of +120° suggests:

(a) Mild right ventricular hypertrophy
(b) Moderate right ventricular hypertrophy
(c) Severe right ventricular hypertrophy
(d) Isolated left ventricular hypertrophy

Q16. In the presence of electrocardiographic signs of right ventricular hypertrophy, which of the following electrocardiographic findings suggest biventricular enlargement?

(a) Left axis deviation
(b) Broad and deep Q in leads V_5, V_6
(c) Prominent narrow q in leads V_5, V_6
(d) Deep S in leads V_1, V_2

Q17. In the presence of electrocardiographic signs of left ventricular pressure overload, which of the following electrocardiographic findings suggest concomitant right ventricular hypertrophy?

(a) Mean frontal plane QRS axis of +120°
(b) Broad and notched P waves
(c) Shift of transition zone to left in the precordial leads
(d) Peaked P waves

Q18. In the presence of electrocardiographic findings of left ventricular volume overload, which of the following electrocardiographic findings suggest biventricular enlargement?

(a) Deep S wave in lead V_2
(b) Tall R in led V_1
(c) R in lead V_6 taller than R in V_5
(d) Mean frontal plane QRS axis of +120°

Q19. In the presence of electrocardiographic findings of right ventricular volume overload, which of the following electrocardiographic findings suggest concomitant left ventricular overload?

(a) Tall R in lead V_5
(b) RsR' in lead V_1
(c) T wave inversion in lead V_1
(d) T wave inversion in lead V_5

Q20. In the presence of electrocardiographic findings of left ventricular pressure overload, which of the following electrocardiographic findings would suggest biventricular enlargement?

(a) Mean frontal plane QRS axis of +60°
(b) P wave amplitude of 3 mm in lead II
(c) Equal R and S in lead V_5
(d) None

Q21. In the presence of electrocardiographic findings of right ventricular pressure overload, which of the following electrocardiographic findings would suggest, biventricular enlargement?

(a) Left axis deviation
(b) R wave of more than 25 mm in lead V_5
(c) T wave inversion in lead V_1
(d) T wave inversion in lead V_5

Q22. Which combination of electrocardiographic findings do not suggest biventricular enlargement?

(a) Prominent q waves, tall R wave, and inverted T waves in leads II, III, aVF, and V_4 to V_6 with mean frontal plane QRS axis of +120°
(b) Giant P waves in lead II, negative P wave in lead V_1, tall monophasic R of 10 mm in lead V_1 with T wave inversion in leads V_1 to V_4
(c) Giant P wave in lead II, broad negative terminal deflection of P wave in lead V_1, rR' in lead V_1, qR with T wave inversion in lead V_6
(d) qR in lead V_1 and aVR, mean frontal plane QRS axis of +140° and RS in V_6

Answers

(1) b,c (2) c (3)b (4) d (5) a,b (6) b (7) a,b,d (8) a(9) b (10) c (11) d (12) d (13) a,b (14) a,b,c (15) c (16) c (17) a,c,d (18) b,d (19) a,d (20) b,c (21) b,d (22) b,d

References

1. Mirvis DM, Goldberger AL. Electrocardiography. In: Mann DL, Zipes DP, Libby P, Bonow RO, editors. Braunwald's heart disease. Philadelphia: Elsevier; 2015. p. 114–52.
2. De luna AB, Goldwasser D, Fiol M, Bayes-Genis A. Surface electrocardiography. In: Fuster V, Walsh RA, Harrington RA, editors. Hurst's the heart. New York: McGraw Hill; 2011. p. 307–70.
3. Hancock EW, Deal BJ, Mirvis DM, Okin P, Kligfield P, Gettes LS. AHA/ACCF/HRS recommendations for the standardization and interpretation of the electrocardiogram. Part V: electrocardiogram changes associated with cardiac chamber hypertrophy. J Am Coll Cardiol. 2009;53:992–1002.
4. Mirvis DM, Goldberger AL. Electrocardiography. In: Bonow RO, Mann DL, Zipes DP, Libby P, editors. Braunwald's heart disease. St Louis: Elsevier; 2012. p. 126–65.
5. Schamroth L. Right ventricular hypertrophy. In: Schamroth C, editor. An introduction to electrocardiography. France: Blackwell Science; 1982. p. 78–85.
6. Wagner GS, Lim TH. Chamber enlargement. In: Wagner GS, editor. Marriott's practical electrocardiography. Philadelphia: Wolters Kluwer; 2008. p. 71–96.

Hypertrophic Cardiomyopathy (HCM) 37

1 Genesis of ECG Changes

1.1 Genetic Factors

ECG abnormalities are more common in carriers of genetic defect related to HCM than in noncarriers [1]. Less severe phenotypes can have normal ECG [2]. However, particular ECG findings do not correlate with particular gene mutation [3].

1.2 Site of Left Ventricular Hypertrophy

Hypertrophy is usually concentric but some part of the ventricles may show asymmetrically more hypertrophy than the other parts. Hypertrophy can be localized to the basal interventricular septum (IVS), whole IVS, left ventricular apex, or left ventricular lateral wall [4]. ECG findings differ depending on the site of hypertrophy [5].

(a) *Asymmetrical hypertrophy of IVS* [5].

Increased magnitude of septal depolarization vector results in prominent, relatively deep but narrow Q waves in left sided and inferior leads. Representative electrocardiogram are shown in Figs. 37.1, 37.2, and 37.3.

(b) *Asymmetrical Hypertrophy of the lateral wall of the left ventricle* [5]

It presents with tall R waves with inverted T waves in left oriented leads (I, aVL, V_5 and V_6) (Fig. 37.4). Right oriented leads (V_1 and V_2) show deep S waves.

Representative electrocardiograms are shown in Figs. 37.5 and 37.6.

(c) *Asymmetrical hypertrophy of the apex of left ventricle* [5].

QRS vector is directed toward leads V_2 to V_4 and lead II. This results in tall R waves in these leads. R wave is tallest in lead V_3 and V_4 (Fig. 37.7). R wave in lead V_4 is taller than R wave in lead V_5.

Representative electrocardiogram is shown in Fig. 37.8.

(d) *Right ventricular hypertrophy* [5].

It presents as right axis deviation and tall R waves and inverted T waves in right precordial leads, right atrial enlargement, and incomplete or complete RBBB. Representative electrocardiogram is shown in Fig. 37.9.

Caveats

1. Classical electrocardiograms have been shown above. However, there may be many shades having different combinations.
2. Further, electrocardiographic findings may not be classical in children. Classical electro-

S. Mittal, *Insights into Electrocardiograms with MCQs*,
https://doi.org/10.1007/978-981-99-0127-2_37

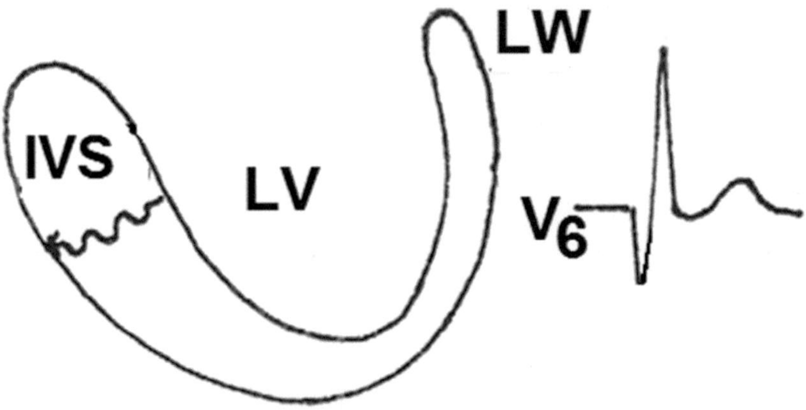

Fig. 37.1 Diagram showing asymmetrical hypertrophy of the interventricular septum (IVS) producing prominent septal q waves. *LV* left ventricle, *LW* left ventricular lateral wall. (Reproduced with permission from: Mittal SR. Hypertrophic Cardiomyopathy. Cardiology Today-2016;20:141–146)

Fig. 37.2 Electrocardiogram from a case of asymmetrical hypertrophy of the interventricular septum producing prominent septal q waves in leads I, II, V_5, V_6 (marked ⇨). (Reproduced with permission from: Mittal SR. Hypertrophic Cardiomyopathy. Cardiology Today-2016;20:141–146)

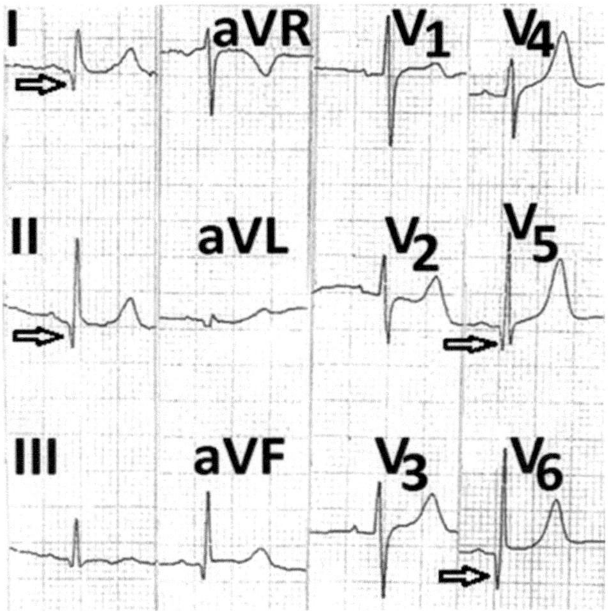

cardiographic finding may appear only with advancing age.

3. Echocardiography is needed for confirmation of diagnosis.
4. Some middle-aged patients may have concomitant systemic hypertension. However HCM should be suspected in any patient showing electrocardiographic evidence of gross left ventricular hypertrophy which is disproportionate to the severity of any concomitant illness, that is, left ventricular outflow tract obstruction, systemic hypertension, or coarctation of aorta.

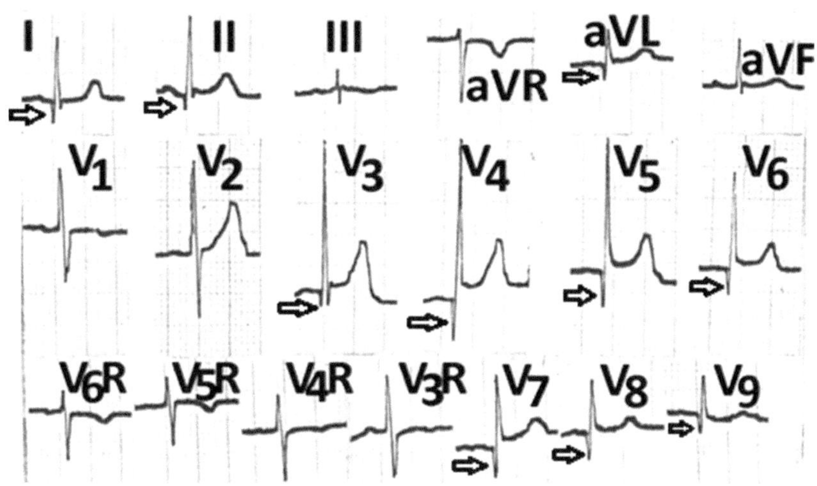

Fig. 37.3 Electrocardiogram from another case of asymmetrical hypertrophy of the interventricular septum producing prominent septal q waves in leads I, II, aVL, V_3 to V_9 (marked ⇨). (Reproduced with permission from: Mittal SR. Hypertrophic Cardiomyopathy. Cardiology Today-2016;20:141–146)

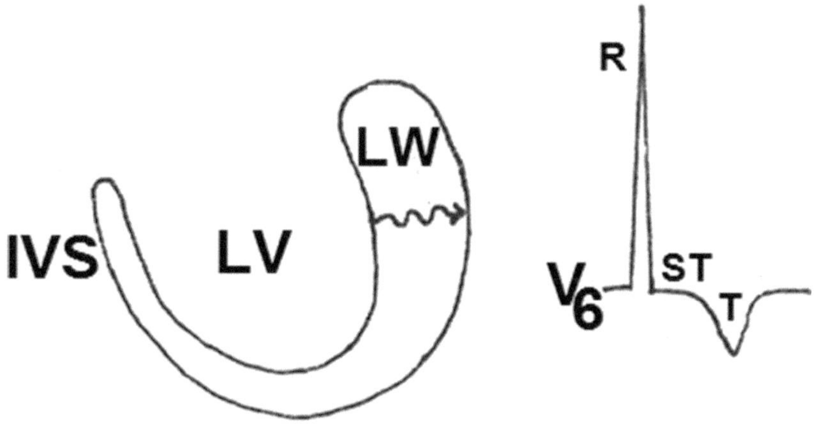

Fig. 37.4 Diagram showing asymmetrical hypertrophy of the left ventricular lateral wall (LW) producing prominent R waves with "shelf" like ST-segment and symmetrical T wave inversion in lead V_6. *IVS* interventricular septum, *LV* Left ventricle, *LW* left ventricular lateral wall. (Reproduced with permission from: Mittal SR Hypertrophic Cardiomyopathy. Cardiology Today-2016;20:141–146)

5. Electrocardiographic findings are also affected by gene mutation.
6. Some patients may have normal electrocardiogram for reason not yet known.
7. Electrocardiographic findings may not correlate with the echocardiographic findings.
8. Electrocardiographic finding cannot differentiate obstructive from nonobstructive hypertrophic cardiomyopathy.

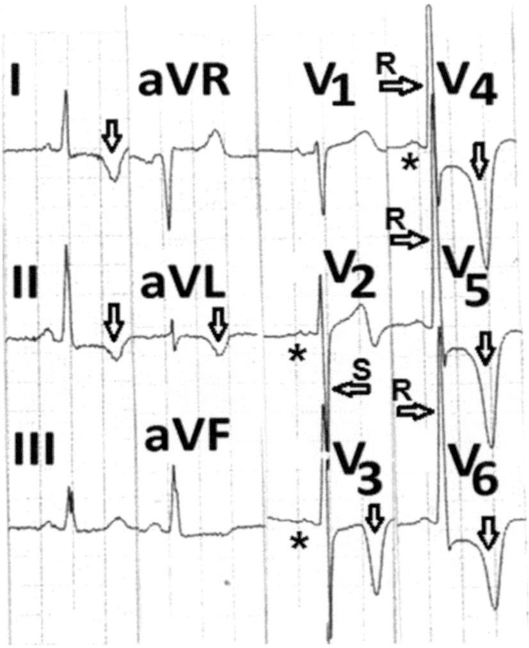

Fig. 37.5 Electrocardiogram from a case with asymmetrical hypertrophy of the left ventricular lateral wall producing tall R waves in leads V_4 to V_6 (marked ⇨) and deep symmetrical T wave inversion in leads I, II, aVL, and V_3 to V_6 (marked ⇩). Lead V_2 shows deep S wave (marked ⇦). P waves are broad and notched (P-mitrale) in leads V_2 to V_4 (marked *). These P waves suggest concomitant left atrial overload. (Reproduced with permission from: Mittal SR. Hypertrophic Cardiomyopathy. Cardiology Today-2016;20:141–146)

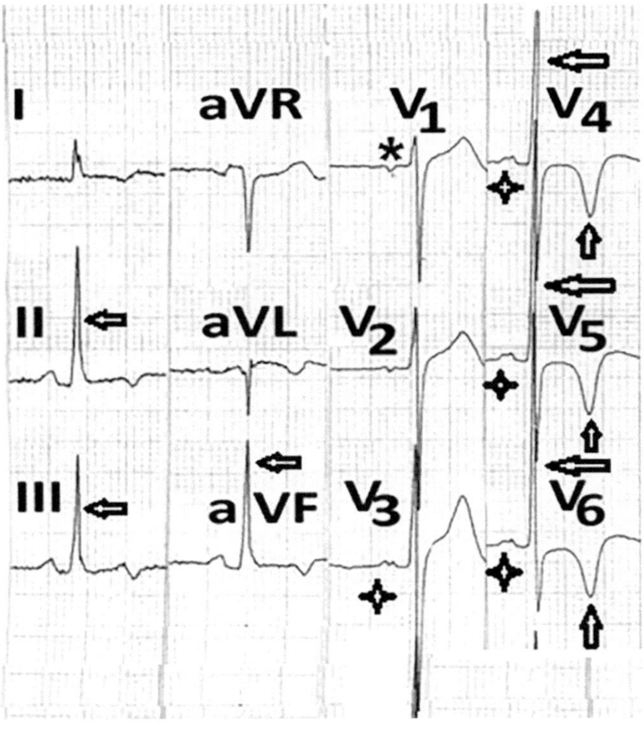

Fig. 37.6 Electrocardiogram from another case with asymmetrical hypertrophy of left ventricular lateral wall showing tall R waves in leads II, III, aVF, and V_4 to V_6 (marked ⇦) with deep symmetrical T wave inversion in leads V_4 to V_6 (marked ⇧). T waves are also inverted in leads I, II, III, aVL, and aVF. Suggesting some degree of concomitant concentric left ventricular hypertrophy. P wave in lead V_1 is negative (marked *) and P waves in leads V_3 to V_6 are broad and notched (P-mitrale) (marked ✧). These P waves suggest concomitant left atrial overload. (Reproduced with permission from: Mittal SR. Hypertrophic Cardiomyopathy. Cardiology Today-2016;20:141–146)

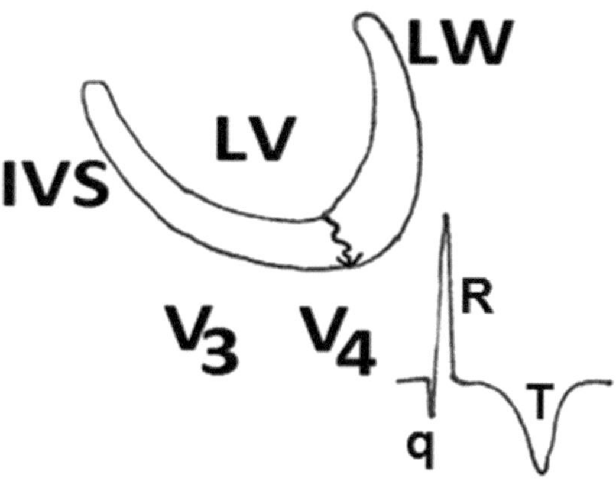

Fig. 37.7 Diagram showing asymmetrical hypertrophy of the left ventricular apex producing deep symmetrical T wave inversion in mid precordial leads (V_3 and V_4). *IVS* interventricular septum, *LV* left ventricle, *LW* left ventricular lateral wall. (Reproduced with permission from: Mittal SR. Hypertrophic Cardiomyopathy. Cardiology Today-2016;20:141–146)

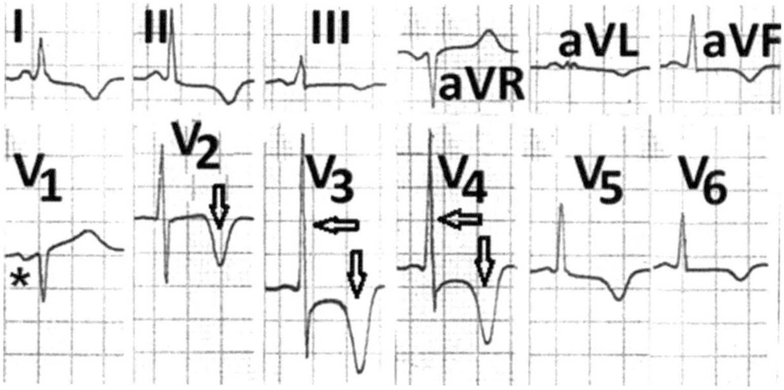

Fig. 37.8 Electrocardiogram from a case of hypertrophy of the left ventricular apex showing tall R waves (marked ⇐) with deep symmetrical inversion of T waves (marked ⇩) in leads V_2 to V_4. T waves are also inverted in leads I, II, aVL, aVF, V_5, and V_6. These suggest some degree of concomitant concentric hypertrophy of the left ventricle. R wave in leads V_3 and V_4 are taller than R wave in lead V_5. P wave in lead V_1 is negative (marked *) suggestive of left atrial overload. (Reproduced with permission from: Mittal SR. Hypertrophic Cardiomyopathy. Cardiology Today-2016;20:141–146)

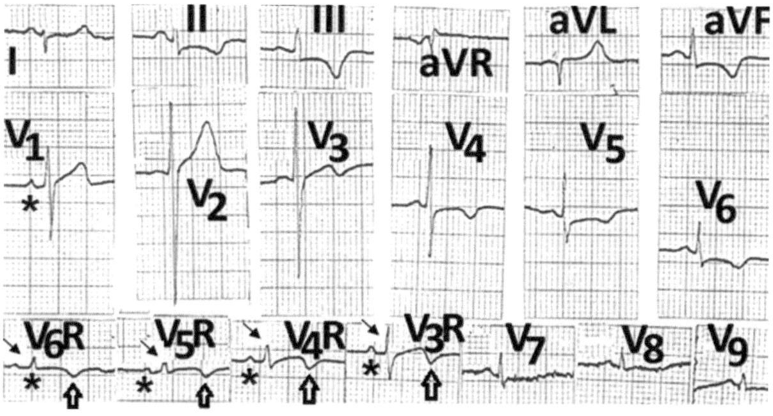

Fig. 37.9 Electrocardiogram from a case of right ventricular hypertrophy showing right axis deviation (prominent S wave in lead I) with prominent R wave (marked ↘) and T wave inversion (marked ⇧) in right-sided chest leads (V_3R to V_6R). T waves are also inverted in leads II, III, aVF, and V_4 to V_6. These suggest that there is also some degree of left ventricular hypertrophy. P wave in leads V_1 to V_6R are peaked (marked *) suggestive of right atrial overload. (Reproduced with permission from: Mittal SR. Hypertrophic Cardiomyopathy. Cardiology Today-2016;20:141–146)

2 Other Abnormalities

2.1 Atrial Enlargement

Left and/or right atrial enlargement is common depending on dominant ventricular involvement (Fig. 37.10).

Uncommon combination of right atrial enlargement with left ventricular hypertrophy or strain is very suggestive of HCM [5]. It is likely to occur in septal hypertrophy which also decreases the compliance of the right ventricle.

2.2 Intraventricular Conduction Defects

Some patients can develop left anterosuperior fascicular block (Fig. 37.10). Absence of small septal q wave in left-sided chest leads (V5, V6) may suggest incomplete left bundle branch block.

2.3 Arrhythmias

Some patients can develop ventricular extrasystoles. Atrial fibrillation can develop due to enlargement of atria and stretching of atrial conduction system (Fig. 37.11).

2.4 ST-Segment and T Wave Changes

Classically, ST-segment shows initial horizontality with sharp ST-T angle. ST-segment appears like a "shelf" [5]. In myocardial infarction, ST-segment is elevated. Uncommonly, patients of

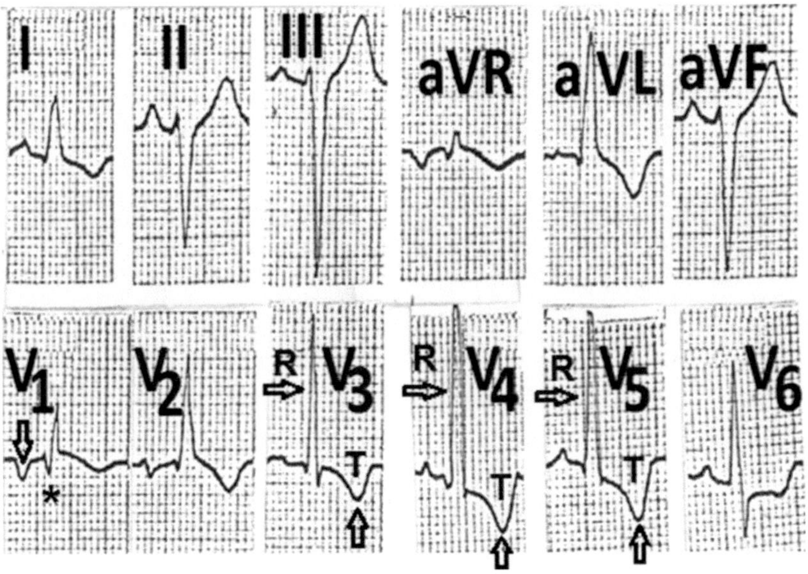

Fig. 37.10 Electrocardiogram from a case of hypertrophic cardiomyopathy showing tall R wave (marked ⇨) with deep symmetrical inversion of T waves (marked ⇧) in leads V_3, V_4, and V_5. There is concomitant incomplete right bundle branch block (narrow rsR' configuration in lead V_1 (marked *)) and left anterosuperior fascicular block (qR in leads I, aVL, and rS in leads II, III, aVF). P wave is negative in lead V_1 (marked ⇩) suggestive of left atrial overload. (Reproduced with permission from: Mittal SR. Hypertrophic Cardiomyopathy. Cardiology Today-2016;20:141–146)

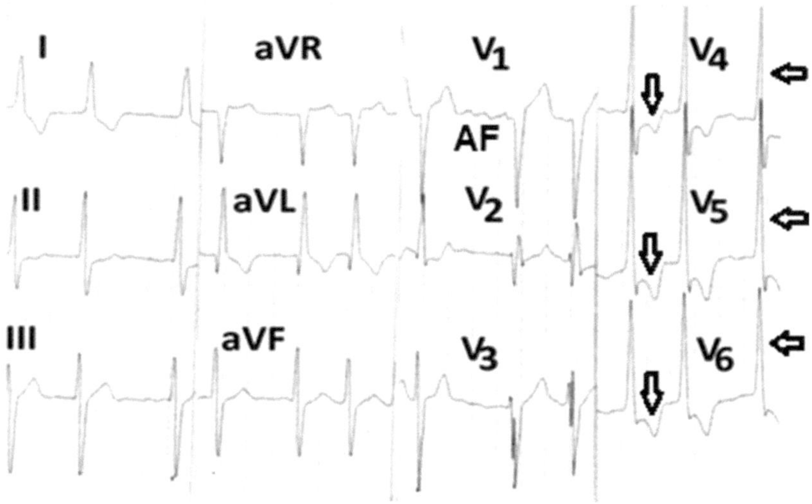

Fig. 37.11 Electrocardiogram from a case of hypertrophic cardiomyopathy showing atrial fibrillation (irregularly irregular rhythm with absence of P waves and wavy baseline in lead V_1 marked AF), left axis deviation with tall R waves (marked ⇦), and inverted T waves (marked ⇩) in leads V_4 to V_6. (Reproduced with permission from: Mittal SR. Hypertrophic Cardiomyopathy. Cardiology Today-2016;20:141–146)

HCM may show persistent ST-segment elevation [6]. In left ventricular hypertrophy due to classical systolic overload or ischemia, ST-segment is down sloping (Figs. 37.12 and 37.13). In HCM, inverted T waves have symmetrical limbs (downslope and upslope). In left ventricular systolic overload, the two limbs of the inverted T waves are asymmetrical. Representative electrocardiograms are shown in.

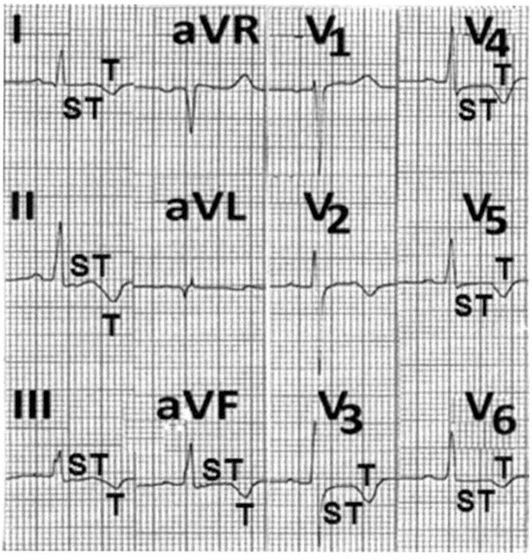

Fig. 37.12 Electrocardiogram from a case of hypertrophic cardiomyopathy showing "shelf" like ST-segment with inversion of T wave in leads I, II, III, aVF, V3–V6. (Reproduced with permission from: Mittal SR. Hypertrophic Cardiomyopathy. Cardiology Today-2016;20:141–146)

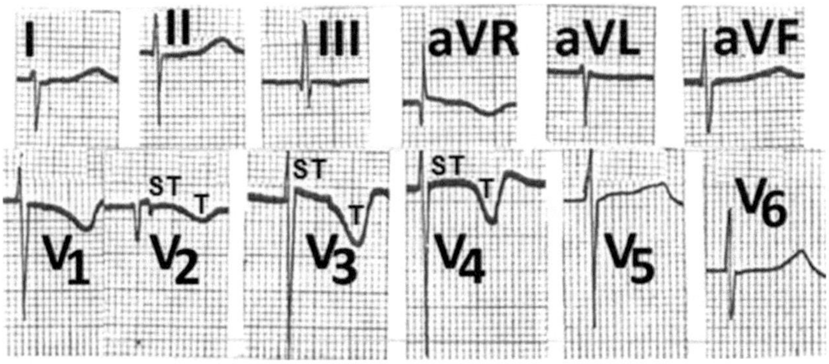

Fig. 37.13 Electrocardiogram from a case of hypertrophy of the LV apex showing "shelf" like ST-segment with symmetrical inversion of T wave in leads V_1 to V_4. (Reproduced with permission from: Mittal SR. Hypertrophic Cardiomyopathy. Cardiology Today-2016;20:141–146)

3 Differential Diagnosis

3.1 Myocardial Infarction

Abnormal Q waves are accompanied by loss of R wave and coved ST segment (Fig. 37.14a).

Non Q wave (subendocardial) infarcts are difficult to differentiate. Absence of symptoms of myocardial infarction, normal serum enzymes and absence of sequential ECG changes support the diagnosis of HCM. However, at times HCM can show ST-segment elevation resembling acute injury pattern [6].

3.2 Left Ventricular Systolic Overload (Fig. 37.14b)

Septal q waves are diminished. Tall R waves are followed by down sloping ST-segment and inversion of T waves with asymmetrical limbs. Hypertrophic cardiomyopathy (Fig. 37.14c) produces deep but narrow q waves. R waves are tall. ST-segment is "shelf like" (horizontal). T waves show deep symmetrical inversion. There are no sequential changes.

3.3 Exercise and Sports Cardiomyopathy [7] (Athlete's Heart Syndrome)

ECG usually shows resting bradycardia. First degree or Mobitz type 1 s-degree atrioventricular block can be present. Early repolarization pattern of the ST-segment supports this possibility.

3.4 Other Causes of Increased Thickness of Left Ventricular Walls on Echocardiography

3.4.1 Infiltrative Cardiomyopathies [8]

Cardiac Amyloidosis [9]

Amyloidosis is characterized by low QRS voltage in limb leads. Unusually rightward deviation of the mean frontal plane QRS axis, first degree atrioventricular block, and Q waves in leads V_1 to V_3 are present. Familial and senile amyloidosis are characterized by normal QRS voltage, nonspecific intraventricular conduction defects (specially left bundle branch block), and ST-segment and T wave changes. Atrioventricular block may occur.

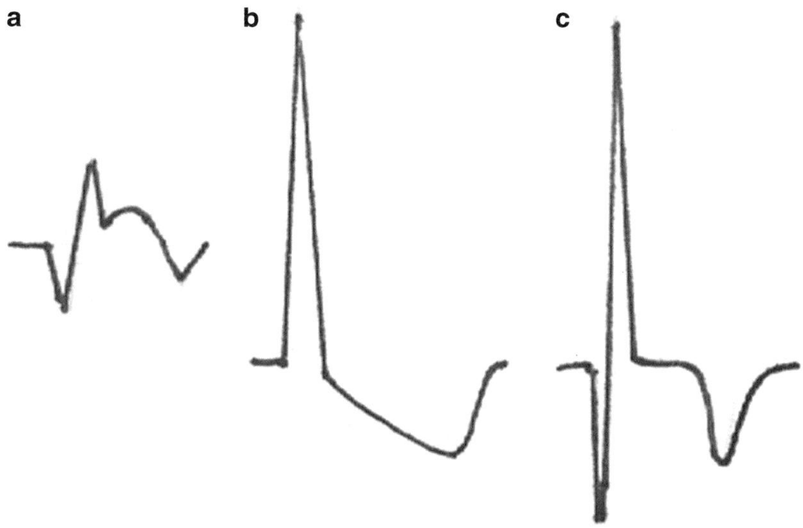

Fig. 37.14 Diagram showing the electrocardiographic configuration in (**a**) Myocardial infarction showing deep and broad Q waves, loss of R wave with coved ST-segment and inversion of T wave. (**b**) Systolic overload showing loss of septal q wave, tall R wave, down sloping ST-segment, and asymmetrical inversion of T wave. (**c**) Hypertrophic cardiomyopathy showing deep but narrow Q wave, tall R wave, "shelf" (horizontal) like ST-segment with sharp ST-T angle, and T wave inversion with symmetrical limbs. (Reproduced with permission from: Mittal SR. Hypertrophic Cardiomyopathy. Cardiology Today-2016;20:141–146)

Fabry Disease

Electrocardiographic findings include left ventricular hypertrophy with nonspecific intraventricular conduction defects.

Hemochromatosis

Common ECG findings include atrioventricular and bundle branch blocks.

3.4.2 Neurological Disorders

Friedreich Ataxia [10]

Widespread T wave inversions are common.

3.4.3 Endocrine Disorders

Acromegaly [11]

Common ECG findings include left axis deviation, prominent septal Q waves, ST-segment depression, inversion of the T wave, and conduction system defects.

4 Clinical Correlation

1. ECG is abnormal in the majority of patients of hypertrophic cardiomyopathy [4]. Left ventricular hypertrophy is present in 70–80% of the cases [4]. Totally normal ECG is against the possibility of significant disease.
2. No ECG change is unique to or characteristic of hypertrophic cardiomyopathy [2]. High voltage QRS with depressed ST-segment and inverted T waves in an asymptomatic patient without any murmur or hypertension should raise suspicion of hypertrophic cardiomyopathy [3].
3. Increased QRS voltage only weakly correlates with the magnitude of left ventricular hypertrophy.
4. Electrocardiogram cannot distinguish between obstructive and non-obstructive forms [2].
5. In some patients with severe hypertension, ECG and echocardiographic findings may

resemble hypertrophic cardiomyopathy. Regression of ECG findings after proper control of blood pressure supports the diagnosis of hypertension-induced ECG changes [12]. It may take a long period of adequate control of blood pressure before there is some regression in ECG changes.

Summary

- Patients with mild hypertrophy may have normal ECG or only ST-segment and T wave changes in leads V_5, V_6.
- Prominent relatively deep but narrow Q waves in inferior and left lateral leads suggest disproportionate hypertrophy of the interventricular septum.
- Hypertrophy of left ventricular lateral wall produces abnormally deep S waves in leads V_1, V_2 and abnormally tall R waves in leads V_5, V_6 with down sloping ST- segment and asymmetrically inverted T waves in leads I, aVL, V_5, and V_6. Increase in QRS voltage, however, may not directly correlate with the magnitude of left ventricular hypertrophy. Electrocardiogram resembles left ventricular pressure overload except that the increase in amplitude of QRS is disproportionately more as compared to the magnitude of systolic overload.
- Hypertrophy localized to the apex of the left ventricle produces prominent R waves in leads V_3 to V_5. ST-segment is horizontal with sharp ST-T angle. The T waves are inverted, deep, symmetrical, and narrow pointed. In myocardial infarction, there is loss of R wave, ST-segment is coved, and T waves are not so deep.
- Some patients can have a short PR interval, preexcitation, left axis deviation, or evidence of left atrial enlargement.
- Electrocardiogram cannot differentiate between obstructive and nonobstructive variants.
- Repolarization abnormalities (ST-segment depression and T wave inversion) are the most frequent ECG changes.
- Amyloidosis may resemble hypertrophic cardiomyopathy on echocardiography. However, in contrast to hypertrophic cardiomyopathy, QRS voltages are diminished in amyloidosis. On echocardiography, myocardial contractility is reduced. Hypertrophic cardiomyopathy shows normal contractility of myocardium.
- Isolated right ventricular hypertrophic cardiomyopathy is rare. It may not produce diagnostic electrocardiographic changes in conventionally recorded 12 leads probably because of significantly greater thickness of the left ventricle.

MCQs

Q1. In a case of hypertrophic cardiomyopathy, ECG can be normal in:

(a) Noncarriers of genetic defect
(b) Asymmetrical septal hypertrophy
(c) Hypertrophy of LV apex
(d) Hypertrophy of lateral wall of LV

Q2. Asymmetrical septal hypertrophy produces:

(a) Broad Q waves
(b) Deep but narrow Q waves
(c) Tall R waves in leads V_5, V_6
(d) Deep T wave inversion in leads V_5, V_6

Q3. Asymmetrical hypertrophy of left ventricular apex produces:

(a) Tall R waves in leads V_5, V_6
(b) Deep T inversion in lead V_6
(c) Deep symmetrical T wave inversion in leads V_2 to V_4
(d) Tall R waves in leads V_3 and V_4

Q4. Asymmetrical hypertrophy of lateral wall of the left ventricle produces:

(a) Tall R waves in leads V_5, V_6
(b) Deep symmetrical T inversion in leads V_5 and V_6

(c) Broad Q waves in leads V_3 and V_4
(d) Tall R waves in leads V_3 and V_4

Q5. Right ventricular hypertrophic cardiomyopathy does not produce:

(a) Right atrial enlargement
(b) Left axis deviation
(c) Prominent R wave in right sided chest leads
(d) Deep T inversion in leads V_3 and V_4

Q6. Hypertrophic cardiomyopathy produces:

(a) Broad Q waves
(b) Tall R waves
(c) Coved ST segment
(d) Asymmetrical T wave inversion

Q7. Hypertrophic cardiomyopathy produces:

(a) Coved ST segment
(b) Horizontal ST segment
(c) Upsloping ST segment
(d) Down sloping ST segment

Q8. Athlete's heart syndrome does not produce:

(a) Sinus tachycardia
(b) Short PR interval
(c) Early repolarization ST pattern
(d) Tall R waves

Q9. Amyloidosis produces:

(a) Tall R wave in limb leads
(b) Tall R wave in leads V_5, V_6
(c) Low voltage in limb leads
(d) LVH on echocardiography

Q10. Acromegaly does not produce:

(a) Right axis deviation
(b) Prominent septal Q waves
(c) Intraventricular conduction defects
(d) ST segment depression in leads V_5, V_6

Q11. Hemochromatosis produces:

(a) Atrioventricular block
(b) Bundle branch block
(c) Diffuse T wave inversion
(d) Abnormal Q waves

Q12. Patients with hypertrophic cardiomyopathy must have:

(a) Increased QRS voltages
(b) ST segment depression
(c) T wave inversion
(d) None of the above

Q13. Deep, narrow Q wave in leads II, III, aVF, V_5, and V_6 without any R wave, ST segment, or T wave abnormality in an asymptomatic patient could suggest:

(a) Normal variation
(b) Cardiac amyloidosis
(c) Hypertrophy of interventricular septum
(d) Inferolateral infarction

Q14. Deep S wave in leads V_1, V_2 and tall R with ST segment depression and T wave inversion in leads V_5, V_6 in an asymptomatic patient with normal clinical examination suggests the possibility of:

(a) Critical fixed subaortic stenosis
(b) Severe coarctation of aorta
(c) Hypertrophic cardiomyopathy
(d) Left ventricular volume overload

Q15. In hypertrophic cardiomyopathy, which of the following ECG findings correlates with magnitude of left ventricular hypertrophy?

(a) Amplitude of R wave in leads V5, V6
(b) Repolarization changes
(c) Both
(d) None

Q16. Which of the following electrocardiographic finding can diagnose the presence of subaortic obstruction in a case of hypertrophic cardiomyopathy?

(a) Amplitude of R wave in leads V5, V6

(b) Repolarization changes
(c) Left atrial enlargement
(d) None

Q17. Which electrocardiographic finding is never seen in hypertrophic cardiomyopathy?

(a) Short PR interval
(b) Preexcitation
(c) Left bundle branch block
(d) None

Q18. Which tachyarrhythmia can occur in hypertrophic cardiomyopathy?

(a) Ventricular tachycardia
(b) Atrioventricular nodal reentry tachycardia
(c) Atrial fibrillation
(d) None

Q19. Prominent R wave with isoelectric ST-segment and deep symmetrical pointed T wave inversion in leads V_3 to V_5 in a case of effort angina with normal clinical examination suggests:

(a) Anterior infarction
(b) Septal infarction
(c) Hypertrophic cardiomyopathy
(d) Dilated cardiomyopathy

Q20. A 40-year-old male complained of breathlessness on moderate effort. Clinical examination and electrocardiogram were normal. Treadmill stress test did not reveal any evidence of exercise-induced ischemia. Possibility is:

(a) Dilated cardiomyopathy
(b) Hypertrophic cardiomyopathy
(c) Infiltrative cardiomyopathy
(d) Myocarditis

Q21. Isolated right ventricular hypertrophic cardiomyopathy produces:

(a) Extreme right axis deviation
(b) Prominent R in lead V1
(c) Upright T wave in lead V1
(d) None

Answers

(1) a (2) b (3) c, d (4) a, b (5) b,d (6) b (7) b (8)a,b (9) c, d (10) a (11) a, b(12) d (13)c (14) c, (15) d (16) d (17) d (18) c (19) c (20) b (21) d

References

1. Mirvis DM, Goldberger AL. Electrocardiography. In: Bonow RO, Mann DL, Zipes DP, Libby P, editors. Braunwald's heart disease. Missouri: Elsevier; 2012. p. 126–65.
2. Maron BJ, Olivatto L. Hypertrophic cardiomyopathy. In: Mann DL, Zipes DP, Libby P, Bonow RO, editors. Braunwald's heart disease. Philadelphia: Elsevier; 2015. p. 1574–88.
3. de Luna AB, Goldwasser D, Fiol M, Bayes-Genis A. Surface electrocardiography. In: Fuster V, Walsh RA, Harrington RA, editors. Hurst's the heart. New York: McGraw Hill; 2011. p. 307–70.
4. Omen SR, Nishimura RA, Tajik AJ. Hypertrophic cardiomyopathy. In: Fuster V, Walsh RA, Harrington RA, editors. Hurst's the heart. New York: McGraw Hill; 2011. p. 837–64.
5. Schamroth L. Hypertrophic cardiomyopathy. In: Schamroth C, editor. An introduction to electrocardiography. France: Blackwell Science; 1982. p. 67–77.
6. Khan IA, Ajatta FO, Ansari AW. Persistent ST segment elevation: a new ECG finding in hypertrophic cardiomyopathy. Am J Emerg Med. 1999;17:296–9.
7. Thompson PD, Baggish A. Exercise and sports cardiology. In: Mann DL, Zipes DP, Libby P, Bonow RO, editors. Braunwald's heart disease. Philadelphia: Elsevier; 2015. p. 1771–8.
8. Falk RH, Hershberger RE. The dilated, restrictive and infiltrative cardiomyopathies. In: Mann DL, Zipes DP, Libby P, Bonow RO, editors. Braunwald's heart disease. Philadelphia: Elsevier; 2015. p. 1551–73.
9. Huang J, Zhao S, Chen Z, Zhang S, Lu M. Contribution of electrocardiogram in the differentiation of cardiac amyloidosis and nonobstructive hypertrophic cardiomyopathy. Int Heart J. 2015;56:522–6.
10. Grals WL, Zipes DP. Neurologic disorders and cardiovascular disease. In: Mann DL, Zipes DP, Libby P, Bonow RO, editors. Braunwald's heart disease. Philadelphia: Elsevier; 2015. p. 1890–908.
11. Klein I. Endocrine disorders and cardiovascular disease. In: Mann DL, Zipes DP, Libby P, Bonow RO, editors. Braunwald's heart disease. Philadelphia: Elsevier; 2015. p. 1793–808.
12. Pennacchini E, Musumeci MB, Fierro S, Francia P, Autore C. Distinguishing hypertension from hypertrophic cardiomyopathy as a cause of left ventricular hypertrophy. J Clin Hypertens. 2015;17:239–41.

38 Low Voltage Electrocardiogram

It is defined as QRS complexes of less than 0.5 mV in frontal plane leads and less than 1.0 mV in precordial leads [1]. In nearly 50% of the patients only limb leads are affected and chest leads have normal voltages resulting in disparity in electrocardiographic voltage criteria [2]. Attenuation of QRS voltage on correctly standardized sequential ECGs is also important even if it does not fulfil the definition [3].

1 Causes of Low Voltage QRS

For the sake of understanding, causes have been divided into different groups. However, in a given case, more than one factor may be contributory.

1.1 Incorrect Standardization

Standard ECG is calibrated to a voltage of 10 mm/mV. Incorrect standardization at 5 mm/mV produces low voltage ECG. All waves (P, QRS, and T) and all leads (limb leads as well as precordial leads) are affected.

1.2 Normal Variant Specially in the Elderly

Low QRS voltage can occur in apparently healthy individuals without any apparent explanation. Representative electrocardiogram is shown in Fig. 38.1.

Low QRS is associated with increased risk of mortality even in patients apparently free of cardiovascular disease [4, 5]. Placing left precordial leads over the breast (rather than under the breast) can produce low voltage in these leads in females with heavy breasts.

1.3 Decreased Electrical Potentials Generated by the Myocardium

1.3.1 Multiple Infarcts [6]

Extensive loss of viable myocardium is responsible for decreased QRS voltage (Fig. 38.2).

Heart failure, pulmonary congestion, and subcutaneous edema may also contribute to low voltage in these cases.

S. Mittal, *Insights into Electrocardiograms with MCQs*,
https://doi.org/10.1007/978-981-99-0127-2_38

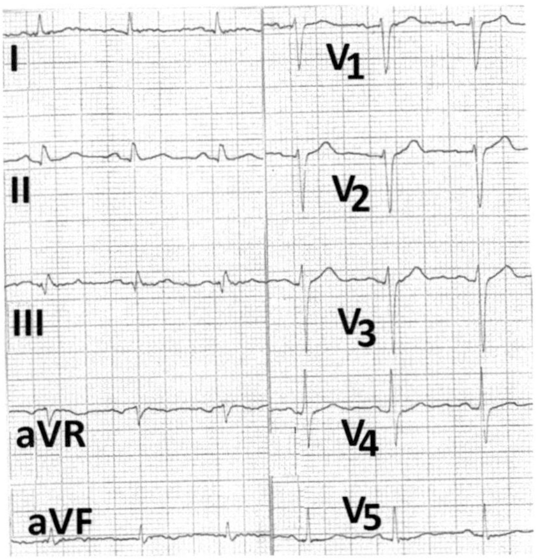

Fig. 38.1 Electrocardiogram from a normal person showing low voltage QRS in limb leads (Reproduced with permission from: Mittal SR. Low Voltage Electrocardiogram. Cardiology Today-2017;21:160–165)

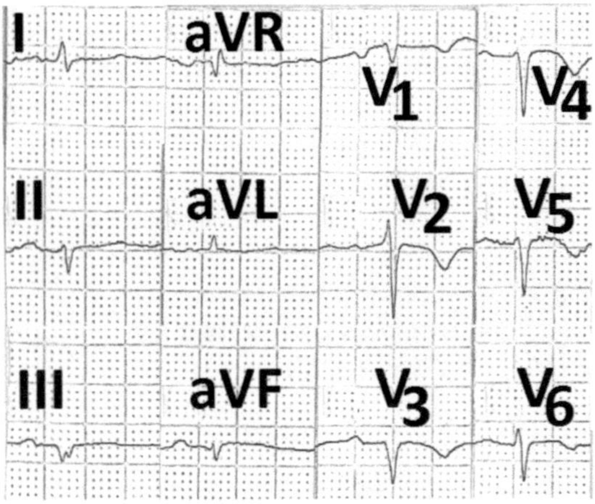

Fig. 38.2 Electrocardiogram from a patient with old inferior and fresh anterior infarction showing low voltage QRS in limb leads and precordial leads (Reproduced with permission from: Mittal SR. Low Voltage Electrocardiogram. Cardiology Today-2017;21:160–165)

1.3.2 Acute Coronary Syndrome

Low QRS voltage is associated with previous myocardial infarction, adverse hemodynamics, and higher in-hospital mortality [7].

1.3.3 Acute Phase of Takotsubo Cardiomyopathy

Reduced voltages are due to underlying myocardial edema [8]. Attenuation of QRS amplitude is transient and improves as the clinical and echocardiographic parameters improve [9]. This is in contrast to acute coronary syndrome in which there is no significant change in the QRS voltages after admission [9, 10]. However, new development of any intraventricular conduction defect can change voltages of the QRS complexes. In acute phase of Takotsubo cardiomyopathy, T waves are inverted in leads V_2 to V_4. Lead aVR may show ST-segment depression with positive T wave as this lead is diagonally opposite to the apex of the left ventricle [10].

1.3.4 Cardiac Amyloidosis

Low QRS voltage is frequent in limb as well as in precordial leads. Paradoxically, the thickness of the myocardium is increased on echocardiography. First-degree atrioventricular block is common [11] and Q waves are frequent in leads V_1 to V_3. Low voltage is associated with adverse outcome [12].

1.3.5 Fulminant Myocarditis

Sinus tachycardia is invariable. In addition to myocardial injury, peripheral edema may also contribute to low voltage of QRS complexes [13]. Ventricular activation time is increased due to slow conduction through the injured myocardium and conduction tissues. Multiple foci of myocardial necrosis produce notching and slurring of the QRS complexes. Pathologic Q waves, ST-segment depression, inversion of T wave, and atrial and/or ventricular extrasystoles can be present [14].

1.3.6 Dilated Cardiomyopathy

Low amplitude with notched and wide QRS is seen particularly in the frontal plane leads [15, 16]. Low amplitude in limb leads with high voltage in precordial leads meeting voltage criteria of LVH is strongly suggestive [2, 17, 18]. Biatrial enlargement is common due to elevated end-

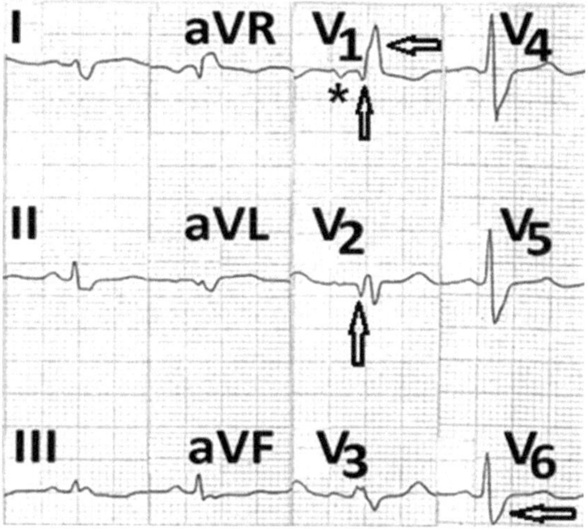

Fig. 38.3 Electrocardiogram from a patient of dilated cardiomyopathy showing low voltage in the limb leads, left atrial overload, (negative P wave in lead V_1 marked *), right bundle branch block (slurred terminal R wave in lead V_1 and slurred terminal S wave in lead V_6 marked ⇦), and q waves in leads V_1 and V_2 (marked ⇧) (Reproduced with permission from: Mittal SR. Low Voltage Electrocardiogram. Cardiology Today-2017;21:160–165)

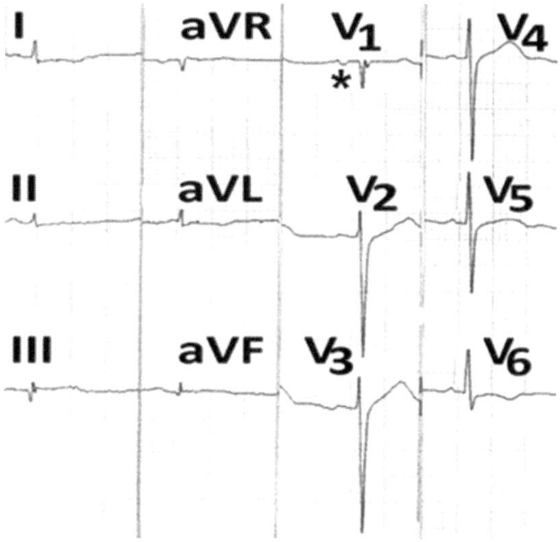

Fig. 38.4 Electrocardiogram from a patient of dilated cardiomyopathy showing low voltage in limb leads and normal to increased voltage in leads V_2 to V_5. P wave in lead V_1 is negative (marked *). It is suggestive of left atrial overload (Reproduced with permission from: Mittal SR. Low Voltage Electrocardiogram. Cardiology Today-2017;21:160–165)

diastolic pressure and associated mitral and tricuspid regurgitation. Left axis deviated is common due to left anterosuperior fascicular block. LBBB is common. Combination of LBBB with right axis deviation is strongly suggestive [19, 20]. Representative ECGs are shown in Figs. 38.3 and 38.4.

1.3.7 Heart Failure

Myocardial edema per se, lung congestion, pleural effusion, and subcutaneous edema contribute to low voltage QRS [21]. Significant decrease in QRS voltage has been observed in cardiomyopathy following Adriamycin chemotherapy [22].

1.3.8 Acute Pulmonary Embolism Complicated by Cardiogenic Shock

Massive pulmonary embolism may be associated with low voltage QRS complexes, right bundle branch block, and ST-segment elevation in lead V_1. These findings are strong predictors of cardiogenic shock in acute pulmonary embolism [23].

1.3.9 Hypothyroidism

Electrocardiogram is characterized by low voltage of all wave forms. Heart rate is usually toward lower limits of normal. Untreated patients will severe hypothyroidism may have resting heart rate less than 60/min. There may be shallow inversion of T waves. There is no deviation of the ST-segment. Hypothyroidism affects all leads [24]. Representative electrocardiogram is shown in Fig. 38.5.

Low voltage correlates with serum level of free T4 and serum albumin [25]. There is no significant relation of QRS voltage to concomitant pericardial effusion [25].

1.4 Impaired Conduction of Electric Potentials to Recording Electrodes

1.4.1 Pericardial Effusion, Cardiac Tamponade

All electrocardiographic deflections (P, QRS, and T waves) have low voltage [26]. T waves can be

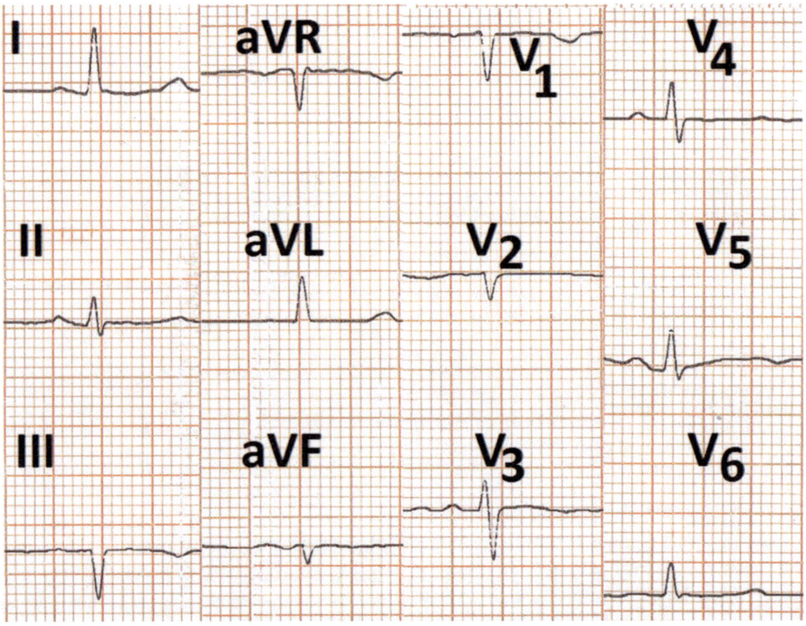

Fig. 38.5 Electrocardiogram showing low voltage of the QRS complexes, and diffuse flattening of the T waves in precordial leads suggestive of hypothyroidism. Heart rate was 58/min (Reproduced with permission from: Mittal SR. Low Voltage Electrocardiogram. Cardiology Today-2017;21:160–165)

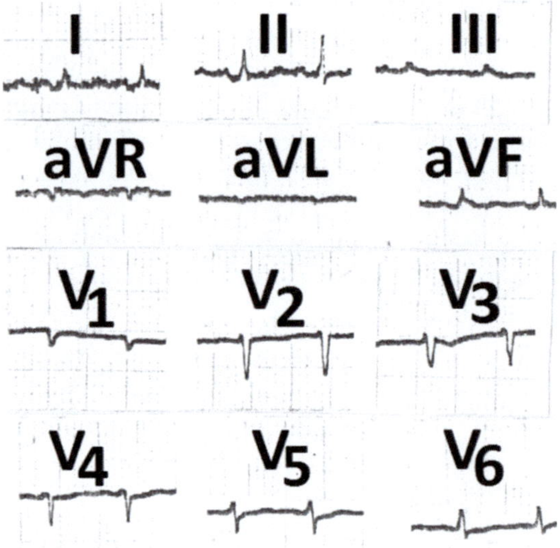

Fig. 38.6 Electrocardiogram from a patient with pericardial effusion showing sinus tachycardia and low voltage of P, QRS, and T waves in all leads (Reproduced with permission from: Mittal SR. Low Voltage Electrocardiogram. Cardiology Today-2017;21:160–165)

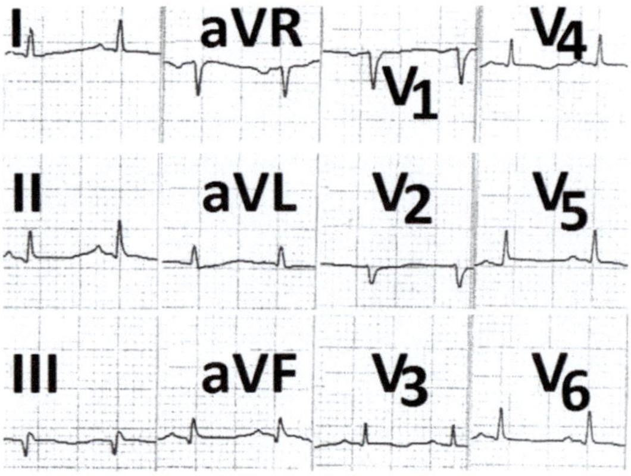

Fig. 38.7 Electrocardiogram from a case of cardiac tamponade showing sinus tachycardia and low voltage P, QRS, and T waves in all leads (Reproduced with permission from: Mittal SR. Low Voltage Electrocardiogram. Cardiology Today-2017;21:160–165)

flat or may show shallow inversion in all leads except lead aVR (Figs. 38.6 and 38.7).

As cardiac tamponade develops, sinus tachycardia appears. Some patients with massive effusion may show electrical alternans (alternate QRS complex show more and less amplitude) (Figs. 38.8 and 38.9). This finding is due to the swinging motion of the heart. Uncommonly alternans may affect all wave forms, that is, the P wave, the QRS complex, and the T wave. Electrical alternans disappear after aspiration of pericardial fluid (Fig. 38.8c).

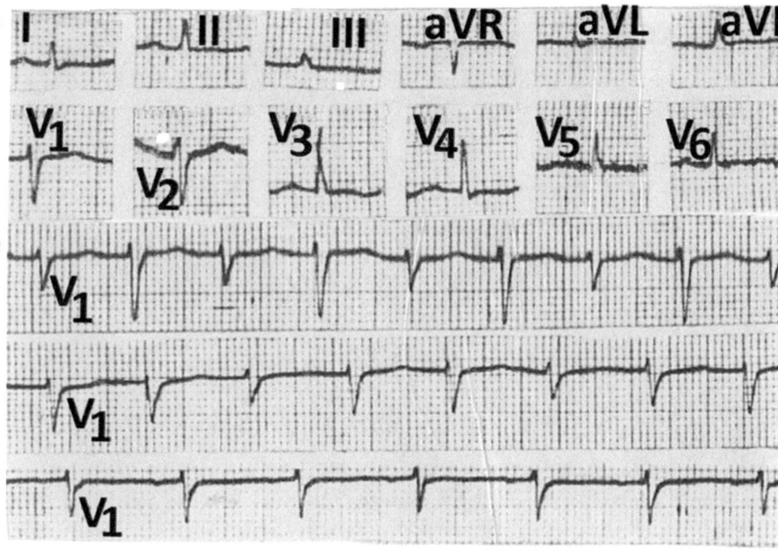

Fig. 38.8 (**a**) Electrocardiogram from a case of cardiac tamponade showing low voltage P, QRS, and T waves in all leads. (**b**) Long lead V_1 showing electrical alternans. (**c**) & (**d**) Showing progressive disappearance of alternans after aspiration of pericardial fluid (Reproduced with permission from: Mittal SR. Low Voltage Electrocardiogram. Cardiology Today-2017;21:160–165)

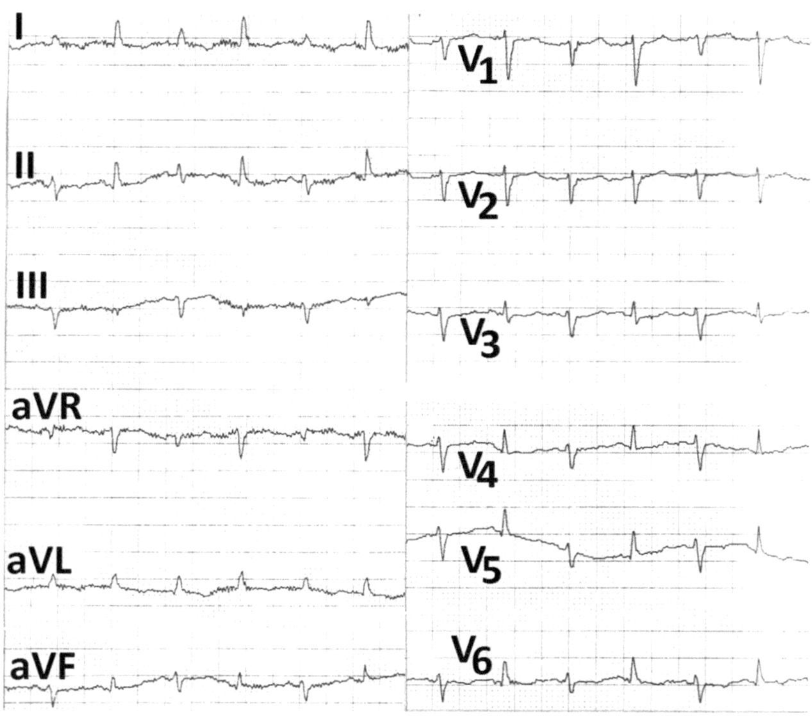

Fig. 38.9 Electrocardiogram from a case of cardiac tamponade showing alternans in most of the leads (Reproduced with permission from: Mittal SR. Low Voltage Electrocardiogram. Cardiology Today-2017;21:160–165)

If acute pericarditis coexists, ST-segment elevation with concavity upward may be present in all leads except in lead aVR (Fig. 38.10). As opposed to myocardial infarction, there are no reciprocal depressions of the ST-segment [27].

1.4.2 Pneumopericardium [29]

Presence of air in the pericardial cavity also hampers conduction of electric potentials to the surface electrodes.

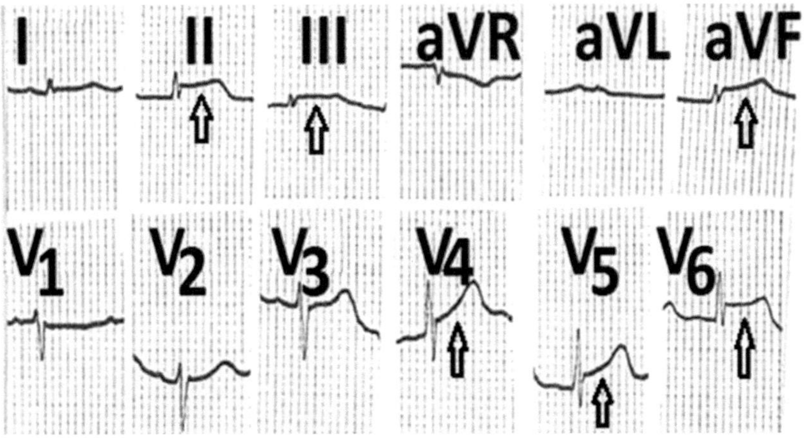

Fig. 38.10 Electrocardiogram from a case of pericardial effusion with pericarditis. It shows low voltage QRS and ST-segment elevation with concavity upward in leads II, III, aVF, V_4 to V_6 (marked ⇧). There are no reciprocal changes and there is no ST-segment elevation in lead aVR (Reproduced with permission from: Mittal SR. Low Voltage Electrocardiogram. Cardiology Today-2017;21:160–165)

1.4.3 Pneumomediastinum [30]

Air in the mediastinum surrounding the heart is also a bad conduction of electrical activity.

1.4.4 Chronic Constrictive Pericarditis

Thickened pericardium prevents transmission of electric potentials to the surface electrodes. However, pericardiectomy only partially restores the QRS amplitude suggesting that fibrosis of the underlying myocardium also contributes to the low voltage of the QRS complexes [28]. Sinus tachycardia is present [26]. Some cases develop atrial fibrillation. It is due to atrial dilatation due to impaired atrial emptying. Thickened pericardium hampers adequate dilatation of the ventricles. This prevents adequate emptying of the atria into the ventricles. Signs of left atrial enlargement may be present in sinus rhythm (Figs. 38.11 and 38.12).

QRS complexes show low amplitude. T waves are of low amplitude or are inverted. Inverted T

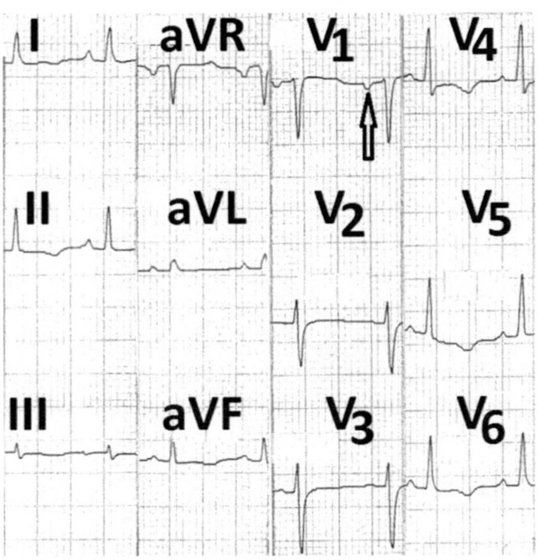

Fig. 38.11 Electrocardiogram from a case of pericardial constriction showing low voltage QRS, T wave inversion in most of the leads, and left atrial overload (negative P wave in lead V_1, marked ⇧) (Reproduced with permission from: Mittal SR. Low Voltage Electrocardiogram. Cardiology Today-2017;21:160–165)

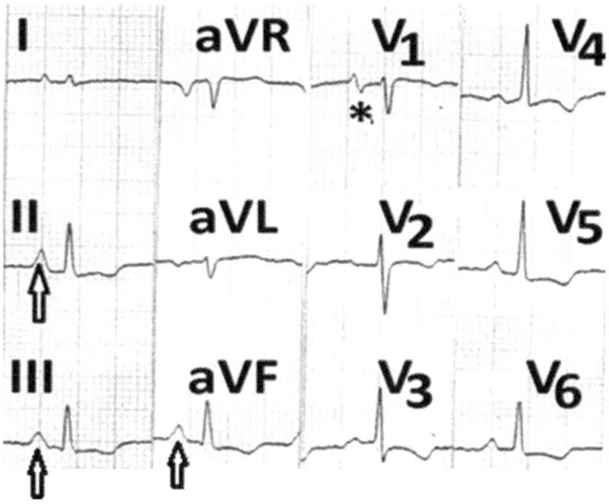

Fig. 38.12 Electrocardiogram from a case of pericardial constriction showing low voltage QRS, T wave inversion in most of the leads, and evidence of biatrial overload (tall P waves in leads II, III, aVF (marked as ⇧) and prominent negative deflection of P wave in lead V_1) (marked *) (Reproduced with permission from: Mittal SR. Low Voltage Electrocardiogram. Cardiology Today-2017;21:160–165)

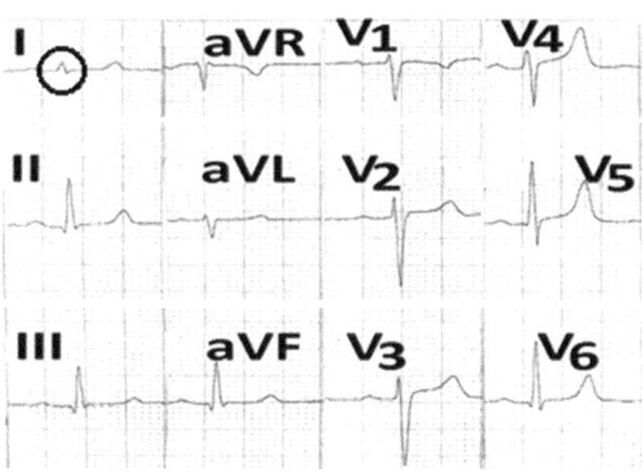

Fig. 38.13 Electrocardiogram from a case of emphysema showing significantly decreased QRS voltage in lead I (marked ○) (Reproduced with permission from: Mittal SR. Low Voltage Electrocardiogram. Cardiology Today-2017;21:160–165)

waves are usually associated with involvement of the underlying myocardium [28].

1.4.5 Chronic Obstructive Pulmonary Disease

Low voltage affects the QRS especially in lead I (Fig. 38.13) and left precordial leads, that is, leads V_5 and V_6 (Fig. 38.14) [24, 31].

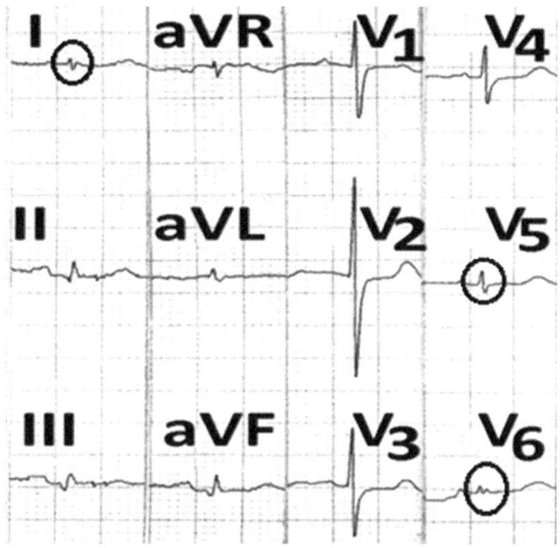

Fig. 38.14 Electrocardiogram from a patient with chronic obstructive airway disease showing markedly diminished QRS amplitude in limb leads (specially Lead I) and leads V_5 and V_6 (marked ○) (Reproduced with permission from: Mittal SR. Low Voltage Electrocardiogram. Cardiology Today-2017;21:160–165)

With the development of pulmonary artery hypertension, P waves become tall and peaked in leads II, III, aVF, and frontal plane QRS axis shifts to the right (Fig. 38.15).

1.4.6 Left-Sided Pleural Effusion [22]

Moderate or massive effusion that covers the inferior and lateral surface of the heart can hamper transmission of electrical impulses to limb lead and left-sided chest leads. Displacement of the heart to the right may also effect the precordial leads.

1.4.7 Left-Sided Pneumothorax [23]

Pneumothorax that covers the lateral surface of the heart can reduce QRS voltage in leads V_5, V_6. Tension pneumothorax can displace the heart to the right. This can also affect the precordial leads.

1.4.8 Pulmonary Edema [32]

Severe pulmonary edema can affect QRS voltage. Exact mechanism is not clear. Presence of fluid in the interested tissue and alveoli of left lung could be the responsible.

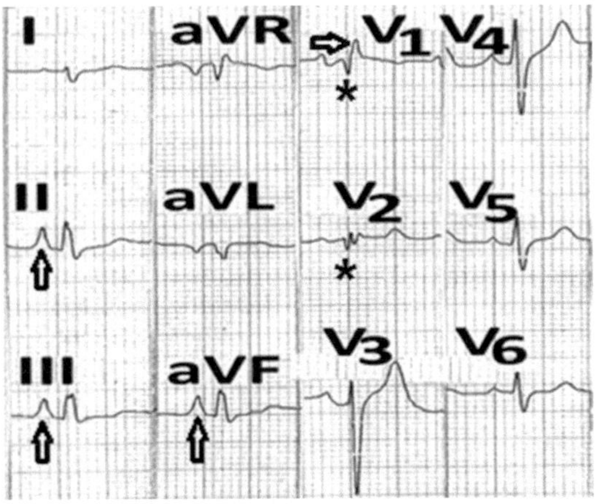

Fig. 38.15 Electrocardiogram from a patient of cor pulmonale showing low QRS voltage in limb leads and leads V_5, V_6, right axis deviation (dominant S wave in lead I), right ventricular hypertrophy (prominent R wave in lead V_1 (marked ⇨) and right atrial overload (tall P waves in leads II, III, aVF (marked ⇧) and qR in V_1 and V_2 (marked *) (Reproduced with permission from Mittal SR. Low Voltage Electrocardiogram. Cardiology Today-2017;21:160–165)

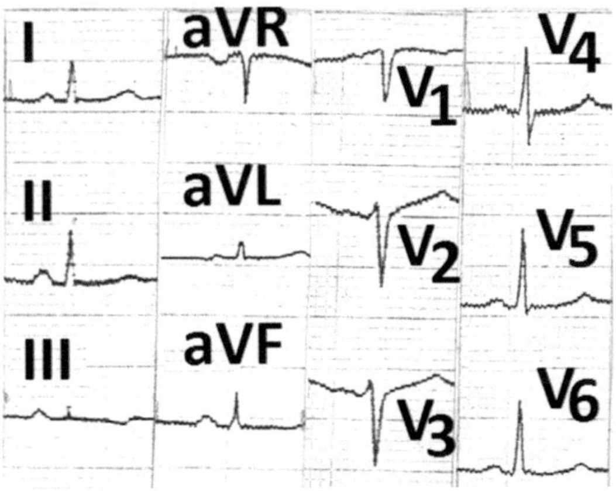

Fig. 38.16 Electrocardiogram from an obese person showing low amplitude of QRS. P and T waves are normal (Reproduced with permission from: Mittal SR. Low Voltage Electrocardiogram. Cardiology Today-2017;21:160–165)

1.4.9 Morbid Obesity

Although obesity is frequently enumerated as one of the causes of low QRS voltage, only a small number of obese patients have low voltage QRS [33]. Representative ECG is shown in Fig. 38.16.

Other electrocardiographic findings include leftward shift of the mean frontal plane QRS axis and prominent q waves in inferior leads presumably because of elevation of the diaphragm.

1.4.10 Ascites

Patients with massive ascites can have low voltage of QRS complexes in precordial leads [34]. Exact mechanism is not known. It is hypothesized that increased intraabdominal pressure could influence the anatomical position and the electrical axis of the heart. Cranial placement of precordial electrodes increases QRS voltage in leads V_4 to V_6. It is considered as one of the causes of low voltage QRS in cirrhosis [35]. Edema and hypoalbuminemia could be contributory. Removal of ascites significantly increases the voltage of the QRS complexes [35].

1.4.11 Edema, Anasarca

In a patient with anasarca, multiple factors, for example, concomitant pericardial effusion, pleural effusion, and ascites, could contribute to low voltage of the QRS complexes [36, 37]. Peripheral edema also affects the voltage of the P wave and the T waves. Changes are reversible.

Peripheral edema secondary to nonsteroidal anti-inflammatory drugs [38], thiazolidinediones, chronic renal failure, cirrhosis, and hypoalbuminemia have also been shown to be associated with reversible reduction in the amplitude of the QRS complexes [3].

Summary

- QRS amplitude less than 5 mm in the limb leads and less than 10 mm in the precordial leads is usually defined as low voltage QRS.
- Incorrect standardization (5 mm/mV) results in low amplitude of all waveforms. Other electrocardiographic parameters are normal.
- Hypothyroidism produces low voltage electrocardiogram with sinus bradycardia and nonspecific diffuse flattening of the T waves. PR interval and QT interval may be prolonged.
- Emphysema also produces low voltage of the QRS complexes. Mean frontal plane QRS axis

is around +90°. P waves are of low amplitude in lead I but are peaked in leads II, III, and aVF in cases of cor pulmonale. QRS voltage is specially diminished in left lateral leads (I, V_5, V_6).
- Moderately severe pericardial effusion also produces low voltage of all waves. Usually there is sinus tachycardia with diffuse flattening of the T waves. T waves may be inverted in all leads other than lead aVR.
- Pericardial constriction also produces low voltage of the QRS complexes with sinus tachycardia and diffuse flattening or inversion of the T waves. ECG evidence of left atrial enlargement or atrial fibrillation support the possibility of pericardial constriction.
- Cardiac amyloidosis causes low voltage in limb leads. Prolonged PR interval and Q waves in leads V_1 to V_3 are frequent.
- Moderate left-sided pneumothorax or pleural effusion can reduce voltage in the left precordial leads.
- Extensive myocardial necrosis (loss of viable myocardium) can produce low amplitude of QRS in most of the leads.
- Low voltage in limb leads with increased voltage in chest leads with slow R wave progression is seen in dilated cardiomyopathy.
- Low QRS voltage is uncommon in isolated obesity.

MCQs

Q1. Low voltage QRS are defined as:

(a) <0.5 mV in limb leads
(b) <1 mV in limb leads
(c) <1 mV in the precordial leads
(d) <2 mV in the precordial leads

Q2. Low voltage QRS mostly affects:

(a) Limb leads
(b) Leads V_1 to V_3
(c) Leads V_4 to V_6
(d) Leads V_7 to V_9

Q3. Low voltage QRS can be due to:

(a) Standardization at 5 mm/mV
(b) Normal variation
(c) Inferior infarction
(d) Acute pericarditis without effusion

Q4. Acute phase of Takotsubo cardiomyopathy produces:

(a) Transient attenuation of QRS complexes
(b) Permanent attenuation of QRS complexes
(c) T wave inversion in leads II, III, aVF
(d) ST-segment depression with positive T wave in lead aVR

Q5. Cardiac amyloidosis produces:

(a) LVH by voltage criteria
(b) LVH on echocardiography
(c) Hypokinesia on echocardiography
(d) First-degree atrioventricular block

Q6. Myocarditis produces:

(a) Atrioventricular block
(b) Sinus bradycardia
(c) Increased ventricular activation time
(d) Pathologic Q waves

Q7. Dilated cardiomyopathy produces:

(a) Attenuation of QRS complexes in limb leads
(b) Tall QRS complexes in precordial leads
(c) Tall QRS complexes in limb leads
(d) Attenuation of QRS in precordial leads

Q8. Which of the following strongly suggests dilated cardiomyopathy?

(a) RBBB + Left anterosuperior fascicular block
(b) RBBB + Left posteroinferior fascicular block
(c) LBBB + RAD
(d) Biatrial enlargement

Q9. Adriamycin cardiomyopathy produces:

(a) Bradycardia
(b) LVH by voltage criteria
(c) Pleural effusion
(d) Pericardial effusion

Q10. Attenuation of QRS voltage can result from:

(a) Pulmonary congestion
(b) Right-sided pleural effusion
(c) Myocardial edema
(d) Pneumomediastinum

Q11. Massive pulmonary embolism produces:

(a) RBBB
(b) ST depression in lead V_1
(c) Tall T wave in leads V_1 to V_3
(d) Q waves in leads II, III, aVF

Q12. Hypothyroidism is characterized by:

(a) Low voltage T waves
(b) ST-segment depression in limb leads
(c) ST-segment elevation in precordial leads
(d) Sinus bradycardia

Q13. In hypothyroidism, QRS voltage correlates with:

(a) Serum T_4
(b) Serum albumin
(c) Pericardial effusion
(d) Myocardial edema

Q14. Acute pericarditis produces:

(a) Concave upward ST-segment elevation in most leads
(b) Reciprocal ST segment depressions
(c) Left atrial enlargement
(d) T wave inversions

Q15. Which ECG finding is specific of cardiac tamponade?

(a) Sinus tachycardia
(b) Electrical alternans
(c) Attenuation of QRS voltage
(d) T wave flattening

Q16. Chronic constrictive pericarditis produces attenuation of QRS voltage due to:

(a) Thick pericardium
(b) Myocardial fibrosis
(c) Pericardial effusion
(d) Right-sided pleural effusion

Q17. In the context of pericardial constriction, T wave inversion correlates with:

(a) Pericardial inflammation
(b) Pericardial effusion
(c) Thickness of pericardium
(d) Myocardial fibrosis

Q18. Which factors can contribute to low voltage QRS?

(a) Edema
(b) Hypoalbuminemia
(c) Pulmonary congestion
(d) All

Q19. Most sensitive ECG finding in emphysema is:

(a) Low voltage QRS in lead I
(b) P- Pulmonale
(c) RVH
(d) RAD

Q20. Which conditions can produce low voltage QRS?

(a) Asthenia
(b) Ascites
(c) Anasarca
(d) All

Q21. Low amplitude of QRS complexes with low normal heart rate, nonspecific diffuse flattening of T wave, and mild prolongation of PR interval and QT interval suggest the possibility of:

(a) Pericardial constriction
(b) Cardiac tamponade
(c) Hypothyroidism
(d) Emphysema

Q22. Low amplitude of the P wave and the QRS complex in lead 1 and low amplitude of QRS in lead V_6 suggest the possibility of:

(a) Emphysema
(b) Hypothyroidism
(c) Left-sided pneumothorax
(d) Cardiac amyloidosis

Q23. The low amplitude of all electrocardiographic wave forms in all leads with normal heart rate, normal configuration of all waves, and normal intervals suggest the possibility of:

(a) Emphysema
(b) Pericardial constriction
(c) Massive pericardial
(d) Incorrect standardization

Q24. Sinus tachycardia, low amplitude of all waves in all leads with diffuse flattening of T waves in all leads suggest the possibility of:

(a) Emphysema
(b) Myxedema
(c) Moderately severe pericardial effusion
(d) Incorrect standardization

Q25. Sinus tachycardia, low amplitude of all waves in all leads, shallow T inversion in all leads other than aVR, and electrical alternans suggest the possibility of:

(a) Pericardial constriction
(b) Cardiac tamponade
(c) Myxedema
(d) Cardiac amyloidosis

Q26. Normal heart rate, low amplitude of QRS in limb leads, prolonged PR interval, and Q waves in leads V_1 to V_3 suggest the possibility of:

(a) Myocarditis
(b) Dilated cardiomyopathy
(c) Cardiac amyloidosis
(d) Pericardial constriction

Q27. The low amplitude of P wave in lead 1, peaked P waves in leads II, III, aVF, S_I, S_{II}, S_{III} pattern, low amplitude of QRS in lead V_6 with slow progressive of r in leads V_2 to V_6, and persistence of S wave in lead V_6 suggest the possibility of:

(a) Cardiac amyloidosis
(b) Emphysema
(c) Emphysema with cor pulmonale
(d) Pericardial constriction

Q28. Normal heart rate, low voltage of all waves, and diffuse flattening of T waves could be due to:

(a) Emphysema
(b) Hypothyroidism
(c) Hypothyroidism with mild pericardial effusion
(d) Cardiac tamponade

Q29. Left ventricular hypertrophy on echocardiogram and low amplitude of QRS in electrocardiogram suggest:

(a) Hypertrophic cardiomyopathy
(b) Amyloidosis
(c) Glycogen storage disease
(d) Cardiac sarcoidosis

Q30. Sinus tachycardia, low voltage of all waves, diffuse shallow inversion of T waves could occur in:

(a) Cor pulmonale
(b) Gross pericardial effusion
(c) Effusive constrictive pericarditis
(d) Cardiac amyloidosis

Q31. Atrial fibrillation with fast ventricular rate, low amplitude of the QRS complex, and shallow T wave inversion in all leads suggest the possibility of:

(a) Acute pericarditis
(b) Pericardial effusion
(c) Pericardial constriction
(d) Mitral regurgitation

Q32. Low amplitude of the QRS complex is seen in:

(a) Systemic hypertension
(b) Left anterosuperior fascicular block
(c) Mitral valve prolapse
(d) None

Q33. Left atrial enlargement, normal QRS axis, low amplitude of QRS, and diffuse flattening of T wave, in all leads, suggest the possibility of:

(a) Tight mitral stenosis with severe pulmonary artery hypertension
(b) Severe MR
(c) Pericardial constriction
(d) None

Q34. The Low amplitude of QRS in leads V_5, V_6 in seen in:

(a) Emphysema
(b) Left anterosuperior fascicular block
(c) Left inferoposterior fascicular block
(d) All

Answers

(1) a, c (2) a (3) a, b (4) a, d (5) b, c, d (6) c, d (7) a, b (8) c (9) c (10) a, c, d (11) a (12) a, d (13) a, b (14) a (15) b (16) a, b (17) d (18) d (19) a (20) b, c (21) c (22) a (23) d (24) c (25) b (26) c (27) c (28) b, c (29) b (30) b c (31) c (32) d (33) c (34) d.

References

1. Kadish AH, Buxton AE, Kennedy HL, et al. ACC/AHA clinical competence statement on electrocardiography and ambulatory electrocardiography. A report of the ACC/AHA/ACP- ASIM task force on clinical competence (ACC/AHA committee to develop a clinical competence statement on electrocardiography and ambulatory electrocardiography). J Am Coll Cardiol. 2001;38:2091–100.
2. Chinitz JS, Cooper JM, Verdino RJ. Electrocardiogram voltage discordance: interpretation of low QRS voltage only in the limb leads. J Electrocardiol. 2008;41:281–6.
3. Madias JE. Low QRS voltage and its causes. J Electrocardiol. 2008;41:498–500.
4. Szewieczek J, Gasior Z, Dulawa J, Francuz T, Legierska K, Batko-Szwaczka A, et al. ECG low QRS voltage and wide QRS complex predictive of centenarian 360 day mortality. Age (Dordr). 2016;38:44–51.
5. Usoro AO, Bradford N, Shah AJ, Soliman EZ. Risk of mortality in individuals with low QRS voltage and free of cardiovascular disease. Am J Cardiol. 2014;113:1514–7.
6. Mirvis DM, Goldberger AL. Electrocardiography. In: Mann DL, Zipes DP, Libby P, Bonow RO, editors. Braunwald's heart disease. Philadelphia: Elsevier; 2015. p. 114–52.
7. Tan NS, Goodman SG, Yan RT, Tan MK, Fox KA, Gore JM, et al. Prognostic significance of low QRS voltage on the admission electrocardiogram in acute coronary syndromes. Int J Cardiol. 2015;190:34–9.
8. Madias JE. Electrocardiographic differentiation between Takotsubo syndrome and acute myocardial infarction. Int J Cardiol. 2016;222:969.
9. Guerra F, Giannini I, Pongetti G, Fabbrizioli A, Rrapaj E, Achieri D, et al. Transient QRS amplitude attenuation is associated with clinical recovery in patients with Takotsubo cardiomyopathy. Int J Cardiol. 2015;187:198–205.
10. Madias JE. Transient attenuation of the amplitude of QRS complexes in the diagnosis of Takotsubo syndrome. Eur Heart J Acute Cardiovasc Care. 2014;3:28–36.
11. Falk RH, Hershberger RE. The dilated, restrictive and infiltrative cardiomyopathies. In: Mann DL, Zipes DP, Libby P, Bonow RO, editors. Braunwald's heart disease. Philadelphia: Elsevier; 2015. p. 1551–73.
12. Cyrille NB, Goldsmith J, Alvarez J, Maurer MS. Prevalence and prognostic significance of low QRS voltage among the three main types of cardiac amyloidosis. Am J Cardiol. 2014;114:1089–93.
13. Madias JE. Low voltage ECG in myocarditis: peripheral oedema as a plausible contributing mechanism. Pacing Clin Electrophysiol. 2007;30:448–52.
14. Schamroth L. Myocarditis. In: Schamroth C, editor. An introduction to electrocardiography. New Delhi: Wiley India; 1982. p. 195–9.
15. De Luna AB, Goldwasser D, Fiol M, Bayes-Genis A. Surface electrocardiography. In: Fuster V, Walsh RA, Harrington RA, editors. Hurst's the heart. New York: McGraw Hill; 2011. p. 307–70.
16. Schamroth L. Congestive cardiomyopathy. In: Schamroth C, editor. An introduction to electrocardiography. New Delhi: Wiley India; 1982. p. 200–1.
17. Glancy DL, Newman WP III. Atrial fibrillation with QRS voltage low in the limb leads and high in the

precordial leads. Baylor Univ Med Centre Proc. 2008;21:437–8.
18. Goldberger AL, Dresselhaus T, Bhargav V. Dilated cardiomyopathy: utility of the transverse frontal plane QRS voltage ratio. J Electrocardiol. 1985;18:35–40.
19. Nickolie G, Marriott HJL. Left bundle branch block with right axis deviation: a marker of congestive cardiomyopathy. J Electrocardiol. 1985;18:395–404.
20. Childers R, Lupovich S, Sochanski M, Konarzewska H. Left bundle branch block and right axis deviation: a report of 36 cases. J Electrocardiol. 2000;33(suppl):93–102.
21. Madias JE. QRS voltage changes in heart failure: a 3-compartment mechanistic model and its implications. Indian Pacing Electrophysiol J. 2010;10:464–73.
22. Ali MK, Soto A, Maroongroge D, Bekheit-Saad S, Buzdar AU, Blumenschein GR, et al. Electrocardiographic changes after Adriamycin chemotherapy. Cancer. 1979;43:465–71.
23. Kukla P, Melntyre WF, Fijorek K, Mirek-Bryniarska E, Bryniarski L, Krupa E, et al. Electrocardiographic abnormalities in patients with acute pulmonary embolism complicated by cardiogenic shock. Am J Emerg Med. 2014;32:507–10.
24. Wagner GS, Wang TY. Miscellaneous conditions. In: Wagner GS, editor. Marriot's practical electrocardiography. New Delhi: Wolters Kluwer; 2008. p. 209–37.
25. Yamanaka S, Kuman Y, Matsumura Y, Kamioka M, Takeuchi H, Sugiura T. Link between pericardial effusion and attenuation of QRS voltage in patients with hypothyroidism. Cardiology. 2010;116:32–6.
26. Schamroth L. Pericarditis. In: Schamroth C, editor. An introduction to electrocardiography. New Delhi: Wiley- India; 1982. p. 189–99.
27. Le Winter MM, Hopkins WE. Pericardial diseases. In: Mann DL, Zipes DP, Libby P, Bonow RO, editors. Braunwald's heart disease. Philadelphia: Elsevier; 2015. p. 1636–57.
28. Vicente T, Valdes M, Garcia A, Soria F, Hernandez A, Castello MJ, et al. The electrocardiographic changes in constrictive pericarditis before and after pericardiectomy. The role of the myocardium in the genesis of preoperative electrocardiographic changes. Ann Med Interna. 1990;7:345–8.
29. Kanijn AJ, Egbers PHM, Kuiper MA. Pneumopericardium should be considered with electrocardiogram changes after blunt chest trauma; a case report. J Med Case Rep. 2008;2:100–5.
30. Brearley WB Jr, Taylor L III, Haley MW, et al. Pneumomediastinum mimicking acute ST elevation Myocardial infarction. Int J Cardiol. 2002;117:e73.
31. Schikawa A, Matsumura Y, Ohnishi H, Kataoka H, Ogura K, Yokoyama A, et al. Identification of electrocardiographic values that indicate chronic obstructive pulmonary disease. Heart Lung. 2016;45:359–62.
32. Dudley SC Jr, Baumgarten CM, Ornato JP. Reversal of low voltage and infarction pattern on the surface electrocardiogram after renal haemodialysis for pulmonary oedema. J Electrocardiol. 1990;23:341–5.
33. Frank S, Colliver JA, Frank A. The electrocardiogram in obesity: statistical analysis of 1029 patients. J Am Coll Cardiol. 1986;7:295–9.
34. Cuculi F, Jamshidi P, Kobza R, Rohacek M, Erne P. Precordial low voltage in patients with ascites. EP Europace. 2008;10:96–8.
35. Cichoz-Lach H, Tomaszewski M, Kowalik A, Lis E, Tomaszewski A, Lach T, Boczkowska S, et al. QT interval prolongation and QRS voltage reduction in patients with liver cirrhosis. Adv Clin Exp Med. 2015;24:615–22.
36. Madias JE. aVR, an index of all ECG limb leads with clinical utility for monitoring patients with oedematous states including heart failure. Pacing Clin Electrophysiol. 2009;32:1567–76.
37. Madias JE, Bazaz R, Agarwal H, Moethu W, Medepalli L. Anasarca mediated attenuation of the amplitude of electrocardiogram complexes: a description of a heretofore unrecognized phenomenon. J Am Coll Cardiol. 2001;38:756–64.
38. Madias JE, Madias NE. Reversible attenuation of the ECG voltage due to peripheral oedema associated with treatment with a cox 2 inhibitor. Heart Fail. 2006;12:46–50.

39 J Point and the J Wave Syndrome

J point is the point of junction of end of depolarization (QRS complex) with onset of repolarization (the ST-segment). This is usually not a sharp point because the end of QRS smoothly merges in the ST-segment. Normally the J point is at the level of the PR segment (base line). If it is elevated above the base line, it forms a smooth dome shaped wave at the end of the QRS complex (the J wave). It is also called "Osborn wave," J wave has the same polarity as that of the preceding QRS complex [1]. J wave should be differentiated from terminal notch produced due to intraventricular conduction defect. Notch due to intraventricular conduction defect is relatively sharp and shows tachycardia or prematurity- dependent augmentation [2]. Structural heart disease is common in these cases. On the other hand J wave is smooth, dome like, and shows bradycardia or pause-dependent augmentation [2]. It is usually seen in the absence of structural heart disease. At times the J wave may not be distinct but may be buried inside the QRS resulting in slurring of terminal part of the QRS [2]. Mostly J wave is a normal variant especially in vagotonic individuals, for example, young athletes or hard manual workers. Normally it is most prominent in left precordial leads (V_4 to V_6), is $\leq$0.2 mV, and is followed by rapidly upsloping ST-segment (Fig. 39.1).

When variations in the J wave are associated with polymorphic VT, idiopathic VF, arrhythmic syncope, or aborted sudden cardiac death, these are called J wave syndromes. Arrhythmic syncope is characterized by absence of prodromes or specific triggers, for example, hot and crowded surrounding, pain, the sight of blood, or prolonged standing [2]. Arrhythmic syncopes can be associated with urinary incontinence [2].

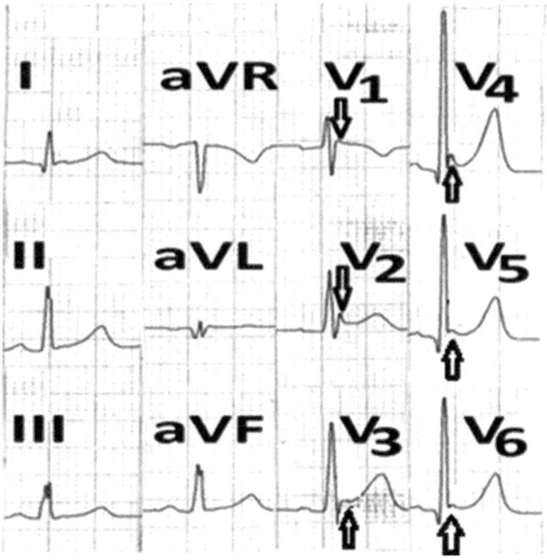

Fig. 39.1 Electrocardiogram is showing J waves (arrows) in leads V_1 to V_6 followed by the rapidly upsloping ST-segment (Reproduced with permission from: Mittal SR. 'J' Point. Cardiology Today-2017;21:121–125)

S. Mittal, *Insights into Electrocardiograms with MCQs*,
https://doi.org/10.1007/978-981-99-0127-2_39

Clinically important J wave syndromes include:

(1) Early repolarization syndrome
(2) Brugada syndrome
(3) Acute myocardial ischemia
(4) Arrhythmogenic right ventricular cardiomyopathy

(1) Early repolarization syndrome

Early repolarization of the endocardium as compared to the epicardium results in the genesis of the J wave [3] (Fig. 39.2a, b).

The term "early repolarization syndrome" is used only when the ECG finding is associated with idiopathic VF [4]. Idiopathic VF is more frequently associated with the following electrocardiographic features:

- J wave > 0.2 mV [5].
- Presence of J wave in inferior (II, III, aVF) as well as lateral (I, aVL, V_4 to V_6) leads, Fig. 39.3 shows J waves with horizontal ST-segment in inferior leads.
- Global (inferior, lateral, and right precordial leads) presence of J wave are more frequently associated with idiopathic VF [6]. Figure 39.4 shows J wave in lead V_2.
- Horizontal or down sloping ST segment [7] (Figs. 39.1, 39.5 and 39.6).
- Transient increase in the amplitude of J point elevation especially during episodes of myocardial ischemia.

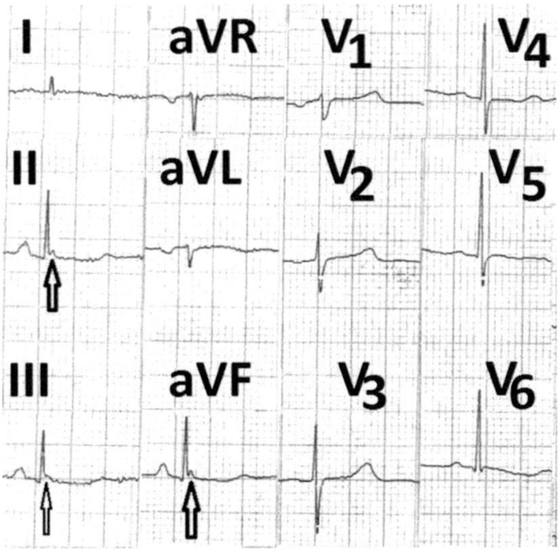

Fig. 39.3 Electrocardiogram is showing J waves (marked ⇧) in leads II, III, and aVF with horizontal ST-segment (Reproduced with permission from: Mittal SR. 'J' Point. Cardiology Today-2017;21:121–125)

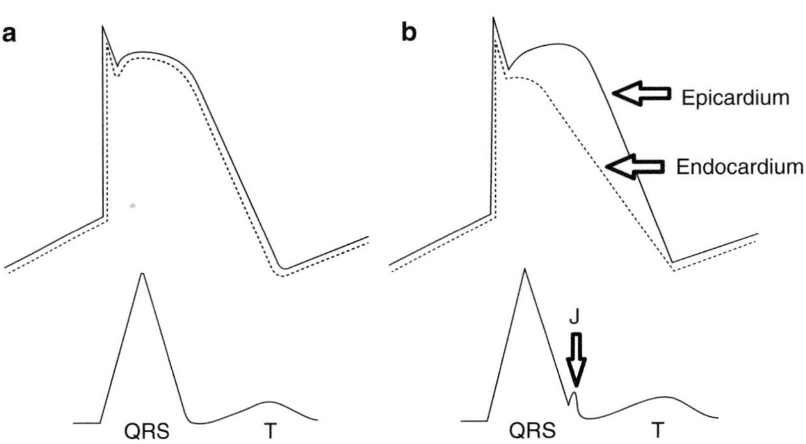

Fig. 39.2 Diagrammatic representation of genesis of the J wave. (**a**) Normal action potentials from epicardium and endocardium with normal QRS. (**b**) Early repolarization of the endocardium producing J wave (marked J) ___ epicardium, - - - endocardium (Reproduced with permission from: Mittal SR. 'J' Point. Cardiology Today-2017;21:121–125)

- Association of prominent J wave with notching (fragmentation) of the QRS [8] (Figs. 39.7 and 39.8).
- Presence of co-existing Brugada ECG pattern or short QT interval [2].

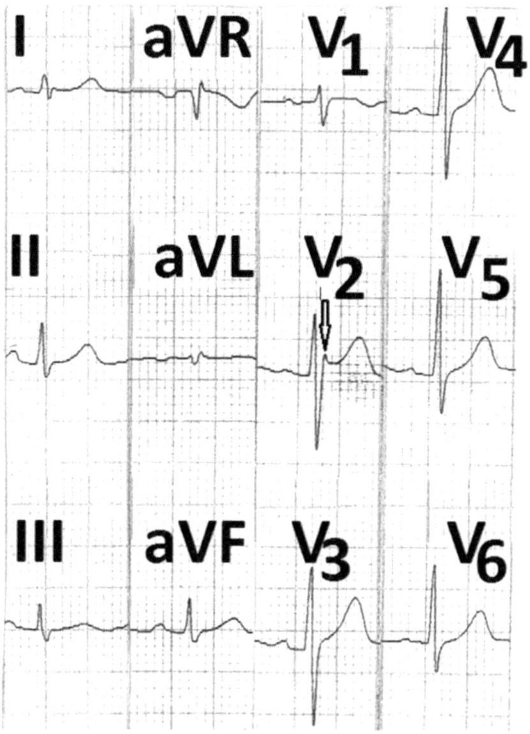

Fig. 39.4 Electrocardiogram is showing J wave (marked ⇩) in lead V_2 (Reproduced with permission from: Mittal SR. 'J' Point. Cardiology Today-2017;21:121–125)

Although these findings are more frequently associated with idiopathic VF, etiological relation is not proven [9]. Further, patients with above-mentioned features alone need not necessarily have the risk of VF. Electrophysiologic study is not justified in asymptomatic patients. Similarly, implantable cardioverter-defibrillator

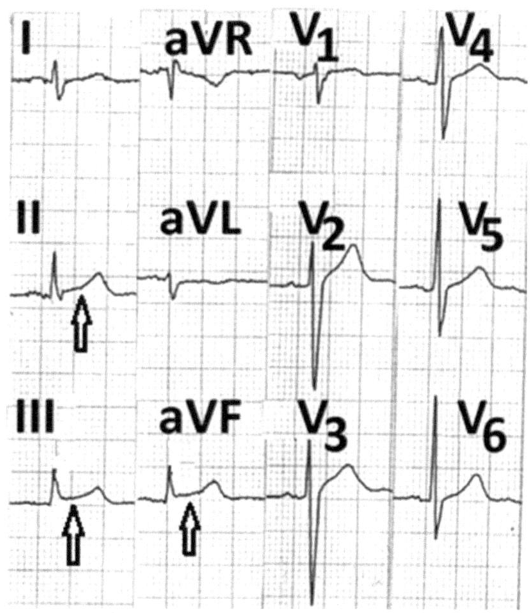

Fig. 39.6 Electrocardiogram shows early repolarization with upsloping ST-segment (marked ⇧) in leads II, III, aVF. There is slurring of terminal QRS in leads III and aVF (Reproduced with permission from: Mittal SR. 'J' Point. Cardiology Today-2017;21:121–125)

Fig. 39.5 Diagrammatic representation of various types of ST-segment following the J wave (**a**) rapidly upsloping, (**b**) horizontal, and (**c**) down sloping (Reproduced with permission from: Mittal SR. 'J' Point. Cardiology Today-2017;21:121–125)

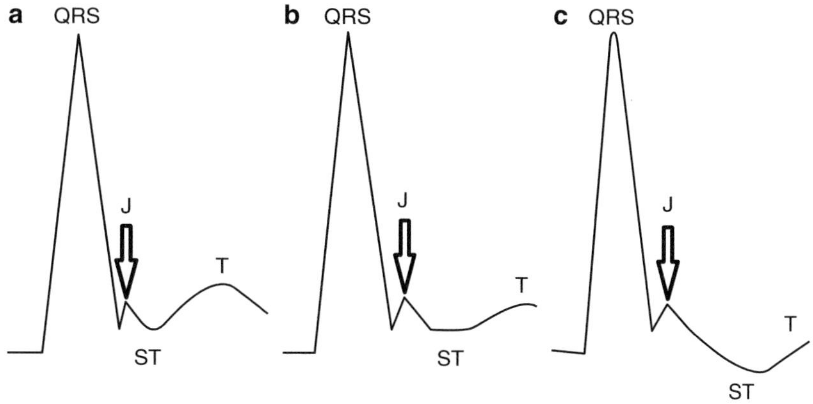

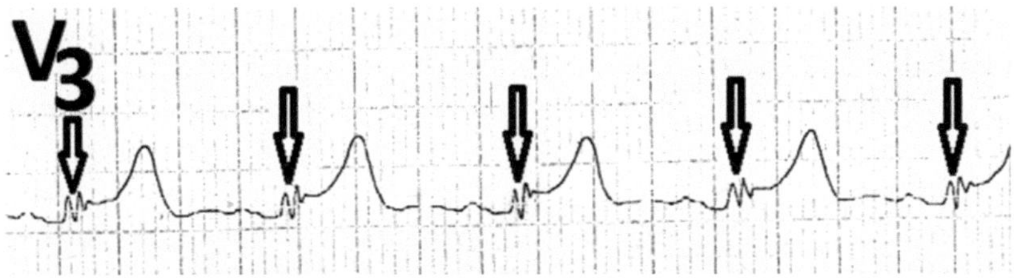

Fig. 39.7 Electrocardiogram is showing fragmented QRS (marked ⇩) with early repolarization (Reproduced with permission from: Mittal SR. 'J' Point. Cardiology Today-2017;21:121–125)

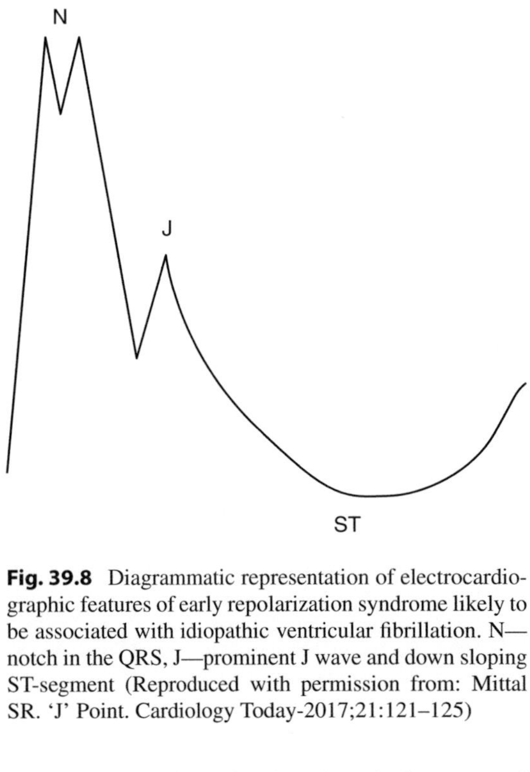

Fig. 39.8 Diagrammatic representation of electrocardiographic features of early repolarization syndrome likely to be associated with idiopathic ventricular fibrillation. N—notch in the QRS, J—prominent J wave and down sloping ST-segment (Reproduced with permission from: Mittal SR. 'J' Point. Cardiology Today-2017;21:121–125)

is not to be considered unless there is documented VF or patient is a survivor of sudden cardiac death. Vagotonia (e.g., during sleep) increases chances of early repolarization associated VF [10]. Sympathetic stimulation suppresses early repolarization and associated arrhythmias [10]. Quinidine is effective in suppressing arrhythmias in early repolarization syndrome [2]. Beta blockers, Verapamil, Mexilitine, Amiodarone, and class IC antiarrhythmic agents are not effective. An idiopathic VF should be considered to be associated with early repolarization only after excluding other etiologies by detailed evaluation. Other causes of VF without structural heart disease also need to be excluded. These include long QT syndrome, short QT syndrome, catecholaminergic arrhythmias, and Brugada syndrome.

(2) Brugada syndrome

It is produced by delayed depolarization over the anterior aspect of the right ventricular outflow tract epicardium [2]. Classical electrocardiographic findings (Type I) include right bundle branch block pattern with ST-segment elevation of >2 mm and inverted T waves in right precordial leads V_1, V_2 [11] (Fig. 39.9a).

Sensitivity is increased by recording these leads in the 2nd and 3rd intercostal spaces [2]. Saddle back type ST-segment elevation of more than 2 mm with a positive or biphasic T wave has been termed as type 2 (Fig. 39.9b). Saddle back type ST-segment elevation of less than 2 mm with a positive T wave is termed type 3 (Fig. 39.9c). Type 2 and 3 are not classical and do not carry the same significance as type I. However, in a given patient all three patterns may be recorded at different times. Type 1 pattern may be transient and may appear spontaneously or after administration of sodium channel blockers, for example, Procainamide, Flecainide, Azmalin, or Pilsicainide [12].

Hypokalemia and oral steroids may also act as proarrhythmic triggers in Brugada syndrome and early repolarization syndromes [13]. Fever can also unmask ECG signs of Brugada syndrome. Several other drugs can produce Brugada-like ECG pattern. These include phenothiazines, tricyclic antidepressants, calcium channel blockers (verapamil, diltiazem), and beta blockers (Propranatal, Nadolol) [2]. The clinical significance of such an ECG change is not clear.

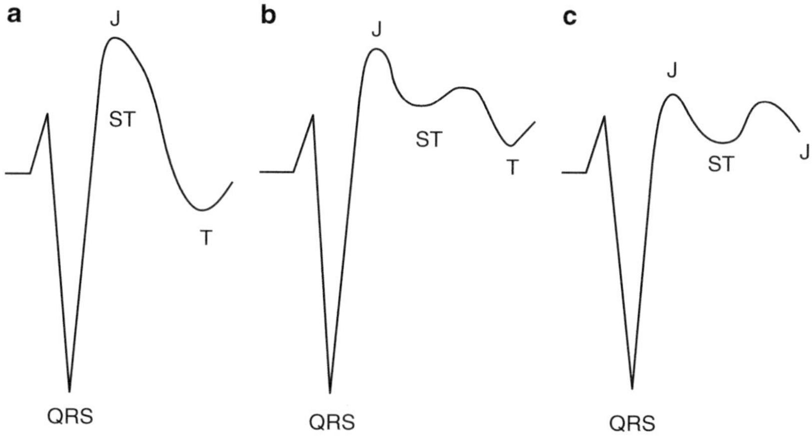

Fig. 39.9 Electrocardiographic types of Brugada syndrome (**a**) Type 1, (**b**) Type 2, and (**c**) Type 3 (Reproduced with permission from: Mittal SR. 'J' Point. Cardiology Today-2017;21:121–125)

Quinidine, Cilostazol, Isoproterenol, and Orciprenaline normalize the ECG and help in controlling arrhythmias. Amiodarone is ineffective [2]. As the classical pattern may appear only transiently, Holter monitoring may be required to detect classical findings in patients with strong clinical suspicion [12].

Brugada like ECG can also occur in RBBB, pectus excavatum, arrhythmogenic right ventricular cardiomyopathy, occlusion of the left anterior descending coronary artery, or conus branch of the right coronary artery which supplies the right ventricular outflow tract. Right bundle branch block pattern of Brugada syndrome may not be accompanied by an S wave in the left precordial leads. It suggests that R' in right precordial leads is an accentuation of the J wave. It can be an asymptomatic ECG finding. The term "Brugada syndrome" should be used only when such electrocardiographic finding is associated with the following:

- Arrhythmogenic syncope or aborted sudden death or documented polymorphic VT or idiopathic VF.
- Family history of sudden cardiac death in young (less than 45 years).
- Presence of classical ECG findings in other family members.
- Inducibility of VT or VF during electrophysiologic study is important only in the presence of clinical findings. Presence of QRS fragmentation correlates with worse prognosis [14]. Other electrocardiographic markers for greater risk of arrhythmias include the following:
- Coexistence of type 1 Brugada pattern with early repolarization in inferolateral leads [2].
- Dynamic changes in the prominence of the J wave and the ST-segment elevation and configuration [15].
- Wide (≥40 ms) and/or large S wave in lead I. It is thought to be due to significant delay in depolarization of the anterolateral right ventricular outflow tract [16].

Other conditions where the J wave can be prominent [2] include:

- Acute myocardial ischemia
- Acidosis
- Hyperkalemia
- Hypercalcemia [17]
- Arrhythmogenic right ventricular cardiomyopathy
- Hypoxia
- Hypothermia

Summary

- J point is the point of junction between the end of the QRS complex and the beginning of the ST-segment.
- Normally it is over the isoelectric line of the electrocardiogram. However, it can be elevated by as much as 2 mm above the base line in leads V_2 and V_3 in persons without any structural heart disease.
- Elevation from J point and its decline to base line can take the form of a wave called the J wave (also called Osborn wave).
- Polarity of the J wave is the same as that of the preceding QRS complex.
- J wave can be prominent in several conditions.
- Brugada syndrome is characterized by:
  - Incomplete or complete right bundle branch block pattern with a coved ST-segment elevation of ≥2 mm followed by a negative T wave in leads V_1 to V_3.
  - Placing leads V_1 to V_3 in higher intercostal space (2nd or 3rd intercostal space) increases sensitivity of detecting Brugada syndrome.
  - There is a high prevalence of atrial fibrillation, polymorphic VT or VF.
- Arrhythmogenic right ventricular dysplasia needs to be differentiated. It is characterized by:
  - Complete or incomplete right bundle branch block.
  - T wave inversion in leads V_1 to V_3 or beyond in an individual ≥14 years.
  - Low amplitude wave between end of the QRS and onset of the T wave (Epsilon wave).
- Other causes of ST-segment elevation in leads V_1 and V_2 and their electrocardiographic differentiating features include:
  - Right bundle branch block—the ST-segment is depressed rather than elevated.
  - Occlusion of a septal branch of the left anterior descending artery-coved ST-segment, abnormal Q wave, and sequential changes in the ST-segment and the T wave help differentiation. Recording leads V_1 and V_2 in higher space do not increase ST-segment elevation.
  - Occlusion of conus branch of the right coronary artery-coved ST-segment, abnormal Q waves, and sequential changes in the ST-segment and the T wave help differentiation. The ST-segment is also elevated in lead V_3R.
  - Pectus excavatum can produce incomplete right bundle branch block. The ST-segment is depressed with the terminal T wave inversion. Recording in a higher space can increase prominence of r' wave.
  - Juvenile pattern—T waves are inverted in leads V_1 to V_4. There are no QRS or ST-segment changes.
- Early repolarization syndrome is characterized by:
  - Notch or slur on terminal part of the R wave, J point elevation, or J wave of 0.1 mV or more in two or more contiguous leads (II, III, aVF, and/or leads V_4, V_5, V_6) (excluding leads V_1 to V_3) with or without. ST-segment elevation with upward concavity and prominent T waves with evidence of syncope, VT/VF, or family history of sudden cardiac death.
- Other conditions that can produce an early repolarization pattern include:
  - Acute pericarditis:

    There is concave upward elevation of the ST-segment in most leads other than leads aVR and V_1.

    There is depression of PR segment due to involvement of pericardium overlying the atria.

    Sequentially the ST-segment settles and the T waves flatten or show shallow inversion.
  - Hypercalcemia

It is characterized by J point/ST segment elevation in leads V_1 and V_2 with short QT interval (due to shortening of the ST-segment).

- Stress cardiomyopathy also presents with diffuse ST-segment elevation simulating transmural infarction.
- Transmural ischemia produces ST-segment elevation which is usually localized to the territory of a coronary artery. There is ST-segment depression in reciprocal leads. Sequentially, there is appearance of abnormal Q waves with deep symmetrically inverted T waves.
- Chronic aortic regurgitation with left ventricular volume overload also produces an early repolarization pattern with prominent narrow q waves and tall R waves in leads V_5, V_6.
- Systemic hypothermia produces prominent J waves with sinus bradycardia and prolongation of the QT interval.

Brugada syndrome and early repolarization syndrome are generally associated with bradycardia or pause. This explains the high incidence of polymorphic VT/VF during sleep or at low levels of physical activity in both conditions.

MCQs

Q1. Normally J point is:

(a) Isoelectric
(b) Elevated
(c) Depressed
(d) Dome shaped

Q2. Polarity of J wave is:

(a) Isoelectric
(b) Same as preceding QRS
(c) Opposite to preceding QRS
(d) Same as following ST segment

Q3. J wave is produced by early repolarization of:

(a) Epicardium
(b) Endocardium
(c) "M"-layer
(d) All layers

Q4. J wave is a normal variant:

(a) In athletes
(b) In hard manual workers
(c) In anxiety
(d) After exercise

Q5. Early repolarization syndrome is diagnosed by:

(a) ECG
(b) Clinical profile
(c) ECG + clinical profile
(d) Electrophysiological study

Q6. Which ECG finding of early repolarization syndrome is more likely to be associated with VF?

(a) J wave in inferior leads
(b) J wave, followed by the rapid upsloping ST segment
(c) J wave in the setting of myocardial ischemia
(d) J wave recorded during sleep

Q7. Which association strongly favors the possibility of early repolarization syndrome?

(a) Presyncope
(b) Documented idiopathic VF
(c) VF in the setting of acute myocardial ischemia
(d) Inducibility of VT during electrophysiological study

Q8. Arrhythmogenic syncopes are characterized by:

(a) Presence of prodromes
(b) Absence of prodromes
(c) Presence of specific triggers
(d) Urinary incontinence

Q9. VF in the absence of structural heart disease can occur in:

(a) Vagotonia
(b) Athletes

(c) Short QT syndrome
(d) Exercise

Q10. Drugs effective in suppressing recurrent VF in a case of early repolarization syndrome:

(a) Quinidine
(b) Beta blockers
(c) Isoproterenol
(d) Amiodarone

Q11. Brugada syndrome is caused by delay in depolarization over:

(a) Anterior aspect of RVOT epicardium
(b) LVOT epicardium
(c) Apex of RV
(d) Interventricular septum

Q12. In classical Brugada syndrome (Type 1), the ST-segment is:

(a) Horizontal
(b) Elevated
(c) Depressed
(d) Saddle back type

Q13. In classical Brugada syndrome T wave is:

(a) Flat
(b) Positive
(c) Inverted
(d) Biphasic

Q14. ECG findings of Brugada syndrome are seen in which leads:

(a) V_1 and V_2
(b) V_3 and V_4
(c) V_3R to V_6R
(d) II, III, and aVF

Q15. Sensitivity of the ECG in diagnosing Brugada syndrome is increased by recording leads V_1 and V_2 in:

(a) 2nd intercostal space
(b) 3rd intercostal space
(c) 4th intercostal space
(d) 5th intercostal space

Q16. Diagnosis of Brugada syndrome is made by:

(a) ECG
(b) Clinical profile
(c) ECG + clinical profile
(d) Electrophysiologic study

Q17. ECG manifestations of Brugada syndrome are unmasked by:

(a) Sodium channel blockers
(b) Fever
(c) Exercise
(d) Beta blockers

Q18. Which of the following is diagnostic of Brugada syndrome?

(a) Type 2 or Type 3 ECG changes
(b) Conversion of type 3 to type 2
(c) Conversion of type 2 to type 1
(d) Spontaneous appearance of type 1 during sleep

Q19. ECG findings of Brugada syndrome are unmasked by:

(a) Sodium channel blockers
(b) Calcium channel blockers
(c) Beta blocker
(d) Isoprenaline

Q20. Arrhythmias of Brugada syndrome are suppressed by:

(a) Quinidine
(b) Procainamide
(c) Ajmaline
(d) Amiodarone

Q21. Which drugs are effective in controlling arrhythmias in Brugada syndrome?

(a) Beta blockers

(b) Cilostazol
(c) Orciprenaline
(d) Amiodarone

Q22. Which drug does not produce Brugada like ECG pattern?

(a) Flecainide
(b) Ajmaline
(c) Quinidine
(d) Procainamide

Q23. Which conditions can produce type 1 Brugada like ECG?

(a) LBBB
(b) Arrhythmogenic right ventricular cardiomyopathy
(c) Pectus excavatum
(d) RBBB

Q24. Which drug can be useful in the management of Brugada syndrome?

(a) Quinidine
(b) Propranolol
(c) Verapamil
(d) Amiodarone

Q25. Which conditions can produce J wave?

(a) Acidosis
(b) Hyperkalemia
(c) Hypercalcemia
(d) All

Q26. Occlusion of which coronary artery can produce type 1 Brugada like ECG?

(a) LAD
(b) LCX
(c) Distal RCA
(d) Conus branch of RCA

Q27. J point is the:

(a) Junction of the P wave and the PR segment
(b) Junction of the PR segment and the Q wave
(c) Junction of the distal part of the QRS complex and the ST-segment
(d) Junction the ST-segment and the T wave

Q28. Normally J point is:

(a) On the isoelectric line
(b) Above the isoelectric line
(c) Below the isoelectric line
(d) At any point

Q29. In normal persons, J point can be elevated in leads:

(a) V_2, V_3
(b) V_3R, V_4R
(c) V_5, V_6
(d) V_7, V_8

Q30. J wave is also called:

(a) Osborn wave
(b) Epsilon wave
(c) Delta wave
(d) None

Q31. Right bundle branch block with prominent J wave, the ST-segment elevation of 3 mm followed by a negative T wave in leads V_1 to V_3 suggest the possibility of:

(a) Occlusion of septal branch of LAD
(b) Brugada syndrome
(c) Occlusion of conus branch of RCA
(d) Arrhythmogenic right ventricular dysplasia

Q32. Incomplete right bundle branch block with T wave inversion in leads V_1 to V_3 in a child of 8 years suggests:

(a) Brugada syndrome
(b) Arrhythmogenic right ventricular cardiomyopathy
(c) Juvenile pattern
(d) Myocarditis

Q33. Incomplete right bundle branch block with T wave inversion in leads V_1 to V_3 in a

26-year-old male with a low amplitude wave between end of QRS and onset of T wave suggests the possibility of:

(a) Brugada syndrome
(b) Juvenile pattern
(c) Arrhythmogenic right ventricular cardiomyopathy
(d) Dilated cardiomyopathy

Q34. In a 30-year-old male with a history of syncope, electrocardiogram showed incomplete right bundle branch block with T wave inversion in leads V_1 and V_2. Broad J wave were present in leads V_1 and V_2 recorded in second intercostal space. What is the most likely possibility?

(a) Brugada syndrome
(b) Pectus excavatum
(c) Juvenile pattern
(d) Arrhythmogenic right ventricular cardiomyopathy

Q35. Patients with Brugada syndrome are prone to:

(a) Atrial fibrillation
(b) Polymorphic VT
(c) AVNRT
(d) VF

Q36. Patients with Brugada syndrome are prone to:

(a) Bradycardia dependent VF
(b) Pause dependent VT
(c) VF during sleep
(d) VF during high levels of exercise

Q37. Coved elevation of the ST-segment with T wave inversion in leads V_3R, V_1, V_2 recorded in the third intercostal space suggests occlusion of:

(a) Septal branch of left anterior descending coronary artery
(b) Conus branch of right coronary artery
(c) Right ventricular branch of RCA
(d) First diagonal

Q38. Complete right bundle branch block with coved ST segment elevation and T wave inversion in leads V_1 and V_2 suggests:

(a) Brugada syndrome
(b) Arrhythmogenic right ventricular cardiomyopathy
(c) Infarction of interventricular septum
(d) Pectus excavatum

Q39. Epsilon wave is seen in:

(a) Brugada syndrome
(b) Arrhythmogenic right ventricular cardiomyopathy
(c) Isolated complete right bundle branch block
(d) Left lateral accessory pathway

Q40. Epsilon wave is seen in leads:

(a) V_1, V_2
(b) V_5, V_6
(c) V_8, V_9
(d) V_3R, V_4R

Q41. rsr' pattern with inverted T waves without any ST-segment deviation in leads V_1, V_2 recorded in third intercostal space can be seen in:

(a) Juvenile pattern
(b) Pectus excavatum
(c) Brugada syndrome
(d) Secundum ASD

Q42. A 28-year-oldmale presented with a history of revival from sudden cardiac death while at rest. One cousin died suddenly at the age of 35 years. There was no history suggestive of coronary artery disease. Electrocardiogram and electrolytes were normal. Possibilities include:

(a) Brugada syndrome
(b) Early repolarization syndrome

(c) Arrhythmogenic right ventricular cardiomyopathy
(d) Catecholaminergic polymorphic VT

Q43. J wave of the Brugada syndrome is usually seen:

(a) Constantly
(b) Intermittently
(c) During bradycardia
(d) During sleep

Q44. Classical ECG findings in Brugada syndrome are:

(a) Transient
(b) Subtle
(c) Spontaneous
(d) None

Q45. J waves of Brugada syndrome are augmented by:

(a) Fever
(b) Following short coupled PVC with long compensatory pause
(c) Sodium channel blockers
(d) None

Q46. Brugada syndrome should be diagnosed only when electrocardiographic findings are associated with:

(a) Documented VT/VF
(b) Syncope of probable arrhythmogenic cause
(c) Family history of sudden cardiac death below 45 years of age
(d) Any of the above

Q47. Drugs that can increase prominence of J wave of Brugada syndrome:

(a) Digoxin
(b) Amiodarone
(c) Ajmaline
(d) Flecainide

Q48. In a case of early repolarization syndrome, the risk of sudden cardiac death is increased in the presence of:

(a) Evidence of spontaneous VT/VF
(b) Suspicion of arrhythmogenic syncope
(c) Family history of sudden cardiac death
(d) Any the of above

Q49. Normally J wave of early repolarization syndrome appears as:

(a) J wave
(b) J point elevation
(c) Notch or slur of terminal part of R
(d) Any the of above

Q50. Which arrhythmia is more frequent in Brugada syndrome than in early repolarization syndrome?

(a) Atrial fibrillation
(b) Polymorphic VT
(c) Ventricular fibrillation
(d) AVRT

Q51. In early repolarization syndrome T wave is:

(a) Flat
(b) Inverted
(c) Prominent positive
(d) Biphasic

Q52. In early repolarization syndrome, the ST-segment shows:

(a) Elevation with concavity upward
(b) Elevation with convexity upward
(c) Depression
(d) Isoelectric line

Q53. Early repolarization syndrome should be diagnosed only when electrocardiographic findings are associated with:

(a) Arrhythmogenic syncope

(b) Documented VT/VF
(c) Family history of sudden cardiac death
(d) Any the of above

Q54. Isolated T wave inversion in leads V_1 to V_3 suggests:

(a) Brugada syndrome
(b) Juvenile pattern
(c) Arrhythmogenic right ventricular cardiomyopathy
(d) Septal infarction

Q55. Diffuse concave upward elevation of the ST-segment, with depression of the PR segment are suggestive of:

(a) Pericarditis
(b) Early repolarization syndrome
(c) Hypercalcemia
(d) Extensive ischemia

Q56. J point and ST-segment elevation in leads V_1 and V_2 with short ST-segment suggests:

(a) Brugada syndrome
(b) Arrhythmogenic right ventricular cardiomyopathy
(c) Septal infarction
(d) Hypercalcemia

Q57. Prominent narrow Q wave, tall R wave, j point elevation, and mild ST-segment elevation with concavity upward in lead V_6 suggests:

(a) Early repolarization syndrome
(b) Lateral infarction
(c) Chronic left ventricular volume overload
(d) Systemic hypothermia

Answers

(1) a (2) b (3) b (4) a, b (5) c (6) c (7) b (8) b, d (9) c (10) a, c (11) a (12) b (13) c (14) a (15) a, b (16) c (17) a, d (18) c, d (19) a, b, c (20) a (21) b, c (22) c (23) b, c, d (24) a (25) d (26) a, d (27) c (28) a (29) a (30) a (31) b (32) c (33) c (34) a (35) a, b, d (36) a, b, c (37) b (38) c (39) b (40) a (41) a, b (42) a (43) b, c, d (44) a, b, c (45) a, b, c (46) d (47) c, d (48) d (49) d (50) a (51) c (52) a (53) d (54) b (55) a (56) d (57) a, c

References

1. Mirvis DM, Goldberger AL. Electrocardiography. In: Mann DL, Zipes DP, Libby P, Bonow RO, editors. Braunwald's heart disease. Philadelphia: Elsevier; 2015. p. 44–152.
2. Antzelevitch C, Yan GX, Ackerman MJ, Borggrefe M, Corrado D, Guo J, et al. J wave syndrome expert consensus conference report: emerging concepts and gaps in knowledge. Heart Rhythm. 2016;13:e296–324.
3. Sethi KK, Sethi K, Chutani SK. Early repolarization and J wave syndromes. Indian Heart J. 2014;66:443–52.
4. Priori SG, Wilde AA, Horie M, Cho Y, Behr ER, Berul C, et al. HRS/EHRA/APHRS expert consensus statement on the diagnosis and management of patients with inherited primary arrhythmia syndromes. Europace. 2013;15:1389–406.
5. Haissaguerre M, Derval N, Sacher F, Jesel L, Deisenhofer I, de Roy L, et al. Sudden cardiac arrest associated with early repolarization. N Engl J Med. 2008;358:2016–23.
6. Antzelevitch C, Yan JX. J wave syndromes – contemporary review. Heart Rhythm. 2010;7:549–58.
7. Rosso R, Glikson E, Belhassen B, Katz A, Halkin A, Steinvil A, et al. Distinguishing "benign" from "malignant early repolarization" the value of ST segment morphology. Heart Rhythm. 2012;9:225–9.
8. Merchant FM, Noseworthy PA, Weiner RB, Singh SM, Ruskin JN, Reddy VY. Ability of terminal QRS notching to distinguish benign from malignant electrocardiographic forms of early repolarization. Am J Cardiol. 2009;104:1402–6.
9. Olgin JE, Zipes DP. Specific arrhythmias: diagnosis and treatment. In: Mann DL, Zipes DP, Libby P, Bonow RO, editors. Braunwald's heart disease. Philadelphia: Elsevier; 2015. p. 748–97.
10. Satish OS, Srivastav KS. Early repolarization changes in ECG: are they benign or malignant? J Assoc Physicians India. 2016;64:51–4.
11. Brugada P, Brugada J. Right bundle branch block, persistent ST segment elevation and sudden cardiac death: a distinct clinical and electrocardiographic syndrome: a multicenter report. J Am Coll Cardiol. 1992;20:1391–6.
12. Priori SG, Napolitano C. Genetics of channelopathies and clinical implications. In: Fuster V, Walsh RA, Harrington RA, editors. Hurst's the heart. New York: McGraw Hill; 2011. p. 897–910.
13. Talib AK, Sato N, Myojo T, Sugiyama E, Nakagawa N, Sakamato N, et al. Insight into specific pro-arrhythmic triggers in Brugada and early repo-

larization syndromes: results of long-term follow up. Heart Vessels. 2016; https://doi.org/10.1007/s00380-016-0828-8.
14. Morita H, Kusano KF, Miura D, et al. Fragmented QRS as a marker of conduction abnormality and a predictor of prognosis of Brugada syndrome. Circulation. 2008;118:1697–704.
15. Postema PG, Wilde AA. Do J waves constitute a syndrome? J Electrocardiol. 2013;46:461–5.
16. Calo L, Giustetto C, Martino A, Sciarra L, Cerrato N, Marziali M, et al. A new electrocardiographic marker of sudden death in Brugada syndrome. The S wave in lead I. J Am Coll Cardiol. 2016;67:1427–40.
17. Sonoda K, Watanabe H, Hisamatsu T, Ashihara T, Ohno S, Hayashi H, et al. High frequency of early repolarization and Brugada – type electrocardiograms in hypercalcemia. Ann Noninvasive Electrocardiol. 2016;21:30–40.

40 ST-Segment Elevation

ST-segment starts from the end of the QRS and ends with beginning of the T wave. Usually it is slightly elevated above the base line and merges smoothly with the T wave so that the exact point of end of the ST-segment is difficult to define [1]. ST-segment elevation is diagnosed by elevation at J point. Criteria for diagnosing ST-segment elevation are 0.2 mV in leads V_2 and V_3, 0.1 mV in all other leads in the absence of LBBB or LVH [2].

It is, however, important that clinical significance of ST-segment elevation of any magnitude should be decided in the context of the total clinical profile [3].

1 Causes of ST-Segment Elevation

1.1 ST-Segment Elevation with Concavity Upward

Positive T wave produces ST-segment elevation with concavity upward.

(i) Early (hyperacute) phase of transmural myocardial injury which is dominantly subepicardial.

Normal electrocardiogram is shown in Fig. 40.1a.

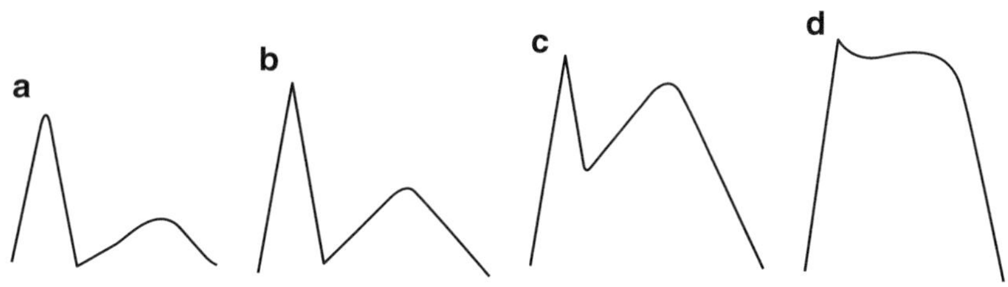

Fig. 40.1 Diagram showing sequential ECG changes during evolution of transmural MI (**a** to **d**) (Reproduced with permission from: Mittal SR. ST Segment Elevation. Cardiology Today-2017;21:203–210)

S. Mittal, *Insights into Electrocardiograms with MCQs*,
https://doi.org/10.1007/978-981-99-0127-2_40

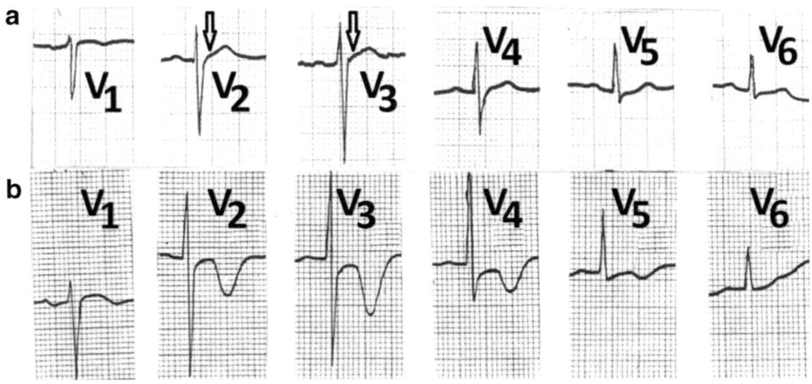

Fig. 40.2 Electrocardiograms (**a**) showing straight elevation of the ST segment in leads V_2 and V_3 (arrow). (**b**) Showing sequential appearance of deep symmetrical T wave inversion in leads V_2 to V_4 (Reproduced with permission from: Mittal SR. ST Segment Elevation. Cardiology Today-2017;21:203–210)

Earliest finding is the straight upward sloping of the ST-segment [4] (Fig. 40.1b). ST-segment is elevated from the J point (early uptake of the ST-segment).

Representative electrocardiogram is shown in Fig. 40.2.

Subsequently, the ST-segment becomes elevated (Fig. 40.1c). Reciprocal ST-segment depressions are seen in opposite leads. Sometimes, reciprocal ST-segment depressions may be more impressive. R wave amplitude is increased due to focal injury block. The T wave is tall and wide. Tall R wave, elevated ST-segment, and tall and widened T wave may merge with each other to form a monophasic deflection (Fig. 40.1d). This constitutes an ominous sign.

There is rapid evolution into appearance of the Q wave and loss of the R wave (Fig. 40.3a, b). Inversion of the T wave results in ST-segment elevation with convexity upward (Fig. 40.3c).

Sometimes during healing, ECG may appear apparently normal before the T waves become inverted. This is known as pseudonormalization (Fig. 40.4). Representative electrocardiograms are shown in Fig. 40.4a, b.

(ii) Coronary vasospasm (Prinzmetal's angina)

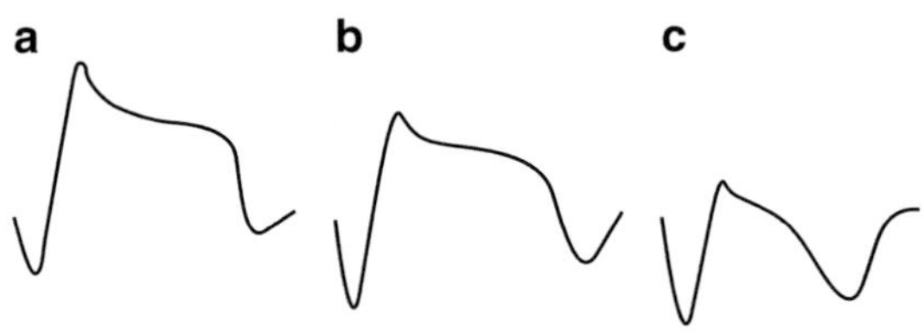

Fig. 40.3 Diagram showing sequential ECG changes during healing of transmural MI (Reproduced with permission from: Mittal SR. ST Segment Elevation. Cardiology Today-2017;21:203–210)

ST-segment elevation is significant. Associated T wave is tall, upright, and widened. Sometimes tall R wave, elevated ST-segment, and tall T wave may fuse to form a single deflection (Figs. 40.5a and 40.6a). The ECG may be indistinguishable from the hyperacute phase of transmural myocardial injury. However, the ECG rapidly normalizes with relief of pain. There are no evolutionary changes like the development of a Q wave, loss of the R wave, or inversion of the T wave (Figs. 40.5 and 40.6).

Leads showing ST-segment elevation may vary between episodes depending on the vessel involved. Patients with ST-segment elevation in one episode may show ST-segment depression in the same or other leads in another episode. ST-segment elevation may be preceded by ST-segment depression.

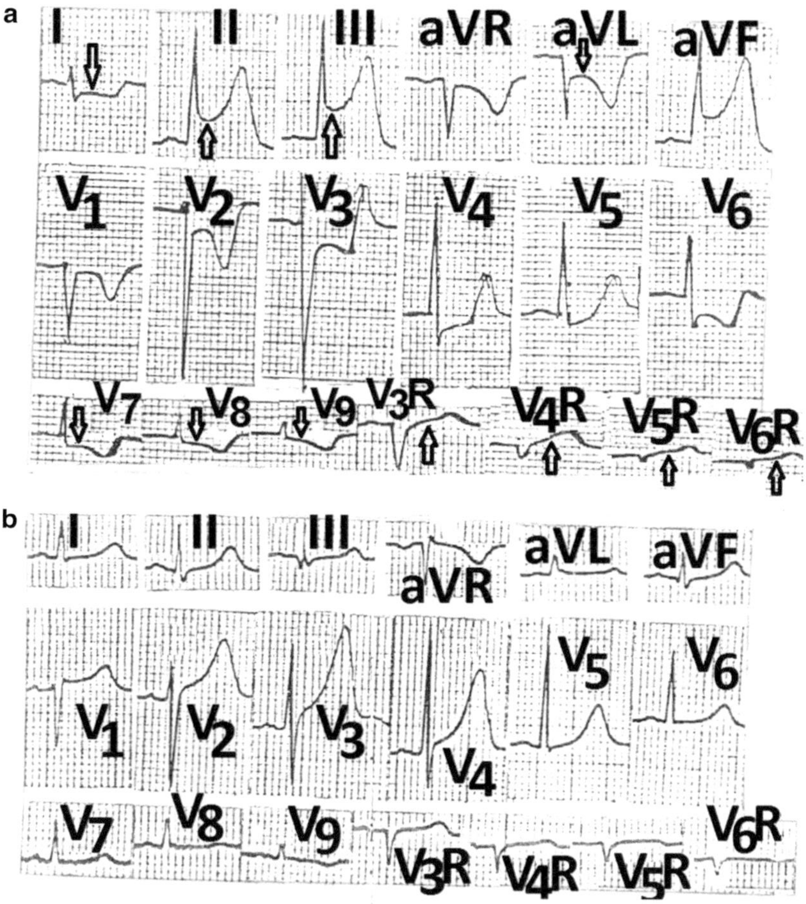

Fig. 40.4 Electrocardiograms (**a**) showing acute inferior and right ventricular infarction. ST-segment is elevated in leads II, III, aVF, and V_3R to V_6R (marked ⇧). There are reciprocal ST-segment depressions in leads I and aVL and V_7 to V_9 (marked ⇩). (**b**) Showing transient pseudo normalization during healing (Reproduced with permission from: Mittal SR. ST Segment Elevation. Cardiology Today-2017;21:203–210)

(iii) Takotsubo cardiomyopathy

ST-segment elevation in lead aVR in association with ST-segment elevation in inferior or anteroseptal leads supports the possibility of Takotsubo cardiomyopathy [5]. There is no ST-segment elevation in lead V_1. In acute myocardial infarction, ST-segment elevation in lead aVR is accompanied by diffuse ST-segment depressions.

(iv) Non-ischemic causes [6, 7]

(a) Early phase of acute pericarditis

ST-segment elevation is mild. PR segment may show mild depression in most of the leads with elevation in lead aVR. This is due to current of atrial injury. Sequentially the ST-segment may settle down and the T waves may become flat or may show shallow inversion. Prolongation of the QRS complex and shortening of the QT interval in the leads with ST-segment elevation supports the pos-

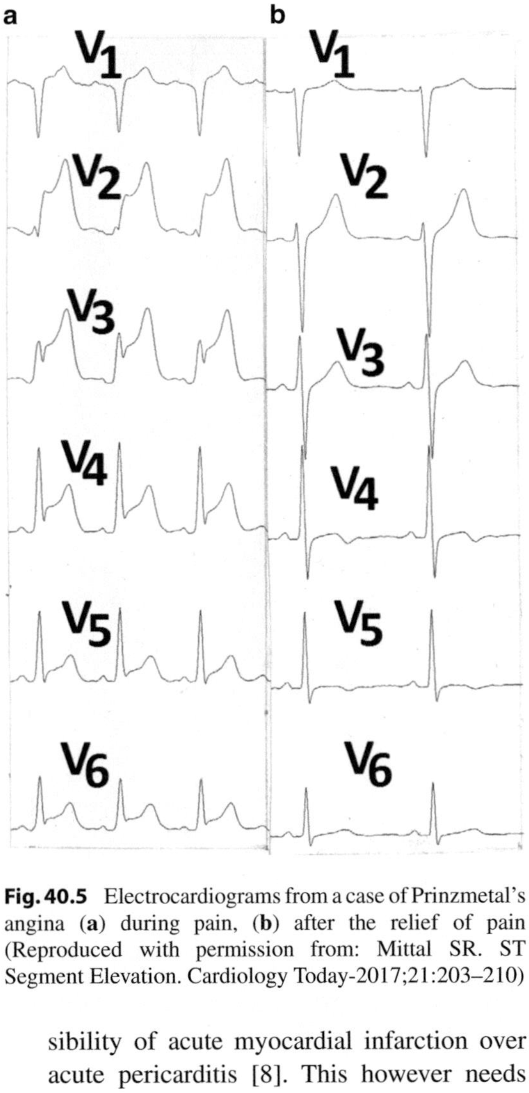

Fig. 40.5 Electrocardiograms from a case of Prinzmetal's angina (**a**) during pain, (**b**) after the relief of pain (Reproduced with permission from: Mittal SR. ST Segment Elevation. Cardiology Today-2017;21:203–210)

sibility of acute myocardial infarction over acute pericarditis [8]. This however needs recording of ECG at 50 mm/s speed. In acute pericarditis, there is no transmural injury and hence QRS complex and QT interval remain normal.

(b) Vagotonia, early repolarization

ST-segment elevation is seen mostly in left lateral or inferior leads.

Elevation is usually mild (rarely exceeds 2 mm). Preceding R wave is usually tall. Sinus bradycardia is common and there are no reciprocal or sequential changes. ST-segment may normalize by sympathetic stimulation, for example, exercise.

(c) Hypothermia

ST-segment elevation is associated with sinus bradycardia, prominent J wave, and prolongation of the QT interval.

(d) Other causes

- LVH—ST-segment is elevated in leads V_1 to V_3.
- LBBB—ST-segment elevation >1 mm concordant with positive QRS or ST-segment elevation >5 mm nonconcordant with negative QRS is defined as ST-segment elevation.
- Pulmonary embolism—ST-segment is elevated in leads V_1 to V_3.
- Brugada syndrome—ST-segment is elevated in leads V_1, V_2.
- Arrhythmogenic right ventricular dysplasia—ST-segment is elevated in leads V_1, V_2.
- Hyperkalemia, Hypercalcemia.
- Left pneumothorax.
- Chest deformities.
- Trauma to myocardium.
- Myocarditis.
- Tumor invading the left ventricular myocardium.
- Pneumomediastinum [9].
- Gastrointestinal causes—ileus, cholecystitis, pancreatitis, gastric dilatation [10].
- Subarachnoid hemorrhage [11].

1.2 ST-Segment Elevation with Convexity Upward

Inversion of T wave produces convex contour of the ST-segment.

(i) Fully evolved myocardial infarction (Fig. 40.3c)

It is characterized by symmetrically inverted, deep, and pointed T waves [12].

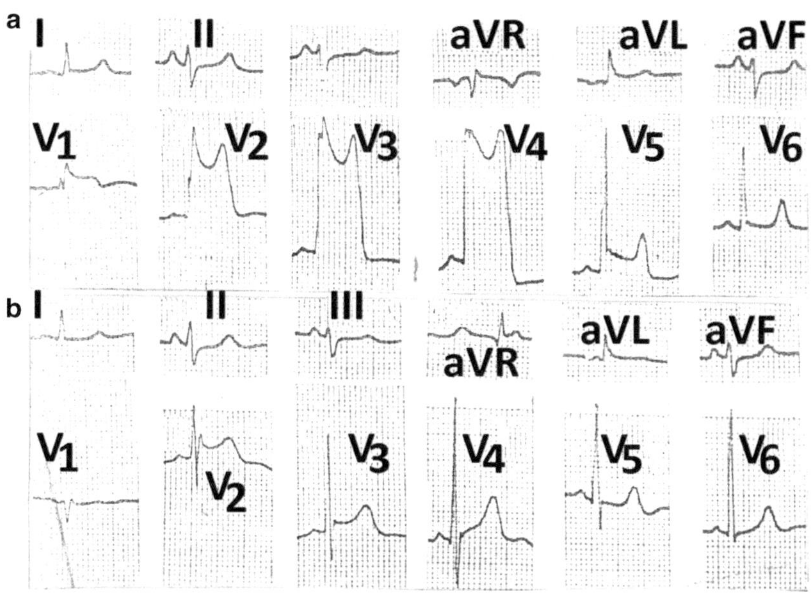

Fig. 40.6 Electrocardiograms from another case of Prinzmetal's angina (**a**) during pain, (**b**) after the relief of pain (Reproduced with permission from: Mittal SR. ST Segment Elevation. Cardiology Today-2017;21:203–210)

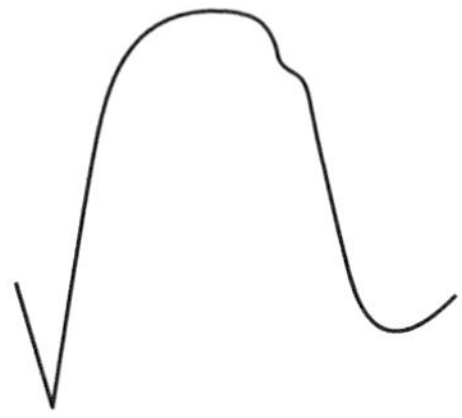

Fig. 40.7 Diagram showing ECG suggestive of ventricular aneurysm (Reproduced with permission from: Mittal SR. ST Segment Elevation. Cardiology Today-2017;21:203–210)

Typically the terminal portion of the T wave is the first to become inverted followed by middle and initial portions. Q waves may or may not be present. ≥70% reduction in the ST-segment elevation in the maximally involved leads is suggestive of successful reperfusion [13].

(ii) Left ventricular aneurysm

Persistent ST-segment elevation 3 months or longer after an acute infarction suggests ventricular aneurysm (Fig. 40.7).

2 Electrocardiographic Localization of Myocardial Infarction According to the Leads Showing ST-Segment Elevation [4] (Fig. 40.8)

(a) Leads V_1 to V_4—anteroseptal and apical (Fig. 40.9).
(b) Leads V_4 toV_6—apical and lateral (Fig. 40.10).
(c) Leads I, aVL—high lateral infarction (Fig. 40.11).
(d) Leads I, aVL, V_4 toV_6—lateral infarction.
(e) Leads I, aVL, V_1 to V_6—extensive anterior infarction (Fig. 40.12).
(f) Leads II, III, and aVF—inferior infarction (Fig. 40.13).
(g) Leads II, III, aVF, V_5, V_6—inferolateral infarction.
(h) Leads V_7 to V_9—posterior infarction. It is usually associated with inferior or lateral

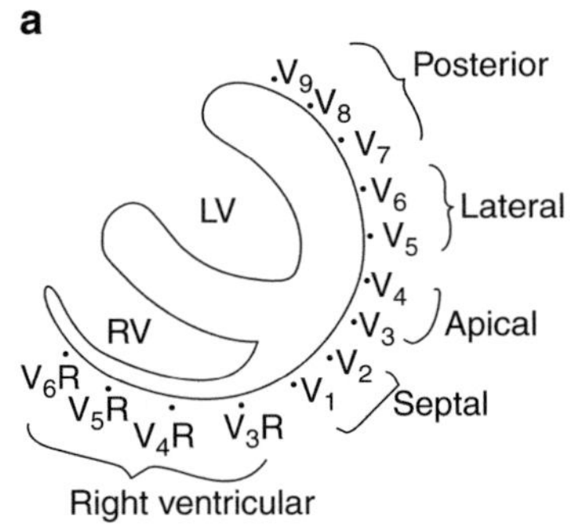

Fig. 40.8 Diagram showing localization of myocardial infarction (**a**) in horizontal plane and (**b**) in frontal plane RV—Right ventricle, LV—left ventricle (Reproduced with permission from: Mittal SR. ST Segment Elevation. Cardiology Today-2017;21:203–210)

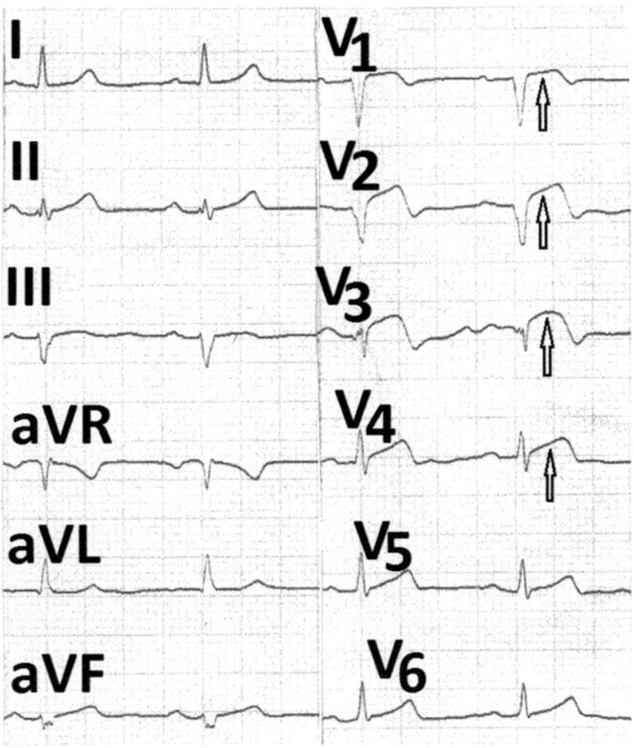

Fig. 40.9 Electrocardiogram is showing anteroseptal and apical infarction. ST-segment is elevated in leads V_1 to V_4 (marked ⇧) (Reproduced with permission from: Mittal SR. ST Segment Elevation. Cardiology Today-2017;21:203–210)

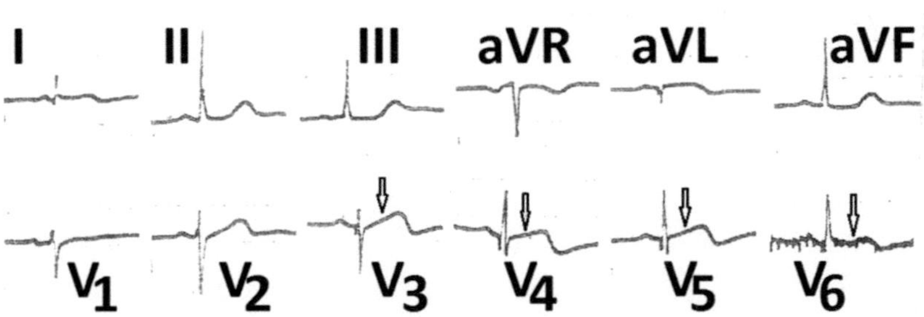

Fig. 40.10 Electrocardiogram is showing apical and lateral infarction. ST-segment elevations in leads V_3 to V_6 are marked ⇧ (Reproduced with permission from: Mittal SR. ST Segment Elevation. Cardiology Today-2017;21:203–210)

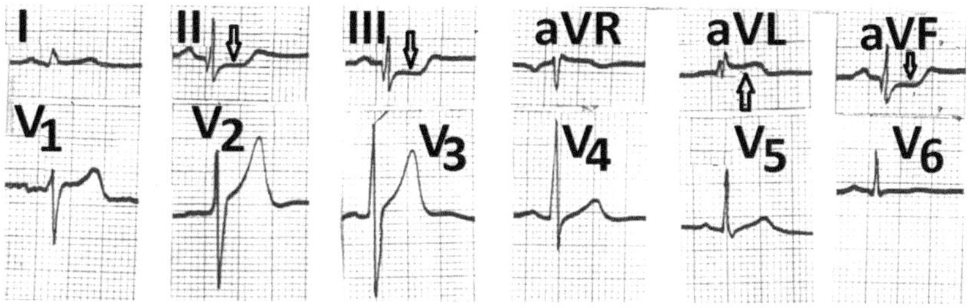

Fig. 40.11 Electrocardiogram is showing high lateral infarction. ST-segment is elevated in lead aVL (marked ⇩). Leads II, III, and aVF show reciprocal ST-segment depressions (marked ⇧) (Reproduced with permission from: Mittal SR. ST Segment Elevation. Cardiology Today-2017;21:203–210)

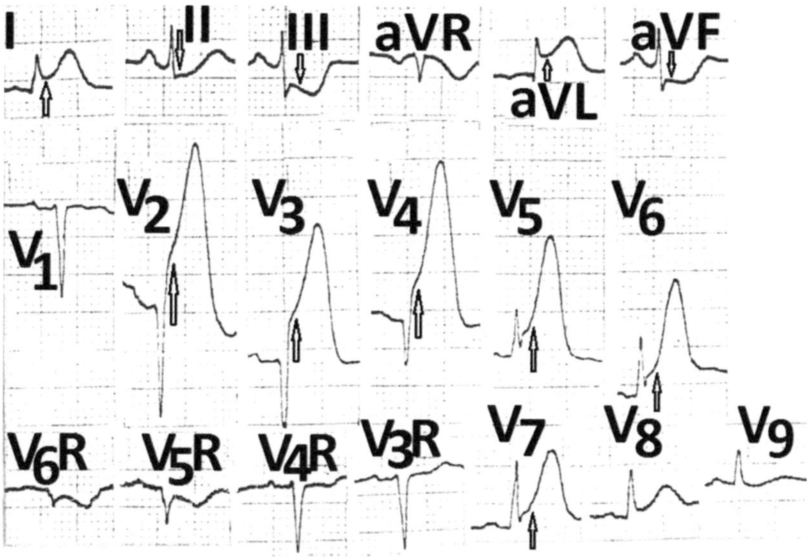

Fig. 40.12 Electrocardiogram is showing extensive anterior infarction. ST-segment is elevated in leads I, aVL, and V_2 to V_7 (marked ⇩). Leads II, III, and aVF show reciprocal ST-segment depressions (marked ⇧) (Reproduced with permission from: Mittal SR. ST Segment Elevation. Cardiology Today-2017;21:203–210)

myocardial infarction. Reciprocal changes are common in leads V_1 to V_3 especially in lead V_2. These include tall and slightly wide R waves and relatively tall T waves (Fig. 40.14).

(i) Leads II, III, aVF, V_4 to V_9—inferoposterolateral infarction (Fig. 40.15)
(j) Leads V1, V_3R to V_6R—right ventricular infarction (Fig. 40.16).

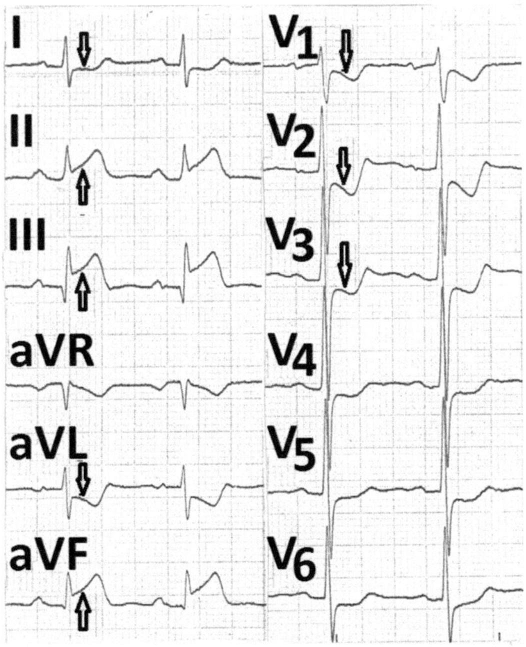

Fig. 40.13 Electrocardiogram is showing inferior infarction. ST-segment is elevated in leads II, III, and aVF (marked ⇩). Reciprocal ST-segment depression is seen in leads I, aVL, and V_1 to V_3 (marked ⇧) (Reproduced with permission from: Mittal SR. ST Segment Elevation. Cardiology Today-2017;21:203–210)

The following ECG findings also suggest right ventricular involvement:

(i) Absence of ST-segment depression in right precordial leads in the setting of inferior myocardial infarction. It is due to ST-segment deviation toward the surface of myocardial injury, that is, right ventricular leads counterbalancing reciprocal ST-segment depression.
(ii) ST-segment depression in lead V_2 less than 50% of the magnitude of ST-segment elevation in lead aVF [14].
(iii) Some cases of right ventricular infarction can have ST-segment elevation extending from lead V_1 to leads V_3 or even lead V_4 [15]. ST-segment elevation in lead $V_1 > V_2$ and V_3 suggests occlusion of right coronary artery. Occlusion of the proximal right coronary artery can hamper the flow to sinoatrial node and produce sinus bradycardia. ST-segment elevation in lead V_1 decreasing toward lead V_5 suggests right ventricular infarction.
(iv) ST-segment elevation in lead V_1 with ST-segment depression in lead V_2.
(v) ST-segment elevation in lead III exceeding ST-segment elevation in lead II (Fig. 40.17), particularly when combined with ST-segment elevation in lead V_1.

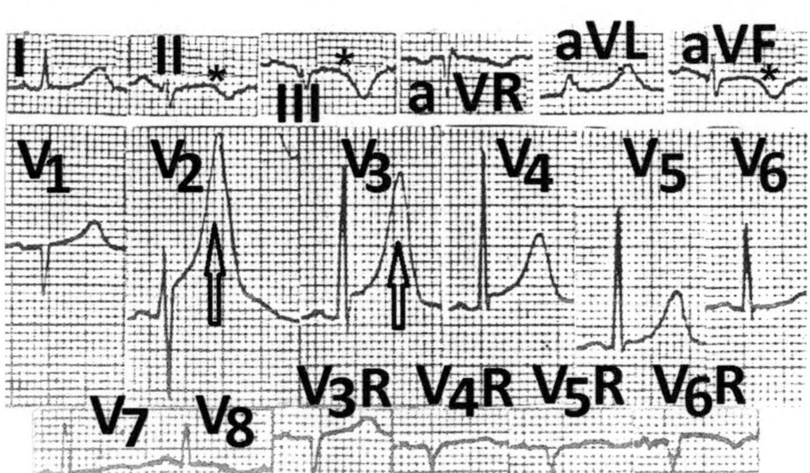

Fig. 40.14 Electrocardiogram is showing inferior infarction (marked *) and reciprocal changes of posterior infarction—Tall and broad T waves in leads V_2 and V_3 (marked ⇩) (Reproduced with permission from: Mittal SR. ST Segment Elevation. Cardiology Today-2017;21:203–210)

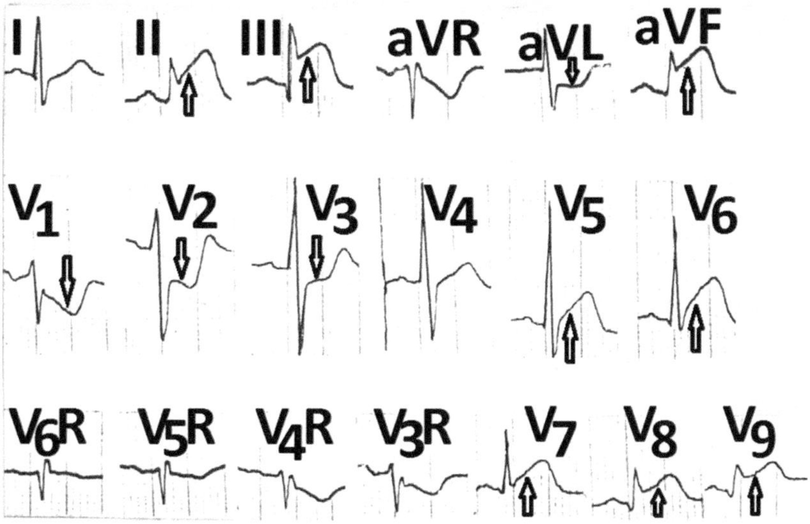

Fig. 40.15 Electrocardiogram is showing inferior and posterolateral infarction. ST-segment is elevated in leads II, III, aVF, and V_5 to V_9 (marked ⇧). Reciprocal ST-segment depression is seen in leads aVL and V_1 to V_3 (marked ⇧) (Reproduced with permission from: Mittal SR. ST Segment Elevation. Cardiology Today-2017;21:203–210)

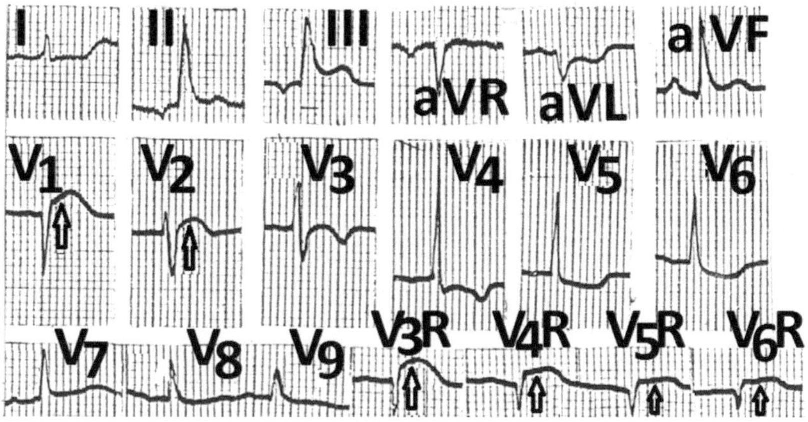

Fig. 40.16 Electrocardiogram is showing right ventricular infarction. ST segment is elevated in leads V_6R to V_2 (marked ⇩) (Reproduced with permission from: Mittal SR. ST Segment Elevation. Cardiology Today-2017;21:203–210)

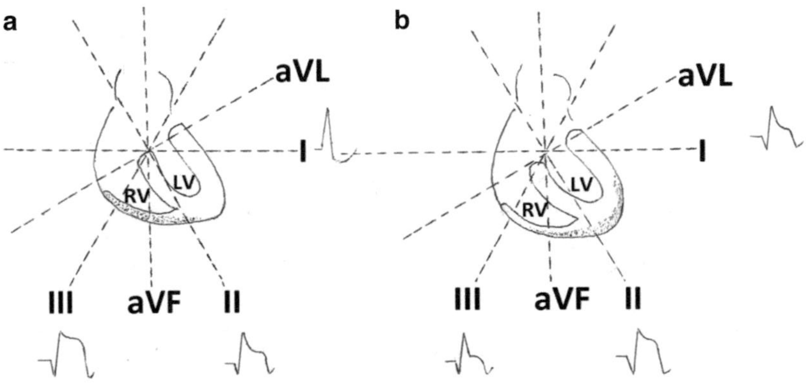

Fig. 40.17 Diagram (**a**) showing ECG changes in inferior infarction due to occlusion of the right coronary artery (**b**) showing ECG changes in inferior infarction due to occlusion of LCX. RV—Right ventricle, LV—Left ventricle (Reproduced with permission from: Mittal SR. ST Segment Elevation. Cardiology Today-2017;21:203–210)

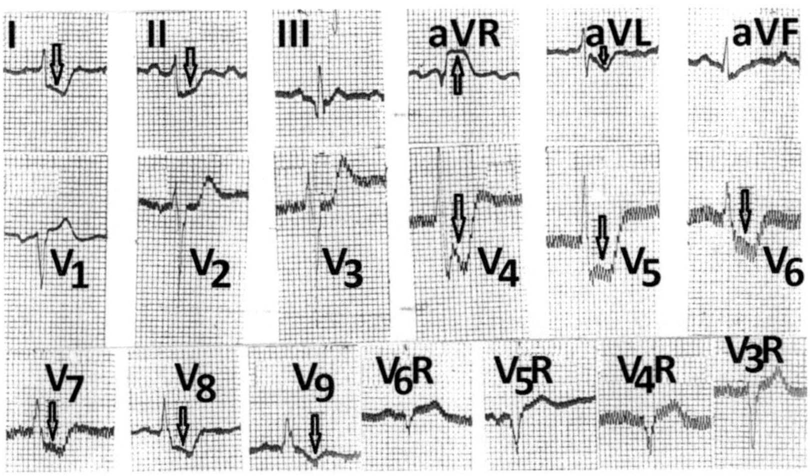

Fig. 40.18 Electrocardiogram due to occlusion of the left main coronary artery showing ST-segment elevation in lead aVR (marked ⇧) and reciprocal ST-segment depression in leads I, II, aVL, and V_3 to V_9 (marked ⇧) (Reproduced with permission from: Mittal SR. ST Segment Elevation. Cardiology Today-2017;21:203–210)

3 Localization of Culprit Lesion [6, 7, 12, 16, 17]

(i) ST-segment elevation in lead aVR more than that in lead V_1 with diffuse ST-segment depression in other leads

Such an ECG suggests occlusion of the left main coronary artery (Fig. 40.18) or its equivalent (significant lesion of LAD and LCX).

Infarction involves bases of the heart. Vector of injury is directed toward lead aVR and away from most of the other leads (Fig. 40.19).

This is a serious situation and needs urgent intervention. Transmural ischemia in the right ventricular outflow tract caused by impaired blood flow in the large conal branch originating from the right coronary artery can also cause ST-segment elevation in lead

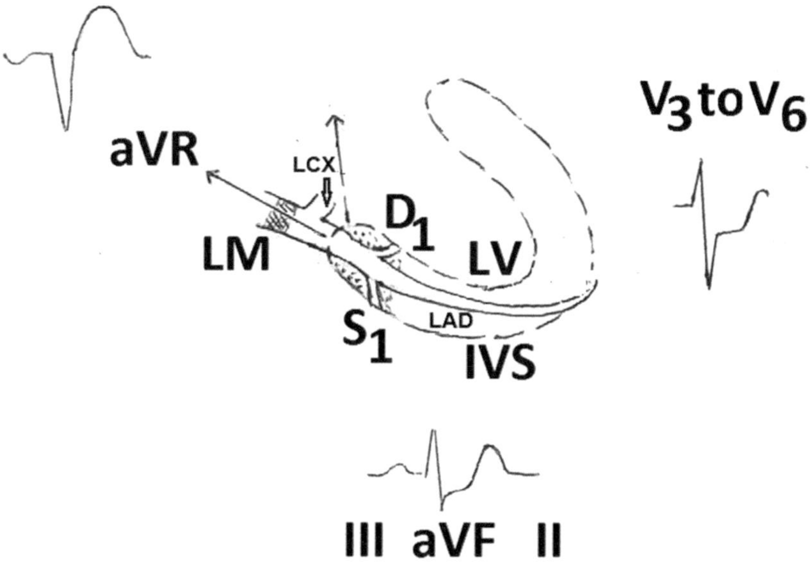

Fig. 40.19 Diagram showing direction of current of injury (marked ⇩) in occlusion of left main coronary artery and its effect on various ECG leads. Lead aVR shows ST-segment elevation. Other leads show ST-segment depression. LM—Left main coronary artery, LCX—Left circumflex, D_1—first diagonal, S_1—first septal, LAD—Left anterior descending coronary artery (Reproduced with permission from: Mittal SR. ST Segment Elevation. Cardiology Today-2017;21:203–210)

aVR [18]. This is, however, not accompanied by diffuse ST-segment depression.

(ii) ST elevation in leads I, aVL, and V_1 to V_6

It suggests occlusion of the left anterior descending coronary artery proximal to the first diagonal (Fig. 40.20).

Inferior leads show reciprocal ST-segment depression. Simultaneous ST-segment elevation in inferior leads suggests that LAD is wrapping around the cardiac apex to supply part of the inferior surface. Occlusion before the first septal branch of LAD results in concomitant fresh appearance of right bundle branch block (Figs. 40.21 and 40.22) as the right bundle branch is supplied by the septal branches of LAD (Fig. 40.23).

ST-segment elevation in leads aVR and V_1 also suggests occlusion before the first septal branch of LAD (Fig. 40.24).

Occlusion of LAD after first diagonal branch spares lead aVL (Fig. 40.25).

(iii) ST-segment elevation in leads V_1 to V_3 (Fig. 40.26)

It suggests occlusion of mid/distal LAD (Fig. 40.26).

Simultaneous elevation of the ST-segment in leads II, III, aVF suggests a wrap around LAD (Fig. 40.27).

(iv) ST-segment elevation in leads I and aVL with ST-segment depression in leads II, III, aVF

It suggests occlusion of proximal diagonal or obtuse marginal coronary artery (Fig. 40.28).

(v) ST-segment elevation in leads II, III, and aVF

It suggests occlusion of the proximal part of posterior descending coronary artery. It is associated with ST depression in leads I, aVL, and V_1 to V_3 (Figs. 40.29 and 40.30).

Prominent R and tall T in leads V_2, V_3 suggest concomitant posterior infarction (Fig. 40.31).

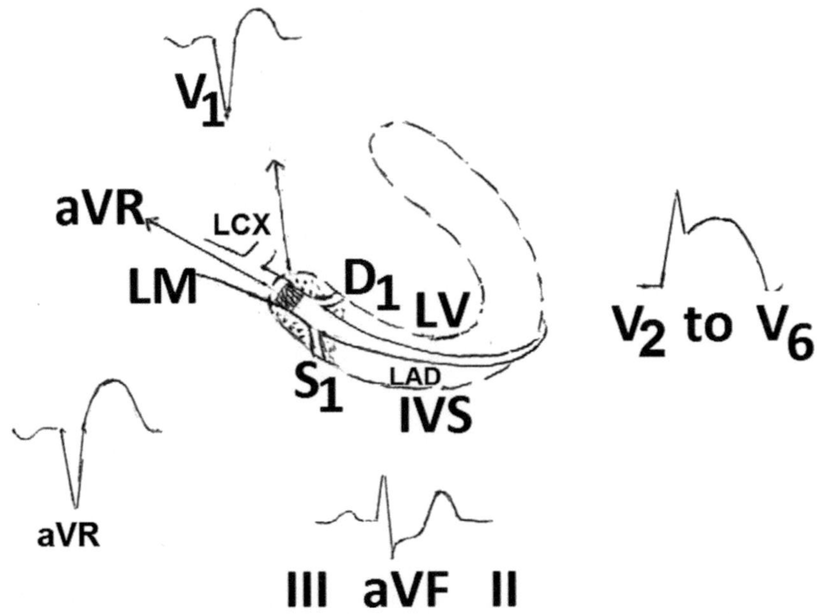

Fig. 40.20 Diagram showing effect of occlusion of proximal LAD on different ECG leads. Leads V_2 to V_6 show ST-segment elevation in addition to lead aVR and V_1. Leads II, III, and aVF show ST-segment depressions. LM—Left main coronary artery, LCX—circumflex coronary artery, D_1—first diagonal, S_1—first septal, LAD—Left anterior descending coronary artery (Arrow showing direction of current and injury) (Reproduced with permission from: Mittal SR. ST Segment Elevation. Cardiology Today-2017;21:203–210)

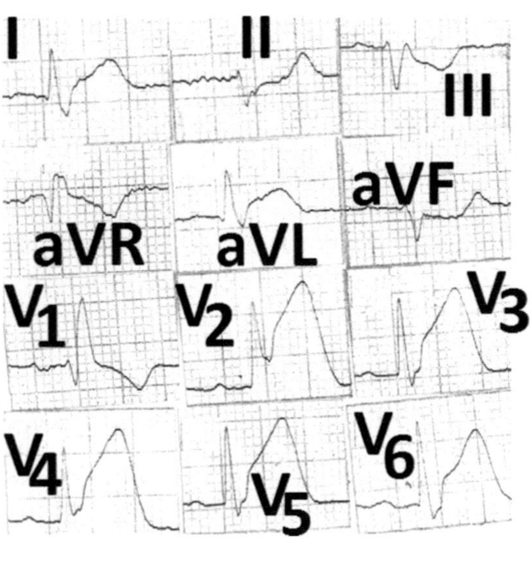

Fig. 40.21 Electrocardiogram shows anterior infarction with RBBB suggesting occlusion of LAD before the first septal branch (Reproduced with permission from: Mittal SR. ST Segment Elevation. Cardiology Today-2017;21:203–210)

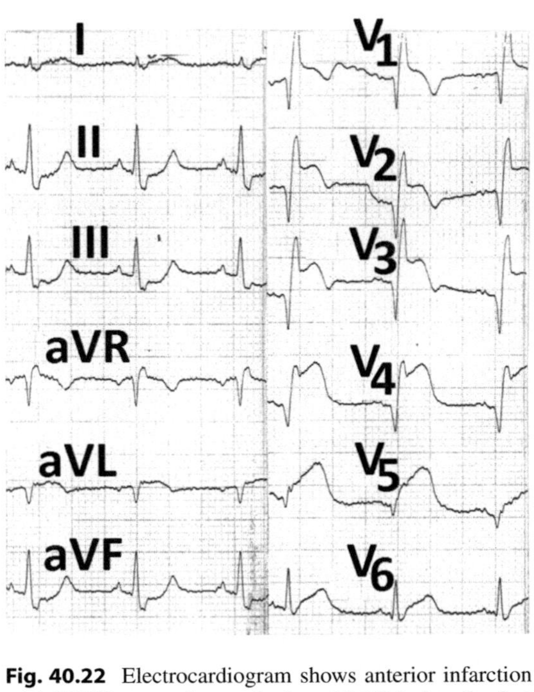

Fig. 40.22 Electrocardiogram shows anterior infarction with RBBB suggesting occlusion of LAD before the first septal branch (Reproduced with permission from: Mittal SR. ST Segment Elevation. Cardiology Today-2017;21:203–210)

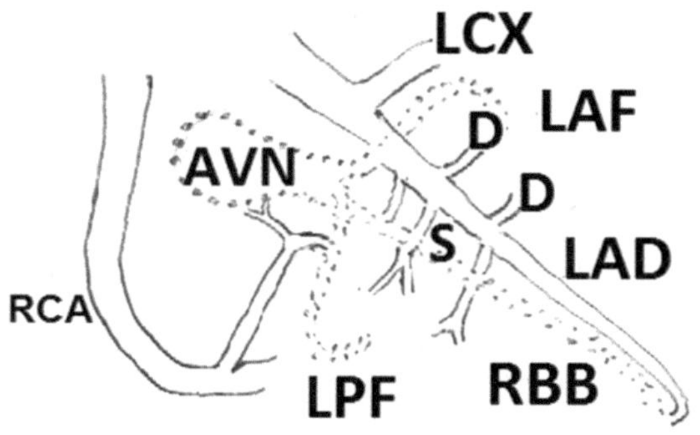

Fig. 40.23 Diagram showing the arterial supply of cardiac conduction tissue. RCA—right coronary artery, LCX—Left circumflex, LAD—Left anterior descending coronary artery, D—Diagonals, S—septal branches of LAD, AVN—Atrioventricular node, LAF—Left anterosuperior fascicle, RBB—Right bundle branch, LPF—Left posteroinferior fascicle. RBB is supplied by septal branches of LAD (Reproduced with permission from: Mittal SR. ST Segment Elevation. Cardiology Today-2017;21:203–210)

Depending on dominance, the posterior descending coronary artery could be arising from the left circumflex or the right coronary artery. Differentiation may be difficult. Some points may be helpful.

- ST-segment elevation in lead I suggests occlusion of LCX whereas ST-segment depression in lead I suggests occlusion of RCA.
- ST-segment elevation in lead II > that in lead III suggests occlusion of LCX. On the other hand, ST-segment elevation in lead III > that in lead II suggests occlusion of RCA.
- ST-segment elevation in leads V_1 and V_4R suggest occlusion of proximal RCA. Positive T wave in lead V_4R without elevation of the ST-segment suggests occlusion of distal RCA. T wave

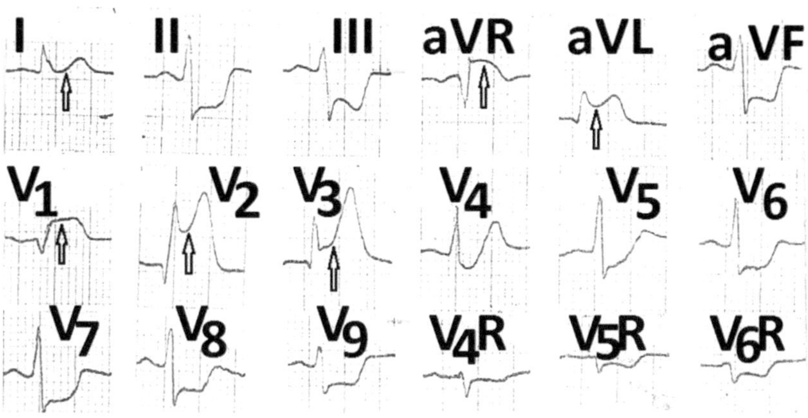

Fig. 40.24 Electrocardiogram from occlusion of proximal LAD. It shows ST-segment is elevated in leads I, aVR, aVL, V_1 to V_3 (marked ↖). Other leads show depression of ST-segment (Reproduced with permission from: Mittal SR. ST Segment Elevation. Cardiology Today-2017;21:203–210)

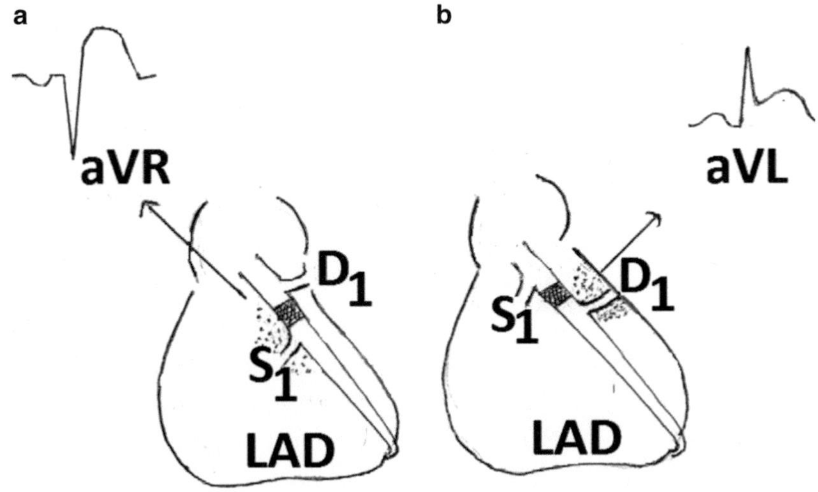

Fig. 40.25 Diagram (**a**) showing the effect of occlusion of left anterior descending coronary artery (LAD) before 1st septal branch (S_1) after first diagonal (D_1), (**b**) showing effect of occlusion of LAD after first septal (S_1) but before first diagonal (D_1) (Arrow—direction of current of injury) (Reproduced with permission from: Mittal SR. ST Segment Elevation. Cardiology Today-2017;21:203–210)

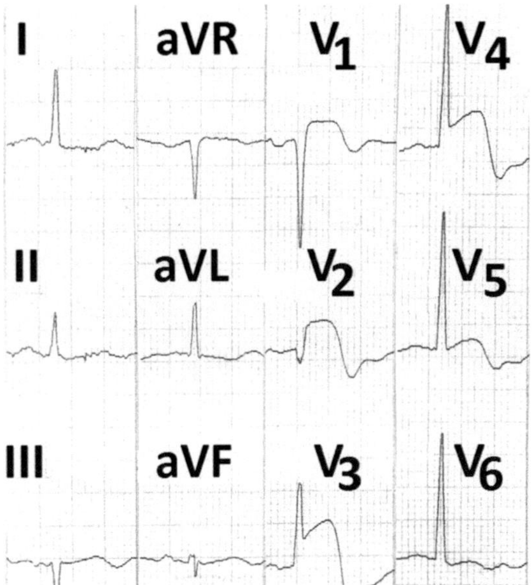

Fig. 40.26 Electrocardiogram is showing anteroseptal and apical myocardial infarction due to occlusion of mid LAD (Reproduced with permission from: Mittal SR. ST Segment Elevation. Cardiology Today-2017;21:203–210)

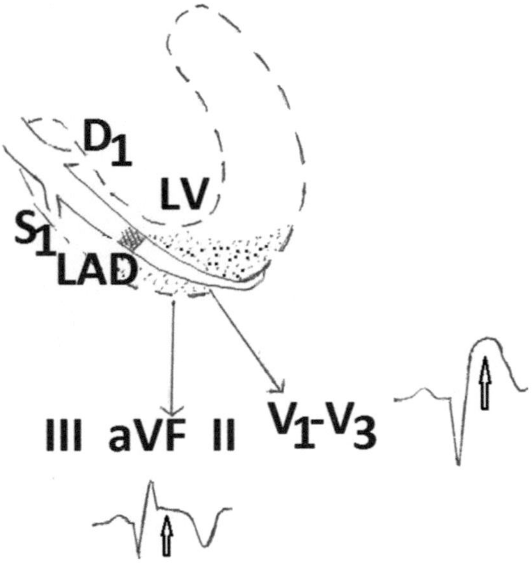

Fig. 40.27 Diagram showing direction of current of injury (marked arrows) in occlusion of the mid segment of the left anterior descending coronary artery (LAD) wrapping around cardiac apex and supplying part of the inferior surface of the heart (Reproduced with permission from: Mittal SR. ST Segment Elevation. Cardiology Today-2017;21:203–210)

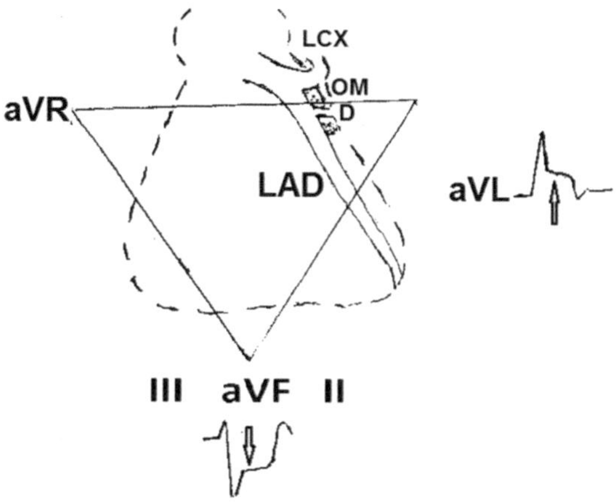

Fig. 40.28 Diagram showing the arterial supply of high lateral surface by diagonal and/or obtuse marginal coronary artery and resulting high lateral infarction on ECG leads. LCX—Left circumflex, OM—Obtuse marginal, LAD—Left anterior descending coronary artery, D—Diagonal branch (Reproduced with permission from: Mittal SR. ST Segment Elevation. Cardiology Today-2017;21:203–210)

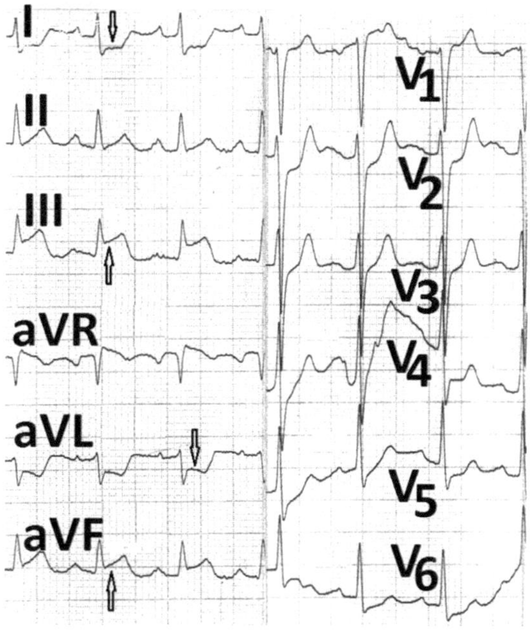

Fig. 40.29 Electrocardiogram shows inferior infarction (marked ⇧) due to occlusion of the posterior descending coronary artery. Reciprocal ST-segment depression is seen in leads I and aVL (marked ⇧) (Reproduced with permission from: Mittal SR. ST Segment Elevation. Cardiology Today-2017;21:203–210)

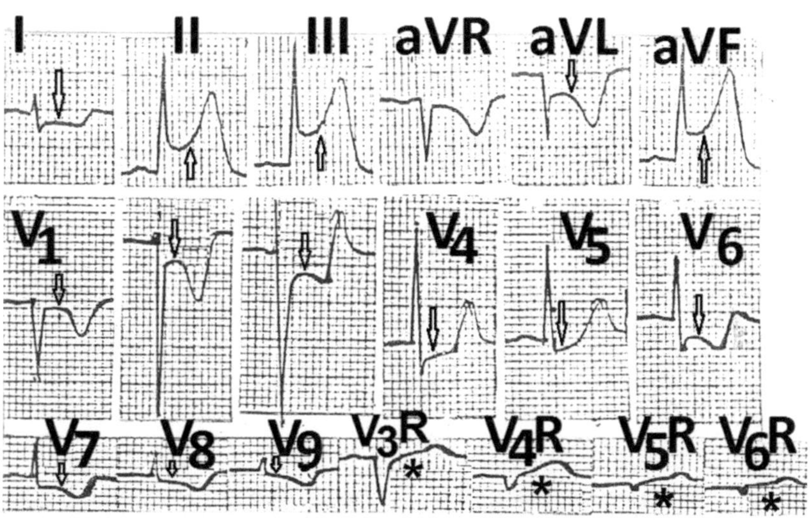

Fig. 40.30 Electrocardiogram shows inferior infarction (marked ⇩) due to occlusion of the posterior descending coronary artery. Reciprocal ST segment depressions are seen in leads I, aVL, and V_1 to V_9 (marked ⇧). Leads V_3R to V_6R also show ST-segment elevation (marked *). This finding suggests the possibility of concomitant right ventricular infarction (Reproduced with permission from: Mittal SR. ST Segment Elevation. Cardiology Today-2017;21:203–210)

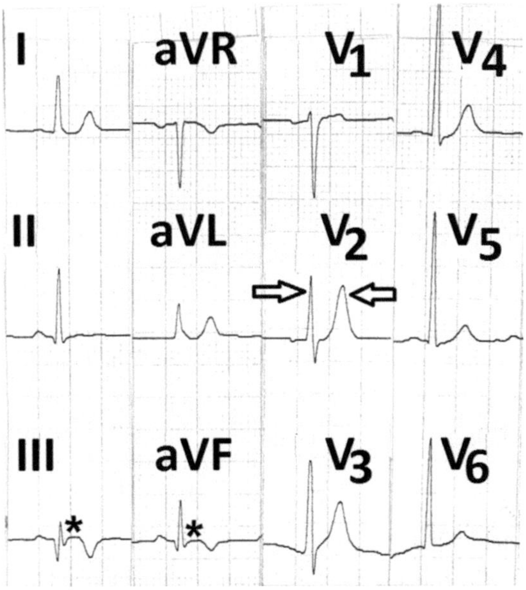

Fig. 40.31 Electrocardiogram showing inferior infarction (marked *). Prominent R wave and tall T wave in lead V_2 (marked arrows) are due to concomitant posterior infarction (Reproduced with permission from: Mittal SR. ST Segment Elevation. Cardiology Today-2017;21:203–210)

inversion in lead V_4R suggests occlusion of LCX.

- ST-segment elevation of >0.1 mV in lead V_6 supports occlusion of LCX [19].

(vi) ST segment elevation in leads V_5 to V_9

In dominant LCA, it is due to occlusion of the LCX (Fig. 40.32) whereas in dominant RCA, it is due to occlusion of RCA (Fig. 40.33).

Posterior leads (V_7 to V_9) are diagonally opposite to the right ventricular leads (V_3R to V_6R). ST-segment elevation in either group of leads can produce ST-segment depression in opposite groups of leads (Fig. 40.34).

Limitations [7, 20]—Above-mentioned criteria may vary depending on:

- Variation in coronary anatomy.
- Presence of multiple lesions in a vessel.
- Presence of multivessel disease.
- Collateral flow.
- Ventricular conduction delay.
- Cancellation of opposing vectorial forces.
- Prior myocardial infarction.

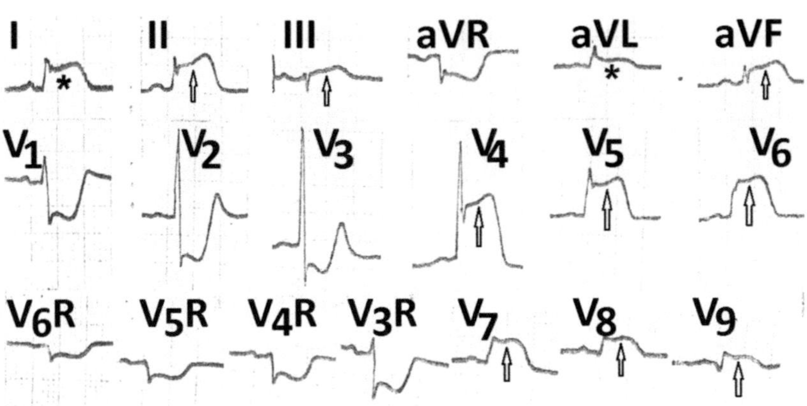

Fig. 40.32 Electrocardiogram is showing inferior and posterolateral infarction (marked ⇩) due to occlusion of the LCX. ST-segment is also elevated in leads I and aVL (marked *). ST-segment elevation in lead II is more than in lead III. These findings suggest occlusion of the left circumflex coronary artery (Reproduced with permission from: Mittal SR. ST Segment Elevation. Cardiology Today-2017;21:203–210)

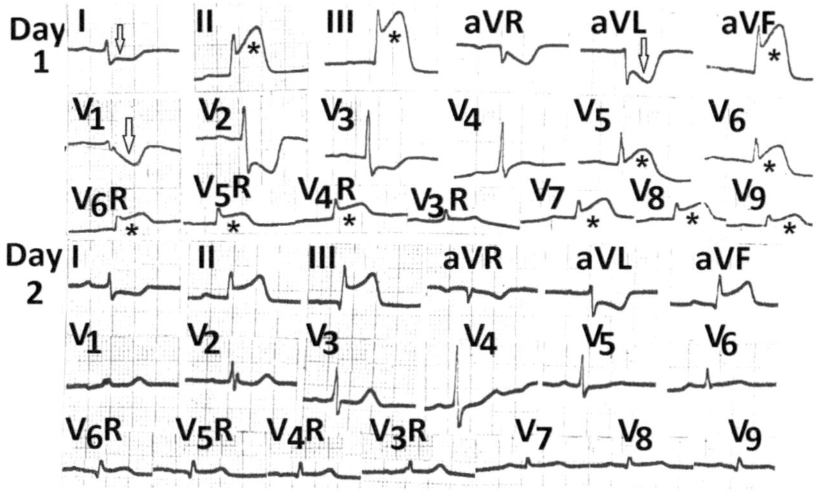

Fig. 40.33 Electrocardiogram (Day 1) showing infero-posterior lateral and right ventricular infarction (marked *). ST-segment elevation in lead III is more than in lead II. Leads I, aVL, and V_1 show depression of the ST-segment (marked ⇧). These findings suggest occlusion of dominant RCA. (Day 2) showing rapid normalization of the ST-segment elevation in right sided chest leads V_3R to V_6R (Reproduced with permission from: Mittal SR. ST Segment Elevation. Cardiology Today-2017;21:203–210)

Fig. 40.34 Diagram showing that right ventricular leads (V_3R to V_6R) and posterior chest leads (V_7 to V_9) are diagonally opposite to each other. ST-segment elevation in one group of leads produces reciprocal changes in the opposite group of leads (Reproduced with permission from: Mittal SR. ST Segment Elevation. Cardiology Today-2017;21:203–210)

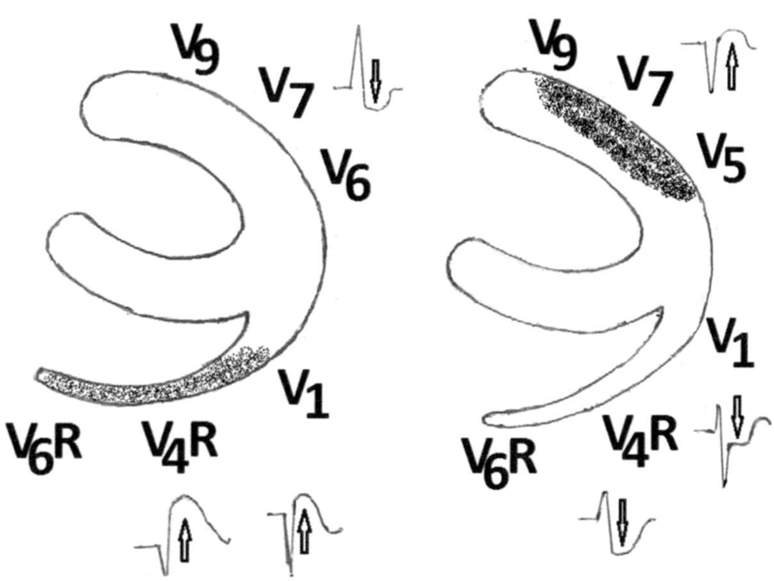

4 Quantification of the Ischemic Burden

Sum of the ST-segment changes in all leads gives an impression about the amount of myocardium at risk. Total of ST-segment deviation >15 mm suggests great area at risk. Sometimes reciprocal changes may be prominent and may result in overestimation of ischemic burden.

Summary

- Beginning of the normal ST-segment is mildly upsloping or isoelectric depending on weather it starts from the end of the S wave or the R wave, respectively. In leads V_1 and V_2, it may show mild upward convexity. In leads V_5 and V_6 it may show early uptake (early repolarization variant). Soon after onset, the ST-segment leaves the baseline and merges smoothly with the proximal limb of the T wave. There is no horizontal or isoelectric portion.
- Diffuse coved ST-segment elevation is seen in coronary vasospasm and Takotsubo cardiomyopathy (stress cardiomyopathy).
- Coved ST-segment localized to the territory of a coronary artery is seen in transmural injury or ventricular aneurysm. ST-segment elevation of more than 1 mm is considered significant in leads V_5 and V_6. Elevation of more than 2 mm is considered significant in leads V_1, V_2, and V_3. ST-segment elevation greater in V_1 than in V_2 suggest occlusion of right coronary artery.
- ST-segment elevation in leads V_1 to V_3 is usual in:
  - Occlusion of a septal branch of the left anterior descending coronary artery. ST-segment is coved. It may accompany right bundle branch block.
  - ST-segment may also be elevated in lead V_3R in occlusion of conus branch of right coronary artery.
  - Acute pulmonary thromboembolism.
  - Other electrocardiographic findings include sinus tachycardia, right axis deviation, T wave inversion in leads V_1 to V_4, S wave in lead I, and at times complete right bundle branch block.
  - Left ventricular hypertrophy and left bundle branch block.

    ST-segment elevation is upsloping. Elevation of more than 5 mm is considered significant is leads V_1 to V_3.
  - Hypercalcemia.

    ST-segment elevation in leads V_1 and V_2 is associated with short ST-segment.
  - Hyperkalemia.

    ST-segment elevation is associated with tall and peaked T wave.
  - Brugada syndrome.

    Incomplete or complete RBBB is associated with prominent J wave simulating elevated ST-segment. Recording in the third intercostal space produces more prominent J wave. Atrial fibrillation can occur. Bradycardia dependent VT/VF can cause sudden death.
  - Cocaine

    It can produce Brugada like pattern.
  - Arrhythmogenic right ventricular dysplasia.

    It is characterized by RBBB with low amplitude Epsilon wave on the early part of the ST-segment.
- ST-segment elevation with concavity upward is usual in the following situations:
  - In leads V_1 to V_3.

    As normal variant in

    Pectus excavatum. rsr' may be seen in leads V_1, V_2.

    Early repolarization.

    Vagal overdrive—Mild ST-segment elevation is associated with bradycardia and tall T wave.
  - Athletes.

    Sinus bradycardia is associated with increased left ventricular voltage and early repolarization pattern in right to mid precordial leads.
  - ST segment elevation in all leads.
  - Hypothermia—It is associated with bradycardia, prolonged PR interval, prolonged QT interval, and prominent J wave (Osborn wave).

- Acute pericarditis/myopericarditis or myopericardial injury.

  Upward concave ST-segment elevation is associated with sinus tachycardia, depression of PR segment, and normal T waves.

  By the time pericardial effusion accumulates, T waves tend to become flat. Pericardial constriction is associated with the inversion of T waves.
- ST-segment elevation in leads V_4 to V_6.
- Early repolarization syndrome—It is characterized by ST-segment elevation with upward concavity and slurring of terminal part of the R wave or prominent J wave in lateral and/or inferior leads. T waves are prominent.
- Left pneumothorax can cause ST-segment elevation with upward concavity in leads V_4 to V_6.

MCQs

Q1. Earliest ECG change in hyperacute transmural myocardial infarction is:

(a) Straight elevation of the ST-segment
(b) ST-segment elevation with concavity upward
(c) ST-segment elevation with convexity upward
(d) Broadening of T wave

Q2. In hyperacute phase of transmural myocardial infarction:

(a) ST-segment is elevated from J point
(b) R wave amplitude is increased
(c) T wave becomes tall and wide
(d) All

Q3. In Prinzmetal's angina:

(a) ECG simulates acute transmural myocardial injury
(b) ECG normalizes with relief of pain
(c) There are wide and deep Q waves
(d) T waves are inverted

Q4. Monophasic upright complexes can occur in:

(a) Hyperacute phase of myocardial infarction
(b) Coronary vasospasm
(c) Myocardial ischemia
(d) Hyperkalemia

Q5. In pseudonormalization:

(a) Q waves are deep and wide
(b) ST-segment is isoelectric
(c) T waves are symmetrically inverted
(d) All

Q6. Takotsubo cardiomyopathy produces:

(a) ST-segment elevation in lead aVR
(b) Extensive ST-segment depression
(c) Extensive T wave inversion
(d) All

Q7. ST-segment elevation with concavity upward is seen in:

(a) Acute pericarditis
(b) Vagotonia
(c) Hypothermia
(d) All

Q8. ST-segment elevation in leads V_1 to V_3 is seen in:

(a) Massive pulmonary embolism
(b) LBBB
(c) RBBB
(d) Infarction of the interventricular septum

Q9. ST-segment elevation is seen in:

(a) LVH
(b) Hypokalemia
(c) Hypocalcemia
(d) Tumors invading myocardium

Q10. Brugada syndrome produces the ST-segment elevation in:

(a) Leads V_1 and V_2
(b) Leads V_3 and V_4
(c) Leads V_5 and V_6
(d) Leads II, III, aVF

Q11. Which of the following conditions can produce ST-segment elevation?

(a) Trauma to myocardium
(b) Myocarditis
(c) Left pneumothorax
(d) All

Q12. Successful reperfusion is suggested by:

(a) ≥70% reduction in ST-segment elevation
(b) ≥50% reduction in ST-segment elevation
(c) ≥40% reduction in ST-segment elevation
(d) ≥30% reduction in ST-segment elevation

Q13. Ventricular aneurysm is suggested by:

(a) Persistent ST-segment elevation
(b) Deep and broad Q wave
(c) Deep and pointed T wave
(d) Loss of R wave

Q14. ST-segment elevation in leads I and aVL suggests:

(a) Anteroseptal infarction
(b) Apical infarction
(c) Lateral infarction
(d) High lateral infarction

Q15. Posterior infarction produces reciprocal changes in:

(a) Leads V_2, V_3
(b) Leads V_5, V_6
(c) Leads I, aVL
(d) Leads II, III, aVF

Q16. Tall R wave and tall T wave in lead V_2 suggest:

(a) Anteroseptal infarction
(b) Apical infarction
(c) Posterior infarction
(d) High lateral infarction

Q17. Inferior infarction produces ST-segment depression in:

(a) Leads V_1 to V_3
(b) Leads V_4 to V_6
(c) Leads V_7 to V_9
(d) Leads V_3R to V_6R

Q18. In the context of inferior infarction, ST-segment elevation in lead V_1 suggests concomitant:

(a) Anteroseptal infarction
(b) Apical infarction
(c) High lateral infarction
(d) Right ventricular infarction

Q19. In the context of inferior infarction, ST-segment elevation in lead III exceeding that in lead II suggests concomitant:

(a) Right ventricular infarction
(b) Apical infarction
(c) Lateral infarction
(d) Posterior infarction

Q20. ST-segment elevation in lead aVR with diffuse ST-segment depression in other leads suggests occlusion of:

(a) Left main coronary artery
(b) Middle part of the LAD
(c) Middle part of LCX
(d) Proximal RCA

Q21. ST-segment elevation in leads I, aVL, and V_1 to V_6 suggests occlusion of:

(a) Proximal LAD before diagonal branch
(b) Mid LAD
(c) Mid circumflex
(d) Proximal RCA

Q22. Extensive anterior infarction with inferior infarction suggests occlusion of:

(a) Proximal LAD wrapping around the cardiac apex
(b) Distal LAD
(c) Proximal RCA
(d) Proximal PDA

Q23. ST-segment elevation only in leads V_2 to V_6 suggests occlusion of:

(a) LAD before D_1
(b) LAD after D_1 and S_1
(c) LAD before S_1
(d) Mid LCX

Q24. In the context of anterior infarction, occlusion of LAD before S_1 is suggested by:

(a) New RBBB
(b) ST-segment elevation in leads V_1 and aVR
(c) New left anterosuperior fascicular block
(d) New left posteroinferior fascicular block

Q25. Left posteroinferior fascicle is supplied by:

(a) LAD
(b) LAD + LCX
(c) LAD + RCA
(d) $S_1 + D_1$

Q26. ST-segment elevation in leads V_1 to V_3 suggests occlusion of:

(a) Proximal LAD
(b) Mid LAD
(c) Distal LAD
(d) Proximal LCX

Q27. Occlusion of proximal diagonal produces ST-segment elevation in leads:

(a) I, aVL
(b) V_1 to V_3
(c) V_4 to V_6
(d) V_7 to V_9

Q28. In the context of inferior infarction, ST-segment elevation in lead I suggests occlusion of:

(a) Proximal LCX
(b) Mid LAD
(c) Proximal RCA
(d) Proximal PDA

Q29. In the context of inferior infarction, ST-segment elevation in lead II > that in lead III suggests occlusion of:

(a) LCX
(b) LAD
(c) RCA
(d) PDA

Q30. In the context of inferior infarction, T wave inversion in lead V_4R suggests occlusion of:

(a) Proximal RCA
(b) Mid RCA
(c) Distal RCA
(d) LCX

Q31. Normally the ST-segment may show mild coved elevation in leads:

(a) V_1, V_2
(b) V_5, V_6
(c) V_7, V_8
(d) V_3R, V_4R

Q32. Normally the ST-segment may show early repolarization pattern in leads:

(a) V_1, V_2
(b) V_3, V_4
(c) V_5, V_6
(d) V_7, V_8

Q33. Normally, which part of the ST-segment is isoelectric?

(a) Initial
(b) Middle
(c) Terminal
(d) None

Q34. Coronary vasospasm usually produces:

(a) Diffuse coved ST-segment elevation

(b) Diffuse concave ST-segment elevation
(c) Convex ST-segment elevation in leads V_5, V_6
(d) Concave ST-segment elevation in leads V_1, V_2

Q35. Stress cardiomyopathy usually produces:

(a) Diffuse coved ST-segment elevation
(b) Coved ST-segment elevation in leads V_5, V_6
(c) Diffuse concave ST-segment elevation
(d) Coved ST-segment elevation in leads V_3R to V_6R

Q36. Normally the ST-segment can show coved elevation of up to 2 mm in leads:

(a) V_1, V_2
(b) V_3, V_4
(c) V_5, V_6
(d) V_7, V_8

Q37. Coved ST-segment elevation greater in V_1 than in V_2 suggests occlusion of:

(a) Right coronary artery
(b) Major septal branch of LAD
(c) D_1
(d) OM_1

Q38. Sinus tachycardia, right axis deviation, ST-segment elevation with T wave inversion in leads V_1 to V_4 and S wave in lead 1 suggest the possibility of:

(a) Occlusion of conus artery
(b) Occlusion of major septal
(c) Acute cor pulmonale
(d) Chronic cor pulmonale

Q39. In left bundle branch block, ST-segment elevation in leads V_1, V_2 suggests:

(a) Occlusion of conus artery
(b) Normal pattern
(c) Left ventricular hypertrophy
(d) Posterior infarction

Q40. In left bundle branch block, what magnitude of ST-segment elevation in lead V_2 is considered abnormal?

(a) More than 1 mm
(b) More than 3 mm
(c) More than 4 mm
(d) More than 5 mm

Q41. Which leads can show upsloping ST-segment elevation in left ventricular hypertrophy with ST-segment depression and T wave inversion in leads V_5, V_6?

(a) V_3, V_4
(b) V_1, V_2
(c) I, aVL
(d) II, III, aVF

Q42. ST segment elevation in leads V_1, V_2 with short ST-segment (short Q-T interval) suggests:

(a) Left ventricular hypertrophy
(b) Hypercalcemia
(c) Left bundle branch block
(d) Hypocalcemia

Q43. ST-segment elevation with tall peaked T wave suggests:

(a) Hypercalcemia
(b) Hyperkalemia
(c) Hypomagnesemia
(d) Hyponatremia

Q44. Right bundle branch block pattern with prominent J wave simulating ST-segment elevation in lead V_1, V_2 suggests:

(a) RBBB with right ventricular hypertrophy
(b) RBBB with left ventricular hypertrophy
(c) RBBB with right ventricular infarction
(d) Brugada syndrome

Q45. Right bundle branch block with ST-segment elevation and additional low voltage wave on the early part of ST-segment suggests:

(a) Brugada syndrome
(b) Arrhythmogenic right ventricular cardiomyopathy
(c) Right lateral accessory pathway
(d) Normal variation

Q46. Pectus excavatum can produce:

(a) rsr' pattern in lead V_1
(b) Mild ST-segment elevation in leads V_1, V_2
(c) Deep S in lead V_6
(d) All

Q47. Electrocardiogram from an athlete can show:

(a) Sinus bradycardia
(b) Increased left ventricular voltages in leads V_5, V_6
(c) Early repolarization pattern in right to mid precordial leads
(d) None

Q48. Bradycardia with prominent J wave suggest the possibility of:

(a) Hypothermia
(b) Myxedema
(c) Athlete heart
(d) All

Q49. Sinus tachycardia with diffuse upward concave ST-segment elevation and elevation of PR segment in lead aVR suggests:

(a) Coronary vasospasm
(b) Stress cardiomyopathy
(c) Acute pericarditis
(d) Right ventricular infarction

Q50. Slurring of the terminal part of prominent R in lead V_5 with prominent T wave and ST segment elevation with upward concavity suggests:

(a) Early repolarization syndrome
(b) Chronic severe aortic regurgitation
(c) Chronic severe mitral regurgitation
(d) Left lateral accessory pathway

Q51. Normal ECG cannot exclude:

(a) Acute pericarditis
(b) Brugada syndrome
(c) Arrhythmogenic right ventricular cardiomyopathy
(d) All

Q52. Diffuse concave upward ST-segment elevation in leads other than lead aVR and V_1, with normal T wave and depression of PR segment suggests:

(a) Acute pericarditis
(b) Early repolarization syndrome
(c) Cardiac tumor
(d) Vagal overdrive

Q53. Sinus tachycardia, low QRS voltage, isoelectric ST-segment with diffuse inverted T wave suggests:

(a) Acute pericarditis
(b) Pericardial constriction
(c) Emphysema
(d) Acute cor pulmonale

Q54. Atrial fibrillation with diffuse concave ST-segment elevation and low amplitude positive T wave suggest:

(a) Mitral stenosis
(b) Mitral regurgitation
(c) Acute pericarditis
(d) Constrictive pericarditis

Q55. Mild pulmonary embolism produces:

(a) ST-segment elevation in leads V_1 to V_3
(b) Deep T inversion in leads V_1 to V_3
(c) S_1 Q_3 and inverted T_3 pattern
(d) None

Q56. Acute pericarditis produces:

(a) Isoelectric ST-segment
(b) Diffuse T wave inversion
(c) T wave amplitude less than 5 mm
(d) T wave amplitude greater than 5 mm

Q57. ST segment elevation in acute pericarditis is due to:

(a) Myocarditis
(b) Pericardial effusion
(c) Early repolarization
(d) All

Q58. Cocaine can produce:

(a) ST-segment elevation with concavity upward
(b) Brugada pattern
(c) Prolongation of QT interval
(d) ST-segment depression

Answers

(1) a (2) d (3) a, b (4) a, b (5) b (6) a (7) d (8) a, b, d (9) a, d (10) a (11) d (12) a (13) a (14) d (15) a (16) c (17) a (18) d (19) a (20) a (21) a (22) a (23) b (24) a, b (25) c (26) b (27) a (28) a (29) a (30) d (31) a (32) c (33) d (34) a (35) a (36) a (37) a (38) c (39) b (40) d (41) b (42) b (43) b (44) d (45) b (46) a, b (47) a, b, c (48) a (49) c (50) a (51) d (52) a (53) b (54) c (55) d (56) c (57) a (58) b, c

References

1. Schamroth L. Coronary insufficiency. In: Schamroth C, editor. An introduction to electrocardiography. New Delhi: Wiley India; 1982. p. 157–85.
2. O'Gara PT, Kushner FG, Ascheim DD, et al. 2013 ACCF/ AHA guideline for the management of ST elevation myocardial infarction. A report of the American College of Cardiology Foundation/ American Heart Association task force on practice guidelines. Circulation. 2013;127:e362–425.
3. Birnbaum Y, Nikus K, Kligfield P, Fiol M, Barrabes JA, Sionis A, et al. The role of the ECG in diagnosis, risk estimation and catheterization laboratory activation in patients with acute coronary syndromes: a consensus document. Ann Noninvasive Electrocardiol. 2014;19:412–25.
4. Schamroth L. Myocardial infarction. In: Schamroth C, editor. An introduction to electrocardiography. New Delhi: Wiley India; 1982. p. 131–56.
5. Frangieh AH, Obeid S, Ghadri JR, Imori Y, D'Ascenzo F, Kovac M, et al. ECG criteria to differentiate between Takotsubo (Stress) cardiomyopathy and myocardial infarction. J Am Heart Assoc. 2016;5:e003418. https://doi.org/10.1161/JAHA116.003418.
6. De Luna AB, Goldwasser D, Fiol M, Bayes Genis A. Surface electrocardiography. In: Fuster V, Walsh RA, Harrington RA, editors. Hurst's the heart. New York: McGraw Hill; 2011. p. 307–70.
7. Mirvis DM, Goldberger AL. Electrocardiography. In: Mann DL, Zipes DP, Libby P, Bonow RO, editors. Braunwald's heart disease. Philadelphia: Saunders; 2015. p. 114–52.
8. Rosello X, Wiegerinck RF, Alguersuari J, Bardaji A, Worner F, Sutil M, et al. New electrocardiographic criteria to differentiate acute pericarditis and myocardial infarction. Am J Med. 2014;127:233–9.
9. Brearley WD Jr, Taylor L 3rd, Haley MW, Littmann L. Pneumomedistinum mimicking acute ST segment elevation myocardial infarction. Int J Cardiol. 2007;117:e73–5.
10. Siddique O, Rasia S, Clark S, Kokkirala A. A case of ileus and ST segment elevation. Ri Med J. 2016;99:44–6.
11. Park I, Kim YJ, Ahn S, Sohn CH, Seo DW, Kim WY. Subarachnoid hemorrhage mimicking ST segment elevation myocardial infarction after return of spontaneous circulation. Clin Exp Emerg Med. 2015;2:260–3.
12. Wagner GS, Lim TH. Myocardial infarction. In: Wagner GS, editor. Marriott's practical electrocardiography. Philadelphia: Wolters Kluwer; 2008. p. 179–207.
13. Schroder R, Dissmann R, Bruggemann T, et al. Extent of early ST segment elevation resolution: a simple but strong predictor of outcome in patients with acute myocardial infarction. J Am Coll Cardiol. 1994;24:384–91.
14. Lew AS, Laramec P, Shah PK, et al. Ratio of ST segment depression in lead V2 to ST segment elevation in lead aVF in evolving acute inferior myocardial infarction: an aid to the early recognition of right ventricular ischemia. Am J Cardiol. 1986;57:1047–51.
15. Zheng J, Lin J, Shen N, Qu B. Proximal complete occlusion of right coronary artery presenting with precordial ST segment elevation. A case report. Medicine. 2016;95:41 (e5113).
16. Wagner GS, Lim TH. Insufficient blood supply. In: Wagner GS, editor. Marriott's practical electrocardiography. Philadelphia: Wolters Kluwer; 2008. p. 163–78.

17. Reddy K, Khaliq A, Henning RJ. Recent advances in the diagnosis and treatment of acute myocardial infarction. World J Cardiol. 2015;7:243–76.
18. Tamura A. Significance of lead aVR in acute coronary syndrome. World J Cardiol. 2014;6:630–7.
19. Noriega FJ, Vives-Borras M, Sole-Gozalez E, Garcia-Picart J, Arzamendi D, Cinca J. Influence of the extent of coronary atherosclerotic disease on ST segment changes induced by ST elevation myocardial infarction. Am J Cardiol. 2014;113:757–64.
20. Thrudeep S, Geofi G, Rupesh G, Abdulkhadar S. Dilemma of localization of culprit vessel by electrocardiography in acute myocardial infarction. Indian Heart J. 2016;68:515–7.

41 ST-Segment Depression

ST-segment starts with the J point and merges smoothly into the T wave. The level of the ST-segment is considered in relation to the preceding TP segment. ST-segment depression of ≥0.5 mm in consecutive leads is considered significant except in leads V_1 to V_3.

1 Causes of ST-Segment Depression

1.1 Subendocardial Ischemia or Injury

Leads V_5 and V_6 are most commonly involved [1]. Depending on the severity of impairment of coronary flow, the ST-segment can take the following shapes (Fig. 41.1).

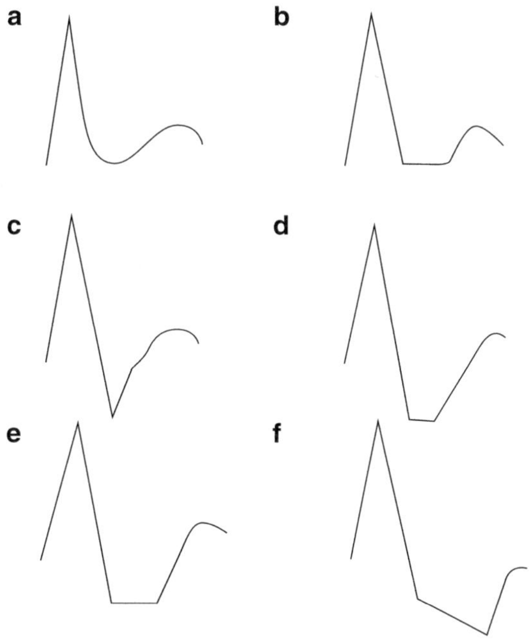

Fig. 41.1 Diagrammatic representation of various shapes of the ST-segment. (**a**) Normal ST-segment. ST-segment in cases of myocardial ischemia. (**b**) Horizontal ST-segment. (**c**) Rapid upsloping ST-segment depression. (**d**) Slow upsloping ST-segment depression. (**e**) Horizontal ST-segment depression. (**f**) Down sloping ST-segment depression (Reproduced with permission from: Mittal SR. ST-Segment Depression. Cardiology Today-2017;21:245–250)

1.1.1 Horizontal ST-Segment

ST-segment is horizontal with sharp angle at the point of junction of the T wave with the ST-segment.

1.1.2 Upsloping ST-Segment

J point is depressed relative to the TP segment. Upsloping depression of the ST-segment can be:

S. Mittal, *Insights into Electrocardiograms with MCQs*,
https://doi.org/10.1007/978-981-99-0127-2_41

- Rapid upsloping
- Slow upsloping

i. Rapid upsloping ST-segment depression

After a depressed J point, the ST-segment rises rapidly with the result that within 0.08 s (two small squares at conventional recording speed of 25 mm/s) after J point, it touches the isoelectric line (Fig. 41.2).

Such a depression can also be present in normal persons due to the effect of atrial repolarization wave. Atrial repolarization wave (Ta wave) is opposite in direction to the P wave. Normally it falls within the QRS complex and does not deform the ST-segment. If the amplitude of the P wave is increased, the Ta wave also becomes prominent and can depress early part of the QRS complex and the J point. In such cases, the line joining the distal limb of the P wave, the PR segment, the J point, the ST-segment, and proximal limb of the T wave form a smooth parabola. In pathological conditions, this line is broken and is step like.

ii. Slow upsloping

In this situation the ST-segment remains 0.2 mV below the isoelectric line after 0.08 s of the J point. This is usually pathological.

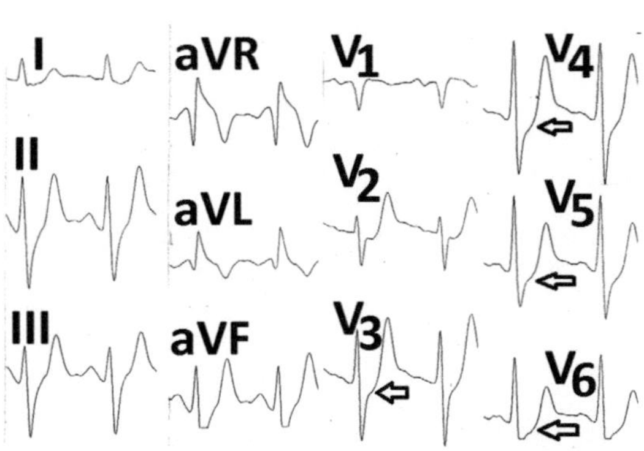

Fig. 41.2 Electrocardiogram is showing rapid upsloping ST segment depression in leads V_3 to V_6 (marked ⇦) (Reproduced with permission from: Mittal SR. ST-Segment Depression. Cardiology Today-2017;21:245–250)

1.1.3 Flat or Horizontal ST-Segment Depression

After a depressed J point, the ST-segment is horizontal with sharp ST-T angle (Fig. 41.3). It is pathological.

1.1.4 Down Sloping

After a depressed J point, the ST segment is further depressed (Fig. 41.4). It is pathological.

In the context of coronary artery disease, greater is the ST-segment depression, greater is the impairment of the coronary flow.

ST-segment depression of 0.5 mm or more that appears during chest discomfort and disappears after relief strongly suggests transient myocardial ischemia. However, if such a change is constantly present with or without chest pain, it is a nonspecific finding.

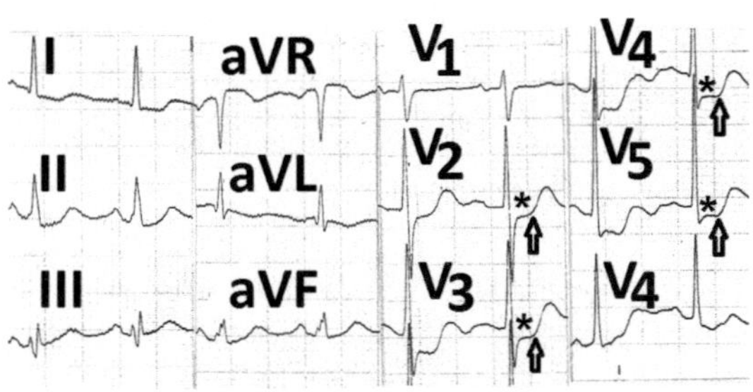

Fig. 41.3 Electrocardiogram shows horizontal ST-segment depression in leads V_2 to V_5 (marked *). Angle between the ST-segment and the T wave is sharp (marked ⇧) (Reproduced with permission from: Mittal SR. ST-Segment Depression. Cardiology Today-2017; 21:245–250)

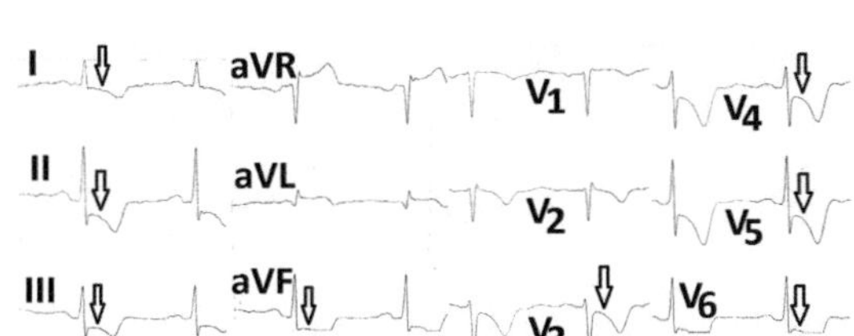

Fig. 41.4 Electrocardiogram is showing down sloping ST-segment depression in leads I, II, III, aVF, and V_3 to V_6 (marked ⇩) (Reproduced with permission from: Mittal SR. ST-Segment Depression. Cardiology Today-2017;21: 245–250)

2 Localization of Ischemia

(i) ST-segment depression in leads V_1 to V_4 suggests anteroseptal and apical subendocardial infarction [2].

It could also be a reciprocal change of inferolateral (posterior) infarction (Fig. 41.5).

In the setting of transmural inferior infarction, such changes may also be due to inferoseptal infarction [3]. Posterolateral infarction can also produce such reciprocal changes (Fig. 41.6).

Right ventricular subendocardial ischemia can also produce similar changes.

(ii) ST-segment depression in leads V_4 to V_6—It suggests lateral subendocardial ischemia. However, such a pattern may also be seen in patients with left ventricular hypertrophy, cardiomyopathy, etc. It can also be a reciprocal change in patients with inferior myocardial infarction. It is, however, important to remember that these changes in leads V_4 to V_6 could also be due to ischemia of anterolateral surface due to concomitant disease of left anterior descending coronary artery (Figs. 41.7 and 41.8).

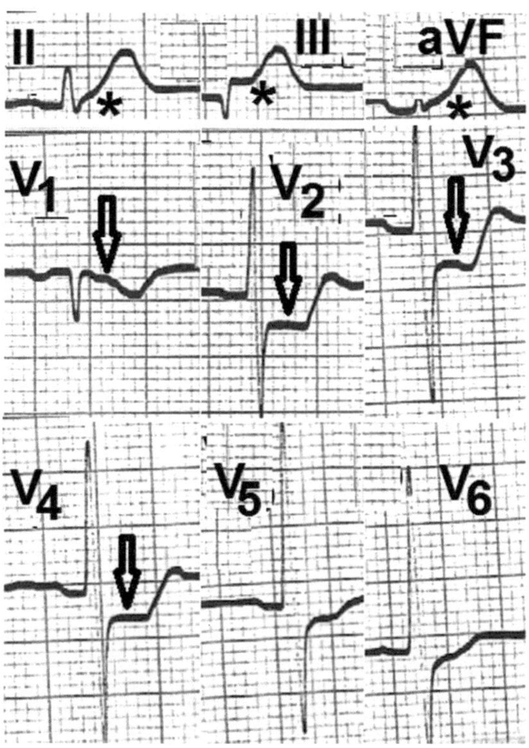

Fig. 41.5 Electrocardiogram is showing ST-segment depression in leads V_1 to V_4 (marked ⇩) in the setting of acute inferior infarction (marked *) (Reproduced with permission from: Mittal SR. ST-Segment Depression. Cardiology Today-2017;21:245–250)

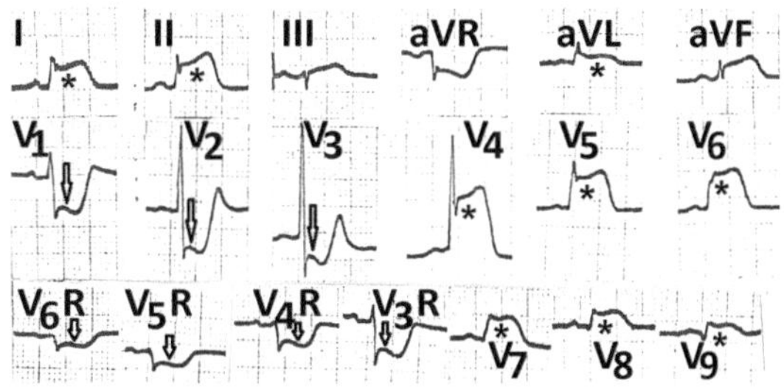

Fig. 41.6 Electrocardiogram from a case of posterolateral infarction (marked *) showing ST-segment depression in leads V_1 to V_3 and V_3R to V_6R (marked ⇩) (Reproduced with permission from: Mittal SR. ST-Segment Depression. Cardiology Today-2017;21:245–250)

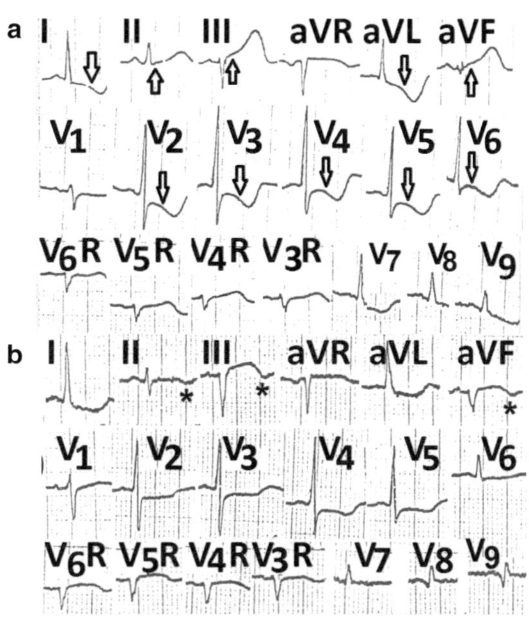

Fig. 41.7 Electrocardiograms from a case of acute inferior infarction (**a**) showing ST-segment depression in leads I, aVL, and V_2 to V_6 (marked ⇩). Leads II, III, and aVF show early uptake of the ST-segment (marked ⇧). (**b**) Electrocardiogram recorded on 2nd day showing sequential T wave inversions in leads II, III, aVF (marked *) with reduction in ST-segment depressions in leads I, aVL, V_2 to V_6 (Reproduced with permission from: Mittal SR. ST-Segment Depression. Cardiology Today-2017;21:245–250)

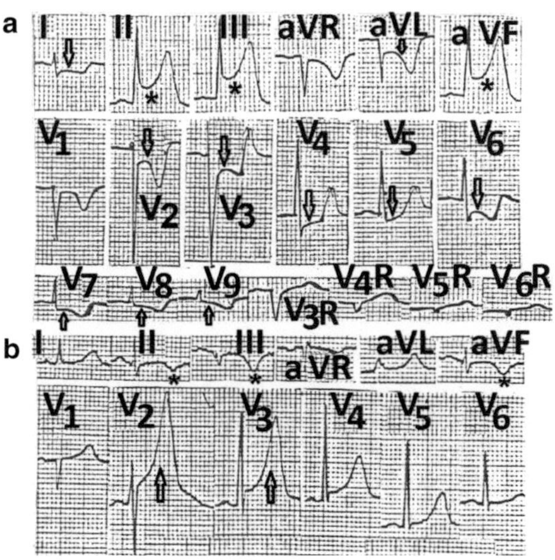

Fig. 41.8 Electrocardiograms from a case of acute inferior infarction (marked *) (**a**) showing ST segment depressions in leads I, aVL, V_1 to V_6 (marked ⇩) and V_7 to V_9 (marked ⇧). (**b**) Electrocardiogram recorded on 2nd day showing sequential loss of R wave and inversion of T wave in leads II, III, and aVF (marked *) with normalization of the ST-segment depressions. Tall and prominent T wave in leads V_2 and V_3 (marked ⇧) suggest the possibility of concomitant posterior infarctions (Reproduced with permission from: Mittal SR. ST-Segment Depression. Cardiology Today-2017;21:245–250)

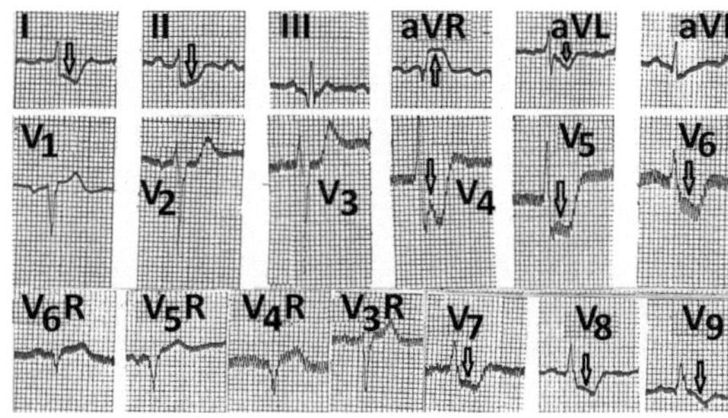

Fig. 41.9 Electrocardiogram is showing ST-segment elevation in lead aVR (marked ⇧) more than that in lead V_1 with ST-segment depression in leads I, II, aVL, and V_4 to V_9 (marked ⇩) (Reproduced with permission from: Mittal SR. ST-Segment Depression. Cardiology Today-2017;21:245–250)

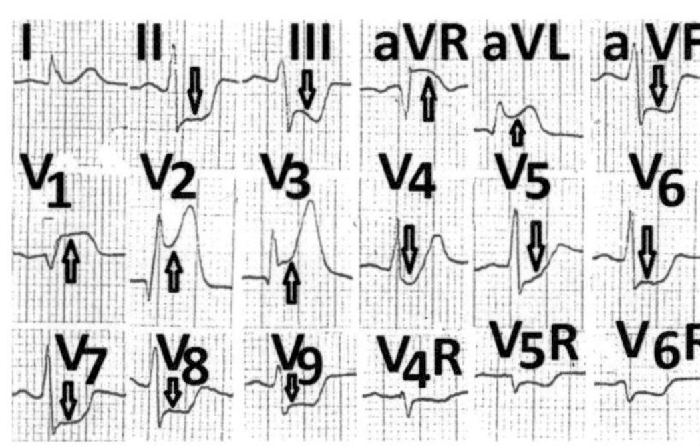

Fig. 41.10 Electrocardiogram is showing ST-segment elevation in leads aVR aVL and V_1 to V_3 (marked ⇧) with ST-segment depression in leads II, III, aVF, V_4 to V_9 (marked ⇩) (Reproduced with permission from: Mittal SR. ST-Segment Depression. Cardiology Today-2017;21: 245–250)

(iii) ST-segment depression in seven or more leads with ST-segment elevation in lead aVR greater than that in lead V_1 suggests occlusion of the left main coronary artery or its equivalent (proximal disease of LAD and LCX) [4] (Figs. 41.9 and 41.10).

(iv) Upsloping ST-segment depression with tall positive T waves in the anterolateral leads (V_4 to V_6) should suggest ischemia due to subtotal occlusion of the LAD or left circumflex artery specially if seen in patients with typical symptoms and at a relatively slow heart rate [5].

(v) ST-segment depression in leads I and aVL may be reciprocal of ST-segment elevation in inferior leads [6] (Fig. 41.11).

ST-segment elevation in inferior leads may not be diagnostic in early stage, especially in patients with low voltage QRS complexes in inferior leads. However, it is important to remember that initial ST-segment depression in leads I and aVL could be additional ischemia in the territory of D_I or OM_I.

(vi) ST-segment depression in leads III and aVF. It suggests inferior subendocardial infarction. It can also be a reciprocal change of infarction involving leads I and aVL (Figs. 41.12 and 41.13).

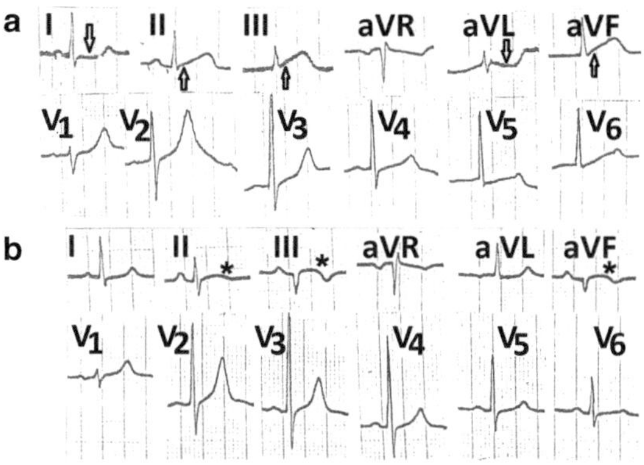

Fig. 41.11 (**a**) Electrocardiogram shows early uptake of the ST-segment in leads II, III, and aVF (marked ⇧). These suggest acute inferior infarction. Leads I and aVL show reciprocal depression of the ST-segment (marked ⇩). (**b**) Electrocardiogram recorded on the second day showing appearance of Q waves with inversion of T waves in leads III and aVF (marked *) with normalization of depression of the ST-segment in leads I and aVL (Reproduced with permission from: Mittal SR. ST-Segment Depression. Cardiology Today-2017;21:245–250)

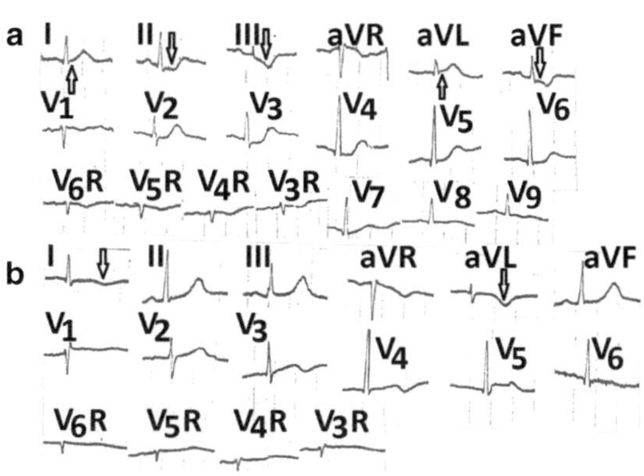

Fig. 41.12 (**a**) Electrocardiogram showing early uptake of ST-segment in leads I and aVL (marked ···) suggestive of acute high lateral infarction. Leads II, III, and aVF show ST-segment depression (marked). (**b**) Electrocardiogram recorded on the second day showing sequential inversion of T waves in leads I and aVL (marked ···). There is normalization of the ST segment depression in leads II, III, and aVF (Reproduced with permission from: Mittal SR. ST-Segment Depression. Cardiology Today-2017;21:245–250)

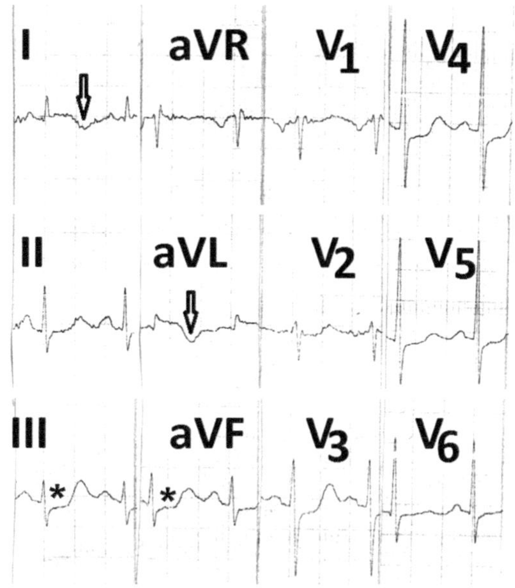

Fig. 41.13 Electrocardiogram from a case of acute high lateral myocardial infarction involving leads I and aVL (marked ⇩) with reciprocal ST-segment depressions in leads III and aVF (marked *) (Reproduced with permission from: Mittal SR. ST-Segment Depression. Cardiology Today-2017;21:245–250)

2.1 Causes Other Than Ischemic Heart Disease

(a) Normal variants [4]
   (i) Sympathetic overdrive
   (ii) Neurocirculatory asthenia
   (iii) Hyperventilation
(b) Secondary to intraventricular conduction defects
   (i) Left bundle branch block
      ST-segment depression is seen in leads I, aVL, and V_5 to V_9 (Fig. 41.14).
   (ii) Right bundle branch block

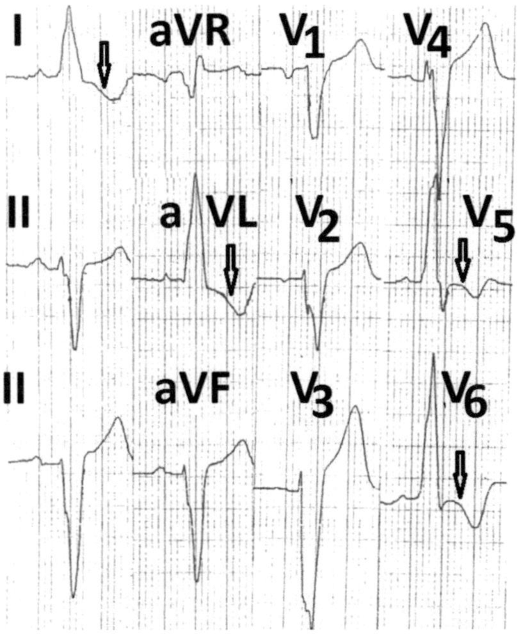

Fig. 41.14 Electrocardiogram from a case of left bundle branch block showing ST-segment depression and T wave inversion in leads I, aVL, V_5, and V_6 (marked ⇩) (Reproduced with permission from: Mittal SR. ST-Segment Depression. Cardiology Today-2017;21:245–250)

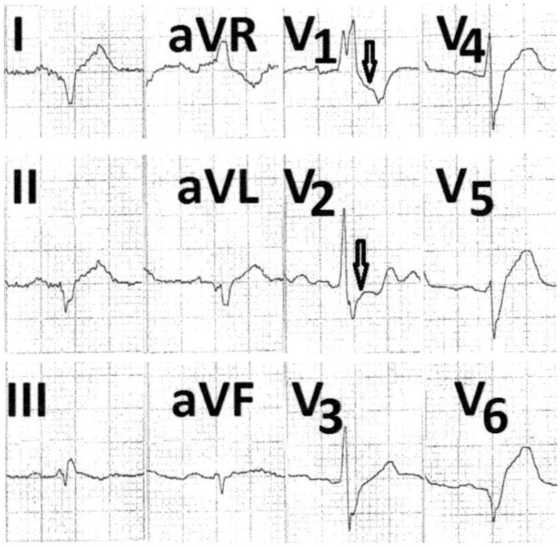

Fig. 41.15 Electrocardiogram from a case of right bundle branch block showing ST-segment depression in leads V_1 and V_2 (marked ⇩). QS pattern in leads I, aVL, and V_6 suggest the possibility of old lateral infarction (Reproduced with permission from: Mittal SR. ST-Segment Depression. Cardiology Today-2017;21:245–250)

ST-segment depression is seen in leads V_1 to V_3 and V_3R to V_6R (Fig. 41.15).

(iii) Preexcitation syndromes

ST-segment is opposite to the delta wave (Fig. 41.16).

(c) Secondary to ventricular hypertrophy

(i) Left ventricular hypertrophy

ST-segment depression is seen in leads I, aVL, V_5 to V_9 (Fig. 41.17).

(ii) Right ventricular hypertrophy (Fig. 41.18).

ST-segment depression is seen in leads II, III, and aVF.

(d) Digitalis [7]

J point is not depressed. ST-segment takes a straight downward course from the J point. T wave amplitude is diminished and there is shortening of the QT interval. Classically it takes the shape of mirror image of correction mark. Changes are most prominent in leads with dominantly positive QRS. Such change suggests an adequate dose of digitalis, but does not suggest digitalis toxicity. First degree atrioventricular block, ventricular

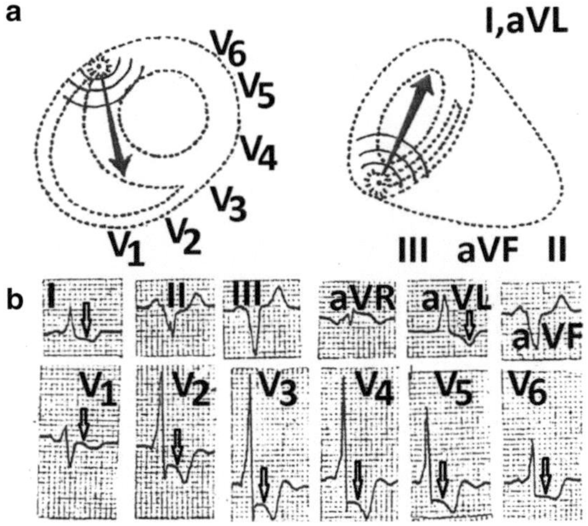

Fig. 41.16 (**a**) Diagram showing direction of the delta wave in a case of posteroseptal accessory pathway. (**b**) Electrocardiogram showing ST-segment depression in leads I, aVL, V_1 to V_6 (marked ⇩) (Reproduced with permission from: Mittal SR. ST-Segment Depression. Cardiology Today-2017;21:245–250)

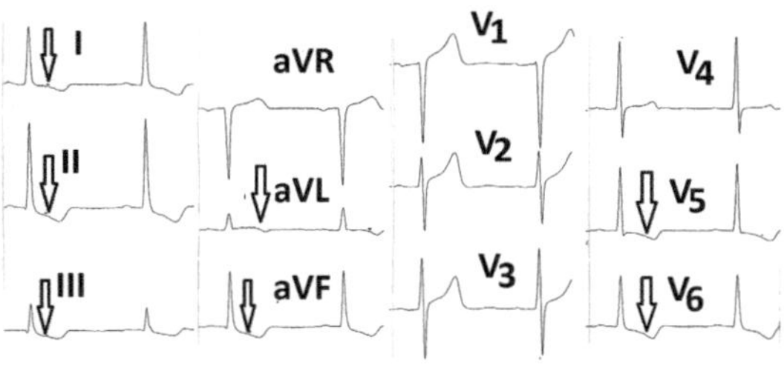

Fig. 41.17 Electrocardiogram from a case of left ventricular hypertrophy showing ST-segment depression and T wave inversion in leads I, II, III, aVL, aVF, V_5 and V_6 (marked ⇩) (Reproduced with permission from: Mittal SR. ST-Segment Depression. Cardiology Today-2017;21: 245–250)

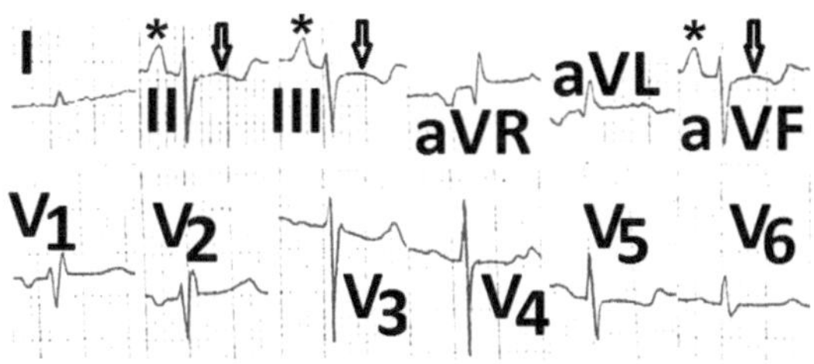

Fig. 41.18 Electrocardiogram from a case of cor pulmonale showing ST-segment depression in leads II, III, and aVF (marked ⇩). Leads II, III, and aVF also show tall and peaked P wave suggestive of right atrial enlargement (marked *). Leads V_1, V_2 show terminal R wave. Lead V_6 shows terminal S wave and lead aVR shows prominent terminal R wave. These findings suggest RVH. Low amplitude of QRS in leads I and V_6 suggest the presence of emphysema (Reproduced with permission from: Mittal SR. ST-Segment Depression. Cardiology Today-2017;21:245–250)

ectopics, multifocal or bidirectional ventricular tachycardia, or junctional tachycardia appearing for the first time after adequate dose of digitalis suggest digitalis overdose (Fig. 41.19).

(e) Hypokalemia

Low amplitude or invisible T waves, prominent U waves, and prolongation of the QT interval suggest hypokalemia.

(f) Mitral valve prolapse

ST-segment depression and/or T wave inversion is seen in leads II, III, aVF, V_5 V_6.

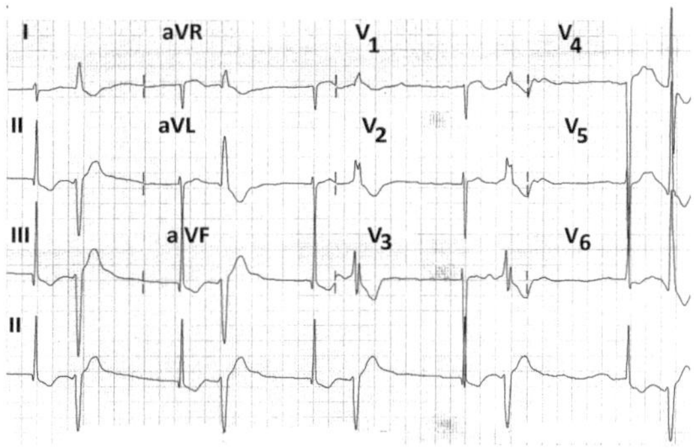

Fig. 41.19 Electrocardiogram from a case of digitalis overdose showing atrial fibrillation with slow ventricular rate (rhythm strip lead II), ST-segment depression in leads II, III, aVF, V_6, and ventricular bigeminy (Reproduced with permission from: Mittal SR. ST-Segment Depression. Cardiology Today-2017;21:245–250)

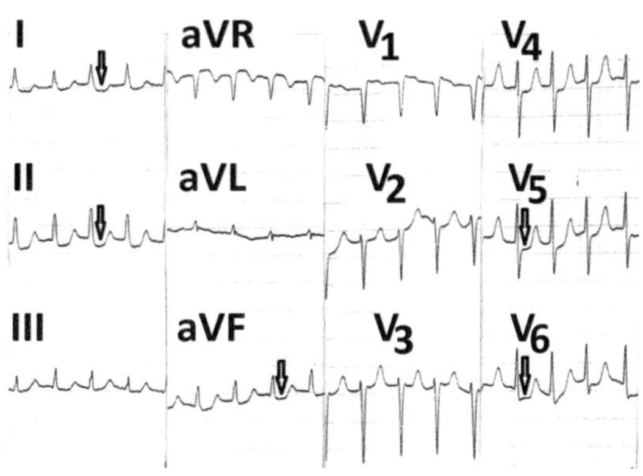

Fig. 41.20 Electrocardiogram from a case of atrioventricular nodal reentrant tachycardia showing ST segment depression in leads I, II, aVF, V_5, and V_6 (marked ⇩) (Reproduced with permission from: Mittal SR. ST-Segment Depression. Cardiology Today-2017;21:245–250)

(g) During tachyarrhythmia

Tachyarrhythmias with fast ventricular rate may show ST-segment depression (Fig. 41.20). Such ST-segment depressions do not suggest myocardial ischemia except in patients who complain of angina during tachyarrhythmia.

(h) Post tachycardia

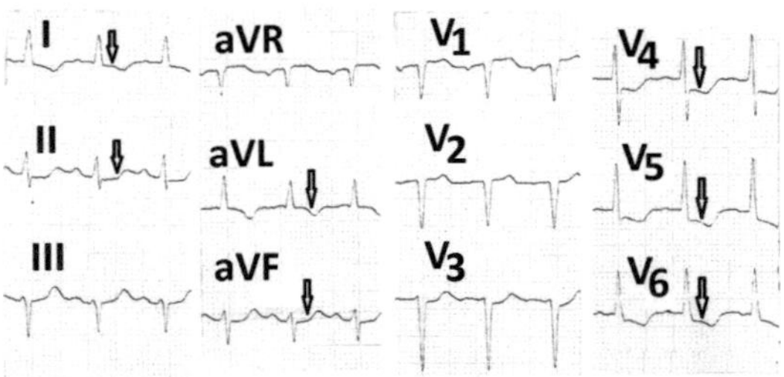

Fig. 41.21 Electrocardiogram from a case of recovery after tachyarrhythmia showing ST-segment depression in leads I, II, aVL, aVF, V_4, and V_6 (marked ⇩) (Reproduced with permission from: Mittal SR. ST-Segment Depression. Cardiology Today-2017;21:245–250)

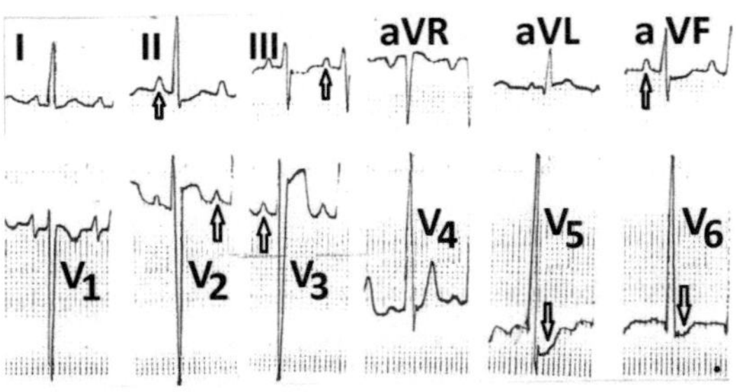

Fig. 41.22 Electrocardiogram from a case of cardiomyopathy showing increased left ventricular voltages and ST segment depression in leads V_5 and V_6 (marked ⇩). Leads V_1 to V_3 show deep S waves suggestive of left ventricular enlargement. P waves are peaked in leads II, III, aVF, V_2, and V_3 (marked ⇧) suggestive of right atrial overload (Reproduced with permission from: Mittal SR.ST-Segment Depression. Cardiology Today-2017;21:245–250)

In some cases, depression of the ST-segment may persist even after cessation of the tachyarrhythmias (Fig. 41.21). Such changes may persist for a few days. Such ST-segment changes also do not suggest myocardial ischemia.

(i) Myocarditis
(j) Cardiomyopathy (Fig. 41.22)

Summary

ST-segment depression is frequently seen in the following situations:

- Normal variations

  Sympathetic overdrive, neurocirculatory asthenia, hyperventilation, ortho static, post meals.
- Subendocardial ischemia

  ST-segment depression of >1 mm at the J point is considered significant. It may be horizontal with sharp angle between the ST-segment and the T wave. It is considered the earliest sign of coronary insufficiency.
- Upsloping ST-segment can be seen in the absence of ischemia due to depressing effect of the Ta wave (atrial repolarization). In this situation, the P wave, PR segment, the J point, the ST-segment, and upstroke of the T wave from a smooth unbroken parabola. If parabola is broken and step like, it is considered suggestive of ischemia.
- Down sloping ST-segment suggests severe ischemia.
- Digitalis overdose produces down sloping (inverted correction mark) ST-segment depression with short QT interval.
- Hypokalemia produces ST-segment depression, flattening of the T wave, increased prominence of the U wave with long QT (U) interval.
- Some patients of mitral valve prolapse can have ST-segment depression in inferior and/or lateral chest leads.
- ST-segment depression may persist for hours to days following termination of any tachyarrhythmia.
- ST-segment depression can be secondary to changes in depolarization, for example, right ventricular hypertrophy, left ventricular hypertrophy, right bundle branch block, left bundle branch block, preexcitation, and aberrant intraventricular conduction. ST-segment depression of more than 1 mm in two or more of leads V_1 to V_3 in the presence of LBBB or LVH is considered suggestive of ischemia.
- Hypothyroidism usually does not produce ST-segment deviation.

MCQs

Q1. During angina, ST-segment depression is considered significant if it is:

(a) ≥0.5 mm
(b) ≥1.0 mm
(c) ≥2.0 mm
(d) ≥2.5 mm

Q2. Subendocardial injury mostly produces ST-segment depression in leads:

(a) V_1 to V_3
(b) V_5 and V_6
(c) V_7 to V_9
(d) V_3R to V_6R

Q3. Severity of impairment of coronary flow is maximum when ST-segment depression is:

(a) Horizontal
(b) Rapid upsloping
(c) Slow upsloping
(d) Down sloping

Q4. In normal persons, upsloping ST-segment depression touches the baseline in:

(a) 0.08 s
(b) 0.10 s
(c) 0.14 s
(d) 0.16 s

Q5. Normally Ta wave falls in:

(a) The P wave
(b) The QRS complex
(c) The ST-segment
(d) The T wave

Q6. ST-segment depression in leads V_1 to V_4 can be seen in:

(a) Anteroseptal subendocardial infarction
(b) Posterior infarction
(c) Right ventricular subendocardial ischemia
(d) All

Q7. ST-segment depression in leads V_4 to V_6 is seen in:

(a) Subendocardial ischemia
(b) Left ventricular hypertrophy
(c) Left bundle branch block
(d) All

Q8. ST-segment depression in leads V_4 to V_6 is seen in:

(a) Cardiomyopathy
(b) Myocarditis
(c) Inferior infarction
(d) All

Q9. ST-segment depression in seven or more leads with ST-segment elevation in lead aVR more than that in lead V_1 suggests:

(a) Left main stem disease
(b) Proximal LCX occlusion
(c) Proximal RCA occlusion
(d) All

Q10. Upsloping ST-segment depression with tall positive T waves in anterolateral leads can occur in:

(a) Subtotal occlusion of LAD
(b) Subtotal occlusion of LCX
(c) Occlusion of mid RCA
(d) All

Q11. ST-segment depression in leads I and aVL can occur in:

(a) High lateral transmural myocardial infarction
(b) Inferior infarction
(c) Posterior infarction
(d) None

Q12. ST-segment depression in leads III and aVF can occur in:

(a) Inferior subendocardial infarction

(b) High lateral transmural myocardial infarction
(c) Right ventricular infarction
(d) All

Q13. Upsloping ST-segment depression can be seen in:

(a) Sympathetic overdrive
(b) Neurocirculatory asthenia
(c) Hyperventilation
(d) All

Q14. ST-segment depression alone in a patient receiving digitalis suggests:

(a) Inadequate dose of digitalis
(b) Digitalis effect
(c) Digitalis toxicity
(d) Hyperkalemia

Q15. Which of the following should suggest digitalis overdose?

(a) New onset of otherwise unexplained ventricular ectopics
(b) Atrial fibrillation with otherwise unexplained slow ventricular rate
(c) Bidirectional ventricular tachycardia
(d) None

Q16. Which of the following does not cause ST-segment depression?

(a) Hyperkalemia
(b) Viral myocarditis
(c) Atrioventricular nodal reentrant tachycardia
(d) Hyperventilation

Q17. ST-segment depression can be seen in:

(a) Sympathetic overdrive
(b) Vagal overtone
(c) Neurocirculatory asthenia
(d) All

Q18. ST-segment can depress:

(a) During hyperventilation
(b) Following a meal
(c) On standing from supine position
(d) None

Q19. ST-segment depression suggests:

(a) Subendocardial ischemia
(b) Subepicardial ischemia
(c) Transmural infarction
(d) Transmural ischemia

Q20. Which is the earliest ST-segment change suggestive of myocardial ischemia?

(a) Horizontality
(b) Rapid upsloping ST-segment depression
(c) Horizontal ST-segment depression
(d) Down sloping ST-segment depression

Q21. Down sloping ST-segment depression with short QT interval suggests:

(a) Digitalis effect
(b) Hypercalcemia
(c) Digitalis toxicity
(d) Hypokalemia

Q22. ST-segment depression with flattening of the T wave, prominent U wave, and prolonged QT (U) interval suggests:

(a) Hypokalemia
(b) Hypomagnesemia
(c) Hypocalcemia
(d) Hyponatremia

Q23. Mitral valve prolapse can produce ST-segment depression in leads:

(a) II, III, aVF
(b) I, aVL
(c) V_1, V_2
(d) V_3, V_4

Q24. How long can post tachycardia ST-segment depression persist?

(a) Seconds
(b) Minutes
(c) Hours
(d) Days

Q25. Post tachycardia ST-segment depression suggests:

(a) Mild ischemia
(b) Severe ischemia
(c) Left ventricular dysfunction
(d) None

Q26. ST-segment depression in leads V_1 to V_6 can occur in:

(a) Right ventricular hypertrophy
(b) Left ventricular hypertrophy
(c) Left lateral accessory pathway
(d) Biventricular hypertrophy

Q27. ST-segment depression in leads I, aVL, V_5, V_6 suggests:

(a) Left ventricular pressure overload
(b) Left bundle branch block
(c) Preexcitation
(d) All

Q28. What is not seen in uncomplicated left bundle branch block?

(a) Down sloping ST-segment depression in leads I and aVL
(b) ST-segment elevation in leads V_5, V_6
(c) ST-segment depression more than 1 mm in leads V_1, V_2
(d) ST-segment elevation in leads V_1, V_2

Q29. What is not seen in uncomplicated RBBB?

(a) Down sloping ST-segment depression in leads V_1, V_2
(b) Down sloping ST-segment depression in leads V_5, V_6
(c) Horizontal ST-segment in leads V_5, V_6
(d) ST-segment depression in leads I and aVL

Q30. In aberrant ventricular conduction ST-segment is:

(a) Isoelectric
(b) Concordant to terminal QRS
(c) Discordant to terminal QRS
(d) Any of the above

Answers

(1) a (2) b (3) d (4) a (5) b (6) d (7) d (8) d (9) a (10) a (11) b (12) a, b (13) d (14) b (15) a, b, c (16) a (17) a, c (18) a, b, c (19) a (20) a (21) a (22) a (23) a (24) d (25) d (26) c (27) a, b (28) b, c (29) b, d (30) c

References

1. Schamroth L. Coronary insufficiency. In: Schamroth C, editor. An introduction to electrocardiography. New Delhi: Wiley India; 1982. p. 157–85.
2. Bimbaum Y, Nikus K, Kligfield P, et al. The role of the ECG in diagnosis, risk estimation and catheterization laboratory activation in patients with acute coronary syndromes: a consensus document. Ann Non-invasive Electrocardiol. 2014;19:412–25.
3. Boden WE, Bough EW, Korr KS, Russo J, Gandsman EJ, Shulman RS. Inferoseptal myocardial infarction: another cause of precordial ST segment depression in transmural inferior wall myocardial infarction? Am J Cardiol. 1984;54:1216–23.
4. De Luna AB, Goldwasser D, Fiol M, Bayes Genis A. Surface electrocardiography. In: Fuster V, Walsh RA, Harrington RA, editors. Hurst's the heart. New York: McGraw Hill; 2011. p. 307–70.
5. Bimbaum Y, Wilson JM, Fiol M, de Luna AB, Eskola M, Nikus K. ECG diagnosis and classification of acute coronary syndromes. Ann Noninvasive Electrocardiol. 2014;19:4–14.
6. Hassen GW, Talebi S, Fernaine G, Kalantari H. Lead aVL on electrocardiogram: emerging as important lead in early diagnosis of myocardial infarction? Am J Emerg Med. 2014;32:785–8.
7. Schamroth L. Drugs and electrolytes. In: Schamroth C, editor. An introduction to electrocardiography. New Delhi: Wiley India; 1982. p. 241–53.

42 T Wave: Normal Variations and Flat T Wave

T wave is produced by repolarization of the ventricles.

The normal T wave is in the same direction as the preceding QRS complex. In leads V_1 and V_2, the T wave may be positive in spite of a dominantly negative QRS complex. Repolarization of the ventricular cells causes a flow of ions in a direction opposite to that of depolarization. It is, therefore, expected that direction of the T wave should be opposite to that of the QRS. However, epicardial cells repolarize earlier than the endocardial cells [1]. This causes the wave of repolarization to spread in the same direction as that of depolarization. Normal T wave is, therefore, in the same direction as that of the preceding QRS complex (Fig. 42.1).

Normally the proximal limb of the T wave is sloping and the distal limb is relatively sharp (Fig. 42.2).

Sloping of the distal limb results in widening of the T wave and prolongation of the QT interval (Fig. 42.3).

Inversion of the proximal limb in the right precordial leads suggest diagnosis of ASD [2, 3]. Inversion of the distal limb with elevation of the ST-segment suggests myocardial infarction.

Normally the apex of the T wave is blunt. Pointed and tall T waves are seen in hyperkalemia (Fig. 42.4).

Asynchronous repolarization of the two ventricles may result in a notch on the apex of the T wave "Dart T wave" (Figs. 42.5 and 42.6).

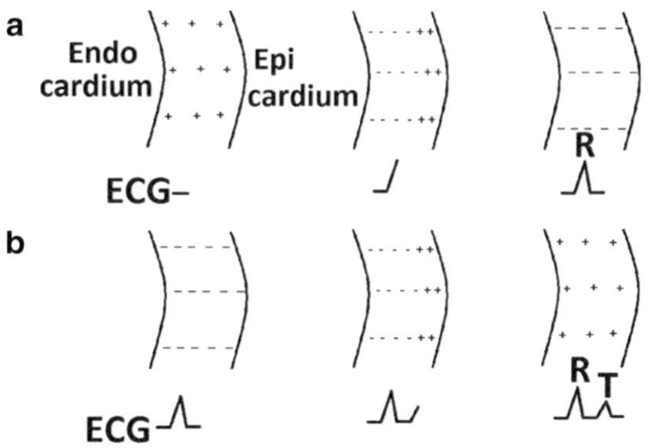

Fig. 42.1 Diagrammatic representation of (**a**) depolarization and (**b**) repolarization of free walls of the ventricles producing T wave in the same direction as that of preceding QRS complex (Reproduced with permission from: Mittal SR. T wave. Cardiology Today-2018;22:30–33)

Prominent U wave can fuse with the T wave and give the impression of a notched T wave (Fig. 42.7). Analysis of complete ECG can help in differentiation.

Normally the T wave is inverted in lead aVR and sometimes in lead V_1 (Fig. 42.8).

Uncommonly lead aVL may record cavity potential. In such a situation all waves (P, QRS, and T) may be negative in lead aVL (Fig. 42.9).

This finding is usually seen in vertical heart. High lateral infarction can also produce QS complex with inverted T wave in lead aVL. However, in this situation P wave is normal, ST-segment is coved, and lead I also shows Q wave and inverted

S. Mittal, *Insights into Electrocardiograms with MCQs*,
https://doi.org/10.1007/978-981-99-0127-2_42

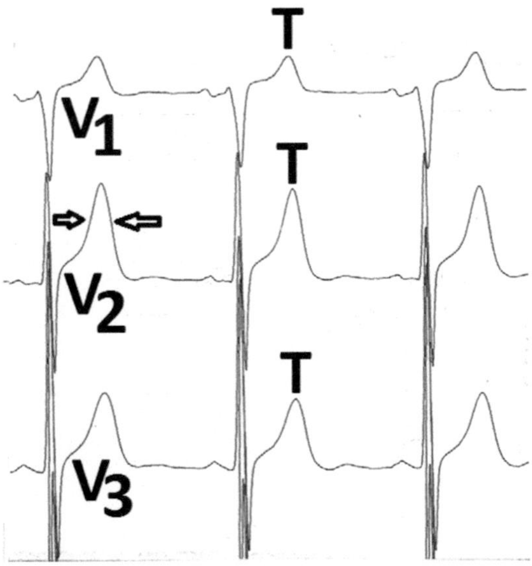

Fig. 42.2 Electrocardiogram showing normal T waves. The proximal limb is sloping (marked ⇨) and the distal limb is sharp (marked ⇦) (Reproduced with permission from: Mittal SR. T wave. Cardiology Today-2018;22:30–33)

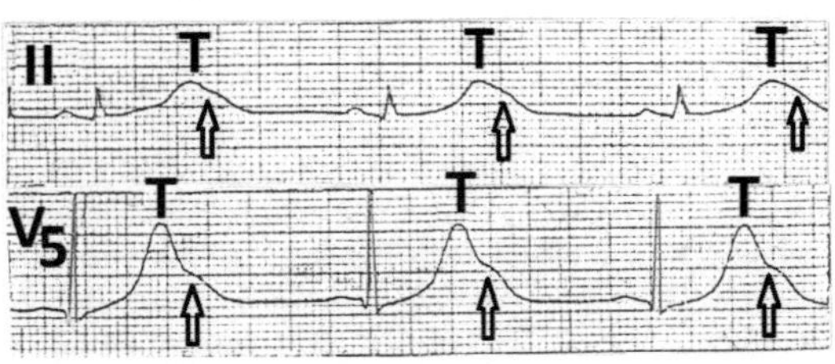

Fig. 42.3 Electrocardiogram is showing a sloping of distal limb of the T wave (arrow) with prolongation of the QT interval (Reproduced with permission from: Mittal SR. T wave. Cardiology Today-2018;22:30–33)

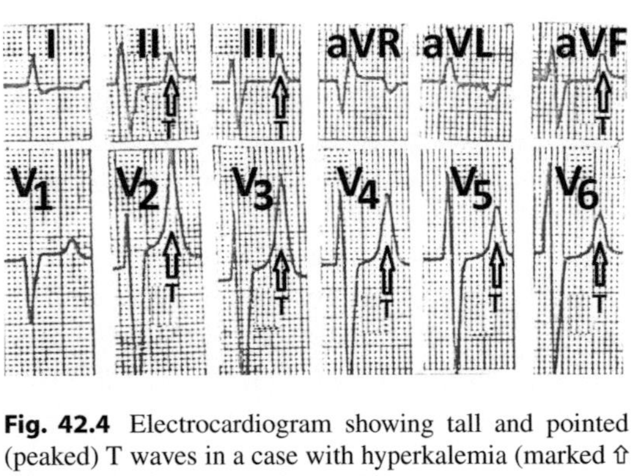

Fig. 42.4 Electrocardiogram showing tall and pointed (peaked) T waves in a case with hyperkalemia (marked ⇧) (Reproduced with permission from: Mittal SR. T wave. Cardiology Today-2018;22:30–33)

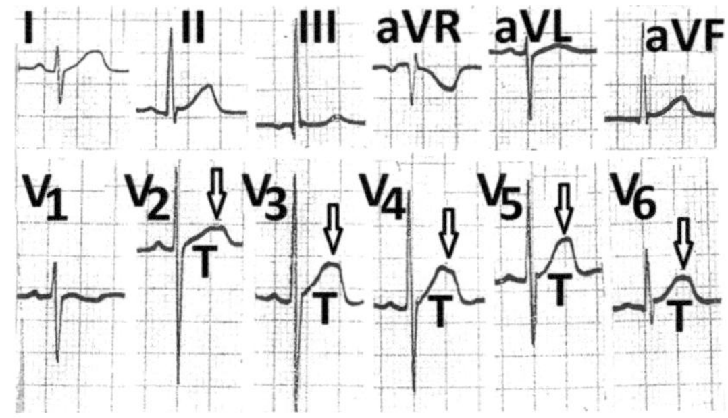

Fig. 42.5 Electrocardiogram is showing notch on the apex of the T wave (marked ⇩) (Reproduced with permission from: Mittal SR. T wave. Cardiology Today-2018; 22:30–33)

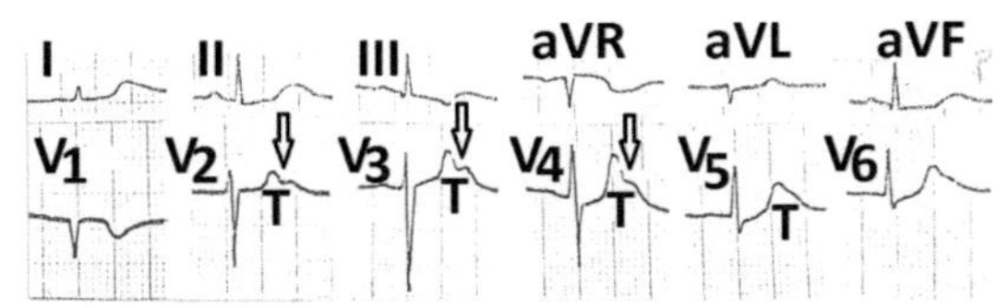

Fig. 42.6 Electrocardiogram is showing notch on the apex of the T wave (arrow) (Reproduced with permission from: Mittal SR. T wave. Cardiology Today-2018;22:30–33 and Flat T wave. Cardiology Today-2018;22:71)

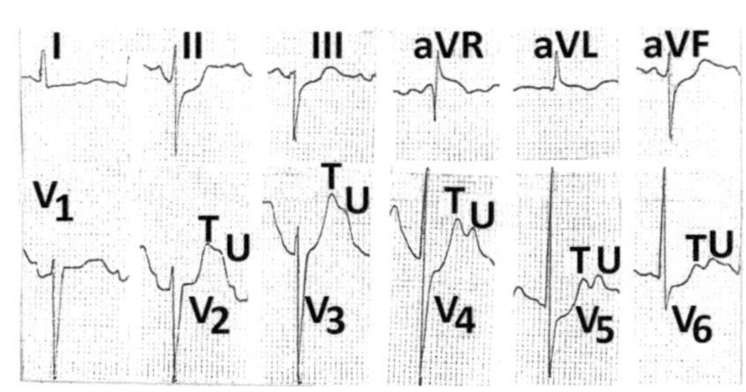

Fig. 42.7 Electrocardiogram is showing fusion of prominent U wave with the T wave (Reproduced with permission from: Mittal SR. T wave. Cardiology Today-2018; 22:30–33)

T wave. Isolated T inversion in lead aVL with positive P wave and QRS is usually abnormal (Fig. 42.10).

In horizontal heart, as in obese persons, the T wave may be inverted in lead III (Fig. 42.11) and inverted or biphasic in lead aVF.

In such a situation, T wave may become less deep or even flat on deep inspiration. On deep inspiration, the diaphragm moves down, and the heart becomes relatively vertical. Symmetrical

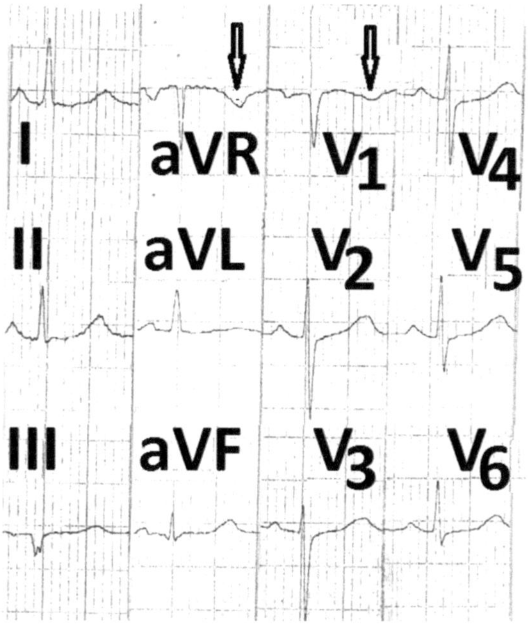

Fig. 42.8 Electrocardiogram is showing normal inversion of the T wave in leads aVR and in lead V_1 (arrow) (Reproduced with permission from: Mittal SR. T wave. Cardiology Today-2018;22:30–33)

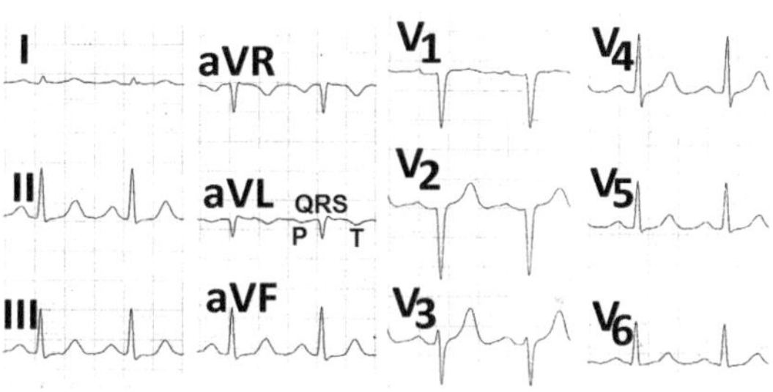

Fig. 42.9 Electrocardiogram from a normal person showing negative P wave, QRS complex, and T wave in lead aVL (Reproduced with permission from: Mittal SR. T wave. Cardiology Today-2018;22:30–33)

inversion of the T wave with broad and deep Q wave is abnormal in lead III (Fig. 42.12).

In dextrocardia or technical dextrocardia (interchange of leads of right and left upper limbs), the P wave, the QRS complex, and the T waves are inverted in leads I and aVL and are positive in lead aVR.

Amplitude of the T wave is usually less than the preceding R wave and more than the amplitude of the following U wave, if the latter is present. T wave equal to or more than the amplitude of the R wave is abnormal (Fig. 42.13).

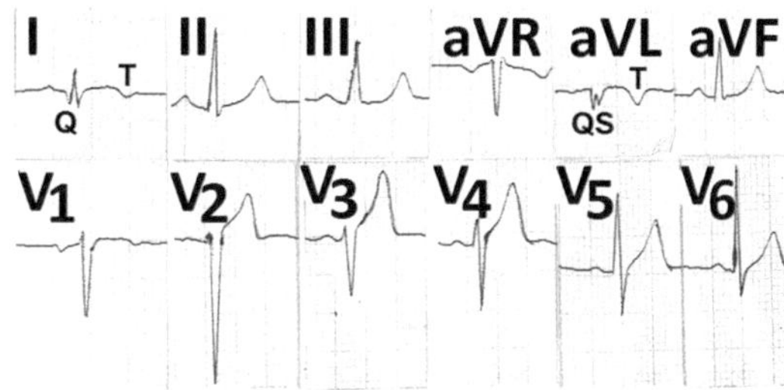

Fig. 42.10 Electrocardiogram from a case of high lateral infarction showing QS configuration with coved ST segment and inverted T wave in leads aVL and Q wave with inverted T wave in lead I (Reproduced with permission from: Mittal SR. T wave. Cardiology Today-2018;22:30–33)

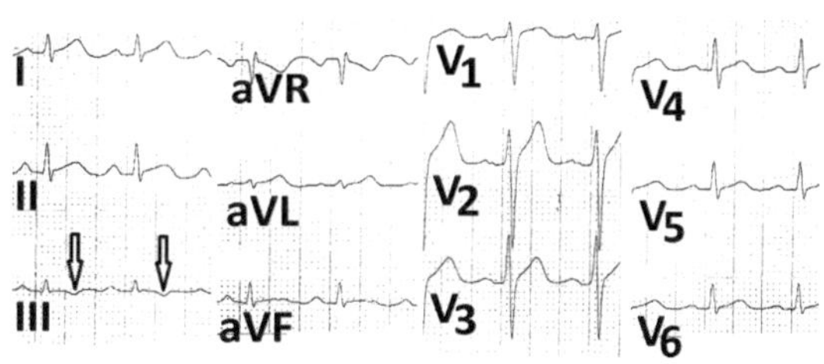

Fig. 42.11 Electrocardiogram is showing normal inversion of the T wave in lead III (marked ⇩) (Reproduced with permission from: Mittal SR. T wave. Cardiology Today-2018;22:30–33)

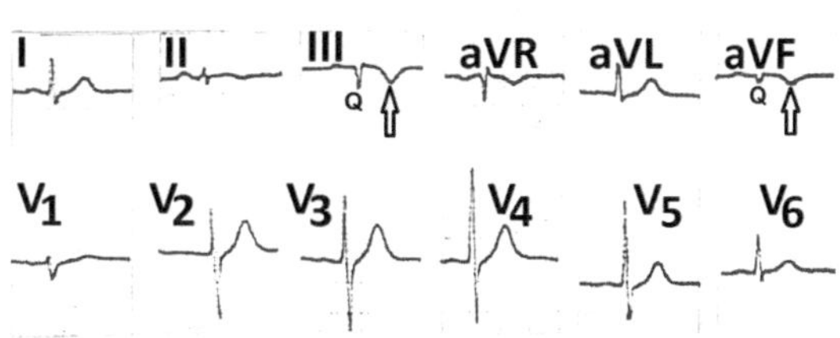

Fig. 42.12 Electrocardiogram is showing Q wave with symmetrical (arrow head) inversion of the T wave in leads III and aVF (marked ⇧) suggestive of inferior infarction (Reproduced with permission from: Mittal SR. T wave. Cardiology Today-2018;22:30–33)

Amplitude of the T wave less than the accompanying U wave is also abnormal (Fig. 42.14). It can occur during acute myocardial ischemia or hypokalemia.

Movement of the electrode (Fig. 42.15) or cable (Fig. 42.16) can produce variation in the configuration of the T wave.

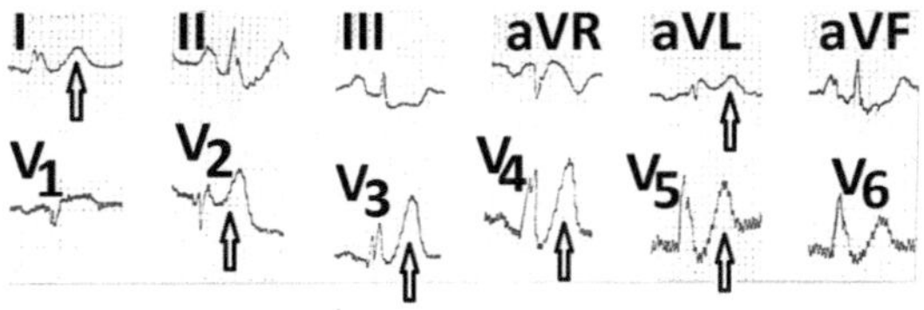

Fig. 42.13 Electrocardiogram is showing T wave amplitude equal to or more than the amplitude of the accompanying "R" wave (marked ⇧). Such T waves suggest hyperacute phase of myocardial infarction in respective leads (Reproduced with permission from: Mittal SR. T wave. Cardiology Today-2018;22:30–33)

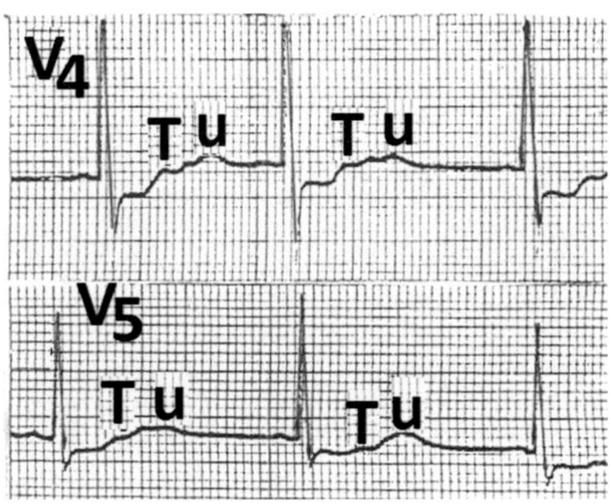

Fig. 42.14 Electrocardiogram is showing T wave amplitude less than the amplitude of the accompanying U wave (Reproduced with permission from: Mittal SR. T wave. Cardiology Today-2018;22:30–33)

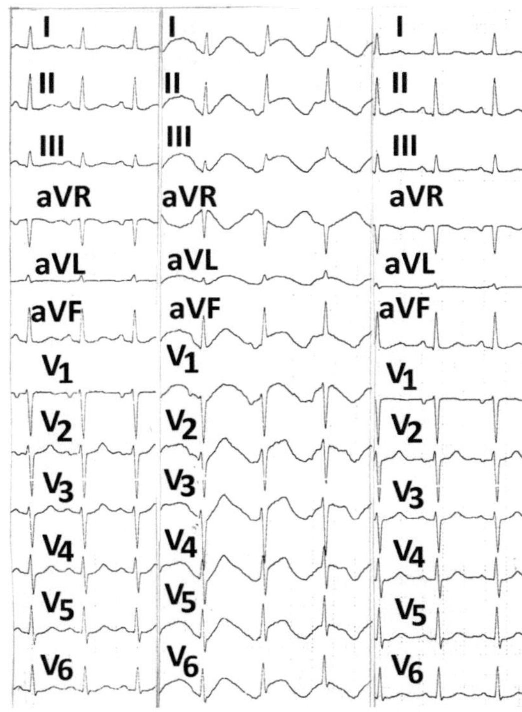

Fig. 42.16 Electrocardiogram showing the effect of movement of cable on the configuration and direction of the T wave (**b**). (**a**) and (**c**) have been recorded after stabilizing the cable (Reproduced with permission from: Mittal SR. T wave. Cardiology Today-2018;22:30–33)

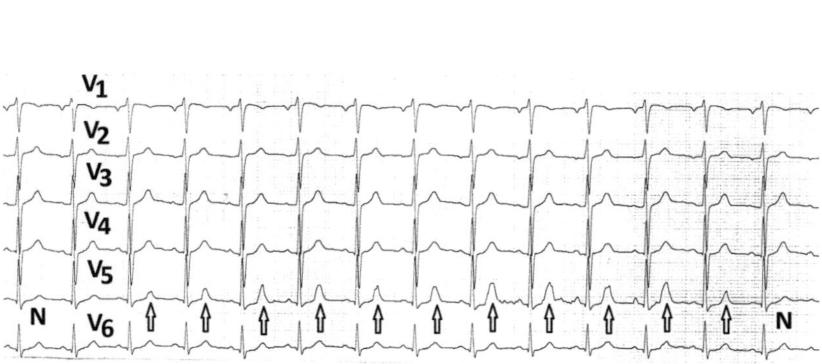

Fig. 42.15 Electrocardiogram showing effect of movement of lead V_5 electrode on configuration of the T wave (arrows). N-normal T wave (Reproduced with permission from: Mittal SR. T wave. Cardiology Today-2018;22:30–33)

1 Flat T Wave

[A] Diffuse flattening of the T wave is observed in

1. Pericardial effusion (Fig. 42.17)
   It is associated with sinus tachycardia and low voltage QRS.
2. Myxedema (Fig. 42.18)
   It is associated with sinus bradycardia and low voltage QRS.
3. Hypokalemia
4. Amiodarone therapy

[B] Flattening of the T wave in leads I, aVL, V_5, and V_6

This finding can be seen in leads I, aVL, V_5 , and V_6 in left ventricular systolic overload (Fig. 42.19). Subsequently the T waves become inverted in these leads.

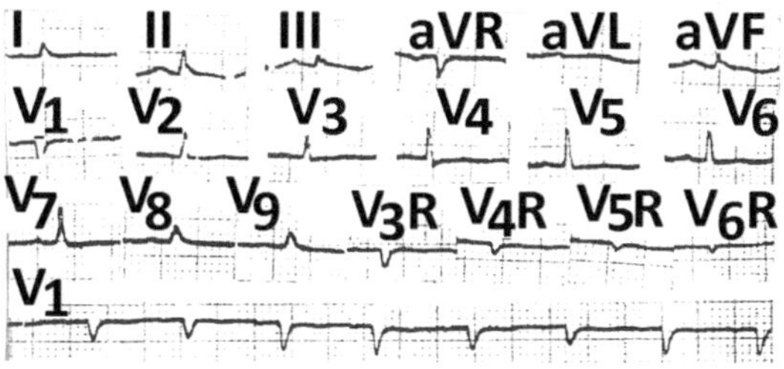

Fig. 42.17 Electrocardiogram from a case of pericardial effusion showing sinus tachycardia, low voltage QRS, and diffuse flattening of the T waves (Reproduced with permission from Mittal SR. Flat T wave. Cardiology Today-2018;22:71)

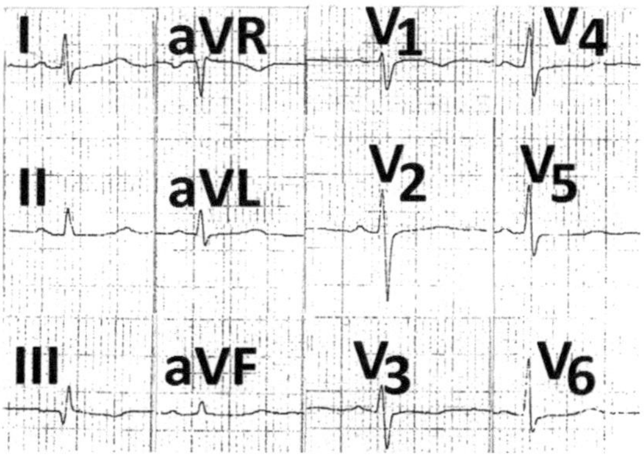

Fig. 42.18 Electrocardiogram from a case of myxedema showing low voltage QRS and diffuse flattening of the T waves in the precordial leads (Reproduced with permission from: Mittal SR. Flat T wave. Cardiology Today-2018;22:71)

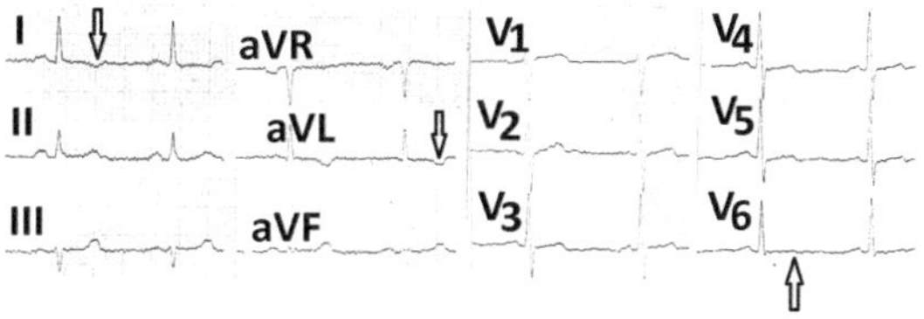

Fig. 42.19 Electrocardiogram from a case of systemic hypertension showing flattening of the T wave in lead V_6 (marked ⇧). T waves are inverted in leads I and aVL (marked ⇩) (Reproduced with permission from Mittal SR. Flat T wave. Cardiology Today-2018;22:71)

Summary

- T wave is produced by repolarization of the ventricular myocardium.
- Polarity is generally the same as the net polarity of the preceding QRS.
- It is smooth and rounded. Proximal limb is shallow. Distal limb is relatively sharp.
- Normally the T wave is positive in leads I, II, aVL, aVF, V_5, and V_6. It is inverted in lead aVR. In lead III it can be inverted. In leads V_1 and V_2 it can be inverted or biphasic (initially positive, terminally negative).
- Amplitude is generally less than 0.5 mV in limb leads and less than 1.5 mV in any precordial lead.
- Amplitude is greater than accompanying U wave, if any.
- T wave may be inverted in right precordial to mid precordial leads (V_1 to V_4) in young (specially in females)—Juvenile pattern.
- In early repolarization variant prominent biphasic T wave (initial positive, terminal negative) with mild elevation of the J point can be seen in precordial leads. T waves normalize during exercise.
- Diffuse flattening of the T waves can be seen in pericardial effusion, myxedema, hypokale-

mia, and Amiodarone therapy. Left ventricular systolic overload can produce flattening of T waves in leads I, aVL, V_5, and V_6.

MCQs

Q1. The T wave is produced by:

(a) Atrial repolarization
(b) Atrial depolarization
(c) Ventricular depolarization
(d) Ventricular repolarization

Q2. In a normal person the T wave can be inverted in:

(a) Lead V_1
(b) Lead III
(c) Lead aVF
(d) All

Q3. In normal persons, the T wave of lead aVR is:

(a) Upright
(b) Biphasic
(c) Inverted
(d) Any shape

Q4. T wave can be inverted in leads I and aVL in:

(a) High lateral myocardial infarction
(b) Dextrocardia
(c) Technical dextrocardia
(d) All

Q5. What is correct for a normal T wave?

(a) Proximal limb is sloping
(b) Apex is blunt
(c) Distal limb is sloping
(d) All

Q6. What suggests an abnormal T wave?

(a) Pointed apex
(b) Notched apex
(c) Terminal inversion of a positive T wave
(d) All

Q7. What suggests an abnormal T wave?

(a) Amplitude greater than accompanying "R" wave
(b) Amplitude lesser than accompanying "U" wave
(c) Inversion of initial portion of the T wave in leads V_1, V_2
(d) All

Q8. Inversion of the terminal portion of the T wave with ST segment elevation suggests:

(a) Myocardial infarction
(b) Left ventricular hypertrophy
(c) Myocarditis
(d) All

Q9. Inverted proximal limb of the T wave in right precordial leads suggests diagnosis of:

(a) ASD
(b) VSD
(c) PDA
(d) TOF

Q10. T wave is produced by repolarization of:

(a) Atrial myocardium
(b) Ventricular myocardium
(c) Purkinje fibers
(d) Midmyocardial (M) cells

Q11. Polarity of a normal T wave is:

(a) Same as net polarity of the preceding QRS complex
(b) Opposite to the direction of the terminal part of the QRS
(c) Opposite to the direction of the initial part of the QRS
(d) Opposite to net polarity of the preceding QRS complex

Q12. Top of normal T wave is:

(a) Rounded
(b) Peaked
(c) Flat
(d) Notched

Q13. In normal T wave:

(a) The two limbs are asymmetrical
(b) Proximal limb has obtuse angle with base line
(c) Distal limb has relatively acute angle with base line
(d) All

Q14. Normally T wave in lead I is:

(a) Positive
(b) Negative
(c) Biphasic
(d) Notched

Q15. Normally T wave in lead V_1 can be:

(a) Flat
(b) Negative
(c) Initial negative, terminal positive
(d) Initial positive, terminal negative

Q16. Normally T wave in lead III can be:

(a) Positive
(b) Negative
(c) Biphasic
(d) Any of the above

Q17. Normal T wave in lead V_5 is:

(a) Positive
(b) Flat
(c) Inverted
(d) Any of the above

Q18. In normal persons, T wave in lead aVL:

(a) Does not exceed 0.5 mV
(b) Exceeds 1 mV
(c) Exceeds 1.5 mV
(d) Exceeds amplitude of preceding R wave:

Q19. In normal persons T wave in lead V_4

(a) Does not exceed 0.5 mV
(b) Does not exceed 1.0 mV
(c) Does not exceed 1.5 mV
(d) Exceeds amplitude of preceding R wave

Q20. In normal persons:

(a) Direction of the T wave is opposite to that of the accompanying U wave
(b) Amplitude of T wave is less than the amplitude of the accompanying U wave
(c) Amplitude of the T wave is equal to the amplitude of the accompanying U wave
(d) Amplitude of the T wave is more than the amplitude of the accompanying U wave

Q21. In juvenile pattern T waves are:

(a) Inverted in leads I, aVL, V_6
(b) Inverted in leads V_1 to V_4
(c) Inverted in leads V_5, V_6
(d) Inverted in leads II, III, aVF

Q22. In early repolarization variant:

(a) J point is mildly elevated
(b) Initial part of T wave is positive
(c) Terminal part of T wave is negative
(d) All of the above

Q23. In early repolarization variant:

(a) Initial part of T wave is negative
(b) Terminal part of T wave in positive
(c) T wave becomes inverted on exercise
(d) T wave becomes normal on exercise

Q24. Early repolarization variant in usually seen in:

(a) Lead III
(b) Lead aVR
(c) Lead V_4R

(d) Leads V_3, V_4

Q25. Combination of sinus bradycardia, low voltage of all wave forms with flat T waves suggests the possibility of:

(a) Acute pericarditis
(b) Pericardial constriction
(c) Cardiac tamponade
(d) Myxedema

Q26. Combination of sinus tachycardia, low voltage of all wave forms with flat T waves suggests possibility of:

(a) Pericardial effusion
(b) Myxedema
(c) Hypokalemia
(d) Amiodarone therapy

Q27. Combination of normal heart rate, flat T waves, prominent U waves, and prolonged QTc interval suggest the possibility of:

(a) Cardiac tamponade
(b) Pericardial constriction
(c) Amiodarone therapy
(d) Hypokalemia

Q28. Electrocardiographic findings of myxedema do not include:

(a) Sinus tachycardia
(b) Prominent P wave
(c) Flat T wave
(d) Prolonged QTc interval

Q29. Electrocardiographic findings of pericardial effusion do not include:

(a) Sinus tachycardia
(b) Prominent P wave
(c) Flat T wave
(d) Prominent "U" wave

Q30. Electrocardiographic findings of hypokalemia do not include:

(a) Flat P wave
(b) Broad QRS
(c) ST segment depression
(d) Flat T wave

Q31. The normal T wave can be flat in:

(a) Lead V_1
(b) Lead III
(c) Lead aVF
(d) All

Q32. In the context of coronary artery disease, flat T wave can be seen:

(a) During evolution of non ST-segment elevation acute coronary syndrome
(b) Residual pattern of old myocardial infarction
(c) During acute ischemia
(d) All

Answers

(1) d (2) d (3) c (4) d (5) a, b (6) d (7) d (8) a (9) a (10) b (11) a (12) a (13) d (14) a (15) a, b, d (16) d (17) a (18) a (19) c (20) d (21) b (22) d (23) d (24) d (25) d (26) a (27) c, d (28) a, b, d (29) b, d (30) a, b, c (31) d (32) a, b

References

1. Wagner GS, Lim TH. Cardiac electrical activity. In: Wagner GS, editor. Marriott's practical electrocardiography. New Delhi: Wolters Kluwer; 2001. p. 1–20.
2. Bayar N, Arstan S, Koklu E, et al. The importance of electrocardiographic findings in the diagnosis of atrial septal defect. Kardiologia Polska. 2015;73:331–6.
3. Wong MX, Wu GF, Gu JL, et al. Defective T wave combined with incomplete right bundle branch block: a new electrocardiographic index for diagnosing atrial septal defect. Chin Med J. 2012;125:1057–62.

Prominent (Tall) T Waves 43

T wave amplitude ≥0.5 mV in limb leads and ≥1.0 mV in the precordial leads is considered as tall [1]. T wave taller than the accompanying R wave is also considered abnormal except in leads V_1 and V_2.

1 Causes of Prominent T Wave

1.1 Vagotonia

It can produce tall symmetrical T waves with minimally elevated ST-segment particularly in leads V_4 to V_6. There is relative bradycardia.

1.2 Early Repolarization

Prominent T waves are accompanied by elevation of the J point (Fig. 43.1).

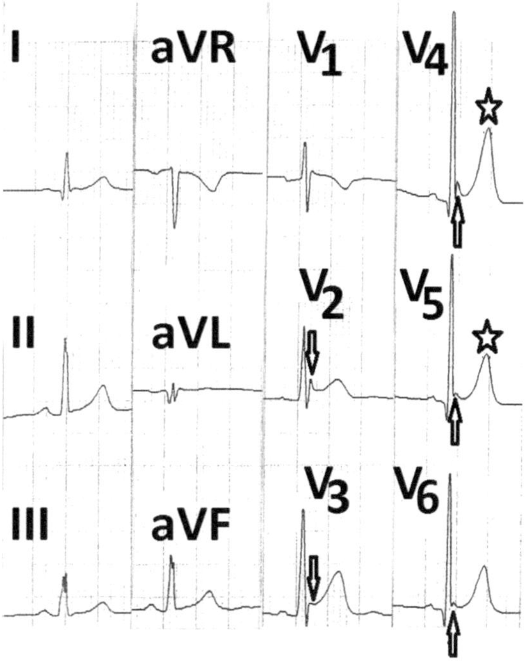

Fig. 43.1 Electrocardiogram from a case of early repolarization showing elevational of the J point (marked arrows) and prominent T waves (marked ☆) (Reproduced with permission from: Mittal SR. Prominent (Tall) T waves. Cardiology Today-2018;22:98–102)

1.3 Ischemic Heart Disease

1.3.1 Acute Subendocardial Ischemia

Prominent T wave is usually accompanied by prolongation of the QT interval (Fig. 43.2).

S. Mittal, *Insights into Electrocardiograms with MCQs*,
https://doi.org/10.1007/978-981-99-0127-2_43

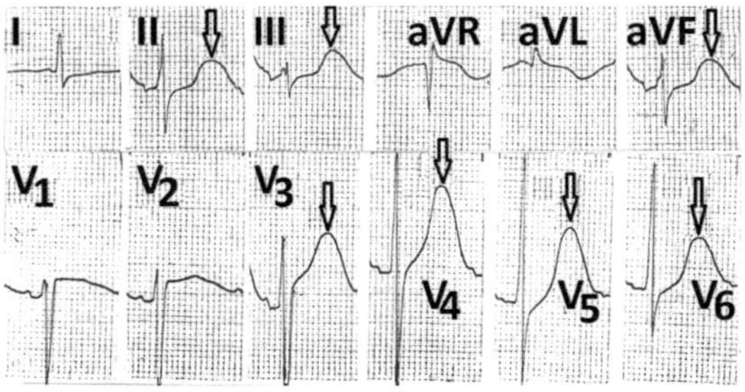

Fig. 43.2 Electrocardiogram from a case of acute subendocardial ischemia showing prominent T waves (arrows) with prolongation of the QT interval (Reproduced with permission from: Mittal SR. Prominent (Tall) T waves. Cardiology Today-2018;22:98–102)

1.3.2 Hyperacute Phase of Myocardial Infarction

There is concomitant elevation of the ST-segment in the affected leads (Figs. 43.3a, b and 43.4a, b).

1.3.3 Recovering Infarction

(i) Recovering true posterior wall myocardial infarction

Prominent T waves are seen in leads V_1 and V_2. These are reciprocal changes of T wave inversion in leads V_7 to V_9. R waves are also prominent in leads V_1 and V_2 as a reciprocal change of prominent Q wave in leads V_7 to V_9. Representative electrocardiogram is shown in Fig. 43.5.

(ii) Recovering inferior infarction

Negative T waves in inferior leads produce reciprocal prominent positive and symmetrical T waves in leads I and aVL [2]. Prominent T waves in leads V_1 to V_3 suggest concomitant posterior infarction (Fig. 43.6).

(iii) Recovering high lateral infarction

It can produce reciprocal prominent T waves in leads II, III, and aVF (Fig. 43.7).

1.4 Left Bundle Branch Block

Tall T waves are present in leads V_1 to V_3. These are associated with deep and broad S waves in

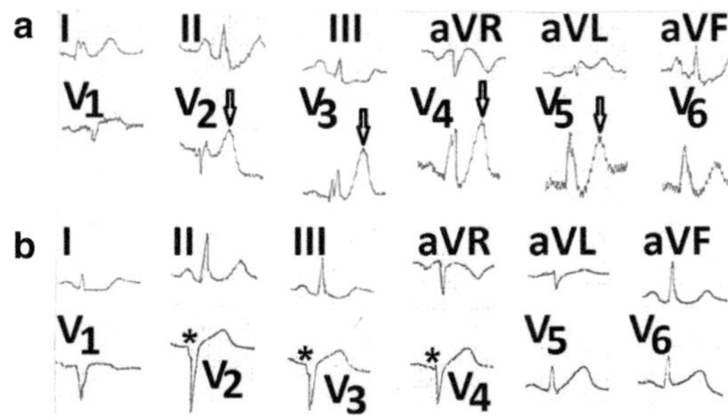

Fig. 43.3 **a** Electrocardiogram during the hyperacute phase of anterior infarction showing prominent T wave in leads V_2 to V_5 (arrows). **b** Electrocardiogram recorded after 24 h showing development of Q waves and loss of the R waves in leads V_2 to V_4 (marked *) suggestive of anterior myocardial infarction (Reproduced with permission from: Mittal SR. Prominent (Tall) T waves. Cardiology Today-2018;22:98–102)

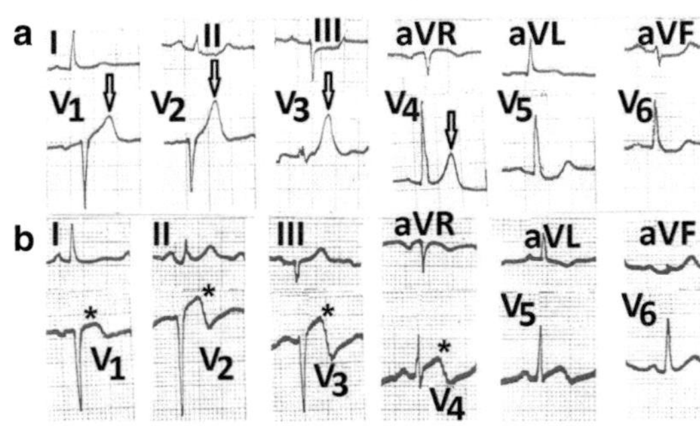

Fig. 43.4 **a** Electrocardiogram during hyperacute phase of anterior infarction showing prominent T waves in leads V_1 to V_4 (arrows). **b** Electrocardiogram recorded after 24 h showing development of coved ST-segment and inverted T waves in leads V_1 to V_4 (marked *) suggestive of anterior infarction (Reproduced with permission from: Mittal SR. Prominent (Tall) T waves. Cardiology Today-2018;22:98–102)

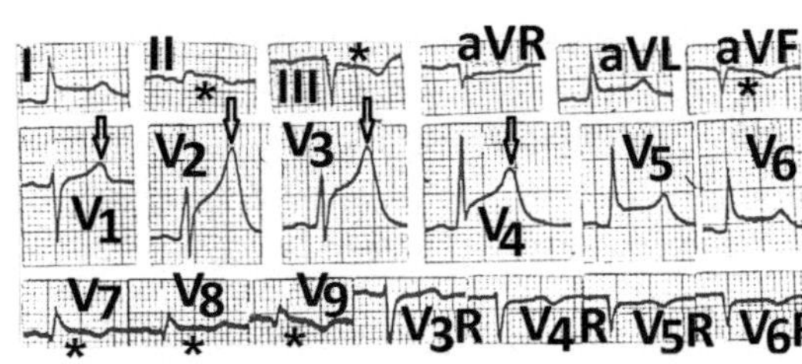

Fig. 43.5 Electrocardiogram is showing inferoposterior infarction (marked *) with reciprocal prominent T waves in leads V_1 to V_4 (arrows) (Reproduced with permission from: Mittal SR. Prominent (Tall) T waves. Cardiology Today-2018;22:98–102)

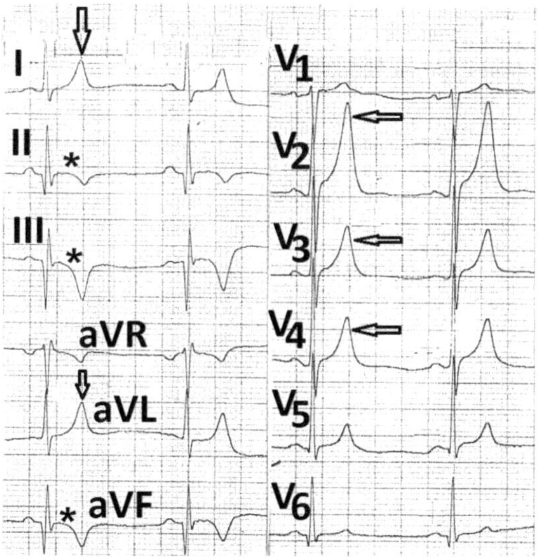

Fig. 43.6 Electrocardiogram shows recovering infero-posterior infarction (marked *) with prominent positive T waves in leads I and aVL (marked ⇩) and tall T waves in leads V_2 to V_4 (arrows) (Reproduced with permission from: Mittal SR. Prominent (Tall) T waves. Cardiology Today-2018;22:98–102)

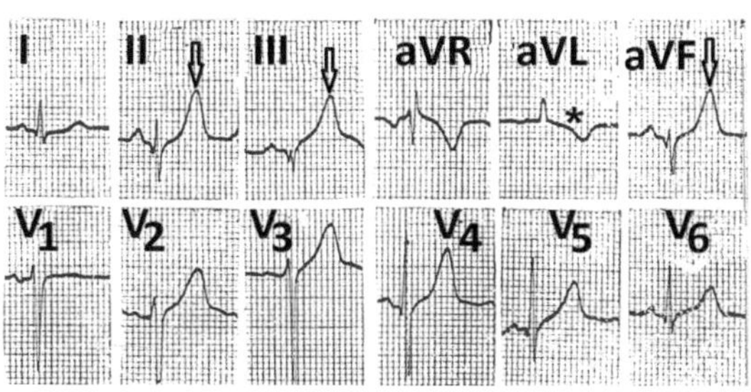

Fig. 43.7 Electrocardiogram shows high lateral infarction (lead aVL-marked *) with prominent T waves in leads II, III, aVF (arrows) (Reproduced with permission from: Mittal SR. Prominent (Tall) T waves. Cardiology Today-2018;22:98–102)

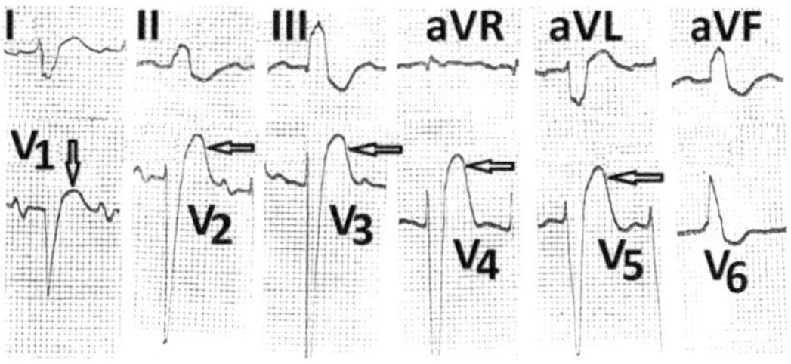

Fig. 43.8 Electrocardiogram shows left bundle branch block with prominent T waves in leads V_1 to V_5 (arrows) (Reproduced with permission from: Mittal SR. Prominent (Tall) T waves. Cardiology Today-2018;22:98–102)

these leads. Leads V_5 to V_9 show broad and notched R waves with inverted T waves. Representative electrocardiogram is shown in Fig. 43.8.

1.5 Moderate LVH with Chronic Diastolic Overload, for Example, Chronic Severe Aortic Regurgitation

Tall T waves are associated with tall R waves and prominent but narrow q waves in leads V_5 and V_6 (Fig. 43.9). Long-standing LVH may produce prominent T waves in leads V_1, V_2.

1.6 Moderate LVH with Chronic Pressure Overload, for Example, Systemic Hypertension

The T waves are prominent in leads V_1 and V_2. T waves in leads V_6 in usually of low amplitude (Fig. 43.10).

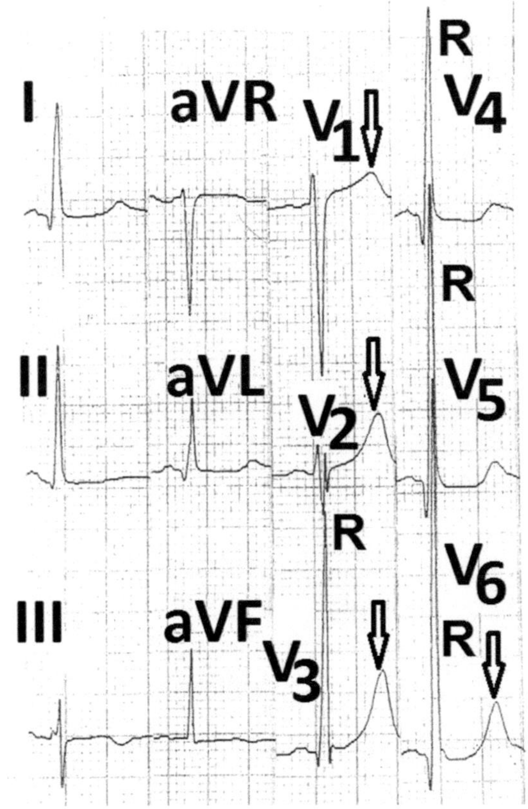

Fig. 43.9 Electrocardiogram from a case of chronic, severe aortic regurgitation showing tall R waves (marked -R) in leads V_3 to V_6 and prominent T waves (arrows) in leads V_1 to V_3 and V_6 (Reproduced with permission from: Mittal SR. Prominent (Tall) T waves. Cardiology Today-2018; 22:98–102)

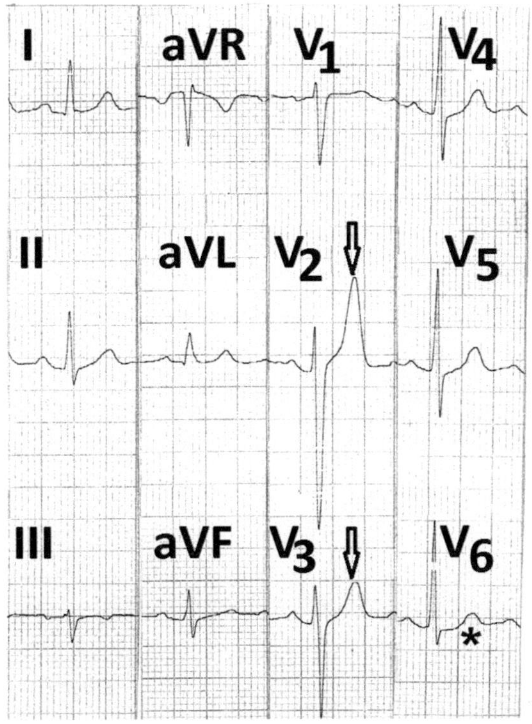

Fig. 43.10 Electrocardiogram from a case of chronic systemic hypertension showing prominent T waves in leads V_2 and V_3 (arrows). T wave of lead V_6 is of low amplitude (marked *) (Reproduced with permission from: Mittal SR. Prominent (Tall) T waves. Cardiology Today-2018;22:98–102)

1.7 Acute Pericarditis

Prominent T waves are associated with diffuse concave upward elevation of the ST-segment in most of the leads (except leads aVR and V_1). There are no reciprocal ST-segment depressions. PR segment depression in leads II, V_5, and V_6 and elevation in lead aVR support the diagnosis of pericarditis. Representative electrocardiogram is shown in Fig. 43.11.

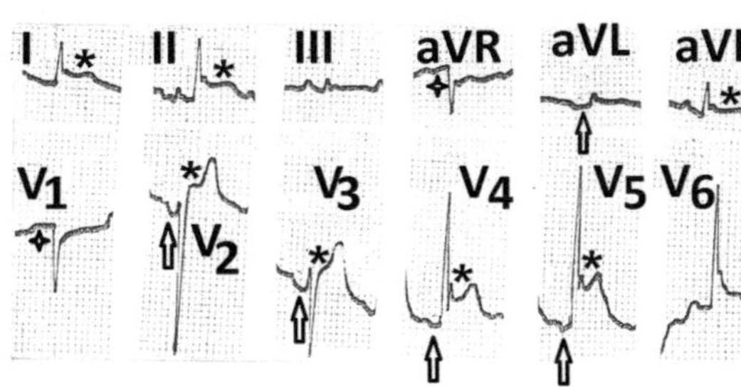

Fig. 43.11 Electrocardiogram from a case of acute pericarditis showing diffuse ST-segment elevation (marked *). The PR segment is depressed in leads aVL and V_2–V_5 (marked as arrows) and is elevated in leads aVR and V_1 (marked ✧) (Reproduced with permission from: Mittal SR. Prominent (Tall) T waves. Cardiology Today-2018;22:98–102)

1.8 Post Stokes–Adams' Attack

T waves are prominent and broad with prolongation of the QT interval. Usually there is high degree atrioventricular block.

1.9 Non-cardiac Causes

1.9.1 Cerebrovascular Accidents

There is widening of the T waves with prolongation of the QT interval (Fig. 43.12). U wave is

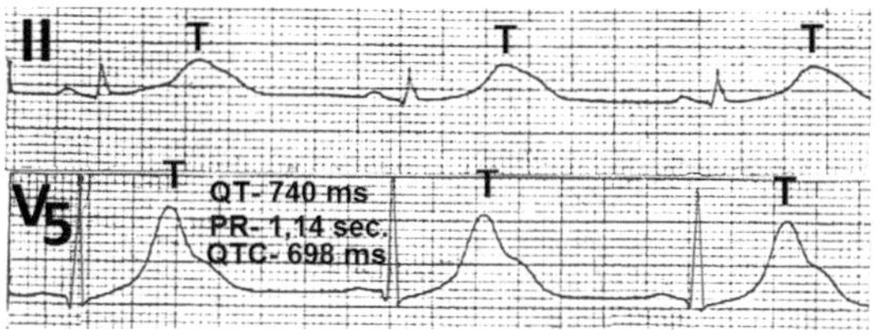

Fig. 43.12 Electrocardiogram shows sinus bradycardia with prominent and broad T waves and prolonged QT interval (Reproduced with permission from: Mittal SR. Prominent (Tall) T waves. Cardiology Today-2018;22: 98–102)

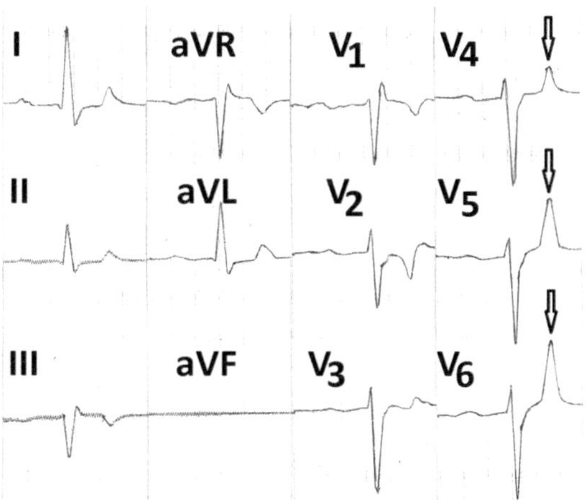

Fig. 43.13 Electrocardiogram from a case of hyperkalemia showing tall tented T waves in leads V_4 to V_6 (arrows) (Reproduced with permission from: Mittal SR. Prominent (Tall) T waves. Cardiology Today-2018;22:98–102)

usually increased in amplitude. Usually there is bradycardia. Representative electrocardiogram is shown in Fig. 43.12.

1.9.2 Hyperkalemia

Earliest change is narrowing and peaking (tenting) of the T waves (Figs. 43.13 and 43.14).

Progressive hyperkalemia results in widening of the QRS complex and decrease in amplitude of the P wave. PR interval may be prolonged (Fig. 43.14). Subsequently, there is complete loss

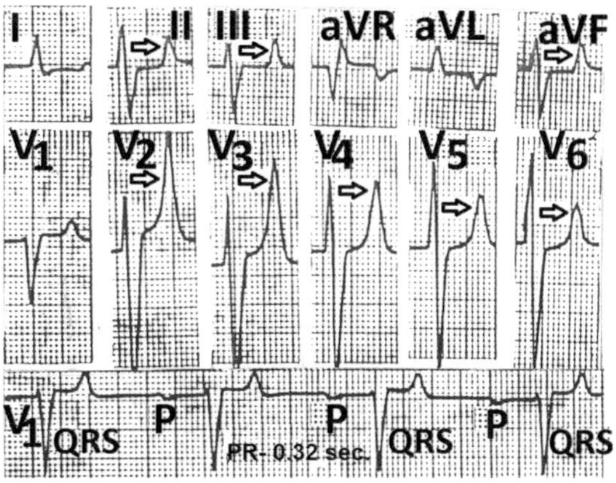

Fig. 43.14 Electrocardiogram from a case of hyperkalemia showing tall, tented T waves in most of the leads (arrows). Rhythm strip of lead V_1 shows prolonged PR interval (Reproduced with permission from: Mittal SR. Prominent (Tall) T waves. Cardiology Today-2018;22:98–102)

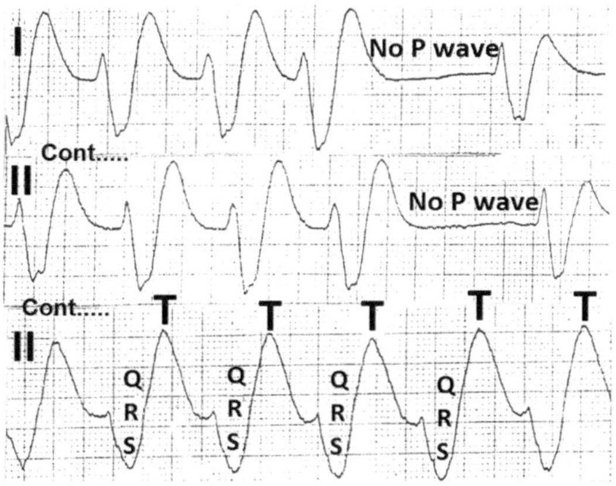

Fig. 43.15 Electrocardiogram (rhythm strip of lead II) from a case of hyperkalemia showing disappearance of the P wave and broad QRS complexes fused with the T waves (Reproduced with permission from: Mittal SR. Prominent (Tall) T waves. Cardiology Today-2018;22:98–102)

of the P wave with further broadening and merging of the QRS complex and the T waves (Fig. 43.15).

Sometimes advanced atrioventricular block may appear (Figs. 43.16 and 43.17).

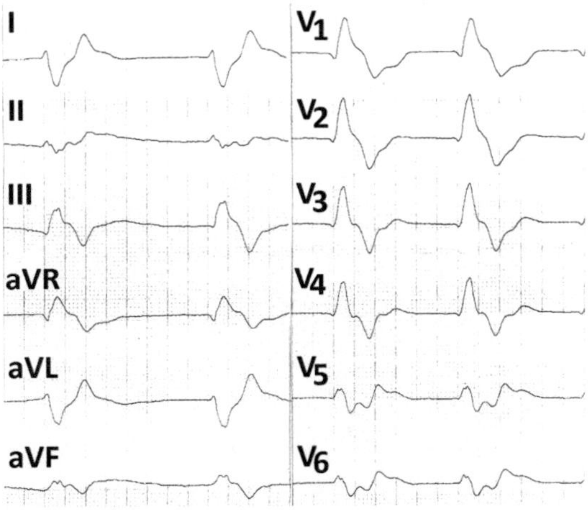

Fig. 43.16 Electrocardiogram from a case of hyperkalemia showing absence of P wave, broad QRS complexes fusing with the T waves and bradycardia (Reproduced with permission from: Mittal SR. Prominent (Tall) T waves. Cardiology Today-2018;22:98–102)

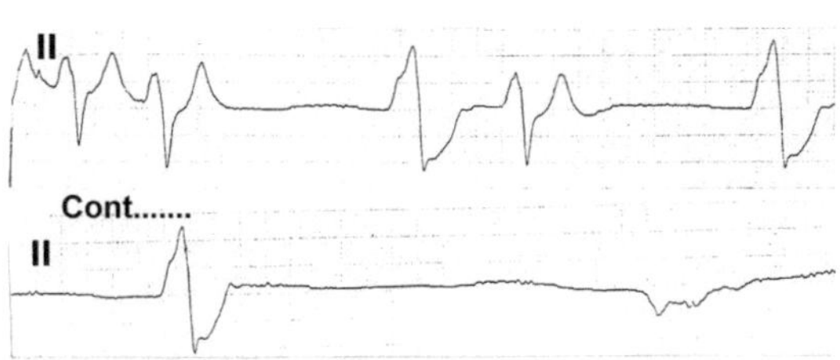

Fig. 43.17 Electrocardiogram from a case of hyperkalemia showing high degree atrioventricular block (Reproduced with permission from: Mittal SR. Prominent (Tall) T waves. Cardiology Today-2018;22:98–102)

Summary

- T waves may be relatively tall as a normal variant and in persons with vagotonia as in athletes.
- Peaked, symmetrical, wider, and taller T wave in leads V_1 to V_4 are present in acute transmural ischemia in left anterior descending coronary artery territory or as a mirror image of the chronic phase of the inferoposterior Q wave myocardial infarction.
- Left ventricular pressure overload and complete left bundle branch block produce tall T waves in leads V_1 and V_2 as reciprocal change of T wave inversions in leads V_5, V_6.
- Peaked, symmetrical, tall T waves with narrow base are seen in hyperkalemia.
- Prominent and broad T waves with a prolonged QTc interval are seen in sub-arachnoid hemorrhage or following Stokes–Adams' syncopal attacks in high-grade atrioventricular block. Such changes are also seen following truncal vagotomy, radical neck dissection, or bilateral carotid endarterectomy.
- Left ventricular diastolic overload (e.g., chronic moderately severe aortic or mitral regurgitation) produce mildly prominent T waves in leads V_5 and V_6.
- In hypercalcemia, proximal limb of T wave acutely slopes to its peak. There is shortening of the ST-segment. ST-segment may not appear.
- Acute pericarditis produces diffuse, mildly prominent T waves with mild elevation of the ST-segment, and mild depression of the PR segment.

MCQs

Q1. T wave in limb leads is considered tall if the amplitude is more than:

(a) 0.5 mV
(b) 1.0 mV
(c) 1.5 mV
(d) Amplitude of the accompanying R wave

Q2. T wave in precordial leads is considered tall if amplitude is more than:

(a) 1.0 mV
(b) 1.5 mV
(c) 2.0 mV
(d) Amplitude of the accompanying R wave

Q3. Normally the T wave can be taller than the r wave in leads:

(a) II, III, aVF
(b) V_1, V_2, V_3
(c) V_5, V_6
(d) None

Q4. Prominent T waves in lead V_5, V_6 can be seen in:

(a) Vagotonia
(b) Anxiety
(c) Hyperventilation
(d) Hypoxia

Q5. In early repolarization syndrome:

(a) J point is elevated
(b) ST-segment is depressed
(c) T wave is prominent
(d) QT interval is prolonged

Q6. Prominent T waves are seen in:

(a) Subendocardial ischemia
(b) Subendocardial infarction
(c) Hyperacute phase of myocardial infarction
(d) Healed myocardial infarction

Q7. Prominent T waves in leads V_1 and V_2 are seen in:

(a) LBBB
(b) True posterior infarction
(c) Anterior subendocardial infarction
(d) Right ventricular infarction

Q8. Prominent T waves in leads I and aVL are seen in:

(a) Recovering inferior infarction
(b) Recovering high lateral infarction
(c) Posterolateral infarction
(d) Acute anterior infarction

Q9. Healing high lateral infarction can produce prominent T waves in:

(a) Leads I and aVL
(b) Leads III and aVF
(c) Leads V_5 and V_6
(d) Leads V_1 to V_4

Q10. Moderate left ventricular hypertrophy due to systemic hypertension can produce prominent T waves in:

(a) Leads V_1, V_2
(b) Leads V_5, V_6
(c) Leads V_7 to V_9
(d) None

Q11. Prominent T waves in leads V_5 and V_6 can be seen in:

(a) Chronic severe aortic regurgitation
(b) Severe aortic stenosis
(c) Hyperacute lateral infarction
(d) Chronic severe mitral regurgitation

Q12. Acute pericarditis produces:

(a) Diffuse ST-segment elevation
(b) Reciprocal ST-segment depression in leads II, III, aVF
(c) PR segment elevation
(d) All

Q13. Prominent and broad T waves with prolonged QT interval are seen in:

(a) Cerebrovascular accident
(b) Following Stokes–Adams' attack
(c) Acute subendocardial ischemia
(d) All

Q14. Hyperkalemia produces:

(a) Tall and pointed T waves
(b) Broad T waves
(c) Prolongation of the QT interval
(d) All

Q15. Hyperkalemia produces:

(a) Prolongation of the PR interval
(b) Disappearance of the P wave
(c) Atrioventricular block

(d) All

Q16. Hyperkalemia produces:

(a) Widening of the QRS complexes
(b) Fusion of the QRS complexes and the T wave
(c) Narrowing of the QRS complexes
(d) Increase in QT interval

Q17. T waves relatively prominent than normal can be present:

(a) As a normal variant
(b) In vagotonia
(c) In athletes
(d) None

Q18. Tall, symmetrical, and wider T waves in leads V_1 to V_4 can be due to:

(a) Acute transmural ischemia in LAD territory
(b) Reciprocal changes of posterior myocardial infarction
(c) LBBB
(d) All

Q19. Complete LBBB produces tall T waves in leads:

(a) V_1, V_2
(b) V_3, V_4
(c) V_5, V_6
(d) V_7, V_8

Q20. Chronic moderately severe aortic regurgitation produces prominent T wave in leads:

(a) V_1, V_2
(b) V_3, V_4
(c) V_5, V_6
(d) V_7, V_8

Q21. Peaked, symmetrical, narrow, tall T waves with absent P wave, and broad QRS are seen in:

(a) Cerebrovascular accident
(b) Hyperkalemia
(c) Hypercalcemia
(d) Truncal vagotomy

Q22. Prominent T waves with broad base, bradycardia, and prolonged QTc interval are seen:

(a) In subarachnoid hemorrhage
(b) Following Stokes–Adams' attacks
(c) Following unilateral carotid endarterectomy
(d) All

Q23. Radical neck dissection can produce:

(a) Symmetrical, narrow, peaked T waves
(b) Symmetrical, broad, prominent T waves with prolonged QTc interval
(c) Prolonged QTc interval
(d) Symmetrical wider and taller T waves

Q24. Diffuse mildly prominent T waves with mild ST-segment elevation and mild PR segment depression without reciprocal changes are seen in:

(a) Hyperkalemia
(b) Acute pericarditis
(c) Pericardial constriction
(d) Vagotonia

Q25. Electrocardiographic findings in athletes do not include:

(a) Sinus tachycardia
(b) Prolonged PR interval
(c) Tall T waves
(d) Prominent U waves

Q26. Electrocardiogram of an athlete can show:

(a) Prominent respirophasic sinus arrhythmia
(b) Early repolarization pattern
(c) Left ventricular hypertrophy
(d) None

Q27. Electrocardiogram of athlete cannot show:

(a) Sinus bradycardia
(b) First degree atrioventricular block
(c) Mobitz type 1 second degree atrioventricular block
(d) Flat T wave

Q28. Which electrocardiographic findings can be seen in athletes?

(a) Incomplete right bundle branch block
(b) Complete right bundle branch block
(c) Complete LBBB
(d) Mobitz type 2 second degree atrioventricular block

Q29. In left ventricular hypertrophy, leads V_1 and V_2 show:

(a) Deep S wave
(b) ST-segment elevation
(c) Symmetrical tall T waves
(d) Asymmetric tall T waves

Q30. In complete LBBB, leads V_1 and V_2 do not show:

(a) QRS duration of less than 120 ms
(b) Absence of initial r wave
(c) Deep S waves
(d) Asymmetric tall T wave

Q31. In initial stage, hyperkalemia produces:

(a) Narrowing of the T wave
(b) Peaking of the T wave
(c) Prolongation of the QT interval
(d) Absence of the P wave

Q32. Cerebrovascular accident can produce:

(a) Bradycardia
(b) Narrow and peaked P wave
(c) Marked prolongation of the QT interval
(d) Prominent U wave

Q33. Chronic moderate aortic regurgitation does not produce:

(a) Tall T waves in leads V_1, V_2
(b) Prominent T waves in leads V_5, V_6
(c) Mild elevation of ST segment in leads V_5, V_6
(d) Narrow deep q wave in leads V_5, V_6

Q34. Acute pericarditis does not produce:

(a) Marked diffuse elevation of the ST-segment
(b) Elevation of the PR segment
(c) Mildly diffuse prominence of the T wave
(d) U wave inversion

Answers

(1) a, d (2) a, d (3) b (4) a (5) a, c (6) a, c (7) a, b (8) a (9) b (10) a (11) a, c (12) a (13) d (14) a (15) d (16) a, b (17) a, b, c (18) a, b (19) a (20) c (21) b (22) a, b (23) b (24) b (25) a, d (26) a, b, c (27) d (28) a (29) a, b, d (30) a (31) a, b (32) a, c (33) a (34) a, b, d

References

1. Wagner GS, Lim TH. Insufficient blood supply. In: Wagner GS, editor. Marriot's practical electrocardiography. New Delhi: Wolters Kluwer; 2001. p. 163–207.
2. de Luna AB, Goldwasser D, Fiol M, Bayes-Genis A. Surface electrocardiography. In: Fuster V, Walsh RA, Harrington RA, editors. Hurst's the heart. New York: McGraw Hill; 2011. p. 307–70.

44 Inverted T Waves

1 Normal Variants

(a) Persistent juvenile pattern.

Inversion of the T wave in right precordial leads is common in children but in some persons, it may persist in adulthood. It is more common in women.

The T waves are inverted in leads V_1 to V_4 (Fig. 44.1). T wave depth usually decreases from leads V_1 to V_4. Changes may normalize on exercise.

(b) Anxiety

Changes may normalize on exercise.

(c) Hyperventilation

Changes may normalize on exercise.

(d) Orthostatic variation

T waves are inverted on standing from supine position. It usually affects the inferior leads. It is common in asthenic individuals.

(e) Postprandial

In some persons the T waves become inverted after meals.

(f) Pregnancy

(g) Idiopathic global T wave inversions

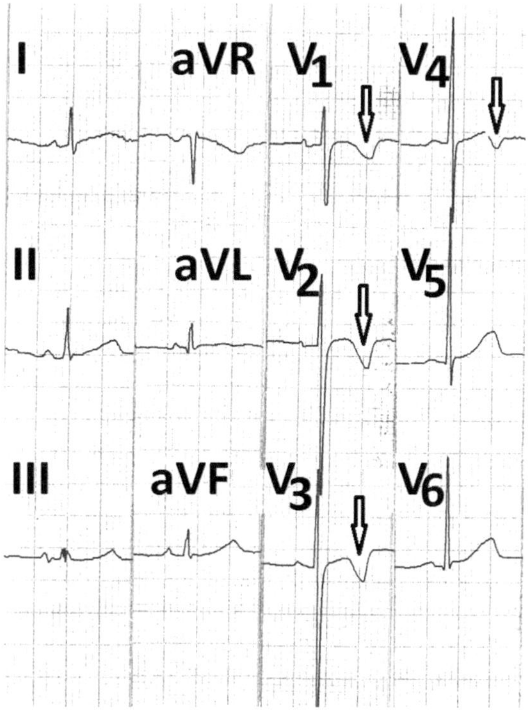

Fig. 44.1 Electrocardiogram from a case of persistent juvenile pattern showing inversion of the T waves (marked ⇩) in leads V_1 to V_4 (Reproduced with permission from: Mittal SR. T wave Inversion. Cardiology Tod ay-2018;22:135–142)

S. Mittal, *Insights into Electrocardiograms with MCQs*,
https://doi.org/10.1007/978-981-99-0127-2_44

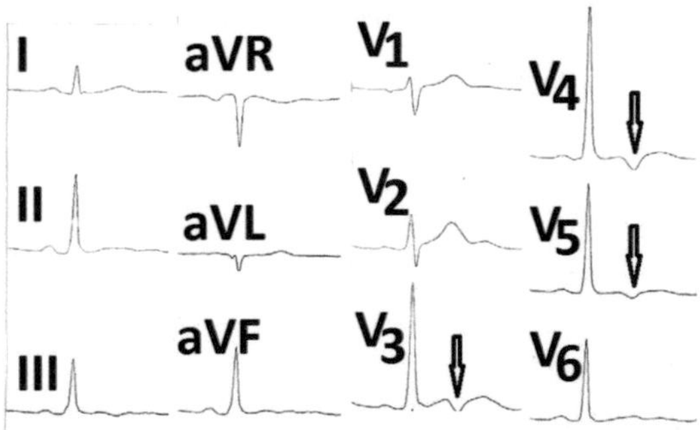

Fig. 44.2 Electrocardiogram from a case of early repolarization variant showing mild ST-segment elevation with shallow T wave inversion (marked ⇩) in leads V_3 to V_5 (Reproduced with permission from: Mittal SR. T wave Inversion. Cardiology Today-2018;22:135–142)

Global T wave inversion means inversion of the T waves in all the standard leads except lead aVR. This may occur in several conditions including myocardial ischemia, central nervous system disorders, myocarditis, apical hypertrophic cardiomyopathy, cardiac tumors, and pheochromocytoma. It is called idiopathic only when no cause is detected even on detailed evaluation. It is common in females [1, 2]. Such T wave changes without any underlying disease per se do not imply a poor prognosis.

(h) Early repolarization variant

Prominent, biphasic T wave inversion with ST-segment elevation is seen in leads V_3 to V_5 [3] (Fig. 44.2). It is usually seen in athletes.

Exercise normalizes such T wave inversions.

2 Noncardiac Causes

(a) Acute abdominal disorders

Pancreatitis (Fig. 44.3), cholecystitis (Fig. 44.4), and peritonitis can produce inversion of the T waves.

(b) Endocrine disorders

(i) Hypothyroidism

There is a generalized diminution of amplitude or shallow inversion of the T waves. Sinus bradycardia and low voltage QRS are other findings.

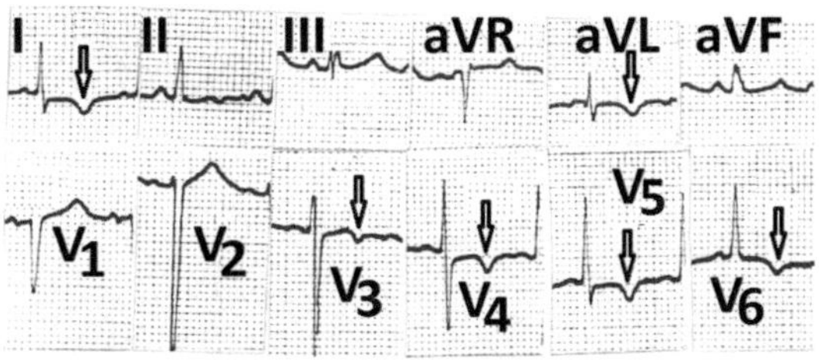

Fig. 44.3 Electrocardiogram from a case of acute pancreatitis showing inversion of the T waves (marked ⇩) in leads I, aVL, V_3 to V_6 (Reproduced with permission from: Mittal SR. T wave Inversion. Cardiology Today-2018;22:135–142)

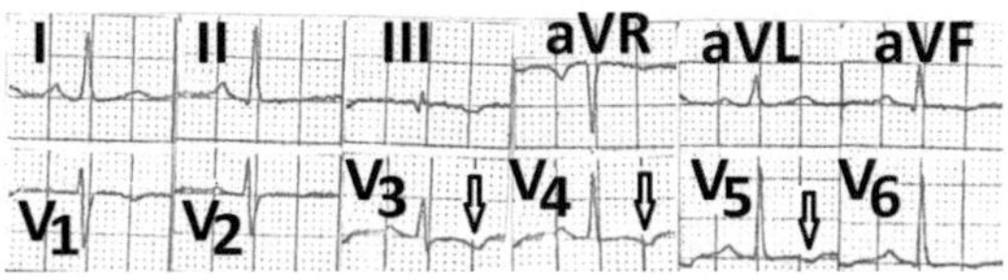

Fig. 44.4 Electrocardiogram from a case of acute cholecystitis showing shallow inversion of the T waves in leads V_3 to V_5 (marked ⇩) (Reproduced with permission from: Mittal SR. T wave Inversion. Cardiology Today-2018;22:135–142)

(ii) Hypoadrenalism

(iii) Hypopituitarism

(iv) Pheochromocytoma

(c) Electrolyte abnormalities

Hypokalemia

Flattening or inversion of the T waves is associated with ST-segment depression, increased prominence of the u wave, and prolongation of the QTc interval (Fig. 44.5).

Concomitant hypomagnesemia also affects the ECG. ECG findings, therefore, may not exactly correlate with the serum levels of potassium.

(d) Acute pulmonary embolism

T waves are inverted in leads III, aVF, and V_1 to V_4. Right ventricular ischemia due to acute increase in afterload is probably responsible. Sinus tachycardia is common right bundle branch block may be present.

(e) Cerebrovascular injury

Stroke [4, 5], intracranial bleed [6], and subarachnoid hemorrhage [7] are associated with electrocardiographic changes even in patients without any history or clinical evidence of primary heart disease. Widening and inversion of the T waves in precordial leads is accompanied by prolongation of the QTc interval and bradycardia. Exact mechanism is not clear. Autonomic neural stimulation from the hypothalamus or elevated levels of catecholamines could be responsible for the ECG changes.

(f) Following truncal vagotomy, radical neck dissection, and bilateral carotid endarterectomy [8].

3 Cardiac Causes

(a) Myocardial infarction

(i) Evolved phase of transmural infarction.

The terminal portion of the T wave is the first to become inverted followed by the middle and initial portions (Fig. 44.6) [9].

T waves may be very deep (Fig. 44.7) but may decrease or even normalize with time [1]. Opposite leads may show prominent T

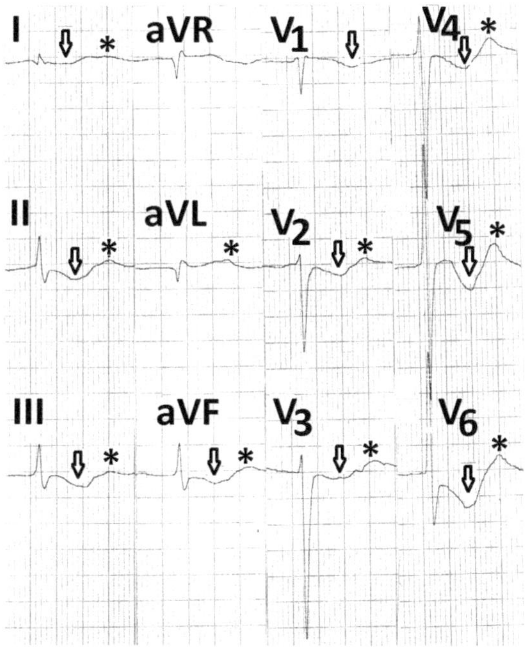

Fig. 44.5 Electrocardiogram from a case of hypokalemia showing inversion of the T waves (marked ⇩), prominent U wave (marked *), and prolonged QTc interval (Reproduced with permission from: Mittal SR. T wave Inversion. Cardiology Today-2018;22:135–142)

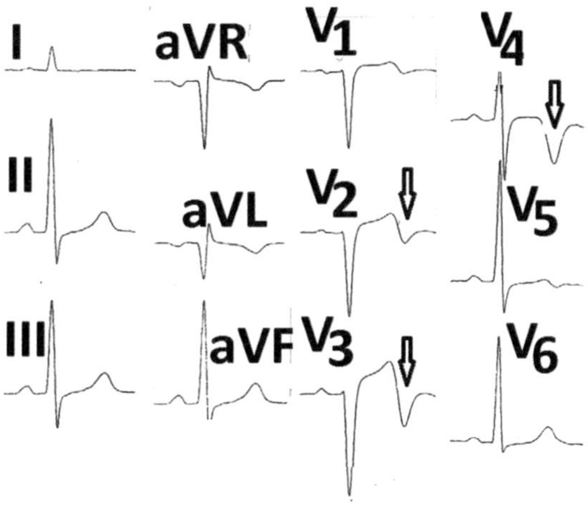

Fig. 44.6 Electrocardiogram from a case of anterior myocardial infarction showing initial inversion of the terminal part of the T wave (marked ⇩) (Reproduced with permission from: Mittal SR. T wave Inversion. Cardiology Today-2018;22:135–142)

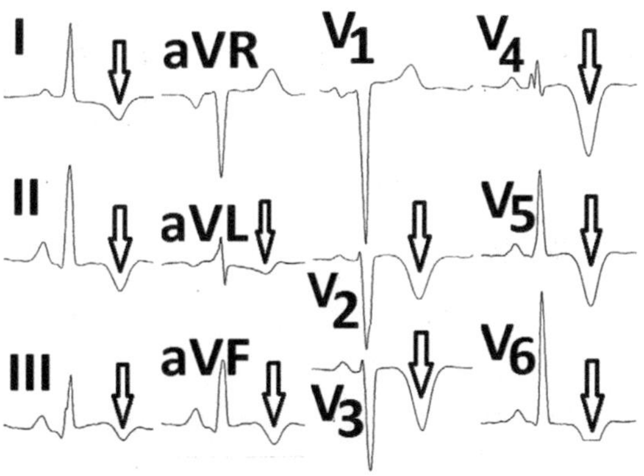

Fig. 44.7 Electrocardiogram from a case of evolved anterior myocardial infarction showing deep inversion of the T waves (marked ⇩) (Reproduced with permission from: Mittal SR. T wave Inversion. Cardiology Today-2018;22:135–142)

wave [1]. Negative T waves in leads II, III, and aVF may produce prominent T waves in leads I and aVL and vice versa. Failure of reperfusion may result in failure of inversion of the T wave.

(ii) Subendocardial infarction (non ST-segment elevation acute coronary syndrome). Inversion of the T waves is usually symmetrical but is usually not deep (Figs. 44.8 and 44.9).

(iii) Wellens' sign

In a case of acute coronary syndrome, biphasic or inverted T waves in leads V_1 to V_4 without Q wave or significant elevation of the ST-segment (Fig. 44.10) suggest tight occlusion of the proximal left anterior descending coronary artery [10, 11]. Precordial R wave progression may be normal.

(iv) T wave inversion in lead aVL (Fig. 44.11). In patients with chronic stable angina, it may predict clinically significant lesion of mid LAD [12, 13].

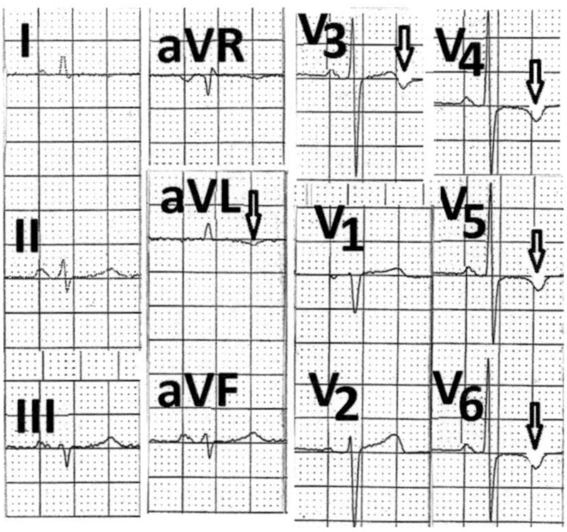

Fig. 44.8 Electrocardiogram from a case of non ST-segment elevation anterior myocardial infarction showing shallow symmetrical T wave inversion in leads aVL, V_3 to V_6 (marked ⇩) (Reproduced with permission from: Mittal SR. T wave Inversion. Cardiology Today-2018;22: 135–142)

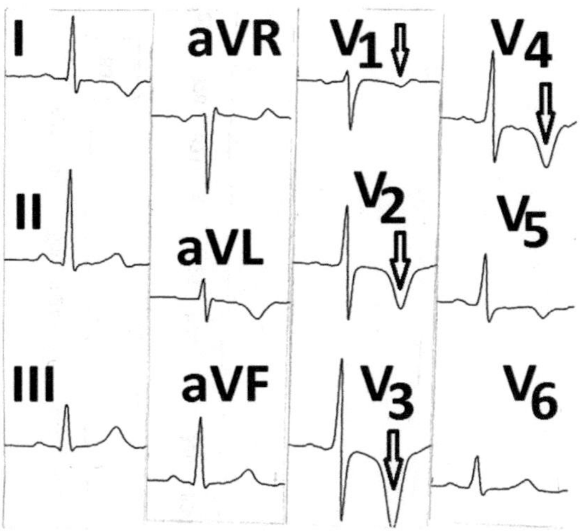

Fig. 44.10 Electrocardiogram shows inverted T waves in leads V_1 to V_4 without Q wave or ST-segment elevation (marked ⇩) (Wellens' sign) (Reproduced with permission from: Mittal SR. T wave Inversion. Cardiology Today-2018;22:135–142)

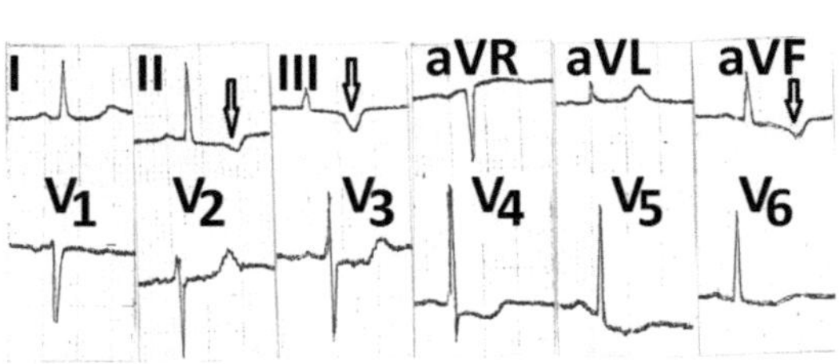

Fig. 44.9 Electrocardiogram from a case of non ST-segment elevation inferior myocardial infarction showing symmetrical inversion of the T waves in leads II, III, and aVF (marked ⇩) (Reproduced with permission from: Mittal SR. T wave Inversion. Cardiology Today-2018;22:135–142)

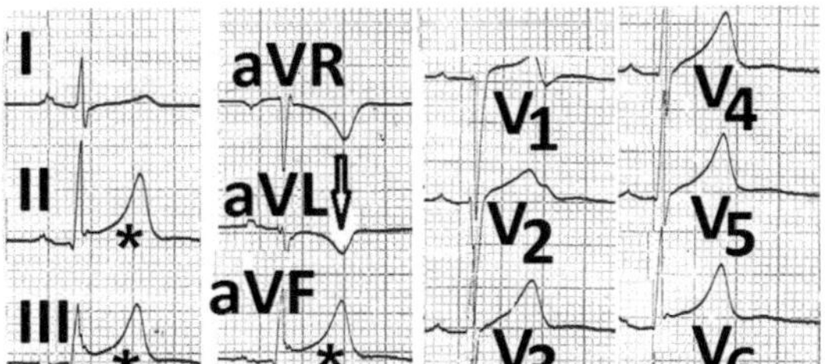

Fig. 44.11 Electrocardiogram showing inversion of the T wave in lead aVL (marked ⇩) with reciprocal prominent T waves in leads II, III, aVF (marked *) (Reproduced with permission from: Mittal SR. T wave Inversion. Cardiology Today-2018;22:135–142)

(b) Mitral valve prolapse

T wave may show shallow inversion in leads II, III, aVF, V_5, and V_6 (Fig. 44.12).

(c) Following a recent Stokes–Adams' attack

The T waves are broad. QTc interval is prolonged (Fig. 44.13). There is high degree atrioventricular block.

(d) Post tachycardia T wave inversion

ST-segment depression may be associated with inversion of the T waves following reversal of supraventricular or ventricular tachycardia (Fig. 44.14). Changes may persist for hours to days.

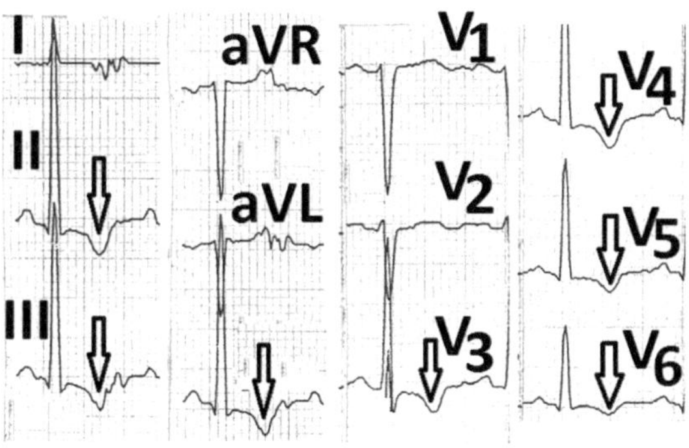

Fig. 44.12 Electrocardiogram from a case of mitral valve prolapse showing inversion of the T waves in leads II, III, aVF, and V_3 to V_6 (marked ⇩) (Reproduced with permission from: Mittal SR. T wave Inversion. Cardiology Today-2018;22:135–142)

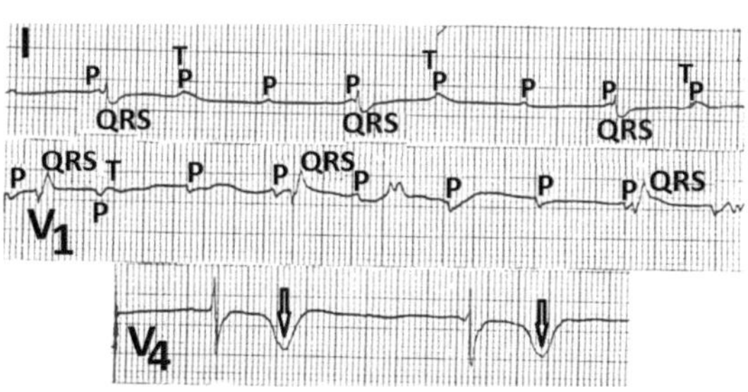

Fig. 44.13 Electrocardiogram from a case of complete atrioventricular block with history of Stokes–Adams' attack showing deep inversion of the T waves in lead V_4 (marked ⇩) (Reproduced with permission from: Mittal SR. T wave Inversion. Cardiology Today-2018;22:135–142)

(e) Rate-dependent T wave inversions

Tachycardia-dependent inversion of the T waves (Fig. 44.15) may be seen in the presence of myocardial ischemia.

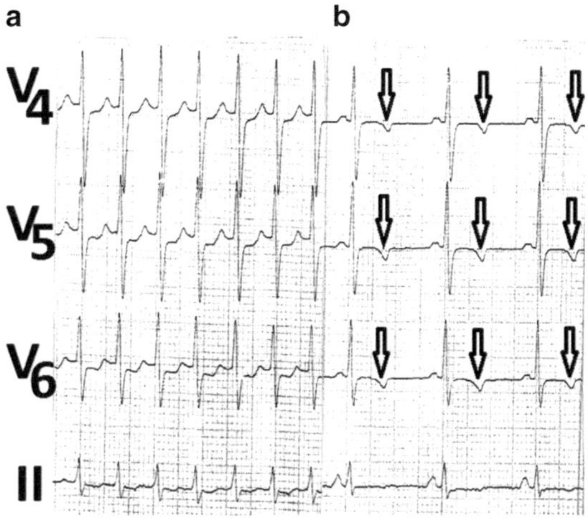

Fig. 44.14 Electrocardiogram (**a**) during supraventricular tachycardia and (**b**) after conversion to sinus rhythm showing post tachycardia T wave inversion (marked ⇩) (Reproduced with permission from: Mittal SR. T wave Inversion. Cardiology Today-2018;22:135–142)

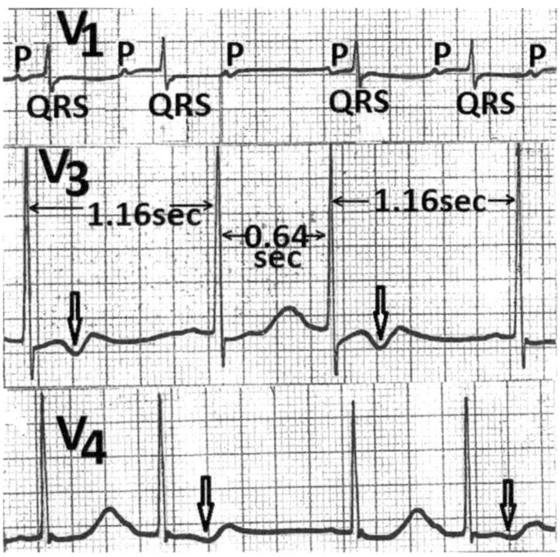

Fig. 44.15 Electrocardiogram from a case of Mobitz type 1 second-degree atrioventricular block showing inversion of the T wave following short R-R cycle in lead V_3 (marked ⇩) (Reproduced with permission from: Mittal SR. T wave Inversion. Cardiology Today-2018;22:135–142)

Bradycardia-dependent inversion of the T waves (Fig. 44.16) may be seen in the setting of prolonged QT syndrome.

(f) Pericardial effusion/constriction

Diffuse low voltage or shallow inversion of the T waves is associated with low voltage QRS complexes (Fig. 44.17).

There are no reciprocal changes. Depth of inversion of the T wave correlates with the degree of pericardial adherence to the myocardium and underlying myocardial damage.

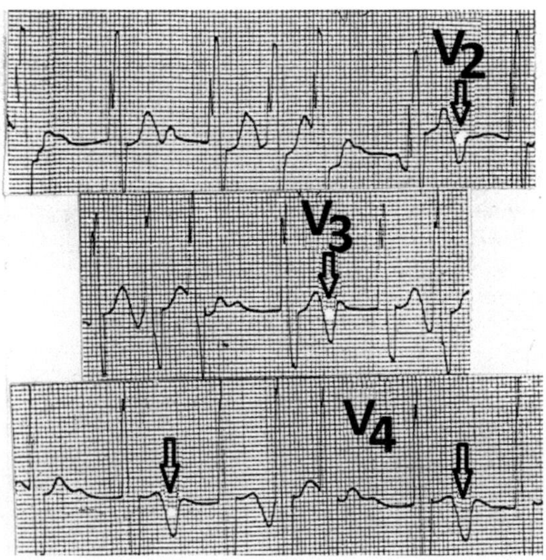

Fig. 44.16 Electrocardiogram from a case of atrial fibrillation with right bundle branch block showing inversion of the T waves in cycles following longer pause (marked ⇩) (Reproduced with permission from: Mittal SR. T wave Inversion. Cardiology Today-2018;22:135–142)

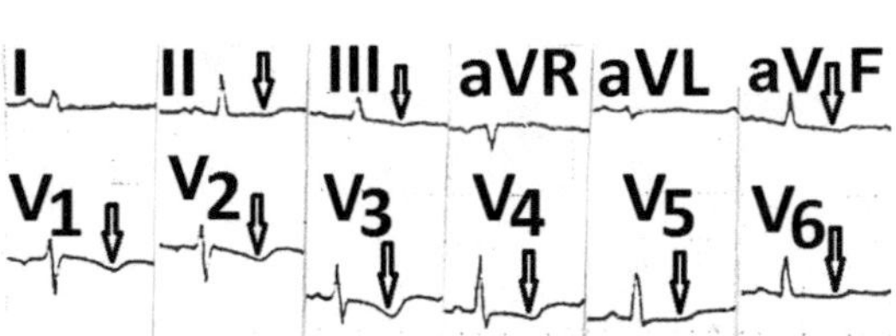

Fig. 44.17 Electrocardiogram from a case of pericardial constriction showing low voltage of QRS and diffuse shallow inversion of the T waves (marked ⇩) (Reproduced with permission from: Mittal SR. T wave Inversion. Cardiology Today-2018;22:135–142)

(g) Hypertrophic cardiomyopathy

Tall QRS complexes are accompanied by diffuse and deep inversion of the T waves [14] (Figs. 44.18 and 44.19).

Normal QRS with diffuse and unexplained inversion of the T waves should also raise suspicion of hypertrophic cardiomyopathy.

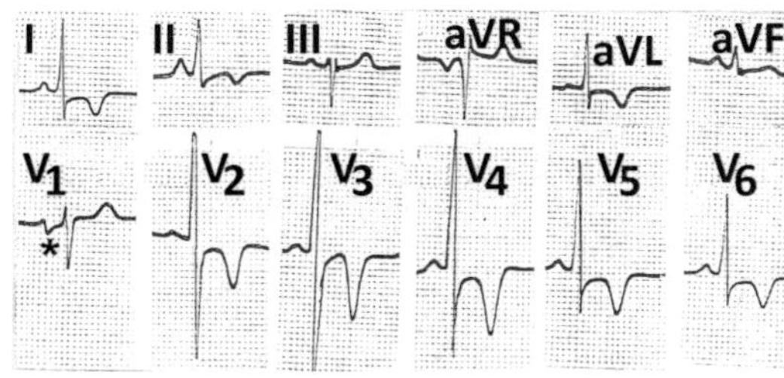

Fig. 44.18 Electrocardiogram from a case of hypertrophic cardiomyopathy showing tall QRS complexes with diffuse inversion of the T wave. Negative P wave in lead V_1 (marked *) suggests concomitant left atrial overload (Reproduced with permission from Mittal SR. T wave Inversion. Cardiology Today-2018;22:135–142)

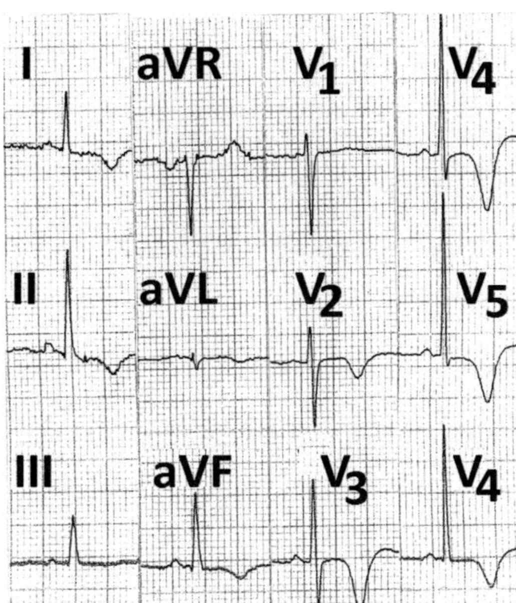

Fig. 44.19 Electrocardiogram from another case of hypertrophic cardiomyopathy showing tall QRS complexes with diffuse and deep inversion of the T waves (Reproduced with permission from: Mittal SR. T wave Inversion. Cardiology Today-2018;22:135–142)

(h) Other cardiomyopathies and myocarditis

Depression of the ST-segments and inversion of the T waves are common in leads V_4 to V_6. Abnormal Q waves, slow progression, or non-progression of the r wave in precordial leads and deep S wave in leads V_2 and V_3 can be other findings (Fig. 44.20).

(i) Arrhythmogenic right ventricular cardiomyopathy

Inversion of the T waves is common in leads V_1 and V_2.

(j) Myocardial tumors

(k) Secondary T wave abnormalities

Intraventricular conduction defects like fascicular blocks, bundle branch block (Fig. 44.21), preexcitation, ventricular ectopics, paced beats, and ventricular hypertrophy (Fig. 44.22) are associated with down sloping depression of the ST-segments with inversion of the T waves in leads showing positive QRS complex.

These changes are secondary to depolarization abnormalities. In the context of left ventricular hypertrophy, these changes suggest increased cardiovascular risk [15, 16]. The T wave is directed away from the latter portion of the QRS complex. In these conditions, if the direction of the T wave is similar to the terminal part of the QRS, it is abnormal and suggests myocardial disease. If the intraventricular conduction defect is intermittent, the T wave changes may persist for hours or days after normalization of ventricular depolarization. Such changes have been termed as "memory T wave" changes [8, 17, 18].

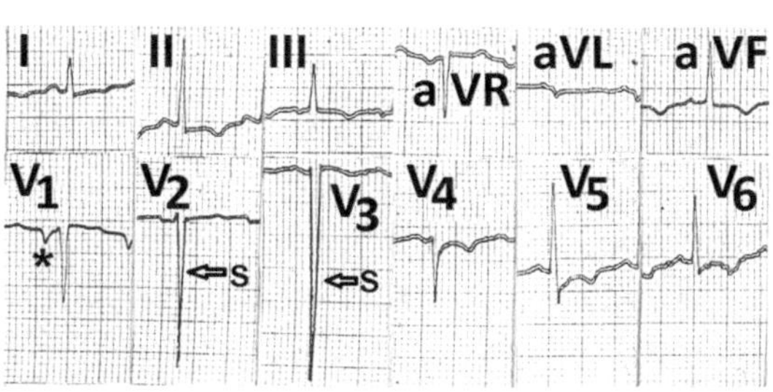

Fig. 44.20 Electrocardiogram from a case of congestive cardiomyopathy showing diffuse inversion of the T wave, nonprogression of the r wave in leads V_1 to V_4, and deep S waves in leads V_2 and V_3 (marked ⇦S). Negative P wave in lead V_1 (marked *) suggests left atrial overload (Reproduced with permission from: Mittal SR. T wave Inversion. Cardiology Today-2018;22:135–142)

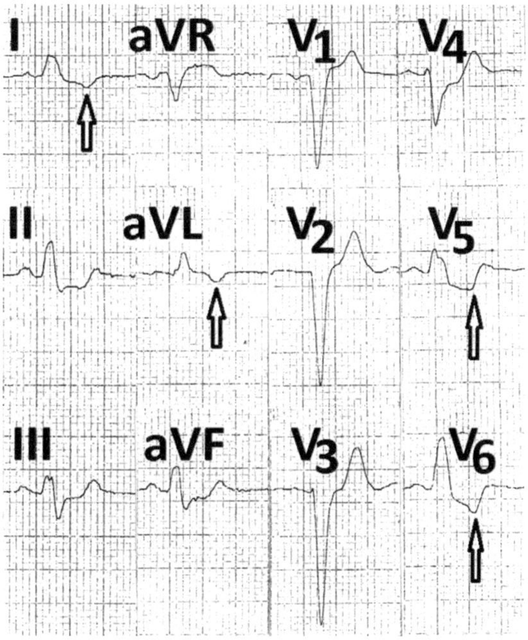

Fig. 44.21 Electrocardiogram from a case of left bundle branch block showing secondary inversion of the T waves in leads I, aVL, V_5, and V_6 (marked ⇧) (Reproduced with permission from: Mittal SR. T wave Inversion. Cardiology Today-2018;22:135–142)

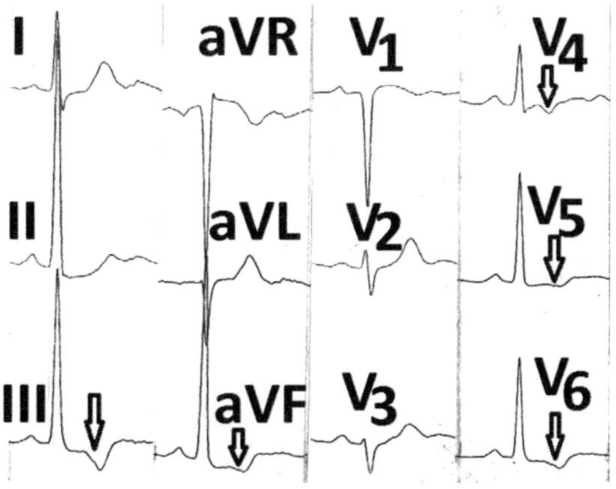

Fig. 44.22 Electrocardiogram from a case of left ventricular hypertrophy showing secondary inversion of the T waves in leads III, aVF, V_4 to V_6 (marked ⇩) (Reproduced with permission from: Mittal SR. T wave Inversion. Cardiology Today-2018;22:135–142)

Summary

- Diffuse and shallow inversion of the T waves are seen in myxedema, pericardial effusion, and pericardial constriction or as normal variant.
- Inversion of the T waves in leads V_1, V_2 (sometimes up to V_4) can be seen in children or even in adults as "juvenile pattern." "Juvenile pattern" is more common in females and at times may be diffuse.
- Right bundle branch block, right ventricular hypertrophy, cor pulmonale, and acute pulmonary embolism produce prominent but asymmetric inversion of the T waves in leads V_1, V_2 (at times up to lead V_4).
- Deep and symmetrically inverted T waves suggest resolution phase of myocardial infarction. At times, inversion of the T wave may not be deep but is symmetrical. Deep and symmetrical inverted T wave in leads V_1 to V_4 in context of acute coronary syndrome suggests high grade stenosis of proximal left anterior descending coronary artery (Wellens' pattern).
- Asymmetrical (not deep) inversion of the T waves in leads V_5 and V_6 can occur in hyperventilation, myocarditis, dilated cardiomyopathy, left ventricular pressure overload, left bundle branch block, pacing from right ventricular apex, or following episode of tachyarrhythmia (post tachycardia). Transient alteration of ventricular depolarization associated with QRS prolongation (e.g., in preexcitation or aberrant ventricular conduction) can also produce T wave changes that can persist for hours or days after normalization of the ventricular depolarization (cardiac memory T waves).
- Deep and symmetrical inversion of the T waves with tall R waves in leads V_3 to V_4 suggest apical hypertrophic cardiomyopathy.
- Mitral valve prolapse can produce shallow T wave inversion in leads II, III, aVF, V_5, and V_6.
- Left anterosuperior fascicular block produces T wave inversion in leads I and aVL.
- Left posteroinferior fascicular block produces T wave inversion in leads II, III, and aVF.
- Preexcitation can produce inversion of the T waves in precordial leads V_2 to V_6.
- Cerebrovascular accident (especially subarachnoid hemorrhage or hemorrhagic stroke) can produce diffuse, deep, and wide T wave inversions.

MCQs

Q1. In children T waves may be normally inverted in leads:

(a) I, aVL
(b) V_5, V_6
(c) V_1 to V_4
(d) V_7 to V_9

Q2. Transient nonischemic T wave inversion can occur during:

(a) Anxiety
(b) Hyperventilation
(c) Standing from supine position
(d) All

Q3. Nonischemic T wave inversion can be seen in:

(a) Postprandial period
(b) Pregnancy
(c) Exercise
(d) Fasting

Q4. Global T wave inversion can be seen in:

(a) Cerebrovascular accidents
(b) Cardiac tumors
(c) Pheochromocytoma
(d) None

Q5. Global T wave inversion is not seen in:

(a) Myocardial ischemia
(b) Myocarditis
(c) Dilated cardiomyopathy
(d) Left bundle branch block

Q6. Early repolarization variant produces changes in leads:

(a) II, III, aVF
(b) V_5 and V_6
(c) V_2 to V_4
(d) V_7 to V_9

Q7. Early repolarization variant produces:

(a) ST-segment elevation
(b) Biphasic T waves
(c) Deep symmetrical inversion of the T waves
(d) Q waves

Q8. Exercise can normalize inversion of the T waves produced by:

(a) Anxiety
(b) Hyperventilation
(c) Early repolarization variant
(d) Myocardial ischemia

Q9. Which of the following conditions can produce inversion of the T waves?

(a) Acute pancreatitis
(b) Cholelithiasis
(c) Renal colic
(d) Nephrotic syndrome

Q10. Which of the following conditions can produce inversion of the T waves?

(a) Acute cholecystitis
(b) Acute peritonitis
(c) Hyperthyroidism
(d) Acute gastritis

Q11. Low voltage QRS, sinus bradycardia, and generalized diminution of the amplitude of the T waves suggest:

(a) Cardiac tamponade
(b) Cor pulmonale
(c) Myxedema
(d) All

Q12. T wave inversion with prolongation of the QTc interval is seen in:

(a) Hypokalemia
(b) Cerebrovascular injury
(c) Recent Stokes–Adams' attack
(d) All

Q13. T wave inversion is seen in:

(a) Hyperkalemia
(b) Hypomagnesemia
(c) Hypocalcemia
(d) Hyponatremia

Q14. Hypokalemia produces:

(a) Flattening of the T wave
(b) Depression of the ST-segment
(c) Prominence of the U wave
(d) Shortening of the QTc interval

Q15. Acute pulmonary embolism produces T wave inversion in leads:

(a) III, aVF
(b) V_1 to V_4
(c) I and aVL
(d) V_5 and V_6

Q16. T wave inversion can be seen in:

(a) Stroke
(b) Intracranial bleeds
(c) Subarachnoid hemorrhage
(d) Subdural hematoma

Q17. Hypothalamic bleed can produce:

(a) Widening of the T wave
(b) Prolongation of the QTc interval
(c) Tachycardia
(d) All

Q18. Which of the following can produce inversion of the T waves?

(a) Truncal vagotomy
(b) Radical neck dissection
(c) Bilateral carotid endarterectomy
(d) None

Q19. In ST-segment elevation myocardial infarction, which portion of the T wave is the first to become inverted?

(a) Initial
(b) Middle
(c) Terminal
(d) Any of the above

Q20. In a case of acute coronary syndrome inverted T waves in leads V_1 to V_4 suggest tight occlusion of proximal:

(a) LAD
(b) LCX
(c) RCA
(d) Left main coronary artery

Q21. In a case of ST-segment elevation myocardial infarction, T wave inversion suggests:

(a) Successful reperfusion
(b) Failure of reperfusion
(c) Ongoing ischemia
(d) Infarction in opposite leads

Q22. In patients with chronic stable angina, T inversion in lead aVL suggests occlusion of:

(a) Proximal LAD
(b) Mid LAD
(c) Proximal D1
(d) Proximal major obtuse marginal

Q23. Mitral valve prolapse can produce T wave inversion in leads:

(a) I and aVL
(b) II, III, and aVF
(c) V_1 to V_3
(d) V_5, V_6

Q24. Broad and inverted T waves, prolongation of the QTc interval, and high degree atrioventricular block suggest:

(a) Severe hypokalemia
(b) Hypothalamic bleed
(c) Recent Stokes–Adams' attack
(d) Severe hypothyroidism

Q25. Post tachycardia T wave inversion suggests:

(a) Nonspecific change
(b) Myocardial ischemia
(c) Myocardial infarction
(d) Electrolyte imbalance

Q26. Combination of sinus tachycardia, low voltage QRS, and diffuse shallow inversion of the T waves should suggest:

(a) Gross pericardial effusion
(b) Pericardial constriction
(c) Myxedema
(d) Cor pulmonale

Q27. Combination of tall QRS, horizontal ST-segment, diffuse and deep inversion of the T waves, absence of Q wave, and absence of any sequential change on repeated ECG should suggest:

(a) Hypertrophic cardiomyopathy
(b) Myocardial infarction
(c) Myocarditis
(d) Cardiomyopathy

Q28. Arrhythmogenic right ventricular dysplasia produces inversion of the T waves in leads:

(a) V_1, V_2
(b) V_5, V_6
(c) I, aVL
(d) II, III, aVF

Q29. Secondary T wave abnormalities are seen in:

(a) Fascicular blocks
(b) Preexcitation
(c) Paced beats
(d) None

Q30. In secondary T wave abnormalities, the T waves are directed:

(a) Away from latter portion of the QRS
(b) Toward the latter portion of the QRS
(c) Away from the initial portion of the QRS
(d) Toward the initial portion of the QRS

Q31. Secondary T waves directed toward the latter portion of the QRS complex suggest:

(a) Nonspecific change
(b) Myocardial damage
(c) Early repolarization
(d) Electrolyte imbalance

Q32. Persistent inversion of the T waves after recovery from transient paced rhythm suggests:

(a) Myocardial ischemia
(b) Myocardial infarction
(c) Memory T wave
(d) Hypokalemia

Q33. Diffuse and shallow inversion of the T waves can be present in:

(a) Myxedema
(b) Pericardial constriction
(c) As a normal variant
(d) None

Q34. Prominent but asymmetrical inversion of the T waves in leads V_1 to V_4 can occur in:

(a) Normal children
(b) Right bundle branch block
(c) Right ventricular hypertrophy
(d) All

Q35. Deep and symmetrical inversion of the T waves in leads V_1 to V_4 suggest:

(a) Resolution phase of anterior myocardial infarction
(b) Acute pulmonary embolism
(c) Chronic cor pulmonale
(d) All

Q36. Juvenile pattern:

(a) Includes prominent inversion of the T waves in leads V_1 to V_2
(b) Is common in females
(c) T wave inversion increases on exercise
(d) All

Q37. Wellens' pattern suggests critical stenosis of:

(a) LAD
(b) LCX
(c) RCA
(d) PDA

Q38. Prominent and asymmetric inversion of the T waves in leads V_1, V_2 occur in:

(a) RBBB
(b) Chronic cor pulmonale
(c) Acute pulmonary embolism
(d) All

Q39. Asymmetric (not deep) inversion of the T waves in leads V_5, V_6 can occur in:

(a) Hyperventilation
(b) Myocarditis
(c) Dilated cardiomyopathy
(d) Apical hypertrophic cardiomyopathy

Q40. LBBB does not produce asymmetric inversion of the T waves in leads:

(a) I
(b) aVL
(c) V_1, V_2
(d) V_5, V_6

Q41. Post tachycardia T wave inversions:

(a) Are commonly seen in leads V_1, V2
(b) Are asymmetric
(c) Last for hours to days after control of tachyarrhythmia
(d) All

Q42. Asymmetric inversion of the T waves in leads V_5, V_6 is seen in:

(a) Left ventricular pressure overload
(b) Pacing from the right ventricular apex
(c) Right ventricular volume overload
(d) All

Q43. T wave changes persisting after recovery from transient alteration in ventricular depolarization is known as:

(a) Wellens' pattern
(b) Cardiac memory T waves
(c) Early repolarization variant
(d) None

Q44. Deep and symmetrical inversion of the T waves with horizontal ST-segment and tall R waves in leads V_3, V_4 suggest:

(a) Severe aortic valve stenosis
(b) Severe systemic hypertension
(c) Severe chronic mitral regurgitation
(d) Apical hypertrophic cardiomyopathy

Q45. Mitral valve prolapse can produce deep and symmetrical inversion of the T waves in:

(a) Leads I, aVL
(b) Leads II, III, aVF
(c) Leads V_5, V_6
(d) None

Q46. Left anterosuperior fascicular block produces T wave inversion in leads:

(a) Leads I, aVL
(b) Leads II, III, aVF
(c) Leads V_1, V_2
(d) Leads V_5, V_6

Q47. Deep and symmetrical inversion of the T waves is classical of:

(a) Left posteroinferior fascicular block
(b) Left bundle branch block
(c) Apical hypertrophic cardiomyopathy
(d) Resolution phase of myocardial infarction

Q48. Asymmetric inversion of the T waves in leads V_1 to V_6 are seen in:

(a) Right lateral accessory pathway
(b) Left lateral accessory pathway
(c) Right posterior paraseptal accessory pathway
(d) Left posterior paraseptal accessory pathway

Q49. Right lateral accessory pathway produces T wave inversion in leads:

(a) Leads I, aVL
(b) Leads II, III, aVF
(c) Leads V_1 to V_4
(d) Leads V_5, V_6

Q50. T wave inversion in subarachnoid hemorrhage is:

(a) Diffuse
(b) Broad
(c) Deep
(d) Peaked

Q51. Bradycardia with diffuse, broad, symmetrical, deep inversion of the T waves, and prolonged QTc interval are seen:

(a) After Stokes–Adams' attack
(b) In hemorrhagic stroke
(c) In myxedema
(d) In athletes

Q52. Deep inversion of the T waves in leads V_1 and V_2 in the presence of complete LBBB suggests:

(a) Normal pattern
(b) Myocardial ischemia
(c) Right ventricular infarction
(d) LVH

Q53. Early repolarization variant produces:

(a) Mild ST-segment elevation
(b) Prominent inversion of the T waves in leads V_1 to V_4
(c) Biphasic T wave (initial positive, terminal negative)
(d) Biphasic T wave (initial negative, terminal positive)

Q54. Global inversion of the T waves:

(a) Suggests triple vessel disease
(b) Can be due to myocarditis
(c) Can be idiopathic
(d) Any one of above

Q55. Shallow inversion of the T waves can occur:

(a) On sudden standing
(b) After meals
(c) After hyperventilation
(d) All

Q56. Normally also, T wave can be inverted in:

(a) Leads V_1 to V_3
(b) Leads III, aVF
(c) Leads I, aVL
(d) All

Q57. Idiopathic global inversion of the T waves:

(a) Is frequent in females
(b) Normalize during exercise
(c) Increase during exercise
(d) Become broad during exercise

Answers

(1) c (2) d (3) a, b (4) a, b, c (5) a, d (6) c (7) a, b (8) a, b, c (9) a (10) a, b (11) c (12) d (13) b (14) a, b, c (15) a, b (16) a, b, c (17) a, b (18) a, b, c (19) c (20) a (21) a (22) b (23) b, d (24) c (25) a (26) a, b (27) a (28) a (29) a, b, c (30) a (31) b (32) c (33) a, b, c (34) d (35) a (36) a, b (37) a (38) d (39) a, b, c (40) c (41) c (42) a, b (43) b (44) d (45) d (46) a (47) c, d (48) b, d (49) c (50) a, b, c (51) a, b (52) b (53) a, c (54) d (55) d (56) a, b (57) a, b

References

1. De Luna AB, Goldwasser D, Fiol M, Bayes Genis A. Surface electrocardiography. In: Fuster V, Walsh RA, Harrington RA, editors. Hurst's the heart. New York: McGraw Hill; 2011. p. 307–70.
2. Walder LA, Spodick DH, et al. Global T wave inversion. Long term follow up. J Am Coll Cardiol. 1993;21:1652.
3. Uberoi A, Jain N, Froelicher V, et al. Early repolarization in an ambulatory clinical population. Circulation. 2011;124:2208–14.
4. Togha M, Sharifpour A, Ashraf H, et al. Electrocardiographic abnormalities in acute cerebrovascular events in patient with/without cardiovascular disease. Ann Indian Acad Neurol. 2013;16:66–71.
5. Murthy SB, Shah S, Venkatasubba Rao CP, et al. Clinical characteristics of myocardial stunning in acute stroke. J Clin Neurosci. 2014;21:1279–52.
6. Qaga AY, Suleiman A, Alsumrain M, et al. Electrocardiographic abnormalities in patients presenting with intracranial parenchymal hemorrhage. Acta Cardiol. 2012;67:635–9.
7. Chatterjee S. ECG changes in subarachnoid hemorrhage: a synopsis. Neth Heart J. 2011;19:31–4.
8. Mirvis DM, Goldberger AL. Electrocardiography. In: Mann DL, Zipes DP, Libby P, Bonow RO, editors. Braunwald's heart disease. Philadelphia: Elsevier; 2015. p. 114–52.
9. Wagner GS, Lim TH. Myocardial infarction. In: Wagner GS, editor. Marriott's practical electrocardiography. Philadelphia: Wolters Kluwer; 2001. p. 178–207.
10. De Zwaan C, Bar FW, Wellens HJ. Characteristic electrocardiographic pattern indicating a critical stenosis high in left anterior descending coronary artery in patients admitted because of impending myocardial infarction. Am Heart J. 1982;103:730–6.

11. Htut Oo SZMW, Khalighi K, Kodali A, May C, Aung TT, Snyder R. Ominous T wave inversions: Wellens' syndrome revisited. J Community Hosp Intern Med Perspect. 2016;6 https://doi.org/10.3402/Jchimp.V6.32011.
12. Farhan HL, Hassan KS, Al Belushi A, et al. The diagnostic value of electrocardiographic T wave inversion in lead aVL in diagnosing coronary artery disease in patients with chronic stable angina. Oman Med J. 2010;25:124–7.
13. Hassen GW, Costea A, Smith T, et al. The neglected lead on electrocardiogram: T wave inversion in lead aVL, nonspecific finding or a sign for left anterior descending artery lesion? J Emerg Med. 2014;46:165–70.
14. Levis J. ECG diagnosis: apical hypertrophic cardiomyopathy. Perm J. 2013;17:84.
15. Larsen CT, Dahlin J, Blackburn H, et al. Prevalence and prognosis of electrocardiographic left ventricular hypertrophy, ST segment depression and negative T wave. The Copenhagen city heart study. Eur Heart J. 2002;23:315–24.
16. Okin PM, Devereux RB, Nieminen MS. Electrocardiographic strain pattern and prediction of cardiovascular morbidity and mortality in hypertensive patients. Hypertension. 2004;44:48–54.
17. Kernohan RJ. Post-paroxysmal tachycardia syndrome. Br Heart J. 1969;31:803–6.
18. Kolb JC. Cardiac memory – persistent T wave changes after ventricular pacing. J Emerg Med. 2002;23:191–7.

The QT Interval 45

1 Definition

The QT interval extends from the beginning of the QRS complex to the end of the T wave.

2 Measurement

Automated measurement can be used for working purpose if the ECG is normal. Automated measurement should be confirmed by manual measurement if the end of the T wave is not well defined or the T wave is notched or biphasic. It should be measured from the lead in which it is longest, usually in leads V_2 to V_4. [1] If the U wave fuses with the end of the T wave, point of notch between the T wave and the U wave should be taken as the end of the T wave [2] (Fig. 45.1).

In tachycardia, the end of the T wave may fuse with the beginning of the next P wave. In such a situation, the junction of the T wave and the P wave should be taken as the end of the T wave [3] (Fig. 45.2).

If the T wave is notched, the end of the terminal wave should be taken as the end of the T wave (Fig. 45.3).

If the end of T wave does not touch the baseline, a tangent should be drawn from the steep portion of the down slope of the T wave. The

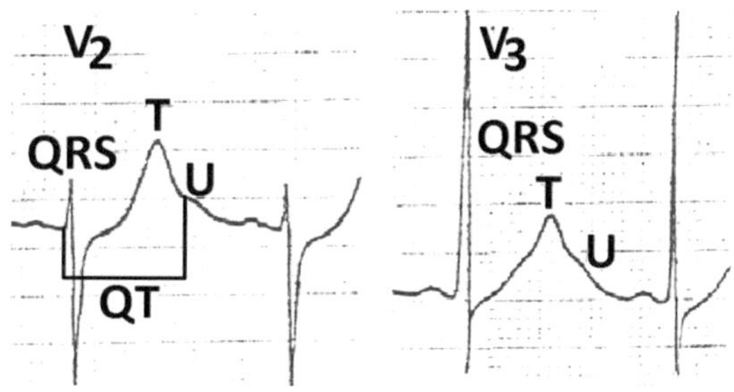

Fig. 45.1 Electrocardiogram (leads V_2 and V_3) showing fusion of the T and the U waves. QT interval is measured from the beginning of the QRS complex to the notch between the T and the U wave. (Reproduced with permission from: Mittal SR. QT Interval. Cardiology Today-2018; 22:179–86)

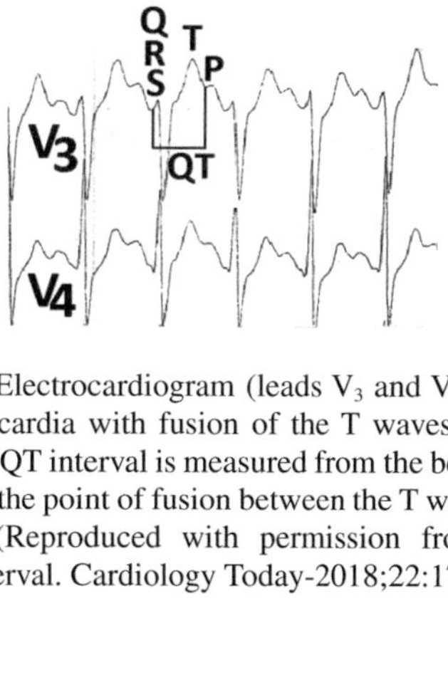

Fig. 45.2 Electrocardiogram (leads V_3 and V_4) showing sinus tachycardia with fusion of the T waves and the P waves. The QT interval is measured from the beginning of the QRS to the point of fusion between the T wave and the P waves. (Reproduced with permission from: Mittal SR. QT Interval. Cardiology Today-2018;22:179–86)

S. Mittal, *Insights into Electrocardiograms with MCQs*,
https://doi.org/10.1007/978-981-99-0127-2_45

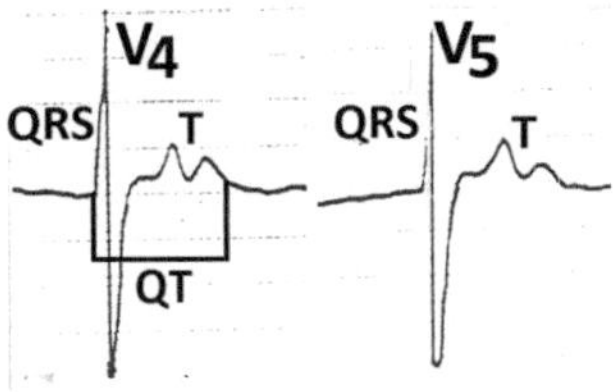

Fig. 45.3 Electrocardiogram (leads V_4 and V_5) showing notched T wave. The QT interval is measured from the beginning of the QRS to the end of the terminal part of the T wave. (Reproduced with permission from: Mittal SR. QT Interval. Cardiology Today-2018;22:179–86)

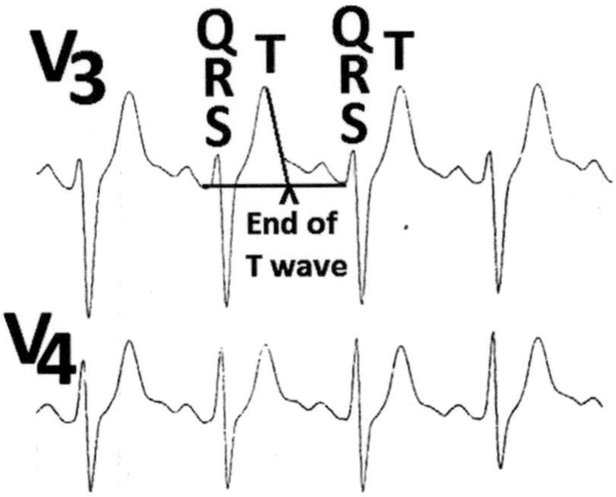

Fig. 45.4 Electrocardiogram (leads V_3 and V_4) showing tangent method to determine point of end of the T wave. (Reproduced with permission from: Mittal SR. QT Interval. Cardiology Today-2018;22:179–86)

point where the tangent touches the baseline should be taken as the end of the T wave (Fig. 45.4).

3 Correction for the Heart Rate

QT interval shortens with increasing heart rate. The exact significance of QT interval can be judged only after correcting it for the heart rate. Several formulae have been proposed. [4] Bazett's formula [5] is most commonly used. It is as under:

$$QTc = \frac{\text{QT interval}\left(\text{in millisecond}\right)}{\sqrt{\text{R} - \text{R}}\ \text{interval}\left(\text{in seconds}\right)}$$

RR is the preceding RR interval.

It is not possible to measure corrected QT interval in irregular rhythm and in patients with wide QRS.

4 Normal Value of QTc (QT Interval Corrected for Heart Rate)

It ranges between 360 milliseconds and 480 milliseconds. [6]

5 Prolongation of the QT Interval

5.1 (A)Causes

(a) Congenital long QT syndromes"
  i. Jervell and Lange-Nielsen syndrome

     Prolonged QT interval is associated with congenital deaf-mutism, fainting attacks, and sudden death. There is no structural heart disease. [7]
  ii. Romano–Ward syndrome

     Prolonged QT interval is associated with fainting attacks and sudden death. However, there is no deaf-mutism. [8, 9]
  iii. Andersen–Tawil syndrome

     It is characterized by a triad of periodic paralysis, skeletal abnormalities (e.g., short stature and scoliosis) and ventricular arrhythmias. [10] Electrocardiogram shows pronounced QTc prolongation, prominent U waves, prolongation of the downslope of the T wave, and ventricular ectopics. [10]
  iv. Timothy syndrome

     Fetal bradycardia and extreme prolongation of the QT interval are often associated with macroscopic T wave alternans and 2:1 atrioventricular block at birth. [10] Cognitive and behavioral problems and immune dysfunction may be seen.

     In addition to these syndromes, mutation of several other genes has been linked to long QT syndrome. In many cases QTc may not be prolonged in the resting ECG (concealed long QT). QTc may be prolonged only during stress or after ectopic beats (Fig. 45.5). It is always prolonged before fainting spell.

A variety of T wave morphologies have been described to be associated with the most common long QT syndromes. [11] LQT 1 is associated with smooth, broad-based T waves. LQT 2 is associated with notched or bifid T waves (Fig. 45.3). LQT 3 is often associated with a prolonged isoelectric ST-segment with a late onset of the T wave (Fig. 45.6) and asymmetrical or biphasic T waves.

However, these electrocardiographic findings have not been found to be specific. [10] Correct diagnosis of the type of long QT syndrome is possible only by study of the gene mutation. However, these T wave morphologies may increase suspicion about the presence of the long QT syndromes. Beat to beat variation in amplitude or polarity of the T wave (T wave alternans) (Figs. 45.7 and 45.8) also suggests increased risk of torsades de pointes.

Torsades de pointes (torsion around a point) is a type of ventricular tachycardia with changing QRS polarity. It is very rapid, usually ill sustained, and associated with syncope.

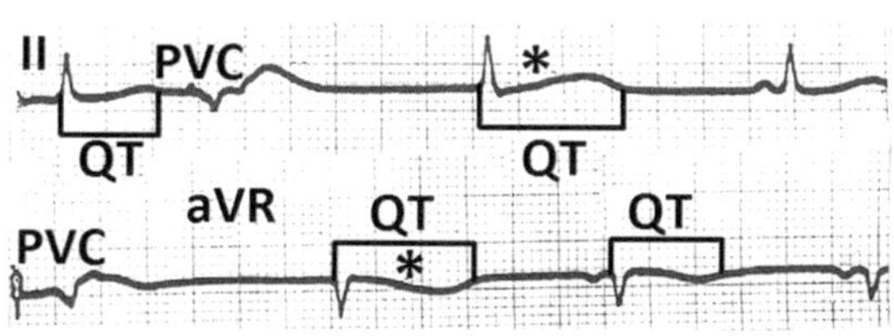

Fig. 45.5 Electrocardiogram (leads II and aVR) showing post ectopic prolongation of the QT interval (*). PVC—premature ventricular contraction. (Reproduced with permission from: Mittal SR. QT Interval. Cardiology Today-2018;22:179–86)

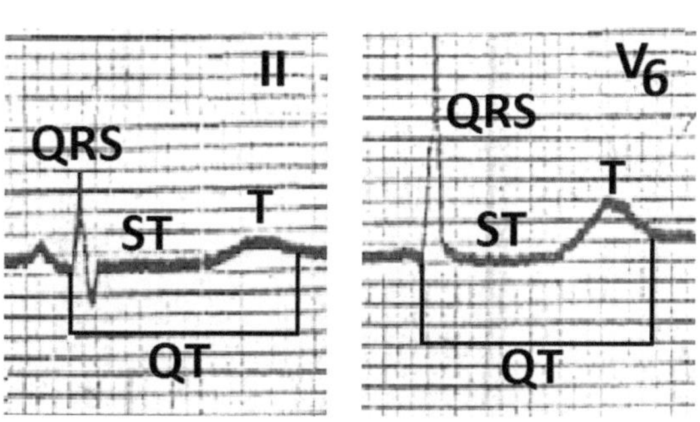

Fig. 45.6 Electrocardiogram (leads II and V_6) showing prolonged isoelectric ST-segment with late onset of the T wave. (Reproduced with permission from: Mittal SR. QT Interval. Cardiology Today-2018;22:179–86)

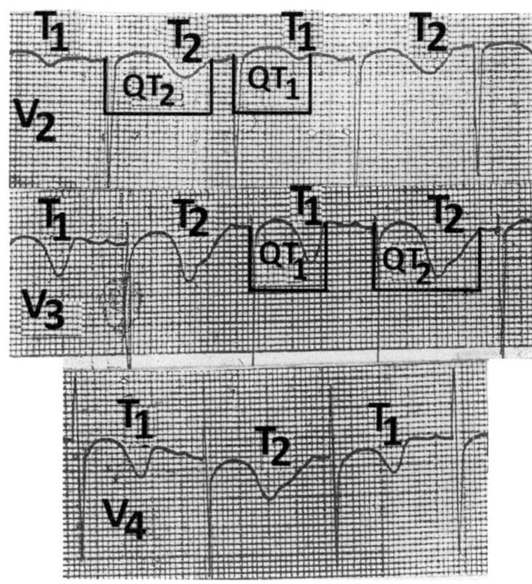

Fig. 45.7 Electrocardiogram (leads V_2, V_3, and V_4) showing alternating change in amplitude of the T waves (T_1 and T_2) and the QT interval (QT_1 and QT_2). (Reproduced with permission from: Mittal SR. QT Interval. Cardiology Today-2018;22:179–86)

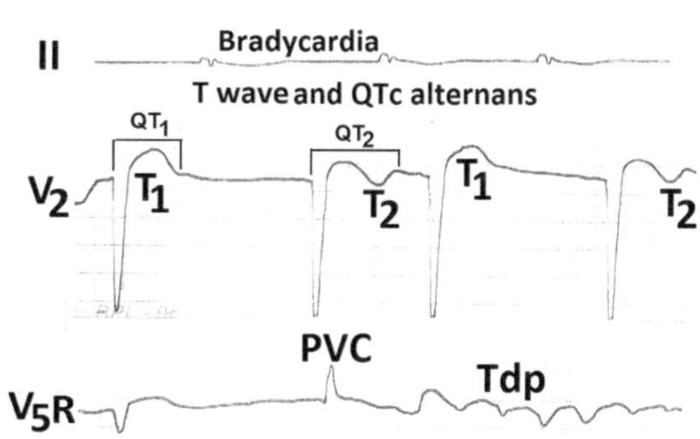

Fig. 45.8 Electrocardiogram (leads II, V_2, and V_5R) from a patient receiving ivabradine, diltiazem, and ranolazine. Lead II is showing bradycardia, lead V_2 is showing alternans in T wave polarity (T_1 and T_2) and QT interval (QT_1 & QT_2), lead V_5R is showing ventricular ectopic (PVC) and torsades de pointes (Tdp). (Reproduced with permission from: Mittal SR. QT Interval. Cardiology Today-2018;22:179–86)

– Significance of congenital prolongation of the corrected QT interval

Patients with congenital long QT syndromes have a high risk of arrhythmias, syncope, and sudden death. Other mutations in long QT susceptibility genes are mostly without consequence. [10] However, triggers like exertion, swimming, emotion, auditory stimuli, and postpartum period can rarely increase electrical instability of the heart resulting in potentially life-threatening arrhythmias. [10, 12]

(b) Acquired prolongation of the QTc interval

(i) Causes

- Drugs: There is a long list of drugs that can prolong QT interval. However, only some drugs predispose to the risk of torsades de pointes.
- Antiarrhythmics

  Class IA: Quinidine, Procainamide, Disopyramide. [13] The QT interval is prolonged. T wave amplitude is decreased. U wave amplitude is increased. QRS prolongation is minimal.

  Class IC: Flecainide. [13]

  There is broadening of the QRS. QT prolongation is minimal.

  Class III: Sotalol, Ibutilide.
- Promotility drugs: Cisapride

  Different drugs prolong different parts of the QT interval. Amiodarone prolongs the interval from the J point to the peak of the T wave. Quinine prolongs the interval between the peak of the T wave to the end of the T wave. [14]
- Risk factors for drug-induced torsades de pointes

  Patients who develop drug-induced torsades de points usually have a QTc interval > 500 milliseconds. [15]

  Torsades de pointes is uncommon with drug-induced QT prolongation alone. Most patients have at least one more identifiable risk factor in addition to prolongation of the QT interval. [14] These include:

– Drug responsible for prolongation of the QT interval

Quinidine, procainamide, dofetilide, and ibutilide have a high risk of inducing torsades de pointes. Amiodarone (Fig. 45.9), dronedarone, ranolazine, lithium, terfenadine, astemizole, antibiotics, antipsychotics, and other noncardiovascular drugs also prolong QT but rarely produce torsades de pointes of their own. [16]

– Dose, route, and rate of administration of the drug

The risk of torsades de pointes with sotalol is dose dependent. Haloperidol can produce torsades de pointes only when given intravenously.

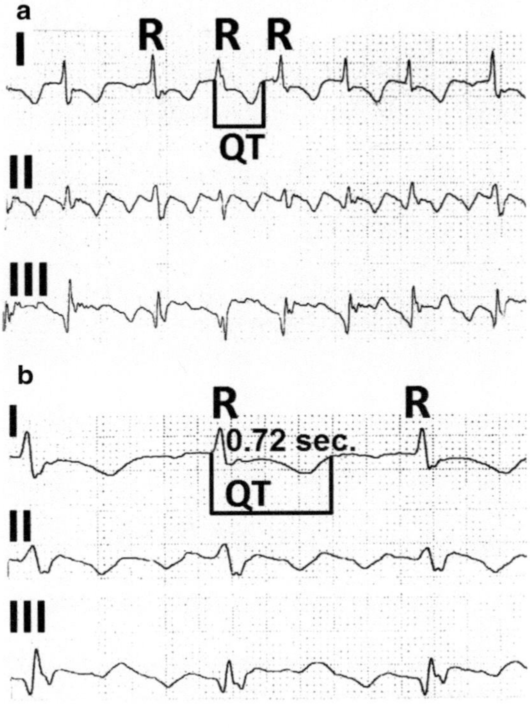

Fig. 45.9 Electrocardiogram (leads I, II, III). (**a**) At admission showing atrial flutter with fast ventricular rate. (**b**) Showing prolongation of the QT interval after amiodarone. (Reproduced with permission from: Mittal SR. QT Interval. Cardiology Today-2018;22:179–86)

- Metabolism and excretion of the drug
  Hepatic and/or renal insufficiency can increase serum level of drugs even when given in recommended dose.
- Concomitant use of other QT prolonging drugs (Fig. 45.10). This can happen due to:
  Additive effect of the drugs
  Drug to drug interaction altering their metabolism
- Electrolyte imbalance
  Hypokalemia, hypocalcemia, hypomagnesemia predispose to drug-induced torsades de pointes.
- Cardiac status
  Gross heart failure, active myocardial ischemia (Fig. 45.11), extreme bradycardia (Fig. 45.12), frequent ventricular ectopics, abrupt slowing of the heart rate, and recent conversion from atrial fibrillation predispose to the drug-induced torsades de pointes.
- Genetic predisposition
  Subclinical mutation in long QT syndrome genes.
  Reduced repolarization reserve.
  Reduced rate of drug metabolism.

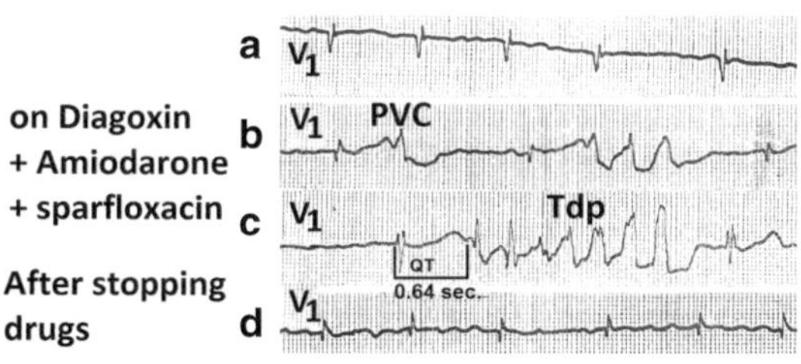

Fig. 45.10 Electrocardiogram (leads V_1) (**a**) showing atrial fibrillation with controlled ventricular rate (**b**) and (**c**) patient on Digoxin + Amiodarone + Sparfloxacin showing prolongation of the QT interval and torsades de pointes (Tdp). (**d**) Disappearance of the ectopics and torsades de pointes, after stopping amiodarone and sparfloxacin. (Reproduced with permission from: Mittal SR. QT Interval. Cardiology Today-2018;22:179–86)

- Electrolyte imbalance
  - Hypokalemia: There is ST-segment depression, flattening of the T wave, and prominence of the U wave. It is the QU interval that is increased (Fig. 45.13).
  - Hypomagnesemia
  - Hypocalcemia—ST-segment is prolonged. T wave remains normal [14]
- Heart diseases
  - Myocarditis
  - Acute coronary syndrome: Terminal part of the T wave is slurred. Interval from peak of the T wave to the end of the T wave (Tp-Te) is increased (Figs. 45.14 and 45.15). QT interval normalizes over time (Fig. 45.16).
- Hypothyroidism
  Electrocardiogram shows sinus bradycardia, low voltage of all wave forms, and flat or slightly inverted T waves without significant deviation of the ST-segment in many or all leads. U waves may be prominent. [13]

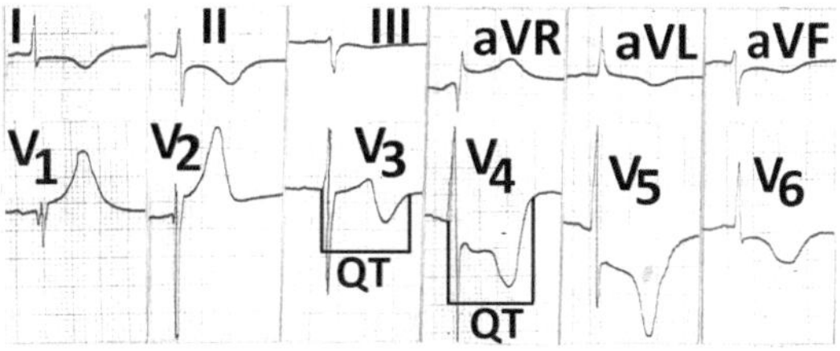

Fig. 45.11 Electrocardiogram is showing prolongation of the QT interval in acute myocardial ischemia. (Reproduced with permission from: Mittal SR. QT Interval. Cardiology Today-2018;22:179–86)

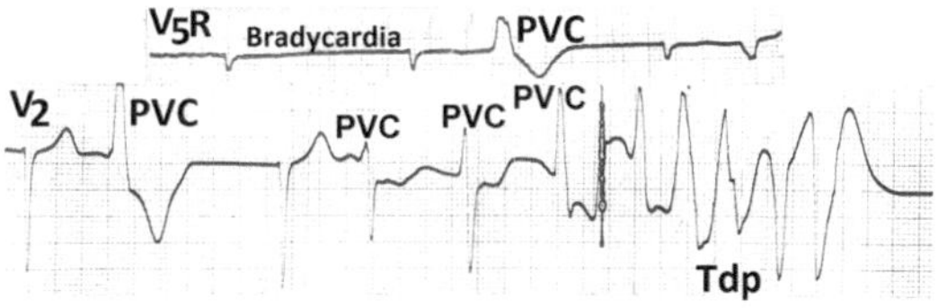

Fig. 45.12 Leads V_5R showing bradycardia and ventricular ectopic (PVC). Lead V_2 showing ventricular ectopics (PVCs) and torsades de pointes (Tdp). (Reproduced with permission from: Mittal SR. QT Interval. Cardiology Today-2018;22:179–86)

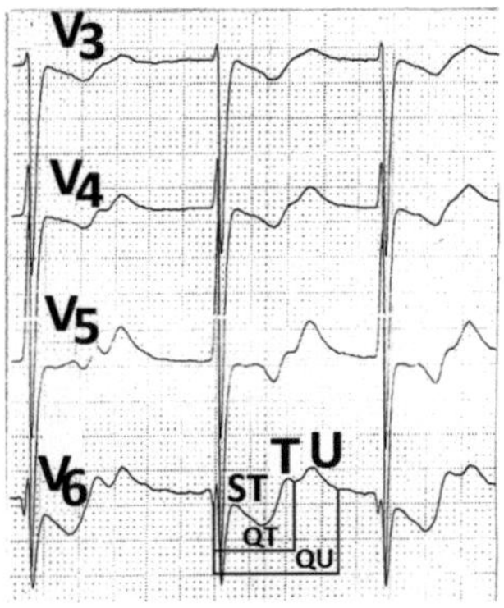

Fig. 45.13 Electrocardiogram (leads V_3 to V_6) from a patient of hypokalemia showing ST-segment depression, reduced amplitude of the T waves, prominent U wave, and prolongation of the Q-U interval. (Reproduced with permission from: Mittal SR. QT Interval. Cardiology Today-2018;22:179–86)

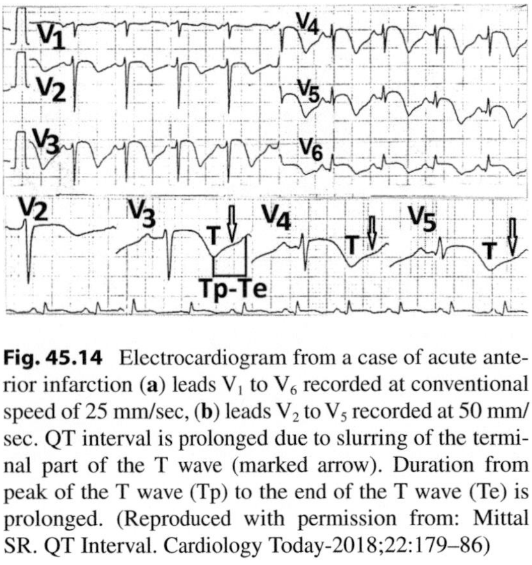

Fig. 45.14 Electrocardiogram from a case of acute anterior infarction (**a**) leads V_1 to V_6 recorded at conventional speed of 25 mm/sec, (**b**) leads V_2 to V_5 recorded at 50 mm/sec. QT interval is prolonged due to slurring of the terminal part of the T wave (marked arrow). Duration from peak of the T wave (Tp) to the end of the T wave (Te) is prolonged. (Reproduced with permission from: Mittal SR. QT Interval. Cardiology Today-2018;22:179–86)

- Head injury, intracerebral hemorrhage

  There is bradycardia, widening, and inversion of the T waves in precordial leads. [13]

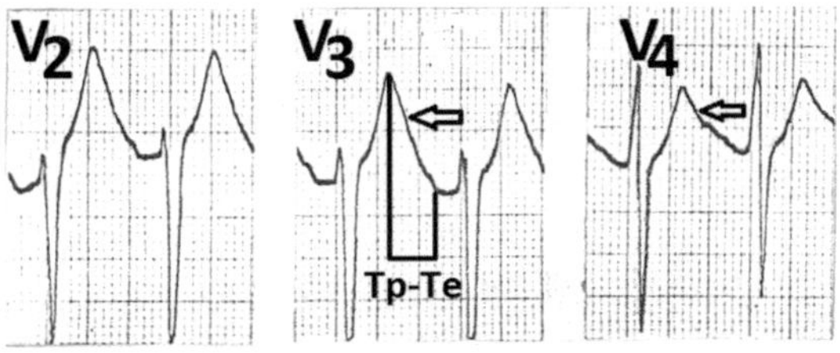

Fig. 45.15 Electrocardiogram (leads V_2 to V_4) from a case of the acute coronary syndrome. QT interval is prolonged due to slurring of the terminal part of the T wave (arrow). Duration from peak of the T wave (Tp) to the end of the T wave (Te) is increased. (Reproduced with permission from: Mittal SR. QT Interval. Cardiology Today-2018;22:179–86)

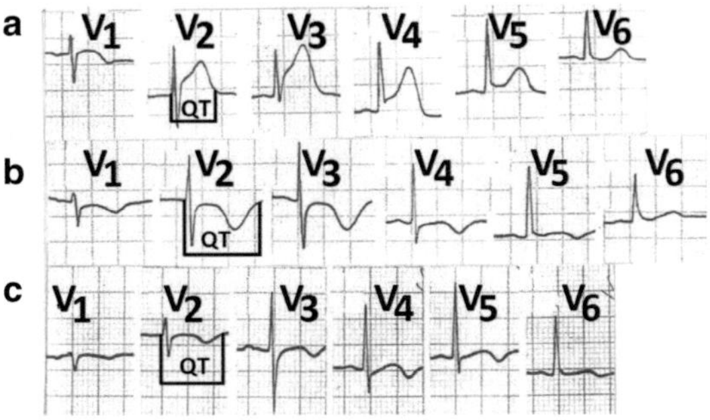

Fig. 45.16 Electrocardiogram (leads V_1 to V_6) (**a**) At the time of admission (**b**) on 2nd day showing prolongation of the QT interval (**c**) on fourth day showing normalization of the QT interval. (Reproduced with permission from: Mittal SR. QT Interval. Cardiology Today-2018;22:179–86)

- High degree or complete atrioventricular block

  It predisposes to torsades de pointes (Fig. 45.17).

  Giant and inverted T waves are seen mostly after a syncopal attack (Stokes–Adams attack).
- Hypothermia

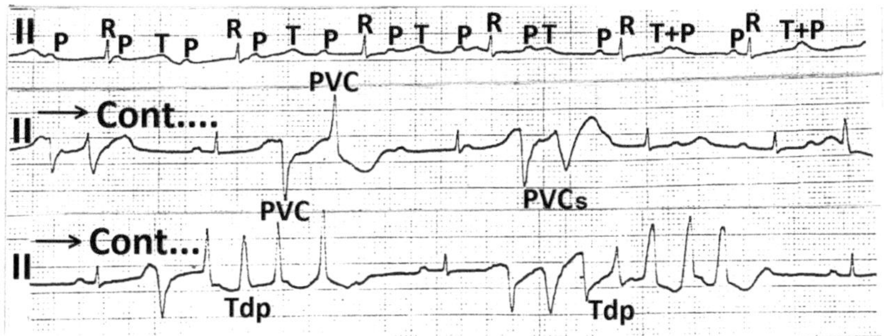

Fig. 45.17 Electrocardiogram (lead II) (**a**) showing complete atrioventricular block, (**b**) showing ventricular ectopics (PVCs), (**c**) showing runs of torsades de pointes (Tdp). (Reproduced with permission from: Mittal SR. QT Interval. Cardiology Today-2018;22:179–86)

6 Short QT Interval

QTc interval less than 320 ms is considered as short [10] (Fig. 45.18).

(a) Causes
  (i) Congenital short QT syndromes
    Patients with specific gene mutations are susceptible to paroxysmal atrial fibrillation, ventricular tachycardia, syncope, or even sudden death. [17, 18, 19] Tall peaked T waves with narrow base are seen in precordial leads, especially in leads V_2 and V_3 with either no or very short ST-segment. [10] J point to peak of the T wave is short (100 ms or less). In the general population, short QTc interval is rare and is not associated with cardiac risk. [20, 21]
  (ii) Acquired
    – Electrolyte imbalance
      Hypercalcemia
      Shortening of the QT interval is due to the shortening of the ST-segment. ST-segment virtually disappears, becoming incorporated into the T wave. [14]
      Hyperkalemia
      Hypermagnesemia
    – Acidosis
    – Drugs

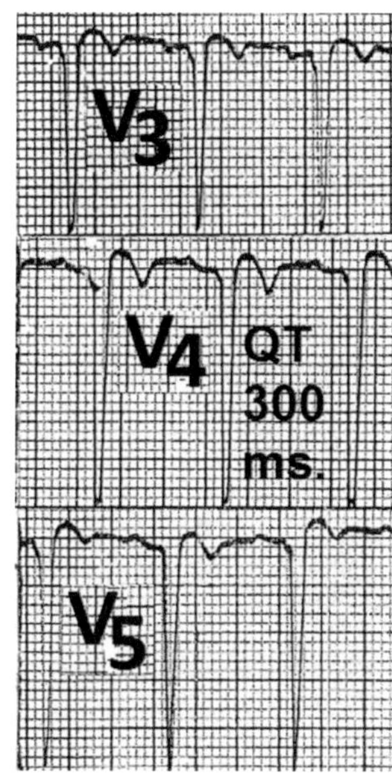

Fig. 45.18 Electrocardiogram (leads V_3, V_4, V_5) showing old anterior myocardial infarction with short QT interval. (Reproduced with permission from: Mittal SR. QT Interval. Cardiology Today-2018;22:179–86)

- Digitalis-induced shortening of the QTc interval is associated with down sloping ST-segment depression. [13] Changes are usually seen in leads with tall R waves.
- Anticonvulsants [22]

- Hyperthermia [22]
- Early repolarization

  It may be associated with a short QT interval. [23] Early repolarization per se may be associated with increased risk of death. [24] However, this increased risk is not related to shortening of the QT interval. [25]
- Carnitine deficiency [26]
- Epilepsy [27]

  It is not clear if shortening of the QT interval is related to epilepsy per se or is an effect of anticonvulsants. [22]

Summary

- QT interval includes the time interval from the beginning of the q wave to the end of the T wave, that is, the period of ventricular depolarization and repolarization.
- Difficulties in measuring correct QT interval.
  - q wave and/or terminal part of the T wave may be isoelectric in some leads. This results in lead to lead variation in QT interval (QT dispersion). It may vary from 50 to 70 milliseconds.
  - End of the T wave is usually not sharp. In such a situation, the point where a tangent line from the steepest portion of the terminal part of the T wave crosses the isoelectric line is taken as the end of the T wave.
  - The end of the T wave may fuse with the U wave or the P wave making the identification of the end of the T wave difficult.
- QT interval decreases with increasing heart rate. Therefore, correction for heart rate is necessary to assess clinical significance. Most commonly used formula is Bazett's formula.

  $$QTc = \frac{\text{QT interval in milliseconds}}{\sqrt{\text{preceding RR interval}}}$$

  Measurement of RR is made in seconds. Normal value ranges from 0.39 to 0.45 seconds.
- Values more than 500 milliseconds or an increase of more than 60 milliseconds above previous value is associated with increased risk of ventricular tachyarrhythmias.
- Congenitally long QT interval may be associated with sensory neural deafness (Jervell and Lange-Nielsen syndrome) or with normal hearing (Romano–Ward syndrome).
- Clinically significant congenital long QT syndrome is divided in three types.
  - Type 1: T waves have an early onset and are broad based. Tachyarrhythmias are usually precipitated by exercise (particularly swimming) and emotional stress.
  - Type 2: It is associated with low amplitude, notched, or biphasic T waves. Lethal events are more frequent at rest or during sleep.
  - Type 3: T waves are narrow based and/or preceded by a long isoelectric ST-segment. Emotional stress or sudden loud noise (e.g., telephone, alarm clock) precipitate lethal arrhythmias.
- Acquired long QT interval can occur with a large number of drugs and clinical conditions. Common being amiodarone, phenothiazines, tricyclic antidepressants, hypokalemia, hypomagnesemia, hemorrhagic stroke, hypocalcemia, acute myocarditis, and acute myocardial infarction.
- QTc interval of 330 milliseconds or less is considered as "short." T waves are peaked and tall, especially in leads V_2 and V_3. The J point to the peak of the T wave is less than 100 milliseconds. U wave is well separated from the T wave. Congenitally short QT interval is associated with increased risk of atrial fibrillation or ventricular tachyarrhythmia. In the general population, congenital short QTc is rare and is not associated with increased cardiac risk. Digitalis, hypercalcemia, hypermagnesemia, and acidosis also produce short QT interval.

MCQs

Q1. QT interval extends from:

(a) Beginning of the q wave to beginning of the T wave
(b) Beginning of the q wave to top of the T wave
(c) Beginning of the q wave to end of the T wave
(d) Beginning of the q wave to beginning of the ST-segment

Q2. The QT interval is usually longest in leads:

(a) I, aVL
(b) II, III, aVF
(c) V_2 to V_4
(d) V_7 to V_9

Q3. Which statement is correct?

(a) The QT interval prolongs with increasing heart rate
(b) The QT interval shortens with increasing heart rate
(c) The QT interval shortens with decreasing heart rate
(d) The QT interval is not affected by heart rate

Q4. According to Bazett's formula corrected QTc is:

(a) $$\frac{\text{QT interval}\left(\text{in second}\right)}{\sqrt{\text{RR}}\ \textit{interval}\left(\textit{in seconds}\right)}$$

(b) $$\frac{\text{QT interval}\left(\text{in millisecond}\right)}{\sqrt{\text{RR}}\ \textit{interval}\left(\textit{in milliseconds}\right)}$$

(c) $$\frac{\text{QT interval}\left(\text{in millisecond}\right)}{\sqrt{\text{R}-\text{R}}\ \textit{interval}\left(\textit{in seconds}\right)}$$

(d) $$\frac{\text{QT interval}\left(\text{in millisecond}\right)}{\sqrt{\text{Heart rate}}}$$

Q5. Normal range of the corrected QT interval is:

(a) 250 to 300 milliseconds
(b) 300 to 350 milliseconds
(c) 360 to 480 milliseconds
(d) 490 to 550 milliseconds

Q6. Congenitally prolonged QT interval with fainting attacks without deaf mutism and any other abnormality suggests:

(a) Jervell and Lange-Nielsen syndrome
(b) Romano–Ward syndrome
(c) Anderson–Tawil syndrome
(d) Timothy syndrome

Q7. In patients with hereditary long QT syndrome the QTc interval is:

(a) Always prolonged at rest
(b) Always prolonged during sleep
(c) May be prolonged during stress
(d) Always prolonged before fainting spell

Q8. Which T wave abnormalities can be associated with long QT syndromes?

(a) Smooth, broad based T wave
(b) Notched or bifid T wave
(c) Biphasic T wave
(d) All

Q9. T wave alternans means beat to beat variation in:

(a) T wave amplitude
(b) T wave duration
(c) T wave polarity
(d) Any of the above

Q10. T wave alternans is associated with increased risk of:

(a) Atrial fibrillation
(b) Atrioventricular nodal reentrant tachycardia
(c) Ventricular tachycardia
(d) None of the above

Q11. Torsades de pointes is a type of:

(a) Atrial flutter
(b) Atrial fibrillation
(c) Atrioventricular reentrant tachycardia
(d) Ventricular tachycardia

Q12. In hereditary long QT syndrome, syncope may be precipitated by:

(a) Exertion
(b) Swimming
(c) Auditory stimuli
(d) All

Q13. Which drugs predispose to risk of torsades de pointes?

(a) Flecainide
(b) Quinine
(c) Ibutilide
(d) Amiodarone

Q14.What magnitude of QTc prolongation predisposes to risk of torsades de pointes?

(a) More than 350msec
(b) More than 400msec
(c) More than 450msec
(d) More than 500msec

Q15. Risk of drug-induced torsades de pointes is increased by:

(a) Extreme tachycardia
(b) Extreme bradycardia
(c) Hyperkalemia
(d) Active myocardial ischemia

Q16. Which factors predispose to risk of drug-induced torsades de pointes?

(a) Hypokalemia
(b) Subclinical mutation in the long QT syndrome genes
(c) Concomitant use of other QT prolonging drugs
(d) Abrupt increase in heart rate

Q17. Which factors can predispose to risk of drug-induced torsades de pointes?

(a) Gross heart failure
(b) Hepatic failure
(c) Renal failure
(d) Hypermagnesemia

Q18. Hypokalemia prolongs:

(a) Duration of the QRS complex
(b) Duration of the ST-segment
(c) Duration of the T wave
(d) Duration of the Q-U interval

Q19. ST-segment is prolonged by:

(a) Hypokalemia
(b) Hypomagnesemia
(c) Hypocalcemia
(d) Hypothyroidism

Q20. Interval from peak of the T wave to its end is increased in:

(a) Acute coronary syndrome
(b) Myocarditis
(c) Heart failure
(d) Acute pericarditis

Q21. Intracerebral hemorrhage produces:

(a) Increase in duration of QRS
(b) Increase in duration of the ST-segment
(c) Widening of the T waves
(d) Inversion of the T waves

Q22. Broad and inverted T waves are seen in:

(a) Head injury
(b) Stokes–Adams' attacks
(c) Hypothyroidism
(d) Hypokalemia

Q23. Risk of torsades de points is increased in:

(a) Hypothyroidism

(b) Head injury
(c) Hypothermia
(d) Complete atrioventricular block

Q24. The QT interval is considered short when the QTc interval is less than:

(a) 400ms
(b) 380ms
(c) 350ms
(d) 320ms

Q25. In congenital short QT syndrome:

(a) QRS in narrow
(b) ST-segment is very short
(c) The T wave is narrow
(d) The T wave is tall and peaked

Q26. In the general population, short QT interval is:

(a) Very frequent
(b) Frequent
(c) Uncommon
(d) Rare

Q27. Patients with congenital short QT syndrome are susceptible to:

(a) Atrial flutter
(b) Paroxysmal atrial fibrillation
(c) Paroxysmal atrial tachycardia
(d) Syncope

Q28. Acquired short QT interval is seen in:

(a) Hyperkalemia
(b) Acidosis
(c) Hypocalcemia
(d) Alkalosis

Q29. Which drugs can shorten QT interval?

(a) Digitalis
(b) Anticonvulsants
(c) Antiarrhythmics
(d) Antipsychotics

Q30. Which conditions are associated with a short QT interval?

(a) Hyperthermia
(b) Early repolarization syndrome
(c) Epilepsy
(d) Raised intracranial pressure

Q31. QT dispersion varies from:

(a) 10 to 20 milliseconds
(b) 30 to 40 milliseconds
(c) 50 to 70 milliseconds
(d) 80 to 100 milliseconds

Q32. QT dispersion is due to:

(a) q wave being isoelectric in some leads
(b) Terminal part of T wave being isoelectric in some leads
(c) Fusion of end of T wave with U wave
(d) All

Q33. The normal upper limit for corrected QTc interval is:

(a) 0.45 seconds
(b) 0.50 seconds
(c) 0.55 seconds
(d) 0.60 seconds

Q34. Jervell and Lange-Nielsen syndrome is associated with:

(a) Normal hearing
(b) Sensory deafness
(c) Neural deafness
(d) Sensorineural deafness

Q35. Romano–Ward syndrome is characterized by:

(a) Congenitally long QT interval
(b) Normal hearing
(c) Episodes of ventricular tachyarrhythmia and syncope
(d) Increased risk of atrial fibrillation

Q36. Type 1 congenital long QT syndrome is characterized by:

(a) Early onset of broad T wave
(b) Low amplitude T wave
(c) Long ST-segment
(d) Ventricular tachyarrhythmias precipitated by exercise

Q37. Type 2 congenital long QT syndrome is characterized by:

(a) Broad base T waves
(b) Notched or biphasic T waves
(c) Lethal events frequent at rest or during sleep
(d) Long ST-segment

Q38. Type 3 congenital long QT syndrome is characterized by:

(a) Broad-based T waves
(b) Biphasic T waves
(c) Narrow-based T waves
(d) Long isoelectric ST-segment

Q39. Which of the following can prolong QT interval?

(a) Beta blockers
(b) Amiodarone
(c) Digitalis
(d) Tricyclic antidepressants

Q40. Which of the following can prolong QT interval?

(a) Hypokalemia
(b) Hypercalcemia
(c) Acidosis
(d) All

Q41. The short QT interval is characterized by:

(a) Peaked and tall T waves
(b) J point to peak of T wave less than 100 msec
(c) U wave well separated from T wave
(d) All

Q42. Which of the following can produce short QT interval?

(a) Hypomagnesemia
(b) Acute myocardial infarction
(c) Hypercalcemia
(d) Digitalis

Q43. Short QT interval increases the risk of:

(a) Sudden cardiac death
(b) Atrial fibrillation
(c) Atrioventricular block
(d) Ventricular tachyarrhythmias

Answers

(1) c (2) c (3) b (4) c (5) c (6) b (7) c, d (8) d, (9) d (10) c (11) d (12) d (13) a,b,c (14) d (15) b,d (16) a,b,c (17) a,b,c (18) d (19) c (20) a (21) c,d (22) a,b (23) d (24) d (25) b,c,d (26) d (27) b,d (28) a,b (29) a,b (30) a,b,c (31) c (32) d (33) a (34) c (35) a,b,c (36) a,d (37) b,c (38) c,d (39) b,d (40) a (41) d (42) c,d (43) a,b,d

References

1. Mirvis DM, Goldberger AL. Electrocardiography. In: Mann DL, Zipes DP, Libby P, Bonow RO, editors. Braunwald's Heart Disease. Philadelphia: Saunders; 2015. p. 114–52.
2. Goldenberg I, Moss AJ, Zareba W. QT interval: How to measure it and what is normal. J Cardiovasc Electrophysiol. 2006;17:333–6.
3. Wagner GS, Lim TH. Interpretation of normal electrocardiogram. In: Wagner GS, editor. Marriott's Practical Electrocardiography. Philadelphia: Wolters Kluwer; 2008. p. 43–70.
4. Luo S, Michler K, Johnston P, Macfarlane PW. A comparison of commonly used QT correction formulae: the effect of heart rate on the QTc of normal ECGs. J Electrocardiol. 2004;37(suppl):81–90.
5. Bazett HC. An analysis of the time relations of electrocardiogram. Heart. 1920;7:353–70.
6. Priori SG, Blomstorm-Lundquist C, Mazzanti A, et al. ESC guidelines for the management of patients with ventricular arrhythmias and prevention of sudden cardiac death. Eur Heart J. 2015;36:2793–867.

7. Jervell A, Lange-Nielsen F. Congenital deaf-mutism, functional heart disease with prolongation of the QT interval and sudden death. Am Heart J. 1954;54(1):59–68.
8. Romano C, Gemme G, Pongiglione R. Rare cardiac arrhythmias of the pediatric age II. Syncopal attacks due to paroxysmal ventricular fibrillation(Italian). Clin Pediatr (Bologna). 1963;45:656–83.
9. Ward OC. A new familial cardiac syndrome in children. J Ir Med Assoc. 1964;54:103–6.
10. Tester DJ, Ackerman MJ. Genetics of cardiac arrhythmias. In: Mann DL, Zipes DP, Libby P, Bonow RO, editors. Braunwald's Heart Disease. Philadelphia: Saunders; 2015. p. 617–23.
11. Zhang I, Timothy KW, Vincent GM, et al. Spectrum of ST-T wave patterns and repolarization parameters in congenital long – QT syndrome. ECG findings identify genotypes. Circulation. 2000;102:2849–55.
12. Goldenberg I, Horr S, Moss AJ, et al. Risk of life threatening cardiac events in patients with genotype-confirmed Long QT syndrome and normal range corrected QT intervals. J Am Coll Cardiol. 2010;57:51–9.
13. Wagner GS, Wang TY. Miscellaneous conditions. In: Wagner GS, editor. Marriott's Practical Electrocardiography. Philadelphia: Wolters Kluwer; 2008. p. 209–37.
14. Yap YG, Camm AJ. Drug induced QT prolongation and torsades de pointes. Heart. 2003;89:1363–7.
15. Kallergis EM, Goudis CA, Simantirakis EN, Kochiadakis GE, Vardas PE. Mechanisms, risk factors and management of acquired long QT syndrome: A comprehensive review. Sci World J. 2012; Article ID: 212178, doi:https://doi.org/10.1100/2012/212178. Epub 2012 Apr 19.
16. Roden DM. Predicting drug induced QT prolongation and torsades de pointes. J Physiol. 2016;594:2459–68.
17. Gaita F, Giustella C, Bianchi F, et al. Short QT syndrome: a familial cause of sudden death. Circulation. 2003;108:965–70.
18. Brugada R, Hong K, Cordeiro JM, Dumaine R. Short QT syndrome. CMAJ. 2005;173:1349–54.
19. Perez-Riera AR, Paixao Almeida A, Barbosa-Barros R, et al. Congenital short QT syndrome: Landmarks of the newest arrhythmogenic cardiac channelopathy. Cardiol J. 2013;20:464–71.
20. Funada A, Hayashi K, Ino H, et al. Assessment of QT interval and prevalence of short QT syndrome in Japan. Clin Cardiol. 2008;31:270–4.
21. Reinig MG, Engel TR. The shortage of short QT intervals. Chest. 2007;132:246–9.
22. Shah RR. Drug- induced QT interval shortening: potential harbinger of pro arrhythmia and regulatory perspectives. British Journal of Pharmacology. 2010;159:58–69.
23. Watanabe H, Makigama T, Koyama T, et al. High prevalence of early repolarization in short QT syndrome. Heart Rhythm. 2010;7:647–52.
24. Antzelevitch C, Yan GX. J wave syndromes. Heart Rhythm. 2010;7:549–58.
25. Tikkanen JT, Anttonen O, Junttila MJ, et al. Long term outcome associated with early repolarization on electrocardiography. N Eng J Med. 2009;361:2529–37.
26. Roussel J, Labarthe F, Thireau J, et al. Carnitine deficiency induces a short QT syndrome. Heart Rhythm. 2015;13:165–74.
27. Surges R, Taggart P, Sander JW, Walker MC. Too long or too short? New insights into abnormal cardiac repolarization in people with chronic epilepsy and its potential role in sudden unexpected death. Epilepsia. 2010;51:738–44.

46 The U Wave

1 Introduction

The U wave is the wave between the T wave and the P wave. Its genesis is not clear. Several hypotheses have been proposed, but there is no consensus of opinion. [1]

2 Physiological U Wave

U wave is a small wave after the end of the T wave. It can be identified in all leads, but is usually most apparent in leads V_2 to V_4 [2] (Fig. 46.1).

It is better visible at slower heart rate (<65 beats/minute) [2] and in elderly. [3] Amplitude is normally less than 0.2mV. Amplitude varies directly with the amplitude of the preceding T wave. [2] Normally it is less than 25% of the amplitude of the preceding T wave. Normally the polarity of the U wave is the same as that of the preceding T wave. [3] Therefore, U wave is normally negative in lead aVR and at times in lead V_1 [2] (Fig. 46.2).

Clear U wave is not recognized in the presence of atrial fibrillation due to wavy base line [2] (Fig. 46.3).

However, it can be visible in patients with straight line fibrillation with controlled ventricular rate (Fig. 46.4).

Sometimes in patients of atrial fibrillation, fibrillary waves may be confused with normal U wave and P wave (Fig. 46.5a). Recording long lead helps in identifying irregularity of the rhythm and true nature of the fibrillary waves (Fig. 46.5b).

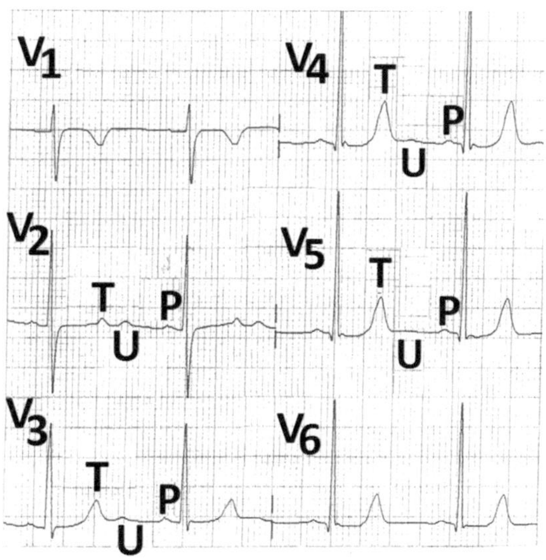

Fig. 46.1 Electrocardiogram showing normal U waves. (Reproduced with permission from: Mittal SR. "U" wave. Cardiology Today-2018;22:218–24)

At times the T wave may encroach on the U wave making a TU complex. In such a situation, it may be difficult to differentiate U wave from the second peak of a notched T wave. Apices of a notched T wave are usually less than 0.15 seconds apart (Fig. 46.6) whereas the interval between the apices of the T wave and the U wave is usually more than 0.15 seconds. [2]

S. Mittal, *Insights into Electrocardiograms with MCQs*,
https://doi.org/10.1007/978-981-99-0127-2_46

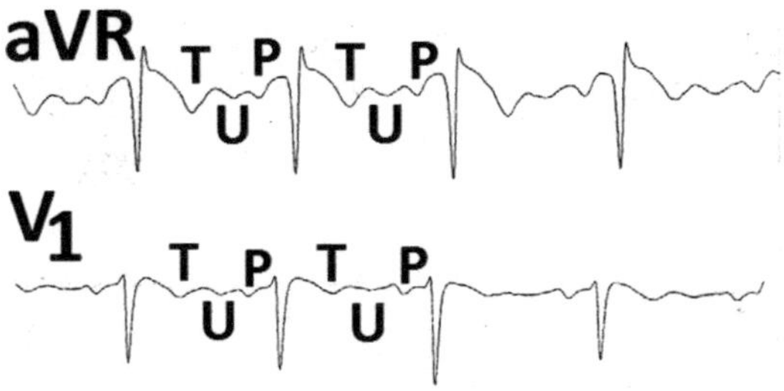

Fig. 46.2 Electrocardiogram showing normally inverted U waves in leads aVR and V_1. (Reproduced with permission from: Mittal SR. "U" wave. Cardiology Today-2018; 22:218–24)

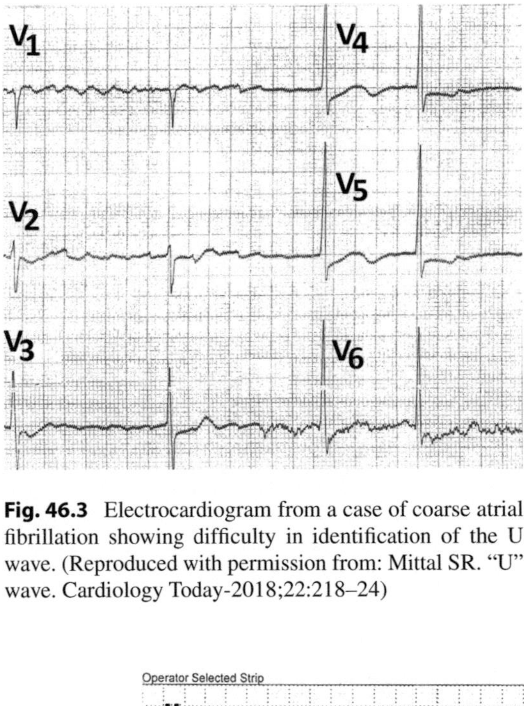

Fig. 46.3 Electrocardiogram from a case of coarse atrial fibrillation showing difficulty in identification of the U wave. (Reproduced with permission from: Mittal SR. "U" wave. Cardiology Today-2018;22:218–24)

Secondly, normal U wave is less than one-third of the amplitude of T wave. Therefore, in a TU complex, the second peak is much smaller than the first peak (Fig. 46.7).

On the contrary, second peak more than 50% of the first peak strongly suggests a bifid T wave.

The U wave is similar to the P wave in appearance. [1] In patients of atrial fibrillation with fast ventricular rate, it may be misinterpreted as a P wave if a QRS complex follows it (Fig. 46.8).

Irregularly irregular rhythm and absence of such pseudo P wave in most of the beats helps correct identification of atrial fibrillation. In patients with sinus bradycardia, U wave may be misinterpreted as additional P wave resulting in a

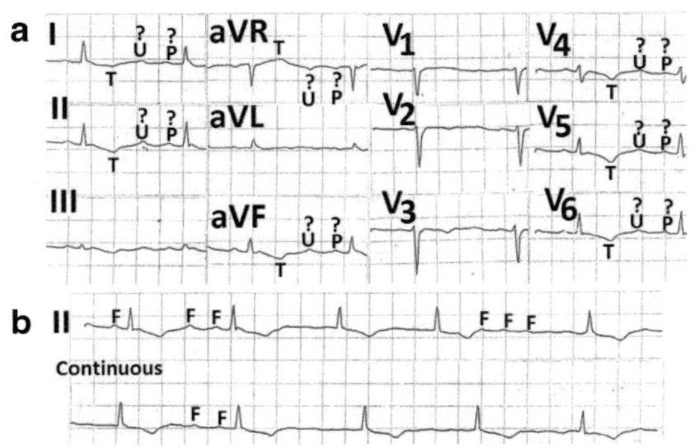

Fig. 46.5 (**a**). Electrocardiogram showing pseudo U waves and pseudo P waves in a case of atrial fibrillation. (**b**). Continuous recording of lead II identifying atrial fibrillary waves (F). (Reproduced with permission from: Mittal SR. "U" wave. Cardiology Today-2018;22:218–24.)

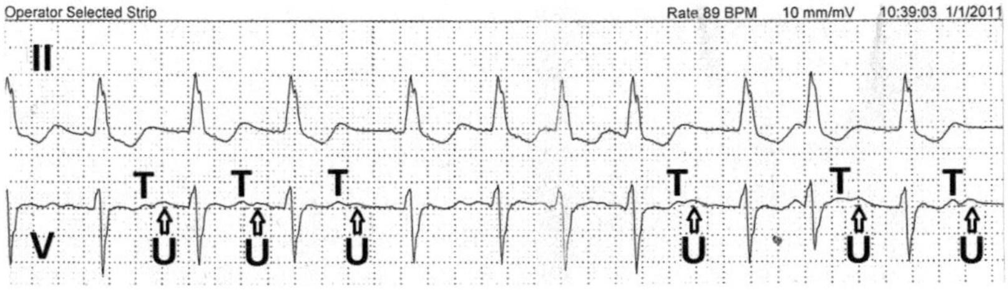

Fig. 46.4 Strip of Holter monitoring from a case of straight line atrial fibrillation with identification of the U wave (marked ⇧). (Reproduced with permission from Mittal SR. "U" wave. Cardiology Today-2018;22:218–24)

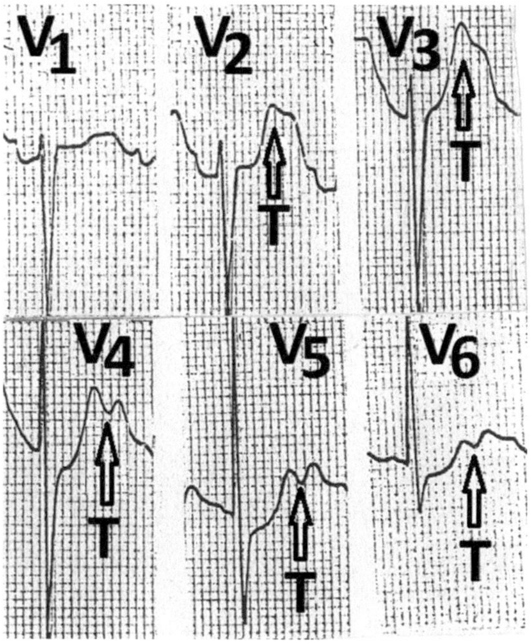

Fig. 46.6 Electrocardiogram is showing bifid (notched) T wave (marked ⇧). (Reproduced with permission from: Mittal SR. "U" wave. Cardiology Today-2018;22:218–24)

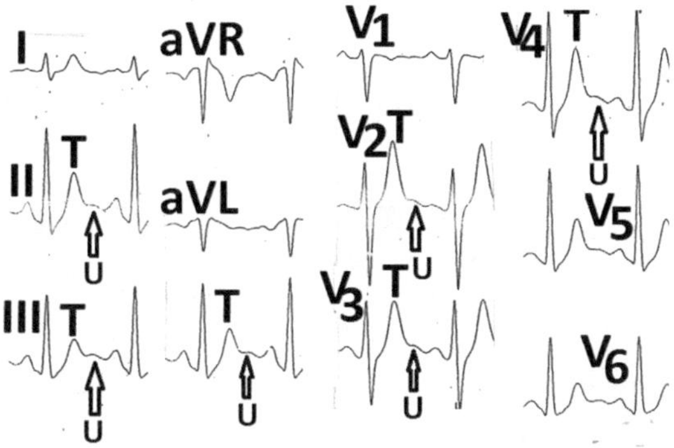

Fig. 46.7 Electrocardiogram shows fusion of T and U waves. (Reproduced with permission from: Mittal SR. "U" wave (marked ⇧). Cardiology Today-2018;22:218–24)

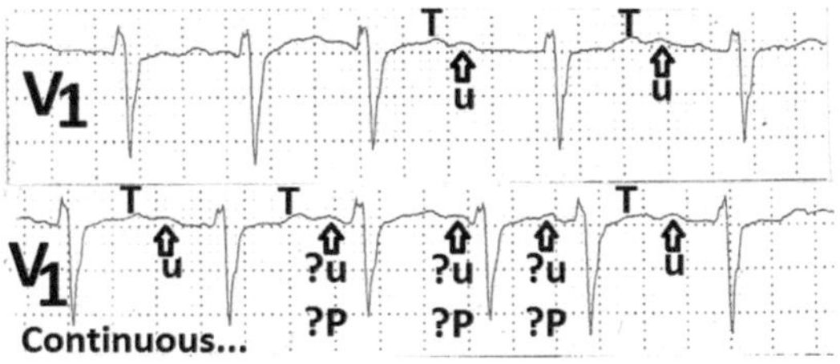

Fig. 46.8 Electrocardiogram showing U wave giving a false impression of the P wave during fast rate in a case of atrial fibrillation. (Reproduced with permission from: Mittal SR. "U" wave. Cardiology Today-2018;22:218–24)

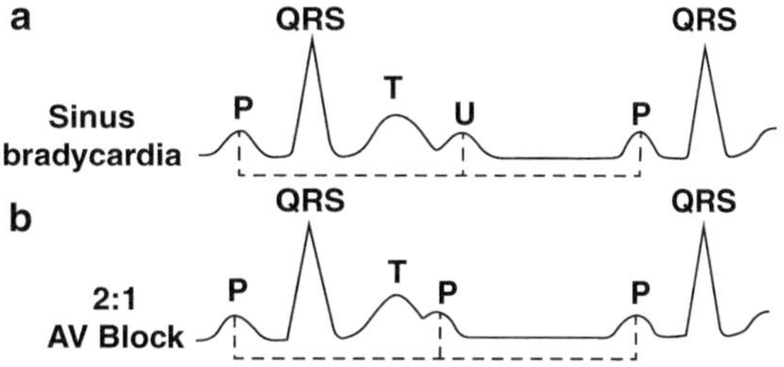

Fig. 46.9 Diagrammatic representation comparing electrocardiographic findings (**a**) in a case of sinus bradycardia with prominent U wave and (**b**) in a case of 2:1 atrioventricular block. (Reproduced with permission from: Mittal SR. "U" wave. Cardiology Today-2018;22:218–24.)

wrong diagnosis of 2:1 atrioventricular block. [1] The time interval between the preceding P wave and the U wave is, however, different from the interval between the U wave and following P wave (Fig. 46.9). This gives a clue to identification of the U wave (Fig. 46.9).

3 Prominent U Wave

U wave is prominent in the following conditions.

(1) Hypokalemia: It is the commonest cause of the prominent U wave. It is associated with depression of the ST-segment and decreased amplitude of the T wave. ECG changes may appear even when the serum potassium concentration is within normal limits [4] (Figs. 46.10 and 46.11).
(2) Hypocalcemia
(3) Hypomagnesemia
(4) Antiarrhythmic drugs, for example, amiodarone, dofetilide

The QT interval is prolonged with fusion of the T wave and the U wave forming a prominent and broad TU fusion wave (Fig. 46.12).

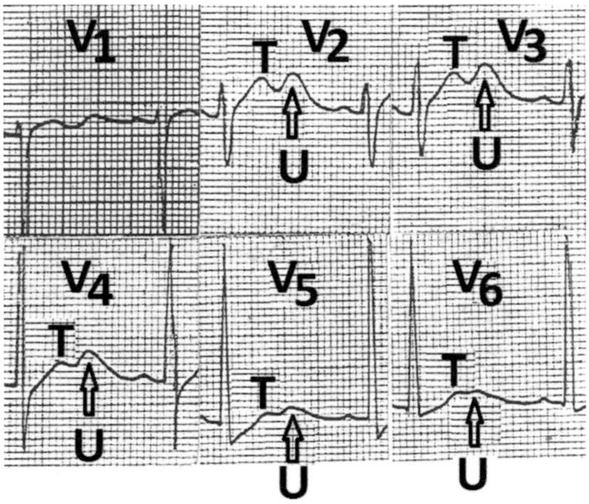

Fig. 46.10 Electrocardiogram from a patient of hypokalemia showing prominent U wave (marked ⇧). (Reproduced with permission from: Mittal SR. "U" wave. Cardiology Today-2018;22:218–24)

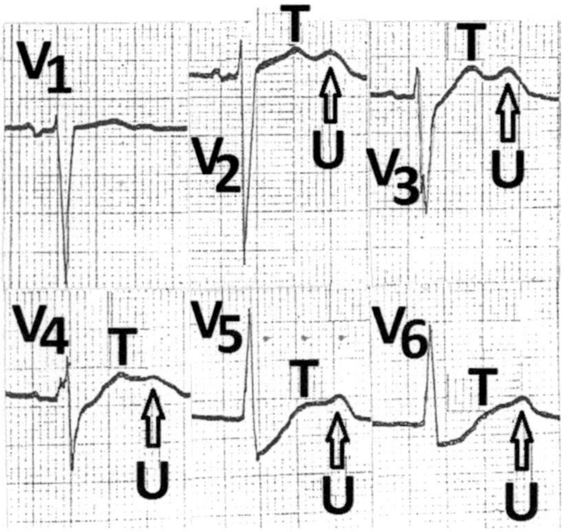

Fig. 46.11 Electrocardiogram from another patient of hypokalemia showing prominent U waves (marked ⇧). (Reproduced with permission from: Mittal SR. "U" wave. Cardiology Today-2018;22:218–24)

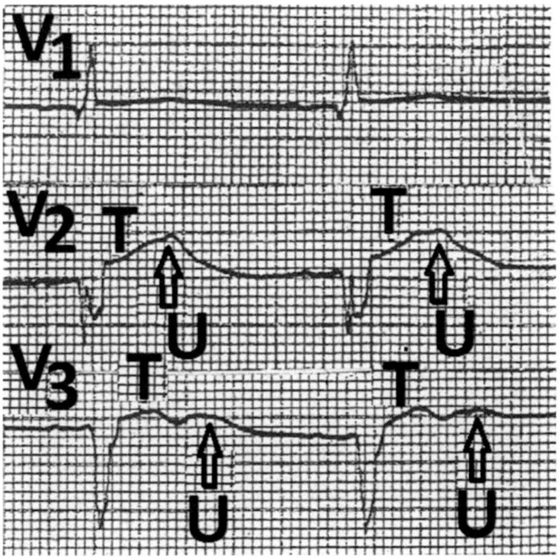

Fig. 46.12 Electrocardiogram is showing sinus bradycardia, fusion of the T wave, and the U wave (marked ⇧) with prolongation of QTU interval. (Reproduced with permission from: Mittal SR. "U" wave. Cardiology Today-2018;22:218–24.)

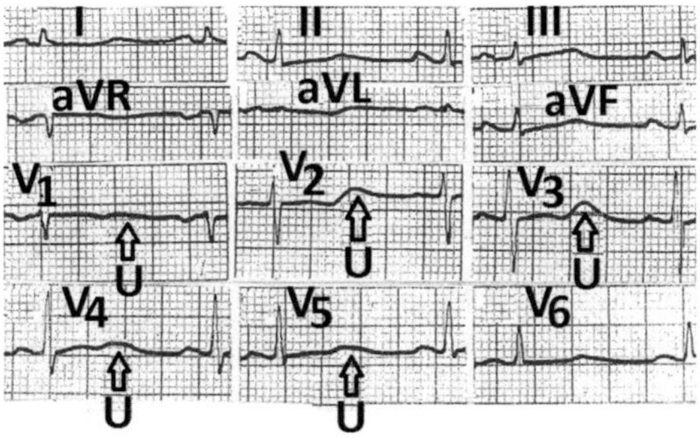

Fig. 46.13 Electrocardiogram from a case of hypothyroidism showing sinus bradycardia, low voltage QRS, decreased amplitude of the T waves, and prominent U waves (marked). (Reproduced with permission from: Mittal SR. "U" wave. Cardiology Today-2018;22:218–24)

(5) Hypothyroidism

Sinus bradycardia, low voltage QRS, and relative flattening of the T waves are other findings (Fig. 46.13). There is no deviation of the ST-segment.

(6) Hypothermia

(7) Inherited long QT syndrome type 3

It is characterized by prolonged QTc interval, late onset of the T wave due to long ST-segment, and frequently prominent U waves [5] (Fig. 46.14).

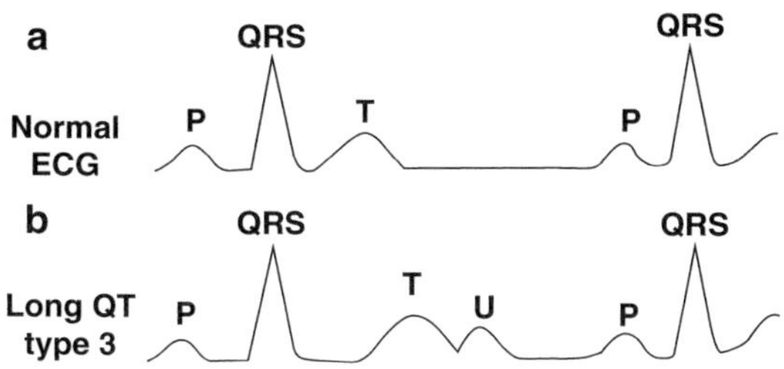

Fig. 46.14 Diagrammatic representation of (**a**) normal electrocardiogram and (**b**) electrocardiographic findings in type 3 inherited long QT syndrome showing prolonged QT interval, late onset of the T wave, and prominent U wave. (Reproduced with permission from: Mittal SR. "U" wave. Cardiology Today-2018;22:218–24)

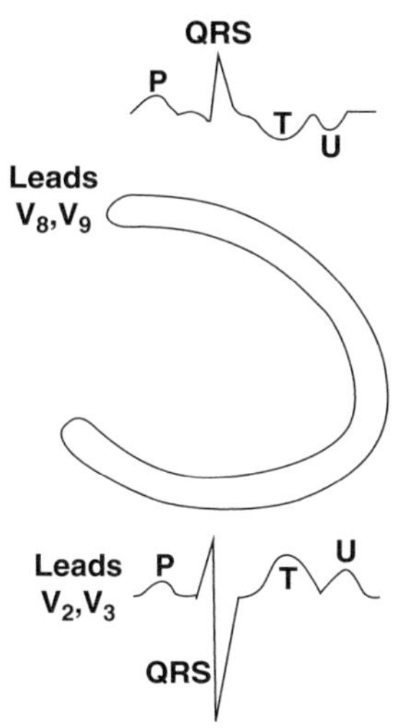

Fig. 46.15 Diagrammatic representation of inverted U wave on the posterior surface of the heart (leads V_8,V_9) producing reciprocal prominent U wave in the anterior precordial leads (V_2, V_3). (Reproduced with permission from: Mittal SR. "U" wave. Cardiology Today-2018;22:218–24)

(8) Significant narrowing of the left circumflex or the right coronary artery

In the clinical context of myocardial ischemia, prominent U wave in the precordial leads may suggest a significant narrowing of the left circumflex or the right coronary artery. [1] The exact mechanism is not known. It could be reciprocal to inversion of the U wave on the posterior surface of the heart (Fig. 46.15).

(9) Right ventricular compression

Yamagata et al. [6] have described a new appearance of abnormal

U waves in leads V_1 to V_3 following a reconstructed stomach tube after surgery for cancer esophagus. Plain chest CT and transthoracic echocardiography demonstrated compression of the right ventricular outflow tract and right ventricular free wall. The exact mechanism of new appearance of such abnormal U wave in this setting is not clear. Mechanical compression of the heart has been reported to produce ECG changes similar to the Brugada syndrome. [7, 8]

4 Inverted U Wave

U wave is normally inverted in lead aVR and sometimes in leads III, aVF, and V_1, if the T wave is inverted in these leads. Inversion in other leads is abnormal.

(1) Inversion in left precordial leads (V_3 to V_6)

It can occur in the following conditions.

(a) Acute myocardial ischemia [4]

It is a marker of significant stenosis of the left anterior descending coronary artery or the left main coronary artery. [9–11]

It may be seen

i. During attack of angina
ii. During coronary spasm (Prinzmetal's angina/variant angina). [12]
iii. During exercise testing, U wave inversion may precede ST-segment depression (Fig. 46.16).
iv. Acute anterior myocardial infarction. U wave inversion may precede other electrocardiographic changes (Figs. 46.17 and 46.18).

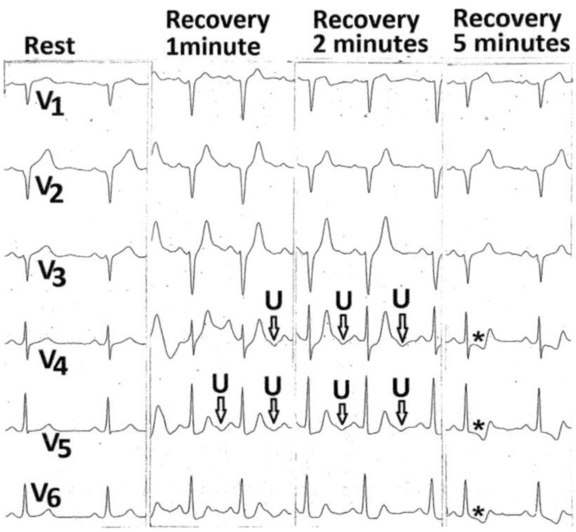

Fig. 46.16 Sequential electrocardiograms during treadmill stress testing of a patient of ischemic heart disease showing inversion of the U wave (recovery 1 and 2 minutes) (marked ⇩) before the appearance of the classical ST-segment depression (recovery 5 minutes, marked *). (Reproduced with permission from: Mittal SR. "U" wave. Cardiology Today-2018;22:218–24)

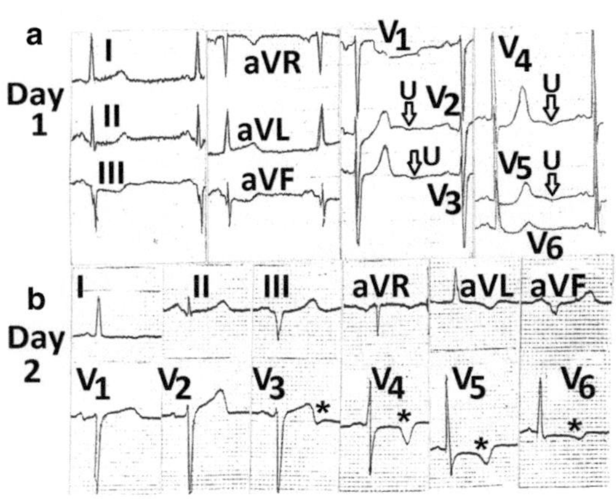

Fig. 46.17 Sequential electrocardiograms from a case of acute myocardial infarction showing (**a**) appearance of the inverted U wave (marked ⇩) before the appearance of classical T wave changes (marked *) (**b**). (Reproduced with permission from: Mittal SR. "U" wave. Cardiology Today-2018;22:218–24)

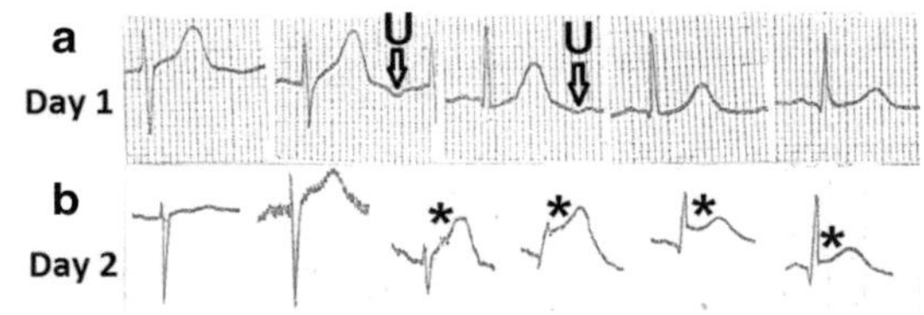

Fig. 46.18 Sequential electrocardiograms from another case of acute myocardial infarction (**a**) showing inversion of the U wave (marked ⇩) preceding the classical ST-segment changes (marked *) (**b**). (Reproduced with permission from: Mittal SR. "U" wave. Cardiology Today-2018;22:218–24)

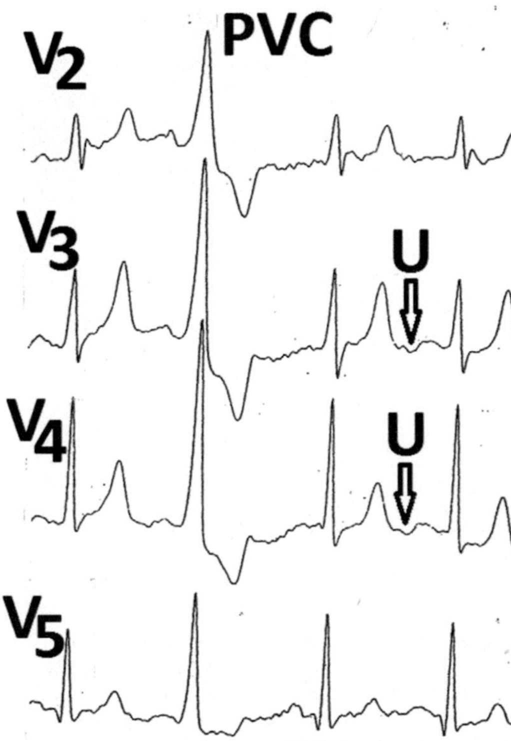

Fig. 46.19 Electrocardiogram is showing inversion of the U wave (marked ⇩) in the sinus beat following a premature ventricular contraction (PVC). (Reproduced with permission from: Mittal SR. "U" wave. Cardiology Today-2018;22:218–24)

v. Post extrasystolic (pause dependent) inversion of the U wave also suggests underlying coronary artery disease [12] (Fig. 46.19).

(b) Left ventricular hypertrophy
   (i) Systemic hypertension (Figs. 46.20 and 46.21).
   (ii) Aortic stenosis.
   (iii) Hypertrophic cardiomyopathy.

Presence of prominent U wave in the presence of electrocardiographic findings of left ventricular hypertrophy should lead to suspicion of underlying hypokalemia (Fig. 46.22).

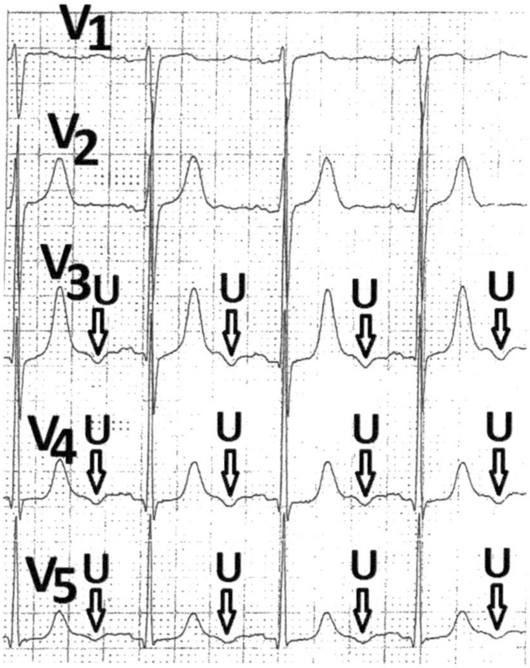

Fig. 46.20 Electrocardiogram from a patient of systemic hypertension showing inversion of the U wave (marked ⇩) in leads V_3 to V_6. (Reproduced with permission from: Mittal SR. "U" wave. Cardiology Today-2018;22:218–24)

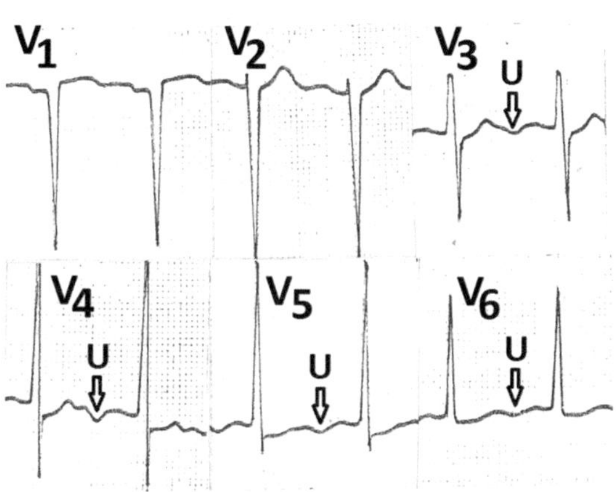

Fig. 46.21 Electrocardiogram from another patient of systemic hypertension showing inversion of the U wave (marked ⇩) in leads V_3 to V_6. (Reproduced with permission from: Mittal SR. "U" wave. Cardiology Today-2018; 22:218–24)

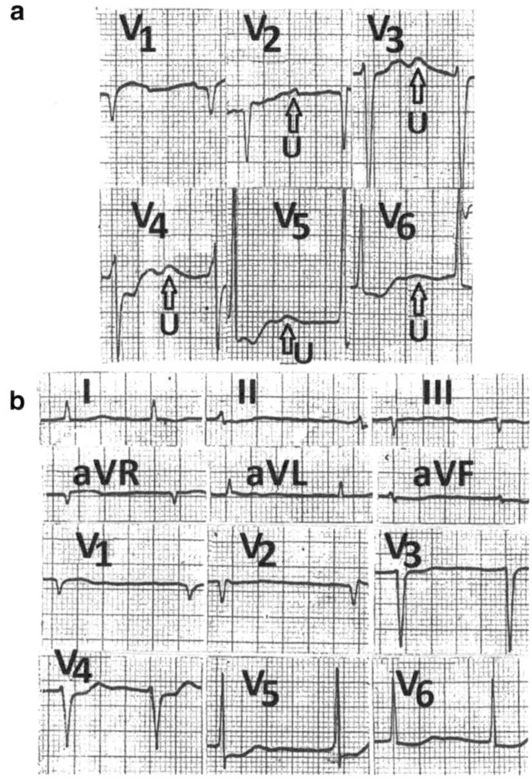

Fig. 46.22 (**a**) Electrocardiogram from a patient of systemic hypertension with hypokalemia showing prominent U waves (marked ⇧) in leads V2 to V6 (**b**) Electrocardiogram recorded after correction of hypokalemia showing normalization of the U waves. (Reproduced with permission from: Mittal SR. "U" wave. Cardiology Today-2018;22:218–24)

(c) Hyperthyroidism [10]

(2) Inversion of the U wave in leads I and aVL

It suggests lateral ischemia. Concomitant prominent U wave in leads V_2 to V_4 (Fig. 46.23) suggests the possibility of additional posterior ischemia.

(3) Inversion in right precordial leads (V_1, V_2)

U wave inversion in leads V_1, V_2 can be seen in cases of right ventricular hypertrophy. [10]

(4) Inversion in the inferior leads (II, III, aVF)

It is suggestive of inferior ischemia.

(5) Broad negative TU complex

Negative TU complex with prolonged QT interval can be seen in the setting of myocardial infarction (Figs. 46.24 and 46.25) or raised intracranial pressure (Fig. 46.26). As a reciprocal change TU fusion waves are positive in lead aVR, right-sided chest leads (V_3R to V_5R), and posterior chest leads (V_7 to V_9).

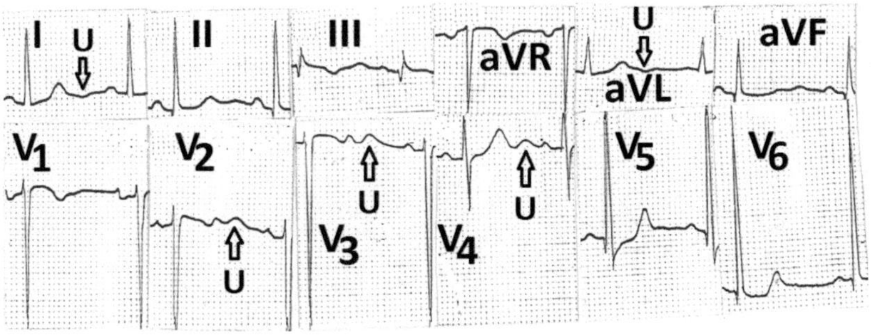

Fig. 46.23 Electrocardiogram is showing inversion of the U wave in leads I and aVL (marked ⇧) with prominent U wave in leads V_2 to V_4 (marked ⇩) (Reproduced with permission from: Mittal SR. "U" wave. Cardiology Today-2018;22:218–24)

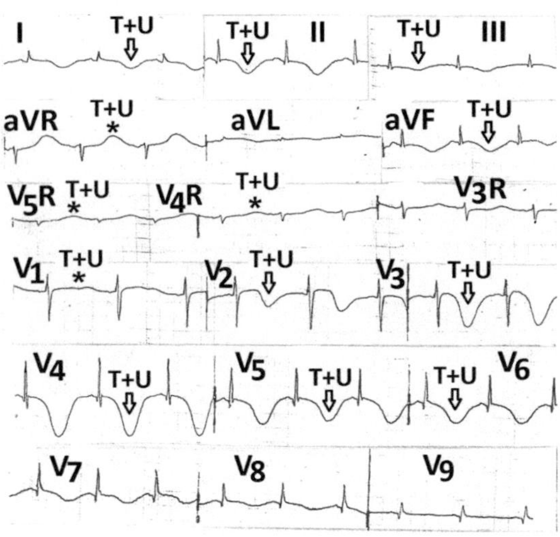

Fig. 46.24 Electrocardiogram is showing prominent negative TU complex (marked ⇩) with prolonged QT interval in leads I, II, III, aVF, and V_2 to V_7 with positive TU complex in leads aVR, V_5R to V_1 (marked *). (Reproduced with permission from: Mittal SR. "U" wave. Cardiology Today-2018;22:218–24)

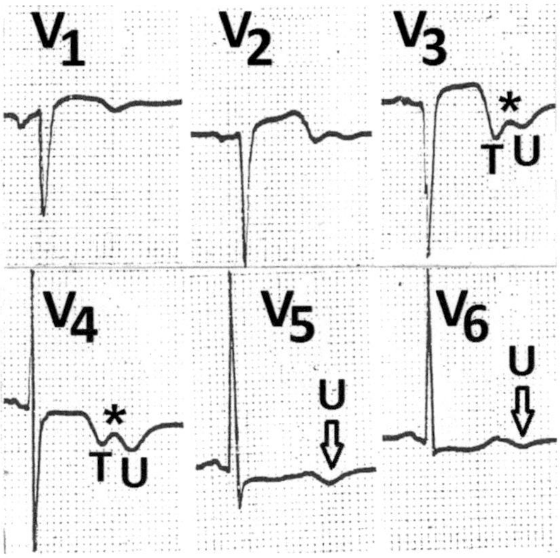

Fig. 46.25 Electrocardiogram is showing anterior myocardial infarction with negative TU complex (marked *) in leads V_3 and V_4 and inverted U wave in leads V_5 and V_6 (marked ⇩). (Reproduced with permission from: Mittal SR. "U" wave. Cardiology Today-2018;22:218–24)

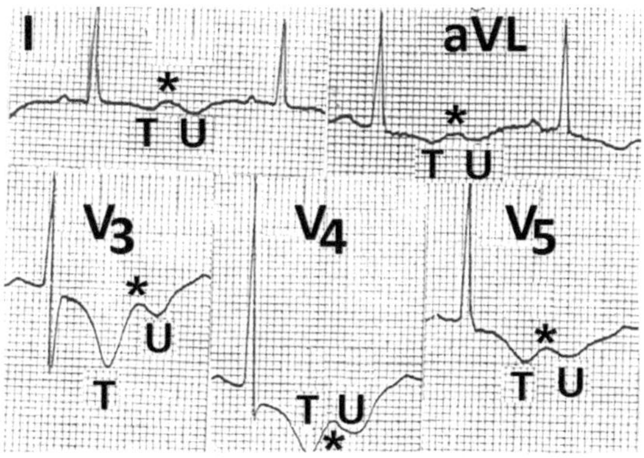

Fig. 46.26 Electrocardiogram shows negative TU complex (marked *) with prolongation of the QT interval in a case with raised intracranial pressure. (Reproduced with permission from: Mittal SR. "U" wave. Cardiology Today-2018;22:218–24)

Summary

- It is a repolarization wave present on the T-P segment.
- It is not seen in all ECGs.
- The mechanism of genesis is not clear. It is probably due to delayed repolarization of part of the myocardium.
- It is positive, low amplitude, smooth wave with broad base and rounded apex.
- Its polarity is the same as that of the preceding T wave.
- It is best seen in leads II, V_2, and V_3. It is isoelectric in lead aVL.
- Beginning of the U wave is usually merged with terminal part of the T wave obscuring exact point of end of the T wave. This interferes in correct estimation of the QT interval.
- During exercise-induced tachycardia, the P wave merges with the T wave leaving no space for the appearance of the U wave. During recovery, as the heart rate declines, the P wave separates from the T wave and the U wave appears between the T waves and P waves.
- U wave may be inverted in left ventricular hypertrophy or critical stenosis of left anterior descending coronary artery.
- U wave becomes more prominent during hypokalemia or use of class 1A antiarrhythmic drugs, for example, quinidine.

MCQs

Q1. U wave is usually most clearly visible in leads:

(a) V_3R to V_6R
(b) V_2,V_3,V_4
(c) V_7,V_8 V_9
(d) II, III, aVF

Q2. U wave is usually best seen at heart rate:

(a) <65/min
(b) 80 to 100/min
(c) 100 to 120/min
(d) >120/min

Q3. U wave is frequently seen in:

(a) Children
(b) Adults
(c) Middle aged
(d) Elderly

Q4. Amplitude of a normal U wave is:

(a) <2mm
(b) 2 to 4mm
(c) 4 to 6mm
(d) >6mm

Q5. Amplitude of a normal U wave is:

(a) <25% of T wave
(b) 30 to 40% of T wave
(c) 40 to 50% of T wave
(d) >50% of T wave

Q6. Amplitude of the U wave:

(a) Increases with decreasing heart rate
(b) Increases with increasing heart rate
(c) Decreases with decreasing heart rate
(d) Is not affected by heart rate

Q7. In the bifid T wave:

(a) Two apices are less than 0.16 seconds apart

(b) Two apices are more than 0.20 seconds apart
(c) Second peak is less than one-third of the amplitude of the first peak
(d) Amplitude of second peak is more than 50% of the amplitude of the first peak

Q8. In fusion of the T wave and the U wave:

(a) Two apices are less than 0.16 seconds apart
(b) Two apices are more than 0.16 seconds apart
(c) Second peak is less than one-third of the amplitude of the first peak
(d) Amplitude of the second peak is more than 50% of the amplitude of the first peak

Q9. Prominent U waves suggests:

(a) Hyperkalemia
(b) Hyponatremia
(c) Hypercalcemia
(d) hypomagnesemia

Q10. Hypokalemia produces:

(a) Tachycardia
(b) ST-segment depression
(c) Tall T wave
(d) Tall U wave

Q11. Prominent U waves are seen in:

(a) Hypokalemia
(b) Hypomagnesemia
(c) Hyperthyroidism
(d) Hyperthermia

Q12. Amiodarone produces:

(a) Prominent U wave
(b) Broad TU fusion wave
(c) Torsades de pointes
(d) All

Q13. Broad T-U fusion waves are produced by:

(a) Amiodarone
(b) Dofetilide
(c) Increased intracranial pressure
(d) Thyrotoxicosis

Q14. Inherited long QT syndrome type 3 is characterized by:

(a) Prolonged QT interval
(b) Early onset T wave
(c) Prominent U wave
(d) Inverted U wave

Q15. In the context of myocardial ischemia, prominent U wave in precordial leads (V_2 toV_4) may suggest significant narrowing of the:

(a) Left anterior descending coronary artery
(b) Ramus artery
(c) Left circumflex artery
(d) Right coronary artery

Q16. U wave is normally inverted in lead:

(a) III
(b) aVR
(c) V_9
(d) V_4R

Q17. U wave inversions in leads V_3 to V_6 suggest:

(a) Myocardial ischemia
(b) Left ventricular hypertrophy
(c) Hypothyroidism
(d) Hypermagnesemia

Q18. U wave inversions in leads V_3 to V_5 suggests significant stenosis of the:

(a) Left main coronary artery
(b) Left anterior descending coronary artery
(c) Left circumflex coronary artery
(d) Posterior descending coronary artery

Q19. U wave inversions in leads II, III, and aVF suggests significant stenosis of:

(a) Left anterior descending coronary artery
(b) Posterior descending coronary artery

(c) Right ventricular branch of right coronary artery
(d) First diagonal artery

Q20. U wave inversions in leads V_1 and V_2 can be seen in:

(a) Dilated cardiomyopathy
(b) Myocarditis
(c) Right ventricular ischemia
(d) Right ventricular compression

Q21. U wave inversions can be seen in:

(a) Systemic hypertension
(b) Aortic stenosis
(c) Hypertrophic cardiomyopathy
(d) Hypothyroidism

Q22. U wave occurs during:

(a) Early part of action potential
(b) Mid part of action potential
(c) Terminal part of action potential
(d) After completion of action potential

Q23. U wave is not:

(a) Of low amplitude
(b) Smooth
(c) Rounded with broad base
(d) Peaked with narrow base

Q24. U wave is best visible during:

(a) Bradycardia
(b) Exercise-induced tachycardia
(c) Recovery from exercise-induced tachycardia
(d) Atrioventricular block

Q25. Polarity of the U wave is:

(a) Same as that of the preceding QRS complex
(b) Same as that of the following QRS complex
(c) Same as that of the preceding T wave
(d) Flat

Q26. Inverted U waves are seen in:

(a) LVH
(b) Critical narrowing of LAD
(c) Critical narrowing of LCX
(d) Atrioventricular block

Q27. U waves are prominent in:

(a) Hypokalemia
(b) Hypermagnesemia
(c) Hypercalcemia
(d) Hyponatremia

Q28. U waves are prominent during use of:

(a) Beta blockers
(b) Calcium channel blockers
(c) Digitalis
(d) Quinidine

Answers

(1) b (2) a (3) d (4) a (5) a (6) a (7) a,d (8) b,c (9) d (10) b,d (11) a,b (12) a,b (13) a,b,c (14) a,c (15) c,d (16) b (17) a,b (18) a,b (19) b (20) c (21) a,b,c (22) d (23) d (24) c (25) c (26) a,b (27) a (28) d

References

1. www.cardiocases.com/en/ecg/traces/normal-and-pathological-ECG-uwave/u-wave.
2. U wave – an overview/Science Direct Topics. https://www.sciencedirect.com/topics/medicine-and-dentistry/u-wave.
3. De Luna AB, Goldwasser D, Fiol M, Bayes GA. Surface electrocardiography. In: Fuster V, Walsh RA, Harrington RA, editors. Hurst's the heart. New York: Mc Graw Hill; 2017. p. 252–317.
4. Wagner GS, Wang TY. Miscellaneous conditions. In: Wagner GS, editor. Marriott's Practical Electrocardiography. New Delhi: Wolters Kluwer; 2001. p. 209–37.
5. Perez-Riera AR, Barbosa Barros R, Daminello Raimundo R, da Costa de Rezende Barbosa MP, Esposito sorpreso IC, de Abreu LC. The congenital long QT syndrome type 3. An update. Indian Pacing Electrophysiol J. 2018;18:25–35.
6. Yamagata K, Uno K, Mori K, Seto Y. Unusual U wave induced by reconstructed retrosternal esophagus. Clin Case Rep. 2015;3:949–51.

7. Tarin N, Farre J, Rubio JM, Tunon J, Castro DJ. Brugada like electrocardiographic pattern in a patient with mediastinal tumor. Pacing Clin Electrophysiol. 1999;22:1264–6.
8. Shimizu W. Acquired forms of Brugada syndrome. J Electrocardiol. 2005;38:22–5.
9. Gerson MC, Mc Henry PL. Resting U wave inversion as a marker of stenosis of the left anterior descending coronary artery. Am J Med. 1980;69:545–50.
10. Burns E. U wave. https!//litfl.com/u-wave-ecg-library/ 2018.
11. Mirvis DM, Goldberger AL. Electrocardiography. In: Zipes DP, Libby P, Bonow RO, Mann DL, Tomaselli GF, editors. Braunwald's Heart Disease. Canada: Elsevier; 2019. p. 117–53.
12. Nice seeing "U" again. Heeps:/ecg.utah.edu/lesson/12.
13. Sovari AA, Farokhi F, Kocheril AG. Inverted U wave, a specific electrocardiographic sign of cardiac ischemia. Am J Emerg Med. 2007;25:235–7.

47 Electrical Alternans

Electrical alternans is an electrocardiographic finding defined as a change in the electrical activity of the heart in alternate beats in the presence of a regular rhythm. [1] When the rhythm is irregular, it is not called electrical alternans (Fig. 47.1).

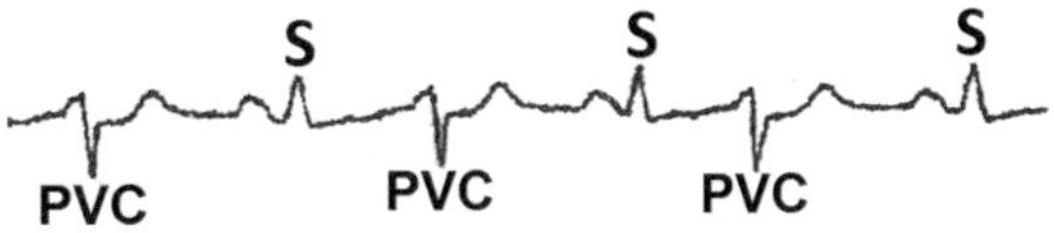

Fig. 47.1 It is not electrical alternans. Rhythm is irregular. 2nd, 4th, and 6th beats are sinus beats (S) with preceding normal P wave and normal PR interval. 1st, 3rd, and 5th beats are premature ventricular beats (PVC). (Reproduced with permission from: Mittal SR. Electrical Alternans. Cardiology Today-2005;9:297)

All components of the electrocardiogram, that is, the P wave, the QRS complex, the ST- segment, and the TU complex, can show alternans.

QRS alternans is the commonest. It is frequently seen in the presence of sinus tachycardia and usually presents as aberrant (slow) conduction in the right bundle branch [2] in alternate beats (Fig. 47.2).

When seen in the presence of tachycardia, it does not have any clinical significance. [2] When this electrocardiographic finding is seen in the absence of tachycardia (Fig. 47.3), there is usually an underlying organic heart disease.

In such a situation, it may be accompanied by pulsus alternans—that is, change in the volume of the arterial pulse in alternate beats. QRS alternans with sinus tachycardia, low voltage, and nonspecific flattening of the T waves is a specific (but not sensitive) marker of cardiac tamponade (Fig. 47.4).

QRS alternans in this setting occurs due to swinging of the heart within a tense pericardial effusion. [3] QRS voltage increases when the heart is close to the chest wall and decreases when the heart is away from the chest wall.

QRS alternans can be seen during supraventricular tachycardia (Fig. 47.5).

At rates less than 180/minute, QRS alternans in supraventricular tachycardia is seen frequently in orthodromic atrioventricular reciprocating tachycardia with an underlying accessory pathway. [4] However, at rates more than 200/minute, QRS alternans can occur in any supraventricular tachycardia.

PR interval duration alternans and QRS duration alternans can be seen when preexcitation is seen in alternate beats [5] (Fig. 47.6).

ST-segment alternans (Fig. 47.7) is a marker of electric instability in acute ischemia where it may precede ventricular tachycardia. [6]

TU wave alternans (Fig. 47.8) can be seen in the presence of long QTU interval. In this setting it is a marker of imminent risk of ventricular tachyarrhythmia. [7]

S. Mittal, *Insights into Electrocardiograms with MCQs*,
https://doi.org/10.1007/978-981-99-0127-2_47

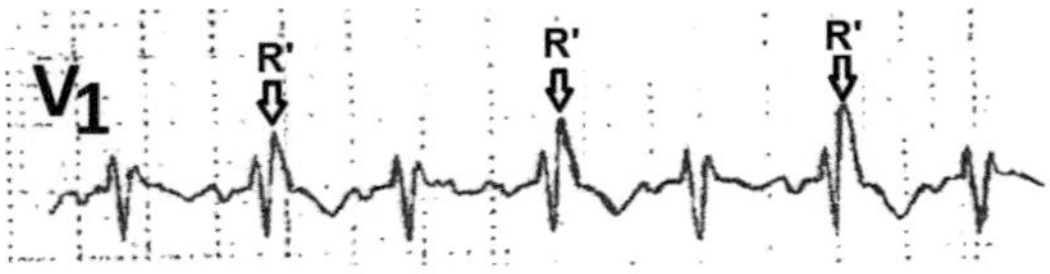

Fig. 47.2 Electrical alternans during tachycardia. 2nd, 4th, and 6th beats show prominent R' (marked) wave due to slow conduction (aberration) in the right bundle branch. (Reproduced with permission from: Mittal SR. Electrical Alternans. Cardiology Today-2005;9:297)

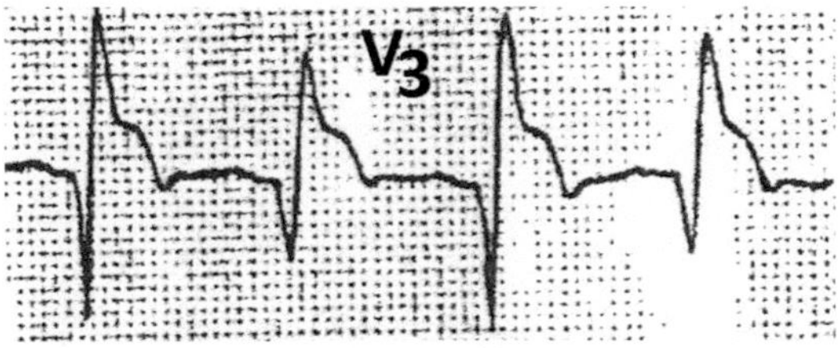

Fig. 47.3 Electrical alternans during normal heart rate in a patient of right bundle branch block and anterior myocardial infarction. (Reproduced with permission from: Mittal SR. Electrical Alternans. Cardiology Today- 2005;9:297)

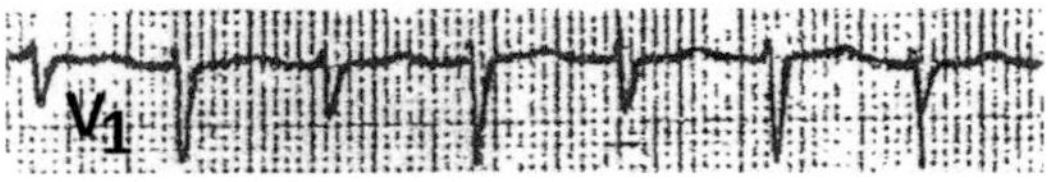

Fig. 47.4 Electrical alternans with sinus tachycardia in a patient of cardiac tamponade. (Reproduced with permission from: Mittal SR. Electrical Alternans. Cardiology Today-2005;9:297)

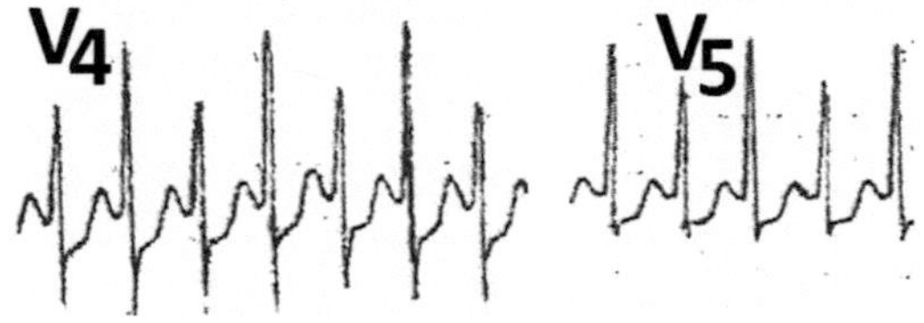

Fig. 47.5 QRS alternans during supraventricular tachycardia. (Reproduced with permission from: Mittal SR. Electrical Alternans. Cardiology Today-2005;9:297)

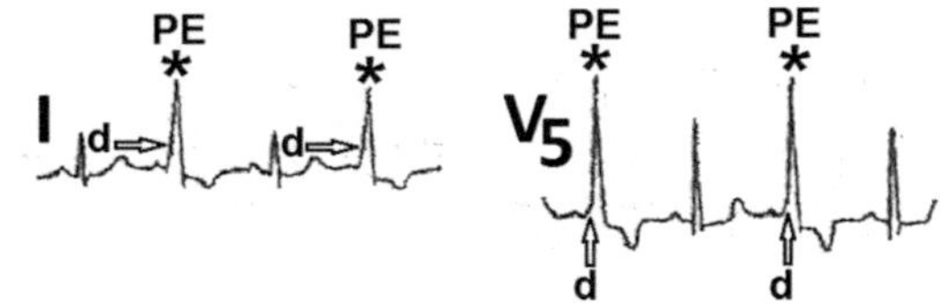

Fig. 47.6 Preexcitation in alternate beat (marked *). Alternate beat shows short PR interval and broad QRS complex with the presence of a delta wave (marked d). (Reproduced with permission from: Mittal SR. Electrical Alternans. Cardiology Today-2005;9:297)

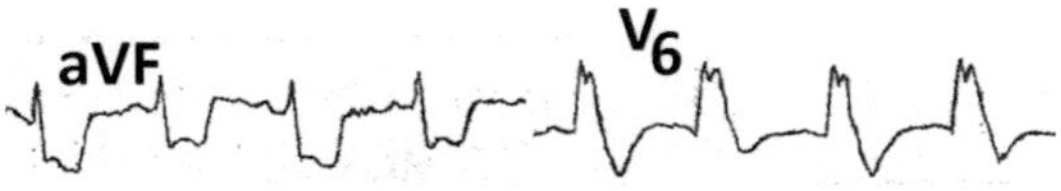

Fig. 47.7 ST-segment alternans. (Reproduced with permission from: Mittal SR. Electrical Alternans. Cardiology Today-2005;9:297)

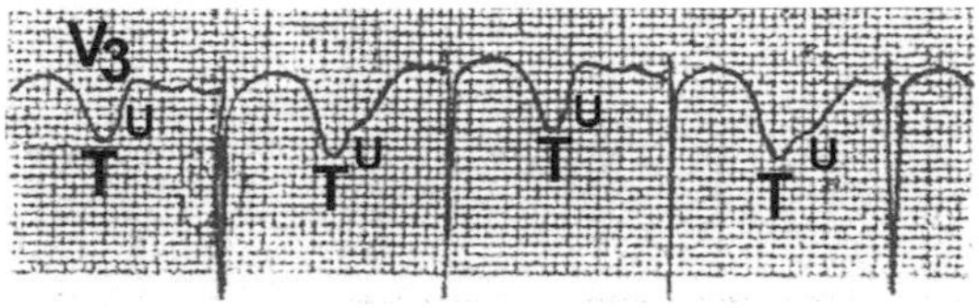

Fig. 47.8 TU wave alternans. (Reproduced with permission from: Mittal SR. Electrical Alternans. Cardiology Today-2005;9:297)

Summary

- Change in the electrical activity of the heart in alternate beats in the presence of a regular rhythm is called electrical alternans.
- QRS alternans can occur due to aberrant ventricular conduction in alternate beats, swinging of the heart within a tense pericardial effusion, or when there is fast ventricular rate during a supraventricular tachycardia.

- Preexcitation in alternate cardiac cycle can produce alternans in PR interval and QRS duration.
- ST-segment alternans is seen in acute ischemia and is a marker of electric instability.
- TU wave alternans is seen in the presence of long QTU interval and is a marker of risk of ventricular tachyarrhythmia.

MCQ

Q1. In electrical alternans, the rhythm is:

(a) Regular
(b) Regularly irregular
(c) Irregularly irregular

Q2. Which condition can produce QRS alternans?

(a) Aberrant ventricular conduction in alternate beat
(b) Ventricular bigeminy
(c) Ventricular trigeminy
(d) Ventricular couplet

Q3. QRS alternans in absence of tachycardia is seen in:

(a) Athletes
(b) Vagal overdrive
(c) Organic heart disease
(d) Atrioventricular block

Q4. QRS alternans, with sinus tachycardia, low voltage of QRS, and nonspecific flattening of T wave is seen in:

(a) Pericardial effusion
(b) Cardiac tamponade
(c) Myxedema
(d) Hypokalemia

Q5. QRS alternans is seen in all cases of:

(a) Pericardial constriction
(b) Pericardial effusion
(c) Cardiac tamponade
(d) None

Q6. QRS alternans is diagnostic of:

(a) Myocarditis
(b) Dilated cardiomyopathy
(c) Hypertrophic cardiomyopathy
(d) None

Q7. QRS alternans is seen in:

(a) Atrial flutter with 2:1 conduction
(b) Atrial fibrillation with controlled ventricular rate
(c) Atrioventricular nodal reentrant tachycardia
(d) Atrial tachycardia

Q8. QRS alternans is frequent in:

(a) Atrial tachycardia
(b) Orthodromic atrioventricular reciprocating tachycardia
(c) Antidromic atrioventricular reciprocating tachycardia
(d) Sinus tachycardia

Q9. QRS alternans is most frequent at ventricular rate between:

(a) 60 to 100/minute
(b) 100 to 150/minute
(c) 150 to 200/minute
(d) 200 to 250/minute

Q10. QRS alternans predisposes to:

(a) Atrial fibrillation
(b) Ventricular tachycardia
(c) Ventricular fibrillation
(d) None

Q11. PR interval alternans with QRS interval alternans can be seen in:

(a) 2:1 Atrioventricular block
(b) Preexcitation in alternate beat

(c) Aberrant ventricular conduction in alternate beats
(d) All

Q12. ST-segment alternans can be seen in:

(a) Hyperventilation
(b) Myocarditis
(c) Exercise
(d) Acute ischemia

Q13. ST-segment alternans may precede:

(a) Atrial flutter
(b) Atrial fibrillation
(c) Atrioventricular block
(d) Ventricular tachycardia

Q14. TU wave alternans can be seen in:

(a) Congenital long QT syndrome
(b) Acquired long QT syndrome
(c) Microvascular angina
(d) Exercise

Q15. TU wave alternans is a marker of imminent risk of:

(a) Atrial flutter
(b) Atrioventricular block
(c) Myocardial infarction
(d) Ventricular tachyarrhythmias

Answers

(1) a (2) a (3) c (4) b (5) d (6) d (7) c (8) b (9) d (10) d (11) b (12) d (13) d (14) a,b (15) d

References

1. Chung DK, Chung EK. Ventricular electrical alternans. WV Med J. 1971;67:99–100.
2. Marriot HJL, Conover MB. Aberrant ventricular conduction. In: Marriot HJL, Conver MB, editors. Advanced concepts in arrhythmias. Mosby: St Louis; 1998. p. 215–36.
3. Rigney DR, Goldberger AL. Nonlinear mechanics of the heart's swinging during pericardial effusion. Am J Physiol. 1989;257:H1292–305.
4. Green M, Heddle B, Dassen W, et al. The value of QRS alternans in diagnosing the site of origin of narrow QRS supraventricular tachycardia. Circulation. 1983;68:368–73.
5. De Pauda F, Pereimho A, Lopes MG. Conduction defects. In: Mc Farlane PW, Veitch Lawrie TD, editors. Comprehensive electrocardiology. Pergamon Press: New York; 1989. p. 459–509.
6. Rosenbaum DS, Jackson LE, Smith JM, et al. Electrical alternans and vulnerability to ventricular arrhythmia. N Engl J Med. 1994;330:235–41.
7. Habbab MA, el -Sherif N. TU alternans, long QTU and torsade de points: clinical and experimental observations. Pacing Clin Electrophysiol. 1992;15:916–31.

48 Electrocardiogram in Asymptomatic Acyanotic Congenital Heart Disease

Nowadays, diagnosis of congenital heart disease is made by detailed echocardiographic evaluation. However, electrocardiogram is still an important part of the clinical diagnosis of congenital heart disease. Clinical examination combined with an electrocardiogram help in arriving at a reasonable differential diagnosis in an asymptomatic, acyanotic patient with congenital heart disease.

The role of surface electrocardiogram in asymptomatic, acyanotic congenital cardiovascular disease can be discussed under the following heads.

(A) **Patient Without Any Clinical Abnormality**

(1) **Normal Electrocardiogram**

The following conditions cannot be ruled out even when clinical examination and electrocardiogram are normal.

(a) **Anomalies of the aorta**
- (i) Unruptured small aneurysm of sinus of Valsalva [1].
- (ii) Uncomplicated right aortic arch.
- (iii) Uncomplicated aneurysm of the thoracic aorta.
- (iv) Mild coarctation of the aorta without significant gradient.

(b) **Anomalies of the systemic veins**
- (i) Persistent left superior vena cava draining into the coronary sinus [2].
- (ii) Congenital absence of the hepatic portion of the inferior vena cava with azygous continuation.

(c) **Narrow small patent ductus arteriosus** [3].

(d) **Congenital partial absence of the pericardium** [4].

(e) **Anomalous drainage of one or two pulmonary veins.**

(f) **Isolated congenital anomalies of the coronary arteries.**
- (i) Anomalies in number, origin, course, or termination (except anomalous origin of the left coronary artery from the pulmonary artery).
- (ii) Small coronary arteriovenous fistula.

(g) **Congenital anomalies of the coronary sinus.**

(h) **Idiopathic dilatation of the right atrium.**

(i) **Isolated inversion of the ventricles.**

(j) **Intracardiac defects.**
- (i) Patent foramen ovale.
- (ii) Small atrial septal defect [5].
- (iii) Nonobstructive left ventricular inflow lesions.
  - • Cor-triatriatum
  - • Supravalvular ring
  - • Congenital mitral stenosis

(k) **Mild prolapse of the atrioventricular valves without significant regurgitation** [6].

(l) **Isolated hypertrophy of the papillary muscle.**

S. Mittal, *Insights into Electrocardiograms with MCQs*,
https://doi.org/10.1007/978-981-99-0127-2_48

Echocardiography and/or angiography is needed for the detection of these anomalies.

This list is likely to increase with advancing knowledge.

(2) **Abnormal electrocardiogram**

(a) **Isolated electrocardiographic abnormalities without clinically detectable abnormalities.**

(i) **Abnormalities of the P wave**

- P waves inverted in leads I and aVL and positive in lead aVR with normal P waves in precordial leads. It suggest situs inversus with levocardia [7] (provided reversal of the arm leads has been excluded) [8] (Fig. 48.1a,b).
- P waves inverted in lead I and isoelectric or negative in leads V_5,V_6 suggest left atrial rhythm [9] (Fig. 48.2) without any other congenital anomaly (exclude sinus venosus ASD). Left atrial rhythm can also produce "dome and dart" shaped P waves in lead V_1 or V_2 [10].

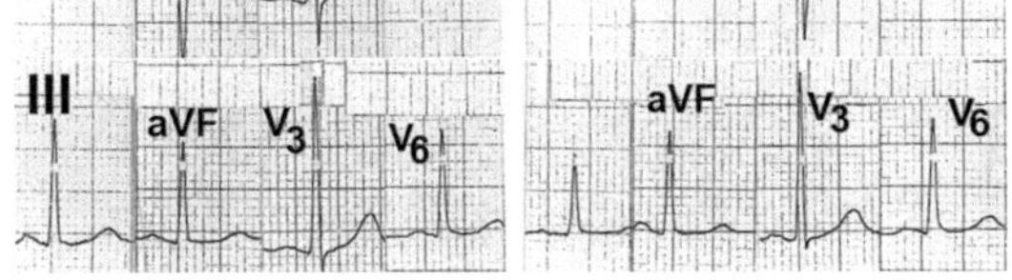

Fig. 48.1 (**a**) Electrocardiogram shows negative P-QRS-T in lead I and positive P-QRS-T in lead aVR. Precordial leads are normal. Findings are suspicious of isolated situs inversus. (**b**) Electrocardiogram recorded after interchanging the leads of the left upper limb and the right upper limb. Normalization of limb leads suggests diagnosis of technical dextrocardia rather than isolated situs inversus. (Reproduced with permission from: Mittal SR. Electrocardiogram in Asymptomatic Acyanotic Congenital Heart Disease. Cardiology Today-2019;23:157–65)

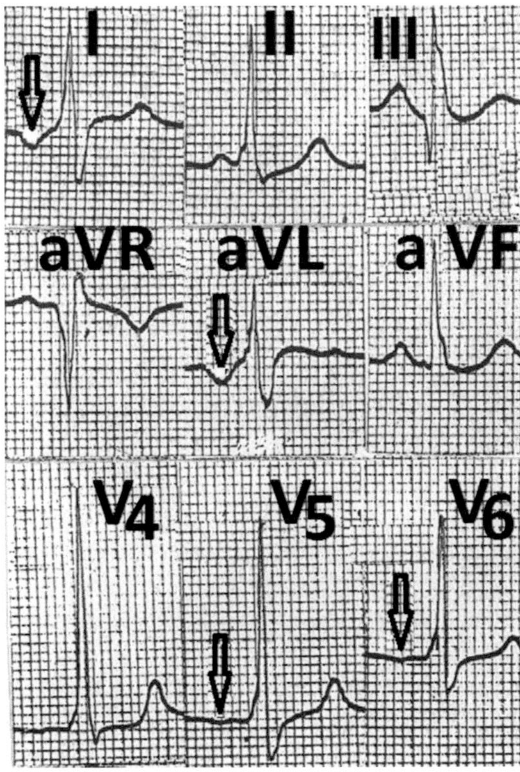

Fig. 48.2 Electrocardiogram is showing inverted P waves in leads I, aVL, V_5, and V_6 (marked ⇩) suggestive of left atrial rhythm. (Reproduced with permission from: Mittal SR. Electrocardiogram in Asymptomatic Acyanotic Congenital Heart Disease. Cardiology Today-2019;23:157–65)

- P waves inverted in leads II, III, and aVF suggest low atrial rhythm (Fig. 48.3) (exclude superior vena cava type sinus venosus ASD) [11].

(ii) **Abnormalities of the PR interval**

- Isolated first or second degree atrioventricular block.
- Isolated preexcitation.

(iii) **Abnormalities of the QRS complexes**

- Inversion of precordial leads—qR in lead V_1 and rS in lead V_6 suggest dextrocardia. It can occur with normal P waves (situs solitus with dextrocardia) [12] (Fig. 48.4) or with inversion of the P waves in lead I and positive P wave in

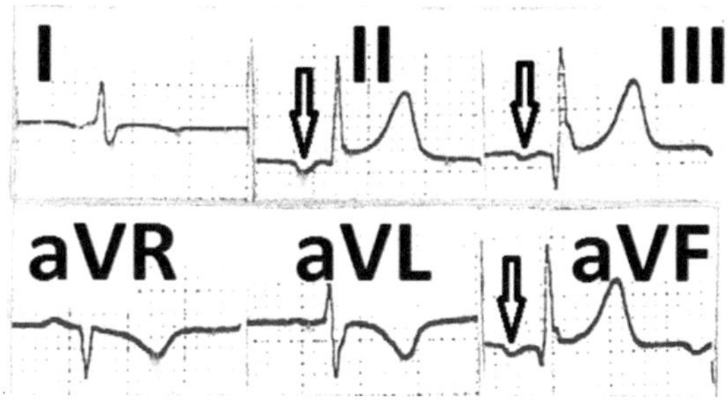

Fig. 48.3 Electrocardiogram is showing inversion of the P waves in leads II, III, and aVF (marked ⇩) suggestive of low atrial rhythm. (Reproduced with permission from Mittal SR. Electrocardiogram in Asymptomatic Acyanotic Congenital Heart Disease. Cardiology Today-2019;23:157–65)

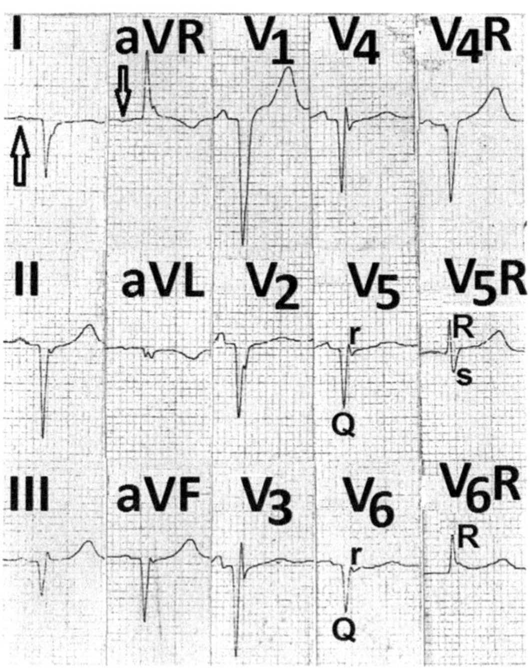

Fig. 48.4 Electrocardiogram is showing normal polarity of the P wave—positive in lead I (marked ⇧) and inverted in lead aVR (marked ⇩). It suggests situs solitus. Leads V_5 and V_6 show Qr configuration and leads V_5R and V_6R show prominent R waves. These findings suggest dextrocardia. (Reproduced with permission from: Mittal SR. Electrocardiogram in Asymptomatic Acyanotic Congenital Heart Disease. Cardiology Today-2019;23:157–65)

lead aVR (situs inversus with dextrocardia (Figs. 48.5 and 48.6).

- Isolated right bundle branch block (Fig. 48.7).

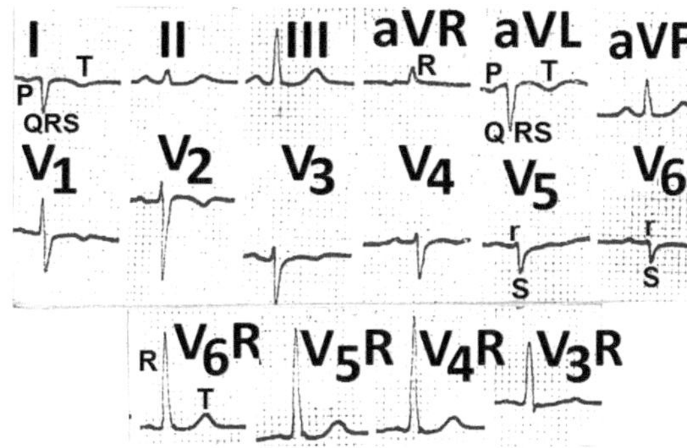

Fig. 48.5 Electrocardiogram is showing negative P-QRS-T in lead I and lead aVL. Lead aVR shows positive QRS. There is a progressive decrease in amplitude of the R wave in leads V_2 to V_6 with rs or QS complex and negative T wave resembling normal right precordial leads. Leads V_1 to V_6R show progressive increase in R wave amplitude and positive T waves resembling normal left precordial leads. These findings are suggestive of situs inversus with dextrocardia. (Reproduced with permission from: Mittal SR. Electrocardiogram in Asymptomatic Acyanotic Congenital Heart Disease. Cardiology Today-2019;23:157–65)

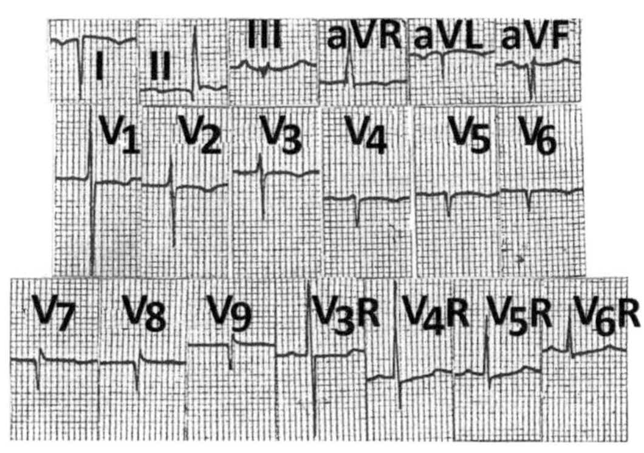

Fig. 48.6 Electrocardiogram is showing another example of situs inversus with dextrocardia. (Reproduced with permission from: Mittal SR. Electrocardiogram in Asymptomatic Acyanotic Congenital Heart Disease. Cardiology Today-2019;23:157–65)

- Isolated pre-excitation (Fig. 48.8)
- Presence of q waves in lead V1 to V4R and the absence of q waves in leads V5 to V9 suggest congenitally corrected transposition of the great arteries (isolated inversion of the ventricles) [13, 14] (Fig. 48.9). Echocardiography is needed for conformation.
- Prominent R wave with deep, inverted T wave in leads V_3 to V_5 suggest isolated

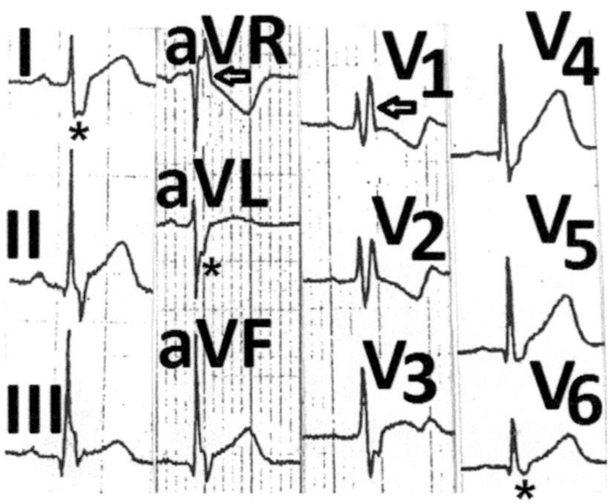

Fig. 48.7 Electrocardiogram is showing isolated right bundle branch block. Leads V_1 and aVR show prominent, terminal, and slurred R wave (marked ⇦). Leads I, aVL, and V_6 show terminal, slurred S wave (marked *) (Reproduced with permission from: Mittal SR. Electrocardiogram in Asymptomatic Acyanotic Congenital Heart Disease. Cardiology Today-2019;23:157–65)

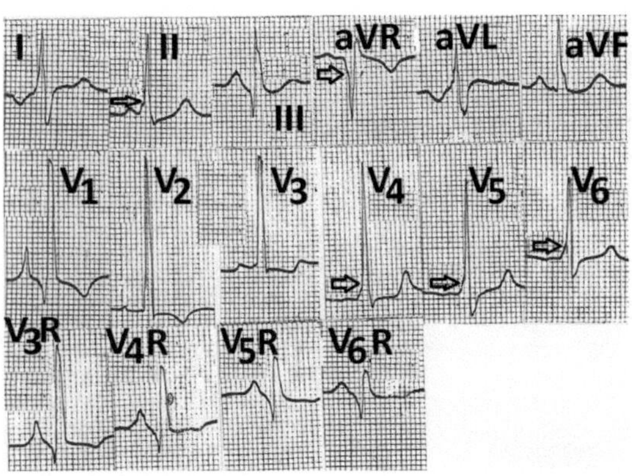

Fig. 48.8 Electrocardiogram is showing short PR interval, broad QRS with slurring of the initial part (delta wave) (marked as ⇨). Findings are diagnostic of preexcitation. (Reproduced with permission from Mittal SR. Electrocardiogram in Asymptomatic Acyanotic Congenital Heart Disease. Cardiology Today-2019;23:157–65)

apical hypertrophic cardiomyopathy. Echocardiography is needed for confirmation.

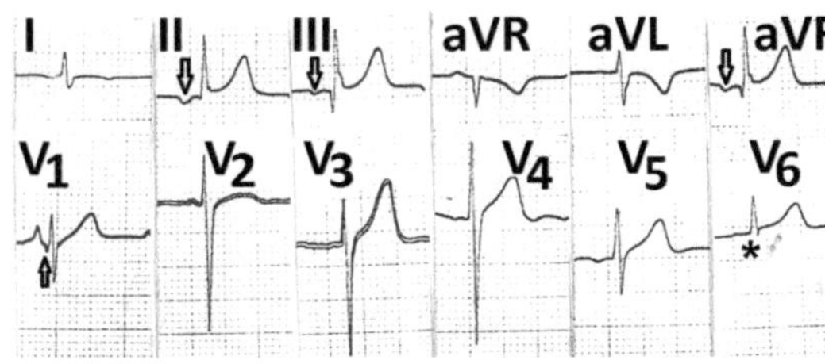

Fig. 48.9 Electrocardiogram is showing inverted P waves in leads II, III, and aVF (marked ⇩) suggestive of low atrial rhythm. Small q waves in lead V_1 (marked ⇧) and no q waves in leads V_5 and V_6 (marked *) suggest inversion of the ventricles. (Reproduced with permission from: Mittal SR. Electrocardiogram in Asymptomatic Acyanotic Congenital Heart Disease. Cardiology Today-2019;23:157–65)

(iv) **Abnormalities of the QT interval**

- Hereditary long QT syndromes
- Hereditary short QT syndrome.

(B) **Patient with Bradycardia**

Bradycardia is defined as slow heart rate for age. Heart rate <100 per minute in newborns and infants, less than 80 per minute in younger children and less than 60 per minute in adolescents and adults is considered bradycardia.

- **Electrocardiogram**

Slow ventricular rate for age, fast atrial rate, narrow QRS, and no relation between the P waves and the QRS complexes suggest congenital complete atrioventricular block.

– Without any other congenital heart disease—Isolated congenital complete atrioventricular block [15] (Fig. 48.10).
– With q waves in lead V_1 to V_4R and the absence of q waves in leads V_6 to V_9 (absence of q wave in left precordial leads is more important) suggest congenital complete atrioventricular block as a part of congenitally corrected transposition of the great arteries

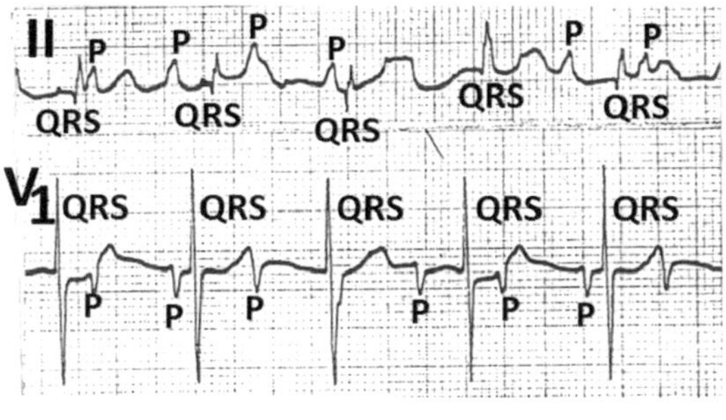

Fig. 48.10 Electrocardiogram is showing that P waves are regular at a rate of 142/minute, QRS complexes are regular at a rate of 83/per minute, QRS is narrow (0.08 second). There is no relation between the P waves and the QRS complexes. These findings suggest congenital complete atrioventricular block. (Reproduced with permission from: Mittal SR. Electrocardiogram in Asymptomatic Acyanotic Congenital Heart Disease. Cardiology Today-2019;23:157–65)

[14] (isolated inversion of the ventricles). Echocardiography is needed for confirmation.

(C) **Patient with Systolic Murmur Along Left Upper Sternal Border**

(1) **Normal electrocardiogram**

Right axis deviation, prominent P waves in lead II, prominent terminal R wave in lead aVR, and prominent R wave in lead V_1 are normal in neonates due to right ventricular dominance [16] (Fig. 48.11).

These changes normalize by the end of one month as left ventricular dominance becomes clear and pulmonary vascular resistance declines.

(a) **No organic congenital heart disease**
- (i) Hemic murmur (vibrations from normal flow in the pulmonary artery)
- (ii) Narrow anteroposterior diameter of the chest
  - – Pectus excavatum.
  - – Straight back.

(b) **Congenital heart diseases**
- (i) Small atrial septal defect.
- (ii) Mild pulmonary valve stenosis [17].

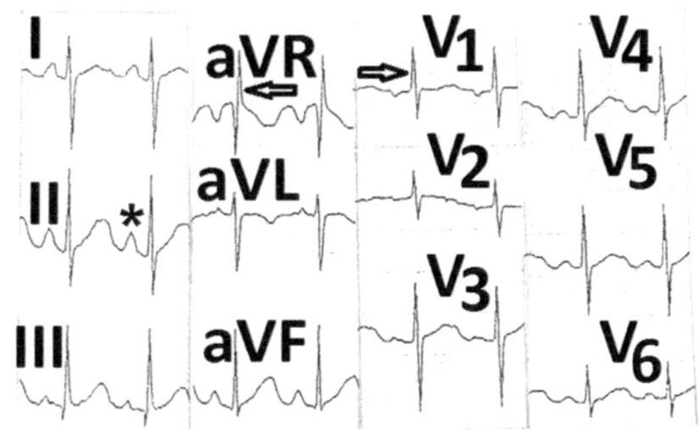

Fig. 48.11 Electrocardiogram from a normal neonate. It shows prominent P wave in lead II (marked *), mean frontal plane QRS axis around +120°, prominent terminal R wave in lead aVR (marked ⇐) and prominent R wave in lead V_1 (marked ⇒). (Reproduced with permission from: Mittal SR. Electrocardiogram in Asymptomatic Acyanotic Congenital Heart Disease. Cardiology Today-2019;23:157–65)

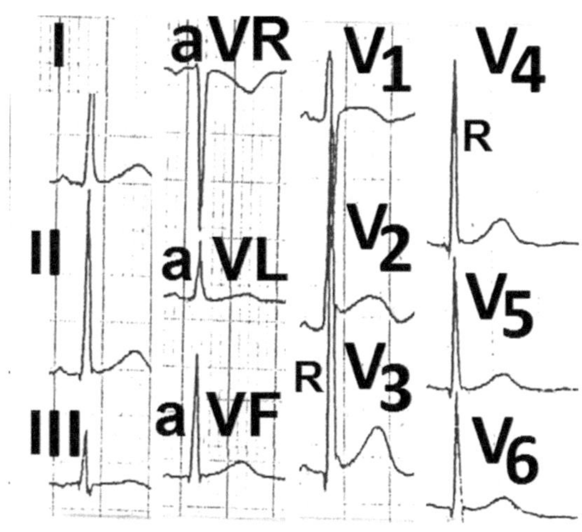

Fig. 48.12 Electrocardiogram from a case of moderate coarctation of the aorta. It shows increased QRS voltage in leads V_3 and V_4 (marked R). QRS axis is normal and there are no ST-segment or T wave changes. (Reproduced with permission from: Mittal SR. Electrocardiogram in Asymptomatic Acyanotic Congenital Heart Disease. Cardiology Today-2019;23:157–65)

- (iii) Idiopathic dilatation of the pulmonary artery [18].
- (iv) Mild to moderate coarctation of the aorta (Fig. 48.12).

Echocardiography is needed for conformation

(2) **rsr' or rsR' in lead V_1**

This electrocardiographic finding suggests right ventricular volume overload [19, 20]. Normal infants can have rsr' in lead V_1 but QRS axis is around +90°. QRS axis of more than +120° is normal in neonates [16].

(a) **Right axis deviation** (around +120°)
- It suggests moderately large secundum atrial septal defect with moderate left to right shunt [5] (Fig. 48.13).
- Nonobstructive anomalous drainage of more than two pulmonary veins. Echocardiography is needed for differentiation.

(b) **Right axis deviation with inverted P waves in leads II, III, and aVF**

These findings suggest the possibility of superior vena cava sinus venosus atrial septal defect [11].

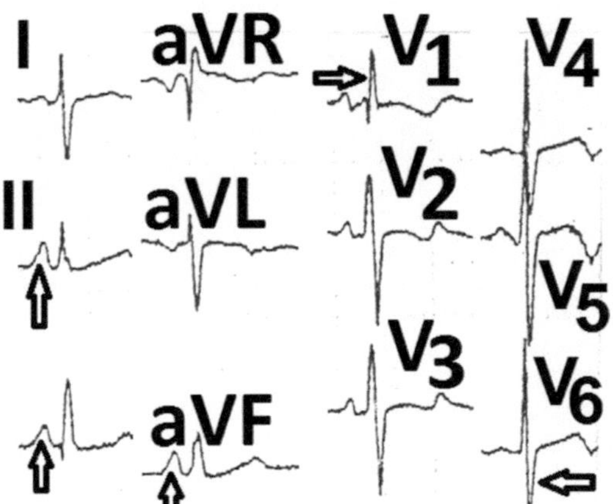

Fig. 48.13 Electrocardiogram from a case of secundum atrial septal defect. Prominent P wave in leads II, III, and aVF (marked ⇧) suggest right atrial overload. The mean frontal plane QRS axis is around +120°. Lead V_1 shows rsR' pattern (marked ⇨) and lead V_6 shows slurred S wave (marked ⇦). These findings suggest right ventricular volume overload. (Reproduced with permission from: Mittal SR. Electrocardiogram in Asymptomatic Acyanotic Congenital Heart Disease. Cardiology Today-2019;23:157–65)

Echocardiography is needed for confirmation.

(c) **Left axis deviation** (−20° to −60°)

It suggests the possibility of isolated ostium primum atrial septal defect without other components of atrioventricular septal defect [21] (Fig. 48.14).

Echocardiography is needed for confirmation.

(3) **Prominent q with tall R wave and positive T wave in leads V_5 & V_6** suggest left ventricular volume overload. Possibilities are as follows:
   - Supracristal (subpulmonic) ventricular septal defect with moderate left to right shunt [22].
   - Patent ductus arteriosus with moderate left to right shunt [23].

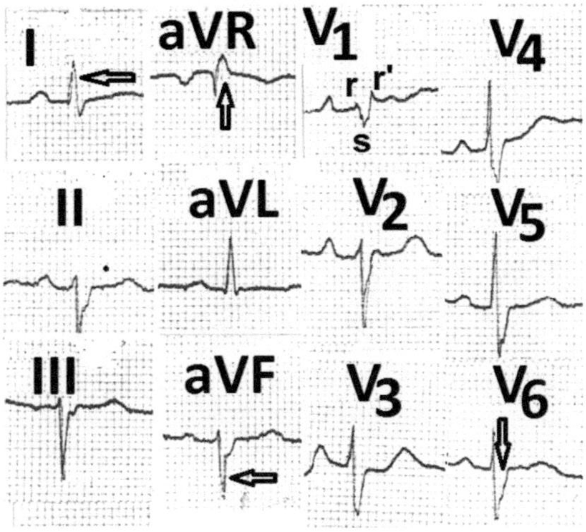

Fig. 48.14 Electrocardiogram from a case of primum atrial septal defect. There is left axis deviation (dominant R wave in lead V_1 and dominant S wave in lead aVF, marked ⇦). Lead V_1 shows rsr' pattern. Lead aVR shows slurred terminal R wave (marked ⇧). Lead V_6 shows slurred S wave (marked ⇩). These findings suggest right ventricular volume overload. (Reproduced with permission from: Mittal SR. Electrocardiogram in Asymptomatic Acyanotic Congenital Heart Disease. Cardiology Today-2019;23:157–65)

(4) **Left axis deviation with prominent R wave in lead V_1**

These findings suggest the possibility of Noonan syndrome with mild to moderate pulmonary valve stenosis [24].

(5) **Prominent R with flat T wave in lead V_6 without prominent q wave** suggest pressure overload of the left ventricle.

It could occur with moderate coarctation of aorta [25] with hypertension in upper limbs.

(D) **Patient with systolic murmur in the left lower parasternal region**
   (1) **Normal electrocardiogram**
      (a) **No organic cardiac disease.** Possibilities include:
         (i) Vibratory murmur (Still's murmur).
         (ii) Transient tricuspid regurgitation in neonates and infants.
         (iii) Mammary soufflé in lactating female.
      (b) **Organic heart disease**
         Small VSD with small left to right shunt [26].
   (2) **Prominent R wave with inversion of the T waves in leads V_5, V_6** suggest left ventricular hypertrophy.
      It could be due to mild to moderate subvalvular left ventricular outflow tract obstruction, for example:
      (a) Hypertrophic obstructive cardiomyopathy.
      (b) Subaortic membrane.
      Echocardiography is needed for differentiation
   (3) **Prominent q wave tall R wave and positive T wave in leads V_5 and V_6** suggest left ventricular volume overload.

It could be due to

– Moderate-sized VSD with moderate left to right shunt [27] (Figs. 48.15 and 48.16).

– Chest and heart of an infant are very small and large chest electrodes meant for adults occupy

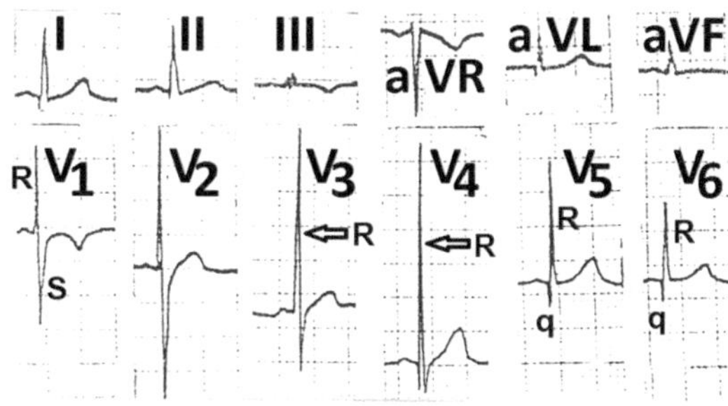

Fig. 48.15 Electrocardiogram from a case of ventricular septal defect with moderate left to right shunt. Leads V_5 and V_6 show qR (suggestive of left ventricular volume overload). Lead V_3 and V_4 show tall R wave (marked ⇦). Lead V_1 shows RS suggestive of right ventricular overload. Thus there is evidence of biventricular overload. (Reproduced with permission from: Mittal SR. Electrocardiogram in Asymptomatic Acyanotic Congenital Heart Disease. Cardiology Today-2019;23:157–65)

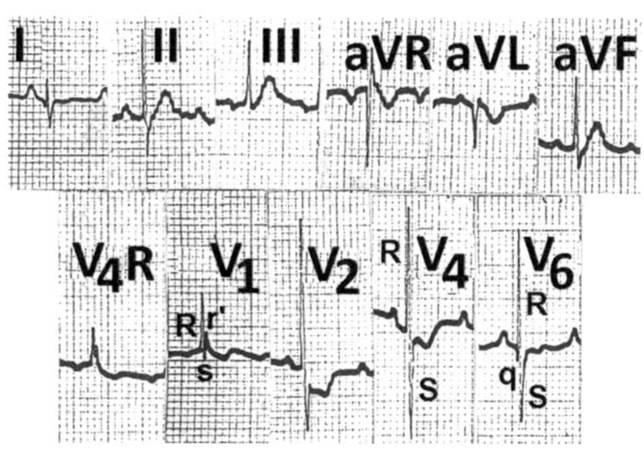

Fig. 48.16 Electrocardiogram from another case of ventricular septal defect with moderate left to right shunt. Lead V_6 shows qRS suggestive of left ventricular volume overload. Leads V_4R and V_1 show Rsr pattern suggestive of right ventricular overload. Lead V_4 shows large equidiphasic QRS (equal R waves and S waves) suggestive of biventricular overload. (Reproduced with permission from: Mittal SR. Electrocardiogram in Asymptomatic Acyanotic Congenital Heart Disease. Cardiology Today-2019;23:157–65)

a greater area. Leads V_5 and V_6, therefore, lie outside the cardiac border of an infant. In an infant changes may be best seen in lead V_4 rather than in leads V_5 and V_6 because usually adult leads are also used for recording ECG in infants.

(4) **Prominent q wave with tall R wave and positive T wave in V_6, rsr' in lead V_1, and terminal r wave in lead aVR** suggest biventricular volume overload.

These findings suggest the possibility of left ventricle to right atrial shunt (Fig. 48.17) [28]. Large shunts may also show right atrial enlargement (tall peaked P waves in lead II).

(5) **QRS axis +90° to +120° and prominent R wave in lead V_1** suggest right ventricular hypertrophy. Possibilities are as follows:
   (a) Mild to moderate right ventricular infundibular stenosis.
   (b) Double chambered right ventricle with mild to moderate obstruction. Right ventricular hypertrophy may not be impressive in leads V_1 to V_3 [29] because only a small portion of the right ventricle (below the obstructive fibrous band) is having high pressure.

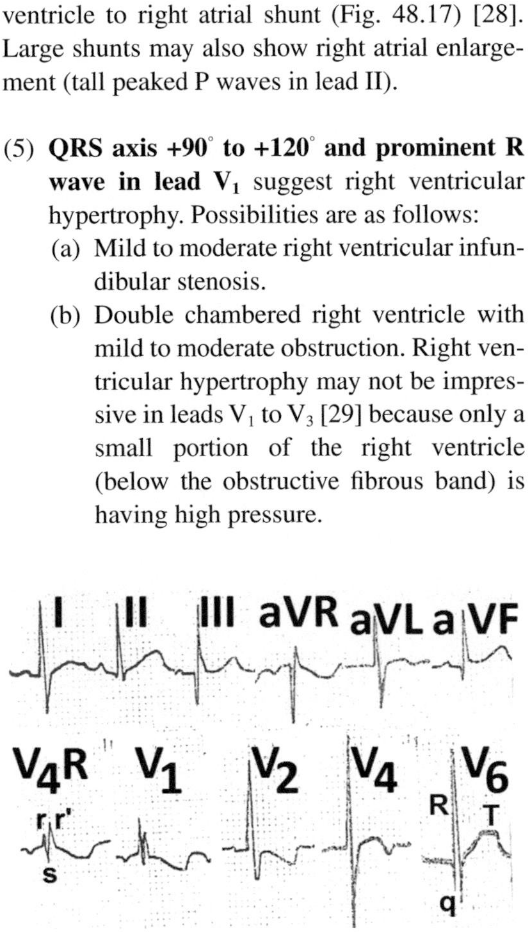

Fig. 48.17 Electrocardiogram from a case of left ventricle to right atrial shunt. Lead V_6 shows prominent q wave with tall R wave and positive T wave suggestive of left ventricular volume overload. Leads V_4R and V_1 show rsr' pattern suggestive of right ventricular volume overload. (Reproduced with permission from: Mittal SR. Electrocardiogram in Asymptomatic Acyanotic Congenital Heart Disease. Cardiology Today-2019;23:157–65)

(6) **Right axis deviation +90° to 120°, prominent R wave in lead V_1, sudden transition from R wave in lead V_1 to rS in lead V_2, and small q wave with well-developed R wave in V_5,V_6.**

These findings suggest the possibility of acyanotic (pink) tetralogy of Fallot (TOF with mild PS) [30].

(7) **Right axis deviation, peaked tall P waves, and splintered small QRS in lead V_1 without RVH** suggest the possibility of mild Ebstein anomaly [31] (Fig. 48.18).

(E) **Patient with systolic murmur along right upper sternal border.**

Normal electrocardiogram—Possibilities include:

(a) Mild to moderate aortic valve stenosis [32] (Fig. 48.19).
(b) Mild to moderate supravalvular aortic stenosis.

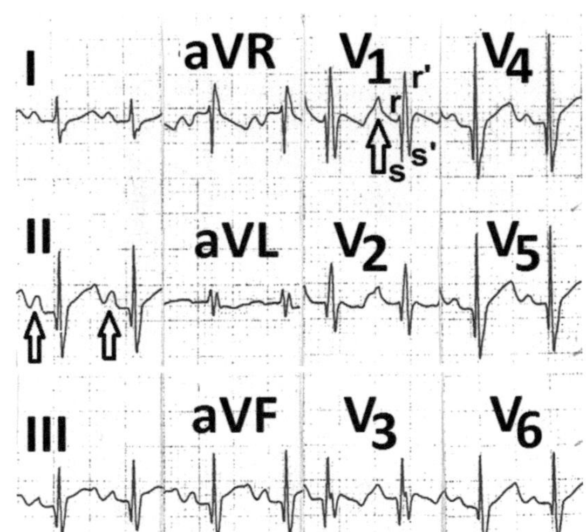

Fig. 48.18 Electrocardiogram from a case of Ebstein anomaly. It shows prominent P wave in leads II and V_1 (marked ⇧) suggestive of right atrial overload. QRS is splintered (rsr's') in lead V_1. (Reproduced with permission from: Mittal SR. Electrocardiogram in Asymptomatic Acyanotic Congenital Heart Disease. Cardiology Today-2019;23:157–65)

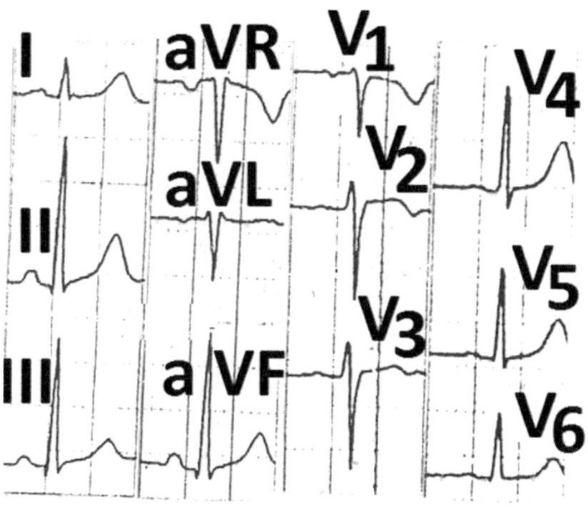

Fig. 48.19 Electrocardiogram from a case of mild to moderate valvular aortic stenosis. It is apparently normal. (Reproduced with permission from: Mittal SR. Electrocardiogram in Asymptomatic Acyanotic Congenital Heart Disease. Cardiology Today-2019;23:157–65)

(F) **Systolic murmur on both sides of upper sternum**

Normal electrocardiogram

It suggests the possibility of mild bilateral pulmonary artery branch stenosis.

(G) **Systolic murmur over the cardiac apex**

It suggests the possibility of congenital mitral regurgitation

Electrocardiogram

- Normal—mild mitral regurgitation.
- Left ventricular volume overload pattern suggests moderately severe mitral regurgitation.
- Additional left axis deviation suggests mitral regurgitation as a part of atrioventricular septal defect (AV canal defect)

(H) **Systolic murmur over lung fields**

Normal ECG suggests possibilities of the following:

- Pulmonary arteriovenous fistula [33].
- Peripheral pulmonary artery branch stenosis.

(I) **Continuous murmur**

(1) **Over left 1st and 2nd intercostal space ECG**

- Normal ECG suggests possibilities of the following:
  - Small patent ductus arteriosus with small left to right shunt [3].
  - Normal venous hum [34]. It is abolished by digital compression of the ipsilateral internal jugular vein.
- Left ventricular volume overload pattern—suggests the possibility of moderate left to right shunt across a patent ductus arteriosus (Fig. 48.20).

(2) **Over left or right sternal border** with normal ECG suggests the possibilities of the following:
  - Coronary arteriovenous fistula [35].
  - Systemic arteriovenous fistula.

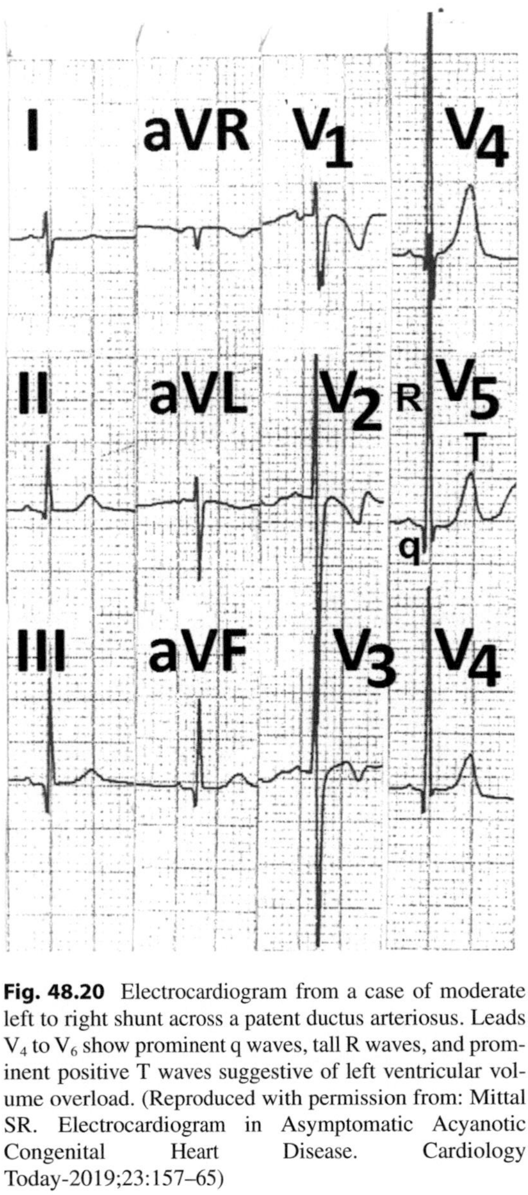

Fig. 48.20 Electrocardiogram from a case of moderate left to right shunt across a patent ductus arteriosus. Leads V_4 to V_6 show prominent q waves, tall R waves, and prominent positive T waves suggestive of left ventricular volume overload. (Reproduced with permission from: Mittal SR. Electrocardiogram in Asymptomatic Acyanotic Congenital Heart Disease. Cardiology Today-2019;23:157–65)

Summary

- Several congenital cardiovascular anomalies may have no appreciable electrocardiographic abnormality. Therefore, a normal electrocardiogram cannot categorically exclude the possibility of a congenital cardiovascular disease.
- Isolated congenital abnormalities of sinus node function, atrioventricular conduction, intraventricular conduction, or QT interval may be present without appreciable clinical findings.
- Congenital complete atrioventricular block may occur as an isolated anomaly or as a part of inversion of the ventricles.
- rsr' or rsR' in lead V_1 with right axis deviation suggests moderately large left to right shunt at the atrial level. Additional inversion of the P waves in leads II, III, and aVF suggest the possibility of superior vena cava sinus venosus atrial septal defect. rsr' or rsR' in lead V_1 with left axis deviation suggests the possibility of ostium primum ASD.
- Prominent q wave with tall R wave and upright T wave in leads V_4/ V_5/V_6 suggest chronic left ventricular volume overload and the possibility of moderate left to right shunt across a ventricular septal defect or patent ductus arteriosus. Additional rsr' in lead V_1 suggests the possibility of left ventricle to right atrial shunt.
- Left axis deviation suggests the possibility of atrioventricular septal defect or Noonan's syndrome. Atrioventricular septal defect can present as mitral regurgitation, ostium primum ASD, or VSD.
- Prominent R wave with flat T wave and no prominence of q wave in leads V_5/V_6 suggest pressure overload of the left ventricle. This can happen in moderate coarctation of the aorta with systemic hypertension, subaortic membrane with moderate gradient, or moderate aortic valve stenosis.
- Prominent R wave with deep symmetrical T wave inversion in leads V_4 to V_6 should raise suspicion of hypertrophic cardiomyopathy.
- Right axis deviation, tall peaked P waves, and prominent R with T wave inversion in leads V_1 to V_3 suggest the possibility of moderately severe valvular pulmonary stenosis with intact interventricular septum.
- Right axis deviation, prominent R wave in lead V_1, sudden transition to RS pattern in lead V_2, and persistence of small q wave and normal R wave in leads V_5 and V_6 should raise suspicion of acyanotic (pink) tetralogy of Fallot.
- Peaked tall P wave and splintered small QRS in lead V_1 should suggest the possibility of mild Ebstein's anomaly.

Flow charts

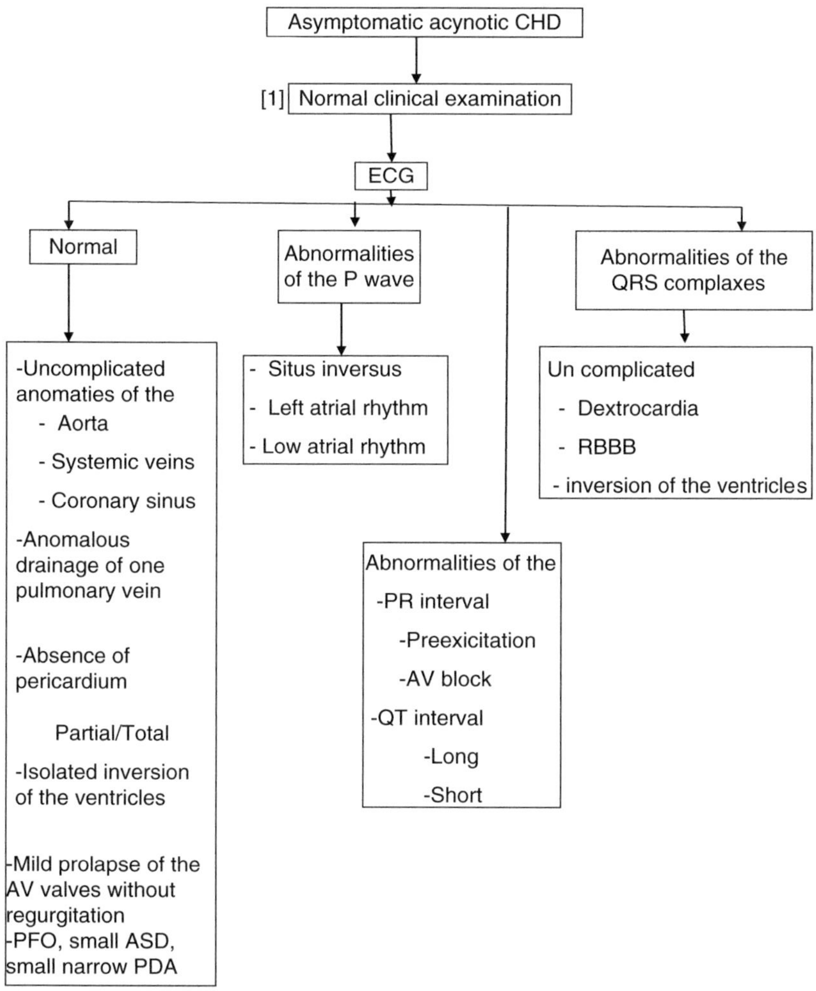
Asymptomatic acynotic CHD
[1] Normal clinical examination
ECG
Normal
-Uncomplicated anomaties of the
- Aorta
- Systemic veins
- Coronary sinus
-Anomalous drainage of one pulmonary vein
-Absence of pericardium
Partial/Total
-Isolated inversion of the ventricles
-Mild prolapse of the AV valves without regurgitation
-PFO, small ASD, small narrow PDA
Abnormalities of the P wave
- Situs inversus
- Left atrial rhythm
- Low atrial rhythm
Abnormalities of the
-PR interval
-Preexicitation
-AV block
-QT interval
-Long
-Short
Abnormalities of the QRS complaxes
Un complicated
- Dextrocardia
- RBBB
- inversion of the ventricles

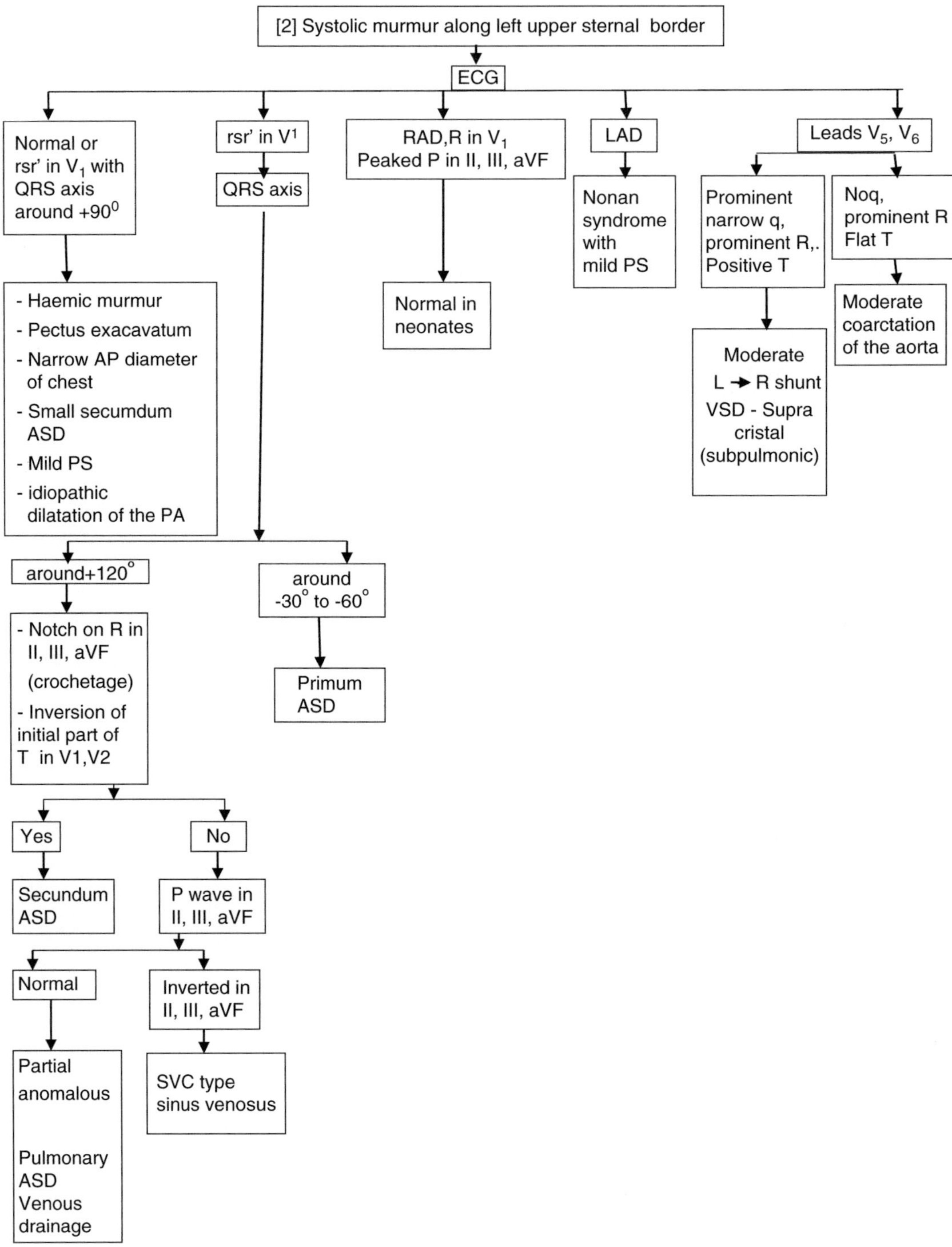
[2] Systolic murmur along left upper sternal border
ECG
Normal or rsr' in V1 with QRS axis around +90⁰
rsr' in V1
RAD,R in V1 Peaked P in II, III, aVF
LAD
Leads V5, V6
QRS axis
Nonan syndrome with mild PS
Prominent narrow q, prominent R,. Positive T
Noq, prominent R Flat T
- Haemic murmur
- Pectus exacavatum
- Narrow AP diameter of chest
- Small secumdum ASD
- Mild PS
- idiopathic dilatation of the PA
Normal in neonates
Moderate coarctation of the aorta
Moderate L → R shunt VSD - Supra cristal (subpulmonic)
around+120°
around -30° to -60°
- Notch on R in II, III, aVF (crochetage)
- Inversion of initial part of T in V1,V2
Primum ASD
Yes
No
Secundum ASD
P wave in II, III, aVF
Normal
Inverted in II, III, aVF
Partial anomalous Pulmonary ASD Venous drainage
SVC type sinus venosus

[3]

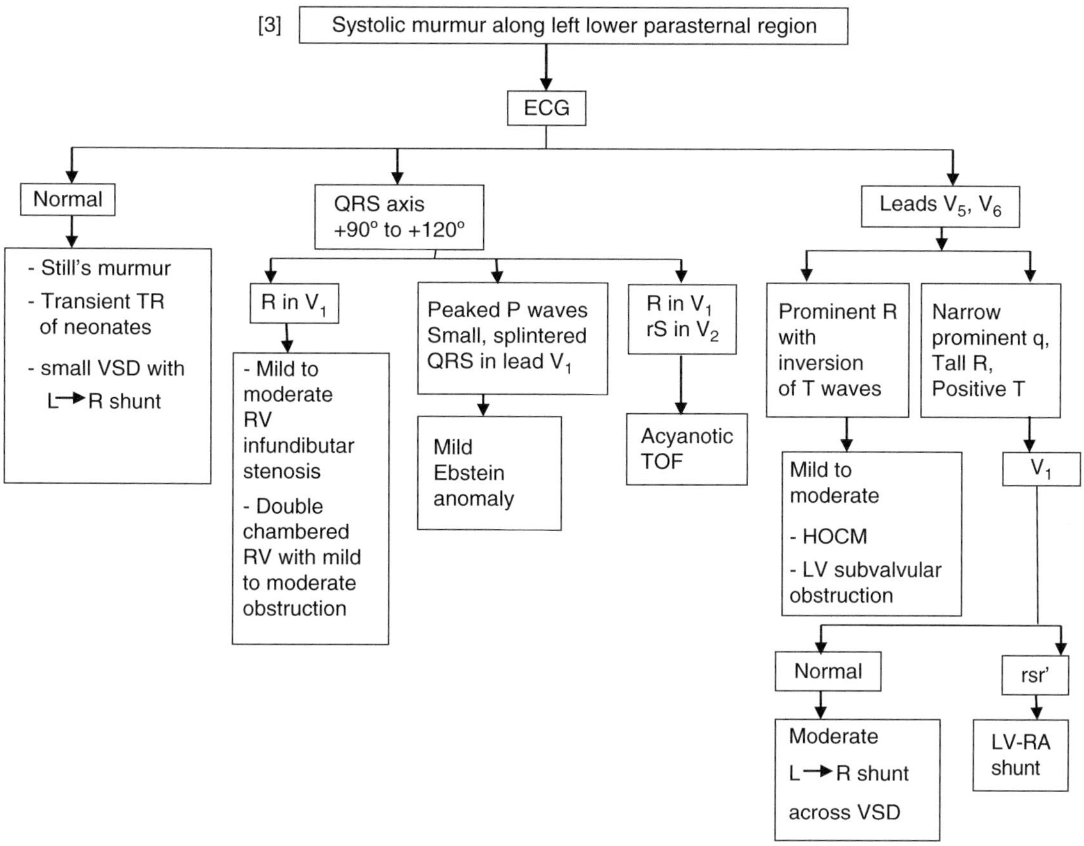
Systolic murmur along left lower parasternal region
ECG
Normal
- Still's murmur
- Transient TR of neonates
- small VSD with L→R shunt
QRS axis +90° to +120°
R in V_1
- Mild to moderate RV infundibutar stenosis
- Double chambered RV with mild to moderate obstruction
Peaked P waves Small, splintered QRS in lead V_1
Mild Ebstein anomaly
R in V_1 rS in V_2
Acyanotic TOF
Leads V_5, V_6
Prominent R with inversion of T waves
Mild to moderate
- HOCM
- LV subvalvular obstruction
Narrow prominent q, Tall R, Positive T
V_1
Normal
Moderate L→R shunt across VSD
rsr'
LV-RA shunt

[4]

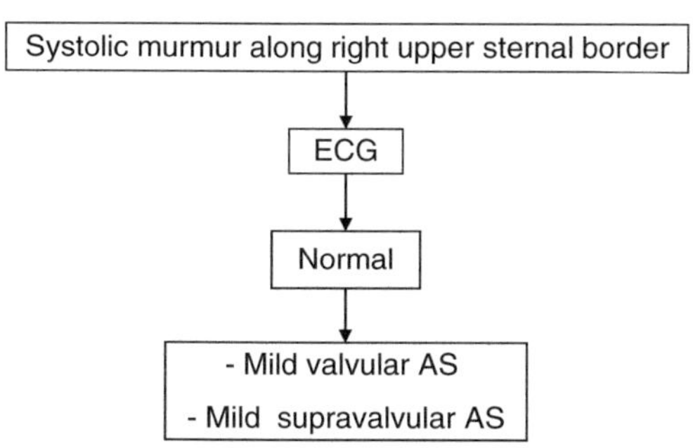
Systolic murmur along right upper sternal border
ECG
Normal
- Mild valvular AS
- Mild supravalvular AS

[5]

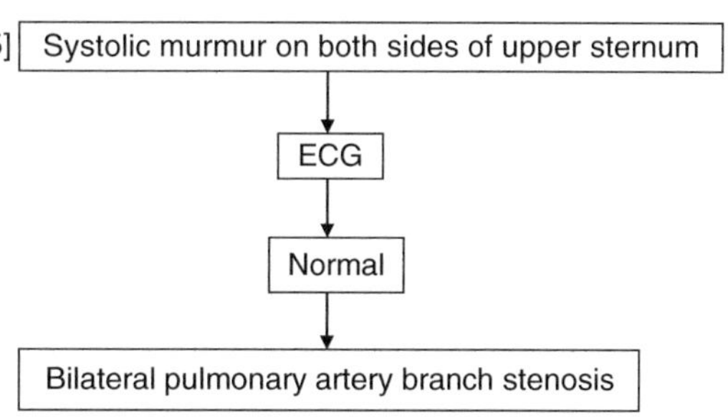
Systolic murmur on both sides of upper sternum
ECG
Normal
Bilateral pulmonary artery branch stenosis

[6]

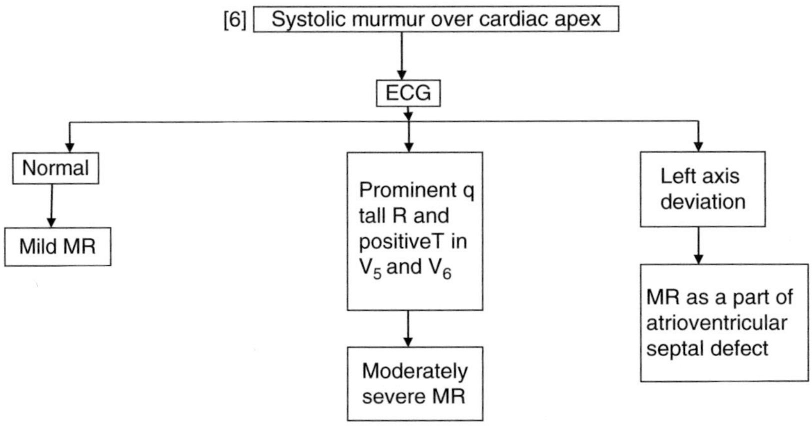
Systolic murmur over cardiac apex
ECG
Normal
Mild MR
Prominent q tall R and positiveT in V_5 and V_6
Moderately severe MR
Left axis deviation
MR as a part of atrioventricular septal defect

[7]

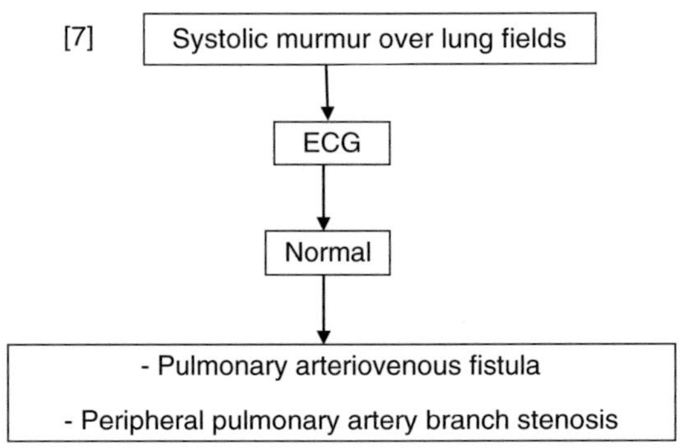
Systolic murmur over lung fields
ECG
Normal
- Pulmonary arteriovenous fistula
- Peripheral pulmonary artery branch stenosis

[8]

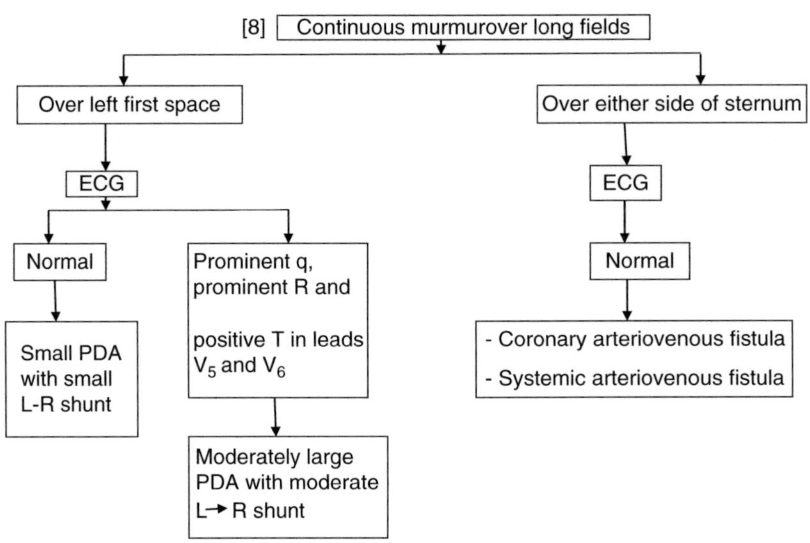
Continuous murmurover long fields
Over left first space
ECG
Normal
Small PDA with small L-R shunt
Prominent q, prominent R and positive T in leads V_5 and V_6
Moderately large PDA with moderate L→R shunt
Over either side of sternum
ECG
Normal
- Coronary arteriovenous fistula
- Systemic arteriovenous fistula

MCQs

Q1. Persistent left superior vena cava draining into the coronary sinus produces:

(a) Right atrial enlargement
(b) Right ventricular enlargement
(c) Right axis deviation
(d) Normal electrocardiogram

Q2. Isolated right aortic arch produces:

(a) No signs
(b) Pulsations in right second intercostal space
(c) Pulsations in the right supraclavicular region
(d) Continuous murmur over right 1st intercostal space

Q3. Congenital absence of hepatic portion of the inferior vena cava with azygos continuation produces:

(a) No signs
(b) Edema in lower limbs
(c) Hepatomegaly
(d) Increased pressure in superior vena cava

Q4. Absence of any abnormal clinical findings excludes:

(a) Narrow patent ductus arteriosus
(b) Patent foramen ovale
(c) Small ventricular septal defect
(d) Single coronary artery

Q5. Congenital complete absence of pericardium produces:

(a) Prominent apical impulse
(b) Prominent right ventricular pulsation
(c) Increased amplitude of QRS in leads V_5 and V_6
(d) Shifting of the heart to the left

Q6. Isolated anomalous drainage of the right upper pulmonary vein in the right atrium produces:

(a) Right atrial enlargement
(b) Right axis deviation
(c) Rsr' in lead V_1
(d) None of the above

Q7. Anomalous origin of the right coronary artery from the left main coronary artery produces:

(a) ST-segment elevation in lead V_3R
(b) ST-segment depression in leads V_5 and V_6
(c) Right ventricular failure
(d) None of the above

Q8. Idiopathic dilatation of the right atrium produces:

(a) Systemic venous congestion
(b) Tall P waves in leads II and V_1
(c) qR in lead V_1
(d) Normal electrocardiogram

Q9. Electrocardiographic findings of isolated inversion of the ventricles include:

(a) Right axis deviation
(b) Tall P waves in leads II and V_1
(c) Left bundle branch block
(d) q wave in lead V_1 and absence of the q wave in lead V_6

Q10. Small secundum atrial septal defect with small left to right shunt produces:

(a) Right axis deviation
(b) Incomplete right bundle branch block
(c) Wide and fixed split of second heart sound
(d) None of the above

Q11. Nonobstructive congenital mitral stenosis produces:

(a) No clinical abnormality
(b) Mid diastolic murmur over cardiac apex
(c) Broad and notched P wave in lead II
(d) Right axis deviation

Q12. Isolated tricuspid valve prolapse can be excluded confidently by:

(a) Thorough clinical examination
(b) Surface electrocardiogram
(c) a+b
(d) None of above

Q13. Normal clinical examination and normal surface electrocardiogram can exclude the presence of:

(a) Isolated tricuspid valve prolapse
(b) Patent foramen ovale
(c) Restrictive ventricular septal defect
(d) Papillary muscle hypertrophy

Q14. Normal clinical examination and normal electrocardiogram cannot exclude the presence of:

(a) Congenital aneurysm of the left ventricle
(b) Persistence of fetal circulation
(c) Ostium primum atrial septal defect
(d) Patent ductus arteriosus with large left to right shunt

Q15. Inversion of the P waves in leads I, V_5, and V_6 suggest:

(a) Situs inversus
(b) Left atrial rhythm
(c) Low atrial rhythm
(d) Situs solitus

Q16. Inversion of the P waves inversion in leads I, aVL, and positive P wave in lead aVR suggest:

(a) Situs inversus
(b) Left atrial rhythm
(c) Reversal of arm leads
(d) Reversal of left upper limb and left lower limb leads

Q17. "Dome and dart" P waves in lead V_1 suggest:

(a) Left atrial rhythm
(b) Low atrial rhythm
(c) Situs inversus
(d) Right ventricular hypertrophy

Q18. qR in lead V_1 and rS in lead V_6 with normal P waves suggest:

(a) Dextrocardia
(b) Right bundle branch block
(c) Right ventricular hypertrophy
(d) Right ventricular volume overload

Q19. q wave in leads V_1 to V_4R and the absence of q wave in leads V_5 to V_9 suggest:

(a) Congenitally corrected transposition of the great arteries
(b) D transposition of the great arteries
(c) Biventricular enlargement
(d) Dextroposition

Q20. Prominent R waves with deep inversion of the T waves in leads V_4 to V_6 without any significant clinical finding in an asymptomatic person suggests the possibility of:

(a) Hypertrophic cardiomyopathy
(b) Severe left ventricular outflow tract obstruction
(c) Severe coarctation
(d) Anterolateral myocardial infarction

Q21. Which conditions may not produce any appreciable clinical findings:

(a) Isolated first degree atrioventricular block
(b) Ebstein anomaly
(c) Hereditary long QT syndrome
(d) None of the above

Q22. Isolated congenital complete atrioventricular block in a neonate produces:

(a) Feeding difficulty
(b) Tachypnoea

(c) Hypotension
(d) None of the above

Q23. Pectus excavatum alone can produce:

(a) Prominent left parasternal lift
(b) Grade 2–3/6 systolic murmur in the left second intercostal space
(c) Prominent systolic pulsations in right second intercostal space
(d) Nonejection click

Q24. The electrocardiogram is normal in:

(a) Primum atrial septal defect
(b) Idiopathic dilatation of the pulmonary artery
(c) Straight back
(d) Origin of left coronary artery from the pulmonary artery

Q25. rSr' configuration in lead V_1 with right axis deviation and inversion of the P waves in leads II, III, and aVF suggest:

(a) Atrioventricular septal defect
(b) Superior vena cava sinus venosus atrial septal defect
(c) Inferior vena cava sinus venosus atrial septal defect
(d) Coronary sinus type atrial septal defect

Q26. Deep S wave in lead V_1 with tall R waves, deep narrow q waves, and positive T waves in leads V_5 and V_6 suggest the possibility of:

(a) Moderate size VSD with moderate left to right shunt
(b) VSD with severe pulmonary stenosis
(c) Left ventricle to right atrial shunt
(d) VSD with severe pulmonary artery hypertension

Q27. Deep S waves in leads V_1,V_2 with tall R waves, small or no q waves, and flat T waves in leads V_5 and V_6 support the possibility of:

(a) Moderate subvalvular membranous aortic stenosis
(b) Apical hypertrophic cardiomyopathy
(c) Supracristal VSD with moderate left to right shunt
(d) Moderate right ventricular infundibular stenosis

Q28. Grade 2/6 short systolic murmur in the left lower parasternal region in an asymptomatic acyanotic neonate with right ventricular dominance in electrocardiogram suggest:

(a) Normal tricuspid regurgitation
(b) Still's murmur
(c) Organic tricuspid regurgitation
(d) Small muscular VSD

Q29. Right axis deviation, prominent R wave in lead V_1, rS in lead V_2, and small q wave with well-developed R wave in lead V_6 in an acyanotic child suggests:

(a) Supracristal VSD with pulmonary artery hypertension
(b) Inlet VSD with pulmonary valve stenosis
(c) PDA with pulmonary artery hypertension
(d) Perimembranous VSD with overriding of the aorta

Q30. Systolic murmur in the left lower parasternal region with right axis deviation, peaked tall P waves, and small, splintered QRS in lead V_1 suggest:

(a) LV to RA shunt
(b) Right ventricular hypertrophy with tricuspid regurgitation
(c) Ebstein anomaly
(d) VSD with prolapse of the septal leaflet of the tricuspid valve in the defect

Q31. Systolic murmur over the cardiac apex in an asymptomatic infant with normal electrocardiogram supports the possibility of mitral regurgitation due to:

(a) Mitral valve prolapse
(b) Atrioventricular septal defect
(c) Anomalous left coronary artery from pulmonary artery
(d) Endocardial fibroelastosis

Q32. Systolic murmur over the right second intercostal space in an asymptomatic, infant with normal electrocardiogram supports the possibility of:

(a) Mild supravalvular aortic stenosis
(b) Right aortic arch
(c) Corrected transposition of great arteries with pulmonary valve stenosis
(d) Right pulmonary artery stenosis

Q33. Continuous murmur over the left interscapular region in an asymptomatic, acyanotic child with normal electrocardiogram is likely to be due to:

(a) Coarctation of aorta
(b) Bronchopulmonary collaterals
(c) Pulmonary arteriovenous fistula
(d) Peripheral pulmonary artery stenosis

Q34. Continuous murmur over the right parasternal region in an asymptomatic child with normal electrocardiogram suggests the possibility of:

(a) Coronary arteriovenous fistula opening in the right atrium
(b) Rupture of sinus of Valsalva aneurysm in right atrium
(c) Pulmonary arteriovenous fistula
(d) Systemic arteriovenous fistula

Q35. Continuous murmur over the left sternal border in an asymptomatic child with normal electrocardiogram suggests the possibility of:

(a) Coronary arteriovenous fistula opening in the right ventricle
(b) Rupture of sinus of Valsalva aneurysm in the right ventricle
(c) Aorto-pulmonary window
(d) Cor triatriatum

Q36. Continuous murmur over the left first intercostal space in an asymptomatic infant with normal electrocardiogram suggests the possibility of:

(a) Large PDA
(b) Venous hum
(c) Stenosis of the left subclavian artery
(d) Left pulmonary artery stenosis

Answers

(1)d (2)a (3)a (4)c (5)d (6)d (7)d (8)d (9)d (10)(d) (11) a (12)d (13)c (14)a (15)b (16)a,c (17)a (18)a (19), (20)a (21)a,b,c (22)d (23)a,b (24)b,c (25)b (26)a (27)a (28)a (29)d (30)c (31)a (32)a (33)c,d (34)a,c (35)a (36)b

References

1. Perloff JK, Marelli AJ. Congenital aneurysms of sinuses of Valsalva. In: Perloff JK, Marelli AJ, editors. Perloff's clinical recognition of congenital heart disease. Philadelphia: Elsevier; 2012. p. 417–28.
2. Mishra BR. Anomalous systemic venous connections. In: Satpathy M, editor. Clinical diagnosis of congenital heart disease. New Delhi: Jaypee; 2008. p. 325–30.
3. Jayarajah M, Satpathy M. Patent ductus arteriosus. In: Satpathy M, editor. Clinical diagnosis of congenital heart disease. New Delhi: Jaypee; 2008. p. 115–23.
4. Perloff JK, Marelli AJ. Congenital abnormalities of the pericardium. In: Perloff JK, Marelli AJ, editors. Perloff's clinical recognition of congenital heart disease. Philadelphia: Elsevier; 2012. p. 51–5.
5. Dash PK, Satpathy M. Atrial septal defect (secundum type). In: Satpathy M, editor. Clinical diagnosis of congenital heart disease. New Delhi: Jaypee; 2008. p. 74–82.
6. Manoria PC, Trivedi SK. Mitral valve prolapse. In: Satpathy M, editor. Clinical diagnosis of congenital heart disease. New Delhi: Jaypee; 2008. p. 60–6.
7. Campbell M, Reynolds G. The significance of the direction of the P wave in dextrocardia and isolated laevocardia. Br Heart J. 1952;14:481–8.
8. Edenbrandt L, Rittner R. Recognition of lead reversals in pediatric electrocardiograms. Am J Cardiol. 1998;82:1290–2.
9. Frankl WS, Soloff LA. Left atrial rhythm. Analysis by intra-atrial electrocardiogram and the vectorcardiogram. Am J Cardiol. 1968;22:645–56.
10. Mirowski M, Neill CA, Taussig HB. Left atrial ectopic rhythm in mirror-image dextrocardia and in normally placed malformed hearts: Report of Twelve cases with "dome and dart" P waves. Circulation. 1963;27:864–77.
11. Maheshwari S, Satpathy M. Atrial septal defect (sinus venosus type). In: Satpathy M, editor. Clinical diagnosis of congenital heart disease. New Delhi: Jaypee; 2008. p. 87–90.

12. Perloff JK, Marelli AJ. Cardiac malpositions. In: Perloff JK, Marelli AJ, editors. Perloff's clinical recognition of congenital heart disease. Philadelphia: Elsevier; 2012. p. 15–40.
13. Schiebler GL, Edwards JE, Burchell HB, Dushane JW, Ongley PA, Wood EH. Congenital corrected transposition of the great vessels: a study of 33 cases. Pediatrics. 1961;27:849–88.
14. Victorica BE, Miller BL, Gessner IH. Electrocardiogram and vectorcardiogram in ventricular inversion (corrected transposition). Am Heart J. 1973;86:733–44.
15. Perloff JK, Marelli AJ. Isolated congenital complete heart block. In: Perloff JK, Marelli AJ, editors. Perloff's clinical recognition of congenital heart disease. Philadelphia: Elsevier; 2012. p. 41–50.
16. Kulkarni ML, Satpathy M. Electrocardiogram - clinically relevant to congenital heart disease. In: Satpathy M, editor. Clinical diagnosis of congenital heart disease. New Delhi: Jaypee; 2008. p. 27–32.
17. Desai N, Kumar R, Mahadevan C, Prakash VS. Congenital pulmonary stenosis. In: Satpathy M, editor. Clinical diagnosis of congenital heart disease. New Delhi: Jaypee; 2008. p. 199–207.
18. Mishra BR. Idiopathic dilatation of pulmonary artery. In: Satpathy M, editor. Clinical diagnosis of congenital heart disease. New Delhi: Jaypee; 2008. p. 212–3.
19. Wagner GS, Lim TH. Chamber enlargement. In: Wagner GS, editor. Marriot's Practical Electrocardiography. New Delhi: Wolters Kluwer; 2008. p. 71–96.
20. Schamroth L. Right ventricular bypertrophy. In: Schamroth C, editor. An introduction to electrocardiography. France: Blackwell Science; 1982. p. 78–85.
21. Pryor R, Woodwork MB, Blount SG. Electrocardiographic changes in atrial septal defects: ostium secundum defect versus ostium primum (endocardial cushion) defect. Am Heart J. 1959;58:689–700.
22. Vince DJ, Keith JD. The electrocardiogram in ventricular septal defect. Circulation. 1961;23:225–40.
23. Perloff JK, Marelli AJ. Patent ductus arteriosus, Aortopulmonary window. In: Perloff JK, Marelli AJ, editors. Perloff's clinical recognition of congenital heart disease. Philadelphia: Elsevier; 2012. p. 368–93.
24. Raaijmakers R, Noordam C, Noonan JA, Croonen EA, Van Der Burgt CJAM, Draaisma JMT. Are ECG abnormalities in Noonan syndrome characteristic for the syndrome? Eur J Pediatr. 2008;167:1363–7.
25. Perloff JK, Marelli AJ. Coarclation of the aorta and interrupted aortic arch. In: Perloff JK, Marelli AJ, editors. Perloff's clinical recognition of congenital heart disease. Philadelphia: Elsevier; 2012. p. 101–28.
26. Toscano-Barboza E, Dushane JW. Ventricular septal defect: correlation of electrocardiographic and hemodynamic findings in 60 proved cases. Am J Cardiol. 1959;3:721–32.
27. Witham AC, Mcdanial JS. Electrocardiogram, vectorcardiogram, and hemodynamics in ventricular septal defect. Am Heart J. 1970;79:335–46.
28. Riemenschneider TA, Moss AJ. Left ventricular-right atrial communication. Am J Cardiol. 1967;19:710–8.
29. Patel R, Astley R. Right ventricular obstruction due to anomalous muscle bands. Br Heart J. 1973;35:890–3.
30. Perloff JK, Marelli AJ. Ventricular septal defect with pulmonary stenosis. In: Perloff JK, Marelli AJ, editors. Perloff's clinical recognition of congenital heart disease. Philadelphia: Elsevier; 2012. p. 316–47.
31. Bhattacharya AK, Satpathy M. Ebstein anomaly. In: Satpathy M, editor. Clinical diagnosis of congenital heart disease. New Delhi: Jaypee; 2008. p. 240–8.
32. Guha S, Satpathy M. Congenital aortic stenosis. In: Satpathy M, editor. Clinical diagnosis of congenital heart disease. New Delhi: Jaypee; 2008. p. 143–56.
33. Perloff JK, Marelli AJ. Congenital pulmonary arteriovenous fistula. In: Perloff JK, Marelli AJ, editors. Perloff's clinical recognition of congenital heart disease. Philadelphia: Elsevier; 2012. p. 513–21.
34. Jones FL Jr. Frequency, characteristics and importance of the cervical venous hum in adults. N Engl J Med. 1982;267:658–60.
35. Bahuleyan CG. Congenital coronary arterial fistula. In: Satpathy M, editor. Clinical diagnosis of congenital heart disease. New Delhi: Jaypee; 2008. p. 182–6.

49 Electrocardiogram in a Neonate Presenting with Failure

Heart failure in the neonatal period is suspected by feeding difficulty, perspiration during feeding, persistent tachycardia (heart rate more than 140/min), and tachypnoea (respiratory rate more than 45/min). [1] Clinical examination correlated with surface electrocardiogram can help in narrowing the differential diagnosis. This correlation helps in more focused echocardiographic evaluation for final diagnosis. It is discussed under the following heads.

1 Normal Examination of the Cardiovascular System

1.1 Electrocardiogram

1. Electrocardiogram normal for age

   Exclude

(i) Noncardiac causes for respiratory distress
   (a) Acute idiopathic respiratory distress syndrome.
   (b) Hyaline membrane disease.
   (c) Meconium aspiration syndrome.
   (d) Bronchopulmonary dysplasia and obstruction.
   (e) Hypoglycemia.

Stridor, cough, and wheezing suggest the possibility of respiratory cause. Skiagram of chest is helpful. Respiratory diseases show ground glass opacity, reticulogranular pattern, coarse nodular opacities, or atelectasis in contrast to pulmonary edema seen in cases of heart failure. Meconium stained liquor supports the possibility of meconium aspiration syndrome. Prematurity and maternal diabetes predispose to idiopathic respiratory distress syndrome. Electrocardiogram is not very useful. Respiratory problems specially idiopathic respiratory distress syndrome and meconium aspiration syndrome are usually associated with pulmonary artery hypertension, right ventricular hypertrophy, and failure. Murmur of additional patent ductus arteriosus may not be audible in preterm infants with respiratory distress.

(ii) Vascular rings

These can present with respiratory distress, dyspnea, stridor, cough, and occasional wheezing [2] and may simulate respiratory disease. There are no diagnostic findings on clinical examination, electrocardiogram, and skiagram of chest. Echocardiography may raise suspicion. However, computerized tomography or magnetic resonance imaging are needed for confirmation.

S. Mittal, *Insights into Electrocardiograms with MCQs*,
https://doi.org/10.1007/978-981-99-0127-2_49

2. Disproportionate tachycardia
   (1) Regular rhythm
      (a) P waves seen before each QRS suggest atrial tachycardia.
      (b) Absence of P waves suggests diagnosis of junctional tachycardia.
      (c) Presence of atrial flutter waves suggest diagnosis of atrial flutter.
   (2) Irregularly irregular rhythm suggests possibility of atrial fibrillation.

Vulnerable neonatal myocardium may not respond adequately to tachycardia. Prolonged duration of tachycardia may per se cause tachycardia-induced cardiomyopathy.

3. Bradycardia: It suggest the possibility of complete atrioventricular block

Block is usually at the level of atrioventricular node. Usually congenital complete atrioventricular block is asymptomatic. However, a heart with bradycardia may not respond adequately to increased circulatory demand of metabolic acidosis and febrile illness. This can result is heart failure.

Electrocardiogram: It shows ventricular rate less than 90/min, faster atrial rate, no relation between P and QRS, normal QRS axis, and narrow QRS. T waves are usually normal, but deep T wave inversion or coving may sometimes occur in right and mid precordial leads.

4. Right axis deviation, peaked P waves, qR configuration in lead V_1, positive T waves in leads V_1 and V_2 beyond third day of life. These findings suggest right ventricular hypertrophy and right atrial enlargement (Fig. 49.1).

Possibilities include:

(a) Severe coarctation of the aorta [3]
(b) Complete interruption of the aortic arch [4]
(c) Hypoplastic left heart syndrome [5]
(d) Severe pulmonary vein stenosis [6]

Echocardiography is needed for differential diagnosis.

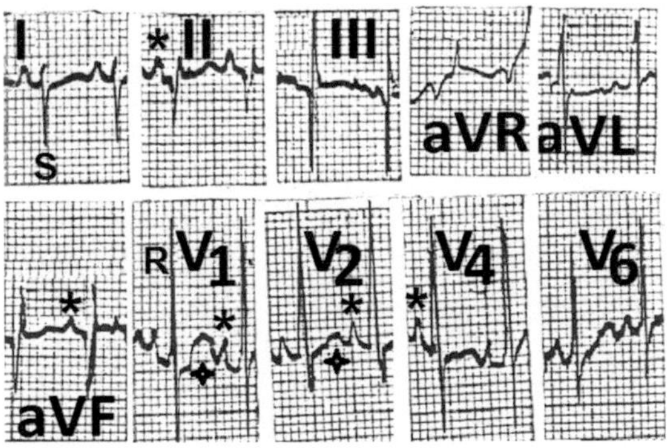

Fig. 49.1 Electrocardiogram from a neonate showing sinus tachycardia (166/minute), right axis deviation (prominent S wave in lead I—marked S), peaked P waves in leads II, aVF, and V_1 to V_4 (marked *) suggestive of right atrial overload, prominent R wave in lead V_1, and upright T waves in leads V_1 and V_2 (marked ✦). These findings are suggestive of right ventricular hypertrophy and right atrial enlargement. (Reproduced with permission from: Mittal SR. Electrocardiogram in Neonate Presenting with Heart Failure. Cardiology Today-2019;23:205–9)

5. Right ventricular hypertrophy, right atrial overload with left atrial overload (broad and notched P waves in lead II or prominent negative deflection of the P waves in lead V_1) (Fig. 49.2).

Possibilities are:

(a) Critical aortic stenosis [7]
(b) Left ventricular inflow obstruction [6]
   i. Congenital mitral stenosis [6]
   ii. Supravalvular stenosing ring [8]
   iii. Cor triatriatum

Echocardiography is needed for differential diagnosis.

6. Q wave with tall R wave, depression of the ST-segment, and inversion of the T waves in leads V_5 and V_6 with broad and notched P wave in lead II and/or prominent negative deflection of the P wave in lead V_1. These findings suggest severe left ventricular hypertrophy with left atrial enlargement (Fig. 49.3).

These findings in a neonate are suggestive of primary endocardial fibroelastosis. [9, 10]

Echocardiography is need for confirmation.

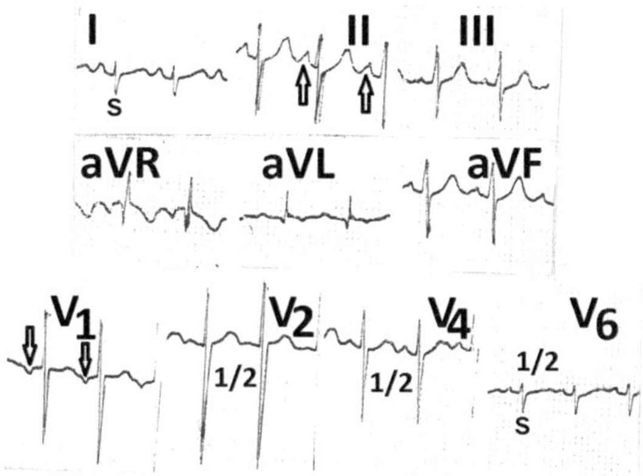

Fig. 49.2 Electrocardiogram from a neonate with heart failure. Leads V_2, V_4, and V_6 have been recorded at half standard, that is, 5 mm/mV. It shows right axis deviation (prominent S wave in lead I—marked S) and clockwise rotation (persistence of S wave up to lead V_5, marked S), suggestive of right ventricular enlargement. Peaked P waves in lead II (marked ⇧) suggesting right atrial enlargement and negative P waves in lead V_1 (marked ⇩) suggest left atrial enlargement. (Reproduced with permission from: Mittal SR. Electrocardiogram in Neonate Presenting with Heart Failure. Cardiology Today-2019;23:205–9)

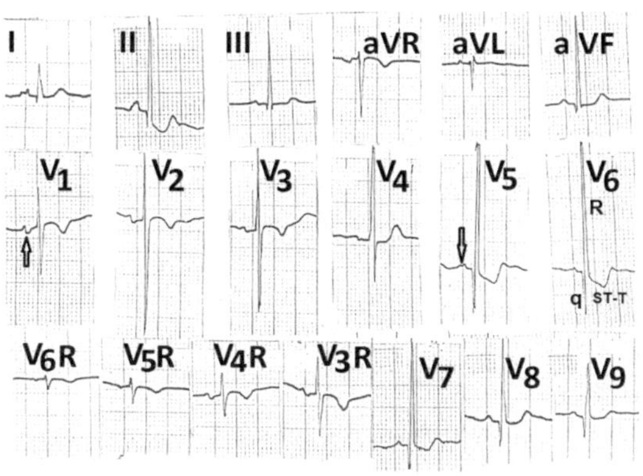

Fig. 49.3 Electrocardiogram from a neonate showing prominent q wave, tall R wave, depression of the ST-segment, and inversion of the T waves in leads V_5 and V_6 suggestive of severe left ventricular hypertrophy. The P wave is broad and notched in lead V_5 (marked ⇩) and is negative in lead V_1 (marked ⇧) suggesting left atrial enlargement. (Reproduced with permission from: Mittal SR. Electrocardiogram in Neonate Presenting with Heart Failure. Cardiology Today-2019;23:205–9)

2 Short Low Pitched Early Diastolic Murmur Along Left Upper Sternal Border

2.1 Electrocardiogram

Right axis deviation, rsr' in lead V_1 without peaked P waves and positive T waves in right precordial leads suggest right ventricular volume overload without right atrial overload. These findings suggest the possibility of severe pulmonary regurgitation due to congenital absence of the pulmonary valve. [11]

Echocardiography is needed for confirmation.

3 Central Cyanosis without Any Murmur

3.1 Electrocardiogram

(1) Right axis deviation and right ventricular hypertrophy.

Possibilities include:

(a) Tetralogy of Fallot with pulmonary atresia (Fig. 49.4).

(b) Transposition of great arteries with no intercirculatory communication. [12]

Echocardiography is needed for differentiation.

(2) Right axis deviation, right ventricular hypertrophy with tall peaked P waves suggestive of right atrial enlargement.
   - These findings suggest the possibility of total anomalous pulmonary venous connection with obstruction (mostly infradiaphragmatic). [13]

Echocardiography is needed for confirmation.

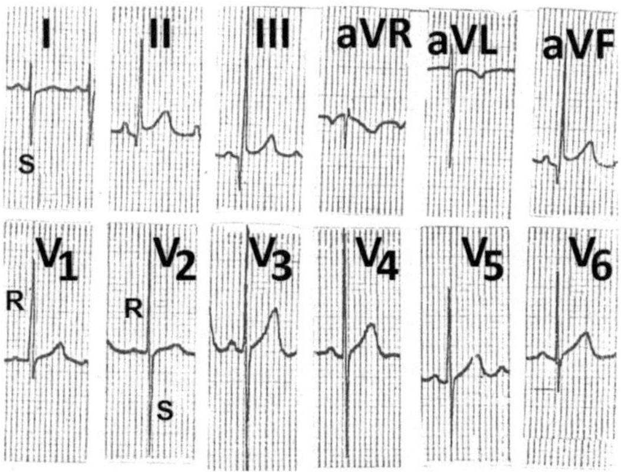

Fig. 49.4 Electrocardiogram from a cyanotic neonate showing right axis deviation (prominent S wave in lead I), prominent R wave in lead V_1, and sudden transition to RS configuration in lead V_2. (Reproduced with permission from: Mittal SR. Electrocardiogram in Neonate Presenting with Heart Failure. Cardiology Today-2019;23:205–9)

(3) Peaked right atrial P wave with prominent R wave with prominent positive T waves in leads V_5 and V_6 suggestive of left ventricular dominance.

(a) Normal QRS axis (+30° to +90°) with clockwise depolarization. [14]

These findings suggest the possibility of pulmonary atresia with intact interventricular septum and significant tricuspid regurgitation.

Echocardiography is needed for confirmation.

(b) Left axis deviation and counterclockwise depolarization (Fig. 49.5).

These findings suggest the possibility of Tricuspid atresia with pulmonary atresia. [15]

Echocardiography is needed for confirmation.

(c) Right upper quadrant axis.

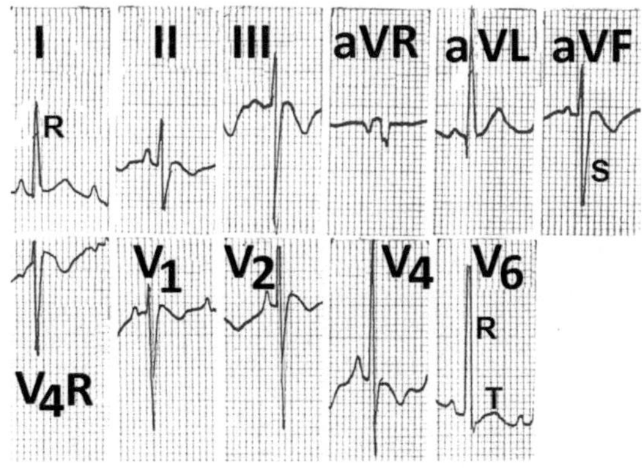

Fig. 49.5 Electrocardiogram from a cyanotic neonate showing leftward deviation of the mean frontal plane QRS axis (around −30°) (dominant R wave in lead I and dominant S wave in lead aVF). Leads V_5 and V_6 show prominent R wave with positive T wave suggestive of left ventricular dominance. (Reproduced with permission from: Mittal SR. Electrocardiogram in Neonate Presenting with Heart Failure. CardiologyToday-2019;23:205–9)

It suggests th3 possibility of critical pulmonary stenosis. [16]

Echocardiography is needed for differentiation from pulmonary atresia.

4 Central Cyanosis with Systolic Murmur Over Left Lower Sternal Border

4.1 Electrocardiogram

Right axis deviation, tall P waves suggestive of right atrial enlargement, low voltage splintered QRS in lead V_1, and absence of right ventricular hypertrophy (Fig. 49.6). These findings suggest the possibility of Ebstein anomaly. [17]

Echocardiography is needed for confirmation.

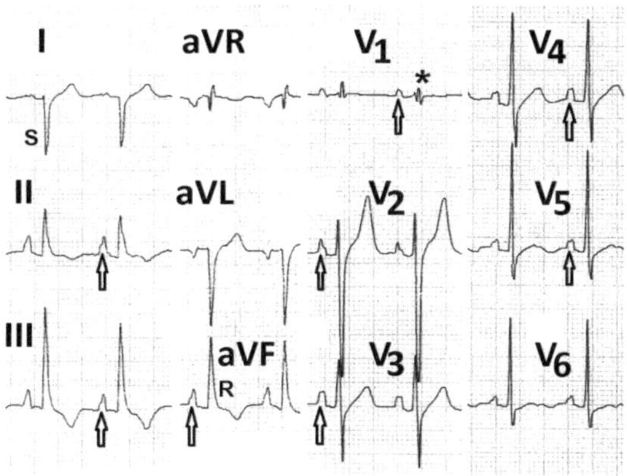

Fig. 49.6 Electrocardiogram from a cyanotic neonate with heart failure. It shows right axis deviation (dominant S wave in lead I and dominant R in lead aVF), peaked P waves in leads II, III, aVF, and V_1 to V_5 (marked ⇧) (suggestive of right atrial enlargement) and low voltage splintered QRS in lead V_1 (marked *). These findings are suggestive of Ebstein anomaly. (Reproduced with permission from: Mittal SR. Electrocardiogram in Neonate Presenting with Heart Failure. Cardiology Today-2019;23:205–9)

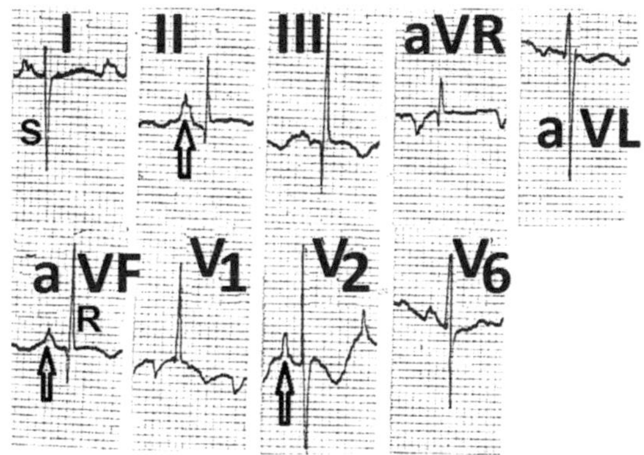

Fig. 49.7 Electrocardiogram from a neonate with heart failure showing right axis deviation (prominent S wave in lead I and prominent R wave in lead aVF), peaked P wave in lead II, aVF, and V_2 (marked ⇧) (suggestive of right atrial enlargement) and prominent R wave in lead V_1 (suggestive of severe right ventricular hypertrophy). (Reproduced with permission from: Mittal SR. Electrocardiogram in Neonate Presenting with Heart Failure. Cardiology Today-2019;23:205–9)

5 Central Cyanosis with Ejection Systolic Murmur Over Left Second Intercostal Space

5.1 Electrocardiogram

– Tall P waves (suggestive of right atrial overload) and right ventricular hypertrophy with
  (a) Right axis deviation (Fig. 49.7)

These findings suggest the possibility of severe pulmonary stenosis with intact interventricular septum. [16]

(b) Left axis deviation

It suggests the possibility of severe stenosis of a dysplastic pulmonary valve (Noonan syndrome). [18]

Clinico-electrocardiographic approach to diagnosis of a neonate presenting with heart failure is shown in Fig. 49.8.

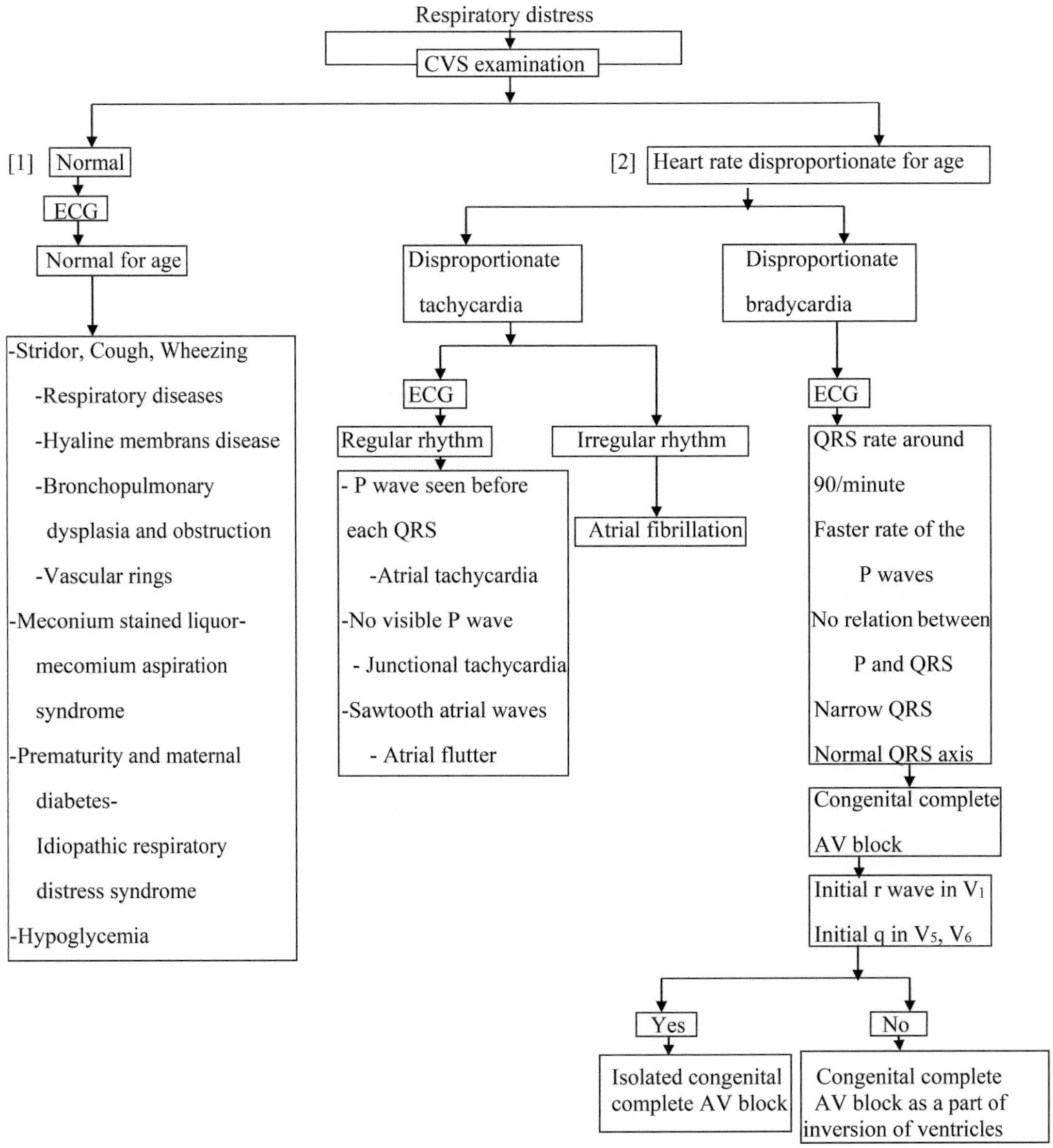

Fig. 49.8 Flowchart of clinico-electrocardiographic approach to diagnosis of a neonate presenting with heart failure

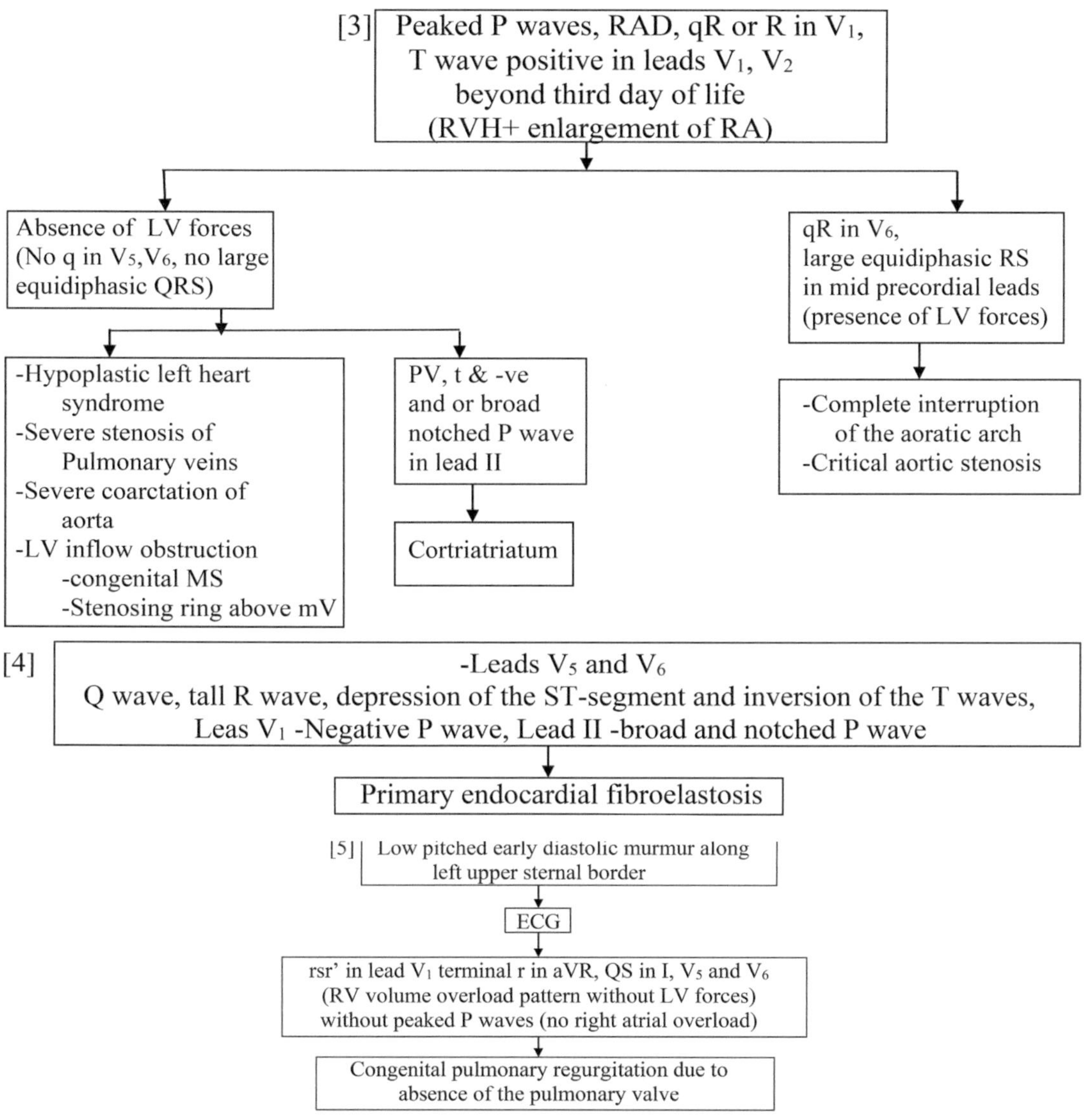

Fig. 49.8 (continued)

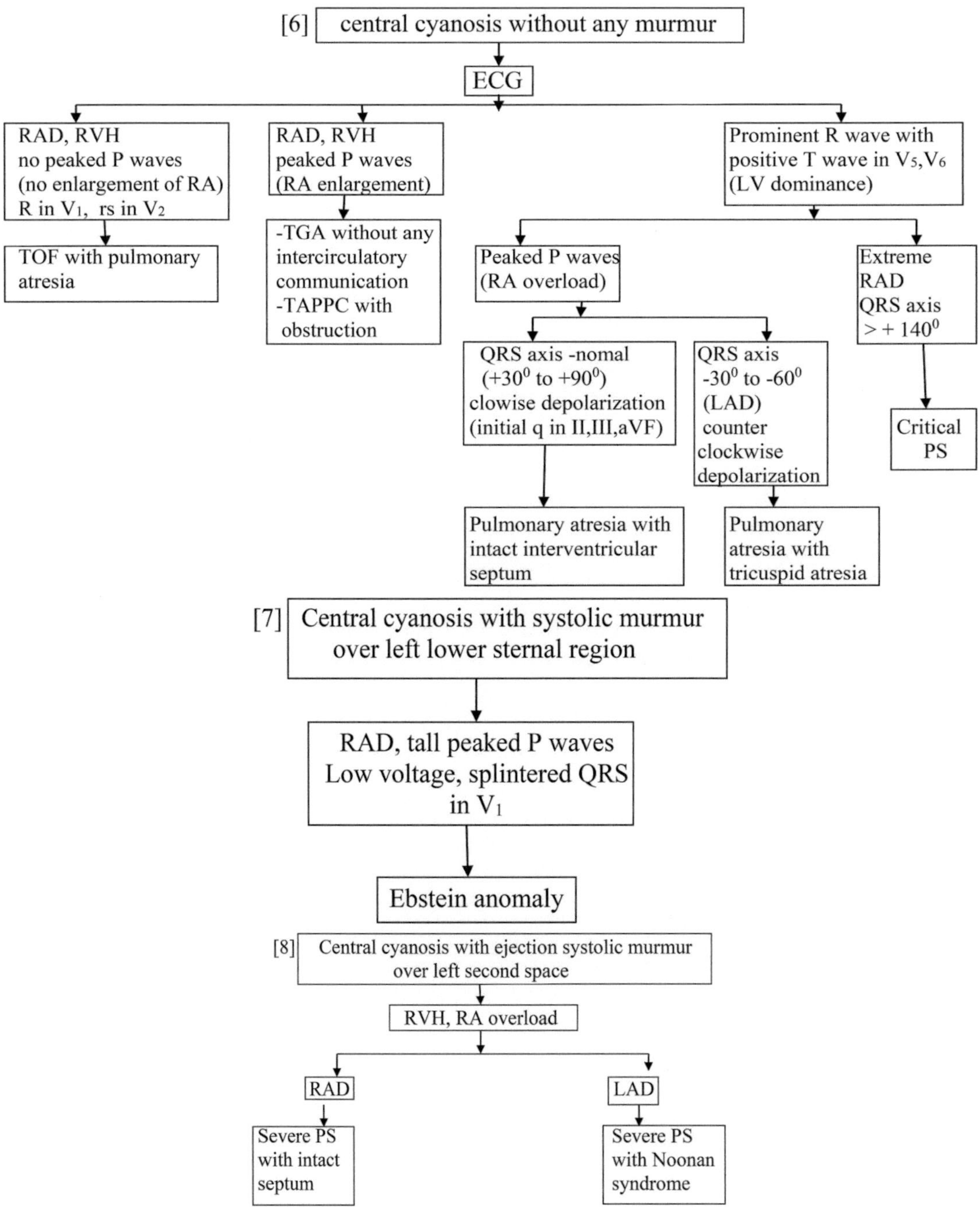

Fig. 49.8 (continued)

Summary

- Tachycardia for age with P wave abnormality suggests the possibility of supraventricular tachyarrhythmia.
- Combination of right axis deviation, right atrial enlargement, and low voltage splintered QRS in lead V_1 in absence of any electrocardiographic sign of RVH suggest the possibility of Ebstein anomaly.
- Combination of right axis deviation and right ventricular hypertrophy without right atrial enlargement and transition from R wave in lead V_1 to RS in lead V_2 suggests the possibility of tetralogy of Fallot with pulmonary atresia.
- Combination of right axis deviation, right ventricular hypertrophy, and right atrial enlargement with absence of left ventricular forces is seen in severe coarctation of aorta, complete interruption of the aortic arch, hypoplastic left heart syndrome, severe pulmonary vein stenosis, total anomalous pulmonary venous drainage with obstruction, severe pulmonary valve stenosis with intact ventricular septum or transposition of the great arteries.
- Combination of right axis deviation, right ventricular hypertrophy, and biatrial enlargement is seen in critical aortic stenosis, congenital mitral stenosis, stenosing ring above the mitral valve or cor triatriatum.
- Combination of right ventricular hypertrophy and right atrial enlargement with left axis deviation suggests the possibility of severe stenosis of dysplastic pulmonary valve with Noonan syndrome.
- Combination of right axis deviation, rsr' in lead V_1 without right atrial enlargement suggests the possibility of severe pulmonary regurgitation due to congenital absence of the pulmonary valve.
- Q wave with tall R wave, ST-segment depression, and T wave inversion in leads V_5 and V_6 and left atrial enlargement suggest the possibility of primary endocardial fibroelastosis.
- Left ventricular dominance with normal QRS axis suggests the possibility of pulmonary atresia with intact interventricular septum.
- Left ventricular dominance with left axis deviation suggests the possibility of tricuspid atresia with pulmonary atresia.
- Left ventricular dominance with right upper quadrant axis suggests the possibility of critical pulmonary stenosis.

MCQs

Q1. What could be the causes of respiratory distress in a neonate?

(a) Hyaline membrane disease
(b) Methemoglobinemia
(c) Hypoglycemia
(d) Right aortic arch

Q2. What could be the cause of heart failure with irregular rhythm in a neonate?

(a) Idiopathic respiratory distress syndrome
(b) Atrial flutter
(c) Atrial fibrillation
(d) All

Q3. In a new born, positive T waves in leads V_1 and V_2 suggest:

(a) Severe right ventricular hypertrophy
(b) Severe left ventricular hypertrophy
(c) Biventricular enlargement
(d) None

Q4. In a new born, T waves in leads V_1 and V_2 become inverted in:

(a) First week of life
(b) After one month
(c) After three months
(d) After one year

Q5. In a new born, positive T waves in leads V_1 and V_2 after seven days of birth suggest:

(a) Right ventricular hypertrophy
(b) Left ventricular hypertrophy
(c) Biventricular hypertrophy
(d) None

Q6. In a neonate with heart failure, qR in V_1, peaked P waves, and positive T waves in leads V_1 and V_2 beyond 3 days of life suggest the possibility of:

(a) Hypoplastic left heart syndrome
(b) Congenital mitral stenosis
(c) Severe aortic stenosis
(d) Severe pulmonary valve stenosis

Q7. In a neonate with heart failure, qR in lead V_1, upright T wave in leads V_1 and V_2 beyond three days of life with biatrial enlargement suggest the possibility of:

(a) Congenital mitral stenosis
(b) Stenosing ring above mitral valve
(c) Hypoplastic left heart syndrome
(d) Severe pulmonary vein stenosis

Q8. In a neonate with heart failure, without cyanosis or murmur, normal QRS axis, left ventricular hypertrophy with depression of the ST-segment, and inversion of the T wave in leads V_5 and V_6, suggest the possibility of:

(a) Critical aortic stenosis
(b) Complete interruption of the aortic arch
(c) Pulmonary atresia
(d) Primary endocardial fibroelastosis

Q9. In a neonate with heart failure, without cyanosis, short low-pitched early diastolic murmur along left upper sternal border, right axis deviation, rsr' in lead V_1 without peaked P wave suggest the possibility of:

(a) Large secundum ASD
(b) Large primum ASD
(c) Congenital absence of pulmonary valve
(d) LV-RA shunt

Q10. In a neonate with central cyanosis and heart failure, the presence of right axis deviation and right ventricular hypertrophy without tall P waves suggests the possibility of:

(a) Tetralogy of Fallot
(b) Pulmonary atresia
(c) Ebstein anomaly
(d) None

Q11. In a neonate with central cyanosis and heart failure, right axis deviation, rsR' in lead V_1, and tall peaked P waves suggest the possibility of:

(a) Tetralogy of Fallot
(b) Tricuspid regurgitation
(c) Total anomalous pulmonary venous connection with obstruction
(d) Tricuspid atresia

Q12. In a neonate with central cyanosis and heart failure, normal QRS axis and prominent R wave with prominent positive T waves in leads V_5 and V_6 suggest the possibility of:

(a) Pulmonary atresia
(b) Tricuspid atresia
(c) Critical pulmonary stenosis
(d) Critical aortic stenosis

Q13. In a neonate with central cyanosis and heart failure, left axis deviation with prominent R and prominent positive T waves in leads V_5 and V_6 suggest the possibility of:

(a) Tricuspid atresia
(b) Pulmonary atresia
(c) Aortic atresia
(d) None

Q14. In a neonate with central cyanosis and heart failure, left axis deviation, right ventricular hypertrophy, and tall P waves suggest the possibility of:

(a) Tricuspid atresia
(b) Pulmonary atresia
(c) Severe pulmonary stenosis
(d) Noonan syndrome with severe pulmonary stenosis

Q15. In a neonate with central cyanosis, systolic murmur over left lower sternal border, right axis deviation, tall P waves, low voltage splintered QRS in lead V_1 but no right ventricular hypertrophy suggest the possibility of:

(a) Severe valvular pulmonary stenosis
(b) Tricuspid atresia
(c) Ebstein anomaly
(d) Tetralogy of Fallot

Answers

(1) a,c (2) c (3) d (4) a (5) a (6) a (7) a,b (8) d, (9) c (10) a (11) c (12) a (13) a (14) d (15) c

References

1. Vijaylakshmi IB, Satpathy M. Bedside diagnosis and classification of congenital heart disease. In: Satpathy M, editor. Clinical diagnosis of congenital heart disease. New Delhi: Jaypee; 2008. p. 14–26.
2. Mishra BR, Satpathy M. Aortic arch anomalies and vascular rings. In: Satpathy M, editor. Clinical diagnosis of congenital heart disease. New Delhi: Jaypee; 2008. p. 192–8.
3. Perloff JK, Marelli AJ. Coarctation of the aorta and interrupted aortic arch. In: Perloff JK, Marelli AJ, editors. Clinical recognition of congenital Heart disease. Philadelphia: Elsevier; 2012. p. 101–28.
4. Roberts WC, Morrow AG, Braunwald E. Complete interruption of the aortic arch. Circulation. 1962;29:39–59.
5. Simpson JM. Hypoplastic left heart syndrome. Ultrasound Obstet Gynecol. 2000;15:271–8.
6. Dalvi B, Venkatesh S, Prabha SS. Left ventricular inflow obstruction. In: Satpathy M, editor. Clinical diagnosis of congenital heart disease. New Delhi: Jaypee; 2008. p. 53–9.
7. Guha S, Satpathy M. Congenital aortic stenosis. In: Satpathy M, editor. Clinical diagnosis of congenital heart disease. New Delhi: Jaypee; 2008. p. 143–56.
8. Perloff JK, Marelli AJ. Congenital obstruction to left atrial flow: mitral stenosis, cor-triatriatum, Pulmonary vein stenosis. In: Perloff JK, Marelli AJ, editors. Clincal recognition of congenital Heart disease. Philadelphia: Elsevier; 2012. p. 129–40.
9. Sellers FJ, Keith JD, Manning JA. The diagnosis of primary endocardial fibroelastosis. Circulation. 1964;29:49–59.
10. Freundlich E, Munk J, Griffel B, Steinlauf J. Primary myocardial disease in infancy. Am J Cardiol. 1963;13:721–33.
11. Ansari A. Isolated pulmonary valvular regurgitation: current perspectives. Prog Cardiovasc Dis. 1991;33:329–44.
12. Manojkumar R, Satpathy M. Complete transposition of the great arteries. In: Satpathy M, editor. Clinical diagnosis of congenital heart disease. New Delhi: Jaypee; 2008. p. 304–11.
13. Gathman GE, Nadas AS. Total anomalous pulmonary venous connection: clinical and physiologic observations of 75 pediatric patients. Circulation. 1970;42:143–54.
14. Gamboa R, Gersony WM, Nadas AS. The electrocardiogram in tricuspid atresia and pulmonary atresia with intact ventricular septum. Circulation. 1966;34:24–37.
15. Devachi F, Lucas RV Jr, Moller JH. The electrocardiogram and vectorcardiogram in tricuspid atresia. Correlation with pathologic anatomy. Am J Cardiol. 1970;25:18–27.
16. Desai N, Kumar R, Mahadevan C, Prakash VS. Congenital pulmonary stenosis. In: Satpathy M, editor. Clinical diagnosis of congenital heart disease. New Delhi: Jaypee; 2008. p. 199–207.
17. Bhattacharyya AK, Satpathy M. Ebstein anomaly. In: Satpathy M, editor. Clinical diagnosis of congenital heart disease. New Delhi: Jaypee; 2008. p. 240–8.
18. Raaijmakers R, Noonan JA, Croonen EA, Van Der Burgt CJAM, Draaisma JMT. Are ECG abnormalities in Noonan syndrome characteristic for the syndrome? Eur J Pediarr. 2008;167:1363–7.

50 Utility of Electrocardiogram in a Case Presenting with Heart Failure in Infancy (After Neonatal Period)

Heart failure in infancy is diagnosed when a previously asymptomatic neonate (first month of life) develops following clinical findings.

- Feeding difficulty (inability to complete feed within 15 to 20 minutes, low volume intake of feeds, frequent demand, and excessive perspiration during feeding).
- Repeated lower respiratory infections.
- Pulse rate >130/minute
- Respiratory rate >30/minute

Good clinical examination and correct interpretation of the electrocardiogram can give a reasonably correct diagnosis. The final diagnosis is, however, made by echocardiographic examination.

Conditions that can produce heart failure in the period of two to four months of infancy can be divided into the following groups.

(A) Non-cardiac causes [1]
1. Severe anemia
2. Overtransfusion
3. Hypoglycemia
4. Sepsis

Electrocardiogram: It is normal for age except sinus tachycardia.

(B) Cardiac causes

Most of these cases have some nonrestrictive left to right shunt. The shunt is not significant soon after birth because of high pulmonary vascular resistance. After birth, pulmonary vascular resistance progressively falls due to thinning of the smooth muscles of the pulmonary arteries and dilatation of the pulmonary vessels which were collapsed during fetal life. Maximum regression of pulmonary vascular resistance occurs by the 2nd to 3rd month of infancy. This results in a progressive increase in left to right shunt and progressive increase in pulmonary flow resulting in a progressive increase in left ventricular volume overload. The infantile left ventricle is still not prepared to tolerate increased volume overload and fails. Asymptomatic infant, therefore, develops heart failure in second to third month of life.

(1) No Cyanosis
  (a) No murmur recorded at birth
    (i) Biventricular enlargement, systolic thrill, and murmur along the left parasternal region.
    Electrocardiogram:

Normal QRS axis, broad and notched P wave in lead II (suggestive of left atrial overload), prominent R wave in lead V_1 (suggestive of right ventricular enlargement), prominent q waves with tall R waves and prominent positive T waves in leads V_5 and V_6 (suggestive of left ventricular overload) and large equidiphasic QRS complexes in mid precordial leads. These findings suggest the possibility of nonrestrictive ventricular septal

S. Mittal, *Insights into Electrocardiograms with MCQs*,
https://doi.org/10.1007/978-981-99-0127-2_50

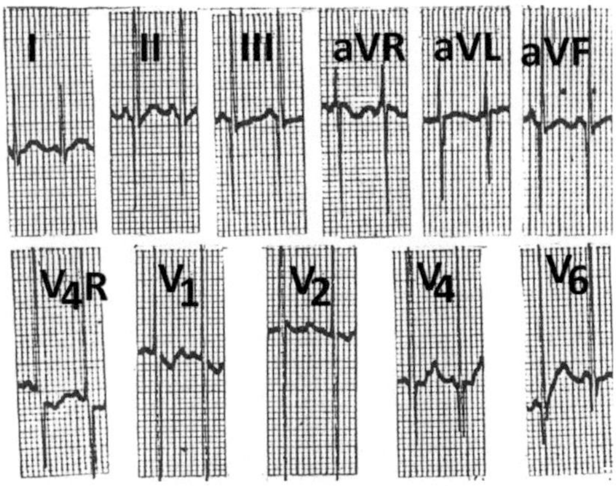

Fig. 50.1 Electrocardiogram is showing normal QRS axis, QR with upright T waves in leads II, III, aVF, and V_4 to V_6 (suggestive of left ventricular volume overload), large equidiphasic QRS complexes in leads V_4R, V_1, and V_2 (suggestive of biventricular overload). (Reproduced with permission from: Mittal SR. Utility of Electrocardiographic Findings in Cases Presenting with Heart Failure in Infancy. Cardiology Today-2019;23:249–57)

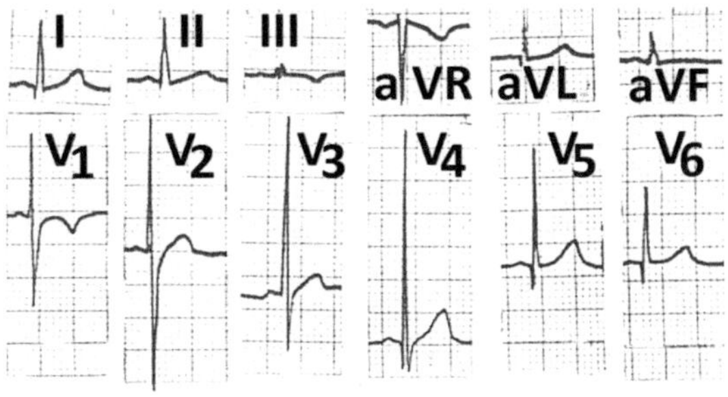

Fig. 50.2 Electrocardiogram showing normal QRS axis, prominent R wave in lead V_1 (suggestive of right ventricular overload), prominent Q waves with tall R waves, and positive T waves in leads V_4 to V_6 (suggestive of left ventricular volume overload), large equidiphasic RS in leads V_1 and V_2 (suggestive of biventricular overload). (Reproduced with permission from Mittal SR. Utility of Electrocardiographic Findings in Cases Presenting with Heart Failure in Infancy. Cardiology Today-2019;23:249–57)

defect with large left to right shunt [2, 3] (Fig. 50.1).

Left axis deviation (−30° to −90°), counter clockwise loop (initial q wave in leads I and aVL), prominent Q waves, R waves, and positive T waves in leads V_5 and V_6, and notch on the upstroke of S wave in leads II, III, and aVF. These findings suggest the possibility of atrioventricular septal defect type ventricular septal defect. [4]

(ii) Bounding pulse, low diastolic pressure, biventricular impulse, harsh ejection systolic murmur with eddy sounds over left third intercostal space.

Electrocardiogram:

Biventricular enlargements. Left ventricular forces dominate in patients with large pulmonary flow. It manifests as qR or qRS in leads III, aVF, and qR in leads V_5,V_6. These findings suggest the possibility of Aorto-pulmonary window with large left to right shunt [5] (Fig. 50.2).

(iii) No murmur or short systolic murmur over the apex.

Electrocardiogram:

Deep narrow Q waves in leads I and aVL with deep, narrow Q waves, elevation of the ST-segment, and inversion of the T waves in anterior chest leads. These findings are suggestive of anomalous origin of the left coronary artery from the pulmonary artery [6] (Fig. 50.3).

(iv) Preceding history of viral illness, displaced apical impulse with loud S_3, and raised levels of Troponins. [7]

Electrocardiogram:

Loss of R waves, diffuse elevation of the ST-segments, inversion of the T waves, atrial or ventricular ectopics. [8]

These findings suggest the possible of myocarditis:

(v) Displaced apical impulse with loud S_3

Electrocardiogram:

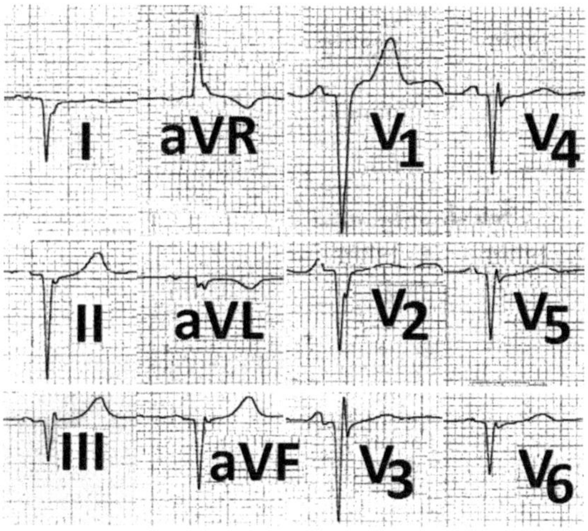

Fig. 50.3 Electrocardiogram is showing deep Q waves with loss of R wave in leads I, aVL, and leads V_2 to V_6 suggestive of anterior myocardial infarction. (Reproduced with permission from : Mittal SR. Utility of Electrocardiographic Findings in Cases Presenting with Heart Failure in Infancy. Cardiology Today-2019;23:249–57)

Low voltage, notched QRS, atypical bundle branch block pattern, but no pathological Q waves.

These findings suggest the possible of dilated cardiomyopathy [9]:

(vi) Left parasternal lift, wide and fixed split of the second sound, ejection systolic murmur along the left upper sternal border, and pansystolic murmur over the cardiac apex.

Electrocardiogram:

Left axis deviation, counterclockwise depolarization (q wave in leads I and aVL), rSR' in lead V_1 with qRS in leads V_5 and V_6, and notch on the ascending limb of the S wave in leads II, III, aVF.

These findings suggest the possibility of ostium primum ASD with significant mitral regurgitation [10, 11]:

(b) Murmur noted at birth
   (i) Pansystolic murmur audible on both sides of the sternum and volume overloaded left and right ventricles.

Electrocardiogram:

Right axis deviation, peaked P waves (suggestive of right atrial enlargement), rSR' in leads V_1 and V_2 (suggestive of right ventricular volume overload), and tall R waves in leads V_5 and V_6 (suggestive of left ventricular volume overload).

These findings suggest the possibility of large left ventricle to right atrial shunt (Gerbode shunt) [12]:

(ii) Bounding pulse with low diastolic pressure, continuous murmur with eddy sounds in the left infraclavicular region with forceful left ventricular apex.

Electrocardiogram:

Broad and notched P wave in lead II and negative P wave in lead V_1 (suggestive of left atrial overload), prominent q waves with tall R waves, and prominent positive T waves in leads V_5 and V_6 (suggestive of left ventricular volume overload) (Fig. 50.4).

These findings suggest the diagnosis of patient ductus arteriosus with large left to right shunt. [13]

(iii) Pansystolic murmur along the left sternal border and right ventricular apical impulse.

Electrocardiogram:

Prolonged PR interval, left axis deviation, counterclockwise depolarization (q waves in leads I and aVL), biventricular enlargement, absence of notching in the ascending limb of the S wave in inferior leads.

These findings suggest the possibility of double outlet right ventricle with large subaortic VSD without pulmonic stenosis. [14]

(iv) Systolic murmur over the cardiac apex radiating to axilla.

Electrocardiogram:

Left atrial enlargement with left ventricular volume overload pattern.

These findings suggest the possibility of severe congenital mitral regurgitation. [15]

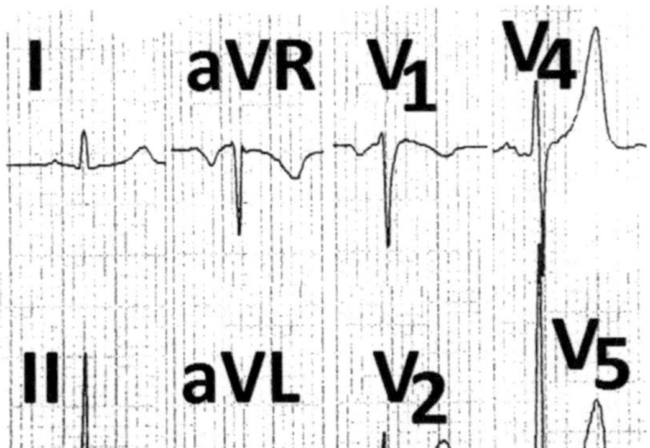

Fig. 50.4 Electrocardiogram showing normal QRS axis, prominent negative P wave in lead V_1 and broad, notched P wave in leads V_3 and V_4 (suggestive of left atrial overload), peaked P waves in leads II, III, aVF (suggestive of right atrial overload), deep S wave in lead V_3, q waves with tall R waves, and prominent positive T waves in leads V_4 and V_5 (suggestive of left ventricular volume overload). (Reproduced with permission from: Mittal SR. Utility of Electrocardiographic Findings in Cases Presenting with Heart Failure in Infancy. Cardiology Today-2019;23:249–57)

Significant congenital mitral regurgitation may be associated with the following:

- Atrial septal defect—ECG will show additional right atrial enlargement (peaked P waves) and right ventricular volume overload (rsR' pattern in lead V_1).
- Ventricular septal defect—ECG will show additional large equidiphasic QRS in mid precordial leads (Fig. 50.5).

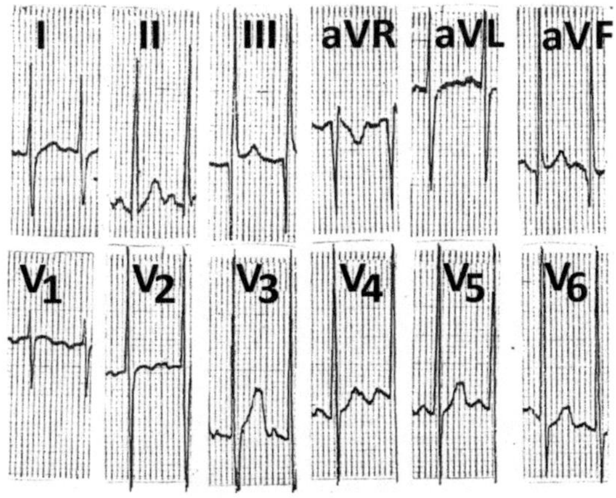

Fig. 50.5 Electrocardiogram is showing normal QRS axis, negative P wave in lead V_1 (suggestive of left atrial enlargement), deep Q waves with tall R waves, and positive T waves in leads III, aVF, and tall R waves with positive T waves in leads V_4 to V_6 (suggestive of left ventricular volume overload). Large equidiphasic QRS in lead V_2 are suggestive of biventricular enlargement. (Reproduced with permission from: Mittal SR. Utility of Electrocardiographic Findings in Cases Presenting with Heart Failure in Infancy. Cardiology Today-2019;23:249–57)

(2) Minimal or Mild Cyanosis

It can be confirmed by pulse oximetry

(a) No murmur detected at birth.
   (i) Systolic murmur over the left parasternal region, single second sound, narrow pedicle of cardiac shadow on skiagram of chest.

Electrocardiogram:
Normal or just rightward QRS axis, tall peaked P waves, large diphasic QRS complexes in mid precordial leads with preserved q waves and R waves in leads V_5 and V_6. These findings suggest the possibility of complete transposition of great arteries with nonrestrictive ventricular septal defect [16] (Fig. 50.6).

   (ii) Systolic murmur in the left parasternal region and normally split second heart sound.

Electrocardiogram:
Vertical or right axis, tall peaked P waves, tall R wave in lead V_1, and deep S waves in leads V_5, V_6, and inverted T waves in right and mid precor-

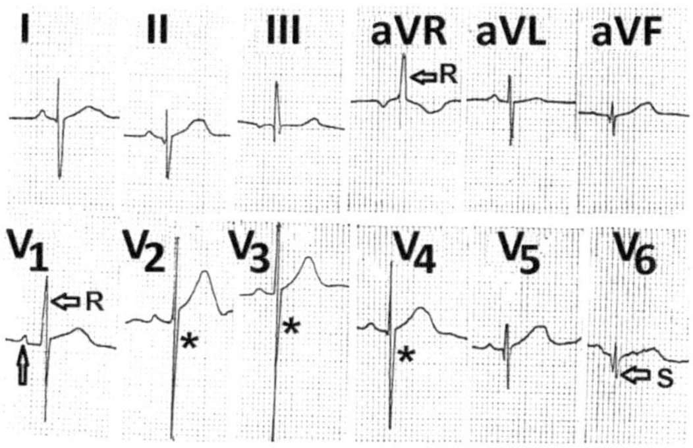

Fig. 50.6 Electrocardiogram showing right axis deviation, peaked P wave in lead V_1 (marked ⇧) (suggestive of right atrial overload), prominent R wave in leads aVR and V_1 and prominent S wave in leads V_5, V_6 (marked ⇦) (suggestive of right ventricular hypertrophy), prominent q wave in leads V_5,V_6 (suggestive of persistent left ventricular forces in spite of RVH), and large diaphic QRS in leads V_2 to V_4 (marked *) (suggestive of biventricular hypertrophy). (Reproduced with permission from: Mittal SR. Utility of Electrocardiographic Findings in Cases Presenting with Heart Failure in Infancy. Cardiology Today-2019;23:249–57)

dial leads (suggestive of right ventricular hypertrophy). Well-developed R waves in leads V_5 and V_6 suggest well-developed left ventricle.

These findings suggest the possibility of double outlet right ventricle with nonrestrictive subpulmonic VSD (Taussig–Bing anomaly) [17].

(iii) Presence of neonatal cyanosis that diminishes or disappears with the appearance of heart failure. Presence of biventricular impulse, single second sound, and systolic murmur over left parasternal region radiating upward and to the right on clinical examination. Presence of truncal regurgitation can produce additional bounding pulse and early diastolic murmur along left sternal border.

Electrocardiogram:

Normal QRS axis, RVH (tall R wave in lead V_1), tall peaked P waves, prominent Q waves with tall R waves and prominent positive T waves in leads V_5 and V_6, and absence of evidence of left atrial enlargement.

These findings suggest the possibility of truncus arteriosus with increased pulmonary flow. [18]

(iv) Presence of left ventricular apical impulse, normal split of second sound, long systolic murmur along left sternal border on clinical examination.

Electrocardiogram:

Large biphasic, monomorphic RS complexes in mid precordial leads with absence of q waves in left precordial leads. These findings suggest the possibility of single morphologic left ventricle without pulmonic stenosis [19] (Fig. 50.7).

Left axis deviation suggests noninverted outlet chamber. Right axis deviation suggests inverted outlet chamber.

(v) Presence of left parasternal systolic lift, wide fixed split of second sound, ejection systolic murmur over the left upper sternal border on clinical examination.

Electrocardiogram:

Left axis deviation (−30° to −120°), counter clockwise depolarization (q waves in leads I and aVL), rsR' or rR' in lead V_1, and peaked P waves in leads II, aVF.

These findings suggest the possibility of single atrium. [20]

(vi) Left parasternal systolic lift, wide fixed split of second sound, and short ejection systolic

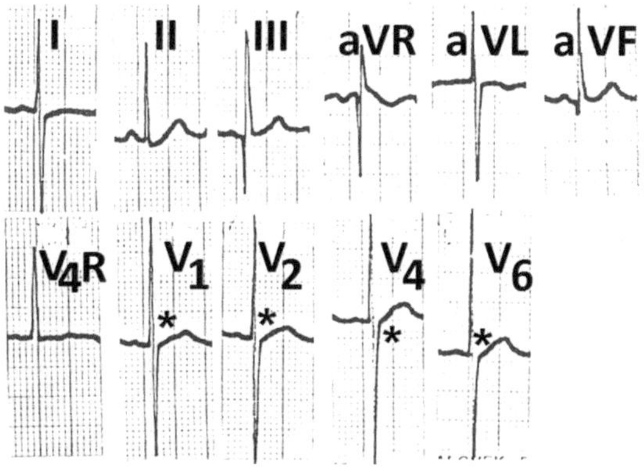

Fig. 50.7 Electrocardiogram is showing large biphasic (monomorphic) RS complexes in precordial leads (marked *) and absence of Q waves in left precordial leads. (Reproduced with permission from : Mittal SR. Utility of Electrocardiographic Findings in Cases Presenting with Heart Failure in Infancy. Cardiology Today-2019;23:249–57)

murmur in left second intercostal space on clinical examination.

Electrocardiogram:

Right axis deviation, peaked P waves suggestive of right atrial enlargement, rsR' in lead V_1 suggestive of right ventricular volume overload (Fig. 50.8). These findings suggest the possibility of total anomalous pulmonary venous drainage without obstruction. [21]

(b) Murmur detected at birth.

Left ventricular apical impulse, no right ventricular pulsation, systolic thrill and murmur in the left parasternal region, single first sound and normally split second heart sound on clinical examination.

Electrocardiogram:

Tall peaked P waves in leads II, III, aVF suggestive of right atrial enlargement, left axis deviation (−45° to −90°), rS configuration in lead V_1, prominent qR in leads V5 and V6 suggestive of left ventricular volume overload. These findings are suggestive of tricuspid atresia with nonrestrictive ventricular septal defect [22] (Fig. 50.9).

Electrocardiographic approach for heart failure in infancy (after neonatal period) is shown in Fig. 50.10.

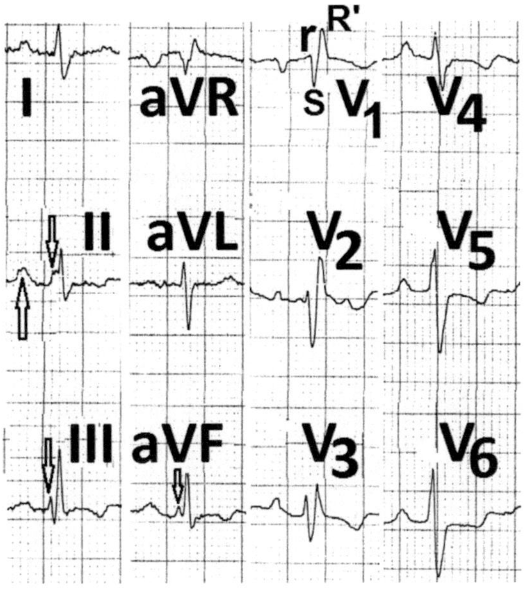

Fig. 50.8 Electrocardiogram showing peaked P wave in lead II (marked ⇧) (suggestive of right atrial enlargement), rSR' pattern in lead V_1 (suggestive of right ventricular volume overload), notch near the apex of R in leads II, III, and aVF—crochetage (marked ⇩) (suggestive of secundum atrial septal defect), and absence of any q wave in leads V_5 and V_6. (Reproduced with permission from: Mittal SR. Utility of Electrocardiographic Findings in Cases Presenting with Heart Failure in Infancy. Cardiology Today-2019;23:249–57)

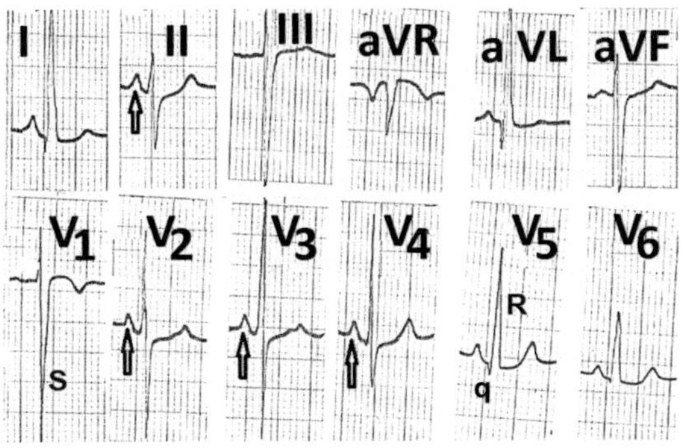

Fig. 50.9 Electrocardiogram showing peaked P waves (marked ⇧) (suggestive of right atrial enlargement), left axis deviation, absence of right ventricular enlargement in lead V_1, deep S wave in lead V_1, and q waves with prominent R waves in leads V_5 and V_6 (suggestive of left ventricular volume overload). (Reproduced with permission from: Mittal SR. Utility of Electrocardiographic Findings in Cases Presenting with Heart Failure in Infancy. Cardiology Today-2019;23:249–57)

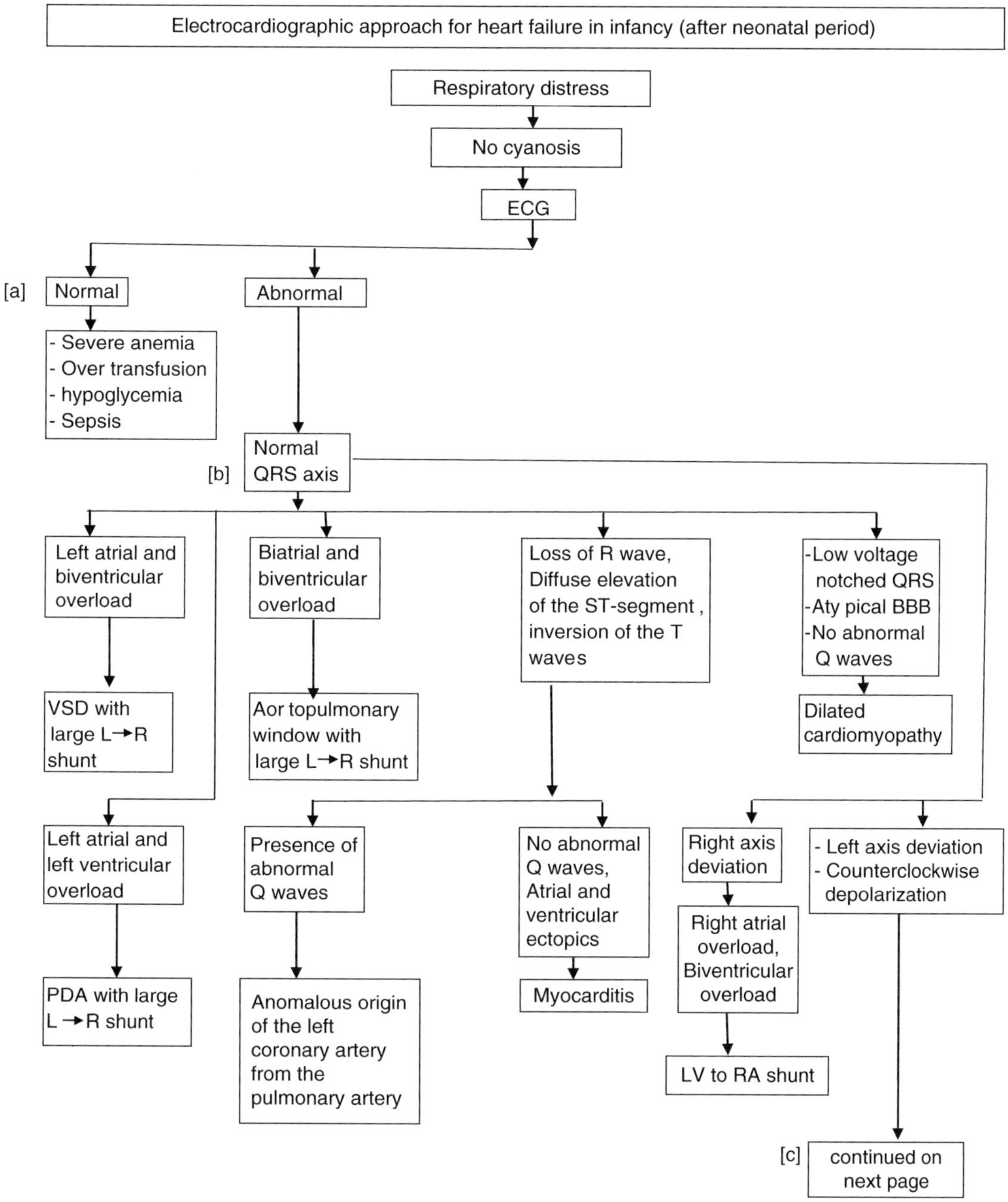

Fig. 50.10 The electrocardiographic approach in a case of heart failure infancy after the neonatal period

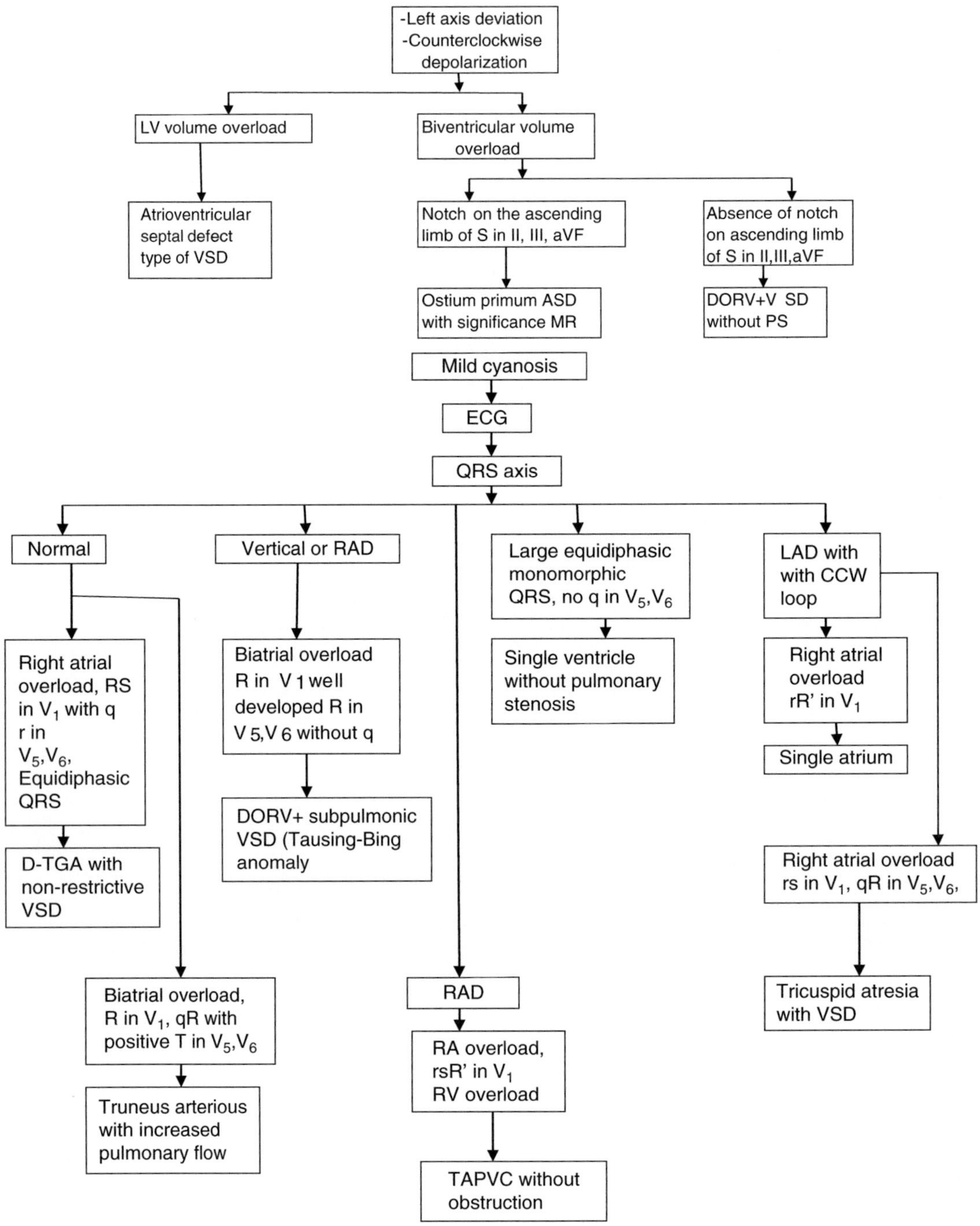

Fig. 50.10 (continued)

Summary

In an acyanotic infant:

- Combination of LV volume overload and left atrial enlargement suggest the possibility of patient ductus arterious with large left to right shunt or severe congenital mitral regurgitation.
- Combination of left ventricular volume overload with right atrial overload suggests the possibility of large left ventricle to right atrial shunt.
- Combination of left ventricular volume overload with right atrial overload and right ventricular volume overload in the presence of left axis deviation suggests the possibility of ostium primum ASD with significant mitral regurgitation.
- Combination of left ventricular volume overload and right ventricular pressure overload (biventricular enlargement) suggests the possibility of ventricular septal defect with large left to right shunt or aortopulmonary window.
- Biventricular enlargement with left axis deviation, counterclockwise depolarization (q in I and aVL), and notch on the upstroke of S wave in leads II, III, and aVF suggest the possibility of atrioventricular septal defect type of VSD with large left to right shunt.
- Biventricular enlargement with left axis deviation, counterclockwise depolarization but without any notch on ascending limb of S in leads II, III, aVF, and without any q wave in leads V_5 and V_6 suggest the possibility of a double outlet right ventricle with large subaortic VSD without pulmonary stenosis.
- ECG suggestive of anterior infarction suggests the diagnosis of anomalous origin of the left coronary artery from the pulmonary artery.
- Loss of the R wave, diffuse elevation of the ST-segments and inversion of the T waves suggest the diagnosis of myocarditis.
- Low voltage QRS with atypical bundle branch block support the possibility of dilated cardiomyopathy.

In an infant with mild to moderate cyanosis:

- Biventricular enlargement with right atrial overload suggests the possibility of transposition of great arteries with nonrestrictive VSD.
- Biventricular enlargement with biatrial overload suggests the possibility of truncus arteriosus with increased pulmonary flow.
- Large biphasic (monomorphic) RS complexes with no q wave in left precordial leads and left axis deviation suggest the possibility of single morphologic left ventricle.
- Left ventricular volume overload with rS complex in lead V_1 and right atrial overload suggest the diagnosis of tricuspid atresia with nonrestrictive VSD.
- Right ventricular volume overload and right atrial overload with right axis deviation suggest the possibility of total anomalous pulmonary venous drainage without obstruction.
- Right ventricular volume overload with left axis deviation suggests the possibility of single atrium.

MCQs

Q1. A four-month-old infant presented with recent onset of respiratory distress. The cardiovascular system examination was normal except a short grade 1 to 2/6 systolic murmur over left second intercostal space. Electrocardiogram was normal. What could be the possibility?

(a) Primum ASD
(b) Muscular VSD
(c) Wide small patent ductus arteriosus
(d) None

Q2. A premature infant of two months presented with respiratory distress. Cardiovascular examination was normal except grade 1–2/6 short ejection systolic murmur over left second intercostal space. Electrocardiogram was normal. What could be the possibility?

(a) Vascular ring
(b) Overtransfusion
(c) Closure of ductus in a case of severe coarctation
(d) All

Q3. A two-month-old infant was admitted for recent onset of excessive perspiration during feeding. Cardiovascular system examination revealed a 3–4/6 systolic murmur along left sternal border. An electrocardiogram revealed frontal plane QRS axis of $+105^{\circ}$, broad notched P wave in lead II, and biventricular enlargement. What could be the possibility?

(a) Nonrestrictive perimembranous VSD
(b) VSD with increased pulmonary vascular resistance
(c) Nonrestrictive VSD with severe valvular pulmonary stenosis
(d) VSD with severe right ventricular infundibular stenosis

Q4. A three-month-old infant was brought with history of recent onset of intake of low volume feeds, frequent demand, and excessive perspiration during feeding. Cardiovascular system examination revealed a 3–4/6 pansystolic murmur over left 3rd and 4th intercostal space. An electrocardiogram revealed QRS axis of $+120^{\circ}$, biatrial overload, and biventricular hypertrophy. What could be the possibility?

(a) Restrictive ventricular septal defect with low pulmonary vascular resistance
(b) Nonrestrictive VSD with normal pulmonary vascular resistance
(c) Nonrestrictive VSD with elevated pulmonary vascular resistance
(d) Nonrestrictive VSD with persistent fetal circulation

Q5. A three-month-old infant was admitted with recent onset of symptoms of heart failure. Cardiovascular system examination revealed a pansystolic murmur along mid left sternal border. An electrocardiogram revealed left axis deviation, counterclockwise depolarization, and biventricular enlargement. What could be the possibility?

(a) Large inlet VSD
(b) Large atrioventricular septal defect type VSD
(c) Large outlet VSD
(d) Large supracristal VSD

Q6. A previously healthy two-month-old infant presented with recent onset of symptoms of heart failure. There was no murmur. Electrocardiogram revealed QRS axis of -15°. Lead 1 showed small narrow q wave. Lead aVL showed narrow Q wave which was deeper than the following R wave. Lead V_1 showed deep S wave and lead V_6 showed tall R wave with inverted T wave. What is the most likely possibility?

(a) Myocarditis
(b) Dilated cardiomyopathy
(c) Anomalous origin of left coronary artery from pulmonary artery
(d) Severe congenital mitral regurgitation as a part of atrioventricular septal defect

Q7. A female infant of three months, from high altitude location presented with heart failure, good volume pulse, dynamic left ventricular apical impulse, and long systolic murmur with eddy sounds over the left infraclavicular region. Electrocardiogram showed normal QRS axis, prominent negative deflection of P wave in lead V_1, and prominent q wave, tall R wave, and positive T wave in leads V_5 and V_6. Most probable diagnosis is:

(a) Restrictive PDA with normal pulmonary vascular resistance
(b) Moderately restrictive PDA with normal pulmonary vascular resistance
(c) Nonrestrictive PDA with normal pulmonary vascular resistance
(d) Nonrestrictive PDA with high pulmonary vascular resistance

Q8. An acyanotic infant presented with heart failure, good volume pulse, forceful left ventricular apical impulse, systolic murmur with eddy sounds over left third intercostal space. Electrocardiogram revealed broad and peaked P waves, prominent R wave in lead V_1, prominent q wave, tall R wave, S wave, and positive T waves in leads V_5 and V_6. Most likely diagnosis is:

(a) Moderately restrictive PDA with normal pulmonary vascular resistance
(b) Nonrestrictive aortopulmonary window with low pulmonary vascular resistance
(c) Truncus arteriosus with truncal incompetence
(d) Perimembranous VSD with aortic regurgitation

Q9. An infant presented with heart failure, good volume pulse, and forceful left ventricular apical impulse. There was no murmur. An electrocardiogram revealed normal QRS axis, prominent negative P wave in lead V_1, and prominent q, tall R, and positive T wave in leads V_5, V_6. Most likely diagnosis is:

(a) Short large ductus with normal pulmonary vascular resistance
(b) Nonrestrictive aortopulmonary window with normal pulmonary vascular resistance
(c) Left ventricle to right atrial shunt
(d) Coronary arteriovenous fistula opening in the right atrium

Q10. A nonrestrictive aortopulmonary window with low pulmonary vascular resistance is associated with:

(a) Biatrial P waves
(b) Combined ventricular hypertrophy
(c) Large equidiphasic RS complexes in mid precordial leads
(d) None

Q11. A three-month-old infant presented with heart failure following respiratory infection. Apical impulse was displaced laterally with a loud S3. No murmur was audible. Serum troponin levels were raised. Electrocardiogram showed loss of R waves, diffuse ST-segment elevation, inverted T waves with atrial and ventricular ectopics. Most likely diagnosis is:

(a) Anomalous origin of the left coronary artery from the pulmonary artery
(b) Myocarditis
(c) Dilated cardiomyopathy
(d) Pericardial effusion

Q12. An infant with heart failure had a prominent systolic murmur over the cardiac apex. An electrocardiogram revealed left axis deviation, counterclockwise depolarization, broad and notched P wave in lead II, prominent q waves with tall R waves, and positive T waves in leads V_5,V_6. Most probable cause of mitral regurgitation was:

(a) Mitral valve prolapse
(b) Endocardial fibroelastosis
(c) Anomalous origin of the left coronary artery from the pulmonary artery
(d) Atrioventricular septal defect

Q13. An acyanotic infant with heart failure had prominent apical impulse, left parasternal systolic lift, wide and fixed split of second sound, an ejection systolic murmur over left second intercostal space, and a pansystolic murmur over the cardiac apex. Electrocardiogram showed left axis deviation, counterclockwise depolarization, peaked P waves, rsR' in lead V_1, and qRS with positive T wave in lead V_6. Most likely possibility is:

(a) Secundum ASD with congenital MR
(b) Ostium primum ASD with congenital MR
(c) Sinus venosus defect with congenital MR
(d) Secundum ASD with mitral valve prolapse

Q14. An infant with heart failure had biventricular pulsations and systolic murmur audible on both sides of the sternum. Electrocardiogram showed right axis deviation, peaked P waves with rsR' pattern in lead V_1, and q waves with tall R waves and prominent positive T waves in leads V_5 and V_6. Most likely diagnosis is:

(a) Moderately restrictive VSD with increased pulmonary vascular resistance
(b) Nonrestrictive VSD with tricuspid regurgitation
(c) Left ventricle to right atrial shunt
(d) Atrioventricular septal defect type VSD

Q15. An infant presented with heart failure. A systolic murmur was detected at birth. Clinical examination revealed prominent right ventricular impulse and a pansystolic murmur over left third and fourth intercostal spaces. Electrocardiogram showed left axis deviation, counterclockwise depolarization, tall R wave in lead V_1 and aVR, RS complex in V_6, and large RS complexes in mid precordial leads. Most likely diagnosis is:

(a) Double outlet right ventricle with nonrestrictive subaortic VSD
(b) Double outlet right ventricle with subpulmonic VSD without pulmonic stenosis
(c) VSD as part of atrioventricular septal defect with increased pulmonary vascular resistance
(d) Left ventricle to right atrial shunt with increased pulmonary vascular resistance

Q16. A two-month-old mildly cyanotic male infant, fifth in birth order, presented with heart failure, left parasternal lift, single second sound, and long systolic murmur in left 3rd and 4th intercostal spaces. No murmur was detected at birth. Electrocardiogram showed biatrial overload, monophasic R wave in lead aVR, qrs complex in lead V_6, and large equidiphasic RS complexes in mid precordial leads. Most likely diagnosis is:

(a) Complete transportation of great arteries with nonrestrictive VSD and low pulmonary vascular resistance
(b) Taussig–Bing anomaly with nonrestrictive VSD
(c) Truncus arteriosus with low pulmonary vascular resistance
(d) Tricuspid atresia with nonrestrictive VSD

Q17. A three-month-old mildly cyanotic infant presented with heart failure, left parasternal systolic lift, single second sound, pansystolic murmur over left 3rd and 4th intercostal spaces. Electrocardiogram showed QRS axis of $+120^{\circ}$, biatrial and biventricular overload. Most likely diagnosis is:

(a) Complete transposition of great arteries with nonrestrictive VSD and low pulmonary vascular resistance
(b) Complete transposition of great arteries with intact ventricular septum
(c) Complete transposition of great arteries with increased pulmonary vascular resistance
(d) Complete transposition of great arteries with atrioventricular septal defect type VSD

Q18. A one-month-old infant presented with heart failure, bounding pulse, single second sound, high-pitched ejection click, an ejection systolic murmur, and high frequency blowing early diastolic murmur over left 3rd and 4th intercostal spaces. No murmur was detected at birth. Electrocardiogram showed normal QRS axis, biatrial and biventricular enlargement. Most likely diagnosis is:

(a) Aortopulmonary window with low pulmonary vascular resistance
(b) Truncus arteriosus with low pulmonary vascular resistance
(c) Patent ductus with low pulmonary vascular resistance
(d) Double outlet right ventricle with subaortic VSD

Q19. A two-month-old infant presented with heart failure, prominent left ventricular apical impulse, normally split second sound and long systolic murmur along left sternal border. Electrocardiogram showed biatrial overload, left axis deviation with counterclockwise depolarization, and remarkably large biphasic (monomorphic) RS complexes in the precordial leads. What is the most likely diagnosis?

(a) Atrioventricular canal type VSD with low pulmonary vascular resistance
(b) Morphologic left single ventricle without pulmonic stenosis with noninverted outlet chamber
(c) Double outlet right ventricle with nonrestrictive subaortic VSD
(d) Large perimembranous VSD with low pulmonary vascular resistance

Q20. A three-month-old infant presented with mild cyanosis, heart failure, left parasternal systolic lift, wide fixed split of second sound, and 2–3/6 ejection systolic murmur over left second intercostal space. An electrocardiogram revealed QRS axis of −45°, counter clockwise depolarization, rsr' configuration in lead V_1, and inverted P waves in leads II, III, and aVF. Most likely diagnosis is:

(a) Lutembacher syndrome
(b) Ostium primum ASD with congenital MR
(c) Single atrium
(d) Total anomalous pulmonary venous drainage without obstruction

Q21. An infant presented with mild cyanosis, heart failure, left ventricular apical impulse, single first sound, normally split second sound with a systolic thrill, and murmur in the left parasternal region. The murmur was also detected at birth. An electrocardiogram revealed tall peaked P waves, left axis deviation, rS configuration in lead V_1, and prominent qR configuration in lead V_6. Most likely diagnosis is:

(a) Tricuspid atresia with nonrestrictive VSD
(b) Atrioventricular septal defect with increased pulmonary vascular resistance
(c) Single morphologic left ventricle with low pulmonary vascular resistance
(d) DORV with nonrestrictive subaortic VSD

Answers

(1)d (2)a,b (3)a (4)b (5)b (6)c (7)b (8)b (9)a (10) a,b,c (11) b (12)d (13)b (14)c (15)a (16)a (17)a (18)b (19)b (20)c (21)a

References

1. Vijayalakshmi IB, Satpathy M. Bedside diagnosis and classification of congenital heart disease. In: Satpathy M, editor. Clinical diagnosis of congenital heart disease. New Delhi: Jaypee; 2008. p. 14–26.
2. Toscano-Barboza E, Dushane JW. Ventricular septal defect: correlation of electrocardiographic and hemodynamic findings in 60 proved cases. Am J Cardiol. 1959;3:721–32.
3. Katz LN, Wachtel S. Diphasic QRS type of electrocardiogram in congenital heart disease. Am Heart J. 1937;13:202–6.
4. Brandenburg RO, Burchell HB, Toscano-Barbosa E. Electrocardiography studies of cases with intracardiac malformations of the atrioventricular canal. Proc Staff Meet Mayo Clin. 1956;31:513–23.
5. Bahuleyan CG. Aorto pulmonary window. In: Satpathy M, editor. Clinical diagnosis of congenital heart disease. New Delhi: Jaypee; 2008. p. 124–8.
6. Perloff JK, Marelli AJ. Anomalous origin of the left coronary artery from the pulmonary trunk. In: Perloff JK, Marelli AJ, editors. Clinical recognition of congenital heart disease. Philadelphia: Elsevier; 2002. p. 394–405.
7. Al-Biltagi M, Issa M, Hagar HA, et al. Circulating cardiac troponins levels and cardiac dysfunction in children with acute fulmimant viral myocarditis. Acta Paediatr. 2010;99:1510–6.
8. Ukena C, Mahfoud F, Kindermann I, et al. Prognotic electrocardiographic parameters in patients with suspected myocarditis. Eur J Heart Fail. 2011;13:398–405.
9. Dec WG, Fuster V. Idiopathic dilated cardiomyopathy. N Engl J Med. 1994;331:1564–75.
10. Perloff JK, Marelli AJ. Atrial septal defect: Simple and complex. In: Perloff JK, Marelli AJ, editors. Clinical recognition of congenital heart disease. Philadelphia: Elsevier; 2002. p. 212–72.
11. Juneja R, Shah S. Complete atrioventricular septal defect. In: Satpathy M, editor. Clinical diagnosis of congenital heart disease. New Delhi: Jaypee; 2008. p. 129–42.

12. Rienenschneider TA, Moff AJ. Left ventricular - right atrial communication. Am J Cardiol. 1967;19:710–8.
13. Perloff JK, Marelli AJ. Patent ductus arteriosus, Aortopulmonary window. In: Perloff JK, Marelli AJ, editors. Clinical recognition of congenital heart disease. Philadelphia: Elsevier; 2002. p. 368–93.
14. Krongard E, Ritter DG, Weidman WH, Dushane JW. Hemodynamic and anatomic correlation of electrocardiogram in double - outlet right ventricle. Circulation. 1972;46:995–1004.
15. Satpathy M. Congenital mitral regurgitation. In: Satpathy M, editor. Clinical diagnosis of congenital heart disease. New Delhi: Jaypee; 2008. p. 67–70.
16. Shaher RM, Deuchar DC. The electrocardiogram in complete transposition of the great vessels. Br. Heart J. 1966;28:265–75.
17. Perloff JK, Marelli AJ. Double outlet ventricle. In: Perloff JK, Marelli AJ, editors. Clinical recognition of congenital heart disease. Philadelphia: Elsevier; 2002. p. 348–67.
18. Perlo FFJK, Marelli AJ. Truncus asteriosus. In: Perloff JK, Marelli AJ, editors. Clinical recognition of congenital heart disease. Philadelphia: Elsevier; 2002. p. 492–504.
19. Davachi F, Moller JH. The electrocardiogram and vectorcardiogram in single ventricle. Anatomic correlations. Am J Cardiol. 1969;23:19–31.
20. Hung JS, Ritter DG, Feldt RH, Kincaid OW. Electrocardiographic and angiographic features of common atrium. Chest. 1973;63:970–5.
21. Gupta CD, Satpathy M. Total anomalous pulmonary venous connections. In: Satpathy M, editor. Clinical diagnosis of congenital heart disease. New Delhi: Jaypee; 2008. p. 317–24.
22. Guller V, Titus JL, Dushane JW. Electrocardiographic diagnosis of malformations associated with tricuspid atresia: correlation with morphologic features. Am Heart J. 1969;78:180–8.

51 Electrocardiogram in Children/ Adolescents/ Adults with Cyanosis Due to Congenital Cardiovascular Disease

Detailed diagnosis of congenital cyanotic cardiovascular diseases in children and adults is possible only by echocardiographic evaluation. Good clinical examination and correct interpretation of the electrocardiogram can, however, significantly narrow down the differential diagnosis. Important clinical findings in relation with classical electrocardiographic features are discussed.

(A) Mild central cyanosis with normal electrocardiogram
   (1) Relative prominence of the left internal jugular venous pulsations suggest the possibility of left superior vena cava draining into the left atrium [1] (Fig. 51.1).
   (2) Abnormal shadow in the skiagram of chest with hereditary hemorrhagic telangiectasia [2] and/or history of recurrent bleeding from nose, lips, and mouth [2] suggests the possibility of pulmonary arteriovenous fistula. Pulmonary artery angiography or contrast-enhanced computerized tomography of the chest is required for confirmation. [3]

(B) Mild to moderate central cyanosis with right-sided accessory bypass tract (Fig. 51.2) suggests the possibility of Ebstein anomaly. [4] Preexcitation may be permanent or intermittent (Fig. 51.3).

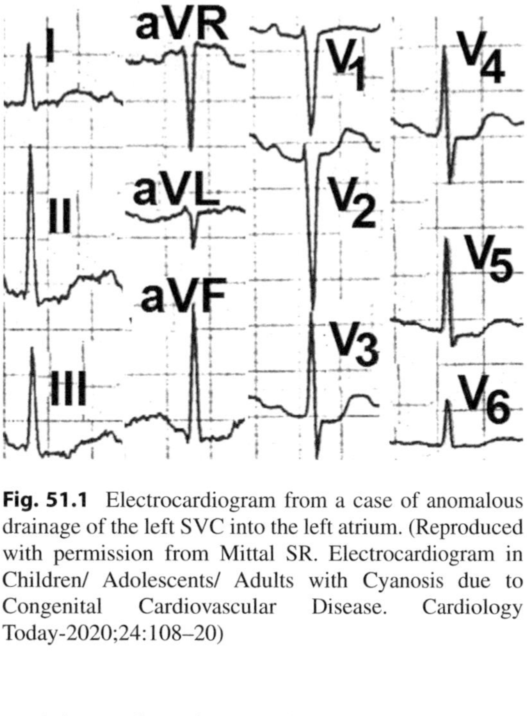

Fig. 51.1 Electrocardiogram from a case of anomalous drainage of the left SVC into the left atrium. (Reproduced with permission from Mittal SR. Electrocardiogram in Children/ Adolescents/ Adults with Cyanosis due to Congenital Cardiovascular Disease. Cardiology Today-2020;24:108–20)

Other salient features include silent precordium on clinical examination and gross cardiomegaly with narrow pedicle of cardiac shadow on the skiagram of chest.

(C) Mild to moderate central cyanosis with supraventricular tachycardia/atrial flutter/

S. Mittal, *Insights into Electrocardiograms with MCQs*,
https://doi.org/10.1007/978-981-99-0127-2_51

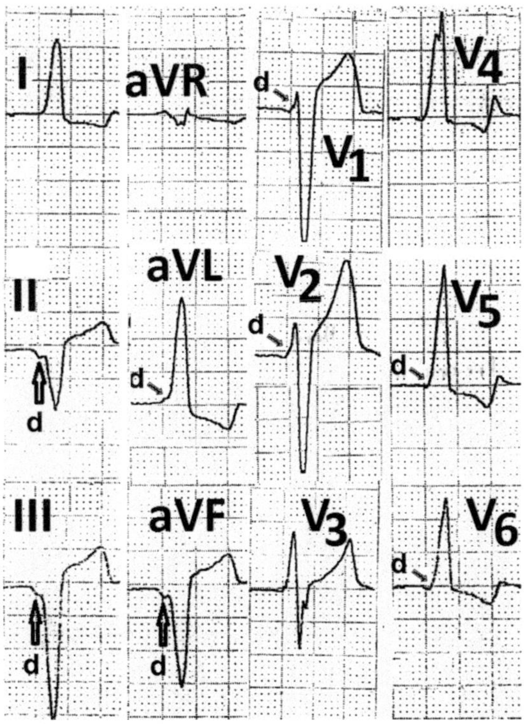

Fig. 51.2 Electrocardiogram from a patient of Ebstein anomaly showing positive QRS in leads V_1 and V_2 (delta wave marked ➘ d) and negative QRS in leads V_5,V_6 (delta wave marked d). Findings are suggestive of right-sided accessory path. Delta waves are also seen in peak II, III, aVL, and aVF (marked d). (Reproduced with permission from: Mittal SR. Electrocardiogram in Children/ Adolescents/ Adults with Cyanosis due to Congenital Cardiovascular Disease. Cardiology Today-2020;24:108–20)

atrial fibrillation should suggest the possibility of underlying condition.

(1) Inferior frontal plane QRS axis with splintered QRS. These findings suggest the possibility of underlying Ebstein anomaly [5] (Figs. 51.4 and 51.5).

(2) Eisenmenger's syndrome.

(a) Right axis deviation, notch on top of the R wave in leads II, III, and aVF, rR' in lead V_1 and rs without any q wave in lead V_6. [6] These ECG findings suggest the possibility of underlying secundum atrial septal defect.

(b) Left axis deviation (−30° to −90°), counterclockwise depolarization

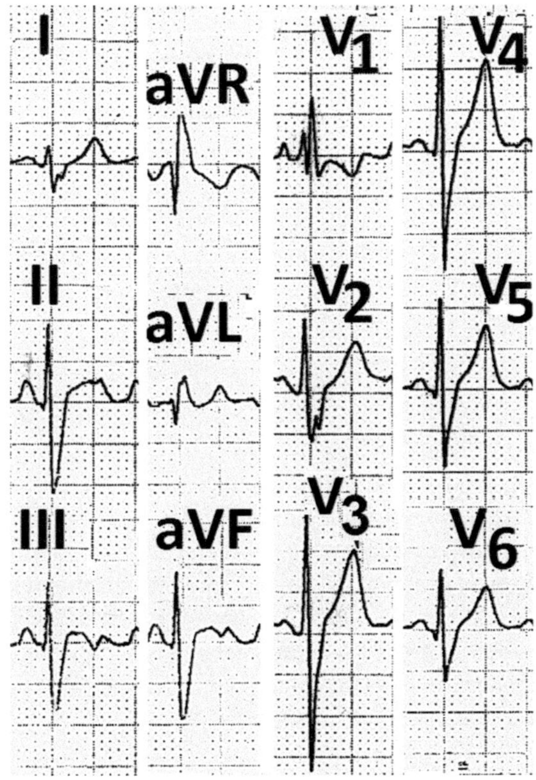

Fig. 51.3 Electrocardiogram during sinus rhythm from the same patient of Ebstein anomaly whose ECG is shown in Fig. 51.2. P waves are peaked in leads II, III, aVF, and V_1 suggestive of right atrial enlargement. The PR interval is normal, suggesting normal atrioventricular conduction. QRS shows right axis deviation and right bundle branch block. (Reproduced with permission from: Mittal SR. Electrocardiogram in Children/ Adolescents/ Adults with Cyanosis due to Congenital Cardiovascular Disease. Cardiology Today-2020;24:108–20)

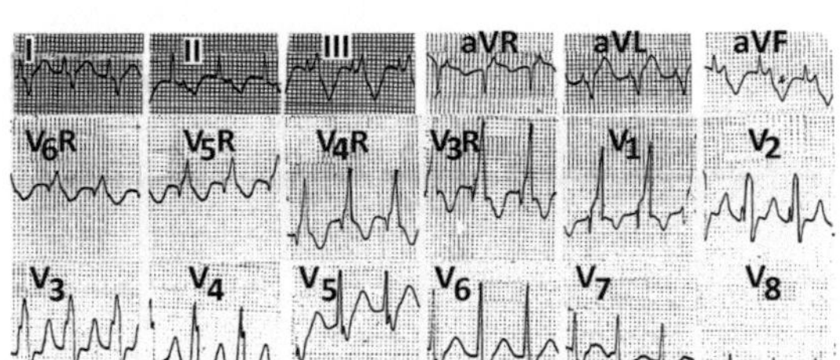

Fig. 51.4 Electrocardiogram from a case of Ebstein anomaly showing supraventricular tachycardia. QRS shows right bundle branch block pattern with Q waves in leads V_6R to V_1 (due to grossly enlarged RA). QRS is fractioned in leads I, II, III, aVL, and aVF. (Reproduced with permission from Mittal SR. Electrocardiogram in Children/ Adolescents/ Adults with Cyanosis due to Congenital Cardiovascular Disease. Cardiology Today-2020;24:108–20)

(initial q wave in leads I and aVL), R wave in lead V_1, and qrs in lead V_6. These ECG findings suggest the possibility of underlying atrioventricular septal defect (Fig. 51.6).

(D) Mild to moderate central cyanosis with atrioventricular block.

1. Ebstein anomaly

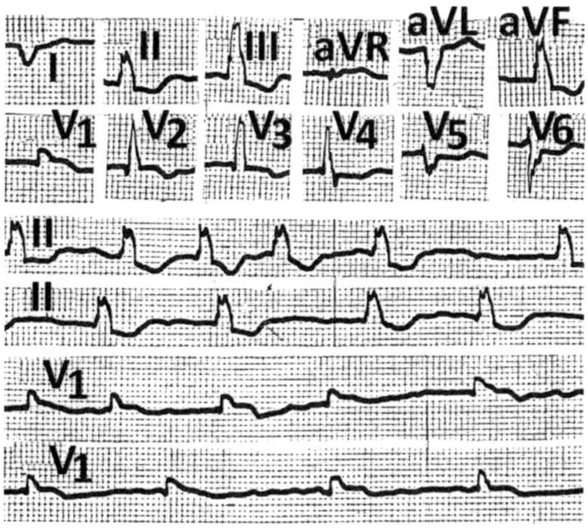

Fig. 51.5 Electrocardiogram from another case of Ebstein' anomaly. Long leads II and V_1 show atrial fibrillation. QRS complex shows right axis deviation, right bundle branch block pattern with q waves in leads V_1 to V_3 (due to enlarged right atrium). (Reproduced with permission from: Mittal SR. Electrocardiogram in Children/ Adolescents/ Adults with Cyanosis due to Congenital Cardiovascular Disease. Cardiology Today-2020;24:108–20)

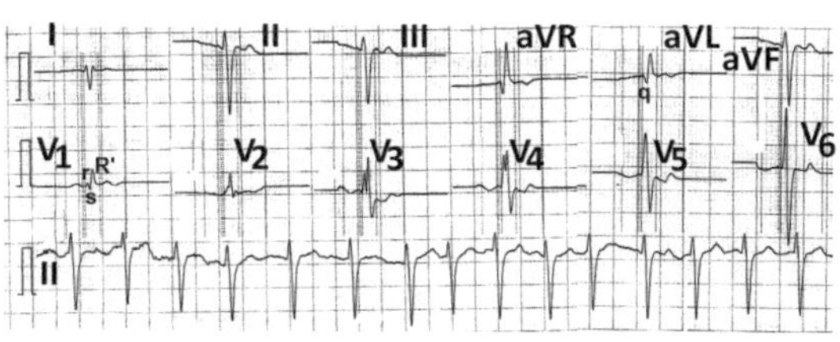

Fig. 51.6 Electrocardiogram from a patient of atrioventricular septal defect showing extreme left axis deviation, counterclockwise depolarization (q wave in aVL), right ventricular hypertrophy (rsR' in lead V_1), and large RS complex in lead V_6. Long lead II shows atrial fibrillation. (Reproduced with permission from: Mittal SR. Electrocardiogram in Children/ Adolescents/Adults with Cyanosis due to Congenital Cardiovascular Disease. Cardiology Today-2020;24:108–20)

Prolonged PR interval, right bundle branch block pattern with q wave in leads V_1 to V_6R. These ECG findings suggest the possibility of underlying Ebstein anomaly (Fig. 51.7). PR interval is prolonged due to intra. His and infra-His delay in conduction. The PR interval can be markedly prolonged [7] but complete atrioventricular block is rare. [8] q waves in right-sided chest leads are due to gross enlargement of the right atrium (Fig. 51.7).

2. Eisenmenger's syndrome
   (a) Presence of q wave in lead V_1 and the absence of q wave in leads V_6 to V_9. These ECG findings should suggest the possibility of underlying congenitally corrected transposition of the great arteries (inversion of the ventricles). [9]
   (b) Left axis deviation (-30° to -90°), counterclockwise depolarization (initial q wave in leads I and aVL) and qrs in leads V_5 and V_6. These findings should suggest the possibility of underlying atrioventricular septal defect. [10] Extreme left axis deviations is more common in the setting of Down syndrome. [11]
   (c) Right axis deviation, qR or R in lead V_1 and rs without any q waves in leads V_5 and V_6. These ECG findings suggest the possibility of underlying secundum atrial septal defect (Fig. 51.8). Atrioventricular blocks are usually seen in the setting of familial secundum atrial septal defect, for example,

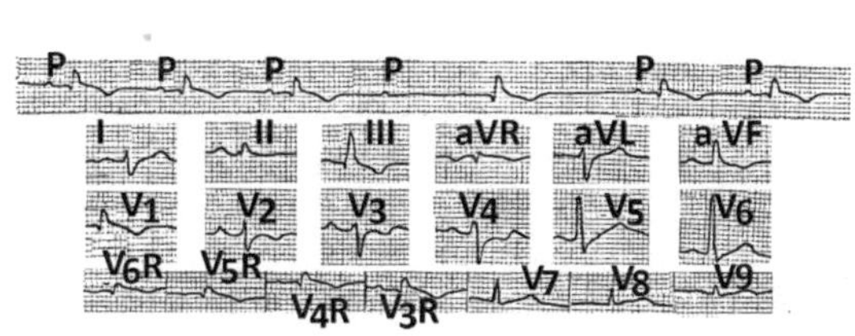

Fig. 51.7 Electrocardiogram from a case of Ebstein anomaly showing P-R interval of 0.20 second with intermittent Mobitz type II atrioventricular block in long lead V_1. QRS shows right bundle branch block pattern with q waves in leads V_1 to V_6R (due to gross right atrial enlargement). (Reproduced with permission from: Mittal SR. Electrocardiogram in Children/ Adolescents/Adults with Cyanosis due to Congenital Cardiovascular Disease. Cardiology Today-2020;24:108–20)

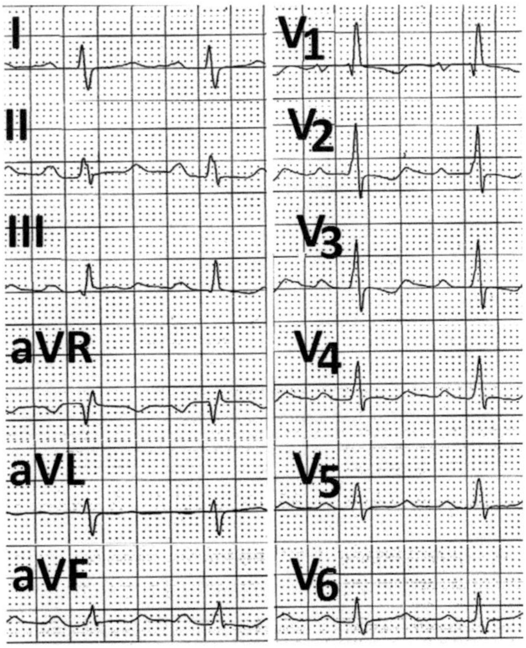

Fig. 51.8 Electrocardiogram from a case of secundum atrial septal defect with prolonged PR interval (0.24 second). QRS shows inferior axis, qR configuration in lead V_1 and Rs configuration in lead V_6. (Reproduced with permission from: Mittal SR. Electrocardiogram in Children/ Adolescents/ Adults with Cyanosis due to Congenital Cardiovascular Disease. Cardiology Today-2020;24:108–20)

Holt-Oram syndrome (hypoplasia of thumb, at times extending to radius). [12]

3. Peaked P waves due to right atrial enlargement, prominent R wave in right precordial leads, large biphasic stereotype RS complexes in mid precordial leads with absence of q wave in leads V_5, V_6. These ECG findings suggest the possibility of single morphologic left ventricle with inversion of outlet chambers [13] (Fig. 51.9).

Right axis deviation, right ventricular hypertrophy (qR in leads V_1, V_2), counterclockwise depolarization (q in I and aVL) with broad and slurred S in leads V_5 and V_6. These ECG findings suggest the possibility of double outlet right ventricle with subaortic ventricular septal defect and pulmonary valve stenosis. [14]

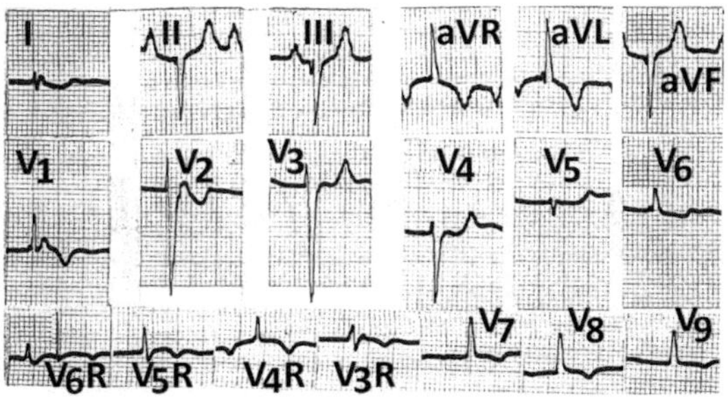

Fig. 51.9 Electrocardiogram from a patient of single morphologic left ventricle with inversion of outlet chambers. P waves are peaked in leads II, III, and aVF and are inverted in lead V_1 (biatrial enlargement), the PR interval is 0.24 second. QRS shows q in leads II, III, aVF (clockwise depolarization), prominent R wave in lead V_1 and aVR, and rs configuration in leads V_2 to V_5 with no q wave in lead V_6. Deep S in leads V_2, V_3, and tall R wave with T wave inversion denotes left ventricular hypertrophy. (Reproduced with permission from Mittal SR. Electrocardiogram in Children/ Adolescents/ Adults with Cyanosis due to Congenital Cardiovascular Disease. Cardiology Today-2020;24:108–20)

(E) Mild to moderate central cyanosis with broad QRS complex simulating right bundle branch block.

These findings should suggest the possibilities of underlying conditions

(1) Ebstein anomaly.

In fact the QRS is fairly small and splintered [15] (Fig. 51.10).

(2) Atrioventricular septal defect.

Left axis deviation and counterclockwise depolarization give a clue to diagnosis [10] (Fig. 51.6).

(3) Double outlet right ventricle with subaortic ventricular septal defect with pulmonary stenosis.

In fact, there is no right bundle branch block. There is prolongation of the terminal forces of the QRS complex resulting in slurred terminal R wave in leads V_1 and aVR. It simulates right bundle branch block. [14]

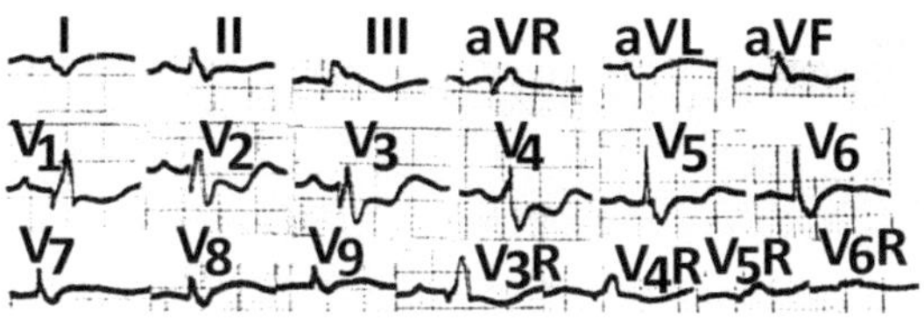

Fig. 51.10 Electrocardiogram from a patient of Ebstein anomaly showing broad and slurred R wave in lead V_1 simulating right bundle branch block. Leads V_1 to V_6R show qR configuration due to right atrial enlargement. Positive P wave in lead V_1 also suggests right atrial enlargement. First normal QRS attached to the second broad QRS is clear in leads V_2 and V_3. This also supports the diagnosis of Ebstein anomaly. (Reproduced with permission from: Mittal SR. Electrocardiogram in Children/ Adolescents/ Adults with Cyanosis due to Congenital Cardiovascular Disease. Cardiology Today-2020;24:108–20)

(F) Moderate cyanosis, systolic pulsations with palpable second sound, left parasternal systolic lift, and short ejection systolic murmur over left second intercostal space.

These findings suggest pulmonary hypertension with right to left shunt (Eisenmenger's syndrome).

(1) QRS axis between $+90^{\circ}$ to $+150^{\circ}$, clockwise depolarization (initial q wave in leads II, III, aVF).
  - (a) Prominent P waves in leads II and V_1, and RS configuration with absence of q wave in leads V_5 and V_6 (absence of left ventricular forces). These electrocardiographic findings suggest the following possibilities:
    - (i) Underlying secundum atrial septal defect [6] (Fig. 51.11)
    - (ii) Primary pulmonary artery hypertension [16] (Fig. 51.12)

(b) Tall monomorphic R wave with inversion of the T wave in lead V_1 with prominent S waves in leads V_5 and V_6. Additional q waves in leads V_5, V_6, and/or large equidiphasic RS complex in leads V_2 to V_5 suggest well developed left ventricle (Figs. 51.13 and 51.14).

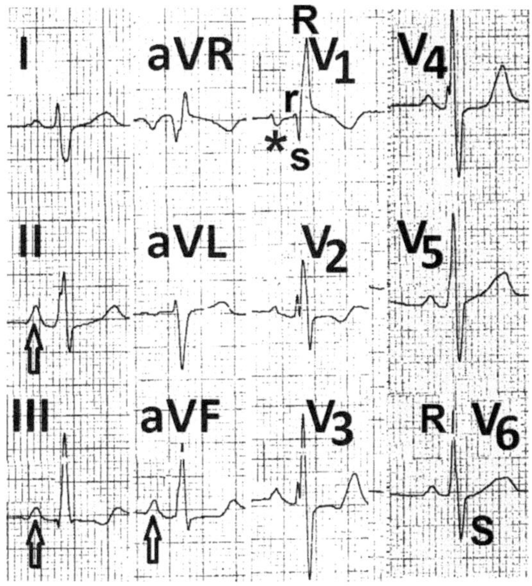

Fig. 51.11 Electrocardiogram from a case of secundum atrial septal defect with Eisenmenger's syndrome. The P waves are peaked in leads II, III, aVF (marked ⇧) (right atrial enlargement). The terminal force of the P wave in lead V_1 is negative (marked *). This is, however, because the terminal vector of the P wave is written by an enlarged right atrium and does not suggest additional left atrial enlargement. QRS axis is $+105^{\circ}$. Lead V_1 shows rsR' and lead V_6 shows the RS configuration without any q wave. (Reproduced with permission from: Mittal SR. Electrocardiogram in Children/ Adolescents/ Adults with Cyanosis due to Congenital Cardiovascular Disease. Cardiology Today-2020;24:108–20)

These findings suggest the possibility of underlying [6]

(ii) Ventricular septal defect
(iii) Patent ductus arteriosus
(iv) Aortopulmonary window
(v) Truncus arteriosus

Differential cyanosis and clubbing (more prominent in the lower limbs than in the upper limbs), confirms the diagnosis of patient ductus arteriosus with reversal of the shunt. [17]

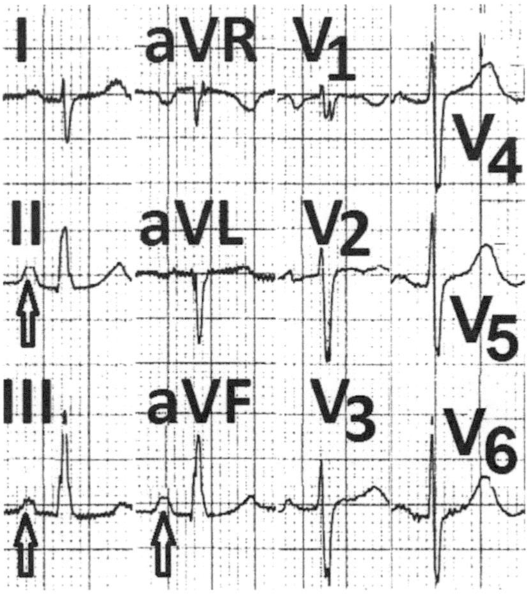

Fig. 51.12 Electrocardiogram from a case of primary pulmonary artery hypertension. The P waves are peaked in leads II, III, aVF (marked ⇧) (suggestive of right atrial enlargement). QRS axis is +105° (right axis deviation). Lead V_6 shows RS without any q wave. (Reproduced with permission from: Mittal SR. Electrocardiogram in Children/ Adolescents/ Adults with Cyanosis due to Congenital Cardiovascular Disease. Cardiology Today-2020;24:108–20)

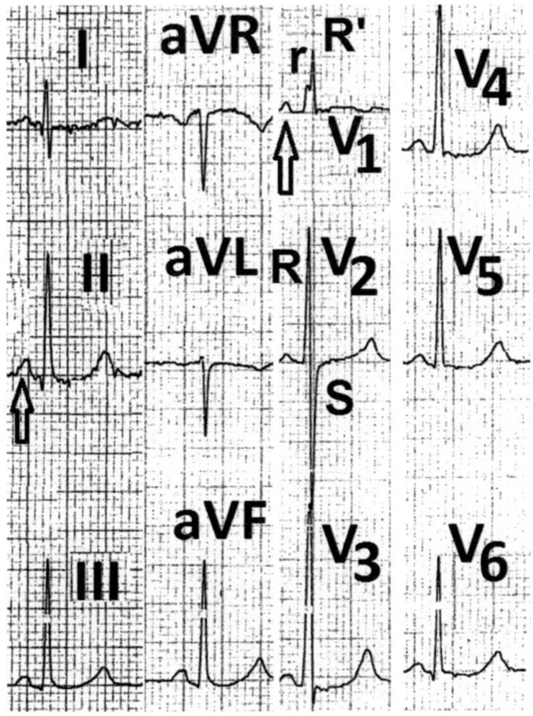

Fig. 51.13 Electrocardiogram from a case of VSD with Eisenmenger's syndrome showing right atrial enlargement (peaked P wave in leads II and V_1, marked ⇧), right ventricular hypertrophy (inferior QRS axis, rR' in lead V_1), and biventricular hypertrophy (large equidiphasic RS complex in lead V_2). (Reproduced with permission from: Mittal SR. Electrocardiogram in Children/ Adolescents/ Adults with Cyanosis due to Congenital Cardiovascular Disease. Cardiology Today-2020;24:108–20)

(c) Long systolic murmur in left 3rd and 4th intercostal spaces with large equidiphasic RS complexes in mid precordial leads should suggest the possibility of single ventricle of left ventricular morphology with ventricular septal defect and inversion of outlet chambers [18] (Fig. 51.15).

(d) Long systolic murmur in the left parasternal region, normal PR interval, and tall stereotyped R waves in precordial leads are features of the single morphologic right ventricle [19] (Fig. 51.16).

(2) Left axis deviation (−30° to −90°).

(a) Counterclockwise depolarization (initial q in leads I and aVL), notch on the ascending limb of S wave in II, III, aVF, rsR' in lead V_1, and RS in leads V_5 and V_6 suggest the possibility of underlying

(i) Ostium primum atrial septal defect [20]

(ii) Complete atrioventricular septal defect [10] (Fig. 51.17)

S waves in leads II, III, and aVF may be notched on upstroke.

(iii) Additional systolic murmur over left 3rd and 4th intercostal spaces without splintering of the S wave in inferior leads should suggest the possibility of double outlet right ventricle with subaortic ventricular septal defect. [14]

(iv) Systolic murmur over the left lower sternal border, left ventricular apical impulse, prominent R wave with an inverted T wave in lead aVL, and large

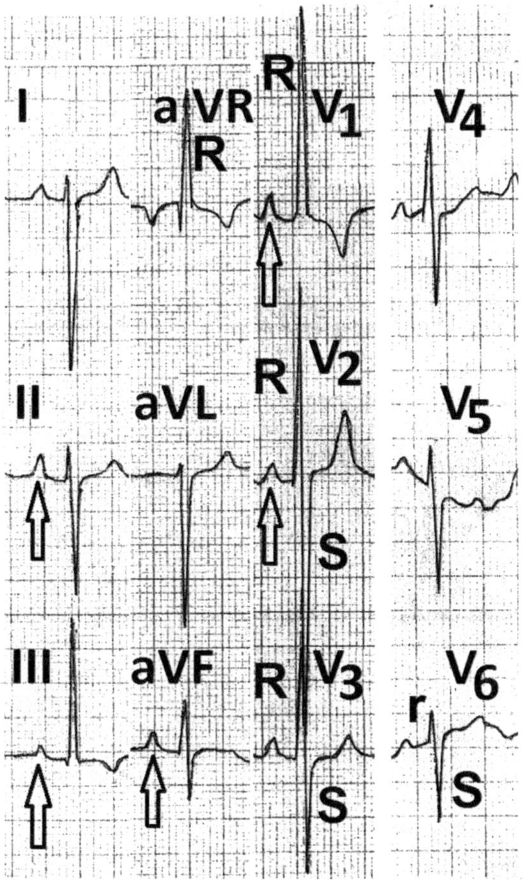

Fig. 51.14 Electrocardiogram from another case of VSD with Eisenmenger's syndrome showing peaked P waves in leads II, III, aVF, V_1, V_2 (marked ⇧) suggestive of right atrial overload, right axis deviation, tall R wave in lead aVR, tall monophasic R wave in lead V_1, rs in lead V_6 suggestive of right ventricular overload. Tall equidiphasic QRS in leads V_2, V_3 suggest biventricular overload. (Reproduced with permission from: Mittal SR. Electrocardiogram in Children/ Adolescents/ Adults with Cyanosis due to Congenital Cardiovascular Disease. Cardiology Today-2020;24:108–20)

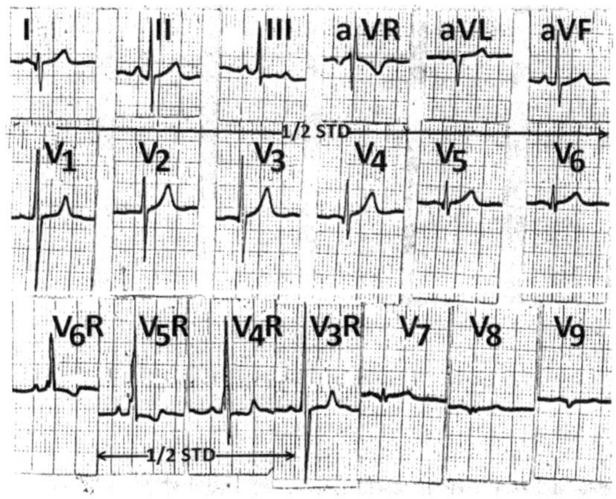

Fig. 51.15 Electrocardiogram from a case of single, morphologic left ventricle with inversion of the outlet chambers and pulmonary artery hypertension. Leads V_1 to V_6 and V_4R and V_5R are recorded at half standard, that is, 5 mm/mV. The P waves are peaked in leads II, III, aVF, V_6R, V_5R (right atrial enlargement). There is right axis deviation with tall RS complexes in precordial leads. (Reproduced with permission from: Mittal SR. Electrocardiogram in Children/ Adolescents/ Adults with Cyanosis due to Congenital Cardiovascular Disease. Cardiology Today-2020;24:108–20)

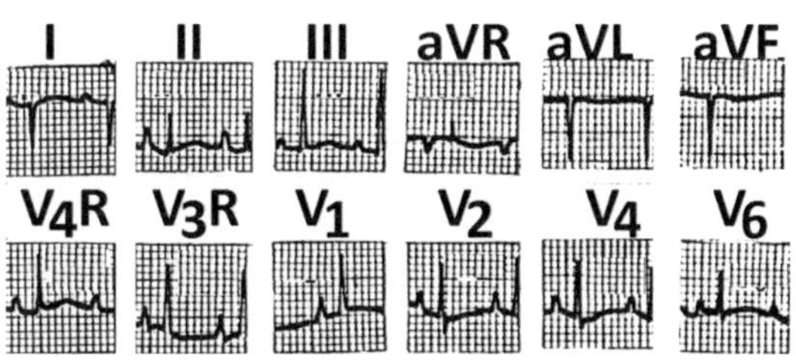

Fig. 51.16 Electrocardiogram shows right axis deviation, peaked P waves in leads V_4R to V_6, and tall uniform R waves in leads V_4R to V_6. (Reproduced with permission from : Mittal SR. Electrocardiogram in Children/ Adolescents/ Adults with Cyanosis due to Congenital Cardiovascular Disease. Cardiology Today-2020;24:108–20)

stereotyped RS complexes in precordial leads should suggest the possibility of single ventricle of left ventricular morphology without inversion of the outlet chambers. [18, 21]

(b) Left axis deviation, clockwise depolarization (deep Q waves in leads III and aVF, deeper in III than in aVF), qR complex in leads V_4R to V_1, no q waves in leads I, aVL with varying degrees of atrioventricular block suggest the possibility of underlying congenitally corrected transposition of the great arteries with nonrestrictive ventricular septal defect and reversal of shunt. [9]

(G) Moderate to marked central cyanosis without systolic pulsations and no palpable second sound in left second intercostal space

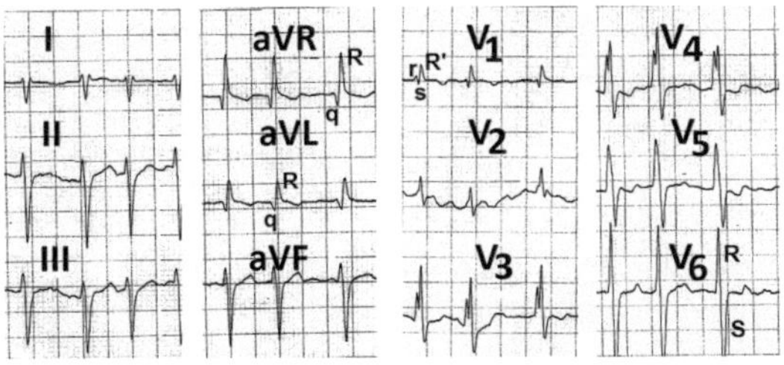

Fig. 51.17 Electrocardiogram from a case of atrioventricular septal defect with Eisenmenger's syndrome showing atrial fibrillation, right upper quadrant QRS axis, counterclockwise depolarization (initial q wave in lead aVL), right ventricular hypertrophy (rsR' pattern in lead V_1, qR in lead aVR, and RS configuration in lead V_6) (Reproduced with permission from: Mittal SR. Electrocardiogram in Children/Adolescents/ Adults with Cyanosis due to Congenital Cardiovascular Disease. Cardiology Today-2020;24:108–20)

suggests congenital cyanotic heart disease with reduced pulmonary flow.

(1) Right axis deviation

(a) Normal jugular venous pulsations, absence of left ventricular impulse and left parasternal systolic lift (right ventricular impulse palpable in subxiphoid area), single second sound, grade2–3/6 ejection systolic murmur along left upper sternal border.

(i) Normal P wave, normal PR interval, normal duration of the QRS complex (neither notched nor slurred), right axis deviation, clockwise depolarization, tall R wave in lead V_1 with inversion of the T wave, sudden change to rS configuration with mildly inverted, flat, or positive T wave in lead V_2, and absence of q wave in leads I, aVL, V_5,V_6 suggest the possibility of tetralogy of Fallot. [22, 23]

– P wave may be peaked in adult unoperated tetralogy (Fig. 51.18).
This is usually due to right ventricular dysfunction.

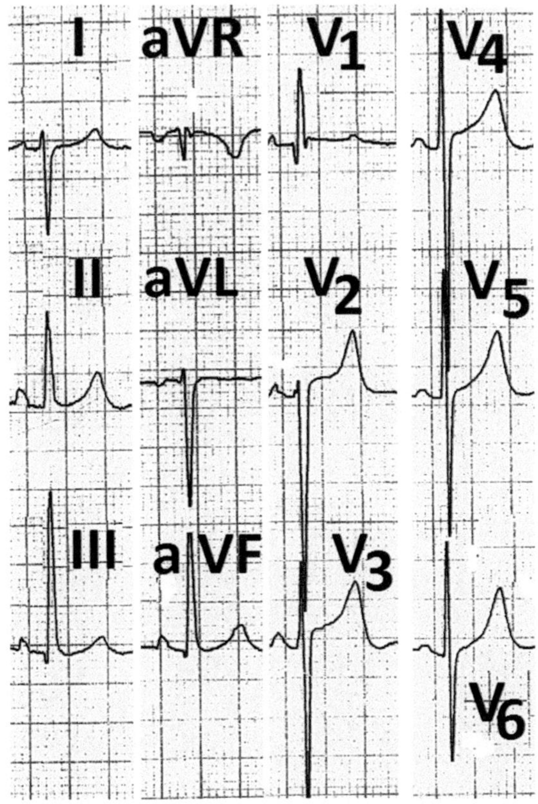

Fig. 51.18 Electrocardiogram from an adult with tetralogy of Fallot showing right ventricular hypertrophy (right axis deviation, clockwise depolarization, prominent R wave in lead V_1) with sudden transition to rS configuration in lead V_2 and absence of left ventricular forces (no q wave in leads V_5,V_6). P waves are peaked in leads II, III, aVF, and V_1 (suggestive of right atrial enlargement). (Reproduced with permission from : Mittal SR. Electrocardiogram in Children/ Adolescents/ Adults with Cyanosis due to Congenital Cardiovascular Disease. Cardiology Today-2020;24:108–20)

– Patients with balanced shunt and reasonable filling of the left ventricle can have q waves in leads V_5 and V_6. [24]

(ii) Prolonged PR interval, right axis deviation, counterclockwise depolarization (q waves in leads I and aVL), slurred S wave in leads I, aVL, V_5 and V_6 with broad R wave in lead aVR suggest the possibility of double outlet right ventricle with subaortic ventricular septal defect and pulmonary stenosis. [25]

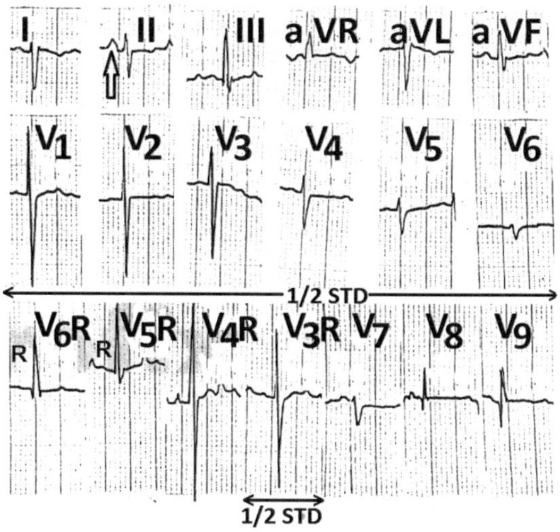

Fig. 51.19 Electrocardiogram from a case of single morphologic left ventricle with inversion of the outlet chamber and pulmonary stenosis showing right atrial enlargement (peaked P wave in lead II, marked ⇧), right axis deviation, prominent R wave in leads V_6R, V_5R, and tall stereotyped RS complexes in leads V_4R to V_4 (leads V_3R to V_2 are recorded at half standard, i.e., 5mm/mV). (Reproduced with permission from: Mittal SR. Electrocardiogram in Children/ Adolescents/ Adults with Cyanosis due to Congenital Cardiovascular Disease. Cardiology Today-2020;24:108–20)

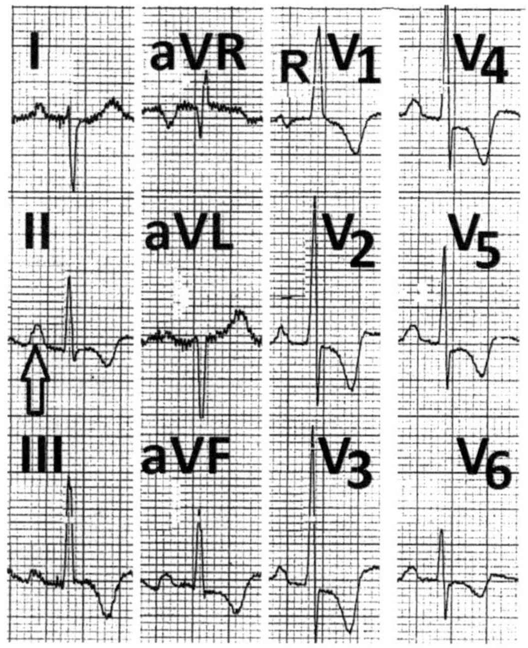

Fig. 51.20 Electrocardiogram from a case of pulmonary valve stenosis with reversal at atrial level showing right atrial enlargement (tall peaked P wave in lead II—marked ⇧), gross right ventricular hypertrophy (right axis deviation, tall monophasic R in lead V_1, and deep T inversion extending from leads V_1 to V_6). (Reproduced with permission from: Mittal SR. Electrocardiogram in Children/ Adolescents/ Adults with Cyanosis due to Congenital Cardiovascular Disease. Cardiology Today-2020;24:108–20)

(iii) Prolonged PR interval, right axis deviation, clockwise depolarization, dominant R wave in lead V_1, and large equidiphasic RS complexes in mid precordial leads (Fig. 51.19) suggest the possibility of single morphologic left ventricle with inversion of the outlet chambers and pulmonary stenosis. [13]

(iv) Normal PR interval, right axis deviation, and tall stereotyped precordial R waves suggest the possibility of single morphologic right ventricle with pulmonary stenosis. [19]

(b) Prominent "a" wave in jugular venous pulse, presence of left parasternal lift, systolic thrill, and long systolic murmur over left upper sternal border in spite of prominent cyanosis.

Peaked P waves in leads II, V_1, and V_2, qR in lead aVR, monophasic R, or qR in lead V_1, inversion of the T waves in leads II, III, aVF, and right precordial leads extending beyond lead V_1 (Fig. 51.20) suggest the possibility of pulmonary valve stenosis with reversal at the atrial level.

(2) Left axis deviation

(a) Prominent "a" wave in the jugular venous pulse, absence of right ventricular impulse in left parasternal or subxiphoid region, left ventricular apical impulse, long systolic murmur (grade 2–3/6) in mid left parasternal region.

(i) Tall peaked P waves in leads II, III, aVF, QRS axis -45° to -90°, counterclockwise depolarization, rs configuration in lead V_1, normal adult progression of QRS in precordial leads with qR configuration in leads V_5,V_6 suggest diagnosis of tricuspid atresia with restrictive ventricular septal defect and pulmonary stenosis [26] (Fig. 51.21). Deep S waves in right precordial leads and tall R wave with inverted T wave in lead aVL also suggest left ventricular hypertrophy.

(b) Peaked P waves, QRS axis around −90° to −180°, counterclockwise depolarization (initial q wave in leads I, aVL, and initial r wave in leads II, III, aVF), prominent R wave in lead V_1 and ST-segment and T wave changes in leads V_1 to V_4 (right ventricular hypertrophy), qRS in leads V_5, V_6 (well-developed left ventricle), notching on the upstroke of the S wave in leads II, III, aVF (due to change in direction of terminal vector of QRS). These findings suggest the possibility of atrioventricular septal defect with pulmonary stenosis. [11]

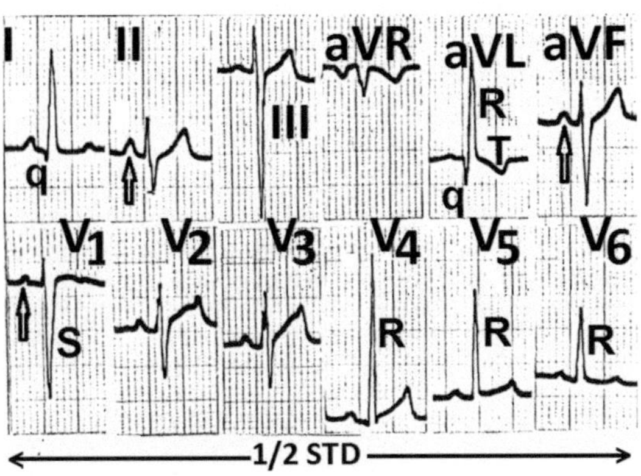

Fig. 51.21 Electrocardiogram from a patient of tricuspid atresia with VSD and PS showing right atrial enlargement (peaked P wave in lead II, aVF, and V_1, marked ⇧), left axis deviation, counterclockwise depolarization (initial q wave in leads I and aVL), and left ventricular hypertrophy (deep S wave in lead V_1, tall R in leads V_4 to V_6, tall R with T inversion in lead aVL). (Reproduced with permission from: Mittal SR. Electrocardiogram in Children/ Adolescents/ Adults with Cyanosis due to Congenital Cardiovascular Disease. Cardiology Today-2020;24:108–20)

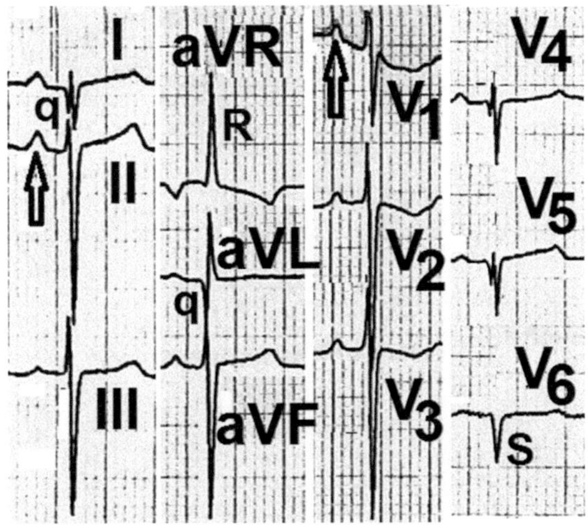

Fig. 51.22 Electrocardiogram from a patient of Noonan's syndrome with stenosis of dysplastic pulmonary valve showing right atrial enlargement (peaked P wave in leads II and V_1, marked ⇧), left axis deviation, counterclockwise depolarization (initial q wave in leads I and aVL), right ventricular hypertrophy (tall monophasic R wave in lead aVR, and dominant S wave in lead V_6). (Reproduced with permission from: Mittal SR. Electrocardiogram in Children/ Adolescents/ Adults with Cyanosis due to Congenital Cardiovascular Disease. Cardiology Today-2020;24:108–20)

(c) Prominent P wave in leads II and V_1, prominent R wave in lead V_1, splintered and prolonged QRS, QS in leads II, III, aVF, V_5,V_6, and inversion of the T waves in inferior and right precordial leads extending beyond lead V_2. These findings suggest the possibility of Noonan's syndrome with stenosis of a dysplastic pulmonary valve [27] and reversal at atrial level (Fig. 51.22).

Electrocardiographic differential diagnosis of an adolescent or adult with cyanosis is shown in Fig. 51.23.

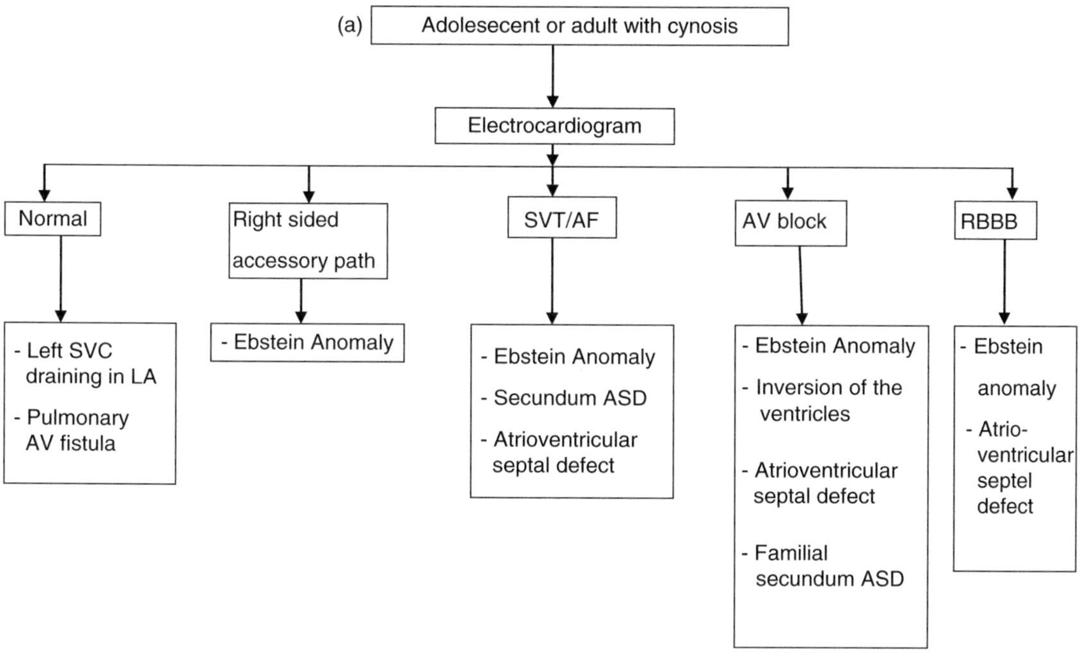

Fig. 51.23 Flow diagram showing electrocardiographic approach in children/adolescent/adults with cyanosis due to congenital cardiovascular

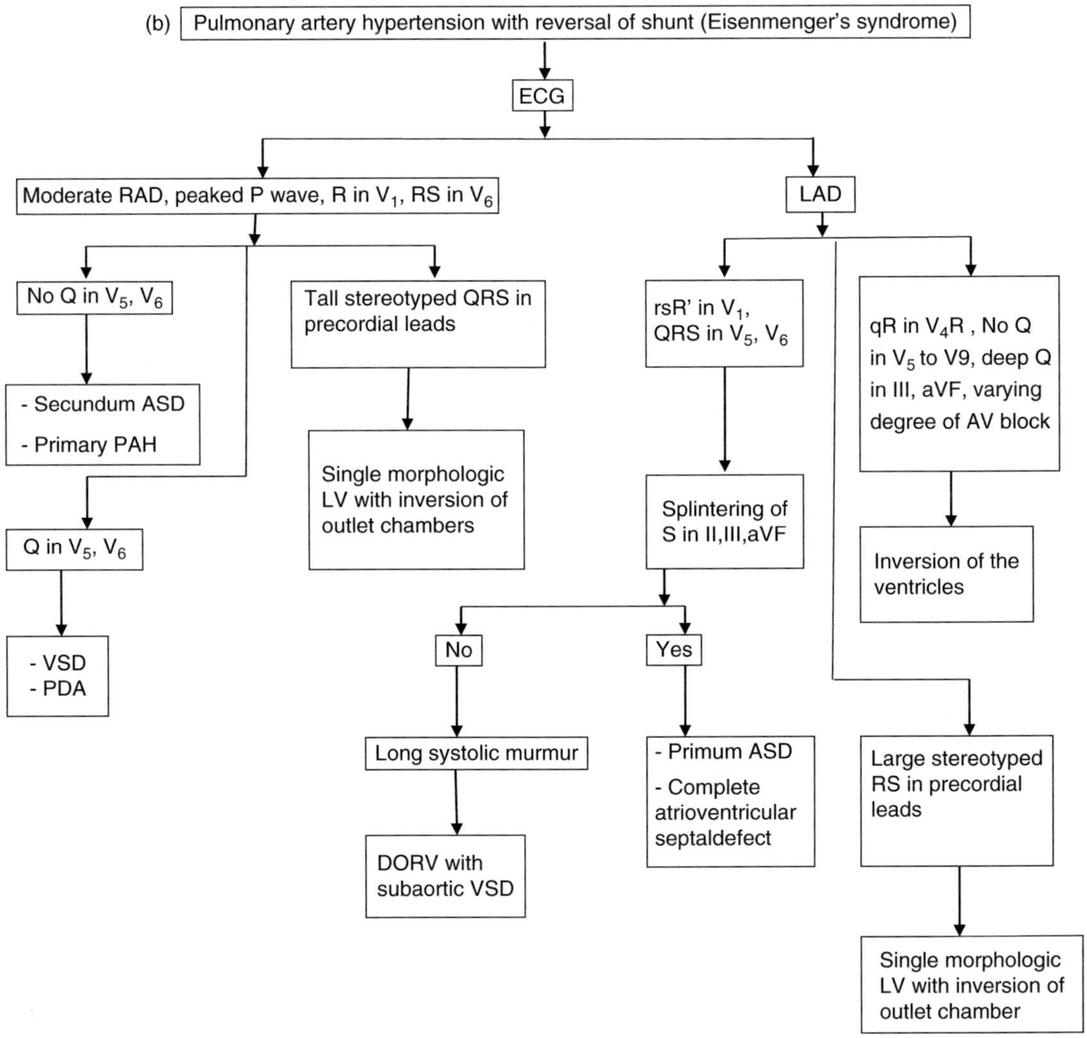

Fig. 51.23 (continued)

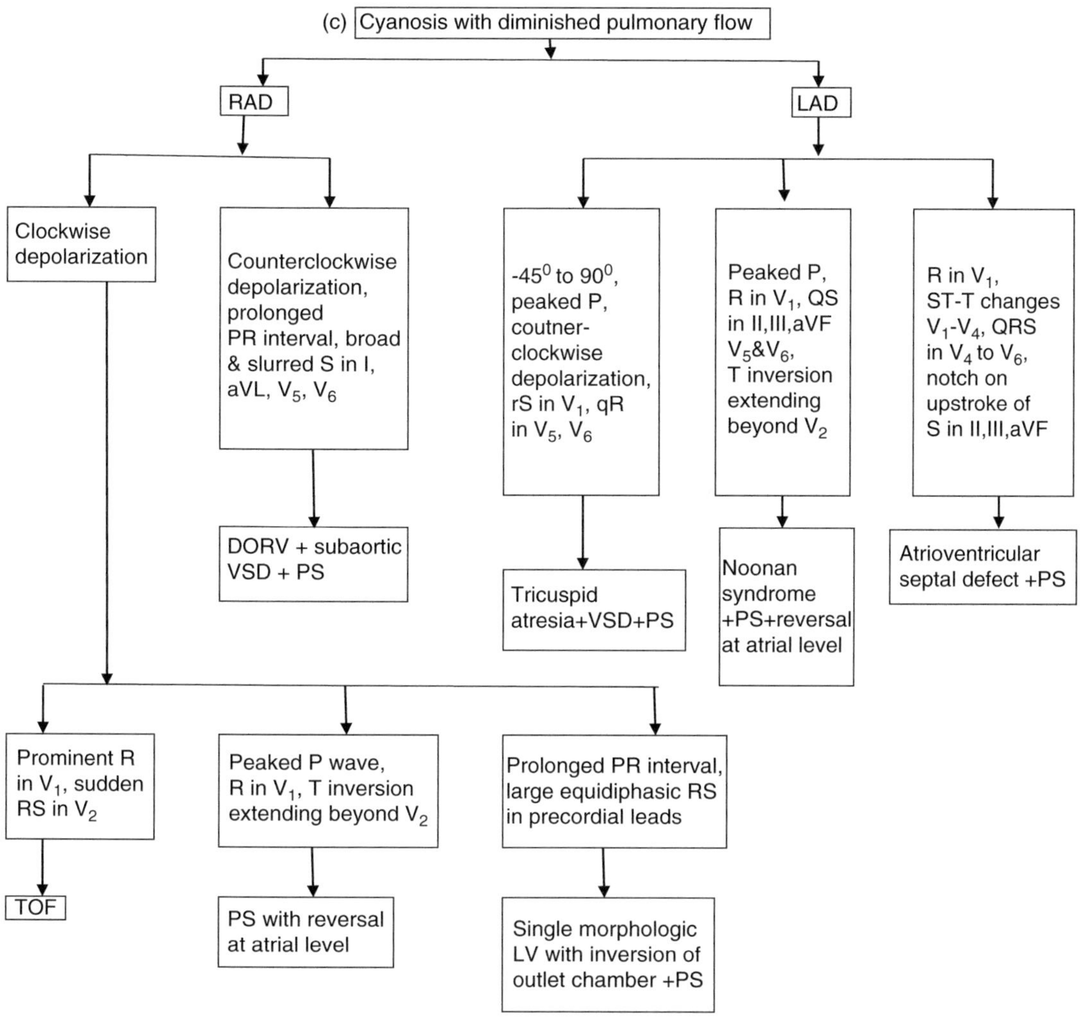

Fig. 51.23 (continued)

Summary

- Uniform central cyanosis with normal electrocardiogram suggests the possibility of left superior vena cava draining in left atrium or pulmonary arteriovenous fistula.
- Preexcitation via a right-sided accessory path in a patient with uniform central cyanosis suggests the possibility of Ebstein anomaly.
- The presence of episodes of supraventricular tachycardia or atrial fibrillation in a patient with uniform central cyanosis should suggest the possibility of Ebstein anomaly, secundum atrial defect, or atrioventricular septal defect.
- In a patient with uniform central cyanosis, the presence of atrioventricular block should raise suspicion of underlying Ebstein anomaly/ inversion of ventricles/atrioventricular septal defect or familial secundum ASD.
- Right bundle branch block pattern is usual in Ebstein anomaly or atrioventricular septal defect.
- In a patient with central cyanosis secondary to pulmonary hypertension with reversal of shunt (Eisenmenger's syndrome), electrocardiogram is useful in deciding the site of shunt.
  - Peaked P waves, moderate right axis deviation, R wave in lead V1, and RS configuration in leads V5 and V6.

    Without any q wave in leads V5 and V6 should suggest the possibility of underlying secundum atrial septal defect.

    With q wave in leads V5 and V6 suggestive of persistence of left ventricular forces. This finding should suggest the possibility of underlying VSD or PDA. Differential cyanosis suggests the possibility of PDA.

    Tall stereotyped QRS complex in precordial leads suggest the possibility of underlying single morphologic LV with inversion of the outlet chambers.
  - Left axis deviation with rsR' in lead V1 and qRS in leads V5 and V6.

    With notch on the upstroke of S wave in leads II, III, and aVF suggests the possibility of primum ASD or complete atrioventricular septal defect.

    Absence of notch on the upstroke of S wave should suggest the possibility of DORV with subaortic VSD.

    With large stereotyped RS complexes in precordial leads should suggest the possibility of single morphologic LV with the inversion of the outlet chamber.

    With qR in leads V1, V4R, and absence of any q wave in leads V5, V6, deep Q wave in leads III, aVF, and varying degree of atrioventricular block suggest the possibility of inversion of ventricles.
- In a patient with central cyanosis and diminished pulmonary flow.
  - Right axis deviation.

    With prominent R wave in lead V1 and sudden transition to RS complex in lead V2 should suggest the possibility of tetralogy of Fallot.

    With peaked P wave, R wave in lead V1, and inversion of T waves beyond lead V2 should suggest the possibility of pulmonary valve stenosis with reversal at atrial level.

    With prolonged PR interval and large equidiphasic RS complexes in precordial leads should suggest the possibility of single morphologic LV with inversion of the outlet chamber and pulmonary stenosis.
  - Left axis deviation.

    With peaked P waves, counterclockwise depolarization, rS in lead V1, and qR in leads V5, V6 should raise the possibility of tricuspid atresia +VSD+PS.

    With peaked P wave, R wave in lead V1, QS in II, III, aVF, V5, V6, and inversion of the T waves extending beyond lead V2 suggest the possibility of Noonan syndrome +PS +reversal at atrial level.

    With R wave in lead V1, depression of the ST-segment and inversion of the T waves in leads V1 to V4, qRS in leads V4 to V6, and notch on the upstroke of S waves in leads II, III, and aVF should suggest the diagnosis of atrioventricular septal defect with pulmonary stenosis.

MCQs

Q1. A completely asymptomatic adult presented for evaluation of uniform central cyanosis. Clinical examination, electrocardiogram, and skiagram of the chest did not reveal any abnormality. Most likely diagnosis is:

(a) Left superior vena cava draining into the coronary sinus
(b) Left superior vena cava draining into the left atrium
(c) Acyanotic tetralogy of Fallot
(d) Mild Ebstein anomaly

Q2. A 10-year-old child presented with uniform central cyanosis. There was history of recurrent epistaxis. Effort tolerance was normal. Clinical examination and electrocardiogram were normal. Most likely diagnosis is:

(a) Left superior vena cava draining into the coronary sinus
(b) Inferior vena cava continuation into azygous vein
(c) Pulmonary arteriovenous fistula
(d) Unroofed coronary sinus

Q3. A 15-year-old male presented with uniform central cyanosis. Effort tolerance was good. Jugular venous pulse was normal. No murmur was audible. An electrocardiogram revealed type preexcitation via right-sided accessory pathway. Most likely diagnosis is:

(a) Ebstein anomaly
(b) Tetralogy of Fallot
(c) Eisenmenger's syndrome
(d) All

Q4. A 20-year-old male presented with a history of episodic palpitation. Effort tolerance was good. There was a rough, scratchy systolic murmur in the left lower parasternal region. Skiagram of chest revealed gross cardiomegaly, narrow pedicle of the cardiac shadow, and normal pulmonary vascularity on skiagram of the chest. Most likely diagnosis is:

(a) Secundum atrial septal defect
(b) Atrioventricular septal defect
(c) Moderate pericardial effusion
(d) Ebstein anomaly

Q5. A child presented with uniform central cyanosis. Review of electrocardiograms revealed varying degrees of atrioventricular block, q wave in lead V_1, and the absence of q wave in leads V_5 to V_9. Most likely diagnosis is:

(a) Ebstein anomaly
(b) Holt–Oram syndrome with reversal of the shunt at atrial level
(c) Down syndrome with reversal of the shunt
(d) Congenitally corrected transposition of the great arteries with reversal of a left to right shunt

Q6. Combination of uniform central cyanosis with prolonged PR interval can be seen in:

(a) Atrioventricular septal defect
(b) Single morphologic left ventricle without inversion of the outlet chambers
(c) Double outlet right ventricle with nonrestrictive subpulmonic VSD
(d) All

Q7. An adult presented with uniform central cyanosis for the last two years. Examination during childhood was reported as normal. Clinical examination revealed a wide split of first and second sounds. An electrocardiogram revealed peaked P waves, right bundle branch block of low voltage, and absence of right ventricular hypertrophy. Most likely diagnosis is:

(a) Eisenmenger's syndrome
(b) Tetralogy of Fallot
(c) Ebstein anomaly
(d) Tricuspid atresia

Q8. A 35-year-old female presented with moderate effort intolerance for the last two years. Clinical examination revealed pulmonary artery hypertension and mild uniform central cyanosis. Electrocardiogram showed peaked P waves, right axis deviation, qR in lead V_1, and rs in lead V_6. Possible diagnosis is:

(a) Large perimembranous VSD with reversal of the shunt
(b) Secundum ASD with reversal of the shunt
(c) Patent ductus arteriosus with reversal of the shunt
(d) Primary pulmonary artery hypertension

Q9. A 5-year-old male child presented with effort intolerance, uniform central cyanosis, and signs of pulmonary artery hypertension. An electrocardiogram revealed peaked P waves, right axis deviation, tall monomorphic R wave in lead V_1, and qRS in lead V_6. Most likely diagnosis is:

(a) Primum ASD with reversal of the shunt
(b) Secundum ASD with reversal of the shunt
(c) Large perimembranous VSD with reversal of the shunt
(d) Patent ductus arteriosus with reversal of the shunt

Q10. A 6-year-old male child presented with effort breathlessness, uniform central cyanosis, and signs of pulmonary artery hypertension. An electrocardiogram revealed normal P wave, right axis deviation, tall monomorphic R in V_1, qRS in lead V_6, and large equidiphasic RS in leads V_2 to V_4. Most likely diagnosis is:

(a) Atrioventricular septal defect with reversal of the shunt
(b) Large perimembranous VSD with reversal of the shunt
(c) Aortopulmonary window with reversal of the shunt
(d) Patent ductus arteriosus with reversal of the shunt

Q11. An 18-year-old female presented with cyanosis and clubbing in both lower limbs and left upper limb. She had signs of pulmonary artery hypertension. An electrocardiogram revealed right axis deviation with qRS in lead V_6. Most likely diagnosis is:

(a) PDA with reversal of the shunt and left subclavian artery arising below the ductus
(b) PDA with reversal of the shunt and left subclavian artery arising above the ductus
(c) PDA with reversal of the shunt and right subclavian artery arising from the descending aorta
(d) Truncus arteriosus with pulmonary artery hypertension

Q12. Eisenmenger's syndrome, extreme left axis deviation, counterclockwise depolarization, and notching on the upstroke of the S wave in leads II, III, and aVF suggest the possibility of:

(a) Ostium primum ASD
(b) Complete atrioventricular septal defect
(c) DORV with PAH
(d) Single morphologic LV with PAH without inversion of the outlet chambers

Q13. Eisenmenger's syndrome with a long systolic murmur, left axis deviation, counterclockwise depolarization without notching of the S wave in leads II, III, aVF should suggest the possibility of:

(a) Large VSD
(b) Complete atrioventricular septal defect
(c) DORV with subaortic VSD
(d) Single morphologic LV without inversion of the outlet chamber

Q14. Eisenmenger's syndrome with a long systolic murmur, right axis deviation, and large equidiphasic RS complexes in mid precordial leads should suggest the possibility of:

(a) VSD with reversal of the shunt
(b) Complete atrioventricular septal defect with reversal of the shunt
(c) Single morphologic LV with inversion of the outlet chamber and PAH
(d) Single morphologic right ventricle with PAH

Q15. Eisenmenger's syndrome with left ventricular apical impulse, long systolic murmur, left axis deviation, prominent R wave with inversion of the T wave in lead aVL, and large stereotyped RS complexes in precordial leads suggests the possibility of underlying:

(a) Ostium primum ASD
(b) Complete atrioventricular septal defect
(c) DORV with PAH
(d) Single morphologic LV with PAH without inversion of the outlet chambers

Q16. A 10-year-old male presented with increasing breathlessness on effort from early childhood. Clinical examination revealed uniform central cyanosis and clubbing, absence of left parasternal systolic lift, no left ventricular apical impulse, and grade 2/6 ejection systolic murmur over left 3rd intercostal space. Electrocardiogram showed peaked P waves, tall monomorphic R wave with an inverted T wave in lead V_1, rS with flat T wave in V_2, rs without any q wave in lead V_6. Most likely diagnosis is:

(a) Classical TOF
(b) Adult TOF
(c) Operated TOF
(d) TOF with left to right shunt

Q17. What is not correct for classical TOF?

(a) P wave and PR interval are normal
(b) Right axis deviation with clockwise depolarization
(c) Q waves in leads I, aVL, V_5, V_6
(d) QRS is neither notched nor slurred

Q18. A combination of uniform central cyanosis and clubbing, normal jugular venous pulse, silent precordium, prolonged PR interval, right axis deviation, counterclockwise depolarization, and broad and slurred S in leads V_5,V_6 suggests the possibility of:

(a) Adult TOF
(b) Atrioventricular septal defect with PS
(c) DORV +VSD +PS
(d) Trilogy of Fallot

Q19. Which electrocardiographic findings are against the diagnosis of congenitally corrected transposition of great arteries with nonrestrictive VSD and PS?

(a) Atrioventricular block
(b) Deep Q wave in lead III
(c) rs in leads V_4R and V_1
(d) QRS in leads V_5, V_6

Q20. Which electrocardiographic finding is against the diagnosis of DORV+VSD+PS?

(a) Prolonged PR interval
(b) Clockwise depolarization
(c) Tall R in leads aVR and V_1
(d) Broad and slurred S in leads V_5, V_6

Q21. Which electrocardiographic finding is against the diagnosis of a single morphologic left ventricle with inversion of the outlet chambers with pulmonary stenosis?

(a) Prolonged PR interval
(b) Right axis deviation
(c) Clockwise depolarization
(d) Low voltage rs in mid precordial leads

Q22. A 20-year-old female presented with uniform central cyanosis and clubbing. On clinical examination, there was left parasternal systolic lift, a long systolic murmur with thrill over left upper sternal border. Electrocardiogram revealed monomorphic R wave in lead V_1, Rs configuration in leads V_2 to V_4. Most likely diagnosis is:

(a) Adult TOF
(b) PS with reversal at atrial level
(c) DORV+VSD+PS
(d) Single morphologic RV with PS

Q23. Which electrocardiographic finding is against the diagnosis of severe valvular PS with reversal at atrial level in an adolescent?

(a) Peaked P wave
(b) Right axis deviation

(c) qR in lead aVR
(d) Positive T waves in leads V_1 to V_6

Q24. A 10-year-old male presented with uniform central cyanosis and clubbing. On clinical examination there was no right ventricular impulse. A long systolic murmur was present in the left mid parasternal region. An electrocardiogram revealed QRS axis of -60°, counterclockwise depolarization, rS in lead V_1, and qR in leads V_5,V_6. Most likely diagnosis is:

(a) TOF with AR
(b) Tricuspid atresia with VSD and PS
(c) Severe PS with reversal at the atrial level
(d) Single morphologic LV with inversion of the outlet chambers and PS

Q25. In atrioventricular septal defect with PS, mean frontal plane QRS axis is between:

(a) $+90^{\circ}$ to $+120^{\circ}$
(b) 0° to $+90^{\circ}$
(c) $0^{\circ}$to -30°
(d) -60° to -180°

Q26. Which electrocardiographic finding is against the diagnosis of Noonan's syndrome with stenosis of dysplastic pulmonary valve?

(a) Right axis deviation
(b) QS in leads II, III, aVF, V_5, V_6
(c) Prominent R in lead V_1
(d) T wave inversion in leads V_1 to V_4

Q27. An adult with history of uniform cyanosis and clubbing from childhood presented with increasing breathlessness on effort. There was no left parasternal lift. Apical impulse was not palpable. An electrocardiogram revealed prominent P waves, right axis deviation, clockwise depolarization, R wave with inversion of the T wave in lead V_1, rS with flat T wave in lead V_2, and rs in lead V_6. Most likely diagnosis is:

(a) DORV+VSD+PS
(b) Adult TOF
(c) Single morphologic LV with inversion of the outlet chambers and PS
(d) Single morphologic RV with PS

Answers

(1)b (2)c (3)a (4)d (5)d(6)a (7)c (8)b (9)c (10)b (11)a (12)a,b (13)c (14)c (15)d (16)a (17)c (18)c, (19)c,d (20)b (21)d (22)b (23)d (24)b (25) d (26) a (27)b

References

1. Perloff JK, Marelli AJ. Congenital anomalies of vena-caval connection. In: Perloff JK, Marelli AJ, editors. Clinical recognition of congenital heart diseases. Philadelphia: Elsevier; 2012. p. 505–12.
2. Hodgson CH, Burchell HB, Good CA, Clagett OT. Hereditary hemorrhagic telangiectasia and pulmonary arteriovenous fistula: survey of a large family. N Engl J Med. 1959;261:625–36.
3. Routray SN, Mishra BR. Congenital pulmonary arteriovenous fistula. In: Satpathy M, editor. Clinical diagnosis of congenital heart disease. New Delhi: Jaypee; 2008. p. 331–4.
4. Perloff JK, Marelli AJ. Ebstein's anomaly of the tricuspid valve. In: Satpathy M, editor. Clinical diagnosis of congenital heart disease. New Delhi: Jaypee; 2008. p. 176–95.
5. Smith WM, Gallagher JJ, Kerr CR, et al. The electrophysiologic basis and management of symptomatic recurrent tachycardia in patients with Ebstein's anomaly of the tricuspid valve. Am J Cardiol. 1982;49:1223–34.
6. Anil SR, Satpathy M. Eisenmenger syndrome. In: Satpathy M, editor. Clinical diagnosis of congenital heart disease. New Delhi: Jaypee; 2008. p. 338–47.
7. Macruz R, Tranchesi J, Ebaid N, Pileggi F, Romero A, Decourt LV, Ebstein's disease. Electrocardiographic and radiologic correlations. Am J Cardiol. 1968;21:653–60.
8. Price JE, Amsterdam EA, Vera Z, Swenson R, Mason DT. Ebstein's disease associated with complete atrioventricular block. Chest. 1978;73:542–4.
9. Victorica BE, Miller BL, Gessner IH. Electrocardiogram and vectorcardiogram in ventricular inversion (corrected transposition). Am Heart J. 1973;86:733–44.
10. Brandenburg BO, Bruchell HB, Toscano-Barbosa E. Electrocardiographic studies of cases with intracardiac malformation of the atrioventricular Canal. Proc Staff Meet Mayo Clin. 1956;31:513–23.

11. PerloFF JK, Marelli AJ. Atrial septal defect: simple or complex. In: Perloff JK, Marelli AJ, editors. Clinical recognition of congenital heart diseases. Philadelphia: Elsevier; 2012. p. 212–72.
12. Silverman ME, Copeland AJ Jr, Hurst JW. The Holt-Oram syndrome: the long and the short of it. Am J Cardiol. 1970;25:11–7.
13. Patnaik AN. Single ventricle. In: Satpathy M, editor. Clinical diagnosis of congenital heart disease. New Delhi: Jaypee; 2008. p. 290–6.
14. Krongrad E, Ritter DG, Weidman WH, Dushane JW. Hemodynamic and anatomic correlation of electrocardiogram in double outlet right ventricle. Circulation. 1972;46:995–1004.
15. Bhattacharyya AK, Satpathy M. Ebstein anomaly. In: Satpathy M, editor. Clinical diagnosis of congenital heart disease. New Delhi: Jaypee; 2008. p. 240–8.
16. Boosone E, Paciocco G, Larussi D, et al. The prognostic role of the ECG in primary pulmonary hypertension. Chest. 2002;121:513–8.
17. Perloff JK, Marelli AJ. Patent ductus arteriosus. Aortopulmonary window. In: Perloff JK, Marelli AJ, editors. Clinical recognition of congenital heart diseases. Philadelphia: Elsevier; 2012. p. 368–93.
18. Elliot LP, Ruttenberg HD, Eliot RS, Anderson RC. Vectorial analysis of the electrocardiogram in common ventricle. Br Heart J. 1964;26:302–11.
19. Perloff JK, Marelli AJ. Univentricular heart. In: Perloff JK, Marelli AJ, editors. Clinical recognition of congenital heart diseases. Philadelphia: Elsevier; 2012. p. 454–71.
20. Maheshwari S. Atrial septal defect (primum type). In: Satpathy M, editor. Clinical diagnosis of congenital heart disease. New Delhi: Jaypee; 2008. p. 83–8.
21. Davachi F, Moller JH. The electrocardiogram and vectorcardiogram in single ventricle. Anatomic correlation. Am J Cardiol. 1969;23:19–31.
22. Bender SR, Dreiful LS, Downing D. Anatomic and electrocardiographic correlation of Fallot's tetralogy: a Study of 100 proved cases. Am J Cardiol. 1961;7:475–80.
23. Pilleggi F, Bocanegra J, Tranchesi J, et al. The electrocardiogram in tetralogy of Fallot: a study of 142 cases. Am Heart J. 1960;59:667–80.
24. Pati PK. Tetralogy of Fallot. In: Satpathy M, editor. Clinical diagnosis of congenital heart disease. New Delhi: Jaypee; 2008. p. 257–67.
25. Titus JL, Neufeld H, Edward JE. The atrioventricular conduction system in hearts with both great vessels originating from the right ventricle. Am Heart J. 1964;67:588–92.
26. Burch GE, Depasquale NP. The electrocardiogram, vectorcardiogram and ventricular gradient in combined pulmonary stenosis and interatrial communication. Am J Cardiol. 1961;7:646–56.
27. Raaimakers R, Noordam C, Noonan JA, Croonen EA, Van Der Burgt CJAM, Draaisma JMT. Are ECG abnormalities in Noonan syndrome characteristic for the syndrome? Eur J Pediar. 2008;167:1363–7.

MIX
Papier aus verantwortungsvollen Quellen
Paper from responsible sources
FSC® C105338

If you have any concerns about our products,
you can contact us on
ProductSafety@springernature.com

In case Publisher is established outside the EU,
the EU authorized representative is:
Springer Nature Customer Service Center GmbH
Europaplatz 3, 69115 Heidelberg, Germany

Printed by Libri Plureos GmbH
in Hamburg, Germany